Introductory Statistics

INTERNATIONAL ADAPTATION

Introductory Statistics

Tenth Edition

INTERNATIONAL ADAPTATION

Prem S. Mann
Former Professor
Eastern Connecticut State University, USA

WILEY

Introductory Statistics

Tenth Edition

INTERNATIONAL ADAPTATION

Contributor Subject Matter Experts: Dr Harmanpreet Singh Kapoor, Central University of Punjab, India; Dr Ravinder Singh, Central University of Haryana, India; Dr Suman Guha, Presidency University, Kolkata, India

Founded in 1807, John Wiley & Sons, Inc. has been a valued source of knowledge and understanding for more than 200 years, helping people around the world meet their needs and fulfill their aspirations. Our company is built on a foundation of principles that include responsibility to the communities we serve and where we live and work. In 2008, we launched a Corporate Citizenship Initiative, a global effort to address the environmental, social, economic, and ethical challenges we face in our business. Among the issues we are addressing are carbon impact, paper specifications and procurement, ethical conduct within our business and among our vendors, and community and charitable support. For more information, please visit our website: www.wiley.com/go/citizenship.

ISBN: 978-1-394-22638-2

ISBN: 978-1-394-22639-9 (ePub)

ISBN: 978-1-394-22640-5 (ePdf)

Printed and bound by CPI Group (UK) Ltd, Croydon, CR0 4YY
C9781394226382_040124

To the memory of my parents

About the Author

PREM S. MANN has a Bachelor's degree in mathematics, economics, and English, and a Master's degree in economics, with a specialization in statistics, from Punjab University, India. Prem taught economics and statistics at a Punjab University college for four years. In 1975 he left India for the United Kingdom, where he studied at the University of Manchester and received his second Master's degree in economics. Then he attended the London School of Economics where he studied operations research. In January 1980, Prem moved to Los Angeles, California, to study for his Ph.D. in Economics at UCLA. One of his fields of specialization in his Ph.D. studies was statistics and quantitative methods. While working on his Ph.D. at UCLA, Prem Mann taught statistics and quantitative methods at the Los Angeles and Fullerton campuses of California State University, first as an adjunct instructor and then as a full-time faculty member. In 1986, he joined Eastern Connecticut State University as a faculty member, where he taught statistics, mathematics for economics, and econometrics for 29 years. He also served as chair of the Department of Economics at Eastern Connecticut State University for 20 years before retiring in 2015. Currently, Prem Mann lives with his wife and close to his children and grandchildren in New York City.

In addition to *Introductory Statistics*, which is currently in its 10th edition, Prem Mann has authored *Introductory Statistics Using Technology* and *Statistics for Business and Economics*. He has also coauthored *Practitioner's Guide to Statistics and Lean Six Sigma for Process Improvements* with Mikel J. Harry, Ofelia C. De Hodgins, Richard L. Hulbert, and Christopher J. Lacke. All these books are Wiley publications. Prem Mann's *Introductory Statistics* book has been translated into Chinese, Serbian, and Portuguese. His *Statistics for Business and Economics* was also translated into Chinese.

Preface

Introductory Statistics is written for a one- or two-semester first course in applied statistics. This book is intended for students who do not have a strong background in mathematics. The only prerequisite for this text is knowledge of elementary algebra.

To quote from pages 8 and 29 of the **GAISE** (Guidelines for Assessment and Instruction in Statistics Education) **College Report 2016**:

"The desired result of all introductory statistics courses is to produce statistically educated students, which means that students should develop the ability to think statistically. ... The democratization of introductory statistics has broadened and diversified the backgrounds, interests, and motivations of those who take the course. Statistics is no longer reserved for future scientists in narrow fields but is now a family of courses, taught to students at many levels, ... with very diverse interests and goals."

In writing and revising this text, the author has kept these comments and all recommendations of the GAISE report in mind to make this text helpful to students in a wide range of areas and at different levels.

As mentioned in the GAISE Report of 2016, college students from almost all fields of study are required to take at least one course in statistics. Consequently, the study of statistical methods has taken on a prominent role in the education of students from a variety of backgrounds and academic pursuits. From the first edition, the goal of ***Introductory Statistics*** has been to make the subject of statistics interesting and accessible to a wide and varied audience. Three major elements of this text support this goal:

1. Realistic content of its examples and exercises, drawing from a comprehensive range of applications from all facets of life
2. Clarity and brevity of presentation
3. Soundness of pedagogical approach

These elements are developed through the interplay of a variety of significant text features. The feedback received from the users of the ninth edition (and earlier editions) of ***Introductory Statistics*** has been very supportive and encouraging. Positive experiences reported by instructors and students have served as evidence that this text offers an interesting and accessible approach to statistics—the author's goal from the very first edition. The author has pursued the same goal through the refinements and updates in this international adaptation, so that ***Introductory Statistics*** can continue to provide a successful experience in statistics to a growing number of students and instructors.

New to the International Adaptation

The following are some of the changes made in this edition:

- New and revised **Learning Objectives** have been added to support each section of the text.
- In Chapter 1, a section on **types of scales** has been added which includes discussion on nominal, ordinal, interval, and ratio scales.
- In Chapter 2, a discussion on **ogive curves** has been added which explores the characteristics of the distributions and provides understating about the scatter, median, percentile, skewness, and central tendency of the data.
- In Chapter 3, the discussion on measures of center for ungrouped data has been updated by adding **geometric and harmonic mean**.
- In Chapter 5, a new section on **negative binomial probability distribution** has been added.
- In Chapter 12, the discussion on analysis of variance has been enhanced by adding a new section on **two-way analysis of variance** (two-way ANOVA).
- A large number of **examples and exercises** throughout the book are new or revised, providing contemporary and varied ways for students to practice statistical concepts.
- The majority of the **case studies** are new or revised, drawing on current uses of statistics in areas of student interest.
- **New data** have been integrated throughout, reinforcing the vibrancy of statistics and the relevance of statistics to student lives right now.
- The **Technology Instructions** sections have been updated to support the use of the latest versions of **TI-84** Color/TI-84, **Minitab** Professional Version 19, and **Excel** Version 19. Comprehensive and **easy-to-download Manuals** have been created for each (including Minitab Express) and can be accessed via the product page on *www.wiley.com*.
- Many **Technology Assignments** at the end of each chapter are either new or have been updated.
- The **data sets** available in the book's product page on *www.wiley.com* have been updated. **Appendix A** contains information about these data sets.
- Some of the **Uses and Misuses of Statistics, Mini-Projects**, and **Decide for Yourself** sections have been updated, revised, or replaced with current and relevant data.

Hallmark Features of the International Adaptation

Clear and Concise Exposition The explanation of statistical methods and concepts is clear and concise. Moreover, the style is user-friendly and easy to understand. Chapter introductions are supported by detailed Learning Objectives which build on the topics and ideas discussed in earlier sections.

Thorough Examples The text contains a wealth of examples. These examples are usually presented in a format showing a problem and its solution. They are well sequenced and thorough, displaying all facets of concepts. Furthermore, the examples capture students' interest because they cover a wide variety of relevant topics. They are based on situations that practicing statisticians encounter every day. Finally, a large number of examples are based on real data taken from sources such as books, government and private data sources, and reports, magazines, newspapers, and professional journals.

Step-by-Step Solutions A clear, concise solution follows each problem presented in an example. When the solution to an example involves many steps, it is presented in a step-by-step format. For instance, examples related to tests of hypothesis contain five steps (four steps for *p*-value approach) that are consistently used to solve such examples in all chapters. Thus, procedures are presented in the concrete settings of applications rather than as isolated abstractions. Frequently, solutions contain highlighted remarks that recall and reinforce ideas critical to the solution of the problem. Such remarks add to the clarity of presentation.

Titles for Examples Each example based on an application of concepts contains a title that describes to what area, field, or concept the example relates.

Example Descriptions Each example contains a note that briefly describes what is being done in that example. Students can use these notes to assist them as they read through sections and to quickly locate appropriate model problems as they work through exercises.

Frequent Use of Diagrams Concepts can often be made more understandable by describing them visually with the help of diagrams. This text uses diagrams frequently to help students understand concepts and solve problems. For example, tree diagrams are used a few times in Chapters 4 and 5 to assist in explaining probability concepts and in computing probabilities. Similarly, solutions to all examples about tests of hypothesis contain diagrams showing rejection regions, nonrejection regions, and critical values.

Highlighting Definitions of important terms, formulas, and key concepts are enclosed in colored boxes so that students can easily locate them.

Cautions Certain items need special attention. These may deal with potential trouble spots that commonly cause errors, or they may deal with ideas that students often overlook. Special emphasis is placed on such items through the heading **Remember**. An icon is used to identify such items.

Real-World Case Studies These case studies, which appear in most of the chapters, provide additional illustrations of the applications of statistics in research and statistical analysis. Most of these case studies are based on articles or data published in journals, magazines, newspapers, or websites. Almost all case studies are based on real data.

Variety of Exercises The text contains a variety of exercises, including technology assignments. Moreover, a large number of these exercises contain several parts. Exercise sets appearing at the end of each section (or sometimes at the end of two or three sections) include problems on the topics of that section. These exercises are divided into two parts: **Concepts and Procedures** that emphasize key ideas and techniques and **Applications** that use these ideas and techniques in concrete settings. **Supplementary exercises** appear toward the end of each chapter and contain exercises on all sections and topics discussed in that chapter. A large number of these exercises are based on real data taken from varied data sources such as books, government and private data sources and reports, magazines, newspapers, and professional journals. Not only do the exercises given in the text provide practice for students, but the real data contained in the exercises provide interesting information and insight into economic, political, social, psychological, and other aspects of life. The exercise sets also contain many problems that demand critical thinking skills. The answers to selected odd-numbered exercises appear in the ***Answers* section** at the back of the book. **Optional exercises** are indicated by an asterisk (*).

Advanced Exercises Almost all chapters have a set of exercises that are of greater difficulty. Such exercises appear under the heading *Advanced Exercises* after the *Supplementary Exercises.*

Uses and Misuses of Statistics This feature toward the end of each chapter (before the Glossary) points out common misconceptions and pitfalls students will encounter in their study of statistics and in everyday life.

Decide for Yourself This feature discusses a real-world problem and raises questions that readers can think about and answer. This feature is available in the book's product page on *www.wiley.com*.

Glossary Each chapter has a glossary that lists the key terms introduced in that chapter, along with a brief explanation of each term.

Self-Review Tests Each chapter contains a *Self-Review Test,* which appears immediately after the *Supplementary* and *Advanced Exercises.* Each *Self-Review Test* has Multiple Choice Questions and Practice Exercises. These problems can help students test their grasp of the concepts and skills presented in respective chapters and monitor their understanding of statistical methods. The problems marked by an asterisk (*) in the *Self-Review Tests* are **optional**. The answers to almost all problems of the *Self-Review Tests* appear in the *Answer* section.

Technology Usage At the end of each chapter is a section covering uses of three major technologies of statistics: the **TI-84** Color/TI-84, **Minitab** Professional Version 19, and **Excel** Version 19. For each technology, students are guided through performing statistical analyses in a step-by-step fashion, showing them how to enter, revise, format, and save data in a spreadsheet, workbook, or named and unnamed lists, depending on the technology used. Illustrations and screenshots demonstrate the use of these technologies. Additional detailed downloadable **technology manuals** (including Minitab Express) are provided in the book's product page on *www.wiley.com*.

Technology Assignments Each chapter contains technology assignments that appear at the end of the chapter. These assignments can be completed using any of the statistical software.

Mini-Projects Associated with each chapter of the text are either very comprehensive exercises or they ask students to perform their own surveys and experiments. They provide practical applications of statistical concepts to real life.

Data Sets A large number of data sets appear in in the book's product page on *www.wiley.com*. These large data sets are collected from various sources, and they contain information on several variables. Many exercises and assignments in the text are based on these data sets. These data sets can also be used for instructor-driven analyses using a wide variety of statistical software packages including TI-84, Minitab Professional Version 19, and Excel Version 19.

Key Formulas Key formulas from all chapters are listed in **Appendix C**.

GAISE Report Recommendations Adopted

In 2003, the American Statistical Association (ASA) funded the Guidelines for Assessment and Instruction in Statistics Education (GAISE) Project to develop ASA-endorsed guidelines for assessment and instruction in statistics for the introductory college statistics course. The report by the committee established in 2003 was submitted in 2005, which made a series of recommendations for a first course in statistics and data analysis. More recently, ASA established another committee for the same purpose. This committee submitted its report in 2016. The **2016 GAISE report** endorsed the same six recommendations that the 2005 GAISE report had endorsed, but in the 2016 report the six recommendations were re-ordered and the language was simplified in some of the recommendations. Here are the six recommendations made in the 2016 report:

1. Teach statistical thinking.
    - Teach statistics as an investigative process of problem-solving and decision-making.
    - Give students experience with multivariable thinking.
2. Focus on conceptual understanding.
3. Integrate real data with a context and purpose.
4. Foster active learning.
5. Use technology to explore concepts and analyze data.
6. Use assessments to improve and evaluate student learning.

This text has integrated all these recommendations in many ways. Here are a few examples of how this text can assist in helping you, the instructor, in meeting the GAISE recommendations.

1. The exercises titled *Concepts and Procedures* and *Applications* help students to understand the concepts. The *Advanced Exercises, Uses and Misuses of Statistics, Decide for Yourself*, and *Mini-Projects* features help to develop statistical thinking and conceptual understanding.
2. Many of the exercises require interpretation, not just answers in terms of numbers. Graphical and numeric summaries are combined in some exercises in order to emphasize looking at the whole picture, as opposed to using just one graph or one summary statistic.
3. All of the **data sets** listed in Appendix A are available in the book's product page on *www.wiley.com* and can be used for statistical analysis. They have been formatted for a variety of statistical software packages. This eliminates the need to enter data into the software. A variety of software instruction manuals also allow the instructor to spend more time on concepts and less time teaching how to use technology.
4. The **Mini-Projects** help students to generate their own data by performing experiments and/or taking random samples from the large data sets mentioned in Appendix A.
5. Many exercises throughout the book and a large number of **Technology Assignments** require students to use real data for statistical analysis.
6. **Chapter 14, Multiple Regression**, is an example of multivariate thinking. Teachers should expose students to multiple regression, even if spending only one or two hours.

We highly recommend that all statistics instructors take the time to read the GAISE report, which is available at **www.amstat.org/education/gaise**. There is a wealth of information in this report that can be used to improve the teaching of statistics.

Instructor and Student Resources

The following resources are available to Instructors and Students in the book's product page on *www.wiley.com*. All these items are available for instructors. The items with * are available to students.

- Test Bank
- PowerPoint Lecture Slides
- Instructor's Solutions Manual
- * Student's Solutions Manual
- * Data Sets (available in the following formats: Comma Delimited, Excel 2019, JMP, Minitab Professional Version 19 and Express, SPSS, and Tab Delimited)
- * Online-only chapter: Chapter 15: Nonparametric Methods
- Correlation Guide
- * Decide for Yourself
- * Technology Resource Manuals:
  - TI-84 Graphing Calculator Manual
  - Minitab Professional Version 19 Manual
  - Minitab Express Manual
  - Excel 2019 Manual

These technology manuals provide step-by-step instructions, screen captures, and examples for using technology in the introductory statistics course.

Acknowledgments

I thank the following reviewers of the tenth edition of this book, whose comments and suggestions were invaluable in improving the text.

D. P. Adhikari, *Marywood University*
Wendy Ahrendsen, *South Dakota State University*
Nan Hutchins Bailey, *The University of Texas at Tyler*
Les Barnhouse, *Athabasca University*
Hossein Behforooz, *Utica College*
Joleen Beltrami, *University of the Incarnate Word*
Bill Burgin, *Gaston College*
David Bush, *Villanova University*
Ferry Butar Butar, *Sam Houston State University*
Chris Chappa, *University of Texas at Tyler*
Jerry Chen, *Suffolk County Community College, Ammerman Campus*
A. Choudhury, *Illinois State University*
Robert C. Forsythe, *Frostburg State University*
Daesung Ha, *Marshall University*
Rhonda Hatcher, *Texas Christian University*
Joanna Jeneralczuk, *University of Massachusetts, Amherst*
Annette Kanko, *University of the Incarnate Word*
Mohammad Kazemi, *University of North Carolina, Charlotte*
Janine Keown-Gerrard, *Athabasca University*
Hoon Kim, *California State Polytechnic University—Pomona*
Matthew Knowlen, *Horry Georgetown Technical College*
Maurice LeBlanc, *Kennesaw State University*
Min-Lin Lo, *California State University, San Bernardino*
Natalya Malakhova, *Johnson County Community College*
Nola McDaniel, *McNeese State University*
Stéphane Mechoulan, *Dalhousie University*
Robert L. Nichols, *Florida Gulf Coast University*
Jonathan Oaks, *Macomb Community College—South Campus*
Marty Rhoades, *West Texas A&M University*
Mary Beth Rollick, *Kent State University—Kent*
Seema Sehgal, *Athabasca University*
Linda Simonsen, *University of Washington, Bothell*
Criselda Toto, *Chapman University*
Viola Vajdova, *Benedictine University*
Daniel Weiner, *Boston University*

I express my thanks to the following for their contributions to earlier editions of this book that made it better in many ways: Doug Tyson, Chris Lacke (Rowan University), Late Gerald Geissert (formerly of Eastern Connecticut State University), Daniel S. Miller (Central Connecticut State University), and David Santana-Ortiz (Rand Organization).

I extend my special thanks to James Bush of Waynesburg University who contributed to this edition in many significant ways—especially his role in authoring the Example Videos and accompanying questions. I would also like to thank Dan Anderson for his work creating the Interactive Applets and James Bush again for developing the questions to support them. I would also like to take this opportunity to thank Julie M. Clark for working on the solutions manuals and preparing the answer section, and Deanna Voehl (Embry-Riddle Aeronautical University – Worldwide) for checking the solutions for accuracy.

It is of utmost importance that a textbook be accompanied by complete and accurate supplements. I take pride in mentioning that the supplements prepared for this text possess these qualities and much more. I thank Sharon Sterken (Emsley A. Laney High School – Wilmington, NC) for her contribution to the Instructor Solutions Manual, Nathan Czuba for his work on the Excel Manual, Dr. Carol Weideman (St. Petersburg College) for creating the Minitab Manuals, and Songfeng Zheng (Missouri State) for updating the Datasets and TI-84 Manuals.

It is my pleasure to thank all the professionals at John Wiley & Sons with whom I enjoyed working during this revision. Among them are Joanna Dingle and Andrew Noble (Executive Editors), Wendy Lai (Senior Creative Product Designer), Valerie Zaborski (Executive Managing Editor), Laura Abrams (Senior Course Production Operations Specialist), Melissa Whelan (Senior Course Content Developer), Thomas Kulesa (Lead Product Designer), and Hannah Larkin (Editorial Assistant). I also want to thank Rajeev Kumar (Course Production Project Manager) for managing the production process and Matt Gervais (Project Manager).

Any suggestions from readers for future revisions would be greatly appreciated. Such suggestions can be sent to me at **premmann@yahoo.com**.

PREM S. MANN

Contents

CHAPTER 1

Introduction

Cathy Yeulet/123RF

What are the opinions of teens ages 13 to 17 years about the anxiety and depression among teens living in their community? Do they think it is a major problem? Or is it a minor problem? Or is it not a problem? A survey conducted by the Pew Research Center in 2018 found that 70% of such U.S. teens said that anxiety and depression is a major problem among teens in the community where they live, 26% said it is a minor problem, and 4% said it is not a problem. (See Case Study 1.2.)

The study of statistics has become more popular than ever over the past four decades. The increasing availability of computers and statistical software packages has enlarged the role of statistics as a tool for empirical research. As a result, statistics is used for research in almost all professions, from medicine to sports. Today, college students in almost all disciplines are required to take at least one statistics course. These days, all print and digital media contain graphs and stories on statistical studies. After you finish reading this book, it should be much easier to understand these graphs and stories.

Every field of study has its own terminology. Statistics is no exception. This introductory chapter explains the basic terms and concepts of statistics. These terms and concepts will bridge our understanding of the concepts and techniques presented in subsequent chapters.

1.1 Statistics and Types of Statistics

LEARNING OBJECTIVES

After completing this section, you should be able to:

- Explain the two meanings of *statistics*.
- Describe in context the difference between descriptive statistics and inferential statistics.

1.1.1 What Is Statistics?

The word **statistics** has two meanings. In the more common usage, *statistics* refers to numerical facts. The numbers that represent the income of a family, the age of a student, the percentage of passes completed by the quarterback of a football team, and the starting salary of a typical college graduate are examples of statistics in this sense of the word. A 1988 article in *U.S. News & World Report* mentioned that "Statistics are an American obsession."[1] During the 1988 baseball World Series between the Los Angeles Dodgers and the Oakland A's, the then NBC commentator Joe Garagiola reported to the viewers numerical facts about the players' performances. In response, fellow commentator Vin Scully said, "I love it when you talk statistics." In these examples, the word *statistics* refers to numbers.

The following examples present some statistics.

1. According to the Institute for College Access and Success, students in the United States who graduated with a bachelor's degree in 2018 had an average student loan debt of $29,200. The same number for 2017 graduates was $28,650. (*USA TODAY*, September 20, 2019.)
2. According to an article published in the journal *Science*, since 1970 there has been a 29% decline in the number of birds in the United States and Canada, which comes out to about 2.9 billion less birds now in these two countries compared to 1970. (*The New York Times*, September 20, 2019.)
3. According to 2019 data by e-importz.com, about 50% of the U.S. population, which comes out to a little more than 150 million people, drink espresso, cappuccino, latte, or iced/cold coffees.
4. According to a 2019 estimate, the total amount of student debt in the United States is $1.56 trillion. Of about 250 million adults in the United States, 45 million have student debt.
5. According to the National Education Association, teachers' average starting salary in the United States was $39,249 in 2017–2018.
6. According to the Nielsen Total Audience Report, U.S. adults age 18 and older spent an average of 4 hours and 46 minutes per day watching TV in the first quarter of 2018 (*USA TODAY*, September 19, 2018).
7. According to a 2019 report, only 2% of the world's population, which comes out to about 140 million people, have green eyes.
8. According to a Deutsche Bank survey of 2018, the average monthly rent of a two-bedroom apartment in San Francisco was $3664.
9. In 2018, the total revenue of Walmart was $514,405 million making it the largest company in the world ranked by revenue (*Fortune*, August 2019).

The second meaning of *statistics* refers to the field or discipline of study. In this sense of the word, *statistics* is defined as follows:

Statistics

Statistics is the science of collecting, analyzing, presenting, and interpreting data, as well as of making decisions based on such analyses.

Every day we make decisions that may be personal, business related, or of some other kind. Usually these decisions are made under conditions of uncertainty. Many times, the situations or problems we face in the real world have no precise or definite solution. Statistical methods help us make scientific and intelligent decisions in such situations. Decisions made by using statistical methods are called *educated guesses*. Decisions made without using statistical (or scientific) methods are *pure guesses* and, hence, may prove to be unreliable. For example, opening a large store in an area with or without assessing the need for it may affect its success.

Like almost all fields of study, statistics has two aspects: theoretical and applied. *Theoretical* or *mathematical statistics* deals with the development, derivation, and proof of statistical theorems, formulas, rules, and laws. *Applied statistics* involves the applications of those theorems, formulas, rules, and laws to solve real-world problems. This text is concerned with applied statistics and not with theoretical statistics. By the time you finish studying this book, you will have learned how to think statistically and how to make educated guesses.

[1]"The Numbers Racket: How Polls and Statistics Lie," *U.S. News & World Report*, July 11, 1988, pp. 44–47.

1.1.2 Types of Statistics

Broadly speaking, applied statistics can be divided into two areas: **descriptive statistics** and **inferential statistics**.

Descriptive Statistics Suppose we have information on the test scores of students enrolled in a statistics class. In statistical terminology, the whole set of numbers that represents the scores of students is called a **data set**, the name of each student is called an **element**, and the score of each student is called an **observation**. (These terms are defined in more detail in Section 1.2.)

Many data sets in their original forms are usually very large, especially those collected by federal and state agencies. Consequently, it is difficult to draw conclusions or make decisions from these large data sets. It is easier to draw conclusions from summary tables and diagrams than from the original version of a data set. So, we summarize data by constructing tables, drawing graphs, and calculating summary measures such as averages and proportions. The portion of statistics that helps us do this type of statistical analysis is called **descriptive statistics**.

Descriptive Statistics

Descriptive statistics consists of methods for organizing, displaying, and describing data by using tables, graphs, and summary measures.

Chapters 2 and 3 discuss descriptive statistical methods. In Chapter 2, we learn how to construct tables and how to graph data. In Chapter 3, we learn how to calculate numerical summary measures, such as averages.

Case Study 1.1 presents an example of descriptive statistics.

Case Study 1.1

Deaths Due to Red-Light Running in the United States

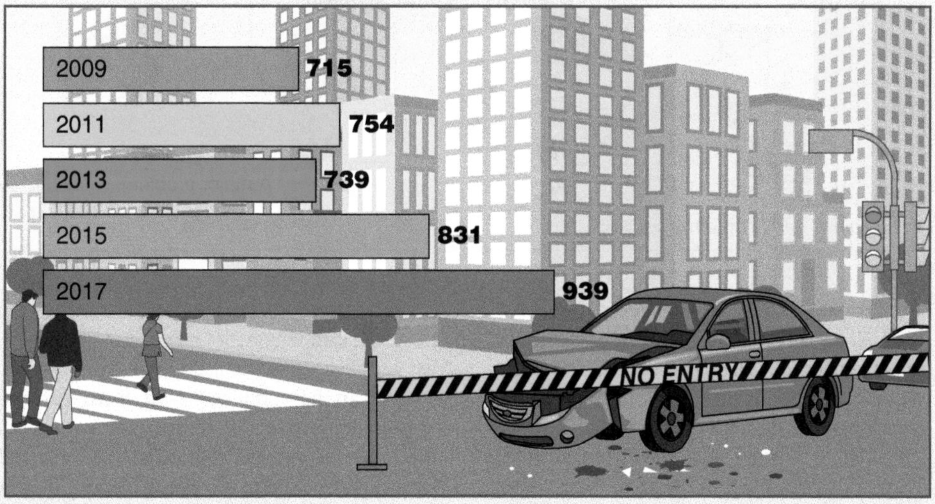

The accompanying chart shows the number of deaths in the United States due to red-light running. As we can observe, the number of such deaths was 715 in 2009 but increased to 939 in 2017. This number has been increasing over the years although the number of such deaths was slightly less in 2013 than in 2011. These numbers simply describe the total number of deaths due to red-light running. We are not drawing any inferences, decisions, or predictions from these data. Hence, this data set and its presentation is an example of descriptive statistics.

Data source: AAA Foundation for Traffic Safety.

Inferential Statistics In statistics, the collection of all elements of interest is called a **population**. The selection of a portion of the elements from this population is called a **sample**. (Population and sample are discussed in more detail in Section 1.6.)

A major portion of statistics deals with making decisions, inferences, predictions, and forecasts about populations based on results obtained from samples. For example, we may make some decisions about the political views of all college and university students based on the political views of 1000 students randomly selected from a few colleges and universities. As another example, we may want to find the starting salary of a typical college graduate. To do so, we may randomly select 2000 recent college graduates, find their starting salaries, and make a decision based on this information. The area of statistics that deals with such decision-making procedures is referred to as **inferential statistics**. This branch of statistics is also called *inductive reasoning* or *inductive statistics*.

Inferential Statistics

Inferential statistics consists of methods that use sample results to help make decisions or predictions about a population.

Case Study 1.2 presents an example of inferential statistics. It shows the results of a survey of U.S. teens ages 13 to 17 conducted by the Pew Research Center in which these teens

Case Study 1.2

Is Anxiety and Depression a Major Problem Among Teens?

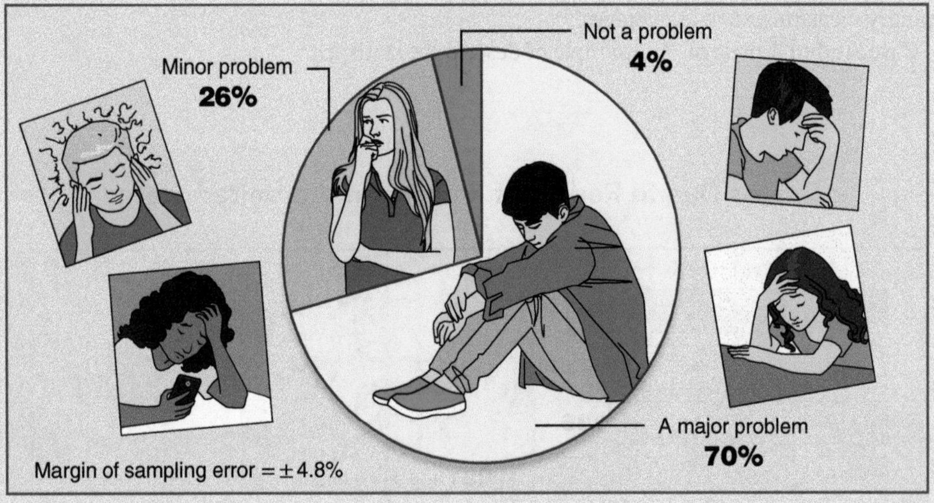

The Pew Research Center conducted a survey of 920 U.S. teens ages 13 to 17 during September 17 to November 25, 2018. These teens were interviewed online or by telephone. They were asked questions on a number of issues related to teen life such as anxiety and depression, bullying, drug addiction, alcohol consumption, poverty, teen pregnancy, gangs, grades, plans to attend college, money, career plans, and so on. On the subject of anxiety and depression among teens, 70% said it is a major problem among teens in the community where they live, 26% said it is a minor problem, and 4% said it is not a problem. As mentioned in the chart, the margin of sampling error was ±4.8%. In Chapter 8, we will discuss the concept of margin of error, which can be combined with these percentages when making inferences. As we notice, the results described in the chart are obtained from a poll of 920 teens. We will learn in later chapters how to apply these results to the entire population of teens. Such decision making about the population based on sample results is called *inferential statistics*.

Source: www.pewresearch.org

were asked numerous questions on a number of issues related to teen life. One of those questions was about anxiety and depression among teens.

Chapters 8 through 15 and parts of Chapter 7 deal with inferential statistics.

Probability, which gives a measurement of the likelihood that a certain outcome will occur, acts as a link between descriptive and inferential statistics. Probability is used to make statements about the likelihood of the occurrence or nonoccurrence of an event under uncertain conditions. Probability and probability distributions are discussed in Chapters 4 through 6 and parts of Chapter 7.

EXERCISES

Concepts and Procedures

1.1 What are the two different meanings of the word *statistics*?

1.2 What are the two major types of statistics?

Applications

1.3 Which of the following is an example of descriptive statistics and which is an example of inferential statistics? Explain.

a. In a Gallup poll conducted April 1–9, 2019, participants were asked to rate their current financial situation as excellent, good, only fair, or poor. The following table lists the summary of the responses of these participants. The results are based on telephone interviews with a random sample of 1012 adults, ages 18 and over, living in all 50 states and the District of Columbia. Assume that the maximum margin of error is ±4%. (www.gallup.com.)

Rating	Percent of Respondents
Excellent	12%
Good	44%
Only fair	29%
Poor	15%

b. The following table gives the earnings of the world's top seven female professional athletes for the year 2019 (www.forbes.com).

Female Professional Athlete	2019 Earnings (millions of dollars)
Serena Williams	29.2
Naomi Osaka	24.3
Angelique Kerber	11.8
Simona Halep	10.2
Sloane Stephens	9.6
Caroline Wozniacki	7.5
Maria Sharapova	7.0

1.2 Basic Terms

LEARNING OBJECTIVES

After completing this section, you should be able to:

- Describe in context the meaning of an element and a variable.
- Describe in context the meaning of an observation and a data set.

It is very important to understand the meaning of some basic terms that will be used frequently in this text. This section explains the meaning of an element (or member), a variable, an observation, and a data set. An element and a data set were briefly defined in Section 1.1. This section defines these terms formally and illustrates them with the help of an example.

Table 1.1 gives information, obtained from *Forbes* magazine, on the annual costs of attending eight universities in the United States. Each university listed in this table is called an **element** or a **member** of this group. Table 1.1 contains information on the eight elements. Note that elements are also called observational units.

TABLE 1.1 Annual Costs of Attending Eight Universities in the United States

University	Annual Cost (dollars) ← Variable
Harvard University	69,600
Princeton University	66,150
An element or member → Stanford University	69,109 ← An observation or measurement
University of California, Berkeley	65,003
University of Chicago	75,735
University of Pennsylvania	71,715
Yale University	71,290
MIT	67,430

Source: *Forbes*, September 30, 2018.

Element or Member

An *element* or *member* of a sample or population is a specific subject or object (for example, a person, firm, item, state, or country) about which the information is collected.

The annual cost of attending a university in our example is called a variable. The annual cost is a characteristic of these universities on which information is collected.

Variable

A *variable* is a characteristic of interest that assumes different values for different elements. In contrast to a variable, the value of a *constant* is fixed.

A few other examples of variables are household incomes, the number of houses built in a city per month during the past year, the makes of cars owned by people, the gross profits of companies, and the number of insurance policies sold by a salesperson per day during the past month.

In general, a variable assumes different values for different elements, as illustrated by the annual costs for the eight universities in Table 1.1. For some elements in a data set, however, the values of the variable may be the same. For example, if we collect information on incomes of households, these households are expected to have different incomes, although some of them may have the same income.

A variable is often denoted by x, y, or z. For instance, in Table 1.1, the annual costs of attending these universities may be denoted by any one of these letters. Starting with Section 1.8, we will begin to use these letters to denote variables.

Each of the values representing the annual costs of attending the eight universities in Table 1.1 is called an **observation** or **measurement**.

Observation or Measurement

The value of a variable for an element is called an *observation* or *measurement*.

From Table 1.1, the annual cost of attending Stanford University is $69,109 per year. The value $69,109 is an observation or a measurement. Table 1.1 contains eight observations, one for each of the eight universities.

The information given in Table 1.1 on the annual costs of attending eight universities is called the data or a **data set**.

Data Set

A *data set* is a collection of observations on one or more variables.

Other examples of data sets are a list of the prices of 25 recently sold homes, test scores of 15 students, opinions of 100 voters, and ages of all employees of a company.

EXERCISES

Concepts and Procedures

1.4 Explain the meaning of an element, a variable, an observation, and a data set.

Applications

1.5 The following table lists the number of deaths by cause in the United States as reported by the Centers for Disease Control and Prevention on September 10, 2017 (www.cdc.gov).

Cause of Death	Number of Deaths
Heart disease	647,457
Cancer	599,108
Accidents	169,936
Chronic lower respiratory diseases	160,201
Stroke	146,383
Alzheimer's disease	121,404
Diabetes	83,564
Influenza and Pneumonia	55,672
Nephritis	50,633

Briefly describe a member, a variable, a measurement, and a data set with reference to the information in this table.

1.6 For Exercise 1.5, answer the following:

a. What is the variable for this data set?

b. How many observations are in this data set?

c. How many elements does this data set contain?

1.3 Types of Variables

LEARNING OBJECTIVES

After completing this section, you should be able to:

- Classify variables as qualitative or quantitative.
- Classify quantitative variables as discrete or continuous.

In Section 1.2, we learned that a variable is a characteristic under investigation that assumes different values for different elements. Family income, height of a person, gross sales of a company, price of a college textbook, make of the car owned by a family, number of accidents, and status (freshman, sophomore, junior, or senior) of a student enrolled at a university are examples of variables.

A variable may be classified as quantitative or qualitative. These two types of variables are explained next.

1.3.1 Quantitative Variables

Some variables (such as the price of a home) can be measured numerically, whereas others (such as hair color) cannot. The price of a home is an example of a **quantitative variable** while hair color is an example of a **qualitative variable.**

Quantitative Variable

A variable that can be measured numerically is called a *quantitative variable.* The data collected on a quantitative variable are called *quantitative data.*

Income, height, gross sales, price of a home, number of cars owned, and number of accidents are examples of quantitative variables because each of them can be expressed numerically. For instance, the income of a family may be $81,520.75 per year, the gross sales for a company may be $567 million for the past year, and so forth. Such quantitative variables may be classified as either *discrete variables* or *continuous variables.*

Discrete Variables The values that a certain quantitative variable can assume may be countable or noncountable. For example, we can count the number of cars owned by a family, but we cannot count the height of a family member, as it is measured on a continuous scale. A variable that assumes countable values is called a **discrete variable.** Note that there are no possible intermediate values between consecutive values of a discrete variable.

Discrete Variable

A variable whose values are countable is called a *discrete variable.* In other words, a discrete variable can assume only certain values with no intermediate values.

For example, the number of cars sold on any given day at a car dealership is a discrete variable because the number of cars sold must be 0, 1, 2, 3, . . . and we can count it. The number of cars sold cannot be between 0 and 1, or between 1 and 2. Other examples of discrete variables are the number of people visiting a bank on any given day, the number of cars in a parking lot, the number of cattle owned by a farmer, and the number of students in a class.

Continuous Variables Some variables assume values that cannot be counted, and they can assume any numerical value between two numbers. Such variables are called **continuous variables.**

Continuous Variable

A variable that can assume any numerical value over a certain interval or intervals is called a *continuous variable.*

The time taken to complete an examination is an example of a continuous variable because it can assume any value, let us say, between 30 and 60 minutes. The time taken may be 42.6 minutes, 42.67 minutes, or 42.674 minutes. (Theoretically, we can measure time as precisely as we want.) Similarly, the height of a person can be measured to the tenth of an inch or to the hundredth of an inch. Neither time nor height can be counted in a discrete fashion. Other examples of continuous variables are the weights of people, the amount of soda in a 12-ounce can (note that a can does not contain exactly 12 ounces of soda), and the yield of potatoes (in pounds) per acre. Note that any variable that involves money and can assume a large number of values is typically treated as a continuous variable.

1.3.2 Qualitative or Categorical Variables

Variables that cannot be measured numerically but can be classified into different categories are called **qualitative** or **categorical variables**.

> **Qualitative or Categorical Variable**
>
> A variable that cannot assume a numerical value but can be classified into two or more (nonnumeric) categories is called a *qualitative* or *categorical variable*. The data collected on such a variable are called *qualitative data*.

For example, the status of an undergraduate college student is a qualitative variable because a student can fall into any one of four categories: freshman, sophomore, junior, or senior. Other examples of qualitative variables are the gender of a person, the make of a computer, the opinions of people, and the make of a car.

Figure 1.1 summarizes the different types of variables.

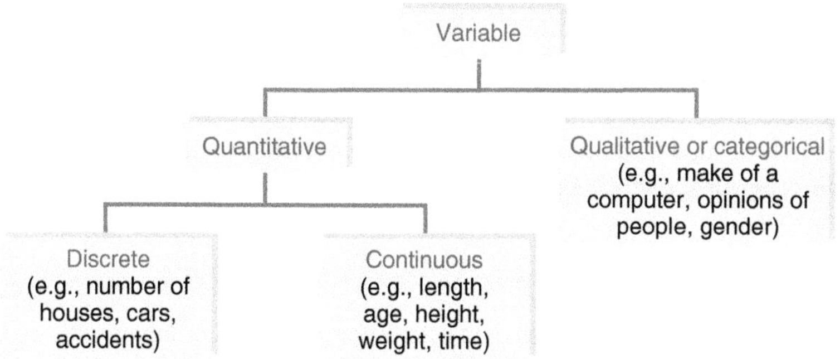

FIGURE 1.1 Types of variables.

EXERCISES

Concepts and Procedures

1.7 Explain the meaning of the following terms:

a. Quantitative variable

b. Qualitative variable

c. Discrete variable

d. Continuous variable

e. Quantitative data

f. Qualitative data

Applications

1.8 Indicate which of the following variables are quantitative and which are qualitative.

a. The number of children under 18 years of age in a family

b. The colors of a car

c. The age of people

d. The amount of time a salesperson requires to commute from home to work

e. The number of errors in a person's credit report

1.9 Classify the following quantitative variables as discrete or continuous.

a. The amount of time a student spent studying for an exam

b. The amount of rain last year in 30 cities

c. The amount of gasoline put into a car at a gas station

d. The number of customers in the line waiting for service at a bank at a given time

1.10 A survey of families living in a certain city was conducted to collect information on the following variables: age of the oldest person in the family, number of family members, number of males in the family, number of females in the family, whether or not they own a house, income of the family, whether or not the family took vacations during the past one year, whether or not they are happy with their financial situation, and the amount of their monthly mortgage or rent.

a. Which of these variables are qualitative variables?

b. Which of these variables are quantitative variables?

c. Which of the quantitative variables of part (b) are discrete variables?

d. Which of the quantitative variables of part (b) are continuous variables?

1.4 Types of Scales

LEARNING OBJECTIVES

After completing this section, you should be able to:

- Understand the concept of measurement scales and their significance in research data analysis.
- Understand the characteristics of different measurement scales which are essential for accurate data collection.

In Section 1.3, we discussed about two types of variables, that is, quantitative and qualitative. This section defines different measurement scales—nominal, ordinal, interval, and ratio—as well as explains how these scales can be implemented on these variables for data analysis, interpretation, and communication. An understanding of these scales is important for making informed decisions in the field of statistics. Each of these scales of measurement possesses some properties, namely, labels, order, differencing, and true zero.

1.4.1 Nominal Scale

A nominal scale deals with categories or labels. These categories or labels are classified to identify the objects or variables that have no numerical value assigned to them. That is why, nominal scale is termed as qualitative in nature. In nominal scale, the aim is to identify the objects. The categories or labels have no inherent numeric values.

Nominal Scale

Nominal scale is used to categorize data into different groups or labels with no inherent order or hierarchy.

An example of a nominal scale measurement is given as follows:

Where is your location?

R – Rural

U – Urban

Here, the data is categorized into two distinct and mutually exclusive categories, that is, rural and urban, where R or U is used to assign categories. Here, the variable location is measured on nominal scale. This scale possesses only one property, that is, label. Some other examples of nominal scales are eye color—black, brown, and green, blood type—A, B, AB, and O, and marital status—single, married, and divorced.

1.4.2 Ordinal Scale

Ordinal scale represents the "order" of categories. It follows a specific order or ranking based on the relative position of the variable with respect to a particular characteristic. In ordinal scale, we cannot quantify the magnitude of the difference between the different ranks. Such type of scales can be used for ranking, Likert scales, etc. For example, consider the class rank of students in a class such as first, second, third, and so on. Here, the students are ordered based on their marks. But their positions do not quantify the magnitude of difference of the marks between the first and second position, second and third position, and so on. This scale possesses the first two properties of the measurement, that is, label and order. Some other examples of ordinal scale are customer satisfaction ratings—very satisfied, satisfied, neutral, dissatisfied, and very dissatisfied and movie ratings—one, two, three, four, or five stars.

Ordinal Scale

Ordinal scale is used to rank the data into a meaningful order but it does not quantify the difference between the values.

1.4.3 Interval Scale

This scale measures the difference between two variables in an exact manner, but there is no true zero point. (A true zero refers to a scale where 0 indicates the absence of something.) Due to no true zero point, ratios cannot be calculated. For example, when temperature is measured in degree Celsius, there are equal intervals as the difference between 20°C and 30°C or between 30°C and 40°C is equal. But it does not have true zero point as 0°C does not signify the complete absence of temperature. It is the most preferred scale in statistics as it helps to quantify the difference between the values. This scale possesses the first three properties of the measurement, that is, label, order, and differencing. Common examples of this scale are temperature scales such as Celsius and Fahrenheit.

Interval Scale

Interval scale is used to measure data with consistent intervals between values but no meaningful ratios are calculated due to the lack of a true zero point.

1.4.4 Ratio Scale

Ratio scale is the most advanced and comprehensive measurement scale. This scale has a unique feature, that is, the data obtained cannot only be categorized and ranked but also have equal intervals. A ratio scale has a true zero point feature, which enables meaningful ratio comparison. This scale does not have negative values. For example, weight measurement scale of objects. It has equal intervals. If the difference of weights between two objects is 10 pounds, this means that one object is 10 pounds heavier than the other. It has a true zero point where zero signifies the absence of any weight. If the weight of one object is 20 pounds and that of another object is 40 pounds, then you can confidently say that the second object is twice as heavy as the first. So, this scale possesses all the four properties of the measurement, that is, label, order, differencing, and true zero point. Some other examples of ratio scales are height, age, and volume.

Ratio Scale

Ratio scale is used to represent data with consistent intervals and a true zero point which allows to find meaningful ratios.

EXERCISES

Concepts and Procedures

1.11 Differentiate between the following scales:

a. Nominal and ordinal scale

b. Interval and ratio scale

Applications

1.12 Indicate which of the following variables follow nominal, ordinal, or ratio scale?

a. Ethnicity of a person

b. Length of an object measured in feet

c. Hair color of a person

d. Level of pain intensity you are experiencing on a scale of 0 to 5

e. Temperature measured in Kelvin unit

1.13 A researcher is conducting a study on the IQ scores of 1000 students studying in a university. The variable "IQ score" is measured on a ratio scale. Is it true? Justify your answer.

1.5 Cross-Section Versus Time-Series Data

LEARNING OBJECTIVE

After completing this section, you should be able to:

- Classify data as cross-section data or time-series data.

Based on the time over which they are collected, data can be classified as either cross-section or time-series data.

1.5.1 Cross-Section Data

Cross-section data contain information on different elements of a population or sample for the *same* period of time. The information on incomes of 100 families for 2020 is an example of cross-section data. All examples of data already presented in this chapter have been cross-section data.

Cross-Section Data

Data collected on different elements at the same point in time or for the same period of time are called *cross-section data.*

Table 1.1, reproduced here as **Table 1.2**, shows the annual costs of attending eight universities in the United States. Because this table presents data on the annual costs of eight universities for the same period, it is an example of cross-section data.

TABLE 1.2 Annual Costs of Attending Eight Universities in the United States

University	Annual Cost (dollars)
Harvard University	69,600
Princeton University	66,150
Stanford University	69,109
University of California, Berkeley	65,003
University of Chicago	75,735
University of Pennsylvania	71,715
Yale University	71,290
MIT	67,430

Source: *Forbes*, September 30, 2018.

1.5.2 Time-Series Data

Time-series data contain information on the same element at *different* points in time. Information on U.S. exports for the years 2001 to 2020 is an example of time-series data.

Time-Series Data

Data collected on the same element for the same variable at different points in time or for different periods of time are called *time-series data*.

The data given in **Table 1.3** are an example of time-series data. This table lists the total population of the United States for five years from 2015 to 2019.

TABLE 1.3 Total Population of the United States

Years	Total Population
2015	320,878,310
2016	323,015,995
2017	325,084,756
2018	327,096,265
2019	329,064,917

Source: www.worldometers.info

EXERCISES

Concepts and Procedures

1.14 Explain the difference between cross-section and time-series data. Give an example of each of these two types of data.

Applications

1.15 Classify the following as cross-section or time-series data.

a. Average prices of houses in 100 cities

b. Salaries of 50 employees

c. Number of cars sold each year by Toyota from 1980 to 2020

d. Number of employees employed by a company each year from 1985 to 2020

e. Number of gyms in 50 cities on December 31, 2020

1.6 Population Versus Sample

LEARNING OBJECTIVES

After completing this section, you should be able to:

- Identify the population and sample in the context of a study.
- Describe the difference between a census and a sample survey.
- Describe the meaning of a representative sample.
- Describe why a sample survey is preferred over a census.
- Identify a sample as a random sample or a nonrandom sample in the context of a study.
- Describe the meaning of sampling with replacement and sampling without replacement.
- Describe different types of samples.
- Describe sampling error and nonsampling errors.
- Provide examples of the four types of nonsampling errors.
- Describe four random sampling techniques and advantages of each technique.

We will encounter the terms *population* and *sample* in every chapter of this text. Consequently, understanding the meaning of each of these two terms and the difference between them is crucial.

Suppose a statistician is interested in knowing the following:

1. The percentage of all voters in a city who will vote for a particular candidate in an election
2. Last year's gross sales of all companies in New York City
3. The prices of all homes in California

In these examples, the statistician is interested in *all* voters in a city, *all* companies in New York City, and *all* homes in California. Each of these groups is called the **population** for the respective example. In statistics, a population does not necessarily mean a collection of people. It can, in fact, be a collection of people or of any kind of item such as houses, books, television sets, or cars. The population of interest is usually called the **target population**.

Population or Target Population

A *population* consists of all elements—individuals, items, or objects—whose characteristics are being studied. The population that is being studied is also called the *target population*.

Most of the time, decisions are made based on portions of populations. For example, the election polls conducted in the United States to estimate the percentages of voters who favor various candidates in any presidential election are based on only a few hundred or a few thousand voters selected from across the country. In this case, the population consists of all registered voters in the United States. The sample is made up of those few hundred or few thousand voters who are included in the poll. Thus, the collection of a number of elements selected from a population is called a **sample**. **Figure 1.2** illustrates the selection of a sample from a population.

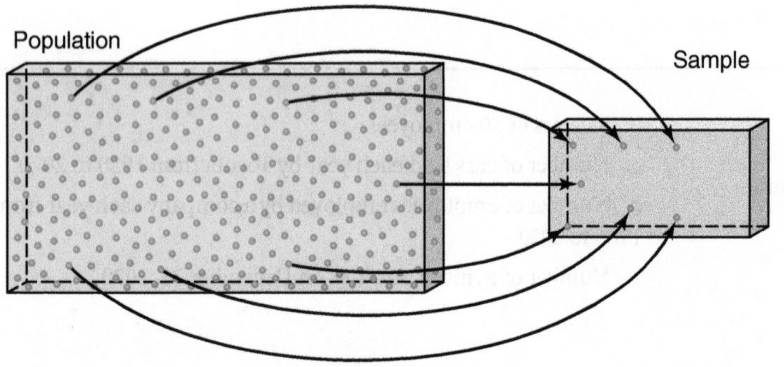

FIGURE 1.2 Population and sample.

Sample

A portion of the population selected for study is referred to as a *sample*.

The collection of information from the elements of a sample is called a **sample survey**. When we collect information on all elements of the target population, it is called a **census**. Often the target population is very large. Hence, in practice, a census is rarely taken because it is expensive and time-consuming. In many cases, it is even impossible to identify each element of the target population. Usually, to conduct a survey, we select a sample and collect the required information from the elements included in that sample. We then make decisions based on this sample information. As an example, collecting information on the 2020 incomes from all families living in Connecticut is referred to as a census. On the other hand, collecting information on the 2020 incomes of 500 families selected from Connecticut is called a sample survey.

Census and Sample Survey

A survey that includes every element of the population is called a *census*. A survey that includes only a portion of the population is called a *sample survey*.

The purpose of conducting a sample survey is to make decisions about the corresponding population. It is important that the results obtained from a sample survey closely match the results that we would obtain by conducting a census. Otherwise, any decision based on a sample survey will not apply to the corresponding population. As an example, to find the average income of families living in Connecticut by conducting a sample survey of 500 families, the sample must contain families who belong to different income groups in almost the same proportion as they exist in the population. Such a sample is called a **representative sample**. Inferences derived from a representative sample will be more reliable.

Representative Sample

A sample that represents the characteristics of the population as closely as possible is called a *representative sample*.

A sample may be selected with or without replacement. In sampling **with replacement**, each time we select an element from the population, we put it back in the population before we select the next element. Thus, in sampling with replacement, the population contains the same number of items each time a selection is made. As a result, we may select the same item more than once in such a sample. Consider a box that contains 25 marbles of different colors. Suppose we draw a marble, record its color, and put it back in the box before drawing the next marble. Every time we draw a marble from this box, the box contains 25 marbles. This is an example of sampling with replacement. The experiment of rolling a die many times is another example of sampling with replacement because every roll has the same six possible outcomes.

Sampling **without replacement** occurs when the selected element is not replaced in the population. In this case, each time we select an item, the size of the population is reduced by one element. Thus, we cannot select the same item more than once in this type of sampling. Most of the time, samples taken in statistics are without replacement. Consider an opinion poll based on 100 voters selected from the population of all eligible voters. In this case, the same voter is not selected more than once. Therefore, this is an example of sampling without replacement.

1.6.1 Why Sample?

Most of the time, surveys are conducted by using samples and not a census of the population. Three of the main reasons for conducting a sample survey instead of a census are listed next.

Time In most cases, the size of the population is quite large. Consequently, conducting a census takes a long time, whereas a sample survey can be conducted quickly. It is time-consuming to interview or contact hundreds of thousands or, sometimes, millions of members of a population. On the other hand, a survey of a sample of a few hundred elements may be completed in a lot less time. In fact, because of the large amount of time needed to conduct a census, by the time the census is completed, the results may be obsolete.

Cost The cost of collecting information from all members of a population may easily fall outside the limited budget of most, if not all, surveys. Consequently, to stay within the available resources, conducting a sample survey may be the best approach.

Impossibility of Conducting a Census Sometimes it is impossible to conduct a census. First, it may not be possible to identify and access each member of the population. For example, if a researcher wants to conduct a survey about homeless people, it is not possible to locate each member of the population and include him or her in the survey. Second, sometimes conducting a survey means destroying the items included in the survey. For example, to estimate the mean life of lightbulbs would necessitate burning out all the bulbs included in the survey. The same is true about finding the average life of batteries. In such cases, only a portion of the population can be selected for the survey.

1.6.2 Random and Nonrandom Samples

Depending on how a sample is drawn, it may be a **random sample** or a **nonrandom sample.**

> **Random and Nonrandom Samples**
>
> A *random sample* is a sample drawn in such a way that each member of the population has some chance of being selected in the sample. In a *nonrandom sample*, some members of the population may not have any chance of being selected in the sample.

Suppose we have a list of 100 students and we want to select 10 of them. If we write the names of these students on 100 pieces of paper, put them in a hat, mix them, and then draw 10 names, the result will be a random sample of 10 students. However, if we arrange the names of these 100 students alphabetically and pick the first 10 names, the result will be a nonrandom sample because the students who are not among the first 10 have no chance of being selected in the sample.

A random sample is usually a representative sample. Note that for a random sample, each member of the population may or may not have the same chance of being included in the sample.

Two types of nonrandom samples are a *convenience sample* and a *judgment sample*. In a **convenience sample**, the most accessible members of the population are selected to obtain the results quickly. For example, an opinion poll may be conducted in a few hours by collecting information from certain shoppers at a single shopping mall. In a **judgment sample**, the members are selected from the population based on the judgment and prior knowledge of an expert. Although such a sample may happen to be a representative sample, the chances of it being so are small. If the population is large, it is not an easy task to select a representative sample based on judgment.

The so-called **pseudo polls** are examples of nonrepresentative samples. For instance, a survey conducted by a magazine that includes only its own readers does not usually involve a representative sample. Similarly, a poll conducted by a television station giving two separate telephone numbers for *yes* and *no* votes is not based on a representative sample. In these two examples, respondents will be only those people who read that magazine or watch that television station, who do not mind paying the postage or telephone charges, or who feel compelled to respond.

Another kind of sample is the **quota sample.** To select such a sample, we divide the target population into different subpopulations based on certain characteristics. Then we select a subsample from each subpopulation in such a way that each subpopulation is represented in the sample in exactly the same proportion as in the target population. As an example of a quota sample, suppose we want to select a sample of 1000 persons from a city whose population has 48% men and 52% women. To select a quota sample, we choose 480 men from the male population and 520 women from the female population. The sample selected in this way will contain exactly 48% men and 52% women. Another way to select a quota sample is to select from the population one person at a time until we have exactly 480 men and 520 women.

Until the 1948 presidential election in the United States, quota sampling was the most commonly used sampling procedure to conduct opinion polls. The voters included in the samples were selected in such a way that they represented the population proportions of voters based on age, sex, education, income, race, and so on. However, this procedure was abandoned after the 1948 presidential election, in which the underdog, Harry Truman, defeated Thomas E. Dewey, who was heavily favored based on several opinion polls. First, the quota samples failed to be representative because the interviewers were allowed to fill their quotas by choosing voters based on their own judgments. This caused the selection of more upper-income and highly educated people, who happened to be Republicans. Thus, the quota samples were not representative of the population because Republicans were overrepresented in these samples. Second, the results of the opinion polls based on quota sampling happened to be false because a large number of factors differentiate voters, but the pollsters considered only a few of those factors. A quota sample based on a few factors will skew the results. A random sample (one that is not based on quotas) has a much better chance of being representative of the population of all voters than a quota sample based on a few factors.

1.6.3 Sampling and Nonsampling Errors

The results obtained from a sample survey may contain two types of errors: sampling and nonsampling errors. The sampling error is also called the chance error, and nonsampling errors are also called the systematic errors.

Sampling or Chance Error Usually, all samples selected from the same population will give different results because they contain different elements of the population. Moreover, the results obtained from any one sample will not be exactly the same as the ones obtained from a census. The difference between a sample result and the result we would have obtained by conducting a census is called the **sampling error**, assuming that the sample is random and no nonsampling error has been made.

Sampling Error

The *sampling error* is the difference between the result obtained from a sample survey and the result that would have been obtained if the whole population had been included in the survey.

The sampling error occurs because of chance, and cannot be avoided. Hence, it is also called **chance error**. A sampling error can occur only in a sample survey. It does not occur in a census. Sampling error is discussed in detail in Chapter 7, and an example of it is given there.

Nonsampling Errors or Biases **Nonsampling errors** or **biases** can occur both in a sample survey and in a census. Such errors occur because of human mistakes and not chance. Nonsampling errors are also called **systematic errors** or **biases**.

Nonsampling Errors or Biases

The errors that occur in the collection, recording, and tabulation of data are called nonsampling errors or biases.

Nonsampling errors can be minimized if questions are prepared carefully and data are handled cautiously. Many types of systematic errors or biases can occur in a survey, including selection error, nonresponse error, response error, and voluntary response error. **Figure 1.3** shows the types of errors.

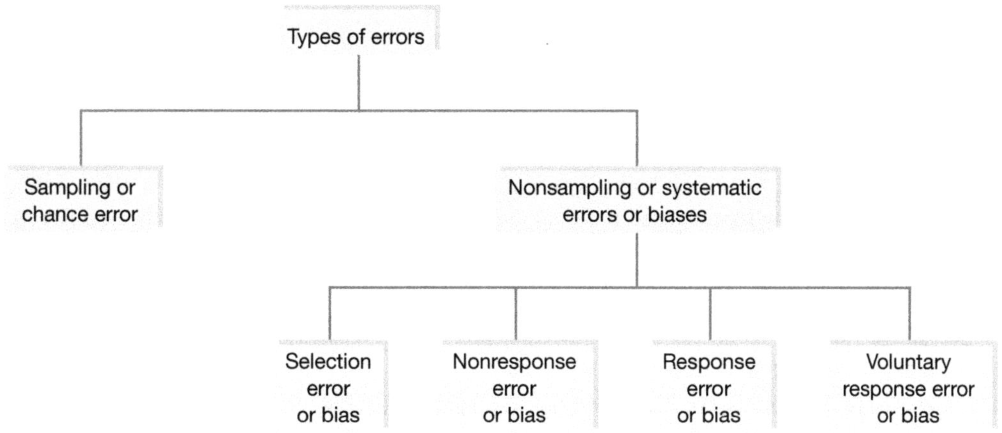

FIGURE 1.3 Types of errors.

(i) Selection Error or Bias When we select a sample, we use a list of elements from which we draw the sample, and this list may not include many members of the target population. Most of the time it is not feasible to include every member of the target population in such a list. This list of members of the population that is used to select a sample is called the **sampling frame**. For example, if we use a telephone directory to select a sample, the list of names that appears in this directory makes the sampling frame. In this case we will miss the people who are not listed in the telephone directory. The people we miss, for example, will be poor people (including homeless people) who do not have telephones and people who do not want to be listed in the directory. These days a large number of people have only cell phones and no landlines. Hence, they are not listed in phone directories. Thus, the sampling frame that is used to select a sample may not be representative of the population. This may cause the sample results to be different from the population results. The error that occurs because the sampling frame is not representative of the population is called the **selection error** or **bias**.

Selection Error or Bias

The list of members of the target population that is used to select a sample is called the sampling frame. The error that occurs because the sampling frame is not representative of the population is called the *selection error* or *bias*.

If a sample is nonrandom (and, hence, nonrepresentative), the sample results may be quite different from the census results.

(ii) Nonresponse Error or Bias Even if our sampling frame and, consequently, the sample are representative of the population, **nonresponse error** may occur because many of the people included in the sample may not respond to the survey.

Nonresponse Error or Bias

The error that occurs because many of the people included in the sample do not respond to a survey is called the *nonresponse error* or *bias*.

This type of error occurs especially when a survey is conducted by mail. A lot of people do not return questionnaires. It has been observed that families with low and high incomes do not respond to surveys by mail. Consequently, such surveys overrepresent middle-income families. This kind of error occurs in other types of surveys, too. For instance, in a face-to-face survey where the interviewer interviews people in their homes, many people may not be home when the interviewer visits their homes. The people who are home at the time the interviewer visits and the ones who are not home at that time may differ in many respects, causing a bias in the survey results. This kind of error may also occur in a telephone survey. Many people may not be home when the interviewer calls. This may distort the results. To avoid nonresponse error, every effort should be made to contact all people included in the sample.

(iii) Response Error or Bias The **response error** or **bias** occurs when the answer given by a person included in the survey is not correct. This may happen for many reasons. One reason is that the respondent may not have understood the question. Thus, the wording of the question may have caused the respondent to answer incorrectly. It has been observed that when the same question is worded differently, many people do not respond the same way. Usually such an error on the part of respondents is not intentional.

Response Error or Bias

The *response error* or *bias* occurs when people included in the survey do not provide correct answers.

Sometimes the respondents do not want to give correct information when answering a question. For example, many respondents will not disclose their true incomes on questionnaires or in interviews. When information on income is provided, it is almost always biased in the upward direction.

Sometimes the race of the interviewer may affect the answers of respondents. This is especially true if the questions asked are about race relations. The answers given by respondents may differ depending on the race of the interviewer.

(iv) Voluntary Response Error or Bias Another source of systematic error is a survey based on a voluntary response sample.

> **Voluntary Response Error or Bias**
>
> *Voluntary response error* or *bias* occurs when a survey is not conducted on a randomly selected sample but on a questionnaire published in a magazine or newspaper and people are invited to respond to that questionnaire.

The polls conducted based on samples of readers of magazines and newspapers suffer from **voluntary response error** or **bias**. Usually only those readers who have very strong opinions about the issues involved respond to such surveys. Surveys in which the respondents are required to call some telephone numbers also suffer from this type of error. Here, to participate, many times a respondent has to pay for the call, and many people do not want to bear this cost. Consequently, the sample is usually neither random nor representative of the target population because participation is voluntary.

1.6.4 Random Sampling Techniques

There are many ways to select a random sample. Five of these techniques are discussed next.

Simple Random Sampling From a given population, we can select a large number of samples of the same size. If each of these samples has the same probability of being selected, then it is called **simple random sampling**.

> **Simple Random Sampling**
>
> In this sampling technique, each sample of the same size has the same probability of being selected from a population. Such a sample is called a *simple random sample*.

One way to select a simple random sample is by a lottery or drawing. For example, if we need to select 5 students from a class of 50, we write each of the 50 names on a separate piece of paper. Then, we place all 50 names in a hat and mix them thoroughly. Next, we draw one name randomly from the hat. We repeat this procedure four more times. The five drawn names make up a simple random sample.

The second procedure to select a simple random sample is to use a table of random numbers, which has become an outdated procedure. In this age of technology, it is much easier to use a graphing calculation or a statistical package, such as Minitab, to select a simple random sample.

Systematic Random Sampling The simple random sampling procedure becomes very tedious if the size of the population is large. For example, if we need to select 150 households from a list of 45,000, it is very time-consuming either to write the 45,000 names on pieces of paper and then select 150 households or to use a table of random numbers. In such cases, it is more convenient to use **systematic random sampling**.

The procedure to select a systematic random sample is as follows: In the example just mentioned, we would arrange all 45,000 households alphabetically (or based on some other characteristic). Since the sample size should equal 150, the ratio of population to sample size is 45,000/150 = 300. Using this ratio, we randomly select one household from the first 300 households in the arranged list using either method. Suppose by using either of the methods, we select the 210th household. We then select the 210th household from every 300 households in the list. In other words, our sample includes the households with numbers 210, 510, 810, 1110, 1410, 1710, and so on.

Systematic Random Sample

In *systematic random sampling,* we first randomly select one member from the first k members of the list of elements arranged based on a given characteristic where k is the number obtained by dividing the population size by the intended sample size. Then every kth member, starting with the first selected member, is included in the sample.

Stratified Random Sampling Suppose we need to select a sample from the population of a city, and we want households with different income levels to be proportionately represented in the sample. In this case, instead of selecting a simple random sample or a systematic random sample, we may prefer to apply a different technique. First, we divide the population into different groups based on income levels. For example, we may form three groups of low-, medium-, and high-income households. We will now have three *subpopulations,* which are usually called **strata.** We then select one sample from each subpopulation or stratum. The collection of all three samples selected from the three strata gives the required sample, called the **stratified random sample**. Usually, the sizes of the samples selected from different strata are proportionate to the sizes of these strata. Note that the elements of each stratum are similar with regard to the possession of the characteristic that defines the strata.

Stratified Random Sample

In a *stratified random sample,* we first divide the population into subpopulations, which are called strata. Then, one sample is selected from each of these strata. The collection of all samples from all strata gives the stratified random sample.

Thus, whenever we observe that a population differs greatly with regard to some characteristic, we may prefer to divide it into different strata and then select one sample from each stratum. We can divide the population on the basis of any characteristic, such as income, expenditure, gender, education, race, employment, or family size.

Cluster Sampling Cluster sampling is a sampling technique which is used when we have relatively homogeneous grouping in a statistical population. In this technique, the whole population is divided into nonoverlapping clusters, and a random sample is taken to select the group. Required information is then collected from each element of each of the selected groups called a cluster. One advantage of using cluster sampling is that it reduces the total number of inspected units and cost to achieve the desired result. The population within each cluster should be as heterogeneous as possible, but there should be homogeneity between the clusters. Cluster sampling is used for a large geographical area. For example, suppose we want to conduct a survey to know about mobile addiction among people living in New York City. Suppose New York City has 800,000 residents. The whole city is divided into 60 nonoverlapping clusters based on neighborhoods, and 6 clusters are selected randomly out of these 60 clusters. Thereafter, we conduct a survey by going door-to-door or approaching people in public places. All individuals in the chosen clusters are surveyed about mobile addiction.

Cluster Sampling

In *cluster sampling,* the whole population is first divided into (geographical) groups called clusters where each cluster is representative of the population. Then a random sample of clusters is selected and each unit of the selected clusters is surveyed.

Two-Stage Sampling Sometimes the target population is scattered over a wide geographical area. Consequently, if a simple random sample is selected, it may be costly to contact each member of the sample. In such a case, we divide the population into different geographical groups or clusters and, as a first step, we select a random sample of certain clusters from all clusters. Then a random sample of certain elements is taken from each selected cluster. For example, suppose we want to conduct a survey of households in the state of New York City. First, the whole state of New York City is divided into, say, 40 regions, which are called clusters or primary units which are similar and, hence, representative of the population. In stage 1, we select at random, say, 5 clusters from 40. Next, in stage 2, we randomly select certain households from each of these five clusters and conduct a survey of these selected households. This is called two-stage sampling.

Two-Stage Sampling

In *two-stage sampling*, the whole population is first divided into (geographical) groups called clusters where each cluster is representative of the population. In stage 1, a random sample of clusters is selected. In stage 2, a random sample of elements from each of the selected clusters is selected.

EXERCISES

Concepts and Procedures

1.16 Briefly explain the terms *population, sample, representative sample, sampling with replacement,* and *sampling without replacement.*

1.17 Differentiate between sampling with replacement and without replacement. Give an example of each of these two types of sampling.

1.18 What are the merits of conducting a sample survey compared to conducting a census?

1.19 Explain the following:

- **a.** Random sample
- **b.** Nonrandom sample
- **c.** Convenience sample
- **d.** Judgment sample
- **e.** Quota sample

1.20 Explain the following four sampling techniques:

- **a.** Simple random sampling
- **b.** Systematic random sampling
- **c.** Stratified random sampling
- **d.** Cluster sampling

1.21 In which sampling technique do all samples of the same size selected from a population have the same chance of being selected?

Applications

1.22 Explain whether each of the following constitutes data collected from a population or a sample.

- **a.** Opinions on a certain issue obtained from all adults living in a city.
- **b.** The price of a gallon of regular unleaded gasoline on a given day at each of 35 gas stations in the Miami Metropolitan Area.
- **c.** Credit card debts of 100 families selected from San Antonio, Texas.
- **d.** The percentage of all U.S. registered voters in each state who voted in the 2016 Presidential election.
- **e.** The number of left-handed students in each of 50 classes selected from a given university.

1.23 A statistics professor wanted to find out the average GPA (grade point average) for all students at her university. She used all students enrolled in her statistics class as a sample and collected information on their GPAs to find the average GPA.

- **a.** Is this sample a random or a nonrandom sample? Explain.
- **b.** What kind of sample is it? In other words, is it a simple random sample, a systematic sample, a stratified sample, a cluster sample, a convenience sample, a judgment sample, or a quota sample? Explain.
- **c.** What kind of systematic error, if any, will be made with this kind of sample? Explain.

1.24 A statistics professor wanted to find the average GPA (grade point average) of all students at her university. The professor obtains a list of all students enrolled at the university from the registrar's office and then selects 250 students at random from this list using a statistical software package such as Minitab.

- **a.** Is this sample a random or a nonrandom sample? Explain.
- **b.** What kind of sample is it? In other words, is it a simple random sample, a systematic sample, a stratified sample, a cluster sample, a convenience sample, a judgment sample, or a quota sample? Explain.
- **c.** Do you think any systematic error will be made in this case? Explain.

1.25 A statistics professor wanted to select 20 students from his class of 300 students to collect detailed information on the profiles of his students. The professor enters the names of all students enrolled in his class on a computer. He then selects a sample of 20 students at random using a statistical software package such as Minitab.

a. Is this sample a random or a nonrandom sample? Explain.

b. What kind of sample is it? In other words, is it a simple random sample, a systematic sample, a stratified sample, a cluster sample, a convenience sample, a judgment sample, or a quota sample? Explain.

c. Do you think any systematic error will be made in this case? Explain.

1.26 A company has 1200 employees, of whom 60% are men and 40% are women. The research department at the company wanted to conduct a quick survey by selecting a sample of 50 employees and asking them about their opinions on an issue. They divided the population of employees into two groups, men and women, and then selected 29 men and 21 women from these respective groups. The interviewers were free to choose any 29 men and 21 women they wanted. What kind of sample is it? Explain.

1.27 A magazine published a questionnaire for its readers to fill out and mail to the magazine's office. In the questionnaire, cell phone owners were asked how much they would have to be paid to do without their cell phones for one month. The magazine received responses from 4568 cell phone owners.

a. What type of sample is this? Explain.

b. To what kind(s) of systematic error, if any, would this survey be subject?

1.28 A researcher wanted to conduct a survey of major companies to find out what benefits are offered to their employees. He mailed questionnaires to 1800 companies and received questionnaires back from 396 companies. What kind of systematic error does this survey suffer from? Explain.

1.29 In a survey, based on a random sample, conducted by an opinion poll agency, the interviewers called the parents included in the sample and asked them the following questions:

i. Do you believe in punishing children?

ii. Have you ever punished your children?

iii. If the answer to the second question is yes, how often?

What kind of systematic error, if any, does this survey suffer from? Explain.

1.30 A survey based on a random sample taken from one borough of Los Angeles City showed that 58% of the people living there would prefer to live somewhere other than Los Angeles City if they had the opportunity to do so. Based on this result, can the researcher say that 58% of people living in Los Angeles City would prefer to live somewhere else if they had the opportunity to do so? Explain.

1.7 Design of Experiments

LEARNING OBJECTIVES

After completing this section, you should be able to:

- Explain the difference between a survey and an experiment.
- Describe the characteristics of an observational study and a designed experiment.
- Identify a study as an observational study or a designed experiment.
- Describe the difference between treatment and control groups.
- Understand the meaning of placebo effect.
- Describe the meaning of a double-blind experiment and identify whether or not an experiment is double-blind.

To use statistical methods to make decisions, we need access to data. Consider the following examples about decision making:

1. A government agency wants to find the average income of households in the United States.
2. A company wants to find the percentage of defective items produced by a machine.
3. A researcher wants to know if there is an association between eating unhealthy food and cholesterol level.
4. A pharmaceutical company has developed a new medicine for a disease and wants to determine if this medicine cures the disease.

All of these cases relate to decision making. We cannot reach a conclusion in these examples unless we have access to data. Data can be obtained from observational studies, experiments, or surveys. This section is devoted mainly to controlled experiments. However, it also explains observational studies and how they differ from surveys.

Suppose two diets, Diet 1 and Diet 2, are being promoted by two different companies, and each of these companies claims that its diet is successful in reducing weight. A research nutritionist wants to compare these diets with regard to their effectiveness for losing weight. Following are two alternatives for the researcher to conduct this research:

1. The researcher contacts the persons who are using these diets and collects information on their weight loss. The researcher may contact as many persons as she has the time and financial resources for. Based on this information, the researcher makes a decision about the comparative effectiveness of these diets.
2. The researcher selects a sample of persons who want to lose weight, divides them randomly into two groups, and assigns each group to one of the two diets. Then she compares these two groups with regard to the effectiveness of these diets.

The first alternative is an example of an **observational study**, and the second is an example of a **controlled experiment**. In an experiment, we exercise control over some factors when we design the study, but in an observational study we do not impose any kind of control.

Treatment

A condition (or a set of conditions) that is imposed on a group of elements by the experimenter is called a *treatment*.

In an observational study the investigator does not impose a condition or **treatment** on subjects or elements included in the study. For instance, in the first alternative mentioned earlier, the researcher simply collects information from the persons who are currently using these diets. In this case, the persons who are included in the study were not assigned to the two diets at random; instead, they chose the diets voluntarily. In this situation the researcher's conclusion about the comparative effectiveness of the two diets may not be valid because the effects of the diets will be **confounded** with many other factors or variables. When the effects of one factor cannot be separated from the effects of some other factors, the effects are said to be *confounded*. The persons who chose Diet 1 may be completely different with regard to age, gender, and eating and exercise habits from the persons who chose Diet 2. Thus, the weight loss may not be due entirely to the diet but to other factors or variables as well. Persons in one group may aggressively manage both diet and exercise, for example, whereas persons in the second group may depend entirely on diet. Thus, the effects of these other variables will get mixed up (confounded) with the effect of the diets.

Under the second alternative, the researcher selects a group of, say, 100 people, and randomly assigns each person to one of the two diets. One way to make random assignments is to write the name of each of these persons on a piece of paper, put them in a hat, mix them, and then randomly draw 50 names from this hat. These 50 persons will be assigned to one of the two diets, say Diet 1. The remaining 50 persons will be assigned to the second diet, Diet 2. This procedure is called **randomization**. Note that random assignments can also be made by using other methods such as a table of random numbers or technology.

Randomization

The procedure in which elements are assigned to different groups at random is called *randomization.*

When people are assigned to one or the other of two diets at random, the other differences among people in the two groups almost disappear. In this case these groups will not differ very much with regard to such factors as age, gender, and eating and exercise habits. The two groups will be very similar to each other. By using the random process to assign people to one or the other of two diets, we have *controlled* for the other factors that can affect the weights of people. Consequently, this is an example of a **designed experiment**.

As mentioned earlier, a condition (or a set of conditions) that is imposed on a group of elements by the experimenter is called a treatment. In the example on diets, each of the two diet types is called a treatment. The experimenter randomly assigns the elements to these two treatments. Again, in such cases the study is called a designed experiment.

Designed Experiment and Observational Study

When the experimenter controls the (random) assignment of elements to different treatment groups, the study is said to be a *designed experiment.* In contrast, in an *observational study* the assignment of elements to different treatments is voluntary, and the experimenter simply observes the results of the study.

The group of people who receive a treatment is called the **treatment group**, and the group of people who do not receive a treatment is called the **control group**. In our example on diets, both groups are treatment groups because each group is assigned to one of the two types of diet. This study does not contain a control group.

Treatment and Control Groups

The group of elements that receives a treatment is called the *treatment group*, and the group of elements that does not receive a treatment is called the *control group*.

EXAMPLE 1.1 Testing a Medicine

An Example of an Observational Study

Suppose a pharmaceutical company has developed a new medicine to cure a disease. To see whether or not this medicine is effective in curing this disease, it will have to be tested on a group of people. Suppose there are 100 persons who have this disease; 50 of them voluntarily decide to take this medicine, and the remaining 50 decide not to take it. The researcher then compares the cure rates for the two groups of patients. Is this an example of a designed experiment or an observational study?

Solution This is an example of an observational study because 50 patients voluntarily joined the treatment group; they were not randomly selected. In this case, the results of the study may not be valid because the effects of the medicine will be confounded with other variables. All of the patients who decided to take the medicine may not be similar to the ones who decided not to take it. It is possible that the persons who decided to take the medicine are in the advanced stages of the disease. Consequently, they do not have much to lose by being in the treatment group. The patients in the two groups may also differ with regard to other factors such as age, gender, and so on.

EXAMPLE 1.2 Testing a Medicine

An Example of a Designed Experiment

Reconsider Example 1.1. Now, suppose that out of the 100 people who have this disease, 50 are selected at random. These 50 people make up one group, and the remaining 50 belong to the second group. One of these groups is the treatment group (people in this group are given the new medicine), and the second is the control group (people in this group are not given the new medicine). The researcher then compares the cure rates for the two groups of patients. Is this an example of a designed experiment or an observational study?

Solution In this case, the two groups are expected to be very similar to each other. Note that we do not expect the two groups to be exactly identical. However, when randomization is used, the two groups are expected to be very similar. After these two groups have been formed, one group will be given the actual medicine. This group is called the treatment group. The other group will be administered a **placebo** (a dummy medicine that looks exactly like the actual medicine). This group is called the control group. This is an example of a designed experiment because the patients are assigned to one of two groups—the treatment or the control group—randomly.

Usually in an experiment like the one in Example 1.2, patients do not know which group they belong to. Most of the time the experimenters do not know which group a patient belongs to. This is done to avoid any bias or distortion in the results of the experiment. When neither patients nor experimenters know who is taking the real medicine and who is taking the placebo, it is called a **double-blind experiment**. For the results of the study to be unbiased and valid, an experiment must be a double-blind designed experiment. Note that if either experimenters or patients or both have access to information regarding which patients belong to treatment or control groups, it will no longer be a double-blind experiment.

The use of a placebo in medical experiments is very important. A placebo is just a dummy pill that looks exactly like the real medicine. Often, patients respond to any kind of medicine. Many studies have shown that even when the patients were given sugar pills (and did not know it), many of them indicated a decrease in pain. Patients respond to a placebo because they have confidence in their physicians and medicines. This is called the **placebo effect**.

Note that there can be more than two groups of elements in an experiment. For example, an investigator may need to compare three diets for people with regard to weight loss. Here, in a designed experiment, the people will be randomly assigned to one of the three diets, which are the three treatments.

In some instances we have to base our research on observational studies because it is not feasible to conduct a designed experiment. For example, suppose a researcher wants to compare the starting salaries of business and psychology majors. The researcher will have to depend on an observational study. She will select two samples, one of recent business majors and another of recent psychology majors. Based on the starting salaries of these two groups, the researcher will make a decision. Note that, here, the effects of the majors on the starting salaries of the two groups of graduates will be confounded with other variables. One of these other factors is that the business and psychology majors may be different in regard to intelligence level, which may affect their salaries. However, the researcher cannot conduct a designed experiment in this case. She cannot select a group of persons randomly and ask them to major in business and select another group and ask them to major in psychology. Instead, persons voluntarily choose their majors.

In a survey we do not exercise any control over the factors when we collect information. This characteristic of a survey makes it very close to an observational study. However, a survey may be based on a probability sample, which differentiates it from an observational study.

If an observational study or a survey indicates that two variables are related, it does not mean that there is a cause-and-effect relationship between them. For example, if an economist takes a sample of families, collects data on the incomes and rents paid by these families, and establishes an association between these two variables, it does not necessarily mean that families with higher incomes pay higher rents. Here the effects of many variables on rents are confounded. A family may pay a higher rent not because of higher income but because of various other factors, such as family size, preferences, or place of residence. We cannot make a statement about the cause-and-effect relationship between incomes and rents paid by families unless we control for these other variables. The association between incomes and rents paid by families may fit any of the following scenarios:

1. These two variables have a cause-and-effect relationship. Families that have higher incomes do pay higher rents. A change in incomes of families causes a change in rents paid.
2. The incomes and rents paid by families do not have a cause-and-effect relationship. Both of these variables have a cause-and-effect relationship with a third variable. Whenever that third variable changes, these two variables change.
3. The effect of income on rent is confounded with other variables, and this indicates that income affects rent paid by families.

If our purpose in a study is to establish a cause-and-effect relationship between two variables, we must control for the effects of other variables. In other words, we must conduct a designed study.

EXERCISES

Concepts and Procedures

1.31 Explain the difference between a survey and an experiment.

1.32 Explain the difference between an observational study and an experiment.

1.33 In March 2005, the *New England Journal of Medicine* published the results of a 10-year clinical trial of low-dose aspirin therapy for the cardiovascular health of women (*Time*, March 21, 2005). The study was based on 40,000 healthy women, most of whom were in their 40s and 50s when the trial began. Half of these women were administered 100 mg of aspirin every other day, and the others were given a placebo. Assume that the women were assigned randomly to these two groups.

a. Is this an observational study or a designed experiment? Explain.

b. From the information given above, can you determine whether or not this is a double-blind study? Explain. If not, what additional information would you need?

1.34 In March 2005, *The New England Journal of Medicine* published the results of a 10-year clinical trial of low-dose aspirin therapy for the cardiovascular health of women (*Time*, March 21, 2005). The study was based on 40,000 healthy women, most of whom were in their 40s and 50s when the trial began. Half of these women were administered 100 mg of aspirin every other day, and the others were given a placebo. The study looked at the incidences of heart attacks in the two groups of women. Overall the study did not find a statistically significant difference in heart attacks between the two groups of women. However, the study noted that among women who were at least 65 years old when the study began, there was a lower incidence of heart attack for those who took aspirin than for those who took a placebo. Suppose that some medical researchers want to study this phenomenon more closely. They recruit 2000 healthy women aged 65 years and older, and randomly divide them into two groups. One group takes 100 mg of aspirin every other day, and the other group takes a placebo. The women do not know to which group they belong, but the doctors who are conducting the study have access to this information.

a. Is this an observational study or a designed experiment? Explain.

b. Is this a double-blind study? Explain.

1.35 In March 2005, *The New England Journal of Medicine* published the results of a 10-year clinical trial of low-dose aspirin therapy for the cardiovascular health of women (*Time*, March 21, 2005). The study was based on 40,000 healthy women, most of whom were in their 40s and 50s when the trial began. Half of these women were administered 100 mg of aspirin every other day, and the others were given a placebo. The study noted that among women who were at least 65 years old when the study began, there was a lower incidence of heart attack for those who took aspirin than for those who took a placebo. Some medical researchers want to study this phenomenon more closely. They recruit 2000 healthy women aged 65 years and older, and randomly divide them into two groups. One group takes 100 mg of aspirin every other day, and the other group takes a placebo. Neither the women nor the doctors know which group the patients belong to.

a. Is this an observational study or a designed experiment? Explain.

b. Is this study a double-blind study? Explain.

1.36 A federal government think tank wanted to investigate whether a job training program helps the families who are on welfare to get off the welfare program. The researchers at this agency selected 5000 volunteer families who were on welfare and offered the adults in those families free job training. The researchers selected another group of 5000 volunteer families who were on welfare and did not offer them such job training. After three years the two groups were compared in regard to the percentage of families who got off welfare. Is this an observational study or a designed experiment? Explain.

1.37 A federal government think tank wanted to investigate whether a job training program helps the families who are on welfare to get off the welfare program. The researchers at this agency selected 10,000 families at random from the list of all families that were on welfare. Of these 10,000 families, the researchers randomly selected 5000 families and offered them free job training. The remaining 5000 families were not offered such job training. After three years the two groups were compared in regard to the percentage of families who got off welfare. Is this an observational study or a designed experiment? Explain.

1.38 A federal government think tank wanted to investigate whether a job training program helps the families who are on welfare to get off the welfare program. The researchers at this agency selected 5000 volunteer families who were on welfare and offered the adults in those families free job training. The researchers selected another group of 5000 volunteer families who were on welfare and did not offer them such job training. After three years the two groups were compared in regard to the percentage of families who got off welfare. Based on that study, the researchers concluded that the job training program causes (helps) families who are on welfare to get off the welfare program. Do you agree with this conclusion? Explain.

1.39 A federal government think tank wanted to investigate whether a job training program helps the families who are on welfare to get off the welfare program. The researchers at this agency selected 10,000 families at random from the list of all families that were on welfare. Of these 10,000 families, the researchers randomly selected 5000 families and offered the adults in those families free job training. The remaining 5000 families were not offered such job training. After three years the two groups were compared in regard to the percentage of families who got off welfare. Based on that study, the researchers concluded that the job training program causes (helps) families who are on welfare to get off the welfare program. Do you agree with this conclusion? Explain.

1.8 Summation Notation

LEARNING OBJECTIVE

After completing this section, you should be able to:

- Use summation notation to calculate sums.

Sometimes mathematical notation helps express a mathematical relationship concisely. This section describes the **summation notation** that is used to denote the sum of values.

Suppose a sample consists of five books, and the prices of these five books are \$175, \$80, \$165, \$97, and \$88, respectively. The variable, *price of a book*, can be denoted by x. The prices of the five books can be written as follows:

Price of the first book = x_1 = \$175

↑ Subscript of x denotes the number of the book

Similarly,

Price of the second book = x_2 = \$80
Price of the third book = x_3 = \$165
Price of the fourth book = x_4 = \$97
Price of the fifth book = x_5 = \$88

In this notation, x represents the price, and the subscript denotes a particular book.

Now, suppose we want to add the prices of all five books. We obtain

$$x_1 + x_2 + x_3 + x_4 + x_5 = 175 + 80 + 165 + 97 + 88 = \$605$$

The uppercase Greek letter $\sum$ (pronounced *sigma*) is used to denote the sum of all values. Using $\sum$ notation, we can write the foregoing sum as follows:

$$\sum x = x_1 + x_2 + x_3 + x_4 + x_5 = \$605$$

The notation $\sum x$ in this expression represents the sum of all values of x and is read as "sigma x" or "sum of all values of x."

EXAMPLE 1.3 | Annual Salaries of Workers

Using Summation Notation: One Variable

Annual salaries (in thousands of dollars) of four workers are 75, 90, 125, and 61, respectively. Find

a. $\sum x$ **b.** $(\sum x)^2$ **c.** $\sum x^2$

iStockphoto.com/Troels Graugaard

Solution Let x_1, x_2, x_3, and x_4 be the annual salaries (in thousands of dollars) of the first, second, third, and fourth worker, respectively. Then,

$$x_1 = 75, \quad x_2 = 90, \quad x_3 = 125, \quad \text{and} \quad x_4 = 61$$

a. $\sum x = x_1 + x_2 + x_3 + x_4 = 75 + 90 + 125 + 61 = \mathbf{351}$

b. Note that $(\sum x)^2$ is the square of the sum of all x values. Thus,

$$(\sum x)^2 = (351)^2 = \mathbf{123{,}201}$$

c. The expression $\sum x^2$ is the sum of the squares of x values. To calculate $\sum x^2$, we first square each of the x values and then sum these squared values. Thus,

$$\sum x^2 = (75)^2 + (90)^2 + (125)^2 + (61)^2 = 5625 + 8100 + 15{,}625 + 3721 = \mathbf{33{,}071}$$

EXAMPLE 1.4

Using Summation Notation: Two Variables

The following table lists four pairs of m and f values.

m	12	15	20	30
f	5	9	10	16

Compute the following:

a. $\sum m$ **b.** $\sum f^2$ **c.** $\sum mf$ **d.** $\sum m^2 f$

Solution We can write

$$m_1 = 12 \quad m_2 = 15 \quad m_3 = 20 \quad m_4 = 30$$
$$f_1 = 5 \quad f_2 = 9 \quad f_3 = 10 \quad f_4 = 16$$

a. $\sum m = 12 + 15 + 20 + 30 = \mathbf{77}$

b. $\sum f^2 = (5)^2 + (9)^2 + (10)^2 + (16)^2 = 25 + 81 + 100 + 256 = \mathbf{462}$

c. To compute $\sum mf$, we multiply the corresponding values of m and f and then add the products as follows:

$$\begin{aligned}\sum mf &= m_1 f_1 + m_2 f_2 + m_3 f_3 + m_4 f_4 \\ &= 12(5) + 15(9) + 20(10) + 30(16) = \mathbf{875}\end{aligned}$$

d. To calculate $\sum m^2 f$, we square each m value, then multiply the corresponding m^2 and f values, and add the products. Thus,

$$\begin{aligned}\sum m^2 f &= (m_1)^2 f_1 + (m_2)^2 f_2 + (m_3)^2 f_3 + (m_4)^2 f_4 \\ &= (12)^2(5) + (15)^2(9) + (20)^2(10) + (30)^2(16) = \mathbf{21{,}145}\end{aligned}$$

The calculations done in parts (a) through (d) to find the values of $\sum m$, $\sum f^2$, $\sum mf$, and $\sum m^2 f$ can be performed in tabular form, as shown in **Table 1.4**.

TABLE 1.4

m	f	f^2	mf	$m^2 f$
12	5	5 × 5 = 25	12 × 5 = 60	12 × 12 × 5 = 720
15	9	9 × 9 = 81	15 × 9 = 135	15 × 15 × 9 = 2025
20	10	10 × 10 = 100	20 × 10 = 200	20 × 20 × 10 = 4000
30	16	16 × 16 = 256	30 × 16 = 480	30 × 30 × 16 = 14,400
$\sum m = 77$	$\sum f = 40$	$\sum f^2 = 462$	$\sum mf = 875$	$\sum m^2 f = 21{,}145$

The columns of Table 1.4 can be explained as follows:

1. The first column lists the values of m. The sum of these values gives $\sum m = 77$.
2. The second column lists the values of f. The sum of this column gives $\sum f = 40$.
3. The third column lists the squares of the f values. For example, the first value, 25, is the square of 5. The sum of the values in this column gives $\sum f^2 = 462$.
4. The fourth column records the product of the corresponding m and f values. For example, the first value, 60, in this column is obtained by multiplying 12 by 5. The sum of the values in this column gives $\sum mf = 875$.
5. Next, the m values are squared and multiplied by the corresponding f values. The resulting products, denoted by $m^2 f$, are recorded in the fifth column. For example, the first value, 720, is obtained by squaring 12 and multiplying this result by 5. The sum of the values in this column gives $\sum m^2 f = 21{,}145$.

EXERCISES

Concepts and Procedures

1.40 The following table lists six pairs of x and y values.

x	7	11	8	4	14	28
y	5	15	7	10	9	19

Compute the value of each of the following:

a. $\sum y$ **b.** $\sum x^2$ **c.** $\sum xy$ **d.** $\sum x^2 y$ **e.** $\sum y^2$

Applications

1.41 Seven randomly selected customers at a local grocery store spent the following amounts on groceries in a single visit: \$137, \$35, \$92, \$105, \$175, \$21, and \$57, respectively. Let y denote the amount spent by a customer on groceries in a single visit. Find:

a. $\sum y$ **b.** $(\sum y)^2$ **c.** $\sum y^2$

1.42 The number of pizzas delivered to a college campus on six randomly selected nights is 48, 103, 95, 188, 286, and 136, respectively. Let x denote the number of pizzas delivered to this college campus on any given night. Find:

a. $\sum x$ **b.** $(\sum x)^2$ **c.** $\sum x^2$

Uses and Misuses of Statistics

Just a Sample Please

Have you ever read an article or seen an advertisement on television that claims to have found some important result based on a scientific study? Or maybe you have seen some results based on a poll conducted by a research firm. Have you ever wondered how samples are selected for such studies, or just how valid the results really are?

One of the most important considerations in conducting a study is to ensure that a *representative sample* has been taken in such a way that the sample represents the underlying population of interest. So, if you see a claim that a recent study found that a new drug was a *magic bullet* for curing cancer, how might we investigate how valid this claim is?

It is very important to consider how the population was sampled in order to understand both how useful the result is and if the result is really *generalizable*. In research, a result that is generalizable means that it applies to the general population of interest, not just to the sample that was taken or to only a small segment of the population.

One of the most common sampling strategies is the *simple random sample* in which each subject in a population has some chance of being chosen for the study. Though there are numerous ways to take a valid statistical sample, simple random samples ensure that there is equal representation of members of the population of interest, and therefore we see this type of sampling strategy employed quite often. But, how many samples taken for studies are truly random? It is not uncommon in survey research, for example, to send people out to a mall or to a busy street corner with clipboard and survey in hand to ask questions of those subjects willing to take the survey. It is also not uncommon for polling firms to call from a random set of telephone numbers—in fact, many of you may have received a call to participate in a poll. In either of these scenarios, it is likely that those subjects willing to take the time to answer a survey may not represent everyone in the population of interest. Additionally, if a telephone poll is used, it is impossible to sample those without a telephone or those who do not answer the call, and hence this segment of the population may not be available to be included in the sample.

One of the major consequences of having a nonrepresentative sample is the risk of creating *bias* in the results. Bias causes the results to potentially deviate from what they would be under ideal sampling circumstances. In the examples just presented, we have an example of *volunteer bias* in which the results are skewed to represent only those willing to participate. There could also be *observer bias* in which the person conducting a poll only approaches certain people—perhaps those who seem friendly or more likely to participate—in which case a large segment of the potential population is ignored. Results from such studies would certainly be suspect.

The take-home message is that, when reading the results of studies or polls, it is very important to stop and consider what population was actually sampled, how representative that sample really is, and if the results may be biased in such a way that the results are not generalizable to the larger population. One of the great things about taking a statistics course is that you will learn many tools that will allow you to look at the world of data more critically so that you can make informed decisions about the studies that you read or hear about in the media. So the next time you hear, "9 out of 10 people say...," consider if they asked only 10 people or if they asked 10,000 people. Learn to be good consumers of data!

Glossary

Census A survey conducted by including every element of the population.

Cluster A subgroup (usually geographical) of the population that is representative of the population.

Cluster sampling A sampling technique in which the population is divided into clusters and a sample is chosen from one or a few clusters.

Continuous variable A (quantitative) variable that can assume any numerical value over a certain interval or intervals.

Control group The group of elements on which no condition is imposed.

Convenience sample A sample that includes the most accessible members of the population.

Cross-section data Data collected on different elements at the same point in time or for the same period of time.

Data or data set Collection of observations or measurements on a variable.

Descriptive statistics Methods for organizing, displaying, and describing data using tables, graphs, and summary measures.

Designed experiment A study in which the experimenter controls the assignment of elements to different treatment groups.

Discrete variable A (quantitative) variable whose values are countable.

Double-blind experiment An experiment in which neither the doctors (or researchers) nor the patients (or members) know to which group a patient (or member) belongs.

Element or member A specific subject or object included in a sample or population.

Experiment A method of collecting data by controlling some or all factors.

Inferential statistics Methods that help make decisions about a population based on sample results.

Interval scale A scale that measures the difference between two variables in an exact manner but there is no true zero point.

Judgment sample A sample that includes the elements of the population selected based on the judgment and prior knowledge of an expert.

Nominal scale A scale which deals with categories or labels only.

Nonrandom sample A sample drawn in such a way that some members of the population have no chance of being selected in the sample.

Nonresponse error The error that occurs because many of the people included in the sample do not respond.

Nonsampling or systematic errors The errors that occur in the collection, recording, and tabulation of data.

Observation or measurement The value of a variable for an element.

Observational study A study in which the assignment of elements to different treatments is voluntary, and the researcher simply observes the results of the study.

Ordinal scale A scale that follows a specific order or ranking based on the relative positions with respect to a particular characteristic.

Population or target population The collection of all elements whose characteristics are being studied.

Qualitative or categorical data Data generated by a qualitative variable.

Qualitative or categorical variable A variable that cannot assume numerical values but is classified into two or more categories.

Quantitative data Data generated by a quantitative variable.

Quantitative variable A variable that can be measured numerically.

Quota sample A sample selected in such a way that each group or subpopulation is represented in the sample in exactly the same proportion as in the target population.

Random sample A sample drawn in such a way that each element of the population has some chance of being included in the sample.

Randomization The procedure in which elements are assigned to different (treatment and control) groups at random.

Ratio scale A scale used to represent data with consistent intervals and a true zero point which allows to find meaningful ratios.

Representative sample A sample that contains the characteristics of the population as closely as possible.

Response error The error that occurs because people included in the survey do not provide correct answers.

Sample A portion of the population of interest.

Sample survey A survey that includes elements of a sample.

Sampling frame The list of elements of the target population that is used to select a sample.

Sampling or chance error The difference between the result obtained from a sample survey and the result that would be obtained from the census.

Selection error The error that occurs because the sampling frame is not representative of the population.

Simple random sampling If all samples of the same size selected from a population have the same chance of being selected, it is called simple random sampling.

Statistics Science of collecting, analyzing, presenting, and interpreting data and making decisions.

Stratified random sampling A sampling technique in which the population is divided into different strata and a sample is chosen from each stratum.

Stratum A subgroup of the population whose members are similar with regard to the possession of a characteristic.

Survey Collecting data from the elements of a population or sample.

Systematic random sampling A sampling method that is used to choose a sample by selecting every *k*th member from the list of elements.

Target population The collection of all subjects of interest.

Time-series data Data that give the values of the same variable for the same element at different points in time or for different periods of time.

Treatment A condition (or a set of conditions) that is imposed on a group of elements by the experimenter. This group is called the **treatment group**.

Variable A characteristic of interest that assumes different values for different elements.

Voluntary response error The error that occurs when a survey is not conducted on a randomly selected sample, but people are invited to respond voluntarily to the survey.

Exercises

Supplementary Exercises

1.43 The following table lists the number of worldwide locations for the 10 largest fast-food restaurant chains.

Fast-Food Restaurant	Number of Locations
Subway	42,998
McDonald's	37,200
Starbucks	30,000
KFC	20,404
Burger King	16,859
Pizza Hut	16,796
Domino's	15,000
Dunkin'	11,300
Baskin-Robbins	7,500
Hunt Brothers Pizza	7,300

a. Explain the meaning of a member, a variable, a measurement, and a data set with reference to this table.

b. Are the data presented in this table an example of cross-section or of time-series data? Explain.

1.44 According to macrotrends.net, Shake Shack is the fastest growing restaurant chain in the United States. The following table gives Shake Shack's annual revenue (in millions of U.S. dollars) from 2012 to 2018.

Year	Annual Revenue (millions of dollars)
2012	459
2013	359
2014	268
2015	191
2016	119
2017	82
2018	57

Are the data presented in this table an example of cross-section or of time-series data? Explain.

1.45 Indicate whether each of the following constitutes data collected from a population or a sample.

a. Salaries of CEOs of all companies in New York City

b. Four hundred houses selected from a city

c. Gross sales of three pizza chains in 2020

d. Annual incomes of all 28 employees of a restaurant

1.46 State which of the following is an example of sampling with replacement and which is an example of sampling without replacement.

a. Selecting 10 cities to market a new shampoo

b. Selecting a high school teacher to drive students to a lecture in September, then selecting a teacher from the same group to chaperone a dance in November

1.47 The number of restaurants in each of five small towns is 4, 14, 8, 10, and 7, respectively. Let y denote the number of restaurants in a small town. Find:

a. $\sum y$ b. $(\sum y)^2$ c. $\sum y^2$

1.48 The following table lists five pairs of m and f values.

m	3	16	11	9	20
f	7	32	17	12	34

Compute the value of each of the following:

a. $\sum m$ b. $\sum f^2$ c. $\sum mf$ d. $\sum m^2 f$ e. $\sum m^2$

1.49 A professor is teaching a large class that has 247 students. He wants to select a sample of 15 students to conduct a study on the habits of his students. For each of the following sampling methods, explain what kind of sampling method is used (e.g., a random sample, nonrandom sample, sample with replacement, sample without replacement, convenience sample, judgment sample, quota sample). Mention all methods that apply.

a. There are 15 sociology majors in his class. He selects all 15 students to include in the sample.

b. He goes through the roster and finds that he knows 30 of the 247 students. Using his knowledge about these students, he selects 15 students from these 30 to include in the sample.

c. He enters names of all 247 students in a spreadsheet on his computer. He uses a statistical software (such as Minitab) to select 15 students.

1.50 A researcher advertised for volunteers to study the relationship between the amount of meat consumed and cholesterol level. In response to this advertisement, 3476 persons volunteered. The researcher collected information on the meat consumption and cholesterol level from each of these persons. Based on these data, the researcher concluded that there is a very strong positive association between these two variables.

a. Is this an observational study or a designed experiment? Explain.

b. Based on this study, can the researcher conclude that consumption of meat increases cholesterol level? Explain why or why not.

1.51 A pharmaceutical company developed a new medicine for compulsive behavior. To test this medicine on humans, the company advertised for volunteers who were suffering from this disease and wanted to participate in the study. As a result, 1820 persons responded. Using their own judgment, the group of physicians who were conducting this study assigned 910 of these patients to the treatment group and the remaining 910 to the control group. The patients in the treatment group were administered the actual medicine, and the patients in the control group were given a placebo. Six months later the conditions of the patients in the two groups were examined and compared. Based on this comparison, the physicians concluded that this medicine improves the condition of patients suffering from compulsive behavior.

a. Comment on this study and its conclusion.

b. Is this an observational study or a designed experiment? Explain.

c. Is this a double-blind study? Explain.

1.52 A pharmaceutical company has developed a new medicine for compulsive behavior. To test this medicine on humans, the physicians conducting this study obtained a list of all patients suffering from compulsive behavior who were being treated by doctors in all hospitals in the country. Further assume that this list is representative of the population of all such patients. The physicians then randomly selected 1820 patients from this list. Of these 1820, randomly selected 910 patients were assigned to the treatment group, and the remaining 910 patients were assigned to the control group. The patients did not know which group they belonged to, but the doctors had access to such information. Six months later the conditions of the patients in the two groups were examined and compared. Based on this comparison, the physicians concluded that this medicine improves the condition of patients suffering from compulsive behavior.

a. Comment on this study and its conclusion.

b. Is this an observational study or a designed experiment? Explain.

c. Is this a double-blind study? Explain.

1.53 A pharmaceutical company has developed a new medicine for compulsive behavior. To test this medicine on humans, the physicians conducting this study obtained a list of all patients suffering from compulsive behavior who were being treated by doctors in all hospitals in the country. Assume that this list is representative of the population of all such patients. The physicians then randomly selected 1820 patients from this list. Of these 1820 patients, randomly selected 910 patients were assigned to the treatment group, and the remaining 910 patients were assigned to the control group. Neither patients nor doctors knew what group the patients belonged to. Six months later the conditions of the patients in the two groups were examined and compared. Based on this comparison, the physicians concluded that this medicine improves the condition of patients suffering from compulsive behavior.

a. Is this an observational study or a designed experiment? Explain.

b. Is this a double-blind study? Explain.

1.54 A researcher wants to understand the vacation spending habits of the middle income group. He prepared a questionnaire with following questions: age; gender; education level; how often do you go on vacation in a year, i.e., one time, two times, three times etc.; about accommodation preference like hotel, resort etc.; how much money do you budget for accommodation per night?; how likely are you to spend on paid activities during your vacation?, i.e., very likely, likely neutral etc. Identify the measurement scales on which the variables are measured.

Advanced Exercises

1.55 A college mailed a questionnaire to all 5432 of its alumni who graduated in the last five years. One of the questions was about the current annual incomes of these alumni. Only 1620 of these alumni returned the completed questionnaires, and 1240 of them answered that question. The current average annual income of these 1240 respondents was $61,200.

a. Do you think $61,200 is likely to be a valid estimate of the current average annual income of all 5432 alumni? If so, explain why.

b. If you think that $61,200 is probably not a valid estimate of the current average annual income of all 5432 alumni, what sources of systematic errors do you think are present here?

c. Do you expect the estimate of $61,200 to be above or below the current average annual income of all 5432 alumni? Explain.

1.56 A group of veterinarians wants to test a new canine vaccine for Lyme disease. (Lyme disease is transmitted by the bite of an infected deer tick.) One hundred dogs are randomly selected to receive the vaccine (with their owners' permission) from an area that has a high incidence of Lyme disease. These dogs are examined by veterinarians for symptoms of Lyme disease once a month for a period of 12 months. During this 12-month period, 10 of these 100 dogs are diagnosed with Lyme disease. During the same 12-month period, 18% of the unvaccinated dogs in the area are found to have contracted Lyme disease.

a. Does this experiment have a control group?

b. Is this a double-blind experiment?

c. Identify any potential sources of bias in this experiment.

d. Explain how this experiment could have been designed to reduce or eliminate the bias pointed out in part (c).

Self-Review Test

Multiple Choice Questions

1. A population in statistics means a collection of all

a. men and women

b. subjects or objects of interest

c. people living in a country

2. A sample in statistics means a portion of the

a. people selected from the population of a country

b. people selected from the population of an area

c. population of interest

3. Under inferential statistics, we study

a. the methods to make decisions about one or more populations based on sample results

b. the methods for organizing, displaying, and describing data

c. how to describe the frequency distribution

4. An independent group wants to determine if the consumption of gasoline has increased due to changes in price. The group randomly selects 320 gas stations from 12 different states and collects data from the month of the year when gas is the cheapest and from the month of the year when gas is the most expensive. The data shows no significant difference in gas consumption between the two months. In this example, what is the variable being studied?

a. The 320 gas stations chosen

b. The 12 different states

c. The consumption of gasoline

5. A continuous variable is a variable that can assume

a. categorical values only

b. a countable set of values only

c. any value in an interval

6. A qualitative variable is a variable that

a. cannot be graphed

b. can assume an uncountable set of values

c. cannot be measured numerically

7. Cross-section data are collected

a. on the same variable for the same variable at different points in time

b. on different elements at the same point in time

c. for a qualitative variable

8. In which scale can you perform meaningful addition and subtraction operations but not meaningful ratios?

a. Ratio scale

b. Ordinal scale

c. Interval scale

9. In a double-blind experiment, who knows which patients are taking the real medicine and which patients are taking the placebo?

a. Only the experimenters

b. Neither the patients nor the experimenters

c. Both the patients and the experimenters

10. In statistics, a representative sample is a sample that

a. contains the characteristics of the population as closely as possible

b. represents the results of a sample exactly

c. contains all people living in an area

11. The following are all valid reasons why a census cannot be conducted except

a. the size of the population is quite large

b. the population is too diverse

c. the cost of collecting information from all members of the population would exceed the budget

12. A random sample is a sample drawn in such a way that

a. all elements of a population are included

b. some members of the population have no chance of being included in the sample

c. each member of the population has some chance of being included in the sample

13. Which of the following is not an example of a nonrandom or nonrepresentative sample?

a. Convenience sample

b. Quota sample

c. Cluster sample

14. Which of the following is not a name for the difference between a sample result and the result we would have obtained by conducting a census that occurs by human mistakes?

a. Nonsampling error

b. Nonsystematic error

c. Systematic error

15. A survey was conducted by mail and 35% of the questionnaires were not returned. This is what type of nonsampling error?

a. Selection error

b. Nonresponse error

c. Response error

16. Realizing that age may affect a person's response, the population under study was divided into different age groups (21–30, 31–40, 41–50, 51–60, and 61–70). Then, one sample of the same size, is selected from each age group. This is what type of sampling technique?

a. Systematic random sampling

b. Stratified random sampling

c. Cluster sampling

17. Consider the following six pairs of x and y values.

x	8	11	15	5	20	21
y	10	16	20	7	28	21

The value of $\sum xy$ is

a. 1482

b. 1392

c. 1592

Practice Exercises

1. Indicate which of the following is an example of a sample with replacement and which is a sample without replacement.

a. Five friends go to a livery stable and select five horses to ride (each friend must choose a different horse).

b. A box contains five marbles of different colors. A marble is drawn from this box, its color is recorded, and it is put back into the box before the next marble is drawn. This experiment is repeated 12 times.

2. Indicate which of the following variables are quantitative and which are qualitative. Classify the quantitative variables as discrete or continuous.

a. Women's favorite TV programs

b. Salaries of football players

c. Number of pets owned by families

d. Favorite breed of dog for each of 20 persons

3. Explain the following types of samples: random sample, nonrandom sample, representative sample, convenience sample, judgment sample, quota sample, sample with replacement, and sample without replacement.

4. Explain the following concepts: sampling or chance error, nonsampling or systematic error, selection bias, nonresponse bias, response bias, and voluntary response bias.

5. Explain the following sampling techniques: simple random sampling, systematic random sampling, stratified random sampling, and cluster sampling.

6. Explain the following concepts: an observational study, a designed experiment, randomization, treatment group, control group, double-blind experiment, and placebo effect.

7. The following table lists the midterm test scores of six students in a class.

Student	Score
John	83
Destiny	97
Colleen	88
Carl	72
Joseph	91
Victoria	68

Explain the meaning of a member, a variable, a measurement, and a data set with reference to this table.

8. The number of types of cereal in the pantries of six households is 6, 11, 3, 5, 6, and 2, respectively. Let x be the number of types of cereal in the pantry of a household. Find:

a. $\sum x$ b. $(\sum x)^2$ c. $\sum x^2$

9. The following table lists the age (rounded to the nearest year) and price (rounded to the nearest thousand dollars) for each of the seven cars of the same model.

Age (x) (years)	Price (y) (thousand dollars)
3	28.4
7	17.2
5	21.6
9	13.9
12	6.3
8	16.8
10	9.4

Let age be denoted by x and price be denoted by y. Calculate

a. $\sum x$ b. $\sum y$ c. $\sum x^2$ d. $\sum xy$

e. $\sum x^2y$ f. $(\sum y)^2$

10. A psychologist needs 10 pigs for a study of the intelligence of pigs. She goes to a pig farm where there are 40 young pigs in a large pen. Assume that these pigs are representative of the population of all pigs. She selects the first 10 pigs she can catch and uses them for her study.

a. Do these 10 pigs make a random sample?

b. Are these 10 pigs likely to be representative of the entire population? Why or why not?

c. If these 10 pigs do not form a random sample, what type of sample is it?

d. Can you suggest a better procedure for selecting a sample of 10 from the 40 pigs in the pen?

11. A newspaper wants to conduct a poll to estimate the percentage of its readers who favor a gambling casino in their city. People register their opinions by placing a phone call that costs them $1.

a. Is this method likely to produce a random sample?

b. Which, if any, of the types of biases are likely to be present and why?

12. A few years ago, the Centre for Nutrition and Food Research at Queen Margaret University College in Edinburgh studied the relationship between sugar consumption and weight gain. All the people who participated in the study were divided into two groups, and both of these groups were put on low-calorie, low-fat diets. The diet of the people in the first group was low in sugar, but the people in the second group received as much as 10% of their calories from sucrose. Both groups stayed on their respective diets for eight weeks. During these eight weeks, participants in both groups lost between 1/2 and 3/4 pound per week.

a. Was this a designed experiment or an observational study?

b. Was there a control group in this study?

c. Was this a double-blind experiment?

13. A researcher wanted to study if regular exercise can control diabetes. He took a sample of 1200 adults who were under treatment for diabetes. He asked them if they exercise regularly or not and whether their diabetes was under control or not. Based on this information he reached his conclusion that regular exercise can keep diabetes under control. Is this an observational study or a designed experiment?

14. A researcher wanted to know if using regular fertilizer or organic fertilizer will affect the yield of potatoes. She divided a large parcel of land into 10 small parcels and then randomly selected five parcels to grow potatoes using regular fertilizer and the other five parcels using organic fertilizer. Is this an observational study or a randomized experiment?

Mini-Project

Mini-Project 1.1

In this mini-project, you are going to obtain a data set of interest to you that you will use for mini-projects in some of the other chapters. The data set should contain at least one qualitative variable and one quantitative variable, although having two of each will be necessary in some cases. Ask your instructor how many variables you should have. A good-size data set to work with should contain somewhere between 50 and 100 observations.

Here are some examples of the procedures to use to obtain data:

1. Take a random sample of used cars and collect data on them. You may use websites like Cars.com, AutoTrader.com, and so forth. Quantitative variables may include the price, mileage, and age of a car. Categorical variables may include the model, drive train (front wheel, rear wheel, and so forth), and type (compact, SUV, minivan, and so forth). You can concentrate on your favorite type of car or look at a variety of different types of cars.
2. Examine the real estate ads in your local newspaper or online and obtain information on houses for sale that may include listed price, number of bedrooms, lot size, living space, town, type of house, number of garage spaces, and number of bathrooms.
3. Use an almanac or go to a government website, such as www.census.gov, www.bls.gov, or www.cdc.gov, to obtain information for each state. Quantitative variables may include income, birth and death rates, crime rates, and the proportion of people living below the poverty level. Categorical variables may include variables such as the region of the country where each state is located and which party won the state governorship in the last election. You can also collect this information on a worldwide level and use the continent or world region as a categorical variable.
4. Take a random sample of students and ask them questions such as these:
   - How much money did you spend on books last semester?
   - How many credit hours did you take?
   - What is your major?
   - Do you work off campus?
   - If you work off campus, how many hours a week do you work?
   - How many hours a week do you spend on studying?
5. If you are a sports fan, you can use an almanac or sports website to obtain statistics on a random sample of athletes. You can look at sport-specific statistics such as home runs, runs batted in, position, left-handed/right-handed, and so forth for baseball players, or you could collect information to compare different sports by gathering information on salary, career length, weight, and so forth.

Once you have collected the information, write a brief report that includes answers to the following tasks/questions:

a. Describe the variables on which you have collected information.

b. Describe a reasonable target population for the sample you used.

c. Is your sample a random sample from this target population?

d. Do you feel that your sample is representative of this population?

e. Is this an example of sampling with or without replacement?

f. For each quantitative variable, state whether it is a continuous or a discrete variable.

g. Describe the meaning of an element, a variable, and a measurement for this data set.

h. Describe any problems you faced in collecting these data.

i. Were any of the data values unusable? If yes, explain why. When you submit your report to your instructor, include a hard copy or electronic copy of your data. Be sure to save your data for use in mini-projects of future chapters.

Your instructor will probably want to see a copy of the data you collected. If you are using statistical software in the class, enter the data into that software and submit a copy of the data file. If you are using a handheld technology calculator, such as a graphing calculator, you will probably have to print out a hard copy version of the data set. Save this data set for projects in future chapters.

Decide for Yourself

Note: The Decide for Yourself feature is located in the book's product page on *www.wiley.com*.

Chapter 1 Technology Instructions

Note: Complete TI-84, Minitab, and Excel manuals are available in the book's product page on *www.wiley.com*.

TI-84 Color/TI-84

The TI-84 Color Technology Instructions feature of this text is written for the TI-84 Plus C color graphing calculator running the 4.0 operating system. Some screens, menus, and functions will be slightly different in older operating systems. The TI-84 and TI-84 Plus can perform all of the same functions but will not have the "Color" option referenced in some of the menus.

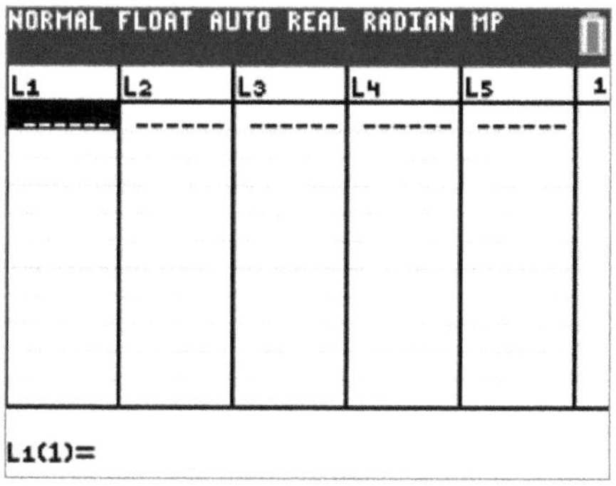

SCREEN 1.1

Entering and Editing Data

1. On the TI-84, variables are referred to as lists.
2. When entering data that does not need to be saved for later use, you can use the default lists, which are named **L1**, **L2**, **L3**, **L4**, **L5**, and **L6**.
3. To set up the editor to view these lists, select **STAT** > **EDIT** > **SetUpEditor**. Press **ENTER**.
4. To view these lists in the editor, select **STAT** > **EDIT** > **Edit.** (See **Screen 1.1.**)
5. To enter data in a list, type each data value and then press **ENTER** after each data value. (See **Screen 1.2.**)
6. To edit data values in a list, use the arrow keys to move to the cell you wish to edit, type the new value, and then press **ENTER**.

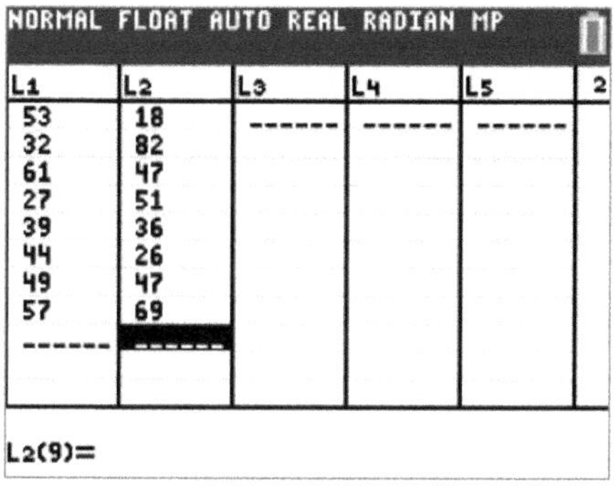

SCREEN 1.2

Creating a Named List

1. When entering data that will be saved for later use, you can create named lists. List names may contain a maximum of five characters, including letters or numbers.
2. Select **STAT** > **EDIT** > **SetUpEditor** and then type the names of your lists, separated by commas. Press **ENTER**. (See **Screen 1.3.**)

   Note: To type list names, select **ALPHA** and the letter (in green on your keypad) or select **2**nd > **ALPHA** to engage the **A-LOCK** (alpha-lock). To disengage the **A-LOCK**, press **ALPHA**.
3. To view these lists in the editor, select **STAT** > **EDIT** > **Edit.** (See **Screen 1.4.**)
4. To return the editor to the default lists, select **STAT** > **EDIT** > **SetUpEditor**. Press **ENTER**.

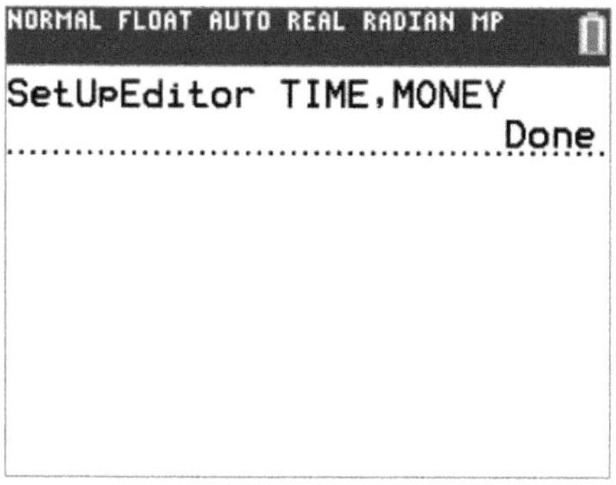

SCREEN 1.3

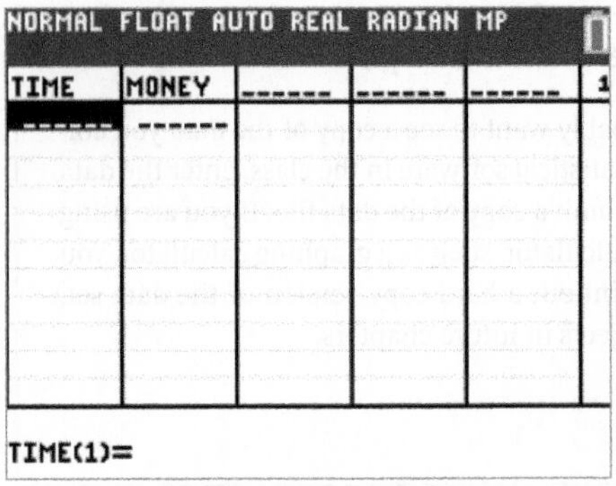

SCREEN 1.4

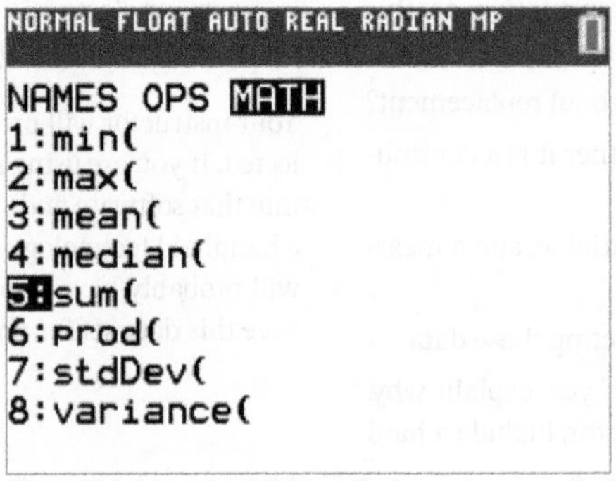

SCREEN 1.5

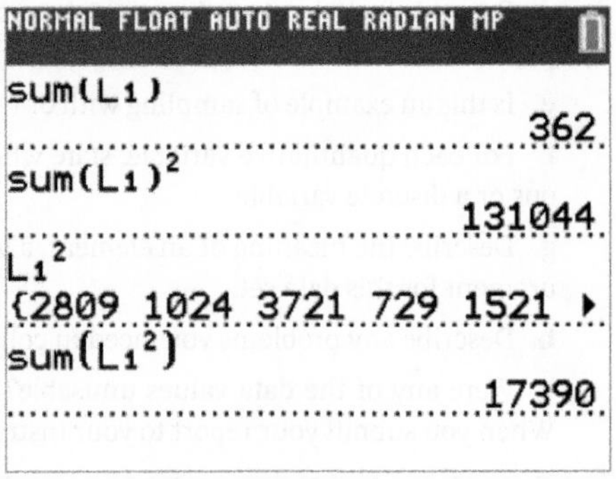

SCREEN 1.6

Operations with Lists

1. To calculate the sum of the values in a list, select 2nd > **STAT** > **MATH** > **sum**, enter the name of the list (e.g., **2nd > 1** for **L1**), and then press the **)** key. Press **ENTER**. (See **Screens 1.5** and **1.6.**)
2. To calculate the square of the sum of the values in a list, denoted by $(\sum x)^2$, use the same instructions as in Step 1, but press the x^2 key before you press **ENTER**. (See **Screen 1.6.**)
3. To calculate the square of each value in a list, select **2nd** > **STAT**, select the name of your list, press the x^2 key, and then press **ENTER**. (See **Screen 1.6.**)
4. To calculate the sum of the squares of the values in a list, denoted by $\sum x^2$, apply the same instructions as in Step 1 to the list created in Step 3. **Screen 1.6** shows the appearance of these two processes. (See **Screen 1.6.**)

Selecting Random Samples

1. The following data represent the ages, in years, at which 10 patients were first diagnosed with diabetes. Enter these data in **L1** in the order they appear below. We will select a random sample of 4 data values from these 10.

   18, 7, 46, 32, 62, 65, 51, 40, 28, 72

2. Select **2nd** > **MODE** to reach the home screen.
3. Select **MATH** > **PROB** > **randIntNoRep**. (*Note:* For older operating systems, use PRB instead of PROB.)
4. Use the following settings in the **randIntNoRep** menu:
   - Type 1 at the **lower** prompt.
   - Type 10 at the **upper** prompt.
   - Type 10 at the **n** prompt.

   Note: Always enter the total number of data values for the **upper** and **n** prompt.
5. Highlight **Paste** and press **ENTER.**
6. Press **STO >.**
7. Select **2nd** > **STAT** > **NAMES** > **L2.**
8. Press **ENTER**. (See **Screen 1.7.**)

   Note: The numbers 1 through 10 may be in a different order on your screen than they appear in **Screen 1.7** because they were randomly sampled.
9. Select **2nd** > **STAT** > **OPS** > **SortA.**
10. Select **2nd** > **2** to display **L2.**
11. Press **,** (the comma key).
12. Select **2nd** > **1** to display **L1.**
13. Press **)** (the right parenthesis key).
14. Press **ENTER.** (See **Screen 1.7.**)
15. Select **STAT** > **EDIT** > **Edit** to view your lists. (See **Screen 1.8.**)
16. The top four numbers in **L1** are the random sample of 4 data values.

    Note: The numbers in **L1** may be in a different order on your screen than they appear in **Screen 1.8** because the data were randomly sorted by the numbers in **L2.**

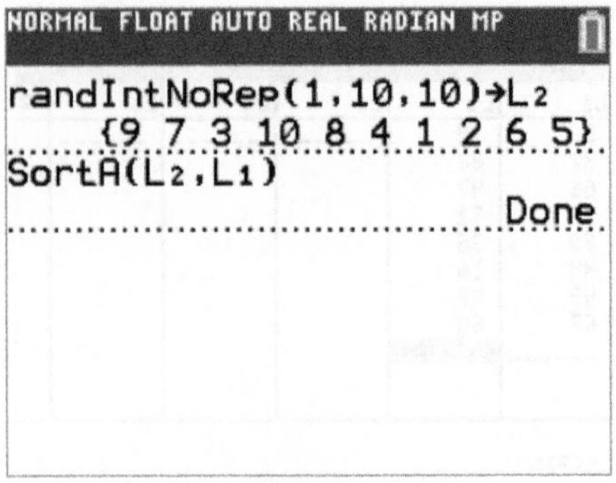

SCREEN 1.7

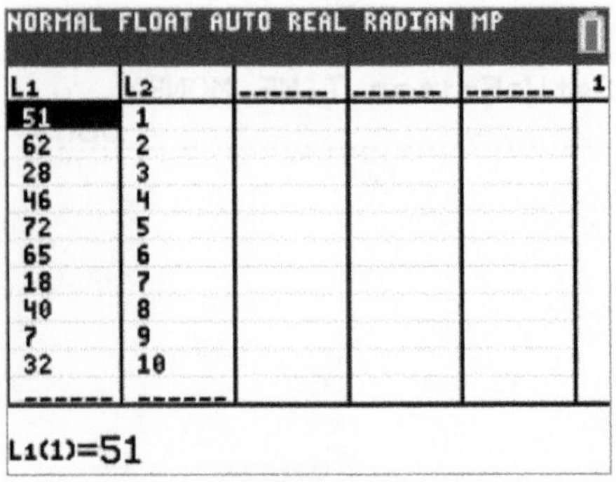

SCREEN 1.8

Minitab

The Minitab Technology Instructions in this text are written for **Minitab Professional Version 19**. Some screens, menus, and functions will be slightly different in older (or in newer) versions of Minitab. If you are using **Minitab Express**, you can download the Minitab Express Manual from the book's product page on *www.wiley.com*.

Entering and Saving Data

1. In Minitab you will see the screen divided into two parts:
   - The top half is called the **Session** window, which will contain numeric output.
   - The bottom half is called the **Worksheet**, which looks similar to a spreadsheet, where you will enter your data. Minitab allows you to have multiple worksheets within a project.
2. Use the mouse or the arrow keys to select where you want to start entering your data in the worksheet. Each column in the worksheet corresponds to a variable, so you can enter only one kind of data into a given column. Data can be numeric, text, or date/time. If a column contains text data, Minitab will add "-T" to the column heading and if a column contains a date/time, Minitab will add "-D" to the column heading. (See **Screen 1.9**.)

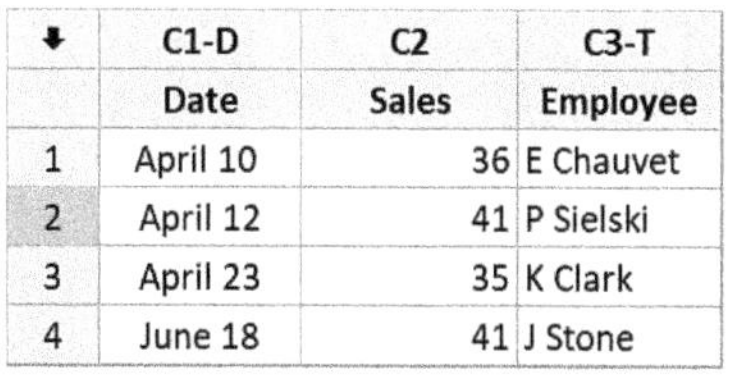

SCREEN 1.9

3. The blank row between the column labels and the first row is for variable names. In these blank cells, you can type the names of variables.
4. The direction in which data is entered can be changed by clicking the direction arrow at the top left of the worksheet. (See **Screen 1.9**.)
5. Click on a cell and begin typing. Press **ENTER** when you are finished an entry.
6. If you need to revise an entry, go to that cell with the mouse or the arrow keys and begin typing. Press **ENTER** to put the revised entry into the cell.
7. When you are finished, select **File > Save Project As** to save your work as a file on your computer. Minitab will automatically assign the file extension *.mpj* to your work after the filename. Or you can also save just the worksheet by selecting **File > Save Current Worksheet**. Minitab will add *.mtb* extensions to your saved file.
8. As an exercise, enter the data shown below into Minitab. Name the columns *Month*, *Cost*, and *Increase*. Save the result as the file *test.mpj*.

Month	Cost	Increase
January	52	0.08
February	48	0.06
March	49	0.07

9. To open a saved file, select **File > Open** and select the file name.
10. If you are already in Minitab and you want to start a new worksheet, select **File > New** and choose Worksheet. Whenever you save a project, Minitab will automatically save all of the worksheets in the project.

Creating New Columns from Existing Columns

Note: This example calculates the squares of values in a column but other computations can be performed, such as computing $\sum x^2$ or $\sum xy$.

1. To calculate a column containing the squares of the values in the column *Sales* as shown in Screen 1.9, select **Calc > Calculator**.
2. In the **Store result in variable** box, type the name of the column to contain the new values (such as C4).
3. Click inside the **Expression** box, click **C2 Sales** in the column to the left of the **Expression** box, and then click **Select**. You can also type **C2** in the **Expression** box.
4. Click on the exponentiation (^) button on the keypad in this dialog box. Type **2** after the ^ in the **Expression** box. Click **OK**. (See **Screen 1.10**.)
5. The numbers 1296, 1681, 1225, and 1681 should appear in column C4 of the worksheet.

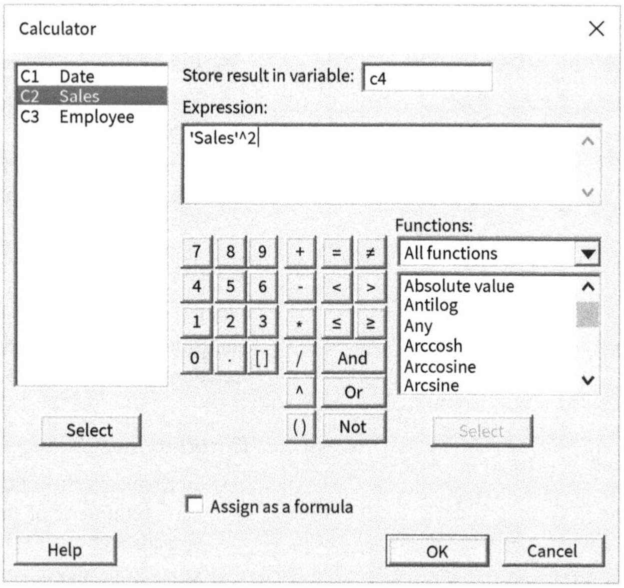

SCREEN 1.10

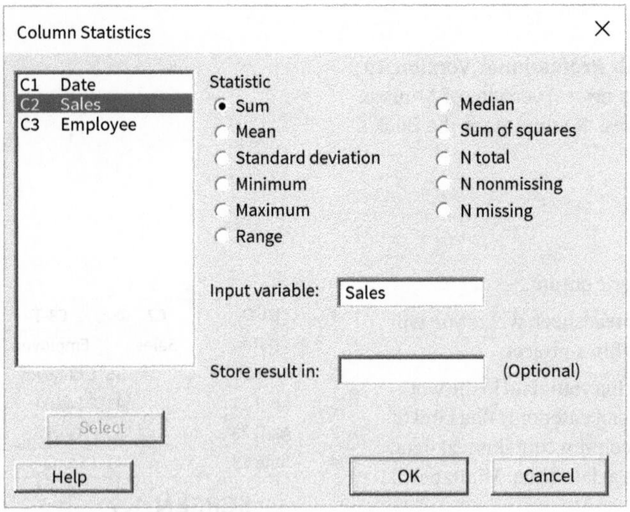

SCREEN 1.11

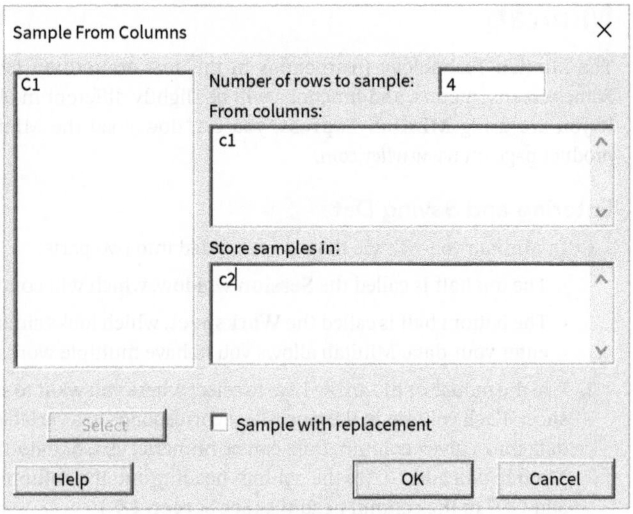

SCREEN 1.12

Calculating the Sum of the Values in a Column

1. To calculate the sum of the values in the *Sales* column from Screen 1.9, select **Calc > Column Statistics**, which will produce a dialog box.
2. From the **Statistic** list, select **Sum**.
3. Click in the **Input Variable** box. The list of variables will appear in the left portion of the dialog box.
4. Click on the column *Sales*, then click **Select**. You can also type **C2** in the **Input variable box**. (See **Screen 1.11**.)
5. Click **OK**. The result will appear in the **Session** window.

Selecting Random Samples

1. The following data represent the ages, in years, at which 10 patients were first diagnosed with diabetes. Enter these data in column **C1** in the order they appear below. We will select a random sample of 4 data values from these 10.

   18, 7, 46, 32, 62, 65, 51, 40, 28, 72

2. Select **Calc > Random Data > Sample from Columns**.
3. Use the following settings in the dialog box that appears on screen (see **Screen 1.12**):
   - Type 4 in the **Number of rows to sample** box.
   - Type C1 in the **From columns** box.
   - Type C2 in the **Store samples in** box.
4. Click **OK**.
5. Four data values from C1 are randomly selected without replacement and stored in C2. (See **Screen 1.13**.)

   Note: The numbers in **C2** may be different on your screen than they appear in **Screen 1.13** because the data were randomly sampled.

↓	C1	C2
1	18	51
2	7	72
3	46	62
4	32	65
5	62	
6	65	
7	51	
8	40	
9	28	
10	72	
11		

SCREEN 1.13

Excel

The Excel Technology Instructions feature of this text is written for Excel 2019. Some screens, menus, and functions will be slightly different in older versions of Excel. To enable some of the advanced Excel functions, you must enable the Data Analysis Add-In for Excel.

Entering and Saving Data

1. In Excel, use the mouse or the arrow keys to select where you want to start entering your data. Data can be numeric or text. The rectangles are called cells, and the cells are collectively known as a spreadsheet.
2. You can format your data by selecting the cells that you want to format, then clicking **HOME** and choosing the formatting option from the **NUMBER** group.

3. If you need to revise an entry, go to that cell with the mouse or the arrow keys. You can retype the entry or you can edit it. To edit it, double-click on the cell and use the arrow keys and the backspace key to help you revise the entry, then press **ENTER** to place the revised entry into the cell.
4. When you are done, select **Save As** from the **FILE** tab to save your work as a file on your computer. Excel will automatically assign the file extension *.xlsx* after the filename.
5. As an exercise, enter the following data into Excel, format the second column as currency, and then format the third column as percent. Save the result as the file *test.xlsx*. (See **Screen 1.14**.)

	A	B	C
1	Month	Cost	Increase
2	January	$ 52.00	8%
3	February	$ 48.00	6%
4	March	$ 49.00	7%

SCREEN 1.14

Month	Cost	Increase
January	$52	8%
February	$48	6%
March	$49	7%

6. To open a saved file, select **Open** from the **FILE** tab and select the filename.

Creating New Columns from Existing Columns

Note: This example calculates the squares of values in a column but other computations can be performed, such as computing $\sum x^2$ or $\sum xy$.

	A	B	C	D
1	Month	Cost	Increase	
2	January	$ 52.00	8%	=B2^2
3	February	$ 48.00	6%	
4	March	$ 49.00	7%	

SCREEN 1.15

1. Open the file *test.xlsx*. Click on cell **D2**.
2. Type **=B2^2** and then press **ENTER**. The value $2704.00 should appear, which is the square of $52.00, the value in cell **B2**. (See **Screen 1.15**.)
3. Highlight cell **D2** and then select **Copy** from the **Clipboard** group on the **HOME** tab. Highlight cells **D3** and **D4** and then select **Paste** from the **Clipboard** group on the **HOME** tab.
4. The numbers $2304.00 and $2401.00 should appear in cells **D3** and **D4**, respectively. (See **Screen 1.16**.)

	A	B	C	D
1	Month	Cost	Increase	
2	January	$ 52.00	8%	$2,704.00
3	February	$ 48.00	6%	$2,304.00
4	March	$ 49.00	7%	$2,401.00
5		=sum(B2:B4)		

SCREEN 1.16

Calculating the Sum of the Values in a Column

1. Open the file *test.xlsx*. Click on cell **B5**.
2. Select **FORMULAS > Function Library >** $\sum$. Excel should automatically enter the Sum function along with the range of cells involved in the sum. (*Note*: If the range of cells is incorrect, you can type any changes.) (See **Screen 1.16**.)
3. Press **ENTER**. The number $149.00 should appear in cell **B5**.

Selecting Random Samples

1. The following data represent the age, in years, at which 10 patients were first diagnosed with diabetes. Enter these data into cells A1 through A10 in the order they appear below. We will select a random sample of 4 data values from these 10.

   18, 7, 46, 32, 62, 65, 51, 40, 28, 72

2. Select **DATA > Data Analysis**.
3. From the dialog box that appears on screen, select **Sampling**.
4. Use the following settings in the dialog box that appears on screen (see **Screen 1.17**):
   - Type A1:A10 in the **Input Range** box.
   - Select Random in the **Sampling Method** box.
   - Type 4 in the **Number of Samples** box.
   - Select New Worksheet Ply in the **Output options** box.
5. Click **OK.**
6. Four data values from A1 to A10 are randomly selected without replacement and stored in a new worksheet.

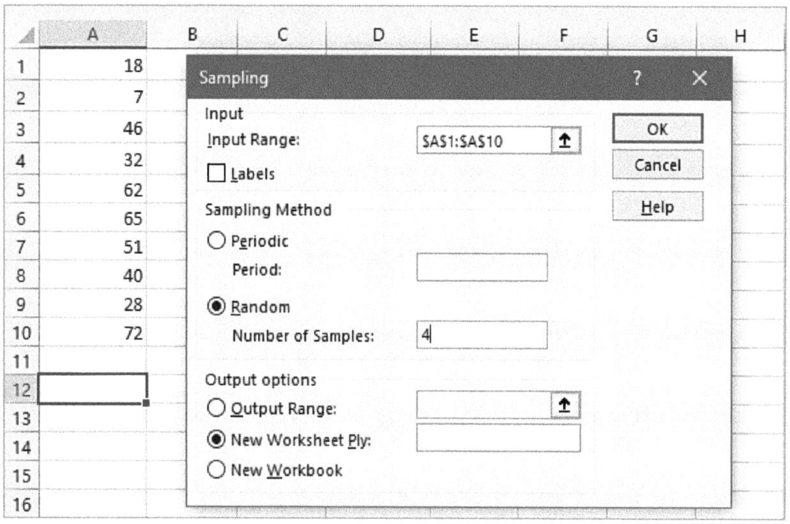

SCREEN 1.17

Enabling the Data Analysis Add-In

1. Click the **FILE > Options**.
2. From the resulting dialog box, click **Add-ins**.
3. In the **Manage** box, select **Excel Add-ins** and then click **Go**.
4. In the **Add-Ins** dialog box, select the **Analysis ToolPak** check box, and then click **OK**.
5. If **Analysis ToolPak** is not listed in the Add-Ins available box, click **Browse** to locate it.
6. If you are prompted that the **Analysis ToolPak** is not currently installed on your computer, click **Yes** to install it.

Technology Assignments

TA1.1 The following table gives the names, hours worked, and salary for the past week for five workers.

Name	Hours Worked	Salary ($)
John	42	6325
Shannon	33	2583
Kathy	28	3255
David	47	5090
Steve	40	4020

a. Enter these data into the spreadsheet. Save the data file as WORKER. Exit the session or program. Then restart the program or software and retrieve the file WORKER.

b. Print a hard copy of the spreadsheet containing the data you entered.

TA1.2 Refer to the data set on the annual revenue for Shake Shack given in Exercise 1.44.

a. Enter these data into Minitab or Excel. Save the data file with the name SHAKE SHACK REVENUE. Close the file, then open it again.

b. Use technology to find the total annual revenue from 2012 to 2018.

c. Print a hard copy or take a screenshot of the data you entered.

TA1.3 Refer to Data Set IV on the Manchester Road Race that accompanies this text (see Appendix A). Using the technology of your choice, select a random sample of 300 from the third column.

TA1.4 Refer to Data Set V on Data on Movies (see Appendix A). This file contains information on the top 200 movies from 2018. Using the technology of your choice, select a random sample of 20 movies from the first column.

CHAPTER 2

Organizing and Graphing Data

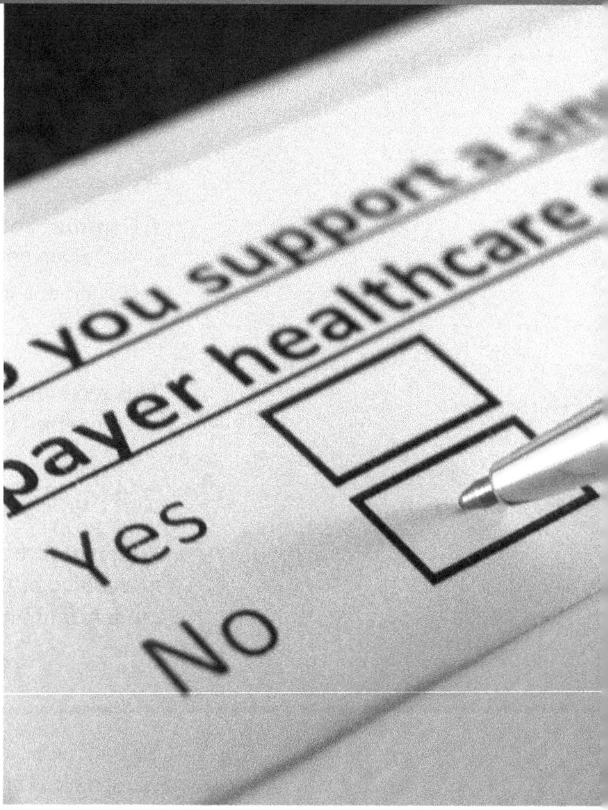

Yeexin Richelle/Shutterstock.com

Are you in favor of a single-payer health-care system where the federal government provides coverage for everyone? Or are you opposed to such a system? Or still, are you undecided? Rasmussen Reports conducted a survey of 1000 likely voters on August 4–5, 2019 to find the answers to this question. To see the results of this survey, see Case Study 2.2.

In addition to thousands of private organizations and individuals, a large number of U.S. government agencies (such as the Bureau of the Census, the Bureau of Labor Statistics, the National Agricultural Statistics Service, the National Center for Education Statistics, the National Center for Health Statistics, and the Bureau of Justice Statistics) conduct hundreds of surveys every year. The data collected from each of these surveys fill hundreds of thousands of pages. In their original form, these data sets may be so large that they do not make sense to most of us. Descriptive statistics, however, supplies the techniques that help to condense large data sets by using tables, graphs, and summary measures. We see such tables, graphs, and summary measures in print and digital media every day. At a glance, these tabular and graphical displays present information on every aspect of life. Consequently, descriptive statistics is of immense importance because it provides efficient and effective methods for summarizing and analyzing information.

This chapter explains how to organize and display data using tables and graphs. We will learn how to prepare frequency distribution tables for qualitative and quantitative data; how to construct bar graphs, pie charts, histograms, and polygons for such data; and how to prepare stem-and-leaf displays.

2.1 Organizing and Graphing Qualitative Data

LEARNING OBJECTIVES

After completing this section, you should be able to:

- Construct and interpret a frequency distribution table, a relative frequency distribution table, and a percentage distribution table for qualitative data.
- Construct and interpret a bar graph, a Pareto chart, and a pie chart.

This section discusses how to organize and display qualitative (or categorical) data. Data sets are organized into tables and displayed using graphs. First we discuss the concept of raw data.

2.1.1 Raw Data

When data are collected, the information obtained from each member of a population or sample is recorded in the sequence in which it becomes available. This sequence of data recording is random and unranked. Such data, before they are grouped or ranked, are called **raw data**.

Raw Data

Data recorded in the sequence in which they are collected and before they are processed or ranked are called *raw data*.

Suppose we collect information on the ages (in years) of 50 students selected from a university. The data values, in the order they are collected, are recorded in **Table 2.1**. For instance, the first student's age is 21, the second student's age is 19 (second number in the first row), and so forth. The data in Table 2.1 are **quantitative raw data**.

TABLE 2.1 Ages of 50 Students

21	19	24	25	29	34	26	27	37	33
18	20	19	22	19	19	25	22	25	23
25	19	31	19	23	18	23	19	23	26
22	28	21	20	22	22	21	20	19	21
25	23	18	37	27	23	21	25	21	24

Suppose we ask the same 50 students about their student status. The responses of the students are recorded in **Table 2.2**. In this table, F, SO, J, and SE are the abbreviations for freshman, sophomore, junior, and senior, respectively. This is an example of **qualitative (or categorical) raw data.**

The data presented in Tables 2.1 and 2.2 are also called **ungrouped data**. An ungrouped data set contains information on each member of a sample or population individually. If we rank the data of Table 2.1 from lowest to the highest age, they will still be ungrouped data but not raw data.

TABLE 2.2 Status of 50 Students

J	F	SO	SE	J	J	SE	J	J	J
F	F	J	F	F	F	SE	SO	SE	J
J	F	SE	SO	SO	F	J	F	SE	SE
SO	SE	J	SO	SO	J	J	SO	F	SO
SE	SE	F	SE	J	SO	F	J	SO	SO

2.1.2 Frequency Distributions

On March 1–10, 2019, the Gallup polling agency surveyed, by telephone interviews, 1039 randomly selected adults aged 18 and over from all 50 U.S. states and the District of Columbia. These adults were asked to answer the following question, "... we'd like to ask you about proposals to dramatically reduce the use of fossil fuels such as gas, oil and coal in the United States within the next 10 or 20 years, in order to reduce greenhouse gas emissions. Would you strongly favor, favor, oppose or strongly oppose establishing policies aimed at achieving this?" The Gallup report contained the percent of adults belonging to each category, which we have converted to numbers and listed in **Table 2.3.** In this table, the variable is "support for proposals to reduce the use of fossil fuels." The categories representing this variable are listed in the first column of the table. Note that these categories are mutually exclusive. In other words, each of the 1039 adults belongs to one and only one of these categories. The number of adults who belong to a certain category is called the frequency of that category. A **frequency distribution** exhibits how the frequencies are distributed over various categories. Table 2.3 is called a *frequency distribution table* or simply a *frequency table.*

TABLE 2.3 Support for Proposals to Reduce the Use of Fossil Fuels

	Response	Number of Adults	
Variable → (Response)			← Frequency column (Number of Adults)
	Strongly favor	280	
	Favor	343	
Category →	Oppose	197	← Frequency
	Strongly oppose	177	
	Other	42	
		Sum = 1039	

Source: Gallup Poll.

Frequency Distribution of a Qualitative Variable

A *frequency distribution* of a qualitative variable lists all categories and the number of elements that belong to each of the categories.

Example 2.1 illustrates how a frequency distribution table is constructed for a qualitative variable.

EXAMPLE 2.1 What Variety of Donuts Is Your Favorite?

Constructing a Frequency Distribution Table for Qualitative Data

A sample of 30 persons who often consume donuts were asked what variety of donuts is their favorite. The responses from these 30 persons are as follows:

glazed	filled	other	plain	glazed	other
frosted	filled	filled	glazed	other	frosted
glazed	plain	other	glazed	glazed	filled
frosted	plain	other	other	frosted	filled
filled	other	frosted	glazed	glazed	filled

Construct a frequency distribution table for these data.

Jack Puccio/Getty Images

Solution Note that the variable in this example is *favorite variety of donut*. This variable has five categories (varieties of donuts): glazed, filled, frosted, plain, and other. To prepare a frequency distribution, we record these five categories in the first column of **Table 2.4**. Then we read each response (each person's favorite variety of donut) from the given information and mark a *tally*, denoted by the symbol |, in the second column of Table 2.4 next to the corresponding category. For example, the first response is *glazed*. We show this in the frequency table by marking a tally in the second column next to the category *glazed*. Note that the tallies are marked in blocks of five for counting convenience. Finally, we record the total of the tallies for each category in the third column of the table. This column is called the *column of frequencies* and is usually denoted by f. The sum of the entries in the frequency column gives the sample size or total frequency. In Table 2.4, this total is 30, which is the sample size.

TABLE 2.4 Frequency Distribution of Favorite Donut Variety

Donut Variety	Tally	Frequency (f)
Glazed	𝍸 |||	8
Filled	𝍸 ||	7
Frosted	𝍸	5
Plain	|||	3
Other	𝍸 ||	7
		Sum = 30

Note that we do not have to have the tally column. It is only for convenience and to avoid errors. We can simply count how many persons in the sample like glazed donuts, how many like filled donuts, and so on. Then we can write these numbers in the frequency column.

2.1.3 Relative Frequency and Percentage Distributions

The **relative frequency** of a category is obtained by dividing the frequency of that category by the sum of all frequencies. Thus, the relative frequency shows what fractional part or proportion of the total frequency belongs to the corresponding category. A *relative frequency distribution* lists the relative frequencies for all categories.

Calculating Relative Frequency of a Category

$$\text{Relative frequency of a category} = \frac{\text{Frequency of that category}}{\text{Sum of all frequencies}}$$

The **percentage** for a category is obtained by multiplying the relative frequency of that category by 100. A *percentage distribution* lists the percentages for all categories.

Calculating Percentage

$$\text{Percentage} = (\text{Relative frequency}) \times 100\%$$

EXAMPLE 2.2 What Variety of Donuts Is Your Favorite?

Constructing Relative Frequency and Percentage Distributions

Determine the relative frequency and percentage distributions for the data of Table 2.4.

Solution The relative frequencies and percentages from Table 2.4 are calculated and listed in **Table 2.5**. Based on this table, we can state that 26.7% of the people in the sample said that glazed donut is their favorite. By adding the percentages for the first two categories, we can state that 50% of the persons included in the sample said that glazed or filled donut is their favorite. The other numbers in Table 2.5 can be interpreted in similar ways.

TABLE 2.5 Relative Frequency and Percentage Distributions of Favorite Donut Variety

Donut Variety	Relative Frequency	Percentage
Glazed	8/30 = .267	.267(100) = 26.7
Filled	7/30 = .233	.233(100) = 23.3
Frosted	5/30 = .167	.167(100) = 16.7
Plain	3/30 = .100	.100(100) = 10.0
Other	7/30 = .233	.233(100) = 23.3
	Sum = 1.000	Sum = 100%

Notice that the sum of the relative frequencies is always 1.00 (or approximately 1.00 if the relative frequencies are rounded), and the sum of the percentages is always 100 (or approximately 100 if the percentages are rounded).

2.1.4 Graphical Presentation of Qualitative Data

All of us have heard the adage "a picture is worth a thousand words." A graphic display can reveal at a glance the main characteristics of a data set. The *bar graph* and the *pie chart* are two types of graphs that are commonly used to display qualitative data.

Bar Graphs To construct a **bar graph** (also called a *bar chart*), we mark the various categories on the horizontal axis as in **Figure 2.1**. Note that all categories are represented by intervals of the same width. We mark the frequencies on the vertical axis. Then we draw one bar for each category such that the height of the bar represents the frequency of the corresponding category. We leave a small gap between adjacent bars. Figure 2.1 gives the bar graph for the frequency distribution of Table 2.4.

Bar Graph

A graph made of bars whose heights represent the frequencies of respective categories is called a *bar graph*.

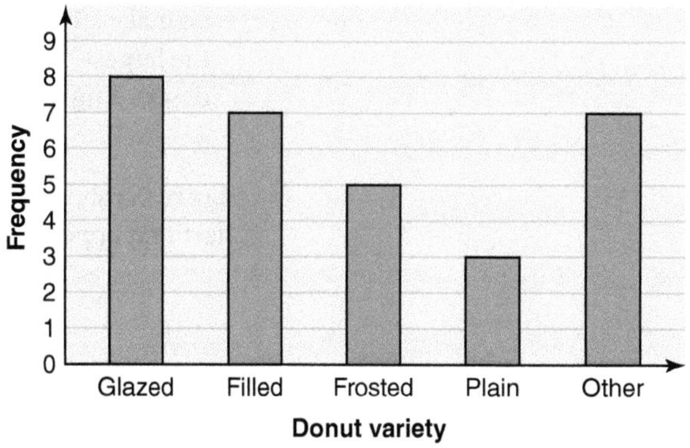

FIGURE 2.1 Bar graph for the frequency distribution of Table 2.4.

The bar graphs for relative frequency and percentage distributions can be drawn simply by marking the relative frequencies or percentages, instead of the frequencies, on the vertical axis.

Sometimes a bar graph is constructed by marking the categories on the vertical axis and the frequencies on the horizontal axis. Case Study 2.1 presents such an example.

Case Study 2.1

Confidence in Charitable and Nongovernmental Organizations

A June 20, 2019 Gallup blog based on Gallup and Wellcome Global Monitor worldwide poll reported that people from different countries were asked whether or not they have confidence in charitable and nongovernmental organizations (NGOs). In the United States, 60% of the adults polled said that they have confidence in charitable organizations and NGOs, 27% said they do not have confidence in such organizations, and 13% said that they do not know. The above chart shows the bar graph for these data. Note that in this chart, the bars are drawn horizontally.

Source: Data from Mohamed Younis and Andrew Rzepa, One in Three Worldwide Lack Confidence in NGOs, JUNE 20, 2019. Gallup, Inc.

Pareto Chart To obtain a Pareto chart, we arrange (in a descending order) the bars in a bar graph based on their heights (frequencies, relative frequencies, or percentages). Thus, the bar with the largest height appears first (on the left side) in a bar graph and the one with the smallest height appears at the end (on the right side) of the bar graph.

Pareto Chart

A *Pareto chart* is a bar graph with bars arranged by their heights in descending order. To make a Pareto chart, arrange the bars according to their heights such that the bar with the largest height appears first on the left side, and then subsequent bars are arranged in descending order with the bar with the smallest height appearing last on the right side.

Figure 2.2 shows the Pareto chart for the frequency distribution of Table 2.4. It is the same bar chart that appears in Figure 2.1 but with bars arranged based on their heights.

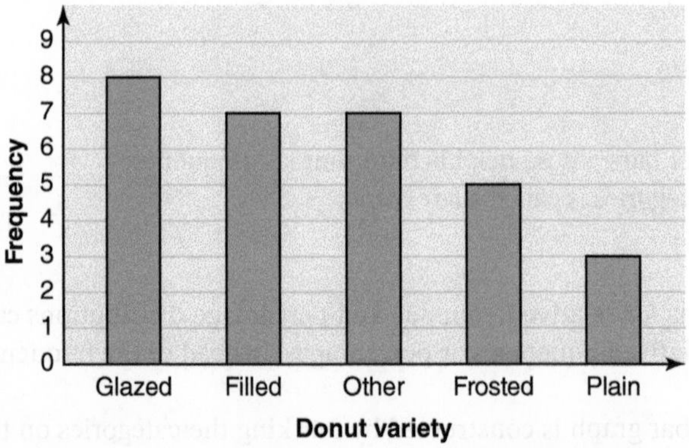

FIGURE 2.2 Pareto chart for the frequency distribution of Table 2.4.

Pie Charts A **pie chart** is more commonly used to display percentages, although it can be used to display frequencies or relative frequencies. The whole pie (or circle) represents the total sample or population. Then we divide the pie into different portions that represent the different categories. **Figure 2.3** shows the pie chart for the percentage distribution of Table 2.5.

Pie Chart

A circle divided into portions that represent the relative frequencies or percentages of a population or a sample belonging to different categories is called a *pie chart.*

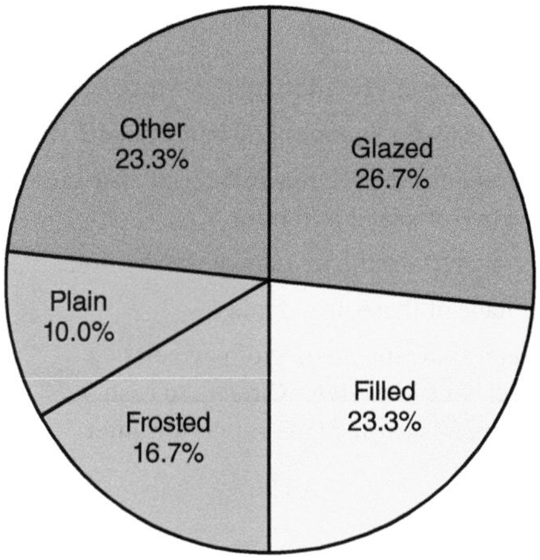

FIGURE 2.3 Pie chart for the percentage distribution of Table 2.5.

Case Study 2.2

Single-Payer Health-Care System Where the Federal Government Provides Coverage for Everyone

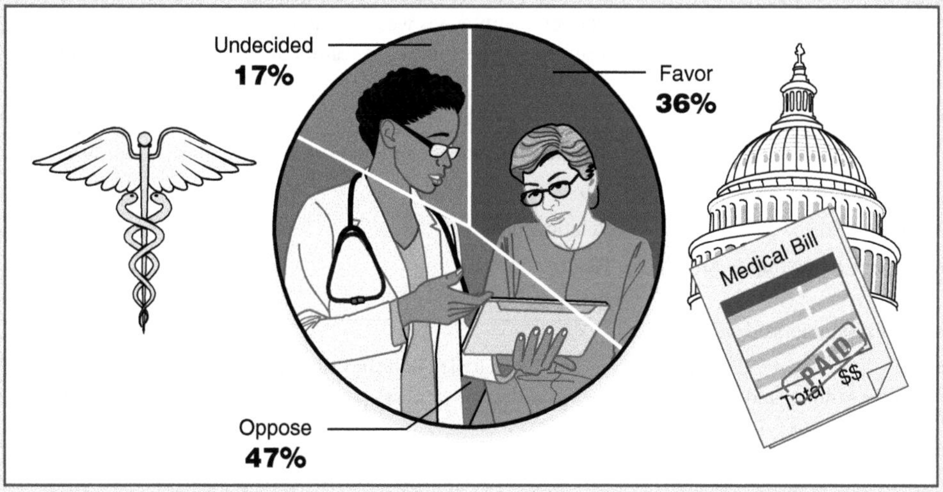

Rasmussen Reports conducted a national telephone and online survey about health care. The survey included 1000 likely voters and was conducted on August 4–5, 2019. Rasmussen Reports use automated polling methodology to conduct surveys, which uses a "single, digitally-recorded, voice to conduct the interview" instead of traditional operator-assisted interviews.

In this survey on health care, one question asked was, "Do you favor or oppose a single-payer health-care system where the federal government provides coverage for everyone?" Of the voters included in the survey, 36% said they favor such a single-payer health-care system, 47% oppose such a system, and 17% were undecided. The above pie chart shows these categories and the percentages belonging to these categories.

Source: Data from Health Care Law, Wednesday, August 07, 2019. Rasmussen Reports, LLC. http://www.rasmussenreports.com/public_content/politics/current_events/healthcare/health_care_law_aug07

EXERCISES

Concepts and Procedures

2.1 What are the benefits of grouping data in the form of a frequency table?

2.2 How are the relative frequencies and percentages of categories obtained from the frequencies of categories? Illustrate with the help of an example.

2.3 The following data give the results of a sample survey. The letters Y, N, and D represent the three categories.

N	N	N	Y	Y	Y	N	Y	D	N
Y	Y	Y	Y	N	Y	Y	N	N	D
D	Y	Y	D	D	N	N	N	Y	N
Y	Y	N	N	Y	Y	N	N	D	Y

a. Prepare a frequency distribution table.

b. Calculate the relative frequencies and percentages for all categories.

c. What percentage of the elements in this sample belong to category Y?

d. What percentage of the elements in this sample belong to category N or D?

e. Draw a pie chart for the percentage distribution.

Applications

2.4 In the past few years, many states have built casinos and many more are in the process of doing so. Forty adults were asked if building casinos is good for society. Following are the responses of these adults, where G stands for good, B indicates bad, and I means indifferent or no answer.

B	G	B	B	I	G	B	I	B	B
G	B	B	G	B	B	B	G	G	I
B	G	B	B	I	G	G	G	B	B
I	G	B	B	B	G	G	B	B	G

a. Prepare a frequency distribution table.

b. Calculate the relative frequencies and percentages for all categories.

c. What percentage of the adults in this sample said building casinos is good?

d. What percentage of the adults in this sample said building casinos is bad or were indifferent?

e. Draw a bar graph for the frequency distribution.

f. Draw a pie chart for the percentage distribution.

g. Make a Pareto chart for the percentage distribution.

2.5 The following data give the political party of each of the first 30 U.S. presidents. In the data, D stands for Democrat, DR for Democratic Republican, F for Federalist, R for Republican, and W for Whig.

F	F	DR	DR	DR	DR	D	D	W	W
D	W	W	D	D	R	D	R	R	R
R	D	R	D	R	R	R	D	R	R

a. Prepare a frequency distribution table for these data.

b. Calculate the relative frequency and percentage distributions.

c. Draw a bar graph for the relative frequency distribution and a pie chart for the percentage distribution.

d. Make a Pareto chart for the frequency distribution.

e. What percentage of these presidents were Whigs?

2.6 The following data show the method of payment by 16 customers in a supermarket checkout line. Here, C refers to cash, CK to check, CC to credit card, D to debit card, and O stands for other.

C	CK	CK	C	CC	D	O	C
CK	CC	D	CC	C	CK	CK	CC

a. Construct a frequency distribution table.

b. Calculate the relative frequencies and percentages for all categories.

c. Draw a pie chart for the percentage distribution.

2.7 In a 2019 survey of employees conducted by PriceWaterhouse-Coopers, employees were asked about their top cause of stress. The following table shows the results of this survey (www.pwc.com).

Top Cause of Stress	Percentage of Responses
Financial or money matters	59
Job	15
Relationships	12
Health concerns	10
Other	4

a. Draw a pie chart for this percentage distribution.

b. Make a Pareto chart for this percentage distribution.

2.2 Organizing and Graphing Quantitative Data

LEARNING OBJECTIVES

After completing this section, you should be able to:

- Describe the midpoint, limits, boundaries, and width of a class.
- Construct and interpret a frequency distribution table, a relative frequency distribution table, and a percentage distribution table for quantitative data.
- Construct and interpret a histogram for frequencies, relative frequencies, and percentages.
- Construct and interpret a polygon.

- Construct and interpret a cumulative frequency distribution table, a cumulative relative frequency distribution table, and a cumulative percentage distribution table and its ogive curve for quantitative data.
- Describe the shape of a histogram.
- Discuss features of misleading graphs.

In the previous section we learned how to group and display qualitative data. This section explains how to group and display quantitative data.

2.2.1 Frequency Distributions

Table 2.6 gives the weekly earnings of 100 employees of a large company. The first column lists the *classes*, which represent the (quantitative) variable *weekly earnings*. For quantitative data, an interval that includes all the values that fall on or within two numbers—the lower and upper limits—is called a **class**. Note that the classes always represent a variable. As we can observe, the classes are nonoverlapping; that is, each value for earnings belongs to one and only one class. The second column in the table lists the number of employees who have earnings within each class. For example, four employees of this company earn $801 to $1000 per week. The numbers listed in the second column are called the **frequencies**, which give the number of data values that belong to different classes. The frequencies are denoted by ***f***.

TABLE 2.6 Weekly Earnings of 100 Employees of a Company

Weekly Earnings (dollars)	Number of Employees f
801 to 1000	4
1001 to 1200	11
1201 to 1400	39
1401 to 1600	24
1601 to 1800	16
1801 to 2000	6

Variable → Weekly Earnings (dollars)
Frequency column → Number of Employees f
Third class → 1201 to 1400
Frequency of the third class → 39
Lower limit of the sixth class → 1801
Upper limit of the sixth class → 2000

For quantitative data, the frequency of a class represents the number of values in the data set that fall in that class. Table 2.6 contains six classes. Each class has a *lower limit* and an *upper limit*. The values 801, 1001, 1201, 1401, 1601, and 1801 give the lower limits, and the values 1000, 1200, 1400, 1600, 1800, and 2000 are the upper limits of the six classes, respectively. The data presented in Table 2.6 are an illustration of a **frequency distribution table** for quantitative data. Whereas the data that list individual values are called ungrouped data, the data presented in a frequency distribution table are called **grouped data**.

Frequency Distribution for Quantitative Data

A *frequency distribution* for quantitative data lists all the classes and the number of values that belong to each class. Data presented in the form of a frequency distribution are called *grouped data*.

The difference between the lower limits of two consecutive classes gives the **class width**. The class width is also called the **class size**.

Finding Class Width

To find the width of a class, subtract its lower limit from the lower limit of the next class. Thus,

$$\text{Width of a class} = \text{Lower limit of the next class} - \text{Lower limit of the current class}$$

Thus, in Table 2.6,

$$\text{Width of the first class} = 1001 - 801 = 200$$

The class widths for the frequency distribution of Table 2.6 are listed in the second column of **Table 2.7**. Each class in Table 2.7 (and Table 2.6) has the same width of 200.

TABLE 2.7 Class Widths and Class Midpoints for Table 2.6

Class Limits	Class Width	Class Midpoint
801 to 1000	200	900.5
1001 to 1200	200	1100.5
1201 to 1400	200	1300.5
1401 to 1600	200	1500.5
1601 to 1800	200	1700.5
1801 to 2000	200	1900.5

The **class midpoint** or **mark** is obtained by dividing the sum of the two limits of a class by 2.

Calculating Class Midpoint or Mark

$$\text{Class midpoint or mark} = \frac{\text{Lower limit} + \text{Upper limit}}{2}$$

Thus, the midpoint of the first class in Table 2.6 or Table 2.7 is calculated as follows:

$$\text{Midpoint of the first class} = \frac{801 + 1000}{2} = 900.5$$

The class midpoints for the frequency distribution of Table 2.6 are listed in the third column of Table 2.7.

2.2.2 Constructing Frequency Distribution Tables

When constructing a frequency distribution table, we need to make the following three major decisions.

(1) Number of Classes Usually the number of classes for a frequency distribution table varies from 5 to 20, depending mainly on the number of observations in the data set.[1] It is preferable to have more classes as the size of a data set increases. The decision about the number of classes is arbitrarily made by the data organizer.

---

[1]One rule to help decide on the number of classes is Sturge's formula:

$$c = 1 + 3.3 \log n$$

where c is the number of classes and n is the number of observations in the data set. The value of $\log n$ can be obtained by using a calculator.

(2) Class Width Although it is not uncommon to have classes of different sizes, most of the time it is preferable to have the same width for all classes. To determine the class width when all classes are the same size, first find the difference between the largest and the smallest values in the data. Then, the approximate width of a class is obtained by dividing this difference by the number of desired classes.

Calculation of Class Width

$$\text{Approximate class width} = \frac{\text{Largest value} - \text{Smallest value}}{\text{Number of classes}}$$

Usually this approximate class width is rounded to a convenient number, which is then used as the class width. Note that rounding this number may slightly change the number of classes initially intended.

(3) Lower Limit of the First Class or the Starting Point Any convenient number that is equal to or less than the smallest value in the data set can be used as the lower limit of the first class.

Example 2.3 illustrates the procedure for constructing a frequency distribution table for quantitative data.

EXAMPLE 2.3 Systolic Blood Pressure of Adults

Constructing a Frequency Distribution Table for Quantitative Data

The following data were collected on the systolic blood pressures of 30 randomly selected adults from a small city.

116	132	123	137	125	110	101	129	147	114
133	118	107	121	143	120	131	109	127	139
111	114	121	130	106	149	124	119	133	127

Construct a frequency table for these data.

Solution In these data, the minimum value is 101, and the maximum value is 149. Suppose we decide to group these data using five classes of equal width. Then,

$$\text{Approximate width of each class} = \frac{149 - 101}{5} = 9.6$$

Now we round this approximate width to a convenient number, say 10. The lower limit of the first class can be taken as 101 or any number less than 101. Suppose we take 101 as the lower limit of the first class. Then our classes will be

101–110, 111–120, 121–130, 131–140, and 141–150

We record these five classes in the first column of **Table 2.8**.

TABLE 2.8 Frequency Distribution of the Systolic Blood Pressures

Systolic Blood Pressure	Tally	Number of Adults (f)
101–110	𝍸	5
111–120	𝍸 ||	7
121–130	𝍸 ||||	9
131–140	𝍸 |	6
141–150	|||	3
		$\Sigma f = 30$

Now we read each value from the given data and mark a tally in the second column of Table 2.8 next to the corresponding class. The first value in our original data set is 116, which belongs to the 111–120 class. To record it, we mark a tally in the second column next to the 111–120 class. We continue this process until all the data values have been read and entered in the tally column. Note that tallies are marked in blocks of five for counting convenience. After the tally column is completed, we count the tally marks for each class and write those numbers in the third column. This gives the column of frequencies. These frequencies represent the number of adults with systolic blood pressures in the corresponding classes. For example, five of the adults have systolic blood pressures in the interval 101–110.

Using the $\sum$ notation (see Section 1.8 of Chapter 1), we can denote the sum of frequencies of all classes by $\sum f$. Hence,

$$\sum f = 5 + 7 + 9 + 6 + 3 = 30$$

The number of observations in a sample is usually denoted by $\boldsymbol{n}$. Thus, for the sample data, $\sum f$ is equal to n. The number of observations in a population is denoted by $\boldsymbol{N}$. Consequently, $\sum f$ is equal to N for population data. Because the data set on the systolic blood pressures in Table 2.8 is for 30 randomly selected adults, it represents a sample. Therefore, in Table 2.8 we can also denote the sum of frequencies by n instead of $\sum f$.

Note that when we present the data in the form of a frequency distribution table, as in Table 2.8, we lose the information on individual observations. We cannot know the exact systolic blood pressure of any adult from Table 2.8. All we know is that five adults have their blood pressures in the interval 101–110, and so on.

2.2.3 Relative Frequency and Percentage Distributions

Using Table 2.8, we can compute the relative frequency and percentage distributions in the same way as we did for qualitative data in Section 2.1.3. The relative frequencies and percentages for a quantitative data set are obtained as follows. Note that relative frequency is the same as proportion.

Calculating Relative Frequency and Percentage

$$\text{Relative frequency of a class} = \frac{\text{Frequency of that class}}{\text{Sum of all frequencies}} = \frac{f}{\sum f}$$

$$\text{Percentage} = (\text{Relative frequency}) \times 100\%$$

Example 2.4 illustrates how to construct relative frequency and percentage distributions.

EXAMPLE 2.4 | Systolic Blood Pressure of Adults

Constructing Relative Frequency and Percentage Distributions

Calculate the relative frequencies and percentages for Table 2.8.

Solution The relative frequencies and percentages for the data in Table 2.8 are calculated and listed in the second and third columns, respectively, of **Table 2.9.**

Using Table 2.9, we can make statements about the percentage of adults with systolic blood pressures within a certain interval. For example, from Table 2.9, we can state that about 16.7% of the adults in this group of 30 adults had blood pressures in the interval 101 to 110 at the time these measurements were taken. By adding the percentages for the first two classes, we can state that 40% of the adults in this group of 30 adults had blood pressures in the interval 101 to 120 at the time these measurements were taken. Similarly, by adding the percentages for the last two classes, we can state that about 30.0% of the adults in this group of 30 adults had blood pressures in the interval 131 to 150 at the time these measurements were taken.

TABLE 2.9 Relative Frequency and Percentage Distributions of the Systolic Blood Pressures

Systolic Blood Pressure	Relative Frequency	Percentage
101–110	5/30 = .167	16.7
111–120	7/30 = .233	23.3
121–130	9/30 = .300	30.0
131–140	6/30 = .200	20.0
141–150	3/30 = .100	10.0
	Sum = 1.000	Sum = 100%

2.2.4 Graphing Grouped Data

Grouped (quantitative) data can be displayed using a *histogram* or a *polygon*. This section describes how to construct such graphs. We can also draw a pie chart to display the percentage distribution for a quantitative data set. The procedure to construct a pie chart is similar to the one for qualitative data explained in Section 2.1.4; it will not be repeated in this section.

Histograms A **histogram** can be drawn for a frequency distribution, a relative frequency distribution, or a percentage distribution. To draw a histogram, we first mark classes on the horizontal axis and frequencies (or relative frequencies or percentages) on the vertical axis. Next, we draw a bar for each class so that its height represents the frequency of that class. The bars in a histogram are drawn adjacent to each other with no gap between them. A histogram is called a **frequency histogram**, a **relative frequency histogram**, or a **percentage histogram** depending on whether frequencies, relative frequencies, or percentages are marked on the vertical axis.

Sometimes we draw a histogram with frequencies marked on the horizontal axis and classes on the vertical axis as in Case Study 2.3.

Histogram

A *histogram* is a graph in which classes are marked on the horizontal axis and the frequencies, relative frequencies, or percentages are marked on the vertical axis. The frequencies, relative frequencies, or percentages are represented by the heights of the bars. In a histogram, the bars are drawn adjacent to each other.

Case Study 2.3

Average Starting Salaries of Teachers in the United States

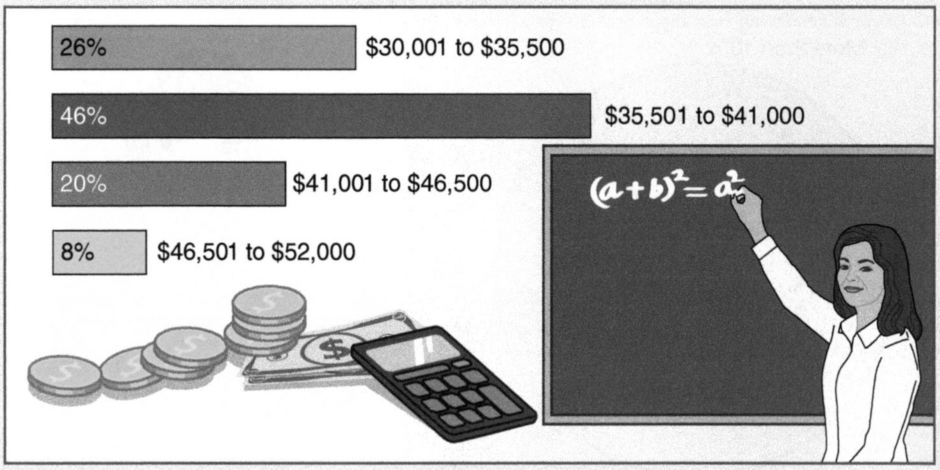

We obtained data on the average starting salaries of teachers in all 50 states for the year 2017–18 from the National Education Association (NEA) website. The latest year for which the data were available at that time was 2017–18. According to National Education Association, these average salaries were based on data obtained from many sources, "including NEA state affiliates, state departments of education, and school district and local affiliate websites." As the above histogram (that is made using these data) shows, in 26% of the states (that is, in 13 states) the average salary of teachers was in the interval $30,001 to $35,500 in 2017–18, 46% of the states (that is 23 states) had the average salary of teachers in the interval $35,501 to $41,000, in 20% of the states the average salary of teachers was in the interval $41,001 to $46,500, and 8% of the states had the average salary of teachers in the interval $46,501 to $52,000.

Source: Data from 2017–2018 Average Starting Teacher Salaries by State, National Education Association. http://www.nea.org/home/2017-2018-average-starting-teacher-salary.html

Figures 2.4 and **2.5** show the frequency and the percentage histograms, respectively, for the data of Tables 2.8 and 2.9 of Sections 2.2.2 and 2.2.3. The two histograms look alike because they represent the same data. A relative frequency histogram can be drawn for the relative frequency distribution of Table 2.9 by marking the relative frequencies on the vertical axis.

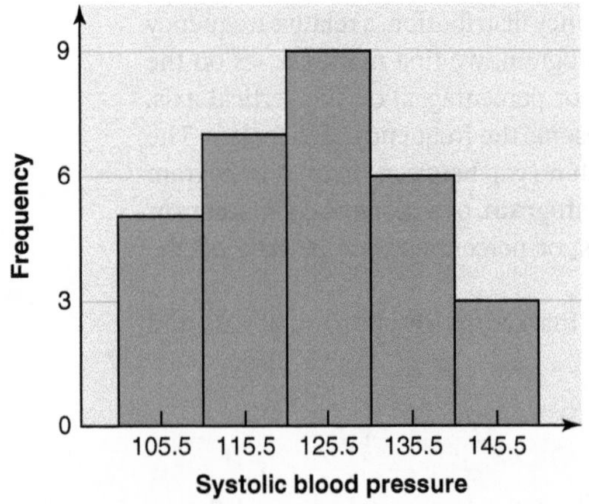

FIGURE 2.4 Frequency histogram for Table 2.8.

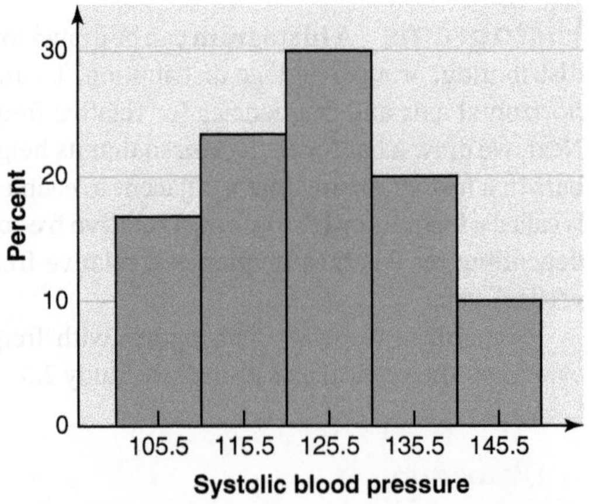

FIGURE 2.5 Percentage histogram for Table 2.9.

In Figures 2.4 and 2.5, we used class midpoints to mark classes on the horizontal axis. However, we can show the intervals on the horizontal axis by using the class limits instead of the class midpoints.

Case Study 2.4

Mom, I Am Hungry

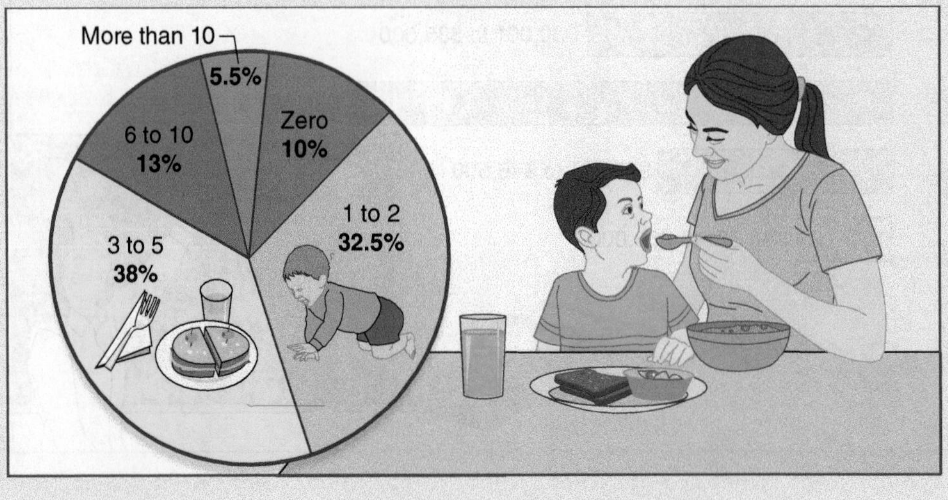

In a 2018 RXBAR and Monkey survey, 1270 parents were asked how many times a day their kids say, "I'm hungry." The above pie chart shows the percentage distribution of the responses of these parents. As the numbers in the pie chart show, 10% of the parents said zero time, 32.5% said 1 to 2 times, 38% said 3 to 5 times, 13% said 6 to 10 times, and 5.5% parents said that it is more than 10 times a day. Note that because of rounding, the percentages do not add to 100% but to 99%. Also note that all classes are not of the same width.

Source: Data from *USA TODAY*, November 27, 2018.

Polygons A **polygon** is another device that can be used to present quantitative data in graphic form. To draw a **frequency polygon**, we first mark a dot above the midpoint of each class at a height equal to the frequency of that class. This is the same as marking the midpoint at the top of each bar in a histogram. Next we include two more classes, one at each end, and mark their midpoints. Note that these two classes have zero frequencies. In the last step, we join the adjacent dots with straight lines. The resulting line graph is called a frequency polygon or simply a polygon.

A polygon with relative frequencies marked on the vertical axis is called a *relative frequency polygon*. Similarly, a polygon with percentages marked on the vertical axis is called a *percentage polygon*.

Polygon

A graph formed by joining the midpoints of the tops of successive bars in a histogram with straight lines is called a *polygon*.

Figure 2.6 shows the frequency polygon for the frequency distribution of Table 2.8.

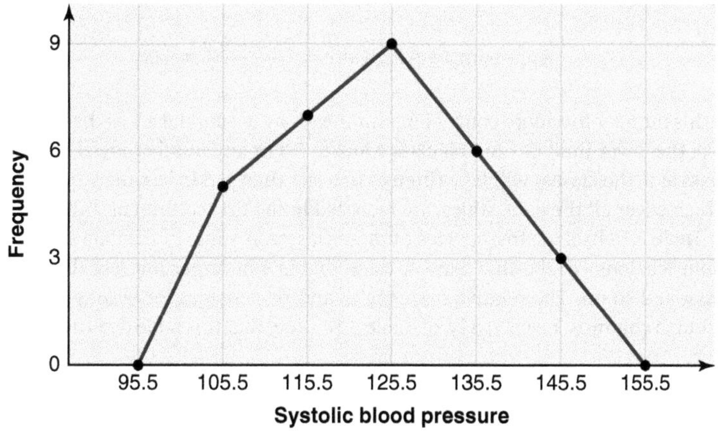

FIGURE 2.6 Frequency polygon for Table 2.8.

For a very large data set, as the number of classes is increased (and the width of classes is decreased), the frequency polygon eventually becomes a smooth curve. Such a curve is called a *frequency distribution curve* or simply a *frequency curve*. **Figure 2.7** shows the frequency curve for a large data set with a large number of classes.

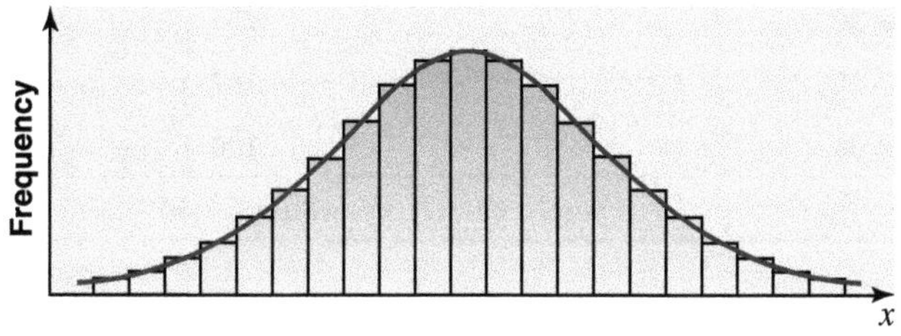

FIGURE 2.7 Frequency distribution curve.

2.2.5 More on Classes and Frequency Distributions

This section presents two alternative methods for writing classes to construct a frequency distribution for quantitative data.

Less-Than Method for Writing Classes The classes in the frequency distribution given in Table 2.8 for the data on systolic blood pressures of adults were written as 101–110, 111–120, and so on. Alternatively, we can write the classes in a frequency distribution table using the *less-than* method. The technique for writing classes shown in Table 2.8 is commonly used for data sets that do not contain fractional values. The *less-than* method is more appropriate when a data set contains fractional values. Example 2.5 illustrates the *less-than* method.

EXAMPLE 2.5 Percentage of Adults with No Bank Account

Constructing a Frequency Distribution Using the Less-than Method

The following data give the percentage of adults from 30 small towns who do not have a bank account.

18.4	9.7	24.3	3.9	14.3	22.8	12.5	7.2	13.9	20.6
26.7	7.7	13.4	15.4	19.3	27.3	4.7	16.3	11.6	17.7
12.9	22.3	6.3	14.8	28.4	13.9	12.4	18.5	24.7	17.2

Construct a frequency distribution table. Calculate the relative frequencies and percentages for all classes.

Solution The minimum value in this data set is 3.9, and the maximum value is 28.4. Suppose we decide to group these data using six classes of equal width. Then,

$$\text{Approximate class width} = \frac{28.4 - 3.9}{6} = 4.08$$

We round this number to a more convenient number—say 5—and take 5 as the width of each class. We can take the lower limit of the first class equal to 3.9 or any number lower than 3.9. If we start the first class at 0, the classes will be written as 0 to less than 5, 5 to less than 10, and so on. The six classes, which cover all the data values, are recorded in the first column of **Table 2.10**. The second column in Table 2.10 lists the frequencies of these classes. A value in the data set that is 0 or larger but less than 5 belongs to the first class, a value that is 5 or larger but less than 10 falls into the second class, and so on. The relative frequencies and percentages for classes are recorded in the third and fourth columns, respectively, of Table 2.10. Note that this table does not contain a column of tallies.

TABLE 2.10 Frequency, Relative Frequency, and Percentage Distributions of the Percent of Adults with No Bank Account

Percentage of Adults with No Bank Account	Frequency	Relative Frequency	Percentage
0 to less than 5	2	.067	6.7
5 to less than 10	4	.133	13.3
10 to less than 15	9	.300	30.0
15 to less than 20	7	.233	23.3
20 to less than 25	5	.167	16.7
25 to less than 30	3	.100	10.0
	Sum = 30	Sum = 1.00	Sum = 100

Note that in Table 2.10, the first column lists the class intervals using the **boundaries**, and not the limits. When we use the *less than method* to write classes, we call the two end-points of a class the

lower and upper boundaries. For example, in the first class, which is *0 to less than 5*, 0 is the lower boundary and 5 is the upper boundary. The difference between the two boundaries gives the width of the class. Thus, the width of the first class is 5 − 0 = 5. All classes in Table 2.10 have the same width, which is 5.

A histogram and a polygon for the data of Table 2.10 can be drawn the same way as was done for the data of Tables 2.8 and 2.9.

Single-Valued Classes If the observations in a data set assume only a few distinct (integer) values, it may be appropriate to prepare a frequency distribution table using *single-valued classes*—that is, classes that are made of single values and not of intervals. This technique is especially useful in cases of discrete data with only a few possible values. Example 2.6 exhibits such a situation.

EXAMPLE 2.6 Number of Vehicles Owned by Households

Constructing a Frequency Distribution Using Single-Valued Classes

The administration in a large city wanted to know the distribution of the number of vehicles owned by households in that city. A sample of 40 randomly selected households from this city produced the following data on the number of vehicles owned.

5	1	1	2	0	1	1	2	1	1
1	3	3	0	2	5	1	2	3	4
2	1	2	2	1	2	2	1	1	1
4	2	1	1	2	1	1	4	1	3

Construct a frequency distribution table for these data using single-valued classes.

Solution The observations in this data set assume only six distinct values: 0, 1, 2, 3, 4, and 5. Each of these six values is used as a class in the frequency distribution in **Table 2.11**, and these six classes are listed in the first column of that table. To obtain the frequencies of these classes, the observations in the data that belong to each class are counted, and the results are recorded in the second column of Table 2.11. Thus, in these data, 2 households own no vehicle, 18 own one vehicle each, 11 own two vehicles each, and so on.

iStockphoto.com/Jorge Salcedo

The data of Table 2.11 can also be displayed in a bar graph, as shown in **Figure 2.8.** To construct a bar graph, we mark the classes, as intervals, on the horizontal axis with a little gap between consecutive intervals. The bars represent the frequencies of respective classes.

The frequencies of Table 2.11 can be converted to relative frequencies and percentages the same way as in Table 2.9. Then, a bar graph can be constructed to display the relative frequency or percentage distribution by marking the relative frequencies or percentages, respectively, on the vertical axis.

TABLE 2.11 Frequency Distribution of the Number of Vehicles Owned

Vehicles Owned	Number of Households (f)
0	2
1	18
2	11
3	4
4	3
5	2
	$\sum f = 40$

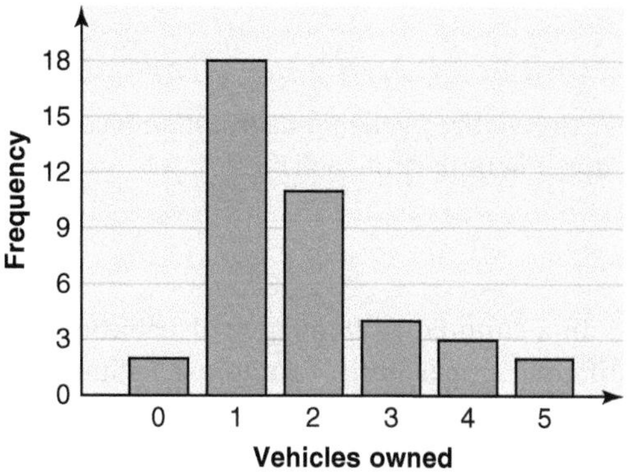

FIGURE 2.8 Bar graph for Table 2.11.

Case Study 2.5

How Many Cups of Coffee Do You Drink a Day?

In a Gallup poll conducted by telephone interviews on July 9–12, 2012, U.S. adults of age 18 years and older were asked, "How many cups of coffee, if any, do you drink on an average day?" According to the results of the poll, shown in the accompanying pie chart, 36% of these adults said that they drink no coffee (represented by zero cups in the chart), 26% said that they drink one cup of coffee per day, and so on. The last class is open-ended class that indicates that 10% of these adults drink four or more cups of coffee a day. This class has no upper limit. Since the values of the variable (cups of coffee) are discrete and the variable assumes only a few possible values, the first four classes are single-valued classes.

Data source: Data from Nearly Half of Americans Drink Soda Daily, Lydia Saad. Gallup, Inc.

2.2.6 Cumulative Frequency Distributions

Consider again Example 2.3 of Section 2.2.2 about the systolic blood pressures of 30 adults. Suppose we want to know how many adults had a systolic blood pressure of 120 or lower. Such a question can be answered by using a **cumulative frequency distribution**. Each class in a cumulative frequency distribution table gives the total number of values that fall below a certain value. A cumulative frequency distribution is constructed for quantitative data only.

Cumulative Frequency Distribution

A *cumulative frequency distribution* gives the total number of values that fall below the upper boundary of each class.

In a cumulative frequency distribution table, each class has the same lower limit but a different upper limit. Example 2.7 illustrates the procedure for preparing a cumulative frequency distribution.

EXAMPLE 2.7 | Systolic Blood Pressure of Adults

Constructing a Cumulative Frequency Distribution Table

Using the frequency distribution of Table 2.8, reproduced here as **Table 2.12**, prepare a cumulative frequency distribution for the systolic blood pressures of those adults.

TABLE 2.12 Systolic Blood Pressures of Adults

Systolic Blood Pressure	Number of Adults (f)
101–110	5
111–120	7
121–130	9
131–140	6
141–150	3

Solution **Table 2.13** gives the cumulative frequency distribution for the systolic blood pressures of adults. As we can observe, 101 (which is the lower limit of the first class in Table 2.12) is taken as the lower limit of each class in Table 2.13. The upper limits of all classes in Table 2.13 are the same as those in Table 2.12. To obtain the cumulative frequency of a class, we add the frequency of that class in Table 2.12 to the frequencies of all preceding classes. The cumulative frequencies are recorded in the second column of Table 2.13.

TABLE 2.13 Cumulative Frequency Distribution of Systolic Blood Pressures

Class Limits	Cumulative Frequency
101–110	5
101–120	$5 + 7 = 12$
101–130	$5 + 7 + 9 = 21$
101–140	$5 + 7 + 9 + 6 = 27$
101–150	$5 + 7 + 9 + 6 + 3 = 30$

From Table 2.13, we can determine the number of observations that fall below the upper limit of each class. For example, 21 adults had a blood pressure of 130 or lower.

The **cumulative relative frequencies** are obtained by dividing the cumulative frequencies by the total number of observations in the data set. The **cumulative percentages** are obtained by multiplying the cumulative relative frequencies by 100.

Calculating Cumulative Relative Frequency and Cumulative Percentage

$$\text{Cumulative relative frequency} = \frac{\text{Cumulative frequency of a class}}{\text{Total observations in the data set}}$$

$$\text{Cumulative percentage} = (\text{Cumulative relative frequency}) \times 100\%$$

Table 2.14 contains both the cumulative relative frequencies and the cumulative percentages for Table 2.13. We can observe, for example, that 70% of adults had a blood pressure between 101 and 130.

TABLE 2.14 Cumulative Relative Frequency and Cumulative Percentage Distributions of Systolic Blood Pressures

Class Limits	Cumulative Relative Frequency	Cumulative Percentage
101–110	5/30 = .1667	16.67
101–120	12/30 = .4000	40.00
101–130	21/30 = .7000	70.00
101–140	27/30 = .9000	90.00
101–150	30/30 = 1.0000	100.00

2.2.7 Ogive Curves

The graphical representation of cumulative distribution or cumulative frequency is known as **ogive curve**. It is also used to envision the cumulative frequency of the data. Ogive curves help to study the characteristics of the distribution and provide understanding about the scatter, median, percentile, skewness, and central tendency of the data.

Ogive Curve

A graph showing the cumulative frequency against the upper boundary of the class is called the *ogive curve*.

Ogive curves can be constructed using two different methods:

1. **Less-than Method** The graphical representation of number of frequencies upto a particular value is known as less-than ogive curve.
2. **More-than Method** The graphical representation of number of frequencies greater than a particular value is known as more-than ogive curve.

Example 2.8 illustrates the concept of ogive curves.

EXAMPLE 2.8 Systolic Blood Pressure of Adults

Using the frequency distribution of Table 2.12, prepare less-than and more-than ogive curves.

Solution **Figures 2.9** and **2.10** show the less-than and more-than ogive curves, respectively. To draw the less-than ogive curve, the upper limits of the classes are shown in **Table 2.15**.

TABLE 2.15

Less-than Systolic Blood Pressure	Less-than Cumulative Frequencies
110	5
120	5 + 7 = 12
130	5 + 7 + 9 = 21
140	5 + 7 + 9 + 6 = 27
150	5 + 7 + 9 + 6 + 3 = 30

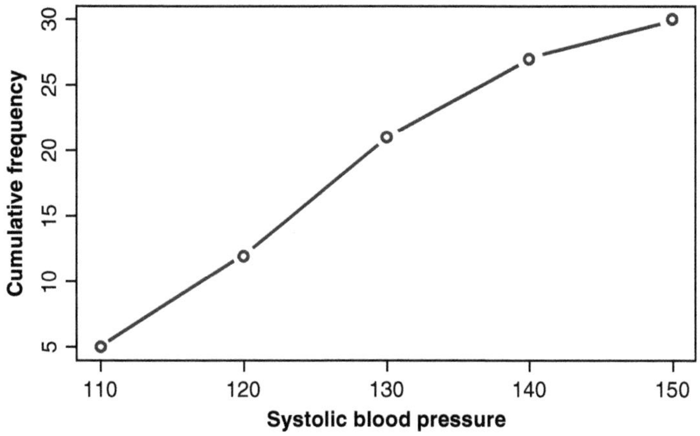

FIGURE 2.9 Less-than ogive curve.

For more-than ogive curve, the lower limits of the classes are shown in **Table 2.16**.

TABLE 2.16

More-than Systolic Blood Pressure	More-than Cumulative Frequencies
101	5 + 7 + 9 + 6 + 3 = 30
111	7 + 9 + 6 + 3 = 25
121	9 + 6 + 3 = 18
131	6 + 3 = 9
141	3

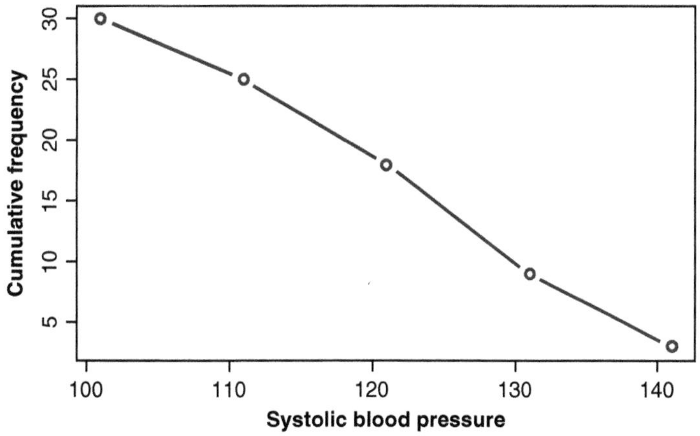

FIGURE 2.10 More-than ogive curve.

2.2.8 Shapes of Histograms

A histogram can assume any one of a large number of shapes. The most common of these shapes are

1. Symmetric
2. Skewed
3. Uniform or rectangular

A **symmetric histogram** is identical on both sides of its central point. The histograms shown in **Figure 2.11** are symmetric around the dashed lines that represent their central points.

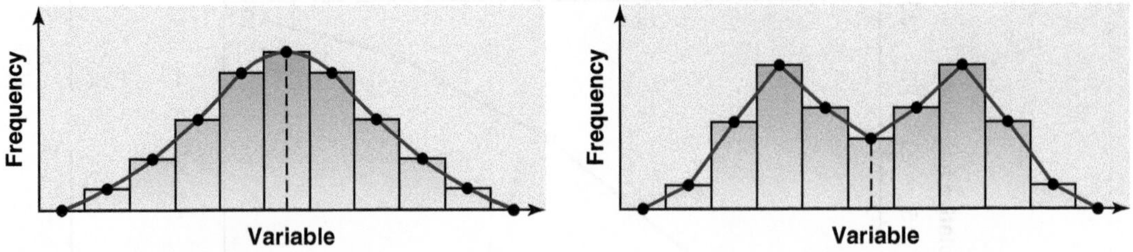

FIGURE 2.11 Symmetric histograms.

A **skewed histogram** is nonsymmetric. For a skewed histogram, the tail on one side is longer than the tail on the other side. A **skewed-to-the-right histogram** has a longer tail on the right side (see **Figure 2.12*a***). A **skewed-to-the-left histogram** has a longer tail on the left side (see **Figure 2.12*b***).

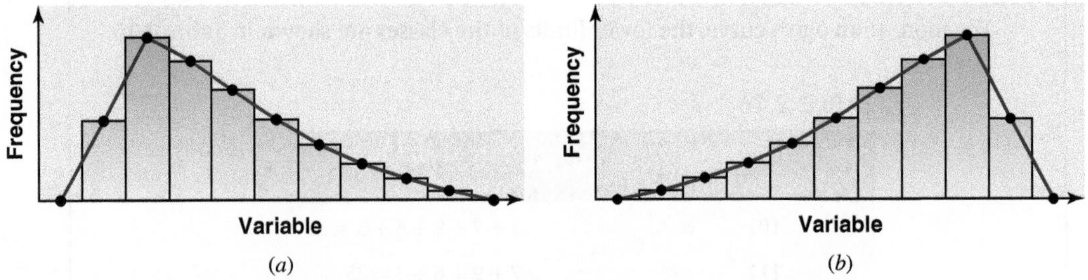

FIGURE 2.12 (*a*) A histogram skewed to the right. (*b*) A histogram skewed to the left.

A **uniform** or **rectangular histogram** has the same frequency for each class. **Figure 2.13** is an illustration of such a case.

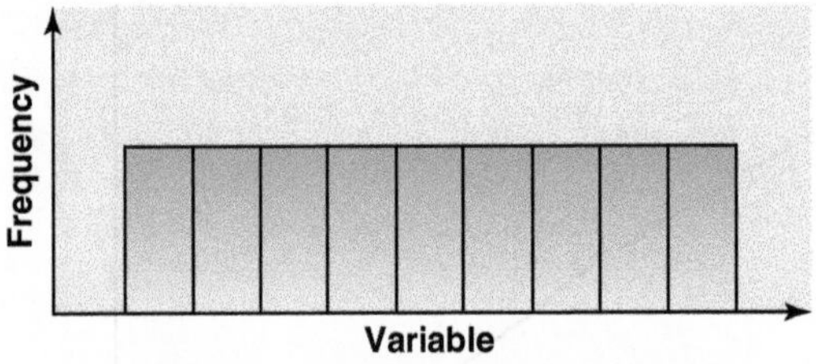

FIGURE 2.13 A histogram with uniform distribution.

Figures 2.14*a* and **2.14*b*** display symmetric frequency curves. **Figures 2.14*c*** and **2.14*d*** show frequency curves skewed to the right and to the left, respectively.

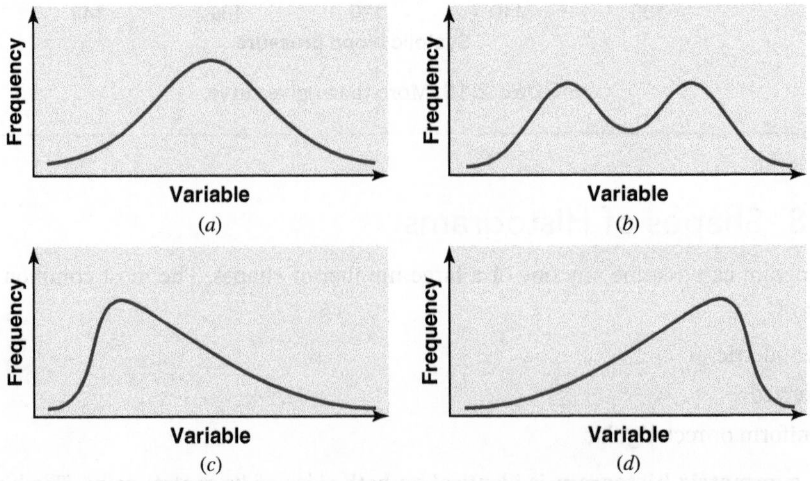

FIGURE 2.14 (*a*), (*b*) Symmetric frequency curves. (*c*) Frequency curve skewed to the right. (*d*) Frequency curve skewed to the left.

2.2.9 Truncating Axes

Describing data using graphs gives us insights into the main characteristics of the data. But graphs, unfortunately, can also be used, intentionally or unintentionally, to distort the facts and deceive the reader. The following are two ways to manipulate graphs to convey a particular opinion or impression.

1. *Changing the scale* either on one or on both axes—that is, shortening or stretching one or both of the axes.
2. *Truncating the frequency axis*—that is, starting the frequency axis at a number greater than zero.

Suppose 400 randomly selected adults were asked whether or not they are happy with their jobs. Of them, 156 said that they are happy, 136 said that they are not happy, and 108 had no opinion. Converting these numbers to percentages, 39% of these adults said that they are happy, 34% said that they are not happy, and 27% had no opinion. Let us denote the three opinions by A, B, and C, respectively. The following table shows the results of this survey.

Opinion	Percentage
A	39
B	34
C	27
	Sum = 100

Now let us make two bar graphs—one showing the complete vertical axis and the second using a truncated vertical axis. **Figure 2.15** shows a bar graph with the complete vertical axis. By looking at this bar graph, we can observe that the opinions represented by three categories in fact differ by small percentages. But now look at **Figure 2.16** in which the vertical axis has been truncated to start at 25%. By looking at this bar chart, if we do not pay attention to the vertical axis, we may erroneously conclude that the opinions represented by three categories vary by large percentages.

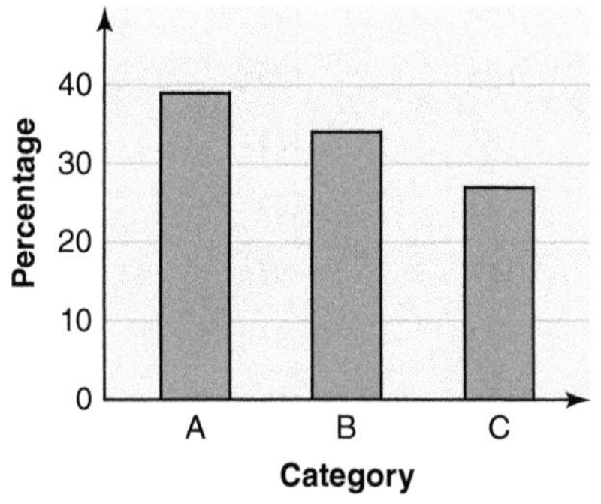

FIGURE 2.15 Bar graph without truncation of the vertical axis.

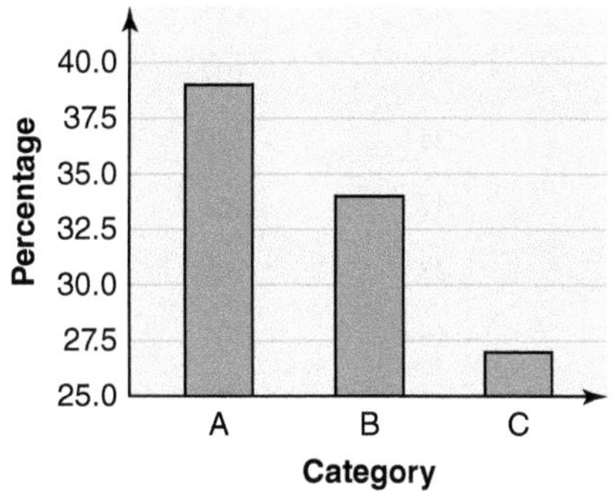

FIGURE 2.16 Bar graph with truncation of the vertical axis.

When interpreting a graph, we should be very cautious. We should observe carefully whether the frequency axis has been truncated or whether any axis has been unnecessarily shortened or stretched.

EXERCISES

Concepts and Procedures

2.8 Briefly explain the three decisions that have to be made to group a data set in the form of a frequency distribution table.

2.9 Explain how to find the relative frequencies and percentages of classes from frequencies of classes with the help of an example.

2.10 Three methods—writing classes using limits, using the *less-than* method, and grouping data using single-valued classes—were discussed to group quantitative data into classes. Explain these three methods and give one example of each.

Applications

2.11 The following data give the number of text messages sent by a high school student on 40 randomly selected days during 2020.

32	33	33	34	35	36	37	37	37	37
38	39	40	41	41	42	42	42	43	44
44	45	45	45	47	47	47	47	47	48
48	49	50	50	51	52	53	54	59	61

a. Construct a frequency distribution table. Take 32 as the lower limit of the first class and 6 as the class width.

b. Calculate the relative frequency and percentage for each class.

c. Construct a histogram for the frequency distribution of part (a).

d. On what percentage of these 40 days did this student send 44 or more text messages?

e. Prepare the cumulative frequency, cumulative relative frequency, and cumulative percentage distributions.

2.12 A data set on money spent on lottery tickets during the past year by 200 households has a lowest value of $1 and a highest value of $983. Suppose we want to group these data into six classes of equal widths.

a. Assuming that we take the lower limit of the first class as $1 and the width of each class equal to $200, write the class limits for all six classes.

b. What are the class boundaries and class midpoints?

2.13 The following data give the one-way commuting times (in minutes) from home to work for a random sample of 50 workers.

23	17	34	26	18	33	46	42	12	37
44	15	22	19	28	32	18	39	40	48
16	11	9	24	18	26	31	7	30	15
18	22	29	32	30	21	19	14	26	37
25	36	23	39	42	46	29	17	24	31

a. Construct a frequency distribution table using the classes 0–9, 10–19, 20–29, 30–39, and 40–49.

b. Calculate the relative frequency and percentage for each class.

c. Construct a histogram for the percentage distribution made in part (b).

d. What percentage of the workers in this sample commute for 30 minutes or more?

e. Prepare the cumulative frequency, cumulative relative frequency, and cumulative percentage distributions using the table of part (a).

f. Construct the more-than ogive curve.

Exercises 2.14 to 2.19 are Based on the Following Data: The following table includes partial data for a health fair held at a local mall by community college nursing students. The data are provided for 30 male and 30 female participants who stopped by the health fair booth. Data includes participant's age to the nearest year, body weight in pounds, and blood glucose level measured in mg/dL.

Participant	Age (Males)	Age (Females)	Weight (Males)	Weight (Females)	Blood Glucose (Males)	Blood Glucose (Females)
1	55	23	197	111	140	124
2	24	38	137	137	106	105
3	45	25	139	136	93	121
4	38	47	261	105	92	109
5	49	23	169	208	117	97
6	41	46	201	117	77	86
7	54	34	176	244	128	131
8	32	35	203	234	95	143
9	30	36	210	258	149	134
10	59	58	168	156	131	90
11	55	58	223	211	130	124
12	21	46	248	163	106	108
13	52	26	151	99	108	110
14	61	28	220	162	143	144

Participant	Age (Males)	Age (Females)	Weight (Males)	Weight (Females)	Blood Glucose (Males)	Blood Glucose (Females)
15	56	53	211	253	76	81
16	55	33	104	119	124	109
17	52	62	262	231	89	81
18	33	35	122	258	107	85
19	34	31	136	138	103	135
20	33	29	156	139	100	105
21	31	20	222	232	98	97
22	27	34	174	140	112	132
23	34	50	230	236	121	113
24	48	47	167	234	105	149
25	25	44	262	149	81	99
26	47	21	234	196	134	149
27	49	55	146	202	113	75
28	22	43	214	229	93	85
29	35	53	130	220	124	126
30	29	31	232	190	76	148

2.14 **a.** Construct a frequency distribution table for ages of male participants using the classes 20–29, 30–39, 40–49, 50–59, and 60–69.

b. Calculate the relative frequency and percentage for each class.

c. Construct a histogram for the frequency distribution of part (a).

d. What percentage of the male participants are younger than 40?

2.15 **a.** Construct a frequency distribution table for ages of female participants using the classes 20–29, 30–39, 40–49, 50–59, and 60–69.

b. Calculate the relative frequency and percentage for each class.

c. Construct a histogram for the frequency distribution of part (a).

d. What percentage of the female participants are younger than 40?

e. Compare the histograms for Exercises 2.14 and 2.15 and mention the similarities and differences.

2.16 **a.** Construct a frequency distribution table for weights of male participants using the classes 91–125, 126–160, 161–195, 196–230, and 231–265.

b. Calculate the relative frequency and percentage for each class.

c. Construct a histogram for the relative frequency distribution of part (b).

d. What percentage of the male participants have weights less than 161 lbs?

2.17 **a.** Construct a frequency distribution table for weights of female participants using the classes 91–125, 126–160, 161–195, 196–230, and 231–265.

b. Calculate the relative frequency and percentage for each class.

c. Construct a histogram for the relative frequency distribution of part (a).

d. What percentage of the female participants have weights less than 161 lbs?

e. Compare the histograms for Exercises 2.16 and 2.17 and mention the similarities and differences.

2.18 **a.** Construct a frequency distribution table for blood glucose levels of male participants using the classes 75–89, 90–104, 105–119, 120–134, and 135–149.

b. Calculate the relative frequency and percentage for each class.

c. Construct a histogram for the percentage distribution of part (b).

d. What percentage of the male participants have blood glucose levels more than 119?

e. Prepare the cumulative frequency, cumulative relative frequency, and cumulative percentage distributions using the table of part (a).

f. Construct the less-than ogive curve using the table of part (a).

2.19 **a.** Construct a frequency distribution table for blood glucose levels of female participants using the classes 75–89, 90–104, 105–119, 120–134, and 135–149.

b. Calculate the relative frequency and percentage for each class.

c. Construct a histogram for the percentage distribution of part (b).

d. What percentage of the female participants have blood glucose levels more than 119?

e. Compare the histograms for Exercises 2.18 and 2.19 and mention the similarities and differences.

f. Prepare the cumulative frequency, cumulative relative frequency, and cumulative percentage distributions using the table of part (a).

g. Construct the more-than ogive curve using the table of part (a).

2.20 The following table lists the number of home runs for each of the 30 Major League baseball teams during the 2019 regular season (https://www.mlb.com).

a. Construct a frequency distribution table. Take 140 as the lower boundary of the first class and 30 as the width of each class.

b. Prepare the relative frequency and percentage distribution columns for the frequency distribution table of part (a).

DATA FOR EXERCISE 2.20

Team	Home Runs	Team	Home Runs	Team	Home Runs
Arizona Diamondbacks	220	Houston Astros	288	Philadelphia Phillies	215
Atlanta Braves	249	Kansas City Royals	162	Pittsburgh Pirates	163
Baltimore Orioles	213	Los Angeles Angels	220	San Diego Padres	219
Boston Red Sox	245	Los Angeles Dodgers	279	San Francisco Giants	167
Chicago Cubs	256	Miami Marlins	146	Seattle Mariners	239
Chicago White Sox	182	Milwaukee Brewers	250	St. Louis Cardinals	210
Cincinnati Reds	227	Minnesota Twins	307	Tampa Bay Rays	217
Cleveland Indians	223	New York Mets	242	Texas Rangers	223
Colorado Rockies	224	New York Yankees	306	Toronto Blue Jays	247
Detroit Tigers	149	Oakland Athletics	257	Washington Nationals	231

2.21 The following data give the number of turnovers (fumbles and interceptions) by a college football team for each game in the past two seasons.

3	1	1	4	0	4	2	3	0	3	2	3
0	2	3	1	4	1	3	2	4	0	0	2

a. Prepare a frequency distribution table for these data using single-valued classes.

b. Calculate the relative frequencies and percentages for all classes.

c. In how many games did the team commit two or more turnovers?

d. Draw a bar graph for the frequency distribution of part (a).

2.22 The following table gives the frequency distribution for the number of parking tickets received on the campus of a university during the past week by 200 students.

Number of Tickets	Number of Students
0	69
1	40
2	37
3	26
4	28

Draw two bar graphs for these data, the first without truncating the frequency axis and the second by truncating the frequency axis. In the second case, mark the frequencies on the vertical axis starting with 25. Briefly comment on the two bar graphs.

2.3 Stem-and-Leaf Displays

LEARNING OBJECTIVES

After completing this section, you should be able to:

- Construct and interpret a stem-and-leaf display for quantitative data.
- Rank the leaves of a stem-and-leaf display.
- Construct a grouped stem-and-leaf display.
- Discuss the advantage a stem-and-leaf display has over a frequency distribution.
- Construct and interpret a stem-and-leaf display with split stems.

Another technique that is used to present quantitative data in condensed form is the **stem-and-leaf display**. An advantage of a stem-and-leaf display over a frequency distribution is that by preparing a stem-and-leaf display we do not lose information on individual observations. A stem-and-leaf display is constructed only for quantitative data.

Stem-and-Leaf Display

In a *stem-and-leaf display* of quantitative data, each value is divided into two portions—a stem and a leaf. The leaves for each stem are shown separately in a display.

Example 2.9 describes the procedure for constructing a stem-and-leaf display.

EXAMPLE 2.9 Scores of Students on a Statistics Test

Constructing a Stem-and-Leaf Display for Two-Digit Numbers

The following are the scores of 30 college students on a statistics test.

75	52	80	96	65	79	71	87	93	95
69	72	81	61	76	86	79	68	50	92
83	84	77	64	71	87	72	92	57	98

Construct a stem-and-leaf display.

Solution To construct a stem-and-leaf display for these scores, we split each score into two parts. The first part contains the first digit of a score, which is called the *stem*. The second part contains the second digit of a score, which is called the *leaf*. Thus, for the score of the first student, which is 75, 7 is the stem and 5 is the leaf. For the score of the second student, which is 52, the stem is 5 and the leaf is 2. We observe from the data that the stems for all scores are 5, 6, 7, 8, and 9 because all these scores lie in the range 50 to 98. To create a stem-and-leaf display, we draw a vertical line and write the stems on the left side of it, arranged in increasing order, as shown in **Figure 2.17**.

```
Stems
  |
  v
  5 | 2 <---- Leaf for 52
  6 |
  7 | 5 <---- Leaf for 75
  8 |
  9 |
```

FIGURE 2.17 Stem-and-leaf display.

After we have listed the stems, we read the leaves for all scores and record them next to the corresponding stems on the right side of the vertical line. For example, for the first score we write the leaf 5 next to stem 7; for the second score we write the leaf 2 next to stem 5. The recording of these two scores in a stem-and-leaf display is shown in Figure 2.17.

Now, we read all scores and write the leaves on the right side of the vertical line in the rows of corresponding stems. The complete stem-and-leaf display for scores is shown in **Figure 2.18**.

```
5 | 2 0 7
6 | 5 9 1 8 4
7 | 5 9 1 2 6 9 7 1 2
8 | 0 7 1 6 3 4 7
9 | 6 3 5 2 2 8
```

FIGURE 2.18 Stem-and-leaf display of test scores.

By looking at the stem-and-leaf display of Figure 2.18, we can observe how the data values are distributed. For example, the scores in the 70s have the highest frequency, followed by the scores in the 80s, 90s, and 60s. Scores in the 50s have the lowest frequency.

Finally, the leaves for each stem of the stem-and-leaf display of Figure 2.18 are *ranked* (in increasing order) and presented in **Figure 2.19**.

```
5 | 0 2 7
6 | 1 4 5 8 9
7 | 1 1 2 2 5 6 7 9 9
8 | 0 1 3 4 6 7 7
9 | 2 2 3 5 6 8
```

FIGURE 2.19 Ranked stem-and-leaf display of test scores.

As already mentioned, one advantage of a stem-and-leaf display is that we do not lose information on individual observations. We can rewrite the individual scores of the 30 college students from the stem-and-leaf display of Figure 2.18 or Figure 2.19. By contrast, the information on individual observations is lost when data are grouped into a frequency table.

EXAMPLE 2.10 Monthly Rents Paid by Households

Constructing a Stem-and-Leaf Display for Three- and Four-Digit Numbers

The following data give the monthly rents paid by a sample of 30 households selected from a small town.

880	1081	721	1075	1023	775	1235	750	965	960
1210	985	1231	932	850	825	1000	915	1191	1035
1151	630	1175	952	1100	1140	750	1140	1370	1280

Construct a stem-and-leaf display for these data.

Solution Each of the values in the data set contains either three or four digits. We will take the first digit for three-digit numbers and the first two digits for four-digit numbers as stems. Then we will use the last two digits of each number as a leaf. Thus, for the first value, which is 880, the stem is 8 and the leaf is 80. The stems for the entire data set are 6, 7, 8, 9, 10, 11, 12, and 13. They are recorded on the left side of the vertical line in **Figure 2.20**. The leaves for the numbers are recorded on the right side.

6	30
7	21 75 50 50
8	80 50 25
9	65 60 85 32 15 52
10	81 75 23 00 35
11	91 51 75 00 40 40
12	35 10 31 80
13	70

FIGURE 2.20 Stem-and-leaf display of rents.

Best practice for a stem-and-leaf plot is always to rank the leaves. That way, we can quickly determine the median of the distribution. Without ranking the leaves, an incorrect median will be obtained. By ranking the leaves of the stem-and-leaf plot of Figure 2.20, we write the following stem-and-leaf plot:

6	30
7	21 50 50 75
8	25 50 80
9	15 32 52 60 65 85
10	00 23 35 75 81
11	00 40 40 51 75 91
12	10 31 35 80
13	70

Sometimes a data set may contain too many stems, with each stem containing only a few leaves. In such cases, we may want to condense the stem-and-leaf display by *grouping the stems*. Example 2.11 describes this procedure.

EXAMPLE 2.11 Number of Hours Spent Working on Computers by Students

Preparing a Grouped Stem-and-Leaf Display

The following stem-and-leaf display is prepared for the number of hours that 25 students spent working on computers during the past month.

0	6
1	1 7 9
2	2 6
3	2 4 7 8
4	1 5 6 9 9
5	3 6 8
6	2 4 4 5 7
7	
8	5 6

Mark Harmel/Getty Images

Prepare a new stem-and-leaf display by grouping the stems.

Solution To condense the given stem-and-leaf display, we can combine the first three rows, the middle three rows, and the last three rows, thus getting the stems 0–2, 3–5, and 6–8. The leaves for each stem of a group are separated by an asterisk (*), as shown in **Figure 2.21**. Thus, the leaf 6 in the first row corresponds to stem 0; the leaves 1, 7, and 9 correspond to stem 1; and leaves 2 and 6 belong to stem 2.

0–2	6 * 1 7 9 * 2 6
3–5	2 4 7 8 * 1 5 6 9 9 * 3 6 8
6–8	2 4 4 5 7 * * 5 6

FIGURE 2.21 Grouped stem-and-leaf display.

If a stem does not contain a leaf, this is indicated in the grouped stem-and-leaf display by two consecutive asterisks. For example, in the stem-and-leaf display of Figure 2.21, there is no leaf for 7; that is, there is no number in the 70s. Hence, in Figure 2.21, we have two asterisks after the leaves for 6 and before the leaves for 8.

Some data sets produce stem-and-leaf displays that have a small number of stems relative to the number of observations in the data set and have too many leaves for each stem. In such cases, it is very difficult to determine if the distribution is symmetric or skewed, as well as other characteristics of the distribution that will be introduced in later chapters. In such a situation, we can create a stem-and-leaf display with *split stems*. To do this, each stem is split into two or five parts. Whenever the stems are split into two parts, any observation having a leaf with a value of 0, 1, 2, 3, or 4 is placed in the first split stem, while the leaves 5, 6, 7, 8, and 9 are placed in the second split stem. Sometimes we can split a stem into five parts if there are too many leaves for one stem. Whenever a stem is split into five parts, leaves with values of 0 and 1 are placed next to the first part of the split stem, leaves with values of 2 and 3 are placed next to the second part of the split stem, and so on. The stem-and-leaf display of Example 2.12 shows this procedure.

EXAMPLE 2.12

Stem-and-Leaf Display with Split Stems

Consider the following stem-and-leaf display, which has only two stems. Using the split stem procedure, rewrite this stem-and-leaf display.

3	1 1 2 3 3 3 4 4 7 8 9 9 9
4	0 0 0 1 1 1 1 1 2 2 2 2 2 3 3 6 6 7

Solution To prepare a split stem-and-leaf display, let us split the two stems, 3 and 4, into two parts each as shown in **Figure 2.22.** The first part of each stem contains leaves from 0 to 4, and the second part of each stem contains leaves from 5 to 9.

3	1 1 2 3 3 3 4 4
3	7 8 9 9 9
4	0 0 0 1 1 1 1 1 1 2 2 2 2 2 3 3
4	6 6 7

FIGURE 2.22 Split stem-and-leaf display.

In the stem-and-leaf display of Figure 2.22, the first part of stem 4 has a substantial number of leaves. So, if we decide to split stems into five parts, the new stem-and-leaf display will look as shown in **Figure 2.23.**

3	1 1
3	2 3 3 3
3	4 4
3	7
3	8 9 9 9
4	0 0 0 1 1 1 1 1 1
4	2 2 2 2 2 3 3
4	
4	6 6 7

FIGURE 2.23 Split stem-and-leaf display.

There are two important properties to note in the split stem-and-leaf display of Figure 2.23. The third part of split stem 4 does not have any leaves. This implies that there are no observations in the data set having a value of 44 or 45. Since there are observations with values larger than 45, we need to leave an empty part of split stem 4 that corresponds to 44 and 45. Also, there are no observations with values of 48 or 49. However, since there are no values larger than 47 in the data, we do not have to write an empty split stem 4 after the largest value.

EXERCISES

Concepts and Procedures

2.23 Illustrate how to construct a stem-and-leaf plot for a given data set using an example.

2.24 When is it more appropriate to use a stem-and-leaf display for grouping a data set rather than a frequency distribution? Give one example.

2.25 Consider the following stem-and-leaf display:

2–3	18 45 56 * 29 67 83 97
4–5	04 27 33 71 * 23 37 51 63 81 92
6–8	22 36 47 55 78 89 * * 10 41

Write the data set that is represented by this display.

Applications

2.26 The U. S. National Highway Traffic Safety Administration collects data on fatal accidents that occur on roads in the United States. The following data represent the number of fatal collisions for all 46 counties in South Carolina for 2017 (*Source*: South Carolina Traffic Collision Fact Book).

8	25	12	4	4	5	17	51
32	33	18	13	61	44	27	18
2	7	18	30	9	3	18	6
42	66	15	17	19	7	49	14
3	20	7	73	17	5	6	25
6	13	15	9	25	7		

Prepare a stem-and-leaf display for these data. Arrange the leaves for each stem in increasing order.

2.27 The following data give the times (in minutes) taken by 50 students to complete a statistics examination that was given a maximum time of 75 minutes to finish.

41	28	45	60	53	69	70	50	63	68
37	44	42	38	74	53	66	65	52	64
26	45	66	35	43	44	39	55	64	54
38	52	58	72	67	65	43	65	68	27
64	49	71	75	45	69	56	73	53	72

a. Prepare a stem-and-leaf display for these data. Arrange the leaves for each stem in increasing order.

b. Prepare a split stem-and-leaf display for the data. Split each stem into two parts. The first part should contain the leaves 0, 1, 2, 3, and 4, and the second part should contain the leaves 5, 6, 7, 8, and 9.

c. Which display (the one in part (a) or the one in part (b)) provides a better representation of the features of the distribution? Explain why you believe this.

2.28 The following data give the taxes paid (rounded to thousand dollars) in 2019 by a random sample of 30 families.

11	17	35	3	15	9	21	13	5	19
5	12	8	16	10	8	12	6	14	18
8	12	5	3	14	28	38	18	22	15

a. Prepare a stem-and-leaf display for these data. Arrange the leaves for each stem in increasing order.

b. Prepare a split stem-and-leaf display for these data. Split each stem into two parts. The first part should contain the leaves 0 through 4, and the second part should contain the leaves 5 through 9.

2.29 The following data give the times (in minutes) taken to commute from home to work for 20 workers.

10	50	65	33	48	5	11	23	39	26
26	32	17	7	15	19	29	43	21	22

Construct a stem-and-leaf display for these data. Arrange the leaves for each stem in increasing order.

2.30 The following data give the times served (in months) by 35 prison inmates who were released recently.

37	6	20	5	25	30	24	10	12	20
24	8	26	15	13	22	72	80	96	33
84	86	70	40	92	36	28	90	36	32
72	45	38	18	9					

a. Prepare a stem-and-leaf display for these data.

b. Condense the stem-and-leaf display by grouping the stems as 0–2, 3–5, and 6–9.

2.4 Dotplots

LEARNING OBJECTIVES

After completing this section, you should be able to:

- Construct and interpret a dotplot for quantitative data.
- Discuss the benefits of a dotplot.

One of the simplest methods for graphing and understanding quantitative data is to create a dotplot. As with most graphs, statistical software should be used to make a dotplot for large data sets. However, Example 2.13 demonstrates how to create a dotplot by hand.

Dotplots can help us detect **outliers** (also called **extreme values**) in a data set. Outliers are the values that are extremely large or extremely small with respect to the rest of the data values.

Outliers or Extreme Values

Values that are very small or very large relative to the majority of the values in a data set are called *outliers* or *extreme values*.

EXAMPLE 2.13 Ages of Students in a Night Class

Creating a Dotplot

A statistics class that meets once a week at night from 7:00 PM to 9:45 PM has 33 students. The following data give the ages (in years) of these students. Create a dotplot for these data.

34	21	49	37	23	22	33	23	21	20	19
33	23	38	32	31	22	20	24	27	33	19
23	21	31	31	22	20	34	21	33	27	21

Solution To make a dotplot, we perform the following steps:

STEP 1 The minimum and maximum values in this data set are 19 and 49 years, respectively. First, we draw a horizontal line (let us call this the *numbers line*) with numbers that cover the given data as shown in **Figure 2.24**. Note that the numbers line in Figure 2.24 shows the values from 19 to 49.

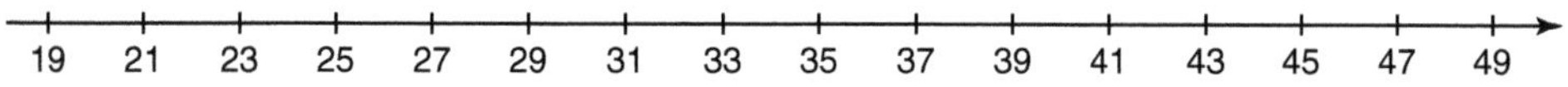

FIGURE 2.24 Numbers line.

STEP 2 Next we place a dot above the value on the numbers line that represents each of the ages listed above. For example, the age of the first student is 34 years. So, we place a dot above 34 on the numbers line as shown in **Figure 2.25**. If there are two or more observations with the same value, we stack dots above each other to represent those values. For example, as shown in the data on ages, two students are 19 years old. We stack two dots (one for each student) above 19 on the numbers line, as shown in Figure 2.25. After all the dots are placed, Figure 2.25 gives the complete dotplot.

As we examine the dotplot of Figure 2.25, we notice that there are two clusters (groups) of data. Eighteen of the 33 students (which is almost 55%) are 19 to 24 years old, and 10 of the 33 students (which is about 30%) are 31 to 34 years old. There is one student who is 49 years old and is an outlier.

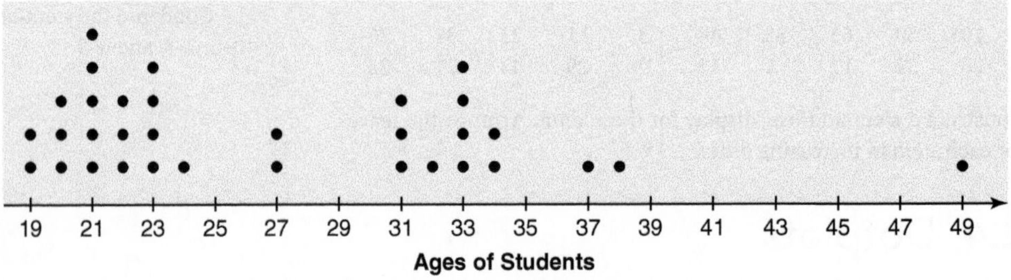

FIGURE 2.25 Dotplot for ages of students in a statistics class.

EXERCISES

Concepts and Procedures

2.31 Briefly explain how to prepare a dotplot for a data set. You may use an example to illustrate.

2.32 What are the benefits of preparing a dotplot? Explain.

2.33 Create a dotplot for the following data set:

1	2	0	5	1	1	3	2	0	5
2	1	2	1	2	0	1	3	1	2

Applications

2.34 The U.S. National Highway Traffic Safety Administration collects data on fatal accidents that occur on roads in the United States. The following data represent the number of fatal collisions for all 46 counties in South Carolina for 2017 (*Source*: South Carolina Traffic Collision Fact Book).

8	25	12	4	4	5	17	51
32	33	18	13	61	44	27	18
2	7	18	30	9	3	18	6
42	66	15	17	19	7	49	14
3	20	7	73	17	5	6	25
6	13	15	9	25	7		

Make a dotplot for these data.

2.35 The following data give the number of text messages sent by a high school student on 40 randomly selected days during 2020.

32	33	33	34	35	36	37	37	37	37
38	39	40	41	41	42	42	42	43	44
44	45	45	45	47	47	47	47	47	48
48	49	50	50	51	52	53	54	59	61

Create a dotplot for these data.

2.36 The following data give the number of orders received for a sample of 30 hours at the Timesaver Mail Order Company.

34	44	31	52	41	47	38	35	32	39
28	24	46	41	49	53	57	33	27	37
30	27	45	38	34	46	36	30	47	50

Create a dotplot for these data.

2.37 The following data represent the ages of the top 25 National Football League quarterbacks as ranked by www.sportingnews.com. Each player's age is as of the first day of the 2019 NFL season (September 5, 2019). Create a dotplot for these data.

23	29	26	27	25
35	37	24	31	22
42	34	23	31	22
40	24	26	30	30
30	37	25	25	23

Uses and Misuses of Statistics

Graphically Speaking...

Imagine a high school in Middle America, if you will. We shall call it Central High School. In Central High there are a fair number of very talented athletes who play soccer, baseball, football, and basketball. At the annual fundraiser for the school, the cheer squad sells advance tickets to the home games so that the community can be entertained by the local school athletes.

As it happens, the cheer squad was able to sell a total of 175 tickets at this year's fundraising event. The ticket sales for the various sports were as follows:

Sport	Number of Tickets Sold
Soccer	23
Baseball	30
Football	63
Basketball	59

In order to show the results of the ticket sales to the Central High community, some of the students got together and decided to make an attractive graph to display the ticket sales in the local paper. As dutiful students, they carefully investigated the various types of charts that might work, and ultimately decided to arrange the graph by sport on the horizontal axis and the number of tickets sold on the vertical axis. To make the graph a bit more vivid, they embellished it using a *pictogram* (that is, a graph made with pictures). They used the following graph to publish in the local paper.

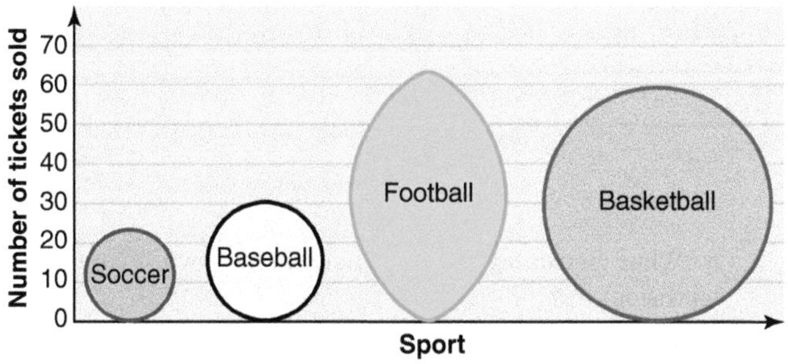

The graph was very well received by the readers of the local paper, and was very attractive to the eye. Further, the football and basketball teams and their supporters were happy to see just how much more popular their respective sports were compared to soccer and baseball. However, there is a major problem with the graph. This graph had the intent of showing the number of tickets sold. This means that we should focus on the *height* of the balls for each sport, which, when read horizontally across to the vertical axis, does in fact match the number of tickets reported to have been sold in the table above. But the eye is easily deceived. What is immediately clear is that we tend to focus on the relative sizes of the balls, not the heights that they represent. In that context, it seems that a huge number of basketball tickets were sold compared to the other sports because the eye focuses immediately on the circumference or the area of the circle represented by the basketball, which appears much larger than the other balls. In fact, comparing soccer to basketball, the area of the basketball in the graph is 6.6 times larger than that of the soccer ball, whereas the tickets sold for basketball were about 2.6 times more than the soccer tickets sold.

A more conservative approach would have been to use a standard traditional *bar chart*. The following bar graph was constructed using the same information as in the pictogram.

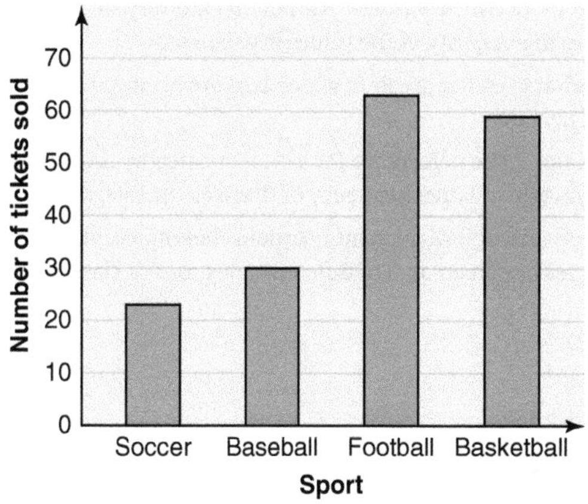

Obviously, this graph does not have the same eye appeal as the one reported in the newspaper; however, it is much clearer to an observer. It is obvious from this graph that the number of soccer tickets and the number of baseball tickets were quite close, as were the number of football and the number of basketball tickets. It is also clear that the number of football tickets is about three times larger than the number of soccer tickets and that the number of football tickets is about two times larger than the number of baseball tickets. These relationships were much more difficult to discern in the pictogram.

In 1954, Darrell Huff wrote a fascinating and entertaining book titled *How to Lie with Statistics* (Huff, Darrell, *How to Lie with Statistics*, 1954, New York: W. W. Norton). This book contains many illustrations of how graphs can be used to deceive—either intentionally or unintentionally—those doing a visual interpretation of data. As Mr. Huff states in his book, "The crooks already know these tricks; honest men must learn them in self-defense." In that regard, it is sometimes useful to learn how *not* to do something in order to understand the importance of doing it correctly.

Glossary

Bar graph A graph made of bars whose heights represent the frequencies of respective categories.

Class An interval that includes all the values in a (quantitative) data set that fall within two numbers, the lower and upper limits of the class.

Class boundary The lower and upper numbers of a class interval in less-than method.

Class frequency The number of values in a data set that belong to a certain class.

Class midpoint or mark The class midpoint or mark is obtained by dividing the sum of the lower and upper limits (or boundaries) of a class by 2.

Class width or size The difference between the two boundaries of a class or the difference between the lower limits of two consecutive classes.

Cumulative frequency The frequency of a class that includes all values in a data set that fall below the upper boundary or limit of that class.

Cumulative frequency distribution A table that lists the total number of values that fall below the upper boundary or limit of each class.

Cumulative percentage The cumulative relative frequency multiplied by 100.

Cumulative relative frequency The cumulative frequency of a class divided by the total number of observations.

Frequency distribution A table that lists all the categories or classes and the number of values that belong to each of these categories or classes.

Grouped data A data set presented in the form of a frequency distribution.

Histogram A graph in which classes are marked on the horizontal axis and frequencies, relative frequencies, or percentages are marked on the vertical axis. The frequencies, relative frequencies, or percentages of various classes are represented by the heights of bars that are drawn adjacent to each other.

Ogive curve The graphical representation of cumulative distribution or cumulative frequency.

Outliers or Extreme values Values that are very small or very large relative to the majority of the values in a data set.

Pareto chart A bar graph in which bars are arranged in decreasing order of heights.

Percentage The percentage for a class or category is obtained by multiplying the relative frequency of that class or category by 100.

Pie chart A circle divided into portions that represent the relative frequencies or percentages of different categories or classes.

Polygon A graph formed by joining the midpoints of the tops of successive bars in a histogram by straight lines.

Raw data Data recorded in the sequence in which they are collected and before they are processed.

Relative frequency The frequency of a class or category divided by the sum of all frequencies.

Skewed-to-the-left histogram A histogram with a longer tail on the left side.

Skewed-to-the-right histogram A histogram with a longer tail on the right side.

Stem-and-leaf display A display of data in which each value is divided into two portions—a stem and a leaf.

Symmetric histogram A histogram that is identical on both sides of its central point.

Ungrouped data Data containing information on each member of a sample or population individually.

Uniform or rectangular histogram A histogram with the same frequency for all classes.

Exercises

Supplementary Exercises

2.38 A whatjapanthinks.com survey asked residents of Japan to name their favorite pizza topping. The possible responses included the following choices: pig-based meats, for example, bacon or ham (PI); seafood, for example, tuna, crab, or cod roe (S); vegetables and fruits (V); poultry (PO); beef (B); and cheese (C). The following data represent the responses of a random sample of 36 people.

V	PI	B	PI	V	PO	S	PI	V	S	V	S
PI	S	V	V	V	PI	S	S	V	PI	C	V
V	V	C	V	S	PO	V	PI	S	PI	PO	PI

a. Prepare a frequency distribution table.

b. Calculate the relative frequencies and percentages for all categories.

c. What percentage of the respondents mentioned vegetables and fruits, poultry, or cheese?

d. Make a Pareto chart for the relative frequency distribution.

2.39 The following data give the number of television sets owned by 40 randomly selected households.

1	1	2	3	2	4	1	3	2	1
3	0	2	1	2	3	2	3	2	2
1	2	1	1	1	3	1	1	1	2
2	4	2	3	1	3	1	2	2	4

a. Prepare a frequency distribution table for these data using single-valued classes.

b. Compute the relative frequency and percentage distributions.

c. Draw a bar graph for the frequency distribution.

d. What percentage of the households own two or more television sets?

2.40 A local gas station collected data from the day's receipts, recording the gallons of gasoline each customer purchased. The following table lists the frequency distribution of the gallons of gas purchased by all customers on this one day at this gas station.

Gallons of Gas	Number of Customers
0 to less than 4	31
4 to less than 8	78
8 to less than 12	49
12 to less than 16	81
16 to less than 20	117
20 to less than 24	13

a. How many customers were served on this day at this gas station?

b. Find the class midpoints. Do all of the classes have the same width? If so, what is this width? If not, what are the different class widths?

c. Prepare the relative frequency and percentage distribution columns.

d. What percentage of the customers purchased 12 gallons or more?

e. Explain why you cannot determine exactly how many customers purchased 10 gallons or less.

f. Prepare the cumulative frequency, cumulative relative frequency, and cumulative percentage distributions using the given table.

g. Construct the less-than ogive curve.

2.41 The following data give the amounts spent on video rentals (in dollars) during 2016 by 36 households randomly selected from those who rented videos in 2016.

595	24	6	100	100	40	622	405	90
55	155	760	405	90	205	70	180	88
808	100	240	127	83	310	350	160	22
111	70	15	22	101	83	121	18	436

a. Construct a frequency distribution table. Take \$1 as the lower limit of the first class and \$200 as the width of each class.

b. Calculate the relative frequencies and percentages for all classes.

c. What percentage of the households in this sample spent more than \$400 on video rentals in 2016?

2.42 The following data give the amounts (in dollars) spent on refreshments by 30 spectators randomly selected from those who patronized the concession stands at a recent Major League Baseball game.

4.95	27.99	8.00	5.80	4.50	2.99	4.85	6.00
9.00	15.75	9.50	3.05	5.65	21.00	16.60	18.00
21.77	12.35	7.75	10.45	3.85	28.45	8.35	17.70
19.50	11.65	11.45	3.00	6.55	16.50		

a. Construct a frequency distribution table using the *less-than* method to write classes. Take \$0 as the lower boundary of the first class and \$6 as the width of each class.

b. Calculate the relative frequencies, and percentages for all classes.

c. Draw a histogram for the frequency distribution.

d. Prepare the cumulative frequency, cumulative relative frequency, and cumulative percentage distributions.

2.43 The following table lists the average one-way commuting times (in minutes) from home to work for a random sample of 30 metropolitan areas in the United States (factfinder.census.gov).

City	Average Commute Length (minutes)	City	Average Commute Length (minutes)
Akron, OH	23.4	Laredo, TX	21.7
Amarillo, TX	18.6	Myrtle Beach, SC	22.5
Beckley, WV	25.7	Odessa, TX	22.0
Bellingham, WA	21.0	Omaha, NE	20.2
Blacksburg, VA	21.1	Ontario, OR	19.0
Bremerton, WA	30.3	Oshkosh, WI	18.2
Columbus, IN	19.6	Providence, RI	25.8
Erie, PA	19.3	Sacramento, CA	26.8
Fairbanks, AK	18.9	San Francisco, CA	32.8
Fort Wayne, IN	21.5	Seattle, WA	29.8
Grand Junction, CO	20.2	Sioux Falls, SD	18.6
Hot Springs, AZ	21.9	St. Joseph, MO	18.8
Iowa City, IA	19.1	Tacoma, WA	31.2
Jackson, TN	19.9	Tyler, TX	23.8
Lake Havasu City, AZ	20.4	Winchester, VA	30.7

a. Prepare a frequency distribution table for these data using the less-than method to write the classes. Use 18 minutes as the lower boundary of the first class and 3 minutes as the class width.

b. Calculate the relative frequencies and percentages for all classes.

c. Draw a histogram for the frequency distribution.

d. Make the cumulative frequency, cumulative relative frequency, and cumulative percentages distributions.

2.44 The following data give the times served (in months) by 35 prison inmates who were released recently.

37	4	20	5	25	28	24	10	12	20
24	8	26	15	13	22	72	80	96	33
84	86	70	40	92	36	24	90	36	30
72	45	38	10	9					

Prepare a stem-and-leaf display for these data.

2.45 The following data give the number of orders received for a sample of 30 hours at the Timesaver Mail Order Company.

34	44	31	52	41	47	38	35	32	39
28	24	46	41	49	53	57	33	27	37
30	27	45	38	34	46	36	30	47	50

Prepare a stem-and-leaf display for these data.

2.46 The following data give the times (in minutes) taken by 50 students to complete a statistics examination that was given a maximum time of 75 minutes to finish.

41	28	45	60	53	69	70	50	63	68
37	44	42	38	74	53	66	65	52	64
26	45	66	35	43	44	39	55	64	54
38	52	58	72	67	65	43	65	68	27
64	49	71	75	45	69	56	73	53	72

Create a dotplot for these data.

2.47 The following data give the one-way commuting time (in minutes) from home to work for a random sample of 50 workers.

23	17	34	26	18	33	46	42	12	37
44	15	22	19	28	32	18	39	40	48
16	11	9	24	18	26	31	7	30	15
18	22	29	32	30	21	19	14	26	37
25	36	23	39	42	46	29	17	24	31

Create a dotplot for these data.

Advanced Exercises

2.48 The following frequency distribution table gives the age distribution of drivers who were at fault in auto accidents that occurred during a one-week period in a city.

Age (years)	*f*
18 to less than 20	7
20 to less than 25	12
25 to less than 30	18
30 to less than 40	14
40 to less than 50	15
50 to less than 60	16
60 and over	35

a. Draw a relative frequency histogram for this table.

b. In what way(s) is this histogram misleading?

c. How can you change the frequency distribution so that the resulting histogram gives a clearer picture?

2.49 Suppose a data set contains the ages of 135 autoworkers ranging from 20 to 53 years.

a. Using Sturge's formula given in footnote 1 in section 2.2.2, find an appropriate number of classes for a frequency distribution for this data set.

b. Find an appropriate class width based on the number of classes in part (a).

2.50 Stem-and-leaf displays can be used to compare distributions for two groups using a back-to-back stem-and-leaf display. In such a display, one group is shown on the left side of the stems, and the other group is shown on the right side. When the leaves are ordered, the leaves increase as one moves away from the stems. The following stem-and-leaf display shows the money earned per tournament entered for the top 30 money winners in the 2008–09 Professional Bowlers Association men's tour and for the top 21 money winners in the 2008–09 Professional Bowlers Association women's tour.

Women's		Men's
8	0	
8871	1	
65544330	2	334456899
840	3	03344678
52	4	011237888
21	5	9
	6	9
5	7	
	8	7
	9	5

The leaf unit for this display is 100. In other words, the data used represent the earnings in hundreds of dollars. For example, for the women's tour, the first number is 08, which is actually 800. The second number is 11, which actually is 1100.

a. Do the top money winners, as a group, on one tour (men's or women's) tend to make more money per tournament played than on the other tour? Explain how you can come to this conclusion using the stem-and-leaf display.

b. What would be a typical earnings level amount per tournament played for each of the two tours?

c. Do the data appear to have similar spreads for the two tours? Explain how you can come to this conclusion using the stem-and-leaf display.

d. Does either of the tours appear to have any outliers? If so, what are the earnings levels for these players?

2.51 Statisticians often need to know the shape of a population to make inferences. Suppose that you are asked to specify the shape of the population of weights of all college students.

a. Sketch a graph of what you think the weights of all college students would look like.

b. The following data give the weights (in pounds) of a random sample of 44 college students (F and M indicate female and male, respectively).

123 F	195 M	138 M	115 F	179 M	119 F	148 F	147 F
180 M	146 F	179 M	189 M	175 M	108 F	193 M	114 F
179 M	147 M	108 F	128 F	164 F	174 M	128 F	159 M
193 M	204 M	125 F	133 F	115 F	168 M	123 F	183 M
116 F	182 M	174 M	102 F	123 F	99 F	161 M	162 M
155 F	202 M	110 F	132 M				

i. Construct a stem-and-leaf display for these data.

ii. Can you explain why these data appear the way they do?

c. Construct a back-to-back stem-and-leaf display for the data on weights, placing the weights of the female students to the left of the stems and those of the male students to the right of the stems. Does one gender tend to have higher weights than the other? Explain how you know this from the display.

Self-Review Test

Multiple Choice Questions

1. Raw data are

a. presented in the form of a frequency table

b. recorded in the sequence in which they are collected

c. arranged in a random order

2. We obtain the relative frequency of a category by dividing the

a. frequency of that category by the sum of all frequencies

b. frequency of that category by 100

c. sum of all frequencies by the frequency of that category

3. The following data gives the results of a sample survey. The letters A, B, and C represent the three categories:

A B A B B C A A C B

B B C A C C C A B B

What percentage of the elements in this sample belong to category B?

a. 40% b. 30% c. 70%

4. The following table gives the frequency distribution of opinions of 50 persons regarding an issue.

Opinion	*f*
In favor	17
Against	18
No opinion	15

The relative frequency, expressed to two decimal places of the *against* category is

a. .30 b. .36 c. .34

5. During a local basketball camp, the participants scored a total of 886 points, of which 485 points were scored by two-point field goals, 179 points by three-point field goals, and 222 points by free-throw goals. Find the pie-chart that better describes the data.

a.

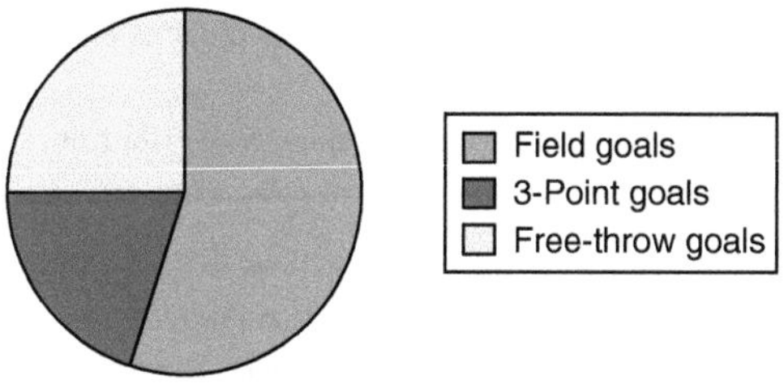

b.

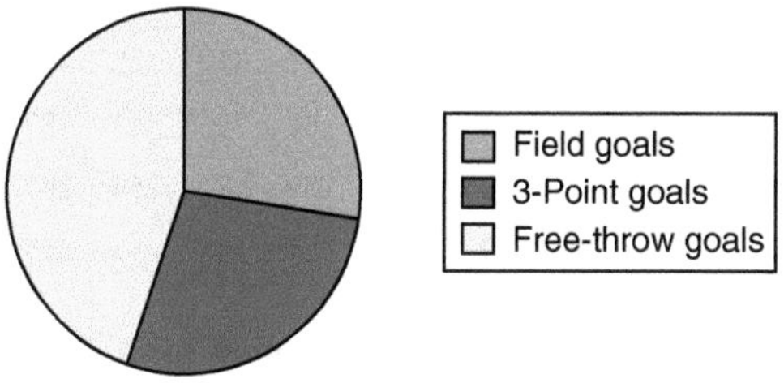

c.

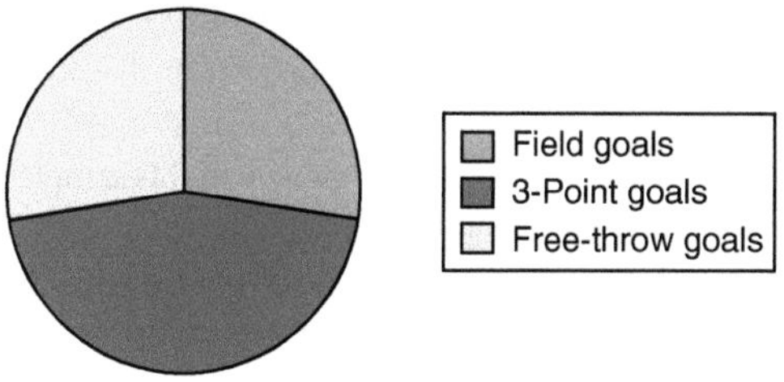

6. The number of classes in a frequency distribution for a quantitative data set depends on the size of the data set. In general, the

a. larger the data set, the larger the number of classes

b. larger the data set, the smaller the number of classes

c. number of classes should be equal to the number of values in the data set divided by 5

7. A distribution curve that is right skewed has

a. a longer tail on the left side

b. a shorter tail on the right side

c. a longer tail on the right side

8. In a frequency histogram, the frequency of a class is given by the

a. height of the corresponding bar

b. width of the corresponding bar

c. height multiplied by the width of the corresponding bar

9. The following table gives the frequency distribution of test scores for a math class of 30 students.

Score	*f*
61 to 70	3
71 to 80	6
81 to 90	10
91 to 100	11

The lower boundary of the first class is

a. 61 b. 60 c. 60.5

10. The following table gives the frequency distribution of the rent paid per month by 500 families selected from a city.

Rent	*f*
901 to 1000	36
1001 to 1100	55
1101 to 1200	73
1201 to 1300	101
1301 to 1400	136
1401 to 1500	99

The midpoint of the second class is

a. 1050 b. 1050.5 c. 1000

11. Here is a dotplot of the daily high temperature (in Fahrenheit) from a sample of 25 U.S. cities.

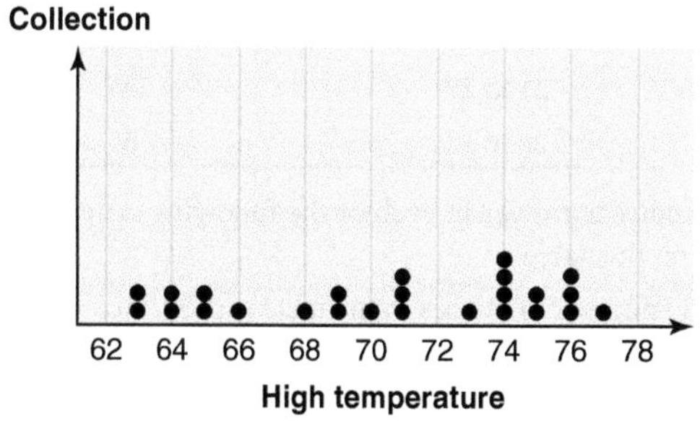

Which daily high temperature has the highest frequency?

a. 76 b. 74 c. 75

12. Consider the following stem-and-leaf display of two-digit numbers.

1	1 3
2	1 2 8
3	5
4	4

What is the sum of the values in the bottom two rows?

a. 9 b. 79 c. 7

13. The following table gives the cumulative frequency distribution of the commuting time (in minutes) from home to work for a sample of 400 persons selected from a city.

Time (minutes)	*f*
0 to less than 10	65
0 to less than 20	155
0 to less than 30	224
0 to less than 40	291
0 to less than 50	356
0 to less than 60	400

The percentage of persons who commute for less than 50 minutes, rounded to two decimal places, is

a. 11.00% b. 72.75% c. 89.00%

14. The following table gives the cumulative frequency distribution of annual incomes (in thousands of dollars) for a sample of 200 families selected from a city.

Income ($1000's)	f
15 to less than 25	30
15 to less than 40	86
15 to less than 55	145
15 to less than 70	171
15 to less than 85	188
15 to less than 100	200

The number of families with an income of $40,000 or less is

a. 86 b. 30 c. cannot be determined

15. The following table gives the frequency distribution of the number of telephones owned by a sample of 50 households selected from a city.

Number of Cellphones Owned	f
0	2
1	19
2	14
3	6
4	9

Would the graph of this distribution be a bar graph or a histogram?

a. Multiple bar b. Histogram c. Bar graph

Practice Exercises

1. Briefly explain the difference between ungrouped and grouped data and give one example of each type.

2. The following table gives the frequency distribution of times (to the nearest hour) that 90 fans spent waiting in line to buy tickets to a rock concert.

Waiting Time (hours)	Frequency
0 to 6	5
7 to 13	27
14 to 20	30
21 to 27	20
28 to 34	8

Circle the correct answer in each of the following statements, which are based on this table.

a. The number of classes in the table is 5, 30, 90.
b. The class width is 6, 7, 34.
c. The midpoint of the third class is 16.5, 17, 17.5.
d. The upper limit of the second class is 12.5, 13, 13.5.
e. The sample size is 5, 90, 11.
f. The relative frequency of the second class is .22, .41, .30.

3. Briefly explain and illustrate with the help of graphs a symmetric histogram, a histogram skewed to the right, and a histogram skewed to the left.

4. Thirty-six randomly selected senior citizens were asked if their net worth is more than or less than $200,000. Their responses are given below, where M stands for more, L represents less, and N means do not know or do not want to tell.

M	M	L	L	L	N	M	N	L	L	M	M
L	L	N	L	L	M	M	L	N	L	L	L
M	M	L	M	L	L	M	M	L	N	L	N

a. Make a frequency distribution for these 36 responses.

b. Using the frequency distribution of part (a), prepare the relative frequency and percentage distributions.

c. What percentage of these senior citizens claim to have net worth less than $200,000?

d. Make a bar graph for the frequency distribution.

e. Make a Pareto chart for the frequency distribution.

f. Draw a pie chart for the percentage distribution.

5. Forty-eight randomly selected car owners were asked about their typical monthly expense (in dollars) on gas. The following data show the responses of these 48 car owners.

210	160	430	255	176	135	221	359	380	405	391	477
333	209	267	121	357	87	167	95	347	487	302	545
351	256	492	277	245	367	159	187	253	287	456	64
76	166	304	444	193	479	188	148	53	327	234	110

a. Construct a frequency distribution table. Use the classes 50–149, 150–249, 250–349, 350–449, and 450–549.

b. Calculate the relative frequency and percentage for each class.

c. Construct a histogram for the percentage distribution made in part (b).

d. What percentage of the car owners in this sample spend $350 or more on gas per month?

e. Prepare the cumulative frequency, cumulative relative frequency, and cumulative percentage distributions using the table of part (a).

6. Thirty customers from all customers who shopped at a large grocery store during a given week were randomly selected and their shopping expenses were noted. The following data show the expenses (in dollars) of these 30 customers.

89.20	145.23	191.54	45.36	67.98	123.67
187.57	56.43	102.45	158.76	134.07	67.49
212.60	165.75	46.50	111.25	64.54	23.67
135.09	193.46	87.65	133.76	156.28	88.64
65.90	120.50	55.45	91.54	153.20	44.39

a. Prepare a frequency distribution table for these data. Use the classes as 20 to less than 60, 60 to less than 100, 100 to less than 140, 140 to less than 180, and 180 to less than 220.

b. What is the width of each class in part (a)?

c. Using the frequency distribution of part (a), prepare the relative frequency and percentage distributions.

d. What percentage of these customers spent less than $140 at this grocery store?

e. Make a histogram for the frequency distribution.

f. Prepare the cumulative frequency, cumulative relative frequency, and cumulative percentage distributions using the table of part (a).

7. Construct a stem-and-leaf display for the following data, which give the times (in minutes) that 24 customers spent waiting to speak to a customer service representative when they called about problems with their Internet service provider.

12	15	7	29	32	16	10	14	17	8	19	21
4	14	22	25	18	6	22	16	13	16	12	20

8. Consider this stem-and-leaf display:

3	0 3 7
4	2 4 6 7 9
5	1 3 3 6
6	0 7 7
7	1 9

Write the data set that was used to construct this display.

9. Make a dotplot for the data on typical monthly expenses on gas for the 48 car owners given in Practice Exercise 5.

Mini-Projects

Mini-Project 2.1

Using the data you gathered for the mini-project in Chapter 1, prepare a summary of that data set that includes the following:

a. Construct a frequency distribution table for one of the quantitative variables and then compute the relative frequencies, percentages, cumulative relative frequencies, and cumulative percentages.

b. Create a histogram, a stem-and-leaf display (by rounding the numbers to integers), and a dotplot of the data and/or frequency distribution of part (a). Comment on any symmetry or skewness and on the presence of clusters and any potential outliers.

c. Construct a frequency distribution table for one of the qualitative variables and then compute the relative frequencies, percentages, cumulative relative frequencies, and cumulative percentages.

d. Create a bar graph and pie chart for the frequency distribution of part (c).

Mini-Project 2.2

Choose 15 of each of two types of magazines (news, sports, fitness, entertainment, and so on) and record the percentage of pages that contain at least one advertisement. Using these percentages and the types of magazines, write a brief report that covers the following:

a. Construct a frequency distribution table for the variable representing the percentage of pages that contain at least one advertisement, and then compute relative frequencies and cumulative relative frequencies.

b. Create a histogram, a stem-and-leaf display (by rounding the percentages to integers), and a dotplot of the frequency distribution or data on the percentage of pages that contain at least one advertisement. Comment on any symmetry or skewness, as well as the presence of clusters and any potential outliers.

Mini-Project 2.3

Have 20 males and 20 females use an online reaction time test like the ones found at www.humanbenchmark.com/tests/reactiontime/ or www.mathsisfun.com/games/reaction-time.html. These tests give an average reaction time for a person after five trials. Record the gender of each person and their average reaction time. Be sure to record the average for each participant and consider it their true reaction time and not the reaction time for each separate trial. Prepare a report on the data that you collected that includes the following elements:

a. Create a frequency distribution of the reaction times for each gender, choosing an appropriate interval width. Also include the relative frequencies and cumulative relative frequencies.

b. Make a histogram, stem-and-leaf display (rounding as needed), and a dotplot of the reaction times for each gender. Comment on any symmetry or skewness, as well as the presence of clusters and potential outliers.

Decide for Yourself

Note: The Decide for Yourself feature is located in the book's product page on *www.wiley.com*.

Chapter 2 Technology Instructions

Note: Complete TI-84, Minitab, and Excel manuals are available in the book's product page on *www.wiley.com.*

TI-84 Color/TI-84

The TI-84 Color Technology Instructions feature of this text is written for the TI-84 Plus C color graphing calculator running the 4.0 operating system. Some screens, menus, and functions will be slightly different in older operating systems. The TI-84 and TI-84 Plus can perform all of the same functions but will not have the "Color" option referenced in some of the menus.

TI-84 calculators do not have the options/capabilities to make a bar graph, pie chart, stem-and-leaf display, and dotplot.

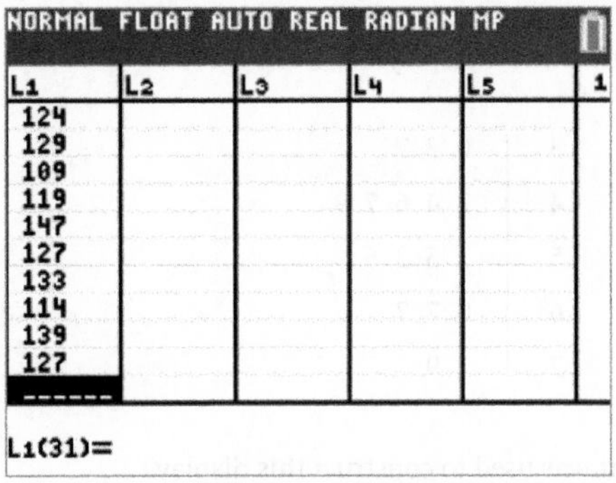

SCREEN 2.1

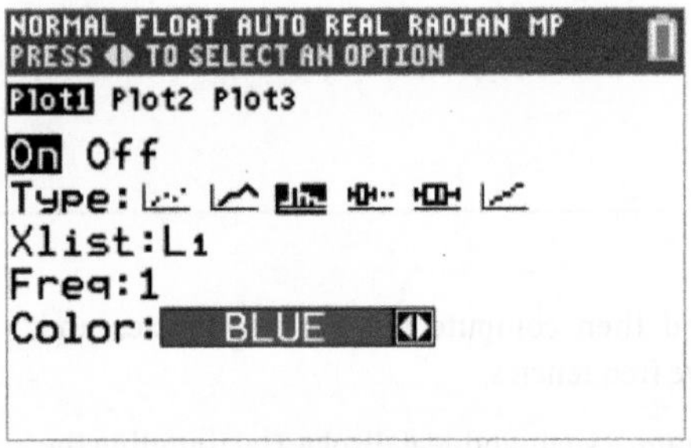

SCREEN 2.2

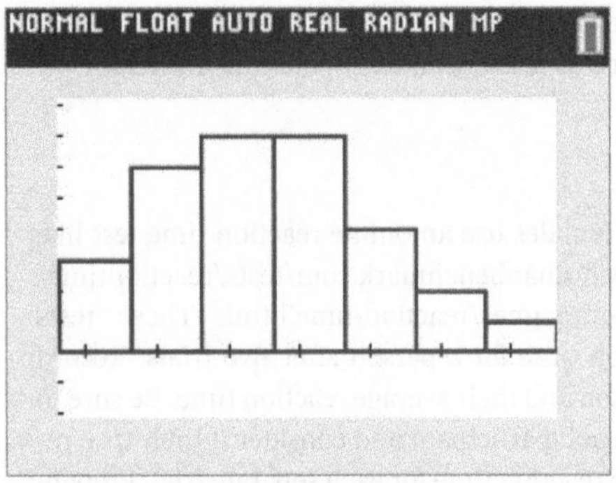

SCREEN 2.3

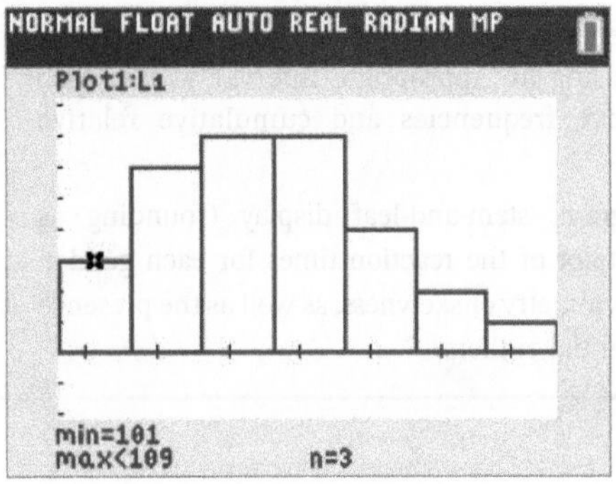

SCREEN 2.4

Creating a Frequency Histogram for Example 2.3 of the Text

1. Enter the data from Example 2.3 of the text into **L1**. (See **Screen 2.1.**)
2. Select **2nd** > **Y** = (the **STAT PLOT** menu).
3. If more than one plot is turned on, select **PlotsOff** and then press **ENTER** to turn off the plots.
4. From the **STAT PLOT** menu, select **Plot 1**. Use the following settings (see **Screen 2.2**):
   - Select **On** to turn the plot on.
   - At the **Type** prompt, select the third icon (histogram).
   - At the **Xlist** prompt, enter L1 by pressing **2nd > 1**.
   - At the **Freq** prompt, enter 1.
   - Select BLUE at the **Color** prompt.

   (Skip this step if you do not have the TI-84 Color calculator.)
5. Select **ZOOM** > **ZoomStat** to display the histogram. This function will automatically choose window settings and class boundaries for the histogram. (See **Screen 2.3.**)
6. Press **TRACE** and use the left and right arrow keys to see the class boundaries and frequencies for each class. (See **Screen 2.4.**)
7. To manually change the window settings and the class boundaries, press **WINDOW** and use the following settings:
   - Type 101 at the **Xmin** prompt. **Xmin** is the extreme left value of the window and the lower boundary value of the first class.
   - Type 110 at the **Xmax** prompt. **Xmax** is the extreme right value of the window.
   - Type 10 at the **Xscl** prompt. **Xscl** is the class width.
   - Type -2 at the **Ymin** prompt. **Ymin** is the extreme bottom value of the window.
   - Type 1 at the **Ymax** prompt. **Ymax** is the extreme top value of the window.
   - Type 2 at the **Yscl** prompt. **Yscl** is the distance between tick marks on the *y*-axis.
   - Press **GRAPH** to see the histogram with the new settings.

Minitab

The Minitab Technology Instructions in this text are written for **Minitab Professional Version 19**. Some screens, menus, and functions will be slightly different in older (or in newer) versions of Minitab. If you are using **Minitab Express**, you can download the Minitab Express Manual from the book's product page on *www.wiley.com.*

Creating a Bar Graph for Example 2.1 of the Text

Bar Graph for Qualitative Ungrouped Data

1. Enter the 30 data values of Example 2.1 of the text into column C1.
2. Name the column *Favorite Donut.*
3. Select **Graph > Bar Chart.**

4. Use the following settings in the resulting dialog box:
   - Select **Counts of unique values**, select **Simple**, and click **OK**.
5. Use the following settings in the new dialog box:
   - Type C1 in the **Categorical variables** box, and click **OK**.
6. The bar graph will appear in a new window. (See **Screen 2.5**.)

Bar Graph for Qualitative Grouped Data

1. Enter the grouped data from Example 2.1 as shown in Table 2.4 of the text. Put the *Donut Variety* into column C1 and the *Frequency* into column C2.
2. Name column C1 *Favorite Donut* and column C2 *Frequency*.
3. Select **Graph > Bar Chart**.
4. Use the following settings in the resulting dialog box:
   - Select **Values from a table**, select **Simple**, and click **OK**.
5. Use the following settings in the new dialog box:
   - Type C2 in the **Graph variables** box, C1 in the **Categorical variable** box, and click **OK**.
6. The bar graph will appear in a new window.

To make a Pareto chart, after Step 4, click on **Chart Options** in the same dialog box. Select **Decreasing Y** in the next dialog box. Click **OK** in both dialog boxes. The Pareto chart will appear in a new window.

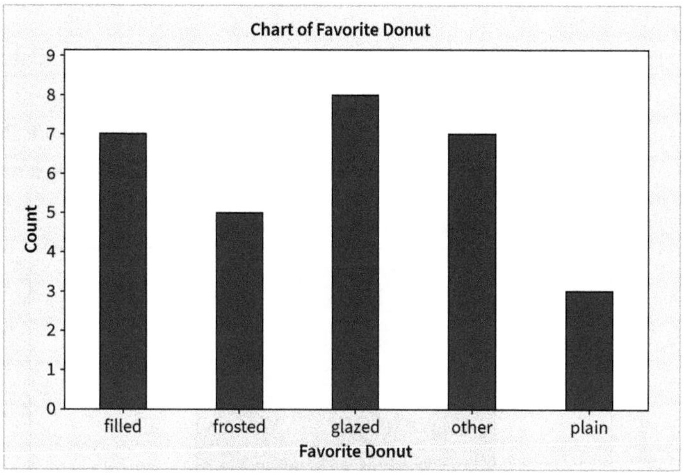

SCREEN 2.5

Creating a Pie Chart for Example 2.1 of the Text

Pie Chart for Qualitative Ungrouped Data

1. Enter the 30 data values of Example 2.1 of the text into column C1.
2. Name the column *Favorite Donut*.
3. Select **Graph > Pie Chart**.
4. Use the following settings in the resulting dialog box:
   - Select **Chart counts of unique values**, and type C1 in the **Categorical variables** box.
   - To label slices of the pie chart with percentages or category names, select **Labels > Slice Labels**. Choose the options you prefer and click **OK**.
5. The pie chart will appear in a new window. (See **Screen 2.6**.)

Pie Chart for Qualitative Grouped Data

1. Enter the grouped data from Example 2.1 as shown in Table 2.4 of the text. Put the *Donut Variety* into column C1 and the *Frequency* into column C2.
2. Select **Graph > Pie Chart**.

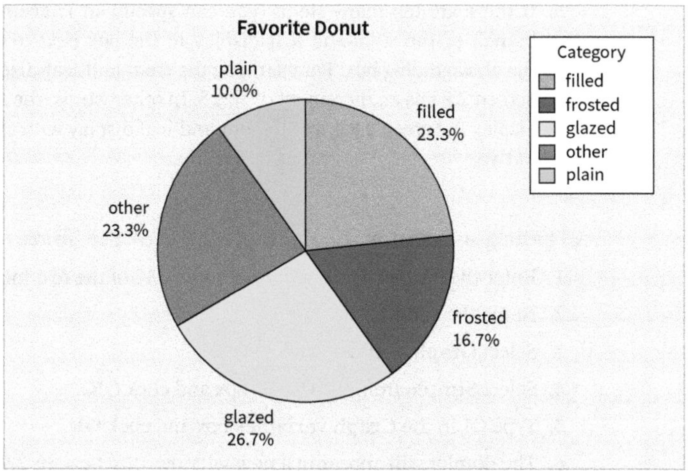

SCREEN 2.6

3. Use the following settings in the resulting dialog box:
   - Select **Chart values from a table**, type C1 in the **Categorical variable** box, and type C2 in the **Summary variables** box.
   - To label slices of the pie chart with percentages or category names, select **Labels > Slice Labels**. Choose the options you prefer, and click **OK**.
4. The pie chart will appear in a new window. (See **Screen 2.6.**)

Creating a Frequency Histogram for Example 2.3 of the Text

1. Enter the 30 data values of Example 2.3 of the text into column C1.
2. Name the column *Systolic Blood Pressure*.
3. Select **Graph > Histogram**, select **Simple**, and click **OK**.

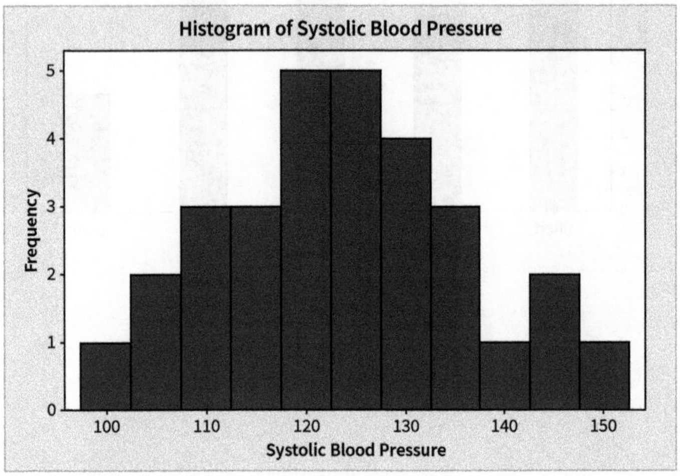

SCREEN 2.7

4. Use the following settings in the resulting dialog box:
   - Type C1 in the **Graph variables** box.
   - If you wish to add labels, titles, or subtitles to the histogram, select **Labels** and choose the options you prefer, and click **OK.**
5. The histogram will appear in a new window. (See **Screen 2.7.**)
6. To change the class boundaries double-click on any of the bars in the histogram and the **Edit Bars** dialog box will appear. Click on the **Binning** tab. Now choose one of the following two options:
   - To set the class midpoints, select **Midpoint** under **Interval Type**, select **Midpoint/Cutpoint positions** under the **Interval Definition** box, and then type the class midpoints (separate each value with a space). Click **OK.**
   - To set the class boundaries, select **Cutpoint** under **Interval Type**, select **Midpoint/Cutpoint positions,** and then type the left boundary for each interval and the right end-point of the last interval (separate each value with a space). Click **OK.**

Creating a Stem-and-Leaf Display for Example 2.9 of the Text

1. Enter the 30 data values of Example 2.9 of the text into column C1.
2. Name the column *Test Scores.*
3. Select **Graph > Stem-and-Leaf.**
4. Type C1 in the **Graph variables** box and click **OK.**
5. The stem-and-leaf display will appear in the Session window. (See **Screen 2.8.**)
6. If there are too many stems, you can specify an **Increment** for each branch of the stem-and leaf display in the box next to Increment of the above dialog box. For example, the stem-and-leaf display shown in Screen 2.8 has an increment of size 5. In other words, the stem-and-leaf display in Screen 2.8 is a split stem-and-leaf display with each stem split in two.

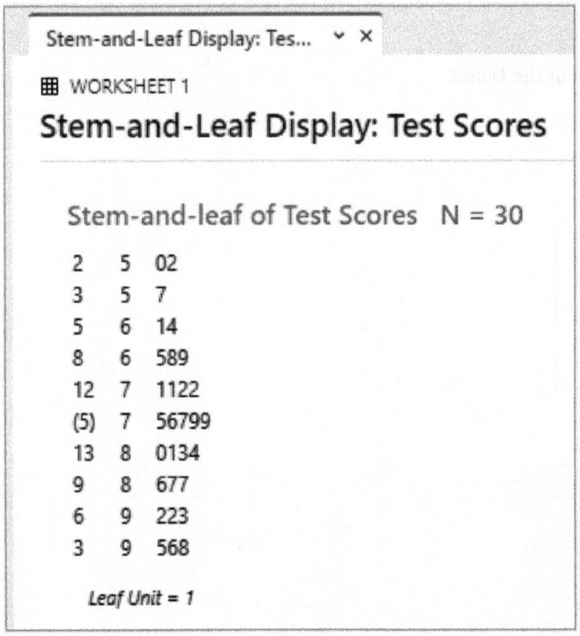

SCREEN 2.8

Creating a Dotplot for Example 2.13 of the Text

1. Enter the 33 data values from Example 2.13 of the text into column C1.
2. Name the column *Age.*
3. Select **Graph > Dotplot.**
4. Select **Simple** from the **One Y** box and click **OK.**
5. Type C1 in the **Graph variables** box and click **OK.**
6. The dotplot will appear in a new window. (See **Screen 2.9.**)

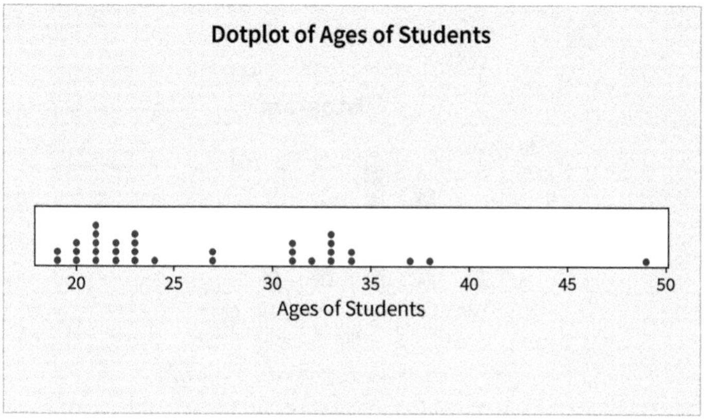

SCREEN 2.9

Excel

The Excel Technology Instructions feature of this text is written for Excel 2019. Some screens, menus, and functions will be slightly different in older versions of Excel. To enable some of the advanced Excel functions, you must enable the Data Analysis Add-In for Excel.

Excel does not have the capability to make a stem-and-leaf display and dotplot.

Creating a Bar Graph for Example 2.1 of the Text

1. Enter the grouped data from Example 2.1 as shown in Table 2.4 of the text.
2. Highlight the cells that contain the donut varieties and the frequencies. (See **Screen 2.10**.)
3. Click **INSERT** and then click the **Insert Column Chart** icon in the **Charts** group.
4. Select the first icon in the **2-D Column** group in the dropdown box.
5. A new graph will appear with the bar graph.

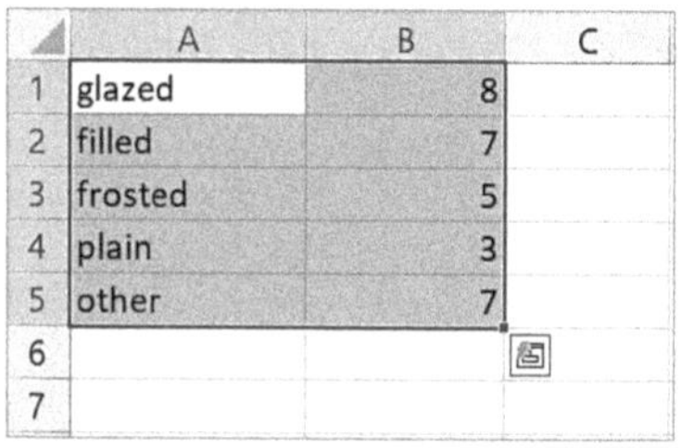

SCREEN 2.10

Creating a Pie Chart for Example 2.1 of the Text

1. Enter the grouped data from Example 2.1 as shown in Table 2.4 of the text.
2. Highlight the cells that contain the donut varieties and the frequencies. (See **Screen 2.10**.)
3. Click **INSERT** and then click the **Insert Pie Chart** icon in the **Charts** group.
4. Select the first icon in the **2-D Pie** group in the dropdown box.
5. A new graph will appear with the pie chart. (See **Screen 2.11**.)

Note that Excel rounds the percentages to the nearest integer.

Note: You may click in the text box and change “Chart Title” to an appropriate title or you may select the text box and delete it.

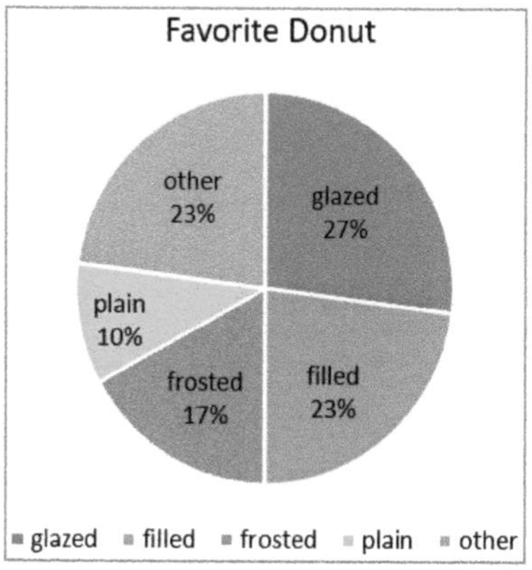

SCREEN 2.11

Creating a Frequency Histogram for Example 2.3 of the Text

1. Enter the data from Example 2.3 of the text into cells A1 to A30.
2. In cell B1, type *Systolic Blood Pressure*.
3. Determine the class boundaries and type the right boundary for each class in cells B2 to B6. In this example, we will use the right boundary values of 110, 120, 130, 140, and 150 from Example 2.3.
4. Click **DATA** and then click **Data Analysis** from the **Analysis** group.
5. Select **Histogram** from the **Data Analysis** dialog box and click **OK**.
6. Click in the **Input Range** box and then highlight cells A1 to A30.
7. Click in the **Bin Range** box and then highlight cells B1 to B6. Then click the box to the left of **Labels**.
8. In the **Output options** box, select **New Worksheet Ply** and check the **Chart Output** check box. Click **OK**.
9. A new worksheet will appear with the frequency distribution and the corresponding histogram. (See **Screen 2.12**.)

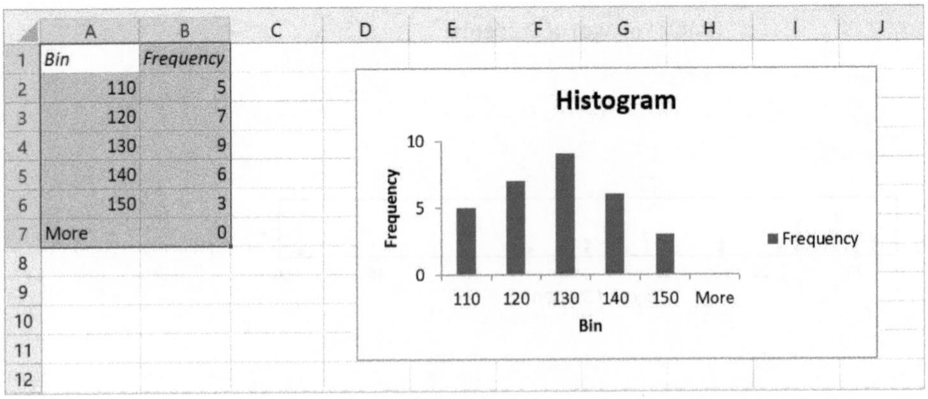

SCREEN 2.12

Technology Assignments

TA2.1 In the past few years, many U.S. states have built casinos, and many more are in the process of doing so. Forty adults were asked if building casinos is good for society. Following are the responses of these adults, where G stands for good, B indicates bad, and I means indifferent or no answer.

B	G	B	B	I	G	B	I	B	B
G	B	B	G	B	B	B	G	G	I
B	G	B	B	I	G	G	G	B	B
I	G	B	B	B	G	G	B	B	G

a. Construct a bar graph and pie chart for these data.

b. Arrange the bars of the bar graph of part (a) in decreasing order to prepare a Pareto chart.

TA2.2 Refer to Data Set I on home prices in Wilmington, NC, that accompanies this text (see Appendix A). Create a bar graph and pie chart for the variable *high school.*

TA2.3 Refer to Data Set V on movies released in 2018 (see Appendix A). Construct a histogram for the opening weekend *gross earnings.*

TA2.4 Refer to Data Set XI that contains information on all airline flights that departed Newark (New Jersey) Liberty International Airport during the month of February 2019 (see Appendix A). Construct a histogram for the variable *departure delay* (minutes) for all *Alaskan Airlines* flights that arrived at the destination airport. Do not include flights that were cancelled or diverted.

TA2.5 Refer to Data Set V. Create a stem-and-leaf display for the variable *length* of these films.

TA2.6 The following data give the one-way commuting times (in minutes) from home to work for a random sample of 50 workers.

23	17	34	26	18	33	46	42	12	37
44	15	22	19	28	32	18	39	40	48
16	11	9	24	18	26	31	7	30	15
18	22	29	32	30	21	19	14	26	37
25	36	23	39	42	46	29	17	24	31

a. Make a histogram for these data.

b. Prepare a stem-and-leaf display for these data.

c. Make a dotplot for these data.

TA2.7 Refer to Data Set VII on McDonald's menu (see Appendix A). Construct a dotplot for the variable *sugar* (g) for the combined burger/chicken sandwiches.

TA2.8 Refer to Data Set VIII on the performance of 31 NFL kickers who had at least one attempt per game during the 2019–2020 NFL (National Football League) regular session (see Appendix A). Construct a dotplot for the variable *longest field goal made* during the season.

TA2.9 Refer to Data Set VII.

a. Construct a histogram for the variable *calories* for the combined Burger/Chicken Sandwiches allowing technology to choose the width of the intervals for you.

b. Construct another histogram for the variable *calories* for the combined Burger/Chicken Sandwiches but use intervals that are half the width of the histogram in part (a).

c. Construct another histogram for the variable *calories* for the combined Burger/Chicken Sandwiches but use intervals that are twice the width of the histogram in part (a).

d. Which of the histograms in parts (a) to (c) do you believe gives the best understanding of the data and why do you feel this way? Explain.

TA2.10 Refer to Data Set IV on the Manchester Road Race (see Appendix A).

a. Create a bar graph for the variable *gender* of the runners.

b. Create a histogram for the *net time* to complete the race.

CHAPTER

3

Numerical Descriptive Measures

Do you drink coffee? If you do, how many cups do you drink a day? Do you drink more or less than a typical American drinks per day? Do you know how much coffee Americans consume per year on average? Do you know in which country people consume the most coffee per year per person? Do you know the average price of coffee in the United States? For answers to all these questions, see Case Study 3.1.

Fabio Balbi/123RF

In Chapter 2 we discussed how to summarize data using different methods and to display data using graphs. Graphs are one important component of statistics; however, it is also important to numerically describe the main characteristics of a data set. The numerical summary measures, such as the ones that identify the center and spread of a distribution, identify many important features of a distribution. For example, the techniques learned in Chapter 2 can help us graph data on family incomes. However, if we want to know the income of a "typical" family (given by the center of the distribution), the spread of the distribution of incomes, or the relative position of a family with a particular income, the numerical summary measures can provide more detailed information (see **Figure 3.1**). The measures that we discuss in this chapter include measures of (1) center, (2) dispersion (or spread), and (3) position.

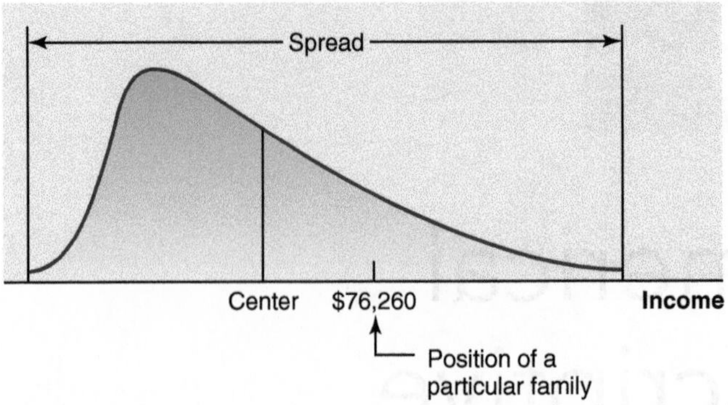

FIGURE 3.1

3.1 Measures of Center for Ungrouped Data

LEARNING OBJECTIVES

After completing this section, you should be able to:

- Calculate and interpret measures of center such as mean, median, mode, trimmed mean, weighted mean, geometric mean, and harmonic mean for quantitative data, and mode for qualitative data.
- Explain the difference between the mean and median calculated for population and sample data.
- Explain the meaning of an outlier.
- Identify the best measure of center for a data set with one or more outliers.

We often represent a data set by numerical summary measures, usually called the *typical values*. A **measure of center** gives the center of a histogram or a frequency distribution curve. This section discusses seven different measures of center: the mean, the median, the trimmed mean, the weighted mean, geometric mean, harmonic mean, and the mode. We will learn how to calculate each of these measures for ungrouped data. Recall from Chapter 2 that the data that give information on each member of the population or sample individually are called *ungrouped data*, whereas *grouped data* are presented in the form of a frequency distribution table.

3.1.1 Mean

The **mean**, also called the *arithmetic mean*, is the most frequently used measure of center. This book will use the words *mean* and *average* synonymously. For ungrouped data, the mean is obtained by dividing the sum of all values by the number of values in the data set:

$$\text{Mean} = \frac{\text{Sum of all values}}{\text{Number of values}}$$

The mean calculated for sample data is denoted by $\bar{x}$ (read as "x bar"), and the mean calculated for population data is denoted by μ (Greek letter *mu*). We know from the discussion in Chapter 2 that the number of values in a data set is denoted by n for a sample and by N for a population. In Chapter 1, we learned that a variable is denoted by x, and the sum of all values of x is denoted by Σx. Using these notations, we can write the following formulas for the mean.

Calculating Mean for Ungrouped Data

The *mean for ungrouped data* is obtained by dividing the sum of all values by the number of values in the data set. Thus,

$$\text{Mean for population data:} \quad \mu = \frac{\Sigma x}{N}$$

$$\text{Mean for sample data:} \quad \bar{x} = \frac{\Sigma x}{n}$$

where Σx is the sum of all values, N is the population size, n is the sample size, μ is the population mean, and $\bar{x}$ is the sample mean.

EXAMPLE 3.1 2018 Revenues of 10 U.S. Companies

Calculating the Sample Mean for Ungrouped Data

Table 3.1 lists the total revenues (rounded to billions of dollars) of 10 U.S. companies for the year 2018 (Fortune.com, August 2019).

TABLE 3.1 2018 Revenues of 10 U.S. Companies

Company	Revenue (billions of dollars)
Apple	266
AT&T	171
Walmart	514
Exxon Mobil	290
Ford Motor	160
General Electric	120
Citigroup	97
Alphabet	137
Amazon.com	233
Microsoft	110

Find the mean of the 2018 revenues for these 10 companies.

Solution The variable in this example is the 2018 revenue of a company. Let us denote this variable by x. The 10 values of x are given in Table 3.1. By adding these 10 values, we obtain the sum of x values, that is

$$\Sigma x = 266 + 171 + 514 + 290 + 160 + 120 + 97 + 137 + 233 + 110$$
$$= 2098$$

Note that the given data include only 10 companies. Hence, it represents a sample with $n = 10$. Substituting the values of Σx and n in the sample formula, we obtain the mean of the 2018 revenues of these 10 companies as follows:

$$\bar{x} = \frac{\Sigma x}{n} = \frac{2098}{10} = 209.8 = \mathbf{\$209.8\ billion}$$

Thus, the average revenue of these 10 companies was $209.8 billion in 2018.

EXAMPLE 3.2 Ages of Employees of a Company

Calculating the Population Mean for Ungrouped Data

The following are the ages (in years) of all eight employees of a small company:

53 32 61 27 39 44 49 57

Find the mean age of these employees.

Solution Because the given data set includes *all* eight employees of the company, it represents the population. Hence, $N = 8$. We have

$$\Sigma x = 53 + 32 + 61 + 27 + 39 + 44 + 49 + 57 = 362$$

The population mean is

$$\mu = \frac{\Sigma x}{N} = \frac{362}{8} = \mathbf{45.25\ years}$$

Thus, the mean age of all eight employees of this company is 45.25 years, or 45 years and 3 months.

Reconsider Example 3.2. If we take a sample of three employees from this company and calculate the mean age of those three employees, this mean will be denoted by $\bar{x}$. Suppose the three values included in the sample are 32, 39, and 57. Then, the mean age for this sample is

$$\bar{x} = \frac{32 + 39 + 57}{3} = \frac{128}{3} = 42.67 \text{ years}$$

If we take a second sample of three employees of this company, the value of $\bar{x}$ will (most likely) be different. Suppose the second sample includes the values 53, 27, and 44. Then, the mean age for this sample is

$$\bar{x} = \frac{53 + 27 + 44}{3} = \frac{124}{3} = 41.33 \text{ years}$$

Consequently, we can state that the value of the population mean μ is constant. However, the value of the sample mean $\bar{x}$ varies from sample to sample. The value of $\bar{x}$ for a particular sample depends on what values of the population are included in that sample.

Sometimes a data set may contain a few very small or a few very large values. As mentioned in Chapter 2, such values are called **outliers** or **extreme values.**

A major shortcoming of the mean as a measure of center is that it is very sensitive to outliers. Example 3.3 illustrates this point.

EXAMPLE 3.3 Prices of Eight Homes

Illustrating the Effect of an Outlier on the Mean

Following are the list prices of eight homes randomly selected from all homes for sale in a city:

\$245,670	176,200	360,280	272,440
450,394	310,160	393,610	3,874,480

Note that the price of the last house is \$3,874,480, which is an outlier. Show how the inclusion of this outlier affects the value of the mean.

Solution If we do not include the price of the most expensive house (the outlier), the mean of the prices of the other seven homes is

$$\text{Mean without the outlier} = \frac{245{,}670 + 176{,}200 + 360{,}280 + 272{,}440 + 450{,}394 + 310{,}160 + 393{,}610}{7}$$

$$= \frac{2{,}208{,}754}{7} = \mathbf{\$315{,}536.29}$$

Now, to see the impact of the outlier on the value of the mean, we include the price of the most expensive home and find the mean price of eight homes. This mean is

Mean with the outlier

$$= \frac{245{,}670 + 176{,}200 + 360{,}280 + 272{,}440 + 450{,}394 + 310{,}160 + 393{,}610 + 3{,}874{,}480}{8}$$

$$= \frac{6{,}083{,}234}{8} = \mathbf{\$760{,}404.25}$$

Thus, when we include the price of the most expensive home, the mean more than doubles, as it increases from \$315,536.29 to \$760,404.25.

The preceding example should encourage us to be cautious. We should remember that the mean is not always the best measure of center because it is heavily influenced by outliers. Sometimes other measures of center give a more accurate impression of a data set and/or typical value. For example, when a data set has outliers, instead of using the mean, we can use either the trimmed mean or the median as a measure of center.

Case Study 3.1

Coffee Consumption Statistics

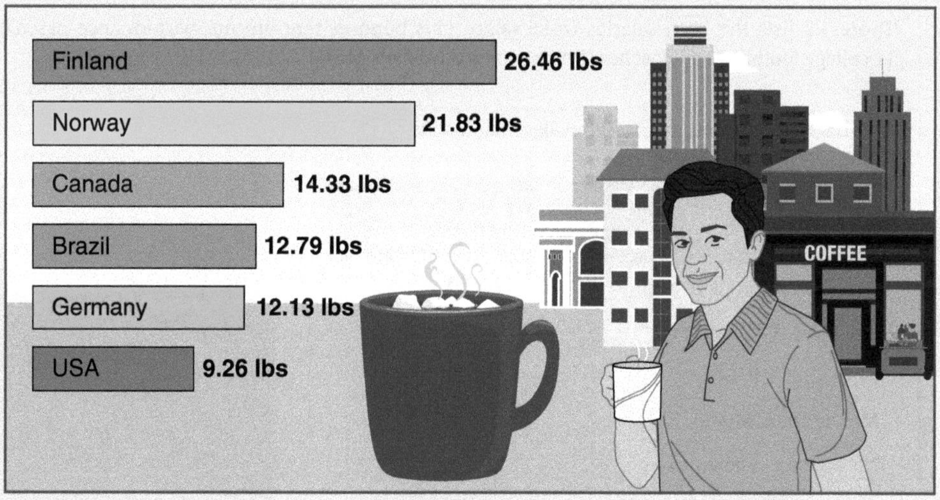

The graph shows the average consumption of coffee per person per year for six countries. According to a worldatlas.com article of January 2018, Finland had the highest consumption of coffee of 26.46 pounds (lbs) per person per year. Norway was ranked second with an average consumption of 21.83 pounds per person per year. Canada was ranked 10th with 14.33 pounds, Brazil was 14th with 12.79 pounds, Germany was 16th with 12.13 pounds, and the United States was ranked 25th with an average consumption of 9.26 pounds per person per year. Here are a few other statistics about coffee consumption in the United States.

Average coffee consumption among coffee drinkers in the United States: 3.1 cups a day
Average coffee cup size: 9 ounces
Average price for an espresso-based drink: \$3.45
Average price for brewed coffee: \$2.38

Data Source: e-importz.com, worldatlas.com

3.1.2 Median

Another important measure of center is the **median**. It is defined as follows:

Median

The *median* is the value that divides a data set that has been ranked in increasing order in two equal halves. If the data set has an odd number of values, the median is given by the middle value in the ranked data set. If the data set has an even number of values, the median is given by the average of the two middle values in the ranked data set.

As is obvious from the definition, the median divides a ranked data set into two equal parts. The calculation of the median consists of the following two steps:

1. Rank the given data set in increasing order.
2. Find the value that divides the ranked data set in two equal parts. This value gives the median.[1]

Note that if the number of observations in a data set is *odd*, then the median is given by the middle value in the ranked data. However, if the number of observations is *even*, then the median is given by the average of the two middle values.

EXAMPLE 3.4 2019 Salaries of College Football Head Coaches

Calculating the Median for Ungrouped Data: Odd Number of Data Values

Table 3.2 lists the 2019 salaries (base salary plus bonuses that are not performance based) of 11 college football head coaches (https://www.boydsbets.com/).

TABLE 3.2 2019 Salaries of 11 College Football Head Coaches

Head Coach and College	2019 Salary (millions of dollars)
Dabo Swinney (Clemson)	10.3
Ryan Day (Ohio State)	4.5
Chip Kelly (UCLA)	3.3
Manny Diaz (Miami, FL)	3.1
Nick Saban (Alabama)	8.7
David Shah (Stanford)	4.3
Mack Brown (North Carolina)	3.5
Lincoln Riley (Oklahoma)	4.8
Tom Herman (Texas)	5.5
James Franklin (Penn State)	4.8
Les Miles (Kansas)	2.8

SW Productions/Stockbyte/Getty Images

Find the median for these data.

[1]Note that the middle value in a data set ranked in *decreasing* order will also give the value of the median.

Solution To calculate the median of this data set, we perform the following two steps:

STEP 1 The first step is to rank the given data. We rank the given data in increasing order as follows:

2.8 3.1 3.3 3.5 4.3 4.5 4.8 4.8 5.5 8.7 10.3

STEP 2 The second step is to find the value that divides this ranked data set in two equal parts. Here there are 11 data values. The sixth value divides these 11 values in two equal parts. Hence, the sixth value gives the median as shown below.

2.8 3.1 3.3 3.5 4.3 4.5 4.8 4.8 5.5 8.7 10.3

↑ Median

Thus, the median of the 2019 salaries of these 11 college football head coaches is \$4.5 million. Note that in this example, there are 11 data values, which is an odd number. Hence, there is one value in the middle that is given by the sixth term and its value is the median. Using the value of the median, we can say that (about) half of these head coaches made less than \$4.5 million and the other half made more than \$4.5 million in 2019. (Note that by the time this book is published, some of these coaches may not be at these colleges.)

EXAMPLE 3.5 Cell Phone Minutes Used

Calculating the Median for Ungrouped Data: Even Number of Data Values

The following data give the cell phone minutes used last month by 12 randomly selected persons.

230 2053 160 397 510 380 263 3864 184 201 326 721

Find the median for these data.

Solution To calculate the median, we perform the following two steps:

STEP 1 In the first step, we rank the given data in increasing order as follows:

160 184 201 230 263 326 380 397 510 721 2053 3864

STEP 2 In the second step, we find the value that divides the ranked data set in two equal parts. This value will be the median. The value that divides 12 data values in two equal parts falls between the sixth and the seventh values. Thus, the median will be given by the average of the sixth and the seventh values as follows:

160 184 201 230 263 326 380 397 510 721 2053 3864

↑ Median = 353

$$\text{Median} = \text{Average of the two middle values} = \frac{326 + 380}{2} = \frac{706}{2} = \mathbf{353\ minutes}$$

Thus, the median cell phone minutes used last month by these 12 persons was 353. We can state that half of these 12 persons used less than 353 cell phone minutes and the other half used more than 353 cell phone minutes last month. Note that this data set has two outliers, 2053 and 3864 minutes, but these outliers do not affect the value of the median.

The median gives the center of a histogram, with half of the data values to the left of the median and half to the right of the median. The advantage of using the median as a measure of center is that it is not influenced by outliers. Consequently, the median is preferred over the mean as a measure of center for data sets that contain outliers.

Sometimes we use the terms resistant and nonresistant with respect to outliers for summary measures. A summary measure that is less affected by outliers is called a **resistant** summary measure, and the one that is affected more by outliers is called a **nonresistant**

Case Study 3.2

Median Prices of Homes in Selected Metro Areas of the United States

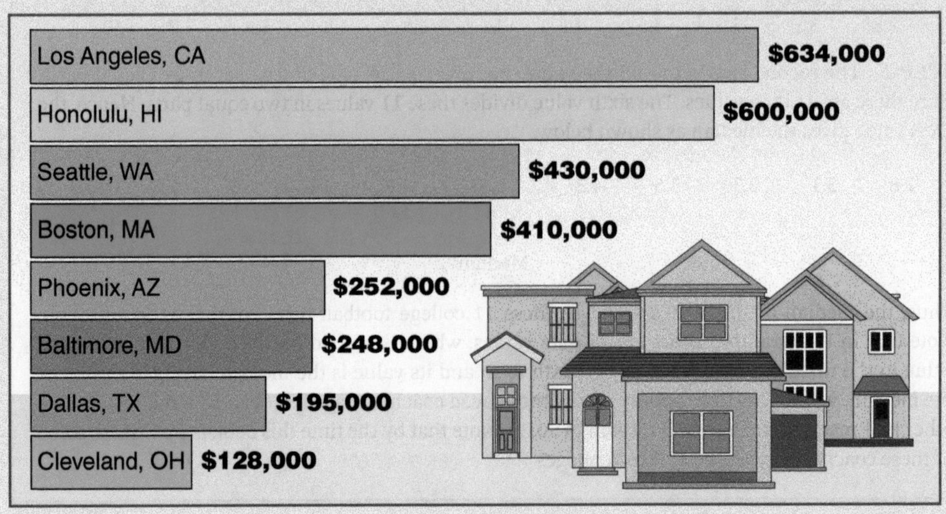

The graph shows the median prices of existing homes (single-family homes and condos) in eight selected cities across the country as of December 31, 2018. These median prices are based on tracking of home prices by Clear Capital. As the graph shows, the median price of homes in Los Angeles was $634,000 at that time, which means that (about) half of the homes in Los Angeles cost less than $634,000 and (about) half cost more than $634,000 at that time. Of these eight cities, the median price of homes was the highest in Los Angeles and lowest in Cleveland, Ohio, which was $128,000. Of the largest 100 metro areas, the highest median price of existing homes was in San Francisco ($860,000) and the lowest median price was in Scranton-Wilkes-Barre, PA ($100,000) as of December 31, 2018.

Source: www.kiplinger.com

summary measure. From the foregoing discussion about the mean and median, it is obvious that the mean is a nonresistant summary measure, as it is influenced by outliers. In contrast, the median is a resistant summary measure because it is not influenced by outliers.

3.1.3 Mode

Mode is a French word that means *fashion*—an item that is most popular or common. In statistics, the mode represents the most common value in a data set.

Mode

The *mode* is the value that occurs with the highest frequency in a data set.

EXAMPLE 3.6 Speeds of Cars

Calculating the Mode for Ungrouped Data

The following data give the speeds (in miles per hour) of eight cars that were stopped on I-95 for speeding violations.

77 82 74 81 79 84 74 78

Find the mode.

Solution In this data set, 74 occurs twice, and each of the remaining values occurs only once. Because 74 occurs with the highest frequency, it is the mode. Therefore,

Mode = **74 miles per hour**

A major shortcoming of the mode is that a data set may have none or may have more than one mode, whereas it will have only one mean and only one median. For instance, a data set with each value occurring only once has no mode. A data set with only one value occurring with the highest frequency has only one mode. The data set in this case is called **unimodal** as in Example 3.6. A data set with two values that occur with the same (highest) frequency has two modes. The distribution, in this case, is said to be **bimodal** as in Example 3.8. If more than two values in a data set occur with the same (highest) frequency, then the data set contains more than two modes and it is said to be **multimodal** as in Example 3.9.

EXAMPLE 3.7 Incomes of Families

Data Set with No Mode

Last year's incomes of five randomly selected families were \$76,150, \$95,750, \$124,985, \$87,490, and \$53,740. Find the mode.

Solution Because each value in this data set occurs only once, this data set contains **no mode**.

EXAMPLE 3.8 Commuting Times of Employees

Data Set with Two Modes

A small company has 12 employees. Their commuting times (rounded to the nearest minute) from home to work are 23, 36, 14, 23, 47, 32, 8, 14, 26, 31, 18, and 28, respectively. Find the mode for these data.

Solution In the given data on the commuting times of these 12 employees, the values 14 and 23 each occurs twice, and each of the remaining values occurs only once. Therefore, this data set has two modes: 14 and 23 minutes.

EXAMPLE 3.9 Ages of Students

Data Set with Three Modes

The ages of 10 randomly selected students from a class at a university are 21, 19, 27, 22, 29, 19, 25, 21, 22, and 30 years, respectively. Find the mode.

Solution This data set has three modes: **19**, **21**, and **22**. Each of these three values occurs with a (highest) frequency of 2.

One advantage of the mode is that it can be calculated for both kinds of data—quantitative and qualitative—whereas the mean and median can be calculated for only quantitative data.

EXAMPLE 3.10 Status of Students

Finding the Mode for Qualitative Data

The status of five students who are members of the student senate at a college are senior, sophomore, senior, junior, and senior, respectively. Find the mode.

Solution Because *senior* occurs more frequently than the other categories, it is the mode for this data set. We cannot calculate the mean and median for this data set.

3.1.4 Trimmed Mean

Earlier in this chapter, we learned that the mean as a measure of center is impacted by outliers. When a data set contains outliers, we can use either median or trimmed mean as a measure of the center of a data set.

Trimmed Mean

After we drop $k\%$ of the values from each end of a ranked data set, the mean of the remaining values is called the $k\%$ trimmed mean.

Thus, to calculate the trimmed mean for a data set, first we rank the given data in increasing order. Then we drop $k\%$ of the values from each end of the ranked data, where k is any positive number, such as 5%, 10%, and so on. The mean of the remaining values is called the $k\%$ trimmed mean. Remember, although we drop a total of $2 \times k\%$ of the values, $k\%$ from each end, it is called the $k\%$ trimmed mean. Example 3.11 illustrates the calculation of the trimmed mean.

EXAMPLE 3.11 Money Spent on Books by Students

Calculating the Trimmed Mean

The following data give the money spent (in dollars) on books during 2020 by 10 students selected from a small college.

890 1354 1861 1644 87 5403 1429 1993 938 2176

Calculate the 10% trimmed mean.

Solution To calculate the trimmed mean, first we rank the given data as below.

87 890 938 1354 1429 1644 1861 1993 2176 5403

To calculate the 10% trimmed mean, we drop 10% of the data values from each end of the ranked data.

$$10\% \text{ of } 10 \text{ values} = 10(.10) = 1$$

Hence, we drop one value from each end of the ranked data. After we drop the two values, one from each end, we are left with the following eight values:

890 938 1354 1429 1644 1861 1993 2176

The mean of these eight values is called the 10% trimmed mean. Since there are eight values, $n = 8$. Adding these eight values, we obtain Σx as follows:

$$\Sigma x = 890 + 938 + 1354 + 1429 + 1644 + 1861 + 1993 + 2176 = 12{,}285$$

The 10% trimmed mean is obtained by dividing 12,285 by 8 as follows:

$$10\% \text{ trimmed mean} = \frac{12{,}285}{8} = 1535.625 = \mathbf{\$1535.63}$$

Thus, by dropping 10% of the values from each end of the ranked data for this example, we can state that students spent an average of $1535.63 on books in 2020.

Since in this data set $87 and $5403 can be considered outliers, it makes sense to drop these two values and calculate the trimmed mean for the remaining values rather than calculating the mean of all 10 values.

3.1.5 Weighted Mean

In many cases, when we want to find the center of a data set, different values in the data set may have different frequencies or different weights. We will have to consider the weights of different values to find the correct mean of such a data set. For example, suppose Maura bought gas for her car four times during June 2019. She bought 10 gallons at a price of $2.60 a gallon, 13 gallons at a price of $2.80 a gallon, 8 gallons at a price of $2.70 a gallon, and 15 gallons at a price of $2.75 a gallon. In such a case, the mean of the four prices as calculated in Section 3.1.1 will not give the actual mean price paid by Maura in June 2019. We cannot add the four prices and divide by four to find the average price. That can be done only if she bought the same amount of gas each time. But because the amount of gas bought each time is different, we will have to calculate the weighted mean in this case. The amounts of gas bought will be considered the weight here.

Weighted Mean

When different values of a data set occur with different frequencies, that is, each value of a data set is assigned different weight, then we calculate the weighted mean to find the center of the given data set.

To calculate the weighted mean for a data set, we denote the variable by x and the weights by w. We add all the weights and denote this sum by Σw. Then we multiply each value of x by the corresponding value of w. The sum of the resulting products gives Σxw. Dividing Σxw by Σw gives the weighted mean.

Weighted Mean

The weighted mean is calculated as

$$\text{Weighted Mean} = \frac{\Sigma xw}{\Sigma w}$$

where x and w denote the variable and the weights, respectively.

Example 3.12 illustrates the calculation of the weighted mean.

EXAMPLE 3.12 | Prices and Amounts of Gas Purchased

Calculating the Weighted Mean

Maura bought gas for her car four times during June 2019. She bought 10 gallons at a price of $2.60 a gallon, 13 gallons at a price of $2.80 a gallon, 8 gallons at a price of $2.70 a gallon, and 15 gallons at a price of $2.75 a gallon. What is the average price that Maura paid for gas during June 2019?

Solution Here the variable is the price of gas per gallon, and we will denote it by x. The weights are the number of gallons bought each time, and we will denote these weights by w. We list the values of x and w in **Table 3.3**, and find Σw. Then we multiply each value of x by the corresponding value of w and obtain Σxw by adding the resulting values. Finally, we divide Σxw by Σw to find the weighted mean.

TABLE 3.3 Prices and Amounts of Gas Purchased

Price (in dollars) x	Gallons of Gas w	xw
2.60	10	26.00
2.80	13	36.40
2.70	8	21.60
2.75	15	41.25
	$\Sigma w = 46$	$\Sigma xw = 125.25$

$$\text{Weighted Mean} = \frac{\Sigma xw}{\Sigma w} = \frac{125.25}{46} = \mathbf{\$2.72}$$

Thus, Maura paid an average of \$2.72 a gallon for the gas she bought in June 2019.

To summarize, we cannot say for sure which of the various measures of center is a better measure overall. Each of them may be better under different situations. Probably the mean is the most-used measure of center, followed by the median. The mean has the advantage that its calculation includes each value of the data set. The median and trimmed mean are better measures when a data set includes outliers. The mode is simple to locate, but it is not of much use in practical applications.

3.1.6 Relationships Among the Mean, Median, and Mode

As discussed in Chapter 2, two of the many shapes that a histogram or a frequency distribution curve can assume are symmetric and skewed. This section describes the relationships among the mean, median, and mode for three such histograms and frequency distribution curves. Knowing the values of the mean, median, and mode can give us some idea about the shape of a frequency distribution curve.

1. For a **symmetric histogram** and frequency distribution curve with one peak (see **Figure 3.2**), the values of the mean, median, and mode are identical, and they lie at the center of the distribution.

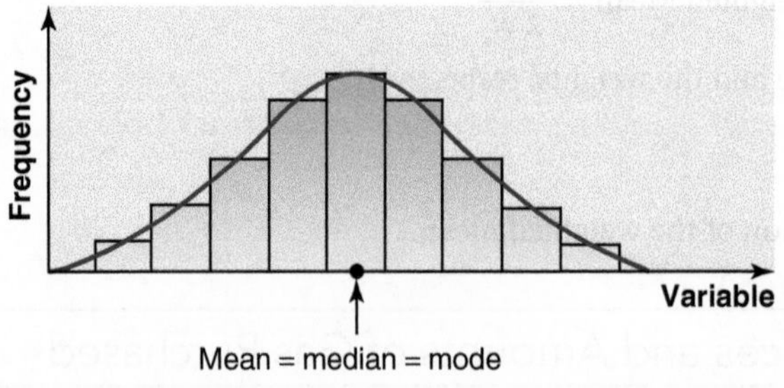

FIGURE 3.2 Mean, median, and mode for a symmetric histogram and frequency distribution curve.

2. For a histogram and a frequency distribution curve **skewed to the right** (see **Figure 3.3**), the value of the mean is the largest, that of the mode is the smallest, and the value of the median lies between these two. (Notice that the mode always occurs at the peak point.) The value of the mean is the largest in this case because it is sensitive to outliers that occur in the right tail. These outliers pull the mean to the right.

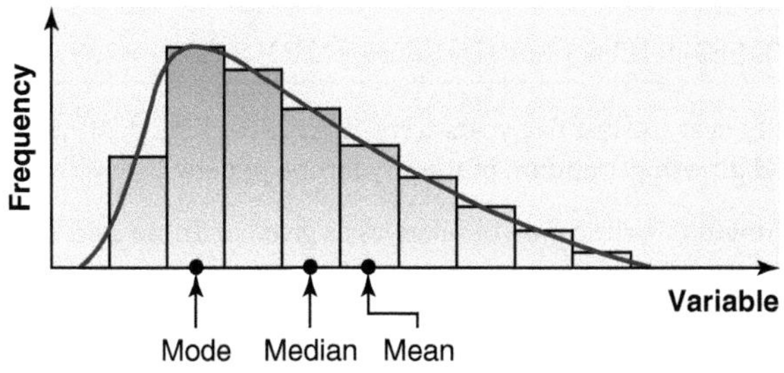

FIGURE 3.3 Mean, median, and mode for a histogram and frequency distribution curve skewed to the right.

3. If a histogram and a frequency distribution curve are **skewed to the left** (see **Figure 3.4**), the value of the mean is the smallest and that of the mode is the largest, with the value of the median lying between these two. In this case, the outliers in the left tail pull the mean to the left.

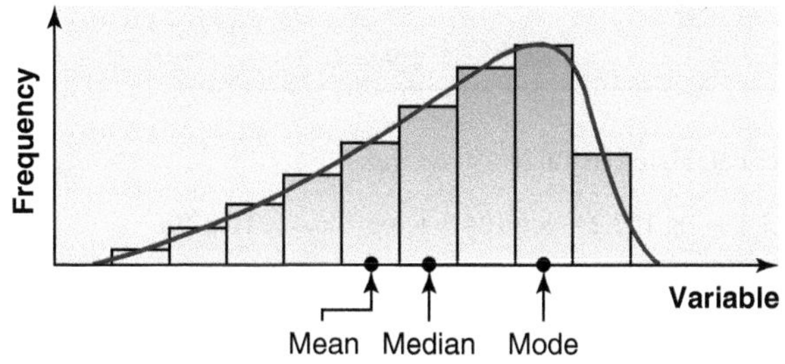

FIGURE 3.4 Mean, median, and mode for a histogram and frequency distribution curve skewed to the left.

3.1.7 Geometric Mean

Geometric mean is one of the methods to find the central tendency of any given data set. It is defined as follows:

Geometric Mean

Geometric mean is defined as the *n*th root of the product of *n* items or observations. If there are two observations, then square root is taken; if there are three observations, then cube root is taken, and so on.

Mathematically, for a set of positive observations $x_1, x_2, x_3, \ldots, x_n$, the geometric mean is calculated as follows:

$$\text{Geometric mean (GM)} = \sqrt[n]{x_1 \cdot x_2 \cdot x_3, \ldots, x_n}$$

Following are the steps to calculate geometric mean:

1. Multiply all the numbers together $x_1, x_2, x_3, \ldots, x_n$.
2. Take the *n*th root of the product, where *n* is the number of terms or observations.

To simplify the calculations, logarithms are used. Geometric mean is then calculated as follows:

$$\log \text{GM} = \frac{\log x_1 + \log x_2 + \log x_3 + \cdots + \log x_n}{n}$$

$$\log \text{GM} = \left(\frac{\Sigma_1^n \log x_i}{n}\right)$$

$$\text{GM} = \text{antilog}\left(\frac{\Sigma_1^n \log x_i}{n}\right)$$

Geometric mean is often used when dealing with average percentage increase in sale, production, population, etc., and dealing with ratios and situations, where values are influenced by compounding effects.

EXAMPLE 3.13 Compound Growth per Annum

Suppose the annual growth rates of a city over the last five years were 1.6%, 2.8%, 3.2%, 4.3%, and 6.4%, respectively. What is the compound growth per annum of the city for the period?

Solution Firstly, convert the annual growth rates into growth relatives as given in **Table 3.4.**

TABLE 3.4 Conversion of Annual Growth Rate into Growth Relatives

Annual Growth Rate	Growth Relatives at the End of the x_i Year
1.6	101.6
2.8	102.8
3.2	103.2
4.3	104.3
6.4	106.4

Taking the logarithm of growth relatives calculated in Table 3.4, we get

$$\Sigma \log x_i = \log 101.6 + \log 102.8 + \log 103.2 + \log 104.3 + \log 106.4 = 10.0778$$

$$\left(\frac{\Sigma_1^n \log x_i}{n}\right) = \left(\frac{10.0778}{5}\right) = 2.01556$$

$$\text{GM} = \text{antilog}\,(2.01556) = 103.6$$

Hence, the compound rate of growth per annum is 103.6 – 100 = 3.6%.

3.1.8 Harmonic Mean

Harmonic mean is another measure of central tendency used in statistics. It is defined as follows:

Harmonic Mean

Harmonic mean is based on the reciprocals of numbers to be averaged. It is calculated by taking the reciprocal of the arithmetic mean of the reciprocals of a given set of observations.

For the set of observations $x_1, x_2, x_3, \ldots, x_n$, the harmonic mean (HM) is defined as

$$\text{Harmonic mean} = \frac{n}{\left(\frac{1}{x_1}+\frac{1}{x_2}+\frac{1}{x_3}+\cdots+\frac{1}{x_n}\right)}$$

Following are the steps to calculate the harmonic mean:

1. Find the reciprocals of all the observations.
2. Sum all the reciprocals.
3. Take the reciprocal of the arithmetic mean of the reciprocals.
4. Multiply by n, the number of observations.

Harmonic mean is useful for computing rates, speed, and ratios. One common application is in calculating average speed when different segments of a journey are covered at different speeds.

EXAMPLE 3.14 Harmonic Mean of Speed of a Car

The following data give the speed (in miles per hour) of a car traveling different segments of a journey.

40 30 60 70 80

Calculate the harmonic mean.

Solution Firstly, calculate the reciprocals of the speed:

$$\frac{1}{40} = .025,\ \frac{1}{30} = .033,\ \frac{1}{60} = .017,\ \frac{1}{70} = .014,\ \frac{1}{80} = .013$$

Now, on summing up all the reciprocals, we obtain:

$$.025 + .033 + .017 + .014 + .013 = .102$$

The reciprocal of the arithmetic mean of the reciprocals is

$$\text{HM} = \frac{5}{.102} = 49.02$$

Hence, the harmonic mean of speed is approximately 49.02 mph.

EXERCISES

Concepts and Procedures

3.1 Explain how the value of the median is determined for a data set that contains an odd number of observations and for a data set that contains an even number of observations.

3.2 Briefly explain the meaning of an outlier. Is the mean or the median a better measure of center for a data set that contains outliers? Illustrate with the help of an example.

3.3 Using an example, show how outliers can affect the value of the mean.

3.4 Which of the five measures of center (the mean, the median, the trimmed mean, the weighted mean, and the mode) can be calculated for quantitative data only, and which can be calculated for both quantitative and qualitative data? Illustrate with examples.

3.5 Which of the five measures of center (the mean, the median, the trimmed mean, the weighted mean, and the mode) can assume more than one value for a data set? Give an example of a data set for which this summary measure assumes more than one value.

3.6 Is it possible for a (quantitative) data set to have no mean, no median, or no mode? Give an example of a data set for which this summary measure does not exist.

3.7 Explain the relationships among the mean, median, and mode for symmetric and skewed histograms. Illustrate these relationships with graphs.

3.8 Prices of mobile phones have a distribution that is skewed to the right with outliers in the right tail. Which of the measures of center is the best to summarize this data set? Explain.

3.9 The following data set belongs to a population.

4 −7 1 0 −9 16 9 8

Calculate the mean, median, and mode.

3.10 Suppose you have invested money in a series of different stocks over a span of six years. The annual returns on your investments are as follows: 12%, 16%, 10%, 18%, 22%, and 17%. To find the annual growth rate of your investment, calculate the geometric mean.

3.11 Suppose the growth rate of a company is as follows: 3%, 2.8%, 3.6%, 4%, 4.8%, 5.3%, 6.5%, 8%, 8.9%, and 10% for the consecutive 10 years. What is the average rate of growth of the company?

3.12 Suppose the incomes of four individuals are $20,000, $60,000, $40,000, and $80,000. Calculate the harmonic mean income for this population.

3.13 A car driver travels form a plain area to a hill station for 100 miles at an average speed of 35 mph. He then makes the return trip at an average speed 25 mph. What is his average speed over the entire distance?

Applications

3.14 The following table lists the number of worldwide locations for the 10 largest fast-food restaurant chains as listed on worldatlas.com on June 10, 2019 (https://www.worldatlas.com).

Name	Number of Locations
Subway	42,998
McDonald's	37,200
Starbucks	30,000
KFC	20,404
Burger King	16,859
Pizza Hut	16,796
Domino's	15,000
Dunkin'	11,300
Baskin-Robbins	7,500
Hunt Brothers Pizza	7,300

Find the mean and median for these data. Do these data have a mode? Explain.

3.15 The following data give the amounts (in dollars) of electric bills for November 2019 for 12 randomly selected households selected from a small town.

211 265 176 314 243 192 297 357 238 281 342 259

Calculate the mean and median for these data. Do these data have a mode? Explain.

3.16 The following data give the numbers of car thefts that occurred in a city during the past 14 days.

8	3	7	11	4	3	5	4	2	6	4	15	8	3

a. Calculate the mean, median, and mode for these data.

b. Calculate the 10% trimmed mean for these data.

3.17 The following data give the monthly incomes (in dollars) of 10 employees in an office.

14,800	15,500	28,600	26,900	23,800
27,500	70,000	29,300	19,000	24,300

a. Calculate the mean and median for the given data.

b. Identify the outlier in this data set. Drop the outlier and recalculate the mean and median. Which of these two summary measures changes by a larger amount when you drop the outlier?

c. Which is the better summary measure for the given data? Explain.

3.18 The following data give the total food expenditures (in dollars) for the past one month for a sample of 20 families.

1125	530	1234	595	427	872	1480	699	1274	1187
933	1127	716	1065	934	930	1046	1199	1353	441

a. Calculate the mean and median for these data.

b. Calculate the 20% trimmed mean for these data.

3.19 The following data give the prices (in thousands of dollars) of 20 houses sold recently in a city.

184	324	365	309	245	387	369	438	195	180
323	578	510	679	307	271	318	795	259	590

a. Calculate the mean, median, and mode for these data.

b. Calculate the 20% trimmed mean for these data.

3.20 The following data give the annual salaries (in thousand dollars) of 20 randomly selected health-care workers.

50	71	57	39	45	64	38	53	35	62
74	40	67	44	77	61	58	55	64	59

a. Calculate the mean, median, and mode for these data.

b. Calculate the 15% trimmed mean for these data.

3.21 The following data give the number of patients who visited a walk-in clinic on each of 20 randomly selected days.

23	37	26	19	33	22	30	42	24	26
28	32	37	29	38	24	35	20	34	38

a. Calculate the mean, median, and mode for these data.

b. Calculate the 15% trimmed mean for these data.

3.22 The following data represent the diastolic blood pressure in (mmHg) for each of the 20 randomly selected middle-aged males.

75	80	72	85	78	82	77	90	70	88
74	76	80	83	81	73	87	86	89	84

a. Calculate the mean, median, and mode for the data given.

b. Calculate the 20% trimmed mean for the given data.

3.23 In a survey of 700 parents of young children, 380 said that they will not want their children to play football because it is a very dangerous sport, 230 said that they will let their children play football, and 90 had no opinion. Considering the opinions of these parents, what is the mode?

3.24 A statistics professor based her final grades on quizzes, in-class group work, homework, a midterm exam, and a final exam. However, not all of the assignments contributed equally to the final grade. John received the scores (out of 100 for each assignment) listed in the table. The instructor weighted each item as shown in the table.

Assignment	John's Score (Total Points)	Percentage of Final Grade Assigned by the Instructor
Quizzes	78	30
In-class group work	55	5
Homework	84	10
Midterm exam	76	15
Final exam	78	40

Calculate John's final grade score (out of 100) in this course. (*Hint:* Note that it is often easier to convert the weights to decimal form for weighting, e.g., 30% = .30; in this way, the sum of weights will equal 1, making the formula easier to apply.)

3.25 A clothing store bought 3000 shirts last year from five different companies. The following table lists the number of shirts bought from each company and the price paid for each shirt.

Company	Number of Shirts Bought	Price per Shirt (dollars)
Best Shirts	200	30
Top Wear	900	45
Smith Shirts	400	40
New Design	1200	35
Radical Wear	300	50

Calculate the weighted mean that represents the average price paid for these 3000 shirts.

***3.26** One property of the mean is that if we know the means and sample sizes of two (or more) data sets, we can calculate the **combined mean** of both (or all) data sets. The combined mean for two data sets is calculated by using the formula

$$\text{Combined mean} = \bar{x} = \frac{n_1\bar{x}_1 + n_2\bar{x}_2}{n_1 + n_2}$$

where n_1 and n_2 are the sample sizes of the two data sets and $\bar{x}_1$ and $\bar{x}_2$ are the means of the two data sets, respectively. Suppose a sample of 10 statistics books gave a mean price of \$140 and a sample of 8 mathematics books gave a mean price of \$160. Find the combined mean. (*Hint:* For this example: $n_1 = 10, n_2 = 8, \bar{x}_1 = \$140, \bar{x}_2 = \$160$.)

***3.27** For any data, the sum of all values is equal to the product of the sample size and mean; that is, $\Sigma x = n\bar{x}$. Suppose the average amount of money spent on shopping by 12 persons during a given week is \$115.50. Find the total amount of money spent on shopping by these 12 persons.

***3.28** The mean 2019 income for five families was \$99,520. What was the total 2019 income of these five families?

***3.29** The mean age of six persons is 52 years. The ages of five of these six persons are 56, 38, 43, 51, and 36 years, respectively. Find the age of the sixth person.

***3.30** Seven airline passengers in economy class on the same flight paid an average of \$354 per ticket. Because the tickets were purchased at different times and from different sources, the prices varied. The first five passengers paid \$425, \$215, \$338, \$600, and \$490. The sixth and seventh tickets were purchased by a couple who paid identical fares. What price did each of them pay?

*3.31 Suppose that the inflation rates for the last five years are 4%, 3%, 5%, 6%, and 8%, respectively. Thus at the end of the first year, the price index will be 1.04 times the price index at the beginning of the year, and so on. Find the mean rate of inflation over the five-year period by finding the geometric mean of the data set 1.04, 1.03, 1.05, 1.06, and 1.08. (*Hint:* Here, $n = 5, x_1 = 1.04, x_2 = 1.03$, and so on. Use the $x^{1/n}$ key on your calculator to find the fifth root. Note that the mean inflation rate will be obtained by subtracting 1 from the geometric mean.)

3.2 Measures of Dispersion for Ungrouped Data

LEARNING OBJECTIVES

After completing this section, you should be able to:

- Identify measures of dispersion for quantitative data.
- Calculate the range for ungrouped quantitative data.
- Explain why the range is influenced by outliers.
- Calculate the variance for ungrouped quantitative data.
- Calculate and interpret the standard deviation for ungrouped quantitative data.
- Explain the difference between the variance and standard deviation calculated for population and sample data.
- Calculate the coefficient of variation for ungrouped quantitative data.

The measures of center, such as the mean, median, and mode, do not provide a complete description of the distribution of a data set. Two data sets with the same mean may have completely different spreads. The variation among the values of observations for one data set may be much larger or smaller than for the other data set. (Note that the words **dispersion**, **spread**, and **variation** have similar meanings.) Consider the following two data sets on the ages (in years) of all workers at each of two small companies.

Company 1:	47	38	35	40	36	45	39
Company 2:	70	33	18	52	27		

The mean age of workers in both these companies is the same, 40 years. If we do not know the ages of individual workers at these two companies and are told only that the mean age of the workers at both companies is the same, we may deduce that the workers at these two companies have a similar age distribution. As we can observe, however, the variation in the workers' ages for each of these two companies is very different. As illustrated in the diagram, the ages of the workers at the second company have a much larger variation than the ages of the workers at the first company.

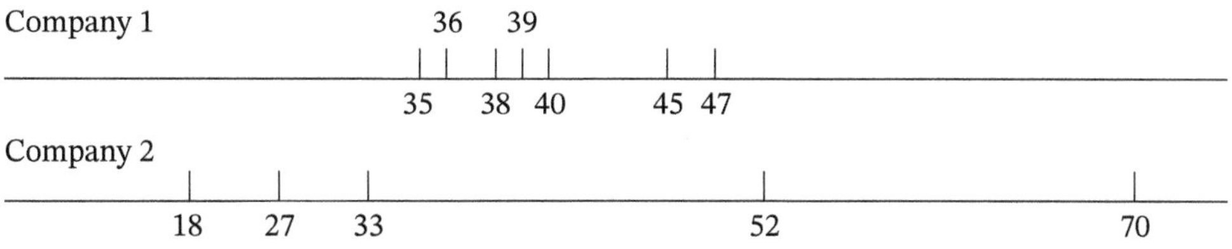

Thus, a measure of center such as the mean, median, or mode by itself is usually not sufficient to fully describe the distribution of a data set. We also need a measure that can provide some information about the variation among data values. The measures that help us learn about the spread of a data set are called the **measures of dispersion**. The measures of center and dispersion taken together give a better picture of a data set than the measures of center alone. This section discusses four measures of dispersion: range, variance, standard deviation, and coefficient of variation.

3.2.1 Range

The **range** is the simplest measure of dispersion to calculate. It is obtained by taking the difference between the largest and the smallest values in a data set.

Finding the Range for Ungrouped Data

Range = Largest value − Smallest value

EXAMPLE 3.15 Total Areas of Four States

Calculating the Range for Ungrouped Data

Table 3.5 gives the total areas in square miles of the four western South-Central states of the United States.

TABLE 3.5

State	Total Area (square miles)
Arkansas	53,182
Louisiana	49,651
Oklahoma	69,903
Texas	267,277

Find the range for this data set.

Solution The largest total area for a state in this data set is 267,277 square miles, and the smallest area is 49,651 square miles. Therefore,

Range = Largest value − Smallest value
= 267,277 − 49,651 = **217,626 square miles**

Thus, the total areas of these four states are spread over a range of 217,626 square miles.

The range, like the mean, has the disadvantage of being influenced by outliers. In Example 3.15, if the state of Texas with a total area of 267,277 square miles is dropped, the range decreases from 217,626 square miles to 69,903 − 49,651 = 20,252 square miles. Consequently, the range is not a good measure of dispersion to use for a data set that contains outliers. This indicates that the range is a **nonresistant measure** of dispersion; that is, the range is affected by the outliers in a data set.

Another disadvantage of using the range as a measure of dispersion is that its calculation is based on two values only: the largest and the smallest. All other values in a data set are ignored when calculating the range. Thus, the range is not a very satisfactory measure of dispersion.

3.2.2 Variance and Standard Deviation

The **standard deviation** is the most-used measure of dispersion. The value of the standard deviation tells how closely the values of a data set are clustered around the mean. In general, a lower value of the standard deviation for a data set indicates that the values of that data set are spread over a relatively smaller range around the mean. In contrast, a larger value of the standard deviation for a data set indicates that the values of that data set are spread over a relatively larger range around the mean.

The *standard deviation is obtained by taking the positive square root of the* **variance**. The variance calculated for population data is denoted by σ^2 (read as *sigma square*),[2] and the variance calculated for sample data is denoted by s^2. Consequently, the standard deviation calculated for population data is denoted by σ and the standard deviation calculated for sample data is denoted by s. Following are what we will call the *basic formulas* that are used to calculate the variance and standard deviation.[3]

$$\sigma^2 = \frac{\Sigma(x-\mu)^2}{N} \quad \text{and} \quad s^2 = \frac{\Sigma(x-\bar{x})^2}{n-1}$$

$$\sigma = \sqrt{\frac{\Sigma(x-\mu)^2}{N}} \quad \text{and} \quad s = \sqrt{\frac{\Sigma(x-\bar{x})^2}{n-1}}$$

where σ^2 is the population variance, s^2 is the sample variance, σ is the population standard deviation, and s is the sample standard deviation.

The quantities $x - \mu$ and $x - \bar{x}$ in the above formulas are called the *deviation* of the x value from the mean. The sum of the deviations of the x values from the mean is always zero; that is, $\Sigma(x-\mu) = 0$ and $\Sigma(x-\bar{x}) = 0$.

For example, suppose the midterm scores of a sample of four students are 82, 95, 67, and 92, respectively. Then, the mean score for these four students is

$$\bar{x} = \frac{82+95+67+92}{4} = \frac{336}{4} = 84$$

The deviations of the four scores from the mean are calculated in **Table 3.6**. As we can observe from the table, the sum of the deviations of the x values from the mean is zero; that is, $\Sigma(x-\bar{x}) = 0$. For this reason we square the deviations to calculate the variance and standard deviation.

TABLE 3.6

x	$(x-\bar{x})$
82	$82 - 84 = -2$
95	$95 - 84 = +11$
67	$67 - 84 = -17$
92	$92 - 84 = +8$
	$\Sigma(x-\bar{x}) = 0$

From the computational point of view, it is easier and more efficient to use *short-cut formulas* to calculate the variance and standard deviation. By using the short-cut formulas, we reduce the computation time and round-off errors. Use of the basic formulas for ungrouped data is illustrated in Section A3.1.1 of Appendix 3.1 of this chapter. The short-cut formulas for calculating the variance and standard deviation are given next.

Short-Cut Formulas for the Variance and Standard Deviation for Ungrouped Data

$$\sigma^2 = \frac{\Sigma x^2 - \left(\frac{(\Sigma x)^2}{N}\right)}{N} \quad \text{and} \quad s^2 = \frac{\Sigma x^2 - \left(\frac{(\Sigma x)^2}{n}\right)}{n-1}$$

where σ^2 is the population variance and s^2 is the sample variance.

The standard deviation is obtained by taking the positive square root of the variance.

$$\text{Population standard deviation:} \quad \sigma = \sqrt{\frac{\Sigma x^2 - \left(\frac{(\Sigma x)^2}{N}\right)}{N}}$$

$$\text{Sample standard deviation:} \quad s = \sqrt{\frac{\Sigma x^2 - \left(\frac{(\Sigma x)^2}{n}\right)}{n-1}}$$

[2]Note that Σ is uppercase sigma and σ is lowercase sigma of the Greek alphabet.

[3]From the formula for σ^2, it can be stated that the population variance is the mean of the squared deviations of x values from the mean. However, this is not true for the variance calculated for a sample data set.

Note that the denominator in the formula for the population variance is N, but that in the formula for the sample variance is $n - 1$.[4]

Thus, to calculate the variance and standard deviation for a data set, we perform the following three steps:

1. In the first step, we add all the values of x in the given data set and denote this sum by Σx.
2. In the second step, we square each value of x and add these squared values, which is denoted by Σx^2.
3. In the third step, we substitute the values of n, Σx, and Σx^2 in the corresponding formula for variance or standard deviation and simplify.

Examples 3.16 and 3.17 show the calculation of the variance and standard deviation.

EXAMPLE 3.16 2019 Salaries of College Football Head Coaches

Calculating the Sample Variance and Standard Deviation for Ungrouped Data

Table 3.2 (reproduced from Example 3.4) lists the 2019 salaries (base salary plus bonuses that are not performance based) of 11 college football head coaches (https://www.boydsbets.com/).

TABLE 3.2 2019 Salaries of 11 College Football Head Coaches

Head Coach and College	2019 Salary (millions of dollars)
Dabo Swinney (Clemson)	10.3
Ryan Day (Ohio State)	4.5
Chip Kelly (UCLA)	3.3
Manny Diaz (Miami, FL)	3.1
Nick Saban (Alabama)	8.7
David Shah (Stanford)	4.3
Mack Brown (North Carolina)	3.5
Lincoln Riley (Oklahoma)	4.8
Tom Herman (Texas)	5.5
James Franklin (Penn State)	4.8
Les Miles (Kansas)	2.8

SW Productions/Stockbyte/Getty Images

Find the variance and standard deviation for these data.

Solution Let x denote the 2019 salaries (in millions of dollars) of the 11 college football head coaches. The calculation of Σx and Σx^2 is shown in **Table 3.7**.

[4]The reason that the denominator in the sample formula is $n - 1$ and not n follows: The sample variance tends to underestimate the population variance when the denominator in the sample formula for variance is n. However, the sample variance does not underestimate the population variance if the denominator in the sample formula for variance is $n - 1$. In Chapter 8 we will learn that $n - 1$ is called the degrees of freedom.

TABLE 3.7 Calculation of Σx and Σx^2

x	x^2
10.3	106.09
4.5	20.25
3.3	10.89
3.1	9.61
8.7	75.69
4.3	18.49
3.5	12.25
4.8	23.04
5.5	30.25
4.8	23.04
2.8	7.84
$\Sigma x = 55.6$	$\Sigma x^2 = 337.44$

Calculation of the variance involves the following three steps:

STEP 1 *Calculate Σx.*

The sum of the values in the first column of Table 3.7 gives the value of Σx, which is 55.6.

STEP 2 *Find Σx^2.*

The value of Σx^2 is obtained by squaring each value of x and then adding the squared values. The results of this step are shown in the second column of Table 3.7. Notice that $\Sigma x^2 = 337.44$.

STEP 3 *Determine the variance.*

Substitute the values of n, Σx, and Σx^2 in the variance formula and simplify. Because the given data are on the 2019 salaries of 11 college football head coaches, we use the formula for the sample variance using $n = 11$.

$$s^2 = \frac{\Sigma x^2 - \left(\frac{(\Sigma x)^2}{n}\right)}{n-1} = \frac{337.44 - \left(\frac{(55.6)^2}{11}\right)}{11-1} = \frac{337.44 - 281.0327}{10} = \mathbf{5.6407}$$

Now to obtain the standard deviation, we take the (positive) square root of the variance.

$$s = \sqrt{5.6407} = 2.3750 = \mathbf{\$2.375\ million}$$

Thus, the standard deviation of the 2019 salaries of these 11 college football head coaches is $2.375 million.

Two Observations

1. **The values of the variance and the standard deviation are never negative.** That is, the numerator in the formula for the variance will never produce a negative value. Usually the values of the variance and standard deviation are positive, but if a data set has no variation, then the variance and standard deviation are both zero. For example, if four persons in a group are the same age—say, 35 years—then the four values in the data set are

   35 35 35 35

   If we calculate the variance and standard deviation for these data, their values will be zero. This is because there is no variation in the values of this data set.

2. **The measurement units of the variance are always the square of the measurement units of the original data.** This is so because the original values are squared to calculate the variance. In Example 3.16, the measurement units of the original data are millions

of dollars. However, the measurement units of the variance are squared million dollars, which, of course, does not make any sense. Thus, the variance of the 2019 salaries of these 11 college football head coaches is 5.6407 squared million dollars. But the standard deviation of these salaries is $2.375 million. The measurement units of the standard deviation are the same as the measurement units of the original data because the standard deviation is obtained by taking the square root of the variance.

Earlier in this chapter, we explained the difference between resistant and nonresistant summary measures. The variance and standard deviation are **nonresistant** summary measures as their values are sensitive to the outliers. The existence of outliers in a data set will increase the values of the variance and standard deviation.

EXAMPLE 3.17 Earnings of Employees

Calculating the Population Variance and Standard Deviation for Ungrouped Data

The following data give the 2019 earnings (in thousands of dollars) before taxes for all six employees of a small company.

88.50 108.40 65.50 52.50 79.80 54.60

Calculate the variance and standard deviation for these data.

Solution Let x denote the 2019 earnings before taxes of an employee of this company. The values of Σx and Σx^2 are calculated in **Table 3.8.**

TABLE 3.8

x	x^2
88.50	7,832.25
108.40	11,750.56
65.50	4,290.25
52.50	2,756.25
79.80	6,368.04
54.60	2,981.16
$\Sigma x = 449.30$	$\Sigma x^2 = 35{,}978.51$

Because the data in this example are on earnings of *all* employees of this company, we use the population formula to compute the variance using $N = 6$. Thus, the variance is

$$\sigma^2 = \frac{\Sigma x^2 - \left(\frac{(\Sigma x)^2}{N}\right)}{N} = \frac{35{,}978.51 - \left(\frac{(449.30)^2}{6}\right)}{6} = \mathbf{388.90}$$

The standard deviation is obtained by taking the (positive) square root of the variance:

$$\sigma = \sqrt{\frac{\Sigma x^2 - \left(\frac{(\Sigma x)^2}{N}\right)}{N}} = \sqrt{388.90} = \mathbf{19.721\ thousand} = \mathbf{\$19{,}721}$$

Thus, the standard deviation of the 2019 earnings of all six employees of this company is $19,721.

Warning

Note that Σx^2 is not the same as $(\Sigma x)^2$. The value of Σx^2 is obtained by squaring the x values and then adding them. The value of $(\Sigma x)^2$ is obtained by squaring the value of Σx.

The uses of the standard deviation are discussed in Section 3.4. Later chapters explain how the mean and the standard deviation taken together can help in making inferences about the population.

3.2.3 Coefficient of Variation

One disadvantage of the standard deviation as a measure of dispersion is that it is a measure of absolute variability and not of relative variability. Sometimes we may need to compare the variability for two different data sets that have different units of measurement. In such cases, a measure of relative variability is preferable. One such measure is the **coefficient of variation.**

Coefficient of Variation

The coefficient of variation, denoted by CV, expresses standard deviation as a percentage of the mean and is computed as follows:

$$\text{For population data:}\quad CV = \frac{\sigma}{\mu} \times 100\%$$

$$\text{For sample data:}\quad CV = \frac{s}{\bar{x}} \times 100\%$$

Note that the coefficient of variation does not have any units of measurement, as it is always expressed as a percent.

Thus, to calculate the coefficient of variation, we perform the following steps: First we calculate the mean and standard deviation for the given data set. Then we divide the standard deviation by the mean and multiply the answer by 100%.

Example 3.18 shows the calculation of the coefficient of variation.

EXAMPLE 3.18 Salaries and Education

Calculating the Coefficient of Variation

The yearly salaries of all employees working for a large company have a mean of $72,350 and a standard deviation of $12,820. The years of schooling (education) for the same employees have a mean of 15 years and a standard deviation of 2 years. Is the relative variation in the salaries higher or lower than that in years of schooling for these employees? Answer the question by calculating the coefficient of variation for each variable.

Solution Because the two variables (salary and years of schooling) have different units of measurement (dollars and years, respectively), we cannot directly compare the two standard deviations. Hence, we calculate the coefficient of variation for each of these data sets.

$$\text{CV for salaries} = \frac{\sigma}{\mu} \times 100\% = \frac{12{,}820}{72{,}350} \times 100\% = \mathbf{17.72\%}$$

$$\text{CV for years of schooling} = \frac{\sigma}{\mu} \times 100\% = \frac{2}{15} \times 100\% = \mathbf{13.33\%}$$

Thus, the standard deviation for salaries is 17.72% of its mean and that for years of schooling is 13.33% of its mean. Since the coefficient of variation for salaries has a higher value than the coefficient of variation for years of schooling, the salaries have a higher relative variation than the years of schooling.

3.2.4 Population Parameters and Sample Statistics

A numerical measure such as the mean, median, mode, range, variance, or standard deviation calculated for a population data set is called a *population parameter*, or simply a **parameter**. A summary measure calculated for a sample data set is called a *sample statistic*, or simply a **statistic**. Thus, μ and σ are population parameters, and $\bar{x}$ and s are sample statistics.

As an illustration, $\bar{x}$ = \$209.8 billion in Example 3.1 is a sample statistic, and μ = 45.25 years in Example 3.2 is a population parameter. Similarly, s = \$2.375 million in Example 3.16 is a sample statistic, whereas σ = \$19,721 in Example 3.17 is a population parameter.

EXERCISES

Concepts and Procedures

3.32 The range, as a measure of spread, has the disadvantage of being influenced by outliers. Illustrate this with an example.

3.33 Can the standard deviation have a negative value? Explain.

3.34 When is the value of the standard deviation for a data set zero? Give one example. Calculate the standard deviation for the example and show that its value is zero.

3.35 Briefly explain the difference between a population parameter and a sample statistic. Give one example of each.

3.36 The following data set belongs to a sample.

14 16 −10 8 8 −18

Calculate the range, variance, and standard deviation.

Applications

3.37 The following data give the prices of seven textbooks randomly selected from a university bookstore.

\$112 \$170 \$93 \$113 \$56 \$161 \$123

a. Find the mean for these data. Calculate the deviations of the data values from the mean. Is the sum of these deviations zero?

b. Calculate the range, variance, and standard deviation.

3.38 The following data give the number of years of employment for all 20 employees of a small company.

23	9	12	21	24	6	33	34	17	3
12	31	5	10	27	9	15	16	30	38

a. Compute the range, variance, and standard deviation for these data.

b. Calculate the coefficient of variation.

c. Are the values of these summary measures population parameters or sample statistics? Explain.

3.39 The following data give the annual salaries (in thousand dollars) of 20 randomly selected health-care workers.

50	71	57	39	45	64	38	53	35	62
74	40	67	44	77	61	58	55	64	59

a. Compute the range, variance, and standard deviation for these data.

b. Calculate the coefficient of variation.

c. Are the values of these summary measures population parameters or sample statistics? Explain.

3.40 The following data give the number of patients who visited a walk-in clinic on each of 20 randomly selected days.

23	37	26	19	33	22	30	42	24	26
28	32	37	29	38	24	35	20	34	38

a. Calculate the range, variance, and standard deviation for these data.

b. Calculate the coefficient of variation.

3.41 The following data give the numbers of hours spent partying by 10 randomly selected college students during the past week.

7 14 5 0 9 7 10 4 0 8

a. Calculate the range, variance, and standard deviation for these data.

b. Calculate the coefficient of variation.

3.42 Following are the temperatures (in Fahrenheit) observed during eight wintry days in a midwestern city:

23 14 6 −7 −2 11 16 19

a. Calculate the range, variance, and standard deviation for these data.

b. Calculate the coefficient of variation.

3.43 The following data give the times (in minutes) that all 10 students took to complete an assignment in a statistics class.

15 26 16 36 31 13 29 18 21 39

a. Calculate the range, variance, and standard deviation for these data.

b. Calculate the coefficient of variation.

c. What does the (large) value of the standard deviation tell you?

3.44 The following data give the amount (in dollars) of electric bills for November 2019 for 12 randomly selected households from a small town.

205 265 176 314 243 192 297 357 238 281 342 259

a. Calculate the range, variance, and standard deviation for these data.

b. Calculate the coefficient of variation.

3.45 The following data give the 2019 bonuses (in thousands of dollars) of 10 randomly selected Wall Street managers.

127 82 45 99 153 3261 77 108 68 278

a. Calculate the range, variance, and standard deviation for these data.

b. Calculate the coefficient of variation.

3.46 The following data give the daily wage rates of eight employees of a company.

\$94 94 94 94 94 94 94 94

Calculate the standard deviation. Is its value zero? If yes, why?

3.47 The yearly salaries of all employees who work for a company have a mean of $61,850 and a standard deviation of $6740. The years of experience for the same employees have a mean of 15 years and a standard deviation of 2 years. Is the relative variation in the salaries larger or smaller than that in years of experience for these employees?

3.48 The SAT scores of 100 students have a mean of 1020 and a standard deviation of 115. The GPAs of the same 100 students have a mean of 3.21 and a standard deviation of .26. Is the relative variation in SAT scores larger or smaller than that in GPAs?

***3.49** Consider the following two data sets:

Data Set I:	4	8	15	9	11
Data Set II:	12	24	45	27	33

Note that each value of the second data set is obtained by multiplying the corresponding value of the first data set by 3. Calculate the standard deviation for each of these two data sets using the formula for population data. Comment on the relationship between the two standard deviations.

3.3 Mean, Variance, and Standard Deviation for Grouped Data

LEARNING OBJECTIVES

After completing this section, you should be able to:

- Calculate and interpret the mean for grouped data.
- Calculate and interpret the variance and standard deviation for grouped data.

In Sections 3.1.1 and 3.2.2, we learned how to calculate the mean, variance, and standard deviation for ungrouped data. In this section, we will learn how to calculate the mean, variance, and standard deviation for grouped data.

3.3.1 Mean for Grouped Data

We learned in Section 3.1.1 that the mean is obtained by dividing the sum of all values by the number of values in a data set. However, if the data are given in the form of a frequency table, we no longer know the values of individual observations. Consequently, in such cases, we cannot obtain the sum of individual values. We find an approximation for the sum of these values using the procedure explained next and in Example 3.19. The formulas used to calculate the mean for grouped data follow:

Calculating Mean for Grouped Data

Mean for population data: $\mu = \frac{\Sigma mf}{N}$

Mean for sample data: $\bar{x} = \frac{\Sigma mf}{n}$

where m is the midpoint and f is the frequency of a class.

To calculate the mean for grouped data, first find the midpoint of each class and then multiply the midpoints by the frequencies of the corresponding classes. The sum of these products, denoted by Σmf, gives an approximation for the sum of all values. To find the value of the mean, divide this sum by the total number of observations in the data.

EXAMPLE 3.19 | Daily Commuting Times for Employees

Calculating the Population Mean for Grouped Data

Table 3.9 gives the frequency distribution of the daily one-way commuting times (in minutes) from home to work for *all* 25 employees of a company.

TABLE 3.9

Daily Commuting Time (minutes)	Number of Employees (f)
0 to less than 10	4
10 to less than 20	9
20 to less than 30	6
30 to less than 40	4
40 to less than 50	2

Calculate the mean of the daily commuting times.

Solution Note that because the data set includes *all* 25 employees of the company, it represents the population. **Table 3.10** shows the calculation of Σmf. Note that in Table 3.10, m denotes the midpoints of the classes.

TABLE 3.10

Daily Commuting Time (minutes)	f	m	mf
0 to less than 10	4	5	20
10 to less than 20	9	15	135
20 to less than 30	6	25	150
30 to less than 40	4	35	140
40 to less than 50	2	45	90
	$N = 25$		$\Sigma mf = 535$

To calculate the mean, we first find the midpoint of each class. The class midpoints are recorded in the third column of Table 3.10. The products of the midpoints and the corresponding frequencies are listed in the fourth column. The sum of the fourth column values, denoted by Σmf, gives the approximate total daily commuting time (in minutes) for all 25 employees. The mean is obtained by dividing this sum by the total frequency. Therefore,

$$\mu = \frac{\Sigma mf}{N} = \frac{535}{25} = \mathbf{21.40\ minutes}$$

Thus, the employees of this company spend an average of 21.40 minutes a day commuting from home to work.

What do the numbers 20, 135, 150, 140, and 90 in the column labeled mf in Table 3.10 represent? We know from this table that four employees spend 0 to less than 10 minutes commuting per day. If we assume that the time spent commuting by these four employees is evenly spread in the interval 0 to less than 10, then the midpoint of this class (which is 5) gives the mean time spent commuting by these four employees. Hence, $4 \times 5 = 20$ is the approximate total time (in minutes) spent commuting per day by these four employees. Similarly, nine employees spend 10 to less than 20 minutes commuting per day, and the total time spent commuting by these nine employees is approximately 135 minutes a day. The other numbers in this column can be interpreted the same way. Note that these numbers give the approximate commuting times for these employees based on the assumption of an even spread within classes. The total commuting time for all 25 employees is approximately 535 minutes. Consequently, 21.40 minutes is an approximate and not the exact value of the mean. We can find the exact value of the mean only if we know the exact commuting time for each of the 25 employees of the company.

EXAMPLE 3.20 Number of Orders Received

Calculating the Sample Mean for Grouped Data

Table 3.11 gives the frequency distribution of the number of orders received each day during the past 50 days at the office of a mail-order company.

TABLE 3.11

Number of Orders	Number of Days (f)
10–12	4
13–15	12
16–18	20
19–21	14

Calculate the mean.

Solution Because the data set includes only 50 days, it represents a sample. The value of Σmf is calculated in **Table 3.12**.

TABLE 3.12

Number of Orders	f	m	mf
10–12	4	11	44
13–15	12	14	168
16–18	20	17	340
19–21	14	20	280
	$n = 50$		$\Sigma mf = 832$

The value of the sample mean is

$$\bar{x} = \frac{\Sigma mf}{n} = \frac{832}{50} = \mathbf{16.64\ orders}$$

Thus, this mail-order company received an average of 16.64 orders per day during these 50 days.

3.3.2 Variance and Standard Deviation for Grouped Data

Following are what we will call the *basic formulas* that are used to calculate the population and sample variances for grouped data:

$$\sigma^2 = \frac{\Sigma f(m - \mu)^2}{N} \quad \text{and} \quad s^2 = \frac{\Sigma f(m - \bar{x})^2}{n - 1}$$

where σ^2 is the population variance, s^2 is the sample variance, and m is the midpoint of a class.

In either case, the standard deviation is obtained by taking the positive square root of the variance.

Again, the *short-cut formulas* are more efficient for calculating the variance and standard deviation. Section A3.1.2 of Appendix 3.1 at the end of this chapter shows how to use the basic formulas to calculate the variance and standard deviation for grouped data.

Short-Cut Formulas for the Variance and Standard Deviation for Grouped Data

$$\sigma^2 = \frac{\Sigma m^2 f - \left(\frac{(\Sigma mf)^2}{N}\right)}{N} \quad \text{and} \quad s^2 = \frac{\Sigma m^2 f - \left(\frac{(\Sigma mf)^2}{N}\right)}{n-1}$$

where σ^2 is the population variance, s^2 is the sample variance, and m is the midpoint of a class.

The standard deviation is obtained by taking the positive square root of the variance.

$$\text{Population standard deviation:} \quad \sigma = \sqrt{\frac{\Sigma m^2 f - \left(\frac{(\Sigma mf)^2}{N}\right)}{N}}$$

$$\text{Sample standard deviation:} \quad s = \sqrt{\frac{\Sigma m^2 f - \left(\frac{(\Sigma mf)^2}{n}\right)}{n-1}}$$

Examples 3.21 and 3.22 illustrate the use of these formulas to calculate the variance and standard deviation.

EXAMPLE 3.21 Daily Commuting Times of Employees

Calculating the Population Variance and Standard Deviation for Grouped Data

The following data, reproduced from Table 3.9 of Example 3.19, give the frequency distribution of the daily one-way commuting times (in minutes) from home to work for all 25 employees of a company.

Daily Commuting Time (minutes)	Number of Employees (f)
0 to less than 10	4
10 to less than 20	9
20 to less than 30	6
30 to less than 40	4
40 to less than 50	2

Calculate the variance and standard deviation.

Solution All four steps needed to calculate the variance and standard deviation for grouped data are shown after **Table 3.13**.

TABLE 3.13

Daily Commuting Time (minutes)	f	m	mf	m^2f
0 to less than 10	4	5	20	100
10 to less than 20	9	15	135	2025
20 to less than 30	6	25	150	3750
30 to less than 40	4	35	140	4900
40 to less than 50	2	45	90	4050
	$N = 25$		$\Sigma mf = 535$	$\Sigma m^2f = 14,825$

STEP 1 *Calculate the value of Σmf.*

To calculate the value of Σmf, first find the midpoint m of each class (see the third column in Table 3.13) and then multiply the corresponding class midpoints and class frequencies (see the fourth column). The value of Σmf is obtained by adding these products. Thus,

$$\Sigma mf = 535$$

STEP 2 *Find the value of $\Sigma m^2 f$.*

To find the value of $\Sigma m^2 f$, square each m value and multiply this squared value of m by the corresponding frequency (see the fifth column in Table 3.13). The sum of these products (that is, the sum of the fifth column) gives $\Sigma m^2 f$. Hence,

$$\Sigma m^2 f = 14{,}825$$

STEP 3 *Calculate the variance.*

Because the data set includes all 25 employees of the company, it represents the population. Therefore, we use the formula for the population variance:

$$\sigma^2 = \frac{\Sigma m^2 f - \left(\frac{(\Sigma mf)^2}{N}\right)}{N} = \frac{14{,}825 - \left(\frac{(535)^2}{25}\right)}{25} = \frac{3376}{25} = \mathbf{135.04}$$

STEP 4 *Calculate the standard deviation.*

To obtain the standard deviation, take the (positive) square root of the variance.

$$\sigma = \sqrt{\sigma^2} = \sqrt{135.04} = \mathbf{11.62\ minutes}$$

Thus, the standard deviation of the daily commuting times for these employees is 11.62 minutes. Note that the variance $\sigma^2 = 135.04$ is in squared minutes, not minutes.

Note that the values of the variance and standard deviation calculated in Example 3.21 for grouped data are approximations. The exact values of the variance and standard deviation can be obtained only by using the ungrouped data on the daily commuting times of these 25 employees.

EXAMPLE 3.22 | Number of Orders Received

Calculating the Sample Variance and Standard Deviation for Grouped Data

The following data, reproduced from Table 3.11 of Example 3.20, give the frequency distribution of the number of orders received each day during the past 50 days at the office of a mail-order company.

Number of Orders	f
10–12	4
13–15	12
16–18	20
19–21	14

Calculate the variance and standard deviation.

Solution All the information required for the calculation of the variance and standard deviation appears in **Table 3.14**.

TABLE 3.14

Number of Orders	f	m	mf	m^2f
10–12	4	11	44	484
13–15	12	14	168	2352
16–18	20	17	340	5780
19–21	14	20	280	5600
	$n = 50$		$\Sigma mf = 832$	$\Sigma m^2f = 14{,}216$

Because the data set includes only 50 days, it represents a sample. Hence, we use the sample formulas to calculate the variance and standard deviation. By substituting the values into the formula for the sample variance, we obtain

$$s^2 = \frac{\Sigma m^2 f - \left(\frac{(\Sigma mf)^2}{n}\right)}{n-1} = \frac{14{,}216 - \left(\frac{(832)^2}{50}\right)}{50-1} = \mathbf{7.5820}$$

Hence, the standard deviation is

$$s = \sqrt{s^2} = \sqrt{7.5820} = \mathbf{2.75\ orders}$$

Thus, the standard deviation of the number of orders received at the office of this mail-order company during the past 50 days is 2.75.

EXERCISES

Concepts and Procedures

3.50 Are the values of the mean and standard deviation that are calculated using grouped data exact or approximate values of the mean and standard deviation, respectively? Explain.

3.51 Using the population formulas, calculate the mean, variance, and standard deviation for the following grouped data:

x	2–4	5–7	8–10	11–13	14–16
f	3	10	14	8	5

3.52 Using the sample formulas, find the mean, variance, and standard deviation for the grouped data displayed in the following table:

x	f
0 to less than 4	17
4 to less than 8	23
8 to less than 12	15
12 to less than 16	11
16 to less than 20	8
20 to less than 24	6

Applications

3.53 The following table gives the frequency distribution of the number of hours spent last week on cell phones (making phone calls and texting) by all 100 students in the tenth grade at a school.

Hours	Number of Students
0 to less than 4	14
4 to less than 8	18
8 to less than 12	25
12 to less than 16	18
16 to less than 20	16
20 to less than 24	9

Find the mean, variance, and standard deviation.

3.54 The following table gives the frequency distribution of the total miles driven during 2019 by 300 car owners.

Miles Driven in 2019 (in thousand)	Number of Car Owners
0 to less than 5	7
5 to less than 10	26
10 to less than 15	59
15 to less than 20	71
20 to less than 25	62
25 to less than 30	39
30 to less than 35	22
35 to less than 40	14

Find the mean, variance, and standard deviation. Give a brief interpretation of the values in the column labeled mf in your table of calculations. What does Σmf represent?

3.55 The following table gives information on the amounts (in dollars) of electric bills for August 2019 for a sample of 50 families.

Amount of Electric Bill (dollars)	Number of Families
0 to less than 60	5
60 to less than 120	16
120 to less than 180	11
180 to less than 240	10
240 to less than 300	8

Find the mean, variance, and standard deviation. Give a brief interpretation of the values in the column labeled mf in your table of calculations. What does Σmf represent?

3.56 The following data give the amounts (in dollars) spent last month on food by 400 randomly selected families.

Food Expenditure (in dollars)	Number of Families
200 to less than 500	38
500 to less than 800	105
800 to less than 1100	130
1100 to less than 1400	60
1400 to less than 1700	42
1700 to less than 2000	18
2000 to less than 2300	7

Calculate the mean, variance, and standard deviation.

3.57 For 50 airplanes that arrived late at an airport during a week, the time by which they were late was observed. In the following table, x denotes the time (in minutes) by which an airplane was late, and f denotes the number of airplanes.

x	f
0 to less than 20	14
20 to less than 40	18
40 to less than 60	9
60 to less than 80	5
80 to less than 100	4

Find the mean, variance, and standard deviation.

3.4 Use of Standard Deviation

LEARNING OBJECTIVES

After completing this section, you should be able to:

- Explain Chebyshev's theorem.
- Calculate the minimum percentage of observations that fall in a given interval using Chebyshev's theorem.
- Explain the empirical rule and when the empirical rule can and cannot be applied.
- Calculate the percentage of observations that fall in a given interval using the empirical rule.

By using the mean and standard deviation, we can find the proportion or percentage of the total observations that fall within a given interval about the mean. This section briefly discusses Chebyshev's theorem and the empirical rule, both of which demonstrate this use of the standard deviation.

3.4.1 Chebyshev's Theorem

Chebyshev's theorem gives a lower bound for the area under a curve between two points that are on opposite sides of the mean and at the same distance from the mean.

Chebyshev's Theorem

For any number k greater than 1, at least $(1 - 1/k^2)$ of the data values lie within k standard deviations of the mean.

Note that k gives the distance between the mean and a point in terms of the number of standard deviations. To find the area under a curve between two points using Chebyshev's theorem, we perform the following steps:

1. First we find the distance between the mean and each of the two given points. Note that both these distances must be the same to apply Chebyshev's theorem.
2. Secondly, we divide the distance calculated in Step 1 by the standard deviation. This gives the value of k.
3. Finally we substitute the value of k in the formula $1 - \frac{1}{k^2}$ and simplify. Multiply this answer by 100 to obtain the percentage. This gives the minimum area (in percent) under the distribution curve between the two points.

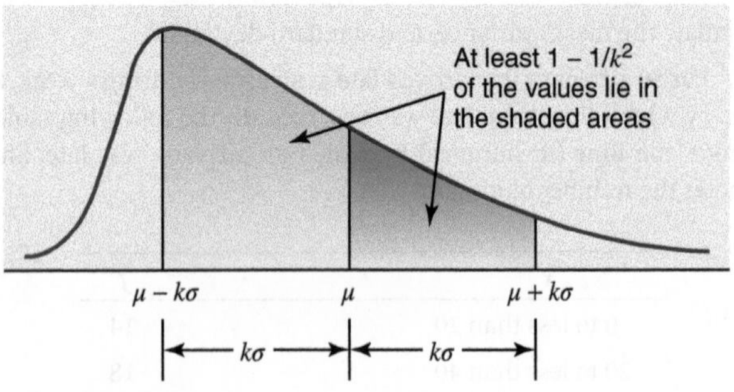

FIGURE 3.5 Chebyshev's theorem.

Figure 3.5 illustrates Chebyshev's theorem.

Table 3.15 lists the areas under a distribution curve for different values of k using Chebyshev's theorem.

TABLE 3.15 Areas Under the Distribution Curve Using Chebyshev's Theorem

k	Interval	$1 - \frac{1}{k^2}$	Minimum Area Within k Standard Deviations
1.5	$\mu \pm 1.5\sigma$	$1 - \frac{1}{1.5^2} = 1 - .44 = .56$	56%
2.0	$\mu \pm 2\sigma$	$1 - \frac{1}{2^2} = 1 - .25 = .75$	75%
2.5	$\mu \pm 2.5\sigma$	$1 - \frac{1}{2.5^2} = 1 - .16 = .84$	84%
3.0	$\mu \pm 3\sigma$	$1 - \frac{1}{3.0^2} = 1 - .11 = .89$	89%

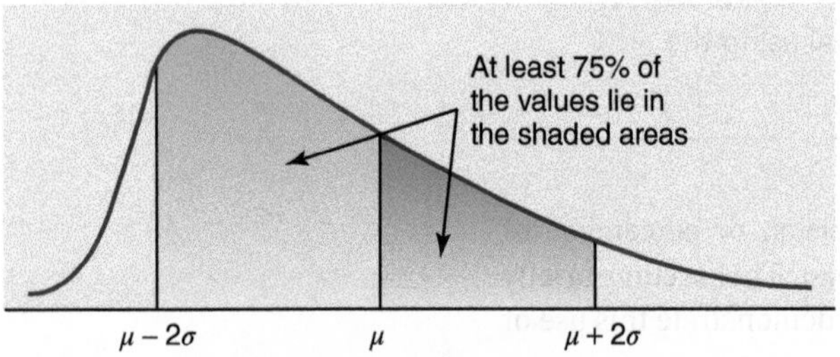

FIGURE 3.6 Percentage of values within two standard deviations of the mean for Chebyshev's theorem.

For example, using Table 3.15, if $k = 2$, then according to Chebyshev's theorem, at least .75, or 75%, of the values of a data set lie within two standard deviations of the mean. This is shown in **Figure 3.6**. Note that this area is at least 75%, which means it can even be 100%.

If $k = 3$, then according to Chebyshev's theorem, at least .89, or 89%, of the values fall within three standard deviations of the mean. This is shown in **Figure 3.7**.

Although in Figures 3.5 through 3.7 we have used the population notation for the mean and standard deviation, the theorem applies to both sample and population data. Note that Chebyshev's theorem is applicable to a distribution of any shape. However, Chebyshev's theorem can be used only for $k > 1$. This is so because when $k = 1$, the value of $1 - 1/k^2$ is zero, and when $k < 1$, the value of $1 - 1/k^2$ is negative.

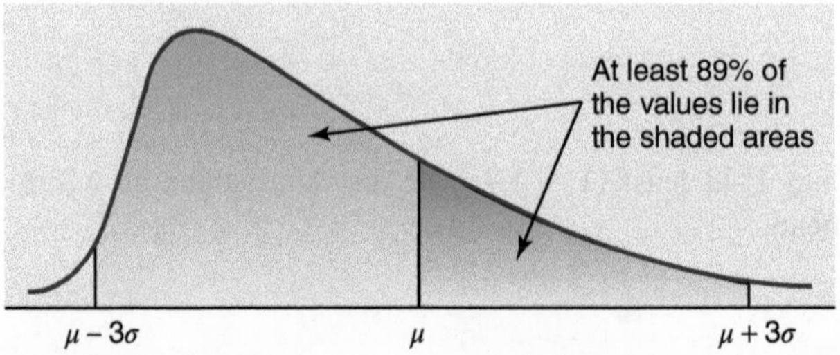

FIGURE 3.7 Percentage of values within three standard deviations of the mean for Chebyshev's theorem.

EXAMPLE 3.23 Blood Pressure of Women

Applying Chebyshev's Theorem

The mean systolic blood pressure for 4000 women who were screened for high blood pressure was found to be 187 mm Hg with a standard deviation of 22. Using Chebyshev's theorem, find the minimum percentage of women in this group who have a systolic blood pressure between 143 and 231 mm Hg.

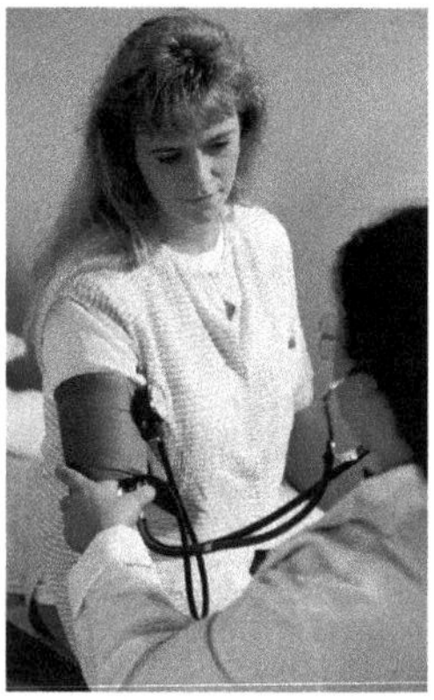

Keith Brofsky/Getty Images

Solution Let μ and σ be the mean and the standard deviation, respectively, of the systolic blood pressures of these women. Then, from the given information,

$$\mu = 187 \quad \text{and} \quad \sigma = 22$$

To find the percentage of women whose systolic blood pressures are between 143 and 231 mm Hg, the first step is to determine k. As shown below, each of the two points, 143 and 231, is 44 units away from the mean.

$$\begin{array}{ccc} |\longleftarrow \; 44 \; \longrightarrow & | \longleftarrow \; 44 \; \longrightarrow & | \\ 143 & \mu = 187 & 231 \end{array}$$

The value of k is obtained by dividing the distance between the mean and each point by the standard deviation. Thus,

$$k = 44/22 = 2$$

From Table 3.15, for $k = 2$, the area under the curve is at least 75%. Hence, according to Chebyshev's theorem, at least 75% of the women have a systolic blood pressure between 143 and 231 mm Hg. This percentage is shown in **Figure 3.8**.

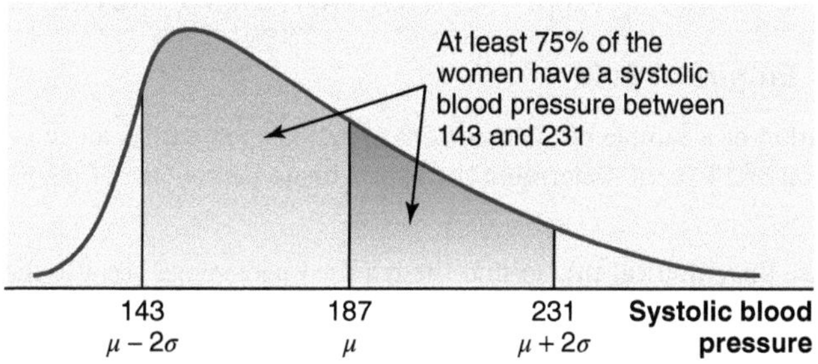

FIGURE 3.8 Percentage of women with systolic blood pressure between 143 and 231.

3.4.2 Empirical Rule

Whereas Chebyshev's theorem is applicable to a distribution of any shape, the **empirical rule** applies only to a specific shape of distribution called a **bell-shaped distribution**, as shown in **Figure 3.9**. More will be said about such a distribution in Chapter 6, where it is called a *normal curve*. In this section, only the following three rules for the bell-shaped curve are given.

Empirical Rule

For a bell-shaped distribution, approximately

1. 68% of the observations lie within one standard deviation of the mean.
2. 95% of the observations lie within two standard deviations of the mean.
3. 99.7% of the observations lie within three standard deviations of the mean.

Table 3.16 lists the areas under a bell-shaped distribution for these three intervals.

TABLE 3.16 Approximate Areas Under a Bell-Shaped Distribution Using the Empirical Rule

Interval	Approximate Area
$\mu \pm 1\sigma$	68%
$\mu \pm 2\sigma$	95%
$\mu \pm 3\sigma$	99.7%

Figure 3.9 illustrates the empirical rule. Again, the empirical rule applies to both population data and sample data.

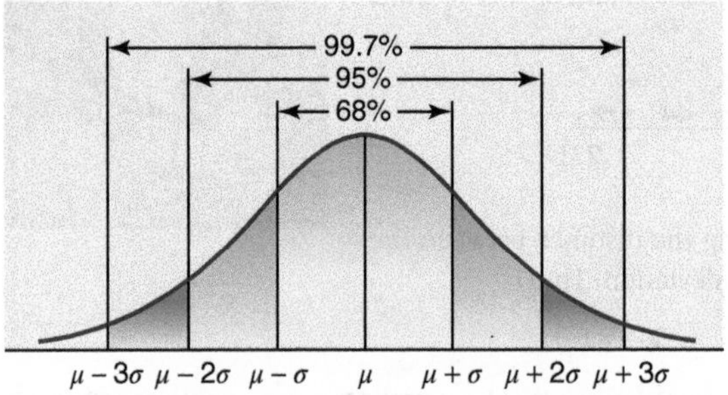

FIGURE 3.9 Illustration of the empirical rule.

EXAMPLE 3.24 Age Distribution of Persons

Applying the Empirical Rule

The age distribution of a sample of 5000 persons is bell-shaped with a mean of 40 years and a standard deviation of 12 years. Determine the approximate percentage of people who are 16 to 64 years old.

Solution We use the empirical rule to find the required percentage because the distribution of ages follows a bell-shaped curve. From the given information, for this distribution,

$$\bar{x} = 40 \text{ years} \quad \text{and} \quad s = 12 \text{ years}$$

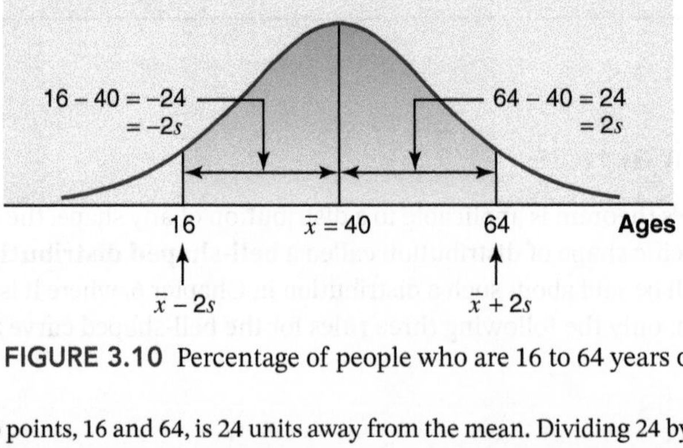

FIGURE 3.10 Percentage of people who are 16 to 64 years old.

Each of the two points, 16 and 64, is 24 units away from the mean. Dividing 24 by 12, we convert the distance between each of the two points and the mean in terms of the number of standard deviations. Thus, the distance between 16 and 40 and that between 40 and 64 is each equal to $2s$. Consequently, as shown in **Figure 3.10**, the area from 16 to 64 is the area from $\bar{x} - 2s$ to $\bar{x} + 2s$.

Because, from Table 3.16, the area within two standard deviations of the mean is approximately 95% for a bell-shaped curve, approximately **95%** of the people in the sample are 16 to 64 years old.

Case Study 3.3

Does Spread Mean the Same as Variability and Dispersion?

In any discipline, there is terminology that one needs to learn in order to become fluent. Accounting majors need to learn the difference between credits and debits, chemists need to know how an ion differs from an atom, and physical therapists need to know the difference between abduction and adduction. Statistics is no different. Failing to learn the difference between the mean and the median makes much of the remainder of this book very difficult to understand.

Another issue with terminology is the use of words other than the terminology to describe a specific concept or scenario. Sometimes the words one chooses to use can be vague or ambiguous, resulting in confusion. One debate in the statistics community involves the use of the word "spread" in place of the words "dispersion" and "variability." In a 2012 article, "Lexical ambiguity: making a case against *spread*," authors Jennifer Kaplan, Neal Rogness, and Diane Fisher point out that the Oxford English Dictionary has more than 25 definitions for the word spread, many of which students know coming into a statistics class. As a result of knowing some of the meanings of spread, students who use the word spread in place of variability or dispersion "do not demonstrate strong statistical meanings of the word spread at the end of a one-semester statistics course."

In order to examine the extent of this issue, the authors of the article designed a study in which they selected 160 undergraduate students taking an introductory statistics course from 14 different professors at three different universities and in the first week of the semester asked them to write sentences and definitions for spread using its primary meaning. Then, at the end of the semester, the same students were asked to write sentences and definitions for spread using its primary meaning in statistics. The authors found that responses of only one-third of the students related spread to the concept of variability, which has to do with how the data vary around the center of a distribution. A slightly larger percentage of students gave responses that "defined spread as 'to cover evenly or in a thin layer,'" while approximately one in eight responded with a definition that was synonymous with the notion of range. Seven other definitions were given by at least three students in the study.

Although more of the definitions and sentences provided at the end of the course had something to do with statistics, the authors did not see an increase in the percentage of definitions that associated spread with the concept of variability. Hence, they suggested that the ambiguity of the term spread is sufficient enough to stop using it in place of the terms variability and dispersion.

Source: Kaplan, J. J., Rogness, N. T., and Fisher, D. G., "Lexical ambiguity: making a case against spread," *Teaching Statistics*, 2011, 34, (2), pp. 56–60. © 2011 Teaching Statistics Trust.

EXERCISES

Concepts and Procedures

3.58 Briefly explain Chebyshev's theorem and its applications.

3.59 Briefly explain the empirical rule. To what kind of distribution is it applied?

3.60 A sample of 2500 observations has a mean of 76 and a standard deviation of 14. Using Chebyshev's theorem, find the minimum percentage of the observations that fall in the intervals $\bar{x} \pm 1.5s$, $\bar{x} \pm 2s$, and $\bar{x} \pm 2.5s$. Note that $\bar{x} \pm 1.5s$ represents the interval $\bar{x} - 1.5s$ to $\bar{x} + 1.5s$, and so on.

3.61 A large population has a mean of 230 and a standard deviation of 41. Using Chebyshev's theorem, find the minimum percentage of the observations that fall in the intervals $\mu \pm 2\sigma$, $\mu \pm 2.5\sigma$, and $\mu \pm 3\sigma$.

3.62 A large population has a bell-shaped distribution with a mean of 280 and a standard deviation of 35. Using the empirical rule, find the approximate percentage of the observations that fall in the intervals $\mu \pm 1\sigma$, $\mu \pm 2\sigma$, and $\mu \pm 3\sigma$.

3.63 A sample of 3000 observations has a bell-shaped distribution with a mean of 82 and a standard deviation of 16. Using the empirical rule, find the approximate percentage of the observations that fall in the intervals $\bar{x} \pm 1s$, $\bar{x} \pm 2s$, and $\bar{x} \pm 3s$.

Applications

3.64 The mean time taken by all participants to run a road race was found to be 250 minutes with a standard deviation of 30 minutes. Using Chebyshev's theorem, find the minimum percentage of runners who completed the race in

- **a.** 190 to 310 minutes
- **b.** 160 to 340 minutes
- **c.** 175 to 325 minutes

3.65 The one-way commuting times from home to work for all employees working at a large company have a mean of 34 minutes and a standard deviation of 8 minutes.

- **a.** Using Chebyshev's theorem, find the minimum percentage of employees at this company who have one-way commuting times in the following intervals:
  - **i.** 14 to 54 minutes
  - **ii.** 18 to 50 minutes
- ***b.** Using Chebyshev's theorem, find the interval that contains one-way commuting times of at least 89% of the employees at this company.

3.66 The mean monthly mortgage paid by all home owners in a town is \$2365 with a standard deviation of \$340.

a. Using Chebyshev's theorem, find the minimum percentage of all home owners in this town who pay a monthly mortgage of

i. \$1685 to \$3045

ii. \$1345 to \$3385

***b.** Using Chebyshev's theorem, find the interval that contains the monthly mortgage payments of at least 84% of all home owners in this town.

3.67 The one-way commuting times from home to work for all employees working at a large company have a bell-shaped curve with a mean of 34 minutes and a standard deviation of 8 minutes. Using the empirical rule, find the approximate percentages of the employees at this company who have one-way commuting times in the following intervals:

a. 10 to 58 minutes **b.** 18 to 50 minutes

3.68 The prices of all college textbooks follow a bell-shaped distribution with a mean of \$180 and a standard deviation of \$30.

a. Using the empirical rule, find the (approximate) percentage of all college textbooks with their prices between

i. \$150 and \$210

ii. \$120 and \$240

***b.** Using the empirical rule, find the interval that contains the prices of (approximate) 99.7% of college textbooks.

3.5 Measures of Position

LEARNING OBJECTIVES

After completing this section, you should be able to:

- Calculate and interpret measures of position (quartiles, percentiles, and percentile rank) for quantitative data.
- Calculate and interpret the interquartile range (IQR) for quantitative data.
- Find the position of an observation in a data set by using its percentile rank.

A **measure of position** determines the position of a single value in relation to other values in a sample or a population data set. There are many measures of position; however, only quartiles, percentiles, and percentile rank are discussed in this section.

3.5.1 Quartiles and Interquartile Range

Quartiles are the summary measures that divide a ranked data set into four equal parts. Three measures will divide any data set into four equal parts. These three measures are the **first quartile** (denoted by Q_1), the **second quartile** (denoted by Q_2), and the **third quartile** (denoted by Q_3). The data must be ranked in increasing order before the quartiles are determined. Note that Q_1 and Q_3 are also called the lower and the upper quartiles, respectively. The quartiles are defined as follows:

Quartiles

Quartiles are three values that divide a ranked data set into four equal parts. The second quartile is the same as the median of a data set. The first quartile is the median of the observations that are less than the median, and the third quartile is the median of the observations that are greater than the median.

Note that if a value in the data set is equal to the median but it falls on the side of the values that are smaller than the median, it is included in these values that are used to calculate the first quartile. Similarly, if a value in the data set is equal to the median but it falls on the side of the values that are larger than the median, it is included in these values that are used to calculate the third quartile.

Figure 3.11 describes the positions of the three quartiles.

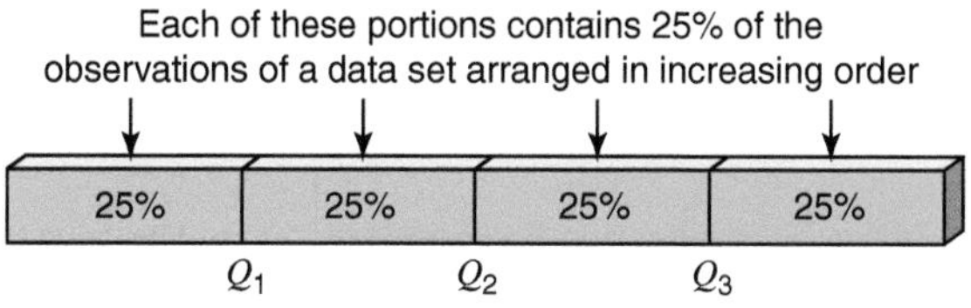

FIGURE 3.11 Quartiles.

Approximately 25% of the values in a ranked data set are less than Q_1 and about 75% are greater than Q_1. The second quartile, Q_2, divides a ranked data set into two equal parts; hence, the second quartile and the median are the same. Approximately 75% of the data values are less than Q_3 and about 25% are greater than Q_3.

The difference between the third quartile and the first quartile for a data set is called the **interquartile range (IQR)**, which is a measure of dispersion.

Calculating Interquartile Range

The difference between the third and the first quartiles gives the *interquartile range*; that is,

$$\text{IQR} = \text{Interquartile range} = Q_3 - Q_1$$

Examples 3.25 and 3.26 show the calculation of the quartiles and the interquartile range.

EXAMPLE 3.25 Commuting Times for College Students

Finding Quartiles and the Interquartile Range

A sample of 12 commuter students was selected from a college. The following data give the typical one-way commuting times (in minutes) from home to college for these 12 students.

29 14 39 17 7 47 63 37 42 18 24 55

a. Find the values of the three quartiles.

b. Where does the commuting time of 47 fall in relation to the three quartiles?

c. Find the interquartile range.

Solution

Finding quartiles for an even number of data values.

a. We perform the following steps to find the three quartiles.

STEP 1 First we rank the given data in increasing order as follows:

7 14 17 18 24 29 37 39 42 47 55 63

STEP 2 We find the second quartile, which is also the median. In a total of 12 data values, the median is between sixth and seventh terms. Thus, the median and, hence, the second quartile is given by the average of the sixth and seventh values in the ranked data set, that is the average of 29 and 37. Thus, the second quartile is

$$Q_2 = \frac{29 + 37}{2} = \mathbf{33}$$

Note that $Q_2 = \mathbf{33}$ is also the value of the median.

STEP 3 We find the median of the data values that are smaller than Q_2, and this gives the value of the first quartile. The values that are smaller than Q_2 are

7 14 17 18 24 29

The value that divides these six data values in two equal parts is given by the average of the two middle values, 17 and 18. Thus, the first quartile is

$$Q_1 = \frac{17 + 18}{2} = \mathbf{17.5}$$

STEP 4 We find the median of the data values that are larger than Q_2, and this gives the value of the third quartile. The values that are larger than Q_2 are

37 39 42 47 55 63

The value that divides these six data values in two equal parts is given by the average of the two middle values, 42 and 47. Thus, the third quartile is

$$Q_3 = \frac{42 + 47}{2} = \mathbf{44.5}$$

Now we can summarize the calculation of the three quartiles in the following figure:

Values smaller than the median | Values larger than the median

7 14 17 ↑ 18 24 29 ↑ 37 39 42 ↑ 47 55 63

$$Q_1 = \frac{17 + 18}{2} = \mathbf{17.5} \qquad Q_2 = \frac{29 + 37}{2} = \mathbf{33} \qquad Q_3 = \frac{42 + 47}{2} = \mathbf{44.5}$$

The value of $Q_1 = 17.5$ minutes indicates that 25% of these 12 students in this sample commute for less than 17.5 minutes and 75% of them commute for more than 17.5 minutes. Similarly, $Q_2 = 33$ indicates that half of these 12 students commute for less than 33 minutes and the other half of them commute for more than 33 minutes. The value of $Q_3 = 44.5$ minutes indicates that 75% of these 12 students in this sample commute for less than 44.5 minutes and 25% of them commute for more than 44.5 minutes.

Position of a value in relation to quartiles.

b. By looking at the position of 47 minutes, we can state that this value lies in the **top 25%** of the commuting times.

Finding the interquartile range.

c. The interquartile range is given by the difference between the values of the third and the first quartiles. Thus,

$$\text{IQR} = \text{Interquartile range} = Q_3 - Q_1 = 44.5 - 17.5 = \mathbf{27\ minutes}$$

EXAMPLE 3.26 Ages of Employees

Finding Quartiles and the Interquartile Range

The following are the ages (in years) of nine employees of an insurance company:

47 28 39 51 33 37 59 24 33

a. Find the values of the three quartiles. Where does the age of 28 years fall in relation to the ages of these employees?

b. Find the interquartile range.

Solution

Finding quartiles for an odd number of data values.

a. First we rank the given data in increasing order. Then we calculate the three quartiles as follows:

Values less than the median | Values greater than the median

24 28 ↑ 33 33 37 39 47 ↑ 51 59
↑

$$Q_1 = \frac{28 + 33}{2} = \mathbf{30.5} \qquad Q_2 = \mathbf{37} \qquad Q_3 = \frac{47 + 51}{2} = \mathbf{49}$$

↑
Also the median

Thus the values of the three quartiles are

$$Q_1 = \mathbf{30.5\ years},\quad Q_2 = \mathbf{37\ years},\quad \text{and}\quad Q_3 = \mathbf{49\ years}$$

The age of 28 falls in the **lowest 25%** of the ages.

Finding the interquartile range.

b. The interquartile range is

$$\text{IQR} = \text{Interquartile range} = Q_3 - Q_1 = 49 - 30.5 = \mathbf{18.5\ years}$$

3.5.2 Percentiles and Percentile Rank

Percentiles are the summary measures that divide a ranked data set into 100 equal parts. Each (ranked) data set has 99 percentiles that divide it into 100 equal parts. The data should be ranked in increasing order to compute percentiles. The kth percentile is denoted by P_k, where k is an integer in the range 1 to 99. For instance, the 25th percentile is denoted by P_{25}. **Figure 3.12** shows the positions of the 99 percentiles.

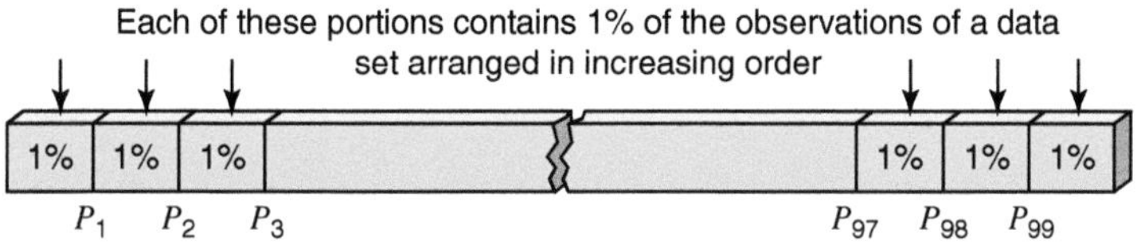

FIGURE 3.12 Percentiles.

Thus, the kth percentile, P_k, can be defined as a value in a data set such that about k% of the measurements are smaller than the value of P_k and about $(100 - k)$% of the measurements are greater than the value of P_k.

The approximate value of the kth percentile is determined as explained next.

Calculating Percentiles

The (approximate) value of the kth *percentile*, denoted by P_k, is

$$P_k = \text{Value of the} \left(\frac{k \times n}{100}\right)\text{th term in a ranked data set}$$

where k denotes the number of the percentile and n represents the sample size.

If the value of $\frac{k \times n}{100}$ is fractional, always round it up to the next higher whole number.

Thus, to calculate the kth percentile, first we rank the given data set in increasing order. Then, to find the kth percentile, we find the $(k \times n/100)$th term, where n is the total number of data values in the given data set. If the value of $(k \times n/100)$ is fractional, we round it up to the next higher whole number. The value of this $(k \times n/100)$th term in the ranked data set gives the kth percentile, P_k.

Example 3.27 describes the procedure to calculate the percentiles.

EXAMPLE 3.27 Commuting Times for College Students

Finding the Percentile for a Data Set

Refer to the data on one-way commuting times (in minutes) from home to college of 12 students given in Example 3.25, which is reproduced below.

29	14	39	17	7	47	63	37	42	18	24	55

Find the value of the 70th percentile. Give a brief interpretation of the 70th percentile.

Solution We perform the following three steps to find the 70th percentile for the given data.

STEP 1 First we rank the given data in increasing order as follows:

7 14 17 18 24 29 37 39 42 47 55 63

STEP 2 We find the $(k \times n/100)$th term. Here $n = 12$ and $k = 70$, as we are to find the 70th percentile.

$$\frac{k \times n}{100} = \frac{70 \times 12}{100} = 8.4 = 9\text{th term}$$

Thus, the 70th percentile, P_{70}, is given by the value of the 9th term in the ranked data set. Note that we rounded 8.4 up to 9, which is always the case when calculating a percentile.

STEP 3 We find the value of the 9th term in the ranked data. This gives the value of the 70th percentile, P_{70}.

$$P_{70} = \text{Value of the 9th term} = \mathbf{42\ minutes}$$

Thus, we can state that approximately 70% of these 12 students commute for less than or equal to 42 minutes.

We can calculate the **percentile rank** for a particular value x_i of a data set by using the formula given below. The percentile rank of x_i gives the percentage of values in the data set that are less than x_i. To find the percentile rank for a data value x_i, we first rank the given data set in increasing order. Then we find the number of data values that are less than x_i. Finally, we divide this number by the number of total values and multiply by 100. This gives the percentile rank for x_i.

Finding Percentile Rank of a Value

$$\text{Percentile rank of } x_i = \frac{\text{Number of values less than } x_i}{\text{Total number of values in the data set}} \times 100\%$$

Example 3.28 shows how the percentile rank is calculated for a data value.

EXAMPLE 3.28 | Commuting Times for College Students

Finding the Percentile Rank for a Data Value

Refer to the data on one-way commuting times (in minutes) from home to college of 12 students given in Example 3.25, which is reproduced here.

29 14 39 17 7 47 63 37 42 18 24 55

Find the percentile rank of 42 minutes. Give a brief interpretation of this percentile rank.

Solution We perform the following three steps to find the percentile rank of 42.

STEP 1 First we rank the given data in increasing order as follows:

7 14 17 18 24 29 37 39 42 47 55 63

STEP 2 Find how many data values are less than 42.

In the above ranked data, there are eight data values that are less than 42.

STEP 3 Find the percentile rank of 42 as follows given that 8 of the 12 values in the given data set are smaller than 42:

$$\text{Percentile rank of } 42 = \frac{8}{12} \times 100\% = \mathbf{66.67\%}$$

Rounding this answer to the nearest integral value, we can state that about 67% of the students in this sample commute for less than 42 minutes.

Most statistical software packages use slightly different methods to calculate quartiles and percentiles. Those methods, while more precise, are beyond the scope of this text.

EXERCISES

Concepts and Procedures

3.69 Briefly describe how the three quartiles are calculated for a data set. Illustrate by calculating the three quartiles for two examples, the first with an odd number of observations and the second with an even number of observations.

3.70 Explain how the interquartile range is calculated. Give one example.

3.71 Briefly describe how the percentiles are calculated for a data set.

3.72 Explain the concept of the percentile rank for an observation of a data set.

Applications

3.73 The following data give the speeds of 14 cars (in mph) measured by radar, traveling on I-84.

73	75	69	68	78	74	74
76	65	79	69	77	71	72

a. Find the values of the three quartiles and the interquartile range.

b. Calculate the (approximate) value of the 35th percentile.

c. Compute the percentile rank of 71.

3.74 The following data give the total food expenditures (in dollars) for the past one month for a sample of 20 families.

1125	530	1234	595	427	872	1480	699	1274	1187
933	1127	716	1065	934	1630	1046	2199	1353	441

a. Calculate the values of the three quartiles and the interquartile range.

b. Find the approximate value of the 57th percentile.

c. Calculate the percentile rank of 1046. Give a brief interpretation of this percentile rank.

3.75 The following data give the number of text messages sent by a high school student on 40 randomly selected days during 2019.

32	33	33	34	35	36	37	37	37	32
38	39	40	41	41	42	42	42	43	44
44	45	45	45	47	58	47	47	41	48
48	60	50	50	51	36	53	54	59	61

a. Calculate the values of the three quartiles and the interquartile range. Where does the value 49 fall in relation to these quartiles?

b. Determine the approximate value of the 91st percentile. Give a brief interpretation of this percentile.

c. For what percentage of the days was the number of text messages sent 40 or higher? Answer by finding the percentile rank of 40.

3.76 Nixon Corporation manufactures computer monitors. The following data give the numbers of computer monitors produced at the company for a sample of 30 days.

24	32	27	23	33	33	29	25	23	36
26	26	31	20	27	33	27	23	28	29
31	35	34	22	37	28	23	35	31	43

a. Calculate the values of the three quartiles and the interquartile range. Where does the value of 31 lie in relation to these quartiles?

b. Find the (approximate) value of the 65th percentile. Give a brief interpretation of this percentile.

c. For what percentage of the days was the number of computer monitors produced 32 or higher? Answer by finding the percentile rank of 32.

3.77 The following data give the annual salaries (in thousand dollars) of 20 randomly selected health-care workers.

50	71	57	39	45	64	38	53	35	62
74	40	67	44	77	61	58	55	64	59

a. Calculate the values of the three quartiles and the interquartile range. Where does the value 57 fall in relation to these quartiles?

b. Find the approximate value of the 30th percentile. Give a brief interpretation of this percentile.

c. Calculate the percentile rank of 61. Give a brief interpretation of this percentile rank.

3.6 Box-and-Whisker Plot

LEARNING OBJECTIVES

After completing this section, you should be able to:

- Determine the five-number summary values for quantitative data.
- Calculate the lower and upper inner fences for quantitative data.
- Determine whether a data set contains any outliers.
- Construct a box-and-whisker plot.
- Describe the shape of a distribution using a box-and-whisker plot.

A **box-and-whisker plot** gives a graphic presentation of data using five measures: the median, the first quartile, the third quartile, and the smallest and the largest values in the data set between the lower and the upper inner fences. (The inner fences are explained in Example 3.29.) A box-and-whisker plot can help us visualize the center, the spread, and the skewness of a data set. It also helps detect outliers. We can compare different distributions by making box-and-whisker plots for each of them.

Box-and-Whisker Plot

A plot that shows the center, spread, and skewness of a data set. It is constructed by drawing a box and two whiskers that use the median, the first quartile, the third quartile, and the smallest and the largest values in the data set between the lower and the upper inner fences.

Note that a box-and-whisker plot is made using the following five-number summary values:

Minimum value $\quad Q_1 \quad$ Median $\quad Q_3 \quad$ Maximum value

The minimum and the maximum values used here must be within the lower and the upper inner fences that will be explained below. Making a box-and-whisker plot involves five steps that are explained in Example 3.29.

EXAMPLE 3.29 Incomes of Households

Constructing a Box-and-Whisker Plot

The following data give the incomes (in thousands of dollars) for a sample of 12 households.

75 69 84 112 74 104 81 90 94 144 79 98

Construct a box-and-whisker plot for these data.

Solution The following five steps are performed to construct a box-and-whisker plot.

STEP 1 First, rank the data in increasing order and calculate the values of the median, the first quartile, the third quartile, and the interquartile range. The ranked data are

69 74 75 79 81 84 90 94 98 104 112 144

For these data,

$$\text{Median} = (84 + 90)/2 = 87$$
$$Q_1 = (75 + 79)/2 = 77$$
$$Q_3 = (98 + 104)/2 = 101$$
$$\text{IQR} = Q_3 - Q_1 = 101 - 77 = 24$$

STEP 2 Find the points that are 1.5 × IQR below Q_1 and 1.5 × IQR above Q_3. These two points are called the **lower** and the **upper inner fences**, respectively.

$$1.5 \times \text{IQR} = 1.5 \times 24 = 36$$
$$\text{Lower inner fence} = Q_1 - 36 = 77 - 36 = 41$$
$$\text{Upper inner fence} = Q_3 + 36 = 101 + 36 = 137$$

STEP 3 Determine the smallest and the largest values in the given data set within the two inner fences. These two values for our example are as follows:

Smallest value within the two inner fences = 69
Largest value within the two inner fences = 112

STEP 4 Draw a horizontal line and mark the income levels on it such that all the values in the given data set are covered. Above the horizontal line, draw a box with its left side at the position of the first quartile and the right side at the position of the third quartile. Inside the box, draw a vertical line at the position of the median. The result of this step is shown in **Figure 3.13.**

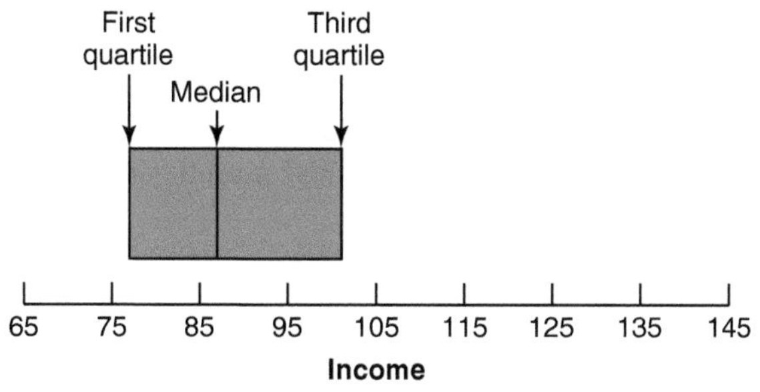

FIGURE 3.13

STEP 5 By drawing two lines, join the points of the smallest and the largest values within the two inner fences to the box. These values are 69 and 112 in this example as listed in Step 3. The two lines that join the box to these two values are called **whiskers**. A value that falls outside the two inner fences is shown by marking an asterisk and is called an **outlier**. This completes the box-and-whisker plot, as shown in **Figure 3.14**.

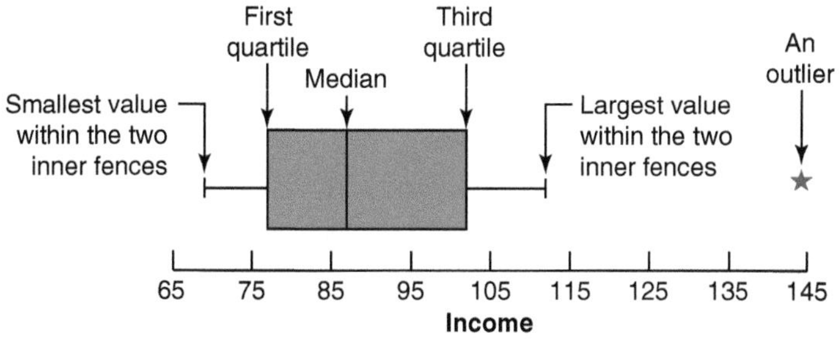

FIGURE 3.14

In Figure 3.14, about 50% of the data values fall within the box, about 25% of the values fall on the left side of the box, and about 25% fall on the right side of the box. Also, 50% of the values fall on the left side of the median and 50% lie on the right side of the median. The data of this example are slightly skewed to the right because the lower 50% of the values are spread over a smaller range than the upper 50% of the values.

The observations that fall outside the two inner fences are called outliers. These outliers can be classified into two kinds of outliers—mild and extreme outliers. To do so, we define two outer fences—a **lower outer fence** at 3.0 × IQR below the first quartile and an **upper outer fence** at 3.0 × IQR above the third quartile. If an observation is outside either of the two inner fences but within the two outer fences, it is called a **mild outlier**. An observation that is outside either of the two outer fences is called an **extreme outlier**. For Example 3.29, the two outer fences are calculated as follows:

$$3 \times \text{IQR} = 3 \times 24 = 72$$
$$\text{Lower outer fence} = Q_1 - 72 = 77 - 72 = 5$$
$$\text{Upper outer fence} = Q_3 + 72 = 101 + 72 = 173$$

Because 144 is outside the upper inner fence but inside the upper outer fence, it is a mild outlier.

Using the box-and-whisker plot, we can conclude whether the distribution of our data is symmetric, skewed to the right, or skewed to the left. If the line representing the median is in the middle of the box and the two whiskers are of about the same length, then the data have a symmetric distribution. If the line representing the median is not in the middle of the box and/or the two whiskers are not of the same length, then the distribution of data values is skewed. The distribution is skewed to the right if the median is to the left of the center of the box with the right-side whisker equal to or longer than the whisker on the left side, or if the median is in the center of the box but the whisker on the right side is longer than the one on the left side. The distribution is skewed to the left if the median is to the right of the center of the box with the left-side whisker equal to or longer than the whisker on the right side, or if the median is in the center of the box but the whisker on the left side is longer than the one on the right side.

EXERCISES

Concepts and Procedures

3.78 Briefly explain what summary measures are used to construct a box-and-whisker plot.

3.79 Prepare a box-and-whisker plot for the following data:

11	8	26	31	62	19	7	3	14	75
33	30	42	15	18	23	29	13	16	6

Does this data set contain any outliers?

Applications

3.80 The following data give the numbers of new cars sold at a dealership during a 20-day period.

8	5	12	3	9	10	6	12	8	8
4	16	10	11	7	7	3	5	9	11

Prepare a box-and-whisker plot. Comment on the skewness of these data.

3.81 The following data give the 2019 bonuses (in thousands of dollars) of 15 randomly selected Wall Street managers.

107	122	175	89	53	361	67	258
61	781	136	208	391	247	71	

Prepare a box-and-whisker plot. Are these data skewed in any direction? Does this data set contain any outliers?

3.82 The following data give the total food expenditures (in dollars) for the past one month for a sample of 20 families.

1125	530	1234	595	427	872	1480	699	1274	1187
933	1127	716	1065	934	1630	1046	2199	1353	441

Prepare a box-and-whisker plot. Is the distribution of these data symmetric or skewed? Are there any outliers? If so, classify them as mild or extreme.

3.83 The following data give the numbers of computer keyboards assembled at the Twentieth Century Electronics Company for a sample of 25 days.

45	52	48	41	56	46	44	42	48	53
51	53	51	48	46	43	52	50	54	47
44	47	50	49	52					

Prepare a box-and-whisker plot. Are these data skewed in any direction?

3.84 The following data represent the credit scores of 22 randomly selected loan applicants.

494	728	468	533	747	639	430	690	604
422	356	805	749	600	797	702	628	625
617	647	772	572					

Prepare a box-and-whisker plot. Are these data skewed in any direction?

Uses and Misuses of Statistics

What Does It Really "Mean"?

People are fascinated by data and, as a result, there are statistical records on just about everything you can imagine. Weather is no exception. Annual precipitation data over many decades are available for most major cities. For example, Washington, DC has reliable precipitation data dating back to 1871 (*Data source*: Capital Climate, capitalclimate.blogspot.com).

One of the most frequently used summary statistics is the *mean* or *arithmetic average*. This statistic has many interesting uses and is one of the most commonly reported summary statistics for quantitative data. The mean allows us to summarize large sets of data into a single number that is readily understandable.

For example, examine the line graph of precipitation data for Washington, DC for the years 1871–2009 presented in **Figure 3.15.** Note that precipitation is measured in inches. The data points on the graph represent the mean precipitation for each decade. At first inspection, we see obvious differences in the mean precipitation from decade to decade. Overall, the points appear to be distributed randomly and uniformly above and below about 40 inches. Hence, we might say that the annual precipitation in Washington, DC is, on average, about 40 inches, with some decades above 40 inches and some decades below 40 inches. However, can we also say that the mean precipitation was unusually high in 1880–1889 and unusually low in 1960–1969? Caution must be taken when making such statements.

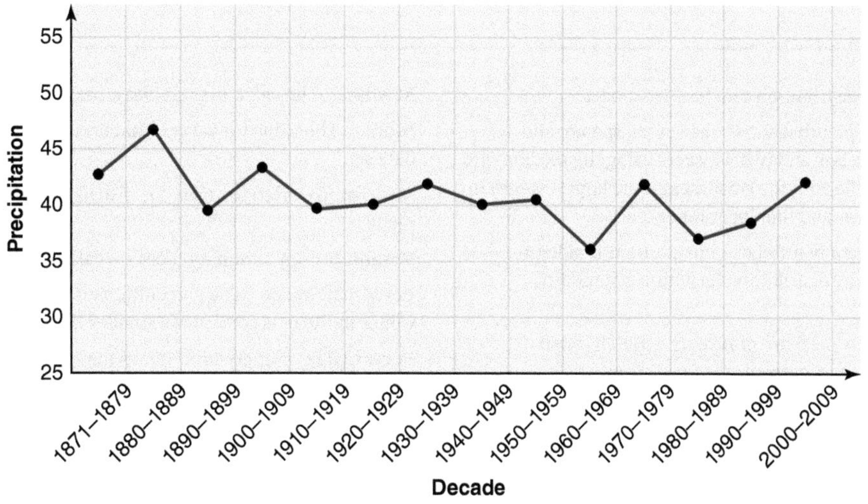

FIGURE 3.15 Mean precipitation by decade for Washington, DC from 1871 to 2009.

Even with the information presented in this graph, we are limited with regards to the interpretation of these data; there is important information that we are missing. The problem is that the mean, by itself, oversimplifies the data. By summarizing the data into a single numeric value, it is easy to be deceived into thinking that the mean tells us everything we need to know about the data. This is an incorrect assumption.

The mean of a data set, of and by itself, provides limited information. As an example, suppose that the average test score for a small class of three students was 50%. If the students' scores were 45%, 50%, and 55%, we might conclude that none of the students scored well on the exam. However, if the students' scores were 0%, 50%, and 100%, we conclude that one student was totally unprepared, one student needs to study more, and one student did extremely well on the test. To provide a complete interpretation of the data, we need to have more information to gain context. We need to know more than just the center of the distribution. We need a measure of the *variability* of the data.

In Figure 3.15, we can see some degree of variability in the mean precipitation from decade to decade, but what about the 10-year pattern of precipitation within each decade? This information is not provided in the graph. Now consider an error bar plot as shown in **Figure 3.16.** This graph shows the mean precipitation for each decade (represented by a solid dot) together with bars representing an interval of values spanning one standard deviation on either side of the mean.

We now have more information to obtain the necessary context. First, even though the distribution of the 10-year mean precipitation varies from decade to decade, the error bars indicate there is overlap in the variability for each decade. Second, even though the mean precipitation appears to be unusually high in 1880–1889 and unusually low in 1960–1969, the variation in the mean shows that, overall, the pattern of precipitation was not significantly different from that of the other decades. Third, we can see that some decades had a relatively consistent pattern of precipitation (1910–1919) while other decades experienced a wide variation in precipitation from year to year (2000–2009 among others).

The take-home message is clear. A measure of center is of limited use without an accompanying measure of variation. Failure to do so easily misleads readers into potentially drawing poor or misinformed conclusions. Additionally, we should be wary when we read results that report only the mean. Without context, we really do not know what the results "mean."

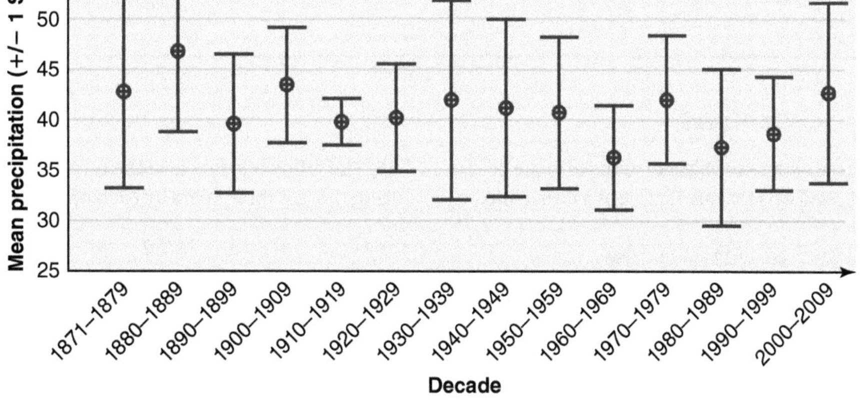

FIGURE 3.16 Mean precipitation in Washington, DC from 1871 to 2009.

Glossary

Bimodal distribution A distribution that has two modes.

Box-and-whisker plot A plot that shows the center, spread, and skewness of a data set with a box and two whiskers using the median, the first quartile, the third quartile, and the smallest and the largest values in the data set between the lower and the upper inner fences.

Chebyshev's theorem For any number k greater than 1, at least $(1 - 1/k^2)$ of the values for any distribution lie within k standard deviations of the mean.

Coefficient of variation A measure of relative variability that expresses standard deviation as a percentage of the mean.

Empirical rule For a specific bell-shaped distribution, about 68% of the observations fall in the interval $(\mu - \sigma)$ to $(\mu + \sigma)$, about 95% fall in the interval $(\mu - 2\sigma)$ to $(\mu + 2\sigma)$, and about 99.7% fall in the interval $(\mu - 3\sigma)$ to $(\mu + 3\sigma)$.

First quartile The first of the three quartiles that divide a ranked data set into four equal parts. The value in a ranked data set such that about 25% of the measurements are smaller than this value and about 75% are larger. It is the median of the values that are smaller than the median of the whole data set.

Geometric mean Calculated by taking the nth root of the product of all values in a data set.

Harmonic mean Calculated by taking the reciprocal of the arithmetic mean of the reciprocals of a given data set.

Interquartile range (IQR) The difference between the third and the first quartiles.

Lower inner fence The value that is 1.5 × IQR units below the first quartile.

Lower outer fence The value that is 3.0 × IQR units below the first quartile.

Mean A measure of center calculated by dividing the sum of all values by the number of values in the data set.

Measures of center Measures that describe the center of a distribution. The mean, median, and mode are three of the measures of center.

Measures of dispersion Measures that give the spread of a distribution. The range, variance, and standard deviation are three such measures.

Measures of position Measures that determine the position of a single value in relation to other values in a data set. Quartiles, percentiles, and percentile rank are examples of measures of position.

Median The value that divides a ranked data set into two equal parts.

Mode The value (or values) that occurs with highest frequency in a data set.

Multimodal distribution A distribution that has more than two modes.

Parameter A summary measure calculated for population data.

Percentile rank The percentile rank of a value gives the percentage of values in the data set that are smaller than this value.

Percentiles Ninety-nine values that divide a ranked data set into 100 equal parts.

Quartiles Three summary measures that divide a ranked data set into four equal parts.

Range A measure of spread obtained by taking the difference between the largest and the smallest values in a data set.

Second quartile The middle or second of the three quartiles that divide a ranked data set into four equal parts. About 50% of the values in the data set are smaller and about 50% are larger than the second quartile. The second quartile is the same as the median.

Standard deviation A measure of spread that is given by the positive square root of the variance.

Statistic A summary measure calculated for sample data.

Third quartile The third of the three quartiles that divide a ranked data set into four equal parts. About 75% of the values in a data set are smaller than the value of the third quartile and about 25% are larger. It is the median of the values that are greater than the median of the whole data set.

Trimmed mean The k% trimmed mean is obtained by dropping k% of the smallest values and k% of the largest values from the given data and then calculating the mean of the remaining $(100 - 2k)$% of the values.

Unimodal distribution A distribution that has only one mode.

Upper inner fence The value that is 1.5 × IQR units above the third quartile.

Upper outer fence The value that is 3.0 × IQR units above the third quartile.

Variance A measure of spread.

Weighted mean Mean of a data set whose values are assigned different weights before the mean is calculated.

Exercises

Supplementary Exercises

3.85 The following data give the odometer mileage (rounded to the nearest thousand miles) for all 20 used cars that are for sale at a dealership.

62	86	58	84	72	40	27	38	50	43
27	40	90	43	94	36	28	48	86	77

a. Calculate the mean and median. Do these data have a mode? Why or why not?

b. Calculate the 10% trimmed mean for these data.

c. Compute the range, variance, standard deviation, and coefficient of variation for these data.

3.86 The following data give the number of driving citations received during the last three years by each of the 12 drivers.

4	8	0	3	11	7	4	14	8	13	7	9

a. Find the mean, median, and mode for these data.

b. Calculate the range, variance, and standard deviation.

c. Are the values of the summary measures in parts (a) and (b) population parameters or sample statistics?

3.87 An electronics store sold 4830 televisions last year. The following table lists the number of different models of televisions sold and the prices for which they were sold.

Television Model	Number of TVs Sold	Price (dollars)
RRV5023	485	1630
TSU7831	1324	625
WEV4920	856	899
HEG7609	633	1178
YOX6938	394	1727
LXT2384	1138	923

Calculate the weighted mean that represents the average price for which these 4830 televisions were sold.

3.88 A company makes five different models of cameras. Last month they sold 13,891 cameras of all five models. The following table lists the number of different models of cameras sold during the last month and the prices for which they were sold.

Camera Model	Number of Cameras Sold	Price (dollars)
B110	3216	425
Z310	1835	1299
N450	4036	369
J150	3142	681
L601	1662	1999

Calculate the weighted mean that represents the average price for which these 13,891 cameras were sold.

3.89 The following table gives the frequency distribution of the times (in minutes) that 50 commuter students at a large university spent looking for parking spaces on the first day of classes in the Fall semester of 2019.

Time	Number of Students
0 to less than 4	1
4 to less than 8	7
8 to less than 12	15
12 to less than 16	18
16 to less than 20	6
20 to less than 24	3

Find the mean, variance, and standard deviation. Are the values of these summary measures population parameters or sample statistics?

3.90 The mean time taken to learn the basics of a software program by all students is 200 minutes with a standard deviation of 20 minutes.

a. Using Chebyshev's theorem, find the minimum percentage of students who learn the basics of this software program in

i. 160 to 240 minutes

ii. 140 to 260 minutes

***b.** Using Chebyshev's theorem, find the interval that contains the times taken by at least 84% of all students to learn this software program.

3.91 The waiting times for patients at a walk-in clinic have a skewed distribution with a mean of 30 minutes and a standard deviation of 6 minutes.

a. Using Chebyshev's theorem, find the minimum percentage of patients at this walk-in clinic who will have to wait for:

i. 15 to 45 minutes

ii. 12 to 48 minutes

***b.** Using Chebyshev's theorem, find the interval that contains waiting times of at least 75% of the patients.

3.92 The mean time taken to learn the basics of a software program by all students has a bell-shaped distribution with a mean of 200 minutes and a standard deviation of 20 minutes.

a. Using the empirical rule, find the (approximate) percentage of students who learn the basics of this software program in

i. 180 to 220 minutes

ii. 160 to 240 minutes

***b.** Using the empirical rule, find the interval that contains the times taken by (approximate) 99.7% of all students to learn this software program.

3.93 Each year the faculty at Metro Business College chooses 10 members from the current graduating class that they feel are most likely to succeed. The data here give the current annual incomes (in thousand dollars) of the 10 members of the class of 2014 who were voted most likely to succeed.

59	68	84	78	107	382	56	74	97	60

a. Determine the values of the three quartiles and the interquartile range. Where does the value of 74 fall in relation to these quartiles?

b. Calculate the (approximate) value of the 70th percentile. Give a brief interpretation of this percentile.

c. Find the percentile rank of 97. Give a brief interpretation of this percentile rank.

3.94 The following data give the number of new cars sold at a dealership during a 20-day period.

8	5	12	3	9	10	6	12	8	8
4	16	10	11	7	7	3	5	9	11

a. Calculate the values of the three quartiles and the interquartile range. Where does the value of 4 lie in relation to these quartiles?

b. Find the (approximate) value of the 25th percentile. Give a brief interpretation of this percentile.

c. Find the percentile rank of 10. Give a brief interpretation of this percentile rank.

3.95 A student washes her clothes at a laundromat once a week. The data here give the time (in minutes) she spent in the laundromat for each of 15 randomly selected weeks. Here, time spent in the laundromat includes the time spent waiting for a machine to become available.

75	62	84	73	107	81	93	72
135	77	85	67	90	83	112	

Prepare a box-and-whisker plot. Is the data set skewed in any direction? If yes, is it skewed to the right or to the left? Does this data set contain any outliers?

Advanced Exercises

3.96 Melissa's grade in her math class is determined by three 100-point tests and a 200-point final exam. To determine the grade for a student in this class, the instructor will add the four scores together and divide this sum by five to obtain a percentage. This percentage must be at least 80 for a grade of B. If Melissa's three test scores are 75, 69, and 87, what is the minimum score she needs on the final exam to obtain a B grade?

3.97 Jeffrey is serving on a six-person jury for a personal-injury lawsuit. All six jurors want to award damages to the plaintiff but cannot agree on the amount of the award. The jurors have decided that each of them will suggest an amount that he or she thinks should be awarded; then they will use the mean of these six numbers as the recommended amount to be awarded to the plaintiff.

a. Jeffrey thinks the plaintiff should receive \$20,000, but he thinks the mean of the other five jurors' recommendations will be about \$12,000. He decides to suggest an inflated amount so that the mean for all six jurors is \$20,000. What amount would Jeffrey have to suggest?

b. How might this jury revise its procedure to prevent a juror like Jeffrey from having an undue influence on the amount of damages to be awarded to the plaintiff?

3.98 On a 300-mile auto trip, Lisa averaged 52 mph for the first 100 miles, 65 mph for the second 100 miles, and 58 mph for the last 100 miles.

a. How long did the 300-mile trip take?

b. Could you find Lisa's average speed for the 300-mile trip by calculating $(52 + 65 + 58)/3$? If not, find the correct average speed for the trip.

3.99 A survey of young people's shopping habits in a small city during the summer months of 2019 showed the following: Shoppers aged 12 to 14 years took an average of 8 shopping trips per month and spent an average of \$14 per trip. Shoppers aged 15 to 17 years took an average of 11 trips per month and spent an average of \$18 per trip. Assume that this city has 1100 shoppers aged 12 to 14 years and 900 shoppers aged 15 to 17 years.

a. Find the total amount spent per month by all these 2000 shoppers in both age groups.

b. Find the mean number of shopping trips per person per month for these 2000 shoppers.

c. Find the mean amount spent per person per month by shoppers aged 12 to 17 years in this city.

3.100 The following table shows the total population and the number of deaths (in thousands) due to heart attack for two age groups (in years) in Countries A and B for 2019.

	Age 30 and Under		Age 31 and Over	
	A	B	A	B
Population	40,000	25,000	20,000	35,000
Deaths due to heart attack	1,000	500	2,000	3,000

a. Calculate the death rate due to heart attack per 1000 population for the 30 years and under age group for each of the two countries. Which country has the lower death rate in this age group?

b. Calculate the death rate due to heart attack per 1000 population for the 31 years and over age group for each of the two countries. Which country has the lower death rate in this age group?

c. Calculate the death rate due to heart attack for the entire population of Country A; then do the same for Country B. Which country has the lower overall death rate?

d. How can the country with lower death rate in both age groups have the higher overall death rate? (This phenomenon is known as Simpson's paradox.)

3.101 The test scores for a large statistics class have an unknown distribution with a mean of 70 and a standard deviation of 10.

a. Find k so that at least 50% of the scores are within k standard deviations of the mean.

b. Find k so that at most 10% of the scores are more than k standard deviations above the mean.

3.102 The test scores for a large statistics class have a bell-shaped distribution with a mean of 70 points.

a. If 16% of all students in the class scored above 85, what is the standard deviation of the scores?

b. If 95% of the scores are between 60 and 80, what is the standard deviation?

3.103 Actuaries at an insurance company must determine a premium for a new type of insurance. A random sample of 40 potential purchasers of this type of insurance were found to have incurred the following losses (in dollars) during the past year. These losses would have been covered by the insurance if it were available.

100	32	0	0	470	50	0	14,589	212	93
0	0	1127	421	0	87	135	420	0	250
12	0	309	0	177	295	501	0	143	0
167	398	54	0	141	0	3709	122	0	0

a. Find the mean, median, and mode of these 40 losses.

b. Which of the mean, median, or mode is largest?

c. Draw a box-and-whisker plot for these data, and describe the skewness, if any. Are there any outliers in this data set?

d. Which measure of center should the actuaries use to determine the premium for this insurance?

3.104 A local golf club has men's and women's summer leagues. The following data give the scores for a round of 18 holes of golf for 17 men and 15 women randomly selected from their respective leagues.

Men	87	68	92	79	83	67	71	92	112
	75	77	102	79	78	85	75	72	
Women	101	100	87	95	98	81	117	107	103
	97	90	100	99	94	94			

a. Make a box-and-whisker plot for each of the two data sets and use them to discuss the similarities and differences between the scores of the men and women golfers.

b. Compute the various descriptive measures you have learned for each sample. How do they compare?

3.105 Answer the following questions:

a. The total weight of all pieces of luggage loaded onto an airplane is 14,124 pounds, which works out to be an average of 51.36 pounds per piece. How many pieces of luggage are on the plane?

b. A group of seven friends, having just gotten back a chemistry exam, discuss their scores. Six of the students reveal that they received grades of 81, 75, 97, 88, 82, and 83, respectively, but the seventh student is reluctant to say what grade she received. After some calculation she announces that the group averaged 86 on the exam. What is her score?

3.106 The following data give the weights (in pounds) of a random sample of 44 college students. (Here F and M indicate female and male, respectively.)

123 F	195 M	138 M	115 F	179 M	119 F
148 F	147 F	180 M	146 F	179 M	189 M
175 M	108 F	193 M	114 F	179 M	147 M
108 F	128 F	164 F	174 M	128 F	159 M
193 M	204 M	125 F	133 F	115 F	168 M
123 F	183 M	116 F	182 M	174 M	102 F
123 F	99 F	161 M	162 M	155 F	202 M
110 F	132 M				

Compute the mean, median, and standard deviation for the weights of all students, of men only, and of women only. Of the mean and median, which is the more informative measure of center? Write a brief note comparing the three measures for all students, men only, and women only.

3.107 The following stem-and-leaf diagram gives the distances (in thousands of miles) driven during the past year by a sample of drivers in a city.

0	3 6 9
1	2 8 5 1 0 5
2	5 1 6
3	8
4	1
5	
6	2

a. Compute the sample mean, median, and mode for the data on distances driven.

b. Compute the range, variance, and standard deviation for these data.

c. Compute the first and third quartiles.

d. Compute the interquartile range. Describe the properties of the interquartile range. When would the IQR be preferable to using the standard deviation when measuring variation?

3.108 Although the standard workweek is 40 hours a week, many people work a lot more than 40 hours a week. The following data give the number of hours worked last week by 50 people.

40.5	41.3	41.4	41.5	42.0	42.2	42.4	42.4	42.6	43.3
43.7	43.9	45.0	45.0	45.2	45.8	45.9	46.2	47.2	47.5
47.8	48.2	48.3	48.8	49.0	49.2	49.9	50.1	50.6	50.6
50.8	51.5	51.5	52.3	52.3	52.6	52.7	52.7	53.4	53.9
54.4	54.8	55.0	55.4	55.4	55.4	56.2	56.3	57.8	58.7

a. The sample mean and sample standard deviation for this data set are 49.012 and 5.080, respectively. Using Chebyshev's theorem, calculate the intervals that contain at least 75%, 88.89%, and 93.75% of the data.

b. Determine the actual percentages of the given data values that fall in each of the intervals that you calculated in part (a). Also calculate the percentage of the data values that fall within one standard deviation of the mean.

c. Do you think the lower endpoints provided by Chebyshev's theorem in part (a) are useful for this problem? Explain your answer.

d. Suppose that the individual with the first number (54.4) in the fifth row of the data is a workaholic who actually worked 84.4 hours last week and not 54.4 hours. With this change, the summary statistics are now $\bar{x} = 49.61$ and $s = 7.10$. Recalculate the intervals for part (a) and the actual percentages for part (b). Did your percentages change a lot or a little?

e. How many standard deviations above the mean would you have to go to capture all 50 data values? Using Chebyshev's theorem, what is the lower bound for the percentage of the data that should fall in the interval?

Appendix 3.1

A3.1.1 Basic Formulas for the Variance and Standard Deviation for Ungrouped Data

Example 3.30 illustrates how to use the basic formulas to calculate the variance and standard deviation for ungrouped data. From Section 3.2.2, the basic formulas for variance for ungrouped data are

$$\sigma^2 = \frac{\Sigma(x-\mu)^2}{N} \quad \text{and} \quad s^2 = \frac{\Sigma(x-\bar{x})^2}{n-1}$$

where σ^2 is the population variance and s^2 is the sample variance.

In either case, the standard deviation is obtained by taking the square root of the variance.

EXAMPLE 3.30 2019 Salaries of 11 College Football Head Coaches

Calculating the Variance and Standard Deviation for Ungrouped Data Using Basic Formulas

Refer to Example 3.16, where the short-cut formula was used to compute the variance and standard deviation for the data on the 2019 salaries of 11 college football head coaches. Calculate the variance and standard deviation for that data set using the basic formula.

Solution Let x denote the 2019 salaries (in millions of dollars) of the 11 college football head coaches. **Table 3.17** shows all the required calculations to find the variance and standard deviation.

TABLE 3.17

x	$x - \bar{x}$	$(x - \bar{x})^2$
10.3	$10.3 - 5.05 = 5.25$	27.5625
4.5	$4.5 - 5.05 = -0.55$	.3025
3.3	$3.3 - 5.05 = -1.75$	3.0625
3.1	$3.1 - 5.05 = -1.95$	3.8025
8.7	$8.7 - 5.05 = 3.65$	13.3225
4.3	$4.3 - 5.05 = -0.75$	.5625
3.5	$3.5 - 5.05 = -1.55$	2.4025
4.8	$4.8 - 5.05 = -0.25$	.0625
5.5	$5.5 - 5.05 = 0.45$	.2025
4.8	$4.8 - 5.05 = -0.25$	.0625
2.8	$2.8 - 5.05 = -2.25$	5.0625
$\Sigma x = 55.6$		$\Sigma(x - \bar{x})^2 = 56.4075$

The following steps are performed to compute the variance and standard deviation.

STEP 1 Find the mean as follows:

$$\bar{x} = \frac{\Sigma x}{n} = \frac{55.6}{11} = 5.05$$

STEP 2 Calculate $x - \bar{x}$, the deviation of each value of x from the mean. The results are shown in the second column of Table 3.17.

STEP 3 Square each of the deviations of x from $\bar{x}$; that is, calculate each of the $(x - \bar{x})^2$ values. These values are called the *squared deviations*, and they are recorded in the third column of Table 3.17.

STEP 4 Add all the squared deviations to obtain $\Sigma(x - \bar{x})^2$; that is, sum all the values given in the third column of Table 3.17. This gives

$$\Sigma(x - \bar{x})^2 = 56.4075$$

STEP 5 Obtain the sample variance by dividing the sum of the squared deviations by $n - 1$. Thus

$$s^2 = \frac{\Sigma(x - \bar{x})^2}{n - 1} = \frac{56.4075}{11 - 1} = 5.64075$$

STEP 6 Obtain the sample standard deviation by taking the positive square root of the variance. Hence,

$$s = \sqrt{5.64075} = \mathbf{2.375 = \$2.375\ million}$$

A3.1.2 Basic Formulas for the Variance and Standard Deviation for Grouped Data

Example 3.31 demonstrates how to use the basic formulas to calculate the variance and standard deviation for grouped data. From Section 3.3.2, the basic formulas for these calculations are

$$\sigma^2 = \frac{\Sigma f(m-\mu)^2}{N} \quad \text{and} \quad s^2 = \frac{\Sigma f(m-\bar{x})^2}{n-1}$$

where σ^2 is the population variance, s^2 is the sample variance, m is the midpoint of a class, and f is the frequency of a class.

In either case, the standard deviation is obtained by taking the square root of the variance.

EXAMPLE 3.31

Calculating the Variance and Standard Deviation for Grouped Data Using Basic Formulas

In Example 3.22, we used the short-cut formula to compute the variance and standard deviation for the data on the number of orders received each day during the past 50 days at the office of a mail-order company. Calculate the variance and standard deviation for those data using the basic formula.

Solution All the required calculations to find the variance and standard deviation appear in **Table 3.18.**

TABLE 3.18

Number of Orders	f	m	mf	$m-\bar{x}$	$(m-\bar{x})^2$	$f(m-\bar{x})^2$
10–12	4	11	44	−5.64	31.8096	127.2384
13–15	12	14	168	−2.64	6.9696	83.6352
16–18	20	17	340	.36	.1296	2.5920
19–21	14	20	280	3.36	11.2896	158.0544
	$n = 50$		$\Sigma mf = 832$			$\Sigma f(m-\bar{x})^2 = 371.5200$

The following steps are performed to compute the variance and standard deviation using the basic formula.

STEP 1 Find the midpoint of each class. Multiply the corresponding values of m and f. Find Σmf. From Table 3.18, $\Sigma mf = 832$.

STEP 2 Find the mean as follows:

$$\bar{x} = \frac{\Sigma mf}{n} = \frac{832}{50} = 16.64$$

STEP 3 Calculate $m - \bar{x}$, the deviation of each value of m from the mean. These calculations are shown in the fifth column of Table 3.18.

STEP 4 Square each of the deviations $m - \bar{x}$; that is, calculate each of the $(m-\bar{x})^2$ values. These are called *squared deviations*, and they are recorded in the sixth column.

STEP 5 Multiply the squared deviations by the corresponding frequencies (see the seventh column of Table 3.18). Adding the values of the seventh column, we obtain

$$\Sigma f(m-\bar{x})^2 = 371.5200$$

STEP 6 Obtain the sample variance by dividing $\Sigma f(m-\bar{x})^2$ by $n-1$. Thus,

$$s^2 = \frac{\Sigma f(m-\bar{x})^2}{n-1} = \frac{371.5200}{50-1} = \mathbf{7.5820}$$

STEP 7 Obtain the standard deviation by taking the positive square root of the variance.

$$s = \sqrt{s^2} = \sqrt{7.5820} = \mathbf{2.75\ orders}$$

Self-Review Test

Multiple Choice Questions

1. The value of the middle term in a ranked data set is called the

a. mean b. median c. mode

2. Which of the following summary measures is/are influenced by extreme values?

a. Mean b. Median c. Mode

3. Which of the following summary measures can be calculated for qualitative data?

a. Mean b. Median c. Mode

4. Which of the following can have more than one value?

a. Mean b. Median c. Mode

5. Which of the following is obtained by taking the difference between the largest and the smallest values of a data set?

a. Variance b. Range c. Mean

6. Which of the following is the mean of the squared deviations of x values from the mean?

a. Standard deviation

b. Population variance

c. Sample variance

7. The values of the variance and standard deviation are

a. never negative

b. always positive

c. never zero

8. A summary measure calculated for the population data is called

a. a population parameter

b. a sample statistic

c. an outlier

9. A summary measure calculated for the sample data is called a

a. population parameter

b. sample statistic

c. box-and-whisker plot

10. Chebyshev's theorem can be applied to

a. any distribution

b. bell-shaped distributions only

c. skewed distributions only

11. The empirical rule can be applied to

a. any distribution

b. bell-shaped distributions only

c. skewed distributions only

12. The first quartile is a value in a ranked data set such that about

a. 75% of the values are smaller and about 25% are larger than this value

b. 50% of the values are smaller and about 50% are larger than this value

c. 25% of the values are smaller and about 75% are larger than this value

13. The third quartile is a value in a ranked data set such that about

a. 75% of the values are smaller and about 25% are larger than this value

b. 50% of the values are smaller and about 50% are larger than this value

c. 25% of the values are smaller and about 75% are larger than this value

14. The 75th percentile is a value in a ranked data set such that about

a. 75% of the values are smaller and about 25% are larger than this value

b. 25% of the values are smaller and about 75% are larger than this value

15. Which of the following is the most influenced summary measure for an outlier?

a. Mean b. Median c. Mode

16. If the histogram of a data set is skewed to the right with one peak, which of the following is true?

a. The values of the mean, median, and mode are the same.

b. The mean is greater than the median, which is greater than the mode.

c. The mean and median are equal, but the mode is different.

17. The scores of 20 students on a mathematical game are 82, 81, 65, 14, 64, 15, 75, 100, 63, 59, 58, 61, 70, 99, 80, 53, 54, 62, 71, and 73. The best measure of central tendency in this case is the

a. weighted mean

b. mean

c. trimmed mean

18. The quantity $x-\mu$ is called the

a. standard deviation

b. deviation of x from the mean

c. variance

19. The coefficient of variation is a measure of

a. average

b. median

c. relative variability

20. Let s_1, s_2, and s_3 be the standard deviations of the bell-shaped graphs I, II, and III, respectively. Place them in increasing order.

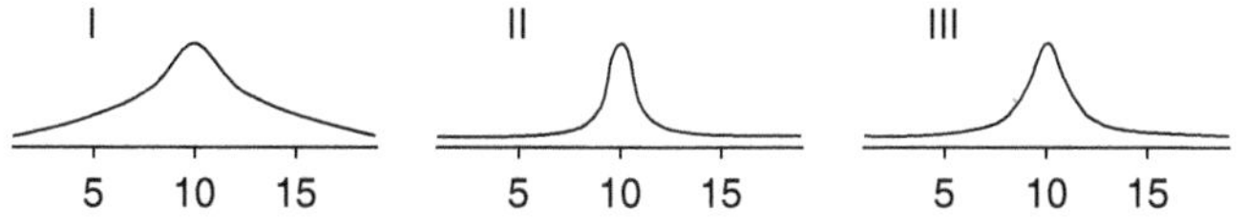

a. $s_2 s_1 s_3$

b. $s_2 s_3 s_1$

c. $s_1 s_3 s_2$

21. You have just recorded the waiting times for a random sample of 50 customers who visited Elmo's Pizza Shop. The following table gives the frequency distribution of waiting times (in minutes) for these customers.

Waiting Time	*f*
10 to less than 16	6
16 to less than 22	10
22 to less than 28	15
28 to less than 34	11
34 to less than 40	7
40 to less than 46	1

The mean waiting time, rounded to two decimal places, is

a. 25.72 minutes

b. 8.33 minutes

c. 30.01 minutes

22. According to Chebyshev's theorem, the minimum percentage of values that fall within 4.4 standard deviations of the mean is

a. 92.83% b. 96.33% c. 94.83%

23. The ages of all high school teachers in a city have a bell-shaped distribution with a mean of 38 years and a standard deviation of 4 years. According to the empirical rule, the percentage of teachers in the city who are 34 to 42 years old is (approximately)

(a) 75% b. 68% c. 95%

24. The value of the 45th percentile for a data set is 52. This means that

a. 55% of the values in that data set are smaller than 52

b. 45% of the values in that data set are greater than 52

c. 45% of the values in that data set are smaller than 52

25. The waiting time (in minutes) for 11 customers at a supermarket is

14 9 15 5 4 7 8 11 14 2 6

The percentile rank for the customer who waited for 11 minutes is

a. 72.72% b. 68.33% c. 63.64%

26. Which of the following is something a box-and-whisker plot does not show?

a. Spread of the data

b. Percent of the data within two standard deviations of the mean

c. Center of the data set

27. The following data represents the age of each of the 15 people buying lift tickets at a ski area.

15 25 26 17 38 16 78 44 30 21 28 40 20 33 31

Are there any outliers in this data set? If so, what are they?

a. Yes: 78

b. Yes: 15

c. Yes: 15 and 78

Practice Exercises

1. Twenty randomly selected persons were asked to keep record of the number of times they used their debit cards during October 2019. The following data show their responses.

5 32 41 14 21 72 19 6 26 7 8 9 18 5 91 14 1 10 8 13

a. Calculate the mean, median, and mode for these data.

b. Compute the 10% trimmed mean for these data.

c. Calculate the range, variance, and standard deviation for these data.

d. Compute the coefficient of variation.

e. Are the values of these summary measures calculated above population parameters or sample statistics? Explain.

2. A company makes five different models of refrigerators. Last year they sold 16,652 refrigerators of all five models. The following table lists the number of different models of refrigerators sold last year and the prices for which they were sold.

Refrigerator Model	Number of Refrigerators Sold	Price (dollars)
VX569	2842	2055
AX255	4364	1165
YU490	3946	1459
MT810	1629	2734
PR920	3871	1672

Calculate the weighted mean that represents the average price for which these 16,652 refrigerators were sold.

3. The mean, as a measure of center, has the disadvantage of being influenced by extreme values. Illustrate this point with an example.

4. The range, as a measure of spread, has the disadvantage of being influenced by extreme values. Illustrate this point with an example.

5. When is the value of the standard deviation for a data set zero? Give one example of such a data set. Calculate the standard deviation for that data set to show that it is zero.

6. The following table gives the frequency distribution of the numbers of computers sold during each of the past 25 weeks at an electronics store.

Computers Sold	Frequency
4 to 9	2
10 to 15	4
16 to 21	10
22 to 27	6
28 to 33	3

a. What does the frequency column in the table represent?

b. Calculate the mean, variance, and standard deviation.

7. The members of a very large health club were observed on a randomly selected day. The distribution of times they spent that day at the health club was found to have a mean of 91.8 minutes and a standard deviation of 16.2 minutes. Suppose these values of the mean and standard deviation hold true for all members of this club.

a. Using Chebyshev's theorem, find the minimum percentage of this health club's members who spend time at this health club between

i. 59.4 and 124.2 minutes

ii. 51.3 and 132.3 minutes

***b.** Using Chebyshev's theorem, find the interval that contains the times spent at this health club by at least 89% of members.

8. The ages of cars owned by all people living in a city have a bell-shaped distribution with a mean of 7.3 years and a standard deviation of 2.2 years.

a. Using the empirical rule, find the (approximate) percentage of cars in this city that are

i. 5.1 to 9.5 years old

ii. .7 to 13.9 years old

***b.** Using the empirical rule, find the interval that contains the ages of (approximate) 95% of the cars owned by all people in this city.

9. The following data give the number of hours worked last week by 18 randomly selected managers working for various Wall Street financial companies.

45	54	63	79	48	50	49	52	74
61	55	56	58	56	77	66	64	70

a. Calculate the three quartiles and the interquartile range. Where does the value of 54 lie in relation to these quartiles?

b. Find the (approximate) value of the 60th percentile. Give a brief interpretation of this value.

c. Calculate the percentile rank of 64. Give a brief interpretation of this value.

10. Make a box-and-whisker plot for the data given in Practice Exercise 9. Comment on the skewness of this data set.

***11.** The mean weekly wage of a sample of 15 employees of a company is $1035. The mean weekly wage of a sample of 20 employees of another company is $1090. Find the combined mean for these 35 employees.

***12.** The mean GPA of five students is 3.21. The GPAs of four of these five students are, respectively, 3.85, 2.67, 3.45, and 2.91. Find the GPA of the fifth student.

***13.** Consider the following two data sets.

Data Set I:	8	16	20	35
Data Set II:	5	13	17	32

Note that each value of the second data set is obtained by subtracting 3 from the corresponding value of the first data set.

a. Calculate the mean for each of these two data sets. Comment on the relationship between the two means.

b. Calculate the standard deviation for each of these two data sets. Comment on the relationship between the two standard deviations.

Mini-Projects

Mini-Project 3.1

Refer to the data you collected for Mini-Project 1.1 of Chapter 1 and analyzed graphically in Mini-Project 2.1 of Chapter 2. Write a report summarizing those data. This report should include answers to at least the following questions:

a. Calculate the summary measures such as the mean, standard deviation, five-number summary (minimum value, first quartile, median, third quartile, and maximum value), and interquartile range for the variable you graphed in part (a) of Mini-Project 2.1.

b. Draw a box-and-whisker plot for the data set used in part (a). Are there any outliers? Does the plot show any skewness?

c. Discuss which measures for the center and spread would be more appropriate to use to describe your data set. Also, discuss your reasons for using those measures.

Mini-Project 3.2

You are employed as a statistician for a company that makes household products, which are sold by part-time salespersons who work during their spare time. The company has four salespersons employed in a small town. Let us denote these salespersons by A, B, C, and D. The sales records (in dollars) for the past six weeks for these four salespersons are shown in the following table:

Week	A	B	C	D
1	1774	2205	1330	1402
2	1808	1507	1295	1665
3	1890	2352	1502	1530
4	1932	1939	1104	1826
5	1855	2052	1189	1703
6	1726	1630	1441	1498

Your supervisor has asked you to prepare a brief report comparing the sales volumes and the consistency of sales of these four salespersons. Use the mean sales for each salesperson to compare the sales volumes, and then choose an appropriate statistical measure to compare the consistency of sales. Make the calculations and write a report.

Mini-Project 3.3

Refer to the data you collected and analyzed graphically for Mini-Project 2.3 of Chapter 2. Write a report summarizing those data. This report should include answers to the following questions:

a. Calculate the summary measures (the mean, standard deviation, five-number summary, and interquartile range) of the reaction times for each gender.

b. Which gender had greater variability in their reaction times? Explain why.

c. Draw a box-and-whisker plot of the reaction time for each gender. Are these plots consistent with your answer in part (b)? Are there any outliers?

d. Discuss which measures for the center and variability would be most appropriate to use to describe the reaction times for each gender.

Decide for Yourself

Note: The Decide for Yourself feature is located in the book's product page on *www.wiley.com*.

Chapter 3 Technology Instructions

Note: Complete TI-84, Minitab, and Excel manuals are available in the book's product page on *www.wiley.com*.

TI-84 Color/TI-84

The TI-84 Color Technology Instructions feature of this text is written for the TI-84 Plus C color graphing calculator running the 4.0 operating system. Some screens, menus, and functions will be slightly different in older operating systems. The TI-84+ can perform all of the same functions but does not have the "Color" option referenced in some of the menus.

The TI-84 calculator does not have functions to calculate the coefficient of variation and trimmed mean or to make a dotplot.

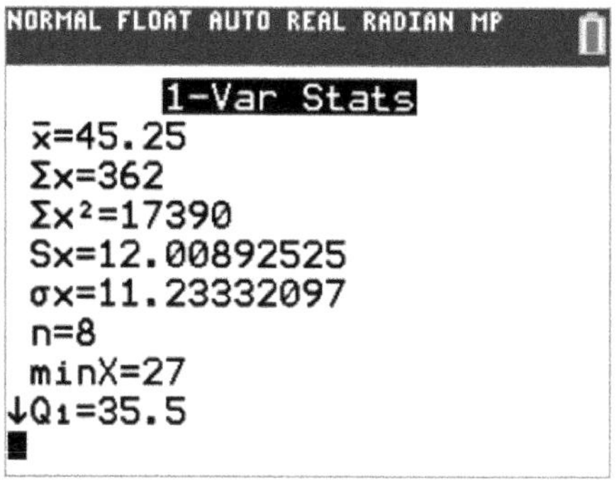

SCREEN 3.1

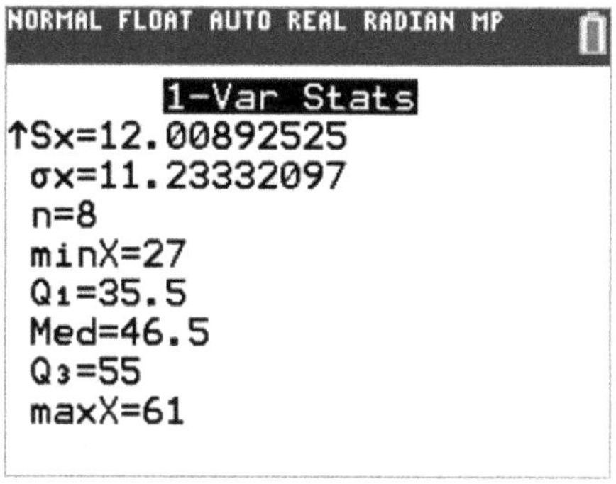

SCREEN 3.2

Calculating Summary Statistics for Example 3.2 of the Text

1. Enter the data from Example 3.2 of the text into L1.
2. Select **STAT > CALC > 1-Var Stats.**
3. Use the following settings in the **1-Var Stats** menu:
   - At the **List** prompt, select the name of the list L1 by pressing 2nd > **STAT** and scroll through the names until you get to your list name.
   - Leave the **FreqList** prompt blank.
4. Highlight **Calculate** and press **ENTER**.
5. The calculator will display the output. (See **Screens 3.1** and **3.2**.)
6. To see all of the output, scroll down by pressing the down arrow. The output will be displayed in the following order:

The mean, the sum of the data values, the sum of the squares of the data values, the sample standard deviation, the population standard deviation (used if your data are from a census), the number of data values (also called the sample size or population size), the minimum data value, the first quartile, the median, the third quartile, and the maximum data value.

Coefficient of Variation

There is no function to calculate the coefficient of variation on the TI. First calculate the sample mean and sample standard deviation as shown above. Then divide the sample standard deviation by the sample mean on the home screen to calculate the coefficient of variation.

Trimmed Mean

There is no function to calculate the trimmed mean on the TI. To calculate the trimmed mean, eliminate the most extreme data values from the data set and then follow the instructions given above to compute the mean from the resulting (trimmed) data.

Weighted Mean

Enter the data into L1. Enter the corresponding weights into L2. Follow the instructions given above, but in Step 3, enter L2 at the **FreqList** prompt.

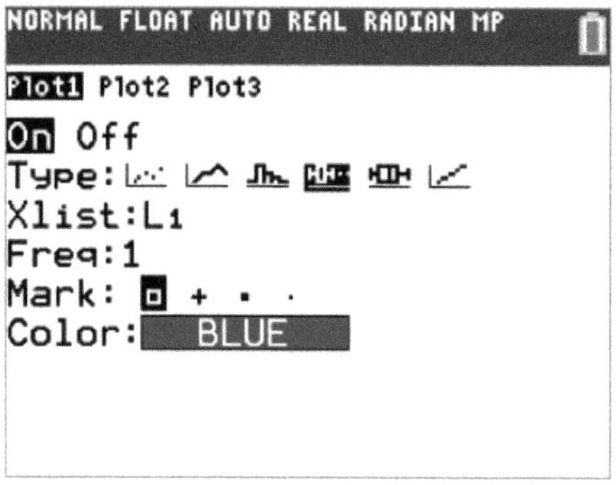

SCREEN 3.3

Creating a Box-and-Whisker Plot for Example 3.29 of the Text

1. Enter the data from Example 3.29 of the text into L1.
2. Select **2nd** > **Y=** (the **STAT PLOT** menu).
3. Select **Plot 1**.
4. Use the following settings in the **Plot 1** menu (see **Screen 3.3**):
   - Select **On** to turn the plot on.

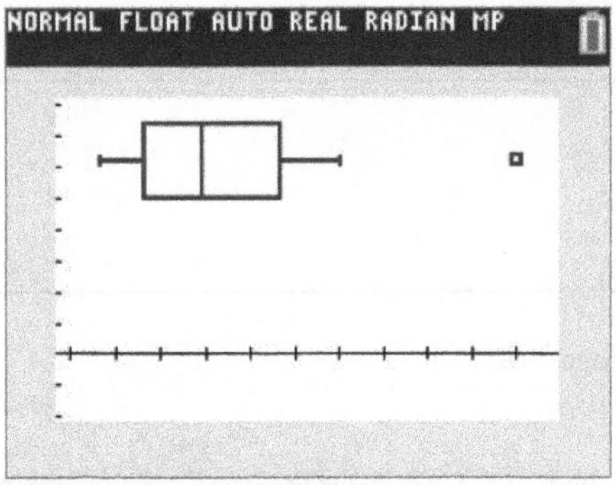

SCREEN 3.4

- At the **Type** prompt, select either the box-and-whisker plot (the fifth choice) or the modified box-and-whisker plot (the fourth choice) which plots outliers as separate points.
- At the **Xlist** prompt, access the name of your list by pressing 2nd > **STAT** and scroll through the names until you get to your list name.
- At the **Freq** prompt, type 1.
- At the **Color** prompt, select BLUE.

  (Skip this step if you do not have the TI-84 Color calculator.)

5. Select **ZOOM** > **ZoomStat** to display the boxplot. This function will automatically choose window settings for the boxplot. (See **Screen 3.4.**)

Minitab

The Minitab Technology Instructions in this text are written for **Minitab Professional Version 19**. Some screens, menus, and functions will be slightly different in older (or in newer) versions of Minitab. If you are using **Minitab Express**, you can download the Minitab Express Manual from the book's product page on *www.wiley.com*.

Calculating Summary Statistics for Example 3.2 of the Text

1. Enter the data from Example 3.2 of the text into column C1. Click below C1 in the blank column heading and name it *Ages*.
2. Select **Stat** > **Basic Statistics** > **Display Descriptive Statistics.**
3. In the dialog box that appears on screen, click inside the **Variables** box. Then either type C1 in this column or double click on C1 Ages on the left.
4. Click the **Statistics** button and check the boxes next to the summary statistics you wish to calculate. You can uncheck the boxes next to statistics that you do not want. (See **Screen 3.5.**)
5. Click **OK** in both dialog boxes. The output will appear in the **Session** window. (See **Screen 3.6.**)

In these summary measures, you can obtain the values of various summary measures including the coefficient of variation and the trimmed mean.

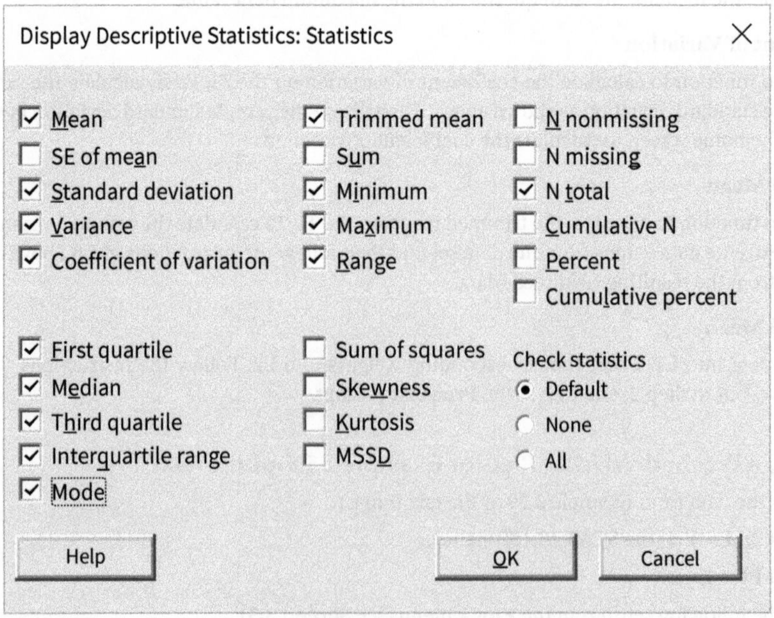

SCREEN 3.5

Descriptive Statistics: Ages

Statistics

Variable	Total Count	Mean	TrMean	StDev	Variance	CoefVar	Minimum	Q1	Median	Q3
Ages	8	45.25	*	12.01	144.21	26.54	27.00	33.75	46.50	56.00

Variable	Maximum	Range	IQR	Mode	N for Mode
Ages	61.00	34.00	22.25	*	0

SCREEN 3.6

Weighted Mean

Enter the data values in C1 and the corresponding weights in C2. Select **Calc > Calculator**. Type C3 in the **Store result in variable** box. Type SUM(C1*C2)/SUM(C2) in the **Expression** box. Click **OK**.

Creating a Box-and-Whisker Plot for Example 3.29 of the Text

1. Enter the data from Example 3.29 of the text into column C1. Name the column *Income*.
2. Select **Graph > Boxplot > Simple** and click **OK**.
3. In the dialog box that appears on screen, click inside the **Graph variables** box. Then either type C1 in this column or double click on C1 Income on the left.
4. Click **OK** and the box-and-whisker plot will appear in a new window. (See **Screen 3.7**.)

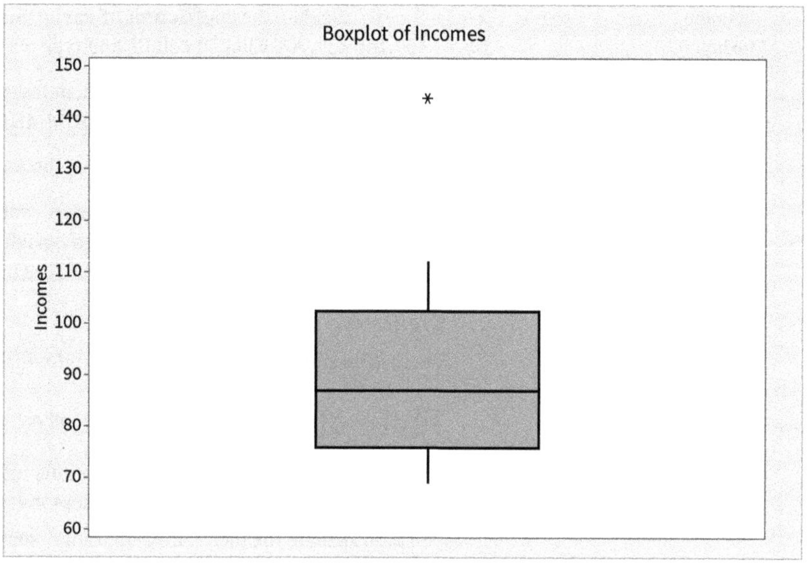

SCREEN 3.7

Excel

The Excel Technology Instructions feature of this text is written for Excel 2019. Some screens, menus, and functions will be slightly different in older versions of Excel. To enable some of the advanced Excel functions, you must enable the Data Analysis Add-In for Excel.

Calculating Summary Statistics for Example 3.2 of the Text

1. Enter the data from Example 3.2 of the text into cells A1 through A8.
2. Click the **Data** tab and then click **Data Analysis** in the **Analysis** group.
3. In the **Data Analysis** dialog box, select **Descriptive Statistics**. Click **OK**.
4. In the **Descriptive Statistics** dialog box, click in the **Input Range** box. Select the cells where your data are located.

   Note: The easiest way to do this is to click and drag from cell A1 to cell A8.
5. Select **Columns** since these data are grouped in columns.
6. Select **New Worksheet Ply**. (See **Screen 3.8**.)

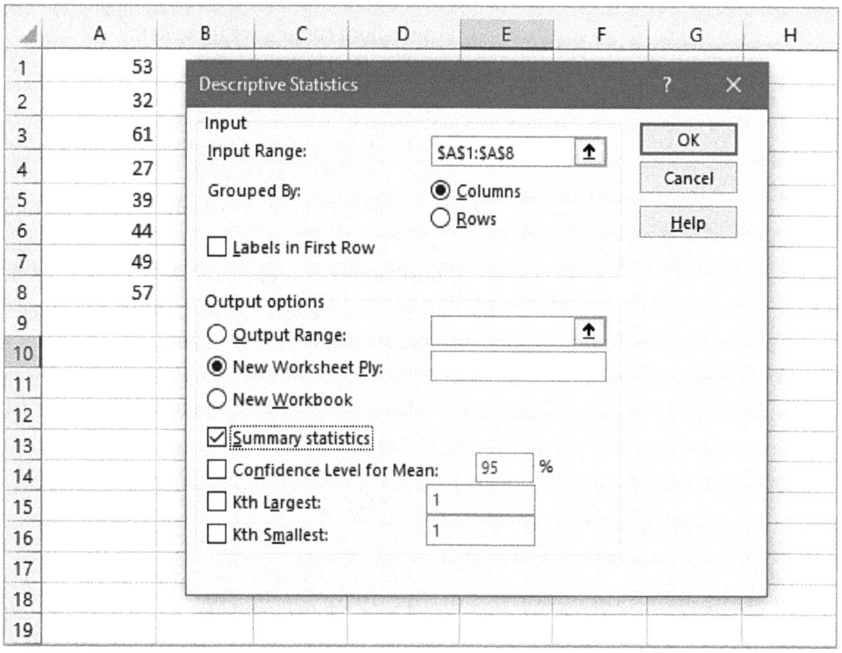

SCREEN 3.8

	A	B
1	COLUMN1	
2		
3	Mean	45.25
4	Standard Error	4.245796
5	Median	46.5
6	Mode	#N/A
7	Standard Deviation	12.00893
8	Sample Variance	144.2143
9	Kurtosis	-1.14659
10	Skewness	-0.28632
11	Range	34
12	Minimum	27
13	Maximum	61
14	Sum	362
15	Count	8

SCREEN 3.9

7. Check **Summary Statistics**. Click **OK**.
8. Excel will place the output, including the mean, median, mode, sample standard deviation, sample variance, range, minimum, and maximum in a new worksheet ply. You may need to make one or both columns wider to reveal all of the output. (See **Screen 3.9.**)
9. The summary statistics do not include the first and third quartiles. To find these quartiles, use the following procedure:
   - Click on an empty cell next to the original data.
   - Type =**quartile(**
   - Select the data by clicking and dragging from cell A1 to cell A8.
   - Type a comma, followed by a 1 for the first quartile (or a 3 for the third quartile).
   - Type a right parenthesis, and then press **Enter**.

To calculate the **coefficient of variation**, enter the data from Example 3.2 of the text into cells A1 through A8. Click in cell B1 and type =STDEV.S(A1:A8)/AVERAGE(A1:A8). Press **Enter**.

To calculate the **trimmed mean**, enter the data from Example 3.2 of the text into cells A1 through A8. Click in cell B1 and type =TRIMMEAN(A1:A8,0.2). Press **Enter**.

Note: The 0.2 indicates that 20% of the total data set will be trimmed, 10% from each tail.

To calculate the **weighted mean**, enter the data from Example 3.2 of the text into cells A1 through A8 and the desired weights in cells B1 through B8. Click in cell C1 and type =SUMPRODUCT(A1:A8,B1:B8)/SUM(B1:B8). Press **Enter**.

Creating a Box-and-Whisker Plot

There is no simple way to make a box-and-whisker plot with Excel.

Technology Assignments

TA3.1 Refer to Data Set III which contains information on all NFL players who were on rosters of all teams on the first day of 2019 football season (see Appendix A).

a. Calculate the mean, median, range, standard deviation, quartiles, and the interquartile range for the weights of these players.

b. Create a box-and-whisker plot for the ages of these players.

c. Calculate the mean and median for the ages of these players.

d. Use your box-and-whisker plot to explain why the mean age is greater than the median age for these players.

TA3.2 Refer to Data Set IX which contains information on Subway's menu items (see Appendix A).

a. Calculate the mean, median, range, standard deviation, quartiles, and the interquartile range for the sodium contents of these items.

b. Create a box-and-whisker plot for the sodium contents of these items.

c. Create a dotplot for the sodium contents of these items.

d. If the technology that you are using lets you calculate the trimmed mean and the coefficient of variation, find these summary measures for the sodium contents of these items.

TA3.3 Refer to Data Set VI which contains information on the S&P 500 companies (see Appendix A).

a. Calculate the mean, median, range, standard deviation, quartiles, and the interquartile range for the closing prices of stocks of the companies listed in column 3.

b. Select a random sample of 20 closing prices. Calculate the mean, median, range, standard deviation, quartiles, and the interquartile range for these 20 closing prices. Are these values of summary measures different from the ones obtained in part (a)?

TA3.4 Refer to Data Set IV which contains information on all participants of the 2019 Manchester Road Race (see Appendix A).

a. Calculate the mean, median, range, standard deviation, quartiles, and the interquartile range for the average pace of these runners.

b. Create a box-and-whisker plot for the average pace of these participants.

c. Create a frequency histogram for the average pace of these participants.

d. Use your frequency histogram to explain why the mean of average pace is greater than the median of average pace for these participants.

e. If the technology that you are using lets you calculate the trimmed mean and the coefficient of variation, find these summary measures for the average pace of these participants.

TA3.5 Refer to Data Set X which contains information on Major League Baseball (MLB) (see Appendix A). Create a dotplot for the number of runs batted in (RBI) listed in column 11.

TA3.6 Refer to Data Set I which contains information on houses sold in Wilmington, NC, between August 25, 2017 and July 17, 2019 (see Appendix A).

a. Calculate the five-number summary (that is used to make a box-and-whisker plot) for the sale prices of these homes (prices at which these homes were sold).

b. Using the values of the summary measures obtained in part (a), calculate the values of the upper and lower inner fences for the sale prices of these homes.

c. Using the values of the summary measures obtained in part (a), calculate the values of the upper and lower outer fences for the sale prices of these homes.

d. Identify any outliers and classify them as mild or extreme.

e. Using technology, make a box-and-whisker plot for the sale prices of these homes.

TA3.7 The following data give the total food expenditures (in dollars) for the past one month for a sample of 20 families.

1125	530	1234	595	427	872	1480	699	1274	1187
933	1127	716	1065	934	1630	1046	2199	1353	441

a. Calculate the mean, median, standard deviation, quartiles, and the interquartile range for these data.

b. Create a box-and-whisker plot for these data.

c. Create a dotplot for these data.

d. If the technology that you are using lets you calculate trimmed mean and coefficient of variation, find these summary measures for these data.

TA3.8 The following data give the odometer mileage (rounded to the nearest thousand miles) for all 20 cars that are for sale at a dealership.

62	86	58	84	72	40	27	38	50	43
27	40	90	43	94	36	28	48	86	77

a. Calculate the mean, median, standard deviation, quartiles, and the interquartile range for these data.

b. Create a box-and-whisker plot for these data.

c. Create a dotplot for these data.

d. If the technology that you are using lets you calculate trimmed mean and coefficient of variation, find these summary measures for these data.

CHAPTER

4

Probability

Are you a vegetarian? If not, do you know someone who is a vegetarian? What percentage of U.S. adults are vegetarians? According to a 2018 Gallup poll, 5% of all U.S. adults, aged 18 and older, are vegetarian. However, from the same poll, only 4% of U.S. adult men are vegetarian while 6% of U.S. adult women are vegetarian. Based on ideology, 2% of conservatives, 3% of moderates, and 11% of liberals are vegetarian. (See Case Study 4.1.)

Monkey Business Images/Shutterstock.com

We often make statements about probability. For example, a weather forecaster may predict that there is an 80% chance of rain tomorrow. A study may predict that a female, compared to a male, has a higher probability of being in favor of gun control. A college student may ask an instructor about the chances of passing a course or getting an A if he or she did not do well on the midterm examination.

Probability, which measures the likelihood that an event will occur, is an important part of statistics. It is the basis of inferential statistics, which will be introduced in later chapters. In inferential statistics, we make decisions under conditions of uncertainty. Probability theory is used to evaluate the uncertainty involved in those decisions. For example, estimating next year's sales for a company is based on many assumptions, some of which may happen to be true and others may not. Probability theory will help us make decisions under such conditions of imperfect information and uncertainty. Combining probability and probability distributions (which are discussed in Chapters 5 through 7) with descriptive statistics will help us make decisions about populations based on information obtained from samples. This chapter presents the basic concepts of probability and the rules for computing probability.

4.1 Experiment, Outcome, and Sample Space

LEARNING OBJECTIVES

After completing this section, you should be able to:

- Explain the terms experiment, outcome, and sample space in the context of probability.
- Explain the difference between a simple event and a compound event.
- Determine whether an event is a simple event or a compound event.
- Determine the number of possible outcomes for a statistical experiment.
- Construct a tree diagram for a statistical experiment.

There are only a few things, if any, in life that have definite, certain, and sure outcomes. Most things in life have uncertain outcomes. For example, will you get an A grade in the statistics class that you are taking this semester? Suppose you recently opened a new business; will it be successful? Your friend will be getting married next month. Will she be happily married for the rest of her life? You just bought a lottery ticket. Will it be a winning ticket? You are playing black jack at a casino. Will you be a winner after 20 plays of the game? In all these situations, the outcomes are random. You know the various outcomes for each of these games/statistical experiments, but you do not know which one of these outcomes will actually occur. As another example, suppose you are working as a quality control manager at a factory where they are making hockey pucks. You randomly select a few pucks from the production line and inspect them for being good or defective. This act of inspecting a puck is an example of a statistical **experiment**. The result of this inspection will be that the puck is either "good" or "defective." Each of these two observations is called an **outcome** (also called a **basic** or **final outcome**) of the experiment, and these outcomes taken together constitute the **sample space** for this experiment. The elements of a sample space are called **sample points.**

Experiment, Outcomes, and Sample Space

An *experiment* is a process that, when performed, results in one and only one of many observations. These observations are called the *outcomes* of the experiment. The collection of all outcomes for an experiment is called a *sample space.*

Table 4.1 lists some examples of experiments, their outcomes, and their sample spaces.

TABLE 4.1 Examples of Experiments, Outcomes, and Sample Spaces

Experiment	Outcomes	Sample Space
Toss a coin once	Head, Tail	{Head, Tail}
Roll a die once	1, 2, 3, 4, 5, 6	{1, 2, 3, 4, 5, 6}
Toss a coin twice	*HH, HT, TH, TT*	{*HH, HT, TH, TT*}
Play lottery	Win, Lose	{Win, Lose}
Take a test	Pass, Fail	{Pass, Fail}
Select a worker	Male, Female	{Male, Female}

The sample space for an experiment can also be illustrated by drawing a tree diagram. In a **tree diagram**, each outcome is represented by a branch of the tree. Tree diagrams help us understand probability concepts by presenting them visually. Examples 4.1 through 4.3 describe how to draw the tree diagrams for statistical experiments.

EXAMPLE 4.1 One Toss of a Coin

Drawing the Tree Diagram: One Toss of a Coin

Draw the tree diagram for the experiment of tossing a coin once.

Solution This experiment has two possible outcomes: head and tail. Consequently, the sample space is given by

$$\text{Sample space} = \{H, T\}, \quad \text{where } H = \text{Head} \quad \text{and} \quad T = \text{Tail}$$

To draw a tree diagram, we draw two branches starting at the same point, one representing the head and the second representing the tail. The two final outcomes are listed at the ends of the branches as shown in **Figure 4.1**.

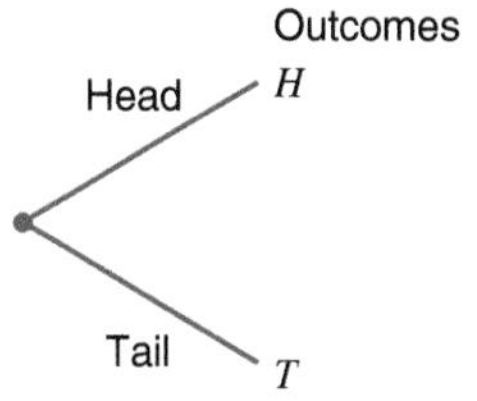

FIGURE 4.1 Tree diagram for one toss of a coin.

EXAMPLE 4.2 Two Tosses of a Coin

Drawing the Tree Diagram: Two Tosses of a Coin

Draw the tree diagram for the experiment of tossing a coin twice.

Solution This experiment can be split into two parts: the first toss and the second toss. Suppose that the first time the coin is tossed, we obtain a head. Then, on the second toss, we can still obtain a head or a tail. This gives us two outcomes: *HH* (head on both tosses) and *HT* (head on the first toss and tail on the second toss). Now suppose that we observe a tail on the first toss. Again, either a head or a tail can occur on the second toss, giving the remaining two outcomes: *TH* (tail on the first toss and head on the second toss) and *TT* (tail on both tosses). Thus, the sample space for two tosses of a coin is

$$\text{Sample space} = \{HH, HT, TH, TT\}$$

The tree diagram is shown in **Figure 4.2**. This diagram shows the sample space for this experiment.

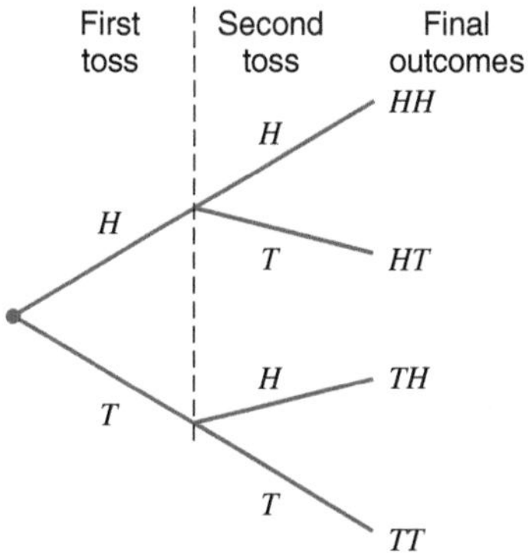

FIGURE 4.2 Tree diagram for two tosses of a coin.

EXAMPLE 4.3 Selecting Two Workers

Drawing the Tree Diagram: Two Selections

Suppose we randomly select two workers from a company and observe whether the worker selected each time is a man or a woman. Write all the outcomes for this experiment. Draw the tree diagram for this experiment.

Solution Let us denote the selection of a man by M and that of a woman by W. We can compare the selection of two workers to two tosses of a coin. Just as each toss of a coin can result in one of two outcomes, head or tail, each selection from the workers of this company can result in one of two outcomes, man or woman. As we can see from the tree diagram of **Figure 4.3**, there are four final outcomes: MM, MW, WM, WW. Hence, the sample space is written as

$$\text{Sample space} = \{MM, MW, WM, WW\}$$

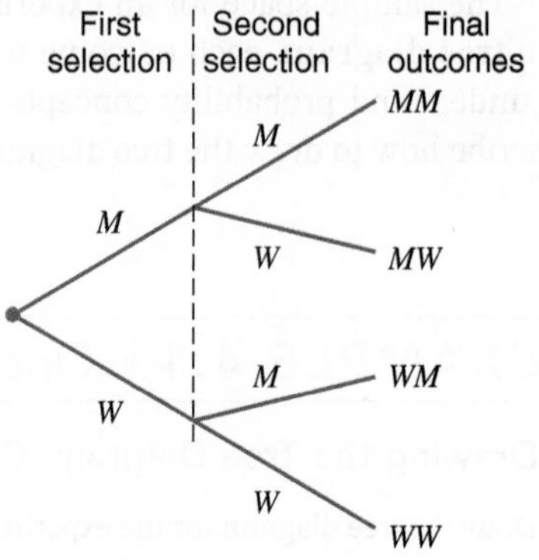

FIGURE 4.3 Tree diagram for selecting two workers.

4.1.1 Simple and Compound Events

An **event** consists of one or more of the outcomes of an experiment.

Event

An *event* is a collection of one or more of the outcomes of an experiment.

An event may be a **simple event** or a **compound event**. A simple event is also called an **elementary event**, and a compound event is also called a **composite event**.

Simple Event Each of the final outcomes for an experiment is called a **simple event**. In other words, a simple event includes one and only one outcome. Usually, simple events are denoted by E_1, E_2, E_3, and so forth. However, we can denote them by any other letter—that is, by A, B, C, and so forth. Many times we denote events by the same letter and use subscripts to distinguish them, as in $A_1, A_2, A_3, \ldots$.

Simple Event

An event that includes one and only one of the (final) outcomes for an experiment is called a *simple event* and is usually denoted by E_i.

Example 4.4 describes simple events.

EXAMPLE 4.4 Selecting Two Workers

Illustrating Simple Events

Reconsider Example 4.3 on selecting two workers from a company and observing whether the worker selected each time is a man or a woman. Each of the final four outcomes (MM, MW, WM, and WW) for this experiment is a simple event. These four events can be denoted by E_1, E_2, E_3, and E_4, respectively. Thus,

$$E_1 = \{MM\}, \quad E_2 = \{MW\}, \quad E_3 = \{WM\}, \quad \text{and} \quad E_4 = \{WW\}$$

Compound Event

A **compound event** consists of more than one outcome.

> **Compound Event**
>
> A *compound event* is a collection of more than one outcome for an experiment.

Compound events are denoted by $A, B, C, D, \ldots$, or by $A_1, A_2, A_3, \ldots$, or by $B_1, B_2, B_3, \ldots$, and so forth. Examples 4.5 and 4.6 describe compound events.

EXAMPLE 4.5 Selecting Two Workers

Illustrating a Compound Event: Two Selections

Reconsider Example 4.3 on selecting two workers from a company and observing whether the worker selected each time is a man or a woman. Let A be the event that at most one man is selected. Is event A a simple or a compound event?

Solution Here *at most one man* means one or no man is selected. Thus, event A will occur if either no man or one man is selected. Hence, the event A is given by

$$A = \text{at most one man is selected} = \{MW, WM, WW\}$$

Because event A contains more than one outcome, it is a compound event. The diagram in **Figure 4.4** gives a graphic presentation of compound event A.

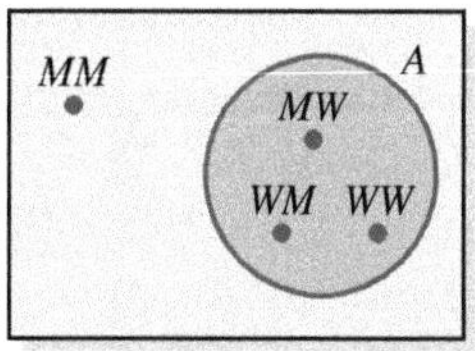

FIGURE 4.4 Diagram showing event A.

EXAMPLE 4.6 Preference for Ice Tea

Illustrating Simple and Compound Events: Two Selections

In a group of college students, some like ice tea and others do not. There is no student in this group who is indifferent or has no opinion. Two students are randomly selected from this group.

a. How many outcomes are possible? List all the possible outcomes.

b. Consider the following events. List all the outcomes included in each of these events. Mention whether each of these events is a simple or a compound event.

- **i.** Both students like ice tea.
- **ii.** At most one student likes ice tea.
- **iii.** At least one student likes ice tea.
- **iv.** Neither student likes ice tea.

Solution Let L denote the event that a student likes ice tea and N denote the event that a student does not like ice tea.

a. This experiment has four outcomes, which are listed below and shown in **Figure 4.5**.

LL = Both students like ice tea

LN = The first student likes ice tea but the second student does not

NL = The first student does not like ice tea but the second student does

NN = Both students do not like ice tea

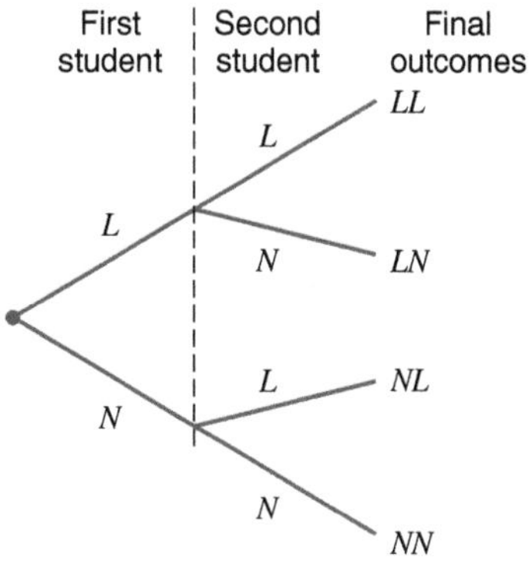

FIGURE 4.5 Tree diagram.

b. **i.** The event *both students like ice tea* will occur if *LL* happens. Thus,

Both students like ice tea = {*LL*}

Since this event includes only one of the four outcomes, it is a **simple** event.

ii. The event *at most one student likes ice tea* will occur if one or none of the two students likes ice tea, which will include the events *LN*, *NL*, and *NN*. Thus,

At most one student likes ice tea = {*LN*, *NL*, *NN*}

Since this event includes three outcomes, it is a **compound** event.

iii. The event *at least one student likes ice tea* will occur if one or two of the two students like ice tea, which will include the events *LN*, *NL*, and *LL*. Thus,

At least one student likes ice tea = {*LN*, *NL*, *LL*}

Since this event includes three outcomes, it is a **compound** event.

iv. The event *neither student likes ice tea* will occur if neither of the two students likes ice tea, which will include the event *NN*. Thus,

Neither student likes ice tea = {*NN*}

Since this event includes one outcome, it is a **simple** event.

EXERCISES

Concepts and Procedures

4.1 Define the following terms: *experiment*, *outcome*, *sample space*, *simple event*, and *compound event*.

4.2 List the simple events for each of the following statistical experiments in a sample space.

a. One roll of a die

b. Three tosses of a coin

c. One toss of a coin and one roll of a die

4.3 A box contains three items that are labeled A, B, and C. Two items are selected at random (without replacement) from this box. List all the possible outcomes for this experiment. Write the sample space.

Applications

4.4 Two students are randomly selected from a statistics class, and it is observed whether or not they suffer from math anxiety. How many total outcomes are possible? Draw a tree diagram for this experiment.

4.5 In a group of adults, some are computer literate and others are computer illiterate. If two adults are randomly selected from this group, how many total outcomes are possible? Draw a tree diagram for this experiment.

4.6 An automated teller machine at a local bank is stocked with $50 and $100 bills. When a customer withdraws $200 from the machine, it dispenses either two $100 bills or four $50 bills. If two customers withdraw $200 each, how many outcomes are possible? Draw a tree diagram for this experiment.

4.7 A box contains a certain number of computer parts, a few of which are defective. Two parts are selected at random from this box and inspected to determine if they are good or defective. How many total outcomes are possible? Draw a tree diagram for this experiment.

4.8 Two students are randomly selected from a statistics class, and it is observed whether or not they suffer from math anxiety. List all the outcomes included in each of the following events. Indicate which are simple and which are compound events.

a. Both students suffer from math anxiety.

b. Exactly one student suffers from math anxiety.

c. The first student does not suffer and the second suffers from math anxiety.

d. None of the students suffers from math anxiety.

4.9 In a group of adults, some own iPads, and others do not. Two adults are randomly selected from this group. List all the outcomes included in each of the following events. Indicate which are simple and which are compound events.

a. One person owns an iPad and the other does not.

b. At least one person owns an iPad.

c. Not more than one person owns an iPad.

d. The first person owns an iPad and the second does not.

4.10 An automated teller machine at a local bank is stocked with $50 and $100 bills. When a customer withdraws $200 from this machine, it dispenses either two $100 bills or four $50 bills. Two customers withdraw $200 each. List all of the outcomes in each of the following events and mention which of these are simple and which are compound events.

a. Exactly one customer receives $100 bills.

b. Both customers receive $50 bills.

c. At most one customer receives $100 bills.

d. The first customer receives $50 bills and the second receives $100 bills.

4.2 Calculating Probability

LEARNING OBJECTIVES

After completing this section, you should be able to:

- Explain the two properties of probability.
- Explain the three approaches to probability.
- Identify which of the three approaches to probability is used to find probability for a specific statistical experiment.
- Calculate the probabilities of simple and compound events.

Probability, which gives the likelihood of occurrence of an event, is denoted by P. The probability that a simple event E_i will occur is denoted by $P(E_i)$, and the probability that a compound event A will occur is denoted by $P(A)$.

Probability

Probability is a numerical measure of the likelihood that a specific event will occur.

4.2.1 Two Properties of Probability

There are two important properties of probability that we should always remember. These properties are mentioned here.

1. **The probability of an event always lies in the range 0 to 1.**

Whether it is a simple or a compound event, the probability of an event is never less than 0 or greater than 1. We can write this property as follows:

First Property of Probability

$$0 \leq P(E_i) \leq 1$$
$$0 \leq P(A) \leq 1$$

An event that cannot occur has zero probability and is called an **impossible (or null) event**. An event that is certain to occur has a probability equal to 1 and is called a **sure (or certain) event**. In the following examples, the first event is an impossible event and the second one is a sure event.

$$P(\text{a tossed coin will stand on its edge}) = 0$$
$$P(\text{a child born today will eventually die}) = 1.0$$

There are very few events in real life that have probability equal to either 0 or 1.0. Most of the events in real life have probabilities that are between 0 and 1.0. In other words, these probabilities are greater than 0 but less than 1.0. A probability closer to 1.0, such as .82, indicates that the event is more likely to occur. On the other hand, an event with a probability closer to 0, such as .12, is less likely to occur. Sometimes events with very low (.05 or lower) probabilities are also called **rare events**.

2. The sum of the probabilities of all simple events (or final outcomes) for an experiment, denoted by $\Sigma P(E_i)$, is always 1.

Second Property of Probability

For an experiment with outcomes $E_1, E_2, E_3, \ldots,$

$$\Sigma P(E_i) = P(E_1) + P(E_2) + P(E_3) + \cdots = 1.0$$

For example, if you buy a lottery ticket, you may either win or lose. The probabilities of these two events must add to 1.0, that is

$$P(\text{you will win}) + P(\text{you will lose}) = 1.0$$

Similarly, for the experiment of one toss of a coin,

$$P(\text{Head}) + P(\text{Tail}) = 1.0$$

For the experiment of two tosses of a coin,

$$P(HH) + P(HT) + P(TH) + P(TT) = 1.0$$

For one game of football by a professional team,

$$P(\text{win}) + P(\text{loss}) + P(\text{tie}) = 1.0$$

4.2.2 Three Conceptual Approaches to Probability

How do we assign probabilities to events? For example, we may say that the probability of obtaining a head in one toss of a coin is .50, or that the probability that a randomly selected family owns a home is .68, or that the Los Angeles Dodgers will win the Major League Baseball championship next year is .14. How do we obtain these probabilities? We will learn the procedures that are used to obtain such probabilities in this section. There are three conceptual approaches to probability: (1) classical probability, (2) the relative frequency concept of probability, and (3) the subjective probability concept. These three concepts are explained next.

Classical Probability Many times, various outcomes for an experiment may have the same probability of occurrence. Such outcomes are called **equally likely outcomes.** The classical probability rule is applied to compute the probabilities of events for an experiment for which all outcomes are equally likely. For example, head and tail are two equally likely outcomes when a fair coin is tossed once. Each of these two outcomes has the same chance of occurrence.

Equally Likely Outcomes

Two or more outcomes that have the same probability of occurrence are said to be *equally likely outcomes.*

Earlier in this section, we learned that the total probability for all simple outcomes of an experiment is 1.0. This total probability is distributed over all the outcomes of the experiment. When all the outcomes of an experiment are equally likely, this total probability will be equally distributed over various outcomes. For example, for a fair coin, there are two equally likely outcomes—a head and a tail. Thus, if we distribute the total probability of 1.0 equally among these two outcomes, then each of these outcomes will have a probability of .50 of occurrence.

According to the **classical probability rule**, to find the probability of a simple event, we divide 1.0 by the total number of outcomes for the experiment. On the other hand, to find the probability of a compound event A, we divide the number of outcomes favorable to event A by the total number of outcomes for the experiment.

Classical Probability Rule to Find Probability

Suppose E_i is a simple event and A is a compound event for an experiment with equally likely outcomes. Then, applying the classical approach, the probabilities of E_i and A are

$$P(E_i) = \frac{1}{\text{Total number of outcomes for the experiment}}$$

$$P(A) = \frac{\text{Number of outcomes favorable to } A}{\text{Total number of outcomes for the experiment}}$$

Examples 4.7 through 4.9 illustrate how probabilities of events are calculated using the classical probability rule.

EXAMPLE 4.7 One Toss of a Coin

Calculating the Probability of a Simple Event

Find the probability of obtaining a head and the probability of obtaining a tail for one toss of a coin.

Solution If we toss a fair coin once, the two outcomes, head and tail, are equally likely outcomes. Hence, this is an example of a classical experiment. Therefore,[1]

$$P(\text{head}) = \frac{1}{\text{Total number of outcomes}} = \frac{1}{2} = \mathbf{.50}$$

Similarly,

$$P(\text{tail}) = \frac{1}{2} = \mathbf{.50}$$

Since this experiment has only two outcomes, a head and a tail, their probabilities add to 1.0.

EXAMPLE 4.8 One Roll of a Die

Calculating the Probability of a Compound Event

Find the probability of obtaining an even number in one roll of a die.

Solution This experiment of rolling a die once has a total of six outcomes: 1, 2, 3, 4, 5, and 6. Given that the die is fair, these outcomes are equally likely. Let A be an event that an even number is observed on the die. Event A includes three outcomes: 2, 4, and 6; that is,

$$A = \text{an even number is obtained} = \{2, 4, 6\}$$

If any one of these three numbers is obtained, event A is said to occur. Since three out of six outcomes are included in the event that an even number is obtained, its probability is

$$P(A) = \frac{\text{Number of outcomes included in } A}{\text{Total number of outcomes}} = \frac{3}{6} = \mathbf{.50}$$

[1]If the final answer for the probability of an event does not terminate within four decimal places, it will be rounded to four decimal places.

EXAMPLE 4.9 | Selecting One Out of Five Homes

Calculating the Probability of a Compound Event

Jim and Kim have been looking for a house to buy in New Jersey. They like five of the homes they have looked at recently and two of those are in West Orange. They cannot decide which of the five homes they should pick to make an offer. They put five balls (of the same size) marked 1 through 5 (each number representing a home) in a box and asked their daughter to select one of these balls. Assuming their daughter's selection is random, what is the probability that the selected home is in West Orange?

Solution With random selection, each home has the same probability of being selected and the five outcomes (one for each home) are equally likely because all five balls are of the same size. Two of the five homes are in West Orange. Hence,

$$P(\text{selected home is in West Orange}) = \frac{2}{5} = \mathbf{.40}$$

Relative Frequency Concept of Probability

Suppose we want to calculate the following probabilities:

1. The probability that the next car that comes out of an auto factory is a "lemon"
2. The probability that a randomly selected family owns a home
3. The probability that a randomly selected woman is an excellent driver
4. The probability that an 80-year-old person will live for at least one more year
5. The probability that a randomly selected adult is in favor of increasing taxes to reduce the national debt
6. The probability that a randomly selected person owns a sport-utility vehicle (SUV)

These probabilities cannot be computed using the classical probability rule because the various outcomes for the corresponding experiments are not equally likely. For example, the next car manufactured at an auto factory may or may not be a lemon. The two outcomes, "the car is a lemon" and "the car is not a lemon," are not equally likely. If they were, then (approximately) half the cars manufactured by this company would be lemons, and this might prove disastrous to the survival of the factory. (Note that in auto manufacturing, a "lemon" is a vehicle with one or more manufacturing defects that affect the vehicle's operation, safety, and value.)

Although the various outcomes for each of these experiments are not equally likely, each of these experiments can be performed again and again to generate data. In such cases, to calculate probabilities, we either use past data or generate new data by performing the experiment a large number of times. Using these data, we calculate the frequencies and relative frequencies for various outcomes. The relative frequency of an event is used as an approximation for the probability of that event. This method of assigning a probability to an event is called the **relative frequency concept of probability**. Because relative frequencies are determined by performing an experiment, the probabilities calculated using relative frequencies may change when an experiment is repeated. For example, every time a new sample of 500 cars is selected from the production line of an auto factory, the number of lemons in those 500 cars is expected to be different. However, the variation in the percentage of lemons will be small if the sample size is large. Note that if we are considering the entire population, the relative frequency will give an exact probability.

Using Relative Frequency as an Approximation of Probability

If an experiment is repeated n times and an event A is observed f times where f is the frequency, then, according to the relative frequency concept of probability:

$$P(A) = \frac{f}{n} = \frac{\text{Frequency of } A}{\text{Sample size}}$$

Examples 4.10 and 4.11 illustrate how the probabilities of events are approximated using the relative frequencies.

EXAMPLE 4.10 Lemons in Car Production

Approximating Probability by Relative Frequency: Sample Data

Ten of the 500 randomly selected cars manufactured at a certain auto factory are found to be lemons. Assuming that lemons are manufactured randomly, what is the probability that the next car manufactured at this auto factory is a lemon?

Solution Let n denote the total number of cars in the sample and f the number of lemons in n. Then, from the given information:

$$n = 500 \quad \text{and} \quad f = 10$$

Using the relative frequency concept of probability, we obtain

$$P(\text{next car is a lemon}) = \frac{f}{n} = \frac{10}{500} = \mathbf{.02}$$

This probability is actually the relative frequency of lemons in 500 cars. **Table 4.2** lists the frequency and relative frequency distributions for this example.

TABLE 4.2 Frequency and Relative Frequency Distributions for the Sample of Cars

Car	f	Relative Frequency
Good	490	490/500 = .98
Lemon	10	10/500 = .02
	$n = 500$	Sum = 1.00

The column of relative frequencies in Table 4.2 is used as the column of approximate probabilities. Thus, from the relative frequency column,

$$P(\text{next car is a lemon}) = \mathbf{.02}$$
$$P(\text{next car is a good car}) = \mathbf{.98}$$

Note that relative frequencies are not exact probabilities but are approximate probabilities unless they are based on a census. However, if the experiment is repeated again and again, this approximate probability of an outcome obtained from the relative frequency will approach the actual probability of that outcome. This is called the **Law of Large Numbers**.

Law of Large Numbers

Law of Large Numbers

If an experiment is repeated again and again, the probability of an event obtained from the relative frequency approaches the actual (or theoretical) probability.

We used Minitab to simulate the tossing of a coin with a different number of tosses. **Table 4.3** lists the results of these simulations.

As we can observe from this table, when the coin was tossed three times, based on relative frequencies, the probability of a head obtained from this simulation was .00 and that of a tail was 1.00. When the coin was tossed eight times, the probability of a head obtained from this simulation was .75 and that of a tail .25. Notice how these probabilities change. As the number of tosses increases, the probabilities of both a head and a tail converge toward .50.

TABLE 4.3 Simulating the Tosses of a Coin

Number of Tosses	Number of Heads	Number of Tails	$P(H)$	$P(T)$
3	0	3	.00	1.00
8	6	2	.75	.25
25	9	16	.36	.64
100	61	39	.61	.39
1000	522	478	.522	.478
10,000	4962	5038	.4962	.5038
1,000,000	500,313	499,687	.5003	.4997

When the simulated tosses of the coin are 1,000,000, the relative frequencies of head and tail are almost equal. This is what is meant by the **Law of Large Numbers**. If the experiment is repeated only a few times, the probabilities obtained may not be close to the actual probabilities. As the number of repetitions increases, the probabilities of outcomes obtained become very close to the actual probabilities. Note that in the example of tossing a fair coin, the actual probability of head and tail is .50 each.

EXAMPLE 4.11 Owning a Home

Approximating Probability by Relative Frequency

Allison wants to determine the probability that a randomly selected family from New York State owns a home. How can she determine this probability?

Solution There are two outcomes for a randomly selected family from New York State: "This family owns a home" and "This family does not own a home." These two outcomes are not equally likely. (Note that these two outcomes will be equally likely if exactly half of the families in New York State own homes and exactly half do not own homes.) Hence, the classical probability rule cannot be applied. However, we can repeat this experiment again and again. In other words, we can select a sample of families from New York State and observe whether or not each of them owns a home. Hence, we will use the relative frequency approach to probability.

Suppose Allison selects a random sample of 1000 families from New York State and observes that 730 of them own homes and 270 do not own homes. Then,

$$n = \text{sample size} = 1000$$
$$f = \text{number of families who own homes} = 730$$

Consequently,

$$P(\text{a randomly selected family owns a home}) = \frac{f}{n} = \frac{730}{1000} = \mathbf{.730}$$

Again, note that .730 is just an approximation of the probability that a randomly selected family from New York State owns a home. Every time Allison repeats this experiment she may obtain a different probability for this event. However, because the sample size ($n = 1000$) in this example is large, the variation in the probability from sample to sample is expected to be relatively small.

Subjective Probability Many times we face experiments that neither have equally likely outcomes nor can be repeated to generate data. In such cases, we cannot compute the probabilities of events using the classical probability rule or the relative frequency concept. For example, consider the following probabilities of events:

1. The probability that Carol, who is taking a statistics course, will earn an A in the course
2. The probability that the Dow Jones Industrial Average will be higher at the end of the next trading day

3. The probability that the New York Giants (football team) will win the Super Bowl next season
4. The probability that Joe will lose the lawsuit he has filed against his landlord

Neither the classical probability rule nor the relative frequency concept of probability can be applied to calculate probabilities for these examples. All these examples belong to experiments that have neither equally likely outcomes nor the potential of being repeated. For example, Carol, who is taking statistics, will take the test (or tests) only once, and based on that she will either earn an A or not. The two events "she will earn an A" and "she will not earn an A" are not equally likely. Also, she cannot take the test (or tests) again and again to calculate the relative frequency of getting or not getting an A grade. The probability assigned to an event in such cases is called **subjective probability**. It is based on the individual's judgment, experience, information, and belief. Carol may be very confident and assign a higher probability to the event that she will earn an A in statistics, whereas her instructor may be more cautious and assign a lower probability to the same event.

Subjective Probability

Subjective probability is the probability assigned to an event based on subjective judgment, experience, information, and belief. There are no definite rules to assign such probabilities.

Subjective probability is assigned arbitrarily. It is usually influenced by the biases, preferences, and experience of the person assigning the probability.

EXERCISES

Concepts and Procedures

4.11 Briefly explain the two properties of probability.

4.12 Briefly describe an impossible event and a sure event. What is the probability of the occurrence of each of these two events?

4.13 Briefly explain the three approaches to probability. Give one example of each approach.

4.14 Briefly explain for what kind of experiments we use the classical approach to calculate probabilities of events and for what kind of experiments we use the relative frequency approach.

4.15 Which of the following values cannot be the probability of an event and why?

.46   2/3   −.09   1.42   .96   9/4   −1/4   .02

Applications

4.16 An economist says that the probability is .47 that a randomly selected adult is in favor of keeping the Social Security system as it is, .32 that this adult is in favor of totally abolishing the Social Security system, and .21 that this adult does not have any opinion or is in favor of other options. Were these probabilities obtained using the classical approach, relative frequency approach, or the subjective probability approach? Explain your answer.

4.17 The president of a company has a hunch that there is a .75 probability that the company will be successful in marketing a new brand of ice cream. Is this a case of classical, relative frequency, or subjective probability? Explain why.

4.18 A financial expert believes that the probability is .13 that the stock price of a specific technology company will double over the next year. Is this a case of classical, relative frequency, or subjective probability? Explain why.

4.19 A hat contains 50 marbles. Of them, 24 are red and 26 are green. If one marble is randomly selected out of this hat, what is the probability that this marble is

a. red?   **b.** green?

4.20 A die is rolled once. What is the probability that

a. a number less than 2 is obtained?

b. a number 3 to 6 is obtained?

4.21 A random sample of 3000 adults showed that 1980 of them have shopped at least once on the Internet. What is the (approximate) probability that a randomly selected adult has shopped on the Internet?

4.22 In a statistics class of 54 students, 34 have volunteered for community service in the past. Find the probability that a randomly selected student from this class has volunteered for community service in the past.

4.23 Jane and Mike are planning to go on a two-week vacation next summer. They have selected six vacation resorts, two of which are in Canada and remaining four are in Caribbean countries. Jane prefers going to a Canadian resort, and Mike prefers to vacation in one of the Caribbean countries. After much argument, they decide that they will put six balls of the same size, each marked with one of the six vacation resorts, in a hat. Then they will ask their eight-year-old son to randomly choose one ball from these six balls. What is the probability that a vacation resort from the Caribbean countries is selected? Is this an example of the classical approach, relative frequency approach, or the subjective probability approach? Explain your answer.

4.24 There are 1320 eligible voters in a town, and 1012 of them are registered to vote. If one eligible voter is selected at random from this town, what is the probability that this voter is

a. registered?

b. not registered?

Do these two probabilities add up to 1.0? If yes, why?

4.25 A sample of 400 large companies showed that 130 of them offer free health fitness centers to their employees on the company premises. If one company is selected at random from this sample, what is the probability that this company offers a free health fitness center to its employees on the company premises? What is the probability that this company does not offer a free health fitness center to its employees on the company premises? Do these two probabilities add to 1.0? If yes, why?

4.26 In a large city, 15,000 workers lost their jobs last year. Of them, 7400 lost their jobs because their companies closed down or moved, 4600 lost their jobs due to insufficient work, and the remainder lost their jobs because their positions were abolished. If one of these 15,000 workers is selected at random, find the probability that this worker lost his or her job

a. because the company closed down or moved

b. due to insufficient work

c. because the position was abolished

Do these probabilities add to 1.0? If so, why?

4.27 A sample of 900 adults showed that 90 of them had no credit cards, 124 had one card each, 103 had two cards each, 77 had three cards each, 51 had four cards each, and 455 had five or more cards each. Write the frequency distribution table for the number of credit cards an adult possesses. Calculate the relative frequencies for all categories. Suppose one adult is randomly selected from these 900 adults. Find the probability that this adult has

a. three credit cards b. five or more credit cards

4.28 Suppose you want to find the (approximate) probability that a randomly selected family from Los Angeles earns more than $175,000 a year. How would you find this probability? What procedure would you use? Explain briefly.

4.3 Marginal Probability, Conditional Probability, and Related Probability Concepts

LEARNING OBJECTIVES

After completing this section, you should be able to:

- Calculate marginal and conditional probabilities.
- Explain the meaning of two mutually exclusive events.
- Determine whether two events are mutually exclusive or mutually nonexclusive events.
- Explain the meaning of two independent events.
- Determine whether two events are independent or dependent events.
- Explain the meaning of the complement of an event.
- Calculate the probability of the complement of an event.

In this section, first we discuss marginal and conditional probabilities, and then we discuss the concepts of mutually exclusive events, independent and dependent events, and complementary events.

4.3.1 Marginal and Conditional Probabilities

Suppose all 100 employees of a company were asked whether they are in favor of or against paying high salaries to CEOs of U.S. companies. **Table 4.4** gives a two-way classification of the responses of these 100 employees. Assume that every employee responds either *in favor* or *against*.

Table 4.4 shows the distribution of 100 employees based on two variables or characteristics: gender (male or female) and opinion (in favor or against). Such a table is called a *contingency table* or a *two-way table*. In Table 4.4, each box that contains a number is called a *cell*. Notice that there are four cells. Each cell gives the frequency for two

TABLE 4.4 Two-Way Classification of Employee Responses

	In Favor	Against
Male	15	45
Female	4	36

characteristics. For example, 15 employees in this group possess two characteristics: "male" and "in favor of paying high salaries to CEOs." We can interpret the numbers in other cells the same way.

By adding the row totals and the column totals in Table 4.4, we write **Table 4.5**.

TABLE 4.5 Two-Way Classification of Employee Responses with Totals

	In Favor	Against	Total
Male	15	45	60
Female	4	36	40
Total	19	81	100

Suppose one employee is selected at random from these 100 employees. This employee may be classified either on the basis of gender alone or on the basis of opinion alone. If only one characteristic is considered at a time, the employee selected can be a male, a female, in favor, or against. The probability of each of these four characteristics or events is called **marginal probability**. These probabilities are called marginal probabilities because they are calculated by dividing the corresponding row margins (totals for the rows) or column margins (totals for the columns) by the grand total.

Marginal Probability

Marginal probability is the probability of a single event without consideration of any other event.

For Table 4.5, the four marginal probabilities are calculated as follows:

$$P(\text{male}) = \frac{\text{Number of males}}{\text{Total number of employees}} = \frac{60}{100} = .60$$

As we can observe, the probability that a male will be selected is obtained by dividing the total of the row labeled "Male" (60) by the grand total (100). Similarly,

$$P(\text{female}) = 40/100 = .40$$
$$P(\text{in favor}) = 19/100 = .19$$
$$P(\text{against}) = 81/100 = .81$$

These four marginal probabilities are shown along the right side and along the bottom of **Table 4.6.**

Now suppose that one employee is selected at random from these 100 employees. Furthermore, assume it is known that this (selected) employee is a male. In other words, the event that the employee selected is a male has already occurred. Given that this selected employee is

TABLE 4.6 Listing the Marginal Probabilities

	In Favor (A)	Against (B)	Total	
Male (M)	15	45	60	$P(M) = 60/100 = .60$
Female (F)	4	36	40	$P(F) = 40/100 = .40$
Total	19	81	100	
	$P(A) = 19/100 = .19$	$P(B) = 81/100 = .81$		

a male, he can be either in favor or against. What is the probability that the employee selected is in favor of paying high salaries to CEOs? This probability is written as follows:

$$P(\text{in favor} \mid \text{male})$$

Read as "given" (the vertical bar)

The event whose probability is to be determined (in favor)

This event has already occurred (male)

This probability, $P(\text{in favor} \mid \text{male})$, is called the **conditional probability** of "in favor" given that the event "male" has already happened. It is read as "the probability that the selected employee is in favor given that this employee is a male."

Conditional Probability

Conditional probability is the probability that an event will occur given that another event has already occurred. If A and B are two events, then the conditional probability of A given B is written as

$$P(A \mid B)$$

and read as "the probability of A given that B has already occurred."

EXAMPLE 4.12 | Opinions of Employees

Calculating the Conditional Probability: Two-Way Table

Compute the conditional probability $P(\text{in favor} \mid \text{male})$ for the data on 100 employees given in Table 4.5.

Solution The probability $P(\text{in favor} \mid \text{male})$ is the conditional probability that a randomly selected employee is in favor given that this employee is a male. It is known that the event "male" has already occurred. Based on the information that the selected employee is a male, we can infer that the selected employee must be one of the 60 males and, hence, must belong to the first row of Table 4.5. Therefore, we are concerned only with the first row of that table.

	In Favor	Against	Total
Male	15	45	60
	↑ Males who are in favor		↑ Total number of males

There are 60 males, and 15 of them are in favor. Hence, the required conditional probability is calculated as follows:

$$P(\text{in favor} \mid \text{male}) = \frac{\text{Number of males who are in favor}}{\text{Total number of males}} = \frac{15}{60} = .25$$

As we can observe from this computation of conditional probability, the total number of males (the event that has already occurred) is written in the denominator and the number of males who are in favor (the event whose probability we are to find) is written in the numerator. Note that we are considering the row of the event that has already occurred.

Case Study 4.1

Vegetarians, Gender, and Ideology

In a Gallup poll of 1033 randomly selected U.S. adults, aged 18 and older, conducted July 1–11, 2018, participants were asked, "In terms of your eating preferences, do you consider yourself to be a vegetarian, or not?" The responses from these adults were classified based on various characteristics of the participants such as gender, race (whites or non-whites), ideology, age, and region of the United States where they lived. The graph shows the percentages of these participants who said that they are vegetarian, classified based on gender and ideology. Of the adults included in the poll, 5% of all adults, 4% of men and 6% of women said that they are vegetarian. Based on ideology, 2% of conservatives, 3% of moderates, and 11% of liberals said that they are vegetarian.

Suppose we randomly select one U.S. adult aged 18 or older. Using the information obtained from the sample, we can estimate the following two marginal probabilities:

$$P(\text{selected person is vegetarian}) = .05$$
$$P(\text{selected person is not vegetarian}) = 1 - .05 = .95$$

Using the same sample information, we can calculate the following conditional probabilities:

$$P(\text{selected person is vegetarian} \mid \text{man}) = .04$$
$$P(\text{selected person is vegetarian} \mid \text{woman}) = .06$$
$$P(\text{selected person is vegetarian} \mid \text{conservative}) = .02$$
$$P(\text{selected person is vegetarian} \mid \text{moderate}) = .03$$
$$P(\text{selected person is vegetarian} \mid \text{liberal}) = .11$$

Note that these probabilities are approximate probabilities because they are based on a sample of 1033 adults. If these probabilities were based on a census, they would be exact probabilities.

Data Source: Adapted from www.gallup.com

EXAMPLE 4.13 | Opinions of Employees

Calculating the Conditional Probability: Two-Way Table

For the data of Table 4.5, calculate the conditional probability that a randomly selected employee is a female given that this employee is in favor of paying high salaries to CEOs.

Solution Here one employee is selected and this employee happens to be *in favor*. Given that this employee is *in favor*, the employee can be either a male or a female. We are to find the probability of a *female* given that the selected employee is *in favor*, that is, we are to compute the probability $P(\text{female} \mid \text{in favor})$. Because it is known that the employee selected is in favor of paying high salaries to CEOs, this employee must belong to the first column (the column labeled "in favor") and must be one of the 19 employees who are in favor. Of these 19 employees who are in favor, four are females.

In Favor	
15	
4	← Females who are in favor
19	← Total number of employees who are in favor

Hence, the required probability is

$$P(\text{female} \mid \text{in favor}) = \frac{\text{Number of females who are in favor}}{\text{Total number of employees who are in favor}} = \frac{4}{19} = \mathbf{.2105}$$

4.3.2 Mutually Exclusive Events

Suppose you toss a coin once. You will obtain either a head or a tail, but not both, because, in one toss, the events head and tail cannot occur together. Such events are called **mutually exclusive events.** Such events do not have any common outcomes. If two or more events are mutually exclusive, then at most one of them will occur every time we repeat the experiment. Thus the occurrence of one event excludes the occurrence of the other event or events.

Mutually Exclusive Events

Events that cannot occur together are said to be *mutually exclusive events.*

For any experiment, the final outcomes are always mutually exclusive because one and only one of these outcomes is expected to occur in one repetition of the experiment. For example, consider tossing a coin twice. This experiment has four outcomes: *HH*, *HT*, *TH*, and *TT*. These outcomes are mutually exclusive because one and only one of them will occur when we toss this coin twice. As another example, suppose one student is selected at random from a statistics class and the gender of this student is observed. This student can be a *male* or a *female*. These are two mutually exclusive events. These events cannot happen together if only one student is selected. But now, if you observe whether the selected student is a math major or a business major, then these two events will not be mutually exclusive if there is at least one student in the class who is a double major in math and business. However, if there is no student with both majors, then these two events will be mutually exclusive events.

EXAMPLE 4.14 One Roll of a Die

Illustrating Mutually Exclusive and Mutually Nonexclusive Events

Consider the following events for one roll of a die:

A = an even number is observed = {2, 4, 6}
B = an odd number is observed = {1, 3, 5}
C = a number less than 5 is observed = {1, 2, 3, 4}

Are events A and B mutually exclusive? Are events A and C mutually exclusive?

Solution **Figures 4.6** and **4.7** show the diagrams of events A and B and of events A and C, respectively.

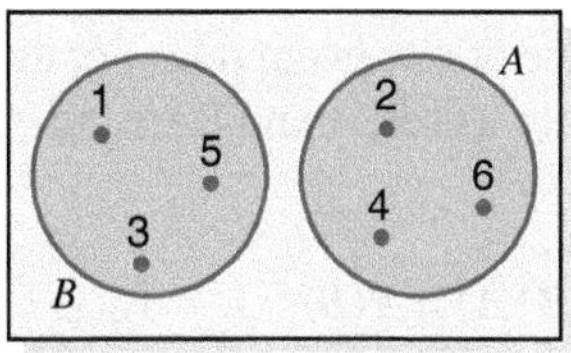

FIGURE 4.6 Mutually exclusive events A and B.

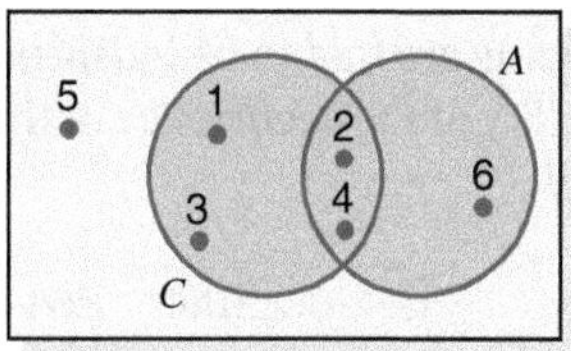

FIGURE 4.7 Mutually nonexclusive events A and C.

As we can observe from the definitions of events A and B and from Figure 4.6, events A and B have no common element. For one roll of a die, only one of the two events A and B can happen. Hence, these are two mutually exclusive events.

We can observe from the definitions of events A and C and from Figure 4.7 that events A and C have two common outcomes: 2-spot and 4-spot. Thus, if we roll a die and obtain either a 2-spot or a 4-spot, then A and C both happen at the same time. Hence, events A and C are not mutually exclusive.

EXAMPLE 4.15 Shopping on the Internet

Illustrating Mutually Exclusive Events

Consider the following two events for a randomly selected adult:

Y = this adult has shopped on the Internet at least once
N = this adult has never shopped on the Internet

Are events Y and N mutually exclusive?

Solution Note that event Y consists of all adults who have shopped on the Internet at least once, and event N includes all adults who have never shopped on the Internet. These two events are illustrated in the diagram in **Figure 4.8**.

As we can observe from the definitions of events Y and N and from Figure 4.8, events Y and N have no common outcome. They represent two distinct sets of adults: the ones who have shopped on the Internet at least once and the ones who have never shopped on the Internet. Hence, these two events are mutually exclusive.

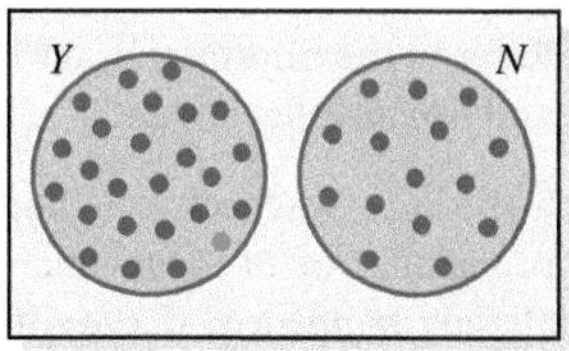

FIGURE 4.8 Mutually exclusive events Y and N.

4.3.3 Independent versus Dependent Events

Consider two tosses of a fair coin. Does the outcome of the first toss affect the outcome of the second toss? In other words, whether we obtain a head or a tail in the first toss, will it change the probability of obtaining a head or a tail in the second toss? The answer is: No. The outcome of the first toss does not affect the outcome of the second toss. Whether we get a head or a tail in the first toss, the probability of obtaining a head in the second toss is still .50 and that of a tail is also .50. This is an example of a statistical experiment where the outcomes of two tosses are independent. In the case of two **independent events**, the occurrence of one event does not change the probability of the occurrence of the other event.

Independent Events

Two events are said to be *independent* if the occurrence of one event does not affect the probability of the occurrence of the other event. In other words, *A* and *B* are *independent events* if

$$\text{either} \quad P(A \mid B) = P(A) \quad \text{or} \quad P(B \mid A) = P(B)$$

Now consider one toss of a coin. Suppose you toss this coin and obtain a head. Did obtaining a head affect the probability of obtaining a tail? Yes, it did. Before you tossed the coin, the probability of obtaining a tail was .50. But once you tossed the coin and obtained a head, the probability of obtaining a tail in that toss becomes zero. In other words, it is obvious that you did not obtain a tail. This is an example of two dependent events or outcomes. Thus, if the occurrence of one event affects the probability of the occurrence of the other event, then the two events are said to be **dependent events**. In probability notation, the two events are dependent if either $P(A \mid B) \neq P(A)$ or $P(B \mid A) \neq P(B)$.

Dependent Events

Two events are said to be *dependent* events if the occurrence of one event affects the probability of the occurrence of the other event. In other words, events *A* and *B* are dependent if

$$\text{either} \quad P(A \mid B) \neq P(A) \quad \text{or} \quad P(B \mid A) \neq P(B)$$

EXAMPLE 4.16 Opinions of Employees

Illustrating Two Dependent Events: Two-Way Table

Refer to the information on 100 employees given in Table 4.5 in Section 4.3.1. Are events *in favor* and *female* independent?

Solution To check whether events *in favor* and *female* are independent, we find the (marginal) probability of *in favor*, and then the conditional probability of *in favor* given that *female* has already happened. If these two probabilities are equal then these two events are independent, otherwise they are dependent.

From Table 4.5, there are 19 employees in favor out of 100 total employees. Hence,

$$P(\text{in favor}) = \frac{19}{100} = .19$$

Now, we find the conditional probability of *in favor* given that the event *female* has already occurred. We know that the person selected is a *female* because this event has already happened. There are a total of 40 employees who are *female* and 4 of them are *in favor*. Hence,

$$P(\text{in favor} \mid \text{female}) = \frac{4}{40} = .10$$

Because these two probabilities are not equal, the two events are dependent. Here, dependence of events means that the percentage of female employees who are in favor of paying high salaries to CEOs is different from the percentage of male employees who are in favor of this issue. This percentage of female employees who are in favor of paying high salaries to CEOs is also different from the overall percentage of employees who are in favor of paying high salaries to CEOs.

EXAMPLE 4.17 Gender and Drinking Coffee with or without Sugar

Illustrating Two Independent Events

In a survey, 500 randomly selected adults who drink coffee were asked whether they usually drink coffee with or without sugar. Of these 500 adults, 240 are men, and 175 drink coffee without sugar. Of the 240 men, 84 drink coffee without sugar. Are the events drinking coffee *without sugar* and *man* independent?

Solution To check if the events drinking coffee *without sugar* and *man* are independent, first we find the probability that a randomly selected adult from these 500 adults drinks coffee *without sugar* and then find the conditional probability that this selected adult drinks coffee *without sugar* given that this adult is a *man*. From the given information:

$$P(\text{drinks coffee without sugar}) = \frac{175}{500} = .35$$

Now to find the conditional probability that a randomly selected adult from these 500 adults drinks coffee *without sugar* given that this adult is a *man*, we divide the number of men in the adults who drink coffee without sugar by the total number of adults who drink coffee without sugar. There are a total of 175 adults who drink coffee without sugar and 84 of them are men. Also, there are a total of 240 men. Hence,

$$P(\text{drinks coffee without sugar} \mid \text{man}) = \frac{84}{240} = .35$$

Since the two probabilities are equal, the two events drinking coffee *without sugar* and *man* are independent. In this example, independence of these events means that men and women have the same preferences in regard to drinking coffee with or without sugar. **Table 4.7** has been written using the given information. The numbers in the shaded cells are given to us. If we calculate the percentages of men and women who drink coffee with and without sugar, they are the same. Sixty-five percent of both groups, men and women, drink coffee with sugar and 35% of both groups drink coffee without sugar. If the percentages of men and women who drink coffee with and without sugar are different, then these events will be dependent.

TABLE 4.7 Gender and Drinking Coffee with or without Sugar

	With Sugar	Without Sugar	Total
Men	156	84	240
Women	169	91	260
Total	325	175	500

Two Important Observations

We can make the following two important observations about mutually exclusive, independent, and dependent events.

1. Two events are either mutually exclusive or independent.[2]
   a. Mutually exclusive events are always dependent.
   b. Independent events are never mutually exclusive.
2. Dependent events may or may not be mutually exclusive.

4.3.4 Complementary Events

Consider one roll of a fair six-sided die. Let A be the event that an even number is obtained and B be the event that an odd number is obtained. Then event A includes outcomes 2, 4, and 6, and event B includes outcomes 1, 3, and 5. As we can notice, events A and B are mutually exclusive events as they do not contain any common outcomes. Taken together, events A and B include all six outcomes of this experiment. In this case, events A and B are called complementary events. Thus, we can state that two mutually exclusive events that taken together include all the outcomes for an experiment are called **complementary events.** Note that two complementary events are always mutually exclusive.

Complementary Events

The *complement of event A*, denoted by $\bar{A}$ and read as "A bar" or "A complement," is the event that includes all the outcomes for an experiment that are not in A.

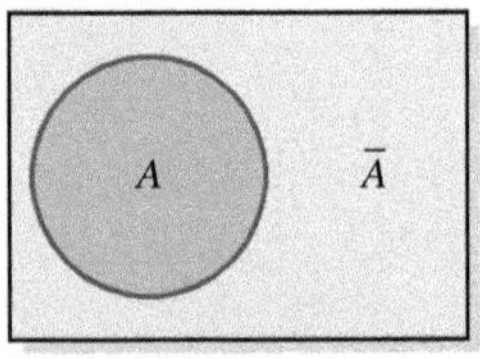

FIGURE 4.9 Two complementary events.

Events A and $\bar{A}$ are complements of each other. The diagram in **Figure 4.9** shows the complementary events A and $\bar{A}$.

Because two complementary events, taken together, include all the outcomes for an experiment and because the sum of the probabilities of all outcomes is 1, it is obvious that

$$P(A) + P(\bar{A}) = 1.0$$

From this equation, we can deduce that

$$P(A) = 1 - P(\bar{A}) \quad \text{and} \quad P(\bar{A}) = 1 - P(A)$$

Thus, if we know the probability of an event, we can find the probability of its complementary event by subtracting the given probability from 1.

EXAMPLE 4.18 Taxpayers Audited by the Taxation Bureau

Calculating Probability of Complementary Events

In a group of 2000 taxpayers, 400 have been audited by the Taxation Bureau at least once. If one taxpayer is randomly selected from this group, what are the two complementary events for this experiment, and what are their probabilities?

[2]The exception to this rule occurs when at least one of the two events has a zero probability.

Solution The two complementary events for this experiment are

A = the selected taxpayer has been audited by the Taxation Bureau at least once
$\bar{A}$ = the selected taxpayer has never been audited by the Taxation Bureau

Note that here event A includes the 400 taxpayers who have been audited by the Taxation Bureau at least once, and $\bar{A}$ includes the 1600 taxpayers who have never been audited by the Taxation Bureau. Hence, the probabilities of events A and $\bar{A}$ are

$$P(A) = \frac{400}{2000} = \mathbf{.20} \quad \text{and} \quad P(\bar{A}) = \frac{1600}{2000} = \mathbf{.80}$$

As we can observe, the sum of these two probabilities is 1. **Figure 4.10** shows the two complementary events A and $\bar{A}$ for this example.

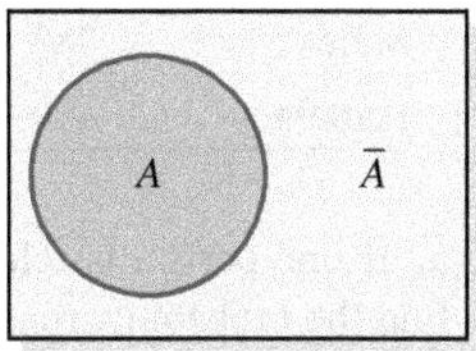

FIGURE 4.10 Complementary events A and $\bar{A}$.

EXAMPLE 4.19 Gender and Drinking Coffee with or without Sugar

Calculating Probabilities of Complementary Events

Refer to the information given in Table 4.7. Of the 500 adults, 325 drink coffee with sugar. Suppose one adult is randomly selected from these 500 adults, and let A be the event that this adult drinks coffee with sugar. What is the complement of event A? What are the probabilities of the two events?

Solution For this experiment, the two complementary events are

A = the selected adult drinks coffee with sugar
$\bar{A}$ = the selected adult drinks coffee without sugar

Note that here $\bar{A}$ includes 175 adults who drink coffee without sugar. The two events A and $\bar{A}$ are complements of each other. The probabilities of the two events are

$$P(A) = P(\text{selected adult drinks coffee with sugar}) = \frac{325}{500} = .65$$
$$P(\bar{A}) = P(\text{selected adult drinks coffee without sugar}) = \frac{175}{500} = .35$$

As we can observe, the sum of these two probabilities is 1, which should be the case with two complementary events. Also, once we find $P(A)$, we can find $P(\bar{A})$ as

$$P(\bar{A}) = 1 - P(A) = 1 - .65 = .35$$

Figure 4.11 shows these two complementary events.

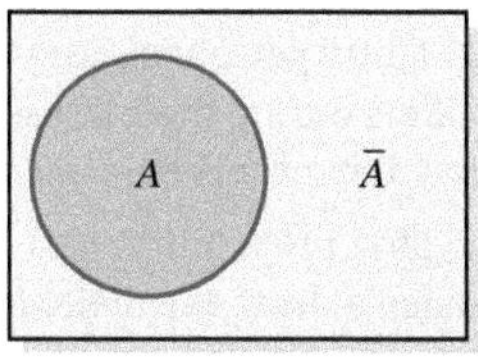

FIGURE 4.11 Complementary events drinking coffee with or without sugar.

EXERCISES

Concepts and Procedures

4.29 Briefly explain the difference between the marginal and conditional probabilities of events. Give one example of each.

4.30 What is meant by two mutually exclusive events? Give one example of two mutually exclusive events and another example of two events that are not mutually exclusive.

4.31 Briefly explain the difference between independent and dependent events. Suppose A and B are two events. What formula will you use to prove whether A and B are independent or dependent?

4.32 What is the complement of an event? What is the sum of the probabilities of two complementary events?

4.33 A statistical experiment has 10 equally likely outcomes that are denoted by 1, 2, 3, 4, 5, 6, 7, 8, 9, and 10. Consider two events: $A = \{3, 4, 6, 9\}$ and $B = \{1, 2, 5\}$.

a. Are events A and B mutually exclusive events?

b. Are events A and B independent events?

c. What are the complements of events A and B, respectively, and what are their probabilities?

Applications

4.34 Two thousand randomly selected adults from a rural area were asked whether or not they have ever shopped on the Internet. The following table gives a two-way classification of the responses.

	Have Shopped	Have Never Shopped
Male	400	800
Female	350	450

a. If one adult is selected at random from these 2000 adults, find the probability that this adult

i. has never shopped on the Internet

ii. is a male

iii. has shopped on the Internet given that this adult is a female

iv. is a male given that this adult has never shopped on the Internet

b. Are the events *male* and *female* mutually exclusive? What about the events *have shopped* and *male*? Why or why not?

c. Are the events *female* and *have shopped* independent? Why or why not?

4.35 Five hundred employees were selected from a city's large private companies, and they were asked whether or not they have any retirement benefits provided by their companies. Based on this information, the following two-way classification table was prepared.

	Have Retirement Benefits	
	Yes	No
Men	240	60
Women	135	65

a. If one employee is selected at random from these 500 employees, find the probability that this employee

i. is a woman

ii. has retirement benefits

iii. has retirement benefits given the employee is a man

iv. is a woman given that she does not have retirement benefits

b. Are the events *man* and *yes* mutually exclusive? What about the events *yes* and *no*? Why or why not?

c. Are the events *woman* and *yes* independent? Why or why not?

4.36 Two thousand randomly selected adults were asked if they are in favor of or against cloning. The following table gives the responses.

	In Favor	Against	No Opinion
Male	380	410	50
Female	400	680	80

a. If one person is selected at random from these 2000 adults, find the probability that this person is

i. in favor of cloning

ii. against cloning

iii. in favor of cloning given the person is a female

iv. a male given the person has no opinion

b. Are the events *male* and *in favor* mutually exclusive? What about the events *in favor* and *against*? Why or why not?

c. Are the events *female* and *no opinion* independent? Why or why not?

4.37 In a survey, 500 randomly selected adults who drink coffee were asked whether they usually drink coffee with or without sugar. Of these 500 adults, 290 are men and 200 drink coffee without sugar. Of the 200 who drink coffee without sugar, 130 are men. Are the events *man* and drinking coffee *without sugar* independent? (*Note*: Compare this exercise to Example 4.17.)

4.38 There are a total of 150 practicing physicians in a city. Of them, 70 are female and 20 are pediatricians. Of the 70 females, 10 are pediatricians. Are the events *female* and *pediatrician* independent? Are they mutually exclusive? Explain why or why not.

4.39 Define the following two events for two tosses of a coin:

A = at least one head is obtained
B = two tails are obtained

a. Are A and B mutually exclusive events? Are they independent? Explain why or why not.

b. Are A and B complementary events? If yes, first calculate the probability of B and then calculate the probability of A using the complementary event rule.

4.40 Let A be the event that a number greater than 2 is obtained if you roll a die once. What is the probability of A? What is the complementary event of A, and what is its probability?

4.41 Thirty percent of last year's graduates from a university received job offers during their last semester in school. What are the two complementary events here and what are their probabilities?

4.42 The probability that a randomly selected college student attended at least one major league baseball game last year is .12. What is the complementary event? What is the probability of this complementary event?

4.4 Intersection of Events and the Multiplication Rule

LEARNING OBJECTIVES

After completing this section, you should be able to:

- Explain the meaning of the intersection of two events.
- Explain the meaning of the joint probability of two or more events.
- Calculate a joint probability using the multiplication rule for independent events.
- Calculate a joint probability using the multiplication rule for dependent events.
- Calculate a joint probability from a two-way table.
- Calculate a conditional probability using the conditional probability formula.

This section discusses the intersection of two events and the application of the multiplication rule to compute the probability of the intersection of events.

4.4.1 Intersection of Events

Refer to the information given in Table 4.7. Suppose one adult is selected at random from this group of 500 adults. Consider the events a *woman* and drinking coffee *without sugar*. Then, 91 women in this group who drink coffee without sugar give the intersection of the events *woman* and drinking coffee *without sugar*. Note that the **intersection of two events** is given by the outcomes that are common to both events.

Intersection of Events

Let A and B be two events defined in a sample space. The *intersection* of A and B represents the collection of all outcomes that are common to both A and B and is denoted by

$$A \text{ and } B$$

The intersection of events A and B is also denoted by either $\boldsymbol{A \cap B}$ or $\boldsymbol{AB}$. Let

A = event that a woman is selected from the 500 adults in Table 4.7
B = event that an adult who drinks coffee without sugar is selected

Figure 4.12 illustrates the intersection of events A and B. The shaded area in this figure gives the intersection of events A and B, and it includes all the women who drink coffee without sugar.

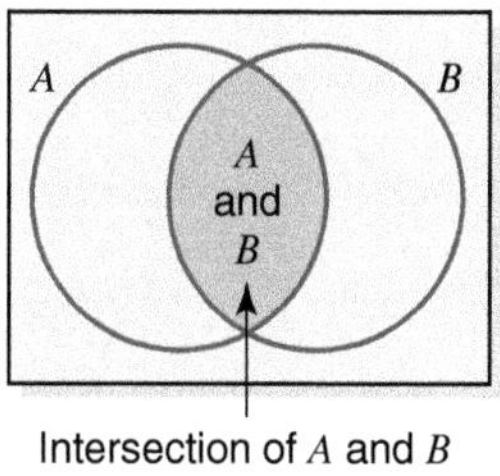

FIGURE 4.12
Intersection of events A and B.

4.4.2 Multiplication Rule for Independent Events

Sometimes we may need to find the probability that two or more events happen together, and this probability is called the **joint probability** of these events.

Joint Probability

The probability of the intersection of two events is called their *joint probability*. It is written as

$$P(A \text{ and } B) \quad \text{or} \quad P(AB)$$

If two events are independent, then they do not affect each other. To find the probability that two independent events happen together, we multiply their marginal probabilities. For example, to find the probability of obtaining two heads in two tosses of a coin, we multiply the probability of obtaining a head in the first toss by the probability of obtaining a head in the second toss since the outcomes of the two tosses are independent of each other.

Multiplication Rule to Calculate the Probability of Independent Events

The probability of the intersection of two independent events A and B is

$$P(A \text{ and } B) = P(A) \times P(B)$$

EXAMPLE 4.20 Two Fire Detectors

Calculating the Joint Probability of Two Independent Events

An office building has two fire detectors. The probability is .02 that any fire detector of this type will fail to go off during a fire. Find the probability that both of these fire detectors will fail to go off in case of a fire. Assume that these two fire detectors are independent of each other.

Solution In this example, the two fire detectors are independent because whether or not one fire detector goes off during a fire has no effect on the second fire detector. We define the following two events:

A = the first fire detector fails to go off during a fire
B = the second fire detector fails to go off during a fire

Then, the joint probability of A and B is

$$P(A \text{ and } B) = P(A) \times P(B) = (.02) \times (.02) = \mathbf{.0004}$$

The multiplication rule can be extended to calculate the joint probability of more than two events. Example 4.21 illustrates such a case for independent events.

EXAMPLE 4.21 Three Students Getting Jobs

Calculating the Joint Probability of Three Independent Events

The probability that a college graduate will find a full-time job within three months after graduation from college is .27. Three college students, who will be graduating soon, are randomly selected. What is the probability that all three of them will find full-time jobs within three months after graduation from college?

Solution Let:

A = the event that the first student finds a full-time job within three months after graduation from college

B = the event that the second student finds a full-time job within three months after graduation from college

C = the event that the third student finds a full-time job within three months after graduation from college

These three events, A, B, and C, are independent events because one student getting a full-time job within three months after graduation from college does not affect the probabilities of the other two students getting full-time jobs within three months after graduation from college. This is a reasonable assumption since the population of students graduating from colleges is very large and the students are randomly selected. Hence, the probability that all three students will find full-time jobs within three months after graduation from college is

$$P(A \text{ and } B \text{ and } C) = P(A) \times P(B) \times P(C) = (.27) \times (.27) \times (.27) = \mathbf{.0197}$$

4.4.3 Multiplication Rule for Dependent Events

We know that when the occurrence of one event affects the occurrence of another event, these events are said to be dependent events. For example, in the two-way Table 4.5, events *female* and *in favor* are dependent. Suppose we randomly select one person from the 100 persons included in Table 4.5 and we want to find the joint probability of *female* and *in favor*. To obtain this probability, we will not multiply the marginal probabilities of these two events. Here, to find the joint probability of these two events, we will multiply the marginal probability of *female* by the conditional probability of *in favor* given that a *female* has been selected.

Thus, the probability of the intersection of two dependent events is obtained by multiplying the marginal probability of one event by the conditional probability of the second event. This rule is called the **multiplication rule** for dependent events.

Multiplication Rule to Find Joint Probability of Two Dependent Events

The probability of the intersection of two dependent events A and B is given by either of these two formulas:

$$P(A \text{ and } B) = P(A) \times P(B \mid A)$$
$$P(A \text{ and } B) = P(B) \times P(A \mid B)$$

EXAMPLE 4.22 Gender and College Degree

Calculating the Joint Probability of Two Events: Two-Way Table

Table 4.8 gives the classification of all employees of a company by gender and college degree.

TABLE 4.8 Classification of Employees by Gender and Education

	College Graduate	Not a College Graduate	Total
Male	7	20	27
Female	4	9	13
Total	11	29	40

If one of these employees is selected at random for membership on the employee–management committee, what is the probability that this employee is a female and a college graduate?

Solution We are to calculate the probability of the intersection of the events *female* and *college graduate*. This probability can be computed using the formula

$$P(\textit{female} \text{ and } \textit{college graduate}) = P(\textit{female}) \times P(\textit{college graduate} \mid \textit{female})$$

The shaded area in **Figure 4.13** shows the intersection of the events *female* and *college graduate*. There are four females who are college graduates in 40 employees.

Notice that there are 13 females among 40 employees. Hence, the probability that a female is selected is

$$P(\text{female}) = \frac{13}{40}$$

To calculate the probability $P(\text{college graduate} \mid \text{female})$, we know that the event *female* has already occurred. Consequently, the selected employee is one of the 13 females. In the table, there are 4 college graduates among 13 female employees. Hence, the conditional probability of the event *college graduate* given that the event *female* has already happened is

$$P(\text{college graduate} \mid \text{female}) = \frac{4}{13}$$

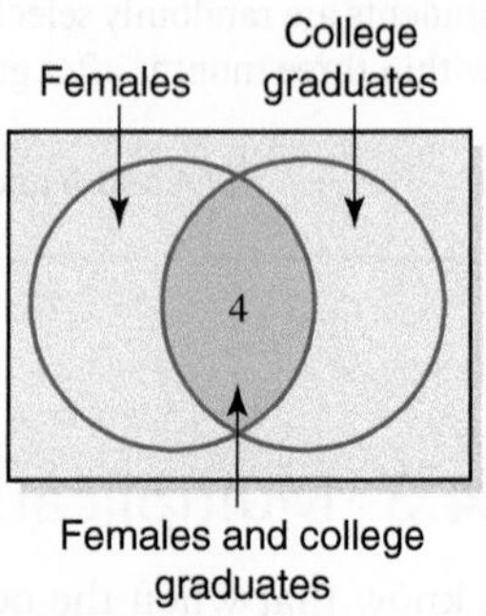

FIGURE 4.13 Intersection of events *female* and *college graduate*.

The joint probability of the events *female* and *college graduate* is

$$P(\textit{female} \text{ and } \textit{college graduate}) = P(\text{female}) \times P(\text{college graduate} \mid \text{female})$$
$$= \frac{13}{40} \times \frac{4}{13} = \mathbf{.10}$$

Thus, the probability is .10 that a randomly selected employee is a *female* and a *college graduate*.

The probability in this example can also be calculated without using the above multiplication rule. As we notice from Table 4.8:

Total employees = 40
Number of employees who are *females* and *college graduates* = 4

Hence, the joint probability of the events *female* and *college graduate* is obtained by dividing 4 by 40:

$$P(\textit{female} \text{ and } \textit{college graduate}) = \frac{4}{40} = \mathbf{.10}$$

EXAMPLE 4.23 Like Ice Tea

Calculating the Joint Probability of Two Events

In a group of 20 college students, 7 like ice tea and the others do not. Two students are randomly selected from this group.

a. Find the probability that both of the selected students like ice tea.
b. Find the probability that the first student selected likes ice tea and the second does not.

Solution In these 20 students, 7 like ice tea and 13 do not. We find the required probabilities as follows:

a. There are 7 students in 20 who like ice tea. Hence, the probability that the first student likes ice tea is

$$P(\textit{first student likes ice tea}) = \frac{7}{20} = .35$$

Now we have 19 students left in the group, of whom 6 like ice tea. Hence,

$$P(\textit{second student likes ice tea} \mid \textit{first student likes ice tea}) = \frac{6}{19} = .3158$$

Thus, the joint probability of the two events is

$$P(\text{both students like ice tea}) =$$
$$P(\textit{first student likes ice tea}) \times P(\textit{second student likes ice tea} \mid \textit{first student likes ice tea})$$
$$= (.35)(.3158) = \mathbf{.1105}$$

b. We find the probability that the first student selected likes ice tea and the second does not as follows:

$$P(\textit{first student likes ice tea}) = \frac{7}{20} = .35$$

Now we have 19 students left in the group, of whom 13 do not like ice tea. Hence,

$$P(\textit{second student does not like ice tea} \mid \textit{first student likes ice tea}) = \frac{13}{19} = .6842$$

Thus, the joint probability of the two events is

$$P(\textit{first student likes ice tea} \text{ and } \textit{second student does not like ice tea}) =$$
$$P(\textit{first student likes ice tea}) \times P(\textit{second student does not like ice tea} \mid \textit{first student likes ice tea})$$
$$= (.35) \times (.6842) = \mathbf{.2395}$$

Conditional probability was discussed in Section 4.3.1. It is obvious from the formula for joint probability that if we know the probability of an event A and the joint probability of events A and B, then we can calculate the conditional probability of B given A.

Calculating Conditional Probability

If A and B are two events, then,

$$P(B \mid A) = \frac{P(A \text{ and } B)}{P(A)} \quad \text{and} \quad P(A \mid B) = \frac{P(A \text{ and } B)}{P(B)}$$

given that $P(A) \neq 0$ and $P(B) \neq 0$.

EXAMPLE 4.24 Major and Status of a Student

Calculating the Conditional Probability of an Event

The probability that a randomly selected student from a college is a senior is .20, and the joint probability that the student is a computer science major and a senior is .03. Find the conditional probability that a student selected at random is a computer science major given that the student is a senior.

Solution From the given information, the probability that a randomly selected student is a senior is

$$P(\textit{senior}) = .20$$

The probability that the selected student is a senior and computer science major is

$$P(\textit{senior} \text{ and } \textit{computer science major}) = .03$$

Hence, the conditional probability that a randomly selected student is a computer science major given that the student is a senior is

$$P(\textit{computer science major} \mid \textit{senior}) = \frac{P(\textit{senior} \text{ and } \textit{computer science major})}{P(\textit{senior})}$$
$$= \frac{.03}{.20} = \mathbf{.15}$$

Thus, the (conditional) probability is .15 that a student selected at random is a computer science major given that the student is a senior.

4.4.4 Joint Probability of Mutually Exclusive Events

We know from an earlier discussion that two mutually exclusive events cannot happen together. Consequently, their joint probability is zero.

Joint Probability of Mutually Exclusive Events

The joint probability of two mutually exclusive events is always zero. If A and B are two mutually exclusive events, then,

$$P(A \text{ and } B) = 0$$

EXAMPLE 4.25 Car Loan Application

Illustrating the Joint Probability of Two Mutually Exclusive Events

Consider the following two events for an application filed by a person to obtain a car loan.

A = event that the loan application is approved
R = event that the loan application is rejected

What is the joint probability of A and R?

Solution The two events A and R are mutually exclusive. Either the loan application will be approved or it will be rejected. Hence,

$$P(A \text{ and } R) = \mathbf{0}$$

EXERCISES

Concepts and Procedures

4.43 Explain the meaning of the intersection of two events. Give one example.

4.44 What is meant by the joint probability of two or more events? Give one example.

4.45 How is the multiplication rule of probability for two dependent events different from the rule for two independent events?

4.46 What is the joint probability of two mutually exclusive events? Give one example.

4.47 Find the joint probability of A and B for the following:

a. $P(A) = .40$ and $P(B \mid A) = .25$

b. $P(B) = .65$ and $P(A \mid B) = .36$

4.48 Given that A and B are two independent events, find their joint probability for the following:

a. $P(A) = .61$ and $P(B) = .27$

b. $P(A) = .39$ and $P(B) = .63$

4.49 Given that A, B, and C are three independent events, find their joint probability for the following:

a. $P(A) = .20$, $P(B) = .46$, and $P(C) = .25$

b. $P(A) = .44$, $P(B) = .27$, and $P(C) = .43$

4.50 Given that $P(B) = .65$ and $P(A \text{ and } B) = .45$, find $P(A \mid B)$.

4.51 Given that $P(A \mid B) = .40$ and $P(A \text{ and } B) = .36$, find $P(B)$.

4.52 Given that $P(B \mid A) = .80$ and $P(A \text{ and } B) = .58$, find $P(A)$.

Applications

4.53 Five hundred employees were selected from a city's large private companies and asked whether or not they have any retirement benefits provided by their companies. Based on this information, the following two-way classification table was prepared.

	Have Retirement Benefits	
	Yes	No
Men	240	60
Women	135	65

a. Suppose one employee is selected at random from these 500 employees. Find the following probabilities:

i. Probability of the intersection of events *woman* and *yes*.

ii. Probability of the intersection of events *no* and *man*.

b. Mention what other joint probabilities you can calculate for this table and then find them. You may draw a tree diagram to find these probabilities.

4.54 In a sample survey, 1800 senior citizens were asked whether or not they have ever been victimized by a dishonest telemarketer. The following table gives the responses by age group (in years).

			Have Been Victimized	Have Never Been Victimized
Age	60–69	(A)	106	698
	70–79	(B)	145	447
	80 or over	(C)	61	343

a. Suppose one person is randomly selected from these senior citizens. Find the following probabilities:

i. P(have been *victimized* and C)

ii. P(have never been *victimized* and A)

b. Find $P(B \text{ and } C)$. Is this probability zero? Explain why or why not.

4.55 A consumer agency randomly selected 1800 flights for two major airlines, *A* and *B*. The following table gives the two-way classification of these flights based on airline and arrival time. Note that *less than 30 minutes late* includes flights that arrived early or on time.

	Less Than 30 Minutes Late	30 Minutes to One Hour Late	More Than One Hour Late
Airline *A*	412	401	87
Airline *B*	385	362	153

a. Suppose one flight is selected at random from these 1800 flights. Find the following probabilities:

i. *P*(*more than one hour late* and *airline A*)

ii. *P*(*airline B* and *less than 30 minutes late*)

b. Find the joint probability of events *30 minutes to one hour late* and *more than one hour late*. Is this probability zero? Explain why or why not.

4.56 In a statistics class of 54 students, 34 have volunteered for community service in the past. If two students are selected at random from this class, what is the probability that both of them have volunteered for community service in the past?

4.57 A company is to hire two new employees. They have prepared a final list of 10 candidates, all of whom are equally qualified. Of these 10 candidates, 7 are women. If the company decides to select two persons randomly from these 10 candidates, what is the probability that both of them are women?

4.58 Of the 35 students in a class, 22 are taking the class because it is a major requirement, and the other 13 are taking it as an elective. If two students are selected at random from this class, what is the probability that the first student is taking the class as an elective and the second is taking it because it is a major requirement? How does this probability compare to the probability that the first student is taking the class because it is a major requirement and the second is taking it as an elective?

4.59 In a group of 10 persons, 4 have a type A personality and 6 have a type B personality. If two persons are selected at random from this group, what is the probability that the first of them has a type A personality and the second has a type B personality?

4.60 Eight percent of all items sold by a mail-order company are returned by customers for a refund. Find the probability that of two items sold during a given hour by this company:

a. both will be returned for a refund

b. neither will be returned for a refund

4.61 In a survey of adults, 40% hold the opinion that there will be another housing bubble in the next four to six years. Three adults are selected at random.

a. What is the probability that all three adults hold the opinion that there will be another housing bubble in the next four to six years?

b. What is the probability that none of the three adults hold the opinion that there will be another housing bubble in the next four to six years?

4.62 The probability that a student graduating from Suburban State University has student loans to pay off after graduation is .70. The probability that a student graduating from this university has student loans to pay off after graduation and is a male is .38. Find the conditional probability that a randomly selected student from this university is a male given that this student has student loans to pay off after graduation.

4.63 The probability that an employee at a company is a female is .36. The probability that an employee is a female and married is .19. Find the conditional probability that a randomly selected employee from this company is married given that she is a female.

4.64 Recent uncertain economic conditions have forced many people to change their spending habits. In a recent telephone poll of 1000 adults, 629 stated that they were cutting back on their daily spending. Suppose that 322 of the 629 people who stated that they were cutting back on their daily spending said that they were cutting back *somewhat* and 97 stated that they were cutting back *somewhat* and *delaying the purchase of a new car by at least six months*. If one of the 629 people who are cutting back on their spending is selected at random, what is the probability that this adult is *delaying the purchase of a new car by at least six months* given that this adult is *cutting back on spending somewhat*?

4.65 Suppose that 20% of all adults in a small town live alone, and 8% of the adults live alone and have at least one pet. What is the probability that a randomly selected adult from this town has at least one pet given that this adult lives alone?

4.5 Union of Events and the Addition Rule

LEARNING OBJECTIVES

After completing this section, you should be able to:

- Explain the meaning of the union of two events.
- Explain the addition rule for two mutually nonexclusive events and two mutually exclusive events.
- Calculate the probability of the union of two mutually nonexclusive events using the addition rule.
- Calculate the probability of the union of two mutually nonexclusive events from a two-way table.
- Calculate the probability of the union of two mutually exclusive events using the addition rule.

This section discusses the union of events and the addition rule that is applied to compute the probability of the union of events.

4.5.1 Union of Events

Suppose one student is randomly selected from a statistics class. Consider the event that this student is a *female* or *junior*. Here, a *female* or *junior* means that this student is either a female or junior or both. This is called the union of events *female* and *junior* and is written as

female or *junior*

Thus, the **union of two events** A and B includes all outcomes that are either in A or in B or in both A and B.

Union of Events

Let A and B be two events defined in a sample space. The *union of events A and B* is the collection of all outcomes that belong either to A or to B or to both A and B and is denoted by

$$A \text{ or } B$$

The union of events A and B is also denoted by $A \cup B$. Example 4.26 illustrates the union of events A and B.

EXAMPLE 4.26 Gender and Hot Drinks

Illustrating the Union of Two Events

All 440 employees of a company were asked what hot drink they like the most. Their responses are listed in **Table 4.9**.

TABLE 4.9 Hot Drinks Employees Like

	Regular Coffee	Regular Tea	Other	Total
Male	110	70	110	290
Female	60	50	40	150
Total	170	120	150	440

Describe the union of events *female* and likes *regular coffee*.

Monkey Business Images/ShutterStock.com

Solution The union of events *female* and likes *regular coffee* includes those employees who are either female or like regular coffee or both. As we can observe from Table 4.9, there are a total of 150 females in these employees, and 170 of all employees like regular coffee. However, 60 females are included in each of these two numbers and, hence, are counted twice. To find the exact number of employees who are either female, or like regular coffee, or are both, we add 150 and 170 and then subtract 60 from this sum to avoid double counting. Thus, the number of employees who are either female or like regular coffee or both is

$$150 + 170 - 60 = 260$$

These 260 employees give the union of events *female* and likes *regular coffee*, which is shown in **Figure 4.14.**

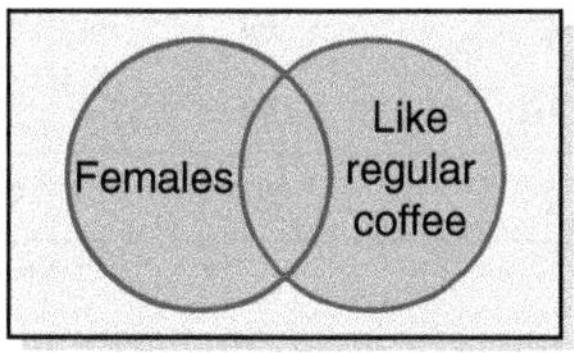

Area shaded in red gives the union of events *female* and likes *regular coffee*, and includes 260 employees who are either female or like regular coffee or both.

FIGURE 4.14 Union of events *female* and likes *regular coffee*.

4.5.2 Addition Rule for Mutually Nonexclusive Events

When two events are not mutually exclusive, which means that they can occur together, then the probability of the union of these events is given by the sum of their marginal probabilities minus their joint probability. We subtract the joint probability of the two events from the sum of their marginal probabilities due to the problem of double counting. For example, refer to Table 4.9 in Example 4.26. Suppose one employee is randomly selected from these 440 employees, and we want to find the probability of the union of events *female* and likes *regular coffee*. In other words, we are to find the probability $P(\textit{female}$ or $\textit{regular coffee})$, which means that the selected employee is a female or likes regular coffee or both. If we add the number of females and the number of employees who like regular coffee, then the employees who are female and like regular coffee (which is 60 in Table 4.9) are counted twice, once in *females* and then in like *regular coffee*. To avoid this double counting, we subtract the joint probability of the two events from the sum of the two marginal probabilities. This method of calculating the probability of the union of events is called the **addition rule.**

Addition Rule to Find the Probability of Union of Two Mutually Nonexclusive Events

The probability of the union of two mutually nonexclusive events A and B is

$$P(A \text{ or } B) = P(A) + P(B) - P(A \text{ and } B)$$

Thus, to calculate the probability of the union of two events A and B, we add their marginal probabilities and subtract their joint probability from this sum. We must subtract the joint probability of A and B from the sum of their marginal probabilities to avoid double counting because of common outcomes in A and B. This is the case where events A and B are not mutually exclusive.

EXAMPLE 4.27 Taking a Course in Ethics

Calculating the Probability of the Union of Two Events: Two-Way Table

A university president proposed that all students must take a course in ethics as a requirement for graduation. Three hundred faculty members and students from this university were asked about their opinions on this issue. **Table 4.10** gives a two-way classification of the responses of these faculty members and students.

TABLE 4.10 Two-Way Classification of Responses

	Favor	Oppose	Neutral	Total
Faculty	45	15	10	70
Student	90	110	30	230
Total	135	125	40	300

Find the probability that if one person is selected at random from these 300 persons, this person is a faculty member or is in favor of this proposal.

Solution To find the probability of the union of events *faculty* and *in favor*, we will use the formula:

$$P(\textit{faculty} \text{ or } \textit{in favor}) = P(\textit{faculty}) + P(\textit{in favor}) - P(\textit{faculty} \text{ and } \textit{in favor})$$

As we can observe from Table 4.10, there are a total of 70 faculty members and a total of 135 in favor of the proposal out of 300 total persons. Hence, the marginal probabilities of the events *faculty* and *in favor* are

$$P(\textit{faculty}) = \frac{70}{300} = .2333$$

$$P(\textit{in favor}) = \frac{135}{300} = .4500$$

We also know from the table that there are a total of 45 faculty members who are in favor out of a total of 300 persons. (Note that these 45 faculty members are counted twice.) Hence, the joint probability of the two events *faculty* and *in favor* is

$$P(\textit{faculty} \text{ and } \textit{in favor}) = \frac{45}{300} = .1500$$

The required probability is

$$\begin{aligned} P(\textit{faculty} \text{ or } \textit{in favor}) &= P(\textit{faculty}) + P(\textit{in favor}) - P(\textit{faculty} \text{ and } \textit{in favor}) \\ &= .2333 + .4500 - .1500 = \mathbf{.5333} \end{aligned}$$

Thus, the probability that a randomly selected person from these 300 persons is a faculty member or is in favor of this proposal is .5333.

The probability in this example can also be calculated without using the addition rule. The total number of persons in Table 4.10 who are either faculty members or in favor of this proposal or both is

$$45 + 15 + 10 + 90 = 160$$

Hence, the required probability is

$$P(\textit{faculty} \text{ or } \textit{in favor}) = \frac{160}{300} = \mathbf{.5333}$$

EXAMPLE 4.28 Gender and Vegetarian

Calculating the Probability of the Union of Two Events

In a group of 2500 persons, 1400 are female, 600 are vegetarian, and 400 are female and vegetarian. What is the probability that a randomly selected person from this group is a male or vegetarian?

Solution From the given information, we know that there are a total of 2500 persons in the group, 1400 of them are female, 600 are vegetarian, and 400 are female and vegetarian. Using this information, we can write **Table 4.11**. In this table, the numbers in the shaded cells are given to us. The remaining numbers are calculated by doing some arithmetic manipulations.

TABLE 4.11 Two-Way Classification Table

	Vegetarian	Nonvegetarian	Total
Female	400	1000	1400
Male	200	900	1100
Total	600	1900	2500

We are to find the probability $P(male$ or $vegetarian)$. To find this probability, we will use the following addition rule formula:

$$P(male \text{ or } vegetarian) = P(male) + P(vegetarian) - P(male \text{ and } vegetarian)$$

We find the marginal probabilities of the two events and their joint probability as follows:

$$P(male) = \frac{1100}{2500} = .44$$

$$P(vegetarian) = \frac{600}{2500} = .24$$

$$P(male \text{ and } vegetarian) = \frac{200}{2500} = .08$$

Thus, the required probability is

$$P(male \text{ or } vegetarian) = P(male) + P(vegetarian) - P(male \text{ and } vegetarian)$$
$$= .44 + .24 - .08 = \mathbf{.60}$$

We can also calculate this probability as follows. From Table 4.11:

Total number of persons who are male or vegetarian or both = 1100 + 600 − 200 = 1500

Hence, the probability of the union of events *male* and *vegetarian* is

$$P(male \text{ or } vegetarian) = \frac{1500}{2500} = \mathbf{.60}$$

4.5.3 Addition Rule for Mutually Exclusive Events

We know from an earlier discussion that the joint probability of two mutually exclusive events is zero. When A and B are mutually exclusive events, the term $P(A$ and $B)$ in the addition rule becomes zero and is dropped from the formula. Thus, the probability of the union of two mutually exclusive events is given by the sum of their marginal probabilities.

Addition Rule to Find the Probability of the Union of Mutually Exclusive Events

The probability of the union of two mutually exclusive events A and B is

$$P(A \text{ or } B) = P(A) + P(B)$$

EXAMPLE 4.29 Taking a Course in Ethics

Calculating the Probability of the Union of Two Mutually Exclusive Events: Two-Way Table

Refer to Example 4.27. A university president proposed that all students must take a course in ethics as a requirement for graduation. Three hundred faculty members and students from this university were asked about their opinion on this issue. The following table, reproduced from Table 4.10 in Example 4.27, gives a two-way classification of the responses of these faculty members and students.

	Favor	Oppose	Neutral	Total
Faculty	45	15	10	70
Student	90	110	30	230
Total	135	125	40	300

What is the probability that a randomly selected person from these 300 faculty members and students is in favor of the proposal or is neutral?

Solution As is obvious from the information given in the table, the events *in favor* and *neutral* are mutually exclusive, as they cannot happen together. Hence, their joint probability is zero. There are a total of 135 persons in favor and 40 neutral in 300 persons. The marginal probabilities of the events *in favor* and *neutral* are

$$P(\textit{in favor}) = \frac{135}{300} = .4500$$

$$P(\textit{neutral}) = \frac{40}{300} = .1333$$

The probability of the union of events *in favor* and *neutral* is

$$P(\textit{in favor} \text{ or } \textit{neutral}) = .4500 + .1333 = \mathbf{.5833}$$

Also, from the information given in the table, we observe that there are a total of 135 + 40 = 175 persons who are in favor or neutral to this proposal. Hence, the required probability is

$$P(\textit{in favor} \text{ or } \textit{neutral}) = \frac{175}{300} = \mathbf{.5833}$$

EXAMPLE 4.30 Rolling a Die

Calculating the Probability of the Union of Two Mutually Exclusive Events

Consider the experiment of rolling a die once. What is the probability that a number less than 3 or a number greater than 4 is obtained?

Solution Here, the event *a number less than 3 is obtained* happens if either 1 or 2 is rolled on the die, and the event *a number greater than 4 is obtained* happens if either 5 or 6 is rolled on the die. Thus, these two events are mutually exclusive as they do not have any common outcome and cannot happen together. Hence, their joint probability is zero. The marginal probabilities of these two events are

$$P(\textit{a number less than 3 is obtained}) = \frac{2}{6}$$

$$P(\textit{a number greater than 4 is obtained}) = \frac{2}{6}$$

The probability of the union of these two events is

$$P(\textit{a number less than 3} \text{ or } \textit{a number greater than 4 is obtained}) = \frac{2}{6} + \frac{2}{6} = \mathbf{.6667}$$

EXERCISES

Concepts and Procedures

4.66 Explain the meaning of the union of two events. Give one example.

4.67 How is the addition rule of probability for two mutually exclusive events different from the rule for two events that are not mutually exclusive?

4.68 Consider the following addition rule to find the probability of the union of two events A and B.

$$P(A \text{ or } B) = P(A) + P(B) - P(A \text{ and } B)$$

When and why is the term $P(A \text{ and } B)$ subtracted from the sum of $P(A)$ and $P(B)$? Give one example where you might use this formula.

4.69 When is the following addition rule used to find the probability of the union of two events A and B?

$$P(A \text{ or } B) = P(A) + P(B)$$

Give one example where you might use this formula.

4.70 Find $P(A \text{ or } B)$ for the following:

a. $P(A) = .66$, $P(B) = .47$, and $P(A \text{ and } B) = .33$

b. $P(A) = .92$, $P(B) = .78$, and $P(A \text{ and } B) = .65$

4.71 Given that A and B are two mutually exclusive events, find $P(A \text{ or } B)$ for the following:

a. $P(A) = .25$ and $P(B) = .27$

b. $P(A) = .58$ and $P(B) = .09$

Applications

4.72 A consumer agency randomly selected 1800 flights for two major airlines, A and B. The following table gives the two-way classification of these flights based on airline and arrival time. Note that *less than 30 minutes late* includes flights that arrived early or on time.

	Less Than 30 Minutes Late	30 Minutes to One Hour Late	More Than One Hour Late
Airline A	412	401	87
Airline B	385	362	153

If one flight is selected at random from these 1800 flights, find the following probabilities:

a. P(*more than one hour late* or *airline A*)

b. P(*airline B* or *less than 30 minutes late*)

c. P(*airline A* or *airline B*)

4.73 Two thousand randomly selected adults were asked whether or not they have ever shopped on the Internet. The following table gives a two-way classification of the responses.

	Have Shopped	Have Never Shopped
Male	400	800
Female	350	450

Suppose one adult is selected at random from these 2000 adults. Find the following probabilities:

a. P(*has never shopped on the Internet* or *is a female*)

b. P(*is a male* or *has shopped on the Internet*)

c. P(*has shopped on the Internet* or *has never shopped on the Internet*)

4.74 Two thousand randomly selected adults were asked if they think they are financially better off than their parents. The following table gives the two-way classification of the responses based on the education levels of the persons included in the survey and whether they are financially better off, the same as, or worse off than their parents.

	Less Than High School	High School	More Than High School
Better off	140	450	420
Same	60	250	110
Worse off	200	300	70

Suppose one adult is selected at random from these 2000 adults. Find the following probabilities:

a. P(*better off* or *high school*)

b. P(*more than high school* or *worse off*)

c. P(*better off* or *worse off*)

4.75 There is an area of free (but illegal) parking near an inner-city sports arena. The probability that a car parked in this area will be ticketed by police is .50, that the car will be vandalized is .30, and that it will be ticketed and vandalized is .15. Find the probability that a car parked in this area will be ticketed or vandalized.

4.76 Jason and Lisa are planning an outdoor reception following their wedding. They estimate that the probability of bad weather is .28, that of a disruptive incident (a fight breaks out, the limousine is late, etc.) is .16, and that bad weather and a disruptive incident will occur is .09. Assuming these estimates are correct, find the probability that their reception will suffer bad weather or a disruptive incident.

4.77 The probability that a randomly selected elementary or secondary school teacher from a city is a female is .66, holds a second job is .33, and is a female and holds a second job is .27. Find the probability that an elementary or secondary school teacher selected at random from this city is a female or holds a second job.

4.78 According to a survey of 2000 home owners, 900 of them own homes with three bedrooms, and 700 of them own homes with four bedrooms. If one home owner is selected at random from these 2000 home owners, find the probability that this home owner owns a home that has three or four bedrooms. Explain why this probability is not equal to 1.0.

4.79 In a Gallup Survey about the U.S. health-care system, conducted November 1–11, 2018, 15% of adults described the U.S. health-care system in a state of crisis and 55% described the U.S. health-care system having major problems. Assuming these percentages are true for the current U.S. population of adults, what is the probability that a randomly selected U.S. adult will describe the U.S. health-care system *in a state of crisis* or *having major problems*? Explain why this probability is not 1.0.

4.80 Twenty percent of a town's voters favor letting a major discount store move into their neighborhood, 63% are against it, and 17% are indifferent. What is the probability that a randomly selected voter from this town will either be against it or be indifferent? Explain why this probability is not equal to 1.0.

***4.81** The probability that a corporation makes charitable contributions is .72. Two corporations are selected at random, and it is noted whether or not they make charitable contributions. Find the probability that at most one corporation makes charitable contributions.

***4.82** The probability that an open-heart operation is successful is .85. What is the probability that in two randomly selected open-heart operations at least one will be successful?

4.6 Counting Rule, Factorials, Combinations, and Permutations

LEARNING OBJECTIVES

After completing this section, you should be able to:

- Calculate the total number of outcomes in a statistical experiment by applying the counting rule.
- Calculate the value of *n factorial* where *n* is an integer and $n \geq 0$.
- Explain the difference between the number of combinations and the number of permutations for selecting *x* elements from *n* distinct elements.
- Calculate the number of combinations by using the combinations formula.
- Calculate the number of permutations by using the permutations formula.

In this section, first we discuss the counting rule that helps us calculate the total number of outcomes for experiments, and then we learn about factorials, combinations, and permutations, respectively.

4.6.1 Counting Rule

The experiments dealt with so far in this chapter have had only a few outcomes, which were easy to list. However, for experiments with a large number of outcomes, it may not be easy to list all outcomes. In such cases, we may use the **counting rule** to find the total number of outcomes.

Counting Rule to Find Total Outcomes

If an experiment consists of three steps, and if the first step can result in *m* outcomes, the second step in *n* outcomes, and the third step in *k* outcomes, then

$$\text{Total outcomes for the experiment} = m \times n \times k$$

The counting rule can easily be extended to apply to an experiment that has fewer or more than three steps.

EXAMPLE 4.31 Three Tosses of a Coin

Applying the Counting Rule: Three Steps

Consider three tosses of a coin. How many total outcomes does this experiment have?

Solution This experiment of tossing a coin three times has three steps: the first toss, the second toss, and the third toss. Each step has two outcomes: a head and a tail. Thus,

$$\text{Total outcomes for three tosses of a coin} = 2 \times 2 \times 2 = \mathbf{8}$$

The eight outcomes for this experiment are *HHH, HHT, HTH, HTT, THH, THT, TTH*, and *TTT*.

EXAMPLE 4.32 | Taking a Car Loan

Applying the Counting Rule: Two Steps

A prospective car buyer can choose between a fixed and a variable interest rate and can also choose a payment period of 36 months, 48 months, or 60 months. How many total outcomes are possible?

Solution This experiment is made up of two steps: choosing an interest rate and selecting a loan payment period. There are two outcomes (a fixed or a variable interest rate) for the first step and three outcomes (a payment period of 36 months, 48 months, or 60 months) for the second step. Hence,

$$\text{Total outcomes} = 2 \times 3 = \mathbf{6}$$

EXAMPLE 4.33 | Outcomes of a Football Game

Applying the Counting Rule: 16 Steps

A National Football League team will play 16 games during a regular season. Each game can result in one of three outcomes: a win, a loss, or a tie. How many total outcomes are possible?

melis/Shutterstock.com

Solution The total possible outcomes for 16 games are calculated as follows:

$$\begin{aligned}\text{Total outcomes} &= 3 \cdot 3 \cdot 3 \cdot 3 \cdot 3 \cdot 3 \cdot 3 \cdot 3 \cdot 3 \cdot 3 \cdot 3 \cdot 3 \cdot 3 \cdot 3 \cdot 3 \cdot 3 \\ &= 3^{16} = \mathbf{43{,}046{,}721}\end{aligned}$$

One of the 43,046,721 possible outcomes is all 16 wins.

4.6.2 Factorials

The symbol ! (read as *factorial*) is used to denote **factorials**. The value of the factorial of a number is obtained by multiplying all the integers from that number to 1. For example, 7! is read as "seven factorial" and is evaluated by multiplying all the integers from 7 to 1.

Factorials

The symbol $n!$, read as "n factorial," represents the product of all the integers from n to 1. In other words,

$$n! = n(n-1)(n-2)(n-3)\cdots 3 \cdot 2 \cdot 1$$

By definition,

$$0! = 1$$

Note that some calculators use $r!$ instead of $n!$ on the factorial key.

Thus, the factorial of a number is given by the product of all integers from that number to 1.

EXAMPLE 4.34	Factorial of a Number

Evaluating a Factorial

Evaluate 7!.

Solution To evaluate 7!, we multiply all the integers from 7 to 1.

$$7! = 7 \cdot 6 \cdot 5 \cdot 4 \cdot 3 \cdot 2 \cdot 1 = \mathbf{5040}$$

Thus, the value of 7! is 5040.

EXAMPLE 4.35	Factorial of a Number

Evaluating a Factorial

Evaluate 10!.

Solution The value of 10! is given by the product of all the integers from 10 to 1. Thus,

$$10! = 10 \cdot 9 \cdot 8 \cdot 7 \cdot 6 \cdot 5 \cdot 4 \cdot 3 \cdot 2 \cdot 1 = \mathbf{3{,}628{,}800}$$

EXAMPLE 4.36	Factorial of the Difference of Two Numbers

Evaluating a Factorial of the Difference Between Two Numbers

Evaluate (12 – 4)!.

Solution The value of (12 – 4)! is

$$(12 - 4)! = 8! = 8 \cdot 7 \cdot 6 \cdot 5 \cdot 4 \cdot 3 \cdot 2 \cdot 1 = \mathbf{40{,}320}$$

EXAMPLE 4.37	Factorial of Zero

Evaluating a Factorial of Zero

Evaluate (5 – 5)!.

Solution The value of (5 – 5)! is 1.

$$(5 - 5)! = 0! = \mathbf{1}$$

Note that 0! is always equal to 1.

Statistical software and most calculators can be used to find the values of factorials. Check if your calculator can evaluate factorials.

4.6.3 Combinations

Quite often we face the problem of selecting a few elements from a larger group of distinct elements in such a way that the order of the selection does not matter. For example, a student may be required to attempt any two questions out of four in an examination. As another example, the faculty in a department may need to select three professors from 20 to form a committee, or a lottery player may have to pick six numbers from 49. The question arises: In how many ways can we make the selections in each of these examples? For instance, how many possible selections exist for the student who is to choose any two questions out of four? The answer is six. Suppose the four questions are denoted by the numbers 1, 2, 3, and 4. Then the six selections are

(1 and 2) (1 and 3) (1 and 4) (2 and 3) (2 and 4) (3 and 4)

The student can choose questions 1 and 2, or 1 and 3, or 1 and 4, and so on. Note that in combinations, all selections are made without replacement.

Each of the possible selections in the given list is called a **combination**. All six combinations are distinct; that is, each combination contains a different set of questions. It is important to remember that the order in which the selections are made is not important in the case of combinations. Thus, whether we write (1 and 2) or (2 and 1), both of these arrangements represent the same combination.

Combinations Notation

Combinations give the number of ways that x elements can be selected from n elements when the order of selection does not matter. The notation used to denote the total number of combinations is

$$_nC_x$$

which is read as "the number of combinations of n elements selected x at a time."

Note that some calculators use r instead of x, so that the combinations notation then reads $_nC_r$.

Suppose there are a total of n elements from which we want to select x elements. Then,

n denotes the total number of elements

$_nC_x$ = the number of combinations of n elements selected x at a time

x denotes the number of elements selected

Number of Combinations

The *number of combinations* for selecting x from n distinct elements is given by the formula

$$_nC_x = \frac{n!}{x!(n-x)!}$$

where $n!$, $x!$, and $(n-x)!$ are read as "n factorial," "x factorial," and "n minus x factorial," respectively.

In the combinations formula, n represents the total items and x is the number of items selected out of n. Note that in combinations, n is always greater than or equal to x. If n is less than x, then we cannot select x distinct elements from n.

EXAMPLE 4.38 Choosing Ice Cream Flavors

Finding the Number of Combinations Using the Formula

An ice cream parlor has six flavors of ice cream. Kristen wants to buy two flavors of ice cream. If she randomly selects two flavors out of six, how many combinations are possible?

Solution For this example,

$$n = \text{total number of ice cream flavors} = 6$$
$$x = \text{number of ice cream flavors to be selected} = 2$$

Therefore, the number of ways in which Kristen can select two flavors of ice cream out of six is

$${}_6C_2 = \frac{6!}{2!\,(6-2)!} = \frac{6!}{2!\,4!} = \frac{6 \cdot 5 \cdot 4 \cdot 3 \cdot 2 \cdot 1}{2 \cdot 1 \cdot 4 \cdot 3 \cdot 2 \cdot 1} = \mathbf{15}$$

Thus, there are 15 ways for Kristen to select two ice cream flavors out of six.

EXAMPLE 4.39 Selecting a Committee

Finding the Number of Combinations and Listing Them

Three members of a committee will be randomly selected from five people. How many different combinations are possible?

Solution There are a total of five persons, and we are to select three of them. Hence,

$$n = 5 \quad \text{and} \quad x = 3$$

Applying the combinations formula, we get

$${}_5C_3 = \frac{5!}{3!\,(5-3)!} = \frac{5!}{3!\,2!} = \frac{120}{6 \cdot 2} = \mathbf{10}$$

If we label the five persons as A, B, C, D, and E, then the 10 possible combinations for the selection of a three-member committee are

ABC ABD ABE ACD ACE ADE BCD BCE BDE CDE

Note that the number of combinations of n items selecting x at a time gives the number of samples of size x that can be selected from a population with n members. For example, in Example 4.39, the population is made of five people, and we are to select a sample of three people to form a committee. There are 10 possible samples of three people each that can be selected from this population of five people.

EXAMPLE 4.40 Selecting Candidates for an Interview

Using the Combinations Formula

Marv & Sons advertised to hire a financial analyst. The company has received applications from 10 candidates who seem to be equally qualified. The company manager has decided to call only three of these candidates for an interview. If he randomly selects 3 candidates from the 10, how many total selections are possible?

Solution The total number of ways to select three applicants from 10 is given by ${}_{10}C_3$. Here, $n = 10$ and $x = 3$. We find the number of combinations as follows:

$${}_{10}C_3 = \frac{10!}{3!\,(10-3)!} = \frac{10!}{3!\,7!} = \frac{3{,}628{,}800}{(6)(5040)} = \mathbf{120}$$

Thus, the company manager can select three applicants from 10 in 120 ways.

Case Study 4.2 describes the probability of winning a Mega Million Lottery jackpot.

Case Study 4.2

Probability of Winning a Mega Millions Lottery Jackpot

Large jackpot lotteries became popular in the United State during the 1970s and 1980s. The introduction of the Powerball lottery in 1992 resulted in the growth of multi-jurisdictional lotteries, as well as the development of a second extensive multi-jurisdictional game called Mega Millions in 2002. Both of these games are offered in many states, the District of Columbia, and the U.S. Virgin Islands resulting in multiple jackpots of more than $300 million.

Both of these games operate on a similar premise. There are two bins—one containing white balls, and one containing red (Powerball) or gold (Mega Millions) balls. When a player fills out a ticket, he or she selects five numbers from the set of white balls (1–69 for Powerball, 1–70 for Mega Millions) and one number from the set of red (1–26) or gold (1–25) balls, depending on the game. Prizes awarded to players are based on how many balls of each color are matched. If all five white ball numbers and the colored ball number are matched, the player wins the jackpot. If more than one player matches all of the numbers, then the jackpot is divided among them. The following table lists the various prizes for the Mega Millions lottery.

Number of White Balls Matched	Number of Gold Balls Matched	Prize	Number of White Balls Matched	Number of Gold Balls Matched	Prize
5	1	Jackpot	3	0	$10
5	0	$1,000,000	2	1	$10
4	1	$10,000	1	1	$4
4	0	$500	0	1	$2
3	1	$200			

The probability of winning each of the various prizes listed in the table for the Mega Millions lottery can be calculated using combinations. To find the probability of winning the jackpot, first we need to calculate the number of ways to draw five white ball numbers from 70 and one gold ball number from 25. These combinations are, respectively,

$${}_{70}C_5 = 12{,}103{,}014 \quad \text{and} \quad {}_{25}C_1 = 25$$

To obtain the total number of ways to select six numbers (five white ball numbers and one gold ball number), we multiply the two numbers obtained above, which gives us 12,103,014 × 25 = 302,575,350. Thus, there are 302,575,350 different sets of five white ball numbers and one gold ball number that can be drawn. Then, the probability that a player with one ticket wins the jackpot is

$$\begin{aligned} P(\text{winning the jackpot}) &= 1/302{,}575{,}350 \\ &= .000000003305 \end{aligned}$$

To calculate the probability of winning each of the other prizes, we calculate the number of ways any prize can be won and divide it by 302,575,350. For example, to win a prize of $10,000, a player needs to match four white ball numbers and the gold ball number. As shown below, there are 350 ways to match four white ball numbers and one gold ball number.

$${}_5C_4 \times {}_{65}C_1 \times {}_1C_1 \times {}_{24}C_0 = 5 \times 65 \times 1 \times 1 = 325$$

Here ${}_5C_4$ gives the number of ways to match four of the five winning white ball numbers, ${}_{65}C_1$ gives the number of ways to match one of the 65 nonwinning white ball numbers, ${}_1C_1$ gives the number of ways to match the winning gold ball number, and ${}_{24}C_0$ gives the number of ways to match none of the 24 nonwinning gold ball numbers. Then, the probability of winning a prize of $10,000 is

$$\begin{aligned} P(\text{winning a 10,000 prize}) &= 325/302{,}575{,}350 \\ &= .000001074 \end{aligned}$$

We can calculate the probabilities of winning the other prizes listed in the table in the same way.

Note: The information about Mega Millions lottery used in this case study is true at the time of revision of this text to the 10th edition. Note that the structure of Mega Millions lottery is changed from time to time.

Source: Adapted from https://www.megamillions.com/How-to-Play.aspx

Statistical software and many calculators can be used to find combinations. Check to see whether your calculator can do so.

Remember

If the total number of elements and the number of elements to be selected are the same, then there is only one combination. In other words,

$$_nC_n = 1$$

Also, the number of combinations for selecting zero items from n is 1; that is,

$$_nC_0 = 1$$

For example,

$$_5C_5 = \frac{5!}{5!\,(5-5)!} = \frac{5!}{5!\,0!} = \frac{120}{(120)\,(1)} = 1$$

$$_8C_0 = \frac{8!}{0!\,(8-0)!} = \frac{8!}{0!\,8!} = \frac{40{,}320}{(1)\,(40{,}320)} = 1$$

4.6.4 Permutations

The concept of permutations is very similar to that of combinations but with one major difference—here the order of selection is important. Suppose there are three marbles in a jar—red, green, and purple—and we select two marbles from these three. When the order of selection is not important, as we know from the previous section, there are three ways (combinations) to do so. Those three ways are RG, RP, and GP, where R represents that a red marble is selected, G means a green marble is selected, and P indicates a purple marble is selected. In these three combinations, the order of selection is not important, and, thus, RG and GR represent the same selection. However, if the order of selection is important, then RG and GR are not the same selections, but they are two different selections. Similarly, RP and PR are two different selections, and GP and PG are two different selections. Thus, if the order in which the marbles are selected is important, then there are six selections—RG, GR, RP, PR, GP, and PG. These are called six **permutations** or **arrangements**.

Permutations Notation

Permutations give the total number of selections of x elements from n (different) elements in such a way that the order of selection is important. The notation used to denote the permutations is

$$_nP_x$$

which is read as "the number of permutations of selecting x elements from n elements." Permutations are also called *arrangements*.

Number of Permutations

The following formula is used to find the number of permutations or arrangements of selecting x items out of n items. Note that here, the n items must all be different.

$$_nP_x = \frac{n!}{(n-x)!}$$

Example 4.41 shows how to apply this formula.

EXAMPLE 4.41 Selecting Office Holders of a Club

Finding the Number of Permutations Using the Formula

A club has 20 members. They are to select three office holders—president, secretary, and treasurer—for next year. They always select these office holders by drawing three names randomly from the names of all members. The first person selected becomes the president, the second is the secretary, and the third one takes over as treasurer. Thus, the order in which three names are selected from the 20 names is important. Find the total number of arrangements of three names from these 20.

Solution For this example,

$$n = \text{total members of the club} = 20$$
$$x = \text{number of names to be selected} = 3$$

Since the order of selections is important, we find the number of permutations or arrangements using the following formula:

$$_{20}P_3 = \frac{20!}{(20-3)!} = \frac{20!}{17!} = \mathbf{6840}$$

Thus, there are 6840 permutations or arrangements for selecting three names out of 20.

Statistical software and many calculators can find permutations. Check to see whether your calculator can do it.

EXERCISES

Concepts and Procedures

4.83 How many different outcomes are possible for four rolls of a die?

4.84 How many different outcomes are possible for 10 tosses of a coin?

4.85 Determine the value of each of the following using the appropriate formula:

$3!$ $(9-3)!$ $9!$ $(14-12)!$ $_5C_3$

$_7C_4$ $_9C_3$ $_4C_0$ $_3C_3$ $_6P_2$ $_8P_4$

Applications

4.86 A small ice cream shop has 12 flavors of ice cream and six kinds of toppings for its sundaes. How many different selections of one flavor of ice cream and one kind of topping are possible?

4.87 A restaurant menu has three kinds of soups, seven kinds of main courses, four kinds of desserts, and five kinds of drinks. If a customer randomly selects one item from each of these four categories, how many different outcomes are possible?

4.88 A ski patrol unit has eight members available for duty, and two of them are to be sent to rescue an injured skier. In how many ways can two of these eight members be selected? Now suppose the order of selection is important. How many arrangements are possible in this case?

4.89 An ice cream shop offers 25 flavors of ice cream. How many ways are there to select 3 different flavors from these 25 flavors? How many permutations are possible?

4.90 A veterinarian assigned to a racetrack has received a tip that one or more of the 12 horses in the third race have been doped. She has time to test only three horses. How many ways are there to randomly select 3 horses from these 12 horses? How many permutations are possible?

4.91 An investor will randomly select 6 stocks from 20 for an investment. How many total combinations are possible? If the order in which stocks are selected is important, how many permutations will there be?

4.92 A company employs a total of 16 workers. The management has asked these employees to select two workers who will negotiate a new contract with management. The employees have decided to select these two workers randomly. How many total selections are possible? Considering that the order of selection is important, find the number of permutations.

4.93 In how many ways can a sample (without replacement) of 9 items be selected from a population of 20 items?

Uses and Misuses of Statistics

Odds and Probability

One of the first things we learn in probability is that the sum of the probabilities of all outcomes for an experiment must equal 1.0. We also learn about the probabilities that are developed from relative frequencies and about subjective probabilities. In the latter case, many of the probabilities involve personal opinions of experts in the field. Still, both scenarios (probabilities obtained from relative frequencies and subjective probabilities) require that all probabilities must be non-negative and the sum of the probabilities of all (simple) outcomes for an experiment must equal 1.0.

Although probabilities and probability models are all around us—in weather prediction, medicine, financial markets, and so forth—they are most obvious in the world of gaming and gambling. Sports betting agencies publish odds of each team winning a specific game or championship. The table here gives the odds as of February 3, 2020 of each National Football League team to win Super Bowl 55 that was to be held in February 2021. These odds were obtained from the website https://www.cbssports.com/nfl.

Note that the odds listed in this table are called the odds in favor of winning the Super Bowl. For example, the defending Champion, Kansas City Chiefs (who won Super Bowl 54 on February 2, 2020), had 1:7 (read as 1 to 7) odds to win Super Bowl 55. If we switch the numbers around, we can state that the odds were 7:1 (read as 7 to 1) against the Kansas City Chiefs to win Super Bowl 55.

How do we convert these odds into probabilities? Let us consider the Kansas City Chiefs. Odds of 1:7 imply that out of eight chances, there is one chance that the Chiefs would win Super Bowl 55 and seven chances that the Chiefs would not win Super Bowl 55. Thus, the probability that the Chiefs would win Super Bowl 55 is $\frac{1}{1+7} = \frac{1}{8} =$ .125 and the probability that Chiefs would not win Super Bowl 55 is $\frac{7}{1+7} = \frac{7}{8} = .875$. Similarly, for the Chicago Bears, the probability of winning Super Bowl 55 is $\frac{1}{1+30} = \frac{1}{31} = .0323$ and the probability of not winning Super Bowl 55 is $\frac{30}{1+30} = \frac{30}{31} = .9677$. We can calculate these probabilities for all teams listed in the table by using this procedure.

Note that here the 32 outcomes (that each team would win Super Bowl 55) are mutually exclusive events because it is impossible for two or more teams to win the Super Bowl during the same year. Hence, if we add the probabilities of winning Super Bowl 55 for all teams, we should obtain a value of 1.0 according to the second property of probability. However, if you calculate the probability of winning Super Bowl 55 for each of the 32 teams using the odds given in the table and then add all these probabilities, the sum is greater than 1.0. So, what happened? Did these odd makers flunk their statistics and probability courses? Probably not.

Casinos and odd makers, who are in the business of making money, are interested in encouraging people to gamble. These probabilities, which seem to violate the basic rule of probability theory, still obey the primary rule for the casinos, which is that, on average, a casino is going to make a profit.

Note: When casinos create odds for sports betting, they recognize that many people will bet on one of their favorite teams, such as the Dallas Cowboys or the Pittsburgh Steelers. To meet the rule that the sum of all of the probabilities is 1.0, the probabilities for the teams more likely to win would have to be lowered. Lowering a probability corresponds to lowering the odds. For example, if the odds for the Kansas City Chiefs had been lowered from 1:7 to 1:20, the probability for them to win would have decreased from .125 to .0476.

Team	Odds	Team	Odds
Arizona Cardinals	1:75	Las Vegas Raiders	1:30
Atlanta Falcons	1:30	Los Angeles Chargers	1:30
Baltimore Ravens	1:7	Los Angeles Rams	1:25
Buffalo Bills	1:30	Miami Dolphins	1:100
Caroline Panthers	1:60	Minnesota Vikings	1:25
Chicago Bears	1:30	New York Giants	1:60
Cincinnati Bengals	1:125	New York Jets	1:60
Cleveland Browns	1:40	New England Patriots	1:12
Dallas Cowboys	1:20	New Orleans Saints	1:10
Denver Broncos	1:50	Philadelphia Eagles	1:15
Detroit Lions	1:60	Pittsburgh Steelers	1:25
Green Bay Packers	1:15	San Francisco 49ers	1:7
Houston Texans	1:25	Seattle Seahawks	1:18
Indianapolis Colts	1:40	Tampa Bay Buccaneers	1:50
Jacksonville Jaguars	1:60	Tennessee Titans	1:25
Kansas City Chiefs	1:7	Washington Redskins	1:100

If the Chiefs had remained as one of the favorites, many people would have bet on them. However, if they had won, the casino would have paid $20, instead of $7, for every $1 bet. The casinos do not want to do this and, hence, they ignore the probability rule in order to make more money. However, the casinos cannot do this with their traditional games, which are bound by the standard rules. From a mathematical standpoint, it is not acceptable to ignore the rule that the probabilities of all final outcomes for an experiment must add to 1.0. (Tampa Bay Buccaneers won Super Bowl 55 in February 2021.)

Glossary

Classical probability rule The method of assigning probabilities to outcomes or events of an experiment with equally likely outcomes.

Combinations The number of ways x elements can be selected from n elements in such a way that the order of selection is not important.

Complementary events Two events that taken together include all the outcomes for an experiment but do not contain any common outcome.

Compound event An event that contains more than one outcome of an experiment. It is also called a **composite event**.

Conditional probability The probability of an event subject to the condition that another event has already occurred.

Dependent events Two events for which the occurrence of one changes the probability of the occurrence of the other.

Equally likely outcomes Two (or more) outcomes or events that have the same probability of occurrence.

Event A collection of one or more outcomes of an experiment.

Experiment A process with well-defined outcomes that, when performed, results in one and only one of the outcomes per repetition.

Factorial Denoted by the symbol !. The product of all the integers from a given number to 1. For example, $n!$ (read as "n factorial") represents the product of all the integers from n to 1.

Impossible event An event that cannot occur.

Independent events Two events for which the occurrence of one does not change the probability of the occurrence of the other.

Intersection of events The intersection of events is given by the outcomes that are common to two (or more) events.

Joint probability The probability that two (or more) events occur together.

Law of Large Numbers If an experiment is repeated again and again, the probability of an event obtained from the relative frequency approaches the actual or theoretical probability.

Marginal probability The probability of one event or characteristic without consideration of any other event.

Mutually exclusive events Two or more events that do not contain any common outcome and, hence, cannot occur together.

Outcome The result of the performance of an experiment.

Permutations Number of arrangements of x items selected from n items in such a way that the order of selection is important.

Probability A numerical measure of the likelihood that a specific event will occur.

Relative frequency as an approximation of probability Probability assigned to an event based on the results of an experiment or based on historical data.

Sample point An outcome of an experiment.

Sample space The collection of all sample points or outcomes of an experiment.

Simple event An event that contains one and only one outcome of an experiment. It is also called an **elementary event**.

Subjective probability The probability assigned to an event based on the information and judgment of a person.

Sure event An event that is certain to occur.

Tree diagram A diagram in which each outcome of an experiment is represented by a branch of a tree.

Union of two events Given by the outcomes that belong either to one or to both events.

Exercises

Supplementary Exercises

4.94 A car rental agency currently has 52 cars available, 38 of which have a GPS navigation system. One of the 52 cars is selected at random. Find the probability that this car

a. has a GPS navigation system

b. does not have a GPS navigation system

4.95 A random sample of 250 adults was taken, and they were asked whether they prefer watching sports or drama shows on television. The following table gives the two-way classification of these adults.

	Prefer Watching Sports	Prefer Watching Drama
Male	104	18
Female	55	73

a. If one adult is selected at random from this group, find the probability that this adult

i. prefers watching drama shows

ii. is a male

iii. prefers watching sports given that the adult is a female

iv. is a male given that he prefers watching sports

v. is a *female* and *prefers watching drama shows*

vi. *prefers watching sports* or *is a male*

b. Are the events *female* and *prefers watching sports* independent? Are they mutually exclusive? Explain why or why not.

4.96 A random sample of 100 lawyers was taken, and they were asked if they are in favor of or against capital punishment. The following table gives the two-way classification of their responses.

	Favors Capital Punishment	Opposes Capital Punishment
Male	44	38
Female	11	7

a. If one lawyer is randomly selected from this group, find the probability that this lawyer

i. favors capital punishment

ii. is a female

iii. opposes capital punishment given that the lawyer is a female

iv. is a male given that he favors capital punishment

v. is a *female* and *favors capital punishment*

vi. *opposes capital punishment* or *is a male*

b. Are the events *female* and *opposes capital punishment* independent? Are they mutually exclusive? Explain why or why not.

4.97 A random sample of 400 college students was asked if college athletes should be paid. The following table gives a two-way classification of the responses.

	Should Be Paid	Should Not Be Paid
Student athlete	110	20
Student nonathlete	170	100

a. If one student is randomly selected from these 400 students, find the probability that this student

i. is in favor of paying college athletes

ii. favors paying college athletes given that the student selected is a nonathlete

iii. is an *athlete* and *favors paying student athletes*

iv. is a *nonathlete* or is *against paying student athletes*

b. Are the events *student athlete* and *should be paid* independent? Are they mutually exclusive? Explain why or why not.

4.98 In a Gallup poll conducted during February 12–28, 2019 and April 1–9, 2019, U.S. adults aged 18 and older were asked to rate their current financial situation as excellent, good, only fair, or poor. Of these adults, 12% rated their current financial situation as excellent. Assume that the results of this survey are true for the current population of U.S. adults aged 18 and over. Suppose two U.S. adults aged 18 or over are selected. Find the following probabilities:

a. Both adults will rate their current financial situation as excellent.

***b.** Exactly one adult will rate his/her current financial situation as excellent.

4.99 A car rental agency currently has 44 cars available, 28 of which have a GPS navigating system. Two cars are selected at random from these 44 cars. Find the probability that both of these cars have GPS navigation systems.

4.100 A company has installed a generator to back up the power in case there is a power failure. The probability that there will be a power failure during a snowstorm is .40. The probability that the generator will stop working during a snowstorm is .08. What is the probability that during a snowstorm the company will lose both sources of power? Assume that the two sources of power are independent.

4.101 Terry & Sons makes bearings for autos. The production system involves two independent processing machines so that each bearing passes through these two processes. The probability that the first processing machine is not working properly at any time is .09, and the probability that the second machine is not working properly at any time is .05. Find the probability that both machines will not be working properly at any given time.

Advanced Exercises

4.102 A player plays a roulette game in a casino by betting on a single number each time. Because the wheel has 38 numbers, the probability that the player will win in a single play is 1/38. Note that each play of the game is independent of all previous plays.

a. Find the probability that the player will win for the first time on the fifth play.

b. Find the probability that it takes the player more than 80 plays to win for the first time.

c. A gambler claims that because she has 1 chance in 38 of winning each time she plays, she is certain to win at least once if she plays 38 times. Does this sound reasonable to you? Find the probability that she will win at least once in 38 plays.

4.103 A certain state's auto license plates have three letters of the alphabet followed by a three-digit number.

a. How many different license plates are possible if all three-letter sequences are permitted and any number from 000 to 999 is allowed?

b. Arnold witnessed a hit-and-run accident. He knows that the first letter on the license plate of the offender's car was a B, that the second letter was an O or a Q, and that the last number was a 5. How many of this state's license plates fit this description?

4.104 Powerball is a game of chance that has generated intense interest because of its large jackpots. To play this game, a player selects five different numbers from 1 through 69, and then picks a Powerball number from 1 through 26. The lottery organization randomly draws five different white balls from 69 balls numbered 1 through 69, and then randomly picks a red Powerball number from 1 through 26.

Note that it is possible for the Powerball number to be the same as one of the first five numbers.

a. If a player's first five numbers match the numbers on the five white balls drawn by the lottery organization and the player's red Powerball number matches the Powerball number drawn by the lottery organization, the player wins the jackpot. Find the probability that a player who buys one ticket will win the jackpot. (Note that the order in which the five white balls are drawn is unimportant.)

b. If a player's first five numbers match the numbers on the five white balls drawn by the lottery organization, the player wins about $1,000,000. Find the probability that a player who buys one ticket will win this prize.

4.105 A box contains 12 red marbles and 12 green marbles.

a. Sampling at random from this box six times with replacement, you have drawn a red marble all six times. What is the probability of drawing a red marble the seventh time?

b. Sampling at random from this box six times without replacement, you have drawn a red marble all six times. Without replacing any of the marbles, what is the probability of drawing a red marble the seventh time?

c. You have tossed a fair coin five times and have obtained heads all five times. A friend argues that according to the law of averages, a tail is due to occur and, hence, the probability of obtaining a head on the sixth toss is less than .50. Is she right? Is coin tossing mathematically equivalent to the procedure mentioned in part (a) or the procedure mentioned in part (b) above? Explain.

4.106 A thief has stolen Roger's automated teller machine (ATM) card. The card has a four-digit personal identification number (PIN). The thief knows that the first two digits are 3 and 5, but he does not know the last two digits. Thus, the PIN could be any number from 3500 to 3599. To protect the customer, the ATM will not allow more than three unsuccessful attempts to enter the PIN. After the third wrong PIN, the machine keeps the card and allows no further attempts.

a. What is the probability that the thief will find the correct PIN within three tries? (Assume that the thief will not try the same wrong PIN twice.)

b. If the thief knew that the first two digits were 3 and 5 and that the third digit was either 1 or 7, what is the probability of the thief guessing the correct PIN in three attempts?

4.107 A gambler has given you two jars and 20 marbles. Of these 20 marbles, 10 are red and 10 are green. You must put all 20 marbles in these two jars in such a way that each jar must have at least one marble in it. Then a friend of yours, who is blindfolded, will select one of the two jars at random and then will randomly select a marble from this jar. If the selected marble is red, you and your friend win $100.

a. If you put five red marbles and five green marbles in each jar, what is the probability that your friend selects a red marble?

b. If you put two red marbles and two green marbles in one jar and the remaining marbles in the other jar, what is the probability that your friend selects a red marble?

c. How should these 20 marbles be distributed among the two jars in order to give your friend the highest possible probability of selecting a red marble?

4.108 A screening test for a certain disease is prone to giving false positives or false negatives. If a patient being tested has the disease, the probability that the test indicates a (false) negative is .14. If the patient does not have the disease, the probability that the test indicates a (false) positive is .08. Assume that 3% of the patients being tested actually have the disease. Suppose that one patient is chosen at random and tested. Find the probability that

a. this patient has the disease and tests positive

b. this patient does not have the disease and tests positive

c. this patient tests positive

d. this patient has the disease given that he or she tests positive

(*Hint:* A tree diagram may be helpful in part (c).)

4.109 A pizza parlor has 15 different toppings available for its pizzas, and 2 of these toppings are pepperoni and anchovies. If a customer picks two toppings at random, find the probability that

a. neither topping is anchovies

b. pepperoni is one of the toppings

4.110 A restaurant chain is planning to purchase 100 ovens from a manufacturer, provided that these ovens pass a detailed inspection. Because of high inspection costs, five ovens are selected at random for inspection. These 100 ovens will be purchased if at most 1 of the 5 selected ovens fails inspection. Suppose that there are 8 defective ovens in this batch of 100 ovens. Find the probability that this batch of ovens is purchased. (*Note*: In Chapter 5 you will learn another method to solve this problem.)

4.111 A production system has two production lines; each production line performs a two-part process, and each process is completed by a different machine. Thus, there are four machines, which we can identify as two first-level machines and two second-level machines. Each of the first-level machines works properly 98% of the time, and each of the second-level machines works properly 96% of the time. All four machines are independent in regard to working properly or breaking down. Two products enter this production system, one in each production line.

a. Find the probability that both products successfully complete the two-part process (i.e., all four machines are working properly).

b. Find the probability that neither product successfully completes the two-part process (i.e., at least one of the machines in each production line is not working properly).

4.112 A Wired Equivalent Privacy (WEP) key is a security code that one must enter in order to access a secure WiFi network. The characters in the key are used from the numbers 0 to 9 and letters from A to F, which gives 16 possibilities for each character of the key. Note that repeats are allowed, that is, the same letter or number can be used more than once in a key. A WEP key for a WiFi network with 64-bit security is 10 characters long.

a. How many different 64-bit WEP keys can be made by using the given numbers and letters?

b. A specific 64-bit network has a WEP key in which the 2nd, 5th, 8th, and 9th characters are numbers and the other six characters are letters. How many different WEP keys are possible for this network?

c. A hacker has determined that the WiFi network mentioned in part (b) will lock him out if he makes 20,000 unsuccessful attempts to break into the network. What is the probability that the hacker will be locked out of the network?

Self-Review Test

Multiple Choice Questions

1. The collection of all outcomes for an experiment is called

   a. a sample space

   b. the intersection of events

   c. joint probability

2. A final outcome of an experiment is called

   a. a compound event

   b. a simple event

   c. a complementary event

3. A compound event includes

   a. all final outcomes

   b. exactly two outcomes

   c. more than one outcome for an experiment

4. Two equally likely events

   a. have the same probability of occurrence

   b. cannot occur together

   c. have no effect on the occurrence of each other

5. Which of the following probability approaches can be applied only to experiments with equally likely outcomes?

   a. Classical probability

   b. Empirical probability

   c. Subjective probability

6. Two mutually exclusive events

   a. have the same probability

   b. cannot occur together

   c. have no effect on the occurrence of each other

7. Two independent events

   a. have the same probability

   b. cannot occur together

   c. have no effect on the occurrence of each other

8. The probability of an event is always

   a. less than 0

   b. in the range 0 to 1.0

   c. greater than 1.0

9. The sum of the probabilities of all final outcomes of an experiment is always

   a. 100 b. 1.0 c. 0

10. The joint probability of two mutually exclusive events is always

    a. 1.0 b. between 0 and 1 c. 0

11. Two independent events are

    a. always mutually exclusive

    b. never mutually exclusive

    c. always complementary

12. A box contains a few red, a few black, and a few white marbles. Two marbles are randomly drawn from this box and the color of these marbles is observed. The total number of outcomes for this experiment is

    a. 3 b. 6 c. 9

13. Friends will be called, one after another, and asked to go on a weekend trip with you. You will call until one agrees to go (A) or four friends are asked. Which of the following is the correct tree diagram for the sample space for this experiment?

    a.

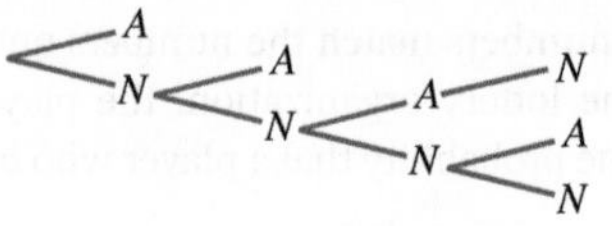

    b.

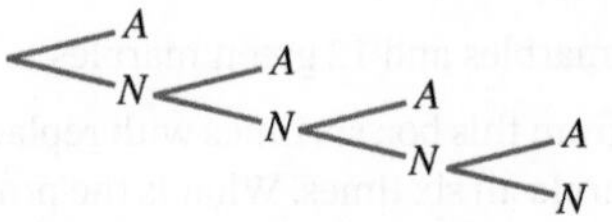

    c.

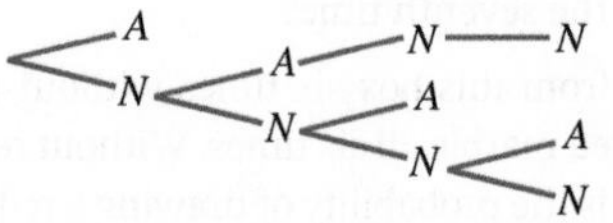

14. When a person makes an educated guess about the likelihood that an event will occur, it is an example of

    a. conditional probability

    b. marginal probability

    c. subjective probability

15. In a group of 72 students, 22 are seniors. If you select one student randomly from this group, the probability (rounded to three decimal places) that this student is a senior is

    a. .306 b. .587 c. .694

16. You roll an unbalanced die 550 times, and a 3-spot is obtained 110 times. The probability (rounded to three decimal places) of not obtaining a 3-spot for this die is (approximately)

    a. .800 b. .200 c. .600

17. Events A and B are independent if

    a. $P(A)$ is equal to $P(B)$

    b. $P(A|B)$ is equal to $P(A)$

    c. $P(A|B)$ is equal to $P(B)$

18. The following table gives the two-way classification of 500 students based on sex and whether or not they suffer from math anxiety.

	Suffer From Math Anxiety	
	Yes	No
Male	166	74
Female	169	91

If you randomly select one student from these 500 students, the probability that this selected student suffers from math anxiety is (round your answer to three decimal places)

    a. .332 b. .338 c. .670

19. A pollster asked 1000 adults whether Republicans or Democrats have better domestic economic policies. The following table gives the two-way classification of their opinions.

	Republicans	Democrats	No Opinion
Male	208	358	24
Female	182	182	46

Which of the following pairs of events are mutually exclusive?

a. Female and no opinion

b. Female and Republican

c. Democrat and no opinion

20. From the probabilities shown in this Venn diagram, determine the probability A does not occur.

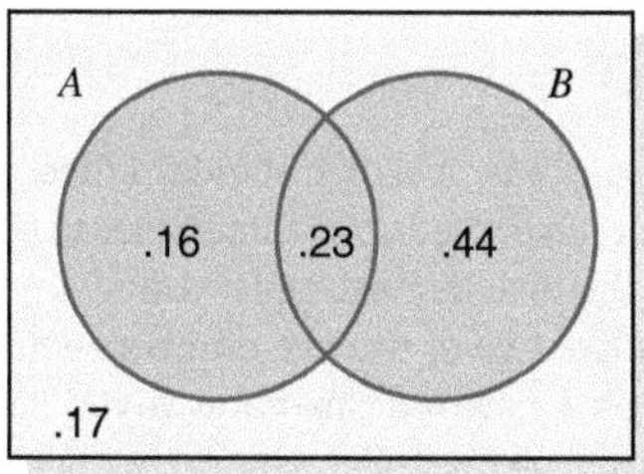

a. .61 b. .84 c. .44

21. The intersection of two events A and B is made up of the outcomes that are

a. either in A or in B or in both A and B

b. common to both A and B

c. either in A or in B, but not both

22. The joint probability of two independent events A and B is

a. $P(A) + P(B) + P(A \text{ or } B)$

b. $P(A) \times P(B)$

c. $P(A) \times P(A|B)$

23. The probability of the union of two events A and B is

a. $P(A) + P(B) + P(A \text{ and } B)$

b. $P(A) + P(B) - P(A \text{ and } B)$

c. $P(A)P(B|A)$

24. A consumer researcher inspects 300 batteries manufactured by two companies for being good or defective. The following table gives the two-way classification of these 300 batteries.

	Good	Defective
Company A	140	10
Company B	130	20

If the researcher selects one battery at random from these 300 batteries, the probability (rounded to three decimal places) that this battery is good or made by company B is

a. .533 b. .933 c. .967

25. You have 10 books on your to-read list. You are selecting three of them to take on the cruise. The first book selected will be read first, the second book selected will be read second, and the third book selected will be read third. Thus, the order in which the three books are selected from the 10 books is important. Find the total number of arrangements of three books from these 10.

a. 3,628,800 b. 6 c. 720

Practice Exercises

1. A couple is planning their wedding reception. The bride's parents have given them a choice of four reception facilities, three caterers, five DJs, and two limo services. If the couple randomly selects one reception facility, one caterer, one DJ, and one limo service, how many different outcomes are possible?

2. Lucia graduated this year with an accounting degree from a university. She has received job offers from an accounting firm, an insurance company, and an airline. She cannot decide which of the three job offers she should accept. Suppose she decides to randomly select one of these three job offers. Find the probability that the job offer selected is

a. from the insurance company

b. not from the accounting firm

3. There are 200 students in a particular graduate program at a state university. Of them, 110 are females and 125 are out-of-state students. Of the 110 females, 70 are out-of-state students.

a. Are the events *female* and *out-of-state student* independent? Are they mutually exclusive? Explain why or why not.

b. If one of these 200 students is selected at random, what is the probability that the student selected is

i. a male?

ii. an out-of-state student given that this student is a female?

4. Reconsider Practice Exercise 3. If 1 of these 200 students is selected at random, what is the probability that the selected student is a *female* or *an out-of-state student*?

5. Reconsider Practice Exercise 3. If 2 of these 200 students are selected at random, what is the probability that both of them are out-of-state students?

6. The probability that an adult has ever experienced a migraine headache is .35. If two adults are randomly selected, what is the probability that neither of them has ever experienced a migraine headache?

7. A hat contains five green, eight red, and seven blue marbles. Let A be the event that a red marble is drawn if we randomly select one marble out of this hat. What is the probability of A? What is the complementary event of A, and what is its probability?

8. The probability that a randomly selected student from a college is a female is .55 and the probability that a student works for more than 10 hours per week is .62. If these two events are independent, find the probability that a randomly selected student is a

a. *male* and *works for more than 10 hours per week*

b. *female* or *works for more than 10 hours per week*

9. A sample was selected of 506 workers who currently receive two weeks of paid vacation per year. These workers were asked if they were willing to accept a small pay cut to get an additional week of paid vacation a year. The following table shows the responses of these workers.

	Yes	No	No Response
Men	77	140	32
Women	104	119	34

a. If one person is selected at random from these 506 workers, find the following probabilities:

i. $P(\text{yes})$

ii. $P(\text{yes} \mid \text{woman})$

iii. $P(\textit{woman} \text{ and } \textit{no})$

iv. $P(\textit{no response} \text{ or } \textit{man})$

b. Are the events *woman* and *yes* independent? Are they mutually exclusive? Explain why or why not.

Mini-Projects

Mini-Project 4.1

Suppose that a small chest contains three drawers. The first drawer contains two \$1 bills, the second drawer contains two \$100 bills, and the third drawer contains one \$1 bill and one \$100 bill. These bills are enclosed in identical envelopes. Suppose that first a drawer is selected at random and then one of the two bills inside that drawer is selected at random. We can define these events as

A = the first drawer is selected, B = the second drawer is selected,

C = the third drawer is selected, D = a \$1 bill is selected

a. Suppose when you randomly select one drawer and then one bill from that drawer, the bill you obtain is a \$1 bill. What is the probability that the second bill in this drawer is a \$100 bill? In other words, find the probability $P(C \mid D)$ because for the second bill to be \$100, it has to be the third drawer. Answer this question intuitively without making any calculations.

b. Use the relative frequency concept of probability to estimate $P(C \mid D)$ as follows: First select a drawer by rolling a die once. If either 1 or 2 occurs, the first drawer is selected; if either 3 or 4 occurs, the second drawer is selected; and if either 5 or 6 occurs, the third drawer is selected. Whenever C occurs, then select a bill by tossing a coin once. (Note that if either A or B occurs, then you do not need to toss the coin because each of these drawers contains both bills of the same denomination.) If you obtain a head, assume that you select a \$1 bill; if you obtain a tail, assume that you select a \$100 bill. Repeat this process 100 times. How many times in these 100 repetitions did the event D occur? What proportion of the time did C occur when D occurred? Use this proportion to estimate $P(C \mid D)$. Does this estimate support your guess of $P(C \mid D)$ in part (a)?

c. Calculate $P(C \mid D)$ using the procedures developed in this chapter (a tree diagram may be helpful). Was your estimate in part (b) close to this value? Explain.

Mini-Project 4.2

There are two families playing in a park, and each of these two families has two children. The Smith family has two daughters, and the Jones family has a daughter and a son. One family is selected at random, and one of the children from this family is chosen at random.

a. Suppose that the selected child is a girl. What is the probability that the second child is also a girl? (*Note*: You need to determine this using conditional probability.)

b. Use the relative frequency concept to estimate the probability that the second child in the family is a girl given that the selected child is a girl. Use the following process to do so. First toss a coin to determine whether the Smith family or the Jones family is chosen. If the Smith family is selected, then record that the second child in this family is a girl, given that the selected child is a girl. This is so because both children in this family are girls. If the Jones family is selected, then toss the coin again to select a child and record the gender of the child selected and that of the second child in this family. Repeat this process 50 times, and then use the results to estimate the required probability. How close is your estimate to the probability calculated in part (a)?

Mini-Project 4.3

The dice game Yahtzee® involves five standard dice. On your turn, you can roll all five or fewer dice up to three times to obtain different sets of numbers on the dice. For example, you will roll all five dice the first time; if you like two of the five numbers obtained, you can roll the other three dice a second time; now if you like three of the five numbers obtained, you can roll the other two dice the third time. Some of the sets of numbers obtained are similar to poker hands (three of a kind, four of a kind, full house, and so forth). However, a few other hands, like five of a kind (called a yahtzee), are not poker hands (or at least they are not the hands you would dare to show anyone).

For the purpose of this project, we will examine the outcomes on the first roll of the five dice. The five scenarios that we will consider are the following:

i. Three of a kind—three numbers are the same and the remaining two numbers are both different, for example, 22254

ii. Four of a kind—four numbers are the same and the fifth number is different, for example, 44442

iii. Full house—three numbers are the same and the other two numbers are the same, for example, 33366

iv. Large straight—five numbers in a row, for example, 12345

v. Yahtzee—all five numbers are the same, for example, 33333

In the first two cases, the dice that are not part of the three or four of a kind must have different values than those in the three or four of a kind. For example, 22252 cannot be considered three of a kind, but 22254 is three of a kind. (Yahtzee players know that this situation differs from the rules of the actual game, but for the purpose of this project, we will change the rules.)

a. Find the probability for each of these five cases for one roll of the five dice.

b. In a regular game, you do not have to roll all five dice on each of the three rolls. You can leave some dice on the table and roll the others in an attempt to improve your score. For example, if you roll 13555 on the first roll of five dice, you can keep the three fives and roll the dice with 1 and 3 outcomes the second time in an attempt to get more fives, or possibly a pair of another number in order to get a full house. After your second roll, you are allowed to pick up any of the dice for your third roll. For example, suppose your first roll is 13644 and you keep the two fours. Then you roll the dice with 1, 3, and 6 outcomes the second time and obtain three fives. Thus, now you have 55544. Although you met your full house requirement before rolling the dice three times, you still need a yahtzee. So, you keep the fives and roll the two dice with fours the third time. Write a paragraph outlining all of the scenarios you will have to consider to calculate the probability of obtaining a yahtzee within your three rolls.

Decide for Yourself

Note: The Decide for Yourself feature is located in the book's product page on *www.wiley.com*.

Chapter 4 Technology Instructions

Note: Complete TI-84, Minitab, and Excel manuals are available in the book's product page on *www.wiley.com*.

TI-84 Color/TI-84

The TI-84 Color Technology Instructions feature of this text is written for the TI-84 Plus C color graphing calculator running the 4.0 operating system. Some screens, menus, and functions will be slightly different in older operating systems. The TI-84 and TI-84 Plus can perform all of the same functions but will not have the "Color" option referenced in some of the menus.

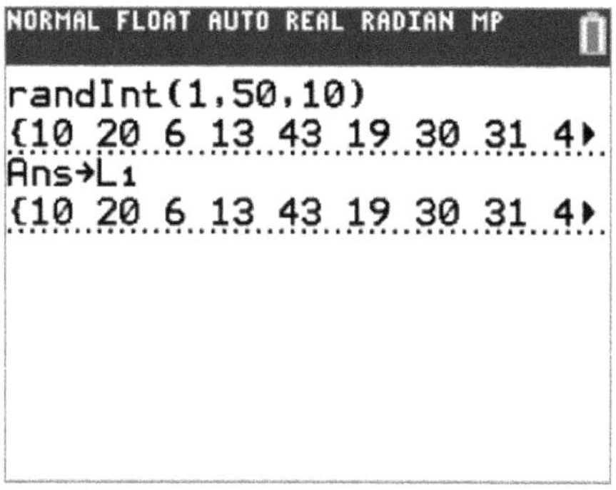

SCREEN 4.1

Generating Random Numbers

To Generate 10 Random Integers Uniformly Distributed Between 1 and 50

1. Select **MATH > PROB > randInt**.
2. Use the following settings in the **randInt** menu:
   - Type 1 at the **lower** prompt.
   - Type 50 at the **upper** prompt.
   - Type 10 at the **n** prompt.
3. Highlight **Paste** and press **ENTER** twice. (See **Screen 4.1.**)
4. The output is 10 randomly chosen integers between 1 and 50. (See **Screen 4.1.**)
5. You may use any lower bound, upper bound, and number of integers.
6. To store these numbers in a list (see **Screen 4.1**):
   - Press **STO.**
   - Press **2nd > STAT** and select the name of the list.
   - Press **ENTER** twice.

Simulating Tosses of a Coin

Follow the instructions shown above, but generate random integers between 0 and 1. Assign 0 to be a tail and 1 to be a head (you can switch these if you prefer). To simulate 500 tosses, type 500 at the **n** prompt. To see a graph of the resulting distribution, store these values in a list and make a histogram of that list.

Simulating Rolls of a Die

Follow the instructions shown above, but generate random integers between 1 and 6. To simulate 500 rolls, type 500 at the **n** prompt. To see a graph of the resulting distribution, store these values in a list and make a histogram of that list.

Calculating 10! for Example 4.35 of the Text

1. Type 10.
2. Select **MATH > PROB > !**.
3. Press **ENTER**. (See **Screen 4.2.**)

Calculating $_5C_3$ for Example 4.39 of the Text

1. Type 5.
2. Select **MATH > PROB > nCr**.
3. Type 3.
4. Press **ENTER**. (See **Screen 4.2.**)

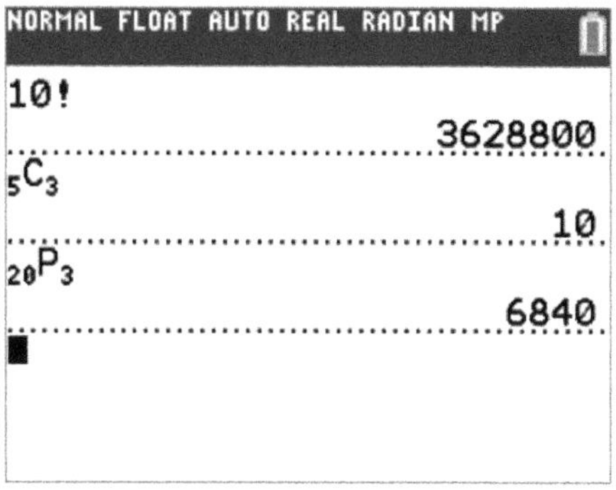

SCREEN 4.2

Calculating $_{20}P_3$ for Example 4.41 of the Text

1. Type 20.
2. Select **MATH > PROB > nPr.**
3. Type 3.
4. Press **ENTER.** (See **Screen 4.2.**)

Minitab

The Minitab Technology Instructions in this text are written for **Minitab Professional Version 19**. Some screens, menus, and functions will be slightly different in older (or in newer) versions of Minitab. If you are using **Minitab Express**, you can download the Minitab Express Manual from the book's product page on *www.wiley.com.*

Generating Random Numbers

To Generate 10 Random Integers Uniformly Distributed Between 1 and 50

1. Select **Calc > Random Data > Integer.**
2. Use the following settings in the dialog box that appears on screen:
   - Type 10 in the **Number of rows of data to generate** box.
   - Type C1 in the **Store in column(s)** box.
   - Type 1 in the **Minimum value** box.
   - Type 50 in the **Maximum value** box.
3. Click **OK.** Minitab will list 10 random integers between 1 and 50 in column C1.

Simulating Tosses of a Coin

Follow the instructions shown earlier, but generate random integers between 0 and 1. Assign 0 to be a tail and 1 to be a head (you can switch these if you prefer). To simulate 500 tosses, type 500 in the **Number of rows of data to generate** box. To see a graph of the resulting distribution, store these values in C1 and make a histogram of C1.

Simulating Rolls of a Die

Follow the instructions shown earlier, but generate random integers between 1 and 6. To simulate 500 rolls, type 500 in the **Number of rows of data to generate** box. To see a graph of the resulting distribution, store these values in C1 and make a histogram of C1.

Calculating 10! for Example 4.35 of the Text

1. Select **CALC > Calculator.**
2. Use the following settings in the dialog box that appears on screen (see **Screen 4.3**):
   - In the **Store result in variable** box, type C1.
   - In the **Expression** box, type factorial(10).
3. Click **OK.** Minitab will list the value of 10! in row 1 of column C1.

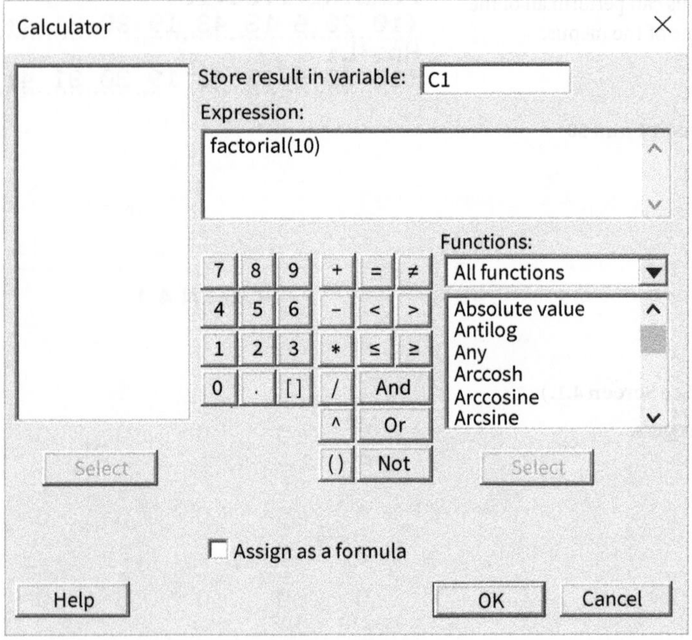

SCREEN 4.3

Calculating $_5C_3$ for Example 4.39 of the Text

1. Select **CALC > Calculator.**
2. Use the following settings in the dialog box that appears on screen (see **Screen 4.4**):
   - In the **Store result in variable** box, type C1.
   - In the **Expression** box, type combinations(5,3).
3. Click **OK.** Minitab will list the value of $_5C_3$ in row 1 of column C1.

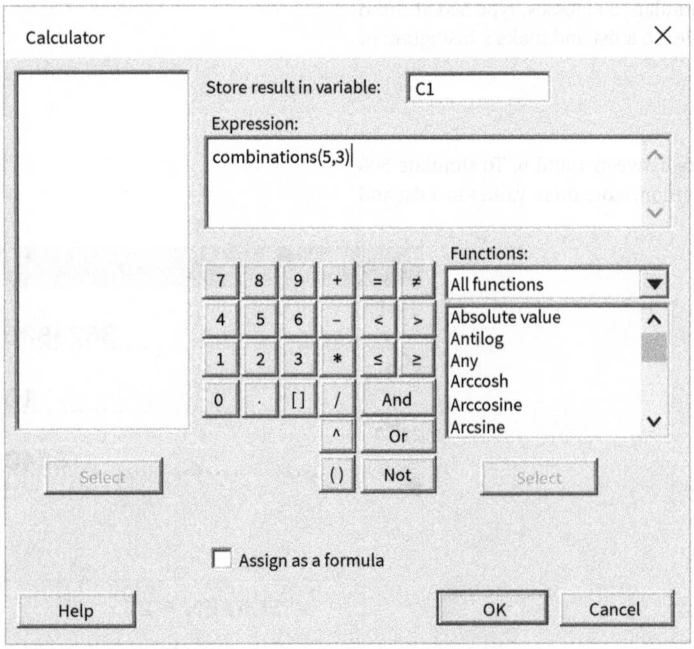

SCREEN 4.4

Calculating ${}_{20}P_3$ for Example 4.41 of the Text

1. Select **CALC > Calculator**.
2. Use the following settings in the dialog box that appears on screen:
   - In the **Store result in variable** box, type C1.
   - In the **Expression** box, type permutations(20,3).
3. Click **OK**. Minitab will list the value of ${}_{20}P_3$ in row 1 of column C1.

Excel

The Excel Technology Instructions feature of this text is written for Excel 2019. Some screens, menus, and functions will be slightly different in older versions of Excel. To enable some of the advanced Excel functions, you must enable the Data Analysis Add-In for Excel.

Generating Random Numbers

To Generate Random Integers Uniformly Distributed Between 1 and 50

1. Click on cell A1.
2. Type **=randbetween(1,50)**. (See **Screen 4.5.**)
3. Press **ENTER.**
4. A randomly chosen integer between 1 and 50 will appear in cell A1.
5. To generate more than one such random number, copy and paste the formula into as many cells as you require.

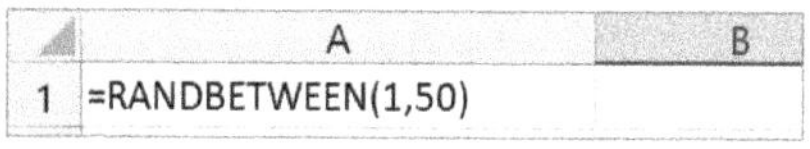

SCREEN 4.5

Simulating Tosses of a Coin

Follow the instructions shown earlier, but generate random integers between 0 and 1. Assign 0 to be a tail and 1 to be a head (you can switch these if you prefer). To simulate 500 tosses, copy the formula to 500 cells. To see a graph of the resulting distribution, make a histogram of these 500 values.

Simulating Rolls of a Die

Follow the instructions shown earlier, but generate random integers between 1 and 6. To simulate 500 rolls, copy the formula to 500 cells. To see a graph of the resulting distribution, make a histogram of these values.

Calculating 10! for Example 4.35 of the Text

1. Click on cell A1.
2. Type **=FACT(10)**.
3. Press **ENTER**.

Calculating ${}_5C_3$ for Example 4.39 of the Text

1. Click on cell A1.
2. Type **=COMBIN(5,3)**.
3. Press **ENTER**. (See **Screen 4.6.**)

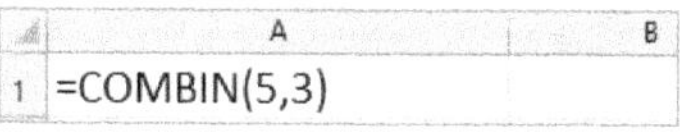

SCREEN 4.6

Calculating ${}_{20}P_3$ for Example 4.41 of the Text

1. Click on cell A1.
2. Type **=PERMUT(20,3)**.
3. Press **ENTER**.

Technology Assignments

TA4.1 You want to simulate the tossing of a coin. Assign a value of 0 (zero) to Head and a value of 1 to Tail.

a. Simulate 50 tosses of the coin by generating 50 random integer numbers between 0 and 1. Then calculate the mean of these 50 numbers. This mean gives you the proportion of 50 tosses that resulted in tails. Using this proportion, calculate the number of heads and tails you obtained in

50 simulated tosses. Prepare the frequency tables for the expected (theoretical) frequencies and for the actual frequencies you obtained.

b. Repeat part (a) by simulating 600 tosses.

c. Repeat part (a) by simulating 4000 tosses.

Comment on the percentage of Tails obtained as the number of tosses is increased.

TA4.2 You want to simulate the rolling of a die. Assign the values 1 through 6 to the outcomes from 1-spot through 6-spot on the die, respectively.

a. Simulate 200 rolls of the die by generating 200 random integer numbers between 1 and 6. Then make a bar graph for these 200 numbers. Prepare the frequency tables for the expected (theoretical) frequencies and for the actual frequencies you obtained.

b. Repeat part (a) by simulating 1000 rolls of the die.

c. Repeat part (a) by simulating 6000 rolls of the die.

Comment on the bar graphs obtained in parts (a) through (c).

TA4.3 Suppose each child born has equal chance of being a boy or a girl. Assign a value of 0 to boy and a value of 1 to girl.

a. Simulate births of 50 children by generating 50 random integers equal to 0 and 1. Make a bar graph for the outcomes of these 50 random numbers. Create a frequency table showing the simulated (actual) frequencies obtained and the expected (theoretical) frequencies for the two outcomes. How many of these 50 children are boys and how many are girls in the simulated data?

b. Repeat part (a), but simulate births of 500 children.

c. Repeat part (a), but simulate births of 5000 children.

d. For which simulation of parts (a) to (c) were the simulated frequencies closest to the theoretical frequencies?

TA4.4 Using a technology of your choice, find 6!, 9!, and 15!.

TA4.5 Using a technology of your choice, find the number of combinations of 16 items selecting 7 at a time.

TA4.6 Using a technology of your choice, find the number of ways to select 9 items out of 24.

TA4.7 Using a technology of your choice, find the number of samples of size 9 each that can be selected from a population of 36 cars.

TA4.8 Using a technology of your choice, find the number of permutations of 16 items selecting 7 at a time.

TA4.9 Using a technology of your choice, find the number of permutations of 24 items selecting 9 at a time.

CHAPTER 5

Discrete Random Variables and Their Probability Distributions

New York Daily News Archive/Getty Images

Now that you know a little about probability, do you feel lucky enough to play the lottery? If you have $20 to spend on lunch today, are you willing to spend it all on four $5 lottery tickets to increase your chance of winning? Do you think you will profit, on average, if you continue buying lottery tickets over time? Can lottery players beat the state, on average? Not a chance! (See Case Study 5.1 for answers.)

Chapter 4 discussed the concepts and rules of probability. This chapter extends the concept of probability to explain probability distributions. As we learned in Chapter 4, any given statistical experiment has more than one outcome. It is impossible to predict with certainty which of the many possible outcomes will occur if an experiment is performed. Consequently, decisions are made under uncertain conditions. For example, a lottery player does not know in advance whether or not he is going to win that lottery. If he knows that he is not going to win, he would definitely not play. It is the uncertainty about winning (some positive probability of winning) that makes him play. This chapter shows that if the outcomes and their probabilities for a statistical experiment are known, we can find out what will happen, on average, if that experiment is performed many times. For the lottery example, we can find out how much a lottery player can expect to win (or lose), on average, if he continues playing this lottery again and again.

In this chapter, random variables and types of random variables are explained. Then, the concept of a probability distribution and its mean and standard deviation for a discrete random variable are discussed. Finally, various special probability distributions for a discrete random variable—the binomial probability distribution, the negative binomial probability distribution, the hypergeometric probability distribution, and the Poisson probability distribution—are developed.

5.1 Random Variables

LEARNING OBJECTIVES

After completing this section, you should be able to:

- Explain the meaning of a random variable.
- Explain the difference between a discrete random variable and a continuous random variable.
- Identify random variables as discrete or continuous.

Randomness plays an important part in our lives. Most things happen randomly. For example, consider the following events: winning a lottery, your car breaking down, getting sick, getting involved in an accident, losing a job, making money in the stock market, and so on. We cannot predict when, where, and to whom these things can or will happen. Who will win a lottery and when? Who will get sick and when? Where, when, and who will get involved in an accident? All these events are uncertain, and they happen randomly to people. In other words, there is no definite time, place, and person for these events to happen. Similarly, consider the following events: how many customers will visit a bank, a grocery store, or a gas station on a given day? How many cars will cross a bridge on a given day? How many students will be absent from a class on a given day? In all these examples, the *number* of customers, cars, and students are random; that is, each of these can assume any value within a certain interval.

Suppose **Table 5.1** gives the frequency and relative frequency distributions of the number of vehicles owned by all 2000 families living in a small town.

TABLE 5.1 Frequency and Relative Frequency Distributions of the Number of Vehicles Owned by Families

Number of Vehicles Owned	Frequency	Relative Frequency
0	30	30/2000 = .015
1	320	320/2000 = .160
2	910	910/2000 = .455
3	580	580/2000 = .290
4	160	160/2000 = .080
	$N = 2000$	Sum = 1.000

Suppose one family is randomly selected from this population. The process of randomly selecting a family is called a *random* or *chance experiment*. Let x denote the number of vehicles owned by the selected family. Then x can assume any of the five possible values (0, 1, 2, 3, and 4) listed in the first column of Table 5.1. The value assumed by x depends on which family is selected. Thus, this value depends on the outcome of a random experiment. Consequently, x is called a **random variable** or a **chance variable**. In general, a random variable is denoted by x or y.

Random Variable

A *random variable* is a variable whose value is determined by the outcome of a random experiment.

As will be explained next, a random variable can be discrete or continuous.

5.1.1 Discrete Random Variable

A **discrete random variable** assumes values that can be counted. In other words, the consecutive values of a discrete random variable are separated by a certain gap.

Discrete Random Variable

A random variable that assumes countable values is called a *discrete random variable.*

In Table 5.1, *the number of vehicles owned by a family* is an example of a discrete random variable because the values of the random variable x are countable: 0, 1, 2, 3, and 4. Here are a few other examples of discrete random variables:

1. The number of cars sold at a dealership during a given month
2. The number of houses in a certain block
3. The number of fish caught on a fishing trip
4. The number of complaints received at the office of an airline on a given day
5. The number of customers who visit a bank during any given hour
6. The number of heads obtained in three tosses of a coin

5.1.2 Continuous Random Variable

A random variable whose values are not countable is called a **continuous random variable**. A continuous random variable can assume any value over an interval or intervals.

Continuous Random Variable

A random variable that can assume any value contained in one or more intervals is called a *continuous random variable.*

Because the number of values contained in any interval is infinite, the possible number of values that a continuous random variable can assume is also infinite. Moreover, we cannot count these values. Consider the life of a battery. We can measure it as precisely as we want. For instance, the life of this battery may be 40 hours, or 40.25 hours, or 40.247 hours. Assume that the maximum life of a battery is 200 hours. Let x denote the life of a randomly selected battery of this kind. Then, x can assume any value in the interval 0 to 200. Consequently, x is a continuous random variable. As shown in the diagram below, every point on the line representing the interval 0 to 200 gives a possible value of x.

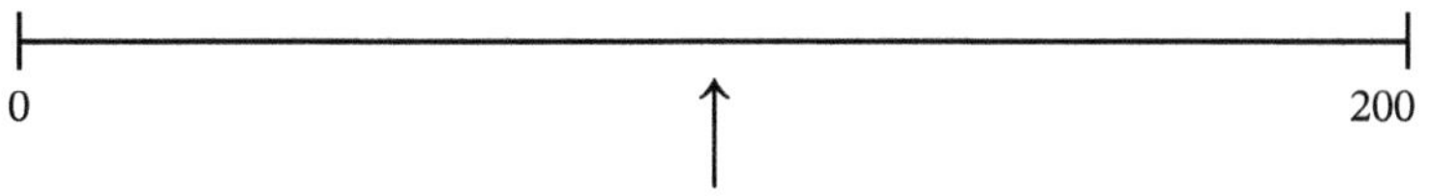

Every point on this line represents a possible value of x that denotes the life of a battery. There is an infinite number of points on this line. The values represented by points on this line are uncountable.

The following are a few other examples of continuous random variables:

1. The length of a room
2. The time to commute from home to work
3. The amount of milk in a gallon (note that we do not expect "a gallon" to contain exactly one gallon of milk but either slightly more or slightly less than one gallon)
4. The weight of a letter
5. The price of a house

Note that money is often treated as a continuous random variable, specifically when there are a large number of unique values.

This chapter is limited to the discussion of discrete random variables and their probability distributions. Continuous random variables will be discussed in Chapter 6.

EXERCISES

Concepts and Procedures

5.1 Explain the meaning of a random variable, a discrete random variable, and a continuous random variable. Give one example each of a discrete random variable and a continuous random variable.

5.2 Classify each of the following random variables as discrete or continuous.

- **a.** The total runs scored by a batsman in a cricket match
- **b.** The duration of a movie in minutes
- **c.** The price of a mobile phone
- **d.** The number of passengers who rode metro train in a given hour
- **e.** The number of accidents on a given road on a given day
- **f.** Health insurance claim amounts that come to an insurance company in a given month
- **g.** The number of persons vaccinated with the COVID-19 vaccine

Applications

5.3 In a hospital, patients visit a dentist for their dental treatment. Let x denote the number of persons who visited the hospital in last six months. Is x a discrete or continuous random variable? What are the possible values that x can assume?

5.2 Probability Distribution of a Discrete Random Variable

LEARNING OBJECTIVES

After completing this section, you should be able to:

- Explain the meaning of the probability distribution of a discrete random variable.
- Explain the two characteristics of the probability distribution of a discrete random variable.
- Verify the conditions of a probability distribution.
- Determine the probability distribution of a discrete random variable.
- Calculate the probability of an event from a probability distribution.

Let x be a discrete random variable. The **probability distribution** of x describes how the probabilities are distributed over all the possible values of x.

Probability Distribution of a Discrete Random Variable

The *probability distribution of a discrete random variable* lists all the possible values that the random variable can assume and their corresponding probabilities.

We learned in Chapter 4 about the classical and the relative frequency approaches to probability. Depending on the type of the statistical experiment, we can use one of these two approaches to write the probability distribution of a random variable. Example 5.1 uses the relative frequency approach to write the probability distribution.

From the probability distribution listed in Table 5.3, we observe two characteristics:

1. Each probability listed in the $P(x)$ column is in the range 0 to 1.
2. The sum of all the probabilities listed in the $P(x)$ column is 1.0.

EXAMPLE 5.1 Number of Vehicles Owned by Families

Writing the Probability Distribution of a Discrete Random Variable

Recall the frequency and relative frequency distributions of the number of vehicles owned by families given in Table 5.1. That table is reproduced here as **Table 5.2**. Let x be the number of vehicles owned by a randomly selected family. Write the probability distribution of x and make a histogram for this probability distribution.

TABLE 5.2 Frequency and Relative Frequency Distributions of the Number of Vehicles Owned by Families

Number of Vehicles Owned	Frequency	Relative Frequency
0	30	.015
1	320	.160
2	910	.455
3	580	.290
4	160	.080

Solution In Chapter 4, we learned that the relative frequencies obtained from an experiment or a sample can be used as approximate probabilities. However, when the relative frequencies represent the population, as in Table 5.2, they give the actual (theoretical) probabilities of outcomes. Using the relative frequencies of Table 5.2, we can write the *probability distribution* of the discrete random variable x in **Table 5.3**. Note that the values of x listed in Table 5.3 are pairwise mutually exclusive events.

TABLE 5.3 Probability Distribution of the Number of Vehicles Owned by Families

Number of Vehicles Owned x	Probability $P(x)$
0	.015
1	.160
2	.455
3	.290
4	.080
	$\Sigma P(x) = 1.000$

Using the probability distribution of Table 5.3, we make the histogram given in **Figure 5.1**. Such a histogram can show whether the probability distribution is symmetric or skewed, and if skewed then whether it is skewed to the left or right.

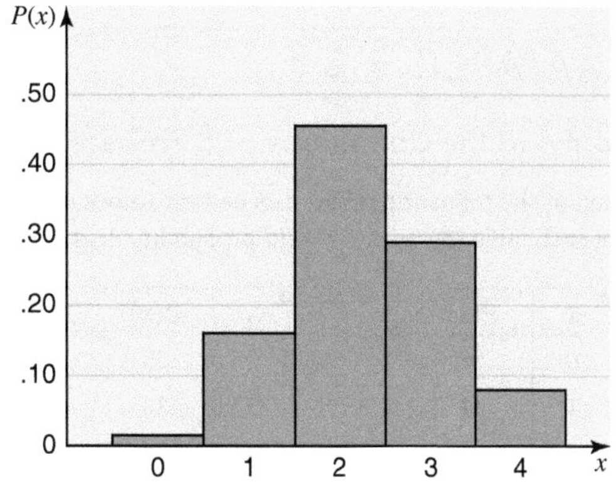

FIGURE 5.1 Histogram for the probability distribution of Table 5.3.

These two characteristics are always true for a valid probability distribution. These two characteristics are the same that we discussed in Chapter 4, and can be written as in the following box:

Two Characteristics of a Probability Distribution

The probability distribution of a discrete random variable possesses the following two characteristics:

1. $0 \le P(x) \le 1$ for each value of x

2. $\Sigma P(x) = 1.0$

These two characteristics are also called the *two conditions* that a probability distribution must satisfy. Notice that in Table 5.3 each probability listed in the column labeled $P(x)$ is between 0 and 1. Also, $\Sigma P(x) = 1.0$. Because both conditions are satisfied, Table 5.3 represents the probability distribution of x.

Using the probability distribution listed in Table 5.3, we can find the probability of any outcome or event. Example 5.2 illustrates this.

EXAMPLE 5.2 Number of Vehicles Owned by Families

Finding Probabilities of Events Using Table 5.3

Using the probability distribution listed in Table 5.3 of Example 5.1, find the following probabilities:

a. The probability that a randomly selected family owns two vehicles

b. The probability that a randomly selected family owns at least two vehicles

c. The probability that a randomly selected family owns at most one vehicle

d. The probability that a randomly selected family owns three or more vehicles

Solution Using the probabilities listed in Table 5.3, we find the required probabilities as follows:

a. $P(\text{selected family owns two vehicles}) = P(2) = .455$

b. $P(\text{selected family owns at least two vehicles}) = P(2 \text{ or } 3 \text{ or } 4) = P(2) + P(3) + P(4)$
$= .455 + .290 + .080 = \mathbf{.825}$

c. $P(\text{selected family owns at most one vehicle}) = P(0 \text{ or } 1) = P(0) + P(1)$
$= .015 + .160 = \mathbf{.175}$

d. $P(\text{selected family owns three or more vehicles}) = P(3 \text{ or } 4) = P(3) + P(4)$
$= .290 + .080 = \mathbf{.370}$

EXAMPLE 5.3

Verifying the Conditions of a Probability Distribution

Each of the following tables lists certain values of x and their probabilities. Determine whether or not each table represents a valid probability distribution.

a.

x	$P(x)$
0	.08
1	.11
2	.39
3	.27

b.

x	$P(x)$
2	.25
3	.34
4	.28
5	.13

c.

x	$P(x)$
7	.70
8	.50
9	−.20

Solution

a. Because each probability listed in this table is in the range 0 to 1, it satisfies the first condition of a probability distribution. However, the sum of all probabilities is not equal to 1.0 because $\Sigma P(x) = .08 + .11 + .39 + .27 = .85$. Therefore, the second condition is not satisfied. Consequently, this table does not represent a valid probability distribution.

b. Each probability listed in this table is in the range 0 to 1. Also, $\Sigma P(x) = .25 + .34 + .28 + .13 = 1.0$. Consequently, this table represents a valid probability distribution.

c. Although the sum of all probabilities listed in this table is equal to 1.0, one of the probabilities is negative. This violates the first condition of a probability distribution. Therefore, this table does not represent a valid probability distribution.

EXAMPLE 5.4 Number of Breakdowns for a Machine

Finding the Probabilities of Events for a Discrete Random Variable

The following table lists the probability distribution of the number of breakdowns per week for a machine based on past data.

Breakdowns per Week	0	1	2	3
Probability	.15	.20	.35	.30

Find the probability that the number of breakdowns for this machine during a given week is

a. exactly 2 **b.** 0 to 2 **c.** more than 1 **d.** at most 1

Solution Let x denote the number of breakdowns for this machine during a given week. **Table 5.4** lists the probability distribution of x.

TABLE 5.4 Probability Distribution of the Number of Breakdowns

x	$P(x)$
0	.15
1	.20
2	.35
3	.30
	$\Sigma P(x) = 1.00$

Using Table 5.4, we can calculate the required probabilities as follows:

a. The probability of exactly two breakdowns is

$$P(\text{exactly 2 breakdowns}) = P(x = 2) = P(2) = \mathbf{.35}$$

Note that we can write the probability $P(x = 2)$ as $P(2)$.

b. The probability of 0 to 2 breakdowns is given by the sum of the probabilities of 0, 1, and 2 breakdowns:

$$\begin{aligned} P(\text{0 to 2 breakdowns}) &= P(0 \le x \le 2) \\ &= P(0 \text{ or } 1 \text{ or } 2) = P(0) + P(1) + P(2) = .15 + .20 + .35 = \mathbf{.70} \end{aligned}$$

c. The probability of more than 1 breakdown is obtained by adding the probabilities of 2 and 3 breakdowns:

$$\begin{aligned} P(\text{more than 1 breakdown}) &= P(x > 1) = P(2 \text{ or } 3) \\ &= P(2) + P(3) \\ &= .35 + .30 = \mathbf{.65} \end{aligned}$$

d. The probability of at most 1 breakdown is given by the sum of the probabilities of 0 and 1 breakdown:

$$P(\text{at most 1 breakdown}) = P(x \leq 1) = P(0 \text{ or } 1)$$
$$= P(0) + P(1)$$
$$= .15 + .20 = \mathbf{.35}$$

EXAMPLE 5.5 Students Suffering from Math Anxiety

Constructing a Probability Distribution

According to a survey, 60% of all students at a large university suffer from math anxiety. Two students are randomly selected from this university. Let x denote the number of students in this sample who suffer from math anxiety. Construct the probability distribution of x.

Solution Let us define the following two events:

N = the student selected does not suffer from math anxiety

M = the student selected suffers from math anxiety

As we can observe from the tree diagram of **Figure 5.2**, there are four possible outcomes for this experiment: NN (neither of the students suffers from math anxiety), NM (the first student does not suffer from math anxiety and the second does), MN (the first student suffers from math anxiety and the second does not), and MM (both students suffer from math anxiety). The probabilities of these four outcomes are listed in the tree diagram. Because 60% of the students suffer from math anxiety and 40% do not, the probability is .60 that any student selected suffers from math anxiety and .40 that he or she does not.

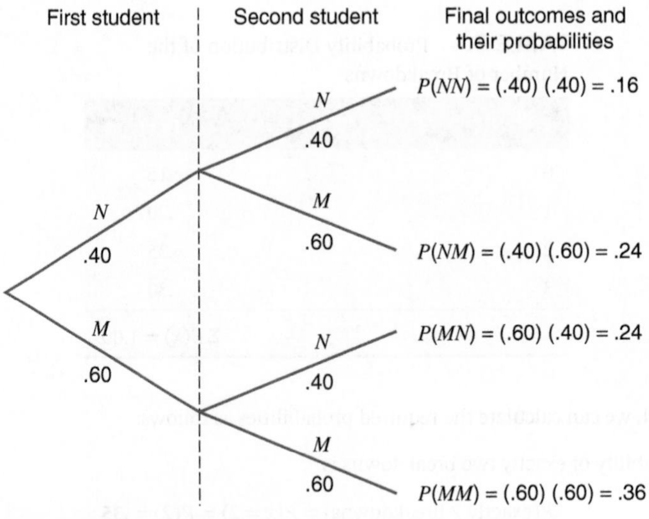

FIGURE 5.2 Tree diagram.

In a sample of two students, the number who suffer from math anxiety can be 0 (given by NN), 1 (given by NM or MN), or 2 (given by MM). Thus, x can assume any of three possible values: 0, 1, or 2. The probabilities of these three outcomes are calculated as follows:

$$P(0) = P(NN) = .16$$
$$P(1) = P(NM \text{ or } MN) = P(NM) + P(MN) = .24 + .24 = .48$$
$$P(2) = P(MM) = .36$$

Using these probabilities, we can write the probability distribution of x as in **Table 5.5**.

TABLE 5.5 Probability Distribution of the Number of Students with Math Anxiety

x	$P(x)$
0	.16
1	.48
2	.36
	$\Sigma P(x) = 1.00$

EXERCISES

Concepts and Procedures

5.4 Explain the meaning of the probability distribution of a discrete random variable. Give one example of such a probability distribution.

5.5 Briefly explain the two characteristics (conditions) of the probability distribution of a discrete random variable.

5.6 Each of the following tables lists certain values of x and their probabilities. Verify whether or not each represents a valid probability distribution and explain why.

a.

x	$P(x)$
0	.45
1	.45
2	.10

b.

x	$P(x)$
0	.25
1	.25
2	.10
3	.25
4	.25

c.

x	$P(x)$
1	−.20
2	.25
3	.10
4	.25
5	.25
6	.35

5.7 The following table gives the probability distribution of a discrete random variable x.

x	1	2	3	4	5	6
$P(x)$	.20	.15	.25	.15	.15	.10

Find the following probabilities:

a. $P(4)$ **b.** $P(x \leq 4)$ **c.** $P(x > 5)$
d. $P(3 < x \leq 4)$ **e.** $P(3 \leq x \leq 5)$

Applications

5.8 A review of emergency room records at rural Millard Fellmore Memorial Hospital was performed to determine the probability distribution of the number of patients entering the emergency room during a one-hour period. The following table lists this probability distribution.

Patients per Hour	0	1	2	3	4	5	6
Probability	.2165	.4183	.2112	.0998	.0314	.0138	.0090

a. Graph the probability distribution.

b. Determine the probability that the number of patients entering the emergency room during a randomly selected one-hour period is

i. two or more **iii.** fewer than three
ii. exactly five **iv.** at most one

5.9 Front camera on the dashboard is considered as one of the most helpful items to record the incidents in case of an accident. Let x denote the number of front cameras installed on a given day at an authorized service center. The following table lists the frequency distribution of x for the last three months.

x	1	2	3	4	5
f	15	8	6	4	2

a. Construct the probability distribution table for the number of front cameras installed on a given day.

b. Find the following probabilities:

i. $P(3)$ **ii.** $P(x \geq 3)$ **iii.** $P(2 \leq x \leq 4)$ **iv.** $P(x < 4)$

5.10 In a poll, 70% of U.S. adults said that CEOs of large U.S. companies are overpaid. Suppose this result is true for the current population of U.S. adults. Two adults are randomly selected from the population of U.S. adults. Let x denote the number of adults in this sample of two adults who hold the opinion that CEOs of large U.S. companies are overpaid. Write the probability distribution of x.

5.11 Suppose that currently 16.1% of motorists in the United States are uninsured. Two motorists are selected at random. Let x denote the number of motorists in this sample of two who are uninsured. Construct the probability distribution table of x. Draw a tree diagram for this problem.

5.12 In a 10 K run, 27.4% of the 5248 participants finished the race in 49 minutes 42 seconds (49:42) or faster, which is equivalent to a pace of less than 8 minutes per mile. Suppose that this result holds true for all people who would participate in and finish a 10 K race. Suppose that two 10 K runners are selected at random. Let x denote the number of runners in these two who would finish a 10 K race in 49:42 or less. Construct the probability distribution table of x. Draw a tree diagram for this problem.

5.13 According to a survey, 35% of employees working at a very large company are happy with their jobs. Suppose that two employees are selected at random from this company. Let x denote the number of employees in this sample of two who are happy with their jobs. Construct the probability distribution table of x. Draw a tree diagram for this problem.

***5.14** In a group of 20 athletes, 6 have used performance-enhancing drugs that are illegal. Suppose that two athletes are randomly selected from this group. Let x denote the number of athletes in this sample who have used such illegal drugs. Write the probability distribution of x. You may draw a tree diagram and use that to write the probability distribution. (*Hint:* Note that the selections are made without replacement from a small population. Hence, the probabilities of outcomes do not remain constant for each selection.)

5.3 Mean and Standard Deviation of a Discrete Random Variable

LEARNING OBJECTIVES

After completing this section, you should be able to:

- Explain the concept of the mean and standard deviation of a discrete random variable.
- Calculate and interpret the mean of a discrete random variable.
- Calculate and interpret the standard deviation of a discrete random variable.

In this section, we will learn how to calculate the mean and standard deviation of a discrete random variable and how to interpret them.

5.3.1 Mean of a Discrete Random Variable

The **mean of a discrete random variable**, denoted by μ, is actually the mean of its probability distribution. The mean of a discrete random variable x is also called its *expected value* and is denoted by $E(x)$. The mean (or expected value) of a discrete random variable is the value that we expect to observe per repetition, on average, if we perform an experiment a large number of times. For example, we may expect a car salesperson to sell, on average, 2.4 cars per week. This does not mean that every week this salesperson will sell exactly 2.4 cars. (Obviously one cannot sell exactly 2.4 cars.) This simply means that if we observe for many weeks, this salesperson will sell a different number of cars during different weeks; however, the average for all these weeks will be 2.4 cars per week.

To calculate the mean of a discrete random variable x, we multiply each value of x by the corresponding probability and sum the resulting products. This sum gives the mean (or expected value) of the discrete random variable x.

Mean of a Discrete Random Variable

The *mean of a discrete random variable* x is the value that is expected to occur per repetition, on average, if an experiment is repeated a large number of times. It is denoted by μ and calculated as

$$\mu = \Sigma x P(x)$$

The mean of a discrete random variable x is also called its expected value and is denoted by $E(x)$; that is,

$$E(x) = \Sigma x P(x)$$

Example 5.6 illustrates the calculation of the mean of a discrete random variable.

EXAMPLE 5.6 Machine Breakdowns per Week

Calculating and Interpreting the Mean of a Discrete Random Variable

The following table lists the number of breakdowns per week and their probabilities for a machine based on past data.

Breakdowns	Probability
0	.15
1	.20
2	.35
3	.30

Find the mean number of breakdowns per week for this machine.

Solution Let x represent the number of breakdowns per week and $P(x)$ be the probability of x breakdowns. To find the mean number of breakdowns per week for this machine, we multiply each value of x by its probability and add these products. This sum gives the mean of the probability distribution of x. The products $xP(x)$ are listed in the third column of **Table 5.6**. The sum of these products gives $\Sigma xP(x)$, which is the mean of x.

TABLE 5.6 Calculating the Mean for the Probability Distribution of Breakdowns

x	$P(x)$	$xP(x)$
0	.15	$0(.15) = .00$
1	.20	$1(.20) = .20$
2	.35	$2(.35) = .70$
3	.30	$3(.30) = .90$
		$\Sigma xP(x) = 1.80$

The mean is

$$\mu = \Sigma xP(x) = \mathbf{1.80}$$

Thus, on average, this machine is expected to break down 1.80 times per week over a period of time. In other words, if this machine is used for many weeks, then for certain weeks we will observe no breakdowns; for some other weeks we will observe one breakdown per week; and for still other weeks we will observe two or three breakdowns per week. The mean number of breakdowns is expected to be 1.80 per week for the entire period.

Note that $\mu = 1.80$ is also the **expected value** of x. It can also be written as

$$E(x) = 1.80$$

Again, this expected value of x means that we will expect an average of 1.80 breakdowns per week for this machine.

Case Study 5.1 illustrates the calculation of the mean amount that an instant lottery player is expected to win.

5.3.2 Standard Deviation of a Discrete Random Variable

The **standard deviation of a discrete random variable**, denoted by σ, measures the spread of its probability distribution. A higher value for the standard deviation of a discrete random variable indicates that x can assume values over a larger range about the mean. In contrast, a smaller value for the standard deviation indicates that most of the values that x can assume are

Case Study 5.1

All State Lottery

A hypothetical country has an instant lottery game called All State Lottery. The cost of each ticket for this lottery game is $5. A player can instantly win $500,000, $10,000, $1000, $100, $50, $20, and $10. Each ticket has seven spots that are covered by latex coating. The top one of these spots, labeled Prize, contains the winning prize. The remaining six spots labeled *Player's Spots* belong to the player. A player will win if any of the *Player's Spots* contains the prize amount that matches the winning prize in the top spot. If none of the six spots contains a dollar amount that matches the winning prize, the player loses.

Initially a total of 20,000,000 tickets are printed for this lottery. The following table lists the prizes and the number of tickets with those prizes. As is obvious from this table, out of a total of 20,000,000 tickets, 18,000,000 are nonwinning tickets (the ones with a prize of $0 in this table). Of the remaining 2,000,000 tickets with prizes, 1,620,000 tickets have a prize of $10 each, 364,000 tickets contain a prize of $20 each, and so forth.

Prize (dollars)	Number of Tickets
0	18,000,000
10	1,620,000
20	364,000
50	10,000
100	5,000
1,000	730
10,000	200
500,000	70
	Total = 20,000,000

The net amount a player wins for each of the winning tickets is equal to the amount of the prize minus $5, which is the cost of the ticket. Thus, the net gain for each of the nonwinning tickets is −$5, which is the cost of the ticket. Let

x = the net amount a player wins by playing this lottery game

The following table shows the probability distribution of x, and all the calculations required to compute the mean of x for this probability distribution. The probability of an outcome (net winnings) is calculated by dividing the number of tickets with that outcome (prize) by the total number of tickets.

x (dollars)	$P(x)$	$xP(x)$
−5	18,000,000/20,000,000 = .9000000	−4.5000000
5	1,620,000/20,000,000 = .0810000	.4050000
15	364,000/20,000,000 = .0182000	.2730000
45	10,000/20,000,000 = .0005000	.0225000
95	5,000/20,000,000 = .0002500	.0237500
995	730/20,000,000 = .0000365	.0363175
9,995	200/20,000,000 = .0000100	.0999500
499,995	70/20,000,000 = .0000035	1.7499825
		$\Sigma xP(x) = -1.8895000$

Hence, the mean or expected value of x is

$$\mu = \Sigma x P(x) = -\$1.8895000 \approx -\$1.89$$

This value of the mean gives the expected value of the random variable x, that is,

$$E(x) = \Sigma x P(x) = -\$1.89$$

Thus, the mean of net winnings for this lottery game is −\$1.89. In other words, all players taken together will lose an average of \$1.89 per ticket. That is, out of every \$5 (the price of a ticket), \$3.11 will be returned to players in the form of prizes and \$1.89 will go to the government, which will cover the costs of operating the lottery, the commission paid to agents, and profit to the government. Note that \$1.89 is 37.8% of \$5. Thus, we can also state that 37.8% of the total money spent by players on this lottery game will go to the government and 100 − 37.8 = 62.2% will be returned to the players in the form of prizes.

clustered closely about the mean. The basic formula to compute the standard deviation of a discrete random variable is

$$\sigma = \sqrt{\Sigma[(x - \mu)^2 \cdot P(x)]}$$

However, it is more convenient to use the following shortcut formula to compute the standard deviation of a discrete random variable.

Standard Deviation of a Discrete Random Variable

The **standard deviation of a discrete random variable** x measures the spread of its probability distribution and is computed as

$$\sigma = \sqrt{\Sigma x^2 P(x) - \mu^2}$$

Note that the variance σ^2 of a discrete random variable is obtained by squaring its standard deviation.

Example 5.7 illustrates how to use the shortcut formula to compute the standard deviation of a discrete random variable.

EXAMPLE 5.7 Defective Computer Parts

Calculating the Standard Deviation of a Discrete Random Variable

Baier's Electronics manufactures computer parts that are supplied to many computer companies. Despite the fact that two quality control inspectors at Baier's Electronics check every part for defects before it is shipped to another company, a few defective parts do pass through these inspections undetected. Let x denote the number of defective computer parts in a shipment of 400. The following table gives the probability distribution of x.

Fuse/Getty Images

x	0	1	2	3	4	5
$P(x)$	.02	.20	.30	.30	.10	.08

Compute the standard deviation of x.

Solution **Table 5.7** shows all the calculations required for the computation of the standard deviation of x.

TABLE 5.7 Computations to Find the Standard Deviation

x	$P(x)$	$xP(x)$	x^2	$x^2P(x)$
0	.02	.00	0	.00
1	.20	.20	1	.20
2	.30	.60	4	1.20
3	.30	.90	9	2.70
4	.10	.40	16	1.60
5	.08	.40	25	2.00
		$\Sigma xP(x) = 2.50$		$\Sigma x^2P(x) = 7.70$

We perform the following steps to compute the standard deviation of x.

STEP 1 *Compute the mean of the discrete random variable.*
The sum of the products $xP(x)$, recorded in the third column of Table 5.7, gives the mean of x.

$$\mu = \Sigma xP(x) = 2.50 \text{ defective computer parts in 400}$$

STEP 2 *Compute the value of $\Sigma x^2P(x)$.*
First we square each value of x and record it in the fourth column of Table 5.7. Then we multiply these values of x^2 by the corresponding values of $P(x)$. The resulting values of $x^2P(x)$ are recorded in the fifth column of Table 5.7. The sum of this column is

$$\Sigma x^2P(x) = 7.70$$

STEP 3 *Substitute the values of μ and $\Sigma x^2P(x)$ in the formula for the standard deviation of x and simplify.*

By performing this step, we obtain

$$\sigma = \sqrt{\Sigma x^2P(x) - \mu^2} = \sqrt{7.70 - (2.50)^2} = \sqrt{1.45}$$
$$= \mathbf{1.204} \text{ defective computer parts}$$

Thus, a given shipment of 400 computer parts is expected to contain an average of 2.50 defective parts with a standard deviation of 1.204 defective parts.

Remember: Because the standard deviation of a discrete random variable is obtained by taking the positive square root, its value is never negative.

EXAMPLE 5.8 Profits from New Makeup Product

Loraine Corporation is planning to market a new makeup product. According to the analysis made by the financial department of the company, it will earn an annual profit of \$4.5 million if this product has high sales, it will earn an annual profit of \$1.2 million if the sales are mediocre, and it will lose \$2.3 million a year if the sales are low. The probabilities of these three scenarios are .32, .51, and .17, respectively.

a. Let x be the profits (in millions of dollars) earned per annum from this product by the company. Write the probability distribution of x.

b. Calculate the mean and standard deviation of x.

Solution

Writing the probability distribution of a discrete random variable.

a. The table below lists the probability distribution of x. Note that because x denotes profits earned by the company, the loss is written as a *negative profit* in the table.

x	$P(x)$
4.5	.32
1.2	.51
−2.3	.17

Calculating the mean and standard deviation of a discrete random variable.

b. **Table 5.8** shows all the calculations that are required for the computation of the mean and standard deviation of x.

TABLE 5.8 Computations to Find the Mean and Standard Deviation

x	$P(x)$	$xP(x)$	x^2	$x^2P(x)$
4.5	.32	1.440	20.25	6.4800
1.2	.51	.612	1.44	.7344
−2.3	.17	−.391	5.29	.8993
		$\Sigma xP(x) = 1.661$		$\Sigma x^2P(x) = 8.1137$

The mean of x is

$$\mu = \Sigma xP(x) = \mathbf{\$1.661\ million}$$

The standard deviation of x is

$$\sigma = \sqrt{\Sigma x^2P(x) - \mu^2} = \sqrt{8.1137 - (1.661)^2} = \mathbf{\$2.314\ million}$$

Thus, it is expected that Loraine Corporation will earn an average of \$1.661 million in profits per year from the new product with a standard deviation of \$2.314 million.

Interpretation of the Standard Deviation

The standard deviation of a discrete random variable can be interpreted or used the same way as the standard deviation of a data set in Section 3.4 of Chapter 3. In that section, we learned that according to Chebyshev's theorem, at least $[1 - (1/k^2)] \times 100\%$ of the total area under a curve lies within k standard deviations of the mean, where k is any number greater than 1. Thus, if $k = 2$, then at least 75% of the area under a curve lies between $\mu - 2\sigma$ and $\mu + 2\sigma$. In Example 5.7,

$$\mu = 2.50 \quad \text{and} \quad \sigma = 1.204$$

Hence,

$$\mu - 2\sigma = 2.50 - 2(1.204) = .092$$
$$\mu + 2\sigma = 2.50 + 2(1.204) = 4.908$$

Using Chebyshev's theorem, we can state that at least 75% of the shipments (each containing 400 computer parts) are expected to contain .092 to 4.908 defective computer parts each.

EXERCISES

Concepts and Procedures

5.15 Briefly explain the concept of the mean and standard deviation of a discrete random variable.

5.16 Find the mean and standard deviation for each of the following probability distributions:

a.

x	$P(x)$
1	.20
2	.15
3	.25
4	.15
5	.15
6	.10

b.

x	$P(x)$
6	.40
7	.26
8	.21
9	.13

Applications

5.17 Let x be the number of errors that appear on a randomly selected page of a book. The following table lists the probability distribution of x.

x	0	1	2	3	4	5
$P(x)$	.60	.15	.12	.08	.04	.01

Find the mean and standard deviation of x.

5.18 The following table gives the probability distribution of the number of microwaves sold on a given day at a department store.

Microwaves Sold	0	1	2	3	4	5	6
Probability	.05	.12	.19	.30	.20	.10	.04

Calculate the mean and standard deviation of this probability distribution. Give a brief interpretation of the value of the mean.

5.19 The following table gives the probability distribution of the number of patients entering the emergency room during a one-hour period at a government hospital.

Number of Patients	0	1	2	3	4	5
Probability	.15	.45	.20	.10	.07	.03

Calculate the mean and standard deviation of this probability distribution.

5.20 Let x be the number of heads obtained in two tosses of a coin. The following table lists the probability distribution of x.

x	0	1	2
$P(x)$	.25	.50	.25

Calculate the mean and standard deviation of this probability distribution and give a brief interpretation of the value of the mean.

5.21 Let x be the number of magazines a person reads every week. Based on a sample survey of adults, the following probability distribution table was prepared.

x	0	1	2	3	4	5
$P(x)$	.38	.27	.15	.09	.06	.05

Find the mean and standard deviation of x.

5.22 The following table lists the probability distribution of the number of exercise machines sold per day at Elmo's Sporting Goods store.

Machines Sold per Day	4	5	6	7	8	9	10
Probability	.08	.11	.14	.19	.20	.16	.12

Find the mean and standard deviation of this probability distribution. Give a brief interpretation of the values of the mean.

5.23 A contractor has submitted bids on three state jobs: an office building, a theater, and a parking garage. State rules do not allow a contractor to be offered more than one of these jobs. If this contractor is awarded any of these jobs, the profits earned from these contracts are \$10 million from the office building, \$5 million from the theater, and \$2 million from the parking garage. His profit is zero if he gets no contract. The contractor estimates that the probabilities of getting the office building contract, the theater contract, the parking garage contract, or nothing are .15, .30, .45, and .10, respectively. Let x be the random variable that represents the contractor's profits in millions of dollars. Write the probability distribution of x. Find the mean and standard deviation of x. Give a brief interpretation of the values of the mean and standard deviation.

5.24 An instant lottery ticket costs \$2. Out of a total of 20,000 tickets printed for this lottery, 2000 tickets contain a prize of \$5 each, 200 tickets have a prize of \$10 each, 10 tickets have a prize of \$1000 each, and one ticket has a prize of \$10,000. Let x be the random variable that denotes the net amount a player wins by playing this lottery. Write the probability distribution of x. Determine the mean and standard deviation of x. How will you interpret the values of the mean and standard deviation of x?

5.25 In a group of 20 athletes, 6 have used performance-enhancing drugs that are illegal. Suppose that two athletes are randomly selected from this group. Let x denote the number of athletes in this sample who have used such illegal drugs. The following table lists the probability distribution of x.

x	0	1	2
$P(x)$	.4789	.4422	.0789

Calculate the mean and standard deviation of x for this distribution.

5.4 The Binomial Probability Distribution

LEARNING OBJECTIVES

After completing this section, you should be able to:

- Explain the four conditions of a binomial experiment.
- Explain the meaning of a trial in the context of a binomial experiment.
- Verify the conditions of a binomial experiment.
- Explain the meaning of a binomial random variable and a binomial probability distribution.
- Identify and explain the parameters of a binomial probability distribution.
- Identify random variables that have a binomial probability distribution.
- Calculate the probability of x successes in n trials using the binomial formula.
- Construct a binomial probability distribution and its graph.
- Use the binomial table to find probabilities.
- Describe how the shape of the binomial probability distribution changes for $p = .50$, $p < .50$, and $p > .50$.
- Calculate the mean and standard deviation of a binomial probability distribution using the formulas that apply only to the binomial probability distribution.

The **binomial probability distribution** is one of the most widely used discrete probability distributions. It is used to find the probability that (under certain conditions) an outcome will occur x times in n repetitions of an experiment. For example, given that 75% of students at a college use Instagram, we may want to find the probability that in a random sample of five students from this college, exactly three students use Instagram. As a second example, we may be interested in finding the probability that a baseball player with a batting average of .250 will have no hits in 10 trips to the plate.

The binomial probability distribution is applied to experiments that satisfy the four conditions of a binomial experiment. (These conditions are discussed next.) There are n repetitions of an experiment where n is a positive integer. Each of the n repetitions of the experiment is called a **trial** or a **Bernoulli trial** (named after Jacob Bernoulli). Each trial must result in exactly one of two possible outcomes. If an experiment has more than two outcomes for each trial, we must be able to combine those outcomes into two outcomes to use the binomial probability distribution. For example, if an experiment is defined as one toss of a coin and this experiment is repeated 10 times, then each repetition (toss) is called a trial. Consequently, there are 10 total trials for this experiment. The random variable that represents the number of successes in n trials for the binomial probability distribution is denoted by x. Remember that x will assume only integer values from 0 to n. In the example of 10 tosses of a coin, $n = 10$. If we define x as the number of heads in 10 trials, then x can assume values from 0 to 10 because in 10 tosses we can obtain 0 to 10 heads.

5.4.1 The Binomial Experiment

An experiment that satisfies the following four conditions is called a **binomial experiment**.

1. There are ***n* identical trials**. In other words, the given experiment is repeated n times, where n is a positive integer. All of these repetitions are performed under identical conditions.
2. Each trial has two and only two outcomes. These outcomes are usually called a ***success*** and a ***failure***, respectively. In case there are more than two outcomes for an experiment, we can combine outcomes into two events and then apply binomial probability distribution.
3. The probability of success is denoted by $\boldsymbol{p}$ and that of failure by $\boldsymbol{q}$, and $p + q = 1$. The probabilities p and q remain constant for each trial.
4. The **trials are independent**. In other words, the outcome of one trial does not affect the outcome of another trial.

Conditions of a Binomial Experiment

A binomial experiment must satisfy the following four conditions:

1. There are n identical trials.
2. Each trial has only two possible outcomes (or events). In other words, the outcomes of a trial are divided into two mutually exclusive events.
3. The probabilities of the two outcomes (or events) remain constant from trial to trial.
4. The trials are independent.

Note that one of the two outcomes (or events) of a trial is called a *success* and the other a *failure*. Notice that a success does not mean that the corresponding outcome is considered favorable or desirable. Similarly, a failure does not necessarily refer to an unfavorable or undesirable outcome. Success and failure are simply the names used to denote the two possible outcomes of a trial. The outcome to which the question refers is usually called a success; the outcome to which it does not refer is called a failure. For example, if we are to find the probability of four heads in 10 tosses of a coin, we will call the head a *success* and the tail a *failure*. But if the question asks to find the probability of six tails in 10 tosses of a coin, we will call the tail a *success* and the head a *failure*.

EXAMPLE 5.9 Ten Tosses of a Coin

Verifying the Conditions of a Binomial Experiment

Consider the experiment consisting of 10 tosses of a coin. Determine whether or not this is a binomial experiment.

Solution The experiment consisting of 10 tosses of a coin satisfies all four conditions of a binomial experiment as explained below.

1. There are a total of 10 trials (tosses), and they are all identical. All 10 tosses are performed under identical conditions. Here, $n = 10$.
2. Each trial (toss) has only two possible outcomes: a head and a tail. Let a head be called a success and a tail be called a failure.
3. The probability of obtaining a head (a success) is 1/2 and that of a tail (a failure) is 1/2 for any toss. That is,

$$p = P(H) = 1/2 \quad \text{and} \quad q = P(T) = 1/2$$

   The sum of these two probabilities is 1.0. Also, these probabilities remain the same for each toss.
4. The trials (tosses) are independent. The result of any preceding toss has no bearing on the result of any succeeding toss.

Since this experiment satisfies all four conditions, it is a binomial experiment.

EXAMPLE 5.10 Students Using Instagram

Verifying the Conditions of a Binomial Experiment

a. Seventy-five percent of students at a college with a large student population use Instagram. A sample of five students from this college is selected, and these students are asked whether or not they use Instagram. Is this experiment a binomial experiment?

b. In a group of 12 students at a college, 9 use Instagram. Five students are selected from this group of 12 and are asked whether or not they use Instagram. Is this experiment a binomial experiment?

Solution

a. Below we check whether all four conditions of a binomial experiment are satisfied.

1. This example consists of five identical trials. A trial represents the selection of a student.
2. Each trial has two outcomes: a student uses Instagram or a student does not use Instagram. Let us call the outcome a *success* if a student uses Instagram and a *failure* if he/she does not use Instagram.
3. Seventy-five percent of students at this college use Instagram. Hence, the probability p that a student uses Instagram is .75, and the probability q that a student does not use Instagram is .25. These probabilities, p and q, remain constant for each selection of a student. Also, these two probabilities add to 1.0.
4. Each trial (student) is independent. In other words, if one student uses Instagram, it does not affect the outcome of another student using or not using Instagram. This is so because the size of the population is very large compared to the sample size.

Since all four conditions are satisfied, this is an example of a binomial experiment.

b. Below we check whether all four conditions of a binomial experiment are satisfied.

1. This example consists of five identical trials. A trial represents the selection of a student.
2. Each trial has two outcomes: a student uses Instagram or a student does not use Instagram. Let us call the outcome a *success* if a student uses Instagram and a *failure* if he/she does not use Instagram.
3. There are a total of 12 students, and 9 of them use Instagram. Let p be the probability that a student uses Instagram and q be the probability that a student does not use Instagram. These two probabilities, p and q, do not remain constant for each selection of a student because of the limited number (12) of students. The probability of each outcome changes with each selection depending on what happened in the previous selections. For example, if the first selected student uses Instagram, then the probability that the second selected student uses Instagram is 8/11. But if the first selected student does not use Instagram, then the probability that the second selected student uses Instagram is 9/11.
4. Because p and q do not remain constant for each selection, the trials are not independent. The outcome of the first selection affects the outcome of the second selection, and so on.

Since the third and fourth conditions of a binomial experiment are not satisfied, this is not an example of a binomial experiment.

5.4.2 The Binomial Probability Distribution and Binomial Formula

The random variable x that represents the number of successes in n trials for a binomial experiment is called a *binomial random variable*. The probability distribution of x in such experiments is called the **binomial probability distribution** or simply the *binomial distribution*. Thus, the binomial probability distribution is applied to find the probability of x successes in n trials for a binomial experiment. The number of successes x in such an experiment is a discrete random variable. Consider Example 5.10a. Let x be the number of college students in a sample of five who use Instagram. Because the number of students who use Instagram in a sample of five can be any number from zero to five, x can assume any of the values 0, 1, 2, 3, 4, and 5. Since the values of x are countable, x is a discrete random variable.

Binomial Formula

For a binomial experiment, the probability of exactly x successes in n trials is given by the binomial formula

$$P(x) = {}_nC_x\, p^x q^{n-x}$$

where

$$\begin{aligned} n &= \text{total number of trials} \\ p &= \text{probability of success} \\ q &= 1 - p = \text{probability of failure} \\ x &= \text{number of successes in } n \text{ trials} \\ n - x &= \text{number of failures in } n \text{ trials} \end{aligned}$$

In the binomial formula, n is the total number of trials and x is the total number of successes. The difference between the total number of trials and the total number of successes, $n - x$, gives the total number of failures in n trials. The value of ${}_nC_x$ gives the number of ways to obtain x successes in n trials. As mentioned earlier, p and q are the probabilities of success and failure, respectively. Again, although it does not matter which of the two outcomes is called a success and which a failure, usually the outcome to which the question refers is called a success.

To solve a binomial problem, we determine the values of n, x, $n - x$, p, and q and then substitute these values in the binomial formula. To find the value of ${}_nC_x$, we can use either the combinations formula from Section 4.6.3 or a calculator.

To find the probability of x successes in n trials for a binomial experiment, the only values needed are those of n and p. These are called the *parameters of the binomial probability distribution* or simply the **binomial parameters**. The value of q is obtained by subtracting the value of p from 1.0. Thus, $q = 1 - p$.

Example 5.11 illustrates how to solve a binomial problem by using the binomial formula.

EXAMPLE 5.11 | Students Using Instagram

Calculating the Probability of One Success in Three Trials Using the Binomial Formula

Seventy-five percent of students at a college with a large student population use the social media site Instagram. Three students are randomly selected from this college. What is the probability that exactly two of these three students use Instagram?

Solution Here, there are three trials (students selected), and we are to find the probability of two successes (two students using Instagram) in these three trials. We can find the probability of two successes in three trials using the procedures learned in Chapter 4, but that will be a time-consuming and complex process. First, we will have to write all the possible outcomes that give two successes in three trials, and then we will have to find the probability that any one of those outcomes occurs using the union of events. But using the binomial probability distribution to find this probability is very easy, as described next. Here, we are given that:

$$\begin{aligned} n &= \text{total number of trials} = \text{number of students selected} = 3 \\ x &= \text{number of successes} = \text{number of students in three who use Instagram} = 2 \\ p &= \text{probability of success} = \text{probability that a student uses Instagram} = .75 \end{aligned}$$

From this given information, we can calculate the values of $n - x$ and q as follows:

$$\begin{aligned} n - x &= \text{number of failures} = \text{number of students not using Instagram} = 3 - 2 = 1 \\ q &= \text{probability of failure} = \text{probability that a student does not use Instagram} = 1 - .75 = .25 \end{aligned}$$

The probability of two successes is denoted by $P(x = 2)$ or simply by $P(2)$. By substituting all of the values in the binomial formula, we obtain

Number of ways to obtain 2 success in 3 trials → $_3C_2$; Number of successes → exponent 2; Number of failures → exponent 1

$$P(2) = {}_3C_2\,(.75)^2\,(.25)^1 = (3)\,(.5625)\,(.25) = \mathbf{.4219}$$

Probability of success → .75; Probability of failure → .25

Note that the value of $_3C_2$ in the formula can either be obtained from a calculator or be computed as follows:

$$_3C_2 = \frac{3!}{2!\,(3-2)!} = \frac{3 \cdot 2 \cdot 1}{2 \cdot 1 \cdot 1} = 3$$

In the above computation, $_3C_2 = 3$ gives the number of ways to select two students who use Instagram in a random sample of three students. Suppose *I* denotes a student who uses Instagram and *N* denotes a student who does not use Instagram. The three ways to select two students who use Instagram are *IIN*, *INI*, and *NII*.

EXAMPLE 5.12 Home Delivery Service

Calculating the Probability Using the Binomial Formula

At the Express House Delivery Service, providing high-quality service to customers is the top priority of the management. The company guarantees a refund of all charges if a package it is delivering does not arrive at its destination by the specified time. It is known from past data that despite all efforts, 2% of the packages mailed through this company do not arrive at their destinations within the specified time. Suppose a corporation mails 10 packages through Express House Delivery Service on a certain day.

Keith Brofsky/Getty Images

a. Find the probability that exactly one of these 10 packages will not arrive at its destination within the specified time.

b. Find the probability that at most one of these 10 packages will not arrive at its destination within the specified time.

Solution Let us call an event a success if a package does not arrive at its destination within the specified time and a failure if a package does arrive within the specified time. Note that because we are to find the probability that one package *will not arrive* within the specified time, we will call *not arriving* a *success*. Then,

$$n = \text{total number of packages mailed} = 10$$
$$p = P(\text{success}) = .02$$
$$q = P(\text{failure}) = 1 - .02 = .98$$

a. For this part,

$$x = \text{number of successes} = 1$$
$$n - x = \text{number of failures} = 10 - 1 = 9$$

Substituting all values in the binomial formula, we obtain

$$P(1) = {}_{10}C_1\,(.02)^1\,(.98)^9 = \frac{10!}{1!\,(10-1)!}\,(.02)^1\,(.98)^9$$
$$= (10)(.02)(.83374776) = \mathbf{.1667}$$

Thus, there is a .1667 probability that exactly one of the 10 packages mailed will not arrive at its destination within the specified time.

b. The probability that at most one of the 10 packages will not arrive at its destination within the specified time is given by the sum of the probabilities of $x = 0$ and $x = 1$. Thus,

$$P(x \le 1) = P(0) + P(1)$$
$$= {}_{10}C_0(.02)^0(.98)^{10} + {}_{10}C_1(.02)^1(.98)^9$$
$$= (1)(1)(.81707281) + (10)(.02)(.83374776)$$
$$= .8171 + .1667 = \mathbf{.9838}$$

Thus, the probability that at most one of the 10 packages will not arrive at its destination within the specified time is .9838.

EXAMPLE 5.13 Employees Changing Their Jobs

Constructing a Binomial Probability Distribution and Its Graph

According to a survey, 33% of U. S. employees do not plan to change their jobs in the near future. Let x denote the number of employees in a random sample of three U. S. employees who do not plan to change their jobs in the near future. Write the probability distribution of x and draw a histogram for this probability distribution.

Solution Let x be the number of employees who do not plan to change their jobs in the near future in a random sample of three U. S. employees. Then, $n - x$ is the number of U. S. employees who do plan to change their jobs in the near future. (We have included the employees who are uncertain or have no opinion in this later group of $n - x$ employees.) From the given information,

n = total employees in the sample = 3
$p = P$(an employee does not plan to change his/her job in the near future) = .33
$q = P$(an employee does plan to change his/her job in the near future) = 1 − .33 = .67

The possible values that x can assume are 0, 1, 2, and 3. In other words, the number of employees in this sample of three who do not plan to change their jobs in the near future can be 0, 1, 2, or 3. The probability of each of these four outcomes is calculated as follows:

If $x = 0$, then $n - x = 3$. Using the binomial formula, we obtain the probability of $x = 0$ as follows:

$$P(0) = {}_3C_0(.33)^0(.67)^3 = (1)(1)(.300763) = .3008$$

Note that ${}_3C_0$ is equal to 1 by definition and $(.33)^0$ is equal to 1 because any (nonzero) number raised to the power zero is always 1.

If $x = 1$, then $n - x = 2$. Using the binomial formula, the probability of $x = 1$ is

$$P(1) = {}_3C_1(.33)^1(.67)^2 = (3)(.33)(.4489) = .4444$$

Similarly, if $x = 2$, then $n - x = 1$, and if $x = 3$, then $n - x = 0$. The probabilities of $x = 2$ and $x = 3$ are, respectively,

$$P(2) = {}_3C_2(.33)^2(.67)^1 = (3)(.1089)(.67) = .2189$$
$$P(3) = {}_3C_3(.33)^3(.67)^0 = (1)(.035937)(1) = .0359$$

These probabilities are written in **Table 5.9. Figure 5.3** shows the histogram for the probability distribution of Table 5.9.

TABLE 5.9 Probability Distribution of x

x	$P(x)$
0	.3008
1	.4444
2	.2189
3	.0359

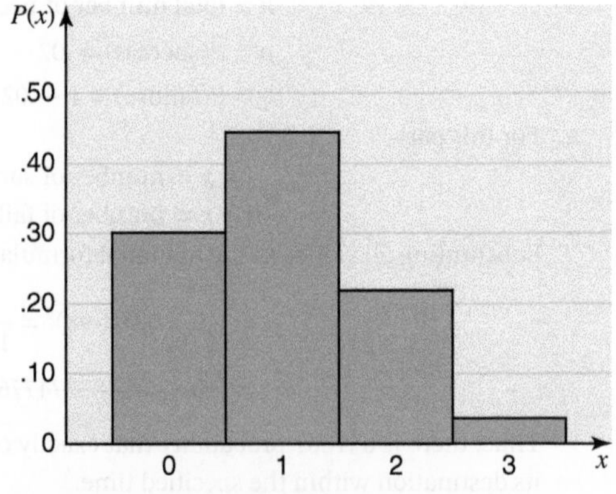

FIGURE 5.3 Histogram for the probability distribution of Table 5.9.

5.4.3 Using the Table of Binomial Probabilities

The probabilities for a binomial experiment can also be read from Table I, the table of binomial probabilities, in Appendix B. This table lists the probabilities of x for $n = 1$ to $n = 25$ and for selected values of p. Example 5.14 illustrates how to read Table I.

EXAMPLE 5.14 Time Spent on Facebook by College Students

Using the Binomial Table to Find Probabilities and to Construct the Probability Distribution and Graph

According to a survey, 30% of college students said that they spend too much time on Facebook. (The remaining 70% said that they do not spend too much time on Facebook or had no opinion.) Suppose this result holds true for the current population of all college students. A random sample of six college students is selected. Using Table I of Appendix B, answer the following:

a. Find the probability that exactly three of these six college students will say that they spend too much time on Facebook.

b. Find the probability that at most two of these six college students will say that they spend too much time on Facebook.

c. Find the probability that at least three of these six college students will say that they spend too much time on Facebook.

d. Find the probability that one to three of these six college students will say that they spend too much time on Facebook.

e. Let x denote the number in a random sample of six college students who will say that they spend too much time on Facebook. Write the probability distribution of x and draw a histogram for this probability distribution.

Solution

a. To read the required probability from Table I of Appendix B, we first determine the values of n, x, and p. For this example,

n = number of students in the sample = 6
x = number of students in this sample who spend too much time on Facebook = 3
$p = P$(a student spends too much time on Facebook) = .30

Then we locate $n = 6$ in the column labeled n in Table I of Appendix B. The relevant portion of Table I with $n = 6$ is reproduced as **Table 5.10** here. Next, we locate 3 in the column for x in the portion of the table for $n = 6$ and locate $p = .30$ in the row for p at the top of the table.

TABLE 5.10 Determining $P(x = 3)$ for $n = 6$ and $p = .30$

$p = .30$

		p					
n	x	.05	.10	.20	.30	. . .	.95
$n = 6$ → 6	0	.7351	.5314	.2621	.1176	. . .	.0000
	1	.2321	.3543	.3932	.3025	. . .	.0000
	2	.0305	.0984	.2458	.3241	. . .	.0001
	$x = 3$ → 3	.0021	.0146	.0819	.1852	. . .	.0021
	4	.0001	.0012	.0154	.0595	. . .	.0305
	5	.0000	.0001	.0015	.0102	. . .	.2321
	6	.0000	.0000	.0001	.0007	. . .	.7351

$P(x = 3) = .1852$

The entry at the intersection of the row for $x = 3$ and the column for $p = .30$ gives the probability of three successes in six trials when the probability of success is .30. From Table I of Appendix B or Table 5.10,

$$P(3) = \mathbf{.1852}$$

Using Table I of Appendix B or Table 5.10, we write **Table 5.11**, which can be used to answer the remaining parts of this example.

b. The event that at most two students will say that they spend too much time on Facebook will occur if x is equal to 0, 1, or 2. From Table I of Appendix B or from Table 5.11, the required probability is

$$P(\text{at most } 2) = P(0 \text{ or } 1 \text{ or } 2) = P(0) + P(1) + P(2)$$
$$= .1176 + .3025 + .3241 = \mathbf{.7442}$$

TABLE 5.11 Portion of Table I for $n = 6$ and $p = .30$

n	x	p .30
6	0	.1176
	1	.3025
	2	.3241
	3	.1852
	4	.0595
	5	.0102
	6	.0007

c. The probability that at least three students will say that they spend too much time on Facebook is given by the sum of the probabilities of 3, 4, 5, or 6. Using Table I of Appendix B or from Table 5.11, we obtain

$$P(\text{at least } 3) = P(3 \text{ or } 4 \text{ or } 5 \text{ or } 6)$$
$$= P(3) + P(4) + P(5) + P(6)$$
$$= .1852 + .0595 + .0102 + .0007 = \mathbf{.2556}$$

d. The probability that one to three students will say that they spend too much time on Facebook is given by the sum of the probabilities of $x = 1$, 2, and 3. Using Table I of Appendix B or from Table 5.11, we obtain

$$P(1 \text{ to } 3) = P(1) + P(2) + P(3)$$
$$= .3025 + .3241 + .1852 = \mathbf{.8118}$$

e. Using Table I of Appendix B or from Table 5.11, we list the probability distribution of x for $n = 6$ and $p = .30$ in **Table 5.12. Figure 5.4** shows the histogram of the probability distribution of x.

TABLE 5.12 Probability Distribution of x for $n = 6$ and $p = .30$

x	$P(x)$
0	.1176
1	.3025
2	.3241
3	.1852
4	.0595
5	.0102
6	.0007

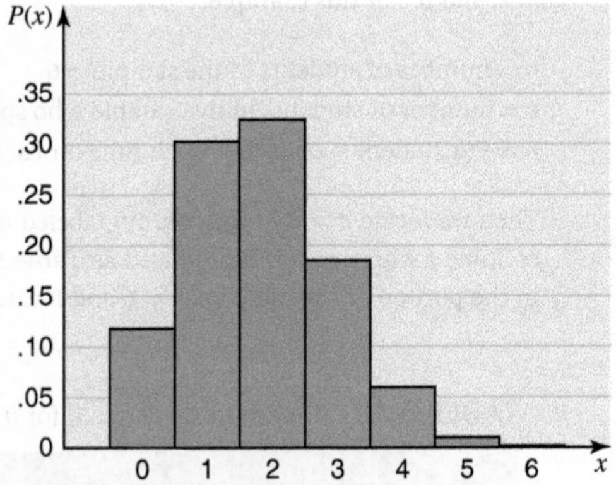

FIGURE 5.4 Histogram for the probability distribution of Table 5.12.

5.4.4 Probability of Success and the Shape of the Binomial Distribution

For any number of trials n:

1. The binomial probability distribution is symmetric if $p = .50$.
2. The binomial probability distribution is skewed to the right if p is less than .50.
3. The binomial probability distribution is skewed to the left if p is greater than .50.

These three cases are illustrated next with examples and graphs.

1. Let $n = 4$ and $p = .50$. Using Table I of Appendix B, we have written the probability distribution of x in **Table 5.13** and plotted the distribution in **Figure 5.5**. As we can observe from Table 5.13 and Figure 5.5, the probability distribution of x is symmetric.

TABLE 5.13 Probability Distribution of x for $n = 4$ and $p = .50$

x	$P(x)$
0	.0625
1	.2500
2	.3750
3	.2500
4	.0625

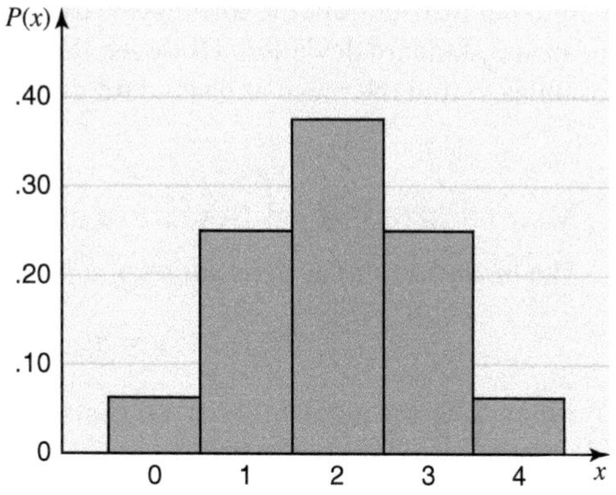

FIGURE 5.5 Histogram for the probability distribution of Table 5.13.

2. Let $n = 4$ and $p = .30$ (which is less than .50). **Table 5.14**, which is written by using Table I of Appendix B, and the graph of the probability distribution in **Figure 5.6** show that the probability distribution of x for $n = 4$ and $p = .30$ is skewed to the right.

TABLE 5.14 Probability Distribution of x for $n = 4$ and $p = .30$

x	$P(x)$
0	.2401
1	.4116
2	.2646
3	.0756
4	.0081

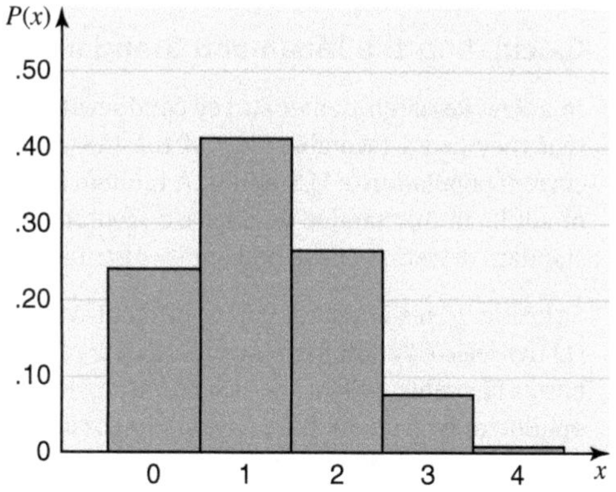

FIGURE 5.6 Histogram for the probability distribution of Table 5.14.

3. Let $n = 4$ and $p = .80$ (which is greater than .50). **Table 5.15**, which is written by using Table I of Appendix B, and the graph of the probability distribution in **Figure 5.7** show that the probability distribution of x for $n = 4$ and $p = .80$ is skewed to the left.

TABLE 5.15 Probability Distribution of x for $n = 4$ and $p = .80$

x	$P(x)$
0	.0016
1	.0256
2	.1536
3	.4096
4	.4096

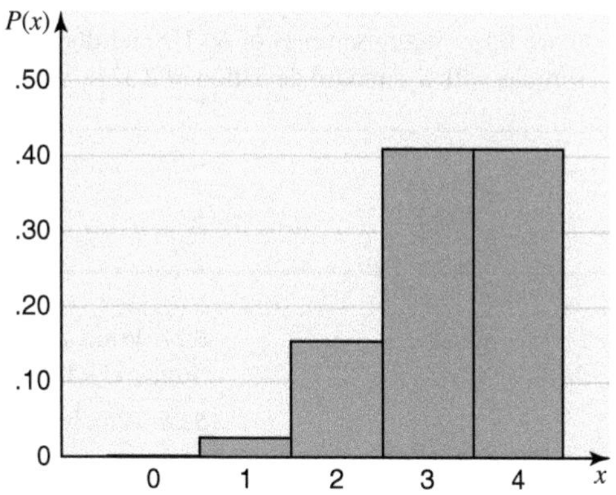

FIGURE 5.7 Histogram for the probability distribution of Table 5.15.

5.4.5 Mean and Standard Deviation of the Binomial Distribution

Section 5.3 explained how to compute the mean and standard deviation, respectively, for a probability distribution of a discrete random variable. When a discrete random variable has a binomial distribution, the formulas learned in Section 5.3 could still be used to compute its mean and standard deviation. However, it is simpler and more convenient to use the following formulas to find the mean and standard deviation of a binomial distribution.

Mean and Standard Deviation of a Binomial Distribution

The *mean* and *standard deviation of a binomial distribution* are, respectively,

$$\mu = np \quad \text{and} \quad \sigma = \sqrt{npq}$$

where n is the total number of trials, p is the probability of success, and q is the probability of failure.

Example 5.15 describes the calculation of the mean and standard deviation of a binomial distribution.

EXAMPLE 5.15 Views of U.S. Adults of the Postal Service

Calculating the Mean and Standard Deviation of a Binomial Distribution

In a Pew Research Center survey conducted during September 5–16, 2019, 90% of U.S. adults said that they had a favorable view of the U.S. postal service. Assume that this result is true for the current population of U.S. adults. A random sample of 60 U.S. adults is selected. Let x be the number of adults in this sample who have a favorable view of the U.S. postal service. Find the mean and standard deviation of the probability distribution of x.

Solution This is a binomial experiment with 60 trials (U.S. adults). Each trial has two outcomes: (1) the selected adult has a favorable view of the U.S. postal service, (2) the selected adult does not have a favorable view of the U.S. postal service (which includes the adults who have an unfavorable opinion or who do not have any opinion or do not want to answer the question). The probabilities p and q for these two outcomes are .90 and .10, respectively. Thus,

$$n = 60, \quad p = .90, \quad \text{and} \quad q = .10$$

Using the formulas for the mean and standard deviation of the binomial distribution, we obtain

$$\mu = np = 60\,(.90) = \mathbf{54}$$
$$\sigma = \sqrt{npq} = \sqrt{(60)(.90)(.10)} = \mathbf{2.3238}$$

Thus, the mean of the probability distribution of x is 54 and the standard deviation is 2.3238. The value of the mean is what we expect to obtain, on average, per repetition of the experiment. In this example, if we select many samples of 60 U.S. adults each, we expect to obtain an average of 54 adults per sample, with a standard deviation of 2.3238, who will view the U.S. postal service favorably.

EXERCISES

Concepts and Procedures

5.26 Briefly explain the following:

a. A binomial experiment

b. A trial

c. A binomial random variable

5.27 What are the parameters of the binomial probability distribution, and what do they mean?

5.28 Which of the following are binomial experiments? Explain why.

a. Rolling a die many times and observing the number of spots

b. Rolling a die many times and observing whether the number obtained is even or odd

c. Selecting a few voters from a very large population of voters and observing whether or not each of them favors a certain proposition in an election when 54% of all voters are known to be in favor of this proposition.

5.29 Which of the following are binomial experiments? Explain why.

a. Drawing 3 balls with replacement from a box that contains 12 balls, 8 of which are red and 4 are blue, and observing the colors of the drawn balls

b. Drawing 3 balls without replacement from a box that contains 10 balls, 6 of which are red and 4 are blue, and observing the colors of the drawn balls

c. Selecting a few households from New York City and observing whether or not they own stocks when it is known that 32% of all households in New York City own stocks

5.30 Let x be a discrete random variable that possesses a binomial distribution. Using the binomial formula, find the following probabilities:

a. $P(5)$ for $n = 6$ and $p = .50$

b. $P(2)$ for $n = 5$ and $p = .60$

c. $P(6)$ for $n = 8$ and $p = .20$

Verify your answers by using Table I of Appendix B.

5.31 Let x be a discrete random variable that possesses a binomial distribution.

a. Using Table I of Appendix B, write the probability distribution of x for $n = 7$ and $p = .30$ and graph it.

b. What are the mean and standard deviation of the probability distribution developed in part (a)?

5.32 The binomial probability distribution is symmetric for $p = .50$, skewed to the right for $p < .50$, and skewed to the left for $p > .50$. Illustrate each of these three cases by writing a probability distribution table and drawing a histogram. Choose any values of n (equal to 4 or higher) and p and use the table of binomial probabilities (Table I of Appendix B) to write the probability distribution tables.

Applications

5.33 According to a survey, 70% of households said that they have never purchased organic fruits or vegetables. Suppose that this result is true for the current population of households.

a. Let x be a binomial random variable that denotes the number of households in a random sample of 10 who have never purchased organic fruits or vegetables. What are the possible values that x can assume?

b. Find the probability that exactly 6 households in a random sample of 10 will say that they have never purchased organic fruits or vegetables. Use the binomial probability distribution formula.

5.34 According to a survey, 18% of the car owners said that they get the maintenance service done on their cars according to the schedule recommended by the auto company. Suppose that this result is true for the current population of car owners.

a. Let x be a binomial random variable that denotes the number of car owners in a random sample of 12 who get the maintenance service done on their cars according to the schedule recommended by the auto company. What are the possible values that x can assume?

b. Find the probability that exactly 3 car owners in a random sample of 12 get the maintenance service done on their cars according to the schedule recommended by the auto company. Use the binomial probability distribution formula.

5.35 In a poll, adult men aged 25–35 were asked, "If you go bald, will you consider hair transplant if you can afford it?" Thirty percent of them said, "Definitely." Suppose that this result is true for the current population of adult men aged 25–35. A random sample of 15 adult men aged 25–35 is selected. Use the binomial probabilities table (Table I of Appendix B) or technology to find the probability that the number of men in this sample of 15 who will say *definitely* in response to the said question is

a. at least 4 b. 1 to 3 c. at most 5

5.36 According to a survey conducted at the local DMV, 60% of the drivers who drive to work stated that they regularly exceed the posted speed limit on their way to work. Suppose that this result is true for the population of drivers who drive to work. A random sample of 15 drivers who drive to work is selected. Use the binomial probabilities table (Table I of Appendix B) or technology to find the probability that the number of drivers in this sample of 15 who regularly exceed the posted speed limit on their way to work is

a. at most 8 b. 5 to 10 c. at least 5

5.37 According to a poll, 63% of U.S. adults believe that climate change and global warming is a real problem that needs to be tackled urgently and immediately. Assume that this result is true of the current population of U.S. adults. Find the probability that in a random sample of four U.S. adults

a. all four will hold the said opinion

b. none of the four will hold the said opinion

5.38 A professional basketball player makes 90% of the free throws he tries. Assuming this percentage holds true for future attempts, use the binomial formula to find the probability that in the next 10 tries, the number of free throws he will make is

a. exactly 7 b. exactly 10

5.39 An office supply company conducted a survey before marketing a new paper shredder designed for home use. In the survey, 80% of the people who used the shredder were satisfied with it. Because of this high acceptance rate, the company decided to market the new shredder. Assume that 80% of all people who will use the new shredder will be satisfied. On a certain day, seven customers bought this shredder.

a. Let x denote the number of customers in this sample of seven who will be satisfied with this shredder. Using the binomial probabilities table (Table I, Appendix B), obtain the probability distribution of x and draw a histogram of the probability distribution. Find the mean and standard deviation of x.

b. Using the probability distribution of part (a), find the probability that exactly four of the seven customers will be satisfied.

5.40 Johnson Electronics makes calculators. Consumer satisfaction is one of the top priorities of the company's management. The company guarantees a refund or a replacement for any calculator that malfunctions within two years from the date of purchase. It is known from past data that despite all efforts, 5% of the calculators manufactured by the company malfunction within a two-year period. The company mailed a package of 20 randomly selected calculators to a store.

a. Let x denote the number of calculators in this package of 20 that will be returned for refund or replacement within a two-year period. Using the binomial probabilities table, obtain the probability distribution of x and draw a graph of the probability distribution. Determine the mean and standard deviation of x.

b. Using the probability distribution of part (a), find the probability that exactly 4 of the 20 calculators will be returned for refund or replacement within a two-year period.

5.5 Negative Binomial Probability Distribution

LEARNING OBJECTIVES

After completing this section, you should be able to:

- Identify the random variables that follow negative binomial distribution.
- Calculate the probability of *r*th success in total trials using negative binomial distribution.
- Calculate the mean and standard deviation of a negative binomial probability distribution using the formulas that apply only to the negative binomial probability distribution.

A negative binomial distribution is a discrete probability distribution. It was first considered by the mathematician B. Pascal in 1679. In 1714, another mathematician, P. R. de Montmort, used negative binomial distribution to conduct an experiment that observed the number of times a coin should be flipped to obtain specific number of heads. Further, researchers have observed the application of negative binomial distribution in different domains, which has led to the study of statistical properties of this distribution.

5.5.1 Characteristics of Negative Binomial Experiment

The experiment must have the following characteristics:

1. There are only two possible outcomes.
2. The trials are identical and independent of each other.
3. The probability (p) of success remains same from trial to trial.

The negative binomial distribution is applicable when we continuously conduct experiments until a desired number of r successes have been achieved, where r is a positive integer which is fixed in advance before experimenting.

For example, a coin is flipped with probability p for heads. The event that head appears on the fourth flip happens only when the coin lands tails three times and falls head at next flip. The probability of this event is $(1-p)^3\,p$. The event that second head appears on the fourth flip happens only when the coin lands exactly two tails and one head in three flips, irrespective of the position, and the last flip results in a head. The probability for this event is $3(1-p)^2\,p$ multiplied by p.

Probability Distribution of a Negative Binomial Distributed Random Variable

The probability distribution of a negative binomial distributed random variable is defined as follows:

Let

r = number of successes
x = number of trials needed to get required success
p = probability of success
$q = 1 - p$ = probability of failure

The probability that rth head appears on $(x + r)$th flips that consist of $r - 1$ heads and x tails in the $x + r - 1$ flips followed by head is defined as

$$P(x) = {}^{x+r-1}C_x(1-p)^x p^r, \quad x = 0, 1, 2, \ldots$$

The distribution having parameters p and r with the above probability function is called negative binomial distribution.

There is another form of negative binomial distribution that is widely used in the literature. The expression is explained further as

Let

r = number of successes
x = number of trials until the rth success
p = probability of success
$q = 1 - p$ = probability of failure

The probability function of x is given by

$$P(x) = {}^{x-1}C_{r-1}(1-p)^{x-r}p^r, \quad x = r, r+1, r+2, \ldots$$

where x denotes the number of trials until the rth success.
Notation: $x \sim NB(r,p)$.

The above form is used further for solving problems related to negative binomial distribution. The following examples illustrate the use of negative binomial distribution for solving the problems.

1. A person tends to buy lottery tickets on weekly basis for a lucky draw. He can win the lottery in any of the coming weeks, and he will buy tickets in a continuous manner, until he wins the lottery. The probability that the person wins the lottery in the third attempt can be derived using probability function of negative binomial distribution.
2. In a city, customers visit stores in a suburban mall to buy groceries and household essentials. Companies are more interested to observe the frequency of items purchased by the customers. A beverage company wants to conduct survey to find out whether customers' opinion can be influenced for second purchase if any reward is attached with that. Arrangements are made so that tokens are allotted to all customers who enter the mall in a queue. Tokenization will help the company to track the customer behavior. The probability that fourth person in the queue will make second purchase can be derived using negative binomial distribution.

5.5.2 Why Is It Called Negative Binomial Distribution?

The reason this distribution is termed as negative binomial distribution is that the cumulative sum of probabilities over range of x is equal to 1, but the final expression consists of negative term in the power.

$$\sum_{x=r}^{\infty} P(x) = \sum_{x=r}^{\infty} {}^{x-1}C_{r-1}p^r q^{x-r} = p^r \sum_{x=r}^{\infty} {}^{x-1}C_{r-1}q^{x-r} = p^r(1-q)^{-r}$$

where $(1-q)^{-r}$ is a binomial expansion with negative term.

EXAMPLE 5.16 Six Hits by a Batsman in a Cricket Match

Let x denote the number of balls played by a batsman until the second six hit by him in a cricket match of 20 overs. If the probability of hitting a six has probability .50, then find the probability that on the sixth ball the player will hit the second six.

Solution The probability of hitting a six on a ball is $p = .50$. The required probability that on the sixth ball the player will hit the second six is

$$P(6) = P(\text{1st six in five balls}) \cdot P(\text{2nd six on sixth ball})$$
$$= {}^5C_1 p^1 q^4 p = {}^5C_1 p^2 q^4 = {}^5C_1(.50)^2(.50)^4 = .0781$$

Thus, the probability that a batsman will score the second six on sixth ball (approximately) is 7% only.

EXAMPLE 5.17 Defective Items in a Manufacturing Unit

In a toothbrush manufacturing company, 5% of the items produced are defective. The company has introduced a new policy in the system, according to which, if there are three defects in the produced items during inspection of more than five items, then the complete lot would be replaced with the new one. This policy helps the company to increase its brand value in the market. What is the probability that at least five units be inspected in order to get three defective items?

Solution Let x be the number of units to be inspected to observe the probability that at least five units be inspected to get three defectives. We have to use probability density function of negative binomial distribution:

$$P(x \geq 5) = 1 - P(x < 5) = 1 - P(3) - P(4)$$

$$= 1 - {}^{2}C_{2}p^{2}q^{0}p - {}^{3}C_{2}p^{2}q^{1}p = 1 - p^{3} - 3p^{3}q$$

$$= 1 - (.05)^{3} - 3(.05)^{3}(.95) = .9995$$

Thus, the required probability is .9995.

5.5.3 Mean and Standard Deviation of the Negative Binomial Distribution

Computation of the mean and the standard deviation of a discrete random variable has been discussed in Section 5.3. As in Section 5.4.5, the following formulas are used to find the mean and standard deviation for a random variable following negative binomial distribution.

Mean and Standard Deviation of a Negative Binomial Distribution

The *mean* and *standard deviation of a negative binomial distribution* are, respectively,

$$\mu = E(x) = \sum_{x=r}^{\infty} x \cdot P(x) = \frac{r}{p}$$

$$\sigma = \sqrt{V(x)} = \sqrt{E(x^2) - E(x)^2} = \sqrt{\frac{rq}{p^2}}$$

where r denotes the number of successes, p denotes the probability of success, and q denotes the probability of failure.

EXAMPLE 5.18 Six Hits by a Batsman in a Cricket Match

Let x denote the number of balls played by a batsman until the second six hit him in a match of 20 overs . If the probability of hitting a six has probability .50, find the mean of the number of balls played by the batsman to score the second six. Also, derive the standard deviation.

Solution The mean of number of balls played by the batsman to score the second six is

$$E(x) = \frac{r}{p} = \frac{2}{.50} = 4$$

Similarly, the standard deviation can be derived as

$$\sigma = \sqrt{\frac{rq}{p^2}} = \sqrt{\frac{2\,(.5)}{(.5)^2}} = 2$$

EXAMPLE 5.19 Defective Items in a Manufacturing Unit

In a toothbrush manufacturing company, 5% of the items produced are defective. The company has introduced a new policy in the system, according to which, if there are three defects in the produced items during inspection of more than five items, then the complete lot would be replaced with the new one. This policy helps the company to increase its brand value in the market. Find the mean and standard deviation of items to be inspected to get three defectives.

Solution The mean of items to get three defectives is

$$E(x) = \frac{r}{p} = \frac{3}{.05} = 60$$

Similarly, the standard deviation can be derived as

$$\sigma = \sqrt{\frac{rq}{p^2}} = \sqrt{\frac{3(.95)}{(.05)^2}} = 33.76$$

5.5.4 Geometric Distribution

The geometric distribution is a special case of negative binomial distribution for $r = 1$. After putting $r = 1$, the probability density function for the negative binomial distribution will be transformed into geometric distribution.

Probability Distribution of a Geometric Distributed Random Variable

The probability distribution of x is given by

$$P(x) = (1 - p)^{x-1} p, \quad x = 1, 2, 3, \ldots$$

where x denotes the number of trials until the first success and p denotes the probability of success.

5.5.5 Mean and Standard Deviation of the Geometric Distribution

The following formulas are used to find the mean and standard deviation for a geometric distributed random variable.

Mean and Standard Deviation of a Geometric Distributed Random Variable

The *mean* and *standard deviation of a geometric distributed random variable* are, respectively,

$$\mu = E(x) = \sum_{x=1}^{\infty} x\,P(x) = \frac{1}{p}$$

$$\sigma = \sqrt{E(x^2) - E(x)^2} = \sqrt{\frac{q}{p^2}}$$

where p and q denote the probability of success and failure, respectively.

EXAMPLE 5.20 Medical Insurance Claims in an Insurance Company

Suppose that a medical insurance company observes claims on a policy in a year with probability .71. Find the probability of the first claim that the company receives on an insurance policy in second year.

Solution The probability of claims on a given policy has a probability .71. The probability of first claim in second year is

$$P(2) = (1 - p)p = (1 - .71)(.71) = .2059$$

The mean of number of claims on a policy in a year can be derived as

$$E(x) = \frac{1}{p} = \frac{1}{.71} = 14.084$$

The standard deviation is given by

$$\sigma = \sqrt{\frac{q}{p^2}} = \sqrt{\frac{.29}{(.71)^2}} = .7584$$

EXERCISES

Concepts and Procedures

5.41 Let $r = 3$ and $p = .25$. Using negative binomial probability distribution formula, find the following:

(a) $P(5)$ (b) $P(x \le 4)$

5.42 Let $r = 2$ and $p = .75$. Using negative binomial probability distribution formula, find the following:

(a) $P(4)$ (b) $P(x \le 3)$

Applications

5.43 In a school, 20% of the students scored 80% marks in a class test. Scholarship will be given to those students who score 80% marks two times in the class test. Find the probability that a student will be able to avail scholarship in exactly four class tests.

5.44 In a college, 10% of the athletes win a match in a season. Sneakers will be given to those athletes who win three matches in a season. Find the probability that an athlete will be awarded sneakers in exactly five matches.

5.6 The Hypergeometric Probability Distribution

LEARNING OBJECTIVES

After completing this section, you should be able to:

- Explain the hypergeometric distribution.
- Explain the similarities and differences between the hypergeometric random variable and the binomial random variable.
- Identify random variables that have a hypergeometric probability distribution.
- Calculate the probability of *x* successes in *n* trials using the hypergeometric formula.

In Section 5.4, we learned that one of the conditions required to apply the binomial probability distribution is that the trials are independent, so that the probabilities of the two outcomes or events (success and failure) remain constant. If the trials are not independent, we cannot apply the binomial probability distribution to find the probability of x successes in n trials. In some cases we can replace the binomial probability distribution by the **hypergeometric probability distribution**. Such a case occurs when a sample is drawn without replacement from a finite population.

As an example, suppose 5% of all auto parts manufactured at a company are defective. Four auto parts are selected at random. What is the probability that three of these four parts are good? Note that we are to find the probability that three of the four auto parts are good and one is defective. In this case, the population is very large and the probability of the first, second, third, and fourth auto parts being defective remains the same at .05. Similarly, the probability of any of the parts being good remains unchanged at .95. Consequently, we will apply the binomial probability distribution to find the probability of three good parts in four.

Now suppose this company shipped 25 auto parts to a dealer. Later on, the company finds out that five of those parts were defective. By the time the company manager contacts the dealer, four auto parts from that shipment have already been sold. What is the probability that three of those four parts were good parts and one part was defective? Here, because the four parts were selected without replacement from a small population, the probability of selecting a good part changes from the first selection to the second selection, to the third selection, and to the fourth selection. In this case we cannot apply the binomial probability distribution. In such instances, we use the hypergeometric probability distribution to find the required probability.

Examples 5.21 and 5.22 provide applications of the hypergeometric probability distribution.

Hypergeometric Probability Distribution

Let

$$N = \text{total number of elements in the population}$$
$$r = \text{number of successes in the population}$$
$$N - r = \text{number of failures in the population}$$
$$n = \text{number of trials (sample size)}$$
$$x = \text{number of successes in } n \text{ trials}$$
$$n - x = \text{number of failures in } n \text{ trials}$$

The probability of x successes in n trials is given by

$$P(x) = \frac{{}_rC_x \times {}_{N-r}C_{n-x}}{{}_NC_n}$$

EXAMPLE 5.21 Defective Auto Parts in a Shipment

Calculating the Probability Using the Hypergeometric Distribution Formula

Brown Manufacturing makes auto parts that are sold to auto dealers. Last week the company shipped 25 auto parts to a dealer. Later on, the company found out that five of those parts were defective. By the time the company manager contacted the dealer, four auto parts from that shipment had already been sold. What is the probability that three of those four parts are good parts and one part is defective?

Solution Let a good part be called a success and a defective part be called a failure. From the given information,

$$N = \text{total number of elements (auto parts) in the population} = 25$$
$$r = \text{number of successes (good parts) in the population} = 20$$
$$N - r = \text{number of failures (defective parts) in the population} = 5$$
$$n = \text{number of trials (sample size)} = 4$$
$$x = \text{number of successes in four trials} = 3$$
$$n - x = \text{number of failures in four trials} = 1$$

Using the hypergeometric formula, we calculate the required probability as follows:

$$P(3) = \frac{{}_rC_x \times {}_{N-r}C_{n-x}}{{}_NC_n} = \frac{{}_{20}C_3 \times {}_5C_1}{{}_{25}C_4} = \frac{\dfrac{20!}{3!\,(20-3)!} \cdot \dfrac{5!}{1!\,(5-1)!}}{\dfrac{25!}{4!\,(25-4)!}}$$

$$= \frac{(1140)\,(5)}{12{,}650} = \mathbf{.4506}$$

Thus, the probability that three of the four parts sold are good and one part is defective is .4506.

In the above calculations, the values of combinations can either be calculated using the formula learned in Section 4.6.3 (as done here) or by using a calculator.

EXAMPLE 5.22 Selecting Employees from a Group

Calculating the Probability Using the Hypergeometric Distribution Formula

Dawn Corporation has 12 employees who hold managerial positions. Of them, seven are females and five are males. The company is planning to send 3 of these 12 managers to a conference. If 3 managers are randomly selected out of 12,

a. find the probability that all 3 selected managers are females

b. find the probability that at most 1 of the 3 selected managers is a female

Solution Let the selection of a female be called a success and the selection of a male be called a failure.

a. From the given information,

$$\begin{aligned} N &= \text{total number of managers in the population} = 12 \\ r &= \text{number of successes (females) in the population} = 7 \\ N - r &= \text{number of failures (males) in the population} = 5 \\ n &= \text{number of selections (sample size)} = 3 \\ x &= \text{number of successes (females) in three selections} = 3 \\ n - x &= \text{number of failures (males) in three selections} = 0 \end{aligned}$$

Using the hypergeometric formula, we calculate the required probability as follows:

$$P(3) = \frac{{}_rC_x \times {}_{N-r}C_{n-x}}{{}_NC_n} = \frac{{}_7C_3 \times {}_5C_0}{{}_{12}C_3} = \frac{(35)(1)}{220} = \mathbf{.1591}$$

Thus, the probability that all three selected managers are females is .1591.

b. The probability that at most one of the selected managers is a female is given by the sum of the probabilities that either none or one of the selected managers is a female.

To find the probability that none of the selected managers is a female, we use

$$\begin{aligned} N &= \text{total number of managers in the population} = 12 \\ r &= \text{number of successes (females) in the population} = 7 \\ N - r &= \text{number of failures (males) in the population} = 5 \\ n &= \text{number of selections (sample size)} = 3 \\ x &= \text{number of successes (females) in three selections} = 0 \\ n - x &= \text{number of failures (males) in three selections} = 3 \end{aligned}$$

Using the hypergeometric formula, we calculate the required probability as follows:

$$P(0) = \frac{{}_rC_x \times {}_{N-r}C_{n-x}}{{}_NC_n} = \frac{{}_7C_0 \times {}_5C_3}{{}_{12}C_3} = \frac{(1)(10)}{220} = .0455$$

To find the probability that one of the selected managers is a female, we use

$$\begin{aligned} N &= \text{total number of managers in the population} = 12 \\ r &= \text{number of successes (females) in the population} = 7 \\ N - r &= \text{number of failures (males) in the population} = 5 \\ n &= \text{number of selections (sample size)} = 3 \\ x &= \text{number of successes (females) in three selections} = 1 \\ n - x &= \text{number of failures (males) in three selections} = 2 \end{aligned}$$

Using the hypergeometric formula, we obtain the required probability as follows:

$$P(1) = \frac{{}_rC_x \times {}_{N-r}C_{n-x}}{{}_NC_n} = \frac{{}_7C_1 \times {}_5C_2}{{}_{12}C_3} = \frac{(7)(10)}{220} = .3182$$

The probability that at most one of the three selected managers is a female is

$$P(x \le 1) = P(0) + P(1) = .0455 + .3182 = \mathbf{.3637}$$

EXERCISES

Concepts and Procedures

5.45 Explain the hypergeometric probability distribution. Under what conditions is this probability distribution applied to find the probability of a discrete random variable x? Give one example of an application of the hypergeometric probability distribution.

5.46 Let $N = 8$, $r = 4$, and $n = 4$. Using the hypergeometric probability distribution formula, find

a. $P(3)$ b. $P(4)$ c. $P(x \leq 2)$

5.47 Let $N = 10$, $r = 3$, and $n = 4$. Using the hypergeometric probability distribution formula, find

a. $P(1)$ b. $P(3)$ c. $P(x \leq 2)$

Applications

5.48 Six jurors are to be selected from a pool of 20 potential candidates to hear a civil case involving a lawsuit between two families. Unknown to the judge or any of the attorneys, 4 of the 20 prospective jurors are potentially prejudiced by being acquainted with one or more of the litigants. They will not disclose this during the jury selection process. If 6 jurors are selected at random from this group of 20, find the probability that the number of potentially prejudiced jurors among the 6 selected jurors is

a. exactly 1 b. none c. at most 2

5.49 A really bad carton of 18 eggs contains 7 spoiled eggs. An unsuspecting chef picks four eggs at random for his "Mega-Omelet Surprise." Find the probability that the number of unspoiled eggs among the four selected eggs is

a. exactly four b. two or fewer c. more than one

5.50 Bender Electronics buys keyboards for its computers from another company. The keyboards are received in shipments of 100 boxes, each box containing 20 keyboards. The quality control department at Bender Electronics first randomly selects one box from each shipment and then randomly selects five keyboards from that box. The shipment is accepted if not more than one of the five keyboards is defective. The quality control inspector at Bender Electronics selected a box from a recently received shipment of keyboards. Unknown to the inspector, this box contains six defective keyboards.

a. What is the probability that this shipment will be accepted?

b. What is the probability that this shipment will not be accepted?

5.7 The Poisson Probability Distribution

LEARNING OBJECTIVES

After completing this section, you should be able to:

- Explain the three conditions required to apply the Poisson probability distribution.
- Identify and explain the parameter of a Poisson probability distribution.
- Identify random variables that have a Poisson probability distribution.
- Calculate the probability of x occurrences of a discrete event using the Poisson probability distribution formula.
- Calculate the probability of x occurrences of a discrete event using the table of Poisson probabilities.
- Calculate the mean and standard deviation of a Poisson probability distribution using the formulas that apply only to the Poisson probability distribution.

The **Poisson probability distribution**, named after the French mathematician Siméon-Denis Poisson, is another important probability distribution of a discrete random variable that has a large number of applications. Suppose a washing machine in a laundromat breaks down an average of three times a month. We may want to find the probability of exactly two breakdowns during the next month. This is an example of a Poisson probability distribution problem. Each breakdown is called an **occurrence** in Poisson probability distribution terminology. The Poisson probability distribution is applied to experiments with random and independent occurrences. The occurrences are random in the sense that they do not follow any pattern, and, hence, they are unpredictable. Independence of occurrences means that one occurrence (or nonoccurrence) of an event does not influence the successive occurrences or nonoccurrences of that event. The occurrences are always considered with respect to an interval. In the example of the washing machine, the interval is one month. The interval may be a time interval, a space interval, or a volume interval. The actual number of occurrences within an interval is random and independent. If the average number of occurrences for a given interval is known, then by using the Poisson probability distribution, we can compute the probability of a certain number of occurrences, x, in that interval. Note that the number of actual occurrences in an interval is denoted by x.

Conditions to Apply the Poisson Probability Distribution

The following three conditions must be satisfied to apply the Poisson probability distribution.

1. x is a discrete random variable.
2. The occurrences are random.
3. The occurrences are independent.

The following are three examples of discrete random variables for which the occurrences are random and independent. Hence, these are examples to which the Poisson probability distribution can be applied.

1. Consider the number of telemarketing phone calls received by a household during a given day. In this example, the receiving of a telemarketing phone call by a household is called an occurrence, the interval is one day (an interval of time), and the occurrences are random (that is, there is no specified time for such a phone call to come in) and discrete. The total number of telemarketing phone calls received by a household during a given day may be 0, 1, 2, 3, 4, and so forth. The independence of occurrences in this example means that the telemarketing phone calls are received individually and none of two (or more) of these phone calls are related.
2. Consider the number of defective items in the next 100 items manufactured on a machine. In this case, the interval is a volume interval (100 items). The occurrences (number of defective items) are random and discrete because there may be 0, 1, 2, 3, . . . , 100 defective items in 100 items. We can assume the occurrence of defective items to be independent of one another.
3. Consider the number of defects in a 5-foot-long iron rod. The interval, in this example, is a space interval (5 feet). The occurrences (defects) are random because there may be any number of defects in a 5-foot iron rod. We can assume that these defects are independent of one another.

The following examples also qualify for the application of the Poisson probability distribution.

1. The number of accidents that occur on a given highway during a one-week period
2. The number of customers entering a grocery store during a one-hour interval
3. The number of television sets sold at a department store during a given week

In contrast, consider the arrival of patients at a physician's office. These arrivals are nonrandom if the patients have to make appointments to see the doctor. The arrival of commercial airplanes at an airport is nonrandom because all planes are scheduled to arrive at certain times, and airport authorities know the exact number of arrivals for any period (although this number may change slightly because of late or early arrivals and cancellations). The Poisson probability distribution cannot be applied to these examples.

In the Poisson probability distribution terminology, the average number of occurrences in an interval is denoted by λ (Greek letter ***lambda***). The actual number of occurrences in that interval is denoted by x. Then, using the Poisson probability distribution, we find the probability of x occurrences during an interval given that the mean number of occurrences during that interval is λ.

Poisson Probability Distribution Formula

According to the *Poisson probability distribution*, the probability of x occurrences in an interval is

$$P(x) = \frac{\lambda^x e^{-\lambda}}{x!}$$

where λ (pronounced ***lambda***) is the mean number of occurrences in that interval and the value of ***e*** is approximately **2.71828**.

The mean number of occurrences in an interval, denoted by λ, is called the *parameter of the Poisson probability distribution* or the **Poisson parameter**. As is obvious from the Poisson probability distribution formula, we need to know only the value of λ to compute the probability of any given value of x. We can read the value of $e^{-\lambda}$ for a given λ from Table II of Appendix B. Examples 5.23 through 5.25 illustrate the use of the Poisson probability distribution formula.

EXAMPLE 5.23 Telemarketing Phone Calls Received

Using the Poisson Formula: x Equals a Specific Value

On average, a household receives 9.5 telemarketing phone calls per week. Using the Poisson probability distribution formula, find the probability that a randomly selected household receives exactly six telemarketing phone calls during a given week.

Solution Let x be the number of telemarketing phone calls received by a household during a given week. Let λ be the mean number of telemarketing phone calls received by a household per week. Then, $\lambda = 9.5$. We are to find the probability of $x = 6$. Substituting all of the values in the Poisson formula, we obtain

$$P(6) = \frac{\lambda^x e^{-\lambda}}{x!} = \frac{(9.5)^6 e^{-9.5}}{6!} = \frac{(735{,}091.8906)(.00007485)}{720} = \mathbf{.0764}$$

To do these calculations, we can find the value of 6! either by using the factorial key on a calculator or by multiplying all integers from 1 to 6, and we can find the value of $e^{-9.5}$ by using the e^x key on a calculator or from Table II in Appendix B.

EXAMPLE 5.24 Washing Machine Breakdowns

Calculating Probabilities Using the Poisson Formula

A washing machine in a laundromat breaks down an average of three times per month. Using the Poisson probability distribution formula, find the probability that during the next month this machine will have

a. exactly two breakdowns

b. at most one breakdown

Solution Let x be the number of breakdowns observed during the next month for this machine. Let λ be the mean number of breakdowns per month. Then,

$$\lambda = 3$$

a. The probability that exactly two breakdowns will be observed during the next month is

$$P(2) = \frac{\lambda^x e^{-\lambda}}{x!} = \frac{(3)^2 e^{-3}}{2!} = \frac{(9)(.04978707)}{2} = \mathbf{.2240}$$

b. The probability that at most one breakdown will be observed during the next month is given by the sum of the probabilities of zero and one breakdown. Thus,

$$\begin{aligned} P(\text{at most 1 breakdown}) &= P(0 \text{ or } 1 \text{ breakdown}) = P(0) + P(1) \\ &= \frac{(3)^0 e^{-3}}{0!} + \frac{(3)^1 e^{-3}}{1!} \\ &= \frac{(1)(.04978707)}{1} + \frac{(3)(.04978707)}{1} \\ &= .0498 + .1494 = \mathbf{.1992} \end{aligned}$$

One important point about the Poisson probability distribution is that *the intervals for λ and x must be equal.* If they are not, the mean λ should be redefined to make them equal. Example 5.25 illustrates this point.

Remember

EXAMPLE 5.25 Returning the Purchased Items

Calculating Probability Using the Poisson Formula

Cynthia's Mail Order Company provides free examination of its products for seven days. If not completely satisfied, a customer can return the product within that period and get a full refund. According to past records of the company, an average of 2 of every 10 products sold by this company are returned for a refund. Using the Poisson probability distribution formula, find the probability that exactly 6 of the 40 products sold by this company on a given day will be returned for a refund.

Solution Let x denote the number of products in 40 that will be returned for a refund. We are to find $P(x = 6)$, which is usually written as $P(6)$. The given mean is defined per 10 products, but x is defined for 40 products. As a result, we should first find the mean for 40 products. Because, on average, 2 out of 10 products are returned, the mean number of products returned out of 40 will be 8. Thus, $\lambda = 8$. Substituting $x = 6$ and $\lambda = 8$ in the Poisson probability distribution formula, we obtain

$$P(6) = \frac{\lambda^x e^{-\lambda}}{x!} = \frac{(8)^6 e^{-8}}{6!} = \frac{(262{,}144)\,(.00033546)}{720} = \mathbf{.1221}$$

Thus, the probability is .1221 that exactly 6 products out of 40 sold on a given day will be returned.

Note that Example 5.25 is actually a binomial problem with $p = 2/10 = .20$, $n = 40$, and $x = 6$. In other words, the probability of success (that is, the probability that a product is returned) is .20 and the number of trials (products sold) is 40. We are to find the probability of six successes (returns). However, we used the Poisson distribution to solve this problem. This is referred to as *using the Poisson distribution as an approximation to the binomial distribution.* We can also use the binomial distribution to find this probability as follows:

$$P(6) = {}_{40}C_6\,(.20)^6\,(.80)^{34} = \frac{40!}{6!\,(40-6)!}(.20)^6\,(.80)^{34}$$

$$= (3{,}838{,}380)\,(.000064)\,(.00050706) = \mathbf{.1246}$$

Thus, the probability $P(6)$ is .1246 when we use the binomial distribution.

As we can observe, simplifying the above calculations for the binomial formula is a little complicated when n is large. It is much easier to solve this problem using the Poisson probability distribution. As a general rule, if the random variable has a binomial probability distribution with $n > 25$ but $\mu \le 25$, then we can use the Poisson probability distribution as an approximation to the binomial distribution. However, if $n > 25$ and $\mu > 25$, we prefer to use the normal distribution as an approximation to the binomial distribution. The latter case will be discussed in Chapter 6. However, if you are using technology, it does not matter how large n is. You can always use the binomial probability distribution to solve a binomial problem.

Case Study 5.2 presents an application of the Poisson probability distribution.

Case Study 5.2

Global Birth and Death Rates

The accompanying graph shows the average global birth and death rates. According to this information, on average, 4.17 children are born per second and 1.8 persons die per second in the world. If we assume that the global birth and death rates follow the Poisson probability distribution, we can find the probability of any given number of global births or deaths for a given time interval. For example, if x is the actual number of global births during a given one-second interval, then x can assume any (nonnegative integer) value, such as 0, 1, 2, 3, ... The same is true for the number of global deaths during a given one-second interval. For example, if y is the actual number of global deaths during a given one-second interval, then y can assume any (nonnegative integer) value, such as 0, 1, 2, 3, ... Here x and y are both discrete random variables.

Using the Poisson formula or technology, we can find the probability of any values of x and y. For example, if we want to find the probability of at most three global births during any given one-second interval, then, using $\lambda = 4.17$, we find this probability using technology as

$$P(x \le 3) = P(0) + P(1) + P(2) + P(3)$$
$$= .0155 + .0644 + .1343 + .1867$$
$$= .4009$$

Now suppose we want to find the probability of exactly six global births using the Poisson formula. This probability is

$$P(6) = \frac{\lambda^x e^{-\lambda}}{x!} = \frac{4.17^6 e^{-4.17}}{6!} = .1128$$

As mentioned earlier, let y be the number of global deaths in a given one-second interval. If we want to find the probability of at most two global deaths during any given one-second interval, then, using $\lambda = 1.8$, we find this probability from Table III as

$$\begin{aligned} P(y \le 2) &= P(0) + P(1) + P(2) \\ &= .1653 + .2975 + .2678 \\ &= .7306 \end{aligned}$$

Now suppose we want to find the probability of exactly three global deaths in a given one-second interval using the Poisson formula. This probability is

$$P(3) = \frac{\lambda^y e^{-\lambda}}{y!} = \frac{1.8^3 e^{-1.8}}{3!} = .1607$$

Using technology, we can prepare the probability distributions of x and y.

Data source: The International Data Base and U.S. Census Bureau.

5.7.1 Using the Table of Poisson Probabilities

The probabilities for a Poisson distribution can also be read from Table III in Appendix B, the table of Poisson probabilities. The following example describes how to read that table.

EXAMPLE 5.26 | New Bank Accounts Opened per Day

Using the Table of Poisson Probabilities

On average, two new accounts are opened per day at an Imperial Savings Bank branch. Using Table III of Appendix B, find the probability that on a given day the number of new accounts opened at this bank will be

a. exactly six **b.** at most three **c.** at least seven

Solution Let

λ = mean number of new accounts opened per day at this bank

x = number of new accounts opened at this bank on a given day

a. The values of λ and x are

$$\lambda = 2 \quad \text{and} \quad x = 6$$

In Table III of Appendix B, we first locate the column that corresponds to $\lambda = 2$. In this column, we then read the value for $x = 6$. The relevant portion of that table is shown here as **Table 5.16**. The probability that exactly six new accounts will be opened on a given day is .0120. Therefore,

$$P(6) = \mathbf{.0120}$$

TABLE 5.16 Portion of Table III for $\lambda = 2.0$

x	1.1	1.2	λ ...	2.0
0				.1353
1				.2707
2				.2707
3				.1804
4				.0902
5				.0361
6				.0120
7				.0034
8				.0009
9				.0002

($\leftarrow \lambda = 2.0$; $x = 6 \rightarrow$; $\leftarrow P(6)$)

Actually, Table 5.16 gives the probability distribution of x for $\lambda = 2.0$. Note that the sum of the 10 probabilities given in Table 5.16 is .9999 and not 1.0. This is so for two reasons. First, these probabilities are rounded to four decimal places. Second, on a given day more than nine new accounts might be opened at this bank. However, the probabilities of 10, 11, 12, . . . new accounts are very small, and they are not listed in the table.

b. The probability that at most three new accounts are opened on a given day is obtained by adding the probabilities of 0, 1, 2, and 3 new accounts. Thus, using Table III of Appendix B or Table 5.16, we obtain

$$\begin{aligned} P(\text{at most } 3) &= P(0) + P(1) + P(2) + P(3) \\ &= .1353 + .2707 + .2707 + .1804 = \mathbf{.8571} \end{aligned}$$

c. The probability that at least seven new accounts are opened on a given day is obtained by adding the probabilities of seven, eight, and nine new accounts. Note that 9 is the last value of x for $\lambda = 2.0$ in Table III of Appendix B or Table 5.16. Hence, 9 is the last value of x whose probability is included in the sum. However, this does not mean that on a given day more than nine new accounts cannot be opened. It simply means that the probability of 10 or more accounts is close to zero. Thus,

$$\begin{aligned} P(\text{at least } 7) &= P(7) + P(8) + P(9) \\ &= .0034 + .0009 + .0002 = \mathbf{.0045} \end{aligned}$$

EXAMPLE 5.27 Cars Sold per Day by a Salesman

Constructing a Poisson Probability Distribution and Graphing It

An auto salesperson sells an average of .9 car per day. Let x be the number of cars sold by this salesperson on any given day. Using the Poisson probability distribution table, write the probability distribution of x. Draw a histogram for the probability distribution.

Solution Let λ be the mean number of cars sold per day by this salesperson. Hence, $\lambda = .9$. Using the portion of Table III of Appendix B that corresponds to $\lambda = .9$, we write the probability distribution of x in **Table 5.17. Figure 5.8** shows the histogram for the probability distribution of Table 5.17.

Note that 6 is the largest value of x for $\lambda = .9$ listed in Table III for which the probability is greater than zero. However, this does not mean that this salesperson cannot sell more than six cars on a given day. What this means is that the probability of selling seven or more cars is very small. Actually, the probability of $x = 7$ for $\lambda = .9$ calculated by using the Poisson formula is .000039. When rounded to four decimal places, this probability is .0000, as listed in Table III.

TABLE 5.17 Probability Distribution of x for $\lambda = .9$

x	$P(x)$
0	.4066
1	.3659
2	.1647
3	.0494
4	.0111
5	.0020
6	.0003

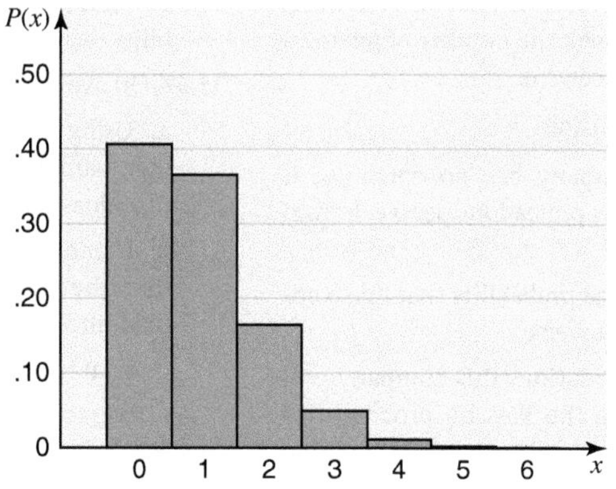

FIGURE 5.8 Histogram for the probability distribution of Table 5.17.

5.7.2 Mean and Standard Deviation of the Poisson Probability Distribution

For the Poisson probability distribution, the mean and variance both are equal to λ, and the standard deviation is equal to $\sqrt{\lambda}$. That is, for the Poisson probability distribution,

$$\mu = \lambda, \quad \sigma^2 = \lambda, \quad \text{and} \quad \sigma = \sqrt{\lambda}$$

For Example 5.27, $\lambda = .9$. Therefore, for the probability distribution of x in Table 5.17, the mean, variance, and standard deviation are, respectively,

$$\mu = \lambda = .9 \text{ car}$$
$$\sigma^2 = \lambda = .9$$
$$\sigma = \sqrt{\lambda} = \sqrt{.9} = .949 \text{ car}$$

EXERCISES

Concepts and Procedures

5.51 What are the conditions that must be satisfied to apply the Poisson probability distribution?

5.52 What is the parameter of the Poisson probability distribution, and what does it mean?

5.53 Using the Poisson formula, find the following probabilities:

a. $P(x \leq 1)$ for $\lambda = 4$

b. $P(3)$ for $\lambda = 3.5$

Verify these probabilities using Table III of Appendix B.

5.54 Let x be a Poisson random variable. Using the Poisson probabilities table, write the probability distribution of x for each of the following. Find the mean, variance, and standard deviation for each of these probability distributions. Draw a histogram for each of these probability distributions.

a. $\lambda = 2.5$ **b.** $\lambda = 3.5$

Applications

5.55 On average, 5.4 shoplifting incidents occur per week at an electronics store. Find the probability that exactly three such incidents will occur during a given week at this store. Use the Poisson probability distribution formula.

5.56 On average, 12.5 rooms stay vacant per day at a large hotel in a city. Find the probability that on a given day exactly four rooms will be vacant. Use the Poisson probability distribution formula.

5.57 A university police department receives an average of 3.7 reports per week of lost student ID cards.

a. Find the probability that at most one such report will be received during a given week by this police department. Use the Poisson probability distribution formula.

b. Using the Poisson probabilities table, find the probability that during a given week the number of such reports received by this police department is

i. one to four **ii.** at least six **iii.** at most three

5.58 A large proportion of small businesses in the United States fail during the first few years of operation. On average, 1.6 businesses file for bankruptcy per day in a particular large city.

a. Using the Poisson probability distribution formula, find the probability that exactly three businesses will file for bankruptcy on a given day in this city.

b. Using the Poisson probabilities table, find the probability that the number of businesses that will file for bankruptcy on a given day in this city is

i. two to three **ii.** more than three **iii.** less than three

5.59 Fred Johnson receives an average of 20 telemarketing calls per week. The number of such calls received by Fred Johnson during any given week follows a Poisson probability distribution.

a. Using the Poisson probability distribution formula, find the probability that during a given week Fred Johnson will receive exactly 13 telemarketing calls.

b. Using the Poisson probability distribution table, determine the probability that during a given week the number of telemarketing calls that Fred Johnson will receive is

i. 12 to 15 **ii.** fewer than nine

5.60 Although Borok's Electronics Company has no openings, it still receives an average of 3.2 unsolicited applications per week from people seeking jobs.

a. Using the Poisson formula, find the probability that this company will receive no applications next week.

b. Let x denote the number of applications this company will receive during a given week. Using the Poisson probabilities table from Appendix B, write the probability distribution table of x.

c. Find the mean, variance, and standard deviation of the probability distribution developed in part (b).

5.61 An average of .6 accident occur per day in a particular large city.

a. Find the probability that no accident will occur in this city on a given day.

b. Let x denote the number of accidents that will occur in this city on a given day. Write the probability distribution of x.

c. Find the mean, variance, and standard deviation of the probability distribution developed in part (b).

***5.62** On average, 30 households in 50 own answering machines.

a. Using the Poisson formula, find the probability that in a random sample of 50 households, exactly 25 will own answering machines.

b. Using the Poisson probabilities table, find the probability that the number of households in 50 who own answering machines is

i. at most 12 **ii.** 13 to 17 **iii.** at least 40

***5.63** Twenty percent of the cars passing through a school zone are exceeding the speed limit by more than 10 mph.

a. Using the Poisson formula, find the probability that in a random sample of 100 cars passing through this school zone, exactly 25 will exceed the speed limit by more than 10 mph.

b. Using the Poisson probabilities table, find the probability that the number of cars exceeding the speed limit by more than 10 mph in a random sample of 100 cars passing through this school zone is

i. at most 8 **ii.** 15 to 20 **iii.** at least 30

Uses and Misuses of Statistics

Batter Up!

In all of sport, there is no other sports that has the history and reputation for the use of statistics than baseball. The book and movie adaption of *Moneyball: The Art of Winning an Unfair Game* by Michael Lewis brought to light the importance of statistics in baseball to the general public. Of course, this is nothing new to true fans of the game who participate in "Fantasy Baseball" and have become experts in *sabermetrics*, the term coined by Bill James for the statistical analysis of baseball in-game activity—the term derives from the acronym SABR (Society for American Baseball Research). In fact, there is quite a following for data published by the Player Empirical Comparison and Optimization Test Algorithm (PECOTA) for projecting baseball player performance.

One of the most commonly reported baseball statistics is the batting average. Even though this is often considered to be a somewhat weak measure of player ability overall because of its singular dependence on one facet of the game, there is still keen interest in this statistic. In baseball, the batting average is defined as the number of hits divided by the number of times at bat. For example, if a batter goes to the plate 10 times and gets a hit on 3 of those occasions, then the batting average is $3/10 = .3$. In baseball parlance, this is translated to three decimal places, so we would report this batting average to be .300.

One would think that most players, inasmuch as batting is such an important part of the game, would have very high batting averages. However, Ty Cobb, who holds the record for highest career batting average, only hit .366; the lowest career average was recorded for Bill Bergen (.170). The highest single season average in the modern era was .426, which was achieved in 1901 by Napoleon Lajoie. Hitting a .400 is a goal of many players, though it is extraordinarily rare even for short periods of time, let alone an entire season or over a career.

To see an example of this, it is important to consider that it is not uncommon for an average player who plays regularly to have over 600 at-bats in a given season. Hence, such a player would have to get 240 hits in order to reach the elusive .400. We can perform some quick calculations to see just how rare it would be for a player of a given batting average to hit .400 or higher. For example, if a player normally bats a .350, how likely would he be able to maintain a batting average of .400 or higher over several at-bats? We can use the binomial distribution to answer this question. For example, we could calculate the probability of hitting .400 or higher over 10 chances at the plate. Out of 10 at-bats, this would be equivalent to calculating $P(x \geq 4) = 1 - P(x \leq 3)$. Using the binomial formula from this chapter, this probability works out to be .4862, or about a 48.62% chance that a regular .350 hitter will bat .400 or higher in 10 batting opportunities. If we repeat this calculation for a .300 average batter, a .250 average batter, and a .200 average batter, we can complete the following table:

Regular Season Batting Average	Probability of Batting at Least .400 in 10 At-Bats
.350	.4862
.300	.3504
.250	.2241
.200	.1209

As can be seen, the probability of hitting a .400 or better for the average batter, even over a relatively small number of bats, is very low. We can easily repeat this procedure for any number of batting opportunities.

In the table below, we examine up to 100 at-bats (or about 1/6 of a standard season) for players with a batting average ranging from .200 to .350.

Number of At-Bats	Probability of Batting at Least .400			
	Regular Season Batting Average			
	.350	.300	.250	.200
5	.572	.472	.367	.263
10	.486	.350	.224	.121
15	.436	.278	.148	.061
20	.399	.228	.102	.032
25	.370	.189	.071	.017
30	.345	.159	.051	.009
35	.324	.135	.036	.005
40	.305	.115	.026	.003
45	.289	.099	.019	.002
50	.274	.085	.014	.001
55	.260	.073	.010	.001
60	.247	.063	.007	<.001
65	.235	.055	.006	<.001
70	.225	.048	.004	<.001
75	.214	.041	.003	<.001
80	.205	.036	.002	<.001
85	.196	.031	.002	<.001
90	.188	.027	.001	<.001
95	.180	.024	.001	<.001
100	.172	.021	.001	<.001

It is clear that reaching .400 happens with very low probability. Even for a very good hitter (.350 average), maintaining a .400 or higher average declines very rapidly as the number of at-bats increases.

Here we have seen that the binomial distribution can be used to make projections of reaching the nearly unattainable .400 batting average. Of course, there are players who manage to exceed a batting average of .400 but just not for very long. There are many other factors that could play into high batting average (e.g., who is pitching, a player having a good batting day just by chance, etc.). But can a player maintain a high batting average for long? We should not bet on it.

Glossary

Bernoulli trial One repetition of a binomial experiment. Also called a **trial**.

Binomial experiment An experiment that contains n identical trials such that each of these n trials has only two possible outcomes (or events), the probabilities of these two outcomes (or events) remain constant for each trial, and the trials are independent.

Binomial parameters The total trials n and the probability of success p for the binomial probability distribution.

Binomial probability distribution The probability distribution that gives the probability of x successes in n trials when the probability of success is p for each trial of a binomial experiment.

Continuous random variable A random variable that can assume any value in one or more intervals.

Discrete random variable A random variable whose values are countable.

Geometric probability distribution The probability distribution that gives the probability of first successes in $(x + 1)$ trials when the probability of success is p for each trial of experiment where x denotes the number of trials needed to get first success.

Hypergeometric probability distribution The probability distribution that is applied to determine the probability of x successes in n trials when the trials are not independent.

Mean of a discrete random variable The mean of a discrete random variable x is the value that is expected to occur per repetition, on average, if an experiment is repeated a large number of times. The mean of a discrete random variable is also called its **expected value**.

Negative Binomial Probability Distribution The probability distribution that gives the probability of r successes in $(x + r)$ trials when the probability of success is p for each trial of experiment where x denotes the number of trials needed to get required success.

Poisson parameter The average occurrences, denoted by λ, during an interval for a Poisson probability distribution.

Poisson probability distribution The probability distribution that gives the probability of x occurrences in an interval when the average number of occurrences in that interval is λ.

Probability distribution of a discrete random variable A list of all the possible values that a discrete random variable can assume and their corresponding probabilities.

Random variable A variable, denoted by x, whose value is determined by the outcome of a random experiment. Also called a **chance variable**.

Standard deviation of a discrete random variable A measure of spread for the probability distribution of a discrete random variable.

Exercises

Supplementary Exercises

5.64 Let x be the number of bikes that a randomly selected auto mechanic repairs on a given day. The following table lists the probability distribution of x.

x	2	3	4	5	6	7
$P(x)$	.10	.25	.40	.15	.08	.02

Find the mean and standard deviation of x. Give a brief interpretation of the value of the mean.

5.65 Let x be the number of accidents happened on a busy route on a given Sunday. The following table lists the probability distribution of x.

x	0	1	2	3	4	5	6	7
$P(x)$	.05	.10	.20	.25	.20	.10	.07	.03

Calculate the mean and standard deviation of x. Give a brief interpretation of the value of the mean.

5.66 Based on its analysis of the future demand for its products, the financial department at Tipper Corporation has determined that there is a .17 probability that the company will lose $1.2 million during the next year, a .21 probability that it will lose $.7 million, a .37 probability that it will make a profit of $.9 million, and a .25 probability that it will make a profit of $2.3 million.

a. Let x be a random variable that denotes the profit earned by this corporation during the next year. Write the probability distribution of x.

b. Find the mean and standard deviation of the probability distribution of part (a). Give a brief interpretation of the value of the mean.

5.67 Spoke Weaving Corporation has eight weaving machines of the same kind and of the same age. The probability is .04 that any weaving machine will break down at any time. Find the probability that at any given time

a. all eight weaving machines will be broken down

b. exactly two weaving machines will be broken down

c. none of the weaving machines will be broken down

5.68 At the Bank of California, past data show that 8% of all credit card holders default at some time in their lives. On one recent day, this bank issued 12 credit cards to new customers. Find the probability that of these 12 customers, eventually

a. exactly 3 will default

b. exactly 1 will default

c. none will default

5.69 Maine Corporation buys motors for electric fans from another company that guarantees that at most 5% of its motors are defective and that it will replace all defective motors at no cost to Maine Corporation. The motors are received in large shipments. The quality control department at Maine Corporation randomly selects 20 motors from each shipment and inspects them for being good or defective. If this sample contains more than two defective motors, the entire shipment is rejected.

a. Using the appropriate probabilities table from Appendix B, find the probability that a given shipment of motors received by Maine Corporation will be accepted. Assume that 5% of all motors received are defective.

b. Using the appropriate probabilities table from Appendix B, find the probability that a given shipment of motors received by Maine Corporation will be rejected.

5.70 In a list of 20 households, 12 households own homes and 8 households do not own homes. Six households are randomly selected from these 20 households. Find the probability that the number of households in these six who own homes is

a. exactly four b. at most one c. exactly five

5.71 Alison Bender works for an accounting firm. To make sure her work does not contain errors, her manager randomly checks on her work. Alison recently filed 15 income tax returns for the company's clients. Alison unintentionally committed minor errors in 4 out of 15 tax returns and is also unaware of that. Alison's manager randomly selects 4 returns from these 15 returns. Find the probability that

a. exactly one of them contains errors

b. none of them contains errors

c. exactly three of them contain errors

5.72 The probability that a child is infected with a communicable disease in a class is .40. What is the probability that the tenth child in the class will be the third child that gets infection?

5.73 An average of 6.3 robberies occur per day in a large city.

a. Using the Poisson formula, find the probability that on a given day exactly three robberies will occur in this city.

b. Using the appropriate probabilities table from Appendix B, find the probability that on a given day the number of robberies that will occur in this city is

i. at least 12 ii. at most three iii. two to six

5.74 An average of 1.4 private airplanes arrive per hour at an airport.

a. Find the probability that during a given hour no private airplane will arrive at this airport.

b. Let x denote the number of private airplanes that will arrive at this airport during a given hour. Write the probability distribution of x. Use the appropriate probabilities table from Appendix B.

5.75 A high school boys' basketball team averages 1.2 technical fouls per game.

a. Using the appropriate formula, find the probability that in a given basketball game this team will commit exactly three technical fouls.

b. Let x denote the number of technical fouls that this team will commit during a given basketball game. Using the appropriate probabilities table from Appendix B, write the probability distribution of x.

Advanced Exercises

5.76 Scott offers you the following game: You will roll two fair dice. If the sum of the two numbers obtained is 2, 3, 4, 9, 10, 11, or 12, Scott will pay you $20. However, if the sum of the two numbers is 5, 6, 7, or 8, you will pay Scott $20. Scott points out that you have seven winning numbers and only four losing numbers. Is this game fair to you? Should you accept this offer? Support your conclusion with appropriate calculations.

5.77 Suppose the owner of a salvage company is considering raising a sunken ship. If successful, the venture will yield a net profit of $10 million. Otherwise, the owner will lose $4 million. Let p denote the probability of success for this venture. Assume the owner is willing to take the risk to go ahead with this project provided the expected net profit is at least $500,000.

a. If $p = .40$, find the expected net profit. Will the owner be willing to take the risk with this probability of success?

b. What is the smallest value of p for which the owner will take the risk to undertake this project?

5.78 York Steel Corporation produces a special bearing that must meet rigid specifications. When the production process is running properly, 10% of the bearings fail to meet the required specifications. Sometimes problems develop with the production process that cause the rejection rate to exceed 10%. To guard against this higher rejection rate, samples of 15 bearings are taken periodically and carefully inspected. If more than 2 bearings in a sample of 15 fail to meet the required specifications, production is suspended for necessary adjustments.

a. If the true rate of rejection is 10% (that is, the production process is working properly), what is the probability that the production will be suspended based on a sample of 15 bearings?

b. What assumptions did you make in part (a)?

5.79 Residents in an inner-city area are concerned about drug dealers entering their neighborhood. Over the past 14 nights, they have taken turns watching the street from a darkened apartment. Drug deals seem to take place randomly at various times and locations on the street and average about three per night. The residents of this street contacted the local police, who informed them that they do not have sufficient resources to set up surveillance. The police suggested videotaping the activity on the street, and if the residents are able to capture five or more drug deals on tape, the police will take action. Unfortunately, none of the residents on this street owns a video camera and, hence, they would have to rent the equipment. Inquiries at the local electronics shop indicated that the best available rate for renting a video camera is $75 for the first night and $40 for each additional night. To obtain this rate, the residents must sign up in advance for a specified number of nights. The residents hold a neighborhood meeting and invite you to help them decide on the length of the rental period. Because it is difficult for them to pay the rental fees, they want to know the probability of taping at least five drug deals on a given number of nights of videotaping.

a. Which of the probability distributions you have studied might be helpful here?

b. What assumption(s) would you have to make?

c. If the residents tape for two nights, what is the probability they will film at least five drug deals?

d. For how many nights must the camera be rented so that there is at least .90 probability that five or more drug deals will be taped?

5.80 A high school history teacher gives a 50-question multiple-choice examination in which each question has four choices. The scoring includes a penalty for guessing. Each correct answer is worth one point, and each wrong answer costs 1/2 point. For example, if a student answers 35 questions correctly, 8 questions incorrectly, and does not answer 7 questions, the total score for this student will be $35 - (1/2)(8) = 31$.

a. What is the expected score of a student who answers 38 questions correctly and guesses on the other 12 questions? Assume that the student randomly chooses one of the four answers for each of the 12 guessed questions.

b. Does a student increase his expected score by guessing on a question if he has no idea what the correct answer is? Explain.

c. Does a student increase her expected score by guessing on a question for which she can eliminate one of the wrong answers? Explain.

5.81 A history teacher has given her class a list of seven essay questions to study before the next test. The teacher announced that she will choose four of the seven questions to give on the test, and each student will have to answer three of those four questions.

a. In how many ways can the teacher choose four questions from the set of seven?

b. Suppose that a student has enough time to study only five questions. In how many ways can the teacher choose four questions from the set of seven so that the four selected questions include both questions that the student did not study?

c. What is the probability that the student in part (b) will have to answer a question that he or she did not study? That is, what is the probability that the four questions on the test will include both questions that the student did not study?

5.82 Consider the following three games. Which one would you be most likely to play? Which one would you be least likely to play? Explain your answer mathematically.

Game I: You toss a fair coin once. If a head appears you receive $3, but if a tail appears you have to pay $1.

Game II: You buy a single ticket for a raffle that has a total of 500 tickets. Two tickets are chosen without replacement from the 500. The holder of the first ticket selected receives $300, and the holder of the second ticket selected receives $150.

Game III: You toss a fair coin once. If a head appears you receive $1,000,002, but if a tail appears you have to pay $1,000,000.

5.83 The number of calls that come into a small mail-order company follows a Poisson distribution. Currently, these calls are serviced by a single operator. The manager knows from past experience that an additional operator will be needed if the rate of calls exceeds 25 per hour. The manager observes that nine calls came into the mail-order company during a randomly selected 10-minute period.

a. If the rate of calls is actually 15 per hour, what is the probability that seven or more calls will come in during a given 10-minute period?

b. If the rate of calls is really 30 per hour, what is the probability that seven or more calls will come in during a given 10-minute period?

c. Based on the calculations in parts (a) and (b), do you think that the rate of incoming calls is more likely to be 15 or 25 per hour?

d. Would you advise the manager to hire a second operator? Explain.

5.84 Many of you probably played the game "Rock, Paper, Scissors" as a child. Consider the following variation of that game. Instead of two players, suppose three players play this game, and let us call these players A, B, and C. Each player selects one of these three items—Rock, Paper, or Scissors—independent of each other. Player A will win the game if all three players select the same item, for example, rock. Player B will win the game if exactly two of the three players select the same item and the third player selects a different item. Player C will win the game if every player selects a different item. If Player B wins the game, he or she will be paid $1. If Player C wins the game, he or she will be paid $3. Assuming that the expected winnings should be the same for each player to make this a fair game, how much should Player A be paid if he or she wins the game?

Self-Review Test

Multiple Choice Questions

1. For the probability distribution of a discrete random variable, the probability of any single value of x is always

a. in the range 0 to 1 **b.** 1.0 **c.** less than 0

2. For the probability distribution of a discrete random variable, the sum of the probabilities of all possible values of x is always

a. greater than 1.0

b. 1.0

c. less than 1.0

3. The parameters of the binomial probability distribution are

a. n, p, and q **b.** n and p **c.** n, p, and x

4. The mean and standard deviation of a binomial probability distribution with $n = 25$ and $p = .20$ are

a. 5 and 2 b. 8 and 4 c. 5 and 3

5. The binomial probability distribution is symmetric if

a. $p < .5$ b. $p = .5$ c. $p > .5$

6. The binomial probability distribution is skewed to the right if

a. $p < .5$ b. $p = .5$ c. $p > .5$

7. The binomial probability distribution is skewed to the left if

a. $p < .5$ b. $p = .5$ c. $p > .5$

8. The parameter(s) of the Poisson probability distribution is(are)

a. λ b. λ and x c. λ and e

9. A continuous random variable is a random variable:

a. that can assume any value in one or more intervals

b. whose set of values is countable

c. that is derived from a random population

10. Which of the following is *not* an example of a discrete random variable?

a. The number of stocks a person owns.

b. The number of persons allergic to penicillin

c. The time spent by a physician with a patient

11. Which of the following is *not* a condition of the binomial experiment?

a. There are only two trials.

b. Each trial has two and only two outcomes.

c. The trials are independent.

12. The mean of a binomial distribution is equal to

a. npq b. np c. $(npq)^2$

13. The standard deviation of a binomial distribution is equal to

a. npq b. np c. $\sqrt{npq}$

14. Which of the following is an example of a binomial experiment?

a. Selecting five persons and observing whether they are in favor of an issue, against it, or have no opinion.

b. Tossing a coin 20 times and observing for a head or tail.

c. Drawing three marbles from a box that contains red, blue, and yellow marbles and observing for a color.

15. Twenty-eight percent of adults did not visit their physicians' offices last year. Let x be the number of adults in a random sample of 32 adults who did not visit their physicians' offices last year. The mean and standard deviation of the probability distribution of x, rounded to two decimal places, are

a. Part A: The mean is 8.96.
Part B: The standard deviation is 2.54.

b. Part A: The mean is 9.68.
Part B: The standard deviation is 2.54.

c. Part A: The mean is 9.68.
Part B: The standard deviation is 2.39.

16. The hypergeometric probability distribution can be used whenever

a. a sample is drawn at random with replacement

b. successive trials are independent of each other

c. the population is finite, and sampling occurs without replacement

17. Which of the following is *not* a condition to apply the Poisson probability distribution?

a. x is a discrete random variable.

b. There are n identical occurrences.

c. The occurrences are random.

Practice Exercises

1. The following table lists the probability distribution of a discrete random variable x.

x	0	1	2	3	4	5	6	7
$P(x)$	.04	.11	.18	.23	.11	.21	.09	.03

The probability that x is less than 5 is ______.

2. The following table lists the probability distribution of a discrete random variable x.

x	0	1	2	3	4	5	6	7
$P(x)$	.04	.11	.18	.22	.13	.20	.09	.03

The mean of the random variable x is ______.

3. The following table lists the probability distribution of a discrete random variable x.

x	2	3	4	5	6	7	8
$P(x)$	.15	.31	.20	.13	.13	.06	.02

The standard deviation of the random variable x, rounded to three decimal places, is ______.

4. Thirty-two percent of adults did not visit their physicians' offices last year. The probability, rounded to four decimal places, that in a random sample of 12 adults, exactly 3 will say they did not visit their physicians' offices last year is ______.

5. Thirty percent of law students who sit for a bar exam pass it in the first time. The probability, rounded to four decimal places, that in a random sample of 15 law students who will sit for the bar examination, maximum 5 will pass it the in the first time is ______.

6. A manufacturer packages bolts in boxes containing 100 each. Each box of 100 bolts contains, on an average, 2 defective bolts. The quality control staff randomly selects a box at the end of the day from an entire production run. What is the probability, rounded to four decimal places, that the box will contain exactly six defective bolts?

7. Historical data indicates that Rickenbacker Airlines receives an average of two complaints per day. What is the probability, rounded to four decimal places, that on a given day, Rickenbacker Airlines will receive at least six complaints?

8. Briefly explain the meaning of a random variable, a discrete random variable, and a continuous random variable. Give one example each of a discrete and a continuous random variable.

9. What name is given to a table that lists all the values that a discrete random variable x can assume and their corresponding probabilities?

10. State the four conditions of a binomial experiment. Give one example of such an experiment.

11. Briefly explain under what conditions a hypergeometric probability distribution is used. Give one example of a hypergeometric probability distribution.

12. Describe the three conditions that must be satisfied to apply the Poisson probability distribution.

13. Let x be the number of houses sold per week by all four real estate agents who work at a realty office. The following table lists the probability distribution of x.

x	0	1	2	3	4	5
$P(x)$	.15	.24	.29	.14	.10	.08

Calculate the mean and standard deviation of x. Give a brief interpretation of the value of the mean.

14. According to a survey, 60% of adults believe that all college students should be required to perform a specified number of hours of community service to graduate. Assume that this percentage is true for the current population of adults.

a. Find the probability that the number of adults in a random sample of 12 who hold this view is

i. exactly 8 (use the appropriate formula)

ii. at least 6 (use the appropriate table from Appendix B)

iii. less than 4 (use the appropriate table from Appendix B)

b. Let x be the number of adults in a random sample of 12 who believe that all college students should be required to perform a specified number of hours of community service to graduate. Using the appropriate table from Appendix B, write the probability distribution of x. Find the mean and standard deviation of x.

15. The Red Cross honors and recognizes its best volunteers from time to time. One of the Red Cross offices has received 12 nominations for the next group of four volunteers to be recognized. Eight of these 12 nominated volunteers are females. If the Red Cross office decides to randomly select 4 names out of these 12 nominated volunteers, find the probability that of these 4 volunteers

a. exactly 3 are females

b. exactly 1 is female

c. at most one is female

16. The police department in a large city has installed a traffic camera at a busy intersection. Any car that runs a red light will be photographed with its license plate visible, and the driver will receive a citation. Suppose that during the morning rush hour of weekdays, an average of 10 drivers are caught running the red light per day by this system.

a. Find the probability that during the morning rush hour on a given weekday this system will catch

i. exactly 14 drivers (use the appropriate formula)

ii. at most seven drivers (use the appropriate table from Appendix B)

iii. 13 to 18 drivers (use the appropriate table from Appendix B)

b. Let x be the number of drivers caught by this system during the morning rush hour on a given weekday. Write the probability distribution of x. Use the appropriate table from Appendix B.

17. The binomial probability distribution is symmetric when $p = .50$, it is skewed to the right when $p < .50$, and it is skewed to the left when $p > .50$. Illustrate these three cases by writing three probability distributions and graphing them. Choose any values of n (4 or higher) and p and use the table of binomial probabilities (Table I of Appendix B).

Mini-Projects

Mini-Project 5.1

Consider Data Set III on the NFL players from the 2019 football season that accompanies this text (see Appendix A).

a. Find the proportion of these players who have 10 or more years of playing experience in the NFL. This gives the probability of success p, which represents the probability that a selected player has 10 or more years of playing experience in the NFL.

b. Suppose a random sample of 22 of these NFL players is taken, and x is the number of players in the sample who have 10 or more years of playing experience in the NFL. Find $P(0)$, $P(1)$, $P(2)$, …, through $P(8)$.

Mini-Project 5.2

Obtain information on the odds and payoffs of one of the instant lottery games in your state or a nearby state. Let the random variable x be the net amount won on one ticket (payoffs minus purchase price). Using the concepts presented in this chapter, find the probability distribution of x. Then calculate the mean and standard deviation of x. What is the player's average net gain (or loss) per ticket purchased?

Mini-Project 5.3

For this project, first collect data by doing the following. Select an intersection in your town that is controlled by a traffic light. For a specific time period (e.g., 9–10 A.M. or 5–6 P.M.), count the number of cars that arrive at that intersection from any one direction during each light cycle. Make sure that you do not count a car twice if it has to sit through two red lights before getting through the intersection. Perform the following tasks with your data.

a. Create a graphical display of your data. Describe the shape of the distribution. Also discuss which of the following graphs is more useful for displaying the data you collected: a dotplot, a bar graph, or a histogram.

b. Calculate the mean and variance for your data for light cycles. Note that your sample size is the number of light cycles you observed. Do you notice a relationship between these two summary measures? If so, explain what it is.

c. For each unique number of arrivals in your data, calculate the proportion of light cycles that had that number of arrivals. For example, suppose you collected these data for 100 light cycles, and you observed eight cars arriving for each of 12 light cycles. Then, $12/100 = .12$ of the light cycles had eight arrivals. Also calculate the theoretical probabilities for each number of arrivals using the Poisson distribution with λ equal to the sample mean that you obtained in part (b). How do the two sets of probabilities compare? Is the Poisson a satisfactory model for your data?

Decide for Yourself

Note: The Decide for Yourself feature is located in the book's product page on *www.wiley.com*.

Chapter 5 Technology Instructions

Note: Complete TI-84, Minitab, and Excel manuals are available in the book's product page on *www.wiley.com*.

TI-84 Color/TI-84

The TI-84 Color Technology Instructions feature of this text is written for the TI-84 Plus C color graphing calculator running the 4.0 operating system. Some screens, menus, and functions will be slightly different in older operating systems. The TI-84 and TI-84 Plus can perform all of the same functions but will not have the "Color" option referenced in some of the menus.

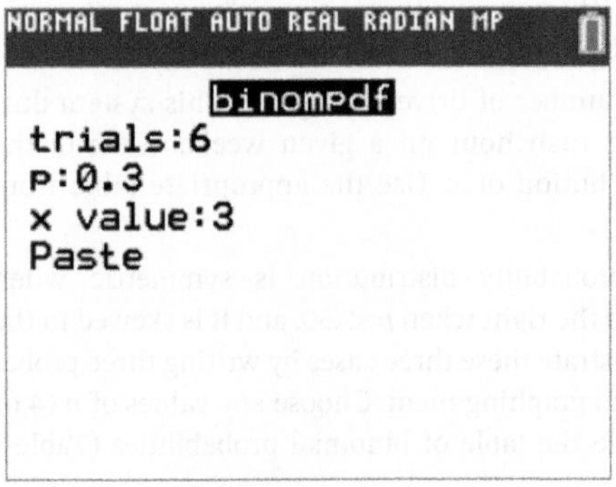

SCREEN 5.1

Calculating a Binomial Probability for Example 5.14(a) of the Text

1. Select 2nd > **VARS** > **binompdf(**.
2. Use the following settings in the **binompdf** menu (see **Screen 5.1**):
   - At the **trials** prompt, type 6.
   - At the **p** prompt, type 0.3.
   - At the **x value** prompt, type 3.
3. Highlight **Paste** and press **ENTER** twice.
4. The output $P(3) = .18522$ will appear on the screen.

Calculating a Cumulative Binomial Probability for Example 5.14(b) of the Text

1. Select 2nd > **VARS** > **binomcdf(**.
2. Use the following settings in the **binomcdf** menu:
   - At the **trials** prompt, type 6.
   - At the **p** prompt, type 0.3.
   - At the **x value** prompt, type 2.
3. Highlight **Paste** and press **ENTER** twice.
4. The output $P(x \leq 2) = .74431$ will appear on the screen.

   Note: The difference between **binompdf** and **binomcdf** is that **binompdf** calculates the probability $P(x = 2)$ while **binomcdf** calculates the cumulative probability $P(x \leq 2)$.

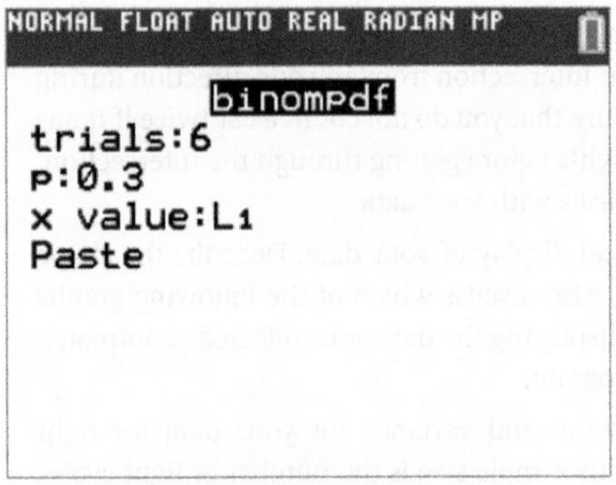

SCREEN 5.2

Preparing a Binomial Probability Distribution for Example 5.14(e) of the Text

1. Select **STAT** > **EDIT** > **Edit.**
2. Type 0, 1, 2, 3, 4, 5, and 6 into **L1.** These are all possible values of x for a binomial probability distribution when $n = 6$. (See **Screen 5.3.**)
3. Select 2nd > **MODE** to return to the home screen.
4. Select 2nd > **VARS** > **binompdf(**.
5. Use the following settings in the **binompdf** menu:
   - At the **trials** prompt, type 6.
   - At the **p** prompt, type 0.3.
   - At the **x value** prompt, select 2nd > **STAT** > **L1.** (See **Screen 5.2.**)
6. Highlight **Paste** and press **ENTER.**
7. Press **STO.**
8. Select 2nd > **STAT** > **L2.**
9. Press **ENTER.**
10. Select **STAT** > **EDIT** > **Edit...** to see the probability distribution. (See **Screen 5.3.**)

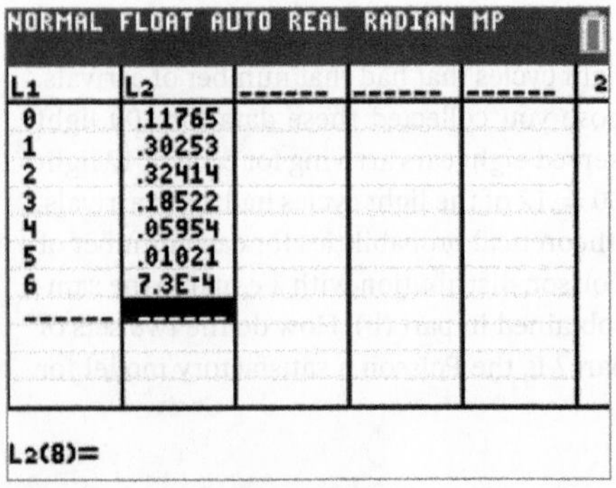

SCREEN 5.3

Calculating a Hypergeometric Probability for Example 5.21 of the Text

1. Access the Home screen.
2. Enter $_{20}C_3$ by following these steps (see **Screen 5.4**):
   - Type (20.
   - Select **MATH > PROB > nCr**.
   - Type 3).
3. Press ×.
4. Enter $_5C_1$ by following these steps (see **Screen 5.4**):
   - Type (5.
   - Select **MATH > PROB > nCr**.
   - Type 1).
5. Press ÷.
6. Enter $_{25}C_4$ by following these steps (see **Screen 5.4**):
   - Type 25.
   - Select **MATH > PROB > nCr**.
   - Type 4.
7. Press **ENTER**. $P(3) = .4505928854$ will appear on the screen.

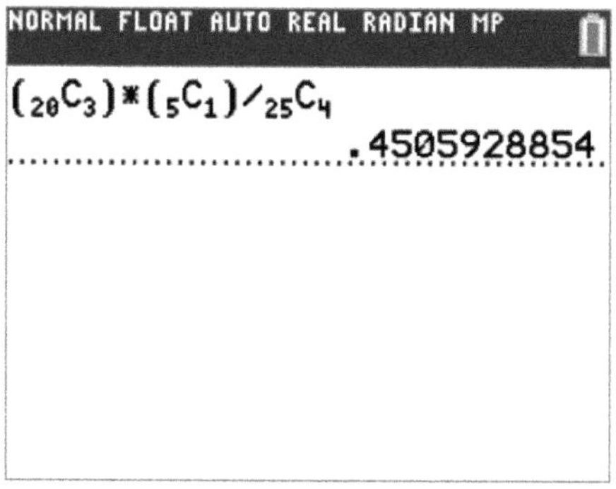

SCREEN 5.4

Calculating a Poisson Probability for Example 5.26(a) of the Text

1. Select **2nd > VARS > poissonpdf(**.
2. Use the following settings in the **poissonpdf** menu (see **Screen 5.5**):
   - At the λ prompt, type 2.
   - At the **x value** prompt, type 6.
3. Highlight **Paste** and press **ENTER** twice.
4. The output $P(6) = .012029803$ will appear on the screen.

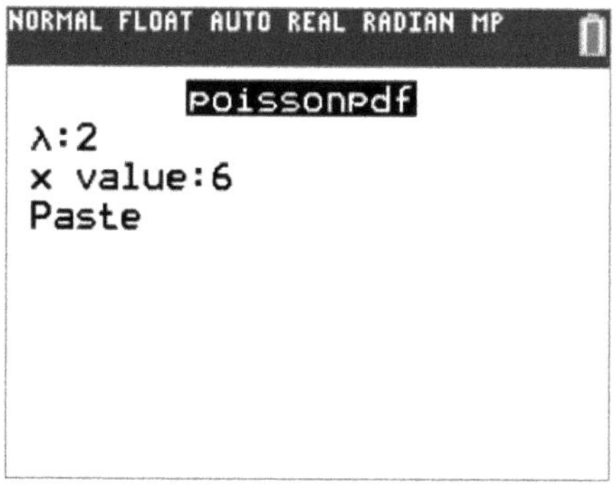

SCREEN 5.5

Calculating a Cumulative Poisson Probability for Example 5.26(b) of the Text

1. Select **2nd > VARS > poissoncdf(**.
2. Use the following settings in the **poissoncdf** menu:
   - At the λ prompt, type 2.
   - At the **x value** prompt, type 3.
3. Highlight **Paste** and press **ENTER** twice.
4. The output $P(x \le 3) = .8571234606$ will appear on the screen.

   Note: The difference between **poissonpdf** and **poissoncdf** is that **poissonpdf** calculates the probability $P(x = 3)$ while **poissoncdf** calculates the cumulative probability $P(x \le 3)$.

Minitab

The Minitab Technology Instructions in this text are written for **Minitab Professional Version 19**. Some screens, menus, and functions will be slightly different in older (or in newer) versions of Minitab. If you are using **Minitab Express**, you can download the Minitab Express Manual from the book's product page on *www.wiley.com*.

Calculating a Binomial Probability for Example 5.14(a) of the Text

1. Select **Calc > Probability Distributions > Binomial**.
2. Use the following settings in the dialog box that appears on screen (see **Screen 5.6**):
   - Select **Probability**.
   - In the **Number of trials** box, type 6.

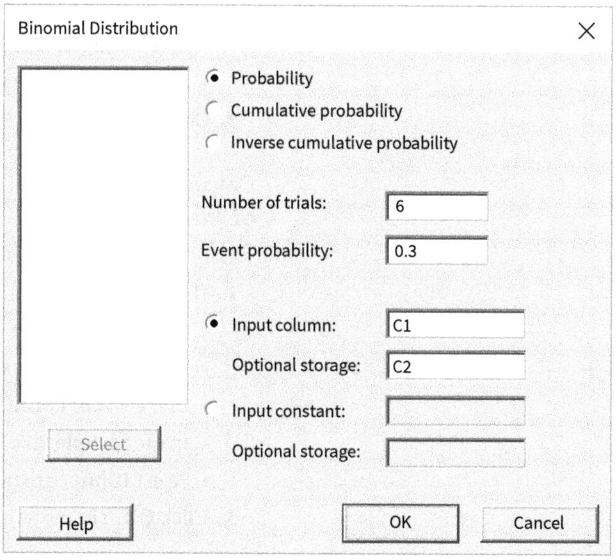

SCREEN 5.6

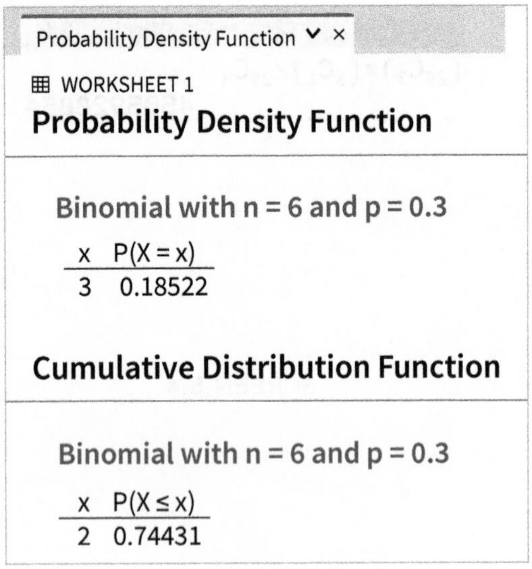

SCREEN 5.7

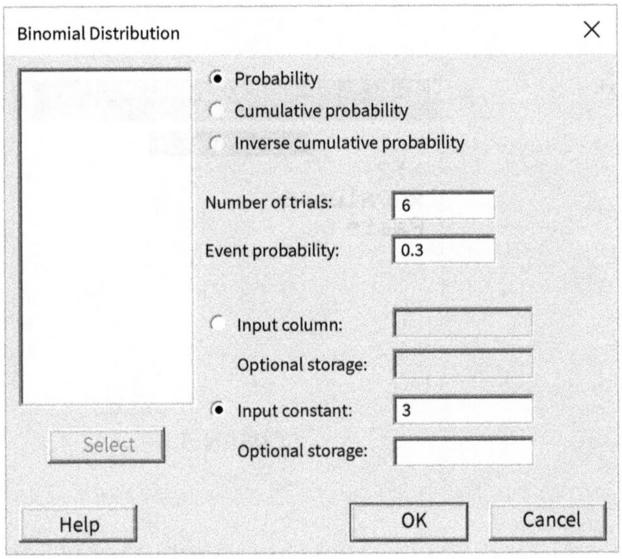

SCREEN 5.8

- In the **Event probability** box, type 0.3
- Select **Input constant** and type 3 in the **input constant** box.

3. Click **OK**. The output $P(3) = .18522$ will appear in the **Session** window. (See **Screen 5.7**.)

Calculating a Cumulative Binomial Probability for Example 5.14(b) of the Text

1. Select **Calc > Probability Distributions > Binomial.**
2. Use the following settings in the dialog box that appears on screen:
   - Select **Cumulative probability**.
   - Type 6 in the **Number of trials** box.
   - Type 0.3 in the **Event probability** box.
   - Select **Input constant** and type 2 in the **Input constant** box.
3. Click **OK**. The output $P(x \leq 2) = .74431$ will appear in the **Session** window. (See **Screen 5.7**.)

Preparing a Binomial Probability Distribution for Example 5.14(e) of the Text

1. Type 0, 1, 2, 3, 4, 5, and 6 into column C1. These are all possible values of x for a binomial probability distribution when $n = 6$.
2. Select **Calc > Probability Distributions > Binomial.**
3. Use the following settings in the dialog box that appears on screen (see **Screen 5.8**):
   - Select **Probability**.
   - Type 6 in the **Number of trials** box.
   - Type 0.3 in the **Event probability** box.
   - Select **Input column** and type C1 in the **Input column** box.
   - Type C2 in the **Optional storage** box.
4. Click **OK**. The output will appear in the **Worksheet** window.

Calculating a Hypergeometric Probability for Example 5.22(a) of the Text

1. Select **Calc > Probability Distributions > Hypergeometric.**
2. Use the following settings in the dialog box that appears on screen:
   - Select **Probability**.
   - In the **Population size** box, type 12.
   - In the **Event count in population** box, type 7.
   - In the **Sample size** box, type 3.
   - Select **Input constant** and type 3 in the **Input constant** box.
3. Click **OK**. The output $P(3) = .159091$ will appear in the **Session** window.

Calculating a Cumulative Hypergeometric Probability for Example 5.22(b) of the Text

1. Select **Calc > Probability Distributions > Hypergeometric.**
2. Use the following settings in the dialog box that appears on screen:
   - Select **Cumulative probability**.
   - In the **Population size** box, type 12.
   - In the **Event count in population** box, type 7.
   - In the **Sample size** box, type 3.
   - Select **Input constant** and type 1 in the **Input constant** box.
3. Click **OK**. The output $P(x \leq 1) = .363636$ will appear in the **Session** window.

Calculating a Poisson Probability for Example 5.26(a) of the Text

1. Select **Calc > Probability Distributions > Poisson**.
2. Use the following settings in the dialog box that appears on screen:
   - Select **Probability**.
   - In the **Mean** box, type 2.
   - Select **Input constant** and type 6 in the **Input constant** box.
3. Click **OK**. The output $P(6) = .0120298$ will appear in the **Session** window.

Calculating a Cumulative Poisson Probability for Example 5.26(b) of the Text

1. Select **Calc > Probability Distributions > Poisson**.
2. Use the following settings in the dialog box that appears on screen:
   - Select **Cumulative probability**.
   - In the **Mean** box, type 2.
   - Select **Input constant** and type 3 in the **Input constant** box.
3. Click **OK**. The output $P(x \leq 3) = .857123$ will appear in the **Session** window.

Excel

The Excel Technology Instructions feature of this text is written for Excel 2019. Some screens, menus, and functions will be slightly different in older versions of Excel. To enable some of the advanced Excel functions, you must enable the Data Analysis Add-In for Excel. For Excel 2007 and older versions of Excel, replace the function **BINOM.DIST** with the function **BINOMDIST**.

Calculating a Binomial Probability for Example 5.14(a) of the Text

1. Click on cell A1.
2. Type =**BINOM.DIST(3, 6, 0.3, 0)**. (See **Screen 5.9**.)

   Note: Entering a 0 as the fourth value in this function causes Excel to calculate the probability $P(x = 3)$.
3. Press **ENTER**. The probability $P(3) = .18522$ will appear in cell A1.

	A	B
1	=BINOM.DIST(3,6,0.3,0)	
2		

SCREEN 5.9

Calculating a Cumulative Binomial Probability for Example 5.14(b) of the Text

1. Click on cell A1.
2. Type =**BINOM.DIST(2, 6, 0.3, 1)**.

   Note: Entering a 1 as the fourth value in this function causes Excel to calculate the cumulative probability $P(x \leq 2)$.
3. Press **ENTER**. The probability $P(x \leq 2) = .74431$ will appear in cell A1.

Preparing a Binomial Probability Distribution for Example 5.14(e) of the Text

1. Type 0, 1, 2, 3, 4, 5, and 6 into cells A1 to A7. These are all possible values of x for a binomial probability distribution when $n = 6$.
2. Click on cell B1.
3. Type =**BINOM.DIST(A1, 6, 0.3, 0)**. (See **Screen 5.10**.)
4. Press **ENTER**.
5. Copy and paste the formula from cell B1 into cells B2 to B7.

	A	B
1	0	=BINOM.DIST(A1,6,0.3,0)
2	1	
3	2	
4	3	
5	4	
6	5	
7	6	

SCREEN 5.10

Calculating a Hypergeometric Probability for Example 5.22(a) of the Text

1. Click on cell A1.
2. Type =**HYPGEOM.DIST(3, 3, 7, 12, 0)**.

   Note: Entering a 0 as the fifth value in this function causes Excel to calculate the probability $P(x = 3)$.
3. Press **ENTER**. The probability $P(3) = .159091$ will appear in cell A1.

Calculating a Cumulative Hypergeometric Probability for Example 5.22(b) of the Text

1. Click on cell A1.
2. Type **=HYPGEOM.DIST(1, 3, 7, 12, 1)**.

   Note: Entering a 1 as the fifth value in this function causes Excel to calculate the cumulative probability $P(x \leq 1)$.
3. Press **ENTER**. The probability $P(x \leq 1) = .363636$ will appear in cell A1.

Calculating a Poisson Probability for Example 5.26(a) of the Text

1. Click on cell A1.
2. Type **=POISSON.DIST(6,2,0)**.

   Note: Entering a 0 as the third value in this function causes Excel to calculate the probability $P(x = 6)$.
3. Press **ENTER**. The probability $P(6) = .01203$ will appear in cell A1.

Calculating a Cumulative Poisson Probability for Example 5.26(b) of the Text

1. Click on cell A1.
2. Type **=POISSON.DIST(3,2,1)**.

   Note: Entering a 1 as the third value in this function causes Excel to calculate the cumulative probability $P(x \leq 3)$.
3. Press **ENTER**. The probability $P(x \leq 3) = .857123$ will appear in cell A1.

Technology Assignments

TA5.1 Forty-five percent of the adult population in a particular large city are women. A court is to randomly select a jury of 12 adults from the population of all adults of this city.

a. Find the probability that none of the 12 jurors is a woman.

b. Find the probability that at most 4 of the 12 jurors are women.

c. Let x denote the number of women in 12 adults selected for this jury. Obtain the probability distribution of x.

d. Using the probability distribution obtained in part (c), find the following probabilities:

i. $P(x > 6)$ **ii.** $P(x \leq 3)$ **iii.** $P(2 \leq x \leq 7)$

TA5.2 In a poll, 70% of U.S. adults said that CEOs of large U.S. companies are overpaid. Suppose this result is true for the current population of U.S. adults.

a. Find the probability that in a random sample of 40 U.S. adults, exactly 25 will say that CEOs of large U.S. companies are overpaid.

b. Find the probability that in a random sample of 80 U.S. adults, at least 60 will say that CEOs of large U.S. companies are overpaid.

c. Find the probability that in a random sample of 100 U.S. adults, at most 62 will say that CEOs of large U.S. companies are overpaid.

d. Find the probability that in a random sample of 70 U.S. adults, 44 to 56 will say that CEOs of large U.S. companies are overpaid.

e. Let x denote the number of adults in a random sample of 14 U.S. adults who will say that CEOs of large U.S. companies are overpaid. Prepare the probability distribution of x.

TA5.3 Suppose that 85% of the employees at a large company drink coffee daily. A manager takes a random sample of 15 employees.

a. Find the probability that all 15 employees drink coffee daily.

b. Find the probability that 10 of the 15 employees drink coffee daily.

c. Create the probability distribution of x. List the probability distribution in table form and display it in a histogram.

d. From the probability distribution in part (c), find the probability that at least 9 of the 15 employees drink coffee daily.

e. From the probability distribution in part (c), find the probability that at most 7 of the 15 employees drink coffee daily.

TA5.4 A friend has a coin that he believes is unfair (not balanced). He suspects that the coin will land heads 70% of the time it is tossed.

a. Assume for a moment that your friend's claim is correct. What is the probability that in 20 tosses, the coin would land heads 12 or fewer times?

b. If your friend actually tossed the coin 20 times and obtained 12 or fewer heads, would you have good reason to doubt that your friend's claim (that the coin lands heads 70% of the time) is true?

c. Let x be the number of heads in 20 tosses of this coin with a 70% chance of landing heads. Obtain the probability distribution of x.

d. Assuming that your friend's claim is true, what is the probability that in 100 tosses, the coin would land heads 60 or fewer times?

e. If your friend actually tossed the coin 100 times and obtained 60 or fewer heads, would you have good reason to doubt that your friend's claim (that the coin lands heads 70% of the time) is true?

TA5.5 The purpose of this assignment is to investigate some of the properties of the binomial probability distribution.

a. Suppose that for a binomial experiment with $n = 10$ trials, the probability of success is $p = .10$, and x is the number of successes. Obtain the probability distribution of x. Make a histogram for the probability distribution and comment on its shape. Find the mean and standard deviation of x.

b. Suppose that for a binomial experiment with $n = 10$ trials, the probability of success is $p = .50$, and x is the number of successes. Obtain the probability distribution of x. Make a histogram for the probability distribution and comment on its shape. Find the mean and standard deviation of x.

c. Suppose that for a binomial experiment with $n = 10$ trials, the probability of success is $p = .80$, and x is the number of successes. Obtain the probability distribution of x. Make a histogram for the probability distribution and comment on its shape. Find the mean and standard deviation of x.

d. Which of the distributions in parts (a–c) had the greatest variability in the number of successes? Why does this make sense?

TA5.6 An Internal Revenue Service inspector is to select 3 corporations from a list of 15 for tax audit purposes. Of the 15 corporations, 6 earned profits and 9 incurred losses during the year for which the tax returns are to be audited. If the IRS inspector decides to select three corporations randomly, find the probability that the number of corporations in these three that incurred losses during the year for which the tax returns are to be audited is

a. exactly two

b. none

c. at most one

TA5.7 A mail-order company receives an average of 40 orders per day.

a. Find the probability that it will receive exactly 55 orders on a certain day.

b. Find the probability that it will receive at most 29 orders on a certain day.

c. Find the following probabilities:

i. $P(x \geq 45)$ **ii.** $P(x < 33)$ **iii.** $P(36 < x < 52)$

TA5.8 A commuter airline receives an average of 13 complaints per week from its passengers. Let x denote the number of complaints received by this airline during a given week.

a. Find $P(x = 0)$. If your answer is zero, does it mean that this cannot happen? Explain.

b. Find $P(x \leq 10)$.

c. Obtain the probability distribution of x.

d. Using the probability distribution obtained in part (c), find the following probabilities:

i. $P(x > 18)$ **ii.** $P(x \leq 9)$ **iii.** $P(10 \leq x \leq 17)$

CHAPTER 6

Continuous Random Variables and the Normal Distribution

mauritius images GmbH/Alamy Stock Photo

Have you ever participated in a road race? If you have, where did you stand in comparison to the other runners? Do you think the time taken to finish a road race varies as much among runners as the runners themselves? See Case Study 6.1 for the distribution of times for runners who completed the Manchester (Connecticut) Road Race in 2019.

Discrete random variables and their probability distributions were presented in Chapter 5. Section 5.1 defined a continuous random variable as a variable that can assume any value in one or more intervals.

The possible values that a continuous random variable can assume are infinite and uncountable. For example, the variable that represents the time taken by a worker to commute from home to work is a continuous random variable. Suppose 5 minutes is the minimum time and 130 minutes is the maximum time taken by all workers to commute from home to work. Let x be a continuous random variable that denotes the time taken to commute from home to work by a randomly selected worker. Then x can assume any value in the interval 5 to 130 minutes. This interval contains an infinite number of values that are uncountable.

A continuous random variable can possess one of many probability distributions. In this chapter, we discuss the normal probability distribution and the use of the normal distribution as an approximation to the binomial distribution.

6.1 Continuous Probability Distribution and the Normal Probability Distribution

LEARNING OBJECTIVES

After completing this section, you should be able to:

- Explain the probability distribution of a continuous random variable.
- Describe the characteristics of a continuous probability distribution.
- Describe the characteristics of a normal distribution.
- Describe the standard normal distribution curve.
- Explain how changing the mean and/or the standard deviation changes the height and spread of a normal distribution.
- Find the area under the standard normal distribution curve for a given interval.
- Calculate a probability for the standard normal distribution for a given interval.

In this section, we will learn about the continuous probability distribution and its properties and then discuss the normal probability distribution.

In Chapter 5, we defined a **continuous random variable** as a random variable whose values are not countable. A continuous random variable can assume any value over an interval or intervals. Because the number of values contained in any interval is infinite, the possible number of values that a continuous random variable can assume is also infinite. Moreover, we cannot count these values. In Chapter 5, it was stated that the life of a battery, heights of people, time taken to complete an examination, amount of milk in a gallon container, weights of babies, and prices of houses are all examples of continuous random variables. Note that although money can be counted, variables involving money are often represented by continuous random variables. This is so because a variable involving money often has a very large number of outcomes.

6.1.1 Continuous Probability Distribution

Suppose 5000 female students are enrolled at a university, and x is the continuous random variable that represents the height of a randomly selected female student. **Table 6.1** lists the frequency and relative frequency distributions of x.

TABLE 6.1 Frequency and Relative Frequency Distributions of Heights of Female Students

Height of a Female Student (inches) x	f	Relative Frequency
60 to less than 61	90	.018
61 to less than 62	170	.034
62 to less than 63	460	.092
63 to less than 64	750	.150
64 to less than 65	970	.194
65 to less than 66	760	.152
66 to less than 67	640	.128
67 to less than 68	440	.088
68 to less than 69	320	.064
69 to less than 70	220	.044
70 to less than 71	180	.036
	$N = 5000$	Sum = 1.0

The relative frequencies given in Table 6.1 can be used as the probabilities of the respective classes. Note that these are exact probabilities because we are considering the population of all female students enrolled at the university.

Figure 6.1 displays the histogram and polygon for the relative frequency distribution of Table 6.1. **Figure 6.2** shows the smoothed polygon for the data of Table 6.1. The smoothed polygon is an approximation of the *probability distribution curve* of the continuous random variable x. Note that each class in Table 6.1 has a width equal to 1 inch. If the width of classes is more (or less) than 1 unit, we first obtain the *relative frequency densities* and then graph these relative frequency densities to obtain the distribution curve. The relative frequency density of a class is obtained by dividing the relative frequency of that class by the class width. The relative frequency densities are calculated to make the sum of the areas of all rectangles in the histogram equal to 1.0. Case Study 6.1, which appears later in this section, illustrates this procedure. The probability distribution curve of a continuous random variable is also called its *probability density function.*

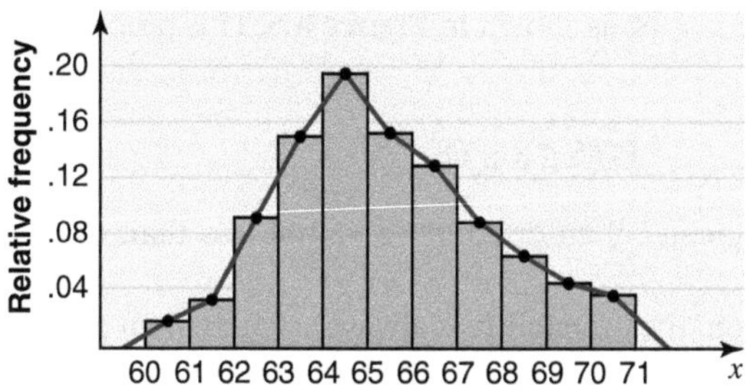

FIGURE 6.1 Histogram and polygon for Table 6.1.

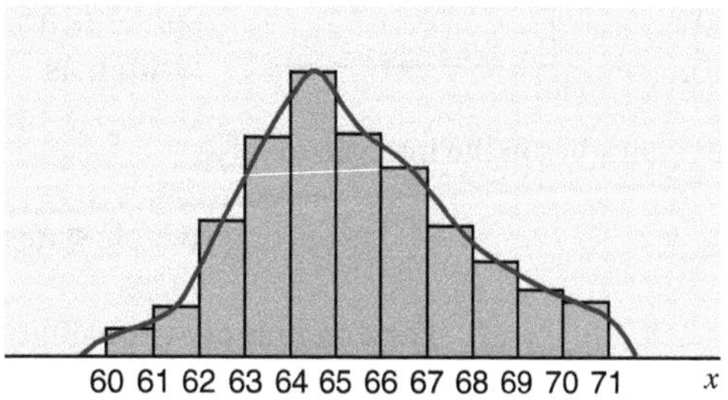

FIGURE 6.2 Probability distribution curve for heights.

The probability distribution of a continuous random variable possesses the following *two characteristics*:

1. The probability that x assumes a value in any interval lies in the range 0 to 1.
2. The total probability of all the (mutually exclusive) intervals within which x can assume a value is 1.0.

The first characteristic states that the area under the probability distribution curve of a continuous random variable between any two points is between 0 and 1, as shown in **Figure 6.3**. The second characteristic indicates that the total area under the probability distribution curve of a continuous random variable is always 1.0, or 100%, as shown in **Figure 6.4**.

The probability that a continuous random variable x assumes a value within a certain interval is given by the area under the curve between the two limits of the interval, as shown in **Figure 6.5**. The shaded area under the curve from a to b in this figure gives the probability that x falls in the interval a to b; that is,

$$P(a \leq x \leq b) = \text{Area under the curve from } a \text{ to } b$$

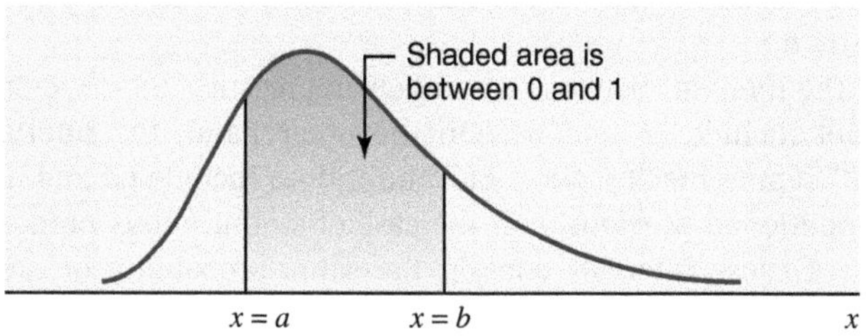

FIGURE 6.3 Area under a curve between two points.

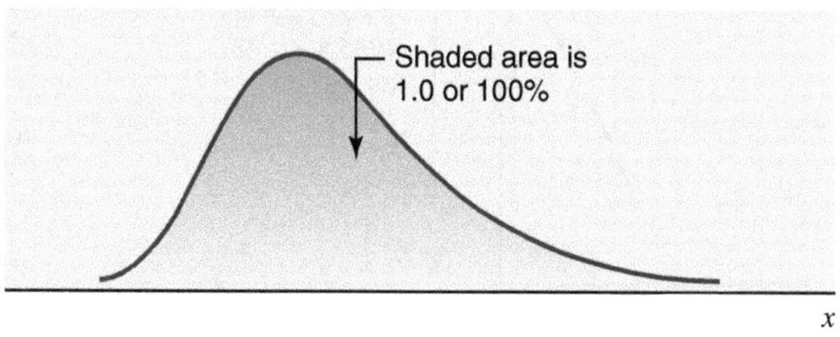

FIGURE 6.4 Total area under a probability distribution curve.

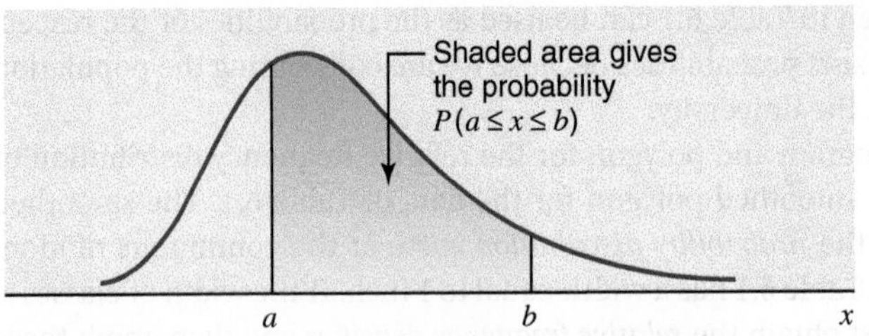

FIGURE 6.5 Area under the curve as probability.

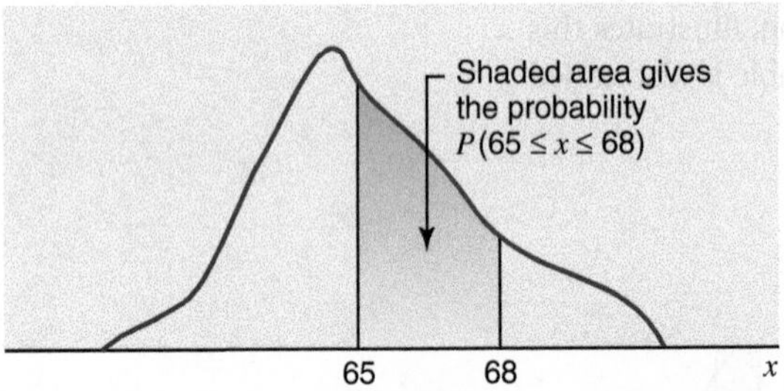

FIGURE 6.6 Probability that x lies in the interval 65 to 68.

Note that the interval $a \leq x \leq b$ states that x is greater than or equal to a but less than or equal to b.

Reconsider the example on the heights of all female students at a university. The probability that the height of a randomly selected female student from this university lies in the interval 65 to 68 inches is given by the area under the distribution curve of the heights of all female students from $x = 65$ to $x = 68$, as shown in **Figure 6.6**. This probability is written as

$$P(65 \leq x \leq 68)$$

which states that x is greater than or equal to 65 but less than or equal to 68.

For a continuous probability distribution, the probability is always calculated for an interval. For example, in Figure 6.6, the interval representing the shaded area is from 65 to 68. Consequently, the shaded area in that figure gives the probability for the interval $65 \leq x \leq 68$. In other words, this shaded area gives the probability that the height of a randomly selected female student from this university is in the interval 65 to 68 inches.

The probability that a continuous random variable x assumes a single value is always zero. This is so because the area of a line, which represents a single point, is zero. For example, if x is the height of a randomly selected female student from that university, then the probability that this student is exactly 66.8 inches tall is zero; that is,

$$P(x = 66.8) = 0$$

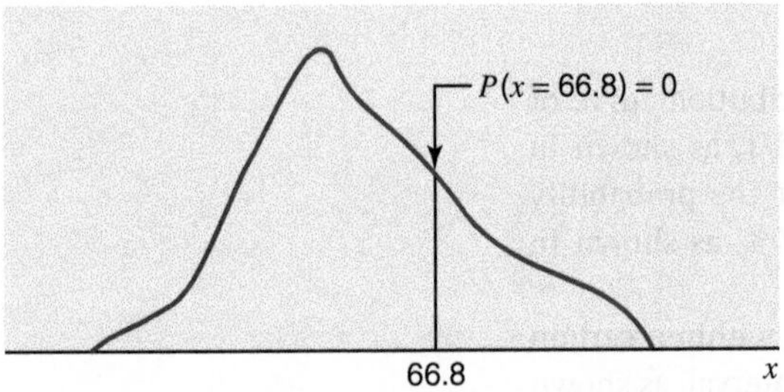

FIGURE 6.7 The probability of a single value of x is zero.

This probability is shown in **Figure 6.7**. Similarly, the probability for x to assume any other single value is zero.

In general, if a and b are two of the values that x can assume, then

$$P(a) = 0 \quad \text{and} \quad P(b) = 0$$

From this we can deduce that for a continuous random variable,

$$P(a \leq x \leq b) = P(a < x \leq b) = P(a \leq x < b) = P(a < x < b)$$

In other words, the probability that x assumes a value in the interval a to b is the same whether or not the values a and b are included in the interval. For the example on the heights of female students, the probability that a randomly selected female student is between 65 and 68 inches tall is the same as the probability that this female is 65 to 68 inches tall. This is shown in **Figure 6.8**.

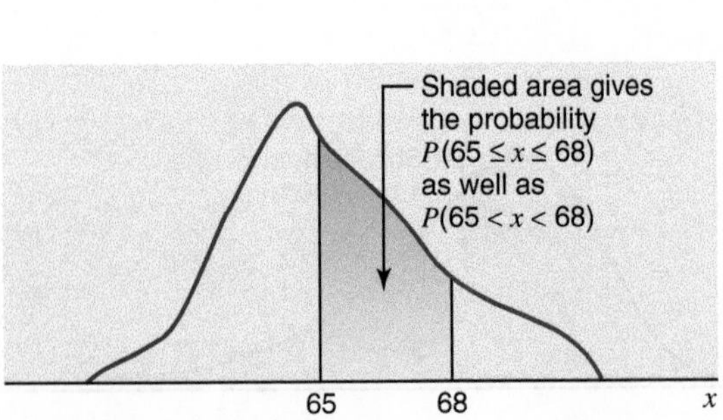

FIGURE 6.8 Probability "from 65 to 68" and "between 65 and 68."

Note that the interval "between 65 and 68" represents "$65 < x < 68$" and it does not include 65 and 68. On the other hand, the interval "from 65 to 68" represents "$65 \leq x \leq 68$" and it does include 65 and 68. However, as mentioned previously, in the case of a continuous random variable, both of these intervals contain the same probability or area under the curve.

Case Study 6.1 describes how we obtain the probability distribution curve of a continuous random variable.

Case Study 6.1

Distribution of Time Taken to Run a Road Race

The following table gives the frequency and relative frequency distributions for the time (in minutes) taken to complete the Manchester Road Race (held on November 28, 2019) by a total of 9956 participants who finished that race. This event is held every year on Thanksgiving Day in Manchester, Connecticut. The total distance of the course is 4.748 miles. The relative frequencies in the following table are used to construct the histogram and polygon in **Figure 6.9**.

Class	Frequency	Relative Frequency
20 to less than 25	49	.00492
25 to less than 30	205	.02059
30 to less than 35	565	.05675
35 to less than 40	1133	.11380
40 to less than 45	1337	.13429
45 to less than 50	1606	.16131
50 to less than 55	1744	.17517
55 to less than 60	1314	.13198
60 to less than 65	615	.06177
65 to less than 70	343	.03445
70 to less than 75	258	.02591
75 to less than 80	166	.01667
80 to less than 85	166	.01667
85 to less than 90	141	.01416
90 to less than 95	132	.01326
95 to less than 100	102	.01025
100 to less than 105	63	.00633
105 to less than 110	17	.00171
	$\Sigma f = 9956$	Sum = .99999

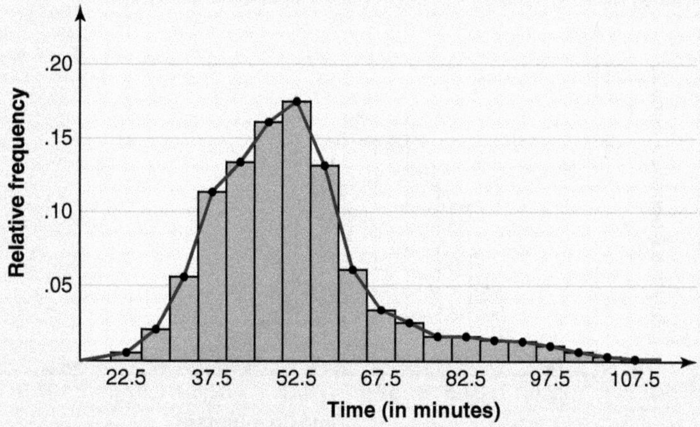

FIGURE 6.9 Histogram and polygon for the road race data.

To derive the probability distribution curve for these data, we calculate the relative frequency densities by dividing the relative frequencies by the class widths. The width of each class in the table is five. By dividing the relative frequencies by five, we obtain the relative frequency densities, which are recorded in the next table. Using the relative frequency densities, we draw a histogram and a smoothed polygon, as shown in **Figure 6.10**. The curve in this figure is the probability distribution curve for the road race data.

Note that the areas of the rectangles in Figure 6.9 do not give probabilities (which are approximated by relative frequencies). Rather here the heights of the rectangles give the probabilities. This is so because the base of each rectangle is five in this histogram. Consequently, the area of any rectangle is given by its height multiplied by five. Thus, the total area of all the rectangles in Figure 6.9 is 5.0, not 1.0. However, in Figure 6.10, it is the areas, not the heights, of rectangles that give the probabilities of the respective classes. Thus, if we add the areas of all the rectangles in Figure 6.10, we obtain the sum of all probabilities equal to 1.0. Consequently the total area under the probability distribution curve is equal to 1.0.

Class	Relative Frequency Density
20 to less than 25	.00098
25 to less than 30	.00412
30 to less than 35	.01135
35 to less than 40	.02276
40 to less than 45	.02686
45 to less than 50	.03226
50 to less than 55	.03503
55 to less than 60	.02640
60 to less than 65	.01235
65 to less than 70	.00689
70 to less than 75	.00518
75 to less than 80	.00333
80 to less than 85	.00333
85 to less than 90	.00283
90 to less than 95	.00265
95 to less than 100	.00205
100 to less than 105	.00127
105 to less than 110	.00034

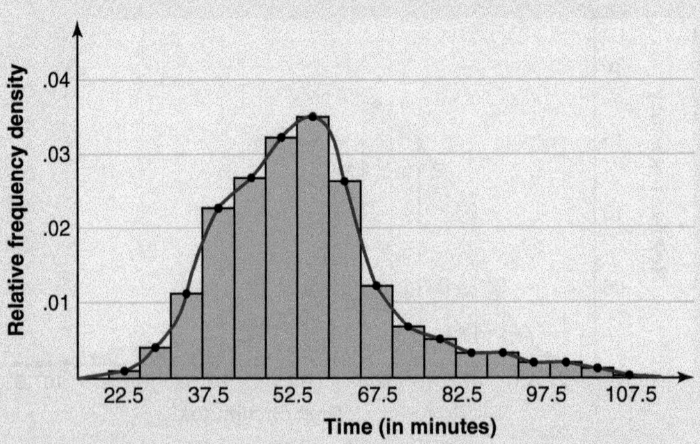

FIGURE 6.10 Probability distribution curve for the road race data.

The probability distribution of a continuous random variable has a mean and a standard deviation, which are denoted by μ and σ, respectively. The mean and standard deviation of the probability distribution given in the previous table and Figure 6.10 are 51.996 and 14.669 minutes, respectively. These values of μ and σ are calculated by using the data on all 9956 participants.

Source: This case study is based on data published on the official website of the Manchester Road Race.

6.1.2 The Normal Distribution

The normal distribution is one of the many probability distributions that a continuous random variable can possess. The normal distribution is the most important and most widely used of all continuous probability distributions. A large number of phenomena in the real world are approximately normally distributed. The continuous random variables representing heights and weights of people, scores on an examination, weights of packages (e.g., cereal boxes, boxes of cookies), amount of milk in a gallon container, life of an item (such as a lightbulb or a battery), and time taken to complete a certain job have all been observed to have an approximate normal distribution.

A **normal probability distribution** or a *normal curve* is a bell-shaped (symmetric) curve. Such a curve is shown in **Figure 6.11**. Its mean is denoted by μ and its standard deviation by σ. A continuous random variable x that has a normal distribution is called a *normal random variable.* Note that not all bell-shaped curves represent a normal distribution curve. Only a specific kind of bell-shaped curve represents a normal curve.

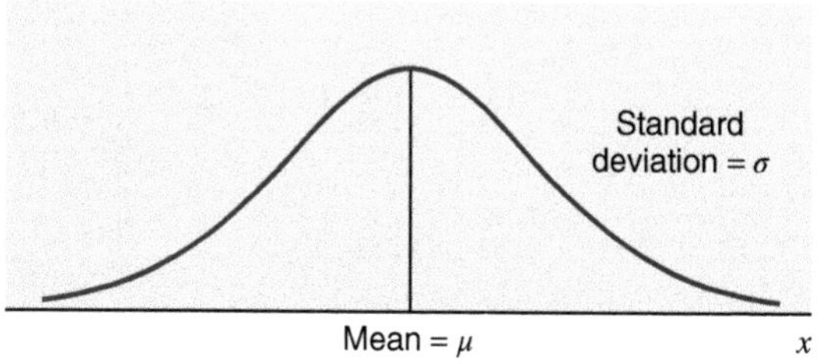

FIGURE 6.11 Normal distribution with mean μ and standard deviation σ.

Normal Probability Distribution

A *normal probability distribution,* when plotted, gives a bell-shaped curve such that:

1. The total area under the curve is 1.0.
2. The curve is symmetric about the mean.
3. The two tails of the curve extend indefinitely.

A normal distribution possesses the following three characteristics:

1. The total area under a normal distribution curve is 1.0, or 100%, as shown in **Figure 6.12.**

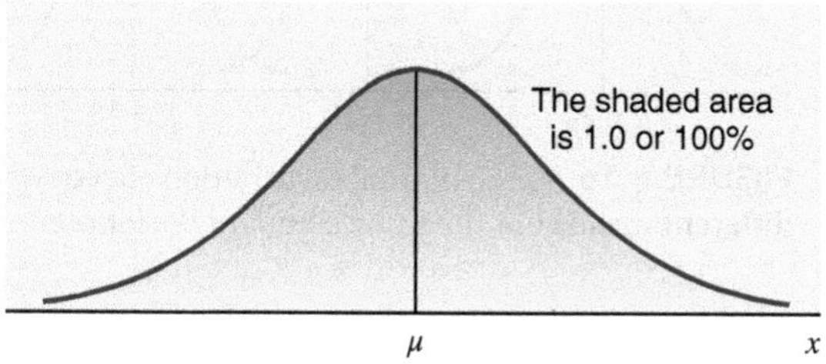

FIGURE 6.12 Total area under a normal curve.

2. A normal distribution curve is symmetric about the mean, as shown in **Figure 6.13**. Consequently, 50% of the total area under a normal distribution curve lies on the left side of the mean, and 50% lies on the right side of the mean.

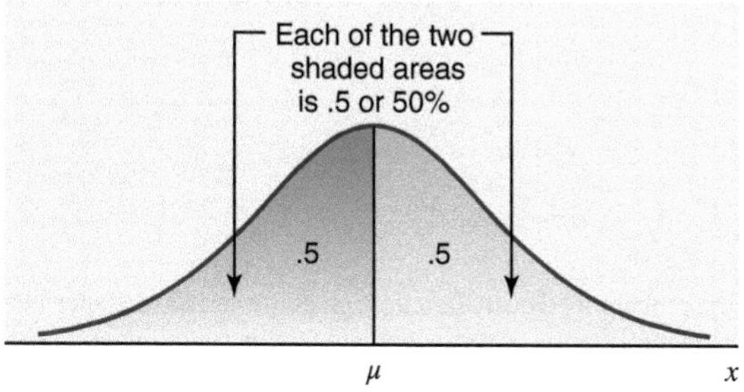

FIGURE 6.13 A normal curve is symmetric about the mean.

3. The tails of a normal distribution curve extend indefinitely in both directions without touching or crossing the horizontal axis. Although a normal distribution curve never meets the horizontal axis, beyond the points represented by $\mu - 3\sigma$ and $\mu + 3\sigma$ it becomes so close to this axis that the area under the curve beyond these points in both directions is very small and can be taken as very close but not equal to zero. The actual area in each tail of the standard normal distribution curve beyond three standard deviations of the mean is .0013. These areas are shown in **Figure 6.14**.

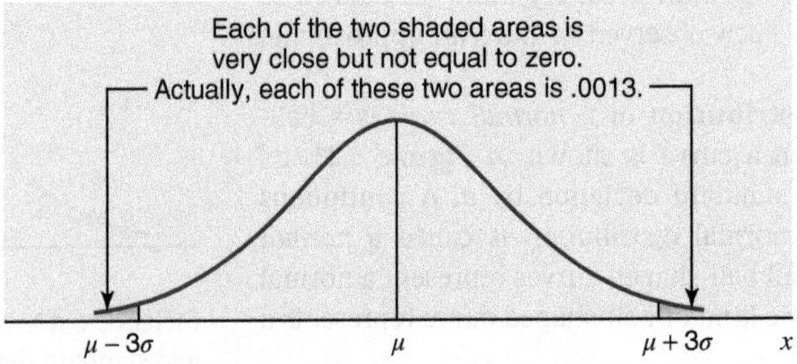

FIGURE 6.14 Areas of the normal curve beyond $\mu \pm 3\sigma$.

The mean, μ, and the standard deviation, σ, are the **parameters** of the normal distribution. Given the values of these two parameters, we can find the area under a normal distribution curve for any interval. Remember, there is not just one normal distribution curve but a *family* of normal distribution curves. Each different set of values of μ and σ gives a normal distribution curve with different height and/or spread. The value of μ determines the center of a normal distribution curve on the horizontal axis, and the value of σ gives the spread of the normal distribution curve. The three normal distribution curves shown in **Figure 6.15** have the same mean but different standard deviations. By contrast, the three normal distribution curves in **Figure 6.16** have different means but the same standard deviation.

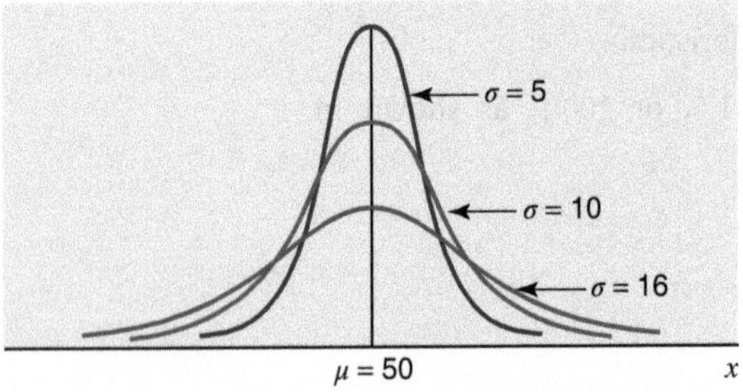

FIGURE 6.15 Three normal distribution curves with the same mean but different standard deviations.

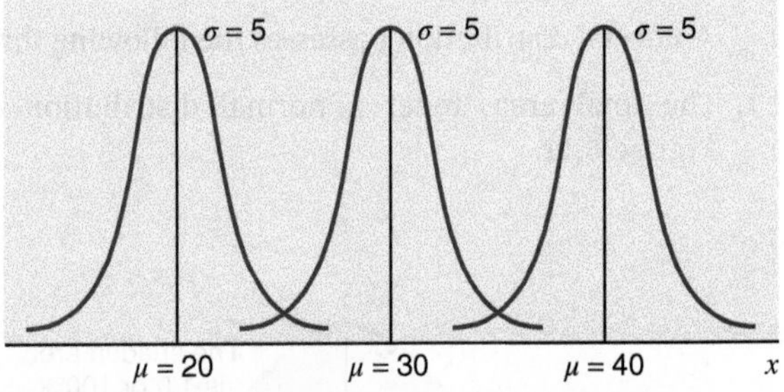

FIGURE 6.16 Three normal distribution curves with different means but the same standard deviation.

Like the binomial and Poisson probability distributions discussed in Chapter 5, the normal probability distribution can also be expressed by a mathematical equation.[1] However, we will not use this equation to find the area under a normal distribution curve. Instead, we will use Table IV of Appendix B.

6.1.3 The Standard Normal Distribution

The **standard normal distribution** is a special case of the normal distribution. For the standard normal distribution, the value of the mean is equal to zero and the value of the standard deviation is equal to 1.

[1]The equation of the normal distribution is

$$f(x) = \frac{1}{\sigma\sqrt{2\pi}} e^{-(1/2)[(x-\mu)/\sigma]^2}$$

where $e \approx 2.71828$ and $\pi \approx 3.14159$; $f(x)$, called the probability density function, gives the vertical distance between the horizontal axis and the curve at point x. For the information of those who are familiar with integral calculus, the definite integral of this function from a to b gives the probability that x assumes a value between a and b.

Standard Normal Distribution

The normal distribution with $\mu = 0$ and $\sigma = 1$ is called the *standard normal distribution.*

Figure 6.17 displays the standard normal distribution curve. The random variable that possesses the standard normal distribution is denoted by z. In other words, the units for the standard normal distribution curve are denoted by z and are called the **z values** or **z scores**. They are also called *standard units* or *standard scores.*

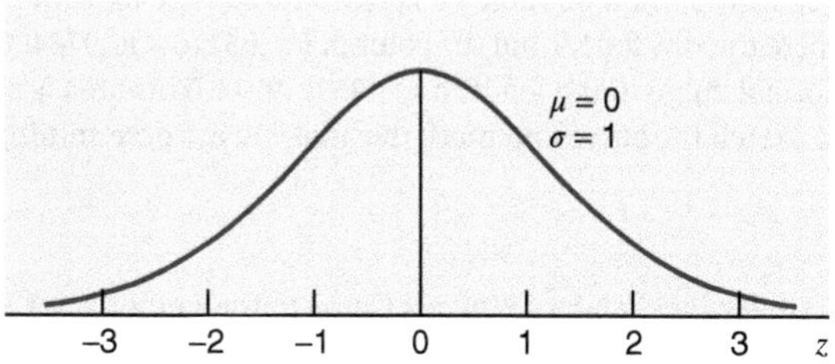

FIGURE 6.17 The standard normal distribution curve.

z Values or z Scores

The units marked on the horizontal axis of the standard normal curve are denoted by z and are called the *z values* or *z scores*. A specific value of z gives the distance between the mean and the point represented by z in terms of the standard deviation.

In Figure 6.17, the horizontal axis is labeled z. The z values on the right side of the mean are positive and those on the left side are negative. **The z value for a point on the horizontal axis gives the distance between the mean and that point in terms of the standard deviation.** For example, a point with a value of $z = 2$ is two standard deviations to the right of the mean. Similarly, a point with a value of $z = -2$ is two standard deviations to the left of the mean.

The standard normal distribution table, Table IV of Appendix B, lists the areas under the standard normal curve to the left of z values from −3.49 to 3.49. To read the standard normal distribution table, we look for the given z value in the table and record the area corresponding to that z value. As shown in **Figure 6.18**, Table IV gives what is called the cumulative probability to the left of any z value.

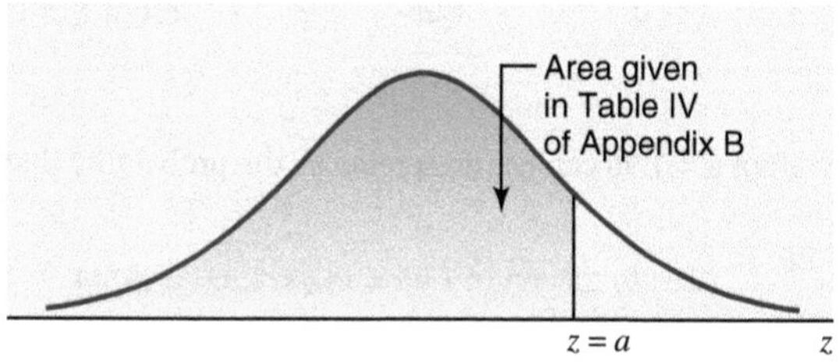

FIGURE 6.18 Area under the standard normal curve.

Although the values of z on the left side of the mean are negative, the area under the curve is always positive.

Remember

The area under the standard normal curve between any two points can be interpreted as the probability that z assumes a value within that interval. Examples 6.1 through 6.4 describe how to read Table IV of Appendix B to find areas under the standard normal curve.

EXAMPLE 6.1

Finding the Area to the Left of a Positive z

Find the area under the standard normal curve to the left of $z = 1.95$.

Solution We divide the given number 1.95 into two portions: 1.9 (the digit before the decimal and one digit after the decimal) and .05 (the second digit after the decimal). (Note that $1.95 = 1.9 + .05$.) To find the required area under the standard normal curve, we locate 1.9 in the column for z on the left side of Table IV of Appendix B and .05 in the row for z at the top of Table IV. The entry where the row for 1.9 and the column for .05 intersect gives the area under the standard normal curve to the left of $z = 1.95$. The relevant portion of Table IV is reproduced here as **Table 6.2.** From Table IV or Table 6.2, the entry where the row for 1.9 and the column for .05 cross is .9744. Consequently, the area under the standard normal curve to the left of $z = 1.95$ is .9744. This area is shown in **Figure 6.19.** (It is always helpful to sketch the curve and mark the area you are determining.)

TABLE 6.2 Area Under the Standard Normal Curve to the Left of $z = 1.95$

z	.00	.01	...	.05	...	.09
−3.4	.0003	.0003	...	.0003	...	.0002
−3.3	.0005	.0005	...	.0004	...	.0003
−3.2	.0007	.0007	...	.0006	...	.0005
.	.	.	...	.	...	.
.	.	.	...	.	...	.
1.9	.9713	.9719	...	.9744 ←	...	.9767
.	.	.	...	.	...	.
.	.	.	...	.	...	.
3.4	.9997	.9997	...	.9997	...	.9998

Required area

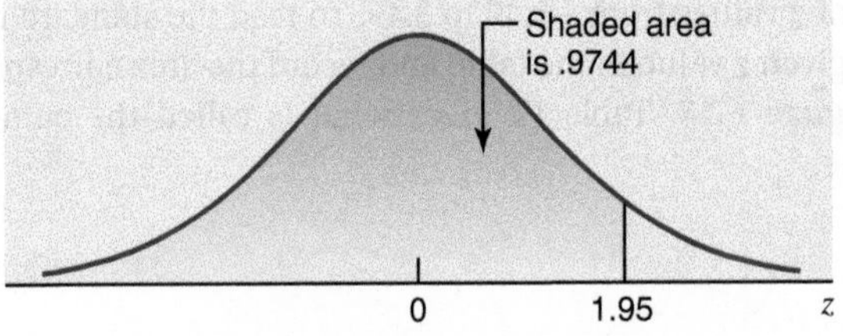

FIGURE 6.19 Area to the left of $z = 1.95$.

The area to the left of $z = 1.95$ can be interpreted as the probability that z assumes a value less than 1.95; that is,

$$\text{Area to the left of } 1.95 = P(z < 1.95) = \mathbf{.9744}$$

As mentioned earlier in Section 6.1, the probability that a continuous random variable assumes a single value is zero. Therefore, the probability that z is exactly equal to 1.95 is zero, that is

$$P(z = 1.95) = 0$$

Hence, $$P(z < 1.95) = P(z \leq 1.95) = \mathbf{.9744}$$

EXAMPLE 6.2

Finding the Area Between a Negative z and $z = 0$

Find the area under the standard normal curve from $z = -2.17$ to $z = 0$.

Solution To find the area from $z = -2.17$ to $z = 0$, which is the shaded area in **Figure 6.20**, first we find the areas to the left of $z = 0$ and to the left of $z = -2.17$ in the standard normal distribution table (Table IV). As shown in **Table 6.3**, these two areas are .5000 and .0150, respectively. Next, we subtract .0150 from .5000 to find the required area.

TABLE 6.3 Area Under the Standard Normal Curve

z	00	.01	...	.07	...	.09
−3.4	.0003	.0003	...	.0003	...	.0002
−3.3	.0005	.0005	...	.0004	...	.0003
−3.2	.0007	.0007	...	.0005	...	.0005
.	.	.	...	.	...	.
.	.	.	...	.	...	.
−2.1	.0179	.0174	...	.0150	...	.0143
.	.	.	...	.	...	.
.	.	.	...	.	...	.
0.0	.5000	.5040	...	.5279	...	.5359
.	.	.	...	.	...	.
.	.	.	...	.	...	.
3.4	.9997	.9997	...	.9997	...	.9998

Area to the left of $z = 0$ Area to the left of $z = -2.17$

The area from $z = -2.17$ to $z = 0$ gives the probability that z lies in the interval −2.17 to 0 (see Figure 6.20); that is,

$$\text{Area from } -2.17 \text{ to } 0 = P(-2.17 \le z \le 0) = \mathbf{.5000} - \mathbf{.0150} = \mathbf{.4850}$$

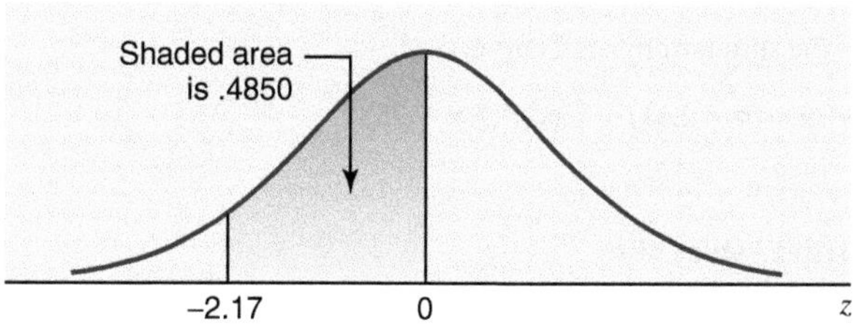

FIGURE 6.20 Area from $z = -2.17$ to $z = 0$.

EXAMPLE 6.3

Finding the Area in the Right and Left Tails

Find the following areas under the standard normal curve.

a. Area to the right of $z = 2.32$

b. Area to the left of $z = -1.54$

Solution

a. Here, we are to find the area that is shaded in **Figure 6.21**. As mentioned earlier, the normal distribution table gives the area to the left of a z value. To find the area to the right of $z = 2.32$, first we find the area to the left of $z = 2.32$. Then we subtract this area from 1.0, which is the total area under the curve. From Table IV, the area to the left of $z = 2.32$ is .9898. Consequently, the required area is $1.0 - .9898 = .0102$, as shown in Figure 6.21.

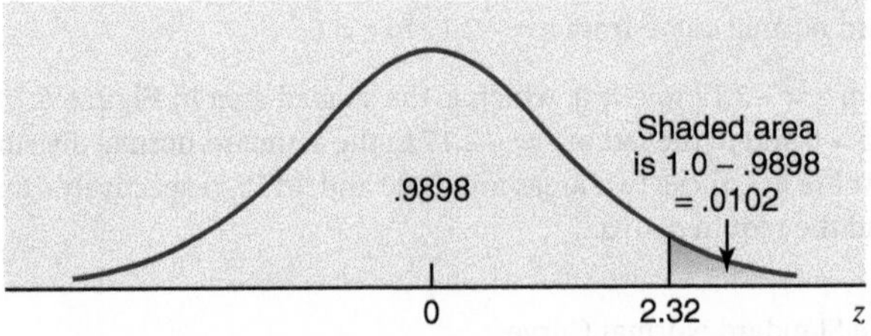

FIGURE 6.21 Area to the right of $z = 2.32$.

Note that the area to the right of $z = 2.32$ gives the probability that z is greater than 2.32. Thus,

$$\text{Area to the right of } 2.32 = P(z > 2.32) = 1.0 - .9898 = \mathbf{.0102}$$

b. To find the area under the standard normal curve to the left of $z = -1.54$, we find the area in Table IV that corresponds to -1.5 in the z column and .04 in the top row. This area is .0618. This area is shown in **Figure 6.22**.

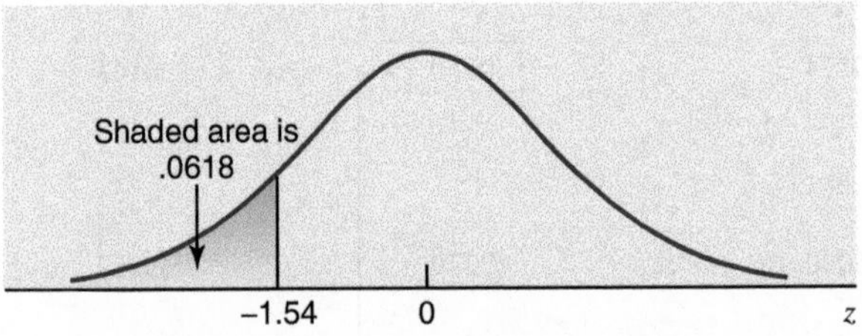

FIGURE 6.22 Area to the left of $z = -1.54$

The area to the left of $z = -1.54$ gives the probability that z is less than -1.54. Thus,

$$\text{Area to the left of } -1.54 = P(z < -1.54) = \mathbf{.0618}$$

EXAMPLE 6.4

Finding Areas Under the Normal Curve

Find the following probabilities for the standard normal curve.

a. $P(1.19 < z < 2.12)$ **b.** $P(-1.56 < z < 2.31)$ **c.** $P(z > -.75)$

Solution

Finding the area between two positive values of z

a. The probability $P(1.19 < z < 2.12)$ is given by the area under the standard normal curve between $z = 1.19$ and $z = 2.12$, which is the shaded area in **Figure 6.23**.

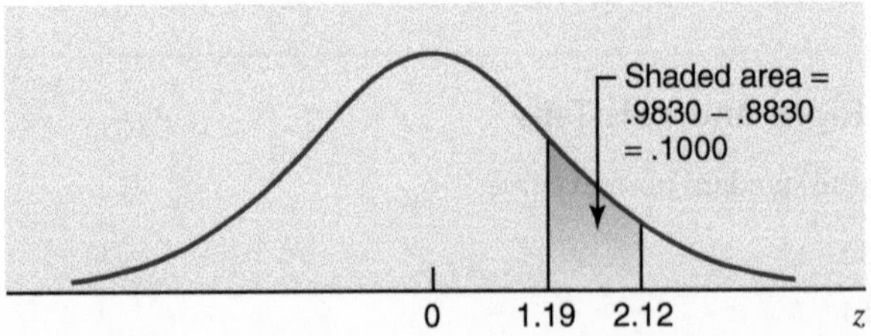

FIGURE 6.23 Finding $P(1.19 < z < 2.12)$.

To find the area between $z = 1.19$ and $z = 2.12$, first we find the areas to the left of $z = 1.19$ and $z = 2.12$. Then we subtract the smaller area (to the left of $z = 1.19$) from the larger area (to the left of $z = 2.12$).

From Table IV (of Appendix B) for the standard normal distribution, we find

Area to the left of 2.12 = .9830
Area to the left of 1.19 = .8830

Then, the required probability is

$$P(1.19 < z < 2.12) = \text{Area between } 1.19 \text{ and } 2.12 = .9830 - .8830 = \mathbf{.1000}$$

Finding the area between a negative and a positive value of *z*

b. The probability $P(-1.56 < z < 2.31)$ is given by the area under the standard normal curve between $z = -1.56$ and $z = 2.31$, which is the shaded area in **Figure 6.24**.

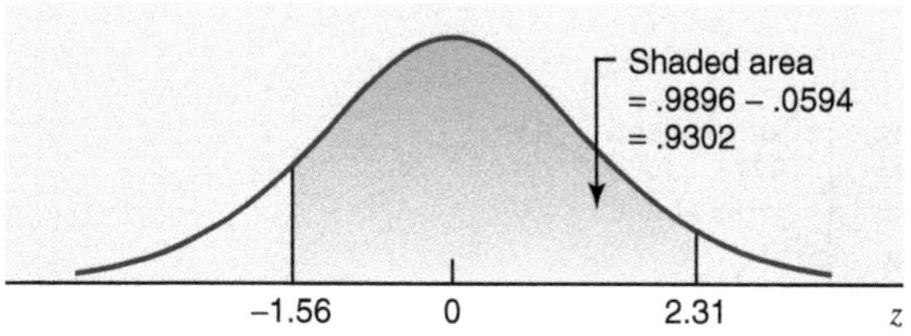

FIGURE 6.24 Finding $P(-1.56 < z < 2.31)$.

From Table IV for the standard normal distribution, we have

Area to the left of 2.31 = .9896
Area to the left of −1.56 = .0594

The required probability is

$$P(-1.56 < z < 2.31) = \text{Area between } -1.56 \text{ and } 2.31 = .9896 - .0594 = \mathbf{.9302}$$

Finding the area to the right of a negative value of *z*

c. The probability $P(z > -.75)$ is given by the area under the standard normal curve to the right of $z = -.75$, which is the shaded area in **Figure 6.25**.

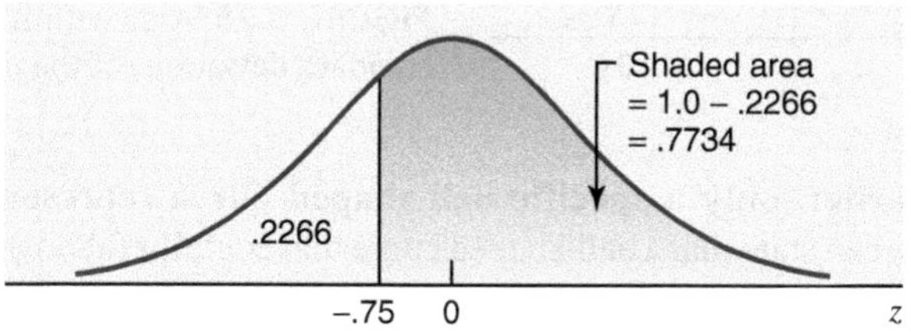

FIGURE 6.25 Finding $P(z > -.75)$.

From Table IV for the standard normal distribution,

Area to the left of −.75 = .2266

The required probability is

$$P(z > -.75) = \text{Area to the right of } -.75 = 1.0 - .2266 = \mathbf{.7734}$$

In the discussion in Section 3.4 of Chapter 3 on the use of the standard deviation, we discussed the empirical rule for a bell-shaped curve. That empirical rule comes from the standard normal distribution. By using the normal distribution table (Table IV of Appendix B), we can now verify the empirical rule as follows:

1. The total area within one standard deviation of the mean is 68.26%. This area is given by the difference between the area to the left of $z = 1.0$ and the area to the left of $z = -1.0$. As shown in **Figure 6.26**, this area is $.8413 - .1587 = .6826$, or 68.26%.

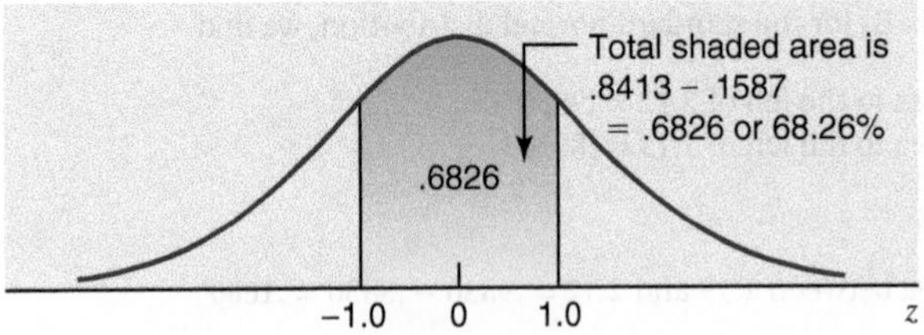

FIGURE 6.26 Area within one standard deviation of the mean.

2. The total area within two standard deviations of the mean is 95.44%. This area is given by the difference between the area to the left of $z = 2.0$ and the area to the left of $z = -2.0$. As shown in **Figure 6.27**, this area is .9772 – .0228 = .9544, or 95.44%.

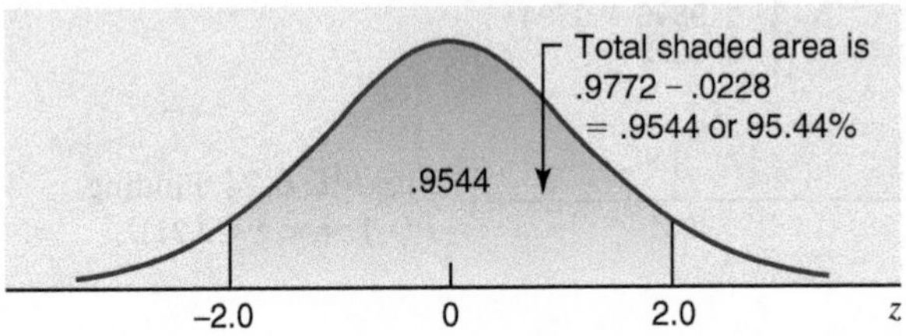

FIGURE 6.27 Area within two standard deviations of the mean.

3. The total area within three standard deviations of the mean is 99.74%. This area is given by the difference between the area to the left of $z = 3.0$ and the area to the left of $z = -3.0$. As shown in **Figure 6.28**, this area is .9987 – .0013 = .9974, or 99.74%.

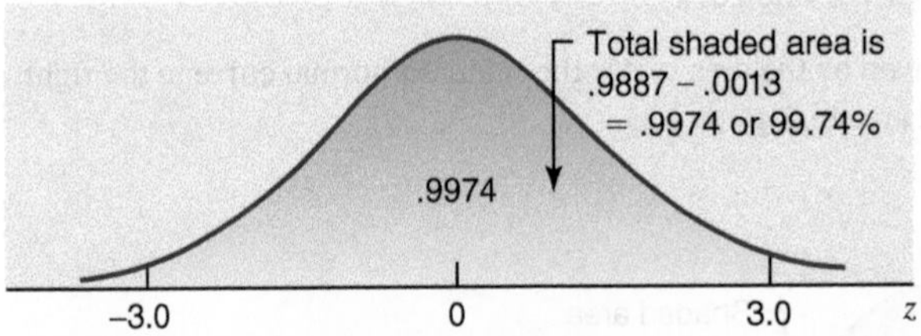

FIGURE 6.28 Area within three standard deviations of the mean.

Again, as mentioned earlier, only a specific bell-shaped curve represents the normal distribution. Now, we can state that a bell-shaped curve that contains (about) 68.26% of the total area within one standard deviation of the mean, (about) 95.44% of the total area within two standard deviations of the mean, and (about) 99.74% of the total area within three standard deviations of the mean represents a normal distribution curve.

The standard normal distribution table, Table IV of Appendix B, goes from $z = -3.49$ to $z = 3.49$. Consequently, if we need to find the area to the left of $z = -3.50$ or a smaller value of z, we can assume it to be approximately .0. If we need to find the area to the left of $z = 3.50$ or a larger number, we can assume it to be approximately 1.0. Example 6.5 illustrates this procedure.

EXAMPLE 6.5

Finding Areas Under the Normal Curve

Find the following probabilities for the standard normal curve.

a. $P(0 < z < 5.67)$

b. $P(z < -5.35)$

Solution

Finding the area between $z = 0$ and a value of z greater than 3.49

a. The probability $P(0 < z < 5.67)$ is given by the area under the standard normal curve between $z = 0$ and $z = 5.67$. Because $z = 5.67$ is greater than 3.49 and is not in Table IV, the area under the standard normal curve to the left of $z = 5.67$ can be approximated by 1.0. Also, the area to the left of $z = 0$ is .5. Hence, the required probability is

$$P(0 < z < 5.67) = \text{Area between 0 and 5.67} = 1.0 - .5 = \mathbf{.5} \text{ approximately}$$

Note that the area between $z = 0$ and $z = 5.67$ is not exactly .5 but very close to .5. This area is shown in **Figure 6.29**.

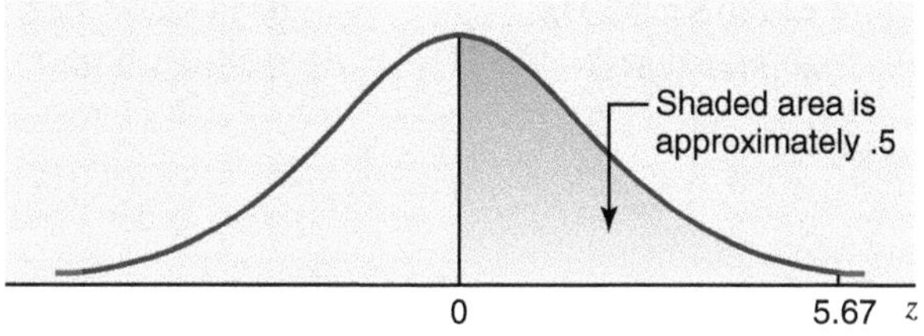

FIGURE 6.29 Area between $z = 0$ and $z = 5.67$.

Finding the area to the left of a z that is less than −3.49

b. The probability $P(z < -5.35)$ represents the area under the standard normal curve to the left of $z = -5.35$. Since $z = -5.35$ is not in the table, we can assume that this area is approximately .00. This is shown in **Figure 6.30**.

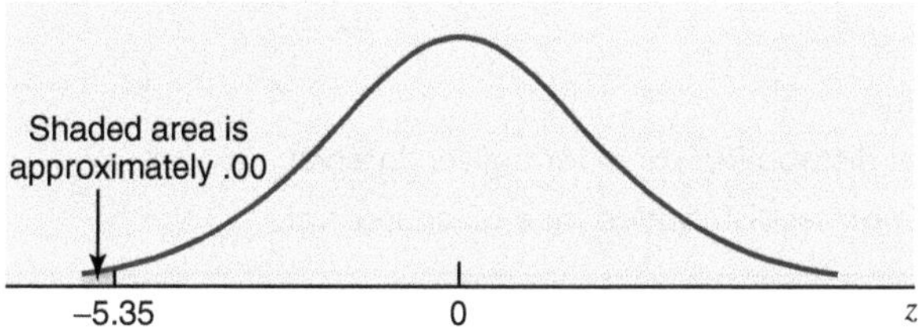

FIGURE 6.30 Area to the left of $z = -5.35$.

The required probability is

$$P(z < -5.35) = \text{Area to the left of } -5.35 \approx \mathbf{.00}$$

Again, note that the area to the left of $z = -5.35$ is not exactly .00 but very close to .00.

We can find the exact areas for parts (a) and (b) of this example by using technology. The reader should review the instructions in the technology section of this chapter.

EXERCISES

Concepts and Procedures

6.1 What is the difference between the probability distribution of a discrete random variable and that of a continuous random variable? Explain.

6.2 Let x be a continuous random variable. What is the probability that x assumes a single value, such as a?

6.3 For a continuous probability distribution, explain why the following holds true.

$$P(a < x < b) = P(a < x \le b) = P(a \le x < b) = P(a \le x \le b)$$

6.4 Briefly explain the main characteristics of a normal distribution. Illustrate with the use of graphs.

6.5 Briefly explain the relationship between standard normal distributed variables with mean 0 and variance 1 and normal distributed variables with mean μ and variance σ^2.

6.6 What are the parameters of the normal distribution?

6.7 How do the spread and height of a normal distribution change when its mean remains the same but its standard deviation decreases?

6.8 Do the spread and height of a normal distribution change when its standard deviation remains the same but its mean increases?

6.9 For the standard normal distribution, what does z represent?

6.10 For the standard normal distribution, find the area within 1.5 standard deviation of the mean—that is, the area between $\mu - 1.5\sigma$ and $\mu + 1.5\sigma$.

6.11 For the standard normal distribution, what is the area within one standard deviation of the mean?

6.12 For the standard normal distribution, what is the area within two standard deviations of the mean?

6.13 Find the area under the standard normal curve:

a. between $z = 0$ and $z = 1.85$

b. between $z = 0$ and $z = -1.43$

c. between $z = .84$ and $z = 1.95$

d. from $z = -.40$ to $z = -2.49$

e. from $z = -2.15$ to $z = 1.34$

6.14 Find the area under the standard normal curve:

a. to the right of $z = 1.36$

b. to the left of $z = -1.76$

c. to the right of $z = -2.48$

d. to the left of $z = 1.89$

6.15 Obtain the area under the standard normal curve:

a. to the right of $z = 1.44$ **b.** to the left of $z = -1.645$

c. to the right of $z = -.25$ **d.** to the left of $z = .53$

6.16 Determine the following probabilities for the standard normal distribution.

a. $P(-2.26 \le z \le 1.68)$ **b.** $P(0 \le z \le 1.65)$

c. $P(-1.645 \le z \le 0)$ **d.** $P(z < 2.58)$

6.17 Find the following probabilities for the standard normal distribution.

a. $P(z < -1.33)$ **b.** $P(.35 \le z \le 2.45)$

c. $P(-1.15 \le z \le -.74)$ **d.** $P(z < 1.75)$

6.18 Obtain the following probabilities for the standard normal distribution.

a. $P(z > -1.645)$ **b.** $P(-1.96 \le z \le 1.96)$

c. $P(0 \le z \le 2.33)$ **d.** $P(-2.575 \le z \le 0)$

e. $P(z > 1.645)$ **f.** $P(z < -2.76)$

6.2 Standardizing a Normal Distribution

LEARNING OBJECTIVES

After completing this section, you should be able to:

- Explain a z value.
- Convert an x value to a z value.
- Find the area under a normal distribution curve for a given interval.
- Calculate a probability for a normal distribution for a given interval.

As was shown in the previous section, Table IV of Appendix B can be used to find areas under the standard normal curve. However, in real-world applications, a (continuous) random variable may have a normal distribution with values of the mean and standard deviation that are different from 0 and 1, respectively. The first step in such a case is to convert the given normal distribution to the standard normal distribution. This procedure is called **standardizing a normal distribution**. The units of a normal distribution (which is not the standard normal distribution) are denoted by x. We know from Section 6.1.3 that units of the standard normal distribution are denoted by z.

Converting an x Value to a z Value

For a normal random variable x, a particular value of x can be converted to its corresponding z value by using the formula

$$z = \frac{x - \mu}{\sigma}$$

where μ and σ are the mean and standard deviation, respectively, of the normal distribution of x. When x follows a normal distribution, z follows the standard normal distribution.

Thus, to find the z value for an x value, we calculate the difference between the given x value and the mean, μ, and divide this difference by the standard deviation, σ. If the value of x is equal to μ, then its z value is equal to zero. Note that we will always round z values to two decimal places.

The z value for the mean of a normal distribution is always zero. The z value for an x value greater than the mean is positive and the z value for an x value smaller than the mean is negative.

Remember

Examples 6.6 through 6.10 describe how to convert x values to the corresponding z values and how to find areas under a normal distribution curve.

EXAMPLE 6.6

Converting x Values to the Corresponding z Values

Let x be a continuous random variable that has a normal distribution with a mean of 50 and a standard deviation of 10. Convert the following x values to z values and find the probability to the left of these points.

a. $x = 55$ **b.** $x = 35$

Solution For the given normal distribution, $\mu = 50$ and $\sigma = 10$.

a. The z value for $x = 55$ is computed as follows:

$$z = \frac{x - \mu}{\sigma} = \frac{55 - 50}{10} = \mathbf{.50}$$

Thus, the z value for $x = 55$ is .50. The z values for both $\mu = 50$ and $x = 55$ are shown in **Figure 6.31**. Note that the z value for $\mu = 50$ is zero. The value $z = .50$ for $x = 55$ indicates that the distance between $\mu = 50$ and $x = 55$ is 1/2 of the standard deviation of 10. Consequently, we can state that the z value represents the distance between μ and x in terms of the standard deviation. Because $x = 55$ is greater than $\mu = 50$, its z value is positive.

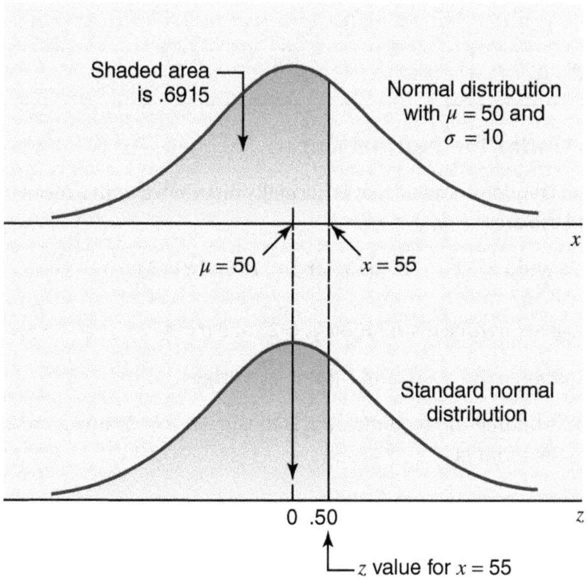

FIGURE 6.31 Area to the left of $x = 55$.

From this point on, we will usually show only the z axis below the x axis and not the standard normal curve itself.

To find the probability to the left of $x = 55$, we find the area to the left of $z = .50$ from Table IV. This area is .6915. Therefore,

$$P(x < 55) = P(z < .50) = \mathbf{.6915}$$

Note that every normal distribution with different μ and σ will have a different bell-shaped curve. But we are drawing similar curves for convenience.

b. The z value for $x = 35$ is computed as follows and is shown in **Figure 6.32**:

$$z = \frac{x - \mu}{\sigma} = \frac{35 - 50}{10} = \mathbf{-1.50}$$

Because $x = 35$ is on the left side of the mean (i.e., 35 is less than $\mu = 50$), its z value is negative. As a general rule, whenever an x value is less than the value of μ, its z value is negative.

To find the probability to the left of $x = 35$, we find the area under the normal curve to the left of $z = -1.50$. This area from Table IV is .0668. Hence,

$$P(x < 35) = P(z < -1.50) = \mathbf{.0668}$$

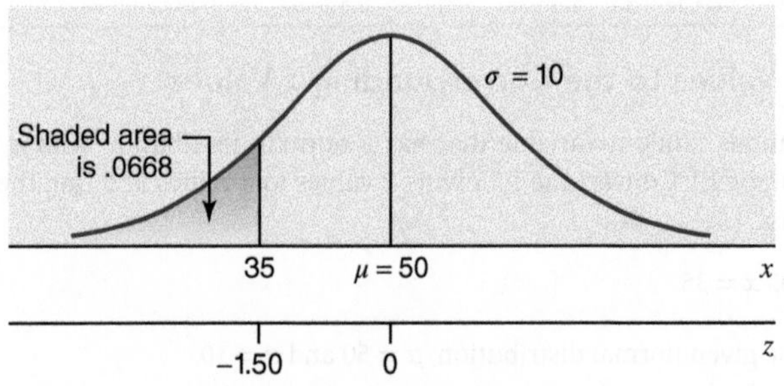

FIGURE 6.32 Area to the left of $x = 35$.

To find the area between two values of x for a normal distribution, we first convert both values of x to their respective z values. Then we find the area under the standard normal curve between those two z values. The area between the two z values gives the area between the corresponding x values. Example 6.7 illustrates this case.

EXAMPLE 6.7

Finding Areas Under the Normal Curve

Let x be a continuous random variable that is normally distributed with a mean of 25 and a standard deviation of 4. Find the area

a. between $x = 25$ and $x = 32$ **b.** between $x = 18$ and $x = 34$

Solution For the given normal distribution, $\mu = 25$ and $\sigma = 4$.

Finding the area between the mean and a point to its right

a. The first step in finding the required area is to convert $x = 25$ and $x = 32$ to their respective z values using the formula

$$z = \frac{x - \mu}{\sigma}$$

The z value for $x = 25$ is zero because it is the mean of the normal distribution. The z value for $x = 32$ is

$$z = \frac{32 - 25}{4} = 1.75$$

The area between $x = 25$ and $x = 32$ under the given normal distribution curve is equivalent to the area between $z = 0$ and $z = 1.75$ under the standard normal curve. From Table IV, the area to the left of $z = 1.75$ is .9599, and the area to the left of $z = 0$ is .50. Hence, the required area is $.9599 - .50 = .4599$, which is shown in **Figure 6.33**.

The area between $x = 25$ and $x = 32$ under the normal curve gives the probability that x assumes a value between 25 and 32. This probability can be written as

$$P(25 < x < 32) = P(0 < z < 1.75) = .9599 - .50 = \mathbf{.4599}$$

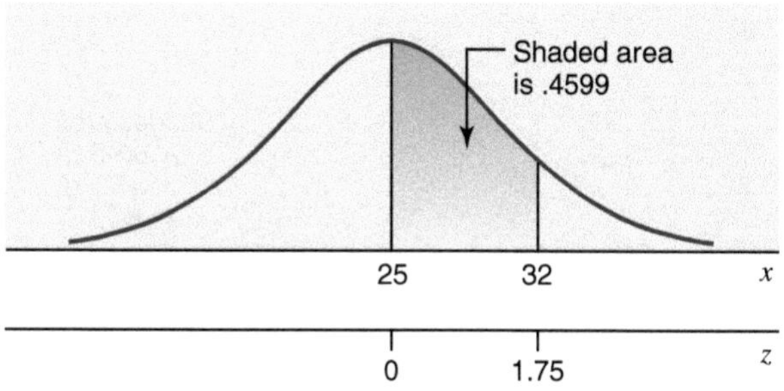

FIGURE 6.33 Area between $x = 25$ and $x = 32$.

Finding the area between two points on different sides of the mean

b. First, we calculate the z values for $x = 18$ and $x = 34$ as follows:

$$\text{For } x = 18: \quad z = \frac{18 - 25}{4} = -1.75$$

$$\text{For } x = 34: \quad z = \frac{34 - 25}{4} = 2.25$$

The area under the given normal distribution curve between $x = 18$ and $x = 34$ is given by the area under the standard normal curve between $z = -1.75$ and $z = 2.25$. From Table IV, the area to the left of $z = 2.25$ is .9878, and the area to the left of $z = -1.75$ is .0401. Hence, the required area is

$$P(18 < x < 34) = P(-1.75 < z < 2.25) = .9878 - .0401 = \mathbf{.9477}$$

This area is shown in **Figure 6.34**.

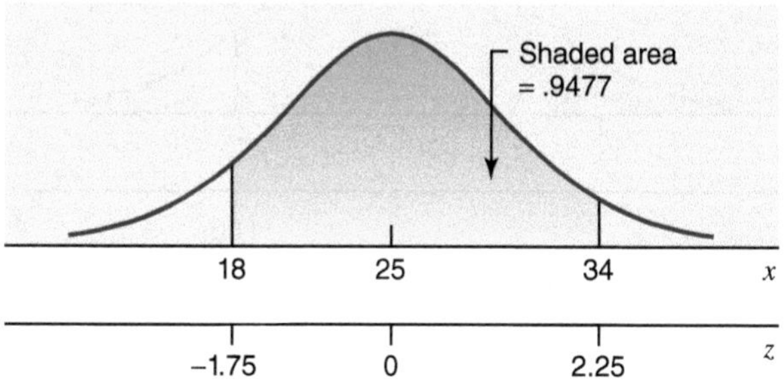

FIGURE 6.34 Area between $x = 18$ and $x = 34$.

EXAMPLE 6.8

Calculating Probabilities for the Normal Distribution

Let x be a normal random variable with its mean equal to 40 and standard deviation equal to 5. Find the following probabilities for this normal distribution.

a. $P(x > 55)$ **b.** $P(x < 49)$

Solution For the given normal distribution, $\mu = 40$ and $\sigma = 5$.

Calculating the probability to the right of a value of x

a. The probability that x assumes a value greater than 55 is given by the area under the normal distribution curve to the right of $x = 55$, as shown in **Figure 6.35**. This area is calculated by subtracting the area to the left of $x = 55$ from 1.0, which is the total area under the curve.

$$\text{For } x = 55: \quad z = \frac{55 - 40}{5} = 3.00$$

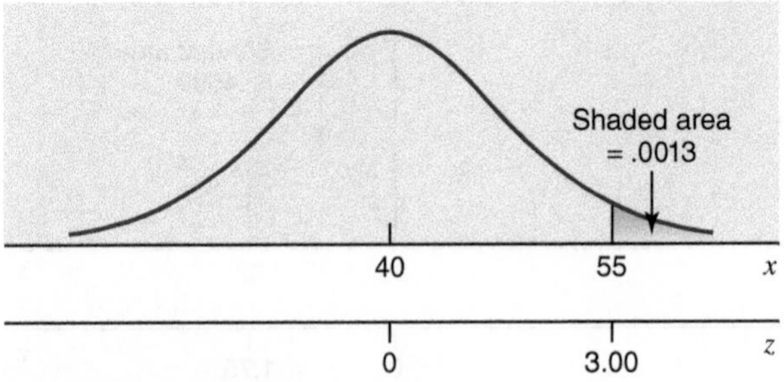

FIGURE 6.35 Finding $P(x > 55)$.

The required probability is given by the area to the right of $z = 3.00$. To find this area, first we find the area to the left of $z = 3.00$, which is .9987. Then we subtract this area from 1.0. Thus,

$$P(x > 55) = P(z > 3.00) = 1.0 - .9987 = \mathbf{.0013}$$

Calculating the probability to the left of a value of *x*

b. The probability that x will assume a value less than 49 is given by the area under the normal distribution curve to the left of 49, which is the shaded area in **Figure 6.36**. This area is obtained from Table IV as follows:

$$\text{For } x = 49\text{:} \quad z = \frac{49 - 40}{5} = 1.80$$

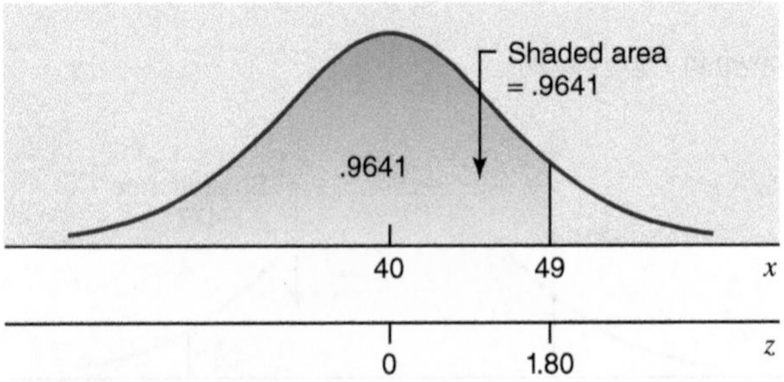

FIGURE 6.36 Finding $P(x < 49)$.

The required probability is given by the area to the left of $z = 1.80$. This area from Table IV is .9641. Therefore, the required probability is

$$P(x < 49) = P(z < 1.80) = \mathbf{.9641}$$

EXAMPLE 6.9

Finding the Area Between Two x Values That Are Less Than the Mean

Let x be a continuous random variable that has a normal distribution with $\mu = 50$ and $\sigma = 8$. Find the probability $P(30 \leq x \leq 39)$.

Solution For this normal distribution, $\mu = 50$ and $\sigma = 8$. The probability $P(30 \leq x \leq 39)$ is given by the area from $x = 30$ to $x = 39$ under the normal distribution curve. As shown in **Figure 6.37**, this area is given by the difference between the area to the left of $x = 39$ and the area to the left of $x = 30$.

$$\text{For } x = 30\text{:} \quad z = \frac{30 - 50}{8} = -2.50$$

$$\text{For } x = 39\text{:} \quad z = \frac{39 - 50}{8} = -1.38$$

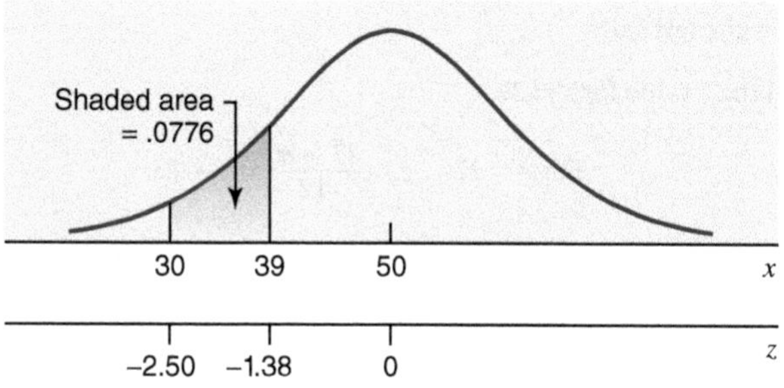

FIGURE 6.37 Finding $P(30 \leq x \leq 39)$.

To find the required area, we first find the area to the left of $z = -2.50$, which is .0062. Then, we find the area to the left of $z = -1.38$, which is .0838. Subtracting the smaller area from the larger area gives the required probability, which is

$$P(30 \leq x \leq 39) = P(-2.50 \leq z \leq -1.38) = .0838 - .0062 = \mathbf{.0776}$$

EXAMPLE 6.10

Calculating Probabilities for the Normal Distribution

Let x be a continuous random variable that has a normal distribution with a mean of 80 and a standard deviation of 12. Find the area under the normal distribution curve

a. from $x = 70$ to $x = 135$ **b.** to the left of $x = 27$

Solution For the given normal distribution, $\mu = 80$ and $\sigma = 12$.

Finding the area between two *x* values that are on different sides of the mean

a. The z values for $x = 70$ and $x = 135$ are

$$\text{For } x = 70: \quad z = \frac{70 - 80}{12} = -.83$$

$$\text{For } x = 135: \quad z = \frac{135 - 80}{12} = 4.58$$

Thus, to find the required area we find the areas to the left of $z = -.83$ and to the left of $z = 4.58$ under the standard normal curve. From Table IV, the area to the left of $z = -.83$ is .2033 and the area to the left of $z = 4.58$ is approximately 1.0. Note that $z = 4.58$ is not in Table IV. See **Figure 6.38**.

Hence,

$$P(70 \leq x \leq 135) = P(-.83 \leq z \leq 4.58)$$
$$= 1.0 - .2033 = \mathbf{.7967} \text{ approximately}$$

Figure 6.38 shows this area.

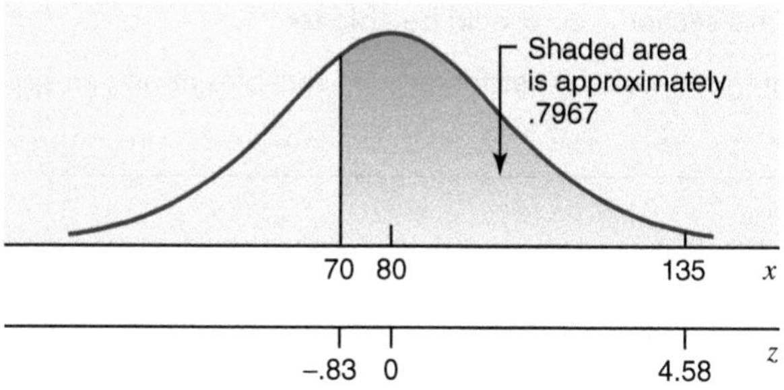

FIGURE 6.38 Area between $x = 70$ and $x = 135$.

Finding an area in the left tail

b. First we find the z value for $x = 27$.

$$\text{For } x = 27: \quad z = \frac{27 - 80}{12} = -4.42$$

As shown in **Figure 6.39**, the required area is given by the area under the standard normal distribution curve to the left of $z = -4.42$. This area is approximately zero.

$$P(x < 27) = P(z < -4.42) = \mathbf{.00} \text{ approximately}$$

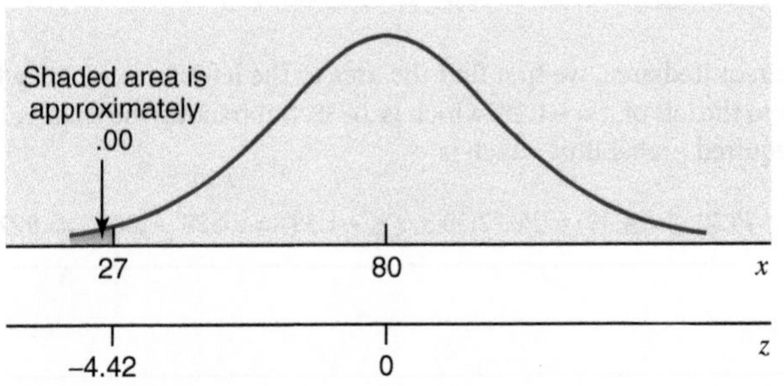

FIGURE 6.39 Area to the left of $x = 27$.

EXERCISES

Concepts and Procedures

6.19 Find the z value for each of the following x values for a normal distribution with $\mu = 20$ and $\sigma = 4$.

a. $x = 21$ **b.** $x = 10$ **c.** $x = 25$ **d.** $x = 15$

6.20 Find the following areas under a normal distribution curve with $\mu = 14$ and $\sigma = 2$.

a. Area between $x = 7.76$ and $x = 14$

b. Area from $x = 14.48$ to $x = 16.54$

c. Area between $x = 8.22$ and $x = 10.06$

6.21 Determine the area under a normal distribution curve with $\mu = 22.5$ and $\sigma = 4.5$

a. to the left of $x = 12.5$ **b.** to the right of $x = 25.2$

c. to the left of $x = 18.7$ **d.** to the right of $x = 30$

6.22 Let x be a continuous random variable that is normally distributed with a mean of 16 and a standard deviation of 5. Find the probability that x assumes a value

a. between 12 and 20 **b.** between 22 and 28

6.23 Let x be a continuous random variable that has a normal distribution with a mean of 15 and a standard deviation of 4. Find the probability that x assumes a value

a. between 14 and 20 **b.** between 19 and 30

6.24 Let x be a continuous random variable that is normally distributed with a mean of 60 and a standard deviation of 10. Find the probability that x assumes a value

a. greater than 49 **b.** less than 55

c. greater than 75 **d.** less than 80

6.3 Applications of the Normal Distribution

LEARNING OBJECTIVE

After completing this section, you should be able to:

- Calculate and interpret probabilities for random variables having an approximate normal distribution.

Sections 6.1 and 6.2 discussed the normal distribution, how to convert a normal distribution to the standard normal distribution, and how to find areas under a normal distribution curve. This section presents examples that illustrate the applications of the normal distribution.

EXAMPLE 6.11 College Debt Among Student Loan Borrowers

Using the Normal Distribution: The Area Between Two Points on Different Sides of the Mean

In the United States, the average student loan debt amount was $37,172 per borrower in 2019 (https://www.nitrocollege.com/research/average-student-loan-debt). Suppose that the 2019 student loan debts for all borrowers are approximately normally distributed with a mean of $37,172 and a standard deviation of $5230. Find the probability that the 2019 student loan debt of a randomly selected borrower was between $30,216 and $39,996.

Solution Let x denote the 2019 student loan debt of a randomly selected borrower. Then, x is approximately normally distributed with

$$\mu = \$37{,}172 \quad \text{and} \quad \sigma = \$5230$$

The probability that the 2019 student loan debt of a randomly selected borrower was between $30,216 and $39,996 is given by the area under the normal distribution curve of x that falls between $x = \$30{,}216$ and $x = \$39{,}996$ as shown in **Figure 6.40**. To find this area, first we find the areas to the left of $x = \$30{,}216$ and $x = \$39{,}996$, respectively, and then take the difference between these two areas. We find the z values for the two x values as follows:

$$\text{For } x = \$30{,}216: \quad z = \frac{30{,}216 - 37{,}172}{5230} = -1.33$$

$$\text{For } x = \$39{,}996: \quad z = \frac{39{,}996 - 37{,}172}{5230} = .54$$

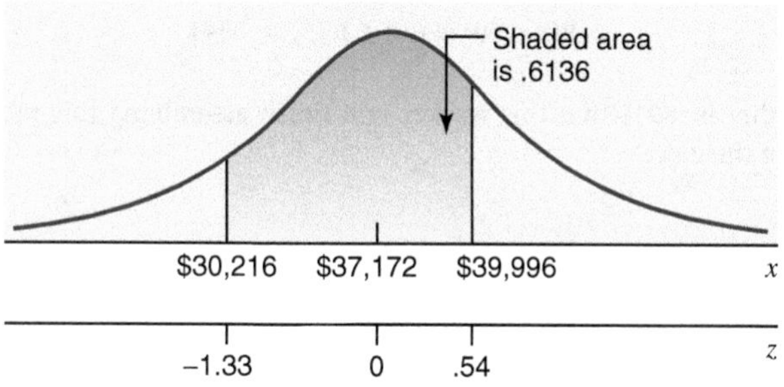

FIGURE 6.40 Area between $x = \$30{,}216$ and $x = 39{,}996$.

Thus, the required probability is given by the difference between the areas under the standard normal curve to the left of $z = .54$ and to the left of $z = -1.33$. From Table IV in Appendix B, the area to the left of $z = -1.33$ is .0918, and the area to the left of $z = .54$ is .7054. Hence, the required probability is

$$P(\$30{,}216 < x < \$39{,}996) = P(-1.33 < z < .54) = .7054 - .0918 = \mathbf{.6136}$$

Thus, the probability is .6136 that the 2019 student loan debt of a randomly selected borrower was between $30,216 and $39,996. Converting this probability into a percentage, we can also state that (about) 61.36% of the student loan borrowers had student loan debts between $30,216 and $39,996 in 2019.

EXAMPLE 6.12 Time Taken to Assemble a Toy

Using the Normal Distribution: Probability That x Is Less Than a Value That Is to the Right of the Mean

A racing car is one of the many toys manufactured by the Mack Corporation. The assembly times for this toy follow a normal distribution with a mean of 55 minutes and a standard deviation of 4 minutes. The company closes at 5 P.M. every day. If one worker starts to assemble a racing car at 4 P.M., what is the probability that this worker will finish this job before the company closes for the day?

Nick Koudis/Getty Images

Solution Let x denote the time this worker takes to assemble a racing car. Then, x is normally distributed with

$$\mu = 55 \text{ minutes} \quad \text{and} \quad \sigma = 4 \text{ minutes}$$

We are to find the probability that this worker can assemble this car in 60 minutes or less (between 4 and 5 P.M.). This probability is given by the area under the normal curve to the left of $x = 60$ minutes as shown in **Figure 6.41**. First we find the z value for $x = 60$ as follows:

$$\text{For } x = 60: \quad z = \frac{60 - 55}{4} = 1.25$$

The required probability is given by the area under the standard normal curve to the left of $z = 1.25$, which is .8944 from Table IV of Appendix B. Thus, the required probability is

$$P(x \leq 60) = P(z \leq 1.25) = \mathbf{.8944}$$

Thus, the probability is .8944 that this worker will finish assembling this racing car before the company closes for the day.

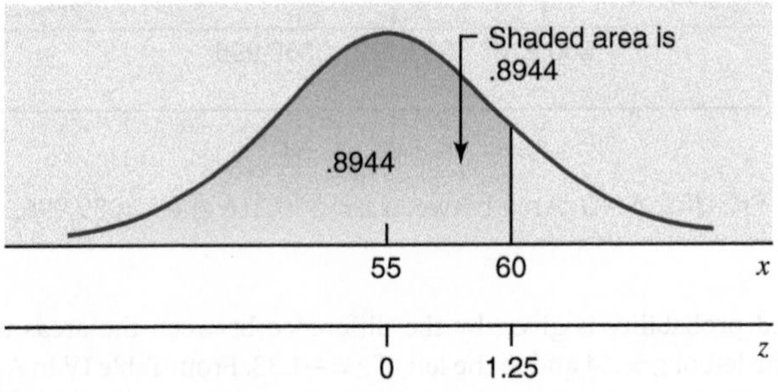

FIGURE 6.41 Area to the left of $x = 60$.

EXAMPLE 6.13 Amount of Soda in a Can

Using the Normal Distribution

Hupper Corporation produces many types of soft drinks, including Orange Cola. The filling machines are adjusted to pour 12 ounces of soda into each 12-ounce can of Orange Cola. However, the actual amount of soda poured into each can is not exactly 12 ounces; it varies from can to can. It has been observed that the net amount of soda in such a can has a normal distribution with a mean of 12 ounces and a standard deviation of .015 ounce.

a. What is the probability that a randomly selected can of Orange Cola contains 11.97 to 11.99 ounces of soda?

b. What percentage of the Orange Cola cans contain 12.02 to 12.07 ounces of soda?

Solution Let x be the net amount of soda in a can of Orange Cola. Then, x has a normal distribution with $\mu = 12$ ounces and $\sigma = .015$ ounce.

Calculating the probability between two points that are to the left of the mean

a. The probability that a randomly selected can contains 11.97 to 11.99 ounces of soda is given by the area under the normal distribution curve from $x = 11.97$ to $x = 11.99$. This area is shown in **Figure 6.42**. First we convert the two values of x to the corresponding z values.

$$\text{For } x = 11.97: \quad z = \frac{11.97 - 12}{.015} = -2.00$$

$$\text{For } x = 11.99: \quad z = \frac{11.99 - 12}{.015} = -.67$$

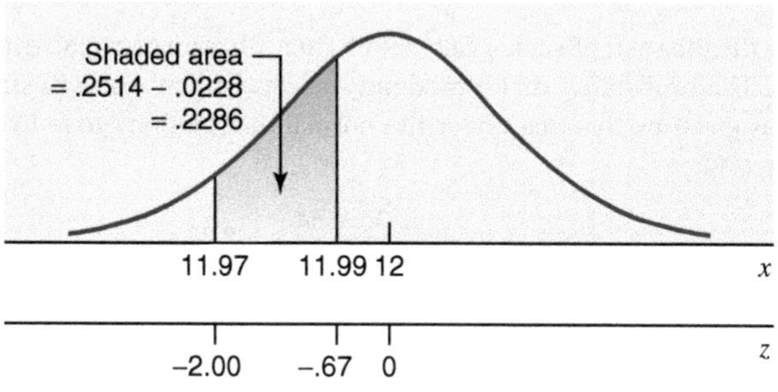

FIGURE 6.42 Area between $x = 11.97$ and $x = 11.99$.

The required probability is given by the area under the standard normal curve between $z = -2.00$ and $z = -.67$. From Table IV of Appendix B, the area to the left of $z = -2.00$ is .0228, and the area to the left of $z = -.67$ is .2514. Hence, the required probability is

$$P(11.97 \le x \le 11.99) = P(-2.00 \le z \le -.67) = .2514 - .0228 = \mathbf{.2286}$$

Thus, the probability is .2286 that a randomly selected can of Orange Cola will contain 11.97 to 11.99 ounces of soda. We can also state that about 22.86% of Orange Cola cans are expected to contain 11.97 to 11.99 ounces of soda.

Calculating the probability between two points that are to the right of the mean

b. The percentage of Orange Cola cans that contain 12.02 to 12.07 ounces of soda is given by the area under the normal distribution curve from $x = 12.02$ to $x = 12.07$, as shown in **Figure 6.43**.

$$\text{For } x = 12.02: \quad z = \frac{12.02 - 12}{.015} = 1.33$$

$$\text{For } x = 12.07: \quad z = \frac{12.07 - 12}{.015} = 4.67$$

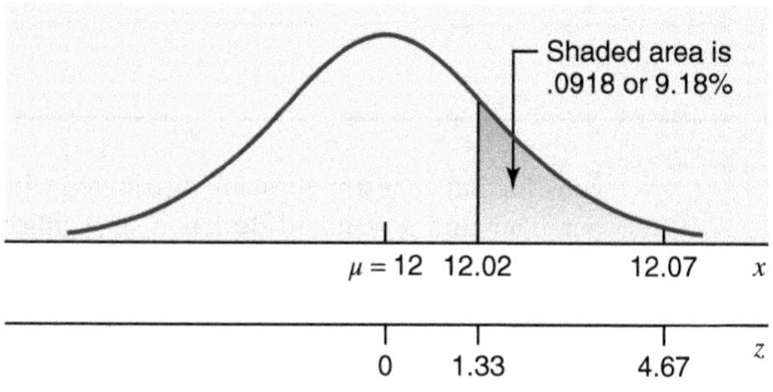

FIGURE 6.43 Area from $x = 12.02$ to $x = 12.07$.

The required probability is given by the area under the standard normal curve between $z = 1.33$ and $z = 4.67$. From Table IV of Appendix B, the area to the left of $z = 1.33$ is .9082, and the area to the left of $z = 4.67$ is approximately 1.0. Hence, the required probability is

$$P(12.02 \le x \le 12.07) = P(1.33 \le z \le 4.67) = 1.0 - .9082 = \mathbf{.0918}$$

Converting this probability to a percentage, we can state that approximately 9.18% of all Orange Cola cans are expected to contain 12.02 to 12.07 ounces of soda.

EXAMPLE 6.14 | Life Span of a Calculator

Finding the Area to the Left of *x* That Is Less Than the Mean

Suppose the life span of a calculator manufactured by the Calculators Corporation has a normal distribution with a mean of 54 months and a standard deviation of 8 months. The company guarantees that any calculator that starts malfunctioning within 36 months of the purchase will be replaced by a new one. About what percentage of calculators made by this company are expected to be replaced?

Solution Let x be the life span of such a calculator. Then x has a normal distribution with $\mu = 54$ and $\sigma = 8$ months. The probability that a randomly selected calculator will start to malfunction within 36 months is given by the area under the normal distribution curve to the left of $x = 36$, as shown in **Figure 6.44**.

$$\text{For } x = 36: \quad z = \frac{36 - 54}{8} = -2.25$$

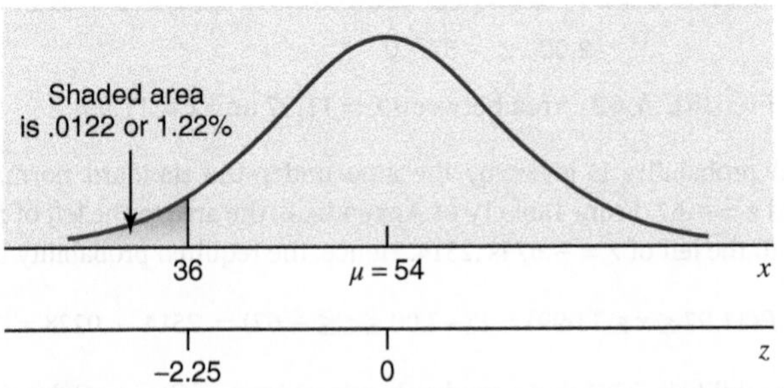

FIGURE 6.44 Area to the left of $x = 36$.

The required percentage is given by the area under the standard normal curve to the left of $z = -2.25$. From Table IV of Appendix B, this area is .0122. Hence, the required probability is

$$P(x < 36) = P(z < -2.25) = \mathbf{.0122}$$

The probability that any randomly selected calculator manufactured by the Calculators Corporation will start to malfunction within 36 months is .0122. Converting this probability to a percentage, we can state that approximately 1.22% of all calculators manufactured by this company are expected to start malfunctioning within 36 months. Hence, 1.22% of the calculators are expected to be replaced.

EXERCISES

Applications

6.25 Let x denote the time taken to run a road race. Suppose x is approximately normally distributed with a mean of 180 minutes and a standard deviation of 14 minutes. If one runner is selected at random, what is the probability that this runner will complete this road race

a. in less than 160 minutes? **b.** in 215 to 245 minutes?

6.26 Tommy Wait, a minor league baseball pitcher, is notorious for taking an excessive amount of time between pitches. In fact, his times between pitches are normally distributed with a mean of 36 seconds and a standard deviation of 2.5 seconds. What percentage of his times between pitches are

a. longer than 40 seconds? **b.** between 25 and 30 seconds?

6.27 A construction zone on a highway has a posted speed limit of 40 miles per hour. The speeds of vehicles passing through this construction zone are normally distributed with a mean of 45 miles per hour and a standard deviation of 4 miles per hour. Find the percentage of vehicles passing through this construction zone that are

a. exceeding the posted speed limit

b. traveling at speeds between 55 and 60 miles per hour

6.28 The Bank of Connecticut issues Visa and MasterCard credit cards. It is estimated that the balances on all Visa credit cards issued by the Bank of Connecticut have a mean of \$920 and a standard deviation of \$270. Assume that the balances on all these Visa cards follow a normal distribution.

a. What is the probability that a randomly selected Visa card issued by this bank has a balance between \$1000 and \$1440?

b. What percentage of the Visa cards issued by this bank have a balance of \$730 or more?

6.29 According to the records of an electric company serving the Boston area, the mean electricity consumption for all households during winter is 1650 kilowatt-hours per month. Assume that the monthly electricity consumptions during winter by all households in this area have a normal distribution with a mean of 1650 kilowatt-hours and a standard deviation of 320 kilowatt-hours.

a. Find the probability that the monthly electricity consumption during winter by a randomly selected household from this area is less than 1950 kilowatt-hours.

b. What percentage of the households in this area have a monthly electricity consumption of 900 to 1300 kilowatt-hours?

6.30 The management of a supermarket wants to adopt a new promotional policy of giving a free gift to every customer who spends more than a certain amount per visit at this supermarket. The expectation of the management is that after this promotional policy is advertised, the expenditures for all customers at this supermarket will be normally distributed with a mean of $110 and a standard deviation of $20. If the management decides to give free gifts to all those customers who spend more than $150 at this supermarket during a visit, what percentage of the customers are expected to receive free gifts?

6.31 One of the cars sold by Walt's car dealership is a very popular subcompact car called the Rhino. The final sale price of the basic model of this car varies from customer to customer depending on the negotiating skills and persistence of the customer. Assume that these sale prices of this car are normally distributed with a mean of $19,800 and a standard deviation of $350.

a. Dolores paid $19,445 for her Rhino. What percentage of Walt's customers paid less than Dolores for a Rhino?

b. Cuthbert paid $20,300 for a Rhino. What percentage of Walt's customers paid more than Cuthbert for a Rhino?

6.32 The average monthly mortgage payment for all homeowners in a city is $2850. Suppose that the distribution of monthly mortgages paid by homeowners in this city follow an approximate normal distribution with a mean of $2850 and a standard deviation of $420. Find the probability that the monthly mortgage paid by a randomly selected homeowner from this city is

a. less than $1200 **b.** between $2300 and $3140

c. more than $3600 **d.** between $3200 and $3700

6.33 Fast Auto Service guarantees that the maximum waiting time for its customers is 20 minutes for oil and lube service on their cars. It also guarantees that any customer who has to wait longer than 20 minutes for this service will receive a 50% discount on the charges. It is estimated that the mean time taken for oil and lube service at this garage is 15 minutes per car and the standard deviation is 2.4 minutes. Suppose the time taken for oil and lube service on a car follows a normal distribution.

a. What percentage of customers will receive a 50% discount on their charges?

b. Is it possible that it may take longer than 25 minutes for oil and lube service for a car? Explain.

6.34 The lengths of 3-inch nails manufactured on a machine are normally distributed with a mean of 3.0 inches and a standard deviation of .009 inch. The nails that are either shorter than 2.97 inches or longer than 3.03 inches are unusable. What percentage of all the nails produced by this machine are unusable?

6.35 The pucks used by the National Hockey League for ice hockey must weigh between 5.5 and 6.0 ounces. Suppose the weights of pucks produced at a factory are normally distributed with a mean of 5.75 ounces and a standard deviation of .11 ounce. What percentage of the pucks produced at this factory cannot be used by the National Hockey League?

6.4 Determining the z and x Values When an Area Under the Normal Distribution Curve Is Known

LEARNING OBJECTIVES

After completing this section, you should be able to:

- Find the z value for a given area under the standard normal distribution curve.
- Convert a z value to an x value.
- Find the x value for a given area under a normal distribution curve.

So far in this chapter, we have discussed how to find the area under a normal distribution curve for an interval of z or x. Now, we reverse this procedure and learn how to find the corresponding value of z or x when an area under a normal distribution curve is known. Examples 6.15 through 6.17 describe this procedure for finding the z value.

EXAMPLE 6.15

Finding z When an Area to the Left of z Is Known

Find the value of z such that the area under the standard normal curve to the left of z is .9251.

Solution As shown in **Figure 6.45**, we are to find the z value such that the area to the left of z is .9251. Since this area is greater than .50, z is positive and lies to the right of zero.

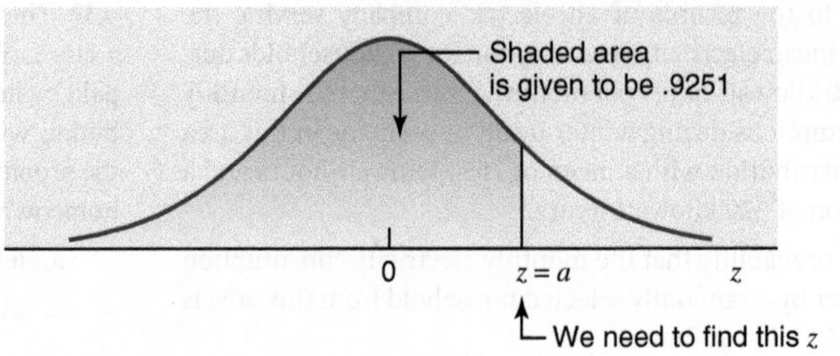

FIGURE 6.45 Finding the z value.

To find the required value of z, we locate .9251 in the body of the normal distribution table, Table IV of Appendix B. The relevant portion of that table is reproduced as **Table 6.4** here. Next, we read the numbers in the column and row for z that correspond to .9251. As shown in Table 6.4, these numbers are 1.4 and .04, respectively. Combining these two numbers, we obtain the required value of $z = 1.44$. Thus, the area to the left of $z = 1.44$ is .9251.

TABLE 6.4 Finding the z Value When Area Is Known

z	.00	.01	...	.04	...	.09
−3.4	.0003	.0003	...		...	.0002
−3.3	.0005	.0005	...		...	.0003
−3.2	.0007	.0007	...		...	.0005
.	.	.	...		...	.
.	.	.	...		...	.
.	.	.	...		...	.
1.4				.9251	...	...
.	.	.	...	.	...	.
.	.	.	...	.	...	.
.	.	.	...	.	...	.
3.4	.9997	.9997	...	.9997	...	.9998

We locate this area in Table IV of Appendix B

EXAMPLE 6.16

Finding z When an Area in the Right Tail Is Known

Find the value of z such that the area under the standard normal curve in the right tail is .0050.

Solution To find the required value of z, we first find the area to the left of z by subtracting .0050 from 1.0. Hence,

$$\text{Area to the left of } z = 1.0 - .0050 = .9950$$

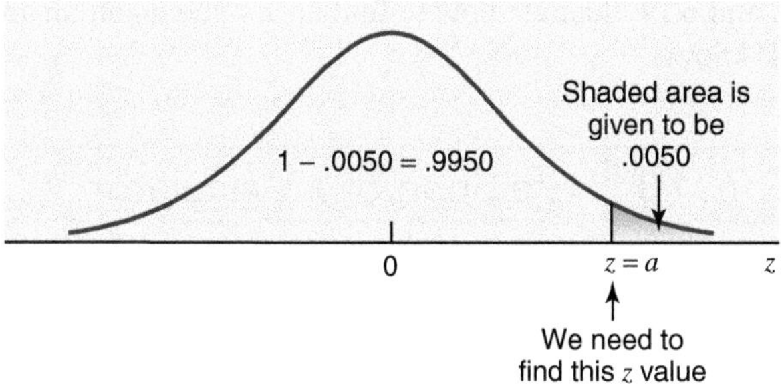

FIGURE 6.46 Finding the z value.

This area is shown in **Figure 6.46**.

Now we look for .9950 in the body of the normal distribution table. Table IV does not contain .9950. So we find the value closest to .9950, which is either .9949 or .9951. We can use either of these two values. If we choose .9951, the corresponding z value is 2.58. Hence, the required value of z is **2.58**, and the area to the right of $z = 2.58$ is approximately .0050. Note that there is no apparent reason to choose .9951 and not to choose .9949. We can use either of the two values. If we choose .9949, the corresponding z value will be 2.57.

EXAMPLE 6.17

Finding z When an Area in the Left Tail Is Known

Find the value of z such that the area under the standard normal curve in the left tail is .05.

Solution Because .05 is less than .5 and it is the area in the left tail, the value of z is negative. This area is shown in **Figure 6.47**.

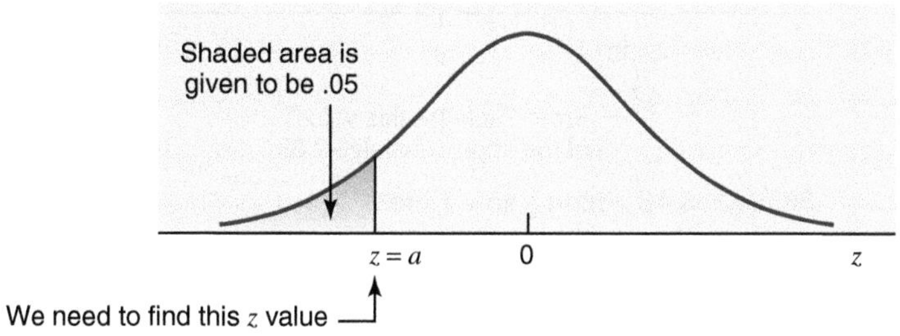

FIGURE 6.47 Finding the z value.

Next, we look for .0500 in the body of the normal distribution table. The value closest to .0500 in the normal distribution table is either .0505 or .0495. Suppose we use the value .0495. The corresponding z value is −1.65. Thus, the required value of z is **−1.65** and the area to the left of $z = -1.65$ is approximately .05.

To find an x value when an area under a normal distribution curve is given, first we find the z value corresponding to that x value from the normal distribution table. Then, to find the x value, we substitute the values of μ, σ, and z in the following formula, which is obtained from $z = (x - \mu)/\sigma$ by doing some algebraic manipulations. Also, if we know the values of x, z, and σ, we can find μ using this same formula.

Finding an x Value for a Normal Distribution

For a normal curve with known values of μ and σ and for a given area under the curve to the left of x, the x value is calculated as

$$x = \mu + z\sigma$$

Examples 6.18 and 6.19 illustrate how to find an x value when an area under a normal distribution curve is known.

EXAMPLE 6.18 | Life Span of a Calculator

Finding x When an Area in the Left Tail Is Known

Recall Example 6.14. It is known that the life of a calculator manufactured by Calculators Corporation has a normal distribution with a mean of 54 months and a standard deviation of 8 months. What should the warranty period be to replace a malfunctioning calculator if the company does not want to replace more than 1% of all the calculators sold?

Solution Let x be the life of a calculator. Then, x follows a normal distribution with $\mu = 54$ months and $\sigma = 8$ months. The calculators that would be replaced are the ones that start malfunctioning during the warranty period. The company's objective is to replace at most 1% of all the calculators sold. The shaded area in **Figure 6.48** gives the proportion of calculators that are replaced. We are to find the value of x so that the area to the left of x under the normal curve is 1%, or .01.

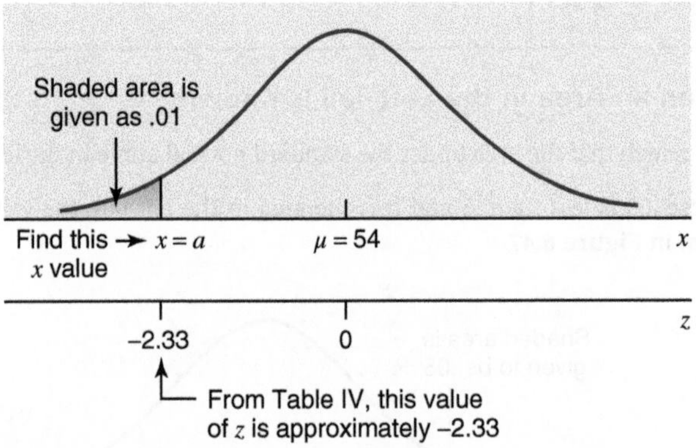

FIGURE 6.48 Finding an x value.

In the first step, we find the z value that corresponds to the required x value.

We find this z value from the normal distribution table for .0100. Table IV of Appendix B does not contain a value that is exactly .0100. The value closest to .0100 in the table is .0099, and the z value for .0099 is −2.33. Hence,

$$z = -2.33$$

In the second step, we substitute the values of $\mu = 54$, $\sigma = 8$, and $z = -2.33$ in the formula $x = \mu + z\sigma$, to obtain the x value as follows:

$$x = \mu + z\sigma = 54 + (-2.33)(8) = 54 - 18.64 = \mathbf{35.36}$$

Thus, the company should replace all the calculators that start to malfunction within 35.36 months (which can be rounded to 35 months) of the date of purchase so that they will not have to replace more than 1% of the calculators.

EXAMPLE 6.19 | College Debt Among Student Loan Borrowers

Finding x When an Area in the Right Tail Is Known

In the United States, the average student loan debt amount was $37,172 per borrower in 2019 (https://www.nitrocollege.com/research/average-student-loan-debt). Suppose that the 2019 student loan debts for all borrowers are approximately normally distributed with a mean of $37,172 and a standard deviation of \$5230. Approximately 10% of all such borrowers had 2019 student loan debts greater than Roger Smith's 2019 student loan debt. What was Roger Smith's 2019 student loan debt?

Solution Let x denote the 2019 student loan debt of a randomly selected borrower. Then, x is normally distributed with

$$\mu = \$37{,}172 \quad \text{and} \quad \sigma = \$5230$$

We are to find the value of x such that the area under the normal distribution curve to the right of x is 10%, as shown in **Figure 6.49**.

First, we find the area under the normal distribution curve to the left of the x value.

$$\text{Area to the left of the } x \text{ value} = 1.0 - .10 = .9000$$

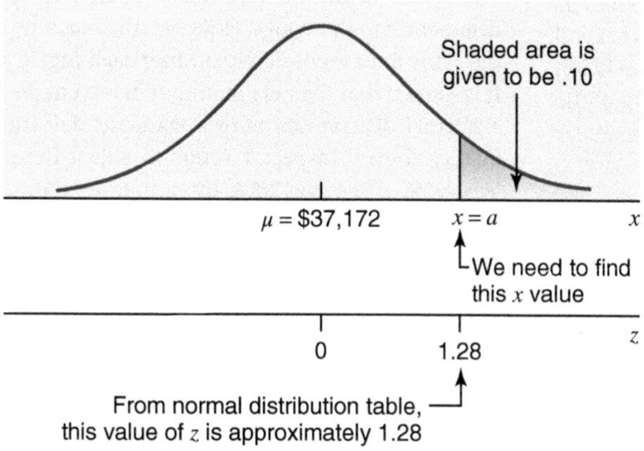

FIGURE 6.49 Finding an x value.

To find the z value that corresponds to the required x value, we look for .9000 in the body of the normal distribution table. The value closest to .9000 in Table IV is .8997, and the corresponding z value is 1.28. Hence, the value of x is computed as

$$x = \mu + z\sigma = 37{,}172 + 1.28(5230) = 37{,}172 + 6694.40 = \mathbf{\$43{,}866.40}$$

Thus, we can state that Roger Smith's student loan debt was approximately \$43,866.40 in 2019.

EXERCISES

Concepts and Procedures

6.36 Find the value of z so that the area under the standard normal curve

a. from 0 to z is (approximately) .1965 and z is positive

b. between 0 and z is (approximately) .2740 and z is negative

c. in the left tail is (approximately) .2050

d. in the right tail is (approximately) .1053

6.37 Determine the value of z so that the area under the standard normal curve

a. in the right tail is .0250

b. in the left tail is .0500

c. in the left tail is .0010

d. in the right tail is .0100

6.38 Determine the value of z so that the area under the standard normal curve

a. in the right tail is .0500

b. in the left tail is .0250

c. in the left tail is .0100

d. in the right tail is .0050

6.39 Let x be a continuous random variable that follows a normal distribution with a mean of 200 and a standard deviation of 25.

a. Find the value of x so that the area under the normal curve to the left of x is approximately .6330.

b. Find the value of x so that the area under the normal curve to the right of x is approximately .050.

c. Find the value of x so that the area under the normal curve to the right of x is .8051.

d. Find the value of x so that the area under the normal curve to the left of x is .0150.

e. Find the value of x so that the area under the normal curve between μ and x is .4525 and the value of x is less than μ.

f. Find the value of x so that the area under the normal curve between μ and x is approximately .4800 and the value of x is greater than μ.

Applications

6.40 As per the record of a private electric company serving in a metro city, the mean electricity consumption during summer for all households is 1800 kilowatt-hours per month. Assume that monthly electric consumption during summer by all households in this area has a normal distribution with a mean of 250 kilowatt-hours. The company sent a notice to one of the households in the city, informing them that about 95% of the households use less electricity per month than they do. What is the household's monthly electricity consumption?

6.41 The management of a supermarket wants to adopt a new promotional policy of giving a free gift to every customer who spends more than a certain amount per visit at this supermarket. The expectation of the management is that after this promotional policy is advertised, the expenditures for all customers at this supermarket will be normally distributed with a mean of $95 and a standard deviation of $20. If the management wants to give free gifts to at most 10% of the customers, what should the amount be above which a customer would receive a free gift?

6.42 HSP Services provides oil and lube services for cars. It is known that the mean time taken for oil and lube service at this garage is 20 minutes per car and the standard deviation is 4 minutes. The management wants to promote the business by guaranteeing a maximum waiting time for its customers. If a customer's car is not serviced within that period, the customer will receive a 50% discount on the charges. The company wants to limit this discount to at most 10% of the customers. What should the maximum guaranteed waiting time be? Assume that the times taken for oil and lube services for all cars have a normal distribution.

***6.43** A study has shown that 20% of all college textbooks have a price of $100 or higher. It is known that the standard deviation of the prices of all college textbooks is $7.50. Suppose the prices of all college textbooks have a normal distribution. What is the mean price of all college textbooks?

***6.44** A machine at Keats Corporation fills 64-ounce detergent jugs. The machine can be adjusted to pour, on average, any amount of detergent into these jugs. However, the machine does not pour exactly the same amount of detergent into each jug; it varies from jug to jug. It is known that the net amount of detergent poured into each jug has a normal distribution with a standard deviation of .35 ounce. The quality control inspector wants to adjust the machine such that at least 95% of the jugs have more than 64 ounces of detergent. What should the mean amount of detergent poured by this machine into these jugs be?

6.5 The Normal Approximation to the Binomial Distribution

LEARNING OBJECTIVES

After completing this section, you should be able to:

- Describe the characteristics of a binomial distribution.
- Verify the conditions for calculating binomial probabilities using the normal approximation.
- Explain correction for continuity and continuity correction factor.
- Calculate and interpret binomial probabilities using the normal approximation with correction for continuity.

Recall from Chapter 5 that:

1. The binomial distribution is applied to a discrete random variable.
2. Each repetition, called a trial, of a binomial experiment results in one of two possible outcomes (or events), either a success or a failure.
3. The probabilities of the two (possible) outcomes (or events) remain the same for each repetition of the experiment.
4. The trials are independent.

The binomial formula, which gives the probability of x successes in n trials, is

$$P(x) = {}_nC_x\, p^x q^{n-x}$$

The use of the binomial formula becomes very tedious when n is large. In such cases, the normal distribution can be used to approximate the binomial probability. Note that for a binomial problem, the exact probability is obtained by using the binomial formula. If we apply the normal distribution to solve a binomial problem, the probability that we obtain is an approximation to the exact probability. The approximation obtained by using the normal distribution is very close to the exact probability when n is large and p is very close to .50. However, this does not mean that we should not use the normal approximation when p is not close to .50. The reason the approximation is closer to the exact probability when p is close to .50 is that the binomial distribution is symmetric when $p = .50$. The normal distribution is always symmetric. Hence, the two distributions are very close to each other when n is large and p is close to .50. However, this does not mean that whenever $p = .50$, the binomial distribution is the same as the normal distribution because not every symmetric bell-shaped curve is a normal distribution curve.

Normal Distribution as an Approximation to Binomial Distribution

Usually, the normal distribution is used as an approximation to the binomial distribution when np and nq are both greater than 5, that is, when

$$np > 5 \quad \text{and} \quad nq > 5$$

Table 6.5 gives the binomial probability distribution of x for $n = 12$ and $p = .50$. This table is written using Table I of Appendix B. **Figure 6.50** shows the histogram and the smoothed polygon for the probability distribution of Table 6.5. As we can observe, the histogram in Figure 6.50 is symmetric, and the curve obtained by joining the upper midpoints of the rectangles is approximately bell shaped.

TABLE 6.5 The Binomial Probability Distribution for $n = 12$ and $p = .50$

x	$P(x)$
0	.0002
1	.0029
2	.0161
3	.0537
4	.1208
5	.1934
6	.2256
7	.1934
8	.1208
9	.0537
10	.0161
11	.0029
12	.0002

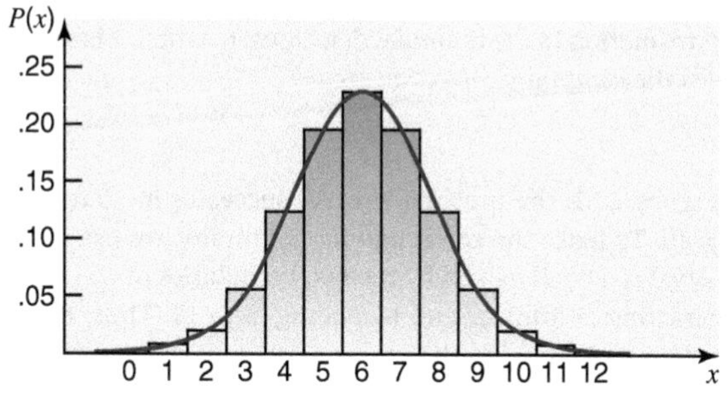

FIGURE 6.50 Histogram for the probability distribution of Table 6.5.

Examples 6.20 through 6.22 illustrate the application of the normal distribution as an approximation to the binomial distribution.

EXAMPLE 6.20 Number of Credit Cards People Have

Using the Normal Approximation to the Binomial Distribution: *x* Equals a Specific Value

According to an estimate, 50% of people in the United States have at least one credit card. If a random sample of 30 persons is selected, what is the probability that 19 of them will have at least one credit card?

Solution Let n be the total number of persons in the sample, x be the number of persons in the sample who have at least one credit card, and p be the probability that a person has at least one credit card. Then, this is a binomial problem with

$$n = 30, \quad p = .50, \quad q = 1 - p = .50,$$
$$x = 19, \quad n - x = 30 - 19 = 11$$

Using the binomial formula, the exact probability that 19 persons in a sample of 30 have at least one credit card is

$$P(19) = {}_{30}C_{19}(.50)^{19}(.50)^{11} = \mathbf{.0509}$$

Now, let us solve this problem using the normal distribution as an approximation to the binomial distribution. For this example,

$$np = 30\,(.50) = 15 \quad \text{and} \quad nq = 30\,(.50) = 15$$

Because np and nq are both greater than 5, we can use the normal distribution as an approximation to solve this binomial problem. We perform the following three steps:

STEP 1. *Compute μ and σ for the binomial distribution.*

To use the normal distribution, we need to know the mean and standard deviation of the distribution. Hence, the first step in using the normal approximation to the binomial distribution is to compute the mean and standard deviation of the binomial distribution. As we know from Chapter 5, the mean and standard deviation of a binomial distribution are given by np and $\sqrt{npq}$, respectively. Using these formulas, we obtain

$$\mu = np = 30(.50) = 15$$

$$\sigma = \sqrt{npq} = \sqrt{30(.50)(.50)} = 2.73861279$$

STEP 2. *Convert the discrete random variable into a continuous random variable.*

The normal distribution applies to a continuous random variable, whereas the binomial distribution applies to a discrete random variable. The second step in applying the normal approximation to the binomial distribution is to convert the discrete random variable to a continuous random variable by making the **correction for continuity.**

Continuity Correction Factor

The addition of .5 and/or subtraction of .5 from the value(s) of x when the normal distribution is used as an approximation to the binomial distribution, where x is the number of successes in n trials, is called the *continuity correction factor.*

As shown in **Figure 6.51**, the probability of 19 successes in 30 trials is given by the area of the rectangle for $x = 19$. To make the correction for continuity, we use the interval 18.5 to 19.5 for 19 persons. This interval is actually given by the two boundaries of the rectangle for $x = 19$, which are obtained by subtracting .5 from 19 and by adding .5 to 19. Thus, $P(x = 19)$ for the binomial problem will be approximately equal to $P(18.5 \le x \le 19.5)$ for the normal distribution.

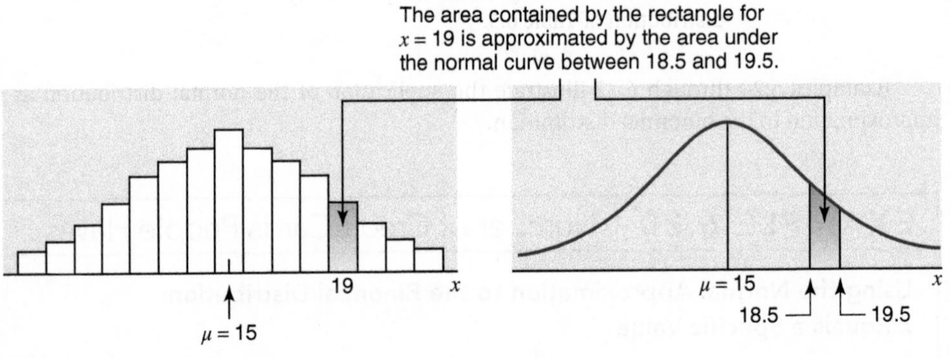

FIGURE 6.51

STEP 3. *Compute the required probability using the normal distribution.*

As shown in **Figure 6.52**, the area under the normal distribution curve between $x = 18.5$ and $x = 19.5$ will give us the (approximate) probability that 19 persons have at least one credit card. We calculate this probability as follows:

$$\text{For } x = 18.5: \quad z = \frac{18.5 - 15}{2.73861279} = 1.28$$

$$\text{For } x = 19.5: \quad z = \frac{19.5 - 15}{2.73861279} = 1.64$$

The required probability is given by the area under the standard normal curve between $z = 1.28$ and $z = 1.64$. This area is obtained by subtracting the area to the left of $z = 1.28$ from the area to the left of $z = 1.64$. From Table IV of Appendix B, the area to the left of $z = 1.28$ is .8997 and the area to the left of $z = 1.64$ is .9495. Hence, the required probability is

$$P(18.5 \le x \le 19.5) = P(1.28 \le z \le 1.64) = .9495 - .8997 = \mathbf{.0498}$$

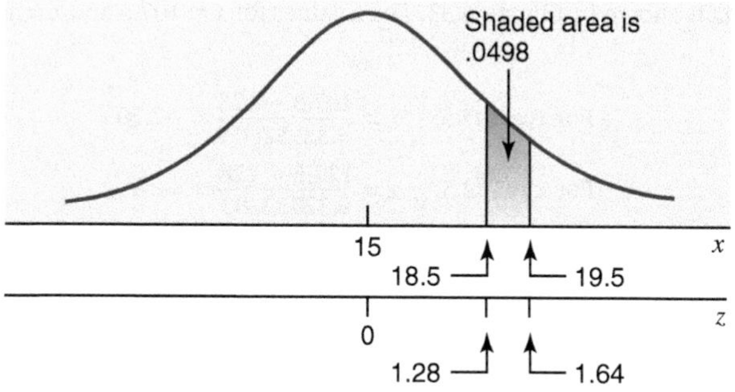

FIGURE 6.52 Area between $x = 18.5$ and $x = 19.5$.

Thus, based on the normal approximation, the probability that 19 persons in a sample of 30 will have at least one credit card is approximately .0498. Earlier, using the binomial formula, we obtained the exact probability .0509. The error due to using the normal approximation is $.0509 - .0498 = .0011$. Thus, the exact probability is underestimated by .0011 if the normal approximation is used.

When applying the normal distribution as an approximation to the binomial distribution, always make a *correction for continuity*. The continuity correction is made by subtracting .5 from the lower limit of the interval and/or by adding .5 to the upper limit of the interval. For example, the binomial probability $P(7 \leq x \leq 12)$ will be approximated by the probability $P(6.5 \leq x \leq 12.5)$ for the normal distribution; the binomial probability $P(x \geq 9)$ will be approximated by the probability $P(x \geq 8.5)$ for the normal distribution; and the binomial probability $P(x \leq 10)$ will be approximated by the probability $P(x \leq 10.5)$ for the normal distribution. Note that the probability $P(x \geq 9)$ has only the lower limit of 9 and no upper limit, and the probability $P(x \leq 10)$ has only the upper limit of 10 and no lower limit.

Remember

EXAMPLE 6.21 Advantage of Working from Home

Using the Normal Approximation to the Binomial Probability: x Assumes a Value in an Interval

According to a survey, 32% of people working from home said that the biggest advantage of working from home is that there is no commute. Suppose that this result is true for the current population of people who work from home. What is the probability that in a random sample of 400 people who work from home, 108 to 122 will say that the biggest advantage of working from home is that there is no commute?

Solution Let n be the number of people in the sample, x be the number of people in the sample who say that the biggest advantage of working from home is that there is no commute, and p be the probability that a person who works from home says that the biggest advantage of working from home is that there is no commute. Then, this is a binomial problem with

$$n = 400, \quad p = .32, \quad \text{and} \quad q = 1 - .32 = .68$$

We are to find the probability of 108 to 122 successes in 400 trials. Because n is large, it is easier to apply the normal approximation than to use the binomial formula. We can check that np and nq are both greater than 5. The mean and standard deviation of the binomial distribution are, respectively,

$$\mu = np = 400(.32) = 128$$
$$\sigma = \sqrt{npq} = \sqrt{400(.32)(.68)} = 9.32952303$$

To make the continuity correction, we subtract .5 from 108 and add .5 to 122 to obtain the interval 107.5 to 122.5. Thus, the probability that 108 to 122 out of a sample of 400 people who work from home will say that the biggest advantage of working from home is that there is no commute is approximated by the area under the normal distribution curve from $x = 107.5$ to

$x = 122.5$. This area is shown in **Figure 6.53.** The z values for $x = 107.5$ and $x = 122.5$ are calculated as follows:

$$\text{For } x = 107.5: \quad z = \frac{107.5 - 128}{9.32952303} = -2.20$$

$$\text{For } x = 122.5: \quad z = \frac{122.5 - 128}{9.32952303} = -.59$$

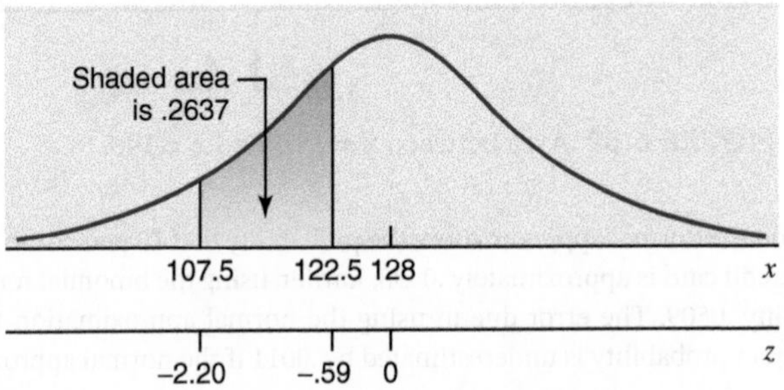

FIGURE 6.53 Area between $x = 107.5$ and $x = 122.5$.

The required probability is given by the area under the standard normal curve between $z = -2.20$ and $z = -.59$. This area is obtained by taking the difference between the areas under the standard normal curve to the left of $z = -2.20$ and to the left of $z = -.59$. From Table IV of Appendix B, the area to the left of $z = -2.20$ is .0139, and the area to the left of $z = -.59$ is .2776. Hence, the required probability is

$$P(107.5 \le x \le 122.5) = P(-2.20 \le z \le -.59) = .2776 - .0139 = \mathbf{.2637}$$

Thus, the probability that 108 to 122 people in a sample of 400 who work from home will say that the biggest advantage of working from home is that there is no commute is approximately .2637.

EXAMPLE 6.22 | Net Neutrality

Using the Normal Approximation to the Binomial Probability: *x* Is Greater Than or Equal to a Value

In a Pew Research Center survey conducted June 3–17, 2019, U.S. adults were asked if they knew what net neutrality was. Of these adults, 45% said that they knew that net neutrality refers to the principle that internet service providers should treat all traffic on their networks equally (https://www.pewinternet.org/2019/10/09/americans-and-digital-knowledge/). Assume that this percentage is true for the current population of U.S. adults. What is the probability that 326 or more adults in a random sample of 700 U.S. adults will know that net neutrality refers to the principle that internet service providers should treat all traffic on their networks equally?

Solution Let n be the sample size, x be the number of adults in the sample who know that net neutrality refers to the principle that internet service providers should treat all traffic on their networks equally, and p be the probability that a randomly selected U.S. adult knows what net neutrality refers to. Then, this is a binomial problem with

$$n = 700, \quad p = .45, \quad \text{and} \quad q = 1 - .45 = .55$$

We are to find the probability of 326 or more successes in 700 trials. The mean and standard deviation of the binomial distribution are

$$\mu = np = 700\,(.45) = 315$$

$$\sigma = \sqrt{npq} = \sqrt{700\,(.45)\,(.55)} = 13.16244658$$

In this example, $np = 700\,(.45) = 315 > 5$ and $nq = 700\,(.55) = 385 > 5$. Hence, in this example, we can use the normal probability distribution as an approximation to the binomial probability distribution.

For the continuity correction, we subtract .5 from 326, which gives 325.5. Thus, the probability that 326 or more of U.S. adults in the sample will know what net neutrality refers to is approximated by the area under the normal distribution curve to the right of $x = 325.5$, as shown in **Figure 6.54**. The z value for $x = 325.5$ is calculated as follows:

$$\text{For } x = 325.5: \quad z = \frac{325.5 - 315}{13.16244658} = .80$$

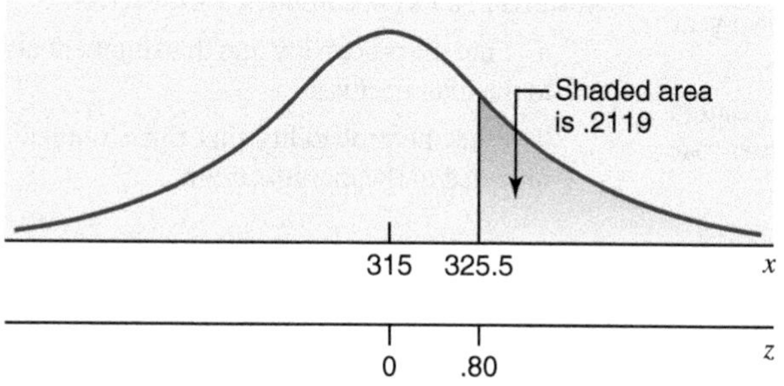

FIGURE 6.54 Area to the right of $x = 325.5$.

To find the required probability, we find the area to the left of $z = .80$ and subtract this area from 1.0. From Table IV of Appendix B, the area to the left of $z = .80$ is .7881. Hence,

$$P(x \geq 325.5) = P(z \geq .80) = 1.0 - .7881 = \mathbf{.2119}$$

Thus, the probability that 326 or more U.S. adults will say that they know that net neutrality refers to the principle that internet service providers should treat all traffic on their networks equally is .2119.

EXERCISES

Concepts and Procedures

6.45 Under what conditions is the normal distribution usually used as an approximation to the binomial distribution?

6.46 For a binomial probability distribution, $n = 20$ and $p = .60$.

a. Find the probability $P(8 \leq x \leq 13)$ using the table of binomial probabilities (Table I of Appendix B).

b. Find the probability $P(x = 12)$ using the normal distribution as an approximation to the binomial distribution. What is the difference between this approximation and the exact probability calculated in part (a)?

6.47 For a binomial probability distribution, $n = 100$ and $p = .50$. Let x be the number of successes in 100 trials.

a. Find the mean and standard deviation of this binomial distribution.

b. Find $P(x \geq 50)$ using the normal approximation.

c. Find $P(51 \leq x \leq 59)$ using the normal approximation.

6.48 Find the following binomial probabilities using the normal approximation.

a. $n = 50$, $p = .50$, $P(x = 20)$

b. $n = 250$, $p = .58$, $P(140 \leq x \leq 150)$

c. $n = 90$, $p = .35$, $P(x \geq 30)$

d. $n = 130$, $p = .65$, $P(x \leq 65)$

Applications

6.49 According to a study by the Haven Life Insurance Agency LLC, approximately 5.2% of all U.S. workers work from home. The city with the highest percentage of telecommuters (workers working from home) is Calsbad, CA, with 15.1% of its population working from home (https://havenlife.com/blog/cities-with-the-most-remote-workers/). Find the probability that in a random sample of 400 workers from Calsbad, CA, the number who work from home is 55 to 70.

6.50 A September 24, 2019, article on news.gallup.com mentioned that according to the Gallup research about climate change issues, 51% of U.S. adults can be classified as "Concerned Believers," meaning they "are highly worried about global warming, think it will pose a serious threat in their lifetime, believe it's the result of human activity, and think news reports about it are accurate or underestimate the problem" (https://news.gallup.com/opinion/polling-matters/267011/environment-climate-change-news.aspx). Assuming that this percentage is true for the current population of U.S. adults, find the probability that in a random sample of 600 U.S. adults, the number of adults who can be classified as "Concerned Believers" is

a. exactly 318 **b.** at most 324 **c.** 90 to 320

6.51 An office supply company conducted a survey before marketing a new paper shredder designed for home use. In the survey, 80% of the people who tried the shredder were satisfied with it. Because of this high satisfaction rate, the company decided to market the new shredder. Assume that 90% of all people are satisfied with this shredder. During a certain month, 100 customers bought this shredder. Find the probability that of these 100 customers, the number who are satisfied is

a. exactly 85 **b.** 80 or fewer **c.** 81 to 95

6.52 COSIO makes calculators. Customer satisfaction is one of the top priorities of the company's management. The company guarantees the refund of money or a replacement for a calculator that malfunctions within two years from the date of purchase. It is known from past data that despite all efforts, 6% of the calculators manufactured by this company malfunction within a two-year period. The company recently mailed 1000 such calculators to its customers.

a. Find the probability that exactly 50 of the 1000 calculators will be returned for refund within a two-year period.

b. What is the probability that 65 or more of the 1000 calculators will be returned for refund or replacement within a two-year period?

c. What is the probability that 60 to 80 of the 1000 calculators will be returned for refund or replacement within a two-year period?

6.53 A sports shop orders several thousand novelty baseball shirts from a manufacturer. Upon delivery of the shirts, the manager of the shop randomly selects 100 shirts and carefully inspects each shirt for any defects. If this sample contains nine or more defective shirts, the entire shipment is returned to the manufacturer. Suppose that 6% of all shirts in a recent shipment are defective.

a. Find the probability that this shipment of shirts will be returned to the manufacturer.

b. Find the probability that this shipment of shirts will not be returned to the manufacturer.

Uses and Misuses of Statistics

1. Don't Lose Your Memory

As discussed in the previous chapter, the Poisson distribution gives the probability of a specified number of events occurring in a time interval. The Poisson distribution provides a model for the number of emails a server might receive during a certain time period or the number of people arriving in line at a bank during lunch hour. These are useful to know for planning purposes, but sometimes we want to know the specific times at which emails or customers arrive. These times are governed by a special continuous probability distribution with certain unusual properties. This distribution is called the *exponential distribution*, and it is derived from the Poisson probability distribution.

Suppose you are a teller at a bank, and a customer has just arrived. You know that the customers arrive according to a Poisson process with a rate of λ customers per hour. Your boss might care how many customers arrive on average during a given time interval to ensure there are enough tellers available to handle the customers efficiently; you are more concerned with the time when the next customer will arrive. Remember that the probability that x customers arrive in an interval of length t is

$$P(x) = \frac{(\lambda t)^x e^{-\lambda t}}{x!}$$

If $t = 1$, this formula converts to the one we learned in Chapter 5. The probability that at least one customer arrives within time t is 1 minus the probability that no customer arrives within time t. Hence,

$$\begin{aligned} P\text{ (at least one customer arrives within time } t) &= 1 - P(0) \\ &= 1 - \frac{(\lambda t)^0 e^{-\lambda t}}{0!} \\ &= 1 - e^{-\lambda t} \end{aligned}$$

If the bank receives an average of 15 customers per hour—an average of one every four minutes—and a customer has just arrived, the probability that a customer arrives within four minutes is $1 - e^{-\lambda t} = 1 - e^{-(15/60)(4)} = .6321$. In the same way, the probability that a customer arrives within eight minutes is .8647.

Let us say that a customer arrived and went to your co-worker's window. No additional customer arrived within the next two minutes—an event with probability .6065—and you dozed off for two more minutes. When you open your eyes, you see that a customer has not arrived yet. What is the probability that a customer arrives within the next four minutes? From the calculation above, you might say that the answer is .8647. After all, you know that a customer arrived eight minutes earlier. But .8647 is not the correct answer.

The exponential distribution, which governs the time between arrivals of a Poisson process, has a property called the *memoryless* property. For you as a bank teller, this means that if you know a customer has not arrived during the past four minutes, then the clock is reset to zero, as if the previous customer had just arrived. So even after your nap, the probability that a customer arrives within four minutes is .6321. This interesting property reminds us again that we should be careful when we use mathematics to model real-world phenomena.

2. Quality Is Job 1

During the early 1980s, Ford Motor Company adopted a new marketing slogan: "Quality Is Job 1." The new slogan coincided with Ford's release of the Taurus and the Mercury Sable, but the groundwork that resulted in Ford's sudden need to improve quality was laid some 30 years earlier by an American statistician—not in Detroit, but in Japan.

W. Edwards Deming is one of the most famous statisticians, if not *the* most famous statistician, in the field of statistical process control, a field that debunked the myth that you could not improve quality and lower costs simultaneously. After World War II, Deming was asked by the U.S. Armed Forces to assist with planning for the 1951 census in Japan. While there, he taught statistical process control and quality management to the managers and engineers at many of Japan's largest companies. After adopting Deming's principles, the quality of and the demand for Japanese products, including Japanese automobiles, increased tremendously.

Ford's interest in Japanese quality resulted from the fact that Ford was having a specific transmission produced simultaneously in Japan and the United States. Ford's U.S. customers were requesting cars with Japanese-produced transmissions, even if it required them to wait longer for the car. Despite the fact that the transmissions were made to the same specifications in the two countries, the parts used in the Japanese transmissions were much closer to the desired size than those used in the transmissions made in America. Knowing that Deming had done a great deal of work with Japanese companies, Ford hired him as a consultant. The result of Deming's work in statistical process control and proper management methods, along with Ford's willingness to implement his recommendations, was the production of Ford's Taurus and Mercury Sable lines of automobiles, which resulted in Ford earning a profit after numerous years of losses.

W. Edwards Deming died in 1993 at the age of 93 years, but he left a legacy. Japan introduced the Deming Prize in 1950. The Deming Prize is awarded annually to individuals and companies whose work has advanced knowledge in the statistical process control area. More information about Deming and The W. Edwards Deming Institute is available at www.deming.org.

Source: The W. Edwards Deming Institute (www.deming.org), and en.wikipedia.org/wiki/W._Edwards_Deming.

Glossary

Continuity correction factor Addition and/or subtraction of .5 from the value(s) of x when the normal distribution is used as an approximation to the binomial distribution, where x is the number of successes in n trials.

Continuous random variable A random variable that can assume any value in one or more intervals.

Normal probability distribution The probability distribution of a continuous random variable that, when plotted, gives a specific bell-shaped curve. The parameters of the normal distribution are the mean μ and the standard deviation σ.

Standard normal distribution The normal distribution with $\mu = 0$ and $\sigma = 1$. A random variable having a standard normal distribution is denoted by z.

z value or z score A random variable having a standard normal distribution is denoted by z.

Exercises

Supplementary Exercises

6.54 A company that has a large number of supermarket grocery stores claims that customers who pay by personal checks spend an average of \$87 on groceries at these stores with a standard deviation of \$22. Assume that the expenses incurred on groceries by all such customers at these stores are normally distributed.

a. Find the probability that a randomly selected customer who pays by check spends more than \$114 on groceries.

b. What percentage of customers paying by check spend between \$40 and \$60 on groceries?

c. What percentage of customers paying by check spend between \$70 and \$105?

d. Is it possible for a customer paying by check to spend more than \$185? Explain.

6.55 At Jen and Perry Ice Cream Company, the machine that fills 1-pound cartons of Top Flavor ice cream is set to dispense 16 ounces of ice cream into every carton. However, some cartons contain slightly less than and some contain slightly more than 16 ounces of ice cream. The amounts of ice cream in all such cartons have a normal distribution with a mean of 16 ounces and a standard deviation of .18 ounce.

a. Find the probability that a randomly selected carton contains 16.20 to 16.50 ounces of ice cream.

b. What percentage of such cartons contain less than 15.70 ounces of ice cream?

c. Is it possible for a carton to contain less than 15.20 ounces of ice cream? Explain.

6.56 A machine at Kasem Steel Corporation makes iron rods that are supposed to be 50 inches long. However, the machine does not make all rods of exactly the same length. It is known that the probability distribution of the lengths of rods made on this machine is normal with a mean of 50 inches and a standard deviation of .06 inch. The rods that are either shorter than 49.85 inches or longer than 50.15 inches are discarded. What percentage of the rods made on this machine are discarded?

6.57 Michel Dolan, who lives in a metropolitan area, commutes by car from home to work. He knows that it takes him an average of 30 minutes for this commute in this morning. However, due to the variability in the traffic situation every morning, the standard deviation of these commutes is 10 minutes. Suppose the population of his morning commute times has a normal distribution with a mean of 30 minutes and a standard deviation of 10 minutes. Dolan has to be at work by 8:30 A.M. every morning. By what time must he leave home in the morning so that he is late for work at most 1% of the time?

6.58 Major League Baseball rules require that the balls used in baseball games must have circumferences between 9 and 9.25 inches. Suppose the balls produced by the factory that supplies balls to Major League Baseball have circumferences normally distributed with a mean of 9.125 inches and a standard deviation of .06 inch. What percentage of these baseballs fail to meet the circumference requirement?

6.59 Mong Corporation makes auto batteries. The company claims that 80% of its LL70 batteries are good for 70 months or longer.

a. What is the probability that in a sample of 100 such batteries, exactly 85 will be good for 70 months or longer?

b. Find the probability that in a sample of 100 such batteries, at most 74 will be good for 70 months or longer.

c. What is the probability that in a sample of 100 such batteries, 75 to 87 will be good for 70 months or longer?

d. Find the probability that in a sample of 100 such batteries, 72 to 77 will be good for 70 months or longer.

6.60 Stress of work is a major concern of a large number of people who go into administrative positions. It is estimated that 60% of the administrators of institutes suffer from work-related stress.

a. What is the probability that in a sample of 150 administrators of institutes, exactly 100 suffer from work-related stress?

b. Find the probability that in a sample of 150 administrators of institutes, at least 110 suffer from work-related stress.

c. What is the probability that in a sample of 150 administrators of institutes, 105 or fewer suffer from work-related stress?

d. Find the probability that in a sample of 150 administrators of institutes, 80 to 110 suffer from work-related stress.

Advanced Exercises

6.61 At Jen and Perry Ice Cream Company, a machine fills 1-pound cartons of Top Flavor ice cream. The machine can be set to dispense, on average, any amount of ice cream into these cartons. However, the machine does not put exactly the same amount of ice cream into each carton; it varies from carton to carton. It is known that the amount of ice cream put into each such carton has a normal distribution with a standard deviation of .18 ounce. The quality control inspector wants to set the machine such that at least 90% of the cartons have more than 16 ounces of ice cream. What should be the mean amount of ice cream put into these cartons by this machine?

6.62 Two companies, A and B, drill wells in a rural area. Company A charges a flat fee of $3500 to drill a well regardless of its depth. Company B charges $1000 plus $12 per foot to drill a well. The depths of wells drilled in this area have a normal distribution with a mean of 250 feet and a standard deviation of 40 feet.

a. What is the probability that Company B would charge more than Company A to drill a well?

b. Find the mean amount charged by Company B to drill a well.

6.63 Lori just bought a new set of four tires for her car. The life of each tire is normally distributed with a mean of 45,000 miles and a standard deviation of 2000 miles. Find the probability that all four tires will last for at least 46,000 miles. Assume that the life of each of these tires is independent of the lives of other tires.

6.64 The Jen and Perry Ice Cream company makes a gourmet ice cream. Although the law allows ice cream to contain up to 50% air, this product is designed to contain only 20% air. Because of variability inherent in the manufacturing process, management is satisfied if each pint contains between 18% and 22% air. Currently, two of Jen and Perry's plants are making gourmet ice cream. At Plant A, the mean amount of air per pint is 20% with a standard deviation of 2%. At Plant B, the mean amount of air per pint is 19% with a standard deviation of 1%. Assuming the amount of air is normally distributed at both plants, which plant is producing the greater proportion of pints that contain between 18% and 22% air?

6.65 Ashley knows that the time it takes her to commute to work is approximately normally distributed with a mean of 45 minutes and a standard deviation of 3 minutes. By what time must she leave home in the morning so that she is 95% sure of arriving at work by 9 A.M.?

6.66 Alpha Corporation is considering two suppliers to secure the large amounts of steel rods that it uses. Company A produces rods with a mean diameter of 8 mm and a standard deviation of .15 mm and sells 10,000 rods for $400. Company B produces rods with a mean diameter of 8 mm and a standard deviation of .12 mm and sells 10,000 rods for $460. A rod is usable only if its diameter is between 7.8 mm and 8.2 mm. Assume that the diameters of the rods produced by each company have a normal distribution. Which of the two companies should Alpha Corporation use as a supplier? Justify your answer with appropriate calculations.

6.67 A gambler is planning to make a sequence of bets on a roulette wheel. Note that a roulette wheel has 38 numbers, of which 18 are red, 18 are black, and 2 are green. Each time the wheel is spun, each of the 38 numbers is equally likely to occur. The gambler will choose one of the following two sequences:

Single-number bet: The gambler will bet $5 on a particular number before each spin. He will win a net amount of $175 if that number comes up and lose $5 otherwise.

Color bet: The gambler will bet $5 on the red color before each spin. He will win a net amount of $5 if a red number comes up and lose $5 otherwise.

a. If the gambler makes a sequence of 25 bets, which of the two betting schemes offers him a better chance of coming out ahead (winning more money than losing) after the 25 bets?

b. Now compute the probability of coming out ahead after 25 single-number bets of $5 each and after 25 color bets of $5 each. Do these results confirm your guess in part (a)? (Before using an approximation to find either probability, be sure to check whether it is appropriate.)

6.68 The amount of time taken by a bank teller to serve a randomly selected customer has a normal distribution with a mean of five minutes and a standard deviation of one minute.

a. What is the probability that both of two randomly selected customers will take less than four minutes each to be served?

b. What is the probability that at least one of four randomly selected customers will need more than six minutes to be served?

6.69 Suppose you are conducting a binomial experiment that has 15 trials and the probability of success of .02. According to the sample size requirements, you cannot use the normal distribution to approximate the binomial distribution in this situation. Use the mean and standard deviation of this binomial distribution and the empirical rule to explain why there is a problem in this situation. (*Note*: Drawing the graph and marking the values that correspond to the empirical rule is a good way to start.)

6.70 A variation of a roulette wheel has slots that are not of equal size. Instead, the width of any slot is proportional to the probability that a standard normal random variable z takes on a value between a and $(a + .1)$, where $a = -3.0, -2.9, -2.8, \ldots, 2.9, 3.0$. In other words, there are slots for the intervals $(-3.0, -2.9)$, $(-2.9, -2.8)$, $(-2.8, -2.7)$ through $(2.9, 3.0)$. There is one more slot that represents the probability that z falls outside the interval $(-3.0, 3.0)$. Find the following probabilities:

a. The ball lands in the slot representing $(.3, .4)$.

b. The ball lands in any of the slots representing $(-.1, .4)$.

c. In at least one out of five games, the ball lands in the slot representing $(-.1, .4)$.

d. In at least 100 out of 500 games, the ball lands in the slot representing $(.4, .5)$.

Appendix 6.1

Normal Quantile Plots

Many of the methods that are used in statistics require that the sampled data come from a normal distribution. While it is impossible to determine if this holds true without taking a census (i.e., looking at the population data), there are statistical tools that can be used to determine if this is a reasonable assumption. One of the simplest tools to use is called a normal quantile plot. The idea of the plot is to compare the values in a data set with the corresponding values one would predict for a standard normal distribution.

Although normal quantile plots are typically created using technology, it is helpful to see an example to understand how they are created and what the various numbers represent. To demonstrate, consider the data in the following table, which contains the 2001 salaries of the mayors of 10 large cities.

City	Mayor's Salary ($)	City	Mayor's Salary ($)
Chicago, IL	170,000	Newark, NJ	147,000
New York, NY	165,000	San Francisco, CA	146,891
Houston, TX	160,500	Jacksonville, FL	127,230
Detroit, MI	157,300	Baltimore, MD	125,000
Los Angeles, CA	147,390	Boston, MA	125,000

Each data point represents 1/10 of the distribution, with the smallest value (salary) representing the smallest 10%, the next representing the 10% to 20% interval, and so on. In each case, we estimate that the data points fall in the middle of their respective intervals. For these 10 data points, these midpoints would be at the 5%, 15%, 25%, and so on, locations, while if we have 20 data points, these locations would be at 2.5%, 7.5%, 12.5%, and so on. Next, we determine the z scores for these locations. The following table shows the z scores for the 10–data point scenario and for the 20–data point scenario.

Ten data points										
Location (%)	5	15	25	35	45	55	65	75	85	95
z Score	−1.645	−1.036	−.674	−.385	−.126	.126	.385	.674	1.036	1.645
Twenty data points										
Location (%)	2.5	7.5	12.5	17.5	22.5	27.5	32.5	37.5	42.5	47.5
z Score	−1.960	−1.440	−1.150	−.935	−.755	−.598	−.454	−.319	−.189	−.063
Location (%)	52.5	57.5	62.5	67.5	72.5	77.5	82.5	87.5	92.5	97.5
z Score	.063	.189	.319	.454	.598	.755	.935	1.150	1.440	1.960

Next, we make a two-dimensional plot that places the data on the horizontal axis and the z scores on the vertical axis.

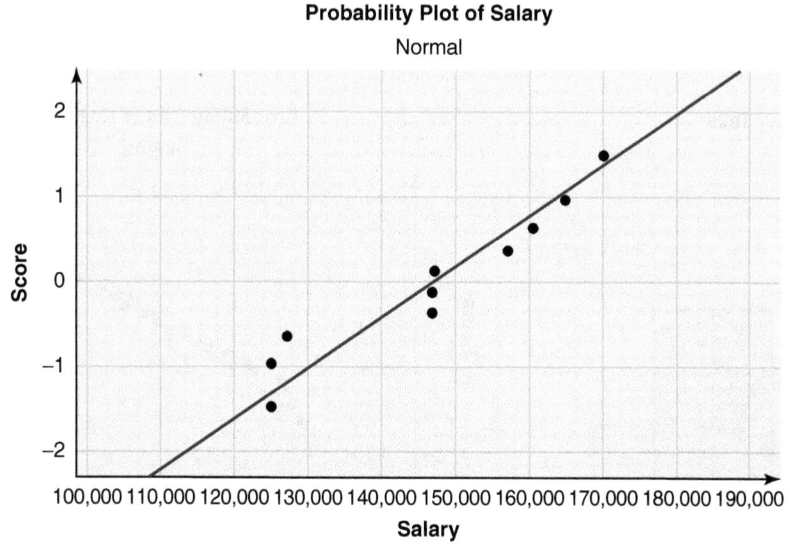

If the data are in complete agreement with a normal distribution, the points will lie on the line displayed in the graph. As the likelihood that the given data come from a normal distribution decreases, the plot of data points will become less linear.

So, how do we interpret the plot of 10 salaries in the graph? There are a couple of features that we can point out. There are two groups of points that are stacked almost vertically (near $125,000 and

$147,000). Depending on the software, multiple data points of the same value will be stacked or will appear as one point. In addition, there is a fairly big gap between these two groups. This is not unusual with small data sets, even if the data come from a normal distribution. Most times, in order to state that a very small data set does not come from a normal distribution, many people will be able to see that the data are very strongly skewed or have an outlier simply by looking at a sorted list of the data.

To understand the correspondence between the shape of a data set and its normal quantile plot, it is useful to look at a dotplot or histogram side by side with the normal quantile plot. We will consider a few common cases here.

1. First, following are the two graphs for 20 data points randomly selected from a normal distribution.

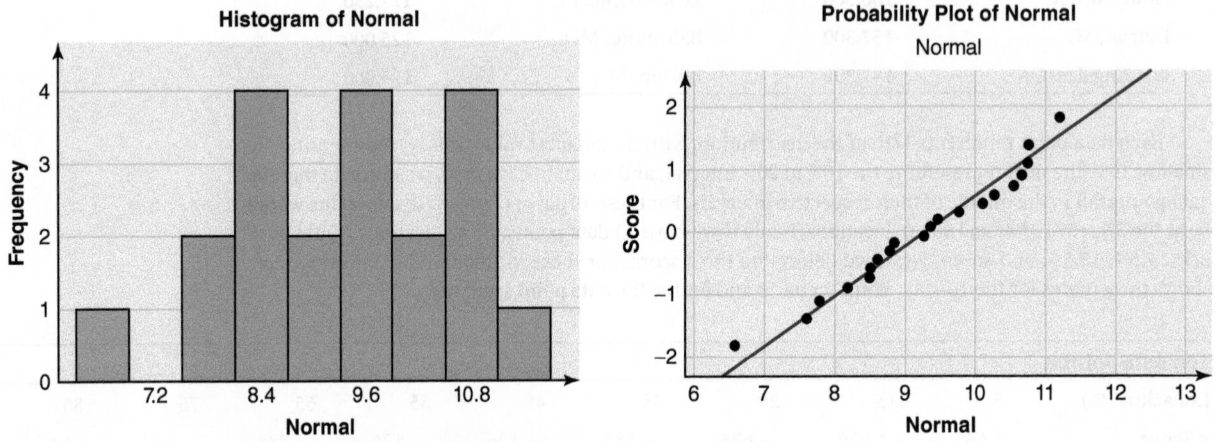

As you can see, just because data come from a normal distribution does not imply that they will be perfectly linear. In general, the points are close to the line, but small patterns such as in the upper right or the gap in the lower left can occur without invalidating the normality assumption.

2. In the next case, we consider data that come from a distribution that is *heavy tailed*, which means that the distribution has higher percentages of values in its tails than one would expect in a normal distribution. For example, the Empirical Rule states that a normal distribution has approximately 2.5% of the observations below $\mu - 2\sigma$ and 2.5% of the observations above $\mu + 2\sigma$. If a data set has 10% of the observations below $\mu - 2\sigma$ and 10% of the observations above $\mu + 2\sigma$, it would be classified as being heavy-tailed. The resulting shape of the normal quantile plot will look somewhat like a playground slide. As the tails get heavier, the ends of the plot become steeper and the middle gets flatter.

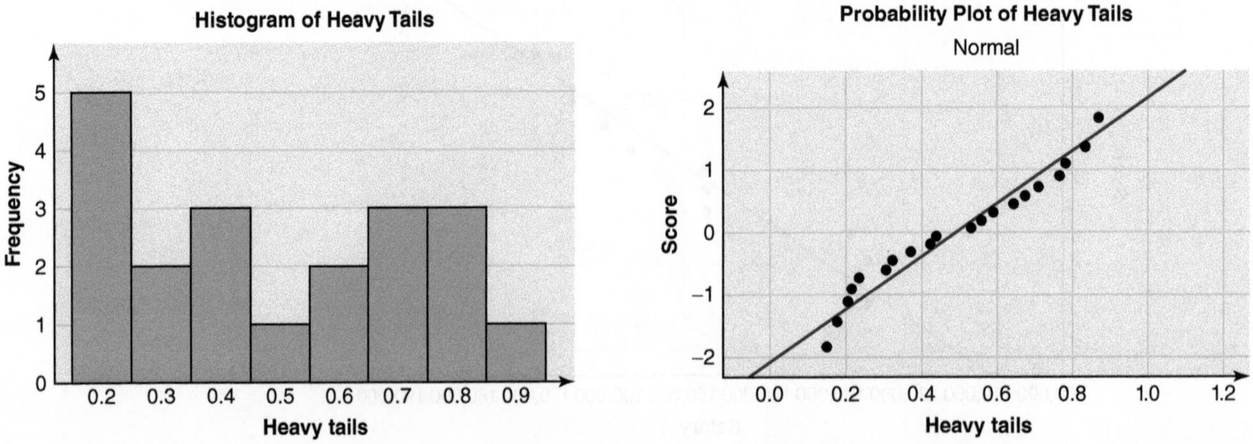

3. Skewed distributions have normal quantile plots that are shaped somewhat like a boomerang, which has a rounded V shape, with one end of the boomerang stretched out more than the other side. Just like all other graphs of skewed distributions, the side that is stretched out identifies the direction of the skew. Again, as the distribution becomes more skewed, the bend in the quantile plot will become more severe.

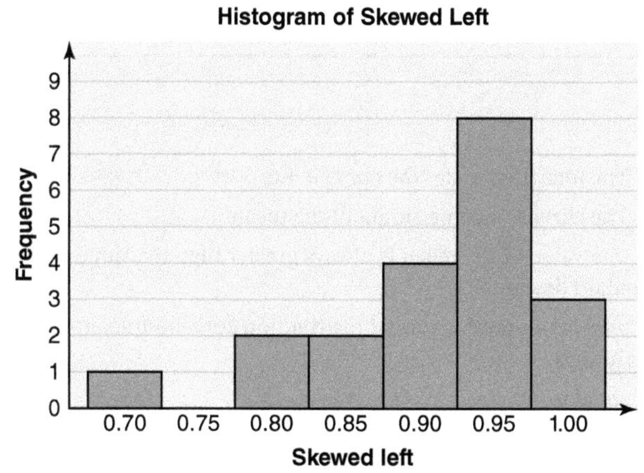

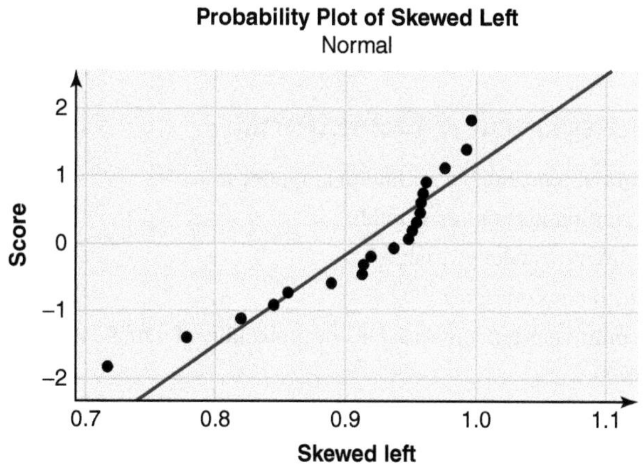

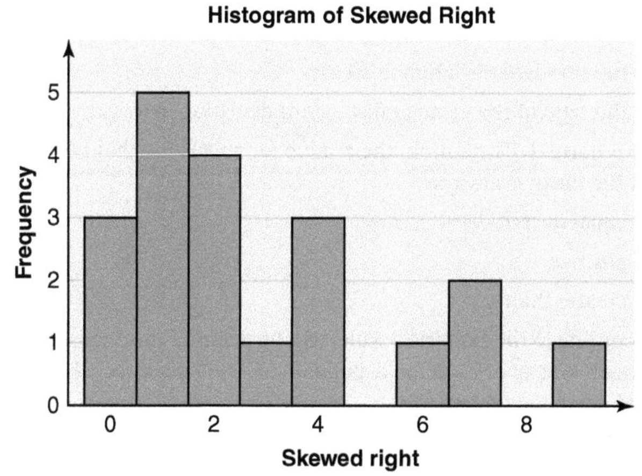

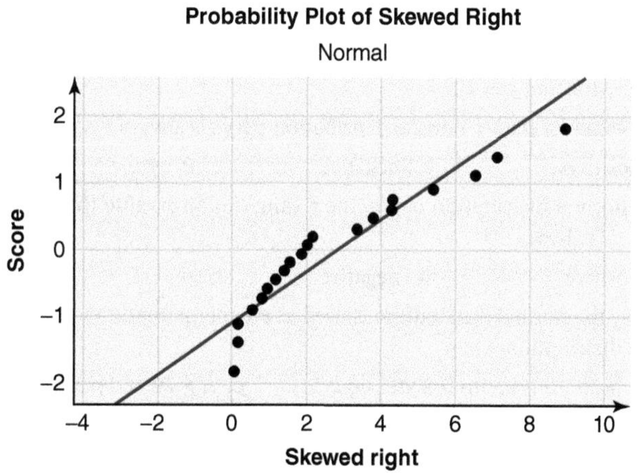

4. Our last example involves an outlier. As in other graphs, potential outliers are fairly easy to identify, basically by finding a large horizontal jump in the left or right tail. However, you need to be careful when distinguishing a skewed distribution from one that has an outlier. In our skewed-to-the-right example, there is an approximate difference of 2 between the two largest values, yet the largest data value is still fairly close to the line in a vertical direction. In our outlier example, there is a substantial vertical distance between the largest data value and the line. Moreover, we do not see the *bow* shape in the latter plot.

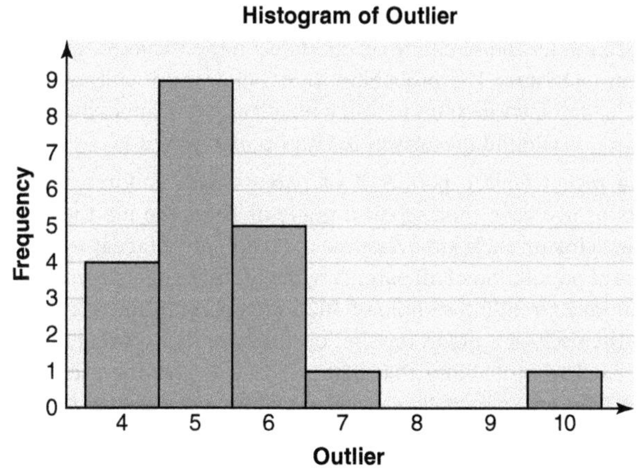

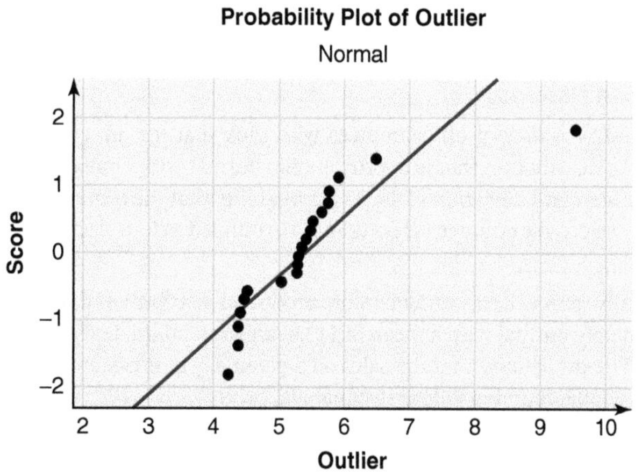

It is important to remember that these plots contain examples of a variety of common features. However, it is also important to remember that some of these features are not mutually exclusive. As one example, it is possible for a distribution to have heavy tails and an outlier. Identifying issues that would reject the notion of normality will be important in determining the types of inference procedures that can be used, which we will begin examining in Chapter 8.

Self-Review Test

Multiple Choice Questions

1. The normal probability distribution is applied to

a. a continuous random variable

b. a discrete random variable

c. any random variable

2. For a continuous random variable, the probability of a single value of x is always

a. zero b. 1.0 c. between 0 and 1

3. The parameters of a normal distribution are

a. μ, z, and σ b. μ and σ c. μ, x, and σ

4. For the standard normal distribution,

a. $\mu = 0$ and $\sigma = 1$

b. $\mu = 1$ and $\sigma = 0$

c. $\mu = 100$ and $\sigma = 10$

5. The z value for μ for a normal distribution curve is always

a. positive b. negative c. 0

6. For a normal distribution curve, the z value for an x value that is less than μ is always

a. positive b. negative c. 0

7. Usually the normal distribution is used as an approximation to the binomial distribution when

a. $n \geq 30$ b. $np > 5$ and $nq > 5$ c. $n > 20$ and $p = .50$

8. Which of the following is *not* a characteristic of the normal distribution?

a. The total area under the curve is 1.0.

b. The curve is symmetric about the mean.

c. The value of the mean is always greater than the value of the standard deviation.

9. The total area under a normal distribution curve to the left of the mean is always

a. equal to 1

b. equal to zero

c. equal to .50

10. For a normal distribution, the spread of the curve decreases and its height increases as

a. the sample size decreases

b. the standard deviation decreases

c. the ratio of the mean and standard deviation increases

11. For a normal distribution, the z value for an x value that is to the right of the mean is always

a. equal to zero

b. positive

c. greater than 1

12. According to the Empirical Rule, what percent of the data should be between 422 and 1342 for a population with mean of 882 and standard deviation of 230?

a. 95.44% b. 47.72% c. 68.26%

Practice Exercises

1. For the standard normal distribution, the area between $z = 0$ and $z = .32$, rounded to four decimal places, is ________.

2. For the standard normal distribution, the area to the right of $z = 1.75$, rounded to four decimal places, is ________.

3. For the standard normal distribution, the area between $z = .47$ and $z = 1.92$, rounded to four decimal places, is ________.

4. Let x have a normal distribution with a mean of 108.2 and a standard deviation of 10.09. The z value for $x = 145.69$, rounded to two decimal places, is ________.

5. The GMAT scores of all examinees who took that test in a year produced a distribution that is approximately normal with a mean of 410 and a standard deviation of 49. The probability that the score of a randomly selected examinee is less than 370, rounded to four decimal places, is ________.

6. The daily sales at a convenience store produce a distribution that is approximately normal with a mean of \$1320 and a standard deviation of \$148. The probability that the sales on a given day at this store are less than \$1305, rounded to four decimal places, is ________.

7. The net weights of all boxes of Top Taste cookies produce a distribution that is approximately normal with a mean of 32.15 and a standard deviation of .80. The probability that the net weight of a randomly selected box of these cookies is less than 31.58 ounces, rounded to four decimal places, is ________.

8. The heights of all female college basketball players produce a distribution that is approximately normal with a mean of 67.59 and a standard deviation of 2.23. The probability that the height of a randomly selected female college basketball player is less than 67.2 inches, rounded to four decimal places, is ________.

9. The area under the standard normal curve from zero to z is .0910 and z is positive. The value of z is ________.

10. We know that football is the favorite sport to watch on television for 38% of adults in the United States. Suppose we take a random sample of 525 adults and obtain their opinions on their favorite sport to watch on television. The probability that 193 or more will say that football is their favorite sport to watch on television is, approximately, ________. (Round the answer to four decimal places.)

11. In a recent Gallup poll, 58% of parents with children under 18 years of age gave themselves a grade of B for the job they are doing bringing up their kids. Assume that this percentage is true for the current population of all parents with children under 18 years of age. You take a random sample of 1021 such parents and ask them to give themselves a grade for the job they are doing bringing up their kids. The probability that 520 to 555 will give themselves a grade of B is, approximately, ________. (Round the answer to four decimal places.)

12. You are given that the area under the standard normal curve to the left of $z = -.78$ is equal to .2177. What is $P(-.78 < z < .78)$?

13. The accountant at a department store is analyzing the credit card purchases of 100 of the store's cardholders over the past year. The following table lists the frequency distribution of the continuous random variable, x, annual credit card purchases.

Annual Credit Card Purchases ($)	Frequency
0 to less than 200	14
200 to less than 400	23
400 to less than 600	25
600 to less than 800	10
800 to less than 1000	15
1000 to less than 1200	13

What is the relative frequency density of the 400 to less than 600 class? (Round your answer to five decimal places.)

14. We know that the length of time required for a student to complete a particular aptitude test has a normal distribution with a mean of 39.5 minutes and a variance of 4.7 minutes. What is the probability, rounded to four decimal places, that a given student will complete the test in more than 36 minutes but less than 43 minutes?

15. Suppose that the distribution for a random variable x is normal with mean 14 and standard deviation σ, and $P(x < 0) = .0864$. Rounded to two decimal places, what is σ?

16. Let x be a continuous random variable that follows a normal distribution with a mean of 254 and a standard deviation of 46. Find the value of x so that the area under the normal curve to the right of x is approximately .0041. (Round your answer to two decimal places.)

17. 94.9% of the parts coming off an assembly line are non-defective. Using the normal approximation to the binomial distribution, what is the probability, rounded to four decimal places, that of 488 parts, fewer than 463 are non-defective?

18. Find the following probabilities for the standard normal distribution.

a. $P(.85 \leq z \leq 2.33)$ **b.** $P(-2.97 \leq z \leq 1.49)$

c. $P(z \leq -1.29)$ **d.** $P(z > -.74)$

19. Find the value of z for the standard normal curve such that the area

a. in the left tail is .1000

b. between 0 and z is .2291 and z is positive

c. in the right tail is .0500

d. between 0 and z is .3571 and z is negative

20. The average length of National Hockey League (NHL) games is 152.5 minutes with a standard deviation of 8.48 minutes. Assume that the lengths of all National Hockey League games have an approximate normal distribution.

a. Find the probability that a randomly selected NHL game takes 142 to 156 minutes to finish.

b. Find the probability that a randomly selected NHL game takes longer than 168 minutes to finish.

c. Find the probability that a randomly selected NHL game is finished within 163 minutes.

d. Find the probability that a randomly selected NHL game takes 159 to 169 minutes to finish.

21. Refer to Practice Exercise 20

a. Suppose that 6% of all NHL games are finished within x minutes. Find the value of x.

b. Suppose that 3% of all NHL games last longer than x minutes. Find the value of x.

22. Estimates show that gluten sensitivity, which is also known as wheat intolerance, affects approximately 15% of people. The condition involves great difficulty in digesting wheat, but is not the same as wheat allergy, which has much more severe reactions. A random sample of 800 individuals is selected.

a. Find the probability that the number of individuals in this sample who have wheat intolerance is

i. exactly 115 **ii.** 103 to 142 **iii.** at least 107

iv. at most 100 **v.** between 111 and 123

b. Find the probability that at least 675 of the individuals in this sample do *not* have wheat intolerance.

c. Find the probability that 682 to 697 of the individuals in this sample do *not* have wheat intolerance.

Mini-Projects

Mini-Project 6.1

Refer to Data Set III on the NFL players from the 2019 football season that accompanies this text (see Appendix A). Consider the heights of the players listed in column 4.

a. Use statistical software to obtain a histogram. Do these heights appear to be symmetrically distributed? If not, in which direction do they seem to be skewed?

b. Compute μ and σ for the heights of all players.

c. What percentage of these heights lie in the interval $\mu - \sigma$ to $\mu + \sigma$? What about in the interval $\mu - 2\sigma$ to $\mu + 2\sigma$? In the interval $\mu - 3\sigma$ to $\mu + 3\sigma$?

d. How do the percentages in part (c) compare to the corresponding percentages for a normal distribution (68.26%, 95.44%, and 99.74%, respectively)?

e. Based on the percentages in the Empirical Rule, approximately 34.13% of data values should fall in each of the intervals $\mu - \sigma$ to μ and μ to $\mu + \sigma$. Similarly, approximately 13.59% of the data values should fall in each of the intervals $\mu - 2\sigma$ to $\mu - \sigma$ and $\mu + \sigma$ to $\mu + 2\sigma$, and approximately 2.15% of the data values should fall in each of the intervals $\mu - 3\sigma$ to $\mu - 2\sigma$ and $\mu + 2\sigma$ to $\mu + 3\sigma$. Calculate the percentages of the values in the NFL data that fall in each of these intervals. How do they compare to the values given by the Empirical Rule?

f. Use statistical software to select three random samples of 20 players each. Create a histogram and a dotplot of heights for each sample, and calculate the mean and standard deviation of heights for each sample. How well do your graphs and summary statistics match up with the corresponding population graph and parameter values obtained in earlier parts? Does it seem reasonable that they might not match up very well?

Mini-Project 6.2

Refer to Data Set III on the NFL players from the 2019 football season (see Appendix A). Consider the weights of the players listed in column 5.

a. Use statistical software to obtain a histogram. Do these weights appear to be symmetrically distributed? If not, in which direction do they seem to be skewed?

b. Compute μ and σ for the weights of all players.

c. What percentage of these weights lie in the interval $\mu - \sigma$ to $\mu + \sigma$? What about in the interval $\mu - 2\sigma$ to $\mu + 2\sigma$? In the interval $\mu - 3\sigma$ to $\mu + 3\sigma$?

d. How do the percentages in part (c) compare to the corresponding percentages for a normal distribution (68.26%, 95.44%, and 99.74%, respectively)?

e. Based on the percentages in the Empirical Rule, approximately 34.13% of data values should fall in each of the intervals $\mu - \sigma$ to μ and μ to $\mu + \sigma$. Similarly, approximately 13.59% of the data values should fall in each of the intervals $\mu - 2\sigma$ to $\mu - \sigma$ and $\mu + \sigma$ to $\mu + 2\sigma$, and approximately 2.15% of the data values should fall in each of the intervals $\mu - 3\sigma$ to $\mu - 2\sigma$ and $\mu + 2\sigma$ to $\mu + 3\sigma$. Calculate the percentages of the values in the NFL data that fall in each of these intervals. How do they compare to the values given by the Empirical Rule?

f. Use statistical software to select three random samples of 20 players each. Create a histogram and a dotplot of weights for each sample, and calculate the mean and standard deviation of weights for each sample. How well do your graphs and summary statistics match up with the corresponding population graph and parameter values obtained in earlier parts? Does it seem reasonable that they might not match up very well?

Mini-Project 6.3

Refer to Data Set VI on the S&P 500 stocks (see Appendix A). Consider these stocks as a sample of all stocks on the U.S. stock market. Consider the high prices of these stocks listed in column 6.

a. Use statistical software to obtain a histogram and a dotplot of these data. Comment on the shape of the distribution as observed from these graphs.

b. Calculate $\bar{x}$ and s.

c. What percentage of the high prices are in the interval $\bar{x} - s$ to $\bar{x} + s$?

d. What percentage are in the interval $\bar{x} - 2s$ to $\bar{x} + 2s$?

e. How do these percentages compare to the corresponding percentages for a normal distribution (68.26% and 95.44%, respectively)?

Decide for Yourself

Note: The Decide for Yourself feature is located in the book's product page on *www.wiley.com*.

Chapter 6 Technology Instructions

Note: Complete TI-84, Minitab, and Excel manuals are available in the book's product page on *www.wiley.com*.

TI-84 Color/TI-84

The TI-84 Color Technology Instructions feature of this text is written for the TI-84 Plus C color graphing calculator running the 4.0 operating system. Some screens, menus, and functions will be slightly different in older operating systems. The TI-84+ can perform all of the same functions but does not have the "Color" option referenced in some of the menus.

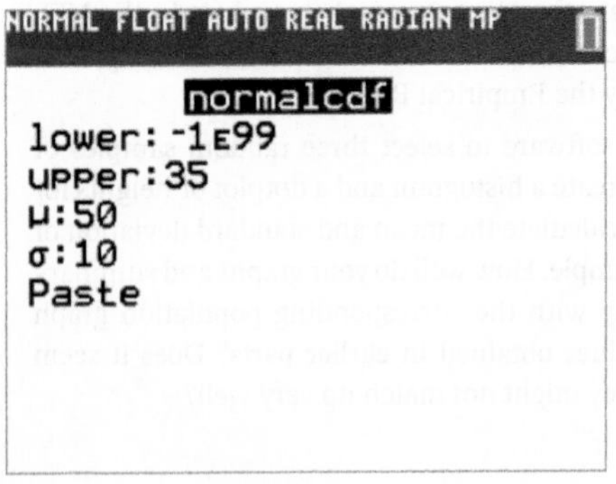

SCREEN 6.1

Calculating a Left-Tail Probability for Example 6.6(b) of the Text

1. Select **2nd > VARS > normalcdf(**.
2. Use the following settings in the **normalcdf** menu (see **Screen 6.1**):
   - At the **lower** prompt, type -1E99. (The "E" symbol means scientific notation. Press **2nd** followed by the comma key to get the "E" symbol. -1E99 is calculator notation for the number -1×10^{99}, which indicates that there is no lower limit of the interval.)
   - At the **upper** prompt, type 35.
   - At the μ prompt, type 50.
   - At the σ prompt, type 10.
3. Highlight **Paste** and press **ENTER** twice.
4. The output gives the probability of .0668072287 that x is less than 35 for a normal distribution with $\mu = 50$ and $\sigma = 10$ (see **Screen 6.2**).

Calculating a Probability Between Two Values for Example 6.7(b) of the Text

1. Select **2nd > VARS > normalcdf(**.
2. Use the following settings in the **normalcdf** menu:
   - At the **lower** prompt, type 18.
   - At the **upper** prompt, type 34.

- At the μ prompt, type 25.
- At the σ prompt, type 4.

3. Highlight **Paste** and press **ENTER** twice.
4. The output gives the probability of .9477164531 that x is between 18 and 34 for a normal distribution with $\mu = 25$ and $\sigma = 4$. (See **Screen 6.2.**)

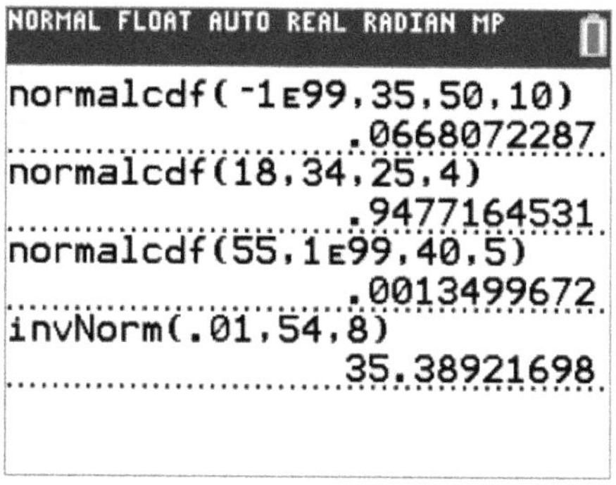

SCREEN 6.2

Calculating a Right-Tail Probability for Example 6.8(a) of the Text

1. Select **2nd > VARS > normalcdf(**.
2. Use the following settings in the **normalcdf** menu:
   - At the **lower** prompt, type 55.
   - At the **upper** prompt, type 1E99. (The "E" symbol means scientific notation. Press **2nd** followed by the comma key to get the "E" symbol. 1E99 is calculator notation for the number 1×10^{99}, which indicates that there is no upper limit of the interval.)
   - At the μ prompt, type 40.
   - At the σ prompt, type 5.
3. Highlight **Paste** and press **ENTER** twice.
4. The output gives the probability of .0013499672 that x is greater than 55 for a normal distribution with $\mu = 40$ and $\sigma = 5$. (See **Screen 6.2.**)

Determining x When a Left-Tailed Probability Is Known for Example 6.18 of the Text

1. Select **2nd > VARS > invNorm(**.
2. Use the following settings in the **invNorm** menu (see **Screen 6.3**):
   - At the **area** prompt, type 0.01.
   - At the μ prompt, type 54.
   - At the σ prompt, type 8.
3. Highlight **Paste** and press **ENTER** twice.
4. The output is 35.38921698, the value of x such that the area under the normal curve in the left tail is 0.01. (See **Screen 6.2.**)

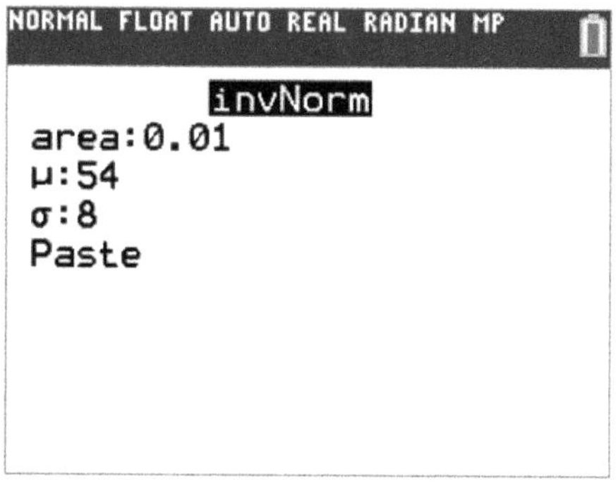

SCREEN 6.3

Minitab

The Minitab Technology Instructions in this text are written for **Minitab Professional Version 19.** Some screens, menus, and functions will be slightly different in older (or in newer) versions of Minitab. If you are using **Minitab Express**, you can download the Minitab Express Manual from the book's product page on *www.wiley.com.*

Calculating a Left-Tail Probability for Example 6.6(b) of the Text

1. Select **Graph > Probability Distribution Plot**.
2. In the dialog box that appears on screen, select **View Probability** and click **OK.**
3. Use the following settings in the dialog box that appears on screen (see **Screen 6.4**):
   - Select Normal in the **Distribution** box.
   - Type 50 in the **Mean** box.
   - Type 10 in the **Standard deviation** box.

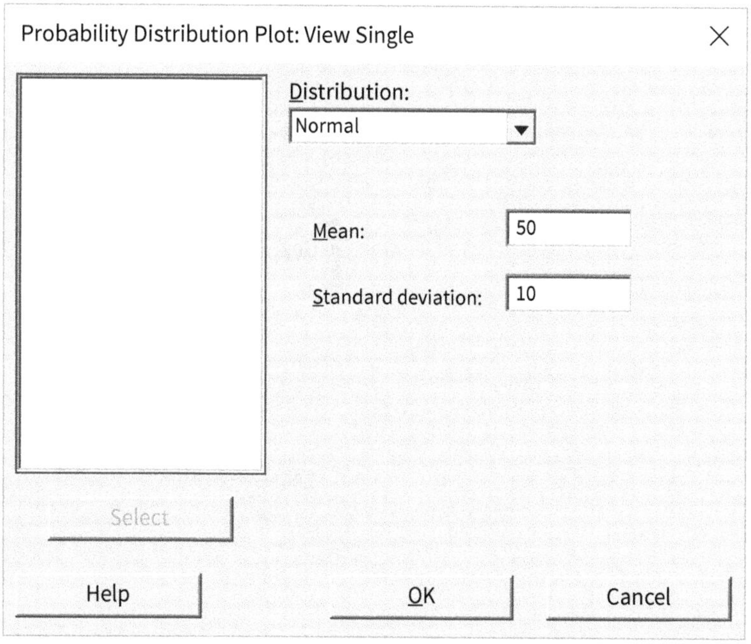

SCREEN 6.4

4. Click on the **Shaded Area** tab, and then use the following settings (see **Screen 6.5**):
   - Select X Value under **Define Shaded Area By.**
   - Select Left Tail.
   - Type 35 in the **X value** box.

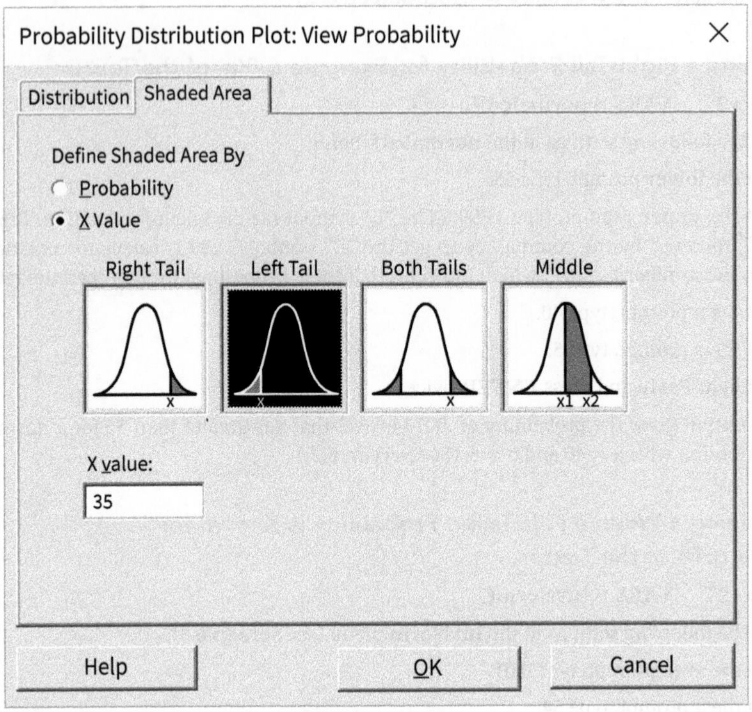

SCREEN 6.5

5. Click **OK.**
6. A new window will appear showing the shaded area of 0.06681 (see **Screen 6.6**).

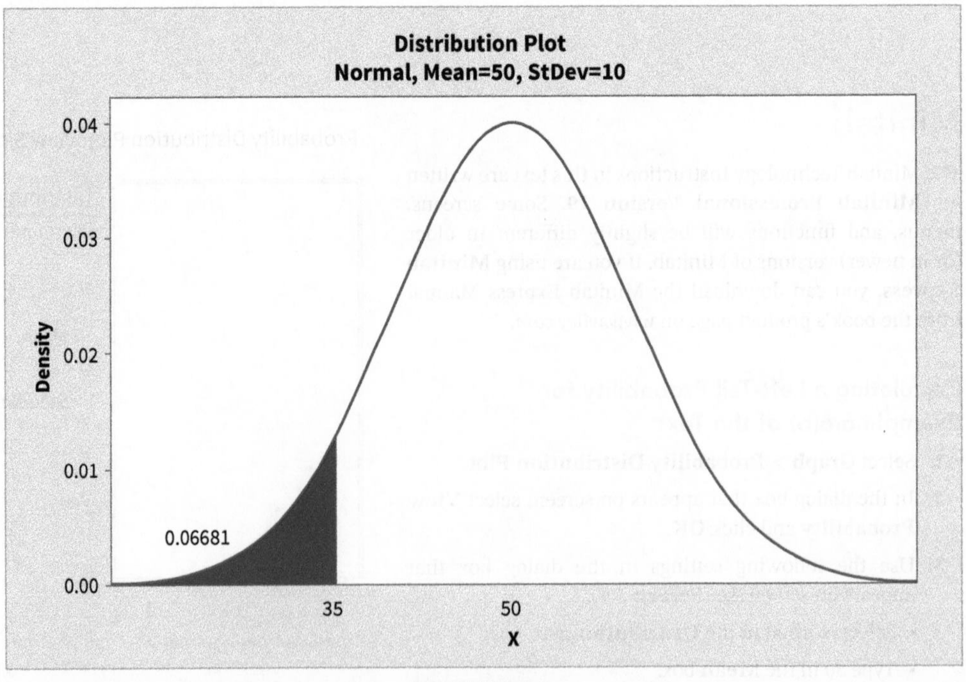

SCREEN 6.6

If you do not want to have a plot of the normal curve in Minitab, select **Calc > Probability Distributions > Normal.** Then select **Cumulative Probability** in the dialog box. Enter the values of the mean and standard deviation. Enter the value of *x* next to **Input constant.** Click ok. The probability to the left of this *x* value will appear on the screen.

Calculating a Probability Between Two Values for Example 6.7(b) of the Text

1. Select **Graph** > **Probability Distribution Plot**.
2. In the dialog box that appears on screen, select **View Probability** and click **OK**.
3. Use the following settings in the dialog box that appears on screen:
   - Select Normal in the **Distribution** box.
   - Type 25 in the **Mean** box.
   - Type 4 in the **Standard deviation** box.
4. Click on the **Shaded Area** tab, and then use the following settings:
   - Select X Value under **Define Shaded Area By**.
   - Select Middle.
   - Type 18 in the **X value 1** box.
   - Type 34 in the **X value 2** box.
5. Click **OK**.
6. A new window will appear showing the shaded area of 0.9477.

Calculating a Right-Tail Probability for Example 6.8(a) of the Text

1. Select **Graph** > **Probability Distribution Plot**.
2. In the dialog box that appears on screen, select **View Probability** and click **OK**.
3. Use the following settings in the dialog box that appears on screen:
   - Select Normal in the **Distribution** box.
   - Type 40 in the **Mean** box.
   - Type 5 in the **Standard deviation** box.
4. Click on the **Shaded Area** tab, and then use the following settings:
   - Select X Value under **Define Shaded Area By**.
   - Select Right Tail.
   - Type 55 in the **X value** box.
5. Click **OK**.
6. A new window will appear showing the shaded area of 0.001350.

Determining *x* When a Left-Tailed Probability Is Known for Example 6.18 of the Text

1. Select **Graph** > **Probability Distribution Plot**.
2. In the dialog box that appears on screen, select **View Probability** and click **OK**.
3. Use the following settings in the dialog box that appears on screen:
   - Select Normal in the **Distribution** box.
   - Type 54 in the **Mean** box.
   - Type 8 in the **Standard deviation** box.
4. Click on the **Shaded Area** tab, then use the following settings (see **Screen 6.7**):
   - Select Probability under **Define Shaded Area By**.
   - Select Left Tail.
   - Type 0.01 in the **Probability** box.
5. Click **OK**.
6. A new window will appear showing the *x* value of 35.39 (see **Screen 6.8**).

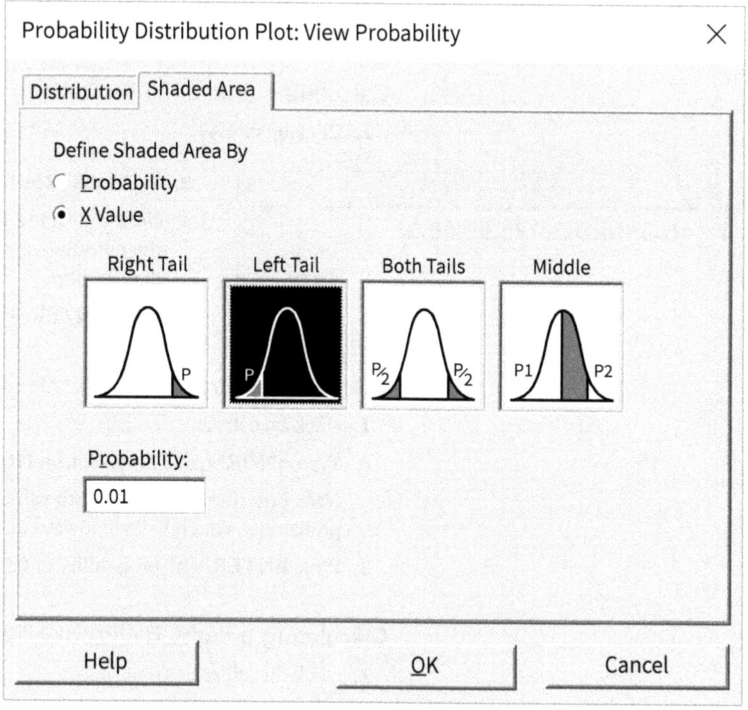

SCREEN 6.7

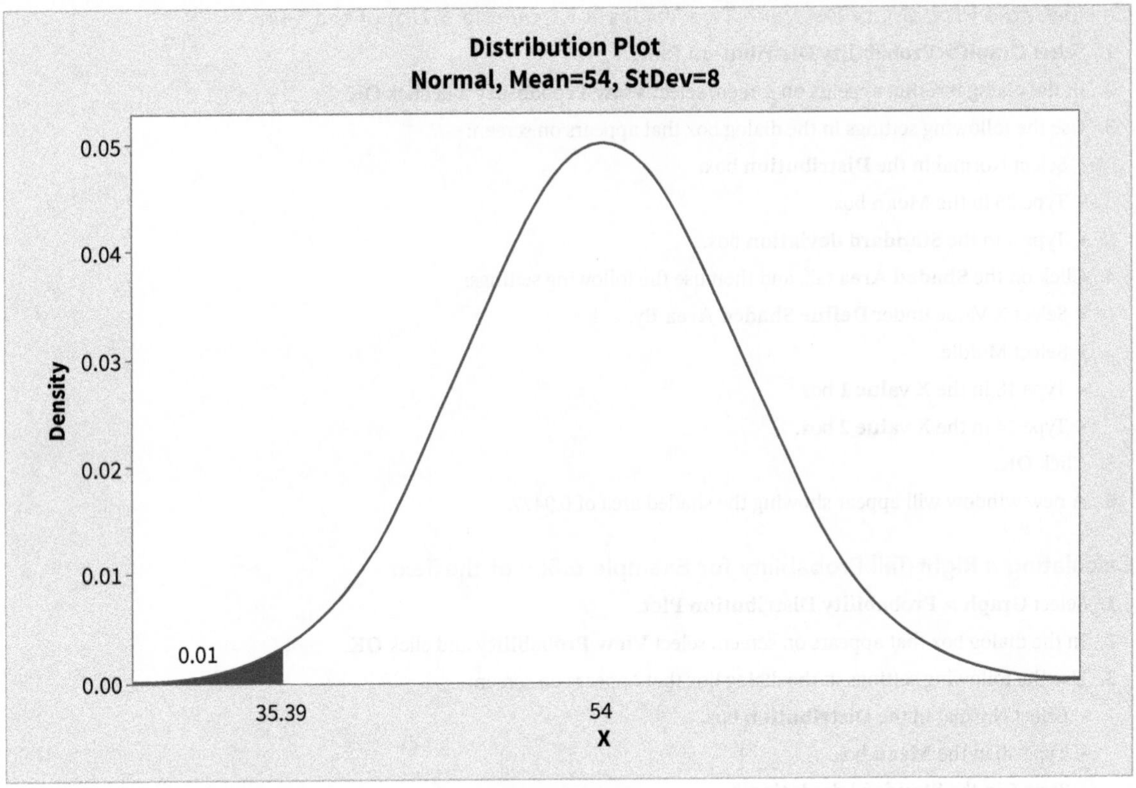

SCREEN 6.8

Excel

The Excel Technology Instructions feature of this text is written for Excel 2019. Some screens, menus, and functions will be slightly different in older versions of Excel. To enable some of the advanced Excel functions, you must enable the Data Analysis Add-In for Excel.

Calculating a Left-Tail Probability for Example 6.6(b) of the Text

1. Click on cell A1.

	A	B
1	=NORM.DIST(35,50,10,1)	
2		

SCREEN 6.9

2. Type **=NORM.DIST(35,50,10,1)** (see **Screen 6.9**).

   Note: Entering a 1 as the fourth value in this function will cause Excel to calculate the cumulative probability, which we will always do for the normal probability distribution.

3. Press **ENTER**. The probability of 0.066807 will appear in cell A1.

Calculating a Probability Between Two Values for Example 6.7(b) of the Text

1. Click on cell A1.
2. Type **=NORM.DIST(34,25,4,1)–NORM.DIST(18,25,4,1)**.

   Note: Entering a 1 as the fourth value in these functions will cause Excel to calculate the cumulative probability, which we will always do for the normal probability distribution.

3. Press **ENTER**. The probability of 0.947716 will appear in cell A1.

Calculating a Right-Tail Probability for Example 6.8(a) of the Text

1. Click on cell A1.
2. Type **=1–NORM.DIST(55,40,5,1)**.

   Note: Entering a 1 as the fourth value in this function will cause Excel to calculate the cumulative probability, which we will always do for the normal probability distribution.

3. Press **ENTER**. The probability of 0.00135 will appear in cell A1.

Determining *x* When a Left-Tailed Probability Is Known for Example 6.18 of the Text

1. Click on cell A1.
2. Type **=NORM.INV(0.01,54,8)** (see **Screen 6.10**).
3. Press **ENTER**. The *x* value of 35.3892 will appear in cell A1.

	A	B
1	=NORM.INV(0.01,54,8)	
2		

SCREEN 6.10

Technology Assignments

TA6.1 Find the area under the standard normal curve

a. to the left of $z = -1.94$

b. to the left of $z = .83$

c. to the right of $z = 1.45$

d. to the right of $z = -1.65$

e. between $z = .75$ and $z = 1.90$

f. between $z = -1.20$ and $z = 1.55$

TA6.2 Find the following areas under a normal curve with $\mu = 86$ and $\sigma = 14$.

a. Area to the left of $x = 71$

b. Area to the left of $x = 96$

c. Area to the right of $x = 90$

d. Area to the right of $x = 75$

e. Area between $x = 65$ and $x = 75$

f. Area between $x = 72$ and $x = 95$

TA6.3 The transmission on a particular model of car has a warranty for 40,000 miles. It is known that the life of such a transmission has a normal distribution with a mean of 72,000 miles and a standard deviation of 12,000 miles. Answer the following questions:

a. What percentage of the transmissions will fail before the end of the warranty period?

b. What percentage of the transmissions will be good for more than 100,000 miles?

c. What percentage of the transmissions will be good for 80,000 to 100,000 miles?

TA6.4 A small company packages and mails gift boxes of fruit and nuts. On average, these boxes weigh 74 ounces with a standard deviation of 3.2 ounces. The weights of these gift boxes are approximately normally distributed.

a. What percentage of these boxes weigh less than 70 ounces?

b. What percentage of these boxes weigh more than 80 ounces?

c. Boxes that weigh more than 80 ounces cost more money to mail. The company has orders for 2400 gift boxes during the holiday season. Based on your answer to part (b), approximately how many boxes weigh more than 80 ounces?

TA6.5 Refer to Data Set X on Major League Baseball (see Appendix A). Consider the data on batting averages given in column 16. The mean and standard deviation of these averages are .251 and .036, respectively. Suppose these batting averages are approximately normally distributed. You randomly select one player from this list of players.

a. What is the probability that the batting average for this player is less than .175?

b. What is the probability that the batting average for this player is more than .310?

c. What is the probability that the batting average for this player is between .200 and .290?

d. What is the probability that the batting average for this player is between .340 and .400?

TA6.6 Shaun is a pretty good tennis player. During match play, his first serve speeds average 92 mph with a standard deviation of 4.2 mph.

a. During match play, what percent of Shaun's first serves are faster than 99 mph?

b. During match play, what percent of Shaun's first serves are slower than 82 mph?

c. During match play, what percent of Shaun's first serves are between 85 and 95 mph?

d. What is the minimum speed of the fastest 10% of Shaun's first serves?

TA6.7 Suppose that systolic blood pressures of adults are normally distributed with a mean of 122 millimeters of mercury (mmHg) and a standard deviation of 6 mmHg.

a. According to one definition, a systolic blood pressure over 140 mmHg means that the person is suffering from hypertension. What percentage of adults have hypertension?

b. What percentage of adults have systolic blood pressure less than 108 mmHg?

c. What percentage of adults have systolic blood pressure between 120 and 130 mmHg?

d. The lowest 5% of adult systolic blood pressures are lower than what mmHg?

CHAPTER

7

Sampling Distributions

Bloomberg/Getty Images

The opinion polls that you read in the newspapers, social media, magazines, and websites are based on sample surveys. These polls do make sampling and nonsampling errors. While reading these opinion polls, you should be aware of sampling and nonsampling errors. Due to these errors, the results and predictions made by these polls may not be valid. See Case Study 7.1 that explains sampling error for the 2016 U.S. presidential election.

Chapters 5 and 6 discussed probability distributions of discrete and continuous random variables. This chapter extends the concept of probability distribution to that of a sample statistic. As we discussed in Chapter 3, a sample statistic is a numerical summary measure calculated for sample data. The mean, median, mode, and standard deviation calculated for sample data are called *sample statistics*. On the other hand, the same numerical summary measures calculated for population data are called *population parameters*. A population parameter is always a constant (at a given point in time), whereas a sample statistic is always a random variable. Because every random variable must possess a probability distribution, each sample statistic possesses a probability distribution. The probability distribution of a sample statistic is more commonly called its *sampling distribution*. This chapter discusses the sampling distributions of the sample mean and the sample proportion. The concepts covered in this chapter are the foundation of the inferential statistics discussed in succeeding chapters.

7.1 Sampling Distribution, Sampling Error, and Nonsampling Errors

LEARNING OBJECTIVES

After completing this section, you should be able to:

- Explain a population probability distribution and a sampling distribution.
- Explain the meaning of the sampling and nonsampling errors.
- Calculate the sampling error for a given value of $\bar{x}$ when μ is known.
- Construct a sampling distribution of $\bar{x}$ when sampling from a small population.

This section introduces the concepts of sampling distribution, sampling error, and nonsampling errors. Before we discuss these concepts, we will briefly describe the concept of a population distribution.

If we have information on all members of a population and we use this information to prepare a probability distribution, this will be called the **population probability distribution**.

Population Probability Distribution

The *population probability distribution* is the probability distribution of the population data.

Suppose there are only five students in an advanced statistics class and the midterm scores of these five students are

70 78 80 80 95

Let x denote the score of a student. Using single-valued classes (because there are only five data values, and there is no need to group them), we can write the frequency distribution of scores as in **Table 7.1** along with the relative frequencies of the classes, which are obtained by dividing the frequencies of the classes by the population size. **Table 7.2**, which lists the probabilities of various x values, presents the probability distribution of the population. Note that these probabilities are the same as the relative frequencies.

TABLE 7.1 Population Frequency and Relative Frequency Distributions

x	f	Relative Frequency
70	1	1/5 = .20
78	1	1/5 = .20
80	2	2/5 = .40
95	1	1/5 = .20
	$N = 5$	Sum = 1.00

TABLE 7.2 Population Probability Distribution

x	$P(x)$
70	.20
78	.20
80	.40
95	.20
	$\Sigma P(x) = 1.00$

The values of the mean and standard deviation calculated for the probability distribution of Table 7.2 give the values of the population parameters μ and σ. These values are $\mu = 80.60$ and $\sigma = 8.09$. The values of μ and σ for the probability distribution of Table 7.2 can be calculated using the formulas given in Section 5.3 of Chapter 5. We can also obtain these values of μ and σ by using the five scores and the appropriate formulas from Chapter 3.

7.1.1 Sampling Distribution

As mentioned at the beginning of this chapter, the value of a population parameter is always constant. For example, for any population data set, there is only one value of the population mean, μ. However, we cannot say the same about the sample mean, $\bar{x}$. We would expect different samples of the same size drawn from the same population to yield different values of the sample mean, $\bar{x}$. The value of the sample mean for any one sample will depend on the elements included in that sample. Consequently, **the sample mean, $\bar{x}$, is a random variable**. Therefore, like other random variables, the sample mean possesses a probability distribution, which is more commonly called the **sampling distribution of $\bar{x}$**. Other sample statistics, such as the median, mode, and standard deviation, also possess sampling distributions.

Sampling Distribution of $\bar{x}$

The probability distribution of $\bar{x}$ is called its sampling distribution. It lists the various values that $\bar{x}$ can assume and the probability of each value of $\bar{x}$.

In general, the probability distribution of a sample statistic is called its *sampling distribution.*

Remember that randomness does not play any role in the calculation of a population parameter, but randomness plays a significant role in obtaining the values of a sample statistic. For example, if we calculate the mean μ of a population data set, there will be no randomness involved, as all the data values in the population are included in the calculation of mean μ. However, if we select a sample from this population and calculate the sample mean $\bar{x}$, it will assume one of the numerous (in most cases millions of) possible values. Assuming we select this sample randomly, the value of $\bar{x}$ depends on which values from the population are included in the sample.

Reconsider the population of midterm scores of five students given in Table 7.1. Consider all possible samples of three scores each that can be selected, without replacement, from that population. The total number of possible samples, given by the combinations formula discussed in Chapter 4, is 10; that is,

$$\text{Total number of samples} = {}_5C_3 = \frac{5!}{3!(5-3)!} = \frac{5 \cdot 4 \cdot 3 \cdot 2 \cdot 1}{3 \cdot 2 \cdot 1 \cdot 2 \cdot 1} = 10$$

Suppose we assign the letters A, B, C, D, and E to the scores of the five students, so that

$$\text{A} = 70, \quad \text{B} = 78, \quad \text{C} = 80, \quad \text{D} = 80, \quad \text{E} = 95$$

Then, the 10 possible samples of three scores each are

ABC, ABD, ABE, ACD, ACE, ADE, BCD, BCE, BDE, CDE

These 10 samples and their respective means are listed in **Table 7.3**. Note that the first two samples have the same three scores. The reason for this is that two of the students (C and D) have the same score of 80, and, hence, the samples ABC and ABD contain the same values. The mean of each sample is obtained by dividing the sum of the three scores included in that sample by 3. For instance, the mean of the first sample is $(70 + 78 + 80)/3 = 76$. Note that the values of the means of the samples in Table 7.3 are rounded to two decimal places.

By using the values of $\bar{x}$ given in Table 7.3, we record the frequency distribution of $\bar{x}$ in **Table 7.4**. By dividing the frequencies of the various values of $\bar{x}$ by the sum of all frequencies, we obtain the relative frequencies of the classes, which are listed in the third column of Table 7.4. These relative frequencies are used as probabilities and listed in **Table 7.5**. This table gives the sampling distribution of $\bar{x}$.

TABLE 7.3 All Possible Samples and Their Means When the Sample Size Is 3

Sample	Scores in the Sample	$\bar{x}$
ABC	70, 78, 80	76.00
ABD	70, 78, 80	76.00
ABE	70, 78, 95	81.00
ACD	70, 80, 80	76.67
ACE	70, 80, 95	81.67
ADE	70, 80, 95	81.67
BCD	78, 80, 80	79.33
BCE	78, 80, 95	84.33
BDE	78, 80, 95	84.33
CDE	80, 80, 95	85.00

TABLE 7.4 Frequency and Relative Frequency Distributions of $\bar{x}$ When the Sample Size Is 3

$\bar{x}$	f	Relative Frequency
76.00	2	2/10 = .20
76.67	1	1/10 = .10
79.33	1	1/10 = .10
81.00	1	1/10 = .10
81.67	2	2/10 = .20
84.33	2	2/10 = .20
85.00	1	1/10 = .10
	$\Sigma f = 10$	Sum = 1.00

TABLE 7.5 Sampling Distribution of $\bar{x}$ When the Sample Size Is 3

$\bar{x}$	$P(\bar{x})$
76.00	.20
76.67	.10
79.33	.10
81.00	.10
81.67	.20
84.33	.20
85.00	.10
	$\Sigma P(\bar{x}) = 1.00$

If we select just one sample of three scores from the population of five scores, we may draw any of the 10 possible samples. Hence, the sample mean, $\bar{x}$, can assume any of the values listed in Table 7.5 with the corresponding probability. For instance, the probability that the mean of a randomly selected sample of three scores is 81.67 is .20. This probability can be written as

$$P(\bar{x} = 81.67) = .20$$

7.1.2 Sampling and Nonsampling Errors

Usually, different samples selected from the same population will give different results because they contain different elements. This is obvious from Table 7.3, which shows that the mean of a sample of three scores depends on which three of the five scores are included in the sample. The result obtained from any one sample will generally be different from the result obtained from the corresponding population. The difference between the value of a sample statistic obtained from a sample and the value of the corresponding population parameter obtained from the population is called the **sampling error.** Note that this difference represents the sampling error only if the sample is random and no nonsampling error has been made. Otherwise, only a part of this difference will be due to the sampling error.

Sampling Error

Sampling error is the difference between the value of a sample statistic and the value of the corresponding population parameter. In the case of the mean,

$$\text{Sampling error} = \bar{x} - \mu$$

assuming that the sample is random and no nonsampling error has been made. The sampling error occurs only in a sample survey, and not in a census.

However, in the real world, it is not possible to find the sampling error because μ is not known. If μ is known then we do not need to find $\bar{x}$. But the concept of sampling error is very important in making inferences in later chapters.

It is important to remember that **a sampling error occurs because of chance.** The errors that occur for other reasons, such as errors made during collection, recording, and tabulation of data, are called **nonsampling errors.** These errors occur because of human mistakes, and not chance. Note that there is only one kind of sampling error—the error that occurs due to chance. However, there is not just one nonsampling error, but there are many nonsampling errors that may occur for different reasons.

Nonsampling Errors

The errors that occur in the collection, recording, and tabulation of data are called *nonsampling errors*. The nonsampling errors can occur both in a sample survey and in a census.

The following paragraph, reproduced from the *Current Population Reports* of the U.S. Bureau of the Census, explains how nonsampling errors can occur.

> **Nonsampling errors can be attributed to many sources, e.g., inability to obtain information about all cases in the sample, definitional difficulties, differences in the interpretation of questions, inability or unwillingness on the part of the respondents to provide correct information, inability to recall information, errors made in collection such as in recording or coding the data, errors made in processing the data, errors made in estimating values for missing data, biases resulting from the differing recall periods caused by the interviewing pattern used, and failure of all units in the universe to have some probability of being selected for the sample (undercoverage).**

The following are the main reasons for the occurrence of nonsampling errors.

1. If a sample is nonrandom (and, hence, most likely nonrepresentative), the sample results may be too different from the census results. Even a randomly selected sample can become nonrandom if some of the members included in the sample cannot be contacted. A very good example of this comes from an article published in a magazine in 1988. As reported in a July 11, 1988, article in *U.S. News & World Report* ("The Numbers Racket: How Polls and Statistics Lie"), during the 1984 presidential election a test poll was conducted in which the only subjects interviewed were those who could be reached on the first try. The results of this poll indicated that the Republican incumbant Ronald Reagan had a 3% point lead over the Democratic challenger Walter Mondale. However, when interviewers made an effort to contact everyone on their lists (calling some households up to 30 times before reaching someone), Reagan's lead increased to 13%. It turned out that this 13% lead was much closer to the actual election results. Apparently, people who planned to vote Republican spent less time at home.
2. The questions may be phrased in such a way that they are not fully understood by the members of the sample or population. As a result, the answers obtained are not accurate.
3. The respondents may intentionally give false information in response to some sensitive questions. For example, people may not tell the truth about their drinking habits, incomes, or opinions about minorities. Sometime, the respondents may give wrong answers because of ignorance. For example, a person may not remember the exact amount he or she spent on clothes last year. If asked in a survey, he or she may give an inaccurate answer.
4. The poll taker may make a mistake and enter a wrong number in the records or make an error while entering the data on a computer.

As mentioned earlier, the nonsampling errors can occur both in a sample survey and in a census, whereas sampling error occurs only when a sample survey is conducted. Nonsampling errors can be minimized by preparing the survey questionnaire carefully and handling the data cautiously. However, it is impossible to avoid sampling error.

Example 7.1 illustrates the sampling and nonsampling errors using the mean.

EXAMPLE 7.1 Scores of Students

Illustrating Sampling and Nonsampling Errors

Reconsider the population of five scores given in Table 7.1. Suppose one sample of three scores is selected from this population, and this sample includes the scores 70, 80, and 95. Find the sampling error.

Solution The scores of the five students are 70, 78, 80, 80, and 95. The population mean is

$$\mu = \frac{70 + 78 + 80 + 80 + 95}{5} = 80.60$$

Now, a random sample of three scores from this population is taken and this sample includes the scores 70, 80, and 95. The mean for this sample is

$$\bar{x} = \frac{70 + 80 + 95}{3} = 81.67$$

Consequently,

$$\text{Sampling error} = \bar{x} - \mu = 81.67 - 80.60 = \mathbf{1.07}$$

That is, the mean score estimated from the sample is 1.07 higher than the mean score of the population. Note that this difference occurred due to chance—that is, because we used a sample instead of the population.

Now suppose, when we select the sample of three scores, we mistakenly record the second score as 82 instead of 80. As a result, we calculate the sample mean as

$$\bar{x} = \frac{70 + 82 + 95}{3} = 82.33$$

Consequently, the difference between this sample mean and the population mean is

$$\bar{x} - \mu = 82.33 - 80.60 = 1.73$$

However, this difference between the sample mean and the population mean does not entirely represent the sampling error. As we calculated earlier, only 1.07 of this difference is due to the sampling error. The remaining portion, which is equal to 1.73 − 1.07 = .66, represents the nonsampling error because it occurred due to the error we made in recording the second score in the sample. Thus, in this case,

$$\text{Sampling error} = \mathbf{1.07}$$
$$\text{Nonsampling error} = \mathbf{.66}$$

Figure 7.1 shows the sampling and nonsampling errors for these calculations.

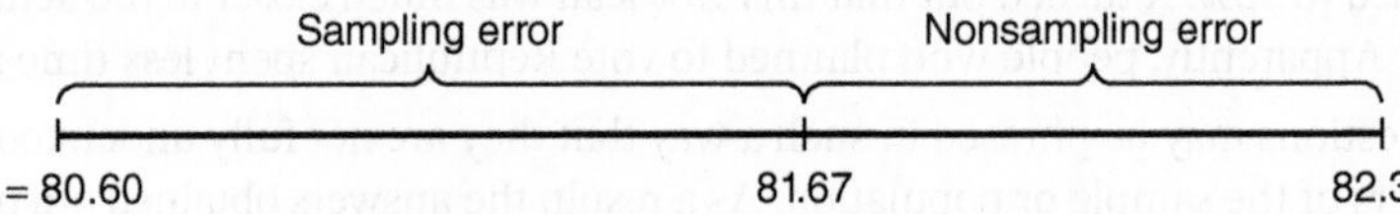

FIGURE 7.1 Sampling and nonsampling errors.

Thus, the sampling error is the difference between the correct value of $\bar{x}$ and the value of μ, where the correct value of $\bar{x}$ is the value of $\bar{x}$ that does not contain any nonsampling errors. In contrast, the nonsampling error(s) is(are) obtained by subtracting the correct value of $\bar{x}$ from the incorrect value of $\bar{x}$, where the incorrect value of $\bar{x}$ is the value that contains the nonsampling error(s). For our example,

$$\text{Sampling error} = \bar{x} - \mu = 81.67 - 80.60 = 1.07$$
$$\text{Nonsampling error} = \text{Incorrect } \bar{x} - \text{Correct } \bar{x} = 82.33 - 81.67 = .66$$

As mentioned earlier, in the real world, we do not know the mean of a population. Hence, we select a sample to use the sample mean as an estimate of the population mean. Consequently, we never know the size of the sampling error.

EXERCISES

Concepts and Procedures

7.1 Briefly explain the meaning of a population probability distribution and a sampling distribution. Give an example of each.

7.2 Explain briefly the meaning of a sampling error. Give an example. Does such an error occur only in a sample survey, or can it occur in both a sample survey and a census?

7.3 Explain briefly the meaning of nonsampling errors. Give an example. Do such errors occur only in a sample survey, or can they occur in both a sample survey and a census?

7.4 Consider the following population of six numbers.

16 13 7 20 9 12

a. Find the population mean.

b. Liza selected one sample of four numbers from this population. The sample included the numbers 13, 7, 9, and 12. Calculate the sample mean and sampling error for this sample.

c. Refer to part (b). When Liza calculated the sample mean, she mistakenly used the numbers 13, 7, 6, and 12 to calculate the sample mean. Find the sampling and nonsampling errors in this case.

d. List all samples of four numbers (without replacement) that can be selected from this population. Calculate the sample mean and sampling error for each of these samples.

Applications

7.5 Using the formulas of Section 5.3 of Chapter 5 for the mean and standard deviation of a discrete random variable, verify that the mean and standard deviation for the population probability distribution of Table 7.2 are 80.60 and 8.09, respectively.

7.6 The following data give the years of teaching experience for all six faculty members of a department in a university.

7 8 14 7 20 12

a. Let x denote the years of teaching experience of a faculty member of this department. Write the population probability distribution of x.

b. List all the possible samples of size five (without replacement) that can be selected from this population. Calculate the mean for each of these samples. Write the sampling distribution of $\bar{x}$.

c. Calculate the mean for the population data. Select one random sample of size five and calculate the sample mean $\bar{x}$. Compute the sampling error.

7.2 Mean and Standard Deviation of $\bar{x}$

LEARNING OBJECTIVES

After completing this section, you should be able to:

- Explain the meaning of an unbiased estimator.
- Explain the meaning of a consistent estimator.
- Explain why $\bar{x}$ is an unbiased estimator of μ.
- Explain why $\bar{x}$ is a consistent estimator.
- Explain when the finite population correction factor is used.
- Calculate the mean and standard deviation of the sampling distribution of $\bar{x}$.

The mean and standard deviation calculated for the sampling distribution of $\bar{x}$ are called the **mean** and **standard deviation of $\bar{x}$**. Actually, the mean and standard deviation of $\bar{x}$ are, respectively, the mean and standard deviation of the means of all samples of the same size selected from a population. The standard deviation of $\bar{x}$ is also called the **standard error of $\bar{x}$**.

Mean and Standard Deviation of $\bar{x}$

The mean and standard deviation of the sampling distribution of $\bar{x}$ are called the *mean* and *standard deviation of* $\bar{x}$ and are denoted by $\mu_{\bar{x}}$ and $\sigma_{\bar{x}}$, respectively.

If we calculate the mean and standard deviation of the 10 values of $\bar{x}$ listed in Table 7.3, we obtain the mean, $\mu_{\bar{x}}$ and the standard deviation, $\sigma_{\bar{x}}$ of $\bar{x}$. Alternatively, we can calculate the mean and standard deviation of the sampling distribution of $\bar{x}$ listed in Table 7.5. These will also be the values of $\mu_{\bar{x}}$ and $\sigma_{\bar{x}}$. From these calculations, we will obtain $\mu_{\bar{x}} = 80.60$ and $\sigma_{\bar{x}} = 3.30$.

The mean of the sampling distribution of $\bar{x}$ is always equal to the mean of the population.

Mean of the Sampling Distribution of $\bar{x}$

The *mean of the sampling distribution of $\bar{x}$* is always equal to the mean of the population. Thus,

$$\mu_{\bar{x}} = \mu$$

Thus, if we select all possible samples (of the same size) from a population and calculate their means, the mean of all these sample means ($\mu_{\bar{x}}$) will be the same as the mean of the population (μ). If we calculate the mean for the population probability distribution of Table 7.2 and the mean for the sampling distribution of Table 7.5 by using the formula learned in Section 5.3 of Chapter 5, we get the same value of 80.60 for μ and $\mu_{\bar{x}}$.

The sample mean, $\bar{x}$, is called an **estimator** of the population mean, μ. When the expected value (or mean) of a sample statistic is equal to the value of the corresponding population parameter, that sample statistic is said to be an **unbiased estimator**. For the sample mean $\bar{x}$, $\mu_{\bar{x}} = \mu$. Hence, $\bar{x}$ is an unbiased estimator of μ. This is a very important property that an estimator should possess.

However, the standard deviation of $\bar{x}$, $\sigma_{\bar{x}}$, is not equal to the standard deviation of the population probability distribution, σ. The standard deviation of $\bar{x}$ is equal to the standard deviation of the population divided by the square root of the sample size; that is,

$$\sigma_{\bar{x}} = \frac{\sigma}{\sqrt{n}}$$

This formula for the standard deviation of $\bar{x}$ holds true only when the sampling is done either with replacement from a finite population or with or without replacement from an infinite population. These two conditions can be replaced by the condition that the above formula holds true if the sample size is small in comparison to the population size. The sample size is considered to be small compared to the population size if the sample size is equal to or less than 5% of the population size; that is, if

$$\frac{n}{N} \leq .05$$

If this condition is not satisfied, we use the following formula to calculate $\sigma_{\bar{x}}$:

$$\sigma_{\bar{x}} = \frac{\sigma}{\sqrt{n}}\sqrt{\frac{N-n}{N-1}}$$

where the factor $\sqrt{\frac{N-n}{N-1}}$ is called the **finite population correction factor.**

In most practical applications, the sample size is small compared to the population size. Consequently, in most cases, the formula used to calculate $\sigma_{\bar{x}}$ is $\sigma_{\bar{x}} = \sigma/\sqrt{n}$.

Standard Deviation of the Sampling Distribution of $\bar{x}$

The *standard deviation of the sampling distribution of $\bar{x}$* is

$$\sigma_{\bar{x}} = \frac{\sigma}{\sqrt{n}}$$

where σ is the standard deviation of the population and n is the sample size. This formula is used when $n/N \leq .05$, where N is the population size.

Following are two important observations regarding the sampling distribution of $\bar{x}$:

1. *The spread of the sampling distribution of $\bar{x}$ is smaller than the spread of the corresponding population distribution.* In other words, $\sigma_{\bar{x}} < \sigma$. This is obvious from the formula for $\sigma_{\bar{x}}$. When n is greater than 1, which is usually true, the denominator in $\sigma/\sqrt{n}$ is greater than 1. Hence, $\sigma_{\bar{x}}$ is smaller than σ.

2. *The standard deviation of the sampling distribution of $\bar{x}$ decreases as the sample size increases.* This feature of the sampling distribution of $\bar{x}$ is also obvious from the formula

$$\sigma_{\bar{x}} = \frac{\sigma}{\sqrt{n}}$$

If the standard deviation of a sample statistic decreases as the sample size is increased, that statistic is said to be a **consistent estimator**. This is another important property that an estimator should possess. It is obvious from the above formula for $\sigma_{\bar{x}}$ that as n increases, the value of $\sqrt{n}$ also increases and, consequently, the value of $\sigma/\sqrt{n}$ decreases. Thus, the sample mean $\bar{x}$ is a consistent estimator of the population mean μ. Example 7.2 illustrates this feature.

EXAMPLE 7.2 | Wages of Employees

Finding the Mean and Standard Deviation of $\bar{x}$

The mean wage per hour for all 5000 employees who work at a large company is \$27.50, and the standard deviation is \$3.70. Let $\bar{x}$ be the mean wage per hour for a random sample of certain employees selected from this company. Find the mean and standard deviation of $\bar{x}$ for a sample size of

a. 30 **b.** 75 **c.** 200

Image Source/Getty Images

Solution From the given information, for the population of all employees,

$$N = 5000, \quad \mu = \$27.50, \quad \text{and} \quad \sigma = \$3.70$$

a. The mean of the sampling distribution of $\bar{x}$, $\mu_{\bar{x}}$, is

$$\mu_{\bar{x}} = \mu = \mathbf{\$27.50}$$

In this case, $n = 30$, $N = 5000$, and $n/N = 30/5000 = .006$. Because n/N is less than .05, the standard deviation of $\bar{x}$ is obtained by using the formula $\sigma/\sqrt{n}$. Hence,

$$\sigma_{\bar{x}} = \frac{\sigma}{\sqrt{n}} = \frac{3.70}{\sqrt{30}} = \mathbf{\$.676}$$

Thus, we can state that if we take all possible samples of size 30 from the population of all employees of this company and prepare the sampling distribution of $\bar{x}$ the mean and standard deviation of this sampling distribution of $\bar{x}$ will be \$27.50 and \$.676, respectively.

b. In this case, $n = 75$ and $n/N = 75/5000 = .015$, which is less than .05. The mean and standard deviation of $\bar{x}$ are

$$\mu_{\bar{x}} = \mu = \mathbf{\$27.50} \quad \text{and} \quad \sigma_{\bar{x}} = \frac{\sigma}{\sqrt{n}} = \frac{3.70}{\sqrt{75}} = \mathbf{\$.427}$$

c. In this case, $n = 200$ and $n/N = 200/5000 = .04$, which is less than .05. Therefore, the mean and standard deviation of $\bar{x}$ are

$$\mu_{\bar{x}} = \mu = \mathbf{\$27.50} \quad \text{and} \quad \sigma_{\bar{x}} = \frac{\sigma}{\sqrt{n}} = \frac{3.70}{\sqrt{200}} = \mathbf{\$.262}$$

From the preceding calculations we observe that the mean of the sampling distribution of $\bar{x}$ is always equal to the mean of the population whatever the size of the sample. However, the value of the standard deviation of $\bar{x}$ decreases from \$.676 to \$.427 and then to \$.262 as the sample size increases from 30 to 75 and then to 200.

Note that in Example 7.2, $\mu_{\bar{x}} = \mu$ in all three parts indicates that $\bar{x}$ is an unbiased estimator of μ. We also observe that $\sigma_{\bar{x}}$ decreases in this example as the sample size increases. This demonstrates that $\bar{x}$ is a consistent estimator of μ.

EXERCISES

Concepts and Procedures

7.7 Let $\bar{x}$ be the mean of a sample selected from a population.

a. What is the mean of the sampling distribution of $\bar{x}$ equal to?

b. What is the standard deviation of the sampling distribution of $\bar{x}$ equal to? Assume that $n/N \le .05$.

7.8 What is an estimator? When is an estimator unbiased? Is the sample mean, $\bar{x}$, an unbiased estimator of μ? Explain.

7.9 When is an estimator said to be consistent? Is the sample mean, $\bar{x}$, a consistent estimator of μ? Explain.

7.10 How does the value of $\sigma_{\bar{x}}$ change as the sample size increases? Explain.

7.11 Consider a large population with $\mu = 100$ and $\sigma = 20$. Assuming $n/N \le .05$, find the mean and standard deviation of the sample mean, $\bar{x}$, for a sample size of

a. 20 **b.** 50

7.12 A population of $N = 100{,}000$ has $\sigma = 60$. In each of the following cases, which formula will you use to calculate $\sigma_{\bar{x}}$ and why? Using the appropriate formula, calculate $\sigma_{\bar{x}}$ for each of these cases.

a. $n = 3000$ **b.** $n = 9000$

***7.13** For a population, $\mu = 50$ and $\sigma = 15$.

a. For a sample selected from this population, $\mu_{\bar{x}} = 50$ and $\sigma_{\bar{x}} = 4$. Find the sample size. Assume that $n/N \le .05$.

b. For a sample selected from this population, $\mu_{\bar{x}} = 50$ and $\sigma_{\bar{x}} = 2.5$. Find the sample size. Assume that $n/N \le .05$.

Applications

7.14 The mean monthly out-of-pocket cost of prescription drugs for all senior citizens in a particular city is \$520 with a standard deviation of \$72. Let $\bar{x}$ be the mean of such costs for a random sample of 25 senior citizens from this city. Find the mean and standard deviation of the sampling distribution of $\bar{x}$.

7.15 According to the 2019 Physician Compensation Report by Medscape (a subsidiary of WebMD), American orthopedists earned an average of \$482,000 in 2019. Suppose that the mean and standard deviation of the 2019 earnings of all American orthopedists were \$482,000 and \$90,000, respectively. Let $\bar{x}$ be the mean 2019 earnings of a random sample of 200 American orthopedists. Find the mean and standard deviation of the sampling distribution of $\bar{x}$. Assume that $n/N \le .05$.

***7.16** Suppose the standard deviation of recruiting costs per player for all female basketball players recruited by all public universities in the Midwestern United States is \$2000. Let $\bar{x}$ be the mean recruiting cost for a sample of a certain number of such players. What sample size will give the standard deviation of $\bar{x}$ equal to \$125? Assume $n/N \le .05$.

***7.17** Consider the sampling distribution of $\bar{x}$ given in Table 7.5.

a. Calculate the value of $\mu_{\bar{x}}$ using the formula $\mu_{\bar{x}} = \Sigma \bar{x} P(\bar{x})$. Is the value of $\mu = 80.60$ calculated earlier the same as the value of $\mu_{\bar{x}}$ calculated here?

b. Calculate the value of $\sigma_{\bar{x}}$ by using the formula

$$\sigma_{\bar{x}} = \sqrt{\Sigma \bar{x}^2 P(\bar{x}) - (\mu_{\bar{x}})^2}$$

c. For the five scores, $\sigma = 8.09$. Also, our sample size is 3, so that $n = 3$. Therefore, $\sigma/\sqrt{n} = 8.09/\sqrt{3} = 4.67$. From part (b), you should get $\sigma_{\bar{x}} = 3.30$. Why does $\sigma/\sqrt{n}$ not equal $\sigma_{\bar{x}}$ in this case?

d. In our example (given in the beginning of Section 7.1) on scores, $N = 5$ and $n = 3$. Hence, $n/N = 3/5 = .60$. Because n/N is greater than .05, the appropriate formula to find $\sigma_{\bar{x}}$ is

$$\sigma_{\bar{x}} = \frac{\sigma}{\sqrt{n}}\sqrt{\frac{N-n}{N-1}}$$

Show that the value of $\sigma_{\bar{x}}$ calculated by using this formula gives the same value as the one calculated in part (b) above.

7.3 Shape of the Sampling Distribution of $\bar{x}$

LEARNING OBJECTIVES

After completing this section, you should be able to:

- Describe the shape of the sampling distribution of $\bar{x}$ when sampling from a normally distributed population.
- Describe the shape of the sampling distribution of $\bar{x}$ when sampling from a population that is not normally distributed.
- Describe the conditions that must hold true for the sampling distribution of $\bar{x}$ to be normal when $n < 30$.
- Explain the central limit theorem.
- Find the mean, standard deviation, and the shape of the sampling distribution of $\bar{x}$ when sampling from a normally distributed population.
- Find the mean, standard deviation, and the shape of the sampling distribution of $\bar{x}$ when sampling from a population that is not normally distributed.

The shape of the sampling distribution of $\bar{x}$ relates to the following two cases:

1. The population from which samples are drawn has a normal distribution.
2. The population from which samples are drawn does not have a normal distribution.

7.3.1 Sampling from a Normally Distributed Population

When the population from which samples are drawn is normally distributed with its mean equal to μ and standard deviation equal to σ, then:

1. The mean of $\bar{x}$, $\mu_{\bar{x}}$, is equal to the mean of the population, μ.
2. The standard deviation of $\bar{x}$, $\sigma_{\bar{x}}$, is equal to $\sigma/\sqrt{n}$, assuming $n/N \leq .05$.
3. The shape of the sampling distribution of $\bar{x}$ is normal, whatever the value of n.

Sampling Distribution of $\bar{x}$ When the Population Has a Normal Distribution

If the population from which the samples are drawn is normally distributed with mean μ and standard deviation σ, then the sampling distribution of the sample mean, $\bar{x}$, will also be normally distributed with the following mean and standard deviation, regardless of the sample size:

$$\mu_{\bar{x}} = \mu \quad \text{and} \quad \sigma_{\bar{x}} = \frac{\sigma}{\sqrt{n}}$$

For $\sigma_{\bar{x}} = \sigma/\sqrt{n}$ to be true, n/N must be less than or equal to .05.

Remember

Figure 7.2*a* shows the probability distribution curve for a population. The distribution curves in Figure 7.2*b* through Figure 7.2*e* show the sampling distributions of $\bar{x}$ for different

(*a*) Population distribution.

(*b*) Sampling distribution of $\bar{x}$ for $n = 5$.

(*c*) Sampling distribution of $\bar{x}$ for $n = 16$.

(*d*) Sampling distribution of $\bar{x}$ for $n = 30$.

(*e*) Sampling distribution of $\bar{x}$ for $n = 100$.

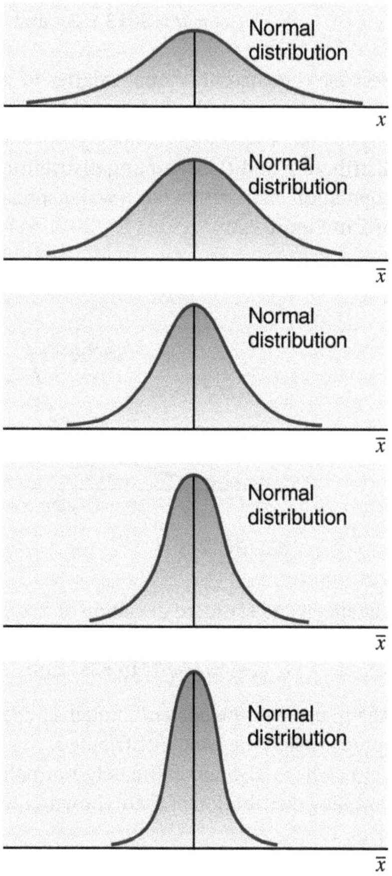

FIGURE 7.2 Population distribution and sampling distributions of $\bar{x}$.

sample sizes taken from the population of Figure 7.2*a*. As we can observe, the population has a normal distribution. Because of this, the sampling distribution of $\bar{x}$ is normal for each of the four cases illustrated in Figure 7.2*b* through Figure 7.2*e*. Also notice from Figure 7.2*b* through Figure 7.2*e* that the spread of the sampling distribution of $\bar{x}$ decreases as the sample size increases.

Example 7.3 illustrates the calculation of the mean and standard deviation of $\bar{x}$ and the description of the shape of its sampling distribution.

EXAMPLE 7.3 Earnings of Internal Medicine Physicians

Finding the Mean, Standard Deviation, and Describing the Shape of the Sampling Distribution of $\bar{x}$: Normally Distributed Population

According to ZipRecruiter, as of October 13, 2019, the average annual salary for internal medicine physicians in the United States was $215,638 a year (https://www.ziprecruiter.com). Suppose that the current annual salaries of all internal medicine physicians in the United States are approximately normally distributed with a mean of $215,638 and a standard deviation of $20,000. Let $\bar{x}$ be the current mean annual salary of a random sample of American internal medicine physicians. Calculate the mean and standard deviation of $\bar{x}$ and describe the shape of its sampling distribution when the sample size is

a. 16 **b.** 50 **c.** 1000

Solution Let μ and σ be the mean and standard deviation of the current annual salaries of all American internal medicine physicians, and $\mu_{\bar{x}}$ and $\sigma_{\bar{x}}$ be the mean and standard deviation of the sampling distribution of $\bar{x}$, respectively. Then, from the given information,

$$\mu = \$215{,}638 \quad \text{and} \quad \sigma = \$20{,}000$$

a. The mean and standard deviation of $\bar{x}$ are, respectively,

$$\mu_{\bar{x}} = \mu = \mathbf{\$215{,}638} \quad \text{and} \quad \sigma_{\bar{x}} = \frac{\sigma}{\sqrt{n}} = \frac{\$20{,}000}{\sqrt{16}} = \mathbf{\$5000}$$

Because the current annual salaries of all American internal medicine physicians are approximately normally distributed, the sampling distribution of $\bar{x}$ for samples of 16 such physicians is also approximately normally distributed. **Figure 7.3** shows the population distribution and the sampling distribution of $\bar{x}$. Note that because σ is greater than $\sigma_{\bar{x}}$, the population distribution has a wider spread but smaller height than the sampling distribution of $\bar{x}$ in Figure 7.3.

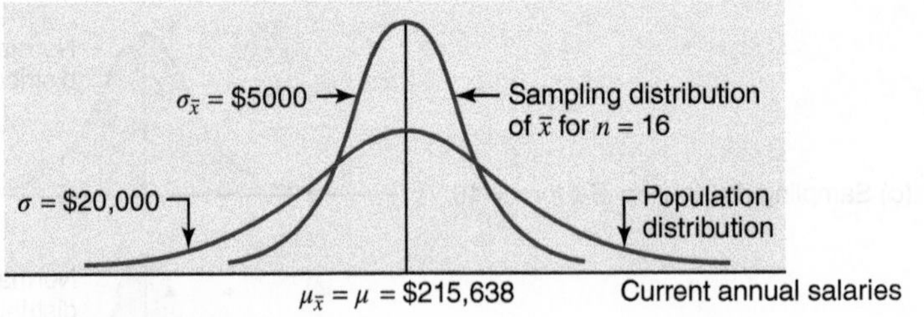

FIGURE 7.3

b. The mean and standard deviation of $\bar{x}$ are, respectively,

$$\mu_{\bar{x}} = \mu = \mathbf{\$215{,}638} \quad \text{and} \quad \sigma_{\bar{x}} = \frac{\sigma}{\sqrt{n}} = \frac{\$20{,}000}{\sqrt{50}} = \mathbf{\$2828.43}$$

Again, because the current annual salaries of all American internal medicine physicians are approximately normally distributed, the sampling distribution of $\bar{x}$ for samples of 50 such physicians is also approximately normally distributed. The population distribution and the sampling distribution of $\bar{x}$ are shown in **Figure 7.4**.

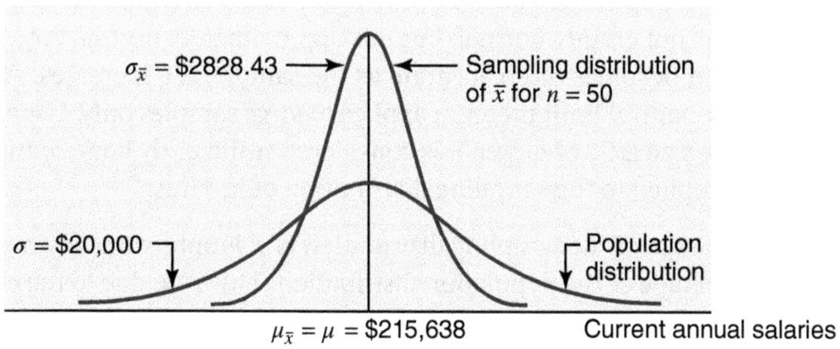

FIGURE 7.4

c. The mean and standard deviation of $\bar{x}$ are, respectively,

$$\mu_{\bar{x}} = \mu = \mathbf{\$215{,}638} \quad \text{and} \quad \sigma_{\bar{x}} = \frac{\sigma}{\sqrt{n}} = \frac{\$20{,}000}{\sqrt{1000}} = \mathbf{\$632.46}$$

Again, because the current annual salaries of all American internal medicine physicians are approximately normally distributed, the sampling distribution of $\bar{x}$ for samples of 1000 such physicians is also approximately normally distributed. The two distributions are shown in **Figure 7.5**.

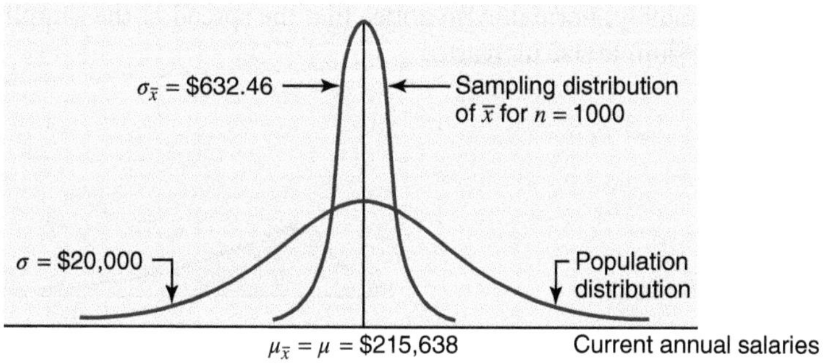

FIGURE 7.5

Thus, whatever the sample size, the sampling distribution of $\bar{x}$ is normal when the population from which the samples are drawn is normally distributed.

7.3.2 Sampling from a Population That Is Not Normally Distributed

Most of the time, the population from which the samples are selected is not normally distributed. In such cases, the shape of the sampling distribution of $\bar{x}$ is inferred from a very important theorem called the **central limit theorem**.

Central Limit Theorem

According to the *central limit theorem*, for a large sample size, the sampling distribution of $\bar{x}$ is approximately normal, irrespective of the shape of the population distribution. The mean and standard deviation of the sampling distribution of $\bar{x}$ are, respectively,

$$\mu_{\bar{x}} = \mu \quad \text{and} \quad \sigma_{\bar{x}} = \frac{\sigma}{\sqrt{n}}$$

In case of the mean, the sample size is usually considered to be large if $n \geq 30$.

Note that when the population does not have a normal distribution, the shape of the sampling distribution is not exactly normal, but it is approximately normal for a large sample size. The approximation becomes more accurate as the sample size increases. Another point to remember is that the central limit theorem applies to *large* samples only. Usually, in case of the mean, if the sample size is 30 or larger, it is considered sufficiently large so that the central limit theorem can be applied to the sampling distribution of $\bar{x}$. Thus:

1. When $n \geq 30$, the shape of the sampling distribution of $\bar{x}$ is approximately normal irrespective of the shape of the population distribution. This is so due to the central limit theorem.
2. The mean of $\bar{x}$, $\mu_{\bar{x}}$, is equal to the mean of the population, μ.
3. The standard deviation of $\bar{x}$, $\sigma_{\bar{x}}$, is equal to $\sigma/\sqrt{n}$ if $n/N \leq .05$.

Again, remember that for $\sigma_{\bar{x}} = \sigma/\sqrt{n}$ to apply, n/N must be less than or equal to .05, otherwise we multiply $\sigma/\sqrt{n}$ by the finite population correction factor explained earlier in this chapter.

Figure 7.6*a* shows the probability distribution curve for a population. The distribution curves in Figure 7.6*b* through Figure 7.6*e* show the sampling distributions of $\bar{x}$ for different sample sizes taken from the population of Figure 7.6*a*. As we can observe, the population is not normally distributed. The sampling distributions of $\bar{x}$ shown in parts (*b*) and (*c*), when $n < 30$, are also not normal. However, the sampling distributions of $\bar{x}$ shown in parts (*d*) and (*e*), when $n \geq 30$, are (approximately) normal. Also notice that the spread of the sampling distribution of $\bar{x}$ decreases as the sample size increases.

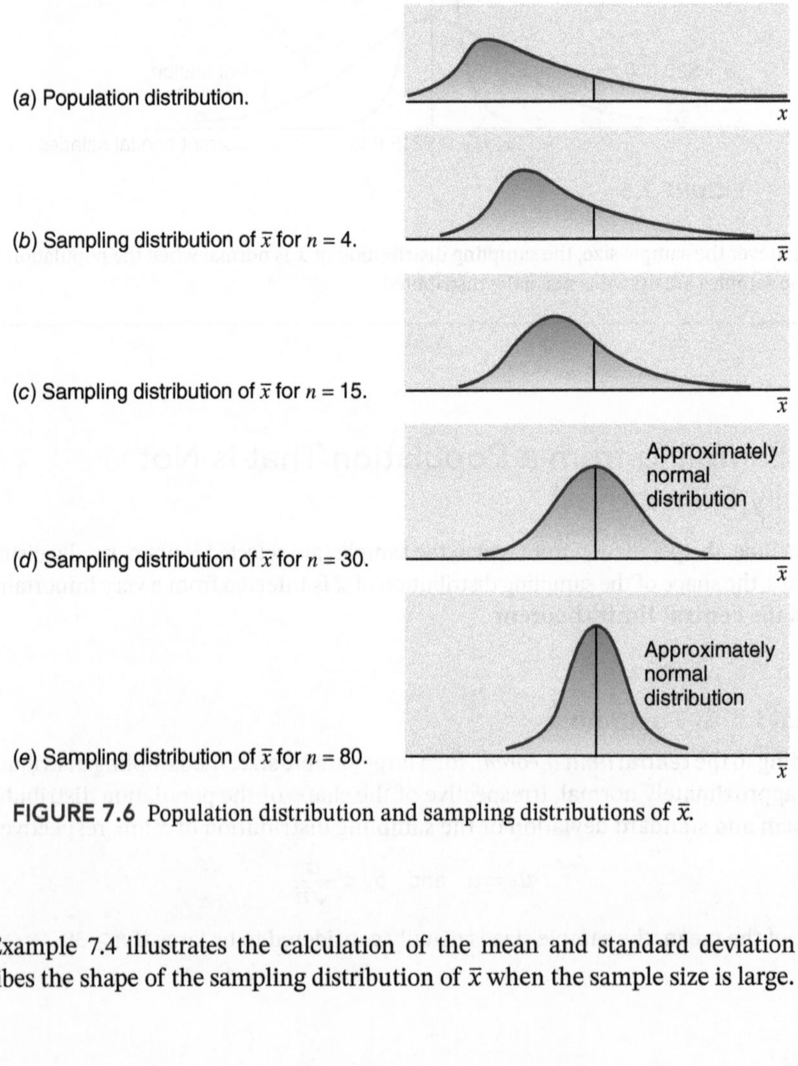

FIGURE 7.6 Population distribution and sampling distributions of $\bar{x}$.

Example 7.4 illustrates the calculation of the mean and standard deviation of $\bar{x}$ and describes the shape of the sampling distribution of $\bar{x}$ when the sample size is large.

EXAMPLE 7.4 | Rents Paid by Tenants in a City

Finding the Mean, Standard Deviation, and Describing the Shape of the Sampling Distribution of $\bar{x}$: Population Not Normally Distributed

The mean rent paid by all tenants in a small city is \$1550 with a standard deviation of \$225. However, the population distribution of rents for all tenants in this city is skewed to the right. Calculate the mean and standard deviation of $\bar{x}$ and describe the shape of its sampling distribution when the sample size is

a. 30 **b.** 100

Solution Although the population distribution of rents paid by all tenants is not normal, in each case the sample size is large ($n \geq 30$). Hence, the central limit theorem can be applied to infer the shape of the sampling distribution of $\bar{x}$. Also, note that the number of total tenants in the city should be large so that $n/N \leq .05$ in both cases.

a. Let $\bar{x}$ be the mean rent paid by a sample of 30 tenants. Then, the sampling distribution of $\bar{x}$ is approximately normal with the values of the mean and standard deviation given as

$$\mu_{\bar{x}} = \mu = \mathbf{\$1550} \quad \text{and} \quad \sigma_{\bar{x}} = \frac{\sigma}{\sqrt{n}} = \frac{225}{\sqrt{30}} = \mathbf{\$41.08}$$

Figure 7.7 shows the population distribution and the sampling distribution of $\bar{x}$.

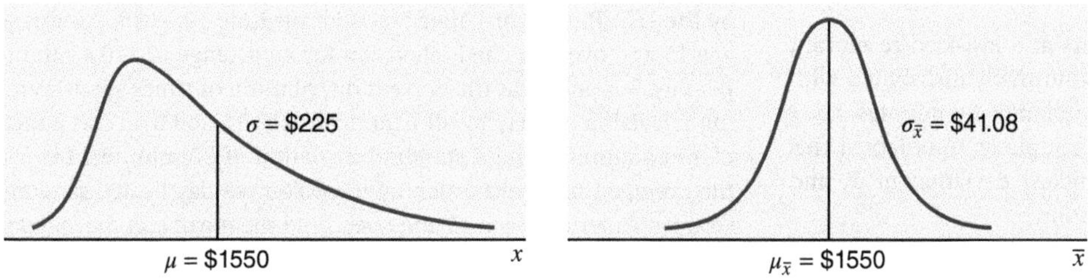

(*a*) Population distribution. (*b*) Sampling distribution of $\bar{x}$ for $n = 30$.

FIGURE 7.7

b. Let $\bar{x}$ be the mean rent paid by a sample of 100 tenants. Then, the sampling distribution of $\bar{x}$ is approximately normal with the values of the mean and standard deviation given as

$$\mu_{\bar{x}} = \mu = \mathbf{\$1550} \quad \text{and} \quad \sigma_{\bar{x}} = \frac{\sigma}{\sqrt{n}} = \frac{225}{\sqrt{100}} = \mathbf{\$22.50}$$

Figure 7.8 shows the population distribution and the sampling distribution of $\bar{x}$.

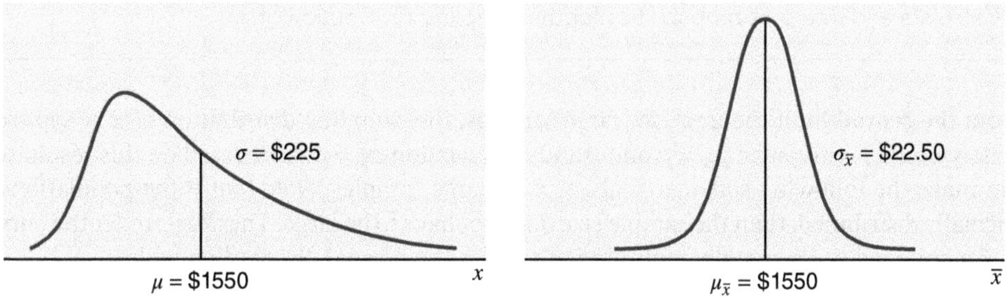

(*a*) Population distribution. (*b*) Sampling distribution of $\bar{x}$ for $n = 100$.

FIGURE 7.8

EXERCISES

Concepts and Procedures

7.18 What condition or conditions must hold true for the sampling distribution of the sample mean to be normal when the sample size is less than 30?

7.19 Explain the central limit theorem.

7.20 A population has a distribution that is skewed to the left. Indicate in which of the following cases the central limit theorem will apply to describe the sampling distribution of the sample mean.

a. $n = 25$ b. $n = 60$ c. $n = 40$

7.21 A population has a distribution that is skewed to the right. A sample of size n is selected from this population. Describe the shape of the sampling distribution of the sample mean for each of the following cases:

a. $n = 35$ b. $n = 100$ c. $n = 26$

7.22 A population has a normal distribution. A sample of size n is selected from this population. Describe the shape of the sampling distribution of the sample mean for each of the following cases:

a. $n = 37$ b. $n = 500$

Applications

7.23 The delivery times for all food orders at a fast-food restaurant during the lunch hour are approximately normally distributed with a mean of 8.4 minutes and a standard deviation of 1.8 minutes. Let $\bar{x}$ be the mean delivery time for a random sample of 16 orders at this restaurant. Calculate the mean and standard deviation of $\bar{x}$, and describe the shape of its sampling distribution.

7.24 Among college students who hold part-time jobs during the school year, the distribution of the time spent working per week is approximately normally distributed with a mean of 20.20 hours and a standard deviation of 2.60 hours. Let $\bar{x}$ be the average time spent working per week for 18 randomly selected college students who hold part-time jobs during the school year. Calculate the mean and the standard deviation of the sampling distribution of $\bar{x}$, and describe the shape of this sampling distribution.

7.25 The weights of all people living in a particular town have a distribution that is skewed to the right with a mean of 150 pounds and a standard deviation of 25 pounds. Let $\bar{x}$ be the mean weight of a random sample of 50 persons selected from this town. Find the mean and standard deviation of $\bar{x}$, and comment on the shape of its sampling distribution.

7.26 According to the National Association of Colleges and Employers Spring 2019 Salary Survey, the average starting salary for college graduates in the United States was $50,004 in 2018. Suppose that the mean starting salary of all 2018 college graduates was $50,004 with a standard deviation of $9200, and that this distribution was strongly skewed to the right. Let $\bar{x}$ be the mean starting salary of 25 randomly selected 2018 college graduates. Find the mean and the standard deviation of the sampling distribution of $\bar{x}$. What are the mean and the standard deviation of the sampling distribution of $\bar{x}$ if the sample size is 100? How do the shapes of the sampling distributions differ for the two sample sizes?

7.27 According to the American Time Use Survey results released by the U.S. Bureau of Labor Statistics on June 19, 2019, Americans age 15 and over watched television for an average of 170.4 minutes per day. Suppose that the current distribution of times spent watching television per day by all Americans age 15 and over has a mean of 170.4 minutes and a standard deviation of 20 minutes. Let $\bar{x}$ be the average time spent watching television per day by 400 randomly selected Americans age 15 and over. Find the mean and the standard deviation of the sampling distribution of $\bar{x}$. What is the shape of the sampling distribution of $\bar{x}$? Do you need to know the shape of the population distribution in order to make this conclusion? Explain why or why not.

7.4 Applications of the Sampling Distribution of $\bar{x}$

LEARNING OBJECTIVE

After completing this section, you should be able to:

- Calculate and interpret probabilities for the sampling distribution of $\bar{x}$.

From the central limit theorem, for large samples, the sampling distribution of $\bar{x}$ is approximately normal with mean $\mu_{\bar{x}} = \mu$ and standard deviation $\sigma_{\bar{x}} = \sigma/\sqrt{n}$. Based on this result, we can make the following statements about $\bar{x}$ for large samples. Note that if the population is normally distributed, then the sample size does not have to be large. The areas under the curve of $\bar{x}$ mentioned in these statements are found from the normal distribution table.

1. *If we take all possible samples of the same (large) size from a population and calculate the mean for each of these samples, then about 68.26% of the sample means will be within one standard deviation ($\sigma_{\bar{x}}$) of the population mean.* Alternatively, we can state that if we take one sample (of $n \geq 30$) from a population and calculate the mean for this sample, the probability that this sample mean will be within one standard deviation ($\sigma_{\bar{x}}$) of the population mean is .6826. That is,

$$P(\mu - 1\sigma_{\bar{x}} \leq \bar{x} \leq \mu + 1\sigma_{\bar{x}}) = P(-1.00 \leq z \leq 1.00) = .8413 - .1587 = .6826$$

This probability is shown in **Figure 7.9**.

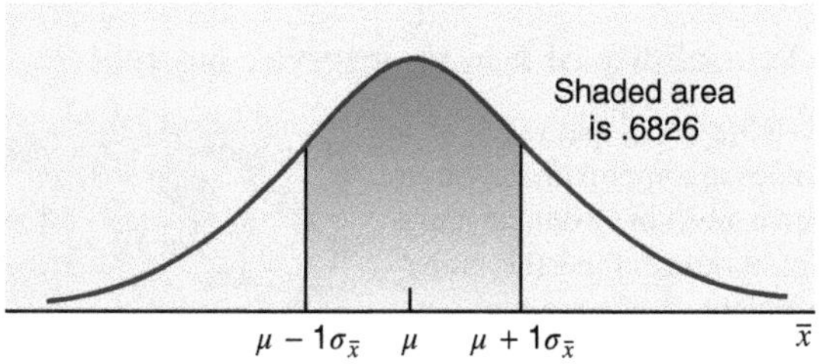

FIGURE 7.9 $P(\mu - 1\sigma_{\bar{x}} \le \bar{x} \le \mu + 1\sigma_{\bar{x}})$.

2. *If we take all possible samples of the same (large) size from a population and calculate the mean for each of these samples, then about 95.44% of the sample means will be within two standard deviations ($\sigma_{\bar{x}}$) of the population mean.* Alternatively, we can state that if we take one sample (of $n \ge 30$) from a population and calculate the mean for this sample, the probability that this sample mean will be within two standard deviations ($\sigma_{\bar{x}}$) of the population mean is .9544. That is,

$$P(\mu - 2\sigma_{\bar{x}} \le \bar{x} \le \mu + 2\sigma_{\bar{x}}) = P(-2.00 \le z \le 2.00) = .9772 - .0228 = .9544$$

This probability is shown in **Figure 7.10**.

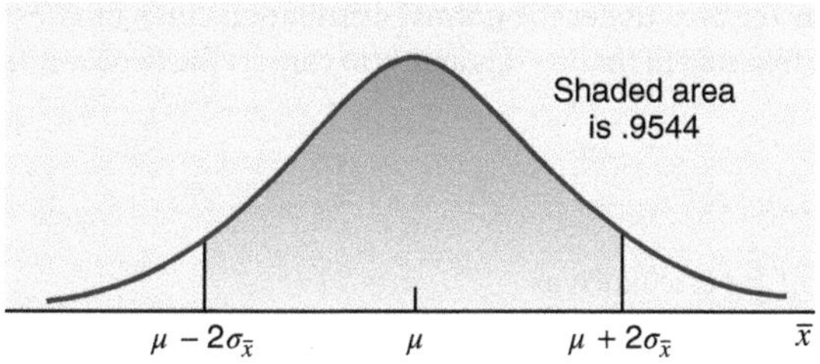

FIGURE 7.10 $P(\mu - 2\sigma_{\bar{x}} \le \bar{x} \le \mu + 2\sigma_{\bar{x}})$.

3. *If we take all possible samples of the same (large) size from a population and calculate the mean for each of these samples, then about 99.74% of the sample means will be within three standard deviations ($\sigma_{\bar{x}}$) of the population mean.* Alternatively, we can state that if we take one sample (of $n \ge 30$) from a population and calculate the mean for this sample, the probability that this sample mean will be within three standard deviations ($\sigma_{\bar{x}}$) of the population mean is .9974. That is,

$$P(\mu - 3\sigma_{\bar{x}} \le \bar{x} \le \mu + 3\sigma_{\bar{x}}) = P(-3.00 \le z \le 3.00) = .9987 - .0013 = .9974$$

This probability is shown in **Figure 7.11**.

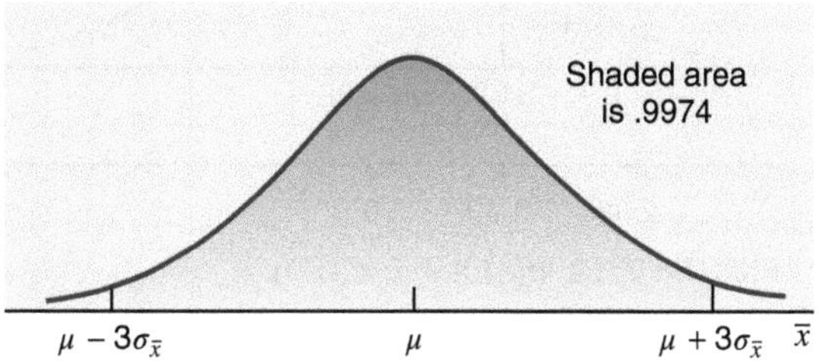

FIGURE 7.11 $P(\mu - 3\sigma_{\bar{x}} \le \bar{x} \le \mu + 3\sigma_{\bar{x}})$.

When conducting a survey, we usually select one sample and compute the value of $\bar{x}$ based on that sample. We never select all possible samples of the same size and then prepare the sampling distribution of $\bar{x}$. Rather, we are more interested in finding the probability that the value of $\bar{x}$ computed from one sample falls within a given interval. Examples 7.5 and 7.6 illustrate this procedure.

EXAMPLE 7.5 Weights of Packages of Cookies

Calculating the Probability of $\bar{x}$ in an Interval: Normal Population

Assume that the weights of all packages of a certain brand of cookies are approximately normally distributed with a mean of 32 ounces and a standard deviation of .3 ounce. Find the probability that the mean weight, $\bar{x}$, of a random sample of 20 packages of this brand of cookies will be between 31.8 and 31.9 ounces.

iStockphoto.com/Burwell and Burwell Photography

Solution Although the sample size is small ($n < 30$), the shape of the sampling distribution of $\bar{x}$ is approximately normal because the population is approximately normally distributed. The mean and standard deviation of $\bar{x}$ are, respectively,

$$\mu_{\bar{x}} = \mu = 32 \text{ ounces} \quad \text{and} \quad \sigma_{\bar{x}} = \frac{\sigma}{\sqrt{n}} = \frac{.3}{\sqrt{20}} = .06708204 \text{ ounce}$$

We are to compute the probability that the value of $\bar{x}$ calculated for one randomly drawn sample of 20 packages is between 31.8 and 31.9 ounces; that is,

$$P(31.8 < \bar{x} < 31.9)$$

This probability is given by the area under the normal distribution curve for $\bar{x}$ between the points $\bar{x} = 31.8$ and $\bar{x} = 31.9$. The first step in finding this area is to convert the two $\bar{x}$ values to their respective z values.

z Value for a Value of $\bar{x}$

The *z value for a value of $\bar{x}$* is calculated as

$$z = \frac{\bar{x} - \mu}{\sigma_{\bar{x}}}$$

The z values for $\bar{x} = 31.8$ and $\bar{x} = 31.9$ are computed below, and they are shown on the z scale below the normal distribution curve for $\bar{x}$ in **Figure 7.12**.

$$\text{For } \bar{x} = 31.8: \quad z = \frac{\bar{x} - \mu}{\sigma_{\bar{x}}} = \frac{31.8 - 32}{.06708204} = -2.98$$

$$\text{For } \bar{x} = 31.9: \quad z = \frac{\bar{x} - \mu}{\sigma_{\bar{x}}} = \frac{31.9 - 32}{.06708204} = -1.49$$

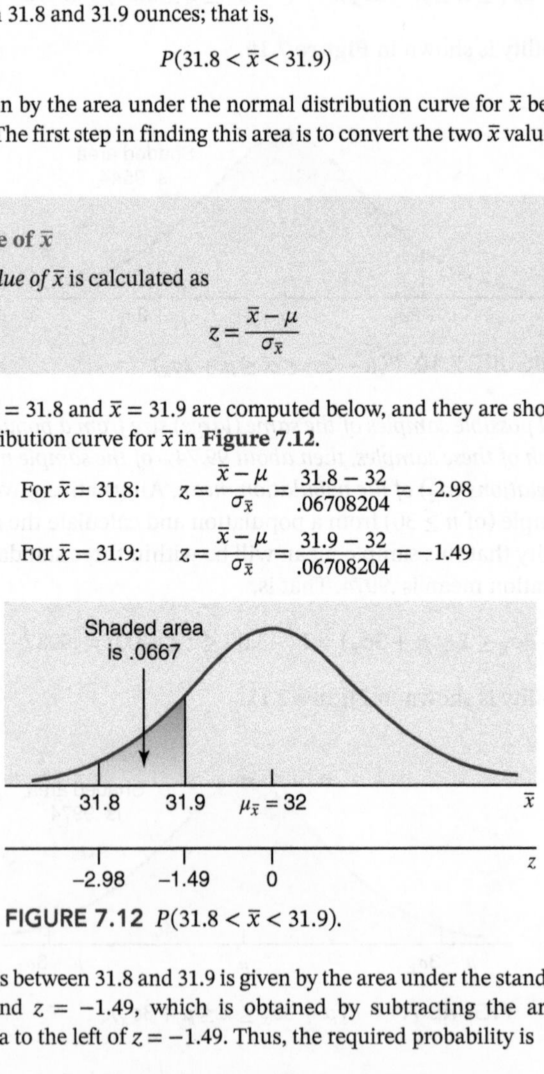

FIGURE 7.12 $P(31.8 < \bar{x} < 31.9)$.

The probability that $\bar{x}$ is between 31.8 and 31.9 is given by the area under the standard normal curve between $z = -2.98$ and $z = -1.49$, which is obtained by subtracting the area to the left of $z = -2.98$ from the area to the left of $z = -1.49$. Thus, the required probability is

$$P(31.8 < \bar{x} < 31.9) = P(-2.98 < z < -1.49)$$
$$= P(z < -1.49) - P(z < -2.98) = .0681 - .0014 = \mathbf{.0667}$$

Therefore, the probability that the mean weight of a sample of 20 packages will be between 31.8 and 31.9 ounces is .0667.

EXAMPLE 7.6 Cost of a Checking Account at U.S. Banks

Calculating the Probability of $\bar{x}$ in an Interval: $n \geq 30$

According to Moebs Services Inc., an individual checking account at major U.S. banks costs the banks between \$350 and \$450 per year. Suppose that the current average cost of all checking accounts at major U.S. banks is \$400 per year with a standard deviation of \$30. Let $\bar{x}$ be the current average annual cost of a random sample of 225 individual checking accounts at major banks in the United States.

a. What is the probability that the average annual cost of the checking accounts in this sample is within \$4 of the population mean?

b. What is the probability that the average annual cost of the checking accounts in this sample is less than the population mean by \$2.70 or more?

Solution From the given information, for the annual costs of all individual checking accounts at major banks in the United States,

$$\mu = \$400 \quad \text{and} \quad \sigma = \$30$$

Although the shape of the probability distribution of the population (annual costs of all individual checking accounts at major U.S. banks) is unknown, the sampling distribution of $\bar{x}$ is approximately normal because the sample is large ($n \geq 30$). Remember that when the sample is large, the central limit theorem applies. The mean and standard deviation of the sampling distribution of $\bar{x}$ are, respectively,

$$\mu_{\bar{x}} = \mu = \$400 \quad \text{and} \quad \sigma_{\bar{x}} = \frac{\sigma}{\sqrt{n}} = \frac{30}{\sqrt{225}} = \$2.00$$

a. The probability that the mean of the annual costs of checking accounts in this sample is within \$4 of the population mean is written as

$$P(396 \leq \bar{x} \leq 404)$$

This probability is given by the area under the normal distribution curve for $\bar{x}$ between $\bar{x} = \$396$ and $\bar{x} = \$404$, as shown in **Figure 7.13**. We find this area as follows:

$$\text{For } \bar{x} = \$396: \quad z = \frac{\bar{x} - \mu}{\sigma_{\bar{x}}} = \frac{396 - 400}{2.00} = -2.00$$

$$\text{For } \bar{x} = \$404: \quad z = \frac{\bar{x} - \mu}{\sigma_{\bar{x}}} = \frac{404 - 400}{2.00} = 2.00$$

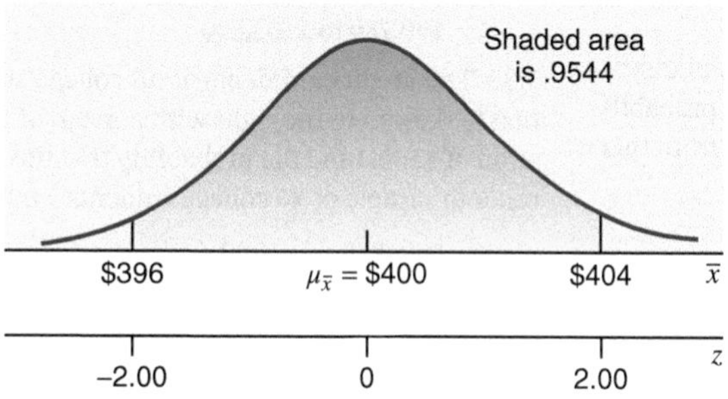

FIGURE 7.13 $P(\$396 \leq \bar{x} \leq \$400)$.

Hence, the required probability is

$$\begin{aligned} P(\$396 \leq \bar{x} \leq \$404) &= P(-2.00 \leq z \leq 2.00) \\ &= P(z \leq 2.00) - P(z \leq -2.00) \\ &= .9772 - .0228 = \mathbf{.9544} \end{aligned}$$

Therefore, the probability is .9544 that the average annual cost of a sample of 225 checking accounts is within \$4 of the population mean.

b. The probability that the average annual cost of the checking accounts in this sample is less than the population mean by \$2.70 or more is written as

$$P(\bar{x} \leq 397.30)$$

This probability is given by the area under the normal curve for $\bar{x}$ to the left of $\bar{x} = \$397.30$, as shown in **Figure 7.14**. We find this area as follows:

$$\text{For } \bar{x} = \$397.30: \quad z = \frac{\bar{x} - \mu}{\sigma_{\bar{x}}} = \frac{397.30 - 400}{2.00} = -1.35$$

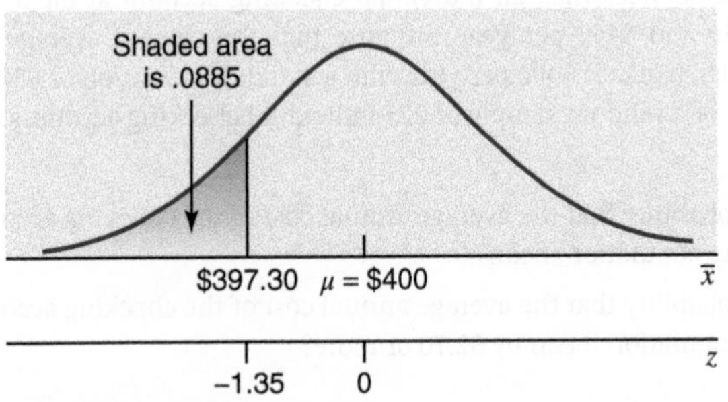

FIGURE 7.14 $P(\bar{x} \le \$397.30)$.

Hence, the required probability is

$$P(\bar{x} \le 397.30) = P(z \le -1.35) = \mathbf{.0885}$$

Thus, the probability that the average annual cost of the checking accounts in this sample is less than the population mean by \$2.70 or more is .0885.

EXERCISES

Concepts and Procedures

7.28 If all possible samples of the same (large) size are selected from a population, what percentage of all the sample means will be within 2.0 standard deviations ($\sigma_{\bar{x}}$) of the population mean?

7.29 If all possible samples of the same (large) size are selected from a population, what percentage of all the sample means will be within 3.5 standard deviations ($\sigma_{\bar{x}}$) of the population mean?

7.30 For a population, $N = 225{,}000$, $\mu = 75$, and $\sigma = 20$. Find the z value for each of the following for $n = 60$.

a. $\bar{x} = 76.33$ **b.** $\bar{x} = 68.25$

c. $\bar{x} = 82.50$ **d.** $\bar{x} = 73.50$

7.31 Let x be a continuous random variable that has a normal distribution with $\mu = 80$ and $\sigma = 15$. Assuming $n/N \le .05$, find the probability that the sample mean $\bar{x}$ for a random sample of 25 taken from this population will be

a. between 75.5 and 82.5 **b.** less than 78

7.32 Let x be a continuous random variable that has a normal distribution with $\mu = 48$ and $\sigma = 8$. Assuming $n/N \le .05$, find the probability that the sample mean $\bar{x}$ for a random sample of 16 taken from this population will be

a. between 49.6 and 52.2 **b.** more than 45.7

7.33 Let x be a continuous random variable that follows a distribution skewed to the left with $\mu = 85$ and $\sigma = 20$. Assuming $n/N \le .05$, find the probability that the sample mean $\bar{x}$ for a random sample of 70 taken from this population will be

a. less than 75 **b.** greater than 80

Applications

7.34 The GPAs of all students enrolled at a large university have an approximate normal distribution with a mean of 3.5 and a standard deviation of .65. Find the probability that the mean GPA of a random sample of 25 students selected from this university is

a. 3.65 or higher **b.** 3.30 or lower **c.** 3.25 to 3.75

7.35 According to a U.S. Bureau of Labor Statistics release of April 29, 2019, financial analysts earn an average of \$100,990 a year. Suppose that the current annual earnings of all financial analysts have the mean and standard deviation of \$100,990 and \$30,000, respectively. Find the probability that the average annual salary of a random sample of 400 financial analysts is

a. more than \$103,090

b. less than \$98,500

c. \$99,780 to \$103,830

7.36 The credit card debts of all college students have a distribution that is skewed to the right with a mean of \$2500 and a standard deviation of \$500. Find the probability that the mean credit card debt for a random sample of 40 college students would be

a. between 2250 and 2750

b. less than 2400

7.37 According to the American Time Use Survey results released by the U.S. Bureau of Labor Statistics on June 14, 2019, Americans age 15 and over watched television for an average of 170.4 minutes per day. Suppose that the current distribution of time spent watching television per day by all Americans age 15 and over has a mean of 170.4 minutes and a standard deviation of 20 minutes. Find the probability that the average time spent per day watching television by a random sample of 400 Americans age 15 and over is

a. 171 to 173 minutes

b. more than 172 minutes

c. at most 168 minutes

7.38 The delivery times for all food orders at a fast-food restaurant during the lunch hour are normally distributed with a mean of 10.5 minutes and a standard deviation of 2.25 minutes. Find the probability that the mean delivery time for a random sample of 20 such orders at this restaurant is

a. between 9 to 12 minutes

b. within 2 minutes of population mean

c. less than the population mean by 2 minutes or more

7.39 According to experian.com, the average household mortgage debt in the United States was \$202,284 for the first quarter of 2019. Suppose that the current distribution of mortgage debts of all U.S. households has a mean of \$202,284 and a standard deviation of \$36,000. Find the probability that the current average mortgage debt for a random sample of 144 U.S. households is

a. less than \$198,00

b. more than \$206,000

c. \$200,000 to \$205,000

7.40 The average length of National Hockey League (NHL) games is 152.5 minutes with a standard deviation of 8.48 minutes. In a random sample of 35 NHL games, what is the probability that the sample mean is within 2.25 minutes of the mean length of all NHL games?

7.41 A machine at Katz Steel Corporation makes 3-inch-long nails. The probability distribution of the lengths of these nails is approximately normal with a mean of 3 inches and a standard deviation of .1 inch. The quality control inspector takes a sample of 25 nails once a week and calculates the mean length of these nails. If the mean of this sample is either less than 2.95 inches or greater than 3.05 inches, the inspector concludes that the machine needs an adjustment. What is the probability, that based on a sample of 25 nails, the inspector will conclude that the machine needs an adjustment?

7.5 Population and Sample Proportions; and the Mean, Standard Deviation, and Shape of the Sampling Distribution of $\hat{p}$

LEARNING OBJECTIVES

After completing this section, you should be able to:

- Describe the difference between a population proportion and a sample proportion.
- Calculate the values of a population proportion and a sample proportion.
- Calculate the sampling error for a given value of $\hat{p}$ when p is known.
- Explain why $\hat{p}$ is an unbiased estimator of p.
- Explain why $\hat{p}$ is a consistent estimator.
- Calculate the mean and standard deviation of the sampling distribution of $\hat{p}$.
- Explain the central limit theorem for a sample proportion.
- Verify the conditions for the central limit theorem for a sample proportion.
- Find the mean, standard deviation, and the shape of the sampling distribution of $\hat{p}$ when the required conditions are met.

The concept of proportion is the same as the concept of relative frequency discussed in Chapter 2 and the concept of probability of success in a binomial experiment. The relative frequency of a category or class gives the proportion of the sample or population that belongs to that category or class. Similarly, the probability of success in a binomial experiment represents the proportion of the sample or population that possesses a given characteristic.

In this section, first we will learn about the population and sample proportions. Then, we will discuss the mean, standard deviation, and shape of the sampling distribution of $\hat{p}$.

7.5.1 Population and Sample Proportions

The **population proportion**, denoted by p, is obtained by taking the ratio of the number of elements in a population with a specific characteristic to the total number of elements in the population. The **sample proportion**, denoted by $\hat{p}$ (pronounced *p hat*), gives a similar ratio for a sample.

Population and Sample Proportions

The *population and sample proportions*, denoted by p and $\hat{p}$, respectively, are calculated as

$$p = \frac{X}{N} \quad \text{and} \quad \hat{p} = \frac{x}{n}$$

where

N = total number of elements in the population
n = total number of elements in the sample
X = number of elements in the population that possess a specific characteristic
x = number of elements in the sample that possess the same specific characteristic

Example 7.7 illustrates the calculation of the population and sample proportions.

EXAMPLE 7.7 Families Owning Homes

Calculating the Population and Sample Proportions

Suppose a total of 789,654 families live in a particular city, and 563,282 of them own homes. A sample of 240 families is selected from this city, and 158 of them own homes. Find the proportion of families who own homes in the population and in the sample.

Solution For the population of this city,

N = population size = 789,654
X = number of families in the population who own homes = 563,282

The population proportion of families in this city who own homes is

$$p = \frac{X}{N} = \frac{563{,}282}{789{,}654} = \mathbf{.71}$$

Now, a sample of 240 families is taken from this city, and 158 of them are home-owners. Then,

n = sample size = 240
x = number of families in the sample who own homes = 158

The sample proportion is

$$\hat{p} = \frac{x}{n} = \frac{158}{240} = \mathbf{.66}$$

As in the case of the mean, the difference between the sample proportion and the corresponding population proportion gives the **sampling error**, assuming that the sample is random and no nonsampling error has been made. Thus, in the case of the proportion,

$$\text{Sampling error} = \hat{p} - p$$

For instance, for Example 7.7,

$$\text{Sampling error} = \hat{p} - p = .66 - .71 = -.05$$

Here, the sampling error being equal to −.05 means that the sample proportion underestimates the population proportion by .05.

Case Study 7.1

2016 U.S. Election and Sampling Error

In the 2016 U.S. Presidential election, the Democratic party candidate, Hillary Clinton, ran against the Republican party candidate, Donald Trump. There were several other candidates from whom voters could choose, but none were a major-party candidate. Most of the major polls leading up to the day before the November 8, 2016 Election Day had predicted that Hillary Clinton would receive the most popular votes and, consequently, would win the election. Although Hillary Clinton did receive the higher number of popular votes than other candidates including Donald Trump, she did not win the 2016 Presidential election. A Presidential candidate wins an election by receiving the highest number of Electoral College votes, and the candidate with the highest number of popular votes is not necessarily the candidate with the highest number of Electoral College votes. With his 304 Electoral College votes, Donald Trump received well over the minimum 270 Electoral College votes needed to win the election. In the history of the United States, there have only been four other elections (1824, 1876, 1888, and 2000) in which the candidate with the majority of the Electoral College votes did not receive the majority of the popular votes.

Hillary Clinton received 48.2% of all the (popular) votes compared to 46.1% received by Donald Trump. The above graph shows the results of five polls that were conducted during the days leading up to November 6, 2016. As we can observe from the graph, the Bloomberg poll, conducted November 4–6, 2016 predicted that Donald Trump would receive 43% of the popular votes while he actually received 46.1% of the popular votes. Thus, for this sample, the sampling error is $\hat{p} - p = .43 - .461 = -.031$ or -3.1%, assuming there were no nonsampling errors made. This means that the Bloomberg poll underestimated the percentage of popular votes received by Donald Trump by 3.1%. The Bloomberg poll also predicted that Hillary Clinton would receive 46% of the popular votes when, in fact, she received 48.2% of the popular votes. Again, assuming there were no nonsampling errors made, the sampling error is $\hat{p} - p = .46 - .482 = -.022$. Thus, the Bloomberg poll underestimated the percentage of popular votes received by Hillary Clinton by 2.2%.

Similarly, we can calculate the sampling errors for the other polls using the information presented in the graph.

Data source: www.yahoo.com and www.realclearpolitics.com

7.5.2 Sampling Distribution of $\hat{p}$

Just like the sample mean $\bar{x}$, the sample proportion $\hat{p}$ is a random variable. In other words, the population proportion p is a contant as it assumes one and only one value. However, the sample proportion $\hat{p}$ can assume one of a large number of possible values depending on which sample is selected. Hence, $\hat{p}$ is a random variable and it possesses a probability distribution, which is called its **sampling distribution**.

Sampling Distribution of the Sample Proportion, $\hat{p}$

The probability distribution of the sample proportion, $\hat{p}$, is called its *sampling distribution*. It gives the various values that $\hat{p}$ can assume and their probabilities.

The value of $\hat{p}$ calculated for a particular sample depends on what elements of the population are included in that sample. Example 7.8 illustrates the concept of the sampling distribution of $\hat{p}$.

EXAMPLE 7.8 Employees' Knowledge of Statistics

Illustrating the Sampling Distribution of $\hat{p}$

Boe Consultant Associates has a total of five employees. **Table 7.6** gives the names of these five employees and information concerning their knowledge of statistics.

TABLE 7.6 Information on the Five Employees of Boe Consultant Associates

Name	Knows Statistics
Ally	Yes
John	No
Susan	No
Lee	Yes
Tom	Yes

If we define the population proportion, p, as the proportion of employees who know statistics, then

$$p = 3/5 = .60$$

Note that this population proportion, $p = .60$, is a constant. As long as the population does not change, this value of p will not change.

Now suppose, we draw all possible samples of three employees each and compute the proportion of employees, for each sample, who know statistics. The total number of samples of size three that can be drawn from the population of five employees is

$$\text{Total number of samples} = {}_5C_3 = \frac{5!}{3!(5-3)!} = \frac{5 \cdot 4 \cdot 3 \cdot 2 \cdot 1}{3 \cdot 2 \cdot 1 \cdot 2 \cdot 1} = 10$$

Table 7.7 lists these 10 possible samples and the proportion of employees who know statistics for each of those samples. Note that we have rounded the values of $\hat{p}$ to two decimal places.

TABLE 7.7 All Possible Samples of Size Three and the Value of $\hat{p}$ for Each Sample

Sample	Proportion Who Know Statistics ($\hat{p}$)
Ally, John, Susan	1/3 = .33
Ally, John, Lee	2/3 = .67
Ally, John, Tom	2/3 = .67
Ally, Susan, Lee	2/3 = .67
Ally, Susan, Tom	2/3 = .67
Ally, Lee, Tom	3/3 = 1.00
John, Susan, Lee	1/3 = .33
John, Susan, Tom	1/3 = .33
John, Lee, Tom	2/3 = .67
Susan, Lee, Tom	2/3 = .67

Using Table 7.7, we prepare the frequency distribution of $\hat{p}$ as recorded in **Table 7.8**, along with the relative frequencies of classes, which are obtained by dividing the frequencies of classes by the total number of possible samples. The relative frequencies are used as probabilities and listed in **Table 7.9**. This table gives the sampling distribution of $\hat{p}$.

TABLE 7.8 Frequency and Relative Frequency Distributions of $\hat{p}$ When the Sample Size Is Three

$\hat{p}$	f	Relative Frequency
.33	3	3/10 = .30
.67	6	6/10 = .60
1.00	1	1/10 = .10
	$\Sigma f = 10$	Sum = 1.00

TABLE 7.9 Sampling Distribution of $\hat{p}$ When the Sample Size Is Three

$\hat{p}$	$P(\hat{p})$
.33	.30
.67	.60
1.00	.10
	$\Sigma P(\hat{p}) = 1.00$

7.5.3 Mean and Standard Deviation of $\hat{p}$

The **mean of $\hat{p}$**, which is the same as the mean of the sampling distribution of $\hat{p}$, is always equal to the population proportion, p, just as the mean of the sampling distribution of $\bar{x}$ is always equal to the population mean, μ.

Mean of the Sample Proportion

The *mean of the sample proportion*, $\hat{p}$, is denoted by $\mu_{\hat{p}}$ and is equal to the population proportion, p. Thus,

$$\mu_{\hat{p}} = p$$

In Example 7.8, $p = .60$. If we calculate the mean of the 10 values of $\hat{p}$ listed in Table 7.7, it will give us $\mu_{\hat{p}} = .60$, which is the mean of $\hat{p}$.

The sample proportion $\hat{p}$ is called an **estimator** of the population proportion p. As mentioned earlier, when the expected value (or mean) of a sample statistic is equal to the value of the corresponding population parameter, that sample statistic is said to be an **unbiased estimator**. Since for the sample proportion $\mu_{\hat{p}} = p$, $\hat{p}$ is an unbiased estimator of p.

The **standard deviation of $\hat{p}$**, denoted by $\sigma_{\hat{p}}$, is given by the following formula. This formula is true only when the sample size is small compared to the population size. As we know from Section 7.2, the sample size is said to be small compared to the population size if $n/N \leq .05$.

Standard Deviation of the Sample Proportion

The *standard deviation of the sample proportion*, $\hat{p}$, is denoted by $\sigma_{\hat{p}}$ and is given by the formula

$$\sigma_{\hat{p}} = \sqrt{\frac{pq}{n}}$$

where p is the population proportion, $q = 1 - p$, and n is the sample size. This formula is used when $n/N \leq .05$, where N is the population size.

However, if n/N is greater than .05, then $\sigma_{\hat{p}}$ is calculated as follows:

$$\sigma_{\hat{p}} = \sqrt{\frac{pq}{n}}\sqrt{\frac{N-n}{N-1}}$$

where the factor $\sqrt{\frac{N-n}{N-1}}$ is called the **finite population correction factor.**

In almost all cases, the sample size is small compared to the population size and, consequently, the formula used to calculate $\sigma_{\hat{p}}$ is $\sqrt{pq/n}$.

As mentioned earlier, if the standard deviation of a sample statistic decreases as the sample size is increased, that statistic is said to be a **consistent estimator**. It is obvious from the above formula for $\sigma_{\hat{p}}$ that as n increases, the value of $\sqrt{pq/n}$ decreases. Thus, the sample proportion, $\hat{p}$, is a consistent estimator of the population proportion, p.

7.5.4 Shape of the Sampling Distribution of $\hat{p}$

The shape of the sampling distribution of $\hat{p}$ is inferred from the central limit theorem.

Central Limit Theorem for Sample Proportion

According to the central limit theorem, the *sampling distribution of* $\hat{p}$ is approximately normal for a sufficiently large sample size. In the case of proportion, the sample size is considered to be sufficiently large if np and nq are both greater than 5; that is, if

$$np > 5 \quad \text{and} \quad nq > 5$$

Note that the sampling distribution of $\hat{p}$ will be approximately normal if $np > 5$ and $nq > 5$. This is the same condition that was required for the application of the normal approximation to the binomial probability distribution in Chapter 6.

Example 7.9 shows the calculation of the mean and standard deviation of $\hat{p}$ and describes the shape of its sampling distribution.

EXAMPLE 7.9 Owning a Home: An American Dream

Finding the Mean and Standard Deviation, and Describing the Shape of the Sampling Distribution of $\hat{p}$

According to a recent poll, 55% of adults polled said that owning a home is a *very* important part of the American Dream. Assume that this result is true for the current population of American adults. Let $\hat{p}$ be the proportion of American adults in a random sample of 2000 who will say that owning a home is a *very* important part of the American Dream. Find the mean and standard deviation of $\hat{p}$ and describe the shape of its sampling distribution.

Solution Let p be the proportion of all American adults who will say that owning a home is a *very* important part of the American Dream. Then,

$$p = .55, \quad q = 1 - p = 1 - .55 = .45, \quad \text{and} \quad n = 2000$$

The mean of the sampling distribution of $\hat{p}$ is

$$\mu_{\hat{p}} = p = \mathbf{.55}$$

The standard deviation of $\hat{p}$ is

$$\sigma_{\hat{p}} = \sqrt{\frac{pq}{n}} = \sqrt{\frac{(.55)(.45)}{2000}} = \mathbf{.0111}$$

The values of np and nq are

$$np = 2000(.55) = 1100 \quad \text{and} \quad nq = 2000(.45) = 900$$

Because np and nq are both greater than 5, we can apply the central limit theorem to make an inference about the shape of the sampling distribution of $\hat{p}$. Thus, the sampling distribution of $\hat{p}$ is approximately normal with a mean of .55 and a standard deviation of .0111, as shown in **Figure 7.15.**

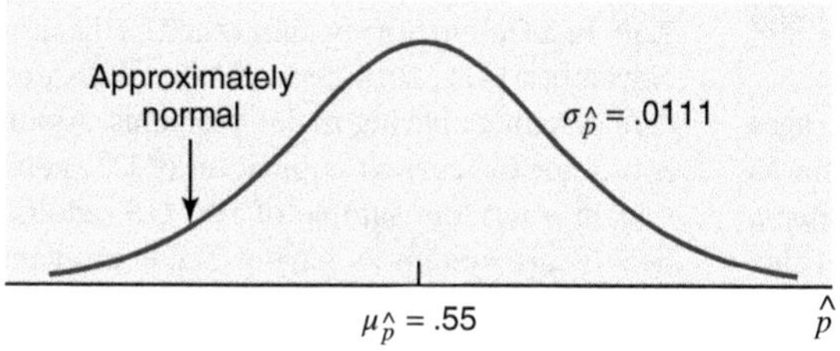

FIGURE 7.15 Sampling distribution of $\hat{p}$.

EXERCISES

Concepts and Procedures

7.42 In a population of 1000 subjects, 580 possess a certain characteristic. In a sample of 50 subjects selected from this population, 32 possess the same characteristic. What are the values of the population and sample proportions?

7.43 In a population of 5000 subjects, 800 possess a certain characteristic. In a sample of 150 subjects selected from this population, 20 possess the same characteristic. What are the values of the population and sample proportions?

7.44 In a population of 16,400 subjects, 30% possess a certain characteristic. In a sample of 200 subjects selected from this population, 25% possess the same characteristic. How many subjects in the population and sample, respectively, possess this characteristic?

7.45 Let $\hat{p}$ be the proportion of elements in a sample that possess a characteristic.

a. What is the mean of $\hat{p}$?

b. What is the formula to calculate the standard deviation of $\hat{p}$? Assume $n/N \leq .05$.

c. What condition(s) must hold true for the sampling distribution of $\hat{p}$ to be approximately normal?

7.46 For a population, $N = 10{,}000$ and $p = .68$. A random sample of 900 elements selected from this population gave $\hat{p} = .64$. Find the sampling error.

7.47 For a population, $N = 2800$ and $p = .28$. A random sample of 100 elements selected from this population gave $\hat{p} = .33$. Find the sampling error.

7.48 What is the estimator of the population proportion? Is this estimator an unbiased estimator of p? Explain why or why not.

7.49 Is the sample proportion a consistent estimator of the population proportion? Explain why or why not.

7.50 How does the value of $\sigma_{\hat{p}}$ change as the sample size increases? Explain. Assume $n/N \leq .05$.

7.51 Consider a large population with $p = .34$. Assuming $n/N \leq .05$, find the mean and standard deviation of the sample proportion $\hat{p}$ for a sample size of

a. 300 **b.** 850

7.52 A population of $N = 4000$ has a population proportion equal to .12. In each of the following cases, which formula will you use to calculate $\sigma_{\hat{p}}$ and why? Using the appropriate formula, calculate $\sigma_{\hat{p}}$ for each of these cases.

a. $n = 800$ **b.** $n = 30$

7.53 According to the central limit theorem, the sampling distribution of $\hat{p}$ is approximately normal when the sample is large. What is considered a large sample in the case of the proportion? Briefly explain.

7.54 Indicate in which of the following cases the central limit theorem will apply to describe the sampling distribution of the sample proportion.

a. $n = 250$ and $p = .32$ **b.** $n = 120$ and $p = .03$

c. $n = 400$ and $p = .008$ **d.** $n = 225$ and $p = .025$

7.55 Indicate in which of the following cases the central limit theorem will apply to describe the sampling distribution of the sample proportion.

a. $n = 30$ and $p = .55$ **b.** $n = 55$ and $p = .15$

c. $n = 65$ and $p = .25$ **d.** $n = 120$ and $p = .04$

Applications

7.56 A company manufactured five television sets on a given day, and these TV sets were inspected for being good or defective. The results of the inspection follow:

Good Good Defective Defective Good

a. What proportion of these TV sets are good?

b. How many total samples (without replacement) of size four can be selected from this population?

c. List all the possible samples of size four that can be selected from this population and calculate the sample proportion, $\hat{p}$, of television sets that are good for each sample. Prepare the sampling distribution of $\hat{p}$.

d. For each sample listed in part (c), calculate the sampling error.

7.57 A September 24, 2019 article on news.gallup.com mentioned that according to the Gallup research about climate change issues, 51% of U.S. adults can be classified as "Concerned Believers," meaning they "are highly worried about global warming, think it will pose a serious threat in their lifetime, believe it's the result of human activity, and think news reports about it are accurate or underestimate the problem" (https://news.gallup.com). Assume that this percentage is true for the current population of U.S. adults. Let $\hat{p}$ be the proportion in a random sample of 200 U.S. adults who can be classified as "Concerned Believers" when it comes to climate change issues. Find the mean and standard deviation of the sampling distribution of $\hat{p}$ and describe its shape.

7.58 In a Gallup poll conducted April 1–9, 2019, U.S. adults, ages 18 and over, were asked to rate their current financial situation as excellent, good, only fair, or poor. In this poll, 12% of the respondents rated their current financial situation as excellent. Assume that this percentage is true for the current population of U.S. adults. Let $\hat{p}$ be the proportion in a random sample of 500 U.S. adults who would rate their current financial situation as excellent. Find the mean and standard deviation of the sampling distribution of $\hat{p}$ and describe its shape.

7.59 In a Gallup Survey about the U.S. health-care system conducted November 1–11, 2018, 55% of U.S. adults described the U.S. health-care system as having major problems. Assume that this percentage is true for the current population of U.S. adults. Let $\hat{p}$ be the proportion in a random sample of 900 U.S. adults who describe the U.S. health-care system as having major problems. Find the mean and standard deviation of the sampling distribution of $\hat{p}$ and describe its shape.

7.6 Applications of the Sampling Distribution of $\hat{p}$

LEARNING OBJECTIVE

After completing this section, you should be able to:

- Calculate and interpret probabilities for the sampling distribution of $\hat{p}$.

As mentioned in Section 7.4, when we conduct a study, we usually take only one sample and make all decisions or inferences on the basis of the results of that one sample. We use the concepts of the mean, standard deviation, and shape of the sampling distribution of $\hat{p}$ to determine the probability that the value of $\hat{p}$ computed from one sample falls within a given interval. Examples 7.10 and 7.11 illustrate this application.

EXAMPLE 7.10 | College Education Too Expensive

Calculating the Probability That $\hat{p}$ Is in an Interval

In a recent Pew Research Center nationwide telephone survey of American adults, 75% of adults said that college education has become too expensive for most people. Suppose that this result is true for the current population of American adults. Let $\hat{p}$ be the proportion in a random sample of 1400 adult Americans who will hold the said opinion. Find the probability that 76.5% to 78% of adults in this sample will hold this opinion.

Solution From the given information,

$$n = 1400, \quad p = .75, \quad \text{and} \quad q = 1 - p = 1 - .75 = .25$$

where p is the proportion of all adult Americans who hold the said opinion.

The mean of the sample proportion $\hat{p}$ is

$$\mu_{\hat{p}} = p = .75$$

The standard deviation of $\hat{p}$ is

$$\sigma_{\hat{p}} = \sqrt{\frac{pq}{n}} = \sqrt{\frac{(.75)(.25)}{1400}} = .01157275$$

The values of np and nq are

$$np = 1400\,(.75) = 1050 \quad \text{and} \quad nq = 1400\,(.25) = 350$$

Because np and nq are both greater than 5, we can infer from the central limit theorem that the sampling distribution of $\hat{p}$ is approximately normal. The probability that $\hat{p}$ is between .765 and .78 is given by the area under the normal curve for $\hat{p}$ between $\hat{p} = .765$ and $\hat{p} = .78$, as shown in **Figure 7.16.**

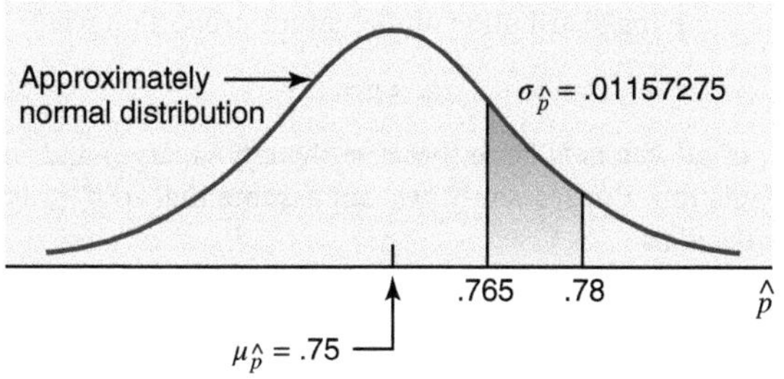

FIGURE 7.16 $P(.765 < \hat{p} < .78)$.

The first step in finding the area under the normal distribution curve between $\hat{p} = .765$ and $\hat{p} = .78$ is to convert these two values to their respective z values. The z value for $\hat{p}$ is computed using the following formula:

z Value for a Value of $\hat{p}$

The *z value for a value of $\hat{p}$* is calculated as

$$z = \frac{\hat{p} - p}{\sigma_{\hat{p}}}$$

The two values of $\hat{p}$ are converted to their respective z values, and then the area under the normal curve between these two points is found using the normal distribution table.

$$\text{For } \hat{p} = .765\text{:} \quad z = \frac{\hat{p} - p}{\sigma_{\hat{p}}} = \frac{.765 - .75}{.01157275} = 1.30$$

$$\text{For } \hat{p} = .78\text{:} \quad z = \frac{\hat{p} - p}{\sigma_{\hat{p}}} = \frac{.78 - .75}{.01157275} = 2.59$$

Thus, the probability that $\hat{p}$ is between .765 and .78 is given by the area under the standard normal curve between $z = 1.30$ and $z = 2.59$. This area is shown in **Figure 7.17**. The required probability is

$$P(.765 < \hat{p} < .78) = P(1.30 < z < 2.59) = P(z < 2.59) - P(z < 1.30)$$
$$= .9952 - .9032 = \mathbf{.0920}$$

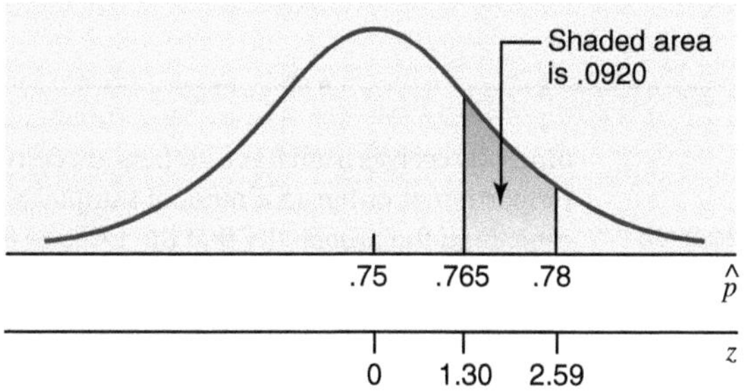

FIGURE 7.17 $P(.765 < \hat{p} < .78)$.

Thus, the probability that 76.5% to 78% of American adults in a random sample of 1400 will say that college education has become too expensive for most people is .0920.

EXAMPLE 7.11 Voting in an Election

Calculating the Probability That $\hat{p}$ Is Less Than a Certain Value

Alexandria Omar, who is running for mayor in a large city, claims that she is favored by 53% of all eligible voters of that city. Assume that this claim is true. What is the probability that in a random sample of 400 registered voters taken from this city, less than 49% will favor Alexandria Omar?

Solution Let p be the proportion of all eligible voters who favor Alexandria Omar. Then,

$$p = .53 \quad \text{and} \quad q = 1 - p = 1 - .53 = .47$$

The mean of the sampling distribution of the sample proportion $\hat{p}$ is

$$\mu_{\hat{p}} = p = .53$$

The population of all voters is large (because the city is large) and the sample size is small compared to the population. Consequently, we can assume that $n/N \leq .05$. Hence, the standard deviation of $\hat{p}$ is calculated as

$$\sigma_{\hat{p}} = \sqrt{\frac{pq}{n}} = \sqrt{\frac{(.53)(.47)}{400}} = .02495496$$

From the central limit theorem, the shape of the sampling distribution of $\hat{p}$ is approximately normal. (The reader should check that $np > 5$ and $nq > 5$ and, hence, the sample size is large.) The probability that $\hat{p}$ is less than .49 is given by the area under the normal distribution curve for $\hat{p}$ to the left of $\hat{p} = .49$ as shown in **Figure 7.18**. The z value for $\hat{p} = .49$ is

$$z = \frac{\hat{p} - p}{\sigma_{\hat{p}}} = \frac{.49 - .53}{.02495496} = -1.60$$

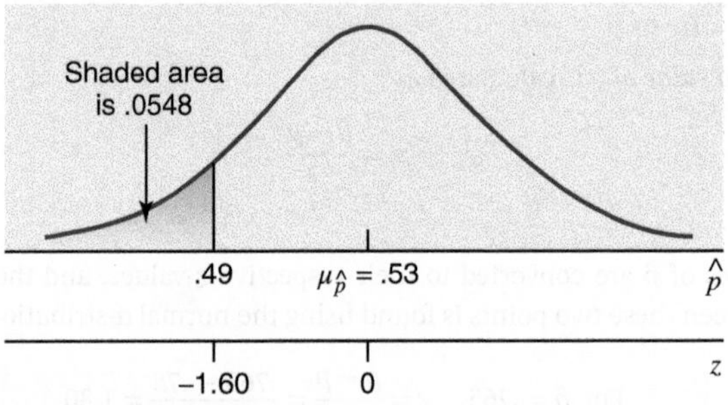

FIGURE 7.18 $P(\hat{p} < .49)$.

Thus, the required probability from Table IV of Appendix B is

$$P(\hat{p} < .49) = P(z < -1.60)$$
$$= \mathbf{.0548}$$

Hence, the probability that less than 49% of the voters in a random sample of 400 will favor Alexandria Omar is .0548.

EXERCISES

Concepts and Procedures

7.60 If all possible samples of the same (large) size are selected from a population, what percentage of all sample proportions will be within 1.5 standard deviations ($\sigma_{\hat{p}}$) of the population proportion?

7.61 If all possible samples of the same (large) size are selected from a population, what percentage of all sample proportions will be within 2.5 standard deviations ($\sigma_{\hat{p}}$) of the population proportion?

7.62 For a population, $N = 30{,}000$ and $p = .62$. Find the z value for each of the following for $n = 100$.

a. $\hat{p} = .56$ **b.** $\hat{p} = .68$

c. $\hat{p} = .53$ **d.** $\hat{p} = .65$

Applications

7.63 A survey of all medium- and large-sized corporations showed that 64% of them offer retirement plans to their employees. Let $\hat{p}$ be the proportion in a random sample of 50 such corporations that offer retirement plans to their employees. Find the probability that the value of $\hat{p}$ will be

a. between .54 and .61 **b.** greater than .71

7.64 Dartmouth Distribution Warehouse makes deliveries of a large number of products to its customers. It is known that 85% of all the orders it receives from its customers are delivered on time. Let $\hat{p}$ be the proportion of orders in a random sample of 100 that are delivered on time. Find the probability that the value of $\hat{p}$ will be

a. between .81 and .88 **b.** less than .87

7.65 Brooklyn Corporation manufactures flash drives. The machine that is used to make these flash drives is known to produce 6% defective flash drives. The quality control inspector selects a sample of 150 flash drives every week and inspects them for being good or defective. If 8% or more of the flash drives in the sample are defective, the process is stopped and the machine is adjusted. What is the probability that based on a sample of 150 flash drives, the process will be stopped to adjust the machine?

7.66 Mong Corporation makes auto batteries. The company claims that 80% of its LL70 batteries are good for 70 months or longer. Assume that this claim is true. Let $\hat{p}$ be the proportion in a sample of 100 such batteries that are good for 70 months or longer.

a. What is the probability that this sample proportion is within .05 of the population proportion?

b. What is the probability that this sample proportion is less than the population proportion by .06 or more?

c. What is the probability that this sample proportion is greater than the population proportion by .07 or more?

Uses and Misuses of Statistics

Beware of Bias

Mathematics tells us that the sample mean, $\bar{x}$, is an unbiased and consistent estimator for the population mean, μ. This is great news because it allows us to estimate properties of a population based on those of a sample; this is the essence of statistics. But statistics always makes a number of assumptions about the sample from which the mean and standard deviation are calculated. Failure to respect these assumptions can introduce bias in your calculations. In statistics, *bias* means a deviation of the expected value of a statistical estimator from the parameter it is intended to estimate.

Let's say you are a quality control manager for a refrigerator parts company. One of the parts that you manufacture has a specification that the length of the part be 2.0 centimeters plus or minus .025 centimeter. The manufacturer expects that the parts it receives have a mean length of 2.0 centimeters and a small variation around that mean. The manufacturing process is to mold the part to something a little bit bigger than necessary—say, 2.1 centimeters—and finish the process by hand. Because the action of cutting material is irreversible, the machinists tend to miss their target by approximately .01 centimeter, so the mean length of the parts is not 2.0 centimeters, but rather 2.01 centimeters. It is your job to catch this.

One of your quality control procedures is to select completed parts randomly and test them against specification. Unfortunately, your measurement device is also subject to variation and might consistently underestimate the length of the parts. If your measurements are consistently .01 centimeter too short, your sample mean will not catch the manufacturing error in the population of parts.

The solution to the manufacturing problem is relatively straightforward: Be certain to calibrate your measurement instrument. Calibration becomes very difficult when working with people. It is known that people tend to overestimate the number of times that they vote and underestimate the time it takes to complete a project. Basing statistical results on this type of data can result in distorted estimates of the properties of your population. It is very important to be careful to weed out bias in your data because once it gets into your calculations, it is very hard to get it out.

Glossary

Central limit theorem The theorem from which it is inferred that, for a large sample size ($n \geq 30$), the shape of the sampling distribution of $\bar{x}$ is approximately normal. Also, by the same theorem, the shape of the sampling distribution of $\hat{p}$ is approximately normal for a sample for which $np > 5$ and $nq > 5$.

Consistent estimator A sample statistic with a standard deviation that decreases as the sample size increases.

Estimator The sample statistic that is used to estimate a population parameter.

Mean of $\hat{p}$ The mean of the sampling distribution of $\hat{p}$, denoted by $\mu_{\hat{p}}$, is equal to the population proportion p.

Mean of $\bar{x}$ The mean of the sampling distribution of $\bar{x}$, denoted by $\mu_{\bar{x}}$, is equal to the population mean μ.

Nonsampling errors The errors that occur during the collection, recording, and tabulation of data.

Population probability distribution The probability distribution of the population data.

Population proportion p The ratio of the number of elements in a population with a specific characteristic to the total number of elements in the population.

Sample proportion $\hat{p}$ The ratio of the number of elements in a sample with a specific characteristic to the total number of elements in that sample.

Sampling distribution of $\hat{p}$ The probability distribution of all the values of $\hat{p}$ calculated from all possible samples of the same size selected from a population.

Sampling distribution of $\bar{x}$ The probability distribution of all the values of $\bar{x}$ calculated from all possible samples of the same size selected from a population.

Sampling error The difference between the value of a sample statistic calculated from a random sample and the value of the corresponding population parameter. This type of error occurs due to chance.

Standard deviation of $\hat{p}$ The standard deviation of the sampling distribution of $\hat{p}$, denoted by $\sigma_{\hat{p}}$, is equal to $\sqrt{pq/n}$ when $n/N \leq .05$.

Standard deviation of $\bar{x}$ The standard deviation of the sampling distribution of $\bar{x}$, denoted by $\sigma_{\bar{x}}$, is equal to $\sigma/\sqrt{n}$ when $n/N \leq .05$.

Unbiased estimator An estimator with an expected value (or mean) that is equal to the value of the corresponding population parameter.

Exercises

Supplementary Exercises

7.67 The print on the package of 100-watt General Electric soft-white lightbulbs claims that these bulbs have an average life of 750 hours. Assume that the lives of all such bulbs have a normal distribution with a mean of 750 hours and a standard deviation of 55 hours. Let $\bar{x}$ be the mean life of a random sample of 25 such bulbs. Find the mean and standard deviation of $\bar{x}$, and describe the shape of its sampling distribution.

7.68 The package of Sylvania CFL 65-watt replacement bulbs that use only 16 watts claims that these bulbs have an average life of 8000 hours. Assume that the lives of all such bulbs have a normal distribution with a mean of 8000 hours and a standard deviation of 400 hours. Find the probability that the mean life of a random sample of 25 such bulbs is

a. less than 7890 hours
b. between 7850 and 7910 hours
c. within 130 hours of the population mean
d. less than the population mean by 150 hours or more

7.69 The Green Company hybrid car is estimated to get an average of 50 miles per gallon (mpg) of gas. However, the gas mileage varies from car to car due to a variety of conditions, driving styles, and other factors and has been reported to be as high as 70 mpg. Suppose that the distribution of miles per gallon for Green Company hybrid cars has a mean of 50 mpg and a standard deviation of 5.9 mpg. Find the probability that the average miles per gallon for 38 randomly selected Green Company hybrid cars is

a. more than 51.5
b. between 48 and 51
c. less than 53
d. greater than the population mean by 2.5 or more

7.70 Suppose that 88% of the cases of car burglar alarms that go off are false. Let $\hat{p}$ be the proportion of false alarms in a random sample of 80 cases of car burglar alarms that go off. Calculate the mean and standard deviation of $\hat{p}$, and describe the shape of its sampling distribution.

7.71 Seventy percent of adults favor some kind of government control on the prices of medicines. Assume that this percentage is true for the current population of all adults. Let $\hat{p}$ be the proportion of adults in a random sample of 400 who favor government control on the prices of medicines. Calculate the mean and standard deviation of $\hat{p}$, and describe the shape of its sampling distribution.

7.72 In a Census-at-School sample of twelfth graders from Pennsylvania, 11.3% said that they walk, bike, skateboard, or rollerblade to school on most days. Assume that this percentage is true for the current population of twelfth graders in Pennsylvania. Let $\hat{p}$ be the proportion of students who walk, bike, skateboard, or rollerblade to school on most days in a random sample of 500 Pennsylvania twelfth graders. Find the mean and standard deviation of $\hat{p}$ and describe its shape.

7.73 AMARAN makes auto batteries. The company claims that 85% of the LL95 batteries are good for 90 months or longer. Assume that this claim is true. Let $\hat{p}$ be the proportion in a sample of 150 such batteries that are good for 90 months or longer.

a. What is the probability that this sample proportion is within .05 of the population proportion?

b. What is the probability that this sample proportion is less than the population proportion by .08 or more?

c. What is the probability that this sample proportion is greater than the population proportion by .075 or more?

Advanced Exercises

7.74 The test scores for 300 students were entered into a computer, analyzed, and stored in a file. Unfortunately, someone accidentally erased a major portion of this file from the computer. The only information that is available is that 30% of the scores were below 65, and 15% of the scores were above 90. Assuming the scores are approximately normally distributed, find their mean and standard deviation.

7.75 A television reporter is covering the election for mayor of a large city and will conduct an exit poll (interviews with voters immediately after they vote) to make an early prediction of the outcome. Assume that the eventual winner of the election will get 60% of the votes.

a. What is the probability that a prediction based on an exit poll of a random sample of 25 voters will be correct? In other words, what is the probability that 13 or more of the 25 voters in the sample will have voted for the eventual winner?

b. How large a sample would the reporter have to take so that the probability of correctly predicting the outcome would be .95 or higher?

7.76 A certain elevator has a maximum legal carrying capacity of 6000 pounds. Suppose that the population of all people who ride this elevator have a mean weight of 160 pounds with a standard deviation of 25 pounds. If 35 of these people board the elevator, what is the probability that their combined weight will exceed 6000 pounds? Assume that the 35 people constitute a random sample from the population.

7.77 Refer to the sampling distribution discussed in Section 7.1. Calculate and replace the sample means in Table 7.3 with the sample medians, and then calculate the average of these sample medians. Does this average of the medians equal the population mean? If yes, why does this make sense? If no, how could you change exactly two of the five data values in this example so that the average of the sample medians equals the population mean?

Self-Review Test

Multiple Choice Questions

1. A sampling distribution is the probability distribution of
   a. a population parameter
   b. a sample statistic
   c. any random variable

2. Nonsampling errors are
   a. the errors that occur because the sample size is too large in relation to the population size
   b. the errors made while collecting, recording, and tabulating data
   c. the errors that occur because an untrained person conducts the survey

3. A sampling error is
   a. the difference between the value of a sample statistic based on a random sample and the value of the corresponding population parameter

b. the error made while collecting, recording, and tabulating data

c. the error that occurs because the sample is too small

4. The mean of the sampling distribution of $\bar{x}$ is always equal to

a. μ b. $\mu - 5$ c. $\sigma/\sqrt{n}$

5. The condition for the standard deviation of the sample mean to be $\sigma/\sqrt{n}$ is that

a. $np > 5$ b. $n/N \leq .05$ c. $n > 30$

6. The standard deviation of the sampling distribution of the sample mean decreases when

a. x increases b. n increases c. n decreases

7. When samples are selected from a normally distributed population, the sampling distribution of the sample mean has a normal distribution

a. if $n \geq 30$ b. if $n/N \leq .05$ c. all the time

8. When samples are selected from a nonnormally distributed population, the sampling distribution of the sample mean has an approximately normal distribution

a. if $n \geq 30$ b. if $n/N \leq .05$ c. always

9. In a sample of 200 customers of a mail-order company, 174 are found to be satisfied with the service they receive from the company. The proportion of customers in this sample who are satisfied with the company's service is

a. .87 b. .174 c. .148

10. The mean of the sampling distribution of $\hat{p}$ is always equal to

a. p b. μ c. $\hat{p}$

11. The condition for the standard deviation of the sampling distribution of the sample proportion to be $\sqrt{pq/n}$ is

a. $np > 5$ and $nq > 5$ b. $n > 30$ c. $n/N \leq .05$

12. The sampling distribution of $\hat{p}$ is approximately normal if

a. $np > 5$ and $nq > 5$ b. $n > 30$ c. $n/N \leq .05$

13. The sampling distribution is the probability distribution of a

a. sample with replacement

b. sample without replacement

c. sample statistic

14. An error that occurs because of chance is called a

a. nonsampling error b. sampling error c. mean error

15. The mean age of all students at a university is 24 years. The mean age of a random sample of 100 students selected from this university is 23.6 years. The difference ($23.6 - 24 = -.4$) is a

a. probability error

b. nonsampling error

c. sampling error

16. A population consists of five elements: 5, 6, 8, 10, and 13. You choose a random sample of three elements from this population (without replacement). Which of the following is *not* a possible value of the sample mean?

a. 6.33 b. 9.33 c. 10.67

17. The mean of the sampling distribution of the sample mean is the mean of

a. the means of all possible samples of the same size taken from the population

b. the frequency distribution of the population

c. the means of all frequency distributions

18. When the sample size is greater than 1, the standard deviation of the sampling distribution of the sample means is always

a. equal to the standard deviation of the population

b. smaller than the standard deviation of the population

c. greater than the standard deviation of the population

19. The standard deviation of the sampling distribution of the sample mean for a sample size of n drawn from a population with a mean of μ and a standard deviation of σ and $n/N \leq .05$ is

a. σ/n b. $\sigma/\sqrt{n}$ c. σ/n^2

20. If the population from which samples are drawn is normally distributed, then the sampling distribution of the sample mean is

a. normally distributed if n is 30 or larger

b. always normally distributed

c. normally distributed if n is less than 30

21. According to the central limit theorem, the sampling distribution of the sample mean is approximately normal, irrespective of the shape of the population distribution, if

a. n is 30 or larger

b. n is less than 30

c. np and nq both are greater than 5

22. A continuous random variable x has a normal distribution with a mean of 90 and a standard deviation of 15. The sampling distribution of the sample mean for a sample of 16 elements taken from this population is

a. normal

b. skewed to the right

c. skewed to the left

23. A continuous random variable x has a left-skewed distribution with a mean of 155 and a standard deviation of 28. The sampling distribution of the sample mean for a sample of 75 elements taken from this population is

a. approximately normal

b. skewed to the right

c. skewed to the left

24. If you divide the number of elements in a population with a specific characteristic by the total number of elements in the population, the result is the population

a. mean b. proportion c. distribution

25. If you divide the number of elements in a sample with a specific characteristic by the total number of elements in the sample, the result is the

a. sample mean

b. sample proportion

c. sample distribution

Practice Exercises

1. For a continuous random variable x, the population mean and the population standard deviation are 75 and 20, respectively. Assuming $n/N \leq .05$, the standard deviation of the sampling distribution of the sample mean for a sample of 16 elements taken from this population, rounded to two decimals, is ________.

2. The weights of all babies born at a hospital have a mean of 7.5 pounds and a standard deviation of .64 pounds. The mean of the sampling distribution of the mean weight of a sample of 31 babies born at this hospital is ________.

3. A population has a mean of 100.5 and a standard deviation of 27.0. Assuming $n/N \leq .05$, the probability, rounded to four decimal places, that the sample mean of a sample of size 67 elements selected from this population will be between 91 and 97 is ________.

4. The number of hours spent per week on household chores by all adults has a mean of 26.5 hours and a standard deviation of 7.8 hours. The probability, rounded to four decimal places, that the mean number of hours spent per week on household chores by a sample of 37 adults will be more than 26.75 is ________.

5. A continuous random variable x has a normal distribution with a mean of 59 and a standard deviation of 17. In the sampling distribution of the sample mean for a sample of 16 elements, what is the z score for a sample mean of 53.05?

6. A company has 506 employees and 196 of them are college graduates. The proportion of employees of this company who are college graduates, rounded to three decimal places, is ________.

7. In a sample of 781 adults, 529 are in favor of banning all advertisements relating to alcohol. The proportion of these adults who are in favor of banning all advertisements relating to alcohol, rounded to three decimal places, is ________.

8. Forty percent of all students at a large university live on campus. Suppose you extract a sample of 104 students from this university. The sample proportion is the proportion of students in this sample who live on campus. The standard deviation of the sampling distribution of this sample proportion, rounded to four decimal places, is ________.

9. Thirty-five percent of all adults read at least one newspaper daily. Suppose you select a sample of 554 adults. The sample proportion is the proportion of adults in this sample who read at least one newspaper daily. The standard deviation of the sampling distribution of this sample proportion, rounded to four decimal places, is ________.

10. There are 2000 bicycles in a community. The owners of 1250 of these bicycles also own locks for the bikes. A sample of 520 bicycles is chosen, and it just so happens that the proportion of bicycles in the sample that have locks is the same as that for the entire community. How many of the bicycles in the sample have locks?

11. A company has 15,000 employees and 9900 of them have completed a particular training course. Management decides to choose a random sample from all employees. What is the minimum number of employees that must be in the sample in order to use the central limit theorem to approximate the distribution of the proportion in the sample who have taken the training course?

12. Suppose that 8.9% or all persons are allergic to penicillin. A sample of 955 persons is selected. The probability that 7.5% to 9% of persons in a sample will be allergic to penicillin, rounded to four decimal places, is approximately ________.

13. A study shows 22.0% of all adult males did not visit their physicians' offices last year. A sample of 467 adult males is selected. The probability that less than 22.5% of adult males in the sample did not visit their physicians' offices last year, rounded to four decimal places, is approximately ________.

14. Briefly state and explain the central limit theorem.

15. According to a Kaiser Family Foundation Health Benefits Survey released in 2019, the total mean cost of employer-sponsored family health coverage was $20,576 per family per year, of which workers were paying an average of $6015. Suppose that currently the distribution of premiums paid by all workers who have employer-sponsored family health coverage is approximately normal with the mean and standard deviation of $6015 and $700, respectively. Let $\bar{x}$ be the average premium paid by a random sample of certain workers who have employer-sponsored family health coverage. Calculate the mean and standard deviation of $\bar{x}$ and describe the shape of its sampling distribution when the sample size is

a. 20 b. 100 c. 800

16. According to a U.S. Bureau of Labor Statistics release of September 4, 2019, statisticians earned an average of $88,190 a year. Suppose that the current annual earnings of all statisticians have the mean and standard deviation of $88,190 and $20,000, respectively, and the shape of this distribution is skewed to the right. Let $\bar{x}$ be the average earnings of a random sample of a certain number of statisticians. Calculate the mean and standard deviation of $\bar{x}$ and describe the shape of its sampling distribution when the sample size is

a. 20 b. 100 c. 800

17. According to a U.S. Bureau of Labor Statistics release of September 4, 2019, statisticians earned an average of $88,190 a year. Suppose that the current annual earnings of all statisticians have the mean and standard deviation of $88,190 and $20,000, respectively, and the shape of this distribution is skewed to the right. Find the probability that the mean earnings of a random sample of 256 statisticians is

a. between $85,600 and $86,800
b. within $1500 of the population mean
c. $89,500 or more
d. not within $1600 of the population mean
e. less than $87,000
f. less than $89,000
g. more than $90,000
h. between $87,000 and 89,500

18. At Jen and Perry Ice Cream Company, the machine that fills 1-pound cartons of Top Flavor ice cream is set to dispense 16 ounces of ice cream into every carton. However, some cartons contain slightly less than and some contain slightly more than 16 ounces of ice cream. The amounts of ice cream in all such cartons have an approximate normal distribution with a mean of 16 ounces and a standard deviation of .18 ounce.

a. Find the probability that the mean amount of ice cream in a random sample of 16 such cartons will be
   i. between 15.90 and 15.95 ounces
   ii. less than 15.95 ounces
   iii. more than 15.97 ounces
b. What is the probability that the mean amount of ice cream in a random sample of 16 such cartons will be within .10 ounce of the population mean?
c. What is the probability that the mean amount of ice cream in a random sample of 16 such cartons will be less than the population mean by .135 ounce or more?

19. According to the Stanford School of Medicine Blood Center, the most common blood type in the United States is O-positive, and approximately 37.4% of Americans have blood type O-positive (https://www.healthline.com/health/rarest-blood-type#rarest-type). Let $\hat{p}$ be the proportion of Americans in a random sample who have

blood type O-positive. Find the mean and standard deviation of the sampling distribution of $\hat{p}$ and describe its shape when the sample size is

a. 25 **b.** 250 **c.** 2500

20. According to a Gallup poll conducted January to June, 2018, 34% of U.S. workers are "engaged" workers, meaning that they are involved in, enthusiastic about, and committed to their work and workplace (https://news.gallup.com/poll/241649/employee-engagement-rise.aspx). The respondents were adults aged 18 and older living in all 50 U.S. states and the District of Columbia and working full or part time for an organization. Assume that this result is true for the current population of American workers.

a. Find the probability that in a random sample of 1150 American workers, the proportion of engaged workers is

i. more than .35 **ii.** between .32 and .37

iii. less than .31 **iv.** between .31 and .36

v. less than .36 **vi.** more than .32

b. What is the probability that in a random sample of 1150 American workers, the proportion of engaged workers is within .025 of the population proportion?

c. What is the probability that in a random sample of 1150 American workers, the proportion of engaged workers is within .03 of the population proportion?

d. What is the probability that in a random sample of 1150 American workers, the proportion of engaged workers is greater than the population proportion by .02 or more?

Mini-Projects

Mini-Project 7.1

Refer to Data Set III on the NFL players from the 2019 football season that accompanies this text (see Appendix A). Consider the weights of the players listed in column 5.

a. Compute μ and σ for these data.

b. Take 20 random samples of five players each, and find $\bar{x}$ for each sample.

c. Compute the mean and standard deviation of the 20 sample means obtained in part (b).

d. Using the formulas given in Section 7.2, find $\mu_{\bar{x}}$ and $\sigma_{\bar{x}}$ for $n = 5$.

e. How do your values of $\mu_{\bar{x}}$ and $\sigma_{\bar{x}}$ in part (d) compare with those in part (c)?

f. What percentage of the 20 sample means found in part (b) lie in the interval $\mu_{\bar{x}} - \sigma_{\bar{x}}$ to $\mu_{\bar{x}} + \sigma_{\bar{x}}$? In the interval $\mu_{\bar{x}} - 2\sigma_{\bar{x}}$ to $\mu_{\bar{x}} + 2\sigma_{\bar{x}}$? In the interval $\mu_{\bar{x}} - 3\sigma_{\bar{x}}$ to $\mu_{\bar{x}} + 3\sigma_{\bar{x}}$?

g. How do the percentages in part (f) compare to the corresponding percentages for a normal distribution (68.26%, 95.44%, and 99.74%, respectively)?

h. Repeat parts (b) through (g) using 20 samples of 10 players each.

Mini-Project 7.2

Refer to Data Set V that contains information on the 200 highest grossing films of 2018 (see Appendix A). Let p denote the proportion of these 200 films that grossed over $100,000,000.

a. Find p.

b. Select 20 random samples of five films each, and find the sample proportion $\hat{p}$ for each sample.

c. Compute the mean and standard deviation of the 20 sample proportions obtained in part (b).

d. Using the formulas given in Section 7.5.3, compute $\mu_{\hat{p}}$ and $\sigma_{\hat{p}}$. Is the finite population correction factor required here?

e. Compare your mean and standard deviation of $\hat{p}$ from part (c) with the values calculated in part (d).

f. Repeat parts (b) through (e) using 30 samples of 25 films each.

Mini-Project 7.3

You are to conduct the experiment of sampling 10 times (with replacement) from the numbers 0, 1, 2, 3, 4, 5, 6, 7, 8, and 9. You can do this in a variety of ways. One way is to write each digit on a separate piece of paper, place all the slips in a hat, and select 10 times from the hat, returning each selected slip before the next pick. As alternatives, you can use a 10-sided die, statistical software, or a calculator that generates random numbers. Perform the experiment using any of these methods, and compute the sample mean $\bar{x}$ for the 10 numbers obtained. Now, repeat the procedure 49 more times. When you are done, you will have 50 sample means.

a. Make a table of the population distribution for the 10 numbers, and display it using a graph.

b. Make a histogram of your 50 sample means. What shape does it have?

c. What does the central limit theorem say about the shape of the sampling distribution of $\bar{x}$? What mean and standard deviation does the sampling distribution of $\bar{x}$ have in this problem?

Decide for Yourself

Note: The Decide for Yourself feature is located in the book's product page on *www.wiley.com*.

Chapter 7 Technology Instructions

Note: Complete TI-84, Minitab, and Excel manuals are available in the book's product page on *www.wiley.com*.

TI-84 Color/TI-84

The TI-84 Color Technology Instructions feature of this text is written for the TI-84 Plus C color graphing calculator running the 4.0 operating system. Some screens, menus, and functions will be slightly different in older operating systems. The TI-84+ can perform all of the same functions but does not have the "Color" option referenced in some of the menus.

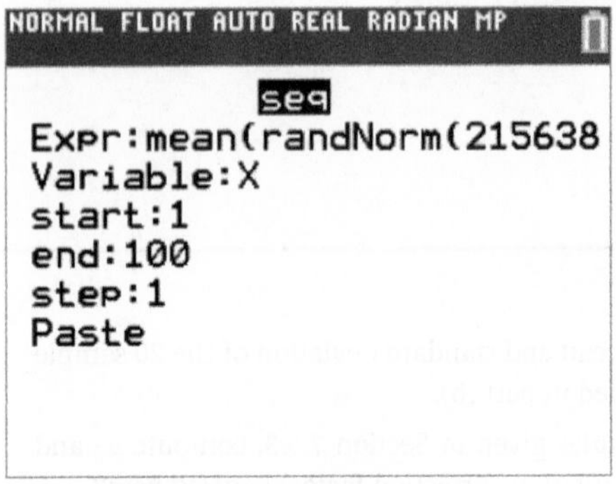

SCREEN 7.1

Approximating the Sampling Distribution of $\bar{x}$ for Example 7.3(a) of the Text by Selecting 100 Samples

1. Select **2nd** > **STAT** > **OPS** > **seq(**.
2. Use the following settings in the **seq** menu (see **Screen 7.1** and **Screen 7.2**):
   - Use the following settings at the **Expr** prompt:
     - Select **2nd** > **STAT** > **MATH** > **mean(**.
     - Select **MATH** > **PROB** > **randNorm(**.
     - Type 215638, 20000, 16)). Be sure to include two right parentheses in this step.
   - At the **Variable** prompt, type X.
   - At the **start** prompt, type 1.
   - At the **end** prompt, type 100.
   - At the **step** prompt, type 1.
   - Highlight **Paste** and press **ENTER**.
3. Select **STO**.
4. Select **2nd** > **1**.
5. Press **ENTER**.

   Note: This process will take a little while to complete. Be patient.
6. There should now be 100 sample means stored in list 1 of your calculator. You can create a histogram of the values in list 1 and compute statistics as described in previous chapters.

   Note: You have only created an approximate sampling distribution. To create the complete sampling distribution, you would need to have each sample mean from every possible sample of 16 American internal medicine physicians.

   Note: You can complete parts (b) and (c) of this example by changing the number 16 in Step 2 to 50 and 1000, respectively.

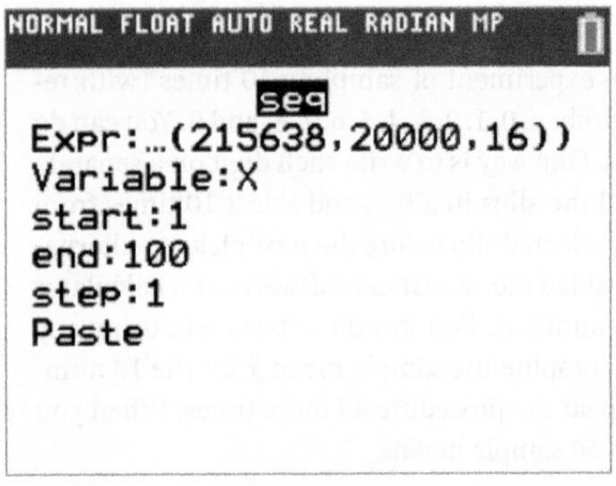

SCREEN 7.2

Approximating the Sampling Distribution of $\hat{p}$ for Example 7.9 of the Text by Selecting 100 Samples

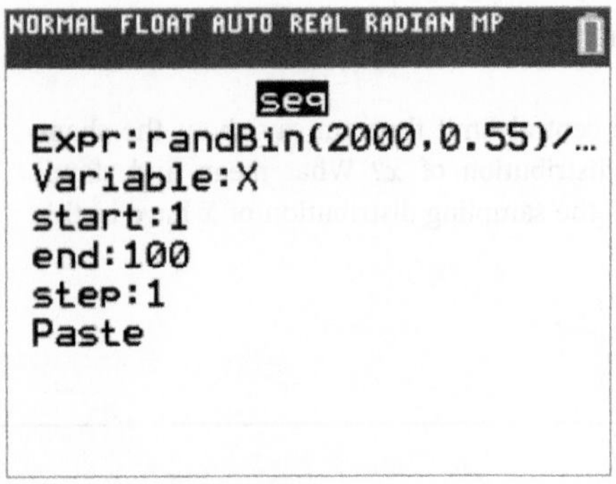

SCREEN 7.3

1. Select **2nd** > **STAT** > **OPS** > **seq(**.
2. Use the following settings in the **seq** menu (see **Screen 7.3**):
   - Use the following settings at the **Expr** prompt:
     - Select **MATH** > **PROB** > **randBin(**.
     - Type 2000,0.55)/2000.
   - At the **Variable** prompt, type X.
   - At the **start** prompt, type 1.
   - At the **end** prompt, type 100.
   - At the **step** prompt, type 1.
   - Highlight **Paste** and press **ENTER**.
3. Select **STO**.
4. Select **2nd** > **1**.

5. Press **ENTER**.

   Note: This process will take a little while to complete. Be patient.

6. There should now be 100 sample proportions stored in list 1 of your calculator. You can create a histogram of the values in list 1 and compute statistics as described in previous chapters.

   Note: You have only created an approximate sampling distribution. To create the complete sampling distribution, you would need to have each sample proportion from every possible sample of size 2000 from this population.

Minitab

The Minitab Technology Instructions in this text are written for **Minitab Professional Version 19**. Some screens, menus, and functions will be slightly different in older (or in newer) versions of Minitab. If you are using **Minitab Express**, you can download the **Minitab Express** Manual from the book's product page on *www.wiley.com*.

Approximating the Sampling Distribution of $\overline{x}$ for Example 7.3(a) of the Text by Selecting 100 Samples

1. Select **Calc** > **Random Data** > **Normal**.
2. Use the following settings in the dialog box that appears on screen (see **Screen 7.4**):
   - In the **Number of rows of data to generate** box, type 100.
   - In the **Store in column(s)** box, type C1-C16.
   - In the **Mean** box, type 215638.
   - In the **Standard deviation** box, type 20000.
3. Click **OK**.
4. Select **Calc** > **Row Statistics**.
5. Use the following settings in the dialog box that appears on screen (see **Screen 7.5**):
   - In the **Statistic** box, select Mean.
   - In the **Input variables** box, type C1-C16.
   - In the **Store result in** box, type C17.
6. Click **OK**.
7. There should now be 100 sample means stored in C17. You can create a histogram of the values in C17 and compute statistics as described in previous chapters.

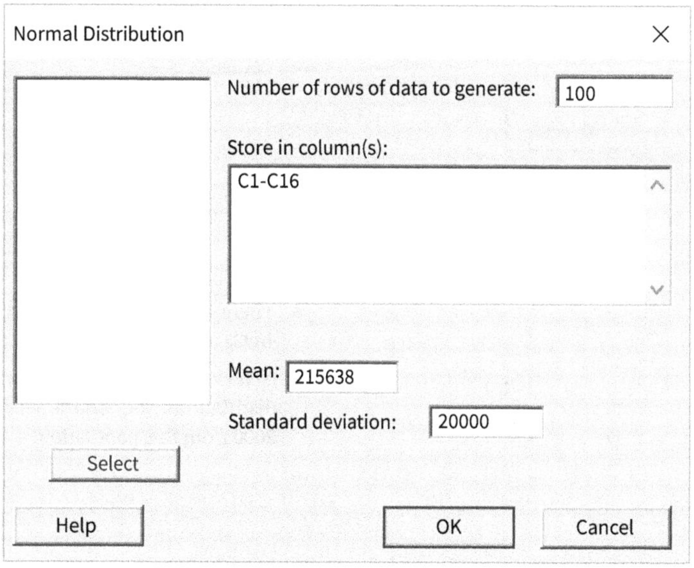

SCREEN 7.4

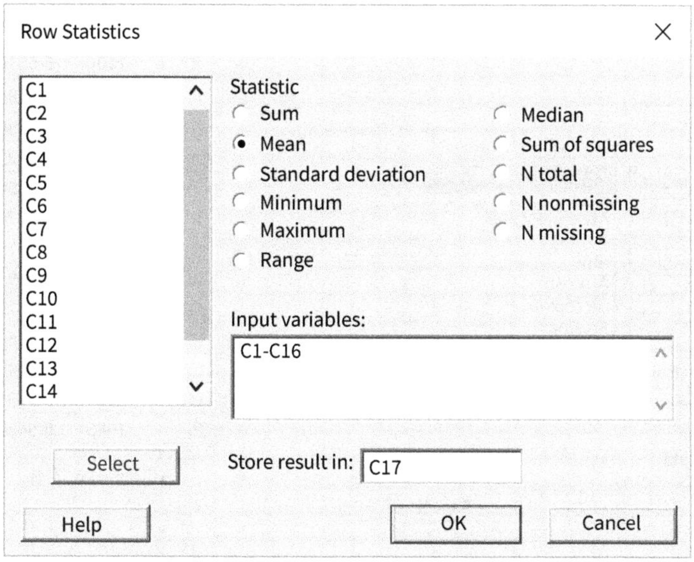

SCREEN 7.5

Note: You have only created an approximate sampling distribution. To create the complete sampling distribution, you would need to have each sample mean from every possible sample of 16 American internal medicine physicians.

Note: You can complete parts (b) and (c) of this example by changing the number 16 in Step 2 to 50 and 1000, respectively, and the number 17 in Step 5 to 51 and 1001, respectively.

Approximating the Sampling Distribution of $\hat{p}$ for Example 7.9 of the Text by Selecting 100 Samples

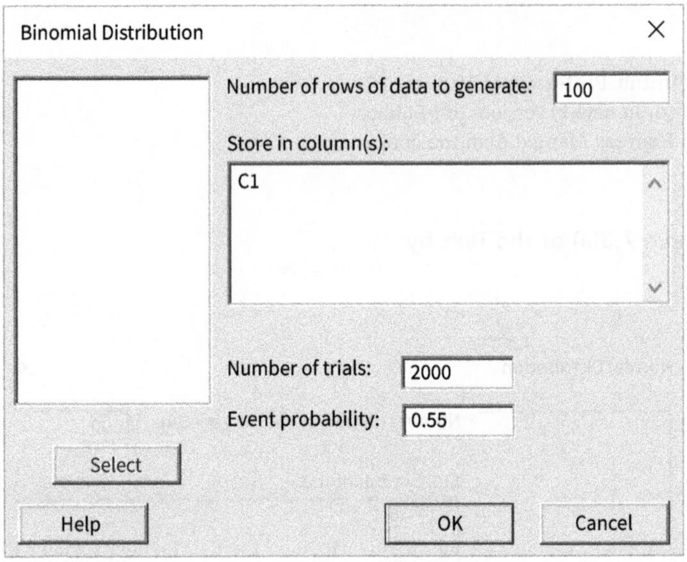

SCREEN 7.6

1. Select **Calc > Random Data > Binomial.**
2. Use the following settings in the dialog box that appears on screen (see **Screen 7.6**):
   - In the **Number of rows of data to generate** box, type 100.
   - In the **Store in column(s)** box, type C1.
   - In the **Number of trials** box, type 2000.
   - In the **Event probability** box, type 0.55.
3. Click **OK.**
4. Each row in C1 should now contain a number of successes from a sample of size 2000. We need to convert these numbers of successes into sample proportions of successes.
5. Select **Calc > Calculator.**
6. Use the following settings in the dialog box that appears on screen (see **Screen 7.7**):
   - In the **Store result in variable** box, type C2.
   - In the **Expression** box, type C1/2000.
7. Click **OK.**
8. There should now be 100 sample proportions stored in C2. You can create a histogram of the values in C2 and compute statistics as described in previous chapters. (See **Screen 7.8.**)

   Note: You have only created an approximate sampling distribution. To create the complete sampling distribution, you would need to have each sample proportion from every possible sample of size 2000 from this population.

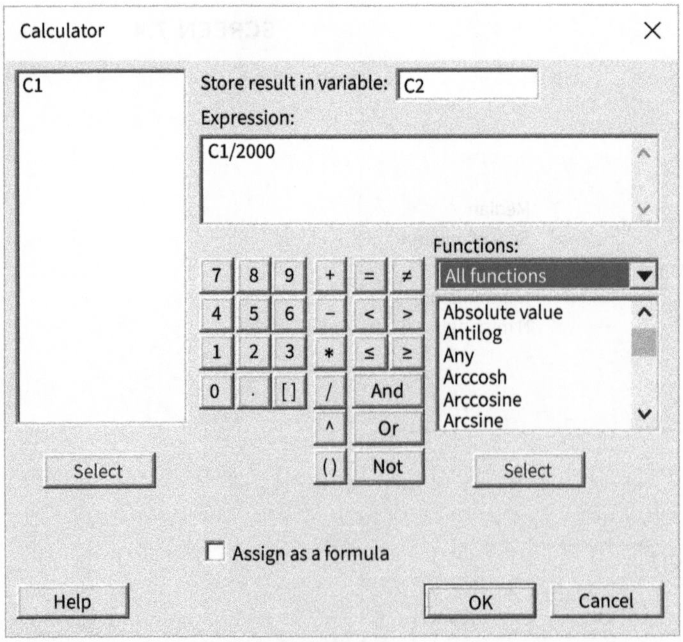

SCREEN 7.7

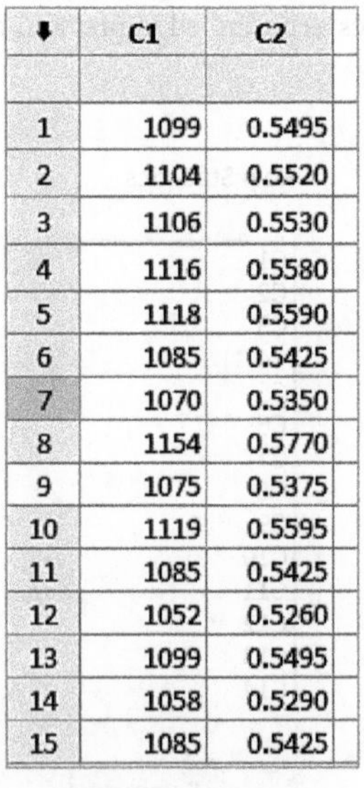

⬇	C1	C2
1	1099	0.5495
2	1104	0.5520
3	1106	0.5530
4	1116	0.5580
5	1118	0.5590
6	1085	0.5425
7	1070	0.5350
8	1154	0.5770
9	1075	0.5375
10	1119	0.5595
11	1085	0.5425
12	1052	0.5260
13	1099	0.5495
14	1058	0.5290
15	1085	0.5425

SCREEN 7.8

Excel

The Excel Technology Instructions feature of this text is written for Excel 2019. Some screens, menus, and functions will be slightly different in older versions of Excel. To enable some of the advanced Excel functions, you must enable the Data Analysis Add-In for Excel.

Approximating the Sampling Distribution of $\bar{x}$ for Example 7.3(a) of the Text by Taking 100 Samples

1. Click on cell A1.
2. Click **DATA** and then click **Data Analysis Tools** from the **Analysis** group.
3. Select **Random Number Generation** from the dialog box that appears on screen.
4. Use the following settings in the dialog box that appears on screen (see **Screen 7.9**):
   - In the **Number of Variables** box, type 100.
   - In the **Number of Random Numbers** box, type 16.
   - From the **Distribution** dropdown menu, select Normal.
   - In the **Mean** box, type 215638.
   - In the **Standard deviation** box, type 20000.
5. Click **OK**.
6. Click on cell A17 and type =AVERAGE(A1:A16).
7. Copy the formula from cell A17 to cells B17 through CV17.
8. There should now be 100 sample means stored in row 17 of the spreadsheet. You can create a histogram of the values in row 17 and compute statistics as described in previous chapters.

   Note: You have only created an approximate sampling distribution. To create the complete sampling distribution, you would need to have each sample mean from every possible sample of 16 American internal medicine physicians.

   Note: You can complete parts (b) and (c) of this example by changing the number 16 in Steps 4 and 6 to 50 and 1000, respectively, and changing the number 17 in Steps 6 and 7 to 51 and 1001, respectively.

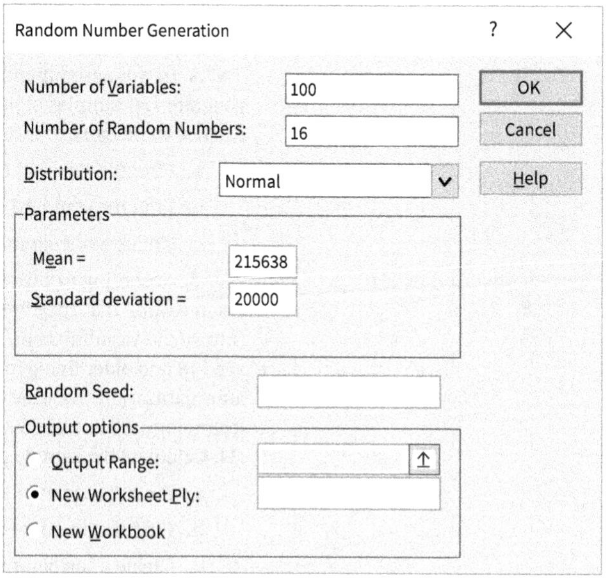

SCREEN 7.9

Approximating the Sampling Distribution of $\hat{p}$ for Example 7.9 of the Text by Taking 100 Samples

1. Click on cell A1.
2. Click **DATA** and then click **Data Analysis Tools** from the **Analysis** group.
3. Select **Random Number Generation** from the dialog box that appears on screen.
4. Use the following settings in the dialog box that appears on screen (see **Screen 7.10**):
   - In the **Number of Variables** box, type 1.
   - In the **Number of Random Numbers** box, type 100.
   - From the **Distribution** drop-down menu, select Binomial.
   - In the **p Value** box, type 0.55.
   - In the **Number of trials** box, type 2000.
5. Click **OK**.
6. Each cell in column A should now contain a number of successes from a sample of size 2000. We need to convert these numbers of successes into sample proportions of successes.
7. Click on cell B1 and type = A1/2000.
8. Copy the formula from cell B1 to cells B2 through B100.

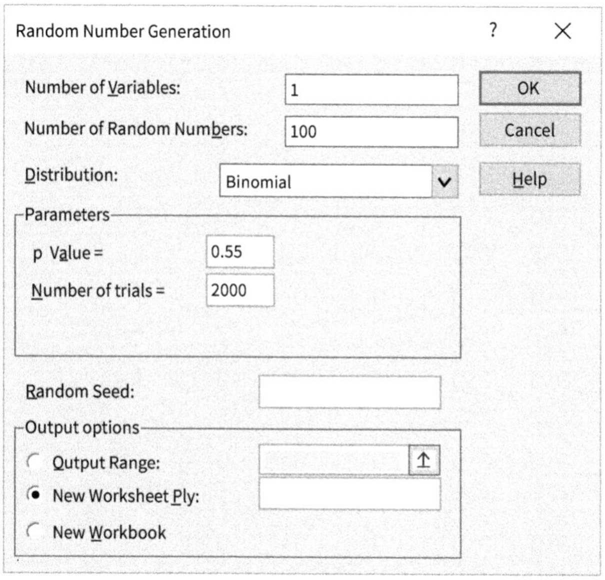

SCREEN 7.10

9. There should now be 100 sample proportions stored in column B. You can create a histogram of the values in column B and compute statistics as described in previous chapters.

   Note: You have only created an approximate sampling distribution. To create the complete sampling distribution, you would need to have each sample proportion from every possible sample of size 2000 from this population.

Technology Assignments

TA7.1 In this assignment, you will simulate rolling a six-sided die. Use technology to generate 200 samples of 30 rolls of a six-sided die. Now, use technology to find the sample mean for each set of 30 rolls.

a. Find the mean of the 200 sample means.

b. Find the standard deviation of the 200 sample means.

c. Create a histogram of the 200 sample means. Describe the shape revealed by the histogram.

TA7.2 In this assignment, you will simulate selecting random integers from 1 to 100. Use technology to generate 150 samples of 40 integers each from 1 to 100. Now, use technology to find the sample mean of each of 40 integers.

a. Find the mean of the 150 sample means.

b. Find the standard deviation of the 150 sample means.

c. Create a histogram of the 150 sample means. Describe the shape revealed by the histogram.

TA7.3 According to a Gallup poll conducted January to June, 2018, 34% of workers are "engaged" workers, meaning that they are involved in, enthusiastic about, and committed to their work and workplace (https://news.gallup.com/poll/241649/employee-engagement-rise.aspx). The respondents were adults aged 18 and older living in all 50 U.S. states and the District of Columbia and working full or part time for an organization. Assume that this result is true for the current population of American workers. Use technology to simulate 500 binomial experiments with 50 trials each and with a probability of success of .34. Calculate the sample proportion of success for each of the 500 experiments.

a. Find the mean of these 500 sample proportions.

b. Find the standard deviation of these 500 sample proportions.

c. Create a histogram of the 500 sample proportions. Describe the shape revealed by the histogram.

TA7.4 It has been conjectured that 10% of all people are left-handed. Use technology to simulate 200 binomial experiments with 20 trials each and a probability of success (observing a left-handed person) of .10. Calculate the sample proportion of successes for each of the 200 experiments.

a. Find the mean of these 200 sample proportions.

b. Find the standard deviation of the 200 sample proportions.

c. Create a histogram of the 200 sample proportions. Describe the shape revealed by the histogram.

CHAPTER

8

Estimation of the Mean and Proportion

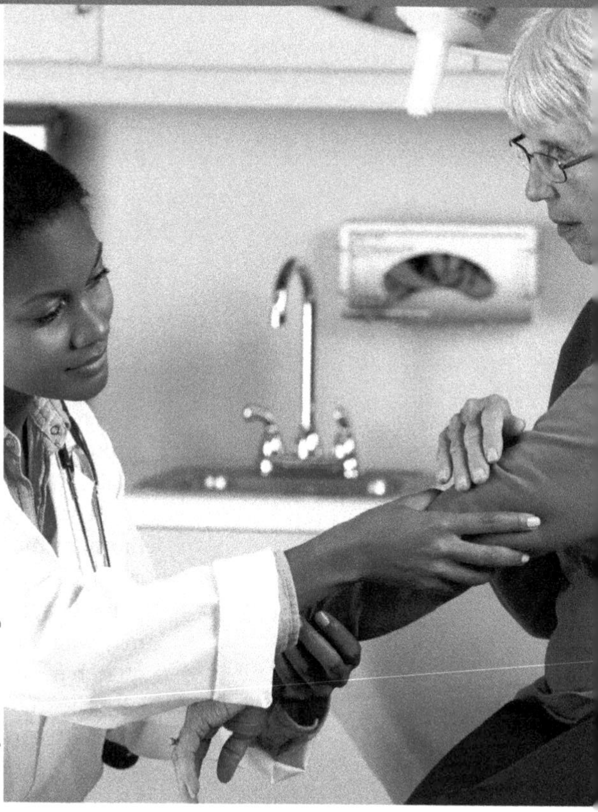
Monkey Business Images/Shutterstock.com

Do you know how much doctors earn? According to a survey by Nomad Health, the average salaries of U.S. doctors vary a lot based on their specialties. According to this survey, whereas the average salary of neurosurgeons in the United States was \$616,823, that of pediatrics was only \$222,942 in 2019. (See Case Study 8.1.)

Now we are entering that part of statistics called *inferential statistics*. In Chapter 1 inferential statistics was defined as the part of statistics that helps us make decisions about some characteristics of a population based on sample information. In other words, inferential statistics uses the sample results to make decisions and draw conclusions about the population from which the sample is drawn. Estimation is the first topic to be considered in our discussion of inferential statistics. Estimation and hypothesis testing (discussed in Chapter 9) taken together are usually referred to as inference making. This chapter explains how to estimate the population mean and population proportion for a single population.

8.1 Estimation, Point Estimate, and Interval Estimate

LEARNING OBJECTIVES

After completing this section, you should be able to:

- Explain the meaning of an estimator and an estimate.
- Explain the meaning of a point estimate.
- Explain the meaning of an interval estimate.
- State the form of a confidence interval.
- Explain the meaning of a confidence level.

In this section, first we discuss the concept of estimation and then the concepts of point and interval estimates.

8.1.1 Estimation: An Introduction

Estimation is a procedure by which a numerical value or values are assigned to a population parameter based on the information collected from a sample.

Estimation

The assignment of value(s) to a population parameter based on a value of the corresponding sample statistic is called *estimation*.

In inferential statistics, μ is called the **true population mean** and p is called the **true population proportion**. There are many other population parameters, such as the median, mode, variance, and standard deviation.

The following are a few examples of estimation: an auto company may want to estimate the mean fuel consumption for a particular model of a car; a manager may want to estimate the average time taken by new employees to learn a job; the U.S. Census Bureau may want to estimate the mean housing expenditure per month incurred by households; and a polling agency may want to estimate the proportion or percentage of adults who are in favor of raising taxes on rich people to reduce the budget deficit.

The examples about estimating the mean fuel consumption, estimating the average time taken to learn a job by new employees, and estimating the mean housing expenditure per month incurred by households are illustrations of estimating the *true population mean*, μ. The example about estimating the proportion (or percentage) of all adults who are in favor of raising taxes on rich people is an illustration of estimating the *true population proportion*, p.

If we can conduct a *census* (a survey that includes the entire population) each time we want to find the value of a population parameter, then the estimation procedures explained in this and subsequent chapters are not needed. For example, if the U.S. Census Bureau can contact every household in the United States to find the mean housing expenditure incurred by households, the result of the survey (which is actually a census) will give the value of μ, and the procedures learned in this chapter will not be needed. However, it is too expensive, very time consuming, or virtually impossible to contact every member of a population to collect information to find the true value of a population parameter. Therefore, we usually take a sample from the population and calculate the value of the appropriate sample statistic. Then we assign a value or values to the corresponding population parameter based on the value of the sample statistic. This chapter (and subsequent chapters) explains how to assign values to population parameters based on the values of sample statistics.

For example, to estimate the mean time taken to learn a certain job by new employees, the manager will take a sample of new employees and record the time taken by each of these employees to learn the job. Using this information, the manager will calculate the sample mean, $\bar{x}$. Then, based on the value of $\bar{x}$, the manager will assign certain values to μ. As another example, to estimate the mean housing expenditure per month incurred by all households in the United States, the Census Bureau will take a sample of households, collect the information on the housing expenditure that each of these households incurs per month, and compute the value of the sample mean, $\bar{x}$. Based on this value of $\bar{x}$, the bureau will then assign values to the population mean, μ. Similarly, the polling agency who wants to find the proportion or percentage of adults who are in favor of raising taxes on rich people to reduce the budget deficit will take a sample of adults and determine the value of the sample proportion, $\hat{p}$, which represents the proportion of adults in the sample who are in favor of raising taxes on rich people to reduce the budget deficit. Using this value of the sample proportion, $\hat{p}$, the agency will assign values to the population proportion, p.

The value(s) assigned to a population parameter based on the value of a sample statistic is called an **estimate** of the population parameter. For example, suppose the manager takes a sample of 40 new employees and finds that the mean time, $\bar{x}$, taken to learn this job for these employees is 5.5 hours. If he or she assigns this value to the population mean, then 5.5 hours is called an estimate of μ. The sample statistic used to estimate a population parameter is called an **estimator**. Thus, the sample mean, $\bar{x}$, is an estimator of the population mean, μ; and the sample proportion, $\hat{p}$, is an estimator of the population proportion, p.

Estimate and Estimator

The value(s) assigned to a population parameter based on the value of a sample statistic is called an *estimate*. The sample statistic used to estimate a population parameter is called an *estimator*.

The estimation procedure involves the following steps:

1. Select a sample.
2. Collect the required information from the members of the sample.
3. Calculate the value of the sample statistic.
4. Assign value(s) to the corresponding population parameter.

Remember, **the procedures to be learned in this chapter assume that the sample taken is a simple random sample**. If the sample is not a simple random sample, then the procedures to be used to estimate a population mean or proportion become more complex. These procedures are outside the scope of this book.

8.1.2 Point and Interval Estimates

An estimate may be a point estimate or an interval estimate. These two types of estimates are described in this section.

A Point Estimate If we select a sample and compute the value of a sample statistic for this sample, then this value gives the **point estimate** of the corresponding population parameter.

Point Estimate

The value of a sample statistic that is used to estimate a population parameter is called a *point estimate*.

Thus, the value of $\bar{x}$ calculated from a random sample is a point estimate of the corresponding population mean, μ. For the example mentioned earlier, suppose the Census Bureau takes a random sample of 10,000 households and determines that the mean housing expenditure per month, $\bar{x}$, for this sample is \$2970. Then, using $\bar{x}$ as a point estimate of μ, the Bureau can state that the mean housing expenditure per month, μ, for all households is about \$2970. Thus,

Point estimate of a population parameter = Value of the corresponding sample statistic

Each sample selected from a population is expected to yield a different value of the sample statistic. Thus, the value assigned to a population mean, μ, based on a point estimate depends on which of the various possible samples is drawn. Consequently, the point estimate assigns a value to μ that differs from sample to sample and is almost always different from the true value of the population mean.

An Interval Estimate In the case of **interval estimation**, instead of assigning a single value to a population parameter, an interval is constructed around the point estimate, and then a probabilistic statement that this interval contains the corresponding population parameter is made.

Interval Estimation

In *interval estimation*, an interval is constructed around the point estimate, and it is stated that this interval contains the corresponding population parameter with a certain confidence level.

For the example about estimating the mean housing expenditure, instead of saying that the mean housing expenditure per month for all households is \$2970, we may obtain an interval by subtracting a number from \$2970 and adding the same number to \$2970. Then we state that this interval contains the population mean, μ. For purposes of illustration, suppose we subtract \$340 from \$2970 and add \$340 to \$2970. Consequently, we obtain the interval (\$2970 − \$340) to (\$2970 + \$340), or \$2630 to \$3310. Then we state that the interval \$2630 to \$3310 is likely to contain the population mean, μ. This procedure is called **interval estimation**. The value \$2630 is called the **lower limit** of the interval, and \$3310 is called the **upper limit** of the interval. The number we add to and subtract from the point estimate is called the **margin of error** or the **maximum error of the estimate.** **Figure 8.1** illustrates the concept of interval estimation.

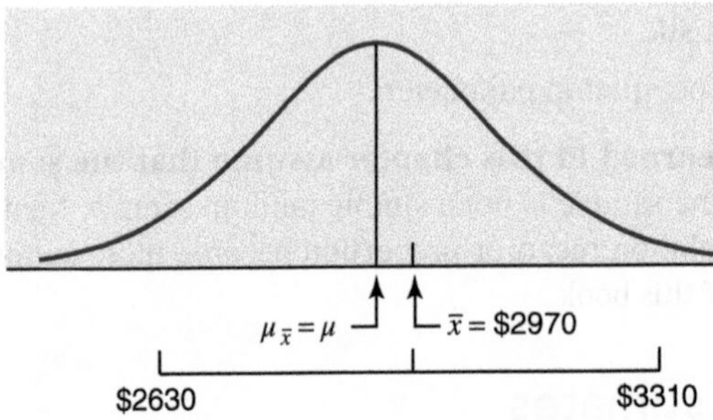

FIGURE 8.1 Interval estimation.

The question arises: What number should we subtract from and add to a point estimate to obtain an interval estimate? The answer to this question depends on two considerations:

1. The standard deviation $\sigma_{\bar{x}}$ of the sample mean, $\bar{x}$
2. The level of confidence to be attached to the interval

First, the larger the standard deviation of $\bar{x}$, the greater is the number subtracted from and added to the point estimate. Thus, it is obvious that if the spread of $\bar{x}$ is larger, then the interval constructed around $\bar{x}$ will be wider to include μ.

Second, the quantity subtracted and added must be larger if we want to have a higher confidence in our interval. We always attach a probabilistic statement to the interval estimation. This probabilistic statement is given by the **confidence level**. An interval constructed based on this confidence level is called a **confidence interval**.

Confidence Level and Confidence Interval

Each interval is constructed with regard to a given *confidence level* and is called a *confidence interval*. The confidence interval is given as

$$\text{Point estimate} \pm \text{Margin of error}$$

The confidence level associated with a confidence interval states how much confidence we have that this interval contains the true population parameter. The confidence level is denoted by $(1 - \alpha)100\%$.

As mentioned above, the confidence level is denoted by $\mathbf{(1 - \alpha)100\%}$, where α is the Greek letter **alpha**. When expressed as probability, it is called the *confidence coefficient* and is denoted by $1 - \alpha$. In passing, note that α is called the *significance level*, which will be explained in detail in Chapter 9.

Although any value of the confidence level can be chosen to construct a confidence interval, the more common values are 90%, 95%, and 99%. The corresponding confidence coefficients are .90, .95, and .99, respectively. The next section describes how to construct a confidence interval for the population mean when the population standard deviation, σ, is known.

Sections 8.2 and 8.3 discuss the procedures that are used to estimate a population mean μ. In Section 8.2 we assume that the population standard deviation σ is known, and in Section 8.3

we do not assume that the population standard deviation σ is known. In the latter situation, we use the sample standard deviation s to approximate the value of σ. In the real world, the population standard deviation σ is almost never known. Consequently, we (almost) always use the sample standard deviation s.

EXERCISES

Concepts and Procedures

8.1 Briefly explain the meaning of an estimator and an estimate.

8.2 Explain the meaning of a point estimate and an interval estimate.

8.2 Estimation of a Population Mean: σ Known

LEARNING OBJECTIVES

After completing this section, you should be able to:

- Determine a point estimate of the population mean μ.
- Verify the conditions for using the normal distribution to make a confidence interval for μ when σ is known.
- Construct a confidence interval for μ when σ is known.
- Identify the margin of error for a confidence interval for μ.
- Interpret a confidence interval for μ.
- Explain the relationship between sample size, confidence level, and the width of a confidence interval for μ.
- Explain various alternatives for decreasing the width of a confidence interval.
- Determine the minimum sample size to produce a confidence interval for μ with a predetermined margin of error.

This section explains how to construct a confidence interval for the population mean μ when the population standard deviation σ is known. Here, there are three possible cases, as follows:

Case I. If the following three conditions are fulfilled:

1. The population standard deviation σ is known
2. The sample size is small (i.e., $n < 30$)
3. The population from which the sample is selected is approximately normally distributed

then we use the normal distribution to make the confidence interval for μ because from Section 7.3.1 of Chapter 7 the sampling distribution of $\bar{x}$ is normal with its mean equal to $\mu_{\bar{x}} = \mu$ and the standard deviation equal to $\sigma_{\bar{x}} = \sigma/\sqrt{n}$, assuming that $n/N \leq .05$.

Case II. If the following two conditions are fulfilled:

1. The population standard deviation σ is known
2. The sample size is large (i.e., $n \geq 30$)

then, again, we use the normal distribution to make the confidence interval for μ because from Section 7.3.2 of Chapter 7, due to the central limit theorem, the sampling distribution of $\bar{x}$ is (approximately) normal with its mean equal to $\mu_{\bar{x}} = \mu$ and the standard deviation equal to $\sigma_{\bar{x}} = \sigma/\sqrt{n}$, assuming that $n/N \leq .05$.

Case III. If the following three conditions are fulfilled:

1. The population standard deviation σ is known
2. The sample size is small (i.e., $n < 30$)

3. The population from which the sample is selected is not normally distributed (or its distribution is unknown)

then we use a nonparametric method to make the confidence interval for μ. Such procedures are covered in Chapter 15.

This section will cover the first two cases. The procedure for making a confidence interval for μ is the same in both these cases. Note that in Case I, the population does not have to be exactly normally distributed. As long as the distribution is close to the normal distribution without any outliers, we can use the normal distribution procedure. In Case II, although 30 is considered a large sample, if the population distribution is very different from the normal distribution, then 30 may not be a large enough sample size for the sampling distribution of $\bar{x}$ to be normal and, hence, to use the normal distribution.

The following chart summarizes the above three cases.

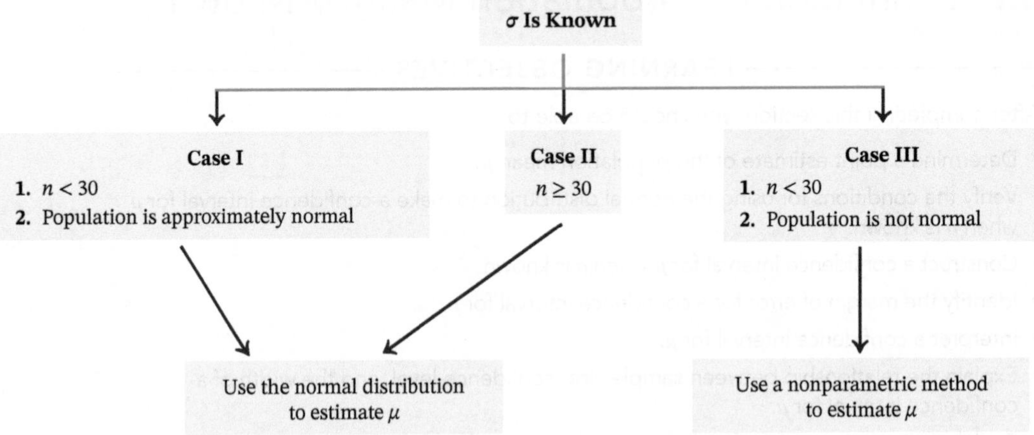

Confidence Interval for μ

The $(1 - \alpha)100\%$ *confidence interval for* μ under Cases I and II is

$$\bar{x} \pm z\sigma_{\bar{x}}$$

where $\sigma_{\bar{x}} = \frac{\sigma}{\sqrt{n}}$ assuming $\frac{n}{N} \le .05$

The value of z used here is obtained from the standard normal distribution table (Table IV of Appendix B) for the given confidence level.

The quantity $z\sigma_{\bar{x}}$ in the confidence interval formula is called the **margin of error** and is denoted by E.

Margin of Error

The margin of error for the estimate of μ, denoted by E, is the quantity that is subtracted from and added to the value of $\bar{x}$ to obtain a confidence interval for μ. Thus,

$$E = z\sigma_{\bar{x}}$$

The value of z in the confidence interval formula is obtained from the standard normal distribution table (Table IV of Appendix B) for the given confidence level. To illustrate, suppose we want to construct a 95% confidence interval for μ. A 95% confidence level means that the total area under the standard normal curve between two points (at the same distance) on different sides of μ is 95%, or .95, as shown in **Figure 8.2.** Note that we have denoted these two points by $-z$ and z in Figure 8.2. To find the value of z for a 95% confidence level, we first find the areas to the left of these two points, $-z$ and z. Then we find the z values for these two areas from the normal distribution table. Note that these two values of z will be the same but with opposite signs. To find these values of z, we perform the following two steps:

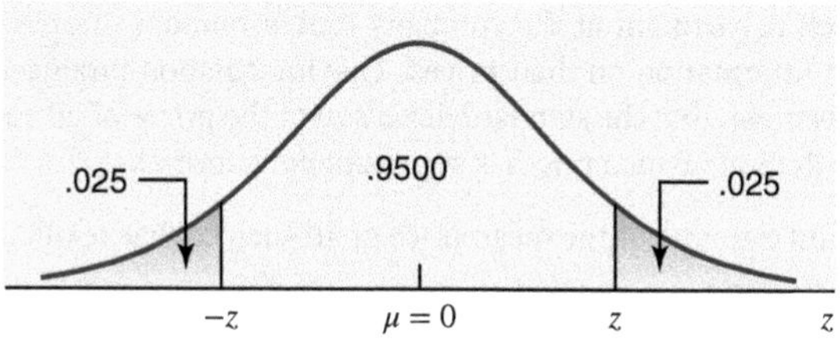

FIGURE 8.2 Finding z for a 95% confidence level.

1. The first step is to find the areas to the left of $-z$ and z, respectively. Note that the area between $-z$ and z is denoted by $1 - \alpha$. Hence, the total area in the two tails is α because the total area under the curve is 1.0. Therefore, the area in each tail, as shown in **Figure 8.3,** is $\alpha/2$. In our example, $1 - \alpha = .95$. Hence, the total area in both tails is $\alpha = 1 - .95 = .05$. Consequently, the area in each tail is $\alpha/2 = .05/2 = .025$. Then, the area to the left of $-z$ is .0250, and the area to the left of z is $.0250 + .95 = .9750$.
2. Next, using Table IV of Appendix B or technology, find the z values such that the areas to the left of $-z$ and z are .0250 and .9750, respectively. These z values are -1.96 and 1.96, respectively.

Thus, for a confidence level of 95%, we will use $z = 1.96$ in the confidence interval formula.

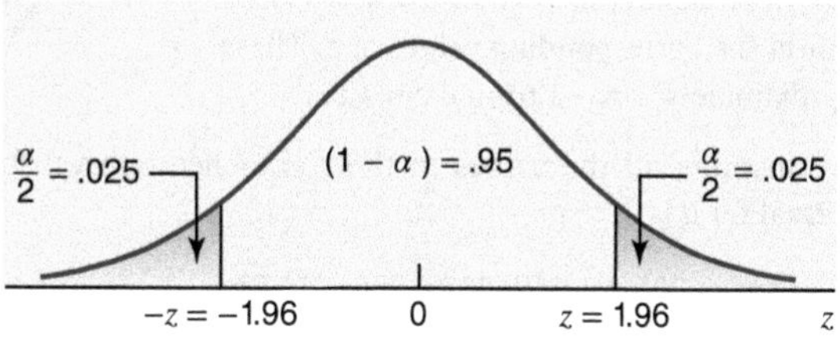

FIGURE 8.3 Area in the tails.

Table 8.1 lists the z values for some of the most commonly used confidence levels. **Note that we always use the positive value of z in the formula.**

TABLE 8.1 z Values for Commonly Used Confidence Levels

Confidence Level	Areas to Look for in Table IV	z Value
90%	.0500 and .9500	1.64 or 1.65
95%	.0250 and .9750	1.96
96%	.0200 and .9800	2.05
97%	.0150 and .9850	2.17
98%	.0100 and .9900	2.33
99%	.0050 and .9950	2.57 or 2.58

Example 8.1 describes the procedure used to construct a confidence interval for μ when σ is known, the sample size is small, but the population from which the sample is drawn is approximately normally distributed.

EXAMPLE 8.1 Prices of Textbooks

Finding the Point Estimate and Confidence Interval for μ: σ Known, $n < 30$, and Population Normal

A publishing company has just published a new college textbook. Before the company decides the price at which to sell this textbook, it wants to know the average price of all such textbooks in the market. The research department at the company took a random sample of 25 comparable textbooks and collected information on their prices. This information produced a mean price of $205 for this sample. It is known that the standard deviation of the prices of all such textbooks is $47 and the population distribution of such prices is approximately normal.

a. What is the point estimate of the mean price of all such college textbooks?

b. Construct a 90% confidence interval for the mean price of all such college textbooks.

Solution Here, σ is known and, although $n < 30$, the population is approximately normally distributed. Hence, we can use the normal distribution. From the given information,

$$n = 25, \quad \bar{x} = \$205, \quad \text{and} \quad \sigma = \$47$$

The standard deviation of $\bar{x}$ is

$$\sigma_{\bar{x}} = \frac{\sigma}{\sqrt{n}} = \frac{47}{\sqrt{25}} = \$9.40$$

iStock.com/Oleg Prikhodko

a. The point estimate of the mean price of all such college textbooks is $205; that is,

$$\text{Point estimate of } \mu = \bar{x} = \mathbf{\$205}$$

b. The confidence level is 90%, or .90. First we find the z value for a 90% confidence level. Here, the area in each tail of the normal distribution curve is $\alpha/2 = (1 - .90)/2 = .05$. Now in Table IV of Appendix B, look for the areas .0500 and .9500 and find the corresponding values of z. These values are (approximately) $z = -1.65$ and $z = 1.65$.[1]

Next, we substitute all the values in the confidence interval formula for μ. The 90% confidence interval for μ is

$$\bar{x} \pm z\sigma_{\bar{x}} = 205 \pm 1.65(9.40) = 205 \pm 15.51$$
$$= (205 - 15.51) \text{ to } (205 + 15.51) = \mathbf{\$189.49 \text{ to } \$220.51}$$

Thus, we are 90% confident that the mean price of all such college textbooks is between $189.49 and $220.51. Note that we cannot say for sure whether the interval $189.49 to $220.51 contains the true population mean or not. Since μ is a constant, we cannot say that the probability is .90 that this interval contains μ because either the interval contains μ or it does not. Consequently, the probability that this interval contains μ is either 1.0 or 0. All we can say is that we are 90% confident that the mean price of all such college textbooks is between $189.49 and $220.51.

In the above estimate, $15.51 is called the margin of error or give-and-take figure.

How do we interpret a 90% confidence level? In terms of Example 8.1, if we take all possible samples of 25 such college textbooks each and construct a 90% confidence interval for μ around each sample mean, we can expect that 90% of these intervals will include μ and 10% will not. In **Figure 8.4** we show means $\bar{x}_1$, $\bar{x}_2$, and $\bar{x}_3$ of three different samples of the same size drawn from the same population. Also shown in this figure are the 90% confidence intervals constructed around these three sample means. As we observe, the 90% confidence intervals constructed around $\bar{x}_1$ and $\bar{x}_2$ include μ, but the confidence interval

[1]Note that there is no apparent reason for choosing .0495 and .9505 and not choosing .0505 and .9495 in Table IV. If we choose .0505 and .9495, the z values will be −1.64 and 1.64. An alternative is to use the average of 1.64 and 1.65, which is 1.645, which we will not do in this text.

constructed around $\bar{x}_3$ does not include μ. We can state for a 90% confidence level that if we take many samples of the same size from a population and construct 90% confidence intervals around the means of these samples, then we expect 90% of these confidence intervals to be like the ones around $\bar{x}_1$ and $\bar{x}_2$ in Figure 8.4, which include μ, and 10% will be like the one around $\bar{x}_3$, which does not include μ.

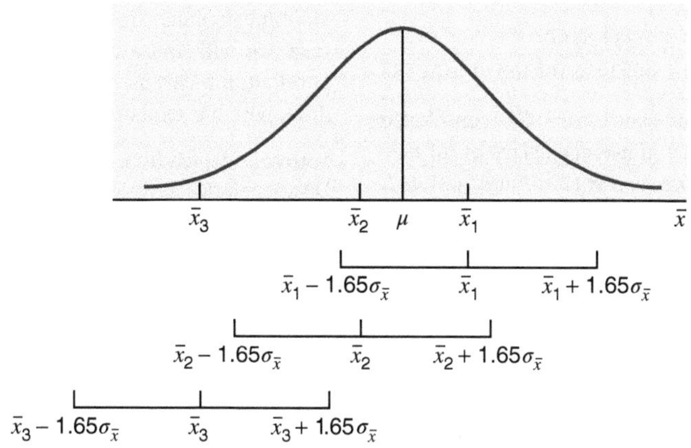

FIGURE 8.4 Confidence intervals.

Case Study 8.1

2019 National Average Salaries of U.S. Doctors

The chart, based on a survey by Nomad Health, lists the 2019 U.S. national average salaries of doctors with six different specialties. The national average salaries of U.S. doctors vary a lot based on their specialties. According to this survey, whereas the 2019 national average salary of neurosurgeons was \$616,823, that of pediatrics was only \$222,942. As we know, such estimates are based on sample surveys. If we know the sample size and the population standard deviation for doctors with a given specialty, we can make a confidence interval for the 2019 national average salary of all doctors with that specialty. For example, if we know the sample size on which the estimate of the 2019 national average salary of neurosurgeons given in the graph is based and the population standard deviation of the 2019 salaries of all U.S. neurosurgeons, we can calculate a confidence interval for the 2019 national average salary of all U.S. neurosurgeons using the formula:

$$\bar{x} \pm z\sigma_{\bar{x}}$$

In this formula, we can substitute the values of $\bar{x}$, z, and $\sigma_{\bar{x}}$ to obtain the confidence interval. Remember that $\sigma_{\bar{x}} = \dfrac{\sigma}{\sqrt{n}}$.

Suppose we want to find a 98% confidence interval for the 2019 national average salary of all neurosurgeons. Further suppose that the estimate of the 2019 national average salary of neurosurgeons (given in the chart) is based on a random sample of 1600 U.S. neurosurgeons and that the population standard deviation for the 2019 salaries of all U.S. neurosurgeons is \$80,000. Then the 98% confidence interval for the corresponding population mean is calculated as follows:

$$\sigma_{\bar{x}} = \frac{\sigma}{\sqrt{n}} = \frac{80,000}{\sqrt{1600}} = \$2000$$

$$\bar{x} \pm z\,\sigma_{\bar{x}} = 616,823 \pm 2.33\,(2000)$$
$$= 616,823 \pm 4660 = \$612,163 \text{ to } \$621,483$$

Thus, we can state with 98% confidence that the 2019 national average salary of all U.S. neurosurgeons is between \$612,163 and \$621,483. Using the same method, we can find a confidence interval for the 2019 national average salary of doctors with each specialty mentioned in the chart. Note that the sample means given in the bar chart are the point estimates of the corresponding population means.

In practice, we typically do not know the value of the population standard deviation, but we do know the value of the sample standard deviation calculated from the sample data. In this case, we will find a confidence interval for the population mean using the *t* distribution procedure, which is explained in the next section.

Source: Adapted from Complete List of Average Doctor Salaries by Specialty. Nomad, 2019. https://blog.nomadhealth.com/complete-list-of-average-doctor-salaries-by-specialty-locum-tenens/

Example 8.2 illustrates how to obtain a confidence interval for μ when σ is known and the sample size is large ($n \geq 30$).

EXAMPLE 8.2 Cost of a Checking Account to Banks

Constructing a Confidence Interval for μ: σ Known and $n \geq 30$

According to a study conducted by Moebs Services Inc., an individual checking account at major U.S. banks costs the banks more than \$380 per year. A recent random sample of 600 such checking accounts produced a mean annual cost of \$500 to major U.S. banks. Assume that the standard deviation of annual costs to major U.S. banks of all such checking accounts is \$40. Make a 99% confidence interval for the current mean annual cost to major U.S. banks of all such checking accounts.

Solution From the given information,

$$n = 600, \quad \bar{x} = \$500, \quad \sigma = \$40,$$
$$\text{Confidence level} = 99\% \text{ or } .99$$

In this example, although the shape of the population distribution is unknown, the population standard deviation is known, and the sample size is large ($n \geq 30$). Hence, we can use the normal distribution to make a confidence interval for μ. To make this confidence interval, first we find the standard deviation of $\bar{x}$. The value of $\sigma_{\bar{x}}$ is

$$\sigma_{\bar{x}} = \frac{\sigma}{\sqrt{n}} = \frac{40}{\sqrt{600}} = 1.63299316$$

To find z for a 99% confidence level, first we find the area in each of the two tails of the normal distribution curve, which is $(1 - .99)/2 = .0050$. Then, we look for areas .0050 and $.0050 + .99 = .9950$ in the normal distribution table to find the two z values. These two z values are (approximately) -2.58 and 2.58. Thus, we will use $z = 2.58$ in the confidence interval formula. Substituting all the values in the formula, we obtain the 99% confidence interval for μ,

$$\bar{x} \pm z\sigma_{\bar{x}} = 500 \pm 2.58\,(1.63299316) = 500 \pm 4.21 = \mathbf{\$495.79 \text{ to } \$504.21}$$

Thus, we can state with 99% confidence that the current mean annual cost to major U.S. banks of all individual checking accounts is between \$495.79 and \$504.21.

The **width of a confidence interval** depends on the size of the margin of error, $z\sigma_{\bar{x}}$, which depends on the values of z, σ, and n because $\sigma_{\bar{x}} = \sigma/\sqrt{n}$. However, the value of σ is not under the control of the investigator. Hence, the width of a confidence interval can be controlled using:

1. The value of z, which depends on the confidence level
2. The sample size n

The confidence level determines the value of z, which in turn determines the size of the margin of error. The value of z increases as the confidence level increases, and it decreases as the confidence level decreases. For example, the value of z is approximately 1.65 for a 90% confidence level, 1.96 for a 95% confidence level, and approximately 2.58 for a 99% confidence level. Hence, the higher the confidence level, the larger the width of the confidence interval as long as σ and n remain the same.

For the same confidence level and the same value of σ, an increase in the sample size decreases the value of $\sigma_{\bar{x}}$, which in turn decreases the size of the margin of error. Therefore, an increase in the sample size decreases the width of the confidence interval.

Thus, if we want to decrease the width of a confidence interval, we have two choices:

1. Lower the confidence level
2. Increase the sample size

Lowering the confidence level is not a good choice because a lower confidence level may give less reliable results. Therefore, we should always prefer to increase the sample size if we want to decrease the width of a confidence interval. Next we illustrate, using Example 8.2, how either a decrease in the confidence level or an increase in the sample size decreases the width of the confidence interval.

1 Confidence Level and the Width of the Confidence Interval

Reconsider Example 8.2. Suppose all the information given in that example remains the same. First, let us decrease the confidence level to 95%. From the normal distribution table, $z = 1.96$ for a 95% confidence level. Then, using $z = 1.96$ in the confidence interval for Example 8.2, we obtain

$$\bar{x} \pm z\sigma_{\bar{x}} = 500 \pm 1.96\,(1.63299316) = 500 \pm 3.20 = \mathbf{\$496.80 \text{ to } \$503.20}$$

Comparing this confidence interval to the one obtained in Example 8.2, we observe that the width of the confidence interval for a 95% confidence level is smaller than the width for a 99% confidence level.

2 Sample Size and the Width of the Confidence Interval

Consider Example 8.2 again. Now suppose the information given in that example is based on a sample size of 1000. Further assume that all other information given in that example, including the confidence level, remains the same. First, we calculate the standard deviation of the sample mean using $n = 1000$:

$$\sigma_{\bar{x}} = \frac{\sigma}{\sqrt{n}} = \frac{40}{\sqrt{1000}} = 1.26491106$$

Then, the 99% confidence interval for μ is

$$\bar{x} \pm z\sigma_{\bar{x}} = 500 \pm 2.58\,(1.26491106) = 500 \pm 3.26 = \mathbf{\$496.74 \text{ to } \$503.26}$$

Comparing this confidence interval to the one obtained in Example 8.2, we observe that the width of the 99% confidence interval for $n = 1000$ is smaller than the width of the 99% confidence interval for $n = 600$.

8.2.1 Determining the Sample Size for the Estimation of Mean

One reason we usually conduct a sample survey and not a census is that almost always we have limited resources at our disposal. In light of this, if a smaller sample can serve our purpose, then we will be wasting our resources by taking a larger sample. For instance, suppose we want to estimate the mean life of a certain auto battery. If a sample of 40 batteries can give us the desired margin of error, then we will be wasting money and time if we take a sample of a much larger size—say, 500 batteries. In such cases, if we know the confidence level and the margin of error that we want, then we can find the (approximate) size of the sample that will produce the required result.

From earlier discussion, we learned that $E = z\sigma_{\bar{x}}$ is called the margin of error of the interval estimate for μ. As we know, the standard deviation of the sample mean is $\sigma_{\bar{x}} = \sigma/\sqrt{n}$. Therefore, we can write the margin of error of the estimate for μ as

$$E = z \cdot \frac{\sigma}{\sqrt{n}}$$

Suppose we predetermine the size of the margin of error, E, and want to find the size of the sample that will yield this margin of error. By manipulating the above expression algebraically, we obtain the following formula to find the required sample size for given values of E, σ, and z.

Determining the Sample Size for the Estimation of μ

Given the confidence level and the standard deviation of the population, the sample size that will produce a predetermined margin of error E of the confidence interval estimate of μ is

$$n = \frac{z^2\sigma^2}{E^2}$$

If we do not know σ, we can take a preliminary sample (of any arbitrarily determined size) and find the sample standard deviation, s. Then we can use s for σ in the formula. However, note that using s for σ may give a sample size that eventually may produce an error much larger (or smaller) than the predetermined margin of error. This will depend on how close s and σ are.

Example 8.3 illustrates how we determine the sample size that will produce the margin of error of estimate for μ to within a certain limit.

EXAMPLE 8.3 Student Loan Debt of College Graduates

Determining the Sample Size for the Estimation of μ

An alumni association wants to estimate the mean debt of this year's college graduates. It is known that the population standard deviation of the debts of this year's college graduates is \$11,800. How large a sample should be selected so that a 99% confidence interval of the estimate is within \$800 of the population mean?

iStock.com/FatCamera

Solution The alumni association wants the 99% confidence interval for the mean debt of this year's college graduates to be

$$\bar{x} \pm 800$$

Hence, the maximum size of the margin of error of estimate is to be \$800; that is,

$$E = \$800$$

The value of z for a 99% confidence level is 2.58. The value of σ is given to be \$11,800. Therefore, substituting all values in the formula and simplifying, we obtain

$$n = \frac{z^2\sigma^2}{E^2} = \frac{(2.58)^2(11{,}800)^2}{(800)^2} = 1448.18 \approx \mathbf{1449}$$

Thus, the minimum required sample size is 1449. If the alumni association takes a random sample of 1449 of this year's college graduates, computes the mean debt for this sample, and then makes a 99% confidence interval around this sample mean, the margin of error of estimate will be approximately \$800. Note that we have rounded the final answer for the sample size to the next higher integer. This is always the case when determining the sample size because rounding down to 1448 will result in a margin of error slightly greater than \$800.

EXERCISES

Concepts and Procedures

8.3 What is the point estimator of the population mean, μ? How would you calculate the margin of error for an estimate of μ?

8.4 Explain the various alternatives for decreasing the width of a confidence interval. Which is the best alternative?

8.5 Briefly explain how the width of a confidence interval decreases with an increase in the sample size. Give an example.

8.6 Briefly explain how the width of a confidence interval decreases with a decrease in the confidence level. Give an example.

8.7 Briefly explain the difference between a confidence level and a confidence interval.

8.8 What is the margin of error of estimate for μ when σ is known? How is it calculated?

8.9 How will you interpret a 99% confidence interval for μ? Explain.

8.10 Find z for each of the following confidence levels:

a. 90% **b.** 95% **c.** 96%

d. 97% **e.** 98% **f.** 99%

8.11 For a data set obtained from a random sample, $n = 20$ and $\bar{x} = 20.4$. It is known that $\sigma = 2.3$. The population is normally distributed.

a. What is the point estimate of μ?

b. Make a 99% confidence interval for μ.

c. What is the margin of error of estimate for part (b)?

8.12 The standard deviation for a population is $\sigma = 12.4$. A random sample of 25 observations selected from this population gave a mean equal to 143.72. The population is known to have a normal distribution.

a. Make a 99% confidence interval for μ.

b. Construct a 95% confidence interval for μ.

c. Determine a 90% confidence interval for μ.

d. Does the width of the confidence intervals constructed in parts (a) through (c) decrease as the confidence level decreases? Explain your answer.

8.13 The standard deviation for a population is $\sigma = 6.35$. A random sample selected from this population gave a mean equal to 81.90.

a. Make a 99% confidence interval for μ assuming $n = 16$.

b. Construct a 99% confidence interval for μ assuming $n = 20$.

c. Determine a 99% confidence interval for μ assuming $n = 25$.

d. Does the width of the confidence intervals constructed in parts (a) through (c) increase as the sample size decreases? Explain.

8.14 For a population, the value of the standard deviation is 3.00. A random sample of 40 observations taken from this population produced the following data:

35	40	42	39	37	44	50	41
25	28	32	36	46	48	49	47
42	51	43	36	36	45	46	37
33	47	41	46	39	48	26	38
46	48	44	41	38	46	42	51

a. What is the point estimate of μ?

b. Make a 98% confidence interval for μ.

c. What is the margin of error of estimate for part (b)?

8.15 For a population data set, $\sigma = 15.5$.

a. How large a sample should be selected so that the margin of error of estimate for a 99% confidence interval for μ is 4.00?

b. How large a sample should be selected so that the margin of error of estimate for a 95% confidence interval for μ is 3.50?

8.16 For a population data set, $\sigma = 18.50$.

a. What should the sample size be for a 98% confidence interval for μ to have a margin of error of estimate equal to 5.50?

b. What should the sample size be for a 95% confidence interval for μ to have a margin of error of estimate equal to 4.00?

8.17 Determine the sample size for the estimate of μ for the following:

a. $E = 3.2$, $\sigma = 14.50$, confidence level = 99%

b. $E = 4.5$, $\sigma = 24.50$, confidence level = 95%

c. $E = 25.5$, $\sigma = 122.25$, confidence level = 90%

8.18 Determine the sample size for the estimate of μ for the following:

a. $E = .25$, $\sigma = .75$, confidence level = 99%

b. $E = 2.25$, $\sigma = 6.75$, confidence level = 95%

c. $E = 5.50$, $\sigma = 17.25$, confidence level = 90%

Applications

8.19 A city planner wants to estimate the average monthly residential water usage in the city. He selected a random sample of 40 households from the city, which gave a mean water usage of 3415.70 gallons over a one-month period. Based on earlier data, the population standard deviation of the monthly residential water usage in this city is 389.60 gallons. Make a 95% confidence interval for the average monthly residential water usage for all households in this city.

8.20 Lazurus Steel Corporation produces iron rods that are supposed to be 40 inches long. The machine that makes these rods does not produce each rod exactly 40 inches long; the lengths of the rods vary slightly. It is known that when the machine is working properly, the mean length of the rods made on this machine is 40 inches. The standard deviation of the lengths of all rods produced on this machine is equal to .15 inch. The quality control department takes a random sample of 25 such rods every week, calculates the mean length of these rods, and makes a 95% confidence interval for the population mean. If either the upper limit of this confidence interval is greater than 40.80 inches or the lower limit of this confidence interval is less than 39.20 inches, the machine is stopped and adjusted. A recent sample of 20 rods produced a mean length of 40.50 inches. Based on this sample, will you conclude that the machine needs an adjustment? Assume that the lengths of all such rods have an approximate normal distribution.

8.21 At Verk's Dairy, a machine is set to fill 1000-milliliter milk bottles. However, this machine does not put exactly 1000 milliliters of milk into each bottle; the amount varies slightly from bottle to bottle. It is known that when the machine is working properly, the mean net weight of these bottles is 1000 milliliters. The standard deviation of the amounts of milk in all such bottles is always equal to 15 milliliters. The quality control department takes a random sample of 30 such bottles every week, calculates the mean net weight of these bottles, and makes a 95% confidence interval for the population mean. If either the upper limit of this confidence interval is greater than 1015 milliliters or the lower limit of this confidence interval is less than 985 milliliters, the machine is stopped and adjusted. A recent sample of

30 such bottles produced a mean net weight of 995 milliliters. Based on this sample, will you conclude that the machine needs an adjustment? Assume that the amounts of milk put in all such bottles have an approximate normal distribution.

8.22 A bank manager wants to know the mean amount of mortgage paid per month by homeowners in an area. A random sample of 120 homeowners selected from this area showed that they pay an average of $1575 per month for their mortgages. The population standard deviation of all such mortgages is $215.

a. Find a 97% confidence interval for the mean amount of mortgage paid per month by all homeowners in this area.

b. Suppose the confidence interval obtained in part (a) is too wide. How can the width of this interval be reduced? Discuss all possible alternatives. Which alternative is the best?

8.23 A company that produces detergents wants to estimate the mean amount of detergent in 64-ounce jugs at a 99% confidence level. The company knows that the standard deviation of the amounts of detergent in all such jugs is .20 ounce. How large a sample should the company select so that the estimate is within .04 ounce of the population mean?

8.24 A department store manager wants to estimate the mean amount spent by all customers at this store at a 98% confidence level. The manager knows that the standard deviation of amounts spent by all customers at this store is $41. What minimum sample size should the manager choose so that the estimate is within $4 of the population mean?

8.25 A city planner wants to estimate the average monthly residential water usage in the city at a 97% confidence level. Based on earlier data, the population standard deviation of the monthly residential water usage in this city is 389.60 gallons. How large a sample should be selected so that the estimate for the average monthly residential water usage in this city is within 100 gallons of the population mean?

***8.26** You are interested in estimating the mean commuting time from home to college for all commuter students at your college. Briefly explain the procedure you will follow to conduct this study. Collect the required data from a sample of 25 or more such students and then estimate the population mean at a 95% confidence level. Assume that the population standard deviation for all such times is 7.5 minutes.

***8.27** You are interested in estimating the mean age of cars owned by all people in the United States. Briefly explain the procedure you will follow to conduct this study. Collect the required data on a sample of 30 or more cars and then estimate the population mean at a 95% confidence level. Assume that the population standard deviation is 2.4 years.

8.3 Estimation of a Population Mean: σ Not Known

LEARNING OBJECTIVES

After completing this section, you should be able to:

- Explain the *t* distribution, and the similarities and differences between the standard normal distribution and the *t* distribution.
- Explain the meaning of the degrees of freedom for a *t* distribution.
- Explain why the *t* distribution is used to construct a confidence interval for μ when σ is unknown and certain other conditions are satisfied.
- Find the value of *t* from the *t* distribution table for the area in the right (or left) tail of a *t* distribution with known degrees of freedom.
- Construct a confidence interval for μ when σ is unknown and certain other conditions are satisfied.
- Identify the point estimate and margin of error for a confidence interval for μ using the *t* distribution.

This section explains how to construct a confidence interval for a population mean μ when the population standard deviation σ is not known. Here, again, there are three possible cases:

Case I. If the following three conditions are fulfilled:

1. The population standard deviation σ is not known
2. The sample size is small (i.e., $n < 30$)
3. The population from which the sample is selected is approximately normally distributed

then we use the *t* distribution (explained in Section 8.3.1) to make the confidence interval for μ.

Case II. If the following two conditions are fulfilled:

1. The population standard deviation σ is not known
2. The sample size is large (i.e., $n \geq 30$)

then again we use the t distribution to make the confidence interval for μ.

Case III. If the following three conditions are fulfilled:

1. The population standard deviation σ is not known
2. The sample size is small (i.e., $n < 30$)
3. The population from which the sample is selected is not normally distributed (or its distribution is unknown)

then we use a nonparametric method to make the confidence interval for μ. Such procedures are covered in Chapter 15.

The following chart summarizes the above three cases.

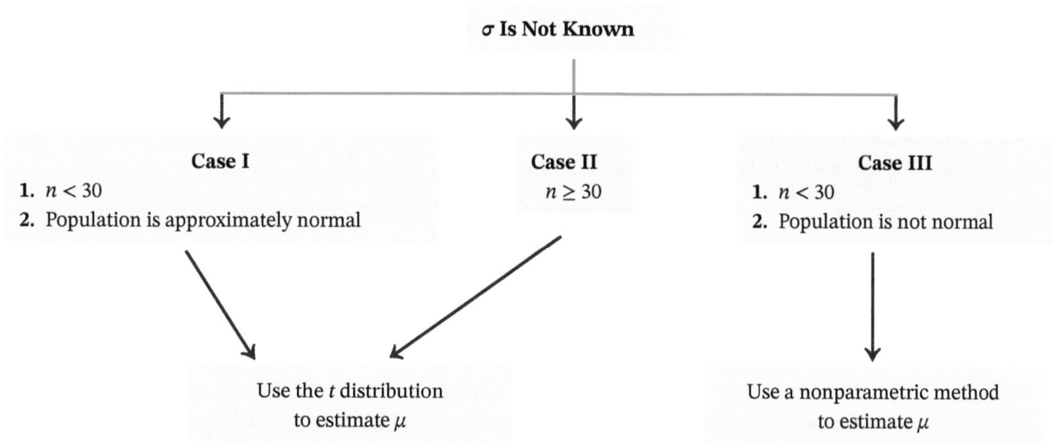

In the next subsection, we discuss the t distribution, and then in Section 8.3.2 we show how to use the t distribution to make a confidence interval for μ when σ is not known and conditions of Cases I or II are satisfied.

8.3.1 The t Distribution

The ***t* distribution** was developed by W. S. Gosset in 1908 and published under the pseudonym *Student*. As a result, the t distribution is also called *Student's t distribution*. The t distribution is similar to the normal distribution in some respects. Like the normal distribution curve, the t distribution curve is symmetric (bell shaped) about the mean and never meets the horizontal axis. The total area under a t distribution curve is 1.0, or 100%. However, the t distribution curve is flatter and wider than the standard normal distribution curve. In other words, the t distribution curve has a lower height and a greater spread (or, we can say, a larger standard deviation) than the standard normal distribution. However, as the sample size increases, the shape of the t distribution gets closer and closer to that of the standard normal distribution. A random variable having a t distribution is denoted by t.

The shape of a particular t distribution curve depends on the number of **degrees of freedom (*df*)**. For the purpose of this chapter and Chapter 9, the number of degrees of freedom for a t distribution is equal to the sample size minus one, that is,

$$df = n - 1$$

The number of degrees of freedom is the only parameter of the t distribution. There is a different t distribution for each number of degrees of freedom. Like the standard normal distribution, the mean of the t distribution is 0. But unlike the standard normal distribution, whose standard deviation is 1, the standard deviation of a t distribution is $\sqrt{df/(df-2)}$ for $df > 2$. Thus, the standard deviation of a t distribution is always greater than 1 and, hence, is larger than the standard deviation of the standard normal distribution.

The *t* Distribution

The *t distribution* is a specific type of bell-shaped distribution with a lower height and a greater spread than the standard normal distribution. As the sample size becomes larger, the t distribution approaches the standard normal distribution. The t distribution has only one parameter, called the degrees of freedom (df). The mean of the t distribution is equal to 0, and its standard deviation is $\sqrt{df/(df-2)}$ for $df > 2$.

Figure 8.5 shows the standard normal distribution and the t distribution for 9 degrees of freedom. The standard deviation of the standard normal distribution is 1.0, and the standard deviation of the t distribution is $\sqrt{df/(df-2)} = \sqrt{9/(9-2)} = 1.134$.

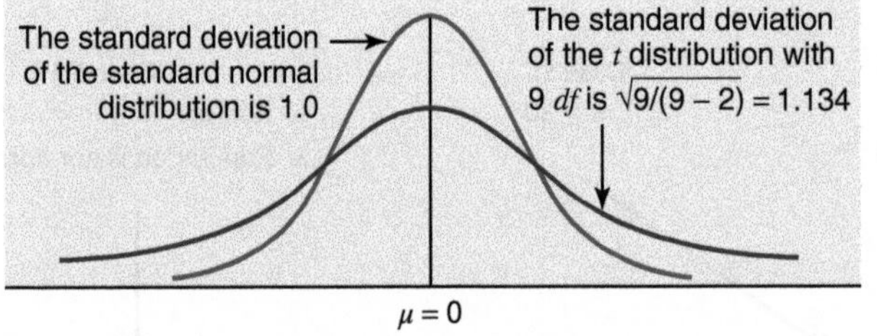

FIGURE 8.5 The t distribution for $df = 9$ and the standard normal distribution.

As stated earlier, the number of degrees of freedom for a t distribution for the purpose of this chapter is $n - 1$. *The number of degrees of freedom is defined as the number of observations that can be chosen freely.* As an example, suppose we know that the mean of four values is 20. Consequently, the sum of these four values is $20(4) = 80$. Now, how many values out of four can we choose freely so that the sum of these four values is 80? The answer is that we can freely choose $4 - 1 = 3$ values. Suppose we choose 27, 8, and 19 as the three values. Given these three values and the information that the sum of the four values is $20 \times 4 = 80$, the fourth value is $80 - 27 - 8 - 19 = 26$. Thus, once we have chosen three values, the fourth value is automatically determined. Consequently, the number of degrees of freedom for this example is

$$df = n - 1 = 4 - 1 = 3$$

We subtract 1 from n because we lose 1 degree of freedom to calculate the mean.

Table V of Appendix B lists the values of t for the given number of degrees of freedom and areas in the right tail of a t distribution. Because the t distribution is symmetric, these are also the values of $-t$ for the same number of degrees of freedom and the same areas in the left tail of the t distribution. Example 8.4 describes how to read Table V of Appendix B.

EXAMPLE 8.4

Reading the t Distribution Table

Find the value of t for 16 degrees of freedom and .05 area in the right tail of a t distribution curve.

Solution In Table V of Appendix B, we locate 16 in the column of degrees of freedom (labeled df) and .05 in the row of *Area in the right tail under the t distribution curve* at the top of the table. The entry at the intersection of the row of 16 and the column of .05, which is 1.746, gives the required value of t. The relevant portion of Table V of Appendix B is shown here as **Table 8.2**. The value of t read from the t distribution table is shown in **Figure 8.6**.

TABLE 8.2 Determining t for 16 df and .05 Area in the Right Tail

	Area in the Right Tail Under the t Distribution Curve				
df	.10	.05	.025	...	.001
1	3.078	6.314	12.706	...	318.309
2	1.886	2.920	4.303	...	22.327
3	1.638	2.353	3.182	...	10.215
.	...	...	...	...	...
.	...	...	...	...	...
.	...	...	...	...	...
16	1.337	1.746	2.120	...	3.686
.	...	...	...	...	...
.	...	...	...	...	...
.	...	...	...	...	...
75	1.293	1.665	1.992	...	3.202
∞	1.282	1.645	1.960	...	3.090

Area in the right tail

df → 16

The required value of t for 16 df and .05 area in the right tail

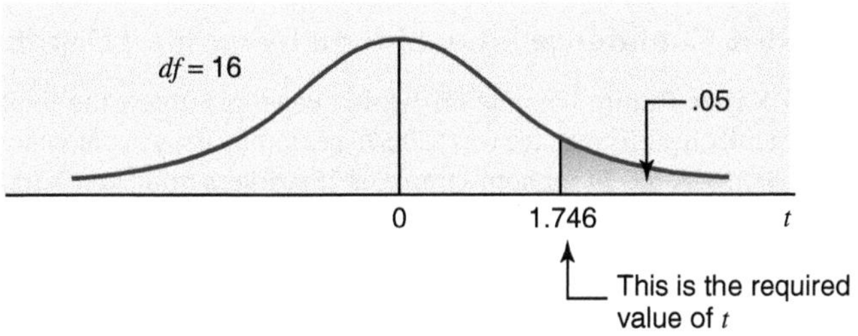

FIGURE 8.6 The value of t for 16 df and .05 area in the right tail.

Because of the symmetric shape of the t distribution curve, the value of t for 16 degrees of freedom and .05 area in the left tail is -1.746. **Figure 8.7** illustrates this case.

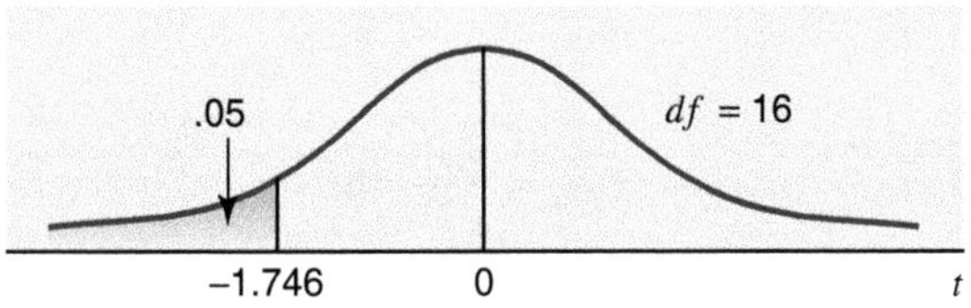

FIGURE 8.7 The value of t for 16 df and .05 area in the left tail.

8.3.2 Confidence Interval for μ Using the t Distribution

To reiterate, when the conditions mentioned under Cases I or II in the beginning of this section hold true, we use the t distribution to construct a confidence interval for the population mean μ.

When the population standard deviation σ is not known, then we replace it by the sample standard deviation s, which is its estimator. Consequently, for the standard deviation of $\bar{x}$, we use

$$s_{\bar{x}} = \frac{s}{\sqrt{n}}$$

for $\sigma_{\bar{x}} = \sigma/\sqrt{n}$. Note that the value of $s_{\bar{x}}$ is a point estimate of $\sigma_{\bar{x}}$.

Confidence Interval for μ Using the t Distribution

The $(1 - \alpha)100\%$ *confidence interval for* μ is

$$\bar{x} \pm t s_{\bar{x}}$$

where

$$s_{\bar{x}} = \frac{s}{\sqrt{n}}$$

The value of t is obtained from the t distribution table for $n - 1$ degrees of freedom and the given confidence level. Here $t s_{\bar{x}}$ is the margin of error or the maximum error of the estimate; that is,

$$E = t s_{\bar{x}}$$

Examples 8.5 and 8.6 describe the procedure of constructing a confidence interval for μ using the t distribution.

EXAMPLE 8.5 Premium for Health Insurance Coverage

Constructing a 95% Confidence Interval for μ Using the t Distribution

According to a 2019 Kaiser Family Foundation Health Benefits Survey, the total mean cost of employer-sponsored family health coverage was $20,576 per family per year, of which workers were paying an average of $6015. A recent random sample of 25 workers from New York City who have employer-provided health insurance coverage paid an average premium of $6600 for family health insurance coverage with a standard deviation of $800. Make a 95% confidence interval for the current average premium paid for family health insurance coverage by all workers in New York City who have employer-provided health insurance coverage. Assume that the distribution of premiums paid for family health insurance coverage by all workers in New York City who have employer-provided health insurance coverage is approximately normally distributed.

Solution Here, σ is not known, $n < 30$, and the population is approximately normally distributed. All conditions mentioned in Case I of the chart given in the beginning of this section are satisfied. Therefore, we will use the t distribution to make a confidence interval for μ. From the given information,

$$n = 25, \quad \bar{x} = \$6600, \quad s = \$800, \quad \text{Confidence level} = 95\% \text{ or } .95$$

The value of $s_{\bar{x}}$ is

$$s_{\bar{x}} = \frac{s}{\sqrt{n}} = \frac{800}{\sqrt{25}} = \$160$$

To find the value of t from Table V of Appendix B, we need to know the degrees of freedom and the area under the t distribution curve in each tail.

$$\text{Degrees of freedom} = n - 1 = 25 - 1 = 24$$

To find the area in each tail, we divide the confidence level by 2 and subtract the number obtained from .5. Thus,

$$\text{Area in each tail} = .5 - (.95/2) = .5 - .4750 = .025$$

From the t distribution table, Table V of Appendix B, the value of t for $df = 24$ and .025 area in the right tail is 2.064. The value of t is shown in **Figure 8.8**.

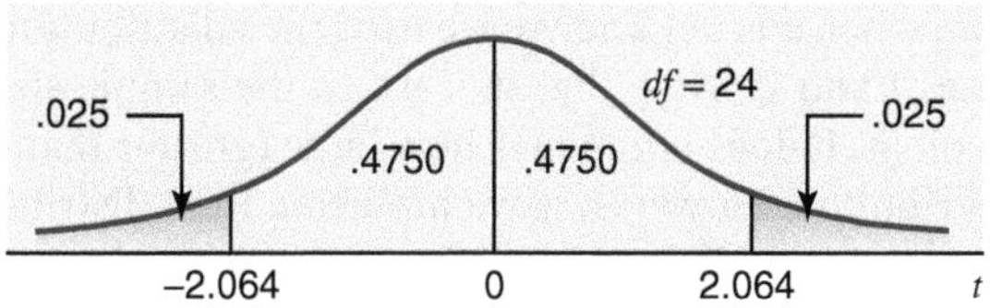

FIGURE 8.8 The value of t.

By substituting all values in the formula for the confidence interval for μ, we obtain the 95% confidence interval as

$$\bar{x} \pm t\, s_{\bar{x}} = 6600 \pm 2.064\,(160) = 6600 \pm 330.24 = \mathbf{\$6269.76 \text{ to } \$6930.24}$$

Thus, we can state with 95% confidence that the current average premium paid for family health insurance coverage by all workers in New York City who have employer-provided health insurance coverage is between \$6269.76 and \$6930.24.

Note that $\bar{x} = \$6600$ is a point estimate of μ in this example, and \$330.24 is the margin of error.

EXAMPLE 8.6 Annual Expenses on Books

Constructing a 99% Confidence Interval for μ Using the t Distribution

Sixty-four randomly selected adults who buy books for general reading were asked how much they usually spend on books per year. This sample produced a mean of \$1450 and a standard deviation of \$300 for such annual expenses. Determine a 99% confidence interval for the corresponding population mean.

Patrick Clark/Getty Images

Solution From the given information,

$$n = 64, \quad \bar{x} = \$1450, \quad s = \$300,$$

and Confidence level = 99% or .99

Here σ is not known, but the sample size is large ($n \geq 30$). Hence, we will use the t distribution to make a confidence interval for μ. First we calculate the standard deviation of $\bar{x}$, the number of degrees of freedom, and the area in each tail of the t distribution.

$$s_{\bar{x}} = \frac{s}{\sqrt{n}} = \frac{300}{\sqrt{64}} = \$37.50$$

$$df = n - 1 = 64 - 1 = 63$$

$$\text{Area in each tail} = .5 - (.99/2) = .5 - .4950 = .005$$

From the t distribution table, $t = 2.656$ for 63 degrees of freedom and .005 area in the right tail. The 99% confidence interval for μ is

$$\bar{x} \pm t\, s_{\bar{x}} = \$1450 \pm 2.656(37.50)$$
$$= \$1450 \pm \$99.60 = \mathbf{\$1350.40 \text{ to } \$1549.60}$$

Thus, we can state with 99% confidence that based on this sample the mean annual expenditure on books by all adults who buy books for general reading is between \$1350.40 and \$1549.60.

Again, we can decrease the width of a confidence interval for μ either by lowering the confidence level or by increasing the sample size, as was done in Section 8.2. However, increasing the sample size is the better alternative.

Note: What If the Sample Size Is Large and the Number of *df* Is Not in the *t* Distribution Table?

Earlier in this section, when σ is not known, we used the *t* distribution to make a confidence interval for μ in Cases I and II. Note that in Case II, the sample size must be large. If we have access to technology, it does not matter how large (greater than 30) the sample size is; we can use the *t* distribution. However, if we are using the *t* distribution table (Table V of Appendix B), this may pose a problem. Usually such a table goes only up to a certain number of degrees of freedom. For example, Table V in Appendix B goes only up to 75 degrees of freedom. Thus, if the sample size is larger than 76, we cannot use Table V to find the *t* value for the given confidence level to use in the confidence interval. In such a situation when *n* is large (for example, 500) and the number of *df* is not included in the *t* distribution table, there are two options:

1. Use the *t* value from the last row (the row of ∞) in Table V.
2. Use the normal distribution as an approximation to the *t* distribution.

Note that the *t* values you will obtain from the last row of the *t* distribution table are the same as obtained from the normal distribution table for the same confidence levels, the only difference being the decimal places. To use the normal distribution as an approximation to the *t* distribution to make a confidence interval for μ, the procedure is exactly like the one in Section 8.2, except that now we replace σ by *s*, and $\sigma_{\bar{x}}$ by $s_{\bar{x}}$.

Again, note that here we can use the normal distribution as a convenience and as an approximation, but if we can, we should use the *t* distribution by using technology.

EXERCISES

Concepts and Procedures

8.28 Briefly explain the similarities and the differences between the standard normal distribution and the *t* distribution.

8.29 What are the parameters of a normal distribution and a *t* distribution? Explain.

8.30 Briefly explain the term degrees of freedom. Explain why the following relationship exists: degrees of freedom ≤ number of observations in a sample. Discuss about the degrees of freedom for a *t* distribution with an example.

8.31 What assumptions must hold true to use the *t* distribution to make a confidence interval for μ?

8.32 Find the value of *t* for the *t* distribution for each of the following:

a. Area in the right tail = .05 and $df = 9$

b. Area in the left tail = .025 and $n = 58$

c. Area in the left tail = .001 and $df = 34$

d. Area in the right tail = .005 and $n = 27$

8.33 a. Find the value of *t* for the *t* distribution with a sample size of 18 and area in the left tail equal to .10.

b. Find the value of *t* for the *t* distribution with a sample size of 22 and area in the right tail equal to .025.

c. Find the value of *t* for the *t* distribution with 45 degrees of freedom and .001 area in the right tail.

d. Find the value of *t* for the *t* distribution with 37 degrees of freedom and .005 area in the left tail.

8.34 For each of the following, find the area in the appropriate tail of the *t* distribution.

a. $t = 2.467$ and $df = 32$ **b.** $t = -1.762$ and $df = 62$

c. $t = -2.670$ and $n = 55$ **d.** $t = 2.819$ and $n = 18$

8.35 a. Find the value of *t* from the *t* distribution table for a sample size of 22 and a confidence level of 95%.

b. Find the value of *t* from the *t* distribution table for 60 degrees of freedom and a 90% confidence level.

c. Find the value of *t* from the *t* distribution table for a sample size of 24 and a confidence level of 99%.

8.36 A random sample of 18 observations taken from a normally distributed population produced the following data:

28.4	25.4	25.5	25.5	31.1	23.0	25.4	24.6	28.4
37.2	23.9	30.4	27.9	25.1	27.2	25.3	23.1	28.9

a. What is the point estimate of μ?

b. Make a 99% confidence interval for μ.

c. What is the margin of error of estimate for μ in part (b)?

8.37 A random sample of 10 observations taken from a normally distributed population produced the following data:

44	52	31	48	46	39	47	36	41	57

a. What is the point estimate of μ?

b. Make a 95% confidence interval for μ.

c. What is the margin of error of estimate for μ in part (b)?

8.38 For a random sample selected from a normally distributed population, suppose $\bar{x} = 25.5$ and $s = 4.9$.

a. Construct a 95% confidence interval for μ assuming $n = 47$.

b. Construct a 99% confidence interval for μ assuming $n = 47$. Is the width of the 99% confidence interval smaller than the width of the 95% confidence interval calculated in part (a)? If yes, explain why.

c. Find a 95% confidence interval for μ assuming $n = 32$. Is the width of the 95% confidence interval for μ with $n = 32$ larger than the width of the 95% confidence interval for μ with $n = 47$ calculated in part (a)? If so, why? Explain.

8.39 a. A random sample of 100 observations taken from a population produced a sample mean equal to 55.32 and a standard deviation equal to 8.4. Make a 90% confidence interval for μ.

b. Another sample of 100 observations taken from the same population produced a sample mean equal to 57.40 and a standard deviation equal to 7.50. Make a 90% confidence interval for μ.

c. A third sample of 100 observations taken from the same population produced a sample mean equal to 56.25 and a standard deviation equal to 7.90. Make a 90% confidence interval for μ.

d. The true population mean for this population is 55.80. Which of the confidence intervals constructed in parts (a) through (c) cover this population mean and which do not?

Applications

8.40 According to the 2019 Physician Compensation Report by Medscape (a subsidiary of WebMD), American orthopedists earned an average of \$482,000 in 2019. Suppose that this mean is based on a random sample of 200 American orthopedists, and the standard deviation for this sample is \$90,000. Make a 90% confidence interval for the population mean μ.

8.41 Almost all employees working for financial companies in New York City receive large bonuses at the end of the year. A random sample of 65 employees selected from financial companies in New York City showed that they received an average bonus of \$55,000 last year with a standard deviation of \$12,000. Construct a 95% confidence interval for the average bonus that all employees working for financial companies in New York City received last year.

8.42 The high price of medicines is a source of major expense for those seniors in the United States who have to pay for these medicines themselves. A random sample of 2000 seniors who pay for their medicines showed that they spent an average of \$4600 last year on medicines with a standard deviation of \$800. Make a 98% confidence interval for the corresponding population mean.

8.43 A random sample of 36 mid-sized cars tested for fuel consumption gave a mean of 26.4 miles per gallon with a standard deviation of 2.3 miles per gallon.

a. Make a 99% confidence interval for the population mean, μ.

b. Suppose the confidence interval obtained in part (a) is too wide. How can the width of this interval be reduced? Describe all possible alternatives. Which alternative is the best and why?

8.44 The mean time taken to design a house plan by 40 architects was found to be 23 hours with a standard deviation of 3.75 hours.

a. Construct a 98% confidence interval for the population mean μ.

b. Suppose the confidence interval obtained in part (a) is too wide. How can the width of this interval be reduced? Describe all possible alternatives. Which alternative is the best and why?

8.45 The following data give the speeds (in miles per hour), as measured by radar, of 10 cars traveling on a highway.

76	74	80	68	79	74	58	78	82	68

Assuming that the speeds of all cars traveling on this highway have a normal distribution, construct a 90% confidence interval for the mean speed of all cars traveling on the highway.

8.46 A dentist wants to find the average time taken by one of her hygienists to take X-rays and clean teeth for patients. She recorded the time to serve 24 randomly selected patients by this hygienist. The data (in minutes) are as follows:

36.80	39.80	38.60	38.30	35.80	32.60	38.70	34.50	37.00
32.70	40.90	33.80	37.10	31.00	35.10	38.20	36.60	38.80
39.60	39.70	35.10	38.20	32.70	40.50			

Assume that such times for this hygienist for all patients are approximately normal.

a. What is the point estimate of the corresponding population mean?

b. Construct a 99% confidence interval for the average time taken by the hygienist to take X-rays and to clean teeth for all patients.

8.47 A random sample of 36 participants in a Zumba dance class had their heart rates measured before and after a moderate 10-minute workout. The following data correspond to the increase in each individual's heart rate (in beats per minute).

59	70	57	42	57	59	41	54	44	36	59	61
52	42	41	32	60	54	52	53	51	47	62	48
62	44	69	50	37	50	54	48	52	61	45	63

a. What is the point estimate of the corresponding population mean?

b. Make a 98% confidence interval for the average increase in a person's heart rate after a moderate 10-minute Zumba workout.

8.48 The following data give the one-way commuting times (in minutes) from home to work for a random sample of 30 workers.

23	17	34	26	18	33	46	42	12	37
44	15	22	19	28	32	18	39	40	48
25	36	23	39	42	46	29	17	24	31

a. Calculate the value of the point estimate of the mean one-way commuting time from home to work for all workers.

b. Construct a 99% confidence interval for the mean one-way commuting time from home to work for all workers.

***8.49** You are working for a supermarket. The manager has asked you to estimate the mean time taken by a cashier to serve customers at this supermarket. Briefly explain how you will conduct this study. Collect data on the time taken by any supermarket cashier to serve each of the 45 customers. Then estimate the population mean. Choose your own confidence level.

***8.50** You are working for a bank. The bank manager wants to know the mean waiting time for all customers who visit this bank. She has asked you to estimate this mean by taking a sample. Briefly explain how you will conduct this study. Collect data on the waiting times for 50 customers who visit this bank. Then estimate the population mean. Choose your own confidence level.

8.4 Estimation of a Population Proportion: Large Samples

LEARNING OBJECTIVES

After completing this section, you should be able to:

- Determine a point estimate of a population proportion *p*.
- Verify the conditions for using the normal distribution to make a confidence interval for *p*.
- Construct a confidence interval for *p* using the normal distribution.
- Identify the margin of error for a confidence interval for *p*.
- Interpret a confidence interval for *p*.
- Explain the relationship between sample size, confidence level, and the width of a confidence interval for *p*.
- Explain various alternatives for decreasing the width of a confidence interval.
- Determine the minimum sample size to produce a confidence interval for *p* with a predetermined margin of error.

Often we want to estimate a population proportion or percentage. (Recall that a percentage is obtained by multiplying the proportion by 100.) For example, the production manager of a company may want to estimate the proportion of defective items produced on a machine. A bank manager may want to find the percentage of customers who are satisfied with the service provided by the bank.

Again, if we can conduct a census each time we want to find the value of a population proportion, there is no need to learn the procedures discussed in this section. However, we usually derive our results from sample surveys. Hence, to take into account the variability in the results obtained from different sample surveys, we need to know the procedures for estimating a population proportion.

Recall from Chapter 7 that the population proportion is denoted by p, and the sample proportion is denoted by $\hat{p}$. This section explains how to estimate the population proportion, p, using the sample proportion, $\hat{p}$. The sample proportion, $\hat{p}$, is a sample statistic, and it possesses a sampling distribution. From Chapter 7, we know that:

1. The sampling distribution of the sample proportion $\hat{p}$ is approximately normal for a large sample. In the case of a proportion, a sample is considered to be large if $np > 5$ and $nq > 5$.
2. The mean of the sampling distribution of $\hat{p}$, $\mu_{\hat{p}}$, is equal to the population proportion, p.
3. The standard deviation of the sampling distribution of $\hat{p}$ is $\sigma_{\hat{p}} = \sqrt{pq/n}$, where $q = 1 - p$, given that $\frac{n}{N} \leq .05$.

Remember

To reiterate, in the case of a proportion, a sample is considered to be large if np and nq are both greater than 5. If p and q are not known, then $n\hat{p}$ and $n\hat{q}$ should each be greater than 5 for the sample to be large.

When estimating the value of a population proportion, we do not know the values of p and q. Consequently, we cannot compute $\sigma_{\hat{p}}$. Therefore, in the estimation of a population proportion, we use the value of $s_{\hat{p}}$ as an estimate of $\sigma_{\hat{p}}$. The value of $s_{\hat{p}}$ is calculated using the formula given next.

Estimator of the Standard Deviation of $\hat{p}$

The value of $s_{\hat{p}}$, which gives a point estimate of $\sigma_{\hat{p}}$, is calculated as follows. Here, $s_{\hat{p}}$ is an estimator of $\sigma_{\hat{p}}$.

$$s_{\hat{p}} = \sqrt{\frac{\hat{p}\hat{q}}{n}}$$

Note that the condition $\frac{n}{N} \leq .05$ must hold true to use this formula.

The sample proportion, $\hat{p}$, is the point estimator of the corresponding population proportion p. Then to find the confidence interval for p, we add to and subtract from $\hat{p}$ a number that is called the **margin of error**, E.

Confidence Interval for the Population Proportion, p

For a large sample, the $(1 - \alpha)100\%$ *confidence interval for the population proportion*, p, is

$$\hat{p} \pm z s_{\hat{p}}$$

The value of z is obtained from the standard normal distribution table for the given confidence level, and $s_{\hat{p}} = \sqrt{\hat{p}\hat{q}/n}$. The term $z s_{\hat{p}}$ is called the *margin of error*, or the maximum error of the estimate, and is denoted by E.

Examples 8.7 and 8.8 illustrate the procedure for constructing a confidence interval for p.

EXAMPLE 8.7 Concerned Believers in Global Warming

Finding the Point Estimate and 99% Confidence Interval for p: Large Sample

A September 24, 2019 article on news.gallup.com mentioned that according to the Gallup research about climate change issues, 51% of U.S. adults can be classified as "Concerned Believers," meaning they "are highly worried about global warming, think it will pose a serious threat in their lifetime, believe it's the result of human activity, and think news reports about it are accurate or underestimate the problem" (https://news.gallup.com). Assume that this poll is based on a random sample of 1600 U.S. adults.

a. What is the point estimate of the corresponding population proportion?

b. Find, with a 99% confidence level, the percentage of all U.S. adults who can be classified as "Concerned Believers" as defined in the Gallup news article.

Solution Let p be the proportion of all U.S. adults who can be classified as "Concerned Believers" as defined in the Gallup news article and let $\hat{p}$ be the corresponding sample proportion. From the given information,

$$n = 1600, \quad \hat{p} = .51, \quad \text{and} \quad \hat{q} = 1 - \hat{p} = 1 - .51 = .49$$

First, we calculate the value of the standard deviation of the sample proportion as follows:

$$s_{\hat{p}} = \sqrt{\frac{\hat{p}\hat{q}}{n}} = \sqrt{\frac{(.51)(.49)}{1600}} = .0124975$$

Note that $n\hat{p}$ and $n\hat{q}$ are both greater than 5. (The reader should check this condition.) Consequently, the sampling distribution of $\hat{p}$ is approximately normal and we will use the normal distribution to make a confidence interval about p.

a. The point estimate of the proportion of all U.S. adults who can be classified as "Concerned Believers" as defined in the Gallup news article is equal to .51; that is,

$$\text{Point estimate of } p = \hat{p} = \mathbf{.51}$$

b. The confidence level is 99%, or .99. To find z for a 99% confidence level, first we find the area in each of the two tails of the normal distribution curve, which is $(1 - .99)/2 = .0050$. Then, we look for .0050 and $.0050 + .99 = .9950$ areas in the normal distribution table to find the two values of z. These two z values are (approximately) -2.58 and 2.58. Thus, we will use $z = 2.58$

in the confidence interval formula. Substituting all the values in the confidence interval formula for p, we obtain

$$\hat{p} \pm z\, s_{\hat{p}} = .51 \pm 2.58\,(.0124975) = .51 \pm .032$$
$$= \mathbf{.478 \text{ to } .542} \quad \text{or} \quad \mathbf{47.8\% \text{ to } 54.2\%}$$

Thus, we can state with a 99% confidence that .478 to .542 or 47.8% to 54.2% of all U.S. adults can be classified as "Concerned Believers" as defined in the Gallup news article.

The margin of error associated with this estimate of p is ±.032 or ±3.2%, that is,

$$\text{Margin of error} = z\, s_{\hat{p}} = \pm .032 \text{ or } \pm 3.2\%$$

EXAMPLE 8.8 | U.S. Workers as Engaged Workers

Constructing a 97% Confidence Interval for p: Large Sample

According to a Gallup poll of U.S. adults, aged 18 and older working full or part time for an organization, conducted January to June, 2018, 34% of U.S. workers were classified as "engaged" workers, meaning that they are involved in, enthusiastic about, and committed to their work and workplace (https://news.gallup.com). Suppose that this poll was based on a random sample of 2000 U.S. adults working full or part time for an organization. Construct a 97% confidence interval for the corresponding population proportion.

Solution Let p be the proportion of all U.S. adults working full or part time for an organization at the time the poll was conducted who could be classified as "engaged" workers, and let $\hat{p}$ be the corresponding sample proportion. From the given information,

$$n = 2000, \quad \hat{p} = .34, \quad \hat{q} = 1 - \hat{p} = 1 - .34 = .66$$

and Confidence level = 97% or .97

The standard deviation of the sample proportion is

$$s_{\hat{p}} = \sqrt{\frac{\hat{p}\hat{q}}{n}} = \sqrt{\frac{(.34)(.66)}{2000}} = .01059245$$

Note that if we check $n\hat{p}$ and $n\hat{q}$, both will be greater than 5. Consequently, we can use the normal distribution to make a confidence interval for p.

From the normal distribution table, the value of z for the 97% confidence level is 2.17. Note that to find this z value, you will look for the areas .0150 and .9850 in Table IV of Appendix B. Substituting all the values in the formula, the 97% confidence interval for p is

$$\hat{p} \pm z\, s_{\hat{p}} = .34 \pm 2.17\,(.01059245) = .34 \pm .023$$
$$= \mathbf{.317 \text{ to } .363} \quad \text{or} \quad \mathbf{31.7\% \text{ to } 36.3\%}$$

Thus, we can state with a 97% confidence that the proportion of all U.S. adults who were working full or part time for an organization at the time the poll was conducted who could be classified as "engaged" workers is between .317 and .363 or 31.7% and 36.3%.

Again, we can decrease the width of a confidence interval for p either by lowering the confidence level or by increasing the sample size. However, lowering the confidence level is not a good choice because it simply decreases the likelihood that the confidence interval contains p. Hence, to decrease the width of a confidence interval for p, we should always choose to increase the sample size.

Case Study 8.2

Is Government, Poor Leadership, or Politicians the Most Important Problem Facing the United States?

The Gallup polling agency conducted a survey of 1526 randomly selected U.S. adults, aged 18 and over, during October 1–13, 2019. One question asked of these adults was, "What do you think is the most important problem facing this country today?" As shown in the table, of the adults included in the survey, 34% (or .34 when converted to proportion) cited "the government, poor leadership, or politicians" as the most important problem currently facing the United States. When classified based on political affiliation, 41% of Democrats, 36% of Republicans, and 27% of Independents in this survey held the this same opinion. As these percentages are based on a sample, we can make a confidence interval to estimate each of the four corresponding population proportions using the procedure learned in this section. Here we assume that the sample selected for this poll was a simple random sample.

Category	Sample Proportion	Confidence Interval
All adults in the sample	.34	$.34 \pm z\, s_{\hat{p}}$
Democrats	.41	$.41 \pm z\, s_{\hat{p}}$
Republicans	.36	$.36 \pm z\, s_{\hat{p}}$
Independents	.27	$.27 \pm z\, s_{\hat{p}}$

For each of the four confidence intervals listed in the table, we can substitute the value of z and of $s_{\hat{p}}$ calculated from $s_{\hat{p}} = \sqrt{\frac{\hat{p}\hat{q}}{n}}$. For example, a 96% confidence interval for the proportion of all adults who hold the opinion that "the government, poor leadership, or politicians" is the most important problem currently facing the United States is determined as follows:

$$s_{\hat{p}} = \sqrt{\frac{\hat{p}\hat{q}}{n}} = \sqrt{\frac{(.34)(.66)}{1526}} = .01212646$$

$$\hat{p} \pm z\, s_{\hat{p}} = .34 \pm 2.05\,(.01212646)$$
$$= .34 \pm .025 = .315 \text{ to } .365$$

Thus, we can state with a 96% confidence that 31.5% to 36.5% of all adults will say that the government, poor leadership, or politicians is the most important problem currently facing the United States.

Using this same method, we can find a confidence interval for the population proportion for each of the other three categories if we know the number of adults in the sample who belong to each of the three political groups.

Source: Adapted from Mentions of Government as Top U.S. Problem Near Record High, Megan Brenan. Gallup 2019. https://news.gallup.com/poll/267581/mentions-govemment-top-problem-near-recordhigh.aspx

8.4.1 Determining the Sample Size for the Estimation of Proportion

Just as we did with the mean, we can also determine the minimum sample size for estimating the population proportion to within a predetermined margin of error, E. By knowing the sample size that can give us the required results, we can save our scarce resources by not taking an unnecessarily large sample. From earlier discussion in this section, the margin of error, E, of the interval estimate of the population proportion is

$$E = z\, s_{\hat{p}} = z \times \sqrt{\frac{\hat{p}\hat{q}}{n}}$$

By manipulating this expression algebraically, we obtain the following formula to find the required sample size for given values of E, $\hat{p}$, $\hat{q}$, and z.

Determining the Sample Size for the Estimation of p

Given the confidence level and the values of $\hat{p}$ and $\hat{q}$, the sample size that will produce a predetermined margin of error E of the confidence interval estimate of p is

$$n = \frac{z^2 \hat{p}\hat{q}}{E^2}$$

We can observe from this formula that to find n, we need to know the values of $\hat{p}$ and $\hat{q}$. However, the values of $\hat{p}$ and $\hat{q}$ are not known to us. In such a situation, we can choose one of the following two alternatives:

1. We make the **most conservative estimate** of the sample size n by using $\hat{p} = .50$ and $\hat{q} = .50$. For a given E, these values of $\hat{p}$ and $\hat{q}$ will give us the largest sample size in comparison to any other pair of values of $\hat{p}$ and $\hat{q}$ because the product of $\hat{p} = .50$ and $\hat{q} = .50$ is greater than the product of any other pair of values for $\hat{p}$ and $\hat{q}$.
2. We take a **preliminary sample** (of arbitrarily determined size) and calculate $\hat{p}$ and $\hat{q}$ for this sample. Then, we use these values of $\hat{p}$ and $\hat{q}$ to find n.

Examples 8.9 and 8.10 illustrate how to determine the sample size that will produce the error of estimation for the population proportion within a predetermined margin of error value. Example 8.9 gives the most conservative estimate of n, and Example 8.10 uses the results from a preliminary sample to determine the required sample size.

EXAMPLE 8.9 Proportion of Parts That Are Defective

Determining the Most Conservative Estimate of n for the Estimation of p

Lombard Electronics Company has just installed a new machine that makes a part that is used in clocks. The company wants to estimate the proportion of these parts produced by this machine that are defective. The company manager wants this estimate to be within .02 of the population proportion for a 95% confidence level. What is the most conservative estimate of the sample size that will limit the margin of error to within .02 of the population proportion?

Solution The company manager wants the 95% confidence interval to be

$$\hat{p} \pm .02$$

Therefore,

$$E = .02$$

The value of z for a 95% confidence level is 1.96. For the most conservative estimate of the sample size, we will use $\hat{p} = .50$ and $\hat{q} = .50$. Hence, the required sample size is

$$n = \frac{z^2 \hat{p}\hat{q}}{E^2} = \frac{(1.96)^2(.50)(.50)}{(.02)^2} = \mathbf{2401}$$

Thus, if the company takes a sample of 2401 parts, there is a 95% chance that the estimate of p will be within .02 of the population proportion.

EXAMPLE 8.10 Proportion of Parts That Are Defective

Determining n for the Estimation of p Using Preliminary Sample Results

Consider Example 8.9 again. Suppose a preliminary sample of 200 parts produced by this machine showed that 7% of them are defective. How large a sample should the company select so that the 95% confidence interval for p is within .02 of the population proportion?

Solution Again, the company wants the 95% confidence interval for p to be

$$\hat{p} \pm .02$$

Hence,

$$E = .02$$

The value of z for a 95% confidence level is 1.96. From the preliminary sample,

$$\hat{p} = .07 \quad \text{and} \quad \hat{q} = 1 - .07 = .93$$

Using these values of $\hat{p}$ and $\hat{q}$, we obtain

$$n = \frac{z^2 \hat{p}\hat{q}}{E^2} = \frac{(1.96)^2(.07)(.93)}{(.02)^2} = \frac{(3.8416)(.07)(.93)}{.0004} = 625.22 \approx \mathbf{626}$$

Note that if the value of n is not an integer, we always round it up.

Thus, if the company takes a sample of 626 items, there is a 95% chance that the estimate of p will be within .02 of the population proportion. However, we should note that this sample size will produce the margin of error within .02 only if $\hat{p}$ is .07 or less for the new sample. If $\hat{p}$ for the new sample happens to be much higher than .07, the margin of error will not be within .02. Therefore, to avoid such a situation, we may be more conservative and take a sample much larger than 626 items.

EXERCISES

Concepts and Procedures

8.51 What assumption(s) must hold true to use the normal distribution to make a confidence interval for the population proportion, p?

8.52 What is the point estimator of the population proportion, p?

8.53 Check if the sample size is large enough to use the normal distribution to make a confidence interval for p for each of the following cases:

a. $n = 50$ and $\hat{p} = .25$ **b.** $n = 160$ and $\hat{p} = .04$

c. $n = 400$ and $\hat{p} = .65$ **d.** $n = 75$ and $\hat{p} = .05$

8.54 Check if the sample size is large enough to use the normal distribution to make a confidence interval for p for each of the following cases:

a. $n = 120$ and $\hat{p} = .04$ **b.** $n = 60$ and $\hat{p} = .08$

c. $n = 40$ and $\hat{p} = .50$ **d.** $n = 900$ and $\hat{p} = .15$

8.55 a. A random sample of 400 observations taken from a population produced a sample proportion of .63. Make a 95% confidence interval for p.

b. Another sample of 400 observations taken from the same population produced a sample proportion of .59. Make a 95% confidence interval for p.

c. A third sample of 400 observations taken from the same population produced a sample proportion of .67. Make a 95% confidence interval for p.

d. The true population proportion for this population is .65. Which of the confidence intervals constructed in parts (a) through (c) cover this population proportion and which do not?

8.56 A random sample of 500 observations selected from a population produced a sample proportion equal to .68.

a. Make a 90% confidence interval for p.

b. Construct a 95% confidence interval for p.

c. Make a 99% confidence interval for p.

d. Does the width of the confidence intervals constructed in parts (a) through (c) increase as the confidence level increases? If yes, explain why.

8.57 A random sample selected from a population gave a sample proportion equal to .73.

a. Make a 99% confidence interval for p assuming $n = 100$.

b. Construct a 99% confidence interval for p assuming $n = 600$.

c. Make a 99% confidence interval for p assuming $n = 1500$.

d. Does the width of the confidence intervals constructed in parts (a) through (c) decrease as the sample size increases? If yes, explain why.

8.58 a. How large a sample should be selected so that the margin of error of estimate for a 99% confidence interval for p is .035 when the value of the sample proportion obtained from a preliminary sample is .29?

b. Find the most conservative sample size that will produce the margin of error for a 99% confidence interval for p equal to .035.

8.59 a. How large a sample should be selected so that the margin of error of estimate for a 95% confidence interval for p is .050 when the value of the sample proportion obtained from a preliminary sample is .55?

b. Find the most conservative sample size that will produce the margin of error for a 98% confidence interval for p equal to .045.

8.60 Determine the most conservative sample size for the estimation of the population proportion for the following:

a. $E = .03$, confidence level = 99%

b. $E = .04$, confidence level = 95%

c. $E = .01$, confidence level = 90%

Applications

8.61 According to a recent Gallup poll, 67% of U.S. adults were dissatisfied with the way income and wealth are distributed in the United States. Assume that this poll is based on a random sample of 1500 U.S. adults.

a. What is the point estimate of the corresponding population proportion?

b. Construct a 98% confidence interval for the proportion of all such U.S. adults who are dissatisfied with the way income and wealth are distributed in the United States. What is the margin of error for this estimate?

8.62 In a Gallup poll conducted April 1–9, 2019, participants were asked to rate their current financial situation as excellent, good, only fair, or poor. From the results based on a random sample of U.S. adults, ages 18 and over, and living in all 50 states and the District of Columbia, 12% of the respondents rated their current financial situation as excellent. Suppose this poll was based on a random sample of 1500 such U.S. adults.

a. What is the point estimate of the corresponding population proportion?

b. Construct a 95% confidence interval for the proportion of all such U.S. adults who will rate their current financial situation as excellent.

8.63 It is said that happy and healthy workers are efficient and productive. A company that manufactures exercising machines wanted to know the percentage of large companies that provide on-site health club facilities. A random sample of 240 such companies showed that 96 of them provide such facilities on site.

a. What is the point estimate of the percentage of all such companies that provide such facilities on site?

b. Construct a 97% confidence interval for the percentage of all such companies that provide such facilities on site. What is the margin of error for this estimate?

8.64 A mail-order company promises its customers that the products ordered will be mailed within 72 hours after an order is placed. The quality control department at the company checks from time to time to see if this promise is fulfilled. Recently the quality control department took a random sample of 50 orders and found that 35 of them were mailed within 72 hours of the placement of the orders.

a. Construct a 98% confidence interval for the percentage of all orders that are mailed within 72 hours of their placement.

b. Suppose the confidence interval obtained in part (a) is too wide. How can the width of this interval be reduced? Discuss all possible alternatives. Which alternative is the best?

8.65 In a Gallup survey about the U.S. health-care system conducted November 1–11, 2018, 55% of U.S. adults described the U.S. health-care system as having major problems. Assume that this poll was based on a random sample of 900 U.S. adults.

a. Construct a 95% confidence interval for the proportion of all such U.S. adults who would describe the U.S. health-care system as having major problems.

b. Explain why we need to construct a confidence interval. Why can we not simply say that 55% of all U.S. adults would describe the U.S. health-care system as having major problems.

8.66 A researcher wants to know the percentage of judges who are in favor of the death penalty. He took a random sample of 16 judges and asked them whether or not they favor the death penalty. The responses of these judges are given here:

Yes	No	Yes	Yes	Yes	No	No	Yes
Yes	Yes	Yes	Yes	Yes	No	Yes	Yes

a. What is the point estimate of the population proportion?

b. Make a 95% confidence interval for the percentage of all judges who are in favor of the death penalty.

8.67 The management of a health insurance company wants to know the percentage of its policyholders who have tried alternative treatments (such as acupuncture, herbal therapy, etc.). A random sample of 24 of the company's policyholders were asked whether or not they have ever tried such treatments. The following are their responses:

Yes	No	No	Yes	No	Yes	No	No
No	Yes	No	No	Yes	No	Yes	No
No	No	Yes	No	No	No	Yes	No

a. What is the point estimate of the corresponding population proportion?

b. Construct a 99% confidence interval for the percentage of this company's policyholders who have tried alternative treatments.

8.68 A consumer agency wants to estimate the proportion of all drivers who wear seat belts while driving. Assume that a preliminary study has shown that 76% of drivers wear seat belts while driving. How large should the sample size be so that the 99% confidence interval for the population proportion has a margin of error of .03?

8.69 A consumer agency wants to estimate the proportion of all drivers who wear seat belts while driving. What is the most conservative estimate of the minimum sample size that would limit the margin of error to within .03 of the population proportion for a 99% confidence interval?

***8.70** You want to estimate the proportion of students at your college who hold off-campus (part-time or full-time) jobs. Briefly explain how you will make such an estimate. Collect data from 35 students at your college on whether or not they hold off-campus jobs. Then calculate the proportion of students in this sample who hold off-campus jobs. Using this information, estimate the population proportion. Select your own confidence level.

***8.71** You want to estimate the percentage of students at your college or university who are satisfied with the campus food services. Briefly explain how you will make such an estimate. Select a sample of 35 students and ask them whether or not they are satisfied with the campus food services. Then calculate the percentage of students in the sample who are satisfied. Using this information, find the confidence interval for the corresponding population percentage. Select your own confidence level.

Uses and Misuses of Statistics

National Versus Local Unemployment Rate

Reading a newspaper article, you learn that the national unemployment rate is 5.1%. The next month you read another article that states that a recent survey in your area, based on a random sample of the labor force, estimates that the local unemployment rate is 4.7% with a margin of error of .5%. Thus, you conclude that the unemployment rate in your area is somewhere between 4.2% and 5.2%.

So, what does this say about the local unemployment picture in your area versus the national unemployment situation? Since a major portion of the interval for the local unemployment rate is below 5.1%, is it reasonable to conclude that the local unemployment rate is below the national unemployment rate? Not really. When looking at the confidence interval, you have some degree of confidence, usually between 90% and 99%. If we use $z = 1.96$ to calculate the margin of error, which is the z value for a 95% confidence level, we can state that we are 95% confident that the local unemployment rate falls in the interval we obtain by using the margin of error. However, since 5.1% is in the interval for the local unemployment rate, the one thing that you can say is that it appears reasonable to conclude that the local and national unemployment rates are not different. However, if the national rate was 5.3%, then a conclusion that the two rates differ is reasonable because we are confident that the local unemployment rate falls between 4.2% and 5.2%.

When making conclusions based on the types of confidence intervals you have learned and will learn in this course, you will only be able to conclude that either there is a difference or there is not a difference. However, the methods you will learn in Chapter 9 will also allow you to determine the validity of a conclusion that states that the local rate is lower (or higher) than the national rate.

Glossary

Confidence interval An interval constructed around the value of a sample statistic to estimate the corresponding population parameter.

Confidence level Confidence level, denoted by $(1 - \alpha)100\%$, states how much confidence we have that a confidence interval contains the true population parameter.

Degrees of freedom (df) The number of observations that can be chosen freely. For the estimation of μ using the t distribution, the degrees of freedom is $n - 1$.

Estimate The value of a sample statistic that is used to estimate the corresponding population parameter.

Estimation A procedure by which a numerical value or values are assigned to a population parameter based on the information collected from a sample.

Estimator The sample statistic that is used to estimate a population parameter.

Interval estimate An interval constructed around the point estimate that is likely to contain the corresponding population parameter. Each interval estimate has a confidence level.

Margin of error The quantity that is subtracted from and added to the value of a sample statistic to obtain a confidence interval for the corresponding population parameter.

Point estimate The value of a sample statistic assigned to the corresponding population parameter.

***t* distribution** A continuous probability distribution with a specific type of bell-shaped curve with its mean equal to 0 and standard deviation equal to $\sqrt{df/(df-2)}$ for $df > 2$.

Exercises

Supplementary Exercises

8.72 A bank manager wants to know the mean amount owed on credit card accounts that become delinquent. A random sample of 100 delinquent credit card accounts taken by the manager produced a mean amount owed on these accounts equal to $2640. The population standard deviation was $578.

a. What is the point estimate of the mean amount owed on all delinquent credit card accounts at this bank?

b. Construct a 97% confidence interval for the mean amount owed on all delinquent credit card accounts for this bank.

8.73 York Steel Corporation produces iron rings that are supplied to other companies. These rings are supposed to have a diameter of 25 inches. The machine that makes these rings does not produce each ring with a diameter of exactly 25 inches. The diameter of each of the rings varies slightly. It is known that when the machine is working properly, the rings made on this machine have a mean diameter of 25 inches. The standard deviation of the diameters of all rings produced on this machine is always equal to .08 inch. The quality control department takes a random sample of 30 such rings every week, calculates the mean of the diameters for these rings, and makes a 95% confidence interval for the population mean. If either the lower limit of this confidence interval is less than 24.975 inches or the upper limit of this confidence interval is greater than 25.025 inches, the machine is stopped and adjusted. A recent such sample of 30 rings produced a mean diameter of 25.010 inches. Based on this sample, can you conclude that the machine needs an adjustment? Explain. Assume that the population distribution is approximately normal.

8.74 Yunan Corporation produces bolts that are supplied to other companies. These bolts are supposed to be 5 inches long. The machine that makes these bolts does not produce each bolt exactly 5 inches long but the length of each bolt varies slightly. It is known that when the machine is working properly, the mean length of the bolts made on this machine is 5 inches. The standard deviation of the lengths of all bolts produced on this machine is always equal to .05 inch. The quality control department takes a random sample of 25 such bolts every week, calculates the mean length of these bolts, and makes a 95% confidence interval for the population mean. If either the upper limit of this confidence interval is greater than 5.03 inches or the lower limit of this confidence interval is less than 4.97 inches, the machine is stopped and adjusted. A recent such sample of 25 bolts produced a mean length of 4.98 inches. Based on this sample, will you conclude that the machine needs an adjustment? Assume that the population distribution is approximately normal.

8.75 A hospital administrator wants to estimate the mean time spent by patients waiting for treatment at the emergency room. The waiting times (in minutes) recorded for a random sample of 32 such patients are given below.

110	42	88	19	35	76	10	151
2	44	27	77	53	102	66	39
20	108	92	55	14	52	3	62
78	15	60	121	40	35	11	72

Construct a 99% confidence interval for the corresponding population mean. Use the *t* distribution.

8.76 A random sample of 25 life insurance policyholders showed that the average premium they pay on their life insurance policies is $685 per year with a standard deviation of $74. Assuming that the life insurance policy premiums for all life insurance policyholders have an approximate normal distribution, make a 99% confidence interval for the population mean, μ.

8.77 A survey of 500 randomly selected adult men showed that the mean time they spend per week watching sports on television is 9.75 hours with a standard deviation of 2.2 hours. Construct a 90% confidence interval for the population mean, μ.

8.78 A random sample of 300 female members of health clubs in Los Angeles showed that they spend, on average, 4.5 hours per week doing physical exercise with a standard deviation of .75 hour. Find a 98% confidence interval for the population mean.

8.79 Salaried workers at a large corporation receive two weeks' paid vacation per year. Sixteen randomly selected workers from this corporation were asked whether or not they would be willing to take a 3% reduction in their annual salaries in return for two additional weeks of paid vacation. The following are the responses of these workers:

No	Yes	No	No	Yes	No	No	Yes
Yes	No	No	No	Yes	No	No	No

Construct a 97% confidence interval for the percentage of all salaried workers at this corporation who would accept a 3% pay cut in return for two additional weeks of paid vacation.

8.80 An insurance company selected a random sample of 50 auto claims filed with it and investigated those claims carefully. The company found that 12% of those claims were fraudulent.

a. What is the point estimate of the percentage of all auto claims filed with this company that are fraudulent?

b. Make a 99% confidence interval for the percentage of all auto claims filed with this company that are fraudulent.

8.81 A random sample of 25 managers was taken, and they were asked whether or not they usually take work home. The responses of these managers are given below, where *yes* indicates they usually take work home and *no* means they do not.

Yes	No	Yes	No	No	Yes	No	Yes	No
Yes	No	No	Yes	Yes	Yes	No	Yes	Yes
No	Yes	No	No	No	No	No		

Make a 95% confidence interval for the percentage of all managers who take work home.

8.82 A researcher wants to determine a 95% confidence interval for the mean number of hours that adults spend per day on social media platforms. How large a sample should the researcher select so that the estimate is within 1.5 hours of the population mean? Assume that the standard deviation for time spent per day spending on social media platforms by all adults is 2.5 hours.

8.83 An economist wants to find a 90% confidence interval for the mean sale price of houses in a state. How large a sample should the economist select so that the estimate is within $3500 of the population mean? Assume that the standard deviation for the sale prices of all houses in this state is $31,500.

8.84 A large city with chronic economic problems is considering construction of a biogas plant. The city council wants to estimate the proportion of all adults in the city who favor construction of the biogas plant. What is the most conservative estimate of the minimum sample size that would limit the margin of error to be within .06 of the population proportion for a 90% confidence interval?

8.85 A large city with chronic economic problems is considering opening a small factory manufacturing toys. The city council wants to estimate the proportion of all adults in the city who favor opening of the factory. Assume that a preliminary sample has shown that 70% of the adults in this city favor opening of the factory. How large should the sample size be so that 90% confidence interval for the population proportion has a margin of error of .06?

Advanced Exercises

8.86 Let μ be the hourly wage (excluding tips) for workers who provide hotel room service in a large city. A random sample of a number (more than 30) of such workers yielded a 95% confidence interval for μ of $8.46 to $9.86 using the normal distribution with a known population standard deviation.

a. Find the value of $\bar{x}$ for this sample.

b. Find a 99% confidence interval for μ based on this sample.

8.87 In a Gallup poll conducted December 4–11, 2017, a random sample of U.S. adults were asked to name their favorite sport to watch on television. Of the respondents, 37% selected football, 11% mentioned basketball, 9% liked baseball, and 7% selected soccer. The remaining adults either did not answer the question or liked other sports (https://news.gallup.com). Using these results, find a 98% confidence interval for the population percentage that corresponds to each response. Assume that sample size is 1049. Write a one-page report to present your results to a group of college students who have not taken statistics. Your report should answer questions such as the following: (1) What is a confidence interval? (2) Why is a range of values (interval) more informative than a single percentage (point estimate)? (3) What does 98% confidence mean in this context? (4) What assumptions, if any, are you making when you construct each confidence interval?

8.88 When one is attempting to determine the required sample size for estimating a population mean, and the information on the population standard deviation is not available, it may be feasible to take a small preliminary sample and use the sample standard deviation to estimate the required sample size, *n*. Suppose that we want to estimate μ, the mean commuting distance for students at a community college, to a margin of error within 1 mile with a confidence level of 95%. A random sample of 20 students yields a standard deviation of 4.1 miles. Use this value of the sample standard deviation, *s*, to estimate the required sample size, n. Assume that the corresponding population has an approximate normal distribution.

8.89 A gas station attendant would like to estimate p, the proportion of all households that own more than two vehicles. To obtain an estimate, the attendant decides to ask the next 200 gasoline customers how many vehicles their households own. To obtain an estimate of p, the attendant counts the number of customers who say there are more than two vehicles in their households and then divides this number by 200. How would you critique this estimation procedure? Is there anything wrong with this procedure that would result in sampling and/or nonsampling errors? If so, can you suggest a procedure that would reduce this error?

8.90 The U.S. Senate just passed a bill by a vote of 55–45 (with all 100 senators voting). A student who took an elementary statistics course last semester says, "We can use these data to make a confidence interval about p. We have $n = 100$ and $\hat{p} = 55/100 = .55$." Hence, according to this student, a 95% confidence interval for p is

$$\hat{p} \pm z\sigma_{\hat{p}} = .55 \pm 1.96\sqrt{\frac{(.55)(.45)}{100}} = .55 \pm .098 = .452 \text{ to } .648$$

Does this make sense? If not, what is wrong with the student's reasoning?

8.91 When calculating a confidence interval for the population mean μ with a known population standard deviation σ, describe the effects of the following two changes on the confidence interval: (1) doubling the sample size, (2) quadrupling (multiplying by 4) the sample size. Give two reasons why this relationship does not hold true if you are calculating a confidence interval for the population mean μ with an unknown population standard deviation.

8.92 Calculating a confidence interval for the proportion requires a minimum sample size. Calculate a confidence interval, using any confidence level of 95% or higher, for the population proportion for each of the following:

a. $n = 250$ and $\hat{p} = .02$

b. $n = 180$ and $\hat{p} = .95$

Explain why these confidence intervals reveal a problem when the conditions for using the normal approximation do not hold.

Self-Review Test

Multiple Choice Questions

1. A 95% confidence interval for μ can be interpreted to mean that if we take 100 samples of the same size and construct 100 such confidence intervals for μ, then

a. 95 of them will not include μ

b. 95 will include μ

c. 95 will include $\bar{x}$

2. The confidence level is denoted by

a. $(1 - \alpha)100\%$ **b.** $100\alpha\%$ **c.** α

3. The margin of error of estimate for μ is

a. $z\sigma_{\bar{x}}$ (or $ts_{\bar{x}}$)

b. $\sigma/\sqrt{n}$ (or $s/\sqrt{n}$)

c. $\sigma_{\bar{x}}$ (or $s_{\bar{x}}$)

4. Which of the following assumptions is not required to use the t distribution to make a confidence interval for μ?

a. Either the population from which the sample is taken is (approximately) normally distributed or $n \geq 30$.

b. The population standard deviation, σ, is not known.

c. The sample size is at least 10.

5. The parameter(s) of the t distribution is (are)

a. n

b. degrees of freedom

c. μ and degrees of freedom

6. Estimation is a procedure by which we assign a numerical value or numerical values to the

a. population parameter based on the information collected from a sample

b. sample statistic based on the information collected from a sample

c. population parameter based on the information collected from a population

7. The values assigned to a population parameter based on the value(s) of a sample statistic are called

a. the probability distribution

b. a sampling distribution

c. estimate(s)

8. The (single) value of a sample statistic that we assign to the population parameter is called a

a. single estimate

b. unique estimate

c. point estimate

9. For most distributions, we can use the normal distribution to make a confidence interval for a population mean provided that the population standard deviation σ is known and the sample size is

a. greater than or equal to 10

b. less than 25

c. greater than or equal to 30

10. Which of the following is *not* part of the procedure for estimating the value of a population parameter?

a. Selecting a sample.

b. Collecting the required information from the members of the sample.

c. Calculating the exact value of the corresponding population parameter.

11. The z value for a 90% confidence interval for the population mean with σ known is

a. 2.05

b. 1.645

c. 2.17

12. The width of a confidence interval depends on the size of the

a. population mean

b. margin of error

c. sample mean

13. To decrease the width of a confidence interval, we should always prefer to

a. lower the confidence level

b. increase the sample size

c. decrease the sample size

14. A sample of size 61 from a population having standard deviation 45 produced a mean of 265. The 95% confidence interval for the population mean (rounded to two decimal places) is

a. 253.71 to 276.29

b. 251.58 to 278.42

c. 250.13 to 279.87

15. The mean IQ score of a sample of 61 students selected from a high school is 87. Suppose the standard deviation of IQs at this school is $\sigma = 8.4$. The 99% confidence interval for the population mean (rounded to two decimal places) is

a. 95.47 to 100.53

b. 95.71 to 100.29

c. 96.08 to 99.92

16. We use the t distribution to make a confidence interval for the population mean if the population from which the sample is drawn is (approximately) normally distributed, the population standard deviation is unknown, and the sample size is at least

a. 30

b. 100

c. 2

17. The value of t for 19 degrees of freedom and a 90% confidence interval is

a. 1.729

b. −1.729

c. 2.539

18. A sample of 20 elements produced a mean of 91.4 and a standard deviation of 13.80. Assuming that the population has a normal distribution, the 90% confidence interval for the population mean is

a. 86.06 to 96.74

b. 86.05 to 96.75

c. 86.08 to 96.72

19. A random sample of 12 life insurance policyholders showed that the mean value of their life insurance policies is $210,000 with a standard deviation of $51,600. Assuming that the values of life insurance policies for all such policyholders are approximately normally distributed, the 99% confidence interval for the mean value of all life insurance policies, rounded to two decimal places, is

a. 163,734.15 to 256,265.85

b. 162,795.73 to 257,204.27

c. 164,493.83 to 255,506.17

20. A random sample of 513 produced a sample proportion of .74. The 95% confidence interval for the population proportion, rounded to four decimal places, is

a. $.7020 to $.7780

b. $.6949 to $.7851

c. $.6900 to $.7900

Practice Exercises

1. Complete the following sentences using the terms *population parameter* and *sample statistic.*

a. Estimation means assigning values to a __________ based on the value of a __________.

b. An estimator is a __________ used to estimate a __________.

c. The value of a __________ is called the point estimate of the corresponding __________.

2. We are using the mean of a sample as a point estimate for the mean of a normal distribution with a standard deviation of 5. The margin of error, with 95% confidence, for this estimate is .834. What is the sample size?

3. A researcher wants to make a 99% confidence interval for a population proportion. The most conservative estimate of the sample size that would limit the margin of error to be within .047 of the population proportion is________.

4. The Labor Bureau wants to estimate, at a 90% confidence level, the proportion of all households that receive welfare. The most conservative estimate of the sample size that would limit the margin of error to be within .024 of the population proportion is__________.

5. A random sample of 36 vacation homes built during the past two years in a coastal resort region gave a mean construction cost of $159,000 with a population standard deviation of $27,000.

a. What is the point estimate of the corresponding population mean?

b. Make a 99% confidence interval for the mean construction cost for all vacation homes built in this region during the past two years. What is the margin of error here?

6. A random sample of 25 malpractice lawsuits filed against doctors showed that the mean compensation awarded to the plaintiffs was $610,425 with a standard deviation of $94,820. Find a 95% confidence interval for the mean compensation awarded to plaintiffs of all such lawsuits. Assume that the compensations awarded to plaintiffs of all such lawsuits are approximately normally distributed.

7. Recently, Harris Interactive conducted an online poll of 2097 U.S. adults on the topic "Who are we lying to?" In response to one of the questions, 37% of U.S. adults said that they have lied to get out of work (www.harrisinteractive.com).

a. What is the point estimate of the corresponding population proportion?

b. Construct a 99% confidence interval for the proportion of all U.S. adults who will say that they have lied to get out of work.

8. A company that makes toaster ovens has done extensive testing on the accuracy of its temperature-setting mechanism. For a previous toaster model of this company, the standard deviation of the temperatures when the mechanism is set for 350°F is 5.78°. Assume that this is the population standard deviation for a new toaster model that uses the same temperature mechanism. How large a sample must be taken so that the estimate of the mean temperature when the mechanism is set for 350°F is within 1.25° of the population mean temperature? Use a 95% confidence level.

9. A college registrar has received numerous complaints about the online registration procedure at her college, alleging that the system is slow, confusing, and error prone. She wants to estimate the proportion of all students at this college who are dissatisfied with the online registration procedure. What is the most conservative estimate of the minimum sample size that would limit the margin of error to be within .05 of the population proportion for a 90% confidence interval?

10. Refer to Practice Problem 9. Assume that a preliminary study has shown that 70% of the students surveyed at this college are dissatisfied with the current online registration system. How large a sample should be taken in this case so that the margin of error is within .05 of the population proportion for a 90% confidence interval?

11. Dr. Garcia estimated the mean stress score before a statistics test for a random sample of 25 students. She found the mean and standard deviation for this sample to be 7.1 (on a scale of 1 to 10) and 1.2, respectively. She used a 97% confidence level. However, she thinks that the confidence interval is too wide. How can she reduce the width of the confidence interval? Describe all possible alternatives. Which alternative do you think is best and why?

***12.** You want to estimate the mean number of hours that students at your college work per week. Briefly explain how you will conduct this study using a small sample. Take a sample of 12 students from your college who hold a job. Collect data on the number of hours that these students spent working last week. Then estimate the population mean. Choose your own confidence level. What assumptions will you make to estimate this population mean?

***13.** You want to estimate the proportion of people who are happy with their current jobs. Briefly explain how you will conduct this study. Take a sample of 35 persons and collect data on whether or not they are happy with their current jobs. Then estimate the population proportion. Choose your own confidence level.

Mini-Projects

Mini-Project 8.1

A study conducted by the Oregon Employment Agency and Bureau of Labor Statistics included information on the average annual salaries for a variety of jobs identified by high school students as being careers of interest. The following table contains this information from that study.

Occupation	Average Salary ($)	Educational Requirement
Accountant	63,018	Bachelor's degree
Programmer	70,644	Bachelor's degree
Applications software designer	88,643	Bachelor's degree
Systems software designer	104,005	Bachelor's degree
Fashion designer	66,544	Bachelor's degree
Firefighter	52,248	Postsecondary training
Forensic scientist	57,887	Bachelor's degree
Photographer	39,534	On-the-job training
Physician or surgeon	180,390	Professional degree
Police officer	59,088	On-the-job training
Preschool teacher	25,017	Associate degree
Elementary school teacher	52,549	Bachelor's degree
High school teacher	52,919	Bachelor's degree
Postsecondary school teacher	77,289	Master's degree
Veterinarian	80,788	Professional degree

Source: Data from Oregon Employment Department and Bureau of Labor Statistics.

The study did not mention sample sizes for the average salary calculations, nor were any standard deviations provided. For the purpose of this mini-project, we will assume that the average salaries were calculated using random samples of 50 workers from each occupation and that the distribution of salaries is approximately normal for each occupation. Furthermore, we will assume that the sample standard deviation for each of the listed occupations is equal to 10% of the average salary. For example, the sample standard deviation of the accountants' salaries will be $6301.80, the sample standard deviation of the programmers' salaries will be $7064.40, and so on. Calculate a 95% confidence interval for the population mean that corresponds to each of the sample average salaries listed in the table. Compare the intervals for the various occupations. Based on these intervals, rank these occupations from highest to lowest salary. Identify the occupations that have overlapping confidence intervals. What do you conclude about the average salaries of the occupations that have overlapping confidence intervals? Explain your reasoning.

Mini-Project 8.2

Refer to Data Set III on the NFL players from the 2019 football season that accompanies this text (see Appendix A). Consider the heights of the players (in inches) listed in column 4.

a. Take a random sample of 15 players and find a 95% confidence interval for the mean height μ. Assume that the heights of these players are normally distributed.

b. Repeat part (a) for samples of size 31 and 60, respectively.

c. Compare the widths of your three confidence intervals.

d. Now calculate the mean, μ, of the heights of all players. Do all of your confidence intervals contain this μ? If not, which ones do not contain μ?

Mini-Project 8.3

In recent years, merchants and advertisers have reacted to the increase in the purchasing power, whether direct or indirect, of tweens, teens, and young adults. In Harris Interactive's Youth Trends survey, a random sample of Americans of age 8 to 24 years old were asked whether they would purchase or impact the purchase of tickets to entertainment or sporting events over the next month. The following table shows the results for the three age groups 8 to 12, 13 to 17, and 18 to 24 year:

Age Group	Percentage Answering Yes
8–12	40
13–17	43
18–24	45

As shown in the table, 45% of the 18- to 24-year olds said that they will purchase or impact the purchase of tickets to entertainment or sporting events over the next month.

a. Using the results given in the table, calculate a 95% confidence interval for the proportion of all Americans of age 18 to 24 years who will say that they will purchase or impact the purchase of tickets to entertainment or sporting events over the next month. Assume that the survey included 434 Americans of age 18 to 24 years.

b. Take a random sample of 50 college students within the specified age group (18 to 24 years) and ask them the same question. Using your results, calculate a 95% confidence interval for the proportion of all Americans of age 18 to 24 years who will say that they will purchase or impact the purchase of tickets to entertainment or sporting events over the next month. If the sample size is not large enough to use the normal approximation, increase the sample size to 75.

c. Compare the confidence intervals calculated in parts (a) and (b). Are the results consistent between the two surveys?

d. Is there any reason to believe that your results are not representative of the population of interest, which is all Americans of age 18 to 24 years? Explain why.

Decide for Yourself

Note: The Decide for Yourself feature is located in the book's product page on *www.wiley.com*.

Chapter 8 Technology Instructions

Note: Complete TI-84, Minitab, and Excel manuals are available in the book's product page on *www.wiley.com*.

TI-84 Color/TI-84

The TI-84 Color Technology Instructions feature of this text is written for the TI-84 Plus C color graphing calculator running the 4.0 operating system. Some screens, menus, and functions will be slightly different in older operating systems. The TI-84 and TI-84 Plus can perform all of the same functions but will not have the "Color" option referenced in some of the menus.

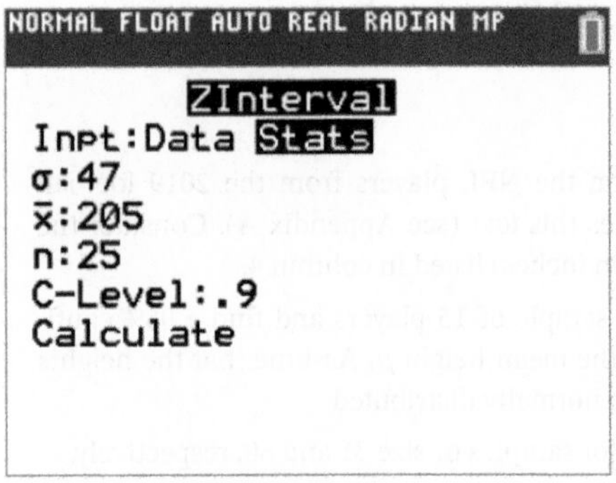

SCREEN 8.1

Estimating the Population Mean, σ, Known for Example 8.1(b) of the Text

1. Select **STAT > TESTS > ZInterval.**
2. Use the following settings in the **Zinterval** menu (see **Screen 8.1**):
   - Select **Stats** at the **Inpt** prompt.

     Note: If you have the data in a list, select **Data** at the **Inpt** prompt.
   - At the σ prompt, type 47.
   - At the $\bar{x}$ prompt, type 205.
   - At the **n** prompt, type 25.
   - At the **C-Level** prompt, type 0.90.
3. Highlight **Calculate** and press **ENTER.**
4. The output includes the confidence interval 189.54 to 220.46. (See **Screen 8.2.**)

   Note: The confidence interval from your calculator may differ slightly from the one in the text since the calculator uses a more precise value for z.

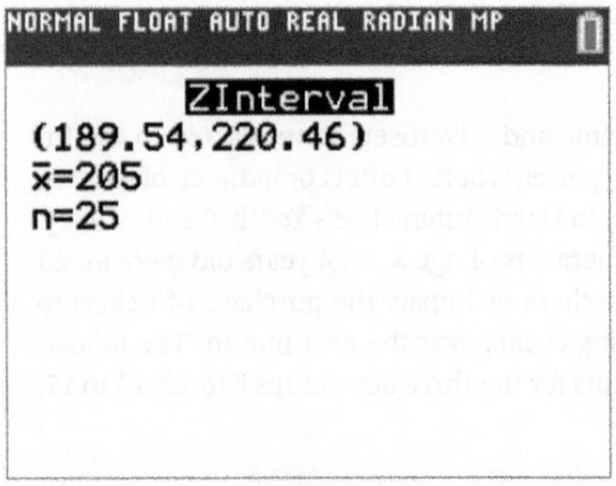

SCREEN 8.2

Estimating the Population Mean, σ, Unknown for Example 8.6 of the Text

1. Select **STAT > TESTS > TInterval.**
2. Use the following settings in the **Tinterval** menu (see **Screen 8.3**):
   - At the **Inpt** prompt, select **Stats.**

     Note: If you have the data in a list, select **Data** at the **Inpt** prompt.
   - At the $\bar{x}$ prompt, type 1450.
   - At the **Sx** prompt, type 300.
   - At the **n** prompt, type 64.
   - At the **C-Level** prompt, type 0.99.
3. Highlight **Calculate** and press **ENTER.**

4. The output includes the confidence interval 1350.4 to 1549.6. (See **Screen 8.4.**)

   Note: The confidence interval from your calculator may differ slightly from the one in the text since the calculator uses a more precise value for *t*.

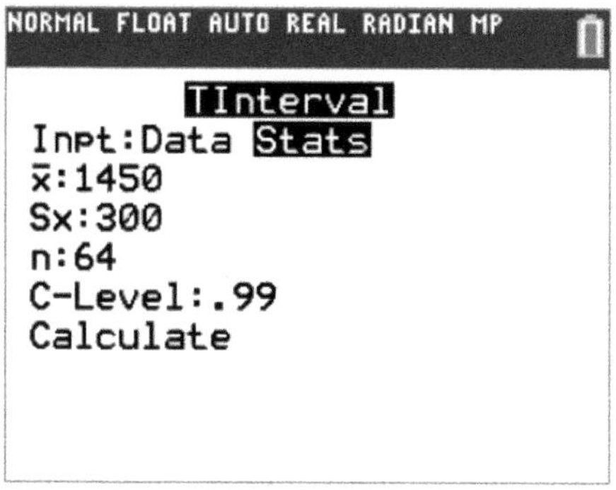

SCREEN 8.3

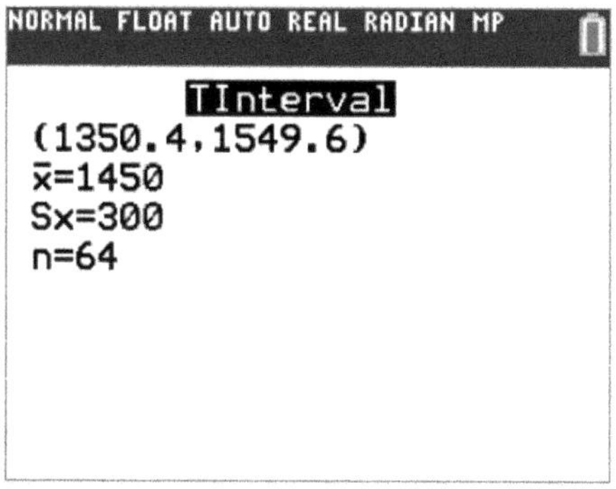

SCREEN 8.4

Estimating the Population Proportion for Example 8.7 of the Text

1. Select **STAT > TESTS > 1-PropZInt.**
2. Use the following settings in the **1-PropZInt** menu (see **Screen 8.5**):
   - At the **x** prompt, type 816.

     Note: The value of **x** is the number of successes in the sample, and it must be a whole number or the calculator will return an error message. If **x** is not given, multiply *n* by $\hat{p}$ to obtain **x** and round the result to the nearest whole number.
   - At the **n** prompt, type 1600.
   - At the **C-Level** prompt, type 0.99.
3. Highlight **Calculate** and press **ENTER.**
4. The output includes the confidence interval .47781 to .54219. (See **Screen 8.6.**)

   Note: The confidence interval from your calculator may differ slightly from the one in the text since the calculator uses a more precise value for *z*.

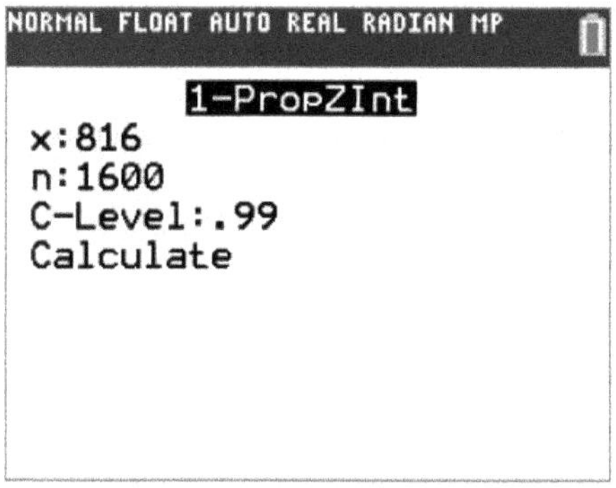

SCREEN 8.5

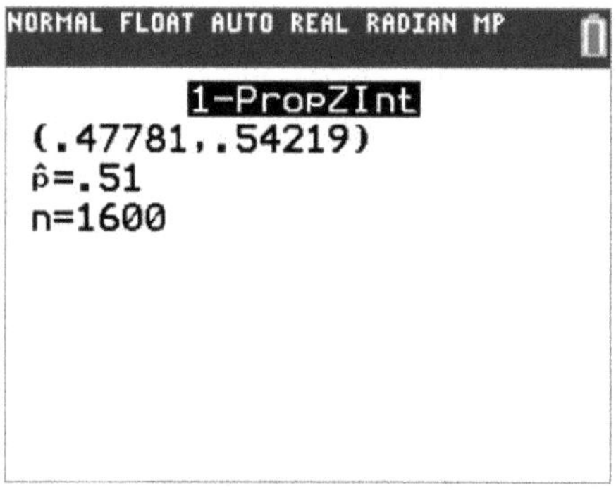

SCREEN 8.6

Minitab

The Minitab Technology Instructions in this text are written for **Minitab Professional Version 19**. Some screens, menus, and functions will be slightly different in older (or in newer) versions of Minitab. If you are using **Minitab Express**, you can download the Minitab Express Manual from the book's product page on *www.wiley.com*.

Estimating the Population Mean, σ, Known for Example 8.1(b) of the Text

1. Select **Stat > Basic Statistics > 1-Sample Z.**
2. Use the following settings in the dialog box that appears on screen (see **Screen 8.7**):
   - From the dropdown box, select **Summarized Data.**

     Note: If you have the data in a column, select **One or more samples, each in a column**, type the column name(s) in the box, and move to Step 3.
   - In the **Sample size** box, type 25.
   - In the **Sample mean** box, type 205.
   - In the **Known standard deviation** box, type 47.

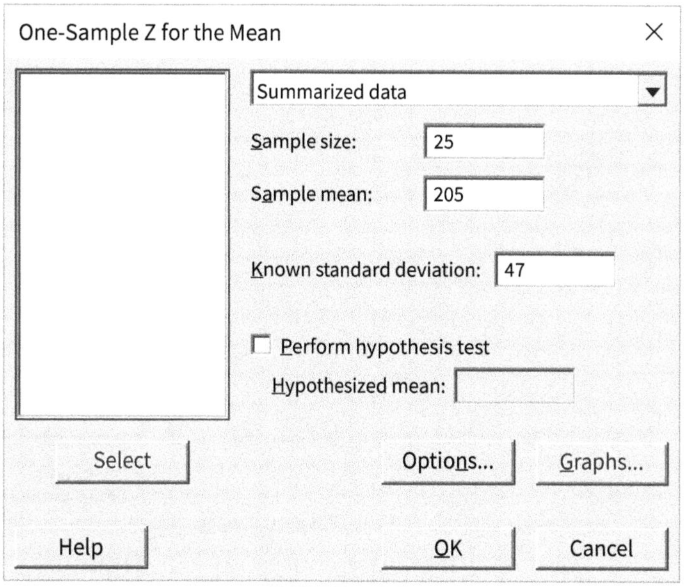

SCREEN 8.7

3. Click on **Options** and type 90 in the **Confidence level** box when the new dialog box appears on screen. (See **Screen 8.8**.)
4. Click **OK** in both dialog boxes.
5. The confidence interval will be displayed in the Session window, which is 189.54 to 220.46. (See **Screen 8.9**.)

   Note: The confidence interval from Minitab may differ slightly from the one in the text since Minitab uses a more precise value for z.

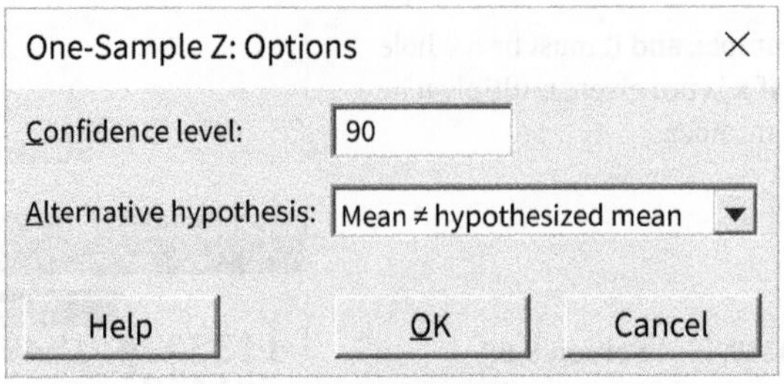

SCREEN 8.8

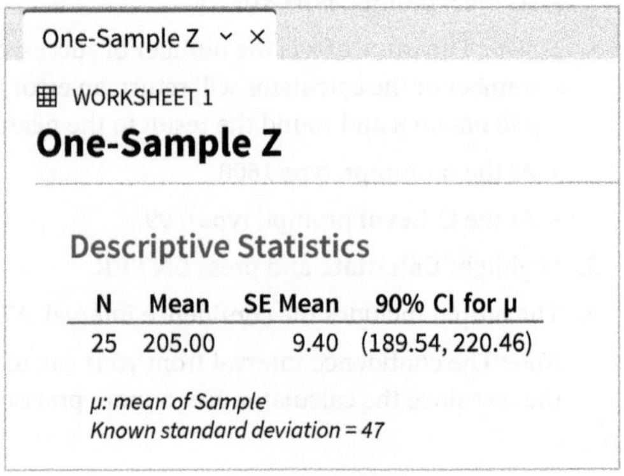

SCREEN 8.9

Estimating the Population Mean, σ, Unknown for Example 8.6 of the Text

1. Select **Stat > Basic Statistics > 1-Sample t.**
2. Use the following settings in the dialog box that appears on screen (see **Screen 8.10**):
   - Select **Summarized Data.**

     Note: If you have the data in a column, select **Samples in columns**, type the column name(s) in the box, and move to Step 3.
   - In the **Sample size** box, type 64.
   - In the **Sample mean** box, type 1450.
   - In the **Standard deviation** box, type 300.
3. Click on **Options** and type 99 in the **Confidence level** box when the new dialog box appears on screen.
4. Click **OK** in both dialog boxes.
5. The confidence interval will be displayed in the Session window.

   Note: The confidence interval from Minitab may differ slightly from the one in the text since Minitab uses a more precise value for t.

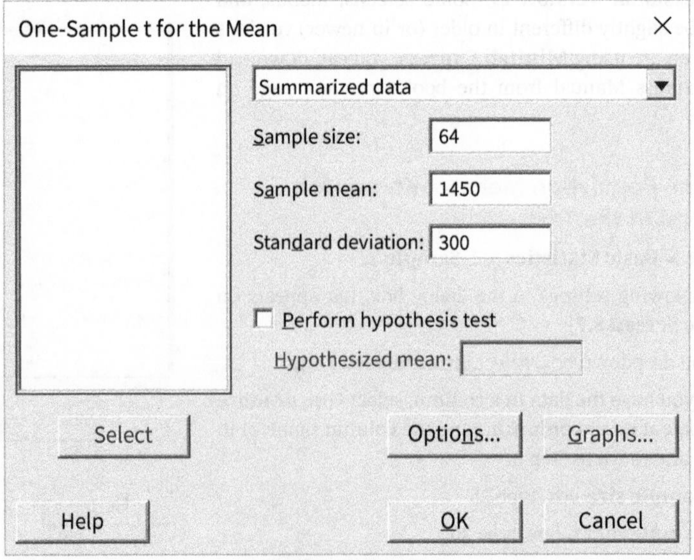

SCREEN 8.10

Estimating the Population Proportion for Example 8.7 of the Text

1. Select **Stat > Basic Statistics > 1 Proportion.**
2. Use the following settings in the dialog box that appears on screen (see **Screen 8.11**):
   - From the drop-down box, select **Summarized Data.**

     Note: If you have the data in a column, select **One or more samples, each in a column**, type the column name(s) in the box, and move to Step 3.
   - In the **Number of events** box, type 816.

     Note: The value 816 is the number of successes in the sample, and it must be a whole number or Minitab will return an error message. If the number of successes is not given, multiply n by $\hat{p}$ to obtain the number of successes and round the result to the nearest whole number.
   - In the **Number of trials** box, type 1600.
3. Click on **Options.** Use the following settings in the new dialog box that appears on screen:
   - In the **Confidence level** box, type 99.
   - In the Method box, select **Normal approximation** from the drop-down menu.
4. Click **OK** in both dialog boxes.
5. The confidence interval will be displayed in the Session window.

   Note: The confidence interval from Minitab may differ slightly from the one in the text since Minitab uses a more precise value for z.

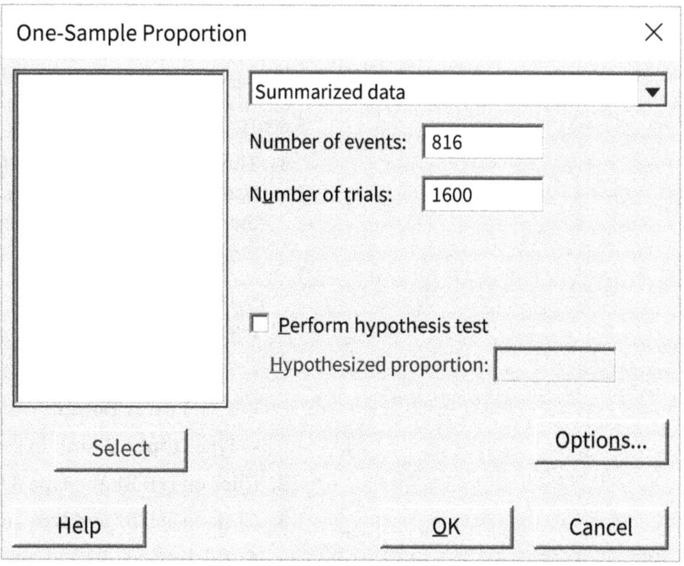

SCREEN 8.11

Excel

The Excel Technology Instructions feature of this text is written for Excel 2019. Some screens, menus, and functions will be slightly different in older versions of Excel. To enable some of the advanced Excel functions, you must enable the Data Analysis Add-In for Excel. For example, in Excel 2007 and older versions of Excel, replace the function **CONFIDENCE.NORM** with the function **CONFIDENCE** and replace the function **NORM.S.INV** with the function **NORMSINV**. For Excel 2007 and older versions of Excel, there is no function that corresponds to **CONFIDENCE.T**.

Estimating the Population Mean, σ, Known for Example 8.1(b) of the Text

1. Click on cell A1.
2. Type **=CONFIDENCE.NORM(0.10,47,25)**. This function calculates the margin of error for the confidence interval. (See **Screen 8.12.**)

   Note: The first number in this function, denoted alpha, is calculated by subtracting the confidence level (as a decimal) from 1. The second number is the population standard deviation, σ, and the third number is the sample size.
3. Click on cell A2 and type **=205 − A1** to calculate the lower bound for the confidence interval. (See **Screen 8.12.**)
4. Click on cell A3 and type **=205 + A1** to calculate the upper bound for the confidence interval. (See **Screen 8.12.**)

   Note: The confidence interval from Excel may differ slightly from the one in the text since Excel uses a more precise value for z.

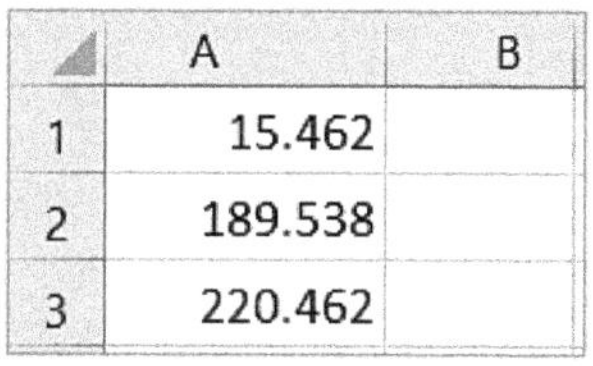

SCREEN 8.12

Estimating the Population Mean, σ, Unknown for Example 8.6 of the Text

1. Click on cell A1.
2. Type **=CONFIDENCE.T(0.01,300,64)**. This function calculates the margin of error for the confidence interval.

   Note: The first number in this function, denoted alpha, is calculated by subtracting the confidence level (as a decimal) from 1. The second number is the sample standard deviation, s, and the third number is the sample size.

3. Click on cell A2 and type **=1450 − A1** to calculate the lower bound for the confidence interval.
4. Click on cell A3 and type **=1450 + A1** to calculate the upper bound for the confidence interval.
   Note: The confidence interval from Excel may differ slightly from the one in the text since Excel uses a more precise value for *t*.

Estimating the Population Proportion for Example 8.7 of the Text

1. There is no Excel function that can be used to create a confidence interval for a population proportion, so we must enter a few calculations in sequence. To make our work easier to understand, type the following text into the corresponding spreadsheet cells (see **Screen 8.13**):
   - Type **p-hat** in cell A1.
   - Type **n** in cell A2.
   - Type **SE of p-hat** in cell A3.
   - Type **Critical z** in cell A4.
   - Type **Lower Bound** in cell A5.
   - Type **Upper Bound** in cell A6.
2. Click on cell B1 and type **0.51.**
3. Click on cell B2 and type **1600.**
4. Click on cell B3 and type **=SQRT(B1*(1-B1)/B2).**
5. Click on cell B4 and type **=NORM.S.INV(0.995).**
   Note: The value 0.995 is calculated by adding the confidence level (as a decimal) to the area in the left tail (as a decimal).
6. Click on cell B5 and type **=B1–B4*B3.**
7. Click on cell B6 and type **=B1+B4*B3.**
8. The lower bound of the interval is in cell B5 and the upper bound of the interval is in cell B6. (See **Screen 8.13.**)
   Note: The confidence interval from Excel may differ slightly from the one in the text since Excel uses a more precise value for *z*.

	A	B
1	p-hat	0.51
2	n	1600
3	SE for p-hat	0.0125
4	Critical z	2.5758
5	Lower Bound	0.4778
6	Upper Bound	0.5422

SCREEN 8.13

Technology Assignments

TA8.1 The following data give the annual incomes (in thousands of dollars) before taxes for a sample of 36 randomly selected families from a city.

61.6	33.0	25.6	37.9	80.0	148.1
50.1	21.5	70.0	72.8	58.2	85.4
91.2	57.0	72.2	45.0	95.0	47.8
92.8	79.4	45.3	76.0	48.6	69.3
40.6	69.0	75.5	57.5	159.7	75.1
96.3	74.5	84.0	43.0	61.7	126.0

Construct a 99% confidence interval for μ assuming that the population standard deviation is \$23.75 thousand.

TA8.2 The following data give the checking account balances (in dollars) on a certain day for a randomly selected sample of 30 households.

500	100	650	1917	2200	500	180	3000	1500	1300
319	1500	1102	405	124	1000	134	2000	150	800
200	750	300	2300	40	1200	500	900	20	160

Construct a 97% confidence interval for μ assuming that the population standard deviation is unknown.

TA8.3 Refer to Data Set I, Home Prices in Wilmington, NC, that accompanies this text (see Appendix A). Using the data on the sale price of these homes, make a 98% confidence interval for the population mean μ.

TA8.4 Refer to Data Set IV on the Manchester Road Race for all participants (see Appendix A). Take a sample of 100 observations from this data set.

a. Using the sample data, make a 95% confidence interval for the mean net time (listed in column 3) taken to complete this race by all participants.

b. Now calculate the mean time taken to run this race by all participants. Does the confidence interval made in part (a) include this population mean?

TA8.5 Repeat Technology Assignment TA8.4 for a sample of 40 observations.

TA8.6 The following data give the prices (in thousands of dollars) of 16 recently sold houses in an area.

341	163	327	204	197	203	313	279
456	228	383	289	533	399	271	381

Construct a 99% confidence interval for the mean price of all houses in this area. Assume that the distribution of prices of all houses in the given area is approximately normal.

TA8.7 A researcher wanted to estimate the mean contributions made to charitable causes by major companies. A random sample of 18 companies produced the following data on contributions (in millions of dollars) made by them.

1.8	.6	1.2	.3	2.6	1.9	3.4	2.6	.2
2.4	1.4	2.5	3.1	.9	1.2	2.0	.8	1.1

Make a 98% confidence interval for the mean contributions made to charitable causes by all major companies. Assume that the contributions made to charitable causes by all major companies have an approximate normal distribution.

TA8.8 A mail-order company promises its customers that their orders will be processed and mailed within 72 hours after an order is placed. The quality control department at the company checks from time to time to see if this promise is kept. Recently the quality control department took a sample of 200 orders and found that 176 of them were processed and mailed within 72 hours of the placement of the orders. Make a 98% confidence interval for the corresponding population proportion.

TA8.9 One of the major problems faced by department stores is a high percentage of returns. The manager of a department store wanted to estimate the percentage of all sales that result in returns. A random sample of 500 sales showed that 95 of them had products returned within the time allowed for returns. Make a 99% confidence interval for the corresponding population proportion.

TA8.10 One of the major problems faced by auto insurance companies is the filing of fraudulent claims. An insurance company carefully investigated 1000 randomly selected auto claims filed with it and found 108 of them to be fraudulent. Make a 96% confidence interval for the corresponding population proportion.

TA 8.11 A manufacturer of contact lenses must ensure that each contact lens is properly manufactured with no flaws since flaws lead to poor vision or even eye damage. In a recent quality control check, 48 of 5000 randomly selected lenses were found to have flaws.

a. Construct a 98% confidence interval for the population proportion of lenses that have flaws.

b. The company advertises that less than 1% of all its lenses have flaws. Based on the interval you constructed in part (a), do you have reason to doubt this claim?

TA 8.12 Chicken eggs are graded AA, A, or B according to strict guidelines set by the USDA. A producer of chicken eggs selects a random sample of 500 eggs and finds that 463 of these are grade A. Construct a 92% confidence interval for the relevant population parameter.

CHAPTER 9

Hypothesis Tests About the Mean and Proportion

ImagesBazaar/Getty Images

Will you graduate from college with debt? If yes, how much money do you think you will owe? Do you know that in the United States students who graduated from public and private nonprofit colleges in 2018 with loans had an average debt of \$29,200? The average debt of the class of 2018 varied a great deal from state to state, with the highest average debt for students who graduated from colleges in Connecticut at \$38,669 and the lowest average for students who graduated from colleges in Utah at \$19,728. (See Case Study 9.1.)

This chapter introduces the second topic in inferential statistics: tests of hypotheses. In a test of hypothesis, we test a certain given theory or belief about a population parameter. We may want to find out, using some sample information, whether or not a given claim (or statement) about a population parameter is true. This chapter discusses how to make such tests of hypotheses about the population mean, μ, and the population proportion, p.

As an example, a soft-drink company may claim that, on average, its cans contain 12 ounces of soda. A consumer agency may want to test whether or not such cans do contain, on average, 12 ounces of soda. As another example, in a Gallup poll conducted a few years ago, U.S. adults were asked, "Do you feel that the distribution of money and wealth in this country today is fair, or do you feel that the money and wealth in this country should be more evenly distributed among a larger percentage of the people?" Of the adults included in the poll, 63% said that money and wealth in this country should be more evenly distributed among a larger percentage of the people. A researcher wants to check if this percentage is still true. In the first of these two examples we are to test a hypothesis about the population mean, μ, and in the second example we are to test a hypothesis about the population proportion, p.

9.1 Hypothesis Tests: An Introduction

LEARNING OBJECTIVES

After completing this section, you should be able to:

- Explain the difference between the null hypothesis and the alternative hypothesis in the context of a test of hypothesis.
- Describe the role of a critical value and significance level in the context of a test of hypothesis.
- Explain the meaning of a rejection region and a nonrejection region in the context of a test of hypothesis.
- Explain the meaning of a Type I error and a Type II error in the context of a test of hypothesis.
- Explain the power of a test.
- Classify a test as a left-tailed test, a right-tailed test, or a two-tailed test.
- Write an appropriate null and alternative hypothesis in the context of a test of hypothesis.

Why do we need to perform a test of hypothesis? Reconsider the example about soft-drink cans. Suppose we take a sample of 100 cans of the soft drink under investigation. We then find out that the mean amount of soda in these 100 cans is 11.89 ounces. Based on this result, can we state that all such cans contain, on average, less than 12 ounces of soda and that the company is lying to the public? Not until we perform a test of hypothesis can we make such an accusation. The reason is that the mean, $\bar{x} = 11.89$ ounces, is obtained from a sample. The difference between 12 ounces (the claimed average amount for the population) and 11.89 ounces (the observed average amount for the sample) may have occurred only because of the sampling error (assuming that no nonsampling errors have been committed). Another sample of 100 cans may give us a mean of 12.04 ounces. Therefore, we perform a test of hypothesis to find out how large the difference between 12 ounces and 11.89 ounces is and whether or not this difference has occurred as a result of chance alone. Now, if we knew that 11.89 ounces is the mean for all cans and not for just 100 cans, then we do not need to make a test of hypothesis. Instead, we can immediately state that the mean amount of soda in all such cans is less than 12 ounces. We perform a test of hypothesis only when we are making a decision about a population parameter based on the value of a sample statistic.

9.1.1 Two Hypotheses

Consider a nonstatistical example of a person who has been indicted for committing a crime and is being tried in a court of law. Based on the available evidence, the judge or jury will make one of two possible decisions:

1. The person is not guilty.
2. The person is guilty.

At the outset of the trial, the person is presumed not guilty. The prosecutor's job is to convince the judge or jury that the person has committed the crime and, hence, is guilty.

In statistics, *the person is not guilty* is called the **null hypothesis** and *the person is guilty* is called the **alternative hypothesis**. The null hypothesis is denoted by $\boldsymbol{H_0}$, and the alternative hypothesis is denoted by $\boldsymbol{H_1}$. In the beginning of the trial it is assumed that the person is not guilty. The null hypothesis is usually the hypothesis that is assumed to be true to begin with. The two hypotheses for the court case are written as follows (notice the colon after H_0 and H_1):

Null hypothesis:	H_0: The person is not guilty
Alternative hypothesis:	H_1: The person is guilty

In a statistics example, the null hypothesis states that a given claim (or statement) about a population parameter is true. Reconsider the example of the soft-drink company's claim that its cans contain, on average, 12 ounces of soda. In reality, this claim may or may not be true. However, we will initially assume that the company's claim is true (that is, the company is not cheating or lying). To test the claim of the soft-drink company, the null hypothesis will be that the company's claim is true. Let μ be the mean amount of soda in all cans. The company's claim will be true if $\mu = 12$ ounces. Thus, the null hypothesis will be written as

$$H_0: \mu = 12 \text{ ounces} \quad \text{(The company's claim is true.)}$$

In this example, the null hypothesis can also be written as $\mu \geq 12$ ounces because the claim of the company will still be true if the cans contain, on average, more than 12 ounces of soda. The company will be accused of cheating the public only if the cans contain, on average, less than 12 ounces of soda. However, it will not affect the test whether we use an = or a $\geq$ sign in the null hypothesis as long as the alternative hypothesis has a < sign. Remember that in the null hypothesis (and also in the alternative hypothesis) we use a population parameter (such as μ or p) and not a sample statistic (such as $\bar{x}$ or $\hat{p}$).

Null Hypothesis

A *null hypothesis* is a claim (or statement) about a population parameter that is assumed to be true until it is declared false.

The alternative hypothesis in our statistics example will be that the company's claim is false and its soft-drink cans contain, on average, less than 12 ounces of soda, that is, $\mu < 12$ ounces. The alternative hypothesis will be written as

$$H_1: \mu < 12 \text{ ounces} \quad \text{(The company's claim is false.)}$$

Alternative Hypothesis

An *alternative hypothesis* is a claim about a population parameter that will be declared true if the null hypothesis is declared to be false.

Let us return to the example of the court trial. The trial begins with the assumption that the null hypothesis is true—that is, the person is not guilty. The prosecutor assembles all the possible evidence and presents it in court to convince the judge or jury that the null hypothesis is false and the alternative hypothesis is true (that is, the person is guilty). In the case of our statistics example, the information obtained from a sample will be used as evidence to decide whether or not the claim of the company is true. In the court case, the decision made by the judge (or jury) depends on the amount and strength of evidence presented by the prosecutor. At the end of the trial, the judge (or jury) will consider whether or not the evidence presented by the prosecutor is sufficient to declare the person guilty. The amount and strength of evidence that will be considered to be sufficient to declare the person guilty depends on the discretion of the judge (or jury).

9.1.2 Rejection and Nonrejection Regions

In **Figure 9.1**, which represents the court case, the point marked 0 indicates that there is no convincing evidence against the person being tried. The farther we move toward the right on the horizontal axis, the more convincing the evidence is that the person has committed the crime. We have arbitrarily marked a point C on the horizontal axis. Let us assume that a judge (or jury) considers any evidence at point C or to the right of point C to be sufficient and any amount of evidence to the left of C to be insufficient to declare the person guilty. Point C is called the **critical value** or **critical point** in statistics. If the amount and strength of evidence presented by the prosecutor falls in the area to the left of point C, the verdict will reflect that there is not enough convincing evidence to declare the person guilty. Consequently, the accused person will be declared *not guilty*. In statistics, this decision is stated as *do not reject* H_0 or failing to reject H_0. It is equivalent to saying that there is not sufficient evidence to declare the null hypothesis false. The area to the left of point C is called the *nonrejection region*; that is, this is the region where the null hypothesis is not rejected. However, if the amount and strength of evidence falls at point *C* or to the right of point *C*, the verdict will be that there is sufficient evidence to declare the person guilty. In statistics, this decision is stated as *reject* H_0 or *the null hypothesis is false*. Rejecting H_0 is equivalent to saying that *the alternative hypothesis is true*. The area to the right of point *C* (including point *C*) is called the *rejection region*; that is, this is the region where the null hypothesis is rejected.

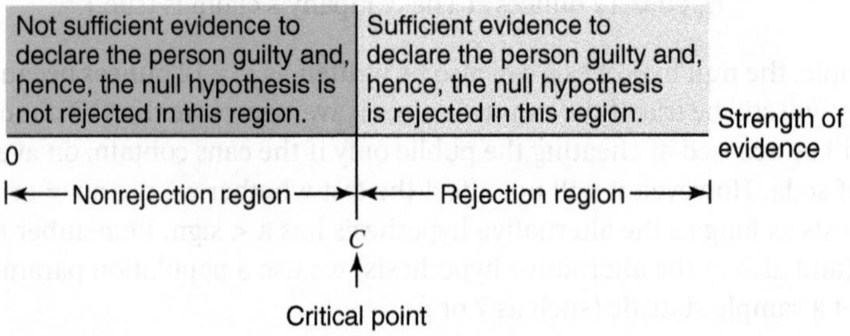

FIGURE 9.1 Nonrejection and rejection regions for the court case.

9.1.3 Two Types of Errors

We all know that a court's verdict is not always correct. If a person is declared guilty at the end of a trial, there are two possibilities:

1. The person has *not* committed the crime but is declared guilty (because of what may be false evidence).
2. The person *has* committed the crime and is correctly declared guilty by the court.

In the first case, the court has made an error by punishing an innocent person. In statistics, this kind of error is called a **Type I** or an **α (*alpha*) error**. In the second case, because the guilty person has been punished, the court has made the correct decision. The second row in **Table 9.1** shows these two cases. The two columns of Table 9.1, corresponding to *the person is not guilty* and *the person is guilty*, give the two actual situations. Which one of these is true is known only to the person being tried. The two rows in this table, corresponding to *the person is not guilty* and *the person is guilty*, show the two possible court decisions.

TABLE 9.1 Four Possible Outcomes for a Court Case

		Actual Situation	
		The Person Is Not Guilty	**The Person Is Guilty**
Court's decision	The person is not guilty	Correct decision	Type II or β error
	The person is guilty	Type I or α error	Correct decision

In our statistics example, a Type I error will occur when H_0 is actually true (that is, the cans do contain, on average, 12 ounces of soda), but it just happens that we draw a sample with a mean that is much less than 12 ounces and wrongfully reject the null hypothesis, H_0. The value of α, called the **significance level** of the test, represents the probability of making a Type I error. In other words, α is the probability of rejecting the null hypothesis, H_0, when in fact H_0 is true.

Type I Error

A *Type I error* occurs when a true null hypothesis is rejected. The value of α represents the probability of committing a Type I error; that is,

$$\alpha = P(H_0 \text{ is rejected} \mid H_0 \text{ is true})$$

The value of α represents the *significance level* of the test.

The size of the **rejection region** in a statistics problem of a test of hypothesis depends on the value assigned to α. In one approach to the test of hypothesis, we assign a value to α before making the test. Although any value can be assigned to α, commonly used values of α are .01, .025, .05, and .10. Usually the value assigned to α does not exceed .10 (or 10%).

Now, suppose that in the court trial case the person is declared not guilty at the end of the trial. Such a verdict does not mean that the person has indeed *not* committed the crime. It is possible that the person is guilty but there is not enough convincing evidence to prove the guilt. Consequently, in this situation there are again two possibilities:

1. The person has *not* committed the crime and is declared not guilty.
2. The person *has* committed the crime but, *because of the lack of enough evidence*, is declared not guilty by the court.

In the first case, the court's decision is correct. In the second case, however, the court has committed an error by setting a guilty person free. In statistics, this type of error is called a **Type II** or a β (the Greek letter *beta*) **error.** These two cases are shown in the first row of Table 9.1.

In our statistics example, a Type II error will occur when the null hypothesis, H_0, is actually false (that is, the soda contained in all cans is, on average, less than 12 ounces), but it happens by chance that we draw a sample with a mean that is close to or greater than 12 ounces and we wrongfully conclude *do not reject* H_0. The value of β represents the probability of making a Type II error. It represents the probability that H_0 is not rejected when actually H_0 is false. The value of $1 - \beta$ is called the **power of the test**. It represents the probability of not making a Type II error. In other words, $1 - \beta$ is the probability that H_0 is rejected when, in fact, H_0 is false.

Type II Error

A *Type II error* occurs when a false null hypothesis is not rejected. The value of β represents the probability of committing a Type II error; that is,

$$\beta = P(H_0 \text{ is not rejected} \mid H_0 \text{ is false})$$

The value of $1 - \beta$ is called the *power of the test*. It represents the probability of rejecting H_0 when it is false.

The two types of errors that occur in tests of hypotheses depend on each other. We cannot lower the values of α and β simultaneously for a test of hypothesis for a fixed sample size. Lowering the value of α will raise the value of β, and lowering the value of β will raise the value of α. However, we can decrease both α and β simultaneously by increasing the sample size. The explanation of how α and β are related and the computation of β are not within the scope of this text.

Table 9.2, which is similar to Table 9.1, is written for the statistics problem of a test of hypothesis. In Table 9.2 *the person is not guilty* is replaced by H_0 *is true, the person is guilty* by H_0 *is false*, and the *court's decision* by *decision.*

TABLE 9.2 Four Possible Outcomes for a Test of Hypothesis

		Actual Situation	
		H_0 Is True	H_0 Is False
Decision	Do not reject H_0	Correct decision	Type II or β error
	Reject H_0	Type I or α error	Correct decision

9.1.4 Tails of a Test

The statistical hypothesis-testing procedure is similar to the trial of a person in court but with two major differences. The first major difference is that in a statistical test of hypothesis, the partition of the total region into rejection and nonrejection regions is not arbitrary. Instead, it depends on the value assigned to α (the probability of a Type I error). As mentioned earlier, α is also called the significance level of the test.

The second major difference relates to the rejection region. In the court case, the rejection region includes the critical point and the area to the right side of the critical point, as shown in Figure 9.1. However, in statistics, the rejection region for a hypothesis-testing problem can be on both sides, with the nonrejection region in the middle, or it can be on the left side or right side of the nonrejection region. These possibilities are explained in the next three parts of this section. A test with two rejection regions is called a **two-tailed test**, and a test with one rejection region is called a **one-tailed test**. The one-tailed test is called a **left-tailed test** if the rejection region is in the left tail of the distribution curve, and is called a **right-tailed test** if the rejection region is in the right tail of the distribution curve.

Tails of the Test

A *two-tailed test* has rejection regions in both tails, a *left-tailed test* has rejection region in the left tail, and a *right-tailed test* has rejection region in the right tail of the distribution curve.

A Two-Tailed Test

According to the U.S. Bureau of Labor Statistics, workers in the United States who were employed earned an average of \$50,400 a year in 2019. Suppose an economist wants to check whether this mean has changed since 2019. The key word here is *changed*. The mean annual earnings of employed Americans has changed if it has either increased or decreased since 2019. This is an example of a two-tailed test. Let μ be the current mean annual earnings of employed Americans. The two possible situations are as follows:

1. The mean annual earnings of employed Americans has not changed since 2019, that is, currently $\mu = \$50{,}400$.
2. The mean annual earnings of employed Americans has changed since 2019, that is, currently $\mu \neq \$50{,}400$.

We will write the null and alternative hypotheses for this test as follows:

$H_0: \mu = \$50{,}400$ (The mean annual earnings of employed Americans has not changed.)

$H_1: \mu \neq \$50{,}400$ (The mean annual earnings of employed Americans has changed.)

To determine whether a test is two-tailed or one-tailed, we look at the sign in the alternative hypothesis. If the alternative hypothesis has a *not equal to* ($\neq$) sign, as in this example, it is a two-tailed test. As shown in **Figure 9.2**, a two-tailed test has two rejection regions, one in each tail of the distribution curve. Figure 9.2 shows the sampling distribution of $\bar{x}$, assuming it has a normal distribution. Assuming H_0 is true, $\bar{x}$ has a normal distribution with its mean equal to \$50,400 (the value of μ in H_0). In Figure 9.2, the area of each of the two

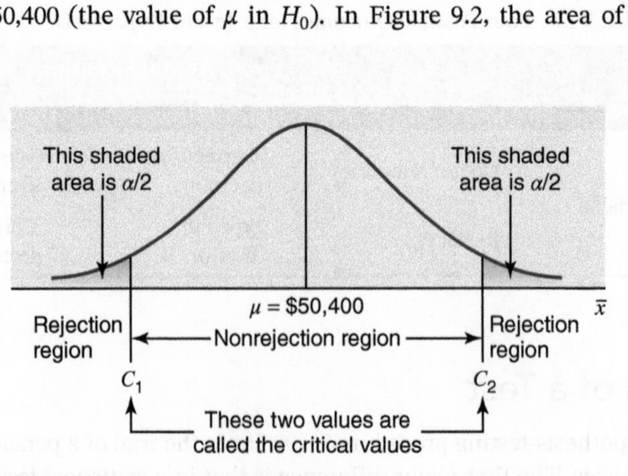

FIGURE 9.2 A two-tailed test.

rejection regions is $\alpha/2$ and the total area of both rejection regions is α (the significance level). As shown in this figure, a two-tailed test of hypothesis has two critical values that separate the two rejection regions from the nonrejection region. We will reject H_0 if the value of $\bar{x}$ obtained from the sample falls in either of the two rejection regions. We will not reject H_0 if the value of $\bar{x}$ lies in the nonrejection region. By rejecting H_0, we are saying that the difference between the value of μ stated in H_0 and the value of $\bar{x}$ obtained from the sample is too large to have occurred because of the sampling error alone. Consequently, this difference appears to be real. By not rejecting H_0, we are saying that the difference between the value of μ stated in H_0 and the value of $\bar{x}$ obtained from the sample is small and may have occurred because of the sampling error alone.

A Left-Tailed Test Reconsider the example of the mean amount of soda in all soft-drink cans produced by a company. The company claims that these cans contain, on average, 12 ounces of soda. However, if these cans contain less than the claimed amount of soda, then the company can be accused of underfilling the cans. Suppose a consumer agency wants to test whether the mean amount of soda per can is less than 12 ounces. Note that the key phrase this time is *less than*, which indicates a left-tailed test. Let μ be the mean amount of soda in all cans. The two possible decisions are as follows:

1. The mean amount of soda in all cans is equal to 12 ounces, that is, $\mu = 12$ ounces.
2. The mean amount of soda in all cans is less than 12 ounces, that is, $\mu < 12$ ounces.

The null and alternative hypotheses for this test are written as

$$H_0: \mu = 12 \text{ ounces} \quad \text{(The mean is equal to 12 ounces.)}$$
$$H_1: \mu < 12 \text{ ounces} \quad \text{(The mean is less than 12 ounces.)}$$

In this case, we can also write the null hypothesis as $H_0: \mu \geq 12$. This will not affect the result of the test as long as the sign in H_1 is *less than* (<).

When the alternative hypothesis has a *less than* (<) sign, as in this case, the test is always left-tailed. In a left-tailed test, the rejection region is in the left tail of the distribution curve, as shown in **Figure 9.3**, and the area of this rejection region is equal to α (the significance level). We can observe from this figure that there is only one critical value in a left-tailed test.

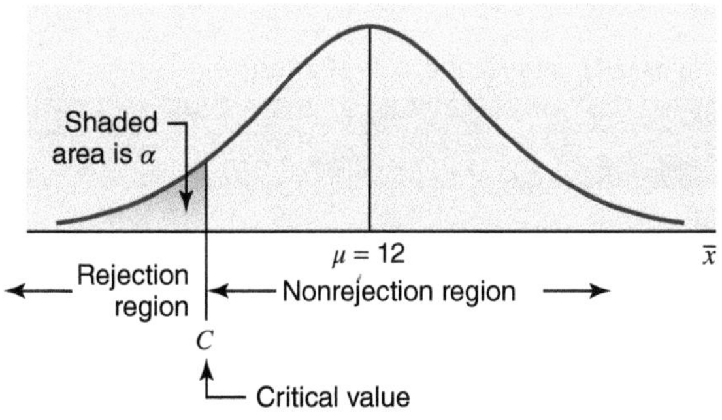

FIGURE 9.3 A left-tailed test.

Assuming H_0 is true, the sampling distribution of $\bar{x}$ has a mean equal to 12 ounces (the value of μ in H_0). We will reject H_0 if the value of $\bar{x}$ obtained from the sample falls in the rejection region; we will not reject H_0 otherwise.

A Right-Tailed Test To illustrate the third case, according to an estimate, the average price of detached homes in Newark, New Jersey, was \$321,257 in 2019. Suppose a real estate researcher wants to check whether the current mean price of homes in this town is higher than \$321,257. The key phrase in this case is *higher than*, which indicates a right-tailed test. Let μ be the current mean price of homes in Newark, New Jersey. The two possible decisions are as follows:

1. The current mean price of detached homes in Newark, NJ, is equal to \$321,257, that is, currently $\mu = \$321{,}257$.
2. The current mean price of detached homes in Newark, NJ, is higher than \$321,257, that is, currently $\mu > \$321{,}257$.

We write the null and alternative hypotheses for this test as follows:

$H_0: \mu = \$321{,}257$ (The current mean price of homes in this town is equal to \$321,257.)
$H_1: \mu > \$321{,}257$ (The current mean price of homes in this town is higher than \$321,257.)

Note that here we can also write the null hypothesis as $H_1: \mu \leq \$321{,}257$, which states that the current mean price of homes in this city is either equal to or less than \$321,257. Again, the result of the test will not be affected by whether we use an *equal to* (=) or a *less than or equal to* (≤) sign in H_0 as long as the alternative hypothesis has a *greater than* (>) sign.

When the alternative hypothesis has a *greater than* (>) sign, the test is always right-tailed. As shown in **Figure 9.4**, in a right-tailed test, the rejection region is in the right tail of the distribution curve. The area of this rejection region is equal to α, the significance level. Like a left-tailed test, a right-tailed test has only one critical value.

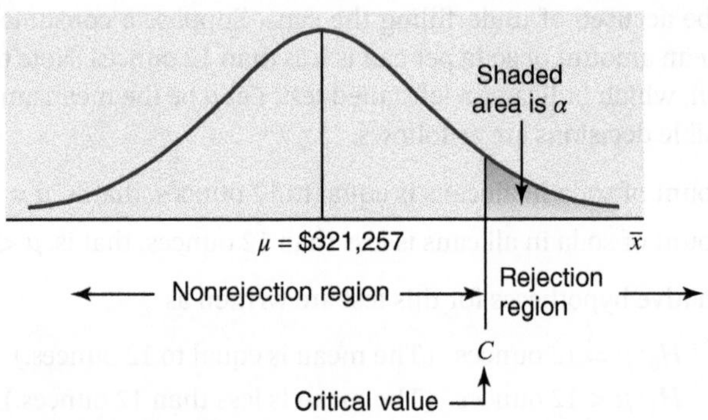

FIGURE 9.4 A right-tailed test.

Again, assuming H_0 is true, the sampling distribution of $\bar{x}$ has a mean equal to \$321,257 (the value of μ in H_0). We will reject H_0 if the value of $\bar{x}$ obtained from the sample falls in the rejection region. Otherwise, we will not reject H_0.

Table 9.3 summarizes the foregoing discussion about the relationship between the signs in H_0 and H_1 and the tails of a test.

TABLE 9.3 Signs in H_0 and H_1 and Tails of a Test

	Two-Tailed Test	Left-Tailed Test	Right-Tailed Test
Sign in the null hypothesis H_0	=	= or ≥	= or ≤
Sign in the alternative hypothesis H_1	≠	<	>
Rejection region	In both tails	In the left tail	In the right tail

Note that the null hypothesis always has an *equal to* (=) or a *greater than or equal to* (≥) or a *less than or equal to* (≤) sign, and the alternative hypothesis always has a *not equal to* (≠) or a *less than* (<) or a *greater than* (>) sign.

9.1.5 Two Procedures to Make Tests of Hypothesis

In this text we will use the following **two procedures** to make tests of hypothesis:

1. **The *p*-value approach.** Under this procedure, we calculate what is called the *p*-value for the observed value of the sample statistic. If we have a predetermined significance level, then we compare the *p*-value with this significance level and make a decision. Note that ***p*-value stands for probability value.**
2. **The critical value approach.** In this approach, we find the critical value(s) from a table (such as the normal distribution table or the *t* distribution table) and find the value

of the test statistic for the observed value of the sample statistic. Then we compare these two values and make a decision.

Remember, the procedures to be learned in this chapter assume that data are obtained from a simple random sample. Also, remember that the critical point is included in the rejection region.

Remember

EXERCISES

Concepts and Procedures

9.1 Briefly explain the meaning of each of the following terms:

a. Null hypothesis **b.** Alternative hypothesis
c. Critical point(s) **d.** Significance level
e. Nonrejection region **f.** Rejection region
g. Tails of a test **h.** Two types of errors

9.2 What are the four possible outcomes for a test of hypothesis? Show these outcomes by writing a table. Briefly describe the Type I and Type II errors.

9.3 Explain how the tails of a test depend on the sign in the alternative hypothesis. Describe the signs in the null and alternative hypotheses for a two-tailed, a left-tailed, and a right-tailed test, respectively.

9.4 Explain which of the following is a two-tailed test, a left-tailed test, or a right-tailed test.

a. $H_0: \mu = 45, H_1: \mu > 45$

b. $H_0: \mu = 23, H_1: \mu \neq 23$

c. $H_0: \mu \geq 75, H_1: \mu < 75$

Show the rejection and nonrejection regions for each of these cases by drawing a sampling distribution curve for the sample mean, assuming that it is normally distributed.

9.5 Which of the two hypotheses (null and alternative) is initially assumed to be true in a test of hypothesis?

9.6 Consider $H_0: \mu = 41$ versus $H_1: \mu < 41$.

a. What type of error would you make if the null hypothesis is actually false and you fail to reject it?

b. What type of error would you make if the null hypothesis is actually true and you reject it?

Applications

9.7 Write the null and alternative hypotheses for each of the following examples. Determine if each is a case of a two-tailed, a left-tailed, or a right-tailed test.

a. To test if the mean number of hours spent working per week by college students who hold jobs is different from 18 hours

b. To test whether or not a bank's ATM is out of service for an average of more than 15 hours per month

c. To test if the mean length of work experience of airport security guards is different from five years

d. To test if the mean credit card debt of college seniors is less than \$800

e. To test if the mean time a customer has to wait on the phone to speak to a representative of a mail-order company about unsatisfactory service is more than 10 minutes

9.8 Write the null and alternative hypotheses for each of the following examples. Determine if each is a case of a two-tailed, a left-tailed, or a right-tailed test.

a. To test if the mean amount of time spent per week watching sports on television by all adult men is different from 6 hours

b. To test if the mean amount of money spent by all customers at a supermarket is more than \$80

c. To test whether the mean starting salary of college graduates is higher than \$39,000 per year

d. To test if the mean waiting time at the drive-through window at a fast food restaurant during rush hour differs from 10 minutes

e. To test if the mean time spent per week on house chores by all housewives is less than 30 hours

9.2 Hypothesis Tests About μ: σ Known

LEARNING OBJECTIVES

After completing this section, you should be able to:

- Describe the five steps of a test of hypothesis using a critical value approach.
- Describe the four steps of a test of hypothesis using a *p*-value approach.
- Use the normal probability distribution to calculate a critical value for a given alternative hypothesis and level of significance.
- Use the normal probability distribution to illustrate the critical value, the rejection region and the nonrejection region for a given alternative hypothesis and level of significance.
- Use the normal probability distribution to calculate a *p*-value for a given alternative hypothesis and set of summary statistics.
- Make a decision based on a comparison of the test statistic and the critical value.
- Make a decision based on a comparison of the *p*-value and the significance level.

- Apply the five-step method using a critical value approach to perform a test of hypothesis about a population mean when the population standard deviation is known.
- Apply the four-step method using a p-value approach to perform a test of hypothesis about a population mean when the population standard deviation is known.
- Describe the Type I and Type II errors in the context of a test of hypothesis about a population mean when the population standard deviation is known.

This section explains how to perform a test of hypothesis for the population mean μ when the population standard deviation σ is known. As in Section 8.2 of Chapter 8, here also there are three possible cases as follows:

Case I. If the following three conditions are fulfilled:

1. The population standard deviation σ is known
2. The sample size is small (i.e., $n < 30$)
3. The population from which the sample is selected is approximately normally distributed

then we use the normal distribution to perform a test of hypothesis about μ because from Section 7.3.1 of Chapter 7 the sampling distribution of $\bar{x}$ is normal with its mean equal to μ and the standard deviation equal to $\sigma_{\bar{x}} = \sigma/\sqrt{n}$, assuming that $n/N \leq .05$.

Case II. If the following two conditions are fulfilled:

1. The population standard deviation σ is known
2. The sample size is large (i.e., $n \geq 30$)

then, again, we use the normal distribution to perform a test of hypothesis about μ because from Section 7.3.2 of Chapter 7, due to the central limit theorem, the sampling distribution of $\bar{x}$ is (approximately) normal with its mean equal to μ and the standard deviation equal to $\sigma_{\bar{x}} = \sigma/\sqrt{n}$, assuming that $n/N \leq .05$.

Case III. If the following three conditions are fulfilled:

1. The population standard deviation σ is known
2. The sample size is small (i.e., $n < 30$)
3. The population from which the sample is selected is not normally distributed (or the shape of its distribution is unknown)

then we use a nonparametric method (explained in Chapter 15) to perform a test of hypothesis about μ.

The following chart summarizes the three cases.

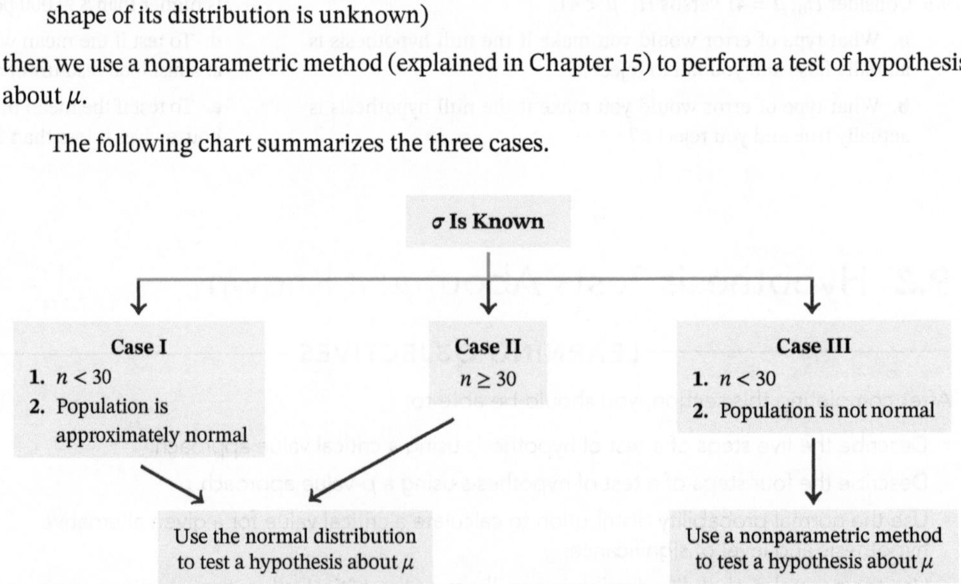

This section will cover the first two cases. The procedure for performing a test of hypothesis about μ is the same in both these cases. Note that in Case I, the population does not have to be exactly normally distributed. As long as it is close to the normal distribution without any outliers, we can use the normal distribution procedure. In Case II, although 30 is considered a large sample, if the population distribution is very different from the normal distribution, then 30 may not be a large enough sample size for the sampling distribution of $\bar{x}$ to be normal and, hence, to use the normal distribution.

Below we explain the two procedures, the *p*-value approach and the critical value approach, to test hypotheses about μ under Cases I and II. We will use the normal distribution to perform such tests.

Note that the two approaches—the *p*-value approach and the critical value approach—are not mutually exclusive. We do not need to use one or the other. We can use both at the same time.

9.2.1 The *p*-Value Approach

In this procedure, we find a probability value such that a given null hypothesis is rejected for any α (significance level) greater than or equal to this value and it is not rejected for any α less than this value. The **probability-value approach**, more commonly called the ***p*-value approach**, gives such a value. In this approach, we calculate the *p-value* for the test, which is defined as the smallest level of significance at which the given null hypothesis is rejected. Using this *p*-value, we state the decision. If we have a predetermined value of α, then we compare the value of p with α and make a decision.

p-Value

Assuming that the null hypothesis is true, the *p*-value can be defined as the probability that a sample statistic (such as the sample mean) is at least as far away from the hypothesized value of the parameter in the direction of the alternative hypothesis as the one obtained from the sample data under consideration. Note that the *p-value* is the smallest significance level at which the null hypothesis is rejected.

Using the *p*-value approach, we reject the null hypothesis if

$$p\text{-value} \leq \alpha \quad \text{which is the same as saying: } \alpha \geq p\text{-value}$$

and we do not reject the null hypothesis if

$$p\text{-value} > \alpha \quad \text{which is the same as saying: } \alpha < p\text{-value}$$

For a one-tailed test, the *p*-value is given by the area in the tail of the sampling distribution curve beyond the observed value of the sample statistic. **Figure 9.5** shows the *p*-value for a right-tailed test about μ. For a left-tailed test, the *p*-value will be the area in the lower tail of the sampling distribution curve to the left of the observed value of $\bar{x}$.

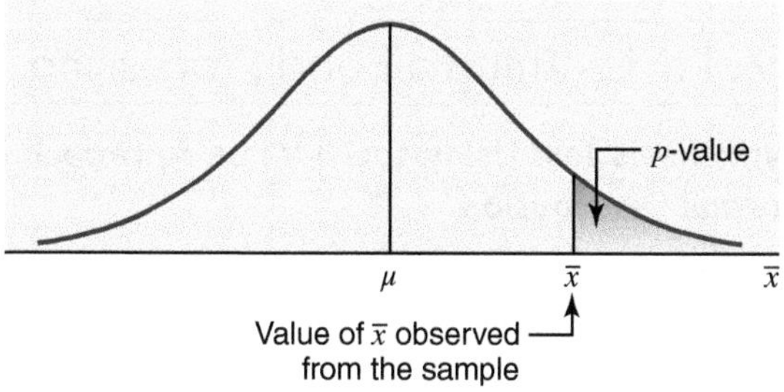

FIGURE 9.5 The *p*-value for a right-tailed test.

For a two-tailed test, the *p*-value is twice the area in the tail of the sampling distribution curve beyond the observed value of the sample statistic. **Figure 9.6** shows the *p*-value for a two-tailed test. Each of the areas in the two tails gives one-half the *p*-value.

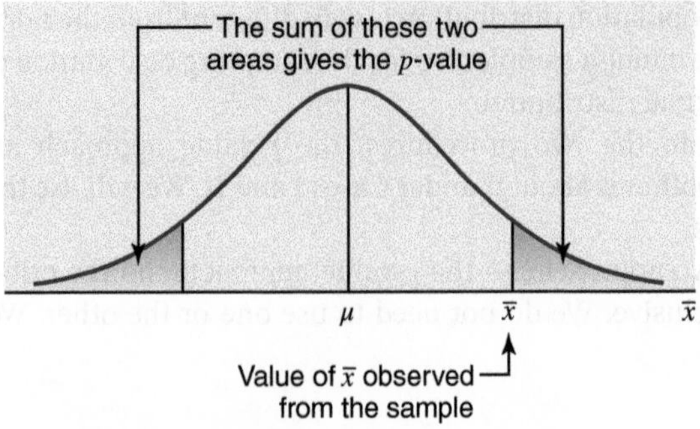

FIGURE 9.6 The p-value for a two-tailed test.

To find the area under the normal distribution curve beyond the sample mean $\bar{x}$, we first find the z value for $\bar{x}$ using the following formula:

Calculating the z Value for $\bar{x}$

When using the normal distribution, *the value of z for $\bar{x}$* for a test of hypothesis about μ is computed as follows:

$$z = \frac{\bar{x} - \mu}{\sigma_{\bar{x}}} \quad \text{where} \quad \sigma_{\bar{x}} = \frac{\sigma}{\sqrt{n}}$$

The value of z calculated for $\bar{x}$ using this formula is also called the *observed value of z*.

Then we find the area under the tail of the normal distribution curve beyond this value of z. This area gives the p-value or one-half the p-value, depending on whether the test is a one-tailed test or a two-tailed test.

A test of hypothesis procedure that uses the p-value approach involves the following four steps:

Steps to Perform a Test of Hypothesis Using the p-Value Approach

1. State the null and alternative hypotheses.
2. Select the distribution to use.
3. Calculate the p-value.
4. Make a decision.

Examples 9.1 and 9.2 illustrate the calculation and use of the p-value to test a hypothesis using the normal distribution.

EXAMPLE 9.1 Learning a Food Processing Job

Performing a Hypothesis Test Using the p-Value Approach for a Two-Tailed Test with the Normal Distribution

At Canon Food Corporation, it used to take an average of 90 minutes for new workers to learn a food processing job. Recently the company installed a new food processing machine. The supervisor at the company wants to find if the mean time taken by new workers to learn the food processing procedure on this new machine is different from 90 minutes. A random sample of 20 workers showed that it took, on average, 85 minutes for them to learn the food processing procedure on the new machine. It is known that the learning times for all new workers are approximately normally distributed with a population standard deviation of 7 minutes. Find the p-value for the test that the mean learning time for new workers for the food processing procedure on the new machine is different from 90 minutes. What will your conclusion be if $\alpha = .01$? Assume that $n/N \le .05$.

Solution Let μ be the mean time (in minutes) taken to learn the food processing procedure on the new machine by all new workers, and let $\bar{x}$ be the corresponding sample mean. From the given information,

$$n = 20, \quad \bar{x} = 85 \text{ minutes}, \quad \sigma = 7 \text{ minutes}, \quad \text{and} \quad \alpha = .01$$

To calculate the *p*-value and perform the test, we apply the following four steps:

STEP 1 *State the null and alternative hypotheses.*

$$H_0: \mu = 90 \text{ minutes}$$
$$H_1: \mu \neq 90 \text{ minutes}$$

Note that the null hypothesis states that the mean time for learning the food processing procedure on the new machine is 90 minutes, and the alternative hypothesis states that this time is different from 90 minutes.

STEP 2 *Select the distribution to use.*

Here, the population standard deviation σ is known, the sample size is small ($n < 30$), but the population distribution is approximately normal. Hence, the sampling distribution of $\bar{x}$ is approximately normal with its mean equal to μ and the standard deviation equal to $\sigma_{\bar{x}} = \sigma/\sqrt{n}$. Consequently, we will use the normal distribution to find the *p*-value and make the test.

STEP 3 *Calculate the p-value.*

The $\neq$ sign in the alternative hypothesis indicates that the test is two-tailed. The *p*-value is equal to twice the area in the tail of the sampling distribution curve of $\bar{x}$ to the left of $\bar{x} = 85$, as shown in **Figure 9.7**. To find this area, we first find the z value for $\bar{x} = 85$ as follows:

$$\sigma_{\bar{x}} = \frac{\sigma}{\sqrt{n}} = \frac{7}{\sqrt{20}} = 1.56524758 \text{ minutes}$$

$$z = \frac{\bar{x} - \mu}{\sigma_{\bar{x}}} = \frac{85 - 90}{1.56524758} = -3.19$$

The area to the left of $\bar{x} = 85$ is equal to the area under the standard normal curve to the left of $z = -3.19$. From the normal distribution table, the area to the left of $z = -3.19$ is .0007.

Consequently, the *p*-value is

$$p\text{-value} = 2(.0007) = \mathbf{.0014}$$

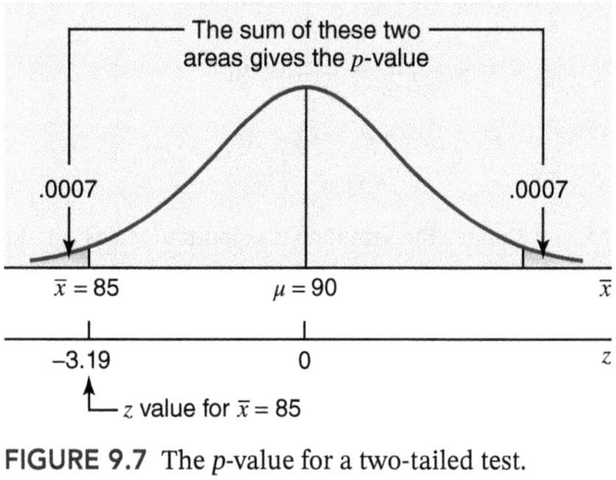

FIGURE 9.7 The *p*-value for a two-tailed test.

STEP 4 *Make a decision.*

Thus, based on the *p*-value of .0014, we can state that for any α (significance level) greater than or equal to .0014, we will reject the null hypothesis stated in Step 1, and for any α less than .0014, we will not reject the null hypothesis.

Because $\alpha = .01$ is greater than the *p*-value of .0014, we reject the null hypothesis at this significance level. Therefore, we conclude that the mean time for learning the food processing procedure on the new machine is different from 90 minutes.

EXAMPLE 9.2 Losing Weight by Joining a Health Club

Performing a Hypothesis Test Using the *p*-Value Approach for a One-Tailed Test with the Normal Distribution

The management of Priority Health Club claims that its members lose an average of 10 pounds or more within the first month after joining the club. A consumer agency that wanted to check this claim took a random sample of 36 members of this health club and found that they lost an average of 9.2 pounds within the first month of membership. The population standard deviation is known to be 2.4 pounds. Find the *p*-value for this test. What will your decision be if $\alpha = .01$? What if $\alpha = .05$? Assume that $n/N \le .05$.

FatCamera/Getty Images

Solution Let μ be the mean weight lost during the first month of membership by all members of this health club, and let $\bar{x}$ be the corresponding mean for the sample. From the given information,

$$n = 36, \quad \bar{x} = 9.2 \text{ pounds}, \quad \text{and} \quad \sigma = 2.4 \text{ pounds}$$

The claim of the club is that its members lose, on average, 10 pounds or more within the first month of membership. To perform the test using the *p*-value approach, we apply the following four steps:

STEP 1 *State the null and alternative hypotheses.*

$H_0: \mu \ge 10$ (The mean weight lost is 10 pounds or more.)
$H_1: \mu < 10$ (The mean weight lost is less than 10 pounds.)

STEP 2 *Select the distribution to use.*

Here, the population standard deviation σ is known, and the sample size is large ($n \ge 30$). Hence, the sampling distribution of $\bar{x}$ is normal (due to the central limit theorem) with its mean equal to μ and the standard deviation equal to $\sigma_{\bar{x}} = \sigma/\sqrt{n}$. Consequently, we will use the normal distribution to find the *p*-value and perform the test.

STEP 3 *Calculate the p-value.*

The < sign in the alternative hypothesis indicates that the test is left-tailed. The *p*-value is given by the area to the left of $\bar{x} = 9.2$ under the sampling distribution curve of $\bar{x}$, as shown in **Figure 9.8**. To find this area, we first find the *z* value for $\bar{x} = 9.2$ as follows:

$$\sigma_{\bar{x}} = \frac{\sigma}{\sqrt{n}} = \frac{2.4}{\sqrt{36}} = .40$$

$$z = \frac{\bar{x} - \mu}{\sigma_{\bar{x}}} = \frac{9.2 - 10}{.40} = -2.00$$

The area to the left of $\bar{x} = 9.2$ under the sampling distribution of $\bar{x}$ is equal to the area under the standard normal curve to the left of $z = -2.00$. From the normal distribution table, the area to the left of $z = -2.00$ is .0228. Consequently,

$$p\text{-value} = \mathbf{.0228}$$

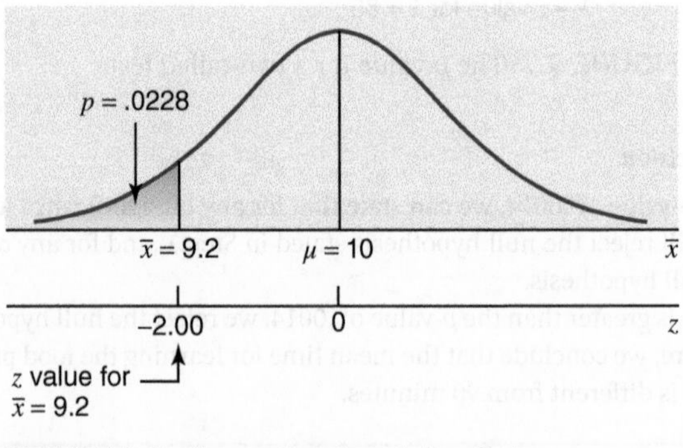

FIGURE 9.8 The *p*-value for a left-tailed test.

STEP 4 *Make a decision.*

Thus, based on the *p*-value of .0228, we can state that for any α (significance level) greater than or equal to .0228 we will reject the null hypothesis stated in Step 1, and for any α less than .0228 we will not reject the null hypothesis.

Since $\alpha = .01$ is less than the *p*-value of .0228, we do not reject the null hypothesis at this significance level. Consequently, we conclude that there is not significant evidence to support the alternative hypothesis that members of this club lose, on average, less than 10 pounds within the first month.

Now, because $\alpha = .05$ is greater than the *p*-value of .0228, we reject the null hypothesis at this significance level. In this case we conclude that there is sufficient evidence to support the consumer agency's claim that the mean weight lost within the first month of membership by the members of this club is less than 10 pounds.

9.2.2 The Critical Value Approach

In this procedure, we have a predetermined value of the significance level α. The value of α gives the total area of the rejection region(s). First we find the critical value(s) of z from the normal distribution table for the given significance level. Then, assuming H_0 is true, we find the value of the test statistic z for the observed value of the sample statistic $\bar{x}$. Finally we compare these two values and make a decision. Remember, if the test is one-tailed, there is only one critical value of z, and it is obtained by using the value of α which gives the area in the left or right tail of the normal distribution curve depending on whether the test is left-tailed or right-tailed, respectively. However, if the test is two-tailed, there are two critical values of z and they are obtained by using $\alpha/2$ area in each tail of the normal distribution curve. The value of the test statistic is obtained as follows:

Test Statistic

In tests of hypotheses about μ using the normal distribution, the random variable

$$z = \frac{\bar{x} - \mu}{\sigma_{\bar{x}}} \quad \text{where} \quad \sigma_{\bar{x}} = \frac{\sigma}{\sqrt{n}}$$

is called the *test statistic*. The test statistic is used to make the decision on whether or not to reject the null hypothesis. Here μ represents the value that appears in the null hypothesis.

A test of hypothesis procedure that uses the critical value approach involves the following five steps:

Steps to Perform a Test of Hypothesis with the Critical Value Approach

1. State the null and alternative hypotheses.
2. Select the distribution to use.
3. Determine the rejection and nonrejection regions.
4. Calculate the value of the test statistic.
5. Make a decision.

Examples 9.3 and 9.4 illustrate the use of these five steps to perform tests of hypotheses about the population mean μ. Example 9.3 is concerned with a two-tailed test, and Example 9.4 describes a one-tailed test.

EXAMPLE 9.3 | Lengths of Long-Distance Calls

Conducting a Two-Tailed Test of Hypothesis About μ Using the Critical Value Approach: σ Known and $n \geq 30$

The TIV Telephone Company provides long-distance telephone service in an area. According to the company's records, the average length of all long-distance calls placed through this company in

2019 was 12.44 minutes. The company's management wanted to check if the mean length of the current long-distance calls is different from 12.44 minutes. A sample of 150 such calls placed through this company produced a mean length of 13.71 minutes. The standard deviation of all such calls is 2.65 minutes. Using a 2% significance level, can you conclude that the mean length of all current long-distance calls is different from 12.44 minutes?

Solution Let μ be the mean length of all current long-distance calls placed through this company and $\bar{x}$ be the corresponding mean for the sample. From the given information,

$$n = 150, \quad \bar{x} = 13.71 \text{ minutes}, \quad \text{and} \quad \sigma = 2.65 \text{ minutes}$$

We are to test whether or not the mean length of all current long-distance calls is different from 12.44 minutes. The significance level α is .02; that is, the probability of rejecting the null hypothesis when it actually is true should not exceed .02. This is the probability of making a Type I error. We perform the test of hypothesis using the five steps as follows:

STEP 1 *State the null and alternative hypotheses.*

Notice that we are testing to find whether or not the mean length of all current long-distance calls is different from 12.44 minutes. We write the null and alternative hypotheses as follows:

$H_0: \mu = 12.44$ (The mean length of all current long-distance calls is 12.44 minutes.)

$H_1: \mu \neq 12.44$ (The mean length of all current long-distance calls is different from 12.44 minutes.)

STEP 2 *Select the distribution to use.*

Here, the population standard deviation σ is known, and the sample size is large ($n \geq 30$). Hence, the sampling distribution of $\bar{x}$ is (approximately) normal (due to the central limit theorem) with its mean equal to μ and the standard deviation equal to $\sigma_{\bar{x}} = \sigma/\sqrt{n}$. Consequently, we will use the normal distribution to perform the test.

STEP 3 *Determine the rejection and nonrejection regions.*

The significance level is .02. The $\neq$ sign in the alternative hypothesis indicates that the test is two-tailed with two rejection regions, one in each tail of the normal distribution curve of $\bar{x}$. Because the total area of both rejection regions is .02 (the significance level), the area of the rejection region in each tail is .01; that is,

$$\text{Area in each tail} = \alpha/2 = .02/2 = .01$$

These areas are shown in **Figure 9.9**. Two critical points in this figure separate the two rejection regions from the nonrejection region. Next, we find the z values for the two critical points using the area of the rejection region. To find the z values for these critical points, we look for .0100 and .9900 areas in the normal distribution table. From Table IV of Appendix B, the z values of the two critical points, as shown in Figure 9.9, are approximately −2.33 and 2.33.

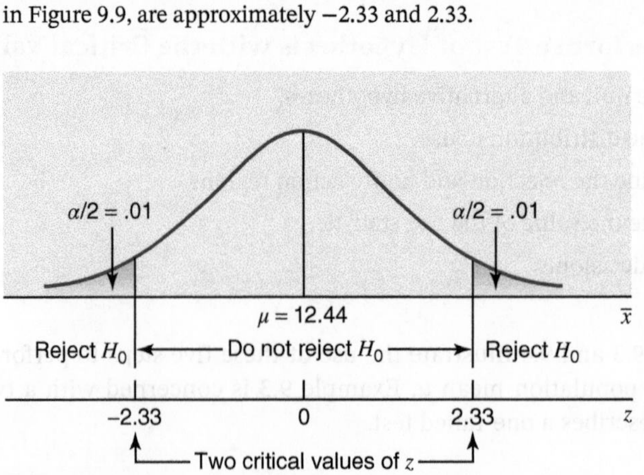

FIGURE 9.9 Rejection and nonrejection regions.

STEP 4 *Calculate the value of the test statistic.*

The decision to reject or not to reject the null hypothesis will depend on whether the evidence from the sample falls in the rejection or the nonrejection region. If the value of $\bar{x}$ falls in either of the two rejection regions, we reject H_0. Otherwise, we do not reject H_0. The value of $\bar{x}$ obtained from the sample is called the *observed value of* $\bar{x}$. To locate the position of $\bar{x} = 13.71$ on the sampling distribution curve of $\bar{x}$ in Figure 9.9, we first calculate the z value for $\bar{x} = 13.71$. This is called the *value of the test statistic.* Then, we compare the value of the test statistic with the two critical values of z, -2.33 and 2.33, shown in Figure 9.9. If the value of the test statistic is between -2.33 and 2.33, we do not reject H_0. If the value of the test statistic is either greater than or equal to 2.33 or less than or equal to -2.33, we reject H_0.

Calculating the Value of the Test Statistic

When using the normal distribution, *the value of the test statistic* z for $\bar{x}$ for a test of hypothesis about μ is computed as follows:

$$z = \frac{\bar{x} - \mu}{\sigma_{\bar{x}}}$$

where $\sigma_{\bar{x}} = \frac{\sigma}{\sqrt{n}}$ given that $\frac{n}{N} \le .05$

This value of z for $\bar{x}$ is also called the *observed value of* z.

The value of $\bar{x}$ from the sample is 13.71. We calculate the z value as follows:

$$\sigma_{\bar{x}} = \frac{\sigma}{\sqrt{n}} = \frac{2.65}{\sqrt{150}} = .21637159$$

$$z = \frac{\bar{x} - \mu}{\sigma_{\bar{x}}} = \frac{13.71 - \overset{\text{From } H_0}{12.44}}{.21637159} = 5.87$$

The value of μ in the calculation of the z value is substituted from the null hypothesis. The value of $z = 5.87$ calculated for $\bar{x}$ is called the *computed value of the test statistic* z. This is the value of z that corresponds to the value of $\bar{x}$ observed from the sample. It is also called the *observed value of* z.

STEP 5 *Make a decision.*

In the final step we make a decision based on the location of the value of the test statistic z computed for $\bar{x}$ in Step 4. This value of $z = 5.87$ is greater than the critical value of $z = 2.33$, and it falls in the right tail of the rejection region in Figure 9.9. Hence, we reject H_0 and conclude that based on the sample information, it appears that the current mean length of all such calls is different from 12.44 minutes.

By rejecting the null hypothesis, we are stating that the difference between the sample mean, $\bar{x} = 13.71$ minutes, and the hypothesized value of the population mean, $\mu = 12.44$ minutes, is too large and may not have occurred because of chance or sampling error alone. This difference seems to be real and, hence, the current mean length of all such calls is most likely different from 12.44 minutes. Note that the rejection of the null hypothesis does not necessarily indicate that the current mean length of all such calls is definitely different from 12.44 minutes. It simply indicates that there is strong evidence (from the sample) that the current mean length of such calls is different from 12.44 minutes. There is a possibility that the current mean length of all such calls is equal to 12.44 minutes, but by the luck of the draw we selected a sample with a mean that is too far from the hypothesized mean of 12.44 minutes. If so, we would have wrongfully rejected the null hypothesis H_0 committing a Type I error.

We can use the p-value approach to perform the test of hypothesis in Example 9.3. In this example, the test is two-tailed. The p-value is equal to twice the area under the sampling distribution of $\bar{x}$ to the right of $\bar{x} = 13.71$. As calculated in Step 4 above, the z value for $\bar{x} = 13.71$ is 5.87. From the normal distribution table, the area to the right of $z = 5.87$ is (approximately) zero. Hence, the p-value is (approximately) zero. (If you use technology, you will obtain the p-value of .000000002.) As we know from earlier discussions, we will reject the null hypothesis for any α (significance level) that is greater than or equal to the p-value. Consequently, in this example, we will reject the null hypothesis for any $\alpha > 0$ approximately. Since $\alpha = .02$ here, which is greater than zero, we reject the null hypothesis.

EXAMPLE 9.4 | Net Worth of Families in a City

Conducting a Left-Tailed Test of Hypothesis About μ Using the Critical Value Approach: σ Known, $n < 30$, and Population Normally Distributed

The mayor of a large city claims that the average net worth of families living in this city is at least \$300,000. A random sample of 25 families selected from this city produced a mean net worth of \$288,000. Assume that the net worths of all families in this city have an approximate normal distribution with a population standard deviation of \$80,000. Using a 2.5% significance level, can you conclude that the mayor's claim is false?

Solution Let μ be the mean net worth of families living in this city and $\bar{x}$ be the corresponding mean for the sample. From the given information,

$$n = 25, \quad \bar{x} = \$288{,}000, \quad \text{and} \quad \sigma = \$80{,}000$$

The significance level is $\alpha = .025$.

STEP 1 *State the null and alternative hypotheses.*

We are to test whether or not the mayor's claim is false. The mayor's claim is that the average net worth of families living in this city is at least \$300,000. Hence, the null and alternative hypotheses are as follows:

$H_0: \mu \geq \$300{,}000$ (The mayor's claim is true. The mean net worth is at least \$300,000.)

$H_1: \mu < \$300{,}000$ (The mayor's claim is false. The mean net worth is less than \$300,000.)

STEP 2 *Select the distribution to use.*

Here, the population standard deviation σ is known, the sample size is small ($n < 30$), but the population distribution is approximately normal. Hence, the sampling distribution of $\bar{x}$ is approximately normal with its mean equal to μ and the standard deviation equal to $\sigma_{\bar{x}} = \sigma/\sqrt{n}$. Consequently, we will use the normal distribution to perform the test.

STEP 3 *Determine the rejection and nonrejection regions.*

The significance level is .025. The < sign in the alternative hypothesis indicates that the test is left-tailed with the rejection region in the left tail of the sampling distribution curve of $\bar{x}$. The critical value of z, obtained from the normal distribution table for .0250 area in the left tail, is −1.96, as shown in **Figure 9.10**.

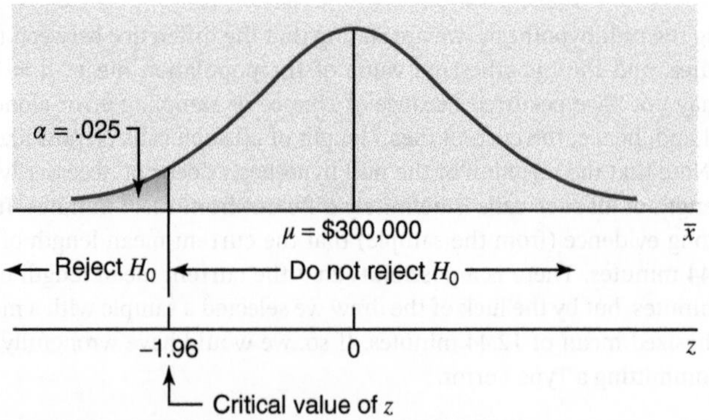

FIGURE 9.10 Rejection and nonrejection regions.

STEP 4 *Calculate the value of the test statistic.*

The value of the test statistic z for $\bar{x} = \$288{,}000$ is calculated as follows:

$$\sigma_{\bar{x}} = \frac{\sigma}{\sqrt{n}} = \frac{80{,}000}{\sqrt{25}} = \$16{,}000$$

$$z = \frac{\bar{x} - \mu}{\sigma_{\bar{x}}} = \frac{288{,}000 - \overset{\text{From } H_0}{300{,}000}}{16{,}000} = -.75$$

STEP 5 *Make a decision.*

The value of the test statistic $z = -.75$ is greater than the critical value of $z = -1.96$, and it falls in the nonrejection region. As a result, we fail to reject H_0. Therefore, we can state that based on the sample information, it appears that the mean net worth of families in this city does not seem to be less than \$300,000. Note that we are not concluding that the mean net worth is definitely not less than \$300,000. By not rejecting the null hypothesis, we are saying that the information obtained from the sample is not strong enough to reject the null hypothesis and to conclude that the mayor's claim is false.

We can use the p-value approach to perform the test of hypothesis in Example 9.4. In this example, the test is left-tailed. The p-value is given by the area under the sampling distribution curve of $\bar{x}$ to the left of $\bar{x} = \$288{,}000$. As calculated in Step 4 above, the z value for $\bar{x} = \$288{,}000$ is $-.75$. From the normal distribution table, the area to the left of $z = -.75$ is .2266. Hence, the p-value is .2266. We will reject the null hypothesis for any α (significance level) that is greater than or equal to the p-value. Consequently, we will reject the null hypothesis in this example for any $\alpha \geq .2266$. Since in this example $\alpha = .025$, which is less than .2266, we fail to reject the null hypothesis.

Case Study 9.1

Class of 2018 Average Loan Debt for U.S. Students

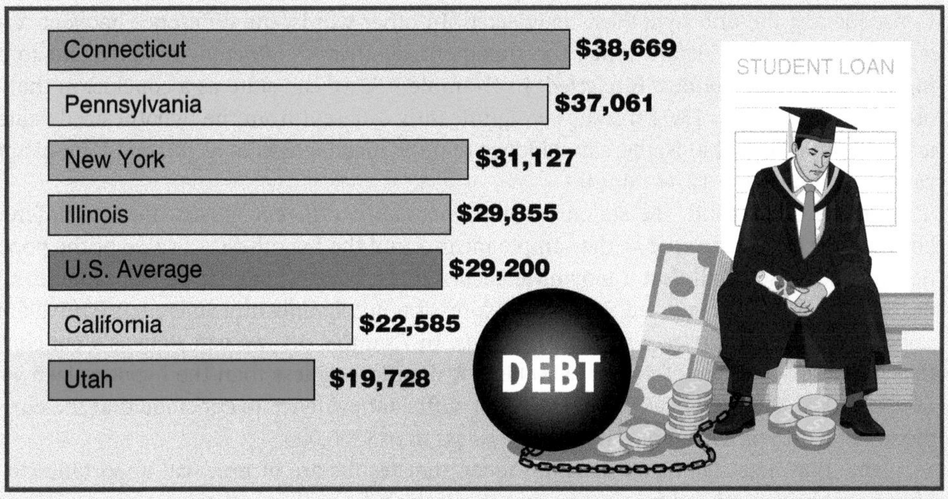

According to estimates by LendEDU, 65% of college seniors who graduated from public and private nonprofit colleges in 2018 had student loan debt. The accompanying chart lists the U.S. average and the averages of such debts for six selected states for students who graduated in 2018 with loan debt. Remember that each of these averages is based on a sample survey of students who graduated in 2018 with student loan debt. According to these estimates, 2018 U.S. college graduates owed an average of \$29,200. Of all 50 states, students who graduated with loan debt from colleges in Connecticut in 2018 had the highest average debt of \$38,669, and those who graduated with loans from colleges in Utah in 2018 had the lowest average debt of \$19,728. Again, note that these averages are based on sample surveys. Suppose we want to find out if the average loan debt for 2018 college graduates in Illinois was higher than the U.S. average of \$29,200. Suppose that the average loan debt for Illinois college students in the class of 2018 is based on a random sample of 900 students who graduated with loans. Assume that the standard deviation of the loans for all Illinois students in the class of 2018 was \$5100 and the significance level is 1%. The test is right-tailed because we are testing the hypothesis that the average student loan debt for the entire class of 2018 in Illinois was higher than \$29,200. The null and alternative hypotheses are as follows where μ is the average student loan debt of all students who graduated from Illinois colleges in 2018:

$$H_0: \mu = \$29{,}200$$
$$H_1: \mu > \$29{,}200$$

Here, $n = 900$, $\bar{x} = \$29{,}855$, $\sigma = \$5100$, and $\alpha = .01$. The population standard deviation is known, and the sample size is large. Hence, we can use the normal distribution to perform this test. From the normal distribution table, we find that the critical value of z is 2.33 for .01 area in the right tail of the normal curve. We find the observed value of z as follows:

$$\sigma_{\bar{x}} = \frac{\sigma}{\sqrt{n}} = \frac{5100}{\sqrt{900}} = \$170$$

$$z = \frac{\bar{x} - \mu}{\sigma_{\bar{x}}} = \frac{29{,}855 - 29{,}200}{170} = 3.85$$

The value of the test statistic $z = 3.85$ for $\bar{x}$ is larger than the critical value of $z = 2.33$, and it falls in the rejection region. Consequently, we reject H_0 and conclude that the average debt of students who graduated in 2018 from colleges in Illinois with student loan debt seems to be higher than \$29,200, which is the average for the United States for the 2018 college graduates. To use the *p*-value approach, we find the area under the normal curve to the right of $z = 3.85$ from the normal distribution table. This area is (approximately) .0000. Therefore, the *p*-value is .0000. Since $\alpha = .01$ is larger than the *p*-value = .0000, we reject the null hypothesis.

Source: Data from Average Student Loan Debt Statistics by School by State 2019. Shop Tutors Inc.

In studies published in various journals, authors usually use the terms **significantly different** and **not significantly different** when deriving conclusions based on hypothesis tests. These terms are short versions of the terms *statistically significantly different* and *statistically not significantly different*. The expression *significantly different* means that the difference between the observed value of the sample mean $\bar{x}$ and the hypothesized value of the population mean μ is so large that it probably did not occur because of the sampling error alone. Consequently, the null hypothesis is rejected. In other words, the difference between $\bar{x}$ and μ is statistically significant. Thus, the statement *significantly different* is equivalent to saying that the *null hypothesis is rejected*. In Example 9.3, we can state as a conclusion that the observed value of $\bar{x} = 13.71$ minutes is significantly different from the hypothesized value of $\mu = 12.44$ minutes. That is, there is evidence that the mean length of all current long-distance calls is different from 12.44 minutes.

On the other hand, the statement *not significantly different* means that the difference between the observed value of the sample mean $\bar{x}$ and the hypothesized value of the population mean μ is so small that it may have occurred just because of chance. Consequently, the null hypothesis is not rejected. Thus, the expression *not significantly different* is equivalent to saying that we *fail to reject the null hypothesis*. In Example 9.4, we can state as a conclusion that the observed value of $\bar{x} = \$288{,}000$ is not significantly less than the hypothesized value of $\mu = \$300{,}000$. In other words, there is not sufficient evidence to conclude that the current mean net worth of households in this city is less than \$300,000.

Note that here *significant* does not mean that results are of practical importance to us. Significant here is related to rejecting or not rejecting the null hypothesis.

EXERCISES

Concepts and Procedures

9.9 What are the five steps of a test of hypothesis using the critical value approach? Explain briefly.

9.10 What does the level of significance represent in a test of hypothesis? Explain.

9.11 By rejecting the null hypothesis in a test of hypothesis example, are you stating that the alternative hypothesis is true?

9.12 What is the difference between the critical value of z and the observed value of z?

9.13 Briefly explain the procedure used to calculate the *p*-value for a two-tailed and for a one-tailed test, respectively.

9.14 Find the *p*-value for each of the following hypothesis tests:

a. $H_0: \mu = 46$, $H_1: \mu \neq 46$, $n = 40$, $\bar{x} = 49.60$, $\sigma = 9.7$

b. $H_0: \mu = 26$, $H_1: \mu < 26$, $n = 33$, $\bar{x} = 24.30$, $\sigma = 4.3$

c. $H_0: \mu = 18$, $H_1: \mu > 18$, $n = 55$, $\bar{x} = 20.50$, $\sigma = 7.8$

9.15 Consider $H_0: \mu = 72$ versus $H_1: \mu > 72$. A random sample of 16 observations taken from this population produced a sample mean of 75.2. The population is normally distributed with $\sigma = 6$.

a. Calculate the *p*-value.

b. Considering the *p*-value of part (a), would you reject the null hypothesis if the test were made at a significance level of .01?

c. Considering the *p*-value of part (a), would you reject the null hypothesis if the test were made at a significance level of .025?

9.16 For each of the following examples of tests of hypotheses about μ, show the rejection and nonrejection regions on the sampling distribution of the sample mean assuming that it is normal.

a. A two-tailed test with $\alpha = .01$ and $n = 100$

b. A left-tailed test with $\alpha = .005$ and $n = 27$

c. A right-tailed test with $\alpha = .025$ and $n = 36$

9.17 Consider the following null and alternative hypotheses:

$$H_0: \mu = 30 \quad \text{versus} \quad H_1: \mu \neq 30$$

Suppose you perform this test at $\alpha = .05$ and reject the null hypothesis. Would you state that the difference between the hypothesized value of the population mean and the observed value of the sample mean is "statistically significant" or would you state that this difference is "statistically not significant"? Explain.

9.18 Consider the following null and alternative hypotheses:

$$H_0: \mu = 75 \quad \text{versus} \quad H_1: \mu < 75$$

Suppose you perform this test at $\alpha = .01$ and fail to reject the null hypothesis. Would you state that the difference between the hypothesized value of the population mean and the observed value of the sample mean is "statistically significant" or would you state that this difference is "statistically not significant"? Explain.

9.19 In a hypothesis test using each of the following significance levels, what is the probability of making a Type I error?

a. $\alpha = .03$ b. $\alpha = .06$ c. $\alpha = .02$ d. $\alpha = .005$

9.20 A random sample of 80 observations produced a sample mean of 88.50. The population standard deviation is known to be 6.40 and the population distribution is normal. Find the critical and observed values of z for each of the following tests of hypothesis using $\alpha = .10$.

a. $H_0: \mu = 91$ versus $H_1: \mu \neq 91$

b. $H_0: \mu = 91$ versus $H_1: \mu < 91$

9.21 Consider the null hypothesis H_0: $\mu = 625$. Suppose that a random sample of 32 observations is taken from a normally distributed population with $\sigma = 29$. Using a significance level of .01, show the rejection and nonrejection regions on the sampling distribution curve of the sample mean and find the critical value(s) of z when the alternative hypothesis is as follows:

a. $H_1: \mu \neq 625$ b. $H_1: \mu > 625$ c. $H_1: \mu < 625$

9.22 Consider $H_0: \mu = 100$ versus $H_1: \mu \neq 100$.

a. A random sample of 64 observations produced a sample mean of 96. Using $\alpha = .01$, would you reject the null hypothesis? The population standard deviation is known to be 12.

b. Another random sample of 64 observations taken from the same population produced a sample mean of 105. Using $\alpha = .01$, would you reject the null hypothesis? The population standard deviation is known to be 12.

c. Comment on the results of parts (a) and (b).

9.23 Consider $H_0: \mu = 45$ versus $H_1: \mu < 45$.

a. A random sample of 25 observations produced a sample mean of 41.8. Using $\alpha = .025$, would you reject the null hypothesis? The population is known to be normally distributed with $\sigma = 6$.

b. Another random sample of 25 observations taken from the same population produced a sample mean of 43.8. Using $\alpha = .025$, would you reject the null hypothesis? The population is known to be normally distributed with $\sigma = 6$.

c. Comment on the results of parts (a) and (b).

9.24 Make the following tests of hypotheses:

a. $H_0: \mu = 25,\ H_1: \mu \neq 25,\ n = 81,\ \bar{x} = 28.5,\ \sigma = 3,\ \alpha = .01$

b. $H_0: \mu = 12,\ H_1: \mu < 12,\ n = 45,\ \bar{x} = 11.25,\ \sigma = 4.5,\ \alpha = .05$

c. $H_0: \mu = 40,\ H_1: \mu > 40,\ n = 100,\ \bar{x} = 47,\ \sigma = 7,\ \alpha = .10$

Applications

9.25 The manufacturer of a certain brand of auto batteries claims that the mean life of these batteries is 45 months. A consumer protection agency that wants to check this claim took a random sample of 20 such batteries and found that the mean life for this sample is 43.05 months. The lives of all such batteries have an approximate normal distribution with the population standard deviation of 6.5 months.

a. Find the p-value for the test of hypothesis with the alternative hypothesis that the mean life of these batteries is less than 45 months. Will you reject the null hypothesis at $\alpha = .025$?

b. Test the hypothesis of part (a) using the critical value approach and $\alpha = .025$.

9.26 According to the U.S. Bureau of Labor Statistics, all workers in the United States who had a bachelor's degree and were employed earned an average of \$1383 a week in 2019. A recent sample of 400 U.S. workers who have a bachelor's degree showed that they earn an average of \$1419 per week. Suppose that the population standard deviation of such earnings is \$160.

a. Find the p-value for the test that the current mean weekly earning of U.S. workers who have a bachelor's degree is higher than \$1383. Will you reject the null hypothesis at $\alpha = .025$?

b. Test the hypothesis of part (a) using the critical value approach and $\alpha = .025$.

9.27 According to the National Association of Colleges and Employers (NACE), the average starting salary for 2019 college graduates was \$55,280 (www.nace.org). This is a substantial increase in the average starting salary from previous years. A random sample of 1000 recent college graduates from one of the U.S. states showed that their average starting salary was \$54,737. Suppose that the population standard deviation for the starting salaries of all recent college graduates from this state is \$7820.

a. Find the p-value for the test that the average starting salary of recent college graduates from this state is less than \$55,280. Will you reject the null hypothesis at $\alpha = .01$? Explain. What if $\alpha = .025$?

b. Test the hypothesis of part (a) using the critical value approach. Will you reject the null hypothesis at $\alpha = .01$? What if $\alpha = .025$?

9.28 A study claims that all adults spend an average of 14 hours or more on chores during a weekend. A researcher wanted to check if this claim is true. A random sample of 200 adults taken by this researcher showed that these adults spend an average of 13.85 hours on chores during a weekend. The population standard deviation is known to be 3.0 hours.

a. Find the p-value for the hypothesis test with the alternative hypothesis that all adults spend more than 14 hours on chores during a weekend. Will you reject the null hypothesis at $\alpha = .01$?

b. Test the hypothesis of part (a) using the critical value approach and $\alpha = .01$.

9.29 Lazurus Steel Corporation produces iron rods that are supposed to be 36 inches long. The machine that makes these rods does not produce each rod exactly 36 inches long. The lengths of the rods are approximately normally distributed and vary slightly. It is known that when the machine is working properly, the mean length of the rods is 36 inches. The standard deviation of the lengths of all rods produced on this machine is .035 inch. The quality control department at the company takes a sample of 20 such rods every week, calculates the mean length of these rods, and tests the null hypothesis, $\mu = 36$ inches, against the alternative hypothesis, $\mu \neq 36$ inches. If the null hypothesis is rejected, the machine is stopped and adjusted. A recent sample of 20 rods produced a mean length of 36.015 inches.

a. Calculate the *p*-value for this test of hypothesis. Based on this *p*-value, will the quality control inspector decide to stop and adjust the machine if he chooses the maximum probability of a Type I error to be .02? What if the maximum probability of a Type I error is .10?

b. Test the hypothesis of part (a) using the critical value approach and $\alpha = .02$. Does the machine need to be adjusted? What if $\alpha = .10$?

9.30 At Farmer's Dairy, a machine is set to fill 42-ounce milk cartons. However, this machine does not put exactly 42 ounces of milk into each carton; the amount varies slightly from carton to carton but has an approximate normal distribution. It is known that when the machine is working properly, the mean net weight of these cartons is 42 ounces. The standard deviation of the milk in all such cartons is .15 ounce. The quality control inspector at this company takes a sample of 50 such cartons every week, calculates the mean net weight of these cartons, and tests the null hypothesis, $\mu = 42$ ounces, against the alternative hypothesis, $\mu \neq 42$ ounces. If the null hypothesis is rejected, the machine is stopped and adjusted. A recent sample of 50 such cartons produced a mean net weight of 41.93 ounces.

a. Calculate the *p*-value for this test of hypothesis. Based on this *p*-value, will the quality control inspector decide to stop and adjust the machine if she chooses the maximum probability of a Type I error to be .01? What if the maximum probability of a Type I error is .05?

b. Test the hypothesis of part (a) using the critical value approach and $\alpha = .01$. Does the machine need to be adjusted? What if $\alpha = .05$?

9.31 According to a MyBankTracker study, the average checking account fee at top U.S. banks was \$9.60 per month in 2018 (https://smartasset.com/checking-account/cost-of-checking-account-fees-and-how-to-avoid). A bank consultant wants to determine whether the current average checking account fee at top U.S. banks is more than \$9.60 per month. A recent random sample of 150 such checking accounts taken from top U.S. banks produced a mean checking account fee of \$10.12 per month. Assume that the standard deviation of all such checking account fees at top U.S. banks is \$3.12.

a. Find the *p*-value for this test of hypothesis. Based on this *p*-value, would you reject the null hypothesis if the maximum probability of Type I error is to be .05? What if the maximum probability of Type I error is to be .01?

b. Test the hypothesis of part (a) using the critical value approach and $\alpha = .05$. Would you reject the null hypothesis? What if $\alpha = .01$? What if $\alpha = 0$?

9.32 According to a survey by the College Board, college students spent an average of \$1290 on books and supplies during the 2018–2019 academic year (www.collegeboard.org). A recent random sample of 200 undergraduate college students from a large college showed that they spent an average of \$1250 on books and supplies during the last academic year. Assume that the standard deviation of annual expenditures on books and supplies by all such students at this college is \$200.

a. Find the *p*-value for the test of hypothesis with the alternative hypothesis that the annual mean expenditure on books by all such students at this college is less than \$1290. Based on this *p*-value, would you reject the null hypothesis if the significance level is .025?

b. Test the hypothesis of part (a) using the critical value approach and $\alpha = .025$. Would you reject the null hypothesis? Explain.

9.33 A company claims that the mean net weight of the contents of its All Taste cereal boxes is at least 18 ounces. Suppose you want to test whether or not the claim of the company is true. Explain briefly how you would conduct this test using a large sample. Assume that $\sigma = .25$ ounce.

9.3 Hypothesis Tests About μ: σ Not Known

LEARNING OBJECTIVES

After completing this section, you should be able to:

- Explain the conditions required to use the *t*-distribution to perform a test of hypothesis about a population mean.
- Use the *t*-distribution to illustrate the critical value, the rejection region and the nonrejection region for a given alternative hypothesis and level of significance.
- Use the *t*-distribution to calculate a *p*-value (or a range for the *p*-value) for a given alternative hypothesis and set of summary statistics.
- Apply the five-step method using a critical value approach to perform a test of hypothesis about a population mean when the population standard deviation is unknown.
- Apply the four-step method using a *p*-value approach to perform a test of hypothesis about a population mean when the population standard deviation is unknown.
- Describe a Type I and Type II error in the context of a test of hypothesis about a population mean when the population standard deviation is unknown.

This section explains how to perform a test of hypothesis about the population mean μ when the population standard deviation σ is not known. Here, again, there are three possible cases as follows:

Case I. If the following three conditions are fulfilled:

1. The population standard deviation σ is not known
2. The sample size is small (i.e., $n < 30$)
3. The population from which the sample is selected is approximately normally distributed

then we use the t distribution to perform a test of hypothesis about μ.

Case II. If the following two conditions are fulfilled:

1. The population standard deviation σ is not known
2. The sample size is large (i.e., $n \geq 30$)

then again we use the t distribution to perform a test of hypothesis about μ.

Case III. If the following three conditions are fulfilled:

1. The population standard deviation σ is not known
2. The sample size is small (i.e., $n < 30$)
3. The population from which the sample is selected is not normally distributed (or the shape of its distribution is unknown)

then we use a nonparametric method to perform a test of hypothesis about μ.

The following chart summarizes these three cases.

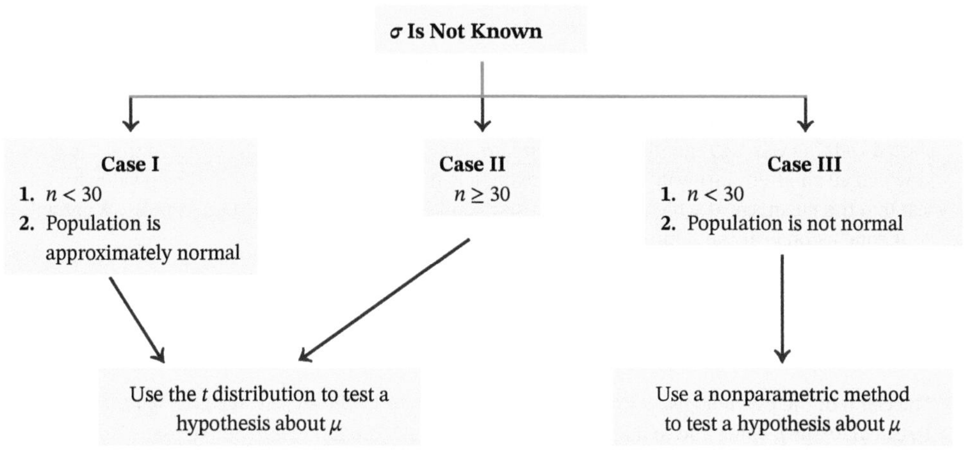

Next we discuss Cases I and II and learn how to use the t distribution to perform a test of hypothesis about μ when σ is not known. When the conditions mentioned for Case I or Case II are satisfied, the random variable

$$t = \frac{\bar{x} - \mu}{s_{\bar{x}}} \quad \text{where} \quad s_{\bar{x}} = \frac{s}{\sqrt{n}}$$

has a t distribution. Here, the t is called the **test statistic** to perform a test of hypothesis about a population mean μ.

Test Statistic

The value of the *test statistic t* for the sample mean $\bar{x}$ is computed as

$$t = \frac{\bar{x} - \mu}{s_{\bar{x}}}$$

where $\quad s_{\bar{x}} = \frac{s}{\sqrt{n}} \quad$ assuming that $\frac{n}{N} \leq .05$

The value of t calculated for $\bar{x}$ by using this formula is also called the *observed value* of t.

In Section 9.2, we discussed two procedures, the *p*-value approach and the critical value approach, to test hypotheses about μ when σ is known. In this section also we will use these two procedures to test hypotheses about μ when σ is not known. The steps used in these procedures are the same as in Section 9.2. The only difference is that we will be using the *t* distribution in place of the normal distribution.

9.3.1 The *p*-Value Approach

To use the *p*-value approach to perform a test of hypothesis about μ using the *t* distribution, we will use the same four steps that we used in Section 9.2.1. Although the *p*-value can be obtained very easily by using technology, we can use Table V of Appendix B to find a **range for the *p*-value** when technology is not available. Note that when using the *t* distribution and Table V, we cannot find the exact *p*-value but only a range within which it falls.

Examples 9.5 and 9.6 illustrate the *p*-value procedure to test a hypothesis about μ using the *t* distribution.

EXAMPLE 9.5 Age at Which Children Start Walking

Finding a p-Value and Making a Decision for a Two-Tailed Test of Hypothesis About μ: σ Not Known, $n < 30$, and Population Normally Distributed

A psychologist claims that the mean age at which children start walking is 12.5 months. Carol wanted to check if the true mean age at which children start walking is different from 12.5 months. She took a random sample of 18 children and found that the mean age at which these children started walking was 12.9 months with a standard deviation of .80 month. It is known that the ages at which all children start walking are approximately normally distributed. Find the *p*-value for the test that the mean age at which all children start walking is different from 12.5 months. What will your conclusion be if the significance level is 1%?

Solution Let μ be the mean age at which all children start walking, and let $\bar{x}$ be the corresponding mean for the sample. From the given information,

$$n = 18, \quad \bar{x} = 12.9 \text{ months}, \quad \text{and} \quad s = .80 \text{ month}$$

The claim of the psychologist is that the mean age at which children start walking is 12.5 months. To calculate the *p*-value and to make the decision, we apply the following four steps:

STEP 1 *State the null and alternative hypotheses.*

We are to test if the mean age at which all children start walking is different from 12.5 months. Hence, the null and alternative hypotheses are

$$H_0: \mu = 12.5 \quad \text{(The mean walking age is 12.5 months.)}$$
$$H_1: \mu \neq 12.5 \quad \text{(The mean walking age is different from 12.5 months.)}$$

STEP 2 *Select the distribution to use.*

In this test of hypothesis, we do not know the population standard deviation σ, the sample size is small ($n < 30$), and the population is approximately normally distributed. Hence, this is an example of Case I mentioned in the beginning of this section. Consequently, we will use the *t* distribution to find the *p*-value for this test.

STEP 3 *Calculate the p-value.*

The $\neq$ sign in the alternative hypothesis indicates that the test is two-tailed. To find the *p*-value, first we find the degrees of freedom and the *t* value for $\bar{x} = 12.9$ months. Then, the *p*-value is equal to twice the area in the tail of the *t* distribution curve beyond this *t* value for $\bar{x} = 12.9$ months. This *p*-value is shown in **Figure 9.11**. We find this *p*-value as follows:

$$s_{\bar{x}} = \frac{s}{\sqrt{n}} = \frac{.80}{\sqrt{18}} = .18856181$$

From H_0

$$t = \frac{\bar{x} - \mu}{s_{\bar{x}}} = \frac{12.9 - 12.5}{.18856181} = 2.121$$

$$df = n - 1 = 18 - 1 = 17$$

Now we can find the range for the *p*-value. To do so, we go to Table V of Appendix B (the *t* distribution table) and find the row for $df = 17$. In this row, we find the two values of t that cover $t = 2.121$. From Table V, for $df = 17$, these two values of t are 2.110 and 2.567. The test statistic $t = 2.121$ falls between these two values. Now look in the top row of this table to find the areas in the tail of the t distribution curve that correspond to 2.110 and 2.567. These two areas are .025 and .01, respectively. In other words, the area in the upper tail of the t distribution curve for $df = 17$ and $t = 2.110$ is .025, and the area in the upper tail of the t distribution curve for $df = 17$ and $t = 2.567$ is .01. Because it is a two-tailed test, the *p*-value for $t = 2.121$ is between $2(.025) = .05$ and $2(.01) = .02$, which can be written as

$$.02 < p\text{-value} < .05$$

Note that by using Table V of Appendix B, we cannot find the exact *p*-value but only a range within which the *p*-value falls. If we have access to technology, we can find the exact *p*-value. Using technology for this example, we obtain a *p*-value of .049.

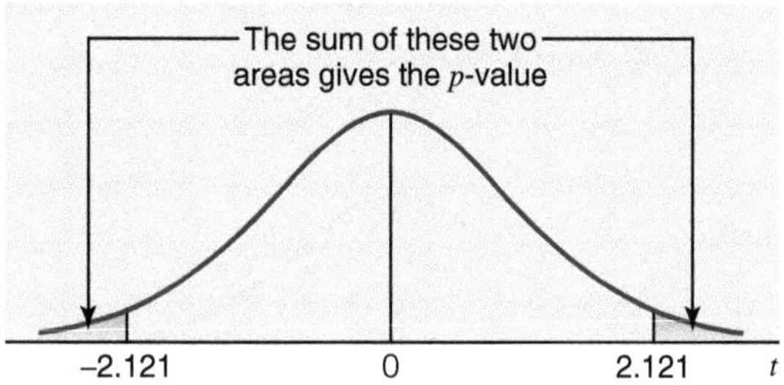

FIGURE 9.11 The required *p*-value.

STEP 4 *Make a decision.*

Thus, we can state that for any α greater than or equal to .05 (the upper limit of the *p*-value range), we will reject the null hypothesis. For any α less than or equal to .02 (the lower limit of the *p*-value range), we will not reject the null hypothesis. However, if α is between .02 and .05, we cannot make a decision. Note that if we use technology, then the *p*-value we will obtain for this example is .049, and we can make a decision for any value of α. For our example, $\alpha = .01$, which is less than the lower limit of the *p*-value range of .02. As a result, we fail to reject H_0 and conclude that the mean age at which all children start walking is not significantly different from 12.5 months. As a result, we can state that the sample data do not support the claim that the mean age at which children start walking is different from 12.5 months.

EXAMPLE 9.6 | Life of Batteries

Finding a *p*-Value and Making a Decision for a Left-Tailed Test of Hypothesis About μ: σ not Known and $n \geq 30$

Grand Auto Corporation produces auto batteries. The company claims that its top-of-the-line Never Die batteries are good, on average, for at least 65 months. A consumer protection agency believes the true mean is less than 65 months and tested 45 such batteries to check the corporation's claim. It found that the mean life of these 45 batteries is 63.4 months, and the standard deviation is 3 months. Find the *p*-value for the test that the mean life of all such batteries is less than 65 months. What will your conclusion be if the significance level is 2.5%?

Solution Let μ be the mean life of all such auto batteries, and let $\bar{x}$ be the corresponding mean life for the sample. From the given information,

$$n = 45, \quad \bar{x} = 63.4 \text{ months}, \quad \text{and} \quad s = 3 \text{ months}$$

The claim of the company is that the mean life of these batteries is at least 65 months. To calculate the *p*-value and make the decision, we apply the following four steps:

STEP 1 *State the null and alternative hypotheses.*

We are to test if the mean life of these batteries is at least 65 months. Hence, the null and alternative hypotheses are

$H_0: \mu \geq 65$ (The mean life of batteries is at least 65 months.)
$H_1: \mu < 65$ (The mean life of batteries is less than 65 months.)

STEP 2 *Select the distribution to use.*

In this example, we do not know the population standard deviation σ, and the sample size is large ($n \geq 30$). Hence, this is an example of Case II mentioned in the beginning of this section. Consequently, we will use the *t* distribution to find the *p*-value for this test.

STEP 3 *Calculate the p-value.*

The < sign in the alternative hypothesis indicates that the test is left-tailed. To find the *p*-value, first we find the degrees of freedom and the *t* value for $\bar{x} = 63.4$ months. Then, the *p*-value is given by the area in the tail of the *t* distribution curve beyond this *t* value for $\bar{x} = 63.4$ months. This *p*-value is shown in **Figure 9.12**. We find this *p*-value as follows:

$$s_{\bar{x}} = \frac{s}{\sqrt{n}} = \frac{3}{\sqrt{45}} = .44721360$$

From H_0

$$t = \frac{\bar{x} - \mu}{s_{\bar{x}}} = \frac{63.4 - 65}{.44721360} = -3.578$$

$$df = n - 1 = 45 - 1 = 44$$

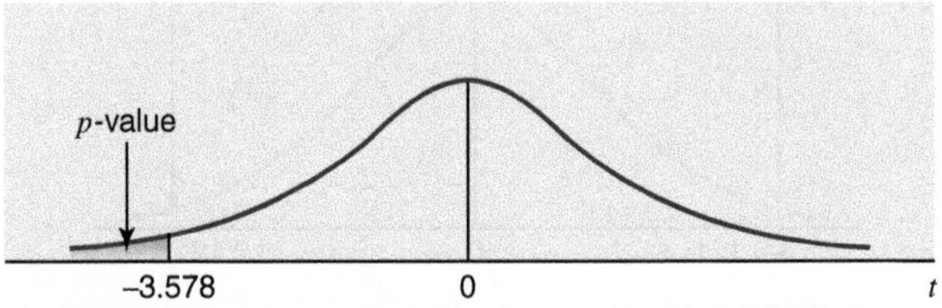

FIGURE 9.12 The required *p*-value.

Now we can find the range for the *p*-value. To do so, we go to Table V of Appendix B (the *t* distribution table) and find the row for $df = 44$. In this row, we find the two values of *t* that cover $t = 3.578$. Note that we use the positive value of the test statistic *t*, although our test statistic has a negative value. From Table V, for $df = 44$, the largest value of *t* is 3.286, for which the area in the tail of the *t* distribution is .001. This means that the area to the left of $t = -3.286$ is .001. Because -3.578 is smaller than -3.286, the area to the left of $t = -3.578$ is smaller than .001. Therefore, the *p*-value for $t = -3.578$ is less than .001, which can be written as

$$p\text{-value} < .001$$

Thus, here the *p*-value has only the upper limit of .001. In other words, the *p*-value for this example is less than .001. If we use technology for this example, we will obtain a *p*-value of .00043.

STEP 4 *Make a decision.*

Thus, we can state that for any α greater than or equal to .001 (the upper limit of the *p*-value range), we will reject the null hypothesis. For our example $\alpha = .025$, which is greater than the upper limit of the *p*-value of .001. As a result, we reject H_0 and conclude that the sample data support the alternative hypothesis that the mean life of such batteries is less than 65 months. Therefore, we can state that the difference between the hypothesized population mean of 65 months and the sample mean of 63.4 is too large to be attributed to sampling error alone.

9.3.2 The Critical Value Approach

In this procedure, as mentioned in Section 9.2.2, we have a predetermined value of the significance level α. The value of α gives the total area of the rejection region(s). First we find the critical value(s) of *t* from the *t* distribution table in Appendix B for the given degrees of freedom and the significance level. Then we find the value of the test statistic *t* for the observed value of the sample statistic $\bar{x}$. Finally we compare these two values and make a decision. Remember, if the test is one-tailed, there is only one critical value of *t*, and it is obtained by using the value of α, which gives the area in the left or right tail of the *t* distribution curve, depending on whether the test is left-tailed or right-tailed, respectively. However, if the test is two-tailed, there are two critical values of *t*, and they are obtained by using $\alpha/2$ area in each tail of the *t* distribution curve. The value of the test statistic *t* is obtained as mentioned earlier in this section.

Examples 9.7 and 9.8 describe the procedure to test a hypothesis about μ using the critical value approach and the *t* distribution.

EXAMPLE 9.7 Age at Which Children Start Walking

Conducting a Two-Tailed Test of Hypothesis About μ Using the Critical Value Approach: σ Unknown, $n < 30$, and Population Normally Distributed

Cohen/Ostrow/Getty Images

Refer to Example 9.5. A psychologist claims that the mean age at which children start walking is 12.5 months. Carol wanted to check if this claim is true. She took a random sample of 18 children and found that the mean age at which these children started walking was 12.9 months with a standard deviation of .80 month. Using a 1% significance level, can Carol conclude that the mean age at which all children start walking is different from 12.5 months? Assume that the ages at which all children start walking have an approximate normal distribution.

Solution Let μ be the mean age at which all children start walking, and let $\bar{x}$ be the corresponding mean for the sample. Then, from the given information,

$$n = 18, \quad \bar{x} = 12.9 \text{ months}, \quad s = .80 \text{ month}, \quad \text{and} \quad \alpha = .01$$

STEP 1 *State the null and alternative hypotheses.*

We are to test if the mean age at which all children start walking is different from 12.5 months. The null and alternative hypotheses are

$H_0: \mu = 12.5$ (The mean walking age is 12.5 months.)
$H_1: \mu \neq 12.5$ (The mean walking age is different from 12.5 months.)

STEP 2 *Select the distribution to use.*

In this example, the population standard deviation σ is not known, the sample size is small ($n < 30$), and the population is approximately normally distributed. Hence, this is an example of Case I mentioned in the beginning of Section 9.3. Consequently, we will use the t distribution to perform the test in this example.

STEP 3 *Determine the rejection and nonrejection regions.*

The significance level is .01. The $\neq$ sign in the alternative hypothesis indicates that the test is two-tailed and the rejection region lies in both tails. The area of the rejection region in each tail of the t distribution curve is

$$\text{Area in each tail} = \alpha/2 = .01/2 = .005$$
$$df = n - 1 = 18 - 1 = 17$$

From the t distribution table, the critical values of t for 17 degrees of freedom and .005 area in each tail of the t distribution curve are −2.898 and 2.898. These values are shown in **Figure 9.13**.

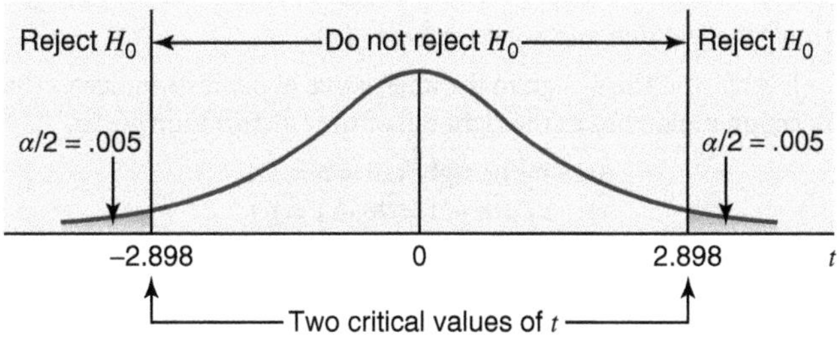

FIGURE 9.13 The critical values of t.

STEP 4 *Calculate the value of the test statistic.*

We calculate the value of the test statistic t for $\bar{x} = 12.9$ as follows:

$$s_{\bar{x}} = \frac{s}{\sqrt{n}} = \frac{.80}{\sqrt{18}} = .18856181$$

From H_0

$$t = \frac{\bar{x} - \mu}{s_{\bar{x}}} = \frac{12.9 - 12.5}{.18856181} = 2.121$$

STEP 5 *Make a decision.*

The value of the test statistic $t = 2.121$ falls between the two critical points, -2.898 and 2.898, which is the nonrejection region. Consequently, we fail to reject H_0. As a result, we consider the difference between the hypothesized population mean and the sample mean to be so small that it may have occurred because of sampling error. Therefore, the sample data do not support the claim that the mean age at which all children start walking is significantly different from 12.5 months.

EXAMPLE 9.8 Waiting Time for Service at a Bank

Conducting a Right-Tailed Test of Hypothesis About μ Using the Critical Value Approach: σ Unknown and $n \geq 30$

Keith Brofsky/Getty Images

The management at Massachusetts Savings Bank is always concerned about the quality of service provided to its customers. With the old computer system, a teller at this bank could serve, on average, 22 customers per hour. The management noticed that with this service rate, the waiting time for customers was too long. Recently the management of the bank installed a new computer system, expecting that it would increase the service rate and consequently make the customers happier by reducing the waiting time. To check if the new computer system is more efficient than the old system, the management of the bank took a random sample of 70 hours and found that during these hours the mean number of customers served by tellers was 27 per hour with a standard deviation of 2.5 customers. Testing at a 1% significance level, would you conclude that the new computer system is more efficient than the old computer system?

Solution Let μ be the mean number of customers served per hour by a teller using the new system, and let $\bar{x}$ be the corresponding mean for the sample. Then, from the given information,

$$n = 70 \text{ hours}, \quad \bar{x} = 27 \text{ customers}, \quad s = 2.5 \text{ customers}, \quad \text{and} \quad \alpha = .01$$

STEP 1 *State the null and alternative hypotheses.*

We are to test whether or not the new computer system is more efficient than the old system. The new computer system will be more efficient than the old system if the mean number of customers served per hour by using the new computer system is significantly more than 22; otherwise, it will not be more efficient. The null and alternative hypotheses are

$$H_0: \mu = 22 \quad \text{(The new computer system is not more efficient.)}$$
$$H_1: \mu > 22 \quad \text{(The new computer system is more efficient.)}$$

STEP 2 *Select the distribution to use.*

In this example, the population standard deviation σ is not known and the sample size is large ($n \geq 30$). Hence, it is Case II mentioned in the beginning of Section 9.3. Consequently, we will use the t distribution to perform the test for this example.

STEP 3 *Determine the rejection and nonrejection regions.*

The significance level is .01. The > sign in the alternative hypothesis indicates that the test is right-tailed and the rejection region lies in the right tail of the t distribution curve.

$$\text{Area in the right tail} = \alpha = .01$$
$$df = n - 1 = 70 - 1 = 69$$

From the t distribution table, the critical value of t for 69 degrees of freedom and .01 area in the right tail is 2.382. This value is shown in **Figure 9.14**.

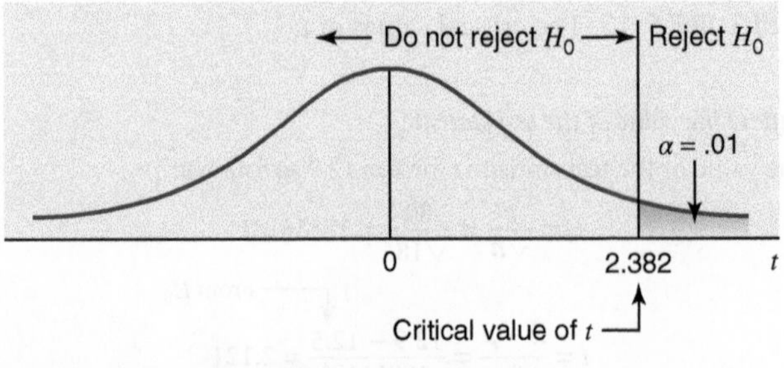

FIGURE 9.14 The critical value of t.

STEP 4 *Calculate the value of the test statistic.*

The value of the test statistic t for $\bar{x} = 27$ is calculated as follows:

$$s_{\bar{x}} = \frac{s}{\sqrt{n}} = \frac{2.5}{\sqrt{70}} = .29880715$$

From H_0

$$t = \frac{\bar{x} - \mu}{s_{\bar{x}}} = \frac{27 - 22}{.29880715} = 16.733$$

STEP 5 *Make a decision.*

The value of the test statistic $t = 16.733$ is greater than the critical value of $t = 2.382$ and falls in the rejection region. Consequently, we reject H_0. As a result, we conclude that the value of the sample mean is very large compared to the hypothesized value of the population mean, and the difference between the two may not be attributed to chance alone. The sample data provide strong evidence that the mean number of customers served per hour using the new computer system is more than 22.

Note: What if the Sample Size Is Large and the Number of *df* Is Not in the *t* Distribution Table?

In this section when σ is not known, we used the t distribution to perform tests of hypothesis about μ in Cases I and II. Note that in Case II, the sample size is large. If we have access to technology, it does not matter how large the sample size is, we can always use the t distribution. However, if we are using the t distribution table (Table V of Appendix B), this may pose a problem. Usually such a table goes only up to a certain number of degrees of freedom. For example, Table V in Appendix B goes only up to 75 degrees of freedom. Thus, if the sample size is larger than 76 (with df more than 75), we cannot use Table V to find the critical value(s) of t to make a decision in this section. In such a situation when n is large and is not included in the t distribution table, there are three options:

1. Use the t value from the last row (the row of ∞) in Table V of Appendix B.
2. Use the normal distribution as an approximation to the t distribution.
3. Use technology to find the t value.

To use the normal distribution as an approximation to the t distribution to make a test of hypothesis about μ, the procedure is exactly like the one in Section 9.2, except that now we will replace σ by s, and $\sigma_{\bar{x}}$ by $s_{\bar{x}}$.

Note that the t values obtained from the last row of the t distribution table are the same as will be obtained from the normal distribution table for the same areas in the upper tail or lower tail of the distribution. Again, note that here we can use the normal distribution as a convenience and as an approximation, but if we can, we should use the t distribution by using technology.

EXERCISES

Concepts and Procedures

9.34 Briefly explain the conditions that must hold true to use the t distribution to make a test of hypothesis about the population mean.

9.35 For each of the following examples of tests of hypothesis about μ, show the rejection and nonrejection regions on the t distribution curve.

a. A two-tailed test with $\alpha = .01$ and $n = 15$

b. A left-tailed test with $\alpha = .005$ and $n = 25$

c. A right-tailed test with $\alpha = .025$ and $n = 22$

9.36 A random sample of 25 observations taken from a population that is normally distributed produced a sample mean of 58.5 and a standard deviation of 7.50. Using $\alpha = .01$, find the critical and observed values of t and the range for the p-value for each of the following tests of hypotheses:

a. $H_0: \mu = 55$ versus $H_1: \mu > 55$

b. $H_0: \mu = 55$ versus $H_1: \mu \neq 55$

9.37 A random sample of 16 observations taken from a population that is normally distributed produced a sample mean of 42.40 and a standard deviation of 8. Using $\alpha = .05$, find the critical and observed values of t and the range for the p-value for each of the following tests of hypotheses:

a. $H_0: \mu = 46$ versus $H_1: \mu < 46$

b. $H_0: \mu = 46$ versus $H_1: \mu \neq 46$

9.38 Consider the null hypothesis $H_0: \mu = 100$. Suppose that a random sample of 30 observations is taken from this population to perform this test. Using a significance level of .01, show the rejection and nonrejection regions and find the critical value(s) of t when the alternative hypothesis is as follows:

a. $H_1: \mu \neq 100$ b. $H_1: \mu > 100$ c. $H_1: \mu < 100$

9.39 Consider $H_0: \mu = 40$ versus $H_1: \mu > 40$.

a. A random sample of 64 observations taken from this population produced a sample mean of 43 and a standard deviation of 5. Using $\alpha = .025$, would you reject the null hypothesis?

b. Another random sample of 64 observations taken from the same population produced a sample mean of 41 and a standard deviation of 7. Using $\alpha = .025$, would you reject the null hypothesis? Explain.

c. Comment on the results of parts (a) and (b).

9.40 Perform the following tests of hypotheses:

a. $H_0: \mu = 94.80$, $H_1: \mu < 94.80$, $n = 12$, $\bar{x} = 92.87$, $s = 5.34$, $\alpha = .10$

b. $H_0: \mu = 18.70$, $H_1: \mu \neq 18.70$, $n = 25$, $\bar{x} = 20.05$, $s = 2.99$, $\alpha = .05$

c. $H_0: \mu = 59$, $H_1: \mu > 59$, $n = 7$, $\bar{x} = 59.42$, $s = .418$, $\alpha = .01$

Applications

9.41 The Knot Worldwide released the results of its annual Real Weddings Study on February 14, 2019. According to this study, the average cost of a wedding in the United States was $33,931 for couples married in 2018. A recent sample of 100 U.S. couples who got married this year produced a mean wedding cost of $36,157 with a standard deviation of $9275. Using a 5% significance level and the critical value approach, can you conclude that the current mean cost of a wedding in the United States is higher than $33,931? Find the range for the p-value for this test. What will your conclusion be using this p-value range and $\alpha = .05$?

9.42 The mean balance of all checking accounts at a bank on December 31, 2019, was $1050. A random sample of 60 checking accounts taken recently from this bank gave a mean balance of $910 with a standard deviation of $230. Using a 1% significance level, can you conclude that the mean balance of such accounts has decreased during this period? Explain your conclusion in words. What if $\alpha = .025$?

9.43 A soft-drink manufacturer claims that its 12-ounce cans do not contain, on average, more than 30 calories. A random sample of 64 cans of this soft drink, which were checked for calories, contained a mean of 32 calories with a standard deviation of 3 calories. Does the sample information support the alternative hypothesis that the manufacturer's claim is false? Use a significance level of 5%. Find the range for the p-value for this test. What will your conclusion be using this p-value and $\alpha = .05$?

9.44 According to Insure.com, the average annual cost of automobile insurance was $1525 in the state of Nevada in 2019 (www.valuepenguin.com). An insurance broker is interested to find if the current mean annual rate of automobile insurance in Nevada is more than $1525. He recently took a random sample of 100 insured automobiles from the state of Nevada and found the mean annual automobile insurance rate of $1550 with a standard deviation of $122.

a. Using a 1% significance level and the critical value approach, can you conclude that the current mean annual automobile insurance rate in Nevada is higher than $1525?

b. Find the range for the p-value for this test. What will your conclusion be using this p-value range and $\alpha = .01$?

9.45 A paint manufacturing company claims that the mean drying time for its paints is not longer than 60 minutes. A random sample of 20 gallons of paints selected from the production line of this company showed that the mean drying time for this sample is 63.50 minutes with a standard deviation of 4 minutes. Assume that the drying times for these paints have a normal distribution.

a. Using a 1% significance level, can you conclude that the company's claim is true?

b. What is the Type I error in this exercise? Explain in words. What is the probability of making such an error?

9.46 A gardener knows from experience that sunflower seeds sprout and grow to an average height of 40 inches in 30 days. The gardener is concerned that a recent drought will affect the growth of the sunflower plants. From a large field of sunflower seeds planted by the gardener, a random sample of 40 sunflower plants was selected 30 days after planting. The average height of these sunflower plants was 38.4 inches with a standard deviation of 2.88 inches. At a 1% significance level, can the gardener conclude that, after 30 days from planting, the mean height of all sunflower plants in the field is less than 40 inches? Use both the p-value and the critical value approaches.

9.47 A past study claimed that adults in the United States spent an average of 18 hours a week on leisure activities. A researcher wanted to test this claim. She took a sample of 12 adults and asked them about the time they spend per week on leisure activities. Their responses (in hours) are as follows:

13.6 14.0 24.5 24.6 22.9 37.7 14.6 14.5 21.5 21.0 17.8 21.4

Assume that the times spent on leisure activities by all U.S. adults are approximately normally distributed. Using a 10% significance level, can you conclude that the average amount of time spent by U.S. adults on leisure activities has changed? (*Hint*: First calculate the sample mean and the sample standard deviation for these data using the formulas learned in Sections 3.1.1 and 3.2.2 of Chapter 3. Then make the test of hypothesis about μ.)

9.48 The past records of a supermarket show that its customers spend an average of $65 per visit at this store. Recently the management of the store initiated a promotional campaign according to which each customer receives points based on the total money spent at the store, and these points can be used to buy products at the store. The management expects that as a result of this campaign, the customers should be encouraged to spend more money at the store. To check whether this is true, the manager of the store took a sample of 12 customers who visited the store. The following data give the money (in dollars) spent by these customers at this supermarket during their visits.

88 69 141 28 106 45 32 51 78 54 110 83

Assume that the money spent by all customers at this supermarket has an approximate normal distribution. Using a 1% significance level, can you conclude that the mean amount of money spent by all customers at this supermarket after the campaign was started is more than $65? (*Hint*: First calculate the sample mean and the sample standard deviation for these data using the formulas learned in Sections 3.1.1 and 3.2.2 of Chapter 3. Then make the test of hypothesis about μ.)

9.49 According to an estimate, U.S. households with credit card debts owed an average of $15,900 on their credit cards last year. A recent random sample of 500 U.S. households with credit card debts produced a mean credit card debt of $16,377 with a standard deviation of $3800. Do these data provide significant evidence at a 1% significance level to conclude that the current mean credit card debt of U.S. households with credit card debts is higher than $15,900? Use both the p-value approach and the critical value approach.

*9.50 The manager of a service station claims that the mean amount spent on gas by its customers is $22.60 per visit. You want to test if the mean amount spent by customers on gas at this station is different from $22.60 per visit. Briefly explain how you would conduct this test when σ is not known.

*9.51 A tool manufacturing company claims that its top-of-the-line machine that is used to manufacture bolts produces an average of 50 or more bolts per hour. A company that is interested in buying this machine wants to check this claim. Suppose you are asked to conduct this test. Briefly explain how you would do so when σ is not known.

9.4 Hypothesis Tests About a Population Proportion: Large Samples

LEARNING OBJECTIVES

After completing this section, you should be able to:

- Explain the conditions required to use the normal distribution to perform a test of hypothesis about a population proportion.
- Determine whether a sample size is large enough to use the normal distribution to perform a test of hypothesis about a population proportion.
- Use the normal distribution to illustrate the critical value, the rejection region and the nonrejection region for a given alternative hypothesis and level of significance.
- Apply the five-step method using a critical value approach to perform a test of hypothesis about a population proportion.
- Apply the four-step method using a *p*-value approach to perform a test of hypothesis about a population proportion.
- Describe the Type I and Type II errors in the context of a test of hypothesis about a population proportion.

Often we want to conduct a test of hypothesis about a population proportion. For example, in a Gallup poll conducted a few years ago, 61% of adults said that upper-income people were paying too little in federal taxes. An agency may want to check if this percentage has changed since then. As another example, a mail-order company claims that 90% of all orders it receives are shipped within 72 hours. The company's management may want to determine from time to time whether or not this claim is true.

This section presents the procedure to perform tests of hypotheses about the population proportion, p, for large samples. The procedures to make such tests are similar in many respects to the ones for the population mean, μ. Again, a test can be two-tailed or one-tailed. We know from Chapter 7 that when the sample size is large, the sampling distribution of the sample proportion, $\hat{p}$, is approximately normally distributed with its mean equal to p and standard deviation equal to $\sqrt{pq/n}$. Hence, we use the normal distribution to perform a test of hypothesis about the population proportion, p, for a large sample. As was mentioned in Chapters 7 and 8, in the case of a proportion, the sample size is considered to be large when np and nq are both greater than 5.

Test Statistic

The value of the *test statistic* z for the sample proportion $\hat{p}$ is computed as

$$z = \frac{\hat{p} - p}{\sigma_{\hat{p}}} \quad \text{where} \quad \sigma_{\hat{p}} = \sqrt{\frac{pq}{n}} \quad \text{assuming } \frac{n}{N} \le .05$$

The value of p that is used in this formula is the one from the null hypothesis. The value of q is equal to $1 - p$.

The value of z calculated for $\hat{p}$ using the above formula is also called the *observed value of z*.

In Section 9.2, we discussed two procedures, the p-value approach and the critical value approach, to test hypotheses about μ. Here too we will use these two procedures to test hypotheses about p. The steps used in these procedures are the same as in Section 9.2. The only difference is that we will be making tests of hypotheses about p rather than about μ.

9.4.1 The p-Value Approach

To use the p-value approach to perform a test of hypothesis about p, we will use the same four steps that we used in Section 9.2. Although the p-value for a test of hypothesis about population proportion p can be obtained very easily by using technology, we can use Table IV of Appendix B to find this p-value when technology is not available.

Examples 9.9 and 9.10 illustrate the p-value procedure to test a hypothesis about p for a large sample.

EXAMPLE 9.9 Emotional Attachment to Alma Mater

Finding a p-Value and Making a Decision for a Two-Tailed Test of Hypothesis About p: Large Sample

According to a Gallup poll conducted a few years ago, 20% of traditional college graduates (those who earned their degree before age 25) were *emotionally attached* to their alma mater (www.gallup.com). Suppose this result is true for the population of traditional college graduates of that year. In a recent random sample of 2000 traditional college graduates, 22% said that they are emotionally attached to their alma mater. Find the p-value to test the hypothesis that the current percentage of traditional college graduates who are emotionally attached to their alma mater is different from 20%. What is your conclusion if the significance level is 5%?

Solution Let p be the current proportion of all traditional college graduates who are emotionally attached to their alma mater, and $\hat{p}$ be the corresponding sample proportion. Then, from the given information,

$$n = 2000, \quad \hat{p} = .22, \quad \text{and} \quad \alpha = .05$$

A few years ago, 20% of traditional college graduates were emotionally attached to their alma mater. Hence,

$$p = .20 \quad \text{and} \quad q = 1 - p = 1 - .20 = .80$$

To calculate the p-value and to make a decision, we apply the following four steps:

STEP 1 *State the null and alternative hypotheses.*

The current percentage of traditional college graduates who are emotionally attached to their alma mater will not be different from 20% if $p = .20$, and the current percentage will be different from 20% if $p \neq .20$. The null and alternative hypotheses are as follows:

$$H_0: p = .20 \quad \text{(The current percentage is not different from 20\%.)}$$
$$H_1: p \neq .20 \quad \text{(The current percentage is different from 20\%.)}$$

STEP 2 *Select the distribution to use.*

To check whether the sample is large, we calculate the values of np and nq.

$$np = 2000(.20) = 400 \quad \text{and} \quad nq = 2000(.80) = 1600$$

Since np and nq are both greater than 5, we can conclude that the sample size is large. Consequently, we will use the normal distribution to find the p-value for this test.

STEP 3 *Calculate the p-value.*

The $\neq$ sign in the alternative hypothesis indicates that the test is two-tailed. The p-value is equal to twice the area in the tail of the normal distribution curve to the right of the test statistic z for $\hat{p} = .22$. This p-value is shown in **Figure 9.15.** To find this p-value, first we find z for $\hat{p} = .22$ as follows:

$$\sigma_{\hat{p}} = \sqrt{\frac{pq}{n}} = \sqrt{\frac{(.20)(.80)}{2000}} = .00894427$$

From H_0

$$z = \frac{\hat{p} - p}{\sigma_{\hat{p}}} = \frac{.22 - .20}{.00894427} = 2.24$$

Now we find the area to the right of $z = 2.24$ from the normal distribution table. This area is $1 - .9875 = .0125$. Consequently, the p-value is

$$p\text{-value} = 2(.0125) = .0250$$

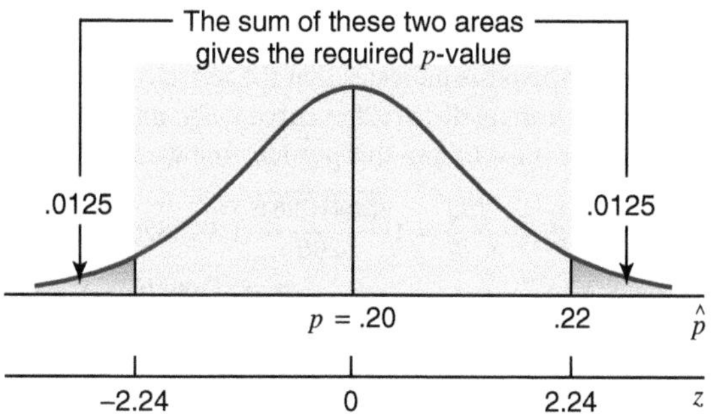

FIGURE 9.15 The required p-value.

STEP 4 *Make a decision.*

Thus, we can state that for any α greater than or equal to .0250 we will reject the null hypothesis, and for any α less than .0250 we will not reject the null hypothesis. In our example, $\alpha = .05$, which is greater than the p-value of .0250. As a result, we reject H_0 and conclude that the sample data provide evidence that the current percentage of traditional college graduates who are emotionally attached to their alma mater is different from .20. Consequently, we can state that the difference between the hypothesized population proportion of .20 and the sample proportion of .22 is most likely large enough to be attributed to sampling error alone when $\alpha = .05$.

EXAMPLE 9.10 Percentage of Defective Chips

Finding a p-Value and Making a Decision for a Right-Tailed Test of Hypothesis About p: Large Sample

When working properly, a machine that is used to make chips for calculators does not produce more than 4% defective chips. Whenever the machine produces more than 4% defective chips, it needs an adjustment. To check if the machine is working properly, the quality control department at the company often takes samples of chips and inspects them to determine if they are good or defective. One such random sample of 200 chips taken recently from the production line contained 12 defective chips. Find the p-value to test whether the machine needs an adjustment. What would your conclusion be if the significance level is 2.5%?

Solution Let p be the proportion of defective chips among all chips produced by this machine, and let $\hat{p}$ be the corresponding sample proportion. Then, from the given information,

$$n = 200, \quad \hat{p} = 12/200 = .06, \quad \text{and} \quad \alpha = .025$$

When the machine is working properly, it does not produce more than 4% defective chips. Hence, assuming that the machine is working properly, we obtain

$$p = .04 \quad \text{and} \quad q = 1 - p = 1 - .04 = .96$$

To calculate the p-value and to make a decision, we apply the following four steps:

STEP 1 *State the null and alternative hypotheses.*

The machine does not need an adjustment if the percentage of defective chips among all chips produced on this machine is 4% or less, and it does need an adjustment if this percentage is greater than 4%. Hence, the null and alternative hypotheses are as follows:

$$H_0: p \le .04 \quad \text{(The machine does not need an adjustment.)}$$
$$H_1: p > .04 \quad \text{(The machine does need an adjustment.)}$$

STEP 2 *Select the distribution to use.*

To check if the sample is large, we calculate the values of np and nq:

$$np = 200(.04) = 8 \quad \text{and} \quad nq = 200(.96) = 192$$

Since np and nq are both greater than 5, we can conclude that the sample size is large. Consequently, we will use the normal distribution to find the p-value for this test.

STEP 3 *Calculate the p-value.*

The > sign in the alternative hypothesis indicates that the test is right-tailed. The *p*-value is given by the area in the upper tail of the normal distribution curve to the right of the test statistic z for $\hat{p} = .06$. This *p*-value is shown in **Figure 9.16.** To find this *p*-value, first we find z for $\hat{p} = .06$ as follows:

$$\sigma_{\hat{p}} = \sqrt{\frac{pq}{n}} = \sqrt{\frac{(.04)(.96)}{200}} = .01385641$$

From H_0

$$z = \frac{\hat{p} - p}{\sigma_{\hat{p}}} = \frac{.06 - .04}{.01385641} = 1.44$$

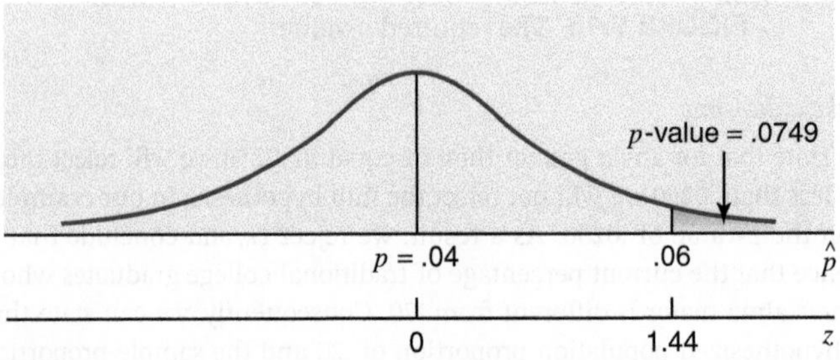

FIGURE 9.16 The required *p*-value.

Now we find the area to the right of $z = 1.44$ from the normal distribution table. This area is $1 - .9251 = .0749$. Consequently, the *p*-value is

$$p\text{-value} = .0749$$

STEP 4 *Make a decision.*

Thus, we can state that for any α greater than or equal to .0749 we will reject the null hypothesis, and for any α less than .0749 we will not reject the null hypothesis. For our example, $\alpha = .025$, which is less than the *p*-value of .0749. As a result, we fail to reject H_0 and conclude that the sample data do not provide evidence that the machine needs an adjustment.

9.4.2 The Critical Value Approach

In this procedure, as mentioned in Section 9.2.2, we have a predetermined value of the significance level α. The value of α gives the total area of the rejection region(s). First we find the critical value(s) of z from the normal distribution table for the given significance level. Then we find the value of the test statistic z for the observed value of the sample statistic $\hat{p}$. Finally we compare these two values and make a decision. Remember, if the test is one-tailed, there is only one critical value of z, and it is obtained by using the value of α, which gives the area in the left or right tail of the normal distribution curve, depending on whether the test is left-tailed or right-tailed, respectively. However, if the test is two-tailed, there are two critical values of z, and they are obtained by using $\alpha/2$ area in each tail of the normal distribution curve. The value of the test statistic z is obtained as mentioned earlier in this section.

Examples 9.11 and 9.12 describe the procedure to test a hypothesis about p using the critical value approach and the normal distribution.

EXAMPLE 9.11 Emotional Attachment to Alma Mater

Making a Two-Tailed Test of Hypothesis About *p* Using the Critical Value Approach: Large Sample

Refer to Example 9.9. According to a Gallup poll conducted a few years ago, 20% of traditional college graduates (those who earned their degree before age 25) were *emotionally attached* to their alma mater (www.gallup.com). Suppose this result is true for the population of traditional college graduates of that year. In a recent random sample of 2000 traditional college graduates, 22% said that they are emotionally attached to their alma mater. Using a 5% significance level, can you conclude that the current percentage of traditional college graduates who are emotionally attached to their alma mater is different from 20%?

Solution Let p be the current proportion of all traditional college graduates who are emotionally attached to their alma mater, and $\hat{p}$ be the corresponding sample proportion. Then, from the given information,

$$n = 2000, \quad \hat{p} = .22, \quad \text{and} \quad \alpha = .05$$

A few years ago, 20% of traditional college graduates were emotionally attached to their alma mater. Hence,

$$p = .20 \quad \text{and} \quad q = 1 - p = 1 - .20 = .80$$

To use the critical value approach to perform a test of hypothesis, we apply the following five steps:

STEP 1 *State the null and alternative hypotheses.*

The current percentage of traditional college graduates who are emotionally attached to their alma mater will not be different from 20% if $p = .20$, and the current percentage will be different from 20% if $p \neq .20$. The null and alternative hypotheses are as follows:

$H_0: p = .20$ (The current percentage is not different from 20%.)
$H_1: p \neq .20$ (The current percentage is different from 20%.)

STEP 2 *Select the distribution to use.*

To check if the sample is large, we calculate the values of np and nq.

$$np = 2000(.20) = 400 \quad \text{and} \quad nq = 2000(.80) = 1600$$

Since np and nq are both greater than 5, we can conclude that the sample size is large. Consequently, we will use the normal distribution to make the test.

STEP 3 *Determine the rejection and nonrejection regions.*

The $\neq$ sign in the alternative hypothesis indicates that the test is two-tailed. The significance level is .05. Therefore, the total area of the two rejection regions is .05, and the area of the rejection region in each tail of the sampling distribution of $\hat{p}$ is $\alpha/2 = .05/2 = .025$. The critical values of z, obtained from the standard normal distribution table, are −1.96 and 1.96, as shown in **Figure 9.17**.

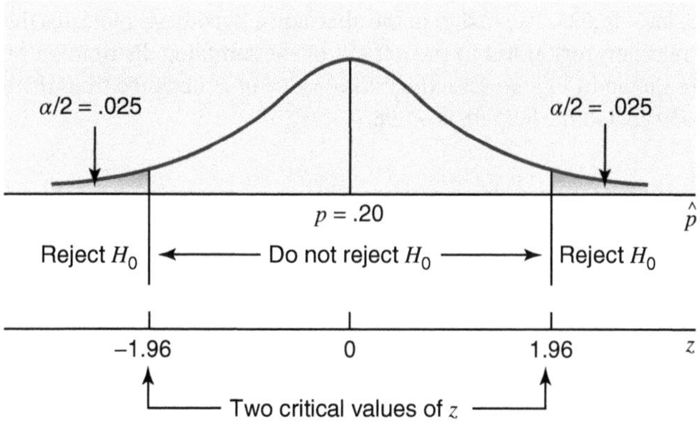

FIGURE 9.17 The critical values of z.

STEP 4 *Calculate the value of the test statistic.*

The value of the test statistic z for $\hat{p} = .22$ is calculated as follows:

$$\sigma_{\hat{p}} = \sqrt{\frac{pq}{n}} = \sqrt{\frac{(.20)(.80)}{2000}} = .00894427$$

From H_0

$$z = \frac{\hat{p} - p}{\sigma_{\hat{p}}} = \frac{.22 - .20}{.00894427} = 2.24$$

STEP 5 *Make a decision.*

The value of the test statistic $z = 2.24$ for $\hat{p}$ falls in the rejection region. As a result, we reject H_0 and conclude that the sample evidence shows that the current percentage of traditional college graduates who are emotionally attached to their alma mater is different from .20. Consequently, we can state that the difference between the hypothesized population proportion of .20 and the sample proportion of .22 is most likely large enough to be attributed to sampling error alone when $\alpha = .05$.

EXAMPLE 9.12 Mailing the Received Orders

Conducting a Left-Tailed Test of Hypothesis About p Using the Critical Value Approach: Large Sample

Direct Mailing Company sells computers and computer parts by mail. The company claims that at least 90% of all orders are mailed within 72 hours after they are received. The quality control department at the company often takes samples to check if this claim is valid. A recently taken sample of 150 orders showed that 129 of them were mailed within 72 hours. Does the sample information provide sufficient evidence to reject the company's claim? Use a 2.5% significance level.

Solution Let p be the proportion of all orders that are mailed by the company within 72 hours, and let $\hat{p}$ be the corresponding sample proportion. Then, from the given information,

$$n = 150, \quad \hat{p} = 129/150 = .86, \quad \text{and} \quad \alpha = .025$$

The company claims that at least 90% of all orders are mailed within 72 hours. Assuming that this claim is true, the values of p and q are

$$p = .90 \quad \text{and} \quad q = 1 - p = 1 - .90 = .10$$

STEP 1 *State the null and alternative hypotheses.*

The null and alternative hypotheses are

$$H_0: p \geq .90 \quad \text{(The company's claim is true.)}$$
$$H_1: p < .90 \quad \text{(The company's claim is false.)}$$

STEP 2 *Select the distribution to use.*

We first check whether np and nq are both greater than 5.

$$np = 150(.90) = 135 > 5 \quad \text{and} \quad nq = 150(.10) = 15 > 5$$

Consequently, the sample size is large. Therefore, we can use the normal distribution to make the hypothesis test about p.

STEP 3 *Determine the rejection and nonrejection regions.*

The significance level is .025. The < sign in the alternative hypothesis indicates that the test is left-tailed, and the rejection region lies in the left tail of the sampling distribution of $\hat{p}$ with its area equal to .025. As shown in **Figure 9.18**, the critical value of z, obtained from the normal distribution table for .0250 area in the left tail, is –1.96.

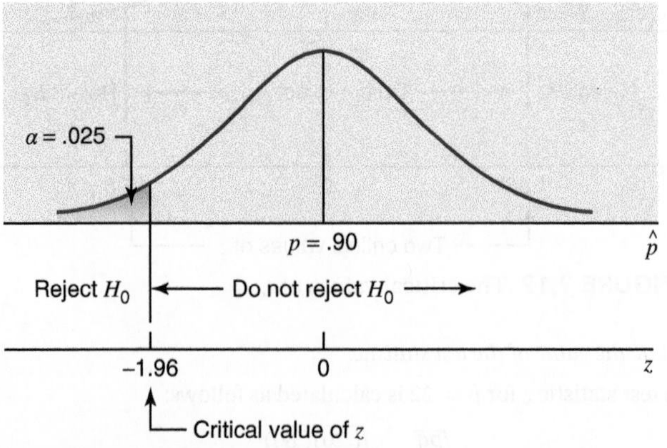

FIGURE 9.18 Critical value of z.

STEP 4 *Calculate the value of the test statistic.*

The value of the test statistic z for $\hat{p} = .86$ is calculated as follows:

$$\sigma_{\hat{p}} = \sqrt{\frac{pq}{n}} = \sqrt{\frac{(.90)(.10)}{150}} = .02449490$$

$$z = \frac{\hat{p} - p}{\sigma_{\hat{p}}} = \frac{.86 - .90}{.02449490} = -1.63$$

(From H_0)

STEP 5 *Make a decision.*

The value of the test statistic $z = -1.63$ is greater than the critical value of $z = -1.96$, and it falls in the nonrejection region. Therefore, we fail to reject H_0. We can state that the difference between the sample proportion and the hypothesized value of the population proportion is small, and most likely this difference may have occurred owing to chance alone. Therefore, the sample information does not provide sufficient evidence to reject the company's claim.

Case Study 9.2

Are Parents Doing Too Much for Their Adult Children?

A Pew Research Center poll of parents of adult children was conducted from June 25 to July 8, 2019. One of the questions that these adults were asked was: Are the parents of adult children ages 18 to 29 doing too much, too little, or about the right amount for these children? The responses of these parents from this poll are shown in the graph. Of these parents, 55% said that parents of adult children ages 18 to 29 do too much for these children, 10% said too little, and 34% said parents do about the right amount for their adult children. Note that these three percentages do not add up to 100 because of rounding.

Suppose we want to check whether the current percentage of parents of adult children who will say that parents do too much for their adult children ages 18 to 29 is different from 55%. Further, suppose we take a sample of 1600 parents of adult children and ask them the same question, and 59% of them say that parents do too much for their adult children ages 18 to 29. Let us choose a significance level of 1%. The test is two-tailed. The null and alternative hypotheses are as follows where p is the current percentage of all parents of adult children who will say that parents do too much for their adult children ages 18 to 29:

$$H_0: p = .55$$
$$H_1: p \neq .55$$

Here, $n = 1600$, $\hat{p} = .59$, $\alpha = .01$, and $\alpha/2 = .005$. The sample is large. (The reader should check that np and nq are both greater than 5.) Using the normal distribution for the test, the critical values of z for .0050 and .9950 areas to the left are −2.58 and 2.58, respectively. We find the observed value of z as follows:

$$\sigma_{\hat{p}} = \sqrt{\frac{pq}{n}} = \sqrt{\frac{(.55)(.45)}{1600}} = .01243734$$

$$z = \frac{\hat{p} - p}{\sigma_{\hat{p}}} = \frac{.59 - .55}{.01243734} = 3.22$$

The value of the test statistic $z = 3.22$ for $\hat{p}$ is larger than the upper critical value of $z = 2.58$, and it falls in the rejection region. Consequently, we reject H_0 and conclude that the current percentage of parents of adult children who will say that parents do too much for their adult children ages 18 to 29 is different from 55%.

We can also use the p-value approach. From the normal distribution table, the area under the normal curve to the right of $z = 3.22$ is .0006. Therefore, the p-value is 2(.0006) = .0012. Since $\alpha = .01$ is larger than .0012, we reject the null hypothesis.

Source: Data from Social and Demographic Trend. Pew Research Center.

EXERCISES

Concepts and Procedures

9.52 Explain when a sample is large enough to use the normal distribution to make a test of hypothesis about the population proportion.

9.53 In each of the following cases, determine whether the sample size is large enough to use the normal distribution to make a test of hypothesis about the population proportion. Explain why or why not.

a. $n = 45$ and $p = .13$

b. $n = 100$ and $p = .79$

c. $n = 80$ and $p = .04$

d. $n = 55$ and $p = .16$

9.54 For each of the following examples of tests of hypothesis about the population proportion, show the rejection and nonrejection regions on the graph of the sampling distribution of the sample proportion assuming that the sample size is large.

a. A two-tailed test with $\alpha = .05$

b. A left-tailed test with $\alpha = .020$

c. A right-tailed test with $\alpha = .025$

9.55 A random sample of 500 observations produced a sample proportion equal to .38. Using $\alpha = .05$, find the critical and observed values of z for each of the following tests of hypotheses:

a. $H_0: p = .40$ versus $H_1: p > .40$

b. $H_0: p = .40$ versus $H_1: p \neq .40$

9.56 A random sample of 200 observations produced a sample proportion equal to .60. Find the critical and observed values of z for each of the following tests of hypotheses using $\alpha = .01$.

a. $H_0: p = .63$ versus $H_1: p < .63$

b. $H_0: p = .63$ versus $H_1: p \neq .63$

9.57 Consider the null hypothesis H_0: $p = .65$. Suppose a random sample of 1000 observations is taken to perform this test about the population proportion. Using $\alpha = .05$, show the rejection and nonrejection regions and find the critical value(s) of z for a

a. left-tailed test

b. two-tailed test

c. right-tailed test

9.58 Consider $H_0: p = .70$ versus $H_1: p \neq .70$.

a. A random sample of 600 observations produced a sample proportion equal to .68. Using $\alpha = .01$, would you reject the null hypothesis?

b. Another random sample of 600 observations taken from the same population produced a sample proportion equal to .76. Using $\alpha = .01$, would you reject the null hypothesis?

c. Comment on the results of parts (a) and (b).

9.59 Consider $H_0: p = .45$ versus $H_1: p < .45$.

a. A random sample of 400 observations produced a sample proportion equal to .42. Using $\alpha = .025$, would you reject the null hypothesis?

b. Another random sample of 400 observations taken from the same population produced a sample proportion of .39. Using $\alpha = .025$, would you reject the null hypothesis?

c. Comment on the results of parts (a) and (b).

9.60 Make the following hypothesis tests about p.

a. $H_0: p = .57$, $H_1: p \neq .57$, $n = 800$, $\hat{p} = .50$, $\alpha = .05$

b. $H_0: p = .26$, $H_1: p < .26$, $n = 400$, $\hat{p} = .23$, $\alpha = .01$

c. $H_0: p = .84$, $H_1: p > .84$, $n = 250$, $\hat{p} = .85$, $\alpha = .025$

Applications

9.61 According to a study conducted in 2019, 18% of shoppers said that they prefer to buy generic instead of name-brand products. Suppose that in a recent sample of 1500 shoppers, 315 stated that they prefer to buy generic instead of name-brand products. At a 5% significance level, can you conclude that the proportion of all shoppers who currently prefer to buy generic instead of name-brand products is higher than .18? Use both the p-value and the critical value approaches.

9.62 According to a recent study, men prefer to shop for clothes in stores as opposed to online. To test this claim, each man in a random sample of 1200 men was asked whether he prefers to shop for clothes in stores or online. Of these 1200 men, 654 said that they prefer to shop for clothes in stores. At a 5% significance level, can you conclude that the proportion of all men who prefer to shop for clothes in stores is higher than .50? Use both the p-value and the critical value approaches.

9.63 According to a CIRP Your First College Year Survey, conducted a few years ago, 88% of the first-year college students said that their college experience exposed them to diverse opinions, cultures, and values (www.heri.ucla.edu). Suppose in a recent poll of 1800 first-year college students, 91% said that their college experience exposed them to diverse opinions, cultures, and values. Perform a hypothesis test to determine if it is reasonable to conclude that the current percentage of all first-year college students who will say that their college experience exposed them to diverse opinions, cultures, and values is higher than 88%. Use a 2% significance level, and use both the p-value and the critical value approaches.

9.64 According to the National Pet Owners Survey, 63.4% of U.S. households own at least one dog. The manager at an animal shelter in a large town believes that this proportion is lower for all households in her town. In a random sample of 400 households from this town, 59% reported owning at least one dog. At a 1% significance level, can you conclude that the percentage of all households in this town who own at least one dog is less than 63.4%? Use both the p-value and the critical value approaches.

9.65 In a Gallup poll conducted April 1–9, 2019, participants were asked to rate their current financial situation as excellent, good, only fair, or poor. In the random sample of U.S. adults included in this survey, ages 18 and over, 12% rated their current financial situation as excellent. The mayor of a midsized city wants to know the opinions of the residents of his city about their financial situation. In a random sample of 250 residents, ages 18 and over, selected from this city, 13.6% rated their current financial situation as excellent. At a 2.5% significance level, can the mayor conclude that more than 12% of all residents, ages 18 and over, of this city would rate their current financial situation as excellent? Use both the p-value and the critical value approaches.

9.66 A food company is planning to market a new type of frozen yogurt. However, before marketing this yogurt, the company wants to find what percentage of the people like it. The company's management has decided that it will market this yogurt only if at least 35% of the people like it. The company's research department selected a random sample of 400 persons and asked them to taste this yogurt. Of these 400 persons, 112 said they liked it.

a. Testing at a 2.5% significance level, can you conclude that the company should market this yogurt?

b. What will your decision be in part (a) if the probability of making a Type I error is zero? Explain.

c. Make the test of part (a) using the p-value approach and $\alpha = .025$.

9.67 Shulman Steel Corporation makes bearings that are supplied to other companies. One of the machines makes bearings that are supposed to have a diameter of 4 inches. The bearings that have a diameter of either more or less than 4 inches are considered defective and are discarded. When working properly, the machine does not produce more than 7% of bearings that are defective. The quality control inspector selects a sample of 200 bearings each week and inspects them for the size of their diameters. Using the sample proportion, the quality control inspector tests the null hypothesis $p \leq .07$ against the alternative hypothesis $p > .07$, where p is the proportion of bearings that are defective. He always uses a 2% significance level. If the null hypothesis is rejected, the machine is stopped to make any necessary adjustments. One sample of 200 bearings taken recently contained 22 defective bearings.

a. Using a 2% significance level, will you conclude that the machine should be stopped to make necessary adjustments?

b. Perform the test of part (a) using a 1% significance level. Is your decision different from the one in part (a)?

c. Comment on the results of parts (a) and (b).

***9.68** Two years ago, 75% of the customers of a bank said that they were satisfied with the services provided by the bank. The manager of the bank wants to know if this percentage of satisfied customers has changed since then. She assigns this responsibility to you. Briefly explain how you would conduct such a test.

***9.69** A study claims that 65% of students at all colleges and universities hold off-campus (part-time or full-time) jobs. You want to check if the percentage of students at your school who hold off-campus jobs is different from 65%. Briefly explain how you would conduct such a test. Collect data from 40 students at your school on whether or not they hold off-campus jobs. Then, calculate the proportion of students in this sample who hold off-campus jobs. Using this information, test the hypothesis. Select your own significance level.

Uses and Misuses of Statistics

Follow the Recipe

Hypothesis testing is one of the most powerful and dangerous tools of statistics. It allows us to make statements about a population and attach a degree of uncertainty to these statements. Read a news article from a newspaper or a credible online news source; rare will be the day when you will not find a news item that does not contain a story featuring a statistical result, often reported with a significance level. Given that the subjects of these reports—public health, the environment, and so on—are important to our lives, it is critical that we perform the statistical calculations and interpretations properly. The first step, one that you should look for when reading statistical results, is proper formulation/specification.

Formulation or *specification*, simply put, is the list of steps you perform when constructing a hypothesis test. In this chapter, these steps are: stating the null and alternative hypotheses; selecting the appropriate distribution; and determining the rejection and nonrejection regions. Once these steps are performed, all you need to do is to calculate the p-value or the test statistic to complete the hypothesis test. It is important to beware of traps in the specification.

Though it might seem obvious, stating the hypothesis properly can be difficult. For hypotheses around a population mean, the null and alternative hypotheses are mathematical statements that do not overlap and also provide no holes. Suppose that a confectioner states that the average mass of his chocolate bars is 100 grams. The null hypothesis is that the mass of the bars is 100 grams, and the alternative hypothesis is that the mass of the bars is not 100 grams. When you take a sample of chocolate bars and measure their masses, all possibilities for the sample mean will fall within one of your decision regions. The problem is a little more difficult for hypotheses based on proportions. Make sure that you only have two categories. For example, if you are trying to determine the percentage of the population that has blonde hair, your groups are "blonde" and "not blonde." You need to decide how to categorize bald people before you conduct this experiment: Do not include bald people in the survey.

Finally, beware of numerical precision. When your sample is large and you assume that the test statistic has a normal distribution, the rejection region for a two-tailed test using the normal distribution with a significance level of 5% will be values of the sample mean that are farther than 1.96 standard deviations from the assumed mean. When you perform your calculations, the sample mean may fall on the border of your decision region. Remember that there is measurement error and sampling error that you cannot account for. In this case, it is probably best to adjust your significance level so that the sample mean falls squarely in a decision region.

The Power of Negative Thinking

In the beginning of this chapter, you learned about Type I and Type II errors. If this is your first statistics class, you might think that this is your first exposure to the concepts of Type I and Type II errors, but that is not the case. As a matter of fact, if you can recall having a medical test, you must have had an interaction with a variety of concepts related to hypothesis testing, including Type I and Type II errors.

In a typical medical test, the assumption (our null hypothesis) is that you do not have the condition for which you are being tested. If the assumption is true, the doctor knows what should happen in the test. If the test results are different from the *normal* range, the doctor has data that would allow the assumption to be rejected. Whenever the null hypothesis is rejected (that is, the test results demonstrate that the person has the condition for which he or she was tested), the medical result is called a *positive* test result. If the doctor fails to reject the null hypothesis (if the test results do not demonstrate that you have the condition), the medical result is called a *negative* test result.

As with other types of hypothesis tests, medical tests are not perfect. Sometimes people are (falsely) diagnosed as having a condition or illness when they actually do not have it. In medical terminology, this Type I error is referred to as a *false positive*. Similarly, the result of a test may (wrongly) indicate that a person does not have a condition or illness when actually he or she has it. This Type II error is called a *false negative* in medical terminology.

Companies that develop medical tests perform intensive research and clinical tests to reduce the risk of making both these types of errors. Specifically, data are collected on the *sensitivity* and the *specificity* of medical tests. In the context of an illness, the sensitivity of a test is the proportion of all people with the illness who are identified by the test as actually having it. For example, suppose 100 students on a college campus have been identified by throat culture as having strep throat. All these 100 students are tested for strep throat using another type of test. Suppose 97 of the 100 tests come back positive with the second test. Then the sensitivity of the (second) test is .97 (or 97%), and the probability of a false negative (Type II error) is one minus the sensitivity, which is .03 here.

The specificity of a test refers to how well a test identifies that a healthy person does not have a given disease. Using the strep throat reference again, suppose that 400 students have been identified by throat culture as not having strep throat. All these 400 students are given a new strep test, and 394 of them are shown to have a negative result (that is, they are identified as not having strep throat). Then the specificity of the test is 394/400 = .985 (or 98.5%), and the probability of a false positive (Type I error) is one minus the specificity, which is .015 here.

Of course, low probabilities for both types of errors are very important in medical testing. A false positive can result in an individual obtaining unnecessary, often expensive, and sometime debilitating treatment, whereas a false negative can allow a disease to progress to an advanced stage when early detection could have helped to save a person's life.

Glossary

α The significance level of a test of hypothesis that denotes the probability of rejecting a null hypothesis when it actually is true. (The probability of committing a Type I error.)

Alternative hypothesis A claim about a population parameter that will be true if the null hypothesis is false.

β The probability of not rejecting a null hypothesis when it actually is false. (The probability of committing a Type II error.)

Critical value or **critical point** One or two values that divide the whole region under the sampling distribution of a sample statistic into rejection and nonrejection regions.

Left-tailed test A test in which the rejection region lies in the left tail of the distribution curve.

Null hypothesis A claim about a population parameter that is assumed to be true until proven otherwise.

Observed value of *z* or *t* The value of *z* or *t* calculated for a sample statistic such as the sample mean or the sample proportion.

One-tailed test A test in which there is only one rejection region, either in the left tail or in the right tail of the distribution curve.

***p*-value** The smallest significance level at which a null hypothesis can be rejected.

Right-tailed test A test in which the rejection region lies in the right tail of the distribution curve.

Significance level The value of α that gives the probability of committing a Type I error.

Test statistic The value of *z* or *t* calculated for a sample statistic such as the sample mean or the sample proportion.

Two-tailed test A test in which there are two rejection regions, one in each tail of the distribution curve.

Type I error An error that occurs when a true null hypothesis is rejected.

Type II error An error that occurs when a false null hypothesis is not rejected.

Exercises

Supplementary Exercises

9.70 Consider the following null and alternative hypotheses:

$$H_0: \mu = 120 \quad \text{versus} \quad H_1: \mu > 120$$

A random sample of 81 observations taken from this population produced a sample mean of 123.5. The population standard deviation is known to be 15.

a. If this test is made at a 2.5% significance level, would you reject the null hypothesis? Use the critical value approach.

b. Calculate the *p*-value for the test. Based on this *p*-value, would you reject the null hypothesis if $\alpha = .01$? What if $\alpha = .05$?

9.71 Consider the following null and alternative hypotheses:

$$H_0: p = .82 \quad \text{versus} \quad H_1: p \neq .82$$

A random sample of 600 observations taken from this population produced a sample proportion of .86.

a. If this test is made at a 2% significance level, would you reject the null hypothesis? Use the critical value approach.

b. What is the probability of making a Type I error in part (a)?

c. Calculate the *p*-value for the test. Based on this *p*-value, would you reject the null hypothesis if $\alpha = .025$? What if $\alpha = .01$?

9.72 Consider the following null and alternative hypotheses:

$$H_0: p = .44 \quad \text{versus} \quad H_1: p < .44$$

A random sample of 450 observations taken from this population produced a sample proportion of .39.

a. If this test is made at a 2% significance level, would you reject the null hypothesis? Use the critical value approach.

b. What is the probability of making a Type I error in part (a)?

c. Calculate the *p*-value for the test. Based on this *p*-value, would you reject the null hypothesis if $\alpha = .01$? What if $\alpha = .025$?

9.73 The mean consumption of water per household in a city was 1245 cubic feet per month. Due to a water shortage because of a drought, the city council campaigned for water use conservation by households. A few months after the campaign was started, the mean consumption of water for a sample of 100 households was found to be 1175 cubic feet per month. The population standard deviation is given to be 250 cubic feet.

a. Find the *p*-value for the hypothesis test that the mean consumption of water per household has decreased due to the campaign by the city council. Would you reject the null hypothesis at $\alpha = .025$?

b. Make the test of part (a) using the critical value approach and $\alpha = .025$.

9.74 According to a U.S. Bureau of Labor Statistics release of April 2, 2019, financial analysts earned an average of $100,990 in 2019. Suppose that the 2019 earnings of all financial analysts had a mean of

$100,990. A recent sample of 400 financial analysts showed that they earn an average of $105,110 a year. Assume that the standard deviation of the annual earnings of all financial analysts is $30,000.

a. Using the critical value approach, can you conclude that the current average annual earnings of financial analysts is higher than $100,990? Use $\alpha = .01$.

b. Describe the Type I error in the context of this test.

c. Will your conclusion of part (a) change if the probability of making a Type I error is zero?

d. Calculate the *p*-value for the test of part (a). What is your conclusion if $\alpha = .01$?

9.75 A real estate agent claims that the mean living area of all single-family homes in his county is at most 2400 square feet. A random sample of 50 such homes selected from this county produced the mean living area of 2540 square feet and a standard deviation of 472 square feet.

a. Using $\alpha = .05$, can you conclude that the real estate agent's claim is true?

b. What will your conclusion be if $\alpha = .01$?

c. Comment on the results of parts (a) and (b).

9.76 The customers at a bank complained about long lines and the time they had to spend waiting for service. It is known that the customers at this bank had to wait 8 minutes, on average, before being served. The management made some changes to reduce the waiting time for its customers. A sample of 60 customers taken after these changes were made produced a mean waiting time of 7.5 minutes with a standard deviation of 2.1 minutes. Using this sample mean, the bank manager displayed a huge banner inside the bank mentioning that the mean waiting time for customers has been reduced by new changes. Do you think the bank manager's claim is justifiable? Use a 2.5% significance level to answer this question. Use both approaches.

9.77 The administrative office of a hospital claims that the mean waiting time for patients to get treatment in its emergency ward is 25 minutes. A random sample of 16 patients who received treatment in the emergency ward of this hospital produced a mean waiting time of 27.5 minutes with a standard deviation of 4.8 minutes. Using a 1% significance level, test whether the mean waiting time at this emergency ward is different from 25 minutes. Assume that the waiting times for all patients at this emergency ward have an approximate normal distribution.

9.78 An earlier study claimed that U.S. adults spent an average of 114 minutes per day with their families. A recently taken sample of 25 adults from a city showed that they spend an average of 109 minutes per day with their families. The sample standard deviation is 11 minutes. Assume that the times spent by adults with their families have an approximate normal distribution.

a. Using a 1% significance level, test whether the mean time spent currently by all adults with their families in this city is different from 114 minutes a day.

b. Suppose the probability of making a Type I error is zero. Can you make a decision for the test of part (a) without going through the five steps of hypothesis testing? If yes, what is your decision? Explain.

9.79 A computer company that recently introduced a new software product claims that the mean time taken to learn how to use this software is not more than 2 hours for people who are somewhat familiar with computers. A random sample of 12 such persons was selected. The following data give the times taken (in hours) by these persons to learn how to use this software.

1.75	2.25	2.40	1.90	1.50	2.75
2.15	2.25	1.80	2.20	3.25	2.60

Test at a 1% significance level whether the company's claim is true. Assume that the times taken by all persons who are somewhat familiar with computers to learn how to use this software are approximately normally distributed.

9.80 According to a U.S. Bureau of Labor Statistics release of a few years ago, 79% of U.S. children under age 18 lived with at least one other sibling at that time. Suppose that in a recent sample of 2000 U.S. children under age 18, 1620 were living with at least one other sibling.

a. Using the critical value approach and $\alpha = .05$, test if the current percentage of all U.S. children under age 18 who live with at least one other sibling is different from 79%.

b. Describe the Type I error in the context of this test.

c. Calculate the *p*-value for the test of part (a). What is your conclusion if $\alpha = .05$?

9.81 PolicyInteractive of Eugene, Oregon, conducted a study of U.S. adults a few years ago for the Center for a New American Dream. Seventy-five percent of the adults included in this study said that having basic needs met is very or extremely important in their vision of the American dream (www.newdream.org). A recent sample of 1500 U.S. adults were asked the same question and 72% of them said that having basic needs met is very or extremely important in their vision of the American dream.

a. Using the critical value approach and $\alpha = .01$, test if the current percentage of U.S. adults who hold the above-mentioned opinion is less than 75%.

b. Describe the Type I error in the context of this test.

c. Calculate the *p*-value for the test of part (a). What is your conclusion if $\alpha = .01$?

9.82 More and more people are abandoning national brand products and buying store brand products to save money. The president of a company that produces national brand coffee claims that 40% of the people prefer to buy national brand coffee. A random sample of 700 people who buy coffee showed that 259 of them buy national brand coffee. Using $\alpha = .01$, can you conclude that the percentage of people who buy national brand coffee is different from 40%? Use both approaches to make the test.

9.83 According to a Pew Research Center survey conducted January 8–February 7, 2019, 81% of Americans own a smartphone. In a recent random sample of 800 Americans age 50–64, 79% own a smartphone. At a 2% significance level, can you conclude that the proportion of Americans age 50–64 who own a smartphone is less than 81%? Use both the *p*-value and the critical value approaches.

9.84 Dartmouth Distribution Warehouse makes deliveries of a large number of products to its customers. To keep its customers happy and satisfied, the company's policy is to deliver on time at least 90% of all the orders it receives from its customers. The quality control inspector at the company quite often takes samples of orders delivered and checks to see whether this policy is maintained. A recent sample of 90 orders taken by this inspector showed that 75 of them were delivered on time.

a. Using a 2% significance level, can you conclude that the company's policy is not maintained?

b. What will your decision be in part (a) if the probability of making a Type I error is zero? Explain.

Advanced Exercises

9.85 Acme Bicycle Company makes derailleurs for mountain bikes. Usually no more than 4% of these parts are defective, but occasionally the machines that make them get out of adjustment and the rate of defectives exceeds 4%. To guard against this, the chief quality control inspector takes a random sample of 130 derailleurs each week and checks each one for defects. If too many of these parts are defective, the machines are shut down and adjusted. To decide how many parts must be defective to shut down the machines, the company's statistician has set up the hypothesis test

$$H_0: p \leq .04 \quad \text{versus} \quad H_1: p > .04$$

where p is the proportion of defectives among all derailleurs being made currently. Rejection of H_0 would call for shutting down the machines. For the inspector's convenience, the statistician would like the rejection region to have the form, "Reject H_0 if the number of defective parts is C or more." Find the value of C that will make the significance level (approximately) .05.

9.86 Alpha Airline claims that only 15% of its flights arrive more than 10 minutes late. Let p be the proportion of all of Alpha's flights that arrive more than 10 minutes late. Consider the hypothesis test

$$H_0: p \leq .15 \quad \text{versus} \quad H_1: p > .15$$

Suppose we take a random sample of 50 flights by Alpha Airline and agree to reject H_0 if 9 or more of them arrive late. Find the significance level for this test.

9.87 Thirty percent of all people who are inoculated with the current vaccine that is used to prevent a disease contract the disease within a year. The developer of a new vaccine that is intended to prevent this disease wishes to test for significant evidence that the new vaccine is more effective.

a. Determine the appropriate null and alternative hypotheses.

b. The developer decides to study 100 randomly selected people by inoculating them with the new vaccine. If 84 or more of them do not contract the disease within a year, the developer will conclude that the new vaccine is superior to the old one. What significance level is the developer using for the test?

c. Suppose 20 people inoculated with the new vaccine are studied and the new vaccine is concluded to be better than the old one if fewer than 3 people contract the disease within a year. What is the significance level of the test?

9.88 In Las Vegas, Nevada, and Atlantic City, New Jersey, tests are performed often on the various gaming devices used in casinos. For example, dice are often tested to determine if they are balanced. Suppose you are assigned the task of testing a die, using a two-tailed test to make sure that the probability of a two spot is 1/6. Using a 5% significance level, determine how many two spots you would have to obtain to reject the null hypothesis when your sample size is

a. 120 b. 1200 c. 12,000

Calculate the value of $\hat{p}$ for each of these three cases. What can you say about the relationship between (1) the difference between $\hat{p}$ and 1/6 that is necessary to reject the null hypothesis and (2) the sample size as the difference gets larger?

9.89 A statistician performs the test $H_0: \mu = 15$ versus $H_1: \mu \neq 15$ and finds the p-value to be .4546.

a. The statistician performing the test does not tell you the value of the sample mean and the value of the test statistic. Despite this, you have enough information to determine the pair of p-values associated with the following alternative hypotheses:

i. $H_1: \mu < 15$ ii. $H_1: \mu > 15$

Note that you will need more information to determine which p-value goes with which alternative. Determine the pair of p-values. Here the value of the sample mean is the same in both cases.

b. Suppose the statistician tells you that the value of the test statistic is negative. Match the p-values with the alternative hypotheses.

Note that the result for one of the two alternatives implies that the sample mean is not on the same side of $\mu = 15$ as the rejection region. Although we have not discussed this scenario in the book, it is important to recognize that there are many real-world scenarios in which this type of situation does occur. For example, suppose the EPA is to test whether or not a company is exceeding a specific pollution level. If the average discharge level obtained from the sample falls below the threshold (mentioned in the null hypothesis), then there would be no need to perform the hypothesis test.

9.90 You read an article that states "50 hypothesis tests of $H_0: \mu = 35$ versus $H_1: \mu \neq 35$ were performed using $\alpha = .05$ on 50 different samples taken from the same population with a mean of 35. Of these, 47 tests failed to reject the null hypothesis." Explain why this type of result is not surprising.

Self-Review Test

Multiple Choice Questions

1. A test of hypothesis is always about
   a. a population parameter
   b. a sample statistic
   c. a test statistic

2. A Type I error is committed when
   a. a null hypothesis is not rejected when it is actually false
   b. a null hypothesis is rejected when it is actually true
   c. an alternative hypothesis is rejected when it is actually true

3. A Type II error is committed when
   a. a null hypothesis is not rejected when it is actually false
   b. a null hypothesis is rejected when it is actually true
   c. an alternative hypothesis is rejected when it is actually true

4. A critical value is the value
   a. calculated from sample data
   b. determined from a table (e.g., the normal distribution table or other such tables)
   c. neither a nor b

5. The computed value of a test statistic is the value
   a. calculated for a sample statistic
   b. determined from a table (e.g., the normal distribution table or other such tables)
   c. neither a nor b

6. The observed value of a test statistic is the value

a. calculated for a sample statistic

b. determined from a table (e.g., the normal distribution table or other such tables)

c. neither a nor b

7. The significance level, denoted by α, is

a. the probability of committing a Type I error

b. the probability of committing a Type II error

c. neither a nor b

8. The value of β gives the

a. probability of committing a Type I error

b. probability of committing a Type II error

c. power of the test

9. The value of $1 - \beta$ gives the

a. probability of committing a Type I error

b. probability of committing a Type II error

c. power of the test

10. A two-tailed test is a test with

a. two rejection regions

b. two nonrejection regions

c. two test statistics

11. A one-tailed test has

a. one rejection region

b. one nonrejection region

c. both a and b

12. The smallest level of significance at which a null hypothesis is rejected is called

a. α b. p-value c. β

13. The sign in the alternative hypothesis in a two-tailed test is always

a. $<$ b. $>$ c. $\neq$

14. The sign in the alternative hypothesis in a left-tailed test is always

a. $<$ b. $>$ c. $\neq$

15. The sign in the alternative hypothesis in a right-tailed test is always

a. $<$ b. $>$ c. $\neq$

16. The null hypothesis is a claim about a

a. parameter, where the claim is assumed to be false until it is declared true

b. parameter, where the claim is assumed to be true until it is declared false

c. statistic, where the claim is assumed to be true until it is declared false

17. In a two-tailed hypothesis test, the two critical points are the points that divide the area under the sampling distribution of a

a. statistic into two rejection regions and one nonrejection region

b. parameter into one rejection region and one nonrejection region

c. statistic into one rejection region and two nonrejection regions

18. In a hypothesis test, the probability of committing a Type I error is called the

a. confidence level

b. confidence interval

c. significance level

19. A researcher wants to test if the mean price of houses in an area is greater than $145,000. The null hypothesis for this example will be that the population mean is

a. less than $145,000

b. less than or equal to $145,000

c. greater than $145,000

20. Which of the following statements describes a Type II error in hypothesis testing?

a. A court declares a defendant guilty when he is actually innocent.

b. A court declares a defendant guilty when he is actually guilty.

c. A statistician determines, through hypothesis testing, that the mean number of televisions per household in a certain community is 1.4, when it is actually greater than 1.4.

21. The power of a hypothesis test is .96. Which of the following statements is true about this test?

a. The probability of a Type II error is .04.

b. The probability of a Type I error is .04.

c. The probability of a Type II error is .96.

22. For a two-tailed test, the p-value is

a. the area under the curve between the mean and the observed value of the sample statistic

b. twice the area under the curve between the mean and the observed value of the sample statistic

c. twice the area in the tail under the curve on the side where the sample statistic lies

23. The following four steps must be taken to perform a hypothesis test using the p-value approach:

1. Calculate the p-value.
2. Select the distribution to use.
3. Make a decision.
4. State the null and alternative hypotheses and determine the significance level.

The correct order for performing these steps is

a. 2, 3, 1, 4 b. 4, 2, 1, 3 c. 3, 2, 1, 4

24. In a hypothesis test with null and alternative hypotheses $H_0: \mu \leq 45$ and $H_1: \mu > 45$, a random sample of 60 elements selected from the population produced a mean of 46.2. Assume that $\sigma = 7.5$ and that the test is to be made at the 2.5% significance level. What is the critical value of z?

a. −1.96 b. 1.65 c. 1.96

25. In a hypothesis test with hypotheses $H_0: \mu = 24$ and $H_1: \mu \neq 24$, a random sample of 37 elements selected from the population produced a mean of 25.2. Assume that $\sigma = 6.1$ and that the test is to be made at the 1% significance level. What is the p-value for this hypothesis test, rounded to four decimal places?

a. .1877 b. .2315 c. .2670

26. In a hypothesis test with hypotheses $H_0: \mu \geq 136$ and $H_1: \mu < 136$, a random sample of 71 elements selected from the population produced a mean of 132.7. Assume that $\sigma = 18.6$ and that the test is to be made at the 5% significance level. What is the value of the test statistic, z, rounded to three decimal places?

a. −1.495 b. −1.459 c. −1.994

27. We use the t distribution to perform a hypothesis test about the population mean when

a. the population from which the sample is drawn is approximately normal and the population standard deviation is known

b. the population from which the sample is drawn is not approximately normal and the population standard deviation is known

c. the population from which the sample is drawn is approximately normal or sample size is 30 or larger and the population standard deviation is unknown

28. A company that manufactures light bulbs claims that its light bulbs last an average of 1150 hours. A sample of 25 light bulbs manufactured by this company gave a mean life of 1071 hours and a standard deviation of 135 hours. A consumer group wants to test the hypothesis that the mean life of light bulbs produced by this company is less than 1150 hours. The significance level is 5%. Assume the population is normally distributed. What is the value of the test statistic, t, rounded to three decimal places?

a. −2.629 b. −1.926 c. −2.926

29. In a hypothesis test with hypotheses $H_0: p \leq .39$ and $H_0: p > .39$, a random sample of size 510 produced a sample proportion of .4475. The test is to be made at the 1% significance level. What is the critical value of z?

a. 2.58 b. 2.33 c. 2.12

30. In a hypothesis test with hypotheses H_0: $p = .26$ and H_0: $p \neq .26$, a random sample of size 757 produced a sample proportion of .2300. The test is to be made at the 5% significance level. What is the p-value for this hypothesis test, rounded to four decimal places?

a. .0592 b. .0299 c. .0005

Practice Exercises

1. According to a Kaiser Family Foundation Health Benefits Survey released in 2019, U.S. workers, who had employer-provided health insurance, had an average annual deductible of $1655 for single (one person) health insurance coverage (https://www.kff.org/health-costs/report/2019-employer-health-benefits-survey/). Suppose that a recent random sample of 100 workers with single coverage by an employer-provided health insurance selected from a large city has an average deductible of $1683. Assume that the deductibles for all such workers in this city have a standard deviation of $125.

a. Using the critical value approach and a 1% significance level, can you conclude that the average deductible for all such workers in this city with single coverage is different from $1655?

b. Using the critical value approach and a 2.5% significance level, can you conclude that the average deductible for all such workers in this city with single coverage is higher than $1655?

c. Describe a Type I error in the context of this test.

d. Calculate the p-value for the test of part (a). What is your conclusion if $\alpha = .01$?

e. Calculate the p-value for the test of part (b). What is your conclusion if $\alpha = .025$?

2. A minor league baseball executive has become concerned about the slow pace of games played in her league, fearing that it will lower attendance. She meets with the league's managers and umpires and discusses guidelines for speeding up the games. Before the meeting, the mean duration of nine-inning games was 3 hours 5 minutes (i.e., 185 minutes). A random sample of 36 nine-inning games after the meeting showed a mean of 179 minutes with a standard deviation of 12 minutes.

a. Testing at a 1% significance level, can you conclude that the mean duration of nine-inning games has decreased after the meeting?

b. Describe a Type I error in the context of this test.

c. What will your decision be in part (a) if the probability of making a Type I error is zero? Explain.

d. Find the range for the p-value for the test of part (a). What is your decision based on this p-value?

3. An editor of a New York publishing company claims that the mean time taken to write a textbook is at least 31 months. A sample of 16 textbook authors found that the mean time taken by them to write a textbook was 25 months with a standard deviation of 7.2 months.

a. Using a 2.5% significance level, would you conclude that the editor's claim is true? Assume that the time taken to write a textbook is approximately normally distributed for all textbook authors.

b. Describe a Type I error in the context of this test.

c. What will your decision be in part (a) if the probability of making a Type I error is .001?

4. A financial advisor claims that less than 50% of adults in the United States have a will. A random sample of 1000 adults showed that 450 of them have a will.

a. At a 5% significance level, can you conclude that the percentage of people who have a will is less than 50%?

b. Describe a Type I error in the context of this test.

c. What would your decision be in part (a) if the probability of making a Type I error were zero? Explain.

d. Find the p-value for the test of hypothesis mentioned in part (a). Using this p-value, will you reject the null hypothesis if $\alpha = .05$? What if $\alpha = .01$?

Mini-Projects

Mini-Project 9.1

Refer to Data Set III on the NFL players from the 2019 football season that accompanies this text (see Appendix A). The mean of heights of players, listed in column 4, of NFL players who were on the rosters of NFL teams during the 2019 NFL season was 74.08 inches.

a. Create the appropriate graph(s) of the heights of all NFL players to determine whether the population of heights is approximately normally distributed. What is your conclusion? (See Appendix 6.1 in Chapter 6.)

b. Take a random sample of 15 players from this NFL data file. Using the heights from this sample, test $H_0: \mu = 74.08$ inches against $H_1: \mu \neq 74.08$ inches, using $\alpha = .05$.

c. Repeat part (b) for samples of 31 and 45 players, respectively.

d. Did any of the three tests in parts (b) and (c) lead to the conclusion that the mean height of all NFL players is different from 74.08 inches? If you were to repeat this process a large number of times using the same sample size, what percentage of samples would be expected to lead to this conclusion?

Mini-Project 9.2

A thumbtack that is tossed on a desk can land in one of the two ways shown in the following illustration:

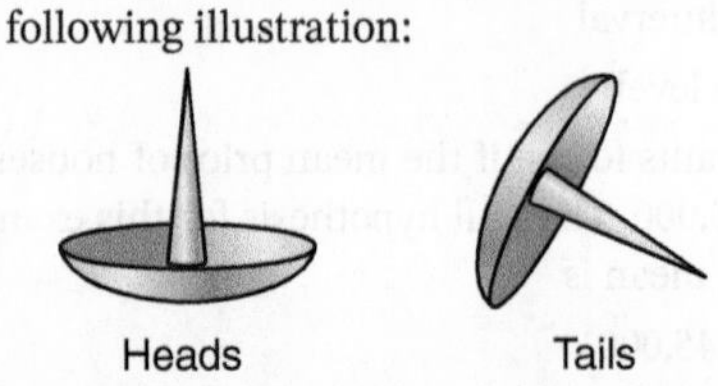

Brad and Dan cannot agree on the likelihood of obtaining a head or a tail. Brad argues that obtaining a tail is more likely than obtaining a head because of the shape of the tack. If the tack had no point at all, it would resemble a coin that has the same probability of coming up heads or tails when tossed. But the longer the point, the less likely it is that the tack will stand up on its head when tossed. Dan believes that as the tack lands tails, the point causes the tack to jump around and come to rest in the heads position. Brad and Dan need you to settle their dispute. Do you think the thumbtack is equally likely to land heads or tails? To investigate this question, find an ordinary thumbtack and toss it a large number of times (say, 100 times).

a. What is the meaning, in words, of the unknown parameter in this problem?

b. Set up the null and alternative hypotheses and compute the p-value based on your results from tossing the tack.

c. How would you answer the original question now? If you decide the tack is not fair, do you side with Brad or Dan?

d. What would you estimate the value of the parameter in part (a) to be? Find a 90% confidence interval for this parameter.

e. After doing this experiment, do you think 100 tosses are enough to infer the nature of your tack? Using your result as a preliminary estimate, determine how many tosses would be necessary to be 95% certain of having 4% accuracy; that is, the margin of error of estimate is 4%. Have you observed enough tosses?

Mini-Project 9.3

Collect pennies in the amount of \$5. Do not obtain rolls of pennies from a bank because many such rolls will consist solely of new pennies. Treat these 500 pennies as your population. Determine the ages, in years, of all these pennies. Calculate the mean and standard deviation of these ages and denote them by μ and σ, respectively.

a. Take a random sample of 10 pennies from these 500. Find the average age of these 10 pennies, which is the value of $\bar{x}$. Perform a test with the null hypothesis that μ is equal to the value obtained for all 500 pennies and the alternative hypothesis that μ is not equal to this value. Use a significance level of .10.

b. Suppose you repeat the procedure of part (a) nine more times. How many times would you expect to reject the null hypothesis? Now actually repeat the procedure of part (a) nine more times, making sure that you put the 10 pennies selected each time back in the population and that you mix all pennies well before taking a sample. How many times did you reject the null hypothesis? Note that you can enter the ages of these 500 pennies in a technology and then use that technology to take samples and make tests of hypothesis.

c. Repeat parts (a) and (b) for a sample size of 25. Did you reject the null hypothesis more often with a sample size of 10 or a sample size of 25?

Mini-Project 9.4

In the article "*Flipping Out—Think a Coin Toss Has a 50–50 Chance? Think Again*," author David E. Adler discusses how a team of Stanford researchers concluded that when you flip a coin, whichever side of the coin is facing up when you place it on your thumb to flip, it is more likely to be the side that faces up after the coin is flipped. To test this concept, you are going to flip a coin 100 times. To simplify the record keeping, you should perform all 100 flips with head facing up when you place the coin on your thumb to flip it or perform all 100 flips with tail facing up when you place the coin on your thumb to flip it. Do your best to use the same amount of force each time you flip the coin.

a. If you had head facing up when you placed the coin on your thumb, calculate the sample proportion of flips in which head occurred. (Or you can perform this experiment with tail facing up.) Perform a test with the null hypothesis that the side that started up will land up 50% of the time versus the alternative hypothesis that the side that started up will land up more than 50% of the time. Use a significance level of 5%.

b. The Stanford research group concluded that the side that started up will land up 51% of the time. Use the data from your 100 flips to test the null hypothesis that the side that started up will land up 51% of the time versus the alternative hypothesis that the side that started up will not land 51% of the time. Use a significance level of 5%.

c. Suppose that you were really bored one day and decided to repeat this experiment four times, using more flips each time, as shown in the following table:

Number of flips	Number of flips in which the side that started up also landed up
500	255
1000	510
5000	2550
10,000	5100

For each of these four cases, calculate the test statistic and p-value for the hypothesis test described in part (a). Based on your results, what can you conclude about the number of repetitions needed to distinguish between a result that occurs 50% of the time and the one that occurs 51% of the time?

Decide for Yourself

Note: The Decide for Yourself feature is located in the book's product page on *www.wiley.com*.

Chapter 9 Technology Instructions

Note: Complete TI-84, Minitab, and Excel manuals are available in the book's product page on *www.wiley.com*.

TI-84 Color/TI-84

The TI-84 Color Technology Instructions feature of this text is written for the TI-84 Plus C color graphing calculator running the 4.0 operating system. Some screens, menus, and functions will be slightly different in older operating systems. The TI-84 and TI-84 Plus can perform all of the same functions but will not have the "Color" option referenced in some of the menus.

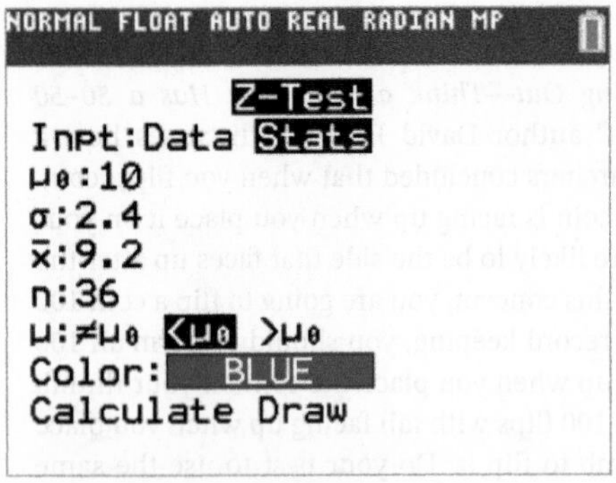

SCREEN 9.1

Testing a Hypothesis About μ, σ Known, for Example 9.2 of the Text

1. Select **STAT > TESTS > Z-Test.**
2. Use the following settings in the **Z-Test** menu (see **Screen 9.1**):
   - At the **Inpt** prompt, select **Stats.**

     Note: If you have the data in a list, select **Data** at the **Inpt** prompt.
   - At the μ_0 prompt, type 10.
   - At the σ prompt, type 2.4.
   - At the $\bar{x}$ prompt, type 9.2.
   - At the n prompt, type 36.
   - At the μ prompt, select $< \mu_0$.
3. Highlight **Calculate** and press **ENTER.**
4. The output includes the test statistic and the *p*-value. (See **Screen 9.2.**)

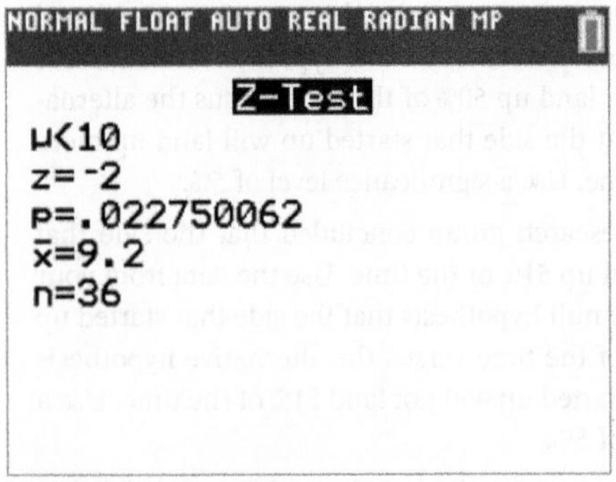

SCREEN 9.2

Compare the test statistic to the critical value of *z* or the *p*-value to the value of α and make a decision.

Testing a Hypothesis About μ, σ Unknown, for Example 9.5 of the Text

1. Select **STAT > TESTS > T-Test.**
2. Use the following settings in the **T-Test** menu (see **Screen 9.3**):
   - At the **Inpt** prompt, select **Stats.**

     Note: If you have the data in a list, select **Data** at the **Inpt** prompt.
   - At the $\mu 0$ prompt, type 12.5.
   - At the $\bar{x}$ prompt, type 12.9.
   - At the **S***x* prompt, type .80.
   - At the n prompt, type 18.
   - At the μ prompt, select $\neq \mu_0$.
3. Highlight **Calculate** and press **ENTER.**
4. The output includes the test statistic and the *p*-value. (See **Screen 9.4.**)

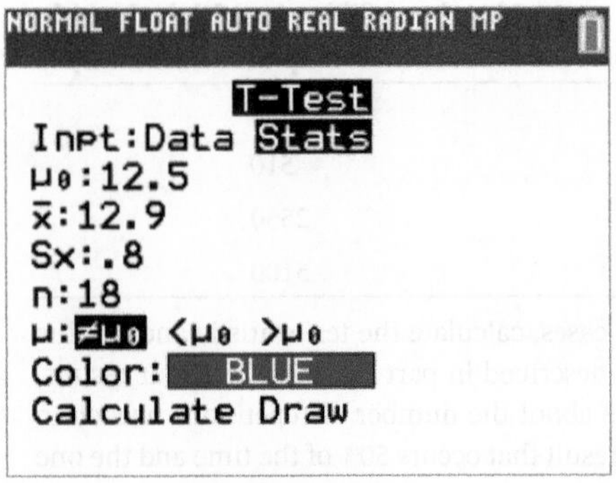

SCREEN 9.3

Compare the test statistic to the critical value of *t* or *p*-value to the value of α and make a decision.

Testing a Hypothesis About p for Example 9.9 of the Text

1. Select **STAT > TESTS > 1-PropZTest.**
2. Use the following settings in the **1-PropZTest** menu (see **Screen 9.5**):
   - At the **p0** prompt, type .20.
   - At the **x** prompt, type 440.

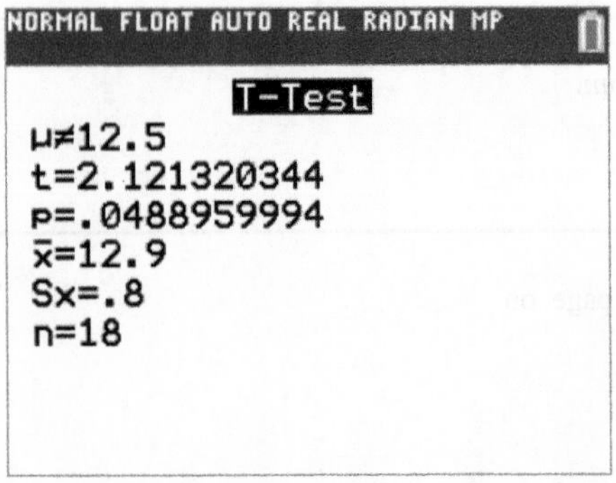

SCREEN 9.4

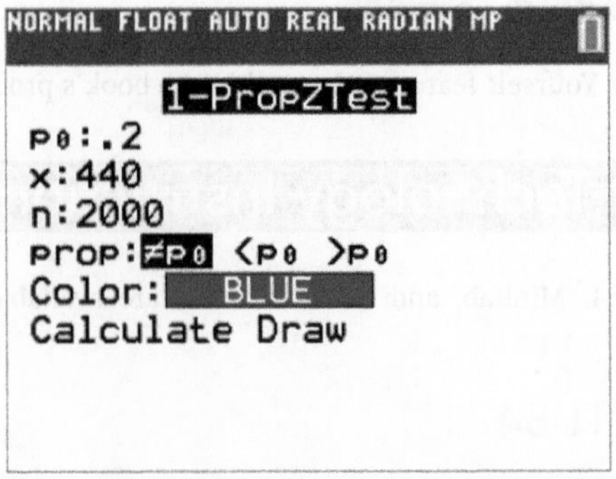

SCREEN 9.5

Note: The value of x is the number of successes in the sample and must be a whole number or the calculator will return an error message. To find x, multiply n by $\hat{p}$ and round the result to the nearest whole number.

- At the **n** prompt, type 2000.
- At the **prop** prompt, select $\neq \mathbf{p_0}$.

3. Highlight **Calculate** and press **ENTER**.
4. The output includes the test statistic and the *p*-value. (See **Screen 9.6.**)

Compare the test statistic to the critical value of z or the p-value to the value of α and make a decision.

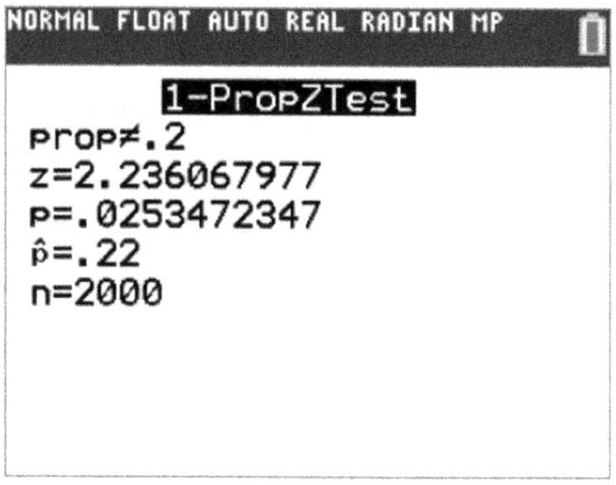

SCREEN 9.6

Minitab

The Minitab Technology Instructions in this text are written for **Minitab Professional Version 19**. Some screens, menus, and functions will be slightly different in older (or in newer) versions of Minitab. If you are using **Minitab Express**, you can download the Minitab Express Manual from the book's product page on *www.wiley.com*.

Testing a Hypothesis About μ, σ Known, for Example 9.2 of the Text

1. Select **Stat > Basic Statistics > 1-Sample Z.**
2. Use the following settings in the dialog box that appears on screen (see **Screen 9.7**):
   - From the drop-down menu, select **Summarized Data.**

     Note: If you have the data in a column, select **One or more samples, each in a column**, type the column name(s) in the box, and move to Step 3.
   - In the **Sample size** box, type 36.
   - In the **Sample mean** box, type 9.2.
   - In the **Known standard deviation** box, type 2.4.
   - Next to **Perform hypothesis test**, check the check box.
   - In the **Hypothesized mean** box, type 10.
3. Select **Options.** When the new dialog box appears on screen, select **Mean < hypothesized mean** from the **Alternative hypothesis** box.
4. Click **OK** in both dialog boxes.
5. The output, including the test statistic and *p*-value, will be displayed in the Session window. (See **Screen 9.8.**)

Compare the test statistic to the critical value of z or the p-value to the value of α and make a decision.

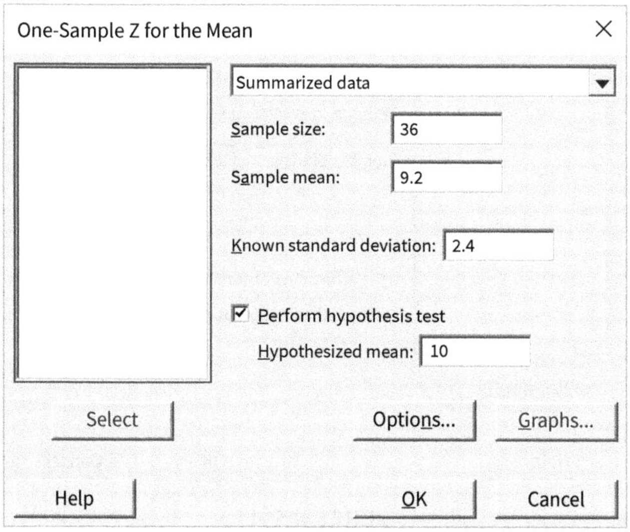

SCREEN 9.7

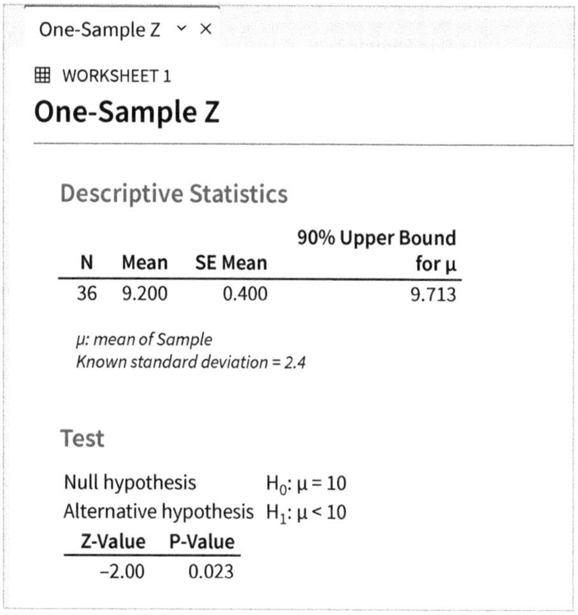

SCREEN 9.8

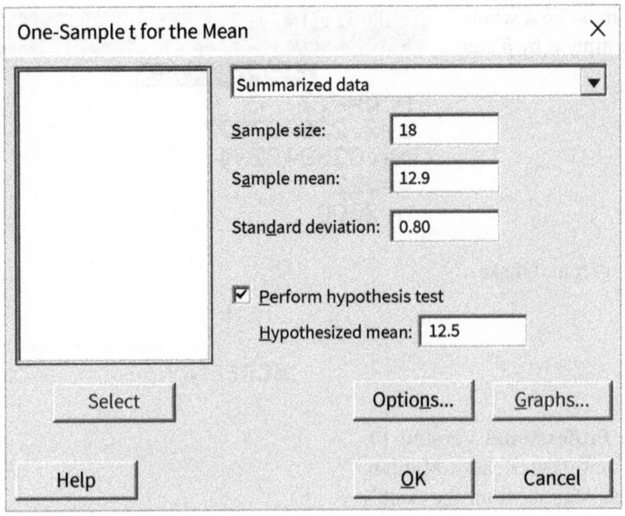

SCREEN 9.9

Testing a Hypothesis About μ, σ Unknown, for Example 9.5 of the Text

1. Select **Stat > Basic Statistics > 1-Sample t.**
2. Use the following settings in the dialog box that appears on screen (see **Screen 9.9**):
   - From the drop-down menu, select **Summarized Data.**
     Note: If you have the data in a column, select **One or more samples, each in a column,** type the column name(s) in the box, and move to Step 3.
   - In the **Sample size** box, type 18.
   - In the **Sample mean** box, type 12.9.
   - In the **Standard deviation** box, type 0.80.
   - Next to **Perform hypothesis test**, check the check box.
   - In the **Hypothesized mean** box, type 12.5.
3. Select **Options**. When the new dialog box appears on screen, select **Mean ≠ hypothesized mean** from the **Alternative hypothesis** box.
4. Click **OK** in both dialog boxes.
5. The output, including the test statistic and *p*-value, will be displayed in the Session window. (See **Screen 9.10.**)

Compare the test statistic to the critical value of *t* or the *p*-value to the value of α and make a decision.

Testing a Hypothesis About *p* for Example 9.9 of the Text

1. Select **Stat > Basic Statistics > 1 Proportion.**
2. Use the following settings in the dialog box that appears on screen (see **Screen 9.11**):
   - From the drop-down menu, select **Summarized Data.**
     Note: If you have the data in a column, select **One or more samples, each in a column,** type the column name(s) in the box, and move to Step 3.
   - In the **Number of events** box, type 440.
     Note: The number of events is the number of successes in the sample, and this must be a whole number or Minitab will return an error message. To find the number of events, multiply *n* by $\hat{p}$ and round the result to the nearest whole number.
   - In the **Number of trials** box, type 2000.
   - Next to **Perform hypothesis test**, check the check box.
   - In the **Hypothesized mean** box, type 0.20.

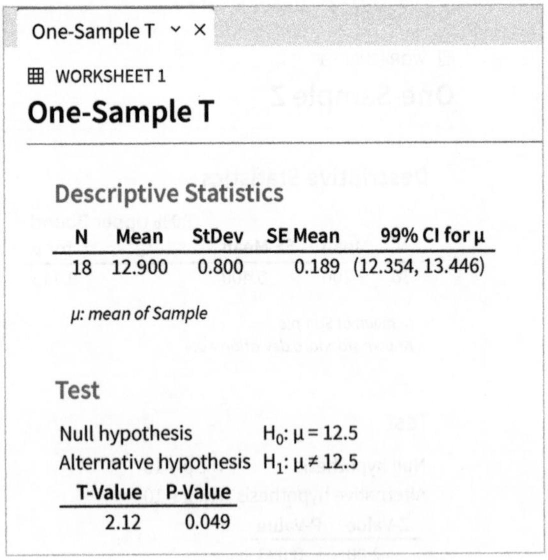

SCREEN 9.10

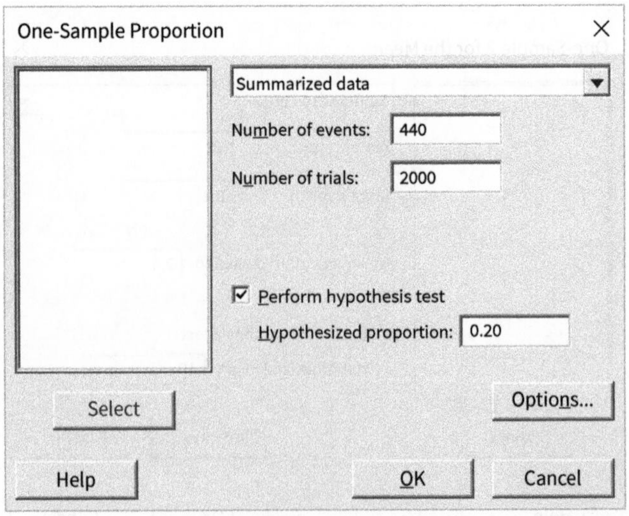

SCREEN 9.11

3. Select **Options**. Use the following settings in the dialog box that appears on screen:
   - Select **Proportion ≠ hypothesized proportion** from the **Alternative hypothesis** box,
   - In the Method box, select **Normal approximation** from the drop-down menu.
4. Click **OK** in both dialog boxes.
5. The output, including the test statistic and *p*-value, will be displayed in the Session window. (See **Screen 9.12**.)

Compare the test statistic to the critical value of z or the *p*-value to the value of α and make a decision.

Test and CI for One Propor...

WORKSHEET 1

Test and CI for One Proportion

Method

p: event proportion
Normal approximation method is used for this analysis.

Descriptive Statistics

N	Event	Sample p	95% CI for p
2000	440	0.220000	(0.201845, 0.238155)

Test

Null hypothesis H_0: p = 0.2
Alternative hypothesis H_1: p ≠ 0.2

Z-Value	P-Value
2.24	0.025

SCREEN 9.12

Excel

The Excel Technology Instructions feature of this text is written for Excel 2019. Some screens, menus, and functions will be slightly different in older versions of Excel. To enable some of the advanced Excel functions, you must enable the Data Analysis Add-In for Excel. For Excel 2007 and older versions of Excel, replace the function **Z.TEST** with the function **ZTEST**, and replace **NORM.S.DIST** with the function **NORMSDIST**.

Testing a Hypothesis About μ, σ Known

1. To perform this test, Excel must have the raw data, not summary statistics, so enter these 10 data values into cells A1 through A10 (see **Screen 9.13**):

   81, 79, 62, 98, 74, 82, 85, 72, 90, 88
2. Click on cell B1.
3. Suppose we wish to test $H_0: \mu = 75$ and it is known that $\sigma = 12$. Type one of the following commands into cell B1 based on your alternative hypothesis (see **Screen 9.13**):
   - For $H_1: \mu > 75$, type **=Z.TEST(A1:A10,75,12)**.
   - For $H_1: \mu < 75$, type **=1 − Z.TEST(A1:A10,75,12)**.
   - For $H_1: \mu \neq 75$, type **=2*MIN(Z.TEST(A1:A10,75,12),1-Z.TEST(A1:A10,75,12))**.
4. The output is the *p*-value for the test.

Compare the *p*-value to the value of α and make a decision.

	A	B
1	81	=Z.TEST(A1:A10,75,12)
2	79	
3	62	=1-Z.TEST(A1:A10,75,12)
4	98	
5	74	=2*MIN(Z.TEST(A1:A10,75,12),1-Z.TEST(A1:A10,75,12))
6	82	
7	85	
8	72	
9	90	
10	88	
11		

SCREEN 9.13

Testing a Hypothesis About μ, σ Unknown

1. To perform this test, Excel must have the raw data, not summary statistics, so enter these 10 data values into cells A1 through A10:

   81, 79, 62, 98, 74, 82, 85, 72, 90, 88
2. Suppose we wish to test $H_0: \mu = 75$. There is no Excel function for a one-sample *t*-test for μ, but it is possible to trick Excel into performing these calculations for us.
3. Enter 0 into cells B1 through B10.
4. Click **DATA** and then click **Data Analysis Tools** in the **Analysis** group.
5. Select **t-Test: Two Sample Assuming Unequal Variances** from the dialog box that appears on screen.
6. Use the following settings in the dialog box that appears on screen (see **Screen 9.14**):

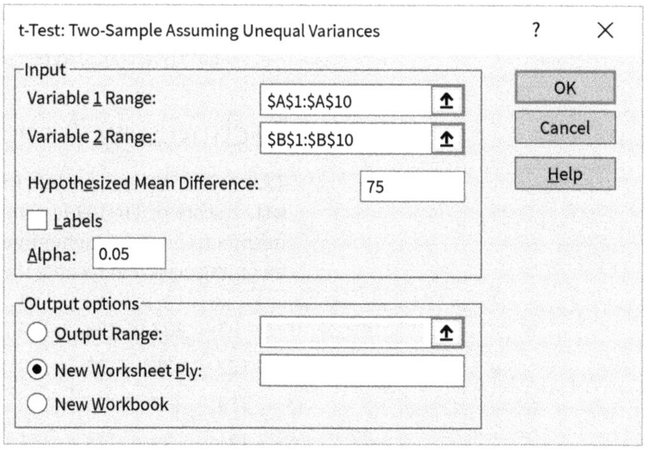

SCREEN 9.14

	A	B	C	D
1	t-Test: Two-Sample Assuming Unequal Variances			
2				
3		*Variable 1*	*Variable 2*	
4	Mean	81.1	0	
5	Variance	103.4333	0	
6	Observations	10	10	
7	Hypothesized Mean Difference	75		
8	df	9		
9	t Stat	1.897		
10	P(T<=t) one-tail	0.045		
11	t Critical one-tail	1.833		
12	P(T<=t) two-tail	0.090		
13	t Critical two-tail	2.262		
14				

SCREEN 9.15

- Type **A1:A10** in the **Variable 1 Range** box.
- Type **B1:B10** in the **Variable 2 Range** box.
- Type **75** in the Hypothesized Mean Difference box.
- Select **New Worksheet Ply** from the **Output Options**.

7. Click **OK**.
8. When the output appears, resize column A so that it is easier to read. (See **Screen 9.15**.)
9. The correct formula for the *p*-value depends on your alternative hypothesis and the *t* statistic.
   - For $H_1: \mu > 75$,
     - If the *t* statistic is positive, use the *p*-value from **P(T<=t) one-tail**.
     - If the *t* statistic is negative, the *p*-value is found by subtracting **P(T<=t) one-tail** from 1.
   - For $H_1: \mu < 75$,
     - If the *t* statistic is positive, the *p*-value is found by subtracting **P(T<=t) one-tail** from 1.
     - If the *t* statistic is negative, use the *p*-value from **P(T<=t) one-tail**.
   - For $H_1: \mu \neq 75$ use the *p*-value from **P(T<=t) two-tail**.

	A	B	C
1	null hyp	0.2	
2	p-hat	0.22	
3	n	2000	
4	SD of p-hat	0.0089	
5	z stat	2.2361	
6	p-value	0.0253	
7			

SCREEN 9.16

Testing a Hypothesis About *p* for Example 9.9 of the Text

1. There is no Excel function that can be used to test a hypothesis for a population proportion, so we must enter a few calculations in sequence. To make our work easier to understand, type the following text into the corresponding spreadsheet cells (see **Screen 9.16**):
   - Type **null hyp** in cell A1.
   - Type ***p*-hat** in cell A2.
   - Type ***n*** in cell A3.
   - Type **SD of *p*-hat** in cell A4.
   - Type ***z* stat** in cell A5.
   - Type ***p*-value** in cell A6.
2. Click on cell B1 and type **0.20**.
3. Click on cell B2 and type **0.22**.
4. Click on cell B3 and type **2000**.
5. Click on cell B4 and type **=SQRT(B1*(1-B1)/B3)**.
6. Click on cell B5 and type **=(B2-B1)/B4**.
7. Click on cell B6. The correct formula for the *p*-value depends on your alternative hypothesis.
   - For $H_1: p > 0.20$, type **=1−NORM.S.DIST(B5,1)**.
   - For $H_1: p < 0.20$, type **=NORM.S.DIST(B5,1)**.
   - For $H_1: p \neq 0.20$, type **=2*MIN(NORM.S.DIST(B5,1),1-NORM.S.DIST(B5,1))** (see **Screen 9.16**).

Technology Assignments

TA9.1 The Wechsler Adult Intelligence Scale-III (WAIS-III) is a widely used intelligence test. On this test, a score of 90–109 is considered *typical*. An administrator at a college believes that the incoming freshmen class is of higher intelligence than typical. She selects a random sample of 36 students from the incoming class and records their intelligence test scores, which are produced below.

95	123	128	114	118	102
122	98	125	120	121	132
113	117	134	100	121	130
121	99	113	111	113	115
116	103	114	107	131	128
115	129	126	118	106	115

Test if this sample data support this administrator's belief that the mean intelligence test score for all freshman is greater than 109. Use $\alpha = .05$.

TA9.2 Some colleges are known for their excellent cafeteria food, so much so that the term "Freshman 15" has been coined to refer to the amount of weight that students gain during their freshman year at college. The following data represent the amount of weight gained by 40 randomly selected students from a college during their freshman year. Note that a negative value implies that a student lost weight.

21.1	17.7	25.1	9.8	25.9	5.3	.3	23.4	22.4	7.6
25.5	15.9	24.2	27.5	−3.0	8.7	13.6	11.2	13.5	7.8
17.8	5.9	−2.4	2.9	.7	−1.0	25.7	18.0	28.7	3.2
2.2	26.7	24.5	10.5	25.5	−3.2	−.5	8.0	5.7	−4.6

Although this college is happy about the reputation of its food service, it is concerned about the health issues of substantial weight gains. As a result, it distributed nutrition pamphlets to students in an attempt to reduce the amount of weight gain. Perform a hypothesis test at a 10% significance level to determine whether the average weight gain by all freshmen at this college during the first year is less than 15 pounds.

TA9.3 General Logs Banana Bombs cereal is sold in 10.40-ounce packages. Because the cereal is sold by weight, the number of pieces of Banana Bombs varies from box to box. The following values represent the number of pieces in 19 boxes of Banana Bombs.

686	695	690	681	683	705	724	701	689	698
715	703	711	676	686	695	697	707	693	

Perform a hypothesis test to determine whether the average number of pieces in all 10.40-ounce boxes of Banana Bombs is different from 700. Assume that the distribution of the number of pieces in a 10.40-ounce box is approximately normal. Use $\alpha = .05$.

TA9.4 According to a basketball coach, the mean height of all male college basketball players is 74 inches. A random sample of 25 such players produced the following data on their heights.

68	76	74	83	77	76	69	67	71	74	79	85	69
78	75	78	68	72	83	79	82	76	69	70	81	

Test at a 2% significance level whether the mean height of all male college basketball players is different from 74 inches. Assume that the heights of all male college basketball players are (approximately) normally distributed.

TA9.5 A past study claimed that adults in the United States spent an average of 18 hours a week on leisure activities. A researcher took a sample of 10 adults from a town and asked them about the time they spend per week on leisure activities. Their responses (in hours) follow:

14	25	22	38	16	26	19	23	41	33

Assume that the times spent on leisure activities by all adults are approximately normally distributed and the population standard deviation is 3 hours. Using a 5% significance level, can you conclude that the claim of the earlier study is false?

TA9.6 Refer to Data Set IV on the Manchester Road Race (see Appendix A). Take a random sample of 100 from the third column that lists net times (in minutes) to complete this race. It is known that $\sigma = 14.669$ minutes for this variable on net times.

a. Using the sample information and a 5% significance level, test the hypothesis that the mean time for all participants to complete this race was less than 54 minutes.

b. Use technology to compute the mean time for all participants, which will be 51.996 minutes. This population mean is less than 54 minutes. Did your test of part (a) give you the correct decision?

TA9.7 Refer to Data Set IV. Let p be the proportion of all participants that are female. Select a random sample of 100 from the sixth column that lists gender of the participants. Using this sample information and 1% significance level, test the hypothesis $H_0: p = .44$ versus $H_1: p > .44$. In this population of participants, 45.75% were females. Did your hypothesis test give you the correct decision?

TA9.8 A manufacturer of contact lenses must ensure that each contact lens is properly manufactured with no flaws since flaws lead to poor vision or even eye damage. In a recent quality control check, 48 of 5000 lenses were found to have flaws. Using this sample information and 1% significance level, test the hypothesis $H_0: p = .01$ versus $H_1: p < .01$.

TA9.9 A mail-order company claims that at least 60% of all orders it receives are mailed within 48 hours. From time to time the quality control department at the company checks if this promise is kept. Recently, the quality control department at this company took a sample of 400 orders and found that 224 of them were mailed within 48 hours of the placement of the orders. Test at a 1% significance level whether the company's claim is false.

CHAPTER 10

Estimation and Hypothesis Testing: Two Populations

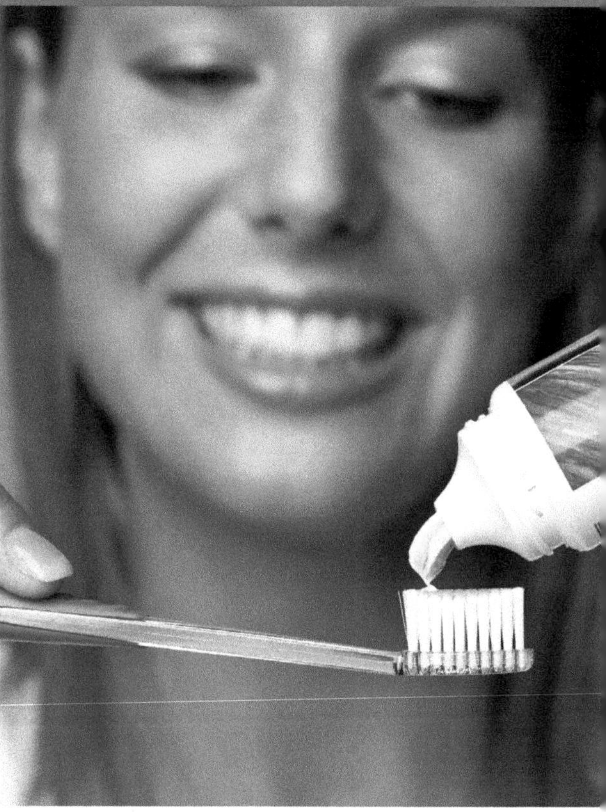

Westend61 GmbH/Alamy Stock Photo

Which toothpaste do you use? Are you a loyal user of this toothpaste? Will you ever switch to another toothpaste? Many of us never switch from one brand to another brand of a product. So, how do we determine which of two toothpastes has more loyal users? In other words, how do we test that one toothpaste has a higher percentage of its users who will not switch to the other toothpaste? (See Example 10.14.)

Chapters 8 and 9 discussed the estimation and hypothesis-testing procedures for μ and p involving a single population. This chapter extends the discussion of estimation and hypothesis-testing procedures to the difference between two population means and the difference between two population proportions. For example, we may want to make a confidence interval for the difference between the mean prices of houses in California and in New York, or we may want to test the hypothesis that the mean price of houses in California is more than that in New York. As another example, we may want to make a confidence interval for the difference between the proportions of all male and female adults who abstain from drinking, or we may want to test the hypothesis that the proportion of all adult men who abstain from drinking is different from the proportion of all adult women who abstain from drinking. Constructing confidence intervals and testing hypotheses about population parameters are referred to as *making inferences*.

10.1 Inferences About the Difference Between Two Population Means for Independent Samples: σ_1 and σ_2 Known

LEARNING OBJECTIVES

After completing this section, you should be able to:

- Explain the meaning of independent and dependent random samples.
- Describe the sampling distribution of the difference between two sample means.
- Determine a point estimate of $\mu_1 - \mu_2$ from sample data.
- Verify the conditions for making an inference for $\mu_1 - \mu_2$ when σ_1 and σ_2 are known.
- Construct and interpret a confidence interval for $\mu_1 - \mu_2$ when σ_1 and σ_2 are known.
- Apply the five-step procedure using the critical value approach to perform a test of hypothesis about $\mu_1 - \mu_2$ when σ_1 and σ_2 are known.
- Apply the four-step procedure using the p-value approach to perform a test of hypothesis about $\mu_1 - \mu_2$ when σ_1 and σ_2 are known.

Let μ_1 be the mean of the first population and μ_2 be the mean of the second population. Suppose we want to make a confidence interval and test a hypothesis about the difference between these two population means, that is, $\mu_1 - \mu_2$. Let $\bar{x}_1$ be the mean of a sample taken from the first population and $\bar{x}_2$ be the mean of a sample taken from the second population. Then, $\bar{x}_1 - \bar{x}_2$ is the sample statistic that is used to make an interval estimate and to test a hypothesis about $\mu_1 - \mu_2$. This section discusses how to make confidence intervals and test hypotheses about $\mu_1 - \mu_2$ when certain conditions (to be explained later in this section) are satisfied. First we explain the concepts of independent and dependent samples.

10.1.1 Independent Versus Dependent Samples

Two samples are **independent** if they are drawn from two different populations and the elements of one sample have no relationship to the elements of the second sample. If the elements of the two samples are somehow related, then the samples are said to be **dependent**. Thus, in two independent samples, the selection of one sample has no effect on the selection of the second sample.

Independent Versus Dependent Samples

Two samples drawn from two populations are *independent* if the selection of one sample from one population does not affect the selection of the second sample from the second population. Otherwise, the samples are *dependent*.

Examples 10.1 and 10.2 illustrate independent and dependent samples, respectively.

EXAMPLE 10.1 | Salaries of Male and Female Executives

Illustrating Two Independent Samples

Suppose we want to estimate the difference between the mean salaries of all male and all female executives. To do so, we draw two samples, one from the population of male executives and another from the population of female executives. These two samples are *independent* because they are drawn from two different populations, and the samples have no effect on each other.

EXAMPLE 10.2 Weights Before and After a Weight Loss Program

Illustrating Two Dependent Samples

Suppose we want to estimate the difference between the mean weights of all participants before and after a weight loss program. To accomplish this, suppose we take a sample of 40 participants and measure their weights before and after the completion of this program. Note that these two samples include the same 40 participants. This is an example of two *dependent* samples.

This section and Sections 10.2, 10.3, and 10.5 discuss how to make confidence intervals and test hypotheses about the difference between two population parameters when samples are independent. Section 10.4 discusses how to make confidence intervals and test hypotheses about the difference between two population means when samples are dependent.

10.1.2 Mean, Standard Deviation, and Sampling Distribution of $\bar{x}_1 - \bar{x}_2$

Suppose we select two (independent) samples from two different populations that are referred to as population 1 and population 2. Let

μ_1 = the mean of population 1

μ_2 = the mean of population 2

σ_1 = the standard deviation of population 1

σ_2 = the standard deviation of population 2

n_1 = the size of the sample drawn from population 1

n_2 = the size of the sample drawn from population 2

$\bar{x}_1$ = the mean of the sample drawn from population 1

$\bar{x}_2$ = the mean of the sample drawn from population 2

Then, as we discussed in Chapters 8 and 9, if

1. The standard deviation σ_1 of population 1 is known
2. At least one of the following two conditions is fulfilled:
   i. The sample is large (i.e., $n_1 \geq 30$)
   ii. If the sample size is small, then the population from which the sample is drawn is approximately normally distributed

then the sampling distribution of $\bar{x}_1$ is approximately normal with its mean equal to μ_1 and the standard deviation equal to $\sigma_1/\sqrt{n_1}$, assuming that $n_1/N_1 \leq .05$.

Similarly, if

1. The standard deviation σ_2 of population 2 is known
2. At least one of the following two conditions is fulfilled:
   i. The sample is large (i.e., $n_2 \geq 30$)
   ii. If the sample size is small, then the population from which the sample is drawn is approximately normally distributed

then the sampling distribution of $\bar{x}_2$ is approximately normal with its mean equal to μ_2 and the standard deviation equal to $\sigma_2/\sqrt{n_2}$, assuming that $n_2/N_2 \leq .05$.

Using these results, we can make the following statements about the mean, the standard deviation, and the shape of the sampling distribution of $\bar{x}_1 - \bar{x}_2$.

If the following conditions are satisfied:

1. The two samples are independent
2. The standard deviations σ_1 and σ_2 of the two populations are known
3. At least one of the following two conditions is fulfilled:
   i. Both samples are large (i.e., $n_1 \geq 30$ and $n_2 \geq 30$)
   ii. If either one or both sample sizes are small, then both populations from which the samples are drawn are approximately normally distributed

then the sampling distribution of $\bar{x}_1 - \bar{x}_2$ is approximately normally distributed with its mean and standard deviation,[1] respectively,

$$\mu_{\bar{x}_1 - \bar{x}_2} = \mu_1 - \mu_2$$

and

$$\sigma_{\bar{x}_1 - \bar{x}_2} = \sqrt{\frac{\sigma_1^2}{n_1} + \frac{\sigma_2^2}{n_2}}$$

In these cases, we can use the normal distribution to make a confidence interval and test a hypothesis about $\mu_1 - \mu_2$. **Figure 10.1** shows the sampling distribution of $\bar{x}_1 - \bar{x}_2$ when the above conditions are fulfilled.

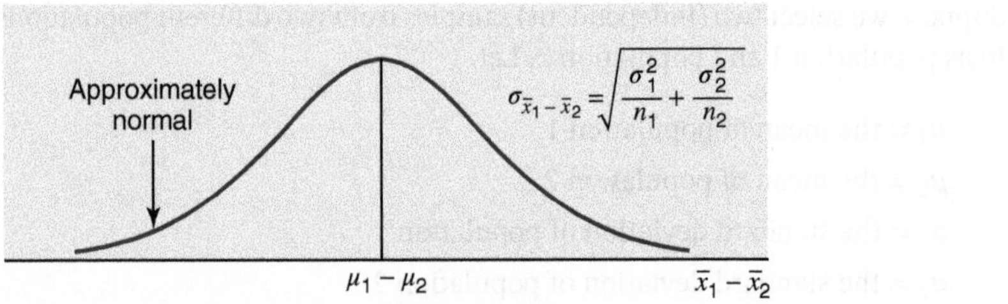

FIGURE 10.1 The sampling distribution of $\bar{x}_1 - \bar{x}_2$.

Sampling Distribution, Mean, and Standard Deviation of $\bar{x}_1 - \bar{x}_2$

When the conditions listed earlier in this section are satisfied, the *sampling distribution of $\bar{x}_1 - \bar{x}_2$* is (approximately) normal with its *mean and standard deviation* as, respectively,

$$\mu_{\bar{x}_1 - \bar{x}_2} = \mu_1 - \mu_2 \quad \text{and} \quad \sigma_{\bar{x}_1 - \bar{x}_2} = \sqrt{\frac{\sigma_1^2}{n_1} + \frac{\sigma_2^2}{n_2}}$$

Note that to apply the procedures learned in this chapter, the samples selected must be random samples.

10.1.3 Interval Estimation of $\mu_1 - \mu_2$

By constructing a confidence interval for $\mu_1 - \mu_2$, we are estimating the difference between the means of two populations. For example, we may want to find the difference between the mean heights of male and female adults. The difference between the two sample means, $\bar{x}_1 - \bar{x}_2$, is the point estimator of the difference between the two population means, $\mu_1 - \mu_2$. When the conditions mentioned earlier in this section hold true, we use the normal distribution to make a confidence interval for the difference between the two population means. The following formula gives the interval estimation for $\mu_1 - \mu_2$.

[1]The formula for the standard deviation of $\bar{x}_1 - \bar{x}_2$ can also be written as

$$\sigma_{\bar{x}_1 - \bar{x}_2} = \sqrt{\sigma_{\bar{x}_1}^2 + \sigma_{\bar{x}_2}^2}$$

where $\sigma_{\bar{x}_1} = \sigma_1/\sqrt{n_1}$ and $\sigma_{\bar{x}_2} = \sigma_2/\sqrt{n_2}$.

Confidence Interval for $\mu_1 - \mu_2$

When using the normal distribution, the $(1 - \alpha) \times 100\%$ *confidence interval for* $\mu_1 - \mu_2$ is

$$(\bar{x}_1 - \bar{x}_2) \pm z\sigma_{\bar{x}_1 - \bar{x}_2}$$

The value of z is obtained from the normal distribution table for the given confidence level. The value of $\sigma_{\bar{x}_1 - \bar{x}_2}$ is calculated as explained earlier. Here, $\bar{x}_1 - \bar{x}_2$ is the point estimator of $\mu_1 - \mu_2$.

Note that in the real world, σ_1 and σ_2 are never known. Consequently we will never use the procedures of this section, but we are discussing these procedures in this book for the information of the readers.

Example 10.3 illustrates the procedure to construct a confidence interval for $\mu_1 - \mu_2$ using the normal distribution.

EXAMPLE 10.3 Average Ages of Male and Female CEOs of U.S. Companies

Constructing a Confidence Interval for $\mu_1 - \mu_2$: σ_1 and σ_2 Known, and Samples Are Large

An economist wanted to know the mean ages of current male and female CEOs of U.S. companies with at least 200 employees. She took a random sample of 60 such male CEOs and found that their mean age was 52 years. Another random sample of 45 such female CEOs selected by her produced their mean age of 56 years. Assume that the population standard deviations for the ages of such male and female CEOs are 6 and 5 years, respectively. Let μ_1 and μ_2 be the mean ages of all such male and female CEOs, respectively.

a. What is the point estimate of $\mu_1 - \mu_2$?

b. Construct a 97% confidence interval for $\mu_1 - \mu_2$.

Solution Let us refer to the male CEOs of all U.S. companies with at least 200 employees as population 1 and the female CEOs of all such companies as population 2. Then the respective samples are samples 1 and 2. Let $\bar{x}_1$ and $\bar{x}_2$ be the means of the two samples, respectively. From the given information,

For male CEOs: $n_1 = 60,\ \bar{x}_1 = 52$ years, $\sigma_1 = 6$ years

For female CEOs: $n_2 = 45,\ \bar{x}_2 = 56$ years, $\sigma_2 = 5$ years

a. The point estimate of $\mu_1 - \mu_2$ is given by the value of $\bar{x}_1 - \bar{x}_2$. Thus,

Point estimate of $\mu_1 - \mu_2 = 52 - 56 = \mathbf{-4}$ **years**

This means that the average age of male CEOs seems to be approximately 4 years less than the average age of female CEOs of such companies.

b. The confidence level is $1 - \alpha = .97$. From the normal distribution table, the values of z for .0150 and .9850 areas to the left are -2.17 and 2.17, respectively. Hence, we will use $z = 2.17$ in the confidence interval formula. First we calculate the standard deviation of $\bar{x}_1 - \bar{x}_2$, $\sigma_{\bar{x}_1 - \bar{x}_2}$, as follows:

$$\sigma_{\bar{x}_1 - \bar{x}_2} = \sqrt{\frac{\sigma_1^2}{n_1} + \frac{\sigma_2^2}{n_2}} = \sqrt{\frac{(6)^2}{60} + \frac{(5)^2}{45}} = 1.0749677 \text{ years}$$

Next, substituting all the values in the confidence interval formula, we obtain a 97% confidence interval for $\mu_1 - \mu_2$ as

$$(\bar{x}_1 - \bar{x}_2) \pm z\sigma_{\bar{x}_1 - \bar{x}_2} = (52 - 56) \pm 2.17(1.0749677)$$
$$= -4 \pm 2.33 = \mathbf{-6.33 \text{ to } -1.67 \text{ years}}$$

Thus, with 97% confidence we can state that the difference between the mean ages of male and female CEOs of all U.S. companies with at least 200 employees is between −6.33 and −1.67 years, which means that all such male CEOs may be younger than all such female CEOs by 1.67 to 6.33 years. The value $z\sigma_{\bar{x}_1-\bar{x}_2} = 2.33$ years is called the margin of error for this estimate.

Note that in Example 10.3 both sample sizes were large and the population standard deviations were known. If the standard deviations of the two populations are known, at least one of the sample sizes is small, and both populations are approximately normally distributed, we use the normal distribution to make a confidence interval for $\mu_1 - \mu_2$. The procedure in this case is exactly the same as in Example 10.3.

10.1.4 Hypothesis Testing About $\mu_1 - \mu_2$

It is often necessary to compare the means of two populations. For example, we may want to know if the mean price of houses in Chicago is different from that in Los Angeles. Similarly, we may be interested in knowing if American children spend, on average, fewer hours in school than Japanese children do. In both these cases, we will perform a test of hypothesis about $\mu_1 - \mu_2$. The alternative hypothesis in a test of hypothesis may be that the means of the two populations are different, or that the mean of the first population is greater than the mean of the second population, or that the mean of the first population is less than the mean of the second population. These three situations are described next.

1. Testing an alternative hypothesis that the means of two populations are different is equivalent to $\mu_1 \neq \mu_2$, which is the same as $\mu_1 - \mu_2 \neq 0$.
2. Testing an alternative hypothesis that the mean of the first population is greater than the mean of the second population is equivalent to $\mu_1 > \mu_2$, which is the same as $\mu_1 - \mu_2 > 0$.
3. Testing an alternative hypothesis that the mean of the first population is less than the mean of the second population is equivalent to $\mu_1 < \mu_2$, which is the same as $\mu_1 - \mu_2 < 0$.

The procedure that is followed to perform a test of hypothesis about the difference between two population means is similar to the one that is used to test hypotheses about single-population parameters in Chapter 9. The procedure involves the same five steps for the critical value approach that were used in Chapter 9 to test hypotheses about μ and p. Here, again, if the following conditions are satisfied, we will use the normal distribution to make a test of hypothesis about $\mu_1 - \mu_2$.

1. The two samples are independent.
2. The standard deviations σ_1 and σ_2 of the two populations are known.
3. At least one of the following two conditions is fulfilled:
   i. Both samples are large (i.e., $n_1 \geq 30$ and $n_2 \geq 30$)
   ii. If either one or both sample sizes are small, then both populations from which the samples are drawn are approximately normally distributed

Test Statistic z for $\bar{x}_1 - \bar{x}_2$

When using the normal distribution, the value of the *test statistic z for $\bar{x}_1 - \bar{x}_2$* is computed as

$$z = \frac{(\bar{x}_1 - \bar{x}_2) - (\mu_1 - \mu_2)}{\sigma_{\bar{x}_1 - \bar{x}_2}}$$

The value of $\mu_1 - \mu_2$ is substituted from H_0. The value of $\sigma_{\bar{x}_1 - \bar{x}_2}$ is calculated as earlier in this section.

Example 10.4 shows how to make a test of hypothesis about $\mu_1 - \mu_2$.

EXAMPLE 10.4 Average Ages of Male and Female CEOs of U.S. Companies

Making a Two-Tailed Test of Hypothesis About $\mu_1 - \mu_2$: σ_1 and σ_2 Are Known, and Samples Are Large

An economist wanted to know the mean ages of current male and female CEOs of U.S. companies with at least 200 employees. She took a random sample of 60 such male CEOs and found that their mean age was 52 years. Another random sample of 45 such female CEOs selected by her produced their mean age of 56 years. Assume that the population standard deviations for the ages of such male and female CEOs are 6 and 5 years, respectively. Test at a 1% significance level if the population means for such male and female CEOs are different.

Solution Let us refer to the male CEOs of all U.S. companies with at least 200 employees as population 1 and the female CEOs of all such companies as population 2. Then the respective samples are samples 1 and 2. Let μ_1 and μ_2 be the mean ages of all such male and female CEOs, respectively, and let $\bar{x}_1$ and $\bar{x}_2$ be the means of the two samples, respectively. From the given information,

For male CEOs: $n_1 = 60$, $\bar{x}_1 = 52$ years, $\sigma_1 = 6$ years

For female CEOs: $n_2 = 45$, $\bar{x}_2 = 56$ years, $\sigma_2 = 5$ years

STEP 1 *State the null and alternative hypotheses.*

We are to test if the two population means are different. The two possibilities are

i. The mean ages of all male and female CEOs of U.S. companies with at least 200 employees are not different. In other words, $\mu_1 = \mu_2$, which can be written as $\mu_1 - \mu_2 = 0$.

ii. The mean ages of all male and female CEOs of U.S. companies with at least 200 employees are different. That is, $\mu_1 \neq \mu_2$, which can also be written as $\mu_1 - \mu_2 \neq 0$.

Considering these two possibilities, the null and alternative hypotheses are, respectively,

$H_0: \mu_1 - \mu_2 = 0$ (The two population means are not different.)

$H_1: \mu_1 - \mu_2 \neq 0$ (The two population means are different.)

STEP 2 *Select the distribution to use.*

Here, the population standard deviations, σ_1 and σ_2, are known, and both samples are large ($n_1 \geq 30$ and $n_2 \geq 30$). Therefore, the sampling distribution of $\bar{x}_1 - \bar{x}_2$ is approximately normal, and we use the normal distribution to perform the hypothesis test.

STEP 3 *Determine the rejection and nonrejection regions.*

The significance level is given to be .01. The $\neq$ sign in the alternative hypothesis indicates that the test is two-tailed. The area in each tail of the normal distribution curve is $\alpha/2 = .01/2 = .005$. The critical values of z for .0050 and .9950 areas to the left are (approximately) −2.58 and 2.58 from Table IV of Appendix B. These values are shown in **Figure 10.2**.

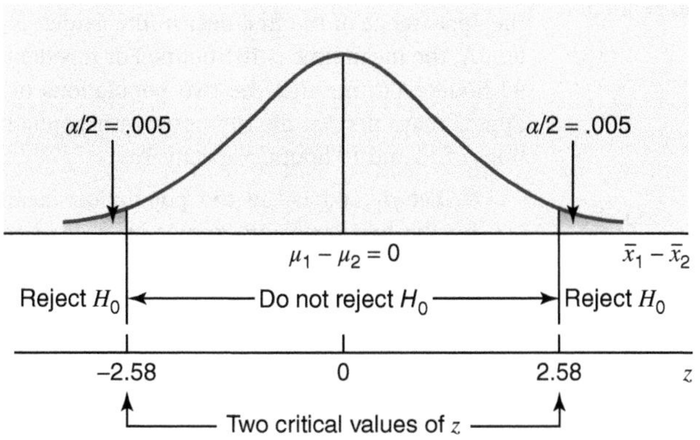

FIGURE 10.2 Rejection and nonrejection regions.

STEP 4 *Calculate the value of the test statistic.*

The value of the test statistic z for $\bar{x}_1 - \bar{x}_2$ is computed as follows:

$$\sigma_{\bar{x}_1 - \bar{x}_2} = \sqrt{\frac{\sigma_1^2}{n_1} + \frac{\sigma_2^2}{n_2}} = \sqrt{\frac{(6)^2}{60} + \frac{(5)^2}{45}} = 1.0749677 \text{ years}$$

From H_0

$$z = \frac{(\bar{x}_1 - \bar{x}_2) - (\mu_1 - \mu_2)}{\sigma_{\bar{x}_1 - \bar{x}_2}} = \frac{(52 - 56) - 0}{1.0749677} = -3.72$$

STEP 5 *Make a decision.*

Because the value of the test statistic $z = -3.72$ falls in the rejection region, we reject the null hypothesis H_0. Therefore, we conclude that the evidence from the sample data supports the decision that the mean ages of male and female CEOs of U.S. companies with at least 200 employees are different.

Using the *p*-Value to Make a Decision for Example 10.4

We can use the *p*-value approach to make the above decision. To do so, we keep Steps 1 and 2. Then in Step 3 we calculate the value of the test statistic z (as done in Step 4) and find the *p*-value for this z from the normal distribution table. In Step 4, the z value for $\bar{x}_1 - \bar{x}_2$ was calculated to be -3.72. In this example, the test is two-tailed. The *p*-value is equal to twice the area under the sampling distribution of $\bar{x}_1 - \bar{x}_2$ to the left of $z = -3.72$. From the normal distribution table (Table IV in Appendix B), the area to the left of $z = -3.72$ is (approximately) zero. Therefore, the *p*-value is zero. As we know from Chapter 9, we will reject the null hypothesis for any α (significance level) that is greater than or equal to the *p*-value. Consequently, in this example, we will reject the null hypothesis for (almost) any $\alpha > 0$. Since $\alpha = .01$ in this example, which is greater than zero, we reject the null hypothesis.

EXERCISES

Concepts and Procedures

10.1 Briefly explain the meaning of independent and dependent samples. Give one example of each.

10.2 Describe the sampling distribution of $\bar{x}_1 - \bar{x}_2$ for two independent samples when σ_1 and σ_2 are known and either both sample sizes are large or both populations are normally distributed. What are the mean and standard deviation of this sampling distribution?

10.3 The following information is obtained from two independent samples selected from two normally distributed populations.

$n_1 = 20 \quad \bar{x}_1 = 8.34 \quad \sigma_1 = 1.89$
$n_2 = 15 \quad \bar{x}_2 = 4.85 \quad \sigma_2 = 2.72$

a. What is the point estimate of $\mu_1 - \mu_2$?

b. Construct a 99% confidence interval for $\mu_1 - \mu_2$. Find the margin of error for this estimate.

10.4 The following information is obtained from two independent samples selected from two populations.

$n_1 = 750 \quad \bar{x}_1 = 1.05 \quad \sigma_1 = 4.85$
$n_2 = 800 \quad \bar{x}_2 = 1.54 \quad \sigma_2 = 7.60$

a. What is the point estimate of $\mu_1 - \mu_2$?

b. Construct a 95% confidence interval for $\mu_1 - \mu_2$. Find the margin of error for this estimate.

10.5 The following information is obtained from two independent samples selected from two normally distributed populations.

$n_1 = 24 \quad \bar{x}_1 = 5.56 \quad \sigma_1 = 1.65$
$n_2 = 27 \quad \bar{x}_2 = 4.80 \quad \sigma_2 = 1.58$

Test at a 5% significance level if the two population means are different.

10.6 The following information is obtained from two independent samples selected from two populations.

$n_1 = 300 \quad \bar{x}_1 = 22.0 \quad \sigma_1 = 4.90$
$n_2 = 250 \quad \bar{x}_2 = 27.6 \quad \sigma_2 = 4.50$

Test at a 5% significance level if μ_1 is less than μ_2.

Applications

10.7 In parts of the eastern United States, whitetail deer are a major nuisance to farmers and homeowners, frequently damaging crops, gardens, and landscaping. A consumer organization arranges a test of two of the leading deer repellents A and B on the market. Fifty-six unfenced gardens in areas having high concentrations of deer are used for the test. Twenty-nine gardens are chosen at random to receive repellent A, and the other 27 receive repellent B. For each of the 56 gardens, the time elapsed between application of the repellent and the appearance of the first deer in the garden is recorded. For repellent A, the mean time is 101 hours. For repellent B, the mean time is 92 hours. Assume that the two populations of elapsed times have approximate normal distributions with population standard deviations of 15 and 10 hours, respectively.

a. Let μ_1 and μ_2 be the population means of elapsed times for the two repellents, respectively. Find the point estimate of $\mu_1 - \mu_2$.

b. Find a 97% confidence interval for $\mu_1 - \mu_2$.

c. Test at a 2% significance level whether the mean elapsed times for repellents A and B are different. Use both approaches, the critical value and *p*-value, to perform this test.

10.8 A local college cafeteria has a self-service soft ice cream machine. The cafeteria provides bowls that can hold up to 16 ounces of ice cream.

The food service manager is interested in comparing the average amount of ice cream dispensed by male students to the average amount dispensed by female students. A measurement device was placed on the ice cream machine to determine the amounts dispensed. Random samples of 90 male and 67 female students who got ice cream were selected. The sample averages were 8.32 and 6.49 ounces for the male and female students, respectively. Assume that the population standard deviations are 1.22 and 1.17 ounces, respectively.

a. Let μ_1 and μ_2 be the population means of ice cream amounts dispensed by all male and all female students at this college, respectively. What is the point estimate of $\mu_1 - \mu_2$?

b. Construct a 95% confidence interval for $\mu_1 - \mu_2$.

c. Using a 1% significance level, can you conclude that the average amount of ice cream dispensed by all male college students is larger than the average amount dispensed by all female college students? Use both approaches to make this test.

10.9 A car magazine is comparing the total repair costs incurred during the first three years on two sports cars, the T-999 and the XPY. Random samples of 45 T-999s and 51 XPYs are taken. All 96 cars are three years old and have similar mileages. The mean of repair costs for the 45 T-999 cars is $3300 for the first three years. For the 51 XPY cars, this mean is $3850. Assume that the standard deviations for the two populations are $800 and $1000, respectively.

a. Construct a 99% confidence interval for the difference between the two population means.

b. Using a 1% significance level, can you conclude that such mean repair costs are different for these two types of cars?

c. What would your decision be in part (b) if the probability of making a Type I error were zero? Explain.

10.10 The management at New Century Bank claims that the mean waiting time for all customers at its branches is less than that at the Public Bank, which is its main competitor. A business consulting firm took a sample of 200 customers from the New Century Bank and found that they waited an average of 4.7 minutes before being served. Another sample of 300 customers taken from the Public Bank showed that these customers waited an average of 5.0 minutes before being served. Assume that the standard deviations for the two populations are 1.2 and 1.5 minutes, respectively.

a. Make a 97% confidence interval for the difference between the two population means.

b. Test at a 2.5% significance level whether the claim of the management of the New Century Bank is true.

c. Calculate the *p*-value for the test of part (b). Based on this *p*-value, would you reject the null hypothesis if $\alpha = .01$? What if $\alpha = .05$?

10.11 Maine Mountain Dairy claims that its 8-ounce low-fat yogurt cups contain, on average, fewer calories than the 8-ounce low-fat yogurt cups produced by a competitor. A consumer agency wanted to check this claim. A sample of 27 such yogurt cups produced by this company showed that they contained an average of 141 calories per cup. A sample of 25 such yogurt cups produced by its competitor showed that they contained an average of 144 calories per cup. Assume that the two populations are approximately normally distributed with population standard deviations of 5.5 and 6.4 calories, respectively.

a. Make a 98% confidence interval for the difference between the mean number of calories in the 8-ounce low-fat yogurt cups produced by the two companies.

b. Test at a 1% significance level whether Maine Mountain Dairy's claim is true.

c. Calculate the *p*-value for the test of part (b). Based on this *p*-value, would you reject the null hypothesis if $\alpha = .05$? What if $\alpha = .025$?

10.2 Inferences About the Difference Between Two Population Means for Independent Samples: σ_1 and σ_2 Unknown but Equal

LEARNING OBJECTIVES

After completing this section, you should be able to:

- Explain the conditions required to use the *t*-distribution to perform an inference about $\mu_1 - \mu_2$ for two independent random samples selected from two populations with unknown but equal population standard deviations.
- Verify the conditions for making an inference for $\mu_1 - \mu_2$ when σ_1 and σ_2 are unknown but equal.
- Construct and interpret a confidence interval for $\mu_1 - \mu_2$ when σ_1 and σ_2 are unknown but equal.
- Apply the five-step procedure using a critical value approach to perform a test of hypothesis about $\mu_1 - \mu_2$ when σ_1 and σ_2 are unknown but equal.
- Apply the four-step procedure using a *p*-value approach to perform a test of hypothesis about $\mu_1 - \mu_2$ when σ_1 and σ_2 are unknown but equal.

This section discusses making a confidence interval and testing a hypothesis about the difference between the means of two populations, $\mu_1 - \mu_2$, assuming that the standard deviations, σ_1 and σ_2, of these populations are not known but are assumed to be equal.

There are some other conditions, explained next, that must be fulfilled to use the procedures discussed in this section.

If the following conditions are satisfied,

1. The two samples are independent
2. The standard deviations σ_1 and σ_2 of the two populations are unknown, but they can be assumed to be equal, that is, $\sigma_1 = \sigma_2$
3. At least one of the following two conditions is fulfilled:
   i. Both samples are large (i.e., $n_1 \geq 30$ and $n_2 \geq 30$)
   ii. If either one or both sample sizes are small, then both populations from which the samples are drawn are approximately normally distributed

then we use the t distribution to make a confidence interval and test a hypothesis about the difference between the means of two populations, $\mu_1 - \mu_2$.

When the standard deviations of the two populations are equal, we can use σ for both σ_1 and σ_2. Because σ is unknown, we replace it by its point estimator s_p, which is called the **pooled sample standard deviation** (hence, the subscript p). The value of s_p is computed by using the information from the two samples as follows:

Pooled Standard Deviation for Two Samples

The *pooled standard deviation for two samples* is computed as

$$s_p = \sqrt{\frac{(n_1 - 1)s_1^2 + (n_2 - 1)s_2^2}{n_1 + n_2 - 2}}$$

where n_1 and n_2 are the sizes of the two samples and s_1^2 and s_2^2 are the variances of the two samples, respectively. Here s_p is an estimator of σ.

In this formula, $n_1 - 1$ are the degrees of freedom for sample 1, $n_2 - 1$ are the degrees of freedom for sample 2, and $n_1 + n_2 - 2$ are the **degrees of freedom for the two samples taken together**. Note that s_p is an estimator of the standard deviation, σ, of each of the two populations.

When s_p is used as an estimator of σ, the standard deviation of $\bar{x}_1 - \bar{x}_2$, $\sigma_{\bar{x}_1 - \bar{x}_2}$, is estimated by $s_{\bar{x}_1 - \bar{x}_2}$. The value of $s_{\bar{x}_1 - \bar{x}_2}$ is calculated by using the following formula:

Estimator of the Standard Deviation of $\bar{x}_1 - \bar{x}_2$

The *estimator of the standard deviation of* $\bar{x}_1 - \bar{x}_2$ is

$$s_{\bar{x}_1 - \bar{x}_2} = s_p \sqrt{\frac{1}{n_1} + \frac{1}{n_2}}$$

Now we are ready to discuss the procedures that are used to make confidence intervals and test hypotheses about $\mu_1 - \mu_2$ for independent samples selected from two populations with unknown but equal standard deviations.

10.2.1 Interval Estimation of $\mu_1 - \mu_2$

As was mentioned earlier in this chapter, the difference between the two sample means, $\bar{x}_1 - \bar{x}_2$, is the point estimator of the difference between the two population means, $\mu_1 - \mu_2$. The formula given next gives the confidence interval for $\mu_1 - \mu_2$ when the t distribution is used and the conditions mentioned earlier in this section are fulfilled.

Confidence Interval for $\mu_1 - \mu_2$

The $(1 - \alpha) \times 100\%$ *confidence interval for* $\mu_1 - \mu_2$ is

$$(\bar{x}_1 - \bar{x}_2) \pm t s_{\bar{x}_1 - \bar{x}_2}$$

where the value of t is obtained from the t distribution table for the given confidence level and $n_1 + n_2 - 2$ degrees of freedom, and $s_{\bar{x}_1 - \bar{x}_2}$ is calculated as explained earlier.

Example 10.5 describes the procedure to make a confidence interval for $\mu_1 - \mu_2$ using the t distribution.

EXAMPLE 10.5 | Caffeine in Two Brands of Coffee

Constructing a Confidence Interval for $\mu_1 - \mu_2$: Two Independent Samples, Unknown but Equal σ_1 and σ_2

ansonsaw/Getty Images

A consumer agency wanted to estimate the difference in the mean amounts of caffeine in two brands of coffee. The agency took a sample of 15 one-pound jars of Brand I coffee that showed the mean amount of caffeine in these jars to be 80 milligrams per jar with a standard deviation of 5 milligrams. Another sample of 12 one-pound jars of Brand II coffee gave a mean amount of caffeine equal to 77 milligrams per jar with a standard deviation of 6 milligrams. Construct a 95% confidence interval for the difference between the mean amounts of caffeine in one-pound jars of these two brands of coffee. Assume that the two populations are approximately normally distributed and that the standard deviations of the two populations are equal.

Solution Let μ_1 and μ_2 be the mean amounts of caffeine per jar in all one-pound jars of Brand I and II coffee, respectively, and let $\bar{x}_1$ and $\bar{x}_2$ be the means of the two respective samples. From the given information,

Brand I coffee:	$n_1 = 15$	$\bar{x}_1 = 80$ milligrams	$s_1 = 5$ milligrams
Brand II coffee:	$n_2 = 12$	$\bar{x}_2 = 77$ milligrams	$s_2 = 6$ milligrams

The confidence level is: $1 - \alpha = .95$

Here, σ_1 and σ_2 are unknown but assumed to be equal, the samples are independent (taken from two different populations), and the sample sizes are small but the two populations are approximately normally distributed. Hence, we will use the t distribution to make the confidence interval for $\mu_1 - \mu_2$ as all conditions mentioned in the beginning of this section are satisfied.

First we calculate the standard deviation of $\bar{x}_1 - \bar{x}_2$ as follows. Note that since it is assumed that σ_1 and σ_2 are equal, we will use s_p to calculate $s_{\bar{x}_1 - \bar{x}_2}$.

$$s_p = \sqrt{\frac{(n_1 - 1)s_1^2 + (n_2 - 1)s_2^2}{n_1 + n_2 - 2}} = \sqrt{\frac{(15 - 1)(5)^2 + (12 - 1)(6)^2}{15 + 12 - 2}} = 5.46260011$$

$$s_{\bar{x}_1 - \bar{x}_2} = s_p\sqrt{\frac{1}{n_1} + \frac{1}{n_2}} = (5.46260011)\sqrt{\frac{1}{15} + \frac{1}{12}} = 2.11565593$$

Next, to find the t value from the t distribution table, we need to know the area in each tail of the t distribution curve and the degrees of freedom.

$$\text{Area in each tail} = \alpha/2 = (1 - .95)/2 = .025$$
$$\text{Degrees of freedom} = n_1 + n_2 - 2 = 15 + 12 - 2 = 25$$

The t value for $df = 25$ and .025 area in the right tail of the t distribution curve is 2.060. The 95% confidence interval for $\mu_1 - \mu_2$ is

$$(\bar{x}_1 - \bar{x}_2) \pm t s_{\bar{x}_1 - \bar{x}_2} = (80 - 77) \pm 2.060(2.11565593)$$
$$= 3 \pm 4.36 = \mathbf{-1.36 \text{ to } 7.36 \text{ milligrams}}$$

Thus, with 95% confidence we can state that based on these two sample results, the difference in the mean amounts of caffeine in one-pound jars of these two brands of coffee lies between −1.36 and 7.36 milligrams. Because the lower limit of the interval is negative, it is possible that the mean amount of caffeine is greater in the second brand than in the first brand of coffee.

Note that the value of $\bar{x}_1 - \bar{x}_2$, which is 80 − 77 = 3, gives the point estimate of $\mu_1 - \mu_2$. The value of $ts_{\bar{x}_1 - \bar{x}_2}$, which is 4.36, is the margin of error.

10.2.2 Hypothesis Testing About $\mu_1 - \mu_2$

When the conditions mentioned in the beginning of Section 10.2 are satisfied, the *t* distribution is applied to make a hypothesis test about the difference between two population means. The test statistic in this case is *t*, which is calculated as follows:

Test Statistic *t* for $\bar{x}_1 - \bar{x}_2$

The value of the *test statistic t for* $\bar{x}_1 - \bar{x}_2$ is computed as

$$t = \frac{(\bar{x}_1 - \bar{x}_2) - (\mu_1 - \mu_2)}{s_{\bar{x}_1 - \bar{x}_2}}$$

The value of $\mu_1 - \mu_2$ in this formula is substituted from the null hypothesis, and $s_{\bar{x}_1 - \bar{x}_2}$ is calculated as explained earlier in Section 10.2.1.

Examples 10.6 and 10.7 illustrate how to use the *t* distribution to perform a test of hypothesis about the difference between two population means for independent samples that are selected from two populations with equal standard deviations.

EXAMPLE 10.6 | Calories in Two Brands of Diet Soda

Making a Two-Tailed Test of Hypothesis About $\mu_1 - \mu_2$: Two Independent Samples, and Unknown but Equal σ_1 and σ_2

A sample of 14 cans of Brand I diet soda gave the mean number of calories of 23 per can with a standard deviation of 3 calories. Another sample of 16 cans of Brand II diet soda gave the mean number of calories of 25 per can with a standard deviation of 4 calories. At a 1% significance level, can you conclude that the mean numbers of calories per can are different for these two brands of diet soda? Assume that the calories per can of diet soda are approximately normally distributed for each of the two brands and that the standard deviations for the two populations are equal.

Solution Let μ_1 and μ_2 be the mean number of calories for all cans of Brand I and Brand II diet soda, respectively, and let $\bar{x}_1$ and $\bar{x}_2$ be the means of the respective samples. From the given information,

Brand I diet soda:	$n_1 = 14$	$\bar{x}_1 = 23$	$s_1 = 3$
Brand II diet soda:	$n_2 = 16$	$\bar{x}_2 = 25$	$s_2 = 4$

The significance level is: $\alpha = .01$

STEP 1 *State the null and alternative hypotheses.*

We are to test for the difference in the mean number of calories for the two brands. The null and alternative hypotheses are, respectively,

H_0: $\mu_1 - \mu_2 = 0$ (The mean number of calories are not different.)

H_1: $\mu_1 - \mu_2 \neq 0$ (The mean number of calories are different.)

STEP 2 *Select the distribution to use.*

Here, the two samples are independent, σ_1 and σ_2 are unknown but equal, and the sample sizes are small but both populations are approximately normally distributed. Hence, all conditions mentioned in the beginning of Section 10.2 are fulfilled. Consequently, we will use the *t* distribution.

STEP 3 *Determine the rejection and nonrejection regions.*

The $\neq$ sign in the alternative hypothesis indicates that the test is two-tailed. The significance level is .01. Hence,

$$\text{Area in each tail} = \alpha/2 = .01/2 = .005$$
$$\text{Degrees of freedom} = n_1 + n_2 - 2 = 14 + 16 - 2 = 28$$

The critical values of t for $df = 28$ and .005 area in each tail of the t distribution curve are -2.763 and 2.763, as shown in **Figure 10.3**.

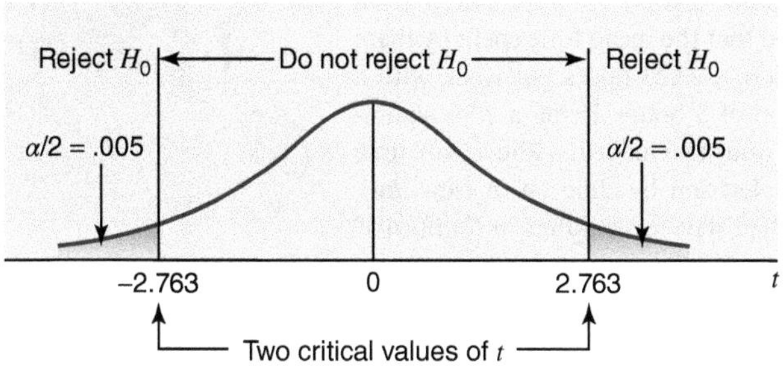

FIGURE 10.3 Rejection and nonrejection regions.

STEP 4 *Calculate the value of the test statistic.*

The value of the test statistic t for $\bar{x}_1 - \bar{x}_2$ is computed as follows:

$$s_p = \sqrt{\frac{(n_1 - 1)s_1^2 + (n_2 - 1)s_2^2}{n_1 + n_2 - 2}} = \sqrt{\frac{(14-1)(3)^2 + (16-1)(4)^2}{14 + 16 - 2}} = 3.57071421$$

$$s_{\bar{x}_1 - \bar{x}_2} = s_p\sqrt{\frac{1}{n_1} + \frac{1}{n_2}} = (3.57071421)\sqrt{\frac{1}{14} + \frac{1}{16}} = 1.30674760$$

$$t = \frac{(\bar{x}_1 - \bar{x}_2) - (\mu_1 - \mu_2)}{s_{\bar{x}_1 - \bar{x}_2}} = \frac{(23 - 25) - 0}{1.30674760} = -1.531$$

(The 0 in the numerator is marked "From H_0".)

STEP 5 *Make a decision.*

Because the value of the test statistic $t = -1.531$ for $\bar{x}_1 - \bar{x}_2$ falls in the nonrejection region, we fail to reject the null hypothesis. Consequently we cannot conclude that the mean number of calories per can are different for these two brands of diet soda. The difference in $\bar{x}_1$ and $\bar{x}_2$ observed for the two samples may have occurred due to sampling error only.

Using the *p*-Value to Make a Decision for Example 10.6

We can use the p-value approach to make the above decision. To do so, we keep Steps 1 and 2 of this example. Then in Step 3, we calculate the value of the test statistic t (as done in Step 4) and then find the p-value for this t from the t distribution table (Table V of Appendix B) or by using technology. In Step 4, the t-value for $\bar{x}_1 - \bar{x}_2$ was calculated to be -1.531. In this example, the test is two-tailed. Therefore, the p-value is equal to twice the area under the t distribution curve to the left of $t = -1.531$. If we have access to technology, we can find the exact p-value, which will be .137. If we use the t distribution table, we can only find a range for the p-value. From Table V of Appendix B, for $df = 28$, the two values that include 1.531 are 1.313 and 1.701. (Note that we use the positive value of t, although our t is negative.) Thus, the test statistic $t = -1.531$ falls between -1.313 and -1.701. The areas in the t distribution table that correspond to 1.313 and 1.701 are .10 and .05, respectively. Because it is a two-tailed test, the p-value for $t = -1.531$ is between $2(.10) = .20$ and $2(.05) = .10$, which can be written as

$$.10 < p\text{-value} < .20$$

As we know from Chapter 9, we will reject the null hypothesis for any α (significance level) that is greater than or equal to the p-value. Consequently, in this example, we will reject the null hypothesis for any $\alpha \geq .20$ using the above range and not reject it for $\alpha < .10$. If we use technology, we will reject the null hypothesis for $\alpha \geq .137$. Since $\alpha = .01$ in this example, which is smaller than both .10 and .137, we fail to reject the null hypothesis.

EXAMPLE 10.7 Time Spent Watching Television by Children

Making a Right-Tailed Test of Hypothesis About $\mu_1 - \mu_2$: Two Independent Samples, σ_1 and σ_2 Unknown but Equal, and Both Samples Are Large

A sample of 40 children from New York State showed that the mean time they spend watching television is 28.50 hours per week with a standard deviation of 4 hours. Another sample of 35 children from California showed that the mean time spent by them watching television is 23.25 hours per week with a standard deviation of 5 hours. Using a 2.5% significance level, can you conclude that the mean time spent watching television by children in New York State is greater than that for children in California? Assume that the standard deviations for the two populations are equal.

Solution Let all children from New York State be referred to as population 1 and those from California as population 2. Let μ_1 and μ_2 be the mean time spent watching television by children in populations 1 and 2, respectively, and let $\bar{x}_1$ and $\bar{x}_2$ be the mean time spent watching television by children in the respective samples. From the given information,

New York:	$n_1 = 40$	$\bar{x}_1 = 28.50$ hours	$s_1 = 4$ hours
California:	$n_2 = 35$	$\bar{x}_2 = 23.25$ hours	$s_2 = 5$ hours

The significance level is $\alpha = .025$.

STEP 1 *State the null and alternative hypotheses.*

The two possible decisions are

1. The mean time spent watching television by children in New York State is not greater than that for children in California. This can be written as $\mu_1 = \mu_2$ or $\mu_1 - \mu_2 = 0$.
2. The mean time spent watching television by children in New York State is greater than that for children in California. This can be written as $\mu_1 > \mu_2$ or $\mu_1 - \mu_2 > 0$.

Hence, the null and alternative hypotheses are, respectively,

$$H_0: \mu_1 - \mu_2 = 0$$
$$H_1: \mu_1 - \mu_2 > 0$$

Note that the null hypothesis can also be written as $\mu_1 - \mu_2 \leq 0$.

STEP 2 *Select the distribution to use.*

Here, the two samples are independent (taken from two different populations), σ_1 and σ_2 are unknown but assumed to be equal, and both samples are large. Hence, all conditions mentioned in the beginning of Section 10.2 are fulfilled. Consequently, we use the t distribution to make the test.

STEP 3 *Determine the rejection and nonrejection regions.*

The > sign in the alternative hypothesis indicates that the test is right-tailed. The significance level is .025.

$$\text{Area in the right tail of the } t \text{ distribution} = \alpha = .025$$
$$\text{Degrees of freedom} = n_1 + n_2 - 2 = 40 + 35 - 2 = 73$$

From the t distribution table, the critical value of t for $df = 73$ and .025 area in the right tail of the t distribution is 1.993. This value is shown in **Figure 10.4**.

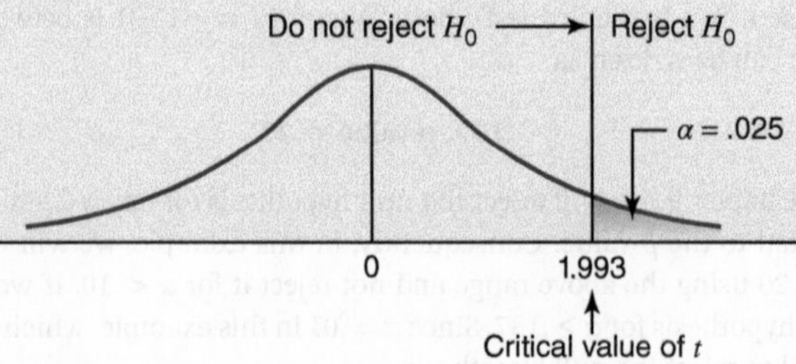

FIGURE 10.4 Rejection and nonrejection regions.

STEP 4 *Calculate the value of the test statistic.*

The value of the test statistic t for $\bar{x}_1 - \bar{x}_2$ is computed as follows:

$$s_p = \sqrt{\frac{(n_1 - 1)s_1^2 + (n_2 - 1)s_2^2}{n_1 + n_2 - 2}} = \sqrt{\frac{(40 - 1)(4)^2 + (35 - 1)(5)^2}{40 + 35 - 2}} = 4.49352655$$

$$s_{\bar{x}_1 - \bar{x}_2} = s_p\sqrt{\frac{1}{n_1} + \frac{1}{n_2}} = (4.49352655)\sqrt{\frac{1}{40} + \frac{1}{35}} = 1.04004930$$

From H_0

$$t = \frac{(\bar{x}_1 - \bar{x}_2) - (\mu_1 - \mu_2)}{s_{\bar{x}_1 - \bar{x}_2}} = \frac{(28.50 - 23.25) - 0}{1.04004930} = 5.048$$

STEP 5 *Make a decision.*

Because the value of the test statistic $t = 5.048$ for $\bar{x}_1 - \bar{x}_2$ falls in the rejection region (see Figure 10.4), we reject the null hypothesis H_0. Hence, we conclude that children in New York State seem to spend more time, on average, watching TV than children in California.

Using the *p*-Value to Make a Decision for Example 10.7

To use the *p*-value approach to make the same decision, we keep Steps 1 and 2 of this example. Then in Step 3, we calculate the value of the test statistic t (as done in Step 4) and then find the *p*-value for this t from the t distribution table (Table V of Appendix B) or by using technology. In Step 4, the t-value for $\bar{x}_1 - \bar{x}_2$ was calculated to be 5.048. In this example, the test is right-tailed. Therefore, the *p*-value is equal to the area under the t distribution curve to the right of $t = 5.048$. If we have access to technology, we can find the exact *p*-value, which will be .000. If we use the t distribution table, for $df = 73$, the value of the test statistic $t = 5.048$ is larger than 3.206. Therefore, the *p*-value for $t = 5.048$ is less than .001, which can be written as

$$p\text{-value} < .001$$

Since we will reject the null hypothesis for any α (significance level) greater than or equal to the *p*-value, here we reject the null hypothesis because $\alpha = .025$ is greater than both the *p*-values, .001 obtained above from the table and .000 obtained by using technology. Note that obtaining the *p*-value = .000 from technology does not mean that the *p*-value is zero. It means that when the *p*-value is rounded to three digits after the decimal, it is .000.

Note: What if the Sample Sizes Are Large and the Number of *df* Are Not in the *t* Distribution Table?

In this section, we used the t distribution to make confidence intervals and perform tests of hypothesis about $\mu_1 - \mu_2$. When both sample sizes are large, it does not matter how large the sample sizes are if we are using technology. However, if we are using the t distribution table (Table V of Appendix B), this may pose a problem if samples are too large. Table V in Appendix B goes up to only 75 degrees of freedom. Thus, if the degrees of freedom are larger than 75, we cannot use Table V to find the critical value(s) of t. As mentioned in Chapters 8 and 9, in such a situation, there are three options:

1. Use the t value from the last row (the row of ∞) in Table V.
2. Use the normal distribution as an approximation to the t distribution.
3. Use technology to find the t value.

EXERCISES

Concepts and Procedures

10.12 Explain what conditions must hold true to use the t distribution to make a confidence interval and to test a hypothesis about $\mu_1 - \mu_2$ for two independent samples selected from two populations with unknown but equal standard deviations.

10.13 The following information was obtained from two independent samples selected from two normally distributed populations with unknown but equal standard deviations.

$n_1 = 21 \quad \bar{x}_1 = 12.82 \quad s_1 = 4.12$

$n_2 = 20 \quad \bar{x}_2 = 16.75 \quad s_2 = 3.46$

a. What is the point estimate of $\mu_1 - \mu_2$?

b. Construct a 95% confidence interval for $\mu_1 - \mu_2$.

10.14 The following information was obtained from two independent samples selected from two normally distributed populations with unknown but equal standard deviations.

$$n_1 = 21 \quad \bar{x}_1 = 12.82 \quad s_1 = 4.12$$
$$n_2 = 20 \quad \bar{x}_2 = 16.75 \quad s_2 = 3.46$$

Test at a 5% significance level if the two population means are different.

10.15 The following information was obtained from two independent samples selected from two populations with unknown but equal standard deviations.

$$n_1 = 50 \quad \bar{x}_1 = 90.40 \quad s_1 = 10.40$$
$$n_2 = 55 \quad \bar{x}_2 = 86.30 \quad s_2 = 9.75$$

Test at a 1% significance level if the two population means are different.

10.16 The following information was obtained from two independent samples selected from two populations with unknown but equal standard deviations.

$$n_1 = 50 \quad \bar{x}_1 = 90.40 \quad s_1 = 10.40$$
$$n_2 = 55 \quad \bar{x}_2 = 86.30 \quad s_2 = 9.75$$

Test at a 5% significance level if μ_1 is greater than μ_2.

10.17 The following information was obtained from two independent samples selected from two normally distributed populations with unknown but equal standard deviations.

Sample 1:	47.7	42.8	51.9	34.1	65.8	61.5	52.1	40.8
	53.1	46.1	47.9	45.7	46.7			
Sample 2:	50.0	47.4	32.7	38.6	54.0	46.3	42.5	40.8
	39.0	68.2						

a. Let μ_1 be the mean of population 1 and μ_2 be the mean of population 2. What is the point estimate of $\mu_1 - \mu_2$?

b. Construct a 98% confidence interval for $\mu_1 - \mu_2$.

c. Test at a 1% significance level if μ_1 is greater than μ_2.

10.18 The following information was obtained from two independent samples selected from two normally distributed populations with unknown but equal standard deviations.

Sample 1:	2.18	2.23	1.96	2.24	2.84	1.87	2.68
	2.15	2.49	1.95				
Sample 2:	1.82	1.86	2.00	1.89	1.73	1.34	1.43
	2.05	1.54	2.50	2.08	2.13		

a. Let μ_1 be the mean of population 1 and μ_2 be the mean of population 2. What is the point estimate of $\mu_1 - \mu_2$?

b. Construct a 99% confidence interval for $\mu_1 - \mu_2$.

c. Test at a 2.5% significance level if μ_1 is lower than μ_2.

Applications

10.19 The standard recommendation for automobile oil changes is once every 3000 miles. A local mechanic is interested in determining whether people who drive more expensive cars are more likely to follow the recommendation. Independent random samples of 45 customers who drive luxury cars and 40 customers who drive compact lower-price cars were selected. The average distance driven between oil changes was 3187 miles for the luxury car owners and 3214 miles for the compact lower-price cars. The sample standard deviations were 42.40 and 50.70 miles for the luxury and compact groups, respectively. Assume that the two population distributions of the distances between oil changes have the same standard deviation.

a. Construct a 95% confidence interval for the difference in the mean distances between oil changes for all luxury cars and all compact lower-price cars.

b. Using a 1% significance level, can you conclude that the mean distance between oil changes is less for all luxury cars than that for all compact lower-price cars?

10.20 A town that recently started a single-stream recycling program provided 60-gallon recycling bins to 30 randomly selected households and 75-gallon recycling bins to 22 randomly selected households. The total volume of recycling over a 10-week period was measured for each of the households. The average total volumes were 365 and 423 gallons for the households with the 60- and 75-gallon bins, respectively. The sample standard deviations were 52.5 and 43.8 gallons, respectively. Assume that the 10-week total volumes of recycling are approximately normally distributed for both groups and that the population standard deviations are equal.

a. Construct a 98% confidence interval for the difference in the mean volumes of 10-week recycling for the households with the 60- and 75-gallon bins.

b. Using the 2% significance level, can you conclude that the average 10-week recycling volume of all households having 60-gallon containers is different from the average volume of all households that have 75-gallon containers?

10.21 A consumer organization tested two paper shredders, the Piranha and the Crocodile, designed for home use. Each of 10 randomly selected volunteers shredded 100 sheets of paper with the Piranha, and then another sample of 10 randomly selected volunteers each shredded 100 sheets with the Crocodile. The Piranha took an average of 218 seconds to shred 100 sheets with a standard deviation of 8 seconds. The Crocodile took an average of 163 seconds to shred 100 sheets with a standard deviation of 4 seconds. Assume that the shredding times for both machines are approximately normally distributed with equal but unknown standard deviations.

a. Construct a 99% confidence interval for the difference between the two population means.

b. Using a 1% significance level, can you conclude that the mean time taken by the Piranha to shred 100 sheets is higher than that for the Crocodile?

c. What would your decision be in part (b) if the probability of making a Type I error were zero? Explain.

10.22 Quadro Corporation has two supermarket stores in a city. The company's quality control department wanted to check if the customers are equally satisfied with the service provided at these two stores. A sample of 380 customers selected from Supermarket I produced a mean satisfaction index of 7.6 (on a scale of 1 to 10, 1 being the lowest and 10 being the highest) with a standard deviation of .75. Another sample of 370 customers selected from Supermarket II produced a mean satisfaction index of 8.1 with a standard deviation of .59. Assume that the customer satisfaction index for each supermarket has unknown but same population standard deviation.

a. Construct a 98% confidence interval for the difference between the mean satisfaction indexes for all customers for the two supermarkets.

b. Test at a 1% significance level whether the mean satisfaction indexes for all customers for the two supermarkets are different.

10.23 A consumer group sampled 40 refrigerator models with side-by-side freezers and 45 refrigerator models with top freezers. For each refrigerator, the annual energy cost of running (based on the price of a kilowatt-hour at the time of testing) was determined. For the 40 models with side-by-side freezers, the average annual energy cost was \$74.72 with a standard deviation of \$9.24. For the 45 models with top freezers, the average annual energy cost was \$62.50 with a standard deviation of \$7.54. Assume that the standard deviations for the two populations are unknown but equal.

a. Let μ_1 and μ_2 be the average annual energy costs for all models of refrigerators with side-by-side freezers and top freezers, respectively. What is the point estimate of $\mu_1 - \mu_2$?

b. Construct a 98% confidence interval for $\mu_1 - \mu_2$.

c. Using a 1% significance level, can you conclude that the average annual energy cost is higher for all refrigerator models with side-by-side freezers than for all refrigerator models with top freezers? Use both the p-value and the critical value approaches.

10.3 Inferences About the Difference Between Two Population Means for Independent Samples: σ_1 and σ_2 Unknown and Unequal

LEARNING OBJECTIVES

After completing this section, you should be able to:

- Explain the conditions required to use the t-distribution to perform an inference about $\mu_1 - \mu_2$ for two independent random samples selected from two populations with unknown and unequal population standard deviations.
- Verify the conditions for making an inference for $\mu_1 - \mu_2$ when σ_1 and σ_2 are unknown and unequal.
- Construct and interpret a confidence interval for $\mu_1 - \mu_2$ when σ_1 and σ_2 are unknown and unequal.
- Apply the five-step procedure using a critical value approach to perform a test of hypothesis about $\mu_1 - \mu_2$ when σ_1 and σ_2 are unknown and unequal.
- Apply the four-step procedure using a p-value approach to perform a test of hypothesis about $\mu_1 - \mu_2$ when σ_1 and σ_2 are unknown and unequal.

Section 10.2 explained how to make inferences about the difference between two population means using the t distribution when the standard deviations of the two populations are unknown but equal and certain other assumptions hold true. Now, what if all other assumptions of Section 10.2 hold true, but the population standard deviations are not only unknown but also unequal? In this case, the procedures used to make confidence intervals and to test hypotheses about $\mu_1 - \mu_2$ remain similar to the ones we learned in Sections 10.2.1 and 10.2.2, except for two differences. When the population standard deviations are unknown and not equal, the degrees of freedom are no longer given by $n_1 + n_2 - 2$, and the standard deviation of $\bar{x}_1 - \bar{x}_2$ is not calculated using the pooled standard deviation s_p.

Because the standard deviations of the two populations are not known, we use $s_{\bar{x}_1 - \bar{x}_2}$ as a point estimator of $\sigma_{\bar{x}_1 - \bar{x}_2}$. The formulas given next are used to calculate the standard deviation $s_{\bar{x}_1 - \bar{x}_2}$ of $\bar{x}_1 - \bar{x}_2$ and the degrees of freedom.

Estimate of the Standard Deviation of $\bar{x}_1 - \bar{x}_2$

The value of $s_{\bar{x}_1 - \bar{x}_2}$ is calculated as

$$s_{\bar{x}_1 - \bar{x}_2} = \sqrt{\frac{s_1^2}{n_1} + \frac{s_2^2}{n_2}}$$

Degrees of Freedom

If

1. The two samples are independent
2. The standard deviations σ_1 and σ_2 of the two populations are unknown and unequal, that is, $\sigma_1 \neq \sigma_2$
3. At least one of the following two conditions is fulfilled:
   i. Both samples are large (i.e., $n_1 \geq 30$ and $n_2 \geq 30$)
   ii. If either one or both sample sizes are small, then both populations from which the samples are drawn are approximately normally distributed

then the t distribution is used to make inferences about $\mu_1 - \mu_2$, and the *degrees of freedom* for the t distribution are given by

$$df = \frac{\left(\frac{s_1^2}{n_1} + \frac{s_2^2}{n_2}\right)^2}{\frac{\left(\frac{s_1^2}{n_1}\right)^2}{n_1 - 1} + \frac{\left(\frac{s_2^2}{n_2}\right)^2}{n_2 - 1}}$$

The number given by this formula is always rounded down for df.

10.3.1 Interval Estimation of $\mu_1 - \mu_2$

Again, the difference between the two sample means, $\bar{x}_1 - \bar{x}_2$, is the point estimator of the difference between the two population means, $\mu_1 - \mu_2$. The following formula gives the confidence interval for $\mu_1 - \mu_2$ when the t distribution is used and the conditions mentioned earlier in this section are satisfied.

Confidence Interval for $\mu_1 - \mu_2$

The $(1 - \alpha)100\%$ *confidence interval for* $\mu_1 - \mu_2$ is

$$(\bar{x}_1 - \bar{x}_2) \pm t s_{\bar{x}_1 - \bar{x}_2}$$

where the value of t is obtained from the t distribution table for a given confidence level and the degrees of freedom that are given by the formula mentioned earlier, and $s_{\bar{x}_1 - \bar{x}_2}$ is also calculated as explained earlier.

Example 10.8 describes how to construct a confidence interval for $\mu_1 - \mu_2$ when the standard deviations of the two populations are unknown and unequal.

EXAMPLE 10.8 Caffeine in Two Brands of Coffee

Constructing a Confidence Interval for $\mu_1 - \mu_2$: Two Independent Samples, σ_1 and σ_2 Unknown and Unequal

According to Example 10.5 of Section 10.2.1, a sample of 15 one-pound jars of coffee of Brand I showed that the mean amount of caffeine in these jars is 80 milligrams per jar with a standard deviation of 5 milligrams. Another sample of 12 one-pound coffee jars of Brand II gave a mean amount of caffeine equal to 77 milligrams per jar with a standard deviation of 6 milligrams. Construct a 95% confidence interval for the difference between the mean amounts of caffeine in one-pound coffee jars of these two brands. Assume that the two populations are approximately normally distributed and that the standard deviations of the two populations are not equal.

Solution Let μ_1 and μ_2 be the mean amounts of caffeine per jar in all one-pound jars of Brands I and II, respectively, and let $\bar{x}_1$ and $\bar{x}_2$ be the means of the two respective samples.

From the given information,

Brand I coffee:	$n_1 = 15$	$\bar{x}_1 = 80$ milligrams	$s_1 = 5$ milligrams
Brand II coffee:	$n_2 = 12$	$\bar{x}_2 = 77$ milligrams	$s_2 = 6$ milligrams

The confidence level is: $1 - \alpha = .95$

First, we calculate the standard deviation of $\bar{x}_1 - \bar{x}_2$ as follows:

$$s_{\bar{x}_1 - \bar{x}_2} = \sqrt{\frac{s_1^2}{n_1} + \frac{s_2^2}{n_2}} = \sqrt{\frac{(5)^2}{15} + \frac{(6)^2}{12}} = 2.16024690$$

Next, to find the t value from the t distribution table, we need to know the area in each tail of the t distribution curve and the degrees of freedom.

$$\text{Area in each tail} = \alpha/2 = (1 - .95)/2 = .025$$

$$df = \frac{\left(\frac{s_1^2}{n_1} + \frac{s_2^2}{n_2}\right)^2}{\frac{\left(\frac{s_1^2}{n_1}\right)^2}{n_1 - 1} + \frac{\left(\frac{s_2^2}{n_2}\right)^2}{n_2 - 1}} = \frac{\left(\frac{(5)^2}{15} + \frac{(6)^2}{12}\right)^2}{\frac{\left(\frac{(5)^2}{15}\right)^2}{15 - 1} + \frac{\left(\frac{(6)^2}{12}\right)^2}{12 - 1}} = 21.42 \approx 21$$

Note that the degrees of freedom are always rounded down as in this calculation. From the t distribution table, the t value for $df = 21$ and .025 area in the right tail of the t distribution curve is 2.080. The 95% confidence interval for $\mu_1 - \mu_2$ is

$$(\bar{x}_1 - \bar{x}_2) \pm t s_{\bar{x}_1 - \bar{x}_2} = (80 - 77) \pm 2.080(2.16024690)$$
$$= 3 \pm 4.49 = \mathbf{-1.49 \text{ to } 7.49}$$

Thus, with 95% confidence we can state that based on these two sample results, the difference in the mean amounts of caffeine in one-pound jars of these two brands of coffee is between −1.49 and 7.49 milligrams.

Comparing this confidence interval with the one obtained in Example 10.5, we observe that the two confidence intervals are very close. From this we can conclude that even if the standard deviations of the two populations are not equal and we use the procedure of Section 10.2.1 to make a confidence interval for $\mu_1 - \mu_2$, the margin of error ($t s_{\bar{x}_1 - \bar{x}_2}$) will be small as long as the difference between the two population standard deviations is not too large.

10.3.2 Hypothesis Testing About $\mu_1 - \mu_2$

When the standard deviations of the two populations are unknown and unequal along with the other conditions of Section 10.3 holding true, we use the t distribution to make a test of hypothesis about $\mu_1 - \mu_2$. This procedure differs from the one in Section 10.2.2 only in the calculation of degrees of freedom for the t distribution and the standard deviation of $\bar{x}_1 - \bar{x}_2$. The df and the standard deviation of $\bar{x}_1 - \bar{x}_2$ in this case are given by the formulas used in Section 10.3.1.

Test Statistic t for $\bar{x}_1 - \bar{x}_2$

The value of the *test statistic t for* $\bar{x}_1 - \bar{x}_2$ is computed as

$$t = \frac{(\bar{x}_1 - \bar{x}_2) - (\mu_1 - \mu_2)}{s_{\bar{x}_1 - \bar{x}_2}}$$

The value of $\mu_1 - \mu_2$ in this formula is substituted from the null hypothesis, and $s_{\bar{x}_1 - \bar{x}_2}$ is calculated using the formula given earlier in this section.

Example 10.9 illustrates the procedure used to conduct a test of hypothesis about $\mu_1 - \mu_2$ when the standard deviations of the two populations are unknown and unequal.

EXAMPLE 10.9 Calories in Two Brands of Diet Soda

Making a Two-Tailed Test of Hypothesis About $\mu_1 - \mu_2$: Two Independent Samples, and Unknown and Unequal σ_1 and σ_2

According to Example 10.6 of Section 10.2.2, a sample of 14 cans of Brand I diet soda gave the mean number of calories per can of 23 with a standard deviation of 3 calories. Another sample of 16 cans of Brand II diet soda gave the mean number of calories of 25 per can with a standard deviation of 4 calories. Test at a 1% significance level whether the mean number of calories per can of diet soda are different for these two brands. Assume that the calories per can of diet soda are approximately normally distributed for each of these two brands and that the standard deviations for the two populations are not equal.

Solution Let μ_1 and μ_2 be the mean number of calories for all cans of Brand I and Brand II diet soda, respectively, and let $\bar{x}_1$ and $\bar{x}_2$ be the means of the respective samples. From the given information,

Brand I diet soda:	$n_1 = 14$	$\bar{x}_1 = 23$	$s_1 = 3$
Brand II diet soda:	$n_2 = 16$	$\bar{x}_2 = 25$	$s_2 = 4$

The significance level is $\alpha = .01$.

STEP 1 *State the null and alternative hypotheses.*

We are to test for the difference in the mean number of calories for the two brands. The null and alternative hypotheses are, respectively,

$$H_0: \mu_1 - \mu_2 = 0 \quad \text{(The mean number of calories are not different.)}$$
$$H_1: \mu_1 - \mu_2 \neq 0 \quad \text{(The mean number of calories are different.)}$$

STEP 2 *Select the distribution to use.*

Here, the two samples are independent, σ_1 and σ_2 are unknown and unequal, the sample sizes are small, but both populations are approximately normally distributed. Hence, all conditions mentioned in the beginning of Section 10.3 are fulfilled. Consequently, we use the t distribution to make the test.

STEP 3 *Determine the rejection and nonrejection regions.*

The $\neq$ sign in the alternative hypothesis indicates that the test is two-tailed. The significance level is .01. Hence,

$$\text{Area in each tail} = \alpha/2 = .01/2 = .005$$

The degrees of freedom are calculated as follows:

$$df = \frac{\left(\frac{s_1^2}{n_1} + \frac{s_2^2}{n_2}\right)^2}{\frac{\left(\frac{s_1^2}{n_1}\right)^2}{n_1 - 1} + \frac{\left(\frac{s_2^2}{n_2}\right)^2}{n_2 - 1}} = \frac{\left(\frac{(3)^2}{14} + \frac{(4)^2}{16}\right)^2}{\frac{\left(\frac{(3)^2}{14}\right)^2}{14 - 1} + \frac{\left(\frac{(4)^2}{16}\right)^2}{16 - 1}} = 27.41 \approx 27$$

From the t distribution table, the critical values of t for $df = 27$ and .005 area in each tail of the t distribution curve are −2.771 and 2.771. These values are shown in **Figure 10.5**.

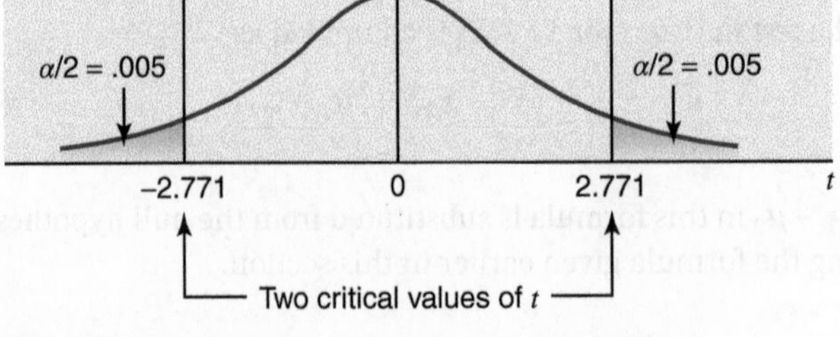

FIGURE 10.5 Rejection and nonrejection regions.

STEP 4 *Calculate the value of the test statistic.*

The value of the test statistic t for $\bar{x}_1 - \bar{x}_2$ is computed as follows:

$$s_{\bar{x}_1 - \bar{x}_2} = \sqrt{\frac{s_1^2}{n_1} + \frac{s_2^2}{n_2}} = \sqrt{\frac{(3)^2}{14} + \frac{(4)^2}{16}} = 1.28173989$$

$$t = \frac{(\bar{x}_1 - \bar{x}_2) - (\mu_1 - \mu_2)}{s_{\bar{x}_1 - \bar{x}_2}} = \frac{(23 - 25) - 0}{1.28173989} = -1.560$$

(The 0 is from H_0.)

STEP 5 *Make a decision.*

Because the value of the test statistic $t = -1.560$ for $\bar{x}_1 - \bar{x}_2$ falls in the nonrejection region, we fail to reject the null hypothesis. Hence, we cannot conclude that the mean number of calories per can are different for the two brands of diet soda. The difference in $\bar{x}_1$ and $\bar{x}_2$ observed for the two samples may have occurred due to sampling error only.

Using the *p*-Value to Make a Decision for Example 10.9

We can use the *p*-value approach to make the same decision. To do so, we keep Steps 1 and 2 of this example. Then in Step 3 we calculate the value of the test statistic t (as done in Step 4) and then find the *p*-value for this t from the t distribution table (Table V of Appendix B) or by using technology. In Step 4, the t-value for $\bar{x}_1 - \bar{x}_2$ was calculated to be -1.560. In this example, the test is two-tailed. Therefore, the *p*-value is equal to twice the area under the t distribution curve to the left of $t = -1.560$. If we have access to technology, we can use it to find the exact *p*-value, which will be .130. If we use the t distribution table, we can only find a range for the *p*-value. From Table V of Appendix B, for $df = 27$, the two values that include 1.560 are 1.314 and 1.703. (Note that we use the positive value of t, although our t is negative.) Thus, the test statistic $t = -1.560$ falls between -1.314 and -1.703. The areas in the t distribution table that correspond to 1.314 and 1.703 are .10 and .05, respectively. Because it is a two-tailed test, the *p*-value for $t = -1.560$ is between $2(.10) = .20$ and $2(.05) = .10$, which can be written as

$$.10 < p\text{-value} < .20$$

Since we will reject the null hypothesis for any α (significance level) that is greater than the *p*-value, we will reject the null hypothesis in this example for any $\alpha \geq .20$ using the above range and not reject for $\alpha < .10$. If we use technology, we will reject the null hypothesis for $\alpha \geq .130$. Since $\alpha = .01$ is smaller than both .10 and .130, we fail to reject the null hypothesis.

The degrees of freedom for the procedures to make a confidence interval and to test a hypothesis about $\mu_1 - \mu_2$ learned in Sections 10.3.1 and 10.3.2 are always rounded down.

Remember

EXERCISES

Concepts and Procedures

10.24 Assuming that the two independent populations are normally distributed with unequal and unknown population standard deviations, construct a 95% confidence interval for $\mu_1 - \mu_2$ for the following:

$$n_1 = 24 \quad \bar{x}_1 = 20.50 \quad s_1 = 3.90$$
$$n_2 = 16 \quad \bar{x}_2 = 22.60 \quad s_2 = 5.15$$

10.25 Assuming that the two independent populations have unequal and unknown population standard deviations, construct a 99% confidence interval for $\mu_1 - \mu_2$ for the following:

$$n_1 = 50 \quad \bar{x}_1 = .863 \quad s_1 = .143$$
$$n_2 = 48 \quad \bar{x}_2 = .796 \quad s_2 = .048$$

10.26 The following information was obtained from two independent random samples selected from two normally distributed populations with unequal and unknown population standard deviations.

$$n_1 = 24 \quad \bar{x}_1 = 20.50 \quad s_1 = 3.90$$
$$n_2 = 16 \quad \bar{x}_2 = 22.60 \quad s_2 = 5.15$$

Test at a 5% significance level if the two population means are different.

10.27 The following information was obtained from two independent random samples selected from two populations with unequal and unknown population standard deviations.

$$n_1 = 50 \quad \bar{x}_1 = .863 \quad s_1 = .143$$
$$n_2 = 48 \quad \bar{x}_2 = .796 \quad s_2 = .048$$

Test at a 1% significance level if the two population means are different.

10.28 Refer to Exercise 10.26. Test at a 1% significance level if μ_1 is less than μ_2.

10.29 Refer to Exercise 10.27. Test at a 2.5% significance level if μ_1 is greater than μ_2.

Applications

10.30 A sample of 45 customers who drive luxury cars showed that their average distance driven between oil changes was 3187 miles with a sample standard deviation of 42.4 miles. Another sample of 40 customers who drive compact lower-price cars resulted in an average distance of 3214 miles with a standard deviation of 50.7 miles. Suppose that the standard deviations for the two populations are not equal.

a. Construct a 95% confidence interval for the difference in the mean distance between oil changes for all luxury cars and all compact lower-price cars.

b. Using a 1% significance level, can you conclude that the mean distance between oil changes is lower for all luxury cars than for all compact lower-price cars?

c. Suppose that the sample standard deviations were 28.9 and 61.4 miles, respectively. Redo parts (a) and (b). Discuss any changes in the results.

10.31 An insurance company wants to know if the average speed at which men drive cars is higher than that of women drivers. The company took a random sample of 30 cars driven by men on a highway and found the mean speed to be 74 miles per hour with a standard deviation of 3.1 miles per hour. Another sample of 18 cars driven by women on the same highway gave a mean speed of 68 miles per hour with a standard deviation of 2.5 miles per hour. Assume that the speeds at which all men and all women drive cars on this highway are both approximately normally distributed with unknown and unequal population standard deviations.

a. Construct a 98% confidence interval for the difference between the mean speeds of cars driven by all men and all women on this highway.

b. Test at a 1% significance level whether the mean speed of cars driven by all men drivers on this highway is higher than that of cars driven by all women drivers.

c. Suppose that the sample standard deviations were 1.9 and 3.4 miles per hour, respectively. Redo parts (a) and (b). Discuss any changes in the results.

10.32 A company claims that its medicine, Brand A, provides faster pain relief than another company's medicine, Brand B. A researcher tested both brands of medicine on two groups of randomly selected patients. The results of these tests are given in the following table. The mean and standard deviation of relief times are in minutes.

Brand	Sample Size	Mean of Relief Times	Standard Deviation of Relief Times
A	30	43	12
B	28	48	7

Assume that the two populations are approximately normally distributed with unknown and unequal standard deviations.

a. Construct a 99% confidence interval for the difference between the mean relief times for the two brands of medicine.

b. Test at a 1% significance level whether the mean relief time for Brand A is less than that for Brand B.

10.33 Quadro Corporation has two supermarket stores in a city. The company's quality control department wanted to check if the customers are equally satisfied with the service provided at these two stores. A sample of 380 customers selected from Supermarket I produced a mean satisfaction index of 7.6 (on a scale of 1 to 10, 1 being the lowest and 10 being the highest) with a standard deviation of .75. Another sample of 370 customers selected from Supermarket II produced a mean satisfaction index of 8.1 with a standard deviation of .59. Assume that the customer satisfaction indexes for the two supermarkets have unknown and unequal population standard deviations.

a. Construct a 98% confidence interval for the difference between the mean satisfaction indexes for all customers for the two supermarkets.

b. Test at a 1% significance level whether the mean satisfaction indexes for all customers for the two supermarkets are different.

c. Suppose that the sample standard deviations were .88 and .39, respectively. Redo parts (a) and (b). Discuss any changes in the results.

10.34 A consumer group sampled 40 refrigerator models with side-by-side freezers and 45 refrigerator models with top freezers. For each refrigerator, the annual energy cost of running (based on the price of a kilowatt-hour at the time of testing) was determined. For the 40 models with side-by-side freezers, the average annual energy cost was $74.72 with a standard deviation of $9.24. For the 45 models with top freezers, the average annual energy cost was $62.50 with a standard deviation of $7.54. Assume that the standard deviations for the two populations are unknown and unequal.

a. Let μ_1 and μ_2 be the average annual energy costs for all models of refrigerators with side-by-side freezers and top freezers, respectively. What is the point estimate of $\mu_1 - \mu_2$?

b. Construct a 98% confidence interval for $\mu_1 - \mu_2$.

c. Using a 1% significance level, can you conclude that the average annual energy cost is higher for all refrigerator models with side-by-side freezers than for all refrigerator models with top freezers? Use both the *p*-value and the critical value approaches.

10.4 Inferences About the Mean of Paired Samples (Dependent Samples)

LEARNING OBJECTIVES

After completing this section, you should be able to:

- Explain when to use the paired samples procedure to make an inference.
- Verify the conditions for making an inference using the paired samples procedure.

- Construct and interpret a confidence interval for μ_d using the paired samples procedure.
- Apply the five-step procedure using the critical value approach to perform a test of hypothesis about μ_d for paired samples.
- Apply the four-step procedure using the p-value approach to perform a test of hypothesis about μ_d for the paired samples.

Sections 10.1, 10.2, and 10.3 were concerned with estimation and hypothesis testing about the difference between two population means when the two samples were drawn independently from two different populations. This section describes estimation and hypothesis testing procedures for the mean of paired samples, which are also called two dependent samples.

In a case of two dependent samples, two data values—one for each sample—are collected from the same source (or element) and, hence, these are also called **paired** or **matched samples**. For example, we may want to make inferences about the mean weight loss for members of a health club after they have gone through an exercise program for a certain period of time. To do so, suppose we select a sample of 15 members of this health club and record their weights before and after the program. In this example, both sets of data are collected from the same 15 persons, once before and once after the program. Thus, although there are two sets of data, they come from the same 15 persons. This is an example of paired (or dependent or matched) samples. The procedures to make confidence intervals and test hypotheses in the case of paired samples are different from those for independent samples discussed in earlier sections of this chapter.

Paired or Matched Samples

Two samples are said to be *paired* or *matched samples* when for each data value collected from one sample there is a corresponding data value collected from the second sample, and both these data values are collected from the same source.

As another example of paired samples, suppose an agronomist wants to measure the effect of a new brand of fertilizer on the yield of potatoes. To do so, he selects 10 pieces of land and divides each piece into two portions. Then he randomly assigns one of the two portions from each piece of land to grow potatoes without using fertilizer (or using some other brand of fertilizer). The second portion from each piece of land is used to grow potatoes with the new brand of fertilizer. Thus, he will have 10 pairs of data values. Then, using the procedure to be discussed in this section, he will make inferences about the mean of the difference in the yields of potatoes with and without the new fertilizer.

The question arises, why does the agronomist not choose 10 pieces of land on which to grow potatoes without using the new brand of fertilizer and another 10 pieces of land to grow potatoes by using the new brand of fertilizer? If he does so, the effect of the fertilizer might be confused with the effects due to soil differences at different locations. Thus, he will not be able to isolate the effect of the new brand of fertilizer on the yield of potatoes. Consequently, the results will not be reliable. By choosing 10 pieces of land and then dividing each of them into two portions, the researcher decreases the possibility that the difference in the productivities of different pieces of land affects the results.

In paired samples, the difference between the two data values for each pair is denoted by ***d***. This value of d is called the **paired difference**. We then treat all the values of d as one sample and make inferences applying procedures similar to the ones used for one-sample cases in Chapters 8 and 9. Note that because each source (or element) gives a pair of values (one for each of the two data sets), each sample contains the same number of values. That is, both samples are of the same size. Therefore, we denote the (common) **sample size** by $\boldsymbol{n}$, which gives the number of paired difference values denoted by d. The **degrees of freedom** for the paired samples are $\boldsymbol{n - 1}$. Let

μ_d = the mean of the paired differences for the population

σ_d = the standard deviation of the paired differences for the population, which is usually not known

$\bar{d}$ = the mean of the paired differences for the sample

s_d = the standard deviation of the paired differences for the sample

n = the number of paired difference values

Mean and Standard Deviation of the Paired Differences for Two Samples

The values of the mean and standard deviation, $\bar{d}$ and s_d, respectively, of paired differences for two samples are calculated as[2]

$$\bar{d} = \frac{\Sigma d}{n}$$

$$s_d = \sqrt{\frac{\Sigma d^2 - \frac{(\Sigma d)^2}{n}}{n-1}}$$

In paired samples, instead of using $\bar{x}_1 - \bar{x}_2$ as the sample statistic to make inferences about $\mu_1 - \mu_2$, we use the sample statistic $\bar{d}$ to make inferences about μ_d. Actually, the value of $\bar{d}$ is always equal to $\bar{x}_1 - \bar{x}_2$, and the value of μ_d is always equal to $\mu_1 - \mu_2$.

Sampling Distribution, Mean, and Standard Deviation of $\bar{d}$

If σ_d is known and either the sample size is large ($n \geq 30$) or the population of paired differences is normally distributed, then the *sampling distribution of* $\bar{d}$ is approximately normal with its *mean* and *standard deviation* given as, respectively,

$$\mu_{\bar{d}} = \mu_d \quad \text{and} \quad \sigma_{\bar{d}} = \frac{\sigma_d}{\sqrt{n}}$$

Thus, if the standard deviation σ_d of the population of paired differences is known and either the sample size is large (i.e., $n \geq 30$) or the population of paired differences is approximately normally distributed (with $n < 30$), then the normal distribution can be used to make a confidence interval and to test a hypothesis about μ_d. However, usually σ_d is not known. Then, if the standard deviation σ_d of the population of paired differences is unknown and either the sample size is large (i.e., $n \geq 30$) or the population of paired differences is approximately normally distributed (with $n < 30$), then the t distribution is used to make a confidence interval and to test a hypothesis about μ_d.

Making Inferences About μ_d

If

1. The standard deviation σ_d of the population of paired differences is unknown
2. At least one of the following two conditions is fulfilled:
   - **i.** The sample size is large (i.e., $n \geq 30$)
   - **ii.** The sample size is small but the population of paired differences is approximately normally distributed

then the t distribution is used to make inferences about μ_d. The standard deviation $\sigma_{\bar{d}}$ of $\bar{d}$ is estimated by $s_{\bar{d}}$, which is calculated as

$$s_{\bar{d}} = \frac{s_d}{\sqrt{n}}$$

---

[2]The basic formula used to calculate s_d is

$$s_d = \sqrt{\frac{\Sigma(d - \bar{d})^2}{n-1}}$$

However, we will not use this formula to make calculations in this chapter.
Also note that the formula for s_d is the same that we used in Chapter 3 with x replaced by d.

Sections 10.4.1 and 10.4.2 describe the procedures that are used to make a confidence interval and to test a hypothesis about μ_d under the above conditions. The inferences are made using the t distribution.

10.4.1 Interval Estimation of μ_d

The mean $\bar{d}$ of paired differences for paired samples is the point estimator of μ_d. The following formula is used to construct a confidence interval for μ_d when the t distribution is used.

Confidence Interval for μ_d

The $(1 - \alpha)100\%$ *confidence interval for* μ_d is

$$\bar{d} \pm t s_{\bar{d}}$$

where the value of t is obtained from the t distribution table for the given confidence level and $n - 1$ degrees of freedom, and $s_{\bar{d}}$ is calculated using the formula given earlier in this section.

Example 10.10 illustrates the procedure to construct a confidence interval for μ_d.

EXAMPLE 10.10 Special Diet and Systolic Blood Pressure

Constructing a Confidence Interval for μ_d: Paired Samples, σ_d Unknown, $n < 30$, and Population Normal

A researcher wanted to find the effect of a special diet on systolic blood pressure. She selected a sample of seven adults and put them on this dietary plan for three months. The following table gives the systolic blood pressures (in mm Hg) of these seven adults before and after the completion of this plan.

Subject	1	2	3	4	5	6	7
Before	210	180	195	220	231	199	224
After	193	186	186	223	220	183	233

Let μ_d be the mean reduction in the systolic blood pressures due to this special dietary plan for the population of all adults. Construct a 95% confidence interval for μ_d. Assume that the population of paired differences is approximately normally distributed.

Solution Because the information obtained is from paired samples, we will make the confidence interval for the paired difference mean μ_d of the population using the paired difference mean $\bar{d}$ of the sample. Let d be the difference in the systolic blood pressure of an adult before and after this special dietary plan. Then, d is obtained by subtracting the systolic blood pressure after the plan from the systolic blood pressure before the plan. The third column of **Table 10.1** lists the values of d for the seven adults. The fourth column of the table records the values of d^2, which are obtained by squaring each of the d values.

TABLE 10.1

Before	After	Difference d	d^2
210	193	17	289
180	186	−6	36
195	186	9	81
220	223	−3	9
231	220	11	121
199	183	16	256
224	233	−9	81
		$\Sigma d = 35$	$\Sigma d^2 = 873$

The values of $\bar{d}$ and s_d are calculated as follows:

$$\bar{d} = \frac{\Sigma d}{n} = \frac{35}{7} = 5.00$$

$$s_d = \sqrt{\frac{\Sigma d^2 - \frac{(\Sigma d)^2}{n}}{n-1}} = \sqrt{\frac{873 - \frac{(35)^2}{7}}{7-1}} = 10.78579312$$

Hence, the standard deviation of $\bar{d}$ is

$$s_{\bar{d}} = \frac{s_d}{\sqrt{n}} = \frac{10.78579312}{\sqrt{7}} = 4.07664661$$

Here, σ_d is not known, the sample size is small, but the population is approximately normally distributed. Hence, we will use the t distribution to make the confidence interval. For the 95% confidence interval, the area in each tail of the t distribution curve is

$$\text{Area in each tail} = \alpha/2 = (1 - .95)/2 = .025$$

The degrees of freedom are

$$df = n - 1 = 7 - 1 = 6$$

From the t distribution table, the t value for $df = 6$ and .025 area in the right tail of the t distribution curve is 2.447. Therefore, the 95% confidence interval for μ_d is

$$\bar{d} \pm t s_{\bar{d}} = 5.00 \pm 2.447(4.07664661) = 5.00 \pm 9.98 = \mathbf{-4.98 \text{ to } 14.98}$$

Thus, we can state with 95% confidence that the mean difference between systolic blood pressures before and after the given dietary plan for all adult participants is between −4.98 and 14.98 mm Hg.

10.4.2 Hypothesis Testing About μ_d

A hypothesis about μ_d is tested by using the sample statistic $\bar{d}$. This section illustrates the case of the t distribution only. Earlier in this section we learned what conditions should hold true to use the t distribution to test a hypothesis about μ_d. The following formula is used to calculate the value of the test statistic t when testing a hypothesis about μ_d.

Test Statistic t for $\bar{d}$

The value of the *test statistic t for $\bar{d}$* is computed as follows:

$$t = \frac{\bar{d} - \mu_d}{s_{\bar{d}}}$$

The critical value of t is obtained from the t distribution table for the given significance level and $n - 1$ degrees of freedom.

Examples 10.11 and 10.12 illustrate the hypothesis-testing procedure for μ_d.

EXAMPLE 10.11 How to Be a Successful Salesperson

Conducting a Left-Tailed Test of Hypothesis About μ_d for Paired Samples: σ_d Not Known, Small Sample but Normally Distributed Population

A company wanted to know if attending a course on "how to be a successful salesperson" can increase the average sales of its employees. The company sent six of its salespersons to attend this course. The following table gives the one-week sales of these salespersons before and after they attended this course.

Subject	1	2	3	4	5	6
Before	12	18	25	9	14	16
After	18	24	24	14	19	20

Using a 1% significance level, can you conclude that the mean weekly sales for all salespersons increase as a result of attending this course? Assume that the population of paired differences has an approximate normal distribution.

Solution Because the data are for paired samples, we test a hypothesis about the paired differences mean μ_d of the population using the paired differences mean $\overline{d}$ of the sample.

Let

$$d = (\text{Weekly sales before the course}) - (\text{Weekly sales after the course})$$

In **Table 10.2**, we calculate d for each of the six salespersons by subtracting the sales after the course from the sales before the course. The fourth column of the table lists the values of d^2.

TABLE 10.2

Before	After	Difference d	d^2
12	18	−6	36
18	24	−6	36
25	24	1	1
9	14	−5	25
14	19	−5	25
16	20	−4	16
		$\Sigma d = -25$	$\Sigma d^2 = 139$

The values of $\overline{d}$ and s_d are calculated as follows:

$$\overline{d} = \frac{\Sigma d}{n} = \frac{-25}{6} = -4.17$$

$$s_d = \sqrt{\frac{\Sigma d^2 - \frac{(\Sigma d)^2}{n}}{n-1}} = \sqrt{\frac{139 - \frac{(-25)^2}{6}}{6-1}} = 2.63944439$$

The standard deviation of $\overline{d}$ is

$$s_{\overline{d}} = \frac{s_d}{\sqrt{n}} = \frac{2.63944439}{\sqrt{6}} = 1.07754866$$

STEP 1 *State the null and alternative hypotheses.*

We are to test if the mean weekly sales for all salespersons increase as a result of taking the course. Let μ_1 be the mean weekly sales for all salespersons before the course and μ_2 the mean weekly sales for all salespersons after the course. Then $\mu_d = \mu_1 - \mu_2$. The mean weekly sales for all salespersons will increase due to attending the course if μ_1 is less than μ_2, which can be written as $\mu_1 - \mu_2 < 0$ or $\mu_d < 0$. Consequently, the null and alternative hypotheses are, respectively,

H_0: $\mu_d = 0$ ($\mu_1 - \mu_2 = 0$ or the mean weekly sales do not increase.)
H_1: $\mu_d < 0$ ($\mu_1 - \mu_2 < 0$ or the mean weekly sales do increase.)

Note that we can also write the null hypothesis as $\mu_d \geq 0$.

STEP 2 *Select the distribution to use.*

Here σ_d is unknown, the sample size is small ($n < 30$), but the population of paired differences is approximately normally distributed. Therefore, we use the t distribution to conduct the test.

STEP 3 *Determine the rejection and nonrejection regions.*

The < sign in the alternative hypothesis indicates that the test is left-tailed. The significance level is .01. Hence,

$$\text{Area in left tail} = \alpha = .01$$
$$\text{Degrees of freedom} = n - 1 = 6 - 1 = 5$$

The critical value of t for $df = 5$ and .01 area in the left tail of the t distribution curve is −3.365. This value is shown in **Figure 10.6**.

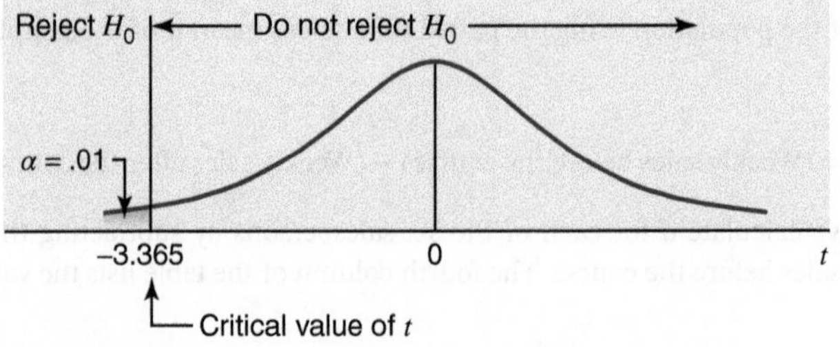

FIGURE 10.6 Rejection and nonrejection regions.

STEP 4 *Calculate the value of the test statistic.*

The value of the test statistic t for $\bar{d}$ is computed as follows:

$$t = \frac{\bar{d} - \mu_d}{s_{\bar{d}}} = \frac{-4.17 - 0}{1.07754866} = -3.870 \quad (\text{0 from } H_0)$$

STEP 5 *Make a decision.*

Because the value of the test statistic $t = -3.870$ falls in the rejection region, we reject the null hypothesis. Consequently, we conclude that the mean weekly sales for all salespersons seems to increase as a result of this course.

Using the *p*-Value to Make the Decision for Example 10.11

We can use the *p*-value approach to make the same decision. To do so, we keep Steps 1 and 2 of this example. Then in Step 3, we calculate the value of the test statistic t for $\bar{d}$ (as done in Step 4) and then find the *p*-value for this test statistic from the t distribution table (Table V of Appendix B) or by using technology. If we have access to technology, we can find the exact *p*-value, which will be .006. By using Table V (of Appendix B), we can find a range for *p*-value. From Table V, for $df = 5$, the test statistic $t = -3.870$ falls between −3.365 and −4.032. The areas in the t distribution table that correspond to −3.365 and −4.032 are .01 and .005, respectively. Because it is a left-tailed test, the *p*-value is between .01 and .005, which can be written as

$$.005 < p\text{-value} < .01$$

Since we will reject the null hypothesis for any α (significance level) that is greater than or equal to the *p*-value, we will reject the null hypothesis in this example for any $\alpha \geq .006$ using the technology and $\alpha \geq .01$ using the above range. Since $\alpha = .01$ in this example, which is larger than .006 obtained from technology, we reject the null hypothesis. Also, because α is equal to .01, using the *p*-value range we reject the null hypothesis.

EXAMPLE 10.12 Special Diet and Systolic Blood Pressure

Making a Two-Tailed Test of Hypothesis About μ_d for Paired Samples: σ_d Not Known, Small Sample but Normally Distributed Population

Refer to Example 10.10. The table that gives the blood pressures (in mm Hg) of seven adults before and after the completion of a special dietary plan is reproduced here.

Subject	1	2	3	4	5	6	7
Before	210	180	195	220	231	199	224
After	193	186	186	223	220	183	233

Let μ_d be the mean of the differences between the systolic blood pressures before and after completing this special dietary plan for the population of all adults. Using a 5% significance level, can you conclude that the mean of the paired differences μ_d is different from zero? Assume that the population of paired differences is approximately normally distributed.

Solution **Table 10.3** gives d and d^2 for each of the seven adults (blood pressure values in mm Hg).

TABLE 10.3

Before	After	Difference d	d^2
210	193	17	289
180	186	−6	36
195	186	9	81
220	223	−3	9
231	220	11	121
199	183	16	256
224	233	−9	81
		$\Sigma d = 35$	$\Sigma d^2 = 873$

The values of $\overline{d}$ and s_d are calculated as follows:

$$\overline{d} = \frac{\Sigma d}{n} = \frac{35}{7} = 5.00$$

$$s_d = \sqrt{\frac{\Sigma d^2 - \frac{(\Sigma d)^2}{n}}{n-1}} = \sqrt{\frac{873 - \frac{(35)^2}{7}}{7-1}} = 10.78579312$$

Hence, the standard deviation of $\overline{d}$ is

$$s_{\overline{d}} = \frac{s_d}{\sqrt{n}} = \frac{10.78579312}{\sqrt{7}} = 4.07664661$$

STEP 1 *State the null and alternative hypotheses.*

H_0: $\mu_d = 0$ (The mean of the paired differences is not different from zero.)
H_1: $\mu_d \neq 0$ (The mean of the paired differences is different from zero.)

STEP 2 *Select the distribution to use.*

Here σ_d is unknown, the sample size is small, but the population of paired differences is approximately normal. Hence, we use the t distribution to make the test.

STEP 3 *Determine the rejection and nonrejection regions.*

The $\neq$ sign in the alternative hypothesis indicates that the test is two-tailed. The significance level is .05.

$$\text{Area in each tail of the curve} = \alpha/2 = .05/2 = .025$$
$$\text{Degrees of freedom} = n - 1 = 7 - 1 = 6$$

The two critical values of t for $df = 6$ and .025 area in each tail of the t distribution curve are −2.447 and 2.447. These values are shown in **Figure 10.7**.

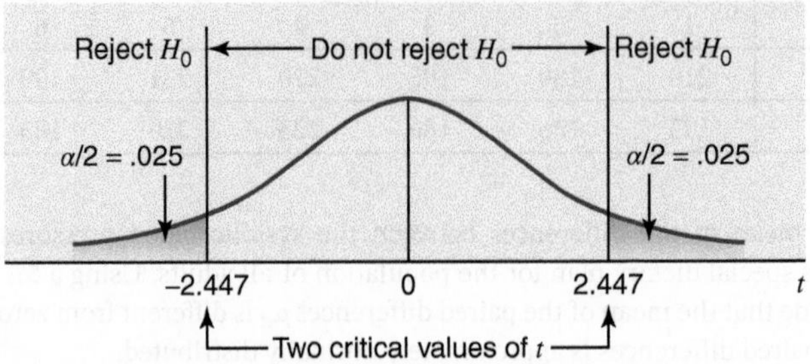

FIGURE 10.7 Rejection and nonrejection regions.

STEP 4 *Calculate the value of the test statistic.*

The value of the test statistic t for $\bar{d}$ is computed as follows:

$$t = \frac{\bar{d} - \mu_d}{s_{\bar{d}}} = \frac{5.00 - 0}{4.07664661} = 1.226 \quad (\leftarrow 0 \text{ from } H_0)$$

STEP 5 *Make a decision.*

Because the value of the test statistic $t = 1.226$ for $\bar{d}$ falls in the nonrejection region, we fail to reject the null hypothesis. Hence, we conclude that the mean of the population of paired differences does not seem to be different from zero. In other words, we can state that the mean of the differences between the systolic blood pressures before and after completing this special dietary plan for the population of all adults is not different from zero.

Using the p-Value to Make the Decision for Example 10.12

We can use the p-value approach to make the same decision. To do so, we keep Steps 1 and 2 of this example. Then in Step 3, we calculate the value of the test statistic t for $\bar{d}$ (as done in Step 4) and then find the p-value for this t from the t distribution table (Table V of Appendix B) or by using technology. If we have access to technology, we can use it to find the exact p-value, which will be .266. By using Table V, we can find a range for the p-value. From Table V, for $df = 6$, the test statistic $t = 1.226$ is less than 1.440. The area in the t distribution table that corresponds to 1.440 is .10. Because it is a two-tailed test, the p-value is greater than 2(.10) = .20, which can be written as

$$p\text{-value} > .20$$

Since $\alpha = .05$ in this example, which is smaller than .20 and also smaller than .266 (obtained from technology), we fail to reject the null hypothesis.

EXERCISES

Concepts and Procedures

10.35 Explain when would you use the paired-samples procedure to make confidence intervals and test hypotheses.

10.36 Find the following confidence intervals for μ_d, assuming that the populations of paired differences are normally distributed.

a. $n = 12$, $\bar{d} = 17.5$, $s_d = 6.3$, confidence level = 99%

b. $n = 27$, $\bar{d} = 55.9$, $s_d = 14.7$, confidence level = 95%

c. $n = 16$, $\bar{d} = 29.3$, $s_d = 8.3$, confidence level = 90%

10.37 Perform the following tests of hypotheses, assuming that the populations of paired differences are normally distributed.

a. $H_0: \mu_d = 0$, $H_1: \mu_d \neq 0$, $n = 26$, $\bar{d} = 9.6$, $s_d = 3.9$, $\alpha = .05$

b. $H_0: \mu_d = 0$, $H_1: \mu_d > 0$, $n = 15$, $\bar{d} = 8.8$, $s_d = 4.7$, $\alpha = .01$

c. $H_0: \mu_d = 0$, $H_1: \mu_d < 0$, $n = 20$, $\bar{d} = -7.4$, $s_d = 2.3$, $\alpha = .10$

Applications

10.38 A company sent seven of its employees to attend a course in building self-confidence. These employees were evaluated for their self-confidence before and after attending this course. The following table gives the scores (on a scale of 1 to 15, 1 being the lowest and 15 being the highest score) of these employees before and after they attended the course.

Before	8	5	4	9	6	9	5
After	10	8	5	11	6	7	9

a. Construct a 95% confidence interval for the mean μ_d of the population of paired differences, where a paired difference is equal to the score of an employee before attending the course minus the score of the same employee after attending the course.

b. Test at a 1% significance level whether attending this course increases the mean score of employees.

Assume that the population of paired differences has an approximate normal distribution.

10.39 A finger-tapping experiment was conducted by a doctoral neuropsychology student. The purpose of this experiment was to determine bilateral nervous system integrity, which provides data about the neuromuscular system and motor control. The experiment requires the subjects to place the palm of the hand on a table and then to tap the index finger on the surface of the table. Ten subjects who had suffered a mild concussion were given a finger-tapping test using the right index finger and then the left index finger. The total number of taps in 30 seconds by each subject are listed in the following table. All subject in the data set are right-handed. Assume that the population of paired differences is approximately normally distributed.

Subject	1	2	3	4	5	6	7	8	9	10
Right	125	131	95	105	137	129	91	112	138	94
Left	127	138	97	107	130	128	85	114	135	85

a. Make a 95% confidence interval for the mean of the population of paired differences, where a paired difference is equal to the number of taps for the right index finger minus the number of taps for the left index finger.

b. Using a 5% significance level, can you conclude that the average number of taps is different for the right and left index fingers?

10.40 The manufacturer of a gasoline additive claims that the use of this additive increases gasoline mileage. A random sample of six cars was selected, and these cars were driven for one week without the gasoline additive and then for one week with the gasoline additive. The following table gives the miles per gallon for these cars without and with the gasoline additive.

Without	23.8	28.3	18.5	23.7	13.6	28.8
With	26.3	30.5	18.2	25.3	16.3	30.9

a. Construct a 99% confidence interval for the mean μ_d of the population of paired differences, where a paired difference is equal to the miles per gallon without the gasoline additive minus the miles per gallon with the gasoline additive.

b. Using a 2.5% significance level, can you conclude that the use of the gasoline additive increases the gasoline mileage?

Assume that the population of paired differences is approximately normally distributed.

10.41 A factory that emits airborne pollutants is testing two different brands of filters for its smokestacks. The factory has two smokestacks. One brand of filter (Filter I) is placed on one smokestack, and the other brand (Filter II) is placed on the second smokestack. Random samples of air released from the smokestacks are taken at different times throughout the day. Pollutant concentrations are measured from both stacks at the same time. The following data represent the pollutant concentrations (in parts per million) for samples taken at 20 different times after passing through the filters. Assume that the differences in concentration levels at all times are approximately normally distributed.

Time	Filter I	Filter II	Time	Filter I	Filter II
1	24	26	11	11	9
2	31	30	12	8	10
3	35	33	13	14	17
4	32	28	14	17	16
5	25	23	15	19	16
6	25	28	16	19	18
7	29	24	17	25	27
8	30	33	18	20	22
9	26	22	19	23	27
10	18	18	20	32	31

a. Make a 95% confidence interval for the mean of the population of paired differences, where a paired difference is equal to the pollutant concentration passing through Filter I minus the pollutant concentration passing through Filter II.

b. Using a 5% significance level, can you conclude that the average paired difference for concentration levels is different from zero?

10.5 Inferences About the Difference Between Two Population Proportions for Large and Independent Samples

LEARNING OBJECTIVES

After completing this section, you should be able to:

- Describe the sampling distribution of the difference between proportions from two large and independent samples.
- Determine a point estimate of $p_1 - p_2$ using data from two large and independent samples.
- Verify the conditions for making an inference about $p_1 - p_2$.
- Construct and interpret a confidence interval for $p_1 - p_2$.
- Apply the five-step procedure using the critical value approach to perform a test of hypothesis about $p_1 - p_2$.
- Apply the four-step procedure using the p-value approach to perform a test of hypothesis about $p_1 - p_2$.

Quite often we need to construct a confidence interval and test a hypothesis about the difference between two population proportions. For instance, we may want to estimate the difference between the proportions of defective items produced on two different machines. If p_1 and p_2 are the proportions of defective items produced on the first and second machine, respectively, then we are to make a confidence interval for $p_1 - p_2$. Alternatively, we may want to test the hypothesis that the proportion of defective items produced on Machine I is different from the proportion of defective items produced on Machine II. In this case, we are to test the null hypothesis $p_1 - p_2 = 0$ against the alternative hypothesis $p_1 - p_2 \neq 0$.

This section discusses how to make a confidence interval and test a hypothesis about $p_1 - p_2$ for two large and independent samples. The sample statistic that is used to make inferences about $p_1 - p_2$ is $\hat{p}_1 - \hat{p}_2$, where $\hat{p}_1$ and $\hat{p}_2$ are the proportions for two large and independent samples. As discussed in Chapter 7, we determine a sample proportion by dividing the number of elements in the sample that possess a given attribute by the sample size. Thus,

$$\hat{p}_1 = \frac{x_1}{n_1} \quad \text{and} \quad \hat{p}_2 = \frac{x_2}{n_2}$$

where x_1 and x_2 are the number of elements that possess a given characteristic in the two samples and n_1 and n_2 are the sizes of the two samples, respectively.

10.5.1 Mean, Standard Deviation, and Sampling Distribution of $\hat{p}_1 - \hat{p}_2$

As discussed in Chapter 7, for a large sample, the sample proportion $\hat{p}$ is approximately normally distributed with mean p and standard deviation $\sqrt{pq/n}$. Hence, for two large and independent samples of sizes n_1 and n_2, respectively, their sample proportions $\hat{p}_1$ and $\hat{p}_2$ are approximately normally distributed with means p_1 and p_2 and standard deviations $\sqrt{p_1q_1/n_1}$ and $\sqrt{p_2q_2/n_2}$, respectively. Using these results, we can make the following statements about the shape of the sampling distribution of $\hat{p}_1 - \hat{p}_2$ and its mean and standard deviation.

Mean, Standard Deviation, and Sampling Distribution of $\hat{p}_1 - \hat{p}_2$

For two large and independent samples, the *sampling distribution* of $\hat{p}_1 - \hat{p}_2$ is approximately normal, with its *mean* and *standard deviation* given as

$$\mu_{\hat{p}_1 - \hat{p}_2} = p_1 - p_2$$

and

$$\sigma_{\hat{p}_1 - \hat{p}_2} = \sqrt{\frac{p_1q_1}{n_1} + \frac{p_2q_2}{n_2}}$$

respectively, where $q_1 = 1 - p_1$ and $q_2 = 1 - p_2$.

Thus, to construct a confidence interval and to test a hypothesis about $p_1 - p_2$ for large and independent samples, we use the normal distribution. As was indicated in Chapter 7, in the case of proportion, the sample is large if np and nq are both greater than 5. In the case of two samples, both sample sizes are large if n_1p_1, n_1q_1, n_2p_2, and n_2q_2 are all greater than 5.

10.5.2 Interval Estimation of $p_1 - p_2$

The difference between two sample proportions $\hat{p}_1 - \hat{p}_2$ is the point estimator for the difference between two population proportions $p_1 - p_2$. Because we do not know p_1 and p_2 when we are making a confidence interval for $p_1 - p_2$, we cannot calculate the value of $\sigma_{\hat{p}_1 - \hat{p}_2}$. Therefore, we use $s_{\hat{p}_1 - \hat{p}_2}$ as the point estimator of $\sigma_{\hat{p}_1 - \hat{p}_2}$ in the interval estimation. We construct the confidence interval for $p_1 - p_2$ using the following formula:

Confidence Interval for $p_1 - p_2$

The $(1 - \alpha)100\%$ *confidence interval for* $p_1 - p_2$ is

$$(\hat{p}_1 - \hat{p}_2) \pm z s_{\hat{p}_1 - \hat{p}_2}$$

where the value of z is obtained from the normal distribution table for the given confidence level, and $s_{\hat{p}_1 - \hat{p}_2}$ is calculated as

$$s_{\hat{p}_1 - \hat{p}_2} = \sqrt{\frac{\hat{p}_1 \hat{q}_1}{n_1} + \frac{\hat{p}_2 \hat{q}_2}{n_2}}$$

Example 10.13 describes the procedure that is used to make a confidence interval for the difference between two population proportions for large samples.

EXAMPLE 10.13 Users of Two Toothpastes

Constructing a Confidence Interval for $p_1 - p_2$: Large and Independent Samples

A researcher wanted to estimate the difference between the percentages of users of two toothpastes who will never switch to another toothpaste. In a sample of 500 users of Toothpaste A taken by this researcher, 100 said that they will never switch to another toothpaste. In another sample of 400 users of Toothpaste B taken by the same researcher, 68 said that they will never switch to another toothpaste.

a. Let p_1 and p_2 be the proportions of all users of Toothpastes A and B, respectively, who will never switch to another toothpaste. What is the point estimate of $p_1 - p_2$?

b. Construct a 97% confidence interval for the difference between the proportions of all users of the two toothpastes who will never switch.

Solution Let p_1 and p_2 be the proportions of all users of Toothpastes A and B, respectively, who will never switch to another toothpaste, and let $\hat{p}_1$ and $\hat{p}_2$ be the respective sample proportions. Let x_1 and x_2 be the number of users of Toothpastes A and B, respectively, in the two samples who said that they will never switch to another toothpaste. From the given information,

Andrey Armyagov/iStockphoto

Toothpaste A: $n_1 = 500$ and $x_1 = 100$

Toothpaste B: $n_2 = 400$ and $x_2 = 68$

The two sample proportions are calculated as follows:

$$\hat{p}_1 = x_1/n_1 = 100/500 = .20$$

$$\hat{p}_2 = x_2/n_2 = 68/400 = .17$$

Then,

$$\hat{q}_1 = 1 - .20 = .80 \quad \text{and} \quad \hat{q}_2 = 1 - .17 = .83$$

a. The point estimate of $p_1 - p_2$ is as follows:

$$\text{Point estimate of } p_1 - p_2 = \hat{p}_1 - \hat{p}_2 = .20 - .17 = \mathbf{.03}$$

b. The values of $n_1\hat{p}_1$, $n_1\hat{q}_1$, $n_2\hat{p}_2$, and $n_2\hat{q}_2$ are

$$n_1\hat{p}_1 = 500(.20) = 100 \qquad n_1\hat{q}_1 = 500(.80) = 400$$

$$n_2\hat{p}_2 = 400(.17) = 68 \qquad n_2\hat{q}_2 = 400(.83) = 332$$

Because each of these values is greater than 5, both sample sizes are large. Consequently we will use the normal distribution to make a confidence interval for $p_1 - p_2$. The standard deviation of $\hat{p}_1 - \hat{p}_2$ is

$$s_{\hat{p}_1 - \hat{p}_2} = \sqrt{\frac{\hat{p}_1\hat{q}_1}{n_1} + \frac{\hat{p}_2\hat{q}_2}{n_2}} = \sqrt{\frac{(.20)(.80)}{500} + \frac{(.17)(.83)}{400}} = .02593742$$

The z value for a 97% confidence level, obtained from the normal distribution table, is 2.17. The 97% confidence interval for $p_1 - p_2$ is

$$(\hat{p}_1 - \hat{p}_2) \pm z s_{\hat{p}_1 - \hat{p}_2} = (.20 - .17) \pm 2.17(.02593742)$$
$$= .03 \pm .056 = \mathbf{-.026 \text{ to } .086}$$

Thus, with 97% confidence we can state that the difference between the two population proportions is between $-.026$ and .086 or -2.6% and 8.6%.

Note that here $\hat{p}_1 - \hat{p}_2 = .03$ gives the point estimate of $p_1 - p_2$ and $z\, s_{\hat{p}_1 - \hat{p}_2} = .056$ is the margin of error of the estimate.

10.5.3 Hypothesis Testing About $p_1 - p_2$

In this section we learn how to test a hypothesis about $p_1 - p_2$ for two large and independent samples. The procedure involves the same five steps we have used previously. Once again, we calculate the standard deviation of $\hat{p}_1 - \hat{p}_2$ as

$$\sigma_{\hat{p}_1 - \hat{p}_2} = \sqrt{\frac{p_1 q_1}{n_1} + \frac{p_2 q_2}{n_2}}$$

When a test of hypothesis about $p_1 - p_2$ is performed, usually the null hypothesis is $p_1 = p_2$ and the values of p_1 and p_2 are not known. Assuming that the null hypothesis is true and $p_1 = p_2$, a common value of p_1 and p_2, denoted by $\bar{p}$, is calculated by using one of the following two formulas:

$$\bar{p} = \frac{x_1 + x_2}{n_1 + n_2} \text{ or } \frac{n_1\hat{p}_1 + n_2\hat{p}_2}{n_1 + n_2}$$

Which of these formulas is used depends on whether the values of x_1 and x_2 or the values of $\hat{p}_1$ and $\hat{p}_2$ are known. Note that x_1 and x_2 are the number of elements in each of the two samples that possess a certain characteristic. This value of $\bar{p}$ is called the **pooled sample proportion**. Using the value of the pooled sample proportion, we compute an estimate of the standard deviation of $\hat{p}_1 - \hat{p}_2$ as follows:

$$s_{\hat{p}_1 - \hat{p}_2} = \sqrt{\bar{p}\bar{q}\left(\frac{1}{n_1} + \frac{1}{n_2}\right)}$$

where $\bar{q} = 1 - \bar{p}$.

Test Statistic z for $\hat{p}_1 - \hat{p}_2$

The value of the *test statistic z for $\hat{p}_1 - \hat{p}_2$* is calculated as

$$z = \frac{(\hat{p}_1 - \hat{p}_2) - (p_1 - p_2)}{s_{\hat{p}_1 - \hat{p}_2}}$$

The value of $p_1 - p_2$ is substituted from H_0, which usually is zero.

Examples 10.14 and 10.15 illustrate the procedure to test hypotheses about the difference between two population proportions for large samples.

EXAMPLE 10.14 Users of Two Toothpastes

Making a Right-Tailed Test of Hypothesis About $p_1 - p_2$: Large and Independent Samples

Reconsider Example 10.13 about the percentages of users of two toothpastes who will never switch to another toothpaste. At a 1% significance level, can you conclude that the proportion of users of Toothpaste A who will never switch to another toothpaste is higher than the proportion of users of Toothpaste B who will never switch to another toothpaste?

Solution Let p_1 and p_2 be the proportions of all users of Toothpastes A and B, respectively, who will never switch to another toothpaste, and let $\hat{p}_1$ and $\hat{p}_2$ be the corresponding sample proportions. Let x_1 and x_2 be the number of users of Toothpastes A and B, respectively, in the two samples who said that they will never switch to another toothpaste. From the given information,

Toothpaste A:	$n_1 = 500$	and	$x_1 = 100$
Toothpaste B:	$n_2 = 400$	and	$x_2 = 68$

The significance level is $\alpha = .01$. The two sample proportions are calculated as follows:

$$\hat{p}_1 = x_1/n_1 = 100/500 = .20$$
$$\hat{p}_2 = x_2/n_2 = 68/400 = .17$$

STEP 1 *State the null and alternative hypotheses.*

We are to test if the proportion of users of Toothpaste A who will never switch to another toothpaste is higher than the proportion of users of Toothpaste B who will never switch to another toothpaste. In other words, we are to test whether p_1 is greater than p_2. This can be written as $p_1 - p_2 > 0$. Thus, the two hypotheses are

$$H_0: p_1 = p_2 \quad \text{or} \quad p_1 - p_2 = 0 \quad (p_1 \text{ is the same as } p_2.)$$
$$H_1: p_1 > p_2 \quad \text{or} \quad p_1 - p_2 > 0 \quad (p_1 \text{ is greater than } p_2.)$$

STEP 2 *Select the distribution to use.*

As shown in Example 10.13, $n_1\hat{p}_1$, $n_1\hat{q}_1$, $n_2\hat{p}_2$, and $n_2\hat{q}_2$ are all greater than 5. Consequently both samples are large, and we use the normal distribution to make the test.

STEP 3 *Determine the rejection and nonrejection regions.*

The > sign in the alternative hypothesis indicates that the test is right-tailed. From the normal distribution table, for a .01 significance level, the critical value of z is 2.33 for .9900 area to the left. This is shown in **Figure 10.8**.

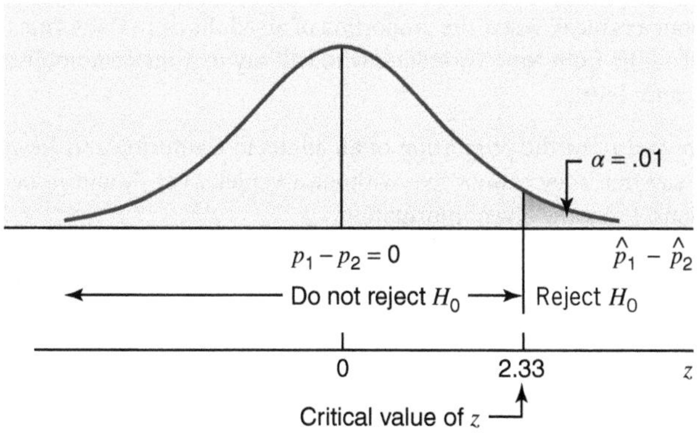

FIGURE 10.8 Rejection and nonrejection regions.

STEP 4 *Calculate the value of the test statistic.*

The pooled sample proportion is

$$\bar{p} = \frac{x_1 + x_2}{n_1 + n_2} = \frac{100 + 68}{500 + 400} = .187 \quad \text{and} \quad \bar{q} = 1 - \bar{p} = 1 - .187 = .813$$

The estimate of the standard deviation of $\hat{p}_1 - \hat{p}_2$ is

$$s_{\hat{p}_1 - \hat{p}_2} = \sqrt{\bar{p}\bar{q}\left(\frac{1}{n_1} + \frac{1}{n_2}\right)} = \sqrt{(.187)(.813)\left(\frac{1}{500} + \frac{1}{400}\right)} = .02615606$$

The value of the test statistic z for $\hat{p}_1 - \hat{p}_2$ is

$$z = \frac{(\hat{p}_1 - \hat{p}_2) - (p_1 - p_2)}{s_{\hat{p}_1 - \hat{p}_2}} = \frac{(.20 - .17) - 0}{.02615606} = 1.15$$

(From H_0: the 0 in the numerator)

STEP 5 *Make a decision.*

Because the value of the test statistic $z = 1.15$ for $\hat{p}_1 - \hat{p}_2$ falls in the nonrejection region (see Figure 10.8), we fail to reject the null hypothesis. Therefore, we cannot conclude that the proportion of users of Toothpaste A who will never switch to another toothpaste is greater than the proportion of users of Toothpaste B who will never switch to another toothpaste.

Using the *p*-Value to Make the Decision for Example 10.14

We can use the *p*-value approach to make the same decision. To do so, we keep Steps 1 and 2 above. Then in Step 3, we calculate the value of the test statistic z (as done in Step 4) and find the *p*-value for test statistic z from the normal distribution table. In Step 4, the z-value for $\hat{p}_1 - \hat{p}_2$ was calculated to be 1.15. In this example, the test is right-tailed. The *p*-value is given by the area under the normal distribution curve to the right of $z = 1.15$. From the normal distribution table (Table IV of Appendix B), this area is $1 - .8749 = .1251$. Hence, the *p*-value is .1251. We reject the null hypothesis for any α (significance level) greater than or equal to the *p*-value; in this example, we will reject the null hypothesis for any $\alpha \geq .1251$ or 12.51%. Because $\alpha = .01$ here, which is less than .1251, we fail to reject the null hypothesis.

EXAMPLE 10.15 Can We Live Without a Vehicle?

Conducting a Two-Tailed Test of Hypothesis About $p_1 - p_2$: Large and Independent Samples

The International Organization of Motor Vehicle Manufacturers (OICA), a group that defends the interests of the vehicle manufacturers, released the findings of a study it conducted a few years ago that included responses to a question that asked people if they can live without a car. Suppose recently a sample of 1000 adults from California and another sample of 1200 adults from New York State were selected and these adults were asked if they can live without a vehicle. Sixty-four percent of these adults from California and 61% from New York State said that they cannot live without a vehicle. Test if the proportion of all adults from California is different from the proportion of all adults from New York State who will say that they cannot live without a vehicle. Use a 1% significance level.

Solution Let p_1 and p_2 be the proportion of all adults in California and New York State, respectively, who will say that they cannot live without a vehicle. Let $\hat{p}_1$ and $\hat{p}_2$ be the corresponding sample proportions. From the given information,

For California: $n_1 = 1000$ and $\hat{p}_1 = .64$
For New York State: $n_2 = 1200$ and $\hat{p}_2 = .61$

The significance level is $\alpha = .01$.

STEP 1 *State the null and alternative hypotheses.*

The null and alternative hypotheses are, respectively,

H_0: $p_1 - p_2 = 0$ (The two population proportions are not different.)
H_1: $p_1 - p_2 \neq 0$ (The two population proportions are different.)

STEP 2 *Select the distribution to use.*

Because the samples are large and independent, we apply the normal distribution to make the test. (The reader should check that $n_1\hat{p}_1$, $n_1\hat{q}_1$, $n_2\hat{p}_2$, and $n_2\hat{q}_2$ are all greater than 5.)

STEP 3 *Determine the rejection and nonrejection regions.*

The ≠ sign in the alternative hypothesis indicates that the test is two-tailed. For a 1% significance level, the critical values of z are −2.58 and 2.58. Note that to find these two critical values, we look for .0050 and .9950 areas in Table IV of Appendix B. These values are shown in **Figure 10.9**.

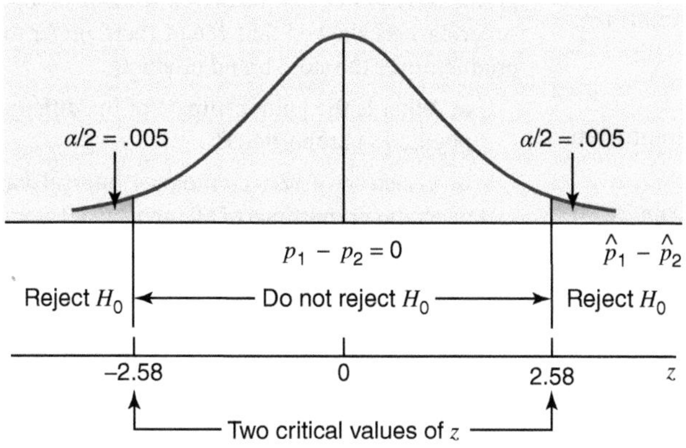

FIGURE 10.9 Rejection and nonrejection regions.

STEP 4 *Calculate the value of the test statistic.*

The pooled sample proportion is

$$\bar{p} = \frac{n_1\hat{p}_1 + n_2\hat{p}_2}{n_1 + n_2} = \frac{1000(.64) + 1200(.61)}{1000 + 1200} = .624$$

and $$\bar{q} = 1 - \bar{p} = 1 - .624 = .376$$

The estimate of the standard deviation of $\hat{p}_1 - \hat{p}_2$ is

$$s_{\hat{p}_1 - \hat{p}_2} = \sqrt{\bar{p}\bar{q}\left(\frac{1}{n_1} + \frac{1}{n_2}\right)} = \sqrt{(.624)(.376)\left(\frac{1}{1000} + \frac{1}{1200}\right)} = .02073991$$

The value of the test statistic z for $\hat{p}_1 - \hat{p}_2$ is

$$z = \frac{(\hat{p}_1 - \hat{p}_2) - (p_1 - p_2)}{s_{\hat{p}_1 - \hat{p}_2}} = \frac{(.64 - .61) - 0}{.02073991} = 1.45$$

STEP 5 *Make a decision.*

Because the value of the test statistic $z = 1.45$ falls in the nonrejection region, we fail to reject the null hypothesis H_0. Therefore, we cannot conclude that the proportions of all adults in California and New York State who will say that they cannot live without a vehicle are different.

Using the *p*-Value to Make the Decision for Example 10.15

We can use the *p*-value approach to make the same decision. To do so, we keep Steps 1 and 2. Then in Step 3, we calculate the value of the test statistic z (as done in Step 4) and find the *p*-value for this z from the normal distribution table. In Step 4, the z-value for $\hat{p}_1 - \hat{p}_2$ was calculated to be 1.45. In this example, the test is two-tailed. The *p*-value is given by twice the area under the normal distribution curve to the right of $z = 1.45$. From the normal distribution table (Table IV of Appendix B), the area to the right of $z = 1.45$ is .0735. Hence, the *p*-value is 2(.0735) = .1470. As we know, we will reject the null hypothesis for any α (significance level) greater than or equal to the *p*-value. Since $\alpha = .01$ in this example, which is smaller than .1470, we fail to reject the null hypothesis.

EXERCISES

Concepts and Procedures

10.42 What is the shape of the sampling distribution of $\hat{p}_1 - \hat{p}_2$ for two large samples? What are the mean and standard deviation of this sampling distribution?

10.43 When are the samples considered large enough for the sampling distribution of the difference between two sample proportions to be (approximately) normal?

10.44 Construct a 95% confidence interval for $p_1 - p_2$ for the following:

$$n_1 = 300,\quad \hat{p}_1 = .55,\quad n_2 = 200,\quad \hat{p}_2 = .62$$

10.45 Construct a 99% confidence interval for $p_1 - p_2$ for the following:

$$n_1 = 100,\quad \hat{p}_1 = .81,\quad n_2 = 150,\quad \hat{p}_2 = .77$$

10.46 Consider the following information obtained from two independent samples:

$$n_1 = 300, \ \hat{p}_1 = .55, \ n_2 = 200, \ \hat{p}_2 = .62$$

Test at a 5% significance level if the two population proportions are different.

10.47 Consider the following information obtained from two independent samples:

$$n_1 = 100, \ \hat{p}_1 = .81, \ n_2 = 150, \ \hat{p}_2 = .77$$

Test at a 5% significance level if $H_0 : p_1 - p_2 = 0$ is different from zero.

10.48 A sample of 500 observations selected from the first population gave $x_1 = 305$. Another sample of 600 observations selected from the second population gave $x_2 = 348$.

a. Find the point estimate of $p_1 - p_2$.

b. Make a 97% confidence interval for $p_1 - p_2$.

c. Show the rejection and nonrejection regions on the sampling distribution of $\hat{p}_1 - \hat{p}_2$ for H_0: $p_1 = p_2$ versus H_1: $p_1 > p_2$. Use a significance level of 2.5%.

d. Find the value of the test statistic z for the test of part (c).

e. Will you reject the null hypothesis mentioned in part (c) at a significance level of 2.5%? Explain.

10.49 A sample of 1000 observations selected from the first population gave $x_1 = 290$. Another sample of 1200 observations selected from the second population gave $x_2 = 396$.

a. Find the point estimate of $p_1 - p_2$.

b. Make a 98% confidence interval for $p_1 - p_2$.

c. Show the rejection and nonrejection regions on the sampling distribution of $\hat{p}_1 - \hat{p}_2$ for H_0: $p_1 = p_2$ versus H_1: $p_1 < p_2$. Use a significance level of 1%.

d. Find the value of the test statistic z for the test of part (c).

e. Will you reject the null hypothesis mentioned in part (c) at a significance level of 1%? Explain.

Applications

10.50 In a survey of U.S. drivers, 79% of women drivers and 85% of men drivers said that they exceeded the speed limit at least once in the past week. Suppose that these percentages are based on random samples of 600 women and 700 men drivers.

a. Let p_1 and p_2 be the proportions of all such women and men drivers, respectively, who will say that they exceeded the speed limit at least once in the past week. Construct a 98% confidence interval for $p_1 - p_2$.

b. Using a 1% significance level, can you conclude that p_1 is lower than p_2? Use both the critical value and the p-value approaches.

10.51 A state that requires periodic emission tests of cars operates two emission test stations, A and B, in one of its towns. Car owners have complained about the lack of uniformity of procedures at the two stations, resulting in different failure rates. A sample of 400 cars at Station A showed that 53 of those failed the test; a sample of 470 cars at Station B found that 51 of those failed the test.

a. What is the point estimate of the difference between the two population proportions?

b. Construct a 95% confidence interval for the difference between the two population proportions.

c. Testing at a 5% significance level, can you conclude that the two population proportions are different? Use both the critical value and the p-value approaches.

10.52 The management of a supermarket chain wanted to investigate if the percentages of men and women who prefer to buy national brand products over the store brand products are different. A sample of 600 men shoppers at the company's supermarkets showed that 246 of them prefer to buy national brand products over the store brand products. Another sample of 700 women shoppers at the company's supermarkets showed that 266 of them prefer to buy national brand products over the store brand products.

a. What is the point estimate of the difference between the two population proportions?

b. Construct a 98% confidence interval for the difference between the proportions of all men and all women shoppers at these supermarkets who prefer to buy national brand products over the store brand products.

c. Testing at a 1% significance level, can you conclude that the proportions of all men and all women shoppers at these supermarkets who prefer to buy national brand products over the store brand products are different?

10.53 The lottery commissioner's office in a state wanted to find if the percentages of men and women who play the lottery often are different. A sample of 500 men taken by the commissioner's office showed that 160 of them play the lottery often. Another sample of 300 women showed that 66 of them play the lottery often.

a. What is the point estimate of the difference between the two population proportions?

b. Construct a 99% confidence interval for the difference between the proportions of all men and all women who play the lottery often.

c. Testing at a 1% significance level, can you conclude that the proportions of all men and all women who play the lottery often are different?

10.54 A mail-order company has two warehouses, one on the West Coast and the second on the East Coast. The company's policy is to mail all orders placed with it within 72 hours. The company's quality control department checks quite often whether or not this policy is maintained at the two warehouses. A recently taken sample of 400 orders placed with the warehouse on the West Coast showed that 364 of them were mailed within 72 hours. Another sample of 300 orders placed with the warehouse on the East Coast showed that 279 of them were mailed within 72 hours.

a. Construct a 97% confidence interval for the difference between the proportions of all orders placed at the two warehouses that are mailed within 72 hours.

b. Using a 2.5% significance level, can you conclude that the proportion of all orders placed at the West Coast warehouse that are mailed within 72 hours is lower than the corresponding proportion for the East Coast warehouse?

10.55 A company that has many department stores in the southern states wanted to find at two such stores the percentage of sales for which at least one of the items was returned. A sample of 800 sales randomly selected from Store A showed that for 280 of them at least one item was returned. Another sample of 900 sales randomly selected from Store B showed that for 279 of them at least one item was returned.

a. Construct a 98% confidence interval for the difference between the proportions of all sales at the two stores for which at least one item is returned.

b. Using a 1% significance level, can you conclude that the proportions of all sales for which at least one item is returned is higher for Store A than for Store B?

Uses and Misuses of Statistics

Statistics and Health-Related Studies

While watching the news or browsing through a newspaper, we are likely to see or read the results of some new medical study. Often, the study result is really eye-catching, such as the following headline from *USA TODAY*: "Chocolate Lowers Heart Stroke Risk" (http://yourlife.usatoday.com/health/healthcare/studies/story/2011-08-29/Chocolate-lowers-heart-stroke-risk/50174422/1). Certainly a headline such as this one can grab your attention, especially if you are a chocolate lover. However, it is important to find out about the type of study that was conducted and determine whether it indicated an association (i.e., it showed a potential link between chocolate consumption and stroke risk) or a causal relationship (i.e., it showed that consuming chocolate actually does lower the risk of a stroke).

Two primary types of medical studies are (1) case–control studies and (b) cohort studies. A case–control study is an observational study that essentially works *backward*. In the case of the aforementioned story, people in the study were classified into two different groups—those who had a stroke, and those who did not have a stroke. The people were interviewed and asked about their chocolate consumption habits. The study revealed that the proportion of people who consumed chocolate was higher in the nonstroke group than in the stroke group. As noted in the *USA TODAY* article, this result regarding the relationship was consistent across a number of independent case–control studies, which were combined into a single study using a process called *meta-analysis*. So, with this information, this set of case–control studies leads to a new question: Does consuming chocolate actually reduce the risk of having a stroke, or is there something else that is the cause of these results?

When a case–control study identifies a potential causal link, the next step is to perform a cohort study (i.e., a *clinical trial*). In a cohort study, individuals are selected to participate in the study, and then these individuals are randomly assigned to different groups, with each group receiving a specific *treatment*. In this case, members in one group would eat chocolate regularly, whereas the members in the other group would not. The participants would be observed over a long period of time, and the proportion of people who have a stroke in each group would be compared. If the result of this study is consistent with the result of the case–control study, more research would be conducted to determine the biological or chemical link between chocolate and stroke prevention.

One question you might ask at this point is: Why would we perform a case–control study instead of commencing with a cohort study from the start? The reason is that cohort studies, especially those involving health-related issues, are expensive and involve a great deal of time. In the chocolate-and-stroke example, the researchers will have to wait to determine whether the individuals in the study have a stroke or not. If you do not know whether there is an association between chocolate consumption and stroke risk, you would probably not want to spend the time and money to perform this study. Case–control studies, on the other hand, are relatively inexpensive and easy to perform because databases are maintained on health events, and, hence, the results of the study can be obtained by accessing databases and performing interviews. The results of a cheaper case–control study would allow us to determine whether it is worth spending the money to perform a cohort study that provides data to help make a more substantial decision.

Glossary

d The difference between two matched values in two samples collected from the same source. It is called the paired difference.

$\bar{d}$ The mean of the paired differences for a sample.

Independent samples Two samples drawn from two populations such that the selection of an observation from one population does not affect the selection of an observation from the other population.

Paired or matched samples Two samples drawn in such a way that they include the same elements and two data values are obtained from each element, one for each sample. Also called **dependent samples**.

μ_d The mean of the paired differences for the population.

s_d The standard deviation of the paired differences for a sample.

σ_d The standard deviation of the paired differences for the population.

Exercises

Supplementary Exercises

10.56 A consulting agency was asked by a large insurance company to investigate if business majors were better salespersons than those with other majors. A sample of 20 salespersons with a business degree showed that they sold an average of 11 insurance policies per week. Another sample of 25 salespersons with a degree other than business showed that they sold an average of nine insurance policies per week. Assume that the two populations are approximately normally distributed with population standard deviations of 1.75 and 1.25 policies per week, respectively.

a. Construct a 99% confidence interval for the difference between the two population means.

b. Using a 1% significance level, can you conclude that persons with a business degree are better salespersons than those who have a degree in another area?

10.57 According to an estimate, the average earnings of female workers who are not union members are \$909 per week and those of female workers who are union members are \$1035 per week. Suppose that these average earnings are calculated based on random samples of 1500 female workers who are not union members and 2000 female workers who are union members. Further assume that the standard deviations for the two corresponding populations are \$70 and \$90, respectively.

a. Construct a 95% confidence interval for the difference between the two population means.

b. Test at a 2.5% significance level whether the mean weekly earnings of female workers who are not union members is less than that of female workers who are union members.

10.58 According to a survey of U.S. doctors by Nomad Health, the national average salary of doctors with specialization in radiology was \$429,000 and that of doctors with specialization in general surgery was \$403,000 in 2019. Assume that the sample sizes for these two groups were 600 and 800, respectively. Further assume that the sample standard deviations of the two groups were \$52,000 and \$50,000, respectively, and that the standard deviations for the two populations are unknown but equal.

a. Let μ_1 and μ_2 be the 2019 national average salaries of doctors with specialization in radiology and of doctors with specialization in general surgery, respectively. What is the point estimate of $\mu_1 - \mu_2$?

b. Construct a 98% confidence interval for $\mu_1 - \mu_2$.

c. Using a 1% significance level, can you conclude that the 2019 national average salary of doctors with specialization in radiology is higher than that of doctors with specialization in general surgery? Use both the *p*-value and the critical value approaches.

10.59 A researcher wants to test if the mean GPAs (grade point averages) of all male and all female college students who actively participate in sports are different. She took a random sample of 28 male students and 24 female students who are actively involved in sports. She found the mean GPAs of the two groups to be 2.62 and 2.74, respectively, with the corresponding standard deviations equal to .43 and .38.

a. Test at a 5% significance level whether the mean GPAs of the two populations are different.

b. Construct a 90% confidence interval for the difference between the two population means. Assume that the GPAs of all male and all female student athletes both are normally distributed with equal but unknown population standard deviations.

10.60 According to estimates by LendEDU, college seniors who graduated from public and private nonprofit colleges in Ohio in 2018 had a mean student loan debt of \$30,323, and those who graduated from such colleges in Michigan in 2018 had a mean student loan debt of \$32,158. Assume that the sample sizes for these estimates for Ohio and Michigan were 900 and 700, respectively. Further assume that the sample standard deviations of student loan debts for these two states were \$4800 and \$4950, respectively, and that the standard deviations for the two populations are unknown and unequal.

a. Let μ_1 and μ_2 be the average student loan debts of all college seniors who graduated from public and private nonprofit colleges in Ohio and Michigan, respectively, in 2018. What is the point estimate of $\mu_1 - \mu_2$?

b. Construct a 95% confidence interval for $\mu_1 - \mu_2$.

c. Using a 2.5% significance level, can you conclude that the average student loan debt of college seniors who graduated from public and private nonprofit colleges in Ohio in 2018 was less than the average student loan debt of college seniors who graduated from public and private nonprofit colleges in Michigan in 2018?

10.61 According to a U.S. Bureau of Labor Statistics report released on May 15, 2018, financial analysts earn an average of \$100,990 a year and engineers earn an average of \$99,230 a year (www.bls.gov). Suppose that these estimates are based on random samples of 1600 financial analysts and 1400 engineers. Further assume that the sample standard deviations of the annual earnings of these two groups are \$12,200 and \$11,500, respectively, and the population standard deviations are unknown and unequal for the two groups.

a. Construct a 98% confidence interval for the difference in the mean annual earnings of the two groups, financial analysts and engineers.

b. Using a 1% significance level, can you conclude that the average annual earnings of financial analysts is higher than that of engineers?

10.62 The manager of a factory has devised a detailed plan for evacuating the building as quickly as possible in the event of a fire or other emergency. An industrial psychologist believes that workers actually leave the factory faster at closing time without following any system. The company holds fire drills periodically in which a bell sounds and workers leave the building according to the system. The evacuation time for each drill is recorded. For comparison, the psychologist also records the evacuation time when the bell sounds for closing time each day. A random sample of 36 fire drills showed a mean evacuation time of 5.1 minutes with a standard deviation of 1.1 minutes. A random sample of 37 days at closing time showed a mean evacuation time of 4.2 minutes with a standard deviation of 1.0 minute.

a. Construct a 99% confidence interval for the difference between the two population means.

b. Test at a 5% significance level whether the mean evacuation time is smaller at closing time than during fire drills.

Assume that the evacuation times at closing time and during fire drills have unknown and unequal population standard deviations.

10.63 Two local post offices are interested in knowing the average number of Christmas cards that are mailed out from the towns that they serve. A random sample of 80 households from Town A showed that they mailed an average of 28.55 Christmas cards with a standard deviation of 10.30. The corresponding values of the mean and standard deviation produced by a random sample of 58 households from Town B were 33.67 and 8.97 Christmas cards. Assume that the distributions of the number of Christmas cards mailed by all households from both these towns have unknown and unequal population standard deviations.

a. Construct a 95% confidence interval for the difference in the average number of Christmas cards mailed by all households in these two towns.

b. Using a 10% significance level, can you conclude that the average number of Christmas cards mailed out by all households in Town A is different from the corresponding average for Town B?

10.64 The owner of a mosquito-infested fishing camp in Alaska wants to test the effectiveness of two rival brands of mosquito repellents, X and Y. During the first month of the season, eight people are chosen at random from those guests who agree to take part in the experiment. For each of these guests, Brand X is randomly applied to one arm and Brand Y is applied to the other arm. These guests fish for four hours, then the owner counts the number of bites on each arm. The table below shows the number of bites on the arm with Brand X and those on the arm with Brand Y for each guest.

Guest	A	B	C	D	E	F	G	H
Brand X	12	23	18	36	8	27	22	32
Brand Y	9	20	21	27	6	18	15	25

a. Construct a 95% confidence interval for the mean μ_d of the population of paired differences, where a paired difference is defined as the number of bites on the arm with Brand X minus the number of bites on the arm with Brand Y.

b. Test at a 5% significance level whether the mean number of bites on the arm with Brand X and the mean number of bites on the arm with Brand Y are different for all such guests.

Assume that the population of paired differences has an approximate normal distribution.

10.65 A random sample of nine students was selected to test the effectiveness of a special course designed to improve memory. The following table gives the scores in a memory test given to these students before and after this course.

Before	43	57	48	65	81	49	38	69	58
After	49	56	55	77	89	57	36	64	69

a. Construct a 95% confidence interval for the mean μ_d of the population of paired differences, where a paired difference is defined as the difference between the memory test scores of a student before and after attending this course.

b. Test at a 1% significance level whether this course makes any statistically significant improvement in the memory of all students.

Assume that the population of paired differences has an approximate normal distribution.

10.66 In a random sample of 800 men aged 25 to 35 years, 24% said they live with one or both parents. In another sample of 850 women of the same age group, 18% said that they live with one or both parents.

a. Construct a 95% confidence interval for the difference between the proportions of all men and all women aged 25 to 35 years who live with one or both parents.

b. Test at a 2% significance level whether the two population proportions are different.

c. Repeat the test of part (b) using the *p*-value approach.

10.67 A drug company has developed a new drug designed to relieve sore throat symptoms. However, some of the volunteers with sore throat symptoms who helped test the new drug developed a mild allergic reaction to the drug. From these volunteers, two samples, one of 60 men and another of 55 women, were selected. Of the 60 men who tested the new drug, 7 had a mild allergic reaction within four hours of taking the drug. Of the 55 women who tested the new drug, 6 had a similar allergic reaction within four hours of taking the drug.

a. Let p_1 and p_2 be the proportions of all such men and women, respectively, who would develop a mild allergic reaction within four hours of taking this drug. Construct a 95% confidence interval for $p_1 - p_2$.

b. Using a 1% significance level, can you conclude that p_1 is different from p_2? Use both the critical value and the *p*-value approaches.

Advanced Exercises

10.68 Manufacturers of two competing automobile models, Gofer and Diplomat, each claims to have the lowest mean fuel consumption. Let μ_1 be the mean fuel consumption in miles per gallon (mpg) for the Gofer and μ_2 the mean fuel consumption in mpg for the Diplomat. The two manufacturers have agreed to a test in which several cars of each model will be driven on a 100-mile test run. The fuel consumption, in mpg, will be calculated for each test run. The average mpg for all 100-mile test runs for each model gives the corresponding mean. Assume that for each model the gas mileages for the test runs are normally distributed with $\sigma = 2$ mpg. Note that each car is driven for one and only one 100-mile test run.

a. How many cars (i.e., sample size) for each model are required to estimate $\mu_1 - \mu_2$ with a 90% confidence level and with a margin of error of estimate of 1.5 mpg? Use the same number of cars (i.e., sample size) for each model.

b. If μ_1 is actually 33 mpg and μ_2 is actually 30 mpg, what is the probability that five cars for each model would yield $\bar{x}_1 \geq \bar{x}_2$?

10.69 Maria and Ellen both specialize in throwing the javelin. Maria throws the javelin a mean distance of 200 feet with a standard deviation of 10 feet, whereas Ellen throws the javelin a mean distance of 210 feet with a standard deviation of 12 feet. Assume that the distances each of these athletes throws the javelin are normally distributed with these population means and standard deviations. If Maria and Ellen each throw the javelin once, what is the probability that Maria's throw is longer than Ellen's?

10.70 A new type of sleeping pill is tested against an older, standard pill. Two thousand insomniacs are randomly divided into two equal groups. The first group is given the old pill, and the second group receives the new pill. The time required to fall asleep after the pill is administered is recorded for each person. The results of the experiment are given in the following table, where $\bar{x}$ and s represent the mean and standard deviation, respectively, for the times required to fall asleep for people in each group after the pill is taken.

	Group 1 (Old Pill)	**Group 2 (New Pill)**
n	1000	1000
$\bar{x}$	15.4 minutes	15.0 minutes
s	3.5 minutes	3.0 minutes

Consider the test of hypothesis $H_0: \mu_1 - \mu_2 = 0$ versus $H_1: \mu_1 - \mu_2 > 0$, where μ_1 and μ_2 are the mean times required for all potential users to fall asleep using the old pill and the new pill, respectively.

a. Find the *p*-value for this test.

b. Does your answer to part (a) indicate that the result is statistically significant? Use $\alpha = .025$.

c. Find a 95% confidence interval for $\mu_1 - \mu_2$.

d. Does your answer to part (c) imply that this result is of great *practical* significance?

10.71 Gamma Corporation is considering the installation of governors on cars driven by its sales staff. These devices would limit the car speeds to a preset level, which is expected to improve fuel economy. The company is planning to test several cars for fuel consumption without governors for one week. Then governors would be installed in the same cars, and fuel consumption will be monitored for another week. Gamma Corporation wants to estimate the mean difference in fuel consumption with a margin of error of estimate of 2 mpg with a 90% confidence level. Assume that the differences in fuel consumption are normally distributed and that previous studies suggest that an estimate of $s_d = 3$ mpg is reasonable. How many cars should be tested? (Note that the critical value of t will depend on n, so it will be necessary to use trial and error.)

10.72 Refer to the previous exercise. Suppose Gamma Corporation decides to test governors on seven cars. However, the management is afraid that the speed limit imposed by the governors will reduce the number of contacts the salespersons can make each day. Thus, both the fuel consumption and the number of contacts made are recorded for each car/salesperson for each week of the testing period, both before and after the installation of governors.

	Number of Contacts		Fuel Consumption (mpg)	
Salesperson	Before	After	Before	After
A	50	49	25	26
B	63	60	21	24
C	42	47	27	26
D	55	51	23	25
E	44	50	19	24
F	65	60	18	22
G	66	58	20	23

Suppose that as a statistical analyst with the company, you are directed to prepare a brief report that includes statistical analysis and interpretation of the data. Management will use your report to help decide whether or not to install governors on all salespersons' cars. Use 90% confidence intervals and .05 significance levels for any hypothesis tests to make suggestions. Assume that the differences in fuel consumption and the differences in the number of contacts are both normally distributed.

10.73 We wish to estimate the difference between the mean scores on a standardized test of students taught by Instructors A and B. The scores of all students taught by Instructor A have a normal distribution with a standard deviation of 15, and the scores of all students taught by Instructor B have a normal distribution with a standard deviation of 10. To estimate the difference between the two means, you decide that the same number of students from each instructor's class should be observed.

a. Assuming that the sample size is the same for each instructor's class, how large a sample should be taken from each class to estimate the difference between the mean scores of the two populations to within 5 points with 90% confidence?

b. Suppose that samples of the size computed in part (a) will be selected in order to test for the difference between the two population mean scores using a .05 level of significance. How large does the difference between the two sample means have to be for you to conclude that the two population means are different?

c. Explain why a paired-samples design would be inappropriate for comparing the scores of Instructor A versus Instructor B.

10.74 Sixty-five percent of all male voters and 40% of all female voters favor a particular candidate. A sample of 100 male voters and another sample of 100 female voters will be polled. What is the probability that at least 10 more male voters than female voters will favor this candidate?

Self-Review Test

Multiple Choice Questions

1. To test the hypothesis that the mean blood pressure of university professors is lower than that of company executives, which of the following would you use?

a. A left-tailed test **b.** A two-tailed test

c. A right-tailed test

2. Two samples drawn from two populations are independent if

a. the selection of one sample from a population is not related to the selection of the second sample from the same population

b. the selection of one sample from one population does not affect the selection of the second sample from the second population

c. the selection of one sample from a population is related to the selection of the second sample from the same population

3. Suppose the following information is obtained from two independent samples selected from normally distributed populations.

$n_1 = 14$ $\bar{x}_1 = 30$ $\sigma_1 = 3.2$

$n_2 = 17$ $\bar{x}_2 = 28$ $\sigma_2 = 3.4$

What is the point estimate of $\mu_1 - \mu_2$?

a. −.2 **b.** 0 **c.** 2

4. A sample of 52 female workers and another sample of 74 male workers from a state produced mean weekly earnings of $735.83 for the females and $771.21 for the males. Suppose that the population standard deviations of the weekly earnings are $80.23 for the females and $84.29 for the males. The null hypothesis is that the mean weekly earnings are the same for females and males, while the alternative hypothesis is that the mean weekly earnings for females is less than the mean weekly earnings for males. Derive the corresponding 95% confidence interval, rounded to two decimal places, for the difference between the mean weekly earnings for all female and male workers in this state.

a. −64.44 to −6.32

b. −69.92 to −.84

c. −73.63 to 2.87

5. A sample of 80 female workers and another sample of 53 male workers from a state produced mean weekly earnings of $730.73 for the females and $775.09 for the males. Suppose that the population standard deviations of the weekly earnings are $84.36 for the females and $100.89 for the males. The null hypothesis is that the mean weekly earnings are the same for females and males, while the alternative hypothesis is that the mean weekly earnings for females is less than the mean weekly earnings for males. The significance level for the test is 1%. What is the critical value of z?

a. −2.58 **b.** −1.96 **c.** −2.33

6. A sample of 80 female workers and another sample of 68 male workers from a state produced mean weekly earnings of $739.50 for the females and $769.95 for the males. Suppose that the population standard deviations of the weekly earnings are $82.28 for the females and $89.02 for the males. The null hypothesis is that the mean weekly earnings are the same for females and males, while the alternative hypothesis is that the mean weekly earnings for females is less than the mean weekly earnings for males. What is the value of the test statistic, rounded to three decimal places?

a. −2.147 **b.** 2.147 **c.** 3.98

7. A sample of 47 female workers and another sample of 73 male workers from a state produced mean weekly earnings of $727.01 for the females and $771.02 for the males. Suppose that the population standard deviations of the weekly earnings are $84.44 for the females and $94.02 for the males. The null hypothesis is that the mean weekly earnings are the same for females and males, while the alternative hypothesis is that the mean weekly earnings for females is less than the mean weekly earnings for males. What is the p-value for this test, rounded to four decimal places?

a. .0046 b. .0022 c. .0039

8. A sample of 27 from a population produced a mean of 74.7 and a standard deviation of 7.2. A sample of 22 from another population produced a mean of 78.5 and a standard deviation of 6.2. Assume that the two populations are normally distributed and the standard deviations of the two populations are equal. Derive the corresponding 90% confidence interval for the difference between the means of these two populations, rounded to three decimal places.

a. −7.063 to −0.537

b. −5.757 to −1.843

c −6.410 to −1.190

9. A sample of 27 from a population produced a mean of 74.7 and a standard deviation of 7.4. A sample of 22 from another population produced a mean of 79.9 and a standard deviation of 6.0. Assume that the two populations are normally distributed and the standard deviations of the two populations are equal. The null hypothesis is that the two population means are equal, while the alternative hypothesis is that the mean of the first population is less than the mean of the second population. The significance level is 2.5%. What is the value of the test statistic, t, rounded to three decimal places?

a. −2.998 b. 3.567 c. −2.659

10. A sample of 27 from a population produced a mean of 74.3 and a standard deviation of 7.6. A sample of 22 from another population produced a mean of 80 and a standard deviation of 7.7. Assume that the two populations are normally distributed and the standard deviations of the two populations are equal. The null hypothesis is that the two population means are equal, while the alternative hypothesis is that the mean of the first population is less than the mean of the second population. The significance level is 2.5%. What is the p-value for this test, rounded to four decimal places?

a. .0063 b. .0021 c. .0079

11. A sample of 14 from a population produced a mean of 56.9 and a standard deviation of 7. A sample of 20 from another population produced a mean of 50.7 and a standard deviation of 10. Assume that the two populations are normally distributed and the standard deviations of the two populations are not equal. The null hypothesis is that the two population means are equal, while the alternative hypothesis is that the two population means are different. The significance level is 10%. What is the number of degrees of freedom of the t distribution to make a confidence interval for the difference between the two population means?

a. 41 b. 31 c. 26

12. A sample of 14 from a population produced a mean of 54.9 and a standard deviation of 7. A sample of 20 from another population produced a mean of 50.9 and a standard deviation of 10. Assume that the two populations are normally distributed and the standard deviations of the two populations are not equal. The null hypothesis is that the two population means are equal, while the alternative hypothesis is that the two population means are different. The significance level is 10%. Derive the corresponding 90% confidence interval for the difference between the means of these two populations, rounded to three decimal places.

a. −0.945 to 8.945 b. −1.948 to 9.948 c. −1.907 to 9.907

13. A sample of 12 from a population produced a mean of 83.0 and a standard deviation of 16. A sample of 16 from another population produced a mean of 71.7 and a standard deviation of 14. Assume that the two populations are normally distributed and the standard deviations of the two populations are not equal. The null hypothesis is that the two population means are equal, while the alternative hypothesis is that the mean of the first population is greater than the mean of the second population. The significance level is 2.5%. What is the critical value of t for the hypothesis test?

a. 2.311 b. 1.997 c. 2.080

14. Two paired or matched samples would imply that

a. data are collected on one variable from the elements of two independent samples

b. two data values are collected from the same source (elements) for two dependent samples

c. two data values are collected from the same source (elements) for two independent samples

15. For two paired samples with sample size of n, the degrees of freedom are

a. $2n - 1$ b. $2n - 2$ c. $n - 1$

16. In a survey regarding job satisfaction, 539 in a sample of 852 female job-holders stated that they are satisfied with their jobs, while 529 in a sample of 799 male job-holders stated that they are satisfied with their jobs. The null hypothesis is that the proportions of all female and male job-holders who are satisfied with their jobs are the same. The alternative hypothesis is that the proportion of female job-holders who are satisfied with their jobs is lower than the proportion of male job-holders stated who are satisfied with their jobs. The significance level is 2.5%. What is the 97% confidence interval for the difference between the proportions of all female and all male job-holders who will say that they are satisfied with their job, rounded to four decimal places?

a. −.0206 to .1006

b. −.0869 to .0233

c. −.0757 to .0203

Practice Exercises

1. Briefly explain the meaning of independent and dependent samples. Give one example of each of these cases.

2. A company psychologist wanted to test if company executives have job-related stress scores higher than those of university professors. He took a sample of 40 executives and 50 professors and tested them for job-related stress. The sample of 40 executives gave a mean stress score of 7.6. The sample of 50 professors produced a mean stress score of 5.4. Assume that the standard deviations of the two populations are .8 and 1.3, respectively.

a. Construct a 99% confidence interval for the difference between the mean stress scores of all executives and all professors.

b. Test at a 2.5% significance level whether the mean stress score of all executives is higher than that of all professors.

3. A sports researcher wanted to find out how long it takes for NFL (National Football League) wide receivers and quarterbacks to run a distance of 40 yards. A random sample of 25 NFL wide receivers produced an average time of 4.48 seconds to run a distance of 40 yards with a standard deviation of .15 seconds. Another random sample of 25 NFL quarterbacks produced an average time of 4.93 seconds to run a distance of 40 yards with a standard deviation of .21 seconds. Assume that the distributions of time to run 40 yards for all quarterbacks and wide receivers are both approximately normal with equal but unknown standard deviations.

a. Construct a 95% confidence interval for the difference between the mean time for wide receivers and quarterbacks to run 40 yards distance.

b. Test at a 1% significance level whether the mean time for all wide receivers to run 40 yards distance is lower than the mean time for all quarterbacks.

4. Repeat Practice Exercise 3 assuming that the distribution of times to run 40 yards by all wide receivers and all quarterbacks are both approximately normally distributed with unequal and unknown standard deviations.

5. Lake City has two shops, Zeke's and Elmer's, that handle the majority of the town's auto body repairs. Seven cars that were damaged in collisions were taken to both shops for written estimates of the repair costs. These estimates (in dollars) are shown in the following table:

Zeke's	1058	544	1349	1296	676	998	1698
Elmer's	995	540	1175	1350	605	970	1520

a. Construct a 99% confidence interval for the mean μ_d of the population of paired differences, where a paired difference is equal to Zeke's estimate minus Elmer's estimate.

b. Test at a 5% significance level whether the mean μ_d of the population of paired differences is different from zero.

Assume that the population of paired differences is (approximately) normally distributed.

6. A sample of 500 male registered voters showed that 57% of them voted in the last presidential election. Another sample of 400 female registered voters showed that 55% of them voted in the same election.

a. Construct a 97% confidence interval for the difference between the proportions of all male and all female registered voters who voted in the last presidential election.

b. Test at a 1% significance level whether the proportion of all male voters who voted in the last presidential election is different from that of all female voters.

Mini-Projects

Mini-Project 10.1

Suppose that a new cold-prevention drug was tested in a randomized, placebo-controlled, doubleblind experiment during the month of January. One thousand healthy adults were randomly divided into two groups of 500 each—a treatment group and a control group. The treatment group was given the new drug, and the control group received a placebo. During the month, 40 people in the treatment group and 120 people in the control group caught a cold. Explain how to construct a 95% confidence interval for the difference between the relevant population proportions. Also describe an appropriate hypothesis test, using the given data, to evaluate the effectiveness of this new drug for cold prevention.

Find a similar article in a journal of medicine, psychology, or other field that lends itself to confidence intervals and hypothesis tests for differences in two means or proportions. First explain how to make the confidence intervals and hypothesis tests; then do so using the data given in the article.

Mini-Project 10.2

Suppose you are planning to move your family to Wilmington, NC, and are concerned about which public high school your children will attend: Eugene Ashley High School or New Hanover High School. You are also concerned about the average sale prices of homes in these two school districts.

Refer to Data Set I on the sale prices of homes in Wilmington, NC (see Appendix A). Using these data, select a random sample of 20 homes from the Eugene Ashley High School district and a second random sample of 20 homes from the New Hanover High School district. For each sample, calculate the mean and standard deviation of the sale prices of the 20 homes you selected. Assume that the standard deviations of the sale prices of all such houses in these two districts are unknown and unequal.

a. Construct a 95% confidence interval for the difference in the mean sale prices for all houses in the Eugene Ashley and New Hanover High School districts.

b. At a 5% level of significance, can you conclude that the mean sale price of houses in the Eugene Ashley High School district is different from the mean sale price of houses in the New Hanover High School district?

Mini-Project 10.3

Many different kinds of analyses have been performed on the salaries of professional athletes. Perform a hypothesis test of whether or not the average salaries of players in two sports are different by taking independent random samples of 35 players, each from any two sports of your choice from the Major League Baseball (MLB), the National Football League (NFL), the National Basketball Association (NBA), and the National Hockey League (NHL). After you take samples, do the following:

a. For each player, calculate the weekly salary. For your information, the approximate length (in weeks) of a season is 32.5 for the MLB, 22.5 for the NFL, 28 for the NBA, and 29.5 for the NHL. This length of a season does not include the playoffs, but it does include training camp and the preseason games because each player is expected to participate in these events. Players may receive bonuses for making the playoffs, but these are not included in their base salaries. You may ignore such bonuses.

b. Perform a hypothesis test to determine if the average weekly salaries are the same for the two sports that you selected. Use a significance level of 5%. Make certain to indicate whether you decide to use the pooled variance assumption or not, and justify your selection.

c. Perform a hypothesis test on the same data to determine if the average annual salaries are the same for the two sports that you selected. Explain why you could get a different answer (with regard to rejecting or failing to reject the null hypothesis) when using the weekly salaries versus the annual salaries.

Mini-Project 10.4

As reported in *USA TODAY* of August 27, 2009, a 3M Privacy Filters survey asked U.S. adults what seat they prefer on a plane when they fly: window, middle, or aisle seat. Obtain random samples of 60 male and 60 female college students and ask the following question: When you fly on a plane, do you prefer to have a window seat or a non-window seat? Perform a hypothesis test to determine if the proportion of female college students who prefer to have a window seat when flying is different from the proportion of male college students who prefer to have a window seat when flying. Use a 5% significance level.

Mini-Project 10.5

The following table gives the average expenditure expected in order to raise a child born in 2008 through the age of 17 years for families in each of three income groups. These estimates are based on a USDA Center for Nutrition Policy and Promotion study reported in *USA Today* of August 21, 2009.

Family Income	Mean Expenditure
Less than $56,870	$159,870
$56,870 to $98,470	$221,190
More than $98,470	$366,660

a. Suppose that the equal variance assumption is reasonable for the lower and middle income groups (the first two groups listed in the table). Using the "Less than $56,870" group as population 1 and the "$56,870 to $98,470" group as population 2, determine the largest possible value of the pooled standard deviation that would cause one to reject the null hypothesis $H_0: \mu_1 = \mu_2$ in favor of the hypothesis $H_1: \mu_1 < \mu_2$ at a 5% significance level when $n_1 = n_2 = 10$.

b. Repeat part (a) for the following sample sizes:

i. $n_1 = n_2 = 15$

ii. $n_1 = n_2 = 20$

iii. $n_1 = n_2 = 30$

c. The sample size for three groups (combined) in the study was more than 3000. Based on your results in parts (a) and (b), do you think that it would have been possible to obtain a pooled standard deviation that would have caused you not to reject $H_0: \mu_1 = \mu_2$ in favor of the hypothesis $H_1: \mu_1 < \mu_2$ at the 5% significance level? Explain why.

Decide for Yourself

Note: The Decide for Yourself feature is located in the book's product page on *www.wiley.com*.

Chapter 10 Technology Instructions

Note: Complete TI-84, Minitab, and Excel manuals are available in the book's product page on *www.wiley.com*.

TI-84 Color/TI-84

The TI-84 Color Technology Instructions feature of this text is written for the TI-84 Plus C color graphing calculator running the 4.0 operating system. Some screens, menus, and functions will be slightly different in older operating systems. The TI-84 and TI-84 Plus can perform all of the same functions, but will not have the "Color" option referenced in some of the menus.

Estimating $\mu_1 - \mu_2$, σ_1 and σ_2 Known, for Example 10.3(b) of the Text

See the graphing calculator manual in the book's product page on *www.wiley.com* for specific directions on creating confidence intervals for $\mu_1 - \mu_2$ when σ_1 and σ_2 are known. The correct menu can be found by selecting **STAT > TESTS > 2-SampZInt.**

Testing a Hypothesis About $\mu_1 - \mu_2$, σ_1 and σ_2 Known, for Example 10.4 of the Text

See the graphing calculator manual in the book's product page on *www.wiley.com* for specific directions on testing a hypothesis for $\mu_1 - \mu_2$ when σ_1 and σ_2 are known. The correct menu can be found by selecting **STAT > TESTS > 2-SampZTest.**

Estimating $\mu_1 - \mu_2$, σ_1 and σ_2 Unknown but Equal, for Example 10.5 of the Text

1. Select **STAT > TESTS > 2-SampTInt.**
2. Use the following settings in the **2-SampTInt** menu:
   - Select **Stats** at the **Inpt** prompt.

     Note: If you have the data in a list, select **Data** at the **Inpt** prompt.
   - Type 80 at the $\bar{x}_1$ prompt.
   - Type 5 at the $\mathbf{Sx_1}$ prompt.
   - Type 15 at the $\mathbf{n_1}$ prompt.
   - Type 77 at the $\bar{x}_2$ prompt.

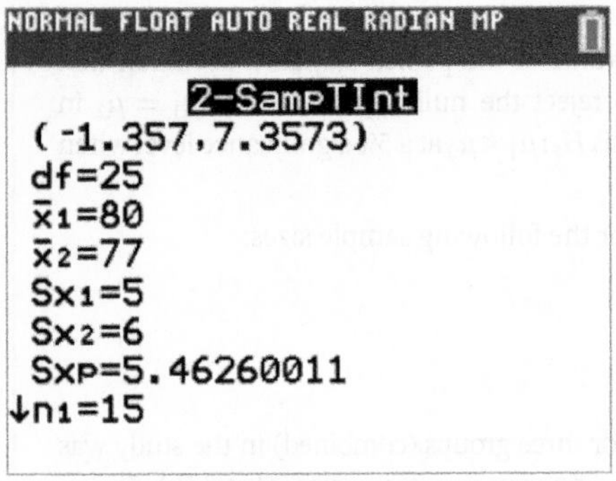

SCREEN 10.1

- Type 6 at the $\mathbf{Sx_2}$ prompt.
- Type 12 at the $\mathbf{n_2}$ prompt.
- Type 0.95 at the **C-Level** prompt.
- Select Yes at the **Pooled** prompt.

3. Scroll down, highlight **Calculate**, and press **ENTER**.
4. The output includes the confidence interval. (See **Screen 10.1**.)

Testing a Hypothesis About $\mu_1 - \mu_2$, σ_1 and σ_2 Unknown but Equal, for Example 10.6 of the Text

1. Select **STAT > TESTS > 2-SampTTest**.
2. Use the following settings in the **2-SampTTest** menu:
   - Select **Stats** at the **Inpt** prompt.

     Note: If you have the data in a list, select **Data** at the **Inpt** prompt.
   - Type 23 at the $\bar{x}_1$ prompt.
   - Type 3 at the $\mathbf{Sx_1}$ prompt.
   - Type 14 at the $\mathbf{n_1}$ prompt.
   - Type 25 at the $\bar{x}_2$ prompt.
   - Type 4 at the $\mathbf{Sx_2}$ prompt.
   - Type 16 at the $\mathbf{n_2}$ prompt.
   - Select $\neq \mu_2$ at the μ_1 prompt.
   - Select Yes at the **Pooled** prompt.
3. Scroll down, highlight **Calculate**, and press **ENTER**.
4. The output includes the test statistic and the *p*-value. (See **Screen 10.2**.)

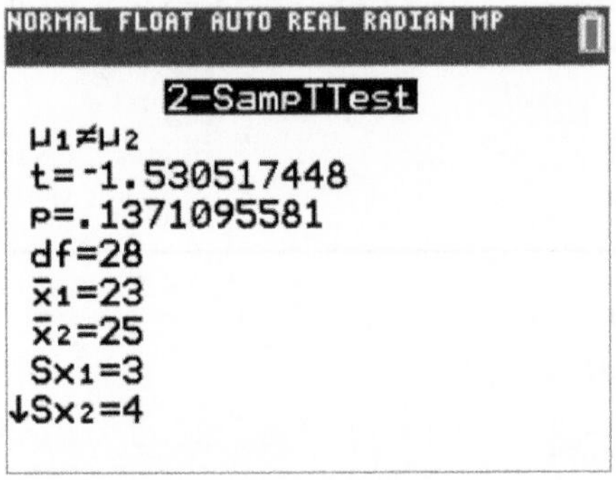

SCREEN 10.2

Compare the value of the test statistic to the critical value of *t* or the *p*-value to α and make a decision.

Estimating $\mu_1 - \mu_2$, σ_1 and σ_2 Unknown and Unequal, for Example 10.8 of the Text

Follow the instructions given previously in the TI-84 section entitled "Estimating $\mu_1 - \mu_2$, σ_1 and σ_2 Unknown but Equal, for Example 10.5 of the Text," but select No at the **Pooled** prompt in Step 2.

Testing a Hypothesis About $\mu_1 - \mu_2$, σ_1 and σ_2 Unknown and Unequal, for Example 10.9 of the Text

Follow the instructions given previously in the TI-84 section entitled "Testing a Hypothesis About $\mu_1 - \mu_2$, σ_1 and σ_2 Unknown but Equal, for Example 10.6 of the Text," but select No at the **Pooled** prompt in Step 2.

Estimating μ_d, σ_d Unknown

Create a list of differences for the paired data and then follow the directions for creating a confidence interval for a single mean when the population standard deviation is unknown. Specific directions can be found in the Chapter 8 Technology Instructions.

Testing a Hypothesis About μ_d, σ_d Unknown

Create a list of differences for the paired data and then follow the directions for testing a hypothesis about a single mean when the population standard deviation is unknown. Specific directions can be found in the Chapter 9 Technology Instructions.

Estimating $p_1 - p_2$ for Example 10.13 of the Text

1. Select **STAT > TESTS > 2-PropZInt**.
2. Use the following settings in the **2-PropZInt** menu:
   - Type 100 at the $\mathbf{x_1}$ prompt.
   - Type 500 at the $\mathbf{n_1}$ prompt.
   - Type 68 at the $\mathbf{x_2}$ prompt.
   - Type 400 at the $\mathbf{n_2}$ prompt.
   - Type 0.97 at the **C-Level** prompt.

3. Highlight **Calculate** and press **ENTER**.
4. The output includes the confidence interval. (See **Screen 10.3**.)

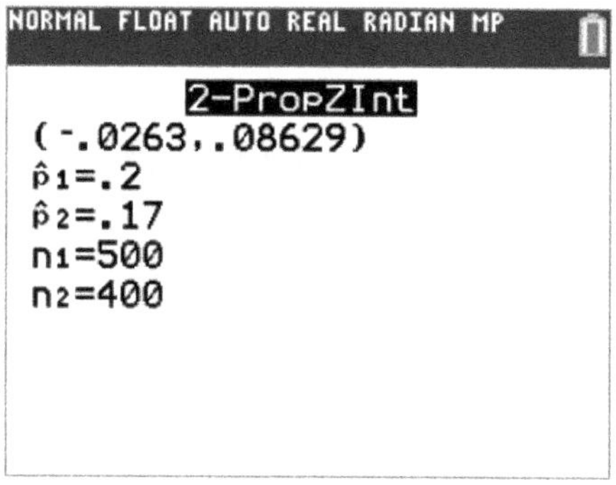

SCREEN 10.3

Testing a Hypothesis About $p_1 - p_2$ for Example 10.14 of the Text

1. Select **STAT > TESTS > 2-PropZTest**.
2. Use the following settings in the **2-PropZTest** menu:
   - Type 100 at the $\mathbf{x_1}$ prompt.
   - Type 500 at the $\mathbf{n_1}$ prompt.
   - Type 68 at the $\mathbf{x_2}$ prompt.
   - Type 400 at the $\mathbf{n_2}$ prompt.
   - Select $>\mathbf{p_2}$ at the $\mathbf{p_1}$ prompt.
3. Highlight **Calculate** and press **ENTER**.
4. The output includes the test statistic and the *p*-value. (See **Screen 10.4**.)

Compare the value of the test statistic to the critical value of z or the *p*-value to α and make a decision.

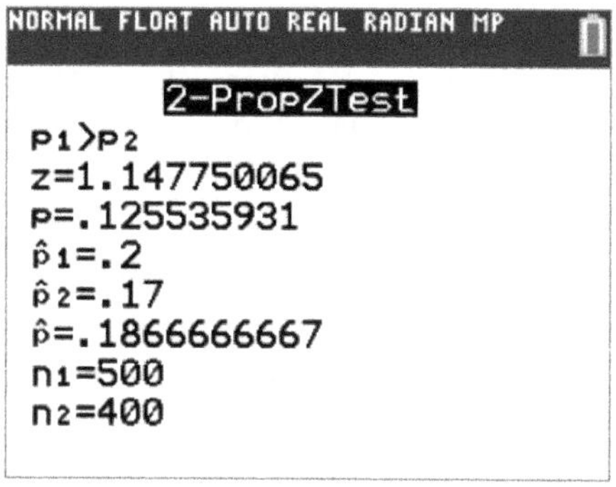

SCREEN 10.4

Minitab

The Minitab Technology Instructions in this text are written for **Minitab Professional Version 19**. Some screens, menus, and functions will be slightly different in older (or in newer) versions of Minitab. If you are using **Minitab Express**, you can download the **Minitab Express** Manual from the book's product page on *www.wiley.com*.

Estimating $\mu_1 - \mu_2$, σ_1 and σ_2 Unknown but Equal, for Example 10.5 of the Text

1. Select **Stat > Basic Statistics > 2-Sample t**.
2. Use the following settings in the dialog box that appears on screen:
   - Select **Summarized Data** from the dropdown menu.

     Note: If you have the data in two different columns, select **Each sample is in its own column**, type the column names in the **Sample 1** and **Sample 2** boxes, and move to Step 3.
   - Type 15 in the **Sample size** box for Sample 1.
   - Type 80 in the **Sample mean** box for Sample 1.
   - Type 5 in the **Standard deviation** box for Sample 1.
   - Type 12 in the **Sample size** box for Sample 2.
   - Type 77 in the **Sample mean** box for Sample 2.
   - Type 6 in the **Standard deviation** box for Sample 2.
3. Select **Options** and use the following settings when the new dialog box appears on screen:
   - Type 95 in the **Confidence level** box.
   - Check the **Assume equal variances** checkbox.
4. Click **OK** in both dialog boxes.
5. The confidence interval will be displayed in the Session window. (See **Screen 10.5**.)

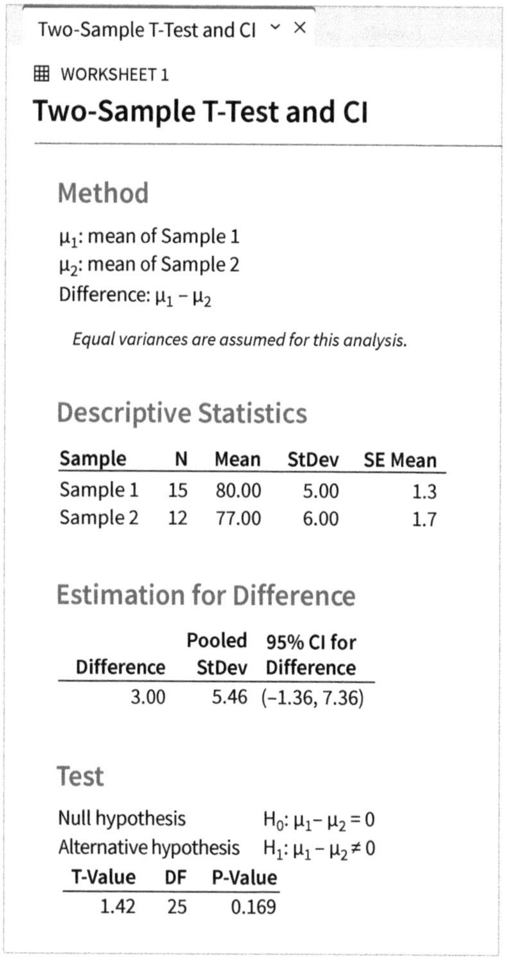

SCREEN 10.5

Testing a Hypothesis About $\mu_1 - \mu_2$, σ_1 and σ_2 Unknown but Equal, for Example 10.6 of the Text

1. Select **Stat > Basic Statistics > 2-Sample t**.
2. Use the following settings in the dialog box that appears on screen:
   - Select **Summarized Data** from the dropdown menu.

     Note: If you have the data in two different columns, select **Each sample is in its own column**, type the column names in the **Sample 1** and **Sample 2** boxes, and move to Step 3.

Two-Sample T-Test and CI

WORKSHEET 1

Two-Sample T-Test and CI

Method

μ_1: mean of Sample 1
μ_2: mean of Sample 2
Difference: $\mu_1 - \mu_2$

Equal variances are assumed for this analysis.

Descriptive Statistics

Sample	N	Mean	StDev	SE Mean
Sample 1	14	23.00	3.00	0.80
Sample 2	16	25.00	4.00	1.0

Estimation for Difference

Difference	Pooled StDev	95% CI for Difference
−2.00	3.57	(−4.68, 0.68)

Test

Null hypothesis $H_0: \mu_1 - \mu_2 = 0$
Alternative hypothesis $H_1: \mu_1 - \mu_2 \neq 0$

T-Value	DF	P-Value
−1.53	28	0.137

SCREEN 10.6

- Type 14 in the **Sample size** box for Sample 1.
- Type 23 in the **Sample mean** box for Sample 1.
- Type 3 in the **Standard deviation** box for Sample 1.
- Type 16 in the **Sample size** box for Sample 2.
- Type 25 in the **Sample mean** box for Sample 2.
- Type 4 in the **Standard deviation** box for Sample 2.

3. Select **Options** and use the following settings in the new dialog box that appears on screen:
   - Type 0 in the **Hypothesized difference** box.
   - Select Difference ≠ hypothesized difference in the **Alternative hypothesis** box.
   - Check the **Assume equal variances** box.
4. Click **OK** in both dialog boxes.
5. The output, including the test statistic and *p*-value, will be displayed in the Session window. (See **Screen 10.6**.)

Compare the value of the test statistic to the critical value of *t* or the *p*-value to α and make a decision.

Estimating $\mu_1 - \mu_2$, σ_1 and σ_2 Unknown and Unequal, for Example 10.8 of the Text

Follow the instructions given previously in the Minitab section entitled "Estimating $\mu_1 - \mu_2$, σ_1 and σ_2 Unknown but Equal, for Example 10.5 of the Text" but do not check the **Assume equal variances** box in Step 3.

Testing a Hypothesis About $\mu_1 - \mu_2$, σ_1 and σ_2 Unknown and Unequal, for Example 10.9 of the Text

Follow the instructions given previously in the Minitab section entitled "Testing a Hypothesis About $\mu_1 - \mu_2$, σ_1 and σ_2 Unknown but Equal, for Example 10.6 of the Text" but do not check the **Assume equal variances** box in Step 3.

Estimating μ_d, σ_d Unknown, for Example 10.10 of the Text

1. Enter the data from Example 10.10 into C1 ("Before") and C2 ("After").
2. Select **Stat > Basic Statistics > Paired t.**
3. Use the following settings in the dialog box that appears on screen:
   - Select **Each sample is in a column** from the dropdown menu.
   - Type C1 in the **Sample 1** box.
   - Type C2 in the **Sample 2** box.
4. Select **Options** and use the following settings when the dialog box appears on screen:
   - Type 95 in the **Confidence level** box.
5. Click **OK** in both dialog boxes.
6. The confidence interval will be displayed in the Session window.

Testing a Hypothesis About μ_d, σ_d Unknown, for Example 10.12 of the Text

1. Enter the data from Example 10.11 into C1 ("Before") and C2 ("After").
2. Select **Stat > Basic Statistics > Paired t.**
3. Use the following settings in the dialog box that appears on screen:
   - Select **Each sample is in a column** from the dropdown menu.
   - Type C1 in the **Sample 1** box.
   - Type C2 in the **Sample 2** box.
4. Select **Options** and use the following settings when the new dialog box appears on screen:
   - Type 0 in the **Hypothesized difference** box.
   - Select Difference < hypothesized difference in the **Alternative hypothesis** box.

Paired T-Test and CI: Befor...

WORKSHEET 1

Paired T-Test and CI: Before, After

Descriptive Statistics

Sample	N	Mean	StDev	SE Mean
Before	7	208.43	18.10	6.84
After	7	203.43	21.08	7.97

Estimation for Paired Difference

Mean	StDev	SE Mean	95% CI for μ_difference
5.00	10.79	4.08	(−4.98, 14.98)

μ_difference: mean of (Before - After)

Test

Null hypothesis H_0: μ_difference = 0
Alternative hypothesis H_1: μ_difference ≠ 0

T-Value	P-Value
1.23	0.266

SCREEN 10.7

5. Click **OK** in both dialog boxes.
6. The output, including the test statistic and *p*-value, will be displayed in the Session window. (See **Screen 10.7**.)

Compare the value of the test statistic to the critical value of t or the *p*-value to α and make a decision.

Estimating $p_1 - p_2$ for Example 10.13 of the Text

1. Select **Stat > Basic Statistics > 2 Proportions**.
2. Use the following settings in the dialog box that appears on screen:
   - Select **Summarized Data** from the dropdown menu.

     Note: If you have the data in two different columns, select **Each sample is in its own column**, type the column names in the **Sample 1** and **Sample 2** boxes, and move to Step 3.
   - Type 100 in the **Number of events** box for Sample 1.
   - Type 500 in the **Number of trials** box for Sample 1.
   - Type 68 in the **Number of events** box for Sample 2.
   - Type 400 in the **Number of trials** box for Sample 2.
3. Select **Options** and use the following settings when the new dialog box appears on screen:
   - Type 97 in the **Confidence level** box.
4. Click **OK** in both dialog boxes.
5. The confidence interval will be displayed in the Session window. (See **Screen 10.8.**)

Testing a Hypothesis About $p_1 - p_2$ for Example 10.14 of the Text

1. Select **Stat > Basic Statistics > 2 Proportions**.
2. Use the following settings in the dialog box that appears on screen:
   - Select **Summarized Data** from the dropdown menu.

Test and CI for Two Propor...

WORKSHEET 1

Test and CI for Two Proportions

Method

p_1: proportion where Sample 1 = Event
p_2: proportion where Sample 2 = Event
Difference: $p_1 - p_2$

Descriptive Statistics

Sample	N	Event	Sample p
Sample 1	500	100	0.200000
Sample 2	400	68	0.170000

Estimation for Difference

Difference	97% CI for Difference
0.03	(−0.026287, 0.086287)

CI based on normal approximation

Test

Null hypothesis H_0: $p_1 - p_2 = 0$
Alternative hypothesis H_1: $p_1 - p_2 \neq 0$

Method	Z-Value	P-Value
Normal approximation	1.16	0.247
Fisher's exact		0.264

SCREEN 10.8

Test and CI for Two Propor...

WORKSHEET 1

Test and CI for Two Proportions

Method

p_1: proportion where Sample 1 = Event
p_2: proportion where Sample 2 = Event
Difference: $p_1 - p_2$

Descriptive Statistics

Sample	N	Event	Sample p
Sample 1	500	100	0.200000
Sample 2	400	68	0.170000

Estimation for Difference

Difference	97% Lower Bound for Difference
0.03	−0.018783

CI based on normal approximation

Test

Null hypothesis H_0: $p_1 - p_2 = 0$
Alternative hypothesis H_1: $p_1 - p_2 > 0$

Method	Z-Value	P-Value
Normal approximation	1.15	0.126
Fisher's exact		0.144

SCREEN 10.9

Note: If you have the data in two different columns, select **Each sample is in its own column**, type the column names in the **Sample 1** and **Sample 2** boxes, and move to Step 3.

- Type 100 in the **Number of events** box for Sample 1.
- Type 500 in the **Number of trials** box for Sample 1.
- Type 68 in the **Number of events** box for Sample 2.
- Type 400 in the **Number of trials** box for Sample 2.

3. Select **Options** and use the following settings in the dialog box that appears on screen:
   - Type 0 in the **Hypothesized difference** box.
   - Select Difference > hypothesized difference from the **Alternative hypothesis** box.
   - Select Use the pooled estimate of the proportion from the **Test method box.**
4. Click **OK** in both dialog boxes.
5. The output, including the test statistic and *p*-value, will be displayed in the Session window. (See **Screen 10.9**.)

Compare the value of the test statistic to the critical value of z or the *p*-value to α and make a decision.

Excel

The Excel Technology Instructions feature of this text is written for Excel 2019. Some screens, menus, and functions will be slightly different in older versions of Excel. To enable some of the advanced Excel functions, you must enable the Data Analysis Add-In for Excel.

Testing a Hypothesis About $\mu_1 - \mu_2$, σ_1 and σ_2 Known

See the Excel manual in the book's product page on *www.wiley.com* for specific directions on creating confidence intervals for $\mu_1 - \mu_2$, when σ_1 and σ_2 are known. The correct menu can be found by clicking on **DATA** and then clicking on **Data Analysis Tools** from the **Analysis** group.

Estimating $\mu_1 - \mu_2$, σ_1 and σ_2 Unknown but Equal, for Example 10.5 of the Text

1. In a new worksheet, enter the **following text** into the indicated cells (see **Screen 10.10**):
   - Type Sample 1 in cell A1, Sample 2 in cell A4, Sample means in cells B1 and B4, Sample std dev in cells B2 and B5, and Sample sizes in cells B3 and B6.

	A	B	C	D
1	Sample 1	Sample mean	80	
2		Sample std dev	5	
3		Sample size	15	
4	Sample 2	Sample mean	77	
5		Sample std dev	6	
6		Sample size	12	
7				
8	xbar1 - xbar2	3		
9	df	25		
10	t*	2.0595		
11	Pooled s	5.4626		
12	Std dev	2.1157		
13				
14	Lower bound	-1.3573		
15	Upper bound	7.3573		

SCREEN 10.10

- Type xbar1 – xbar2 in cell A8, df in cell A9, t* in cell A10, Pooled s in cell A11, and Std dev in cell A12.
- Type Lower bound in cell A14 and Upper bound in cell A15.

2. Enter the values of the summary statistics in the indicated cells (see **Screen 10.10**):
   - Type 80 in cell C1.
   - Type 5 in cell C2.
   - Type 15 in cell C3.
   - Type 77 in cell C4.
   - Type 6 in cell C5.
   - Type 12 in cell C6.
3. Enter the following formulas in the indicated cells (see **Screen 10.10**):
   - Type =C1-C4 in cell B8.
   - Type =C3+C6-2 in cell B9.
   - Type =T.INV(0.975,B9) in cell B10.

     Note: The value 0.975 is found by $(1 - \alpha/2)$.
   - Type =SQRT(((C3-1)*C2^2+(C6-1)*C5^2)/B9) in cell B11.
   - Type =B11*SQRT(1/C3+1/C6) in cell B12.
   - Type =B8-B10*B12 in cell B14.
   - Type =B8+B10*B12 in cell B15.
4. The lower and upper bounds of the confidence interval will be in cells B14 and B15, respectively. (See **Screen 10.10.**)

Testing a Hypothesis About $\mu_1 - \mu_2$, σ_1 and σ_2 Unknown but Equal

1. To perform this test, Excel must have the raw data, not summary statistics. Suppose we wish to test $H_0: \mu_1 = \mu_2$ versus $H_1: \mu_1 > \mu_2$ at the $\alpha = 0.05$ significance level and that the population standard deviations are unknown but equal (see **Screen 10.11**):
   - Enter the first set of 10 data values into cells A1 through A10: 81, 79, 62, 98, 74, 82, 85, 72, 90, 88.
   - Enter the second set of 10 data values into cells B1 through B10: 61, 78, 85, 77, 90, 95, 64, 74, 80, 85.
2. Click **DATA** and then click **Data Analysis Tools** from the **Analysis** group.
3. Select **t-Test: Two-Sample Assuming Equal Variances** from the dialog box that appears on screen and then click OK.
4. Use the following settings in the dialog box that appears on screen (see **Screen 10.11**):
   - Type A1:A10 in the **Variable 1 Range** box.
   - Type B1:B10 in the **Variable 2 Range** box.
   - Type 0 in the **Hypothesized Mean Difference** box.

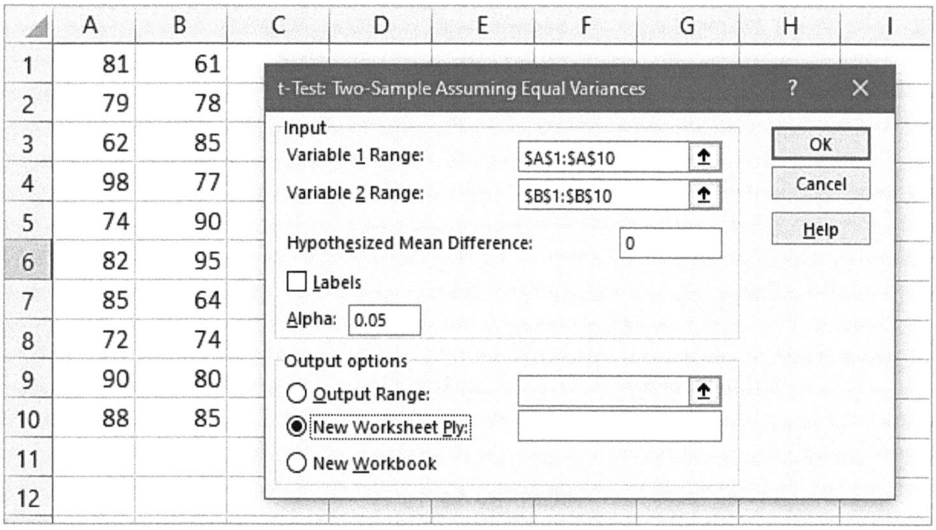

SCREEN 10.11

- Type 0.05 in the **Alpha** box.
- Select New Worksheet Ply from the **Output options.**

5. Click **OK.**
6. When the output appears, resize column A so that it is easier to read.
7. The correct formula for the *p*-value depends on your alternative hypothesis and the *t* statistic.
   - For $H_1: \mu_1 > \mu_2$,
     - If the *t* statistic is positive, use the *p*-value from **P(T<=t) one-tail.**
     - If the *t* statistic is negative, the *p*-value is found by subtracting **P(T<=t) one-tail** from 1.
   - For $H_1: \mu_1 < \mu_2$,
     - If the *t* statistic is positive, the *p*-value is found by subtracting **P(T<=t) one-tail** from 1.
     - If the *t* statistic is negative, use the *p*-value from **P(T<=t) one-tail.**
   - For $H_1: \mu_1 \neq \mu_2$, use the *p*-value from **P(T<=t) two-tail.**

Estimating $\mu_1 - \mu_2$, σ_1 and σ_2 Unknown and Unequal, for Example 10.8 of the Text

1. In a new worksheet, enter the following text into the indicated cells (see **Screen 10.12**):
   - Type Sample 1 in cell A1, Sample 2 in cell A4, Sample means in cells B1 and B4, Sample standard deviations in cells B2 and B5, and Sample size in cells B4 and B6.
   - Type xbar1 – xbar2 in cell A8, df in cell A9, t* in cell A10, and Std dev in cell A11.
   - Type Lower bound in cell A13 and Upper bound in cell A14.
2. Enter the values of the summary statistics in the indicated cells (see **Screen 10.12**):
   - Type 80 in cell C1.
   - Type 5 in cell C2.
   - Type 15 in cell C3.
   - Type 77 in cell C4.
   - Type 6 in cell C5.
   - Type 12 in cell C6.
3. Enter the following formulas in the indicated cells (see **Screen 10.12**):
   - Type =C1-C4 in cell B8.
   - Type =(C2^2/C3+C5^2/C6)^2/((C2^2/C3)^2/(C3-1)+(C5^2/C6)^2/(C6-1)) in cell B9.
   - Type =T.INV(0.975,B9) in cell B10.

	A	B	C	D
1	Sample 1	Sample mean	80	
2		Sample std dev	5	
3		Sample size	15	
4	Sample 2	Sample mean	77	
5		Sample std dev	6	
6		Sample size	12	
7				
8	xbar1 - xbar2	3		
9	df	21.4223		
10	t*	2.0796		
11	Std dev	2.1602		
12				
13	Lower bound	-1.4925		
14	Upper bound	7.4925		
15				

SCREEN 10.12

Note: The value 0.975 is found by $(1 - \alpha/2)$.

- Type =SQRT(C2^2/C3+C5^2/C6) in cell B11.
- Type =B8-B10*B11 in cell B13.
- Type =B8+B10*B11 in cell B14.

4. The lower and upper bounds of the confidence interval will be in cells B13 and B14, respectively. (See **Screen 10.12**.)

Testing a Hypothesis About $\mu_1 - \mu_2$, σ_1 and σ_2 Unknown and Unequal

1. To perform this test, Excel must have the raw data, not summary statistics. Suppose we wish to test $H_0: \mu_1 = \mu_2$ versus $H_1: \mu_1 > \mu_2$ at the $\alpha = 0.05$ significance level and that the population standard deviations are unknown and unequal (see **Screen 10.13**):
   - Enter these 10 data values into cells A1 through A10: 81, 79, 62, 98, 74, 82, 85, 72, 90, 88.
   - Enter these 10 data values into cells B1 through B10: 61, 78, 85, 77, 90, 95, 64, 74, 80, 85.
2. Click **DATA** and then click **Data Analysis Tools** from the **Analysis** group.
3. Select **t-Test: Two-Sample Assuming Unequal Variances** from the dialog box that appears on screen and then click OK.
4. Use the following settings in the dialog box that appears on screen (see **Screen 10.13**):
   - Type A1:A10 in the **Variable 1 Range** box.
   - Type B1:B10 in the **Variable 2 Range** box.
   - Type 0 in the **Hypothesized Mean Difference** box.
   - Type 0.05 in the **Alpha** box.
   - Select New Worksheet Ply from the **Output options**.
5. Click **OK**.
6. When the output appears, resize column A so that it is easier to read.
7. The correct formula for the *p*-value depends on your alternative hypothesis and the *t* statistic.
   - For $H_1: \mu_1 > \mu_2$,
     - If the *t* statistic is positive, use the *p*-value from **P(T<=t) one-tail**.
     - If the *t* statistic is negative, the *p*-value is found by subtracting **P(T<=t) one-tail** from 1.
   - For $H_1: \mu_1 < \mu_2$,
     - If the *t* statistic is positive, the *p*-value is found by subtracting **P(T<=t) one-tail** from 1.
     - If the *t* statistic is negative, use the *p*-value from **P(T<=t) one-tail**.
   - For $H_1: \mu_1 \neq \mu_2$, use the *p*-value from **P(T<=t) two-tail**.

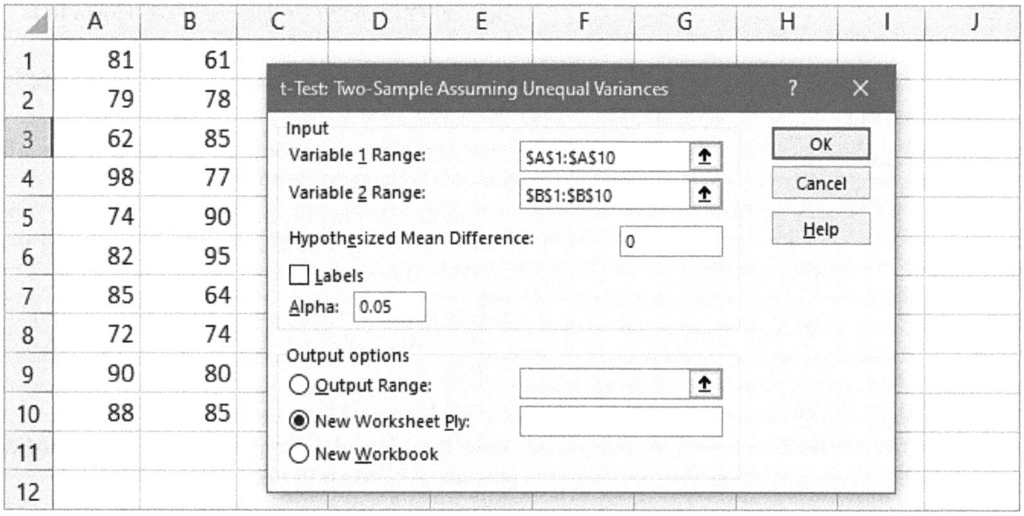

SCREEN 10.13

Estimating μ_d, σ_d Unknown

See the Excel manual in the book's product page on *www.wiley.com* for specific directions on creating confidence intervals for μ_d when σ_d is unknown.

Testing a Hypothesis About μ_d, σ_d Unknown

See the Excel manual in the book's product page on *www.wiley.com* for specific directions on testing a hypothesis about μ_d when σ_d is unknown.

	A	B	C	D
1	Sample 1	Number of successes	100	
2		Number of trials	500	
3	Sample 2	Number of successes	68	
4		Number of trials	400	
5				
6	phat1	0.20		
7	phat2	0.17		
8	phat1 - phat2	0.03		
9	z*	2.1701		
10	Std dev	0.0259		
11				
12	Lower bound	-0.0263		
13	Upper bound	0.0863		
14				

SCREEN 10.14

Estimating $p_1 - p_2$ for Example 10.13 of the Text

1. In a new worksheet, enter the following text into the indicated cells (see **Screen 10.14**):
   - Type Sample 1 in cell A1, Sample 2 in cell A3, Number of successes in cells B1 and B3, and Number of trials in cells B2 and B4.
   - Type phat1 in cell A6, phat2 in cell A7, phat1 – phat2 in cell A8, z* in cell A9, and Std dev in cell A10.
   - Type Lower bound in cell A12 and Upper bound in cell A13.
2. Enter the values of the summary statistics in the indicated cells (see **Screen 10.14**):
   - Type 100 in cell C1.
   - Type 500 in cell C2.
   - Type 68 in cell C3.
   - Type 400 in cell C4.
3. Enter the following formulas in the indicated cells (see **Screen 10.14**):
   - Type =C1/C2 in cell B6.
   - Type =C3/C4 in cell B7.
   - Type =B6-B7 in cell B8.
   - Type =NORM.INV(0.985,0,1) in cell B9.

     Note: The value 0.985 is found by $(1 - \alpha/2)$.
   - Type =SQRT(B6*(1-B6)/C2+B7*(1-B7)/C4) in cell B10.
   - Type =B8-B9*B10 in cell B12.
   - Type =B8+B9*B10 in cell B13.
4. The lower and upper bounds of the confidence interval will be in cells B12 and B13, respectively. (See **Screen 10.14**.)

	A	B	C
1	Sample 1	Number of successes	100
2		Number of trials	500
3	Sample 2	Number of successes	68
4		Number of trials	400
5			
6	phat1	0.20	
7	phat2	0.17	
8	phat1 - phat2	0.03	
9	Pooled sample prop	0.1867	
10	Std dev	0.0261	
11			
12	Test stat	1.1478	
13	p-value	0.1255	

SCREEN 10.15

Testing a Hypothesis About $p_1 - p_2$ for Example 10.14 of the Text

1. In a new worksheet, enter the following text into the indicated cells (see **Screen 10.15**):
   - Type Sample 1 in cell A1, Sample 2 in cell A3, Number of successes in cells B1 and B3, and Number of trials in cells B2 and B4.
   - Type phat1 in cell A6, phat2 in cell A7, phat1 – phat2 in cell A8, Pooled samp prop in cell A9, and Std dev in cell A10.
   - Type Test stat in cell A12 and *p*-value in cell A13.
2. Enter the values of the summary statistics in the indicated cells (see **Screen 10.15**):
   - Type 100 in cell C1.
   - Type 500 in cell C2.
   - Type 68 in cell C3.
   - Type 400 in cell C4.

3. Enter the following formulas in the indicated cells (see **Screen 10.15**):
   - Type =C1/C2 in cell B6.
   - Type =C3/C4 in cell B7.
   - Type =B6-B7 in cell B8.
   - Type =(C1+C3)/(C2+C4) in cell B9.
   - Type =SQRT(B9*(1-B9)/C2+B9*(1-B9)/C4) in cell B10.
   - Type =B8/B10 in cell B12.
   - The formula for cell B13 depends on H_1:
     - For $H_1: \mu_1 > \mu_2$, type =1-NORM.S.DIST(B12,1) in cell B13.
     - For $H_1: \mu_1 < \mu_2$, type =NORM.S.DIST(B12,1) in cell B13.
     - For $H_1: \mu_1 \neq \mu_2$, type =2*(1-NORM.S.DIST(ABS(B12),1)) in cell B13.
4. The test statistic will be in cell B12 and the *p*-value will be in cell B13. (See **Screen 10.15.**)

Technology Assignments

TA10.1 A consumer group sampled 40 refrigerator models with side-by-side freezers and 45 refrigerator models with top freezers. For each refrigerator, the annual energy cost of running (based on the price of a kilowatt-hour at the time of testing) was determined. For the 40 models with side-by-side freezers, the average annual energy cost was $74.72 with a standard deviation of $9.24. For the 45 models with top freezers, the average annual energy cost was $62.50 with a standard deviation of $7.54. Assume that the standard deviations for the two populations are unknown but equal.

a. Construct a 98% confidence interval for $\mu_1 - \mu_2$.

b. Using a 1% significance level, can you conclude that the average annual energy cost is higher for all refrigerator models with side-by-side freezers than for all refrigerator models with top freezers? Use both the *p*-value and the critical value approaches.

TA10.2 A company recently opened two supermarkets in two different areas. The management wants to know if the mean sales per day for these two supermarkets are different. A sample of 10 days for Supermarket A produced the following data on daily sales (in thousand dollars):

47.56	57.66	51.23	58.29	43.71
49.33	52.35	50.13	47.45	53.86

A sample of 12 days for Supermarket B produced the following data on daily sales (in thousand dollars):

56.34	63.55	61.64	63.75	54.78	58.19
55.40	59.44	62.33	67.82	56.65	67.90

Assume that the daily sales of the two supermarkets are both normally distributed with equal but unknown standard deviations.

a. Construct a 99% confidence interval for the difference between the mean daily sales for these two supermarkets.

b. Test at a 1% significance level whether the mean daily sales for these two supermarkets are different.

TA10.3 Refer to Technology Assignment TA10.1. Now do that assignment without assuming that the population standard deviations are equal.

TA10.4 Refer to Technology Assignment TA10.2. Now do that assignment assuming the daily sales of the two supermarkets are both normally distributed with unequal and unknown standard deviations.

TA10.5 A finger-tapping experiment was conducted by a doctoral neuropsychology student. The purpose of this experiment was to determine bilateral nervous system integrity, which provides data about the neuromuscular system and motor control. The experiment requires the subjects to place the palm of the hand on a table and then to tap the index finger on the surface of the table. Ten subjects who had suffered a mild concussion were given a finger tapping test using the right index finger and then the left index finger. The total number of taps in 30 seconds by each subject are listed in the following table. All subject in the data set are right-handed. Assume that the population of paired differences is approximately normally distributed.

Subject	1	2	3	4	5	6	7	8	9	10
Right	125	131	95	105	137	129	91	112	138	94
Left	127	138	97	107	130	128	85	114	135	85

a. Make a 95% confidence interval for the mean of the population of paired differences, where a paired difference is equal to the number of taps for the right index finger minus the number of taps for the left index finger.

b. Using a 5% significance level, can you conclude that the average number of taps is different for the right and left index fingers?

TA10.6 A company is considering installing new machines to assemble its products. The company is considering two types of machines, but it will buy only one type. The company selected eight assembly workers and asked them to use these two types of machines to assemble products. The following table gives the time taken (in minutes) to assemble one unit of the product on each type of machine for each of these eight workers.

Machine I	23	26	19	24	27	22	20	18
Machine II	21	24	23	25	24	28	24	23

a. Construct a 98% confidence interval for the mean μ_d of the population of paired differences, where a paired difference is equal to the time taken to assemble a unit of the product on Machine I minus the time taken to assemble a unit of the product on Machine II by the same worker.

b. Test at a 5% significance level whether the mean of the population of paired differences is different from zero.

Assume that the population of paired differences is (approximately) normally distributed.

TA10.7 A company has two restaurants in two different areas of New York City. The company wants to estimate the percentages of patrons who think that the food and service at each of these restaurants are excellent. A sample of 200 patrons taken from the restaurant in Area A showed that 118 of them think that the food and service are excellent at this restaurant. Another sample of 250 patrons selected from the restaurant in Area B showed that 160 of them think that the food and service are excellent at this restaurant.

a. Construct a 97% confidence interval for the difference between the two population proportions.

b. Testing at a 2.5% significance level, can you conclude that the proportion of patrons at the restaurant in Area A who think that the food and service are excellent is lower than the corresponding proportion at the restaurant in Area B?

TA10.8 The management of a supermarket chain wanted to investigate whether the percentages of all men and all women who prefer to buy national brand products over the store brand products are different. A sample of 600 men shoppers at the company's supermarkets showed that 246 of them prefer to buy national brand products over the store brand products. Another sample of 700 women shoppers at the company's supermarkets showed that 266 of them prefer to buy national brand products over the store brand products.

a. Construct a 99% confidence interval for the difference between the proportions of all men and all women shoppers at these supermarkets who prefer to buy national brand products over the store brand products.

b. Testing at a 1% significance level, can you conclude that the proportions of all men and all women shoppers at these supermarkets who prefer to buy national brand products over the store brand products are different?

CHAPTER 11

Chi-Square Tests

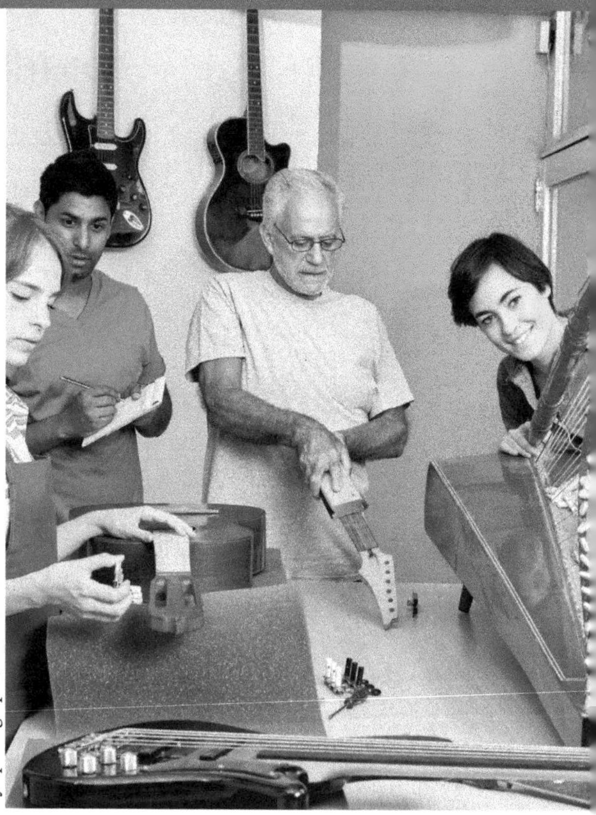
stylephotographs/123 RF

Do you belong to the middle class? If yes, how are the current economic conditions in the country affecting you? Do you think the current economic conditions are hurting you, helping you, or neither helping nor hurting you? In a 2019 Pew Research Center survey of randomly selected U.S. adults, 58% of the adults polled said that the current economic conditions in the country are hurting the middle class a lot or a little, 32% said that they are helping the middle class a lot or a little, and 10% were of the opinion that they neither help nor hurt the middle class. (See Case Study 11.1.)

The tests of hypothesis about a mean, the difference between two means, a proportion, and the difference between two proportions were discussed in Chapters 9 and 10. The tests about proportions dealt with categorical data. In the case of a proportion and the difference between two proportions in Chapters 9 and 10, the tests concerned experiments with only two outcomes. Recall from Chapter 5 that such experiments are called binomial experiments. This chapter describes three types of tests:

1. Tests of hypothesis for experiments with more than two outcomes, called goodness-of-fit tests
2. Tests of hypothesis about contingency tables, called independence and homogeneity tests
3. Tests of hypothesis about the variance and standard deviation of a single population

All of these tests are performed by using the **chi-square distribution**, which is sometimes written as χ^2 *distribution* and is read as "chi-square distribution." The symbol χ is the Greek letter *chi*, pronounced "kī." The values of a chi-square distribution are denoted by the symbol χ^2 (read as "chi-square"), just as the values of the standard normal distribution and the t distribution are denoted by z and t, respectively. Section 11.1 describes the chi-square distribution.

11.1 The Chi-Square Distribution

LEARNING OBJECTIVES

After completing this section, you should be able to:

- Describe the characteristics of a chi-square distribution.
- Identify the parameter of a chi-square distribution.
- Determine the value of χ^2 for a given left-tail or right-tail area of a chi-square distribution with given degrees of freedom.

Like the *t* distribution, the chi-square distribution has only one parameter, called the degrees of freedom (df). The shape of a specific chi-square distribution depends on the number of degrees of freedom.[1] (The degrees of freedom for a chi-square distribution are calculated by using different formulas for different tests. This will be explained when we discuss those tests.) The random variable χ^2 assumes nonnegative values only. Hence, a chi-square distribution curve starts at the origin (zero point) and lies entirely to the right of the vertical axis. **Figure 11.1** shows the chi-square distribution curves for 2, 7, and 12 degrees of freedom, respectively.

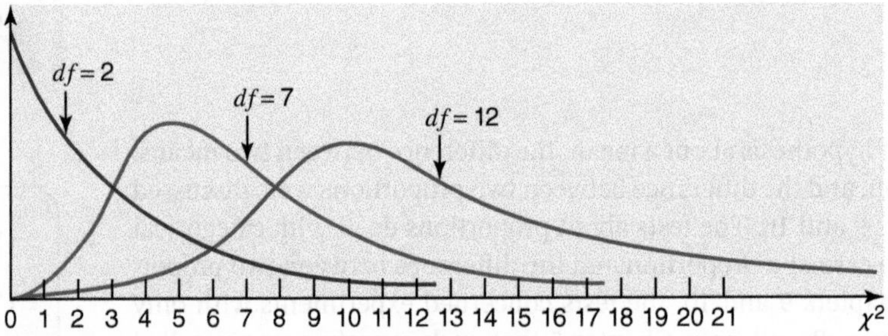

FIGURE 11.1 Three chi-square distribution curves.

As we can see from Figure 11.1, the shape of a chi-square distribution curve is skewed to the right for very small degrees of freedom, and becomes less skewed as the degrees of freedom increase. Eventually, for large degrees of freedom, the chi-square distribution curve looks like a normal distribution curve. The peak (or mode) of a chi-square distribution curve with 1 or 2 degrees of freedom occurs at zero and for a curve with 3 or more degrees of freedom at $df - 2$. For instance, the peak of the chi-square distribution curve with $df = 2$ in Figure 11.1 occurs at zero. The peak for the curve with $df = 7$ occurs at $7 - 2 = 5$. Finally, the peak for the curve with $df = 12$ occurs at $12 - 2 = 10$. Like all other continuous distribution curves, the total area under a chi-square distribution curve is 1.0.

The Chi-Square Distribution

The *chi-square distribution* has only one parameter, called the degrees of freedom. The shape of a chi-square distribution curve is skewed to the right for small *df* and becomes approximately symmetric for large *df*. The chi-square distribution assumes nonnegative values only, and these are denoted by the symbol χ^2 (read as "chi-square").

If we know the degrees of freedom and the area in the right tail of a chi-square distribution curve, we can find the value of χ^2 from Table VI of Appendix B. Examples 11.1 and 11.2 show how to read that table.

[1]The mean of a chi-square distribution is equal to its *df*, and the standard deviation is equal to $\sqrt{2\,df}$.

EXAMPLE 11.1

Reading the Chi-Square Distribution Table: Area in the Right Tail Known

Find the value of χ^2 for 7 degrees of freedom and an area of .10 in the right tail of the chi-square distribution curve.

Solution To find the required value of χ^2, we locate 7 in the column for *df* and .100 in the top row in Table VI of Appendix B. The required χ^2 value is given by the entry at the intersection of the row for 7 and the column for .100. This value is 12.017. The relevant portion of Table VI is presented as **Table 11.1** here.

TABLE 11.1 χ^2 for $df = 7$ and .10 Area in the Right Tail

	Area in the Right Tail Under the Chi-Square Distribution Curve				
df	.995	...	**.100**	...	.005
1	0.000	...	2.706	...	7.879
2	0.010	...	4.605	...	10.597
.	...	...	...	...	...
.	...	...	...	...	...
.	...	...	...	...	...
7	0.989	...	**12.017** ← Required value of χ^2	...	20.278
.	...	...	...	...	...
.	...	...	...	...	...
.	...	...	...	...	...
100	67.328	...	118.498	...	140.169

As shown in **Figure 11.2**, for $df = 7$ and an area of .10 in the right tail of the chi-square distribution curve, the χ^2 value is **12.017**.

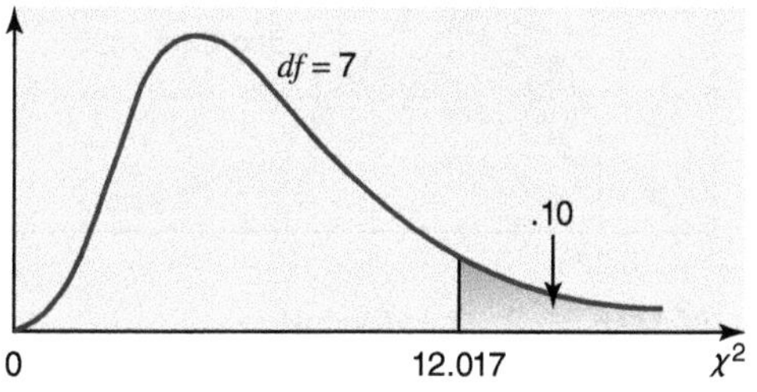

FIGURE 11.2 The χ^2 value.

EXAMPLE 11.2

Reading the Chi-Square Distribution Table: Area in the Left Tail Known

Find the value of χ^2 for 12 degrees of freedom and an area of .05 in the left tail of the chi-square distribution curve.

Solution We can read Table VI of Appendix B only when an area in the right tail of the chi-square distribution curve is known. When the given area is in the left tail, as in this example, the first step is to find the area in the right tail of the chi-square distribution curve as follows:

$$\text{Area in the right tail} = 1 - \text{Area in the left tail}$$

Therefore, for our example,

$$\text{Area in the right tail} = 1 - .05 = .95$$

Next, we locate 12 in the column for df and .950 in the top row in Table VI of Appendix B. The required value of χ^2, given by the entry at the intersection of the row for 12 and the column for .950, is 5.226. The relevant portion of Table VI is presented as **Table 11.2** here.

TABLE 11.2 χ^2 for $df = 12$ and .95 Area in the Right Tail

	Area in the Right Tail Under the Chi-Square Distribution Curve				
df	**.995**	**...**	**.950**	**...**	**.005**
1	0.000	...	0.004	...	7.879
2	0.010	...	0.103	...	10.597
.	...	...	...	...	...
.	...	...	...	...	...
.	...	...	...	...	...
12	3.074	...	5.226	...	28.300
.	...	...	...	...	...
.	...	...	...	...	...
.	...	...	...	...	...
100	67.328	...	77.929	...	140.169

Required value of χ^2

As shown in **Figure 11.3**, for $df = 12$ and .05 area in the left tail, the χ^2 value is **5.226**.

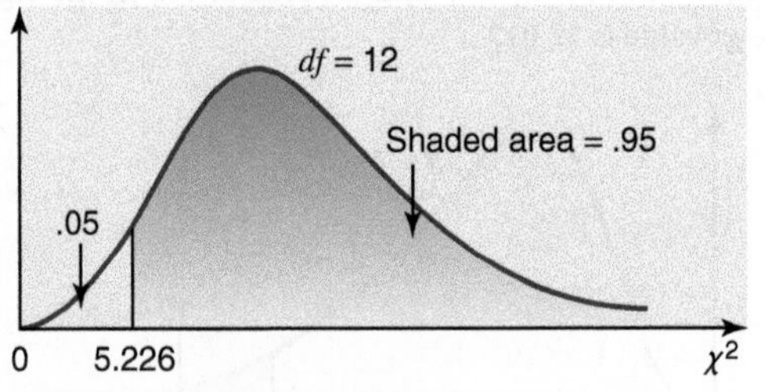

FIGURE 11.3 The χ^2 value.

EXERCISES

Concepts and Procedures

11.1 Describe the characteristics of a chi-square distribution. What is the parameter (parameters) of such a distribution?

11.2 Find the value of χ^2 for 14 degrees of freedom and an area of .025 in the right tail of the chi-square distribution curve.

11.3 Determine the value of χ^2 for 16 degrees of freedom and an area of .10 in the left tail of the chi-square distribution curve.

11.4 Determine the value of χ^2 for 28 degrees of freedom and an area of .990 in the left tail of the chi-square distribution curve.

11.5 Find the value of χ^2 for 12 degrees of freedom and

a. .025 area in the left tail of the chi-square distribution curve

b. .995 area in the right tail of the chi-square distribution curve

11.2 A Goodness-of-Fit Test

LEARNING OBJECTIVES

After completing this section, you should be able to:

- Describe the four characteristics of a multinomial distribution.
- Explain when a goodness-of-fit test is applied.
- Verify the conditions for a goodness-of-fit test.
- Explain the difference between the observed and expected frequencies for a goodness-of-fit test.
- Calculate the expected frequency of each category.
- Calculate the degrees of freedom for a goodness-of-fit test.
- Apply the five-step method using a critical value approach to perform a goodness-of-fit test.

This section explains how to make tests of hypothesis about experiments with more than two possible outcomes (or categories). Such experiments, called **multinomial experiments**, possess four characteristics. Note that a binomial experiment is a special case of a multinomial experiment.

A Multinomial Experiment

An experiment with the following characteristics is called a *multinomial experiment*:

1. The experiment consists of n identical trials (repetitions).
2. Each trial results in one of k possible outcomes (or categories), where $k > 2$.
3. The trials are independent.
4. The probabilities of the various outcomes remain constant for each trial.

An experiment of many rolls of a fair six-sided die is an example of a multinomial experiment. It consists of many identical rolls (trials); each roll (trial) results in one of the six possible outcomes; each roll is independent of the other rolls; and the probabilities of the six outcomes remain constant for each roll.

As a second example of a multinomial experiment, suppose we select a random sample of people and ask them whether or not the quality of American cars is better than that of Japanese cars. The response of a person can be *yes*, *no*, or *no opinion*. Each person included in the sample can be considered as one trial (repetition) of the experiment. There will be as many trials for this experiment as the number of persons selected. Each person can belong to any of the three categories—*yes*, *no*, or *no opinion*. The response of each selected person is independent of the responses of the other persons. Given that the population is large, the probabilities of a person belonging to the three categories remain the same for each trial. Consequently, this is an example of a multinomial experiment.

The frequencies obtained from the actual performance of such an experiment are called the **observed frequencies**. In a **goodness-of-fit test**, we test the null hypothesis that the observed frequencies for an experiment follow a certain pattern or theoretical distribution. The test is called a goodness-of-fit test because the hypothesis tested is how *good* the observed frequencies *fit* a given pattern.

For our first example involving the experiment of many rolls of a die, we may test the null hypothesis that the given die is fair against the alternative hypothesis that the die is not fair. The die will be fair if the observed frequency for each outcome is close to one-sixth of the total number of rolls.

For our second example involving opinions of people on the quality of American cars, suppose such a survey was conducted in 2019, and in that survey 41% of the people said *yes*, 48% said *no*, and 11% had *no opinion*. We want to test if these percentages still hold true. Suppose we take a random sample of 1000 adults and observe that 536 of them think that the quality of American cars is better than that of Japanese cars, 362 say it is worse, and 102 have no opinion. The frequencies 536, 362, and 102 are the observed frequencies. These frequencies are obtained by actually performing the survey. Now, assuming that the 2019 percentages are still true (which will be our null hypothesis), in a sample of 1000 adults we will expect 410 to say *yes*, 480 to say *no*, and 110 to have *no opinion*. These frequencies are obtained by multiplying the sample size (1000) by the 2019 proportions. These frequencies are called the **expected frequencies.** Then, we make a decision to reject or not to reject the null hypothesis based on how large the difference between the observed frequencies and the expected frequencies is. To perform this test, we will use the chi-square distribution. Note that in this case we are testing the null hypothesis that all three percentages (or proportions) are unchanged. However, if we want to make a test for only one of the three proportions, we use the test of a proportion from Chapter 9. For example, if we are testing the hypothesis that the current percentage of people who think the quality of American cars is better than that of the Japanese cars is different from 41%, then we will test the null hypothesis $H_0: p = .41$ against the alternative hypothesis $H_1: p \neq .41$ and use the procedure discussed in Section 9.4 of Chapter 9.

As mentioned earlier, the frequencies obtained from the performance of an experiment are called the observed frequencies. They are denoted by ***O***. To make a goodness-of-fit test, we calculate the expected frequencies for all categories of the experiment. The expected frequency for a category, denoted by ***E***, is given by the product of n and p, where n is the total number of trials and p is the probability for that category.

Observed and Expected Frequencies

The frequencies obtained from the performance of an experiment are called the *observed frequencies* and are denoted by ***O***. The *expected frequencies*, denoted by ***E***, are the frequencies that we expect to obtain if the null hypothesis is true. The expected frequency for a category is obtained as

$$E = np$$

where n is the sample size and p is the probability that an element belongs to that category if the null hypothesis is true.

Degrees of Freedom for a Goodness-of-Fit Test

In a goodness-of-fit test, the *degrees of freedom* are

$$df = k - 1$$

where k denotes the number of possible outcomes (or categories) for the experiment.

The procedure to make a goodness-of-fit test involves the same five steps that we used in the preceding chapters. *The chi-square goodness-of-fit test is always a right-tailed test.*

Test Statistic for a Goodness-of-Fit Test

The *test statistic for a goodness-of-fit test* is χ^2, and its value is calculated as

$$\chi^2 = \sum \frac{(O - E)^2}{E}$$

where O = observed frequency for a category

E = expected frequency for a category = np

Remember that a *chi-square goodness-of-fit test is always a right-tailed test.*

Whether or not the null hypothesis is rejected depends on how much the observed and expected frequencies differ from each other. To find how large the difference between the observed frequencies and the expected frequencies is, we do not look at just $\Sigma(O - E)$, because some of the $O - E$ values will be positive and others will be negative. The net result of the sum of these differences will always be zero. Therefore, we square each of the $O - E$ values to obtain $(O - E)^2$, and then weight them according to the reciprocals of their expected frequencies. The sum of the resulting numbers gives the computed value of the test statistic χ^2.

To make a goodness-of-fit test, the sample size should be large enough so that the expected frequency for each category is at least 5. If there is a category with an expected frequency of less than 5, either increase the sample size or combine two or more categories to make each expected frequency at least 5.

Examples 11.3 and 11.4 describe the procedure for performing goodness-of-fit tests using the chi-square distribution.

EXAMPLE 11.3 Number of People Using an ATM Each Day

Conducting a Goodness-of-Fit Test: Equal Proportions for All Categories

A bank has an ATM installed inside the bank, and it is available to its customers only from 7 A.M. to 6 P.M. Monday through Friday. The manager of the bank wanted to investigate if the number of people who use this ATM is the same for each of the five days (Monday through Friday) of the week. The manager randomly selected one week and counted the number of people who used this ATM on each of the five days during that week. The information obtained is given in the table, where the number of users represents the number of people who used this ATM on these days.

Photodisc/Getty Images, Inc.

Day	Monday	Tuesday	Wednesday	Thursday	Friday
Number of users	253	197	204	279	267

At a 1% level of significance, can you reject the null hypothesis that the number of people who use this ATM each of the five days of the week is the same? Assume that this week is typical of all weeks in regard to the use of this ATM.

Solution To conduct this test of hypothesis, we proceed as follows:

STEP 1 *State the null and alternative hypotheses.*

Because there are five categories (days) as listed in the table, the number of ATM users will be the same for each of these five days if 20% of all users use the ATM each day. The null and alternative hypotheses are as follows:

H_0: The number of people using the ATM is the same for all five days of the week.

H_1: The number of people using the ATM is not the same for all five days of the week.

If the number of people using this ATM is the same for all five days of the week, then .20 of the users will use this ATM on any of the five days of the week. Let p_1, p_2, p_3, p_4, and p_5 be the proportion of people who use this ATM on Monday, Tuesday, Wednesday, Thursday, and Friday, respectively. Then, the null and alternative hypotheses can also be written as

$$H_0: p_1 = p_2 = p_3 = p_4 = p_5 = .20$$
$$H_1: \text{At least two of the five proportions are not equal to .20}$$

STEP 2 *Select the distribution to use.*

Because there are five categories (i.e., five days on which the ATM is used), this is an example of a multinomial experiment. Consequently, we use the chi-square distribution to make this test.

STEP 3 *Determine the rejection and nonrejection regions.*

The significance level is given to be .01, and the goodness-of-fit test is always right-tailed. Therefore, the area in the right tail of the chi-square distribution curve is .01, that is,

$$\text{Area in the right tail} = \alpha = .01$$

The degrees of freedom are calculated as follows:

$$k = \text{number of categories} = 5$$
$$df = k - 1 = 5 - 1 = 4$$

From the chi-square distribution table (Table VI of Appendix B), for $df = 4$ and .01 area in the right tail of the chi-square distribution curve, the critical value of χ^2 is 13.277, as shown in **Figure 11.4.**

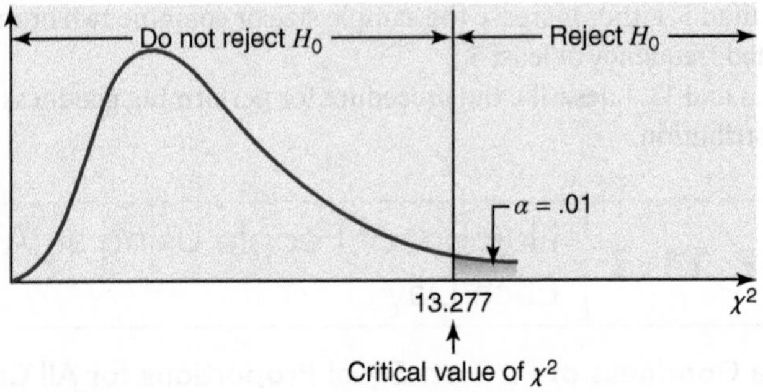

FIGURE 11.4 Rejection and nonrejection regions.

STEP 4 *Calculate the value of the test statistic.*

All the required calculations to find the value of the test statistic χ^2 are shown in **Table 11.3.** Note that the condition that all expected frequencies must be at least 5 is satisfied. The calculations made in Table 11.3 are explained as follows:

1. The first two columns of Table 11.3 list the five categories (days) and the observed frequencies for the 1200 persons who used the ATM during each of the five days of the selected week. The third column contains the probabilities for the five categories assuming that the null hypothesis is true.
2. The fourth column contains the expected frequencies. These frequencies are obtained by multiplying the total users ($n = 1200$) by the probabilities listed in the third column. If the null hypothesis is true (i.e., the ATM users are equally distributed over all five days), then we will expect 240 out of 1200 persons to use the ATM each day. Consequently, each category in the fourth column has the same expected frequency.
3. The fifth column lists the differences between the observed and expected frequencies, that is, $O - E$. These values are squared and recorded in the sixth column.
4. Finally, we divide the squared differences (that appear in the sixth column) by the corresponding expected frequencies (listed in the fourth column) and write the resulting numbers in the seventh column.

TABLE 11.3 Calculating the Value of the Test Statistic

Category (Day)	Observed Frequency O	p	Expected Frequency $E = np$	$(O - E)$	$(O - E)^2$	$\frac{(O - E)^2}{E}$
Monday	253	.20	1200(.20) = 240	13	169	.704
Tuesday	197	.20	1200(.20) = 240	−43	1849	7.704
Wednesday	204	.20	1200(.20) = 240	−36	1296	5.400
Thursday	279	.20	1200(.20) = 240	39	1521	6.338
Friday	267	.20	1200(.20) = 240	27	729	3.038
	$n = 1200$					Sum = 23.184

5. The sum of the seventh column gives the value of the test statistic χ^2. Thus,

$$\chi^2 = \sum \frac{(O-E)^2}{E} = 23.184$$

STEP 5 *Make a decision.*

The value of the test statistic $\chi^2 = 23.184$ is larger than the critical value of $\chi^2 = 13.277$, and falls in the rejection region. Hence, we reject the null hypothesis and state that the number of persons who use this ATM seems not to be the same for each of the five days of the week. In other words, we conclude that a higher number of users of this ATM use this machine on one or more of these days.

If you make this chi-square test using any of the statistical software packages, you will obtain a p-value of .000116 for the test. In this case you can compare the p-value obtained in the computer output with the level of significance and make a decision. As you know from Chapter 9, you will reject the null hypothesis if α (significance level) is greater than or equal to the p-value and not reject it otherwise. In this example, $\alpha = .01$, which is larger than p-value $= .000116$. Hence, we reject the null hypothesis.

EXAMPLE 11.4 Is Climate Change Affecting Your Local Community?

Conducting a Goodness-of-Fit Test: Testing if Results of a Survey Fit a Given Distribution

In a Pew Research Center survey conducted October 1–13, 2019, U.S. adults, aged 18 and older, were asked how much global climate change was affecting their local community. Of the respondents, 22% said a great deal, 39% said some, 38% said not too much or not at all, and 1% gave no answer (www.pewresearch.org). Assume that these percentages hold true for the 2019 population of all U.S. adults aged 18 and older. Recently, 1000 randomly selected U.S. adults, aged 18 and older, were asked the same question. The following table lists the number of adults in this sample who belonged to each response.

Response	Frequency
A great deal	242
Some	406
Not too much/not at all	338
No answer	14

Test at a 2.5% level of significance whether the current distribution of opinions is different from that for 2019.

Solution We perform the following five steps for this test of hypothesis:

STEP 1 *State the null and alternative hypotheses.*

The null and alternative hypotheses are

H_0: The current percentage distribution of opinions is the same as for 2019.

H_1: The current percentage distribution of opinions is different from that for 2019.

STEP 2 *Select the distribution to use.*

Because this experiment has four categories as listed in the table, it is a multinomial experiment. Consequently we use the chi-square distribution to make this test.

STEP 3 *Determine the rejection and nonrejection regions.*

The significance level is given to be .025, and because the goodness-of-fit test is always right-tailed, the area in the right tail of the chi-square distribution curve is .025, that is,

$$\text{Area in the right tail} = \alpha = .025$$

The degrees of freedom are calculated as follows:

$$k = \text{number of categories} = 4$$
$$df = k - 1 = 4 - 1 = 3$$

From the chi-square distribution table (Table VI of Appendix B), for $df = 3$ and .025 area in the right tail of the chi-square distribution curve, the critical value of χ^2 is 9.348, as shown in **Figure 11.5**.

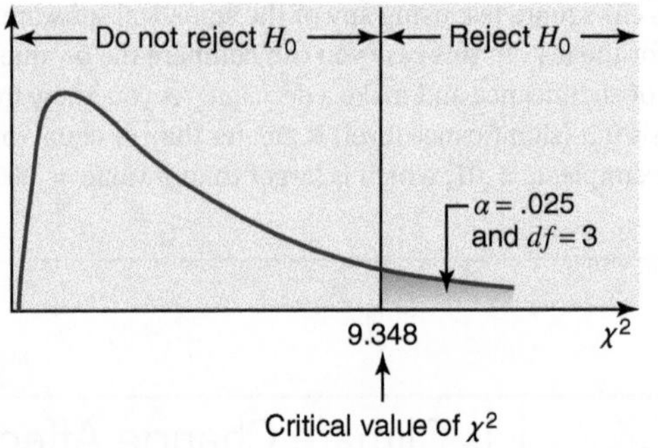

FIGURE 11.5 Rejection and nonrejection regions.

STEP 4 *Calculate the value of the test statistic.*

All the required calculations to find the value of the test statistic χ^2 are shown in **Table 11.4**. Note that the four percentages for 2019 have been converted into probabilities and recorded in the third column of Table 11.4. The value of the test statistic χ^2 is given by the sum of the last column. Thus,

$$\chi^2 = \sum \frac{(O - E)^2}{E} = 9.098$$

TABLE 11.4 Calculating the Value of the Test Statistic

Category (Response)	Observed Frequency O	p	Expected Frequency $E = np$	$(O - E)$	$(O - E)^2$	$\frac{(O - E)^2}{E}$
A great deal	242	.22	1000(.22) = 220	22	484	2.200
Some	406	.39	1000(.39) = 390	16	256	.656
Not too much/ not at all	338	.38	1000(.38) = 380	−42	1764	4.642
No answer	14	.01	1000(.01) = 10	4	16	1.600
	$n = 1000$					Sum = 9.098

Note that the condition that all expected frequencies must be at least 5 is satisfied.

STEP 5 *Make a decision.*

The observed value of the test statistic $\chi^2 = 9.098$ is smaller than the critical value of $\chi^2 = 9.348$, and falls in the nonrejection region. Hence, we fail to reject the null hypothesis, and state that the data from the sample do not provide sufficient evidence that the current percentage distribution of opinions is different from that for 2019.

If you make this chi-square test using any of the statistical software packages, you will obtain a *p*-value of .028 for the test. In this case you can compare the *p*-value obtained from the computer output with the level of significance and make a decision. As you know from Chapter 9, you will reject the null hypothesis if α (significance level) is greater than or equal to the *p*-value and not reject it otherwise. In this example, $\alpha = .025$, which is smaller than *p*-value = .028. Hence, we fail to reject the null hypothesis.

Case Study 11.1

How Are the Economic Conditions in the United States Affecting the Middle Class?

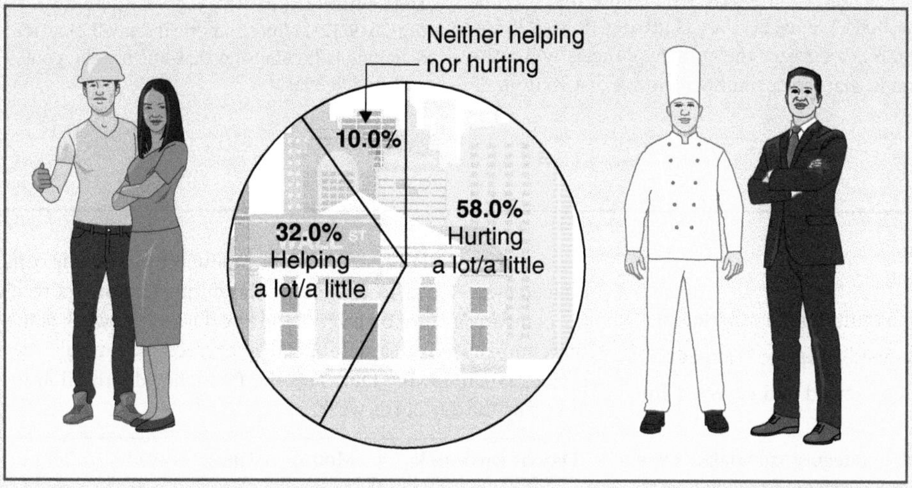

In a Pew Research Center survey of randomly selected U.S. adults conducted September 16–29, 2019, these adults were asked, "How are the economic conditions in the country affecting the middle class?" (https://www.pewsocialtrends.org) The pie chart shows the percentage distribution of the responses of these adults. Fifty-eight percent of the adults polled said that the economic conditions in the country are hurting the middle class a lot or a little, 32% said that they are helping the middle class a lot or a little, and 10% were of the opinion that they neither help nor hurt the middle class. Assume that these percentages were true for the population of U.S. adults in 2019. Suppose that we want to test whether the current percentage distribution for the survey question is different from 2019. Then the two hypotheses are as follows:

H_0: The current percentage distribution of opinions is the same as in 2019

H_1: The current percentage distribution of opinions is not the same as in 2019

To test this hypothesis, suppose we currently take a sample of 2000 U.S. adults and ask them how the economic conditions in the country are affecting the middle class.

Suppose that 1242 of the adults included in the survey say that the current economic conditions are hurting the middle class a lot or a little, 584 say that these conditions are helping the middle class a lot or a little, and 174 said that the current economic conditions in the country are neither helping nor hurting the middle class. Using the given information, we calculate the value of the test statistic as shown in the following table:

Category	Observed Frequency O	p	Expected Frequency $E = np$	$(O - E)$	$(O - E)^2$	$\frac{(O - E)^2}{E}$
Hurting	1242	.58	$2000(.58) = 1160$	82	6724	5.797
Helping	584	.32	$2000(.32) = 640$	−56	3136	4.900
Neither	174	.10	$2000(.10) = 200$	−26	676	3.380
	$n = 2000$					Sum = 14.077

Suppose we use a 1% significance level to perform this test. Then for $df = 3 - 1 = 2$ and .01 area in the right tail, the critical value of χ^2 is 9.210 from Table VI of Appendix B. Since the observed value of χ^2 is 14.077 and it is larger than the critical value of $\chi^2 = 9.210$, we reject the null hypothesis. Thus, based on evidence from the sample information, we conclude that the current percentage distribution of opinions of U.S. adults in response to the survey question is significantly different from the distribution of opinions in 2019.

We can also use the *p*-value approach to make this decision. In Table VI of Appendix B, for

$df = 2$, the largest value of χ^2 is 10.597 and the area to the right of $\chi^2 = 10.597$ is .005. Thus, the p-value for $\chi^2 = 14.077$ will be less than .005. (By using technology, we obtain the p-value of .0009.) Since $\alpha = .01$ is greater than .005 (or .0009), we reject the null hypothesis and conclude that the current percentage distribution of opinions of U.S. adults in response to the survey question seems to be significantly different from the distribution of opinions in 2019.

Data source: https://www.pewsocialtrends.org/2019/12/11/most-americans-say-the-current-economy-is-helping-the-rich-hurting-the-poor-and-middle-class/.

EXERCISES

Concepts and Procedures

11.6 Describe the four characteristics of a multinomial experiment.

11.7 What is a goodness-of-fit test and when is it applied? Explain.

11.8 Explain the difference between the observed and expected frequencies for a goodness-of-fit test.

11.9 How is the expected frequency of a category calculated for a goodness-of-fit test? What are the degrees of freedom for such a test?

11.10 To make a goodness-of-fit test, what should be the minimum expected frequency for each category? What are the alternatives if this condition is not satisfied?

11.11 The following table lists the frequency distribution for 60 rolls of a die.

Outcome	1-spot	2-spot	3-spot	4-spot	5-spot	6-spot
Frequency	7	15	10	16	8	4

Test at a 5% significance level whether we can reject the null hypothesis that the given die is fair.

Applications

11.12 A drug company is interested in investigating whether the color of their packaging has any impact on sales. To test this, they used five different colors (blue, green, orange, red, and yellow) for their packages of an over-the-counter pain reliever, instead of the traditional white package. The following table shows the number of packages of each color sold during the first month.

Package color	Blue	Green	Orange	Red	Yellow
Number of packages sold	340	246	295	189	284

Using a 1% significance level, test the null hypothesis that the number of packages sold of each of these five colors is the same.

11.13 Over the last three years, Art's Supermarket has observed the following distribution of payment methods in the checkout lines: cash (C) 41%, check (CK) 24%, credit or debit card (D) 26%, and other (N) 9%. In an effort to minimize costly credit and debit card fees, Art's has just begun offering a 1% discount for cash payment in the checkout line. The following table lists the frequency distribution of payment methods for a random sample of 600 customers after the discount went into effect.

Payment method	C	CK	D	N
Number of customers	265	129	136	70

Test at a 1% significance level whether the distribution of payment methods in the checkout line changed after the discount went into effect.

11.14 Home Mail Corporation sells products by mail. The company's management wants to find out if the number of orders received at the company's office on each of the five days of the week is the same. The company took a sample of 400 orders received during a four-week period. The following table lists the frequency distribution for these orders by the day of the week.

Day of the week	Mon	Tue	Wed	Thu	Fri
Number of orders received	98	68	52	86	96

Test at a 5% significance level whether we can reject the null hypothesis that the orders are evenly distributed over all days of the week.

11.15 Of all students enrolled at a large undergraduate university, 19% are seniors, 23% are juniors, 27% are sophomores, and 31% are freshmen. A sample of 200 students taken from this university by the student senate to conduct a survey includes 50 seniors, 46 juniors, 55 sophomores, and 49 freshmen. Using a 2.5% significance level, test whether the null hypothesis that this sample is a random sample can be rejected. (*Hint:* This sample will be a random sample if it includes approximately 19% seniors, 23% juniors, 27% sophomores, and 31% freshmen.)

11.16 Chance Corporation produces beauty products. Two years ago the quality control department at the company conducted a survey of users of one of the company's products. The survey revealed that 53% of the users said the product was excellent, 31% said it was satisfactory, 7% said it was unsatisfactory, and 9% had no opinion. Assume that these percentages were true for the population of all users of this product at that time. After this survey was conducted, the company redesigned this product. A recent survey of 800 users of the redesigned product conducted by the quality control department at the company showed that 495 of the users think the product is excellent, 255 think it is satisfactory, 35 think it is unsatisfactory, and 15 have no opinion. Is the percentage distribution of the opinions of users of the redesigned product different from the percentage distribution of users of this product before it was redesigned? Use $\alpha = .025$.

11.17 Henderson Corporation makes metal sheets, among other products. When the process that is used to make metal sheets works properly, 92% of the sheets contain no defects, 5% have one defect each, and 3% have two or more defects each. The quality control inspectors at the company take samples of metal sheets quite often and check them for defects. If the distribution of defects for a sample is significantly different from the above-mentioned percentage distribution, the process is stopped and adjusted. A recent sample of 300 sheets produced the frequency distribution of defects listed in the following table:

Number of defects	None	One	Two or More
Number of metal sheets	262	24	14

Does the evidence from this sample suggest that the process needs an adjustment? Use $\alpha = .01$.

11.3 A Test of Independence or Homogeneity

LEARNING OBJECTIVES

After completing this section, you should be able to:

- Describe the difference between a test of independence and a test of homogeneity.
- Verify the conditions for the tests of independence and homogeneity.
- Write the null and alternative hypotheses for the tests of independence and homogeneity.
- Use a χ^2-distribution to illustrate the critical value, the rejection region and the nonrejection region for the tests of independence and homogeneity.
- Calculate the value of a χ^2 test statistic.
- Make a conclusion based on a comparison of the test statistic and the critical value.
- Apply the five-step method using a critical value approach to perform the tests of hypothesis of independence and homogeneity.

This section is concerned with tests of independence and homogeneity, which are performed using contingency tables. Except for a few modifications, the procedure used to make such tests is almost the same as the one applied in Section 11.2 for a goodness-of-fit test.

11.3.1 A Contingency Table

Often we may have information on more than one categorical variable for each element. Such information can be summarized and presented using a **two-way classification table**, which is also called a *contingency table* or *cross-tabulation*. Suppose a university has a total enrollment of 20,758 students. By classifying these students based on gender and whether these students are full-time or part-time, we can prepare **Table 11.5**, which provides an example of a contingency table. Table 11.5 has two rows (one for males and the second for females) and two columns (one for full-time and the second for part-time students). Hence, it is also called a 2 × 2 (read as "two by two") contingency table.

TABLE 11.5 Total Enrollment at a University

	Full-Time	Part-Time
Male	6768	2615 ← Students who are male and enrolled part time
Female	7658	3717

A contingency table can be of any size. For example, it can be 2 × 3, 3 × 2, 3 × 3, or 4 × 2. Note that in these notations, the first digit refers to the number of rows in the table, and the second digit refers to the number of columns. For example, a 3 × 2 table will contain three rows and two columns. In general, an $R \times C$ table contains R rows and C columns, where R and C represent the number of rows and columns, respectively. For example, in a 3 × 2 table, $R = 3$ and $C = 2$.

Each of the four boxes that contain numbers in Table 11.5 is called a **cell**. The number of cells in a contingency table is obtained by multiplying the number of rows by the number of columns. Thus, Table 11.5 contains 2 × 2 = 4 cells. The subjects that belong to a cell of a contingency table possess two characteristics. For example, the 2615 students listed in the second cell of the first row in Table 11.5 are *male* and enrolled *part time*. The numbers written inside the cells are called the *joint frequencies*. For example, 2615 students belong to the joint category of *male* and *part time*. Hence, 2615 is also referred to as the joint frequency of this category.

11.3.2 A Test of Independence

In a **test of independence** for a contingency table, we test the null hypothesis that the two attributes (characteristics) of the elements of a given population are not related (that is, they are independent) against the alternative hypothesis that the two characteristics are related (that is, they are dependent). For example, we may want to test if the affiliation of people with the Democratic and Republican parties is independent of their income levels. We perform such a test by using the chi-square distribution. As another example, we may want to test if there is an association between being a man or a woman and having a preference for watching sports or soap operas on television.

Degrees of Freedom for a Test of Independence

A test of independence involves a test of the null hypothesis that two attributes of a population are not related against the alternative hypothesis that the two attributes are related. The *degrees of freedom for a test of independence* are

$$df = (R - 1)(C - 1)$$

where R and C are the number of rows and the number of columns, respectively, in the given contingency table.

The value of the test statistic χ^2 in a test of independence is obtained using the same formula as in the goodness-of-fit test described in Section 11.2.

Test Statistic for a Test of Independence

The value of the *test statistic* χ^2 *for a test of independence* is calculated as

$$\chi^2 = \sum \frac{(O - E)^2}{E}$$

where O and E are the observed and expected frequencies, respectively, for a cell.

The null hypothesis in a test of independence always specifies that the two attributes are not related. The alternative hypothesis is that the two attributes are related.

The frequencies obtained from the performance of an experiment for a contingency table are called the **observed frequencies**. The procedure to calculate the **expected frequencies** for a contingency table for a test of independence is different from that for a goodness-of-fit test. Example 11.5 describes this procedure.

EXAMPLE 11.5 Lack of Discipline in Schools

Calculating Expected Frequencies for a Test of Independence

Many adults think that lack of discipline has become a major problem in schools in the United States. A random sample of 300 adults was selected, and these adults were asked if they favor giving more freedom to school teachers to punish students for lack of discipline. The two-way classification of the responses of these adults is presented in the following table:

	In Favor (F)	Against (A)	No Opinion (N)
Men (M)	93	70	12
Women (W)	87	32	6

Calculate the expected frequencies for this table, assuming that the two attributes, gender and opinions on the issue, are independent.

Solution The preceding table is reproduced as **Table 11.6** here. Note that Table 11.6 includes the row and column totals.

TABLE 11.6 Observed Frequencies

	In Favor (F)	Against (A)	No Opinion (N)	Row Totals
Men (M)	93	70	12	175
Women (W)	87	32	6	125
Column Totals	180	102	18	300

The numbers 93, 70, 12, 87, 32, and 6 listed inside Tables 11.5 and 11.6 are the **observed frequencies** of the respective cells. These frequencies are obtained from the sample.

As mentioned earlier, the null hypothesis in a test of independence specifies that the two attributes (or classifications) are independent. In a test of independence, first we assume that the null hypothesis is true and that the two attributes are independent. Assuming that the null hypothesis is true and that gender and opinions are not related, we calculate the expected frequency for the cell corresponding to *Men* and *In Favor* as shown below. From Table 11.6,

$$P(\text{a person is a } Man) = P(M) = 175/300$$
$$P(\text{a person is } In\ Favor) = P(F) = 180/300$$

Because we are assuming that M and F are independent (by assuming that the null hypothesis is true), from the formula learned in Chapter 4, the joint probability of these two events is

$$P(M \text{ and } F) = P(M) \times P(F) = (175/300) \times (180/300)$$

Then, assuming that M and F are independent, the number of persons expected to be *Men* and *In Favor* in a sample of 300 is

$$\begin{aligned} E \text{ for } Men \text{ and } In\ Favor &= 300 \times P(M \text{ and } F) \\ &= 300 \times \frac{175}{300} \times \frac{180}{300} = \frac{175 \times 180}{300} \\ &= \frac{(\text{Row total})(\text{Column total})}{\text{Sample size}} \end{aligned}$$

Thus, the rule for obtaining the expected frequency for a cell is to divide the product of the corresponding row and column totals by the sample size.

Expected Frequencies for a Test of Independence

The expected frequency E for a cell is calculated as

$$E = \frac{(\text{Row total})(\text{Column total})}{\text{Sample size}}$$

Using this rule, we calculate the expected frequencies of the six cells of Table 11.6 as follows:

$$E \text{ for } Men \text{ and } In\ Favor \text{ cell} = (175)(180)/300 = \mathbf{105.00}$$
$$E \text{ for } Men \text{ and } Against \text{ cell} = (175)(102)/300 = \mathbf{59.50}$$
$$E \text{ for } Men \text{ and } No\ Opinion \text{ cell} = (175)(18)/300 = \mathbf{10.50}$$
$$E \text{ for } Women \text{ and } In\ Favor \text{ cell} = (125)(180)/300 = \mathbf{75.00}$$
$$E \text{ for } Women \text{ and } Against \text{ cell} = (125)(102)/300 = \mathbf{42.50}$$
$$E \text{ for } Women \text{ and } No\ Opinion \text{ cell} = (125)(18)/300 = \mathbf{7.50}$$

The expected frequencies are usually written in parentheses below the observed frequencies within the corresponding cells, as shown in **Table 11.7**.

TABLE 11.7 Observed and Expected Frequencies

	In Favor (F)	Against (A)	No Opinion (N)	Row Totals
Men (M)	93 (105.00)	70 (59.50)	12 (10.50)	175
Women (W)	87 (75.00)	32 (42.50)	6 (7.50)	125
Column Totals	180	102	18	300

Like a goodness-of-fit test, **a test of independence is always right-tailed.** To apply a chi-square test of independence, **the sample size should be large enough so that the expected frequency for each cell is at least 5.** If the expected frequency for a cell is less than 5, we either increase the sample size or combine some categories. Examples 11.6 and 11.7 describe the procedure to make tests of independence using the chi-square distribution.

EXAMPLE 11.6 Lack of Discipline in Schools

Making a Test of Independence: 2 × 3 Table

Reconsider the two-way classification table given in Example 11.5. In that example, a random sample of 300 adults was selected, and they were asked if they favor giving more freedom to school teachers to punish students for lack of discipline. Based on the results of the survey, a two-way classification table was prepared and presented in Example 11.5. Does the sample provide sufficient evidence to conclude that the two attributes, gender and opinions of adults, are dependent? Use a 1% significance level.

Solution The test involves the following five steps:

STEP 1 *State the null and alternative hypotheses.*

As mentioned earlier, the null hypothesis must be that the two attributes are independent. Consequently, the alternative hypothesis is that these attributes are dependent.

H_0: Gender and opinions of adults are independent.

H_1: Gender and opinions of adults are dependent.

STEP 2 *Select the distribution to use.*

We use the chi-square distribution to make a test of independence for a contingency table. We will notice in Step 4 that the condition that all expected frequencies are at least 5 is satisfied.

STEP 3 *Determine the rejection and nonrejection regions.*

The significance level is 1%. Because a test of independence is always right-tailed, the area of the rejection region is .01 and falls in the right tail of the chi-square distribution curve. The contingency table contains two rows (*Men* and *Women*) and three columns (*In Favor*, *Against*, and *No Opinion*). Note that we do not count the row and column of totals. The degrees of freedom are

$$df = (R-1)(C-1) = (2-1)(3-1) = 2$$

From Table VI of Appendix B, for $df = 2$ and $\alpha = .01$, the critical value of χ^2 is 9.210. This value is shown in **Figure 11.6**.

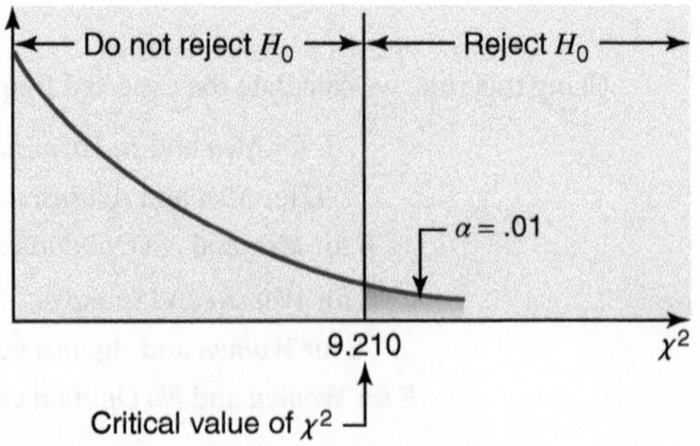

FIGURE 11.6 Rejection and nonrejection regions.

STEP 4 *Calculate the value of the test statistic.*

Table 11.7, with the observed and expected frequencies constructed in Example 11.5, is reproduced as **Table 11.8** here.

TABLE 11.8 Observed and Expected Frequencies

	In Favor (F)	Against (A)	No Opinion (N)	Row Totals
Men (M)	93 (105.00)	70 (59.50)	12 (10.50)	175
Women (W)	87 (75.00)	32 (42.50)	6 (7.50)	125
Column Totals	180	102	18	300

To compute the value of the test statistic χ^2, we take the difference between each pair of observed and expected frequencies listed in Table 11.8, square those differences, and then divide each of the squared differences by the respective expected frequency. The sum of the resulting numbers gives the value of the test statistic χ^2. All these calculations are made as follows:

$$\begin{aligned}\chi^2 &= \Sigma\frac{(O-E)^2}{E}\\ &= \frac{(93-105.00)^2}{105.00}+\frac{(70-59.50)^2}{59.50}+\frac{(12-10.50)^2}{10.50}\\ &\quad+\frac{(87-75.00)^2}{75.00}+\frac{(32-42.50)^2}{42.50}+\frac{(6-7.50)^2}{7.50}\\ &= 1.371+1.853+.214+1.920+2.594+.300=8.252\end{aligned}$$

STEP 5 *Make a decision.*

The value of the test statistic $\chi^2 = 8.252$ is less than the critical value of $\chi^2 = 9.210$, and falls in the nonrejection region. Hence, we fail to reject the null hypothesis and state that there is not enough evidence from the sample to conclude that the two characteristics, *gender* and *opinions of adults*, are dependent for this issue.

The p-value obtained using technology for $df = 2$ and $\chi^2 = 8.252$ is .0161, which is larger than $\alpha = .01$. Hence, we fail to reject the null hypothesis.

EXAMPLE 11.7 Relationship Between Gender and Owning a Smart Phone

Making a Test of Independence: 2 × 2 Table

A researcher wanted to study the relationship between gender and owning smartphones among adults who have cell phones. She took a sample of 2000 adults and obtained the information given in the following table:

	Own a Smartphone	Do Not Own a Smartphone
Men	910	180
Women	714	196

At a 5% level of significance, does the sample provide evidence to conclude that gender and owning a smartphone are related for all adults?

THEGIFT777/iStockphoto

Solution We perform the following five steps to make this test of hypothesis:

STEP 1 *State the null and alternative hypotheses.*

The null and alternative hypotheses are, respectively,

H_0: Gender and owning a smartphone are not related.

H_1: Gender and owning a smartphone are related.

STEP 2 *Select the distribution to use.*

Because we are performing a test of independence, we use the chi-square distribution to make the test. In Step 4, we will notice that all expected frequencies are at least 5.

STEP 3 *Determine the rejection and nonrejection regions.*

With a significance level of 5%, the area of the rejection region is .05 and falls under the right tail of the chi-square distribution curve. The contingency table contains two rows (*men* and *women*) and two columns (*own a smartphone* and *do not own a smartphone*). The degrees of freedom are

$$df = (R-1)(C-1) = (2-1)(2-1) = 1$$

From Table VI of Appendix B, the critical value of χ^2 for $df = 1$ and $\alpha = .05$ is 3.841. This value is shown in **Figure 11.7**.

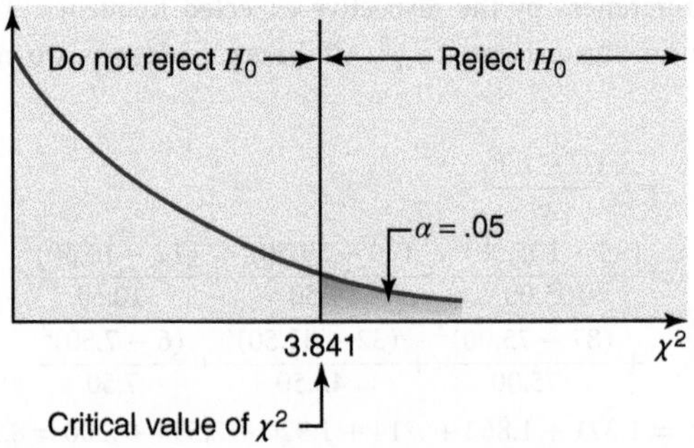

FIGURE 11.7 Rejection and nonrejection regions.

STEP 4 *Calculate the value of the test statistic.*

The expected frequencies for the various cells are calculated as follows and are listed within parentheses in **Table 11.9**.

TABLE 11.9 Observed and Expected Frequencies

	Own a Smartphone (Y)	Do Not Own a Smartphone (N)	Row Totals
Men (M)	910 (885.08)	180 (204.92)	1090
Women (W)	714 (738.92)	196 (171.08)	910
Column Totals	1624	376	2000

E for *men* and *own a smartphone* cell = (1090)(1624)/2000 = 885.08

E for *men* and *do not own a smartphone* cell = (1090)(376)/2000 = 204.92

E for *women* and *own a smartphone* cell = (910)(1624)/2000 = 738.92

E for *women* and *do not own a smartphone* cell = (910)(376)/2000 = 171.08

The value of the test statistic χ^2 is calculated as follows:

$$\chi^2 = \sum \frac{(O-E)^2}{E}$$

$$= \frac{(910-885.08)^2}{885.08} + \frac{(180-204.92)^2}{204.92} + \frac{(714-738.92)^2}{738.92} + \frac{(196-171.08)^2}{171.08}$$

$$= .702 + 3.030 + .840 + 3.630 = 8.202$$

STEP 5 *Make a decision.*

The value of the test statistic $\chi^2 = 8.202$ is larger than the critical value of $\chi^2 = 3.841$ and falls in the rejection region. Hence, we reject the null hypothesis and state that there is sufficient evidence from the sample to conclude that the two characteristics, *gender* and *owning smartphones*, are related for all adults.

From technology, the *p*-value for $df = 1$ and $\chi^2 = 8.202$ is .0042, which is smaller than $\alpha = .05$. Hence, we reject the null hypothesis.

11.3.3 A Test of Homogeneity

In a **test of homogeneity**, we test if two (or more) populations are homogeneous (similar) with regard to the distribution of a certain characteristic. For example, we might be interested in testing the null hypothesis that the proportions of households that belong to different income groups are the same in California and Wisconsin, against the alternative hypothesis that these proportions are different, or we may want to test whether or not the preferences of people in Florida, Arizona, and Vermont are similar with regard to Coke, Pepsi, or 7-Up.

A Test of Homogeneity

A *test of homogeneity* involves testing the null hypothesis that the proportions of elements with certain characteristics in two or more different populations are the same against the alternative hypothesis that these proportions are not the same.

Let us consider the example of testing the null hypothesis that the proportions of households in California and Wisconsin who belong to various income groups are the same. (Note that in a test of homogeneity, the null hypothesis is always that the proportions of elements with certain characteristics are the same in two or more populations. The alternative hypothesis is that these proportions are not the same.) Suppose we define three income strata: high-income group (with an income of more than $400,000), medium-income group (with an income of $70,000 to $400,000), and low-income group (with an income of less than $70,000). Furthermore, assume that we take one sample of 250 households from California and another sample of 150 households from Wisconsin, collect the information on the incomes of these households, and prepare the contingency table as in **Table 11.10**.

TABLE 11.10 Observed Frequencies

	California	Wisconsin	Row Totals
High income	70	34	104
Medium income	80	40	120
Low income	100	76	176
Column Totals	250	150	400

Note that in this example the column totals are fixed. That is, we decided in advance to take samples of 250 households from California and 150 from Wisconsin. However, the row totals (of 104, 120, and 176) are determined randomly by the outcomes of the two samples. If we compare this example to the one about lack of discipline in schools in the previous section, we will notice that neither the column nor the row totals were fixed in that example. Instead, the researcher selected just one sample of 300 adults, collected the information on gender and opinions, and prepared the contingency table. Thus, in that example, the row and column totals were all determined randomly. *Thus, when both the row and column totals are determined randomly, we perform a test of independence. However, when either the column totals or the row totals are fixed, we perform a test of homogeneity.* In the case of income groups in California and Wisconsin, we will perform a test of homogeneity to test for the difference of income groups in the two states.

The procedure to conduct a test of homogeneity is similar to the procedure used to make a test of independence discussed earlier. Like a test of independence, **a test of homogeneity is right-tailed**. Example 11.8 illustrates the procedure to make a homogeneity test.

EXAMPLE 11.8 Distribution of Households in Regard to Income Levels

Performing a Test of Homogeneity

Consider the data on income distributions for households in California and Wisconsin given in Table 11.10. Using a 2.5% significance level, test whether the distribution of households with regard to income levels is different (i.e., not homogeneous) for households in California and Wisconsin.

Solution We perform the following five steps to make this test of hypothesis:

STEP 1 *State the null and alternative hypotheses.*

The two hypotheses are, respectively,[2]

H_0: The proportions of households that belong to different income groups are the same in both states.

H_1: The proportions of households that belong to different income groups are not the same in both states.

STEP 2 *Select the distribution to use.*

We use the chi-square distribution to make a test of homogeneity. We can observe from Step 4 that the condition that all expected frequencies are at least 5 is satisfied.

STEP 3 *Determine the rejection and nonrejection regions.*

The significance level is 2.5%. Because the homogeneity test is right-tailed, the area of the rejection region is .025 and lies in the right tail of the chi-square distribution curve. The contingency table for income groups in California and Wisconsin contains three rows and two columns. Hence, the degrees of freedom are

$$df = (R - 1)(C - 1) = (3 - 1)(2 - 1) = 2$$

From Table VI of Appendix B, the value of χ^2 for $df = 2$ and .025 area in the right tail of the chi-square distribution curve is 7.378. This value is shown in **Figure 11.8**.

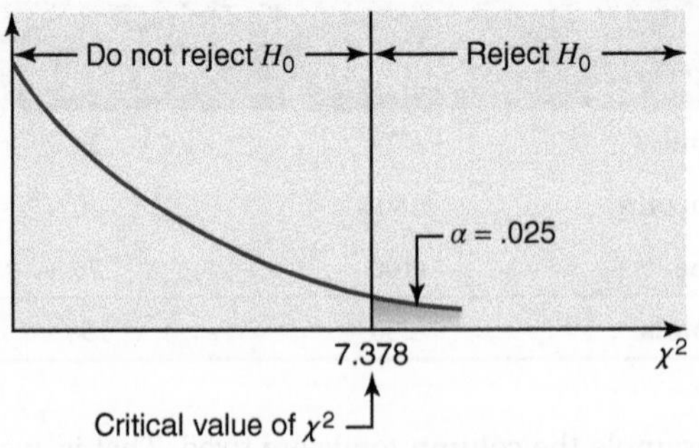

FIGURE 11.8 Rejection and nonrejection regions.

---

[2]Let p_{HC}, p_{MC}, and p_{LC} be the proportions of households in California who belong to high-, middle-, and low-income groups, respectively. Let p_{HW}, p_{MW}, and p_{LW} be the corresponding proportions for Wisconsin. Then we can also write the null hypothesis as

$$H_0: p_{HC} = p_{HW}, p_{MC} = p_{MW}, \text{ and } p_{LC} = p_{LW}$$

and the alternative hypothesis as

H_1: At least one of the equalities mentioned in H_0 is different from the other two.

STEP 4 *Calculate the value of the test statistic.*

To compute the value of the test statistic χ^2, we first need to calculate the expected frequencies. **Table 11.11** lists the observed and the expected frequencies. The numbers in parentheses in this table are the expected frequencies, which are calculated using the formula

$$E = \frac{(\text{Row total})(\text{Column total})}{\text{Total of both samples}}$$

TABLE 11.11 Observed and Expected Frequencies

	California	Wisconsin	Row Totals
High income	70 (65)	34 (39)	104
Medium income	80 (75)	40 (45)	120
Low income	100 (110)	76 (66)	176
Column Totals	250	150	400

Thus, for instance,

$$E \text{ for } \textit{High income} \text{ and } \textit{California} \text{ cell} = \frac{(104)(250)}{400} = 65$$

The remaining expected frequencies are calculated the same way. Note that the expected frequencies in a test of homogeneity are calculated the same way as in a test of independence. The value of the test statistic χ^2 is computed as follows:

$$\chi^2 = \sum \frac{(O - E)^2}{E}$$

$$= \frac{(70-65)^2}{65} + \frac{(34-39)^2}{39} + \frac{(80-75)^2}{75} + \frac{(40-45)^2}{45} + \frac{(100-110)^2}{110} + \frac{(76-66)^2}{66}$$

$$= .385 + .641 + .333 + .556 + .909 + 1.515 = \mathbf{4.339}$$

STEP 5 *Make a decision.*

The value of the test statistic $\chi^2 = 4.339$ is less than the critical value of $\chi^2 = 7.378$ and falls in the nonrejection region. Hence, we fail to reject the null hypothesis and state that there is not sufficient evidence to conclude that the income distribution is different for households in California and Wisconsin.

From technology, the *p*-value for $df = 2$ and $\chi^2 = 4.339$ is .1142, which is larger than $\alpha = .025$. Hence, we fail to reject the null hypothesis.

EXERCISES

Concepts and Procedures

11.18 Describe in your own words a test of independence and a test of homogeneity. Give one example of each.

11.19 Explain how the expected frequencies for cells of a contingency table are calculated in a test of independence or homogeneity. How do you find the degrees of freedom for such tests?

11.20 To perform a test of independence or homogeneity, what should be the minimum expected frequency for each cell? What are the alternatives if this condition is not satisfied?

11.21 Consider the following contingency table, which is based on a sample survey.

	Column 1	Column 2	Column 3
Row 1	143	64	105
Row 2	98	71	58
Row 3	111	92	115

a. Write the null and alternative hypotheses for a test of independence for this table.

b. Calculate the expected frequencies for all cells, assuming that the null hypothesis is true.

c. For $\alpha = .01$, find the critical value of χ^2. Show the rejection and nonrejection regions on the chi-square distribution curve.

d. Find the value of the test statistic χ^2.

e. Using $\alpha = .01$, would you reject the null hypothesis?

11.22 Consider the following contingency table, which records the results obtained for four samples of fixed sizes selected from four populations.

	Sample Selected From			
	Population 1	Population 2	Population 3	Population 4
Row 1	24	81	60	121
Row 2	46	64	91	72
Row 3	20	37	105	93

a. Write the null and alternative hypotheses for a test of homogeneity for this table.

b. Calculate the expected frequencies for all cells assuming that the null hypothesis is true.

c. For $\alpha = .025$, find the critical value of χ^2. Show the rejection and nonrejection regions on the chi-square distribution curve.

d. Find the value of the test statistic χ^2.

e. Using $\alpha = .025$, would you reject the null hypothesis?

Applications

11.23 During the economic recession a few years ago, many families faced hard times financially. Some studies observed that more people stopped buying name brand products and started buying less expensive store brand products instead. Data produced by a recent sample of 700 adults on whether they usually buy store brand or name brand products are recorded in the following table:

	More Often Buy	
	Name Brand	Store Brand
Men	148	170
Women	155	219

Using a 1% significance level, can you reject the null hypothesis that the two attributes, gender and buying name or store brand products, are independent?

11.24 Many students graduate from college deeply in debt from student loans, credit card debts, and so on. A sociologist took a random sample of 401 single persons, classified them by gender, and asked, "Would you consider marrying someone who was $25,000 or more in debt?" The results of this survey are shown in the following table:

	Yes	No	Uncertain
Women	130	59	18
Men	96	82	16

Test at a 1% significance level whether gender and response are related.

11.25 A random sample of 1000 U.S. adults was taken, and these adults were asked if experience in politics was necessary for a candidate to be president of the United States. The following table presents the results of the survey.

	Necessary	Not Necessary	No Opinion
Men	260	220	70
Women	230	180	40

Test at a 1% significance level whether gender and opinions are related.

11.26 One hundred auto drivers, who were stopped by police for some violations, were also checked to see if they were wearing seat belts. The following table records the results of this survey.

	Wearing Seat Belt	Not Wearing Seat Belt
Men	29	22
Women	31	18

Test at a 2.5% significance level whether being a man or a woman and wearing or not wearing a seat belt are related.

11.27 National Electronics Company buys parts from two subsidiaries. The quality control department at this company wanted to check if the distribution of good and defective parts is the same for the supplies of parts received from both subsidiaries. The quality control inspector selected a sample of 300 parts received from Subsidiary A and a sample of 400 parts received from Subsidiary B. These parts were checked for being good or defective. The following table records the results of this investigation.

	Subsidiary A	Subsidiary B
Good	275	360
Defective	25	40

Using a 5% significance level, test the null hypothesis that the distributions of good and defective parts are the same for both subsidiaries.

11.28 Four hundred people were selected from each of the four geographic regions (Midwest, Northeast, South, West) of the United States, and asked which form of camping they prefer. The choices were pop-up camper/trailer, family style (tenting with sanitary facilities), rustic (tenting, no sanitary facilities), or none. The results of the survey are shown in the following table:

	Midwest	Northeast	South	West
Camper/trailer	132	129	129	135
Family style	180	175	168	146
Rustic	46	50	59	68
None	42	46	44	51

Based on the evidence from these samples, can you conclude that the distributions of favorite forms of camping are different for at least two of these regions? Use $\alpha = .01$.

11.29 The following table gives the distributions of grades for three professors for a few randomly selected classes that each of them taught during the last two years.

		Professor		
		Miller	Smith	Moore
Grade	A	16	36	24
	B	25	44	12
	C	85	81	82
	D and F	20	12	8

Using a 2.5% significance level, test whether the grade distributions differ among these three professors.

11.30 Two random samples, one of 95 blue-collar workers and a second of 50 white-collar workers, were taken from a large company. These workers were asked about their views on a certain company issue. The following table gives the results of the survey.

	Opinion		
	Favor	Oppose	Uncertain
Blue-collar workers	44	39	12
White-collar workers	21	26	3

Using a 2.5% significance level, test whether the distributions of opinions differ for the two groups of workers.

11.4 Inferences About the Population Variance

LEARNING OBJECTIVES

After completing this section, you should be able to:

- Construct a confidence interval for the population variance, σ^2, and population standard deviation, σ.
- Write the null and alternative hypotheses for a test about a population variance σ^2.
- Use a χ^2-distribution to illustrate the critical value, the rejection region and the nonrejection region for a test of a population variance σ^2.
- Calculate the value of a χ^2 test statistic.
- Make a conclusion based on a comparison of the test statistic and the critical value.
- Apply the five-step method using a critical value approach to perform a test of hypothesis about a population variance σ^2.

Earlier chapters explained how to make inferences (confidence intervals and hypothesis tests) about the population mean and population proportion. However, we may often need to control the variance (or standard deviation). Consequently, there may be a need to estimate or test a hypothesis about the population variance σ^2. Section 11.4.1 describes how to make a confidence interval for a population variance (or standard deviation). Section 11.4.2 explains how to test a hypothesis about a population variance.

As an example, suppose a machine is set up to fill packages of cookies so that the net weight of cookies per package is 32 ounces. Note that the machine will not put exactly 32 ounces of cookies into each package. Some of the packages will contain less and some will contain more than 32 ounces. However, if the variance (and, hence, the standard deviation) is too large, some of the packages will contain quite a bit less than 32 ounces of cookies, and some others will contain quite a bit more than 32 ounces. The manufacturer will not want a large variation in the amount of cookies put into different packages. To keep this variation within some specified acceptable limit, the machine will be adjusted from time to time. Before the manager decides to adjust the machine at any time, he or she must estimate the variance or test a hypothesis or do both to find out if the variance exceeds the maximum acceptable value.

Like every sample statistic, the sample variance is a random variable and possesses a sampling distribution. If all the possible samples of a given size are taken from a population and their variances are calculated, the probability distribution of these variances is called the **sampling distribution of the sample variance**.

Sampling Distribution of $(n - 1)\, s^2/\sigma^2$

If the population from which the sample is selected is (approximately) normally distributed, then

$$\frac{(n-1)s^2}{\sigma^2}$$

has a chi-square distribution with $n - 1$ degrees of freedom. Note that it is not s^2 but the above expression that has a chi-square distribution.

Thus, the chi-square distribution is used to construct a confidence interval and test a hypothesis about the population variance σ^2.

11.4.1 Estimation of the Population Variance

The value of the sample variance s^2 gives a point estimate of the population variance σ^2. A $(1 - \alpha)100\%$ confidence interval for σ^2 is given by the following formula:

Confidence Interval for the Population Variance σ^2

Assuming that the population from which the sample is selected is (approximately) normally distributed, we obtain a $(1 - \alpha)100\%$ *confidence interval for the population variance* σ^2 as

$$\frac{(n-1)s^2}{\chi^2_{\alpha/2}} \quad \text{to} \quad \frac{(n-1)s^2}{\chi^2_{1-\alpha/2}}$$

where $\chi^2_{\alpha/2}$ and $\chi^2_{1-\alpha/2}$ are obtained from the chi-square distribution table for $\alpha/2$ and $1 - \alpha/2$ areas in the right tail of the chi-square distribution curve, respectively, and for $n - 1$ degrees of freedom.

The confidence interval for the population standard deviation can be obtained by simply taking the positive square roots of the two limits of the confidence interval for the population variance.

The procedure for making a confidence interval for σ^2 involves the following three steps:

1. Take a sample of size n and compute s^2 using the formula learned in Chapter 3. However, if n and s^2 are given, then perform only Steps 2 and 3.
2. Calculate $\alpha/2$ and $1 - \alpha/2$. Find two values of χ^2 from the chi-square distribution table (Table VI of Appendix B): one for $\alpha/2$ area in the right tail of the chi-square distribution curve and $df = n - 1$, and the second for $1 - \alpha/2$ area in the right tail and $df = n - 1$.
3. Substitute all the values in the formula for the confidence interval for σ^2 and simplify.

Example 11.9 illustrates the estimation of the population variance and population standard deviation.

EXAMPLE 11.9 | Weights of Packages of Cookies

Constructing Confidence Intervals for σ^2 and σ

One type of cookie manufactured by Haddad Food Company is Cocoa Cookies. The machine that fills packages of these cookies is set up in such a way that the average net weight of these packages is 32 ounces with a variance of .015 square ounce. From time to time the quality control inspector at the company selects a sample of a few such packages, calculates the variance of the net weights of these packages, and constructs a 95% confidence interval for the population variance. If either one of the two limits of this confidence interval is not in the interval .008 to .030, the machine is stopped and adjusted.

A recent random sample of 25 packages from the production line gave a sample variance of .029 square ounce. Based on this sample information, do you think the machine needs an adjustment? Assume that the net weights of cookies in all packages are approximately normally distributed.

Solution The following three steps are performed to estimate the population variance and to make a decision.

STEP 1 From the given information, $n = 25$ and $s^2 = .029$

STEP 2 Since the population is approximately normally distributed, we will use the chi-square distribution to construct a 95% confidence interval for σ^2. The confidence level is $1 - \alpha = .95$. Hence, $\alpha = 1 - .95 = .05$. Therefore,

$$\alpha/2 = .05/2 = .025$$
$$1 - \alpha/2 = 1 - .025 = .975$$
$$df = n - 1 = 25 - 1 = 24$$

From Table VI of Appendix B,

$$\chi^2 \text{ for 24 } df \text{ and .025 area in the right tail} = 39.364$$
$$\chi^2 \text{ for 24 } df \text{ and .975 area in the right tail} = 12.401$$

These values are shown in **Figure 11.9**.

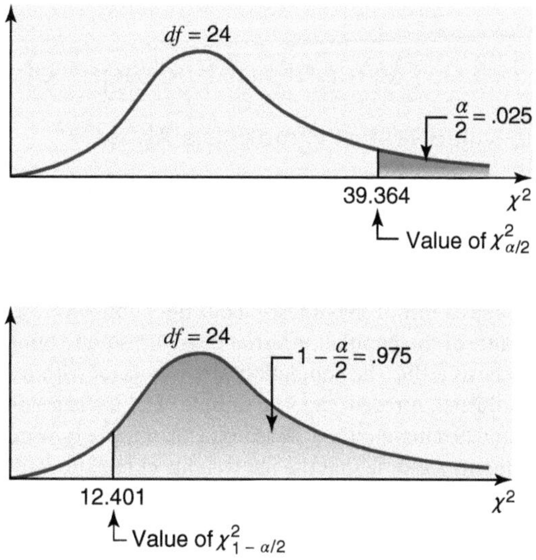

FIGURE 11.9 The values of χ^2.

STEP 3 The 95% confidence interval for σ^2 is

$$\frac{(n-1)s^2}{\chi^2_{\alpha/2}} \text{ to } \frac{(n-1)s^2}{\chi^2_{1-\alpha/2}}$$

$$\frac{(25-1)(.029)}{39.364} \text{ to } \frac{(25-1)(.029)}{12.401}$$

$$\mathbf{.0177 \text{ to } .0561}$$

Thus, with 95% confidence, we can state that the variance for all packages of Cocoa Cookies lies between .0177 and .0561 square ounce. Note that the lower limit (.0177) of this confidence interval is between .008 and .030, but the upper limit (.0561) is larger than .030 and falls outside the interval .008 to .030. Because the upper limit is larger than .030, we can state that, based on the sample information, the machine needs to be stopped and adjusted.

We can obtain the confidence interval for the population standard deviation σ by taking the positive square roots of the two limits of the above confidence interval for the population variance. Thus, a 95% confidence interval for the population standard deviation is

$$\sqrt{.0177} \text{ to } \sqrt{.0561} \quad \text{or} \quad \mathbf{.133 \text{ to } .237}$$

Hence, the standard deviation of all packages of Cocoa Cookies is between .133 and .237 ounce at a 95% confidence level.

11.4.2 Hypothesis Tests About the Population Variance

A test of hypothesis about the population variance can be one-tailed or two-tailed. To make a test of hypothesis about σ^2, we perform the same five steps we used earlier in hypothesis-testing examples. The procedure to test a hypothesis about σ^2 discussed in this section is applied only when the population from which a sample is selected is (approximately) normally distributed.

Test Statistic for a Test of Hypothesis About σ^2

The value of the *test statistic* χ^2 is calculated as

$$\chi^2 = \frac{(n-1)s^2}{\sigma^2}$$

where s^2 is the sample variance, σ^2 is the hypothesized value of the population variance, and $n - 1$ represents the degrees of freedom. The population from which the sample is selected is assumed to be (approximately) normally distributed.

Examples 11.10 and 11.11 illustrate the procedure for making tests of hypothesis about σ^2.

EXAMPLE 11.10 | Weights of Packages of Cookies

Performing a Right-Tailed Test of Hypothesis About σ^2

One type of cookie manufactured by Haddad Food Company is Cocoa Cookies. The machine that fills packages of these cookies is set up in such a way that the average net weight of these packages is 32 ounces with a variance of .015 square ounce. From time to time the quality control inspector at the company selects a sample of a few such packages, calculates the variance of the net weights of these packages, and makes a test of hypothesis about the population variance. She always uses $\alpha = .01$. The acceptable value of the population variance is .015 square ounce or less. If the conclusion from the test of hypothesis is that the population variance is not within the acceptable limit, the machine is stopped and adjusted. A recent random sample of 25 packages from the production line gave a sample variance of .029 square ounce. Based on this sample information, do you think the machine needs an adjustment? Assume that the net weights of cookies in all packages are approximately normally distributed.

Solution From the given information,

$$n = 25, \quad \alpha = .01, \quad \text{and} \quad s^2 = .029$$

The population variance should not exceed .015 square ounce.

STEP 1 *State the null and alternative hypotheses.*

We are to test whether or not the population variance is within the acceptable limit. The population variance is within the acceptable limit if σ^2 is less than or equal to .015; otherwise, it is not within the acceptable limit. Thus, the two hypotheses are

$H_0: \sigma^2 \leq .015$ (The population variance is within the acceptable limit.)
$H_1: \sigma^2 > .015$ (The population variance exceeds the acceptable limit.)

STEP 2 *Select the distribution to use.*

Since the population is approximately normally distributed, we will use the chi-square distribution to test a hypothesis about σ^2.

STEP 3 *Determine the rejection and nonrejection regions.*

The significance level is 1% and, because of the > sign in H_1, the test is right-tailed. The rejection region lies in the right tail of the chi-square distribution curve with its area equal to .01. The degrees of freedom for a chi-square test about σ^2 are $n - 1$; that is,

$$df = n - 1 = 25 - 1 = 24$$

From Table VI of Appendix B, the critical value of χ^2 for 24 degrees of freedom and .01 area in the right tail is 42.980. This value is shown in **Figure 11.10**.

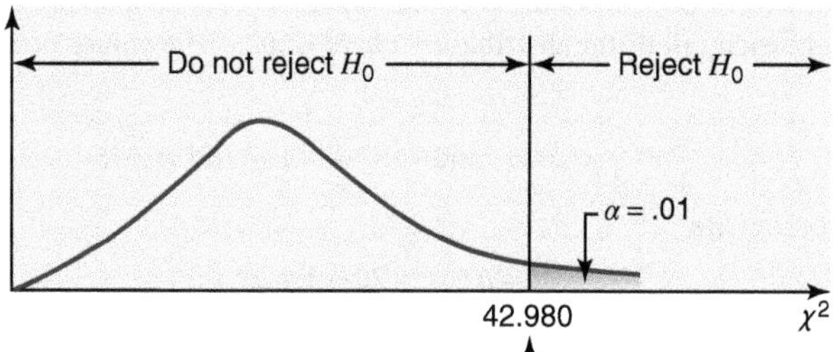

FIGURE 11.10 Rejection and nonrejection regions.

STEP 4 *Calculate the value of the test statistic.*

The value of the test statistic χ^2 for the sample variance is calculated as follows:

$$\chi^2 = \frac{(n-1)s^2}{\sigma^2} = \frac{(25-1)(.029)}{.015} = 46.400$$

(.015: From H_0)

STEP 5 *Make a decision.*

The value of the test statistic $\chi^2 = 46.400$ is greater than the critical value of $\chi^2 = 42.980$ and falls in the rejection region. Consequently, we reject H_0 and conclude that there is strong evidence from the sample information that the population variance is not within the acceptable limit. The machine should be stopped and adjusted.

From technology, the *p*-value for $df = 24$ and $\chi^2 = 46.400$ is .004, which is smaller than $\alpha = .01$. Hence, we reject the null hypothesis.

EXAMPLE 11.11 Variance of GPAs of Students at a University

Conducting a Two-Tailed Test of Hypothesis About σ^2

It is known that the variance of GPAs (with a maximum GPA of 4) of all students at a large university was .24 in 2019. A professor wants to determine whether the variance of the current GPAs of students at this university is different from .24. She took a random sample of 20 students and found that the variance of their GPAs is .27. Using a 5% significance level, can you conclude that the variance of the current GPAs of students at this university is different from .24? Assume that the GPAs of all current students at this university are approximately normally distributed.

skynesher/Getty Images

Solution From the given information,

$$n = 20, \quad \alpha = .05, \quad \text{and} \quad s^2 = .27$$

The population variance was .24 in 2019.

STEP 1 *State the null and alternative hypotheses.*

The null and alternative hypotheses are, respectively,

$$H_0: \sigma^2 = .24 \quad \text{(The population variance is not different from .24.)}$$
$$H_1: \sigma^2 \neq .24 \quad \text{(The population variance is different from .24.)}$$

STEP 2 *Select the distribution to use.*

Since the population of GPAs is approximately normally distributed, we will use the chi-square distribution to test a hypothesis about σ^2.

STEP 3 *Determine the rejection and nonrejection regions.*

The significance level is 5%. The $\neq$ sign in H_1 indicates that the test is two-tailed. The rejection region lies in both tails of the chi-square distribution curve with its total area equal to .05. Consequently, the area in each tail of the distribution curve is .025. The values of $\alpha/2$ and $1 - \alpha/2$ are, respectively,

$$\frac{\alpha}{2} = \frac{.05}{2} = .025 \quad \text{and} \quad 1 - \frac{\alpha}{2} = 1 - .025 = .975$$

The degrees of freedom are

$$df = n - 1 = 20 - 1 = 19$$

From Table VI of Appendix B, the critical values of χ^2 for 19 degrees of freedom and for $\alpha/2$ and $1 - \alpha/2$ areas in the right tail are

$$\chi^2 \text{ for 19 } df \text{ and .025 area in the right tail} = 32.852$$
$$\chi^2 \text{ for 19 } df \text{ and .975 area in the right tail} = 8.907$$

These two values are shown in **Figure 11.11**.

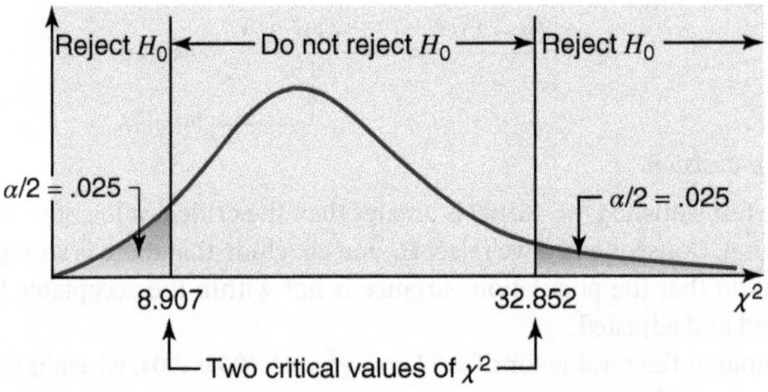

FIGURE 11.11 Rejection and nonrejection regions.

STEP 4 *Calculate the value of the test statistic.*

The value of the test statistic χ^2 for the sample variance is calculated as follows:

$$\chi^2 = \frac{(n-1)s^2}{\sigma^2} = \frac{(20-1)(.27)}{.24} = 21.375$$

From H_0 (applies to .24)

STEP 5 *Make a decision.*

The value of the test statistic $\chi^2 = 21.375$ is between the two critical values of χ^2, 8.907 and 32.852, and falls in the nonrejection region. Consequently, we fail to reject H_0 and state that there is not sufficient evidence to conclude that the population variance of the current GPAs of students at this university is different from .24.

From technology, the *p*-value for $df = 19$ and $\chi^2 = 21.375$ is 2(.3165) = .6330, which is larger than $\alpha = .05$. Hence, we fail to reject the null hypothesis.

Note that we can make a test of hypothesis about the population standard deviation σ using the same procedure as that for the population variance σ^2. To make a test of hypothesis about σ, the only change will be mentioning the values of σ in H_0 and H_1 in place of σ^2. The rest of the procedure remains the same as in the case of σ^2.

EXERCISES

Concepts and Procedures

11.31 A sample of observations selected from a normally distributed population produced a sample variance of 46. Construct a 95% confidence interval for σ^2 for each of the following cases and comment on what happens to the confidence interval of σ^2 when the sample size increases.

a. $n = 10$ **b.** $n = 15$ **c.** $n = 30$

11.32 A sample of 25 observations selected from a normally distributed population produced a sample variance of 35. Construct a confidence interval for σ^2 for each of the following confidence levels and comment on what happens to the confidence interval of σ^2 when the confidence level decreases.

a. $1 - \alpha = .99$

b. $1 - \alpha = .95$

c. $1 - \alpha = .90$

11.33 A sample of 21 observations selected from a normally distributed population produced a sample variance of 1.97.

a. Write the null and alternative hypotheses to test whether the population variance is greater than 1.75.

b. Using $\alpha = .025$, find the critical value of χ^2. Show the rejection and nonrejection regions on a chi-square distribution curve.

c. Find the value of the test statistic χ^2.

d. Using a 2.5% significance level, will you reject the null hypothesis stated in part (a)?

11.34 A sample of 18 observations selected from a normally distributed population produced a sample variance of 4.6.

a. Write the null and alternative hypotheses to test whether the population variance is different from 2.2.

b. Using $\alpha = .05$, find the critical value of χ^2. Show the rejection and nonrejection regions on a chi-square distribution curve.

c. Find the value of the test statistic χ^2.

d. Using a 5% significance level, will you reject the null hypothesis stated in part (a)?

Applications

11.35 Sandpaper is rated by the coarseness of the grit on the paper. Sandpaper that is more coarse will remove material faster. Jobs that involve the final sanding of bare wood prior to painting or sanding in between coats of paint require sandpaper that is much finer. A manufacturer of sandpaper rated 220, which is used for the final preparation of bare wood, wants to make sure that the variance of the diameter of the particles in their 220 sandpaper does not exceed 2.0 square micrometers. Fifty-one randomly selected particles are measured. The variance of the particle diameters is 2.13 square micrometers. Assume that the distribution of particle diameter is approximately normal.

a. Construct the 95% confidence intervals for the population variance and standard deviation.

b. Test at a 2.5% significance level whether the variance of the particle diameters of all particles in 220-rated sandpaper is greater than 2.0 square micrometers.

11.36 An auto manufacturing company wants to estimate the variance of miles per gallon for its auto model AST727. A random sample of 24 cars of this model showed that the variance of miles per gallon for these cars is .64. Assume that the miles per gallon for all such cars are approximately normally distributed.

a. Construct the 95% confidence intervals for the population variance and standard deviation.

b. Test at a 1% significance level whether the sample result indicates that the population variance is different from .30.

11.37 The manufacturer of a certain brand of lightbulbs claims that the variance of the lives of these bulbs is 4200 square hours. A consumer agency took a random sample of 25 such bulbs and tested them. The variance of the lives of these bulbs was found to be 5200 square hours. Assume that the lives of all such bulbs are approximately normally distributed.

a. Make the 99% confidence intervals for the variance and standard deviation of the lives of all such bulbs.

b. Test at a 5% significance level whether the variance of such bulbs is different from 4200 square hours.

Uses and Misuses of Statistics

1. Do Not Feed the Animals

You are a wildlife enthusiast studying African wildlife: gnus, zebras, and gazelles. You know that a herd of each species will visit one of three watering places in a region every day, but you do not know the distribution of choices that the animals make or whether these choices are dependent. You have observed that the animals sometimes drink together and sometimes do not. A statistician offers to help and says that he will perform a test for independence of watering place choices based on your observations of the animals' behavior over the past several months. The statistician performs some calculations and says that he has answered your question because his chi-square test of the independence of watering place choices, at a 5% significance level, told him to reject the null hypothesis. He has also performed a goodness-of-fit test on the hypothesis that the animals are equally likely to choose any watering place, and he has rejected that hypothesis as well.

The statistician barely helped you. In the first case, you know a single piece of information: the choice of a watering place for the three groups of animals is dependent. Another way of stating the result is that your data indicate that the choice of watering places for at least one of the animals is not independent of the others. Perhaps the zebras get up early, and the gnus and gazelles follow, making the gnus and gazelles dependent on the choice of the zebras. Or perhaps the animals choose the watering place of the day independent of the other animals, but always avoid the watering place at which the lions are drinking. Regarding the goodness-of-fit test, all you know is that the hypothesis that the animals equally favor the three watering places was rejected. But you do not know what the expected distribution should be. In short, the rejection of the null hypothesis raises more questions than it answers.

2. Is There a Gender Bias in Admissions?

Categorical data analysis methods, such as a chi-square test for independence, are used quite often in analyzing employment and admissions data in discrimination cases. One of the more famous discrimination cases involved graduate admissions at the University of California, Berkeley, in 1973. The claim was that UC Berkeley was discriminating against women in their admissions decisions, as would seem to be the case based on the data in the following table:

	Applicants	Percentage Admitted
Male	8442	44%
Female	4321	35%

The *p*-value for the corresponding chi-square test of independence is approximately 1.1×10^{-22} so it would seem clear that there is statistical dependence between gender and graduate admission. However, although it is true that admission rates differed by gender, the fact that the University was found not guilty of discrimination against women might shock you.

In order to find out why the University was found not guilty, we need to introduce another variable into the study: program of study. When the various programs were considered separately, the following admission rates for the six largest programs were observed.

	Men		Women	
Department	Applicants	Admitted (%)	Applicants	Admitted (%)
A	825	62	108	82
B	560	63	25	68
C	325	37	593	34
D	417	33	375	35
E	191	28	393	24
F	272	6	341	7

As one can see, four of the six programs had nominally higher admission rates for women than for men. So, why is the overall acceptance rate higher for men than for women? Looking at Departments A and B, which had the highest acceptance rates, we notice that 1385 men (16.4% of all men) applied to those programs, while 133 women (3.1% of all women) applied to these programs. On the other hand, almost twice as many women as men applied to the programs with the lowest acceptance rates. Hence, the overall acceptance rate for each gender was affected by the programs to which they applied and the number who applied to those programs. Since a much higher percentage of men applied to programs with higher (overall) acceptance rates, the overall acceptance rate for men ended up being higher than the overall acceptance rate for women.

This result is an example of what is known as Simpson's Paradox, which occurs when the inclusion of an additional variable (characteristic) reverses the conclusion made without that variable.

Source: P. J. Bickel, E. A. Hammel, and J. W. O'Connell (1975): Sex Bias in Graduate Admissions: Data From Berkeley. *Science* 187(4175): 398–404.

Glossary

Chi-square distribution A distribution, with degrees of freedom as the only parameter, that is skewed to the right for small *df* and approximately a normal curve for large *df*.

Expected frequencies The frequencies for different categories of a multinomial experiment or for different cells of a contingency table that are expected to occur when a given null hypothesis is true.

Goodness-of-fit test A test of the null hypothesis that the observed frequencies for an experiment follow a certain pattern or theoretical distribution.

Multinomial experiment An experiment with *n* trials for which (1) the trials are identical, (2) there are more than two possible outcomes per trial, (3) the trials are independent, and (4) the probabilities of the various outcomes remain constant for each trial.

Observed frequencies The frequencies actually obtained from the performance of an experiment.

Test of homogeneity A test of the null hypothesis that the proportions of elements with certain characteristics in two (or more) populations are the same against the alternative hypothesis that these proportions are not the same.

Test of independence A test of the null hypothesis that two attributes of a population are not related against the alternative hypothesis that the two attributes are related.

Exercises

Supplementary Exercises

11.38 At a County Fair, a local radio station plans to set up a booth with a prize wheel. Each hour, one person is selected to spin the wheel for a cash prize of \$10, \$20, \$50, or \$100. The wheel is partitioned into four uneven *slices*. The radio station claims that the wheel should land on the \$10 amount 40% of the time, the \$20 amount 30% of the time, the \$50 amount 20% of the time, and the \$100 amount 10% of the time. Prior to the start of the fair, the County Game Commission will test the wheel by spinning the wheel 80 times. Suppose the wheel lands on the \$10, \$20, \$50, and \$100 amounts on 41, 25, 9, and 5 of the 80 spins, respectively. Using a 2.5% level of significance, test whether the data from these 80 spins provide evidence to doubt the four percentages as claimed by the radio station.

11.39 In a Gallup poll conducted November 1–14, 2019, randomly selected U.S. adults, aged 18 and older, were asked, "Do you, personally, or does another member of your family living with you, have a long-term medical condition, illness, or disease that would be considered a 'pre-existing condition' by a health insurance company?" The responses of the adults included in the survey are given in the following table:

Person with Pre-existing Condition	Percent Responded
Respondent	15
Respondent and a family member	9
Family member	19
No one in family	57

Assume that these results are true for the population of 2019 U.S. adults. Recently, 2000 randomly selected U.S. adults were asked the same question. The number of adults who belonged to the four categories listed in the table were 280, 210, 350, and 1160, respectively. Test at a 2.5% significance level whether the distribution of current responses is significantly different from the 2019 distribution.

11.40 In a recent poll, U.S. adults were asked, if they have a choice, would they prefer to live in a city, suburb, or countryside. The following table shows the frequencies for the three choices.

Response	City	Suburb	Countryside
Frequency	640	790	570

Test at a 1% significance level if you can reject the null hypothesis that these three places are equally preferred by U.S. adults.

11.41 A randomly selected sample of 100 persons who suffer from allergies were asked during what season they suffer the most. The results of the survey are recorded in the following table:

Season	Fall	Winter	Spring	Summer
Persons allergic	20	3	41	36

Using a 1% significance level, test whether you can reject the null hypothesis that the proportions of all allergic persons are equally distributed over the four seasons.

11.42 All shoplifting cases in the town of Seven Falls are randomly assigned to either Judge Stark or Judge Rivera. A citizens group wants to know whether either of the two judges is more likely to sentence the offenders to jail time. A sample of 180 recent shoplifting cases produced the following two-way summary table:

	Jail	Other Sentence
Judge Stark	29	61
Judge Rivera	34	56

Test at a 5% significance level whether the type of sentence for shoplifting depends on which judge tries the case.

11.43 During a bear market, 150 investors were asked how they were adjusting their portfolios to protect themselves. Some of these investors were keeping most of their money in stocks, whereas others were shifting large amounts of money to bonds, real estate, or cash (such as money market accounts). The results of the survey are shown in the following table:

Favored Choice	Stocks	Bonds	Real Estate	Cash
Number of investors	48	45	35	22

Using a 2.5% significance level, test the null hypothesis that the percentages of investors favoring the four choices are all equal.

11.44 The recession and bad economic conditions of a few years ago forced many people to hold more than one job to make ends meet. A sample of 500 persons who held more than one job produced the following two-way summary table:

	Single	Married	Other
Male	65	231	26
Female	38	120	20

Test at a 10% significance level whether gender and marital status are related for all people who hold more than one job.

11.45 A random sample of 100 persons was selected from each of four regions in the United States. These people were asked whether or not they support a certain farm subsidy program. The results of the survey are summarized in the following table:

	Favor	Oppose	Uncertain
Northeast	56	33	11
Midwest	73	23	4
South	67	28	5
West	59	35	6

Using a 1% significance level, test the null hypothesis that the percentages of people with different opinions are similar for all four regions.

11.46 Construct the 98% confidence intervals for the population variance and standard deviation for the following data, assuming that the respective populations are normally distributed.

a. $n = 21, s^2 = 9.2$ **b.** $n = 17, s^2 = 1.7$

11.47 Construct the 95% confidence intervals for the population variance and standard deviation for the following data, assuming that the respective populations are normally distributed.

a. $n = 10, s^2 = 7.2$ **b.** $n = 18, s^2 = 14.8$

11.48 A sample of 18 units selected from a normally distributed population produced a variance of 14.8. Test at a 5% significance level if the population variance is different from 10.4.

11.49 A sample of 17 units selected from a normally distributed population produced a variance of 1.7. Test at a 2.5% significance level if the population variance is greater than 1.1.

11.50 The variance of the SAT scores for all students who took that test this year is 5000. The variance of the SAT scores for a random sample of 20 students from one school is equal to 3175.

a. Test at a 2.5% significance level whether the variance of the SAT scores for students from this school is lower than 5000. Assume that the SAT scores for all students at this school are (approximately) normally distributed.

b. Construct the 98% confidence intervals for the variance and the standard deviation of SAT scores for all students at this school.

11.51 A company manufactures ball bearings that are supplied to other companies. The machine that is used to manufacture these ball bearings produces them with a variance of diameters of .025 square millimeter or less. The quality control officer selects a sample of these ball bearings quite often and checks, using confidence intervals and tests of hypotheses, whether or not the variance of these bearings is within .025 square millimeter. If it is not, the machine is stopped and adjusted. A recently taken random sample of 23 ball bearings gave a variance of the diameters equal to .034 square millimeter.

a. Using a 5% significance level, can you conclude that the machine needs an adjustment? Assume that the diameters of all ball bearings have an approximate normal distribution.

b. Construct a 95% confidence interval for the population variance.

11.52 A random sample of 25 students taken from a university gave the variance of their GPAs equal to .19.

a. Construct the 99% confidence intervals for the population variance and standard deviation. Assume that the GPAs of all students at this university are approximately normally distributed.

b. The variance of GPAs of all students at this university was .13 two years ago. Test at a 1% significance level whether the variance of current GPAs at this university is different from .13.

11.53 A sample of eight passengers boarding a domestic flight produced the following data on weights (in pounds) of their carry-on bags.

46.3	41.5	40.1	31.0	41.4	35.8	39.8	38.4

a. Using the formula from Chapter 3, find the sample variance, s^2, for these data.

b. Make the 98% confidence intervals for the population variance and standard deviation. Assume that the population from which this sample is selected is normally distributed.

c. Test at a 5% significance level whether the population variance is larger than 20 square pounds.

Advanced Exercises

11.54 A chemical manufacturing company wants to locate a hazardous waste disposal site near a city of 50,000 residents and has offered substantial financial inducements to the city. Two hundred adults (110 women and 90 men) who are residents of this city are chosen at random. Sixty percent of these adults oppose the site, 32% are in favor, and 8% are undecided. Of those who oppose the site, 65% are women; of those in favor, 62.5% are men. Using a 5% level of significance, can you conclude that opinions on the disposal site are dependent on gender?

11.55 A student who needs to pass an elementary statistics course wonders whether it will make a difference if she takes the course with Instructor A rather than Instructor B. Observing the final grades given by each instructor in a recent elementary statistics course, she finds that Instructor A gave 48 passing grades in a class of 52 students and Instructor B gave 44 passing grades in a class of 54 students. Assume that these classes and grades make simple random samples of all classes and grades of these instructors.

a. Compute the value of the standard normal test statistic z of Section 10.5.3 for the data and use it to find the p-value when testing for the difference between the proportions of passing grades given by these instructors.

b. Construct a 2×2 contingency table for these data. Compute the value of the χ^2 test statistic for the test of independence and use it to find the p-value.

c. How do the test statistics in parts (a) and (b) compare? How do the p-values for the tests in parts (a) and (b) compare? Do you think this is a coincidence, or do you think this will always happen?

11.56 Each of five boxes contains a large (but unknown) number of red and green marbles. You have been asked to find if the proportions of red and green marbles are different for each of the five boxes. You sample 50 times, with replacement, from each of the five boxes and observe 20, 14, 23, 30, and 18 red marbles, respectively. Can you conclude that the five boxes have different proportions of red and green marbles? Use a .05 level of significance.

11.57 You have collected data on a variable, and you want to determine if a normal distribution is a reasonable model for these data. The following table shows how many of the values fall within certain ranges of z values for these data.

Category	Count
z score below −2	48
z score from −2 to less than −1.5	67
z score from −1.5 to less than −1	146
z score from −1 to less than −.5	248
z score from −.5 to less than 0	187
z score from 0 to less than .5	125
z score from .5 to less than 1	88
z score from 1 to less than 1.5	47
z score from 1.5 to less than 2	25
z score of 2 or above	19
Total	1000

Perform a hypothesis test to determine if the data provide evidence to doubt the claim that a normal distribution is an appropriate model for these data. Use a significance level of 5%.

Self-Review Test

Multiple Choice Questions

1. The random variable χ^2 assumes only
   a. positive   b. nonnegative   c. nonpositive values

2. The parameter(s) of the chi-square distribution is (are)
   a. degrees of freedom   b. df and n c. χ^2

3. Which of the following is *not* a characteristic of a multinomial experiment?
   a. There are n identical trials.
   b. There are k possible outcomes for each trial and $k > 2$.
   c. The trials are random.
   d. The trials are independent.
   e. The probabilities of outcomes remain constant for each trial.

4. The observed frequencies for a goodness-of-fit test are
   a. the frequencies obtained from the performance of an experiment
   b. the frequencies given by the product of n and p
   c. the frequencies obtained by adding the results of (a) and (b)

5. The expected frequencies for a goodness-of-fit test are
   a. the frequencies obtained from the performance of an experiment
   b. the frequencies given by the product of n and p
   c. the frequencies obtained by adding the results of (a) and (b)

6. The degrees of freedom for a goodness-of-fit test are
   a. $n - 1$ b. $k - 1$ c. $n + k - 1$
   where n is the sample size and k is the number of categories.

7. The chi-square goodness-of-fit test is always
   a. two-tailed   b. left-tailed   c. right-tailed

8. To apply a goodness-of-fit test, the expected frequency of each category must be at least
   a. 10   b. 5   c. 8

9. The degrees of freedom for a test of independence are
   a. $(R - 1)(C - 1)$ b. $n - 2$ c. $(n - 1)(k - 1)$

10. For a chi-square distribution curve with 3 or more degrees of freedom, the peak occurs at the point
    a $x =$ degrees of freedom −1
    b. $x =$ degrees of freedom −2
    c. $n - x$

11. A chi-square distribution assumes
    a. negative values only
    b. positive values only
    c. nonnegative values only

12. For small degrees of freedom, the chi-square distribution is

a. symmetric

b. skewed to the right

c. skewed to the left

13. For very large degrees of freedom, the chi-square distribution becomes

a. symmetric

b. skewed to the right

c. skewed to the left

14. What is the chi-square value for 12 degrees of freedom and a .10 area in the left tail?

a. 6.304

b. −6.304

c. 18.549

15. Find the probability of $\chi^2 < 6.26$ when $df = 15$.

a. .025 b. .05 c. .10

16. If a test of hypothesis involves a multinomial experiment, then which of the following test is used?

a. Test of hypothesis about the variance

b. Goodness-of-fit test

c. χ^2 distribution test

17. Which of the following is *not* a characteristic of a multinomial experiment?

a. It consists of two identical trials.

b. Each trial results in one of k possible outcomes, where k is a number greater than 2.

c. The probabilities of various outcomes are different for each trial.

18. For a goodness-of-fit test, the frequencies obtained from the performance of an experiment are the

a. subjective frequencies

b. objective frequencies

c. observed frequencies

19. For a goodness-of-fit test, when n is the sample size and p is the probability of a category occurring, we calculate the expected frequency for a category by

a. $E = np$ b. $E = xp$ c. $E = nx$

20. For a goodness-of-fit test, the sample size should be large enough so that the

a. expected frequency for each category is at least 10

b. observed frequency for each category is at least 5

c. expected frequency for each category is at least 5

21. If a test of hypothesis involves a contingency table, which of the following test is used?

a. Goodness-of-fit test

b. χ^2 distribution test

c. Independence test or homogeneity test

22. A chi-square test of homogeneity is about

a. testing the null hypothesis that two proportions are homogeneous

b. the independence of two characteristics presented in a contingency table

c. testing the null hypothesis that the proportions of elements with certain characteristics in two or more different populations are the same

23. If a test of hypothesis involves the standard deviation of a single population, which of the following test is used?

a. Test of hypothesis about the variance

b. Goodness-of-fit test

c. χ^2 distribution test

24. A random sample of 28 elements selected from a population produced a variance equal to 8.871. The null hypothesis is that the population variance is greater than or equal to 19.05, and the alternative hypothesis is that the population variance is less than 19.05. Assume that the population is normally distributed. The significance level is 2.5%. The critical value of the chi-square is

a. 14.573 b. 41.923 c. 13.844

Practice Exercises

1. Are birthdays equally distributed across the four seasons (winter, spring, summer, and fall)? A curious statistics student challenges this claim and obtains the birthdays of each student in a random sample of 80 students from his school. In this sample, 14 students have a winter birthday, 13 students were born in spring, 28 students have a summer birthday, and 25 students were born in fall. Using a 1% level of significance, test whether the sample data give reason to doubt the claim that birthdays are equally distributed across the four seasons?

2. The following table gives the two-way classification of 1000 persons who have been married at least once. They are classified by educational level and marital status.

	Educational Level			
	Less Than High School	**High School Degree**	**Some College**	**College Degree**
Divorced	173	158	95	53
Never divorced	162	126	110	123

Test at a 1% significance level whether educational level and ever being divorced are dependent.

3. A researcher wanted to investigate if people who belong to different income groups are homogeneous with regard to playing lotteries. She took a sample of 600 people from the low-income group, another sample of 500 people from the middle-income group, and a third sample of 400 people from the high-income group. All these people were asked whether they play the lottery often, sometimes, or never. The results of the survey are summarized in the following table:

	Income Group		
	Low	**Middle**	**High**
Play often	174	163	90
Play sometimes	286	217	120
Never play	140	120	190

Using a 5% significance level, can you reject the null hypothesis that the percentages of people who play the lottery often, sometimes, and never are the same for each income group?

4. The owner of an ice cream parlor is concerned about consistency in the amount of ice cream his servers put in each cone. He would like the variance of all such cones to be no more than .25 square ounce.

He decides to weigh each cone just before it is given to the customer. For a sample of 20 cones, the weights were found to have a variance of .48 square ounce. Assume that the weights of all such cones are approximately normally distributed.

a. Construct the 99% confidence intervals for the population variance and the population standard deviation.

b. Test at a 1% significance level whether the variance of the weights of all such cones exceeds .25 square ounce.

Mini-Projects

Mini-Project 11.1

In recent years, drivers have become careless about signaling their turns. To study this problem, go to a busy intersection and observe at least 75 vehicles that make left turns. Divide these vehicles into three or four classes. For example, you might use cars, trucks, and others (where "others" include minivans and sport-utility vehicles) as classes. For each left turn made by a vehicle, record the type of vehicle and whether or not the driver used the left turn signal before making this turn. It would be better to avoid intersections that have designated left-turn lanes or green arrows for left turns because drivers in these situations often assume that their intent to turn left is obvious. Carry out an appropriate test at the 1% level of significance to determine if signaling behavior and vehicle type are dependent.

Mini-Project 11.2

One day during lunch, visit your school cafeteria, observe at least 100 people, and write down what they are drinking. Categorize the drinks as soft drink (soda, fruit punch, or lemonade), iced tea, milk or juice, hot drink, and water. Also identify the gender of each person. Perform a hypothesis test to determine if the type of drink and gender are dependent.

Mini-Project 11.3

Many studies have been performed to determine the sources that people use to get their news. Survey at least 50 people at random from your class or dorm and ask the following question:

Which of the following would you classify as being your primary source for news?

a. Network news broadcasts

b. Cable news broadcasts

c. Newspapers

d. Internet-based news sources

e. Radio news broadcasts

Use the data to test the null hypothesis that college students are equally likely to classify the five options as being their primary source for news. Use a 5% significance level.

Mini-Project 11.4

As reported in *USA Today* of August 20, 2009, a Baby Orajel survey of 1004 adults conducted by Kelton Research asked these adults which of a set of four noises (car alarm, jackhammer, baby crying, or dog barking) they find to be the most frustrating to hear. Suppose that the survey also included information on the gender of the respondents as listed in the following table:

Sound That Is Most Frustrating to Hear	Females	Males
Car alarm	225	197
Jackhammer	154	131
Baby crying	79	143
Dog barking	44	31

a. Perform a hypothesis test to determine whether the sound that is most frustrating to hear and gender are dependent. Use a 1% significance level.

b. Suppose that it was noticed that the surveyor incorrectly marked 30 of the survey sheets. Specifically, 30 of the male respondents were mistakenly recorded as saying "baby crying." Furthermore, it was determined that all 30 of the mistakes should have one of the other three responses but all the same response. That is, all these 30 responses should have been "car alarm," or all 30 should have been "jackhammer," or all 30 should have been "dog barking." Determine which of these three changes would result in a different conclusion than the one you obtained in part (a).

Decide for Yourself

Note: The Decide for Yourself feature is located in the book's product page on *www.wiley.com*.

Chapter 11 Technology Instructions

Note: Complete TI-84, Minitab, and Excel manuals are available in the book's product page on *www.wiley.com*.

TI-84 Color/TI-84

The TI-84 Color Technology Instructions feature of this text is written for the TI-84 Plus C color graphing calculator running the 4.0 operating system. Some screens, menus, and functions will be slightly different in older operating systems. The TI-84 and TI-84 Plus can perform all of the same functions but will not have the "Color" option referenced in some of the menus.

Performing a Chi-Square Goodness-of-Fit Test for Example 11.3 of the Text

1. Enter the observed counts from Example 11.3 into list 1 and the corresponding expected counts in list 2.
2. Select **STAT > TESTS >** χ^2 **GOF-Test.**
3. Use the following settings in the χ^2 **GOF-Test** menu (see **Screen 11.1**):
   - Type L1 at the **Observed** prompt.
   - Type L2 at the **Expected** prompt.
   - Type 4 at the **df** prompt.
   - Select any desired color at the **Color** prompt.
4. Highlight **Calculate**, and press **ENTER.**
5. The output includes the test statistic and the *p*-value. The values of $(O-E)^2/E$ are stored in a list named CNTRB. (See **Screen 11.2.**)

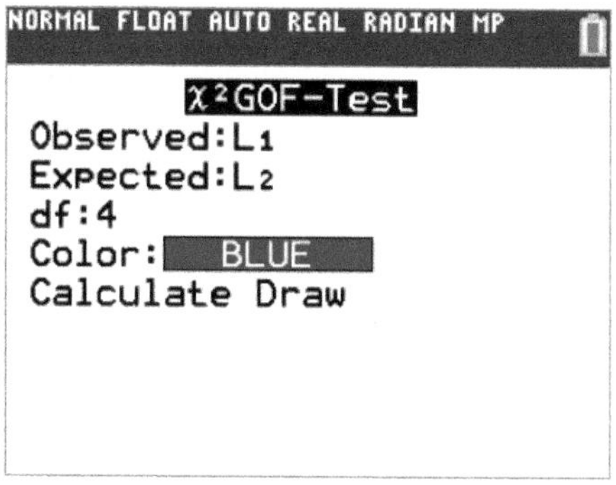

SCREEN 11.1

Now compare the χ^2-value with the critical value of χ^2 or the *p*-value from Screen 11.2 with α and make a decision.

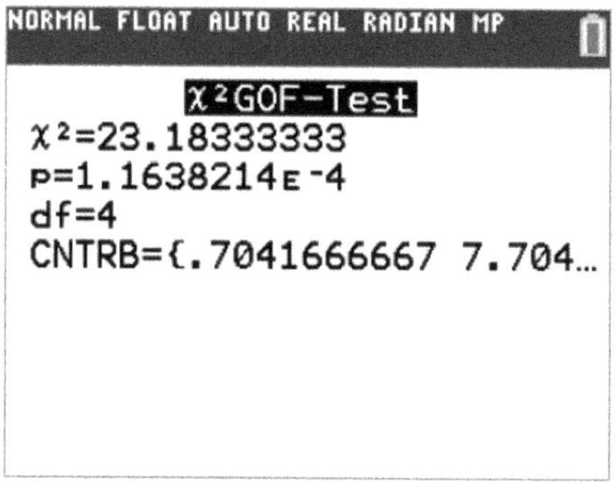

SCREEN 11.2

Performing a Chi-Square Independence/Homogeneity Test for Example 11.6 of the Text

1. Select **2nd >** $\mathbf{X^{-1}}$ **> EDIT > [A].** Press **ENTER.**
2. Use the following settings in the matrix editor (see **Screen 11.3**):
   - At the **Matrix[A]** prompt, type 2 for the number of rows and 3 for the number of columns. Use the arrow keys to navigate between these numbers.
   - Enter the observed counts from Example 11.6 into the matrix. Use **ENTER** and the arrow keys to move between cells.
3. Select **STAT > TESTS >** χ^2**-Test.**
4. Use the following settings in the χ^2**-Test** menu (see **Screen 11.4**):
   - Enter [A] at the **Observed** prompt.

     Note: To enter the name for the matrix [A], select **2nd >** $\mathbf{X^{-1}}$ **> NAMES > [A].**
   - Enter [B] at the **Expected** prompt.

     Note: To enter the name for the matrix [B], select **2nd >** $\mathbf{X^{-1}}$ **> NAMES > [B].**
   - Select any desired color at the **Color** prompt.
5. Highlight **Calculate**, and press **ENTER.**
6. The output includes the test statistic and the *p*-value. (See **Screen 11.5.**)

Now compare the χ^2-value with the critical value of χ^2 or the *p*-value from Screen 11.5 with α and make decision.

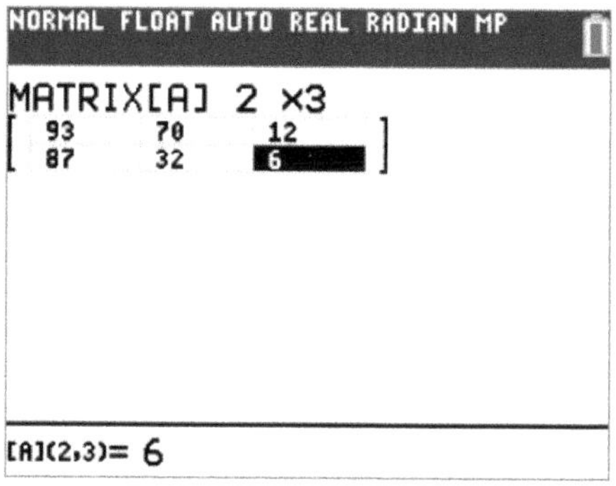

SCREEN 11.3

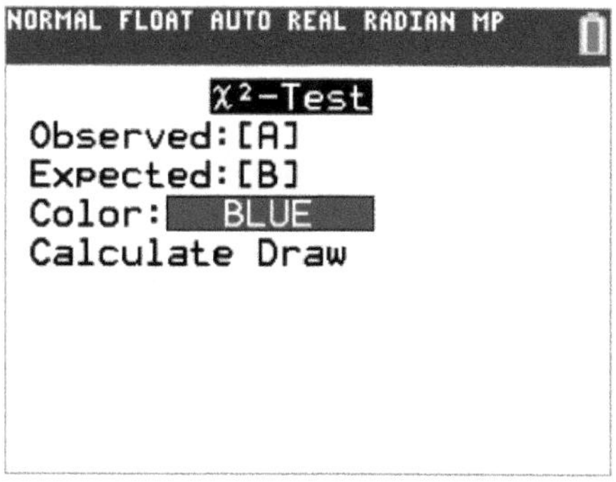

SCREEN 11.4

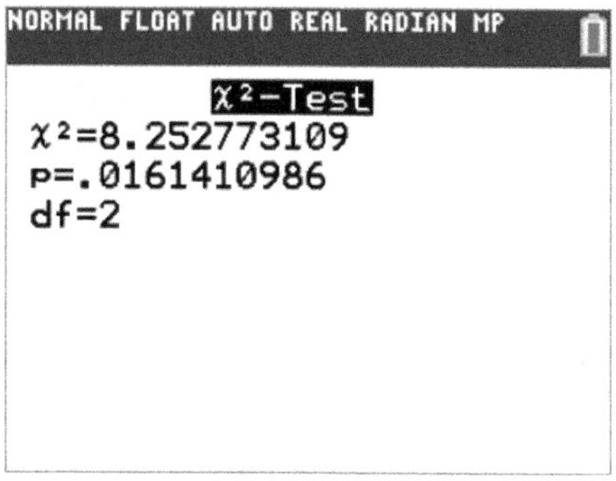

SCREEN 11.5

Minitab

The Minitab Technology Instructions in this text are written for **Minitab Professional Version 19.** Some screens, menus, and functions will be slightly different in older (or in newer) versions of Minitab. If you are using **Minitab Express**, you can download the Minitab Express Manual from the book's product page on *www.wiley.com*.

Performing a Chi-Square Goodness-of-Fit Test for Example 11.3 of the Text

1. Enter the observed counts (frequencies) from Example 11.3 into C1.
2. Select **Stat > Tables > Chi-Square Goodness-of-Fit Test (One Variable).**
3. Use the following settings in the dialog box that appears on screen (see **Screen 11.6**):
   - Select **Observed counts** and type **C1** in the box.
   - Select **Equal proportions** at the **Test** submenu.

     Note: If the alternative hypothesis does not specify equal proportions, go to the worksheet and type the proportions in C2. Return to the dialog box, select **Specific proportions** at the **Test** submenu, and type **C2** in the box.
4. Click **OK.**
5. The output, including the test statistic and *p*-value, will be displayed in the Session window. (See **Screen 11.7.**)

   Note: By default, Minitab will also generate two different bar graphs: one of the observed and expected counts and another of the $(O-E)^2/E$ values, which are called the contributions to the Chi-Square statistic. These graphs are not shown here.

Now compare the χ^2-value with the critical value of χ^2 or the *p*-value from Screen 11.7 with α and make a decision.

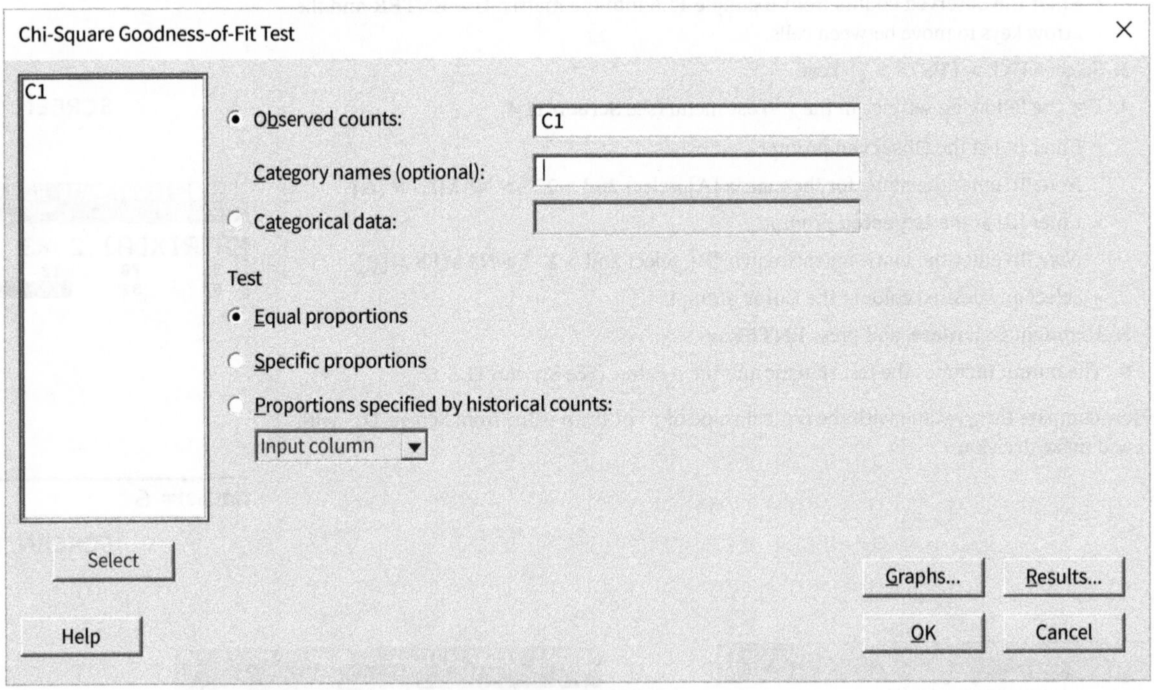

SCREEN 11.6

Performing a Chi-Square Independence/Homogeneity Test for Example 11.6 of the Text

1. Enter the contingency table from Example 11.6 into the first two rows of C1 through C3. (See **Screen 11.8.**)
2. Select **Stat > Tables > Chi-Square Test for Association.**
3. Use the following settings in the dialog box that appears on screen (see **Screen 11.8**):
   - Select **Summarized data in a two-way table** from the drop-down menu.

     Note: For raw data in C1 and C2, select **Raw data (categorical variables)** from the drop-down menu, type C2 in the **Rows** box, and C1 in the **Columns** box. Then go to Step 4.
   - Type C1-C3 in the **Columns containing the table** box.

4. Click **OK**.
5. The output, including the test statistic and *p*-value, will be displayed in the Session window. (See **Screen 11.9**.)

Now compare the χ^2-value with the critical value of χ^2 or the *p*-value from Screen 11.9 with α and make a decision.

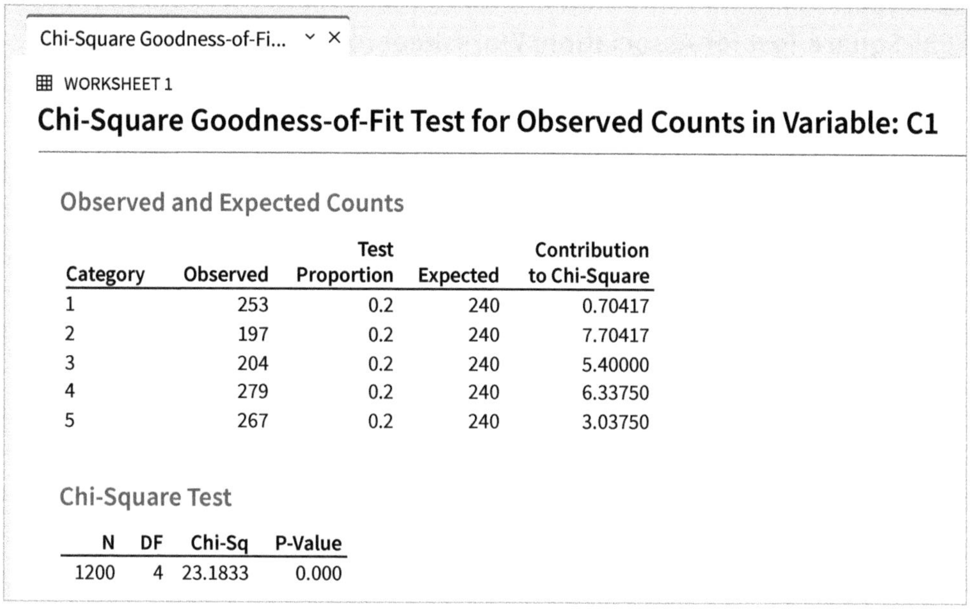
Chi-Square Goodness-of-Fi...

WORKSHEET 1

Chi-Square Goodness-of-Fit Test for Observed Counts in Variable: C1

Observed and Expected Counts

Category	Observed	Test Proportion	Expected	Contribution to Chi-Square
1	253	0.2	240	0.70417
2	197	0.2	240	7.70417
3	204	0.2	240	5.40000
4	279	0.2	240	6.33750
5	267	0.2	240	3.03750

Chi-Square Test

N	DF	Chi-Sq	P-Value
1200	4	23.1833	0.000

SCREEN 11.7

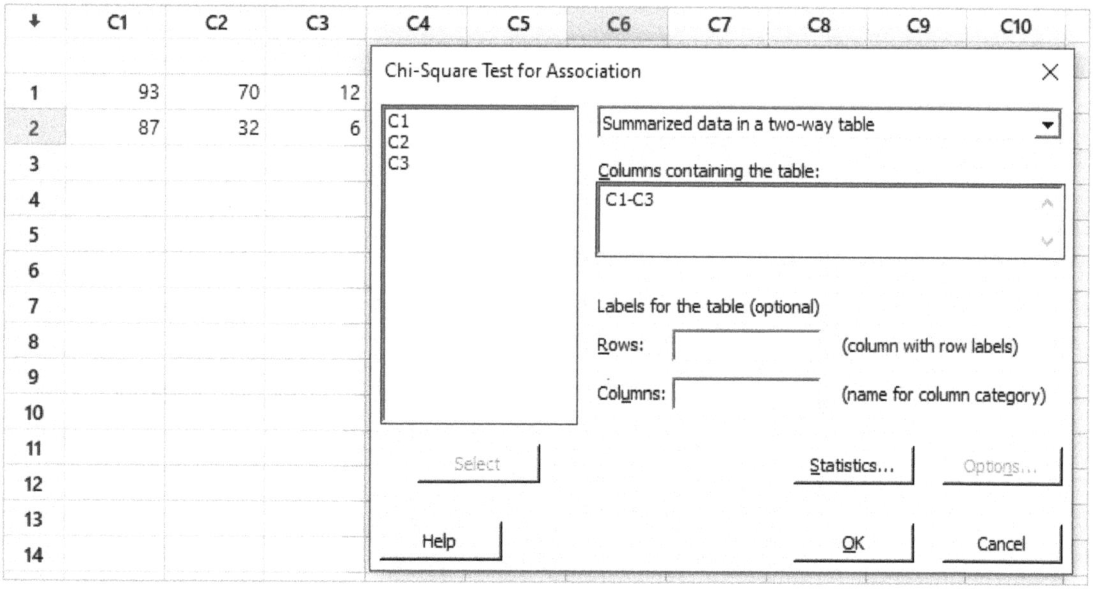

SCREEN 11.8

Testing a Hypothesis About σ^2 for Example 11.10 of the Text

1. Select **Stat > Basic Statistics > 1 Variance**.
2. Use the following settings in the dialog box that appears on screen (see **Screen 11.10**):
   - Select **Sample variance** from the drop-down menu.

     Note: If you have the data in a column, select **One or more samples, each in a column**, type the column name in the next box, and move to Step 3.
   - Type 25 in the **Sample size** box.
   - Type .029 in the **Sample variance** box.

- Check the **Perform hypothesis test** checkbox.
- Select **Hypothesized variance** from the drop-down menu.
- Type .015 in the **Value** box.

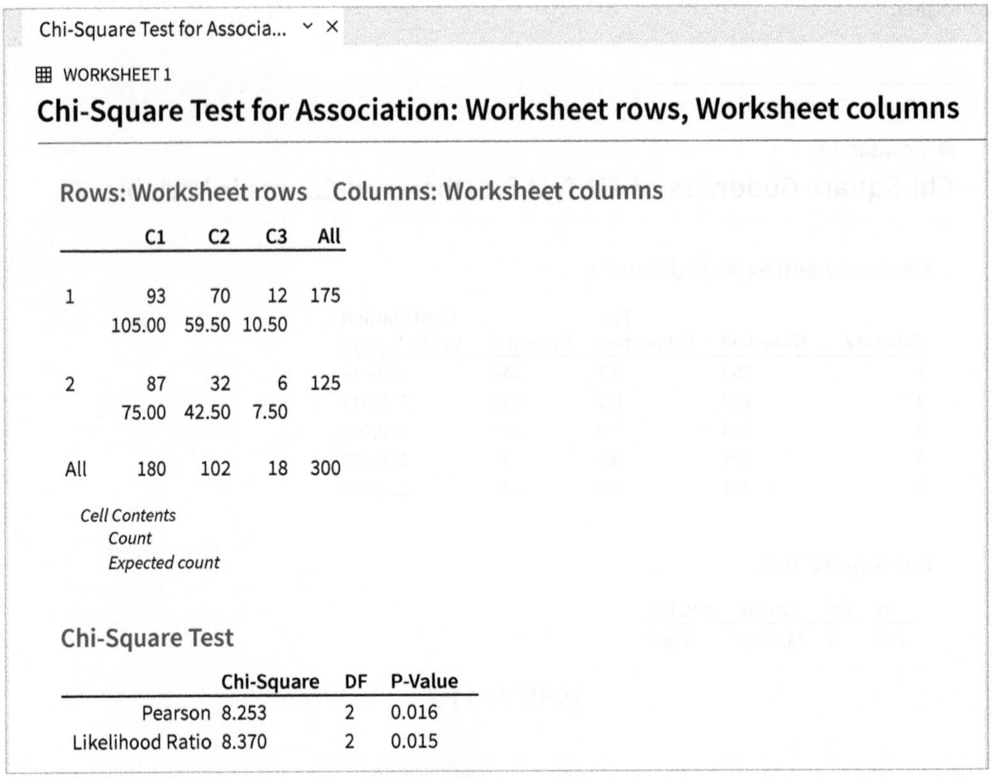

Chi-Square Test for Associa...

WORKSHEET 1

Chi-Square Test for Association: Worksheet rows, Worksheet columns

Rows: Worksheet rows Columns: Worksheet columns

	C1	C2	C3	All
1	93	70	12	175
	105.00	59.50	10.50	
2	87	32	6	125
	75.00	42.50	7.50	
All	180	102	18	300

Cell Contents
Count
Expected count

Chi-Square Test

	Chi-Square	DF	P-Value
Pearson	8.253	2	0.016
Likelihood Ratio	8.370	2	0.015

SCREEN 11.9

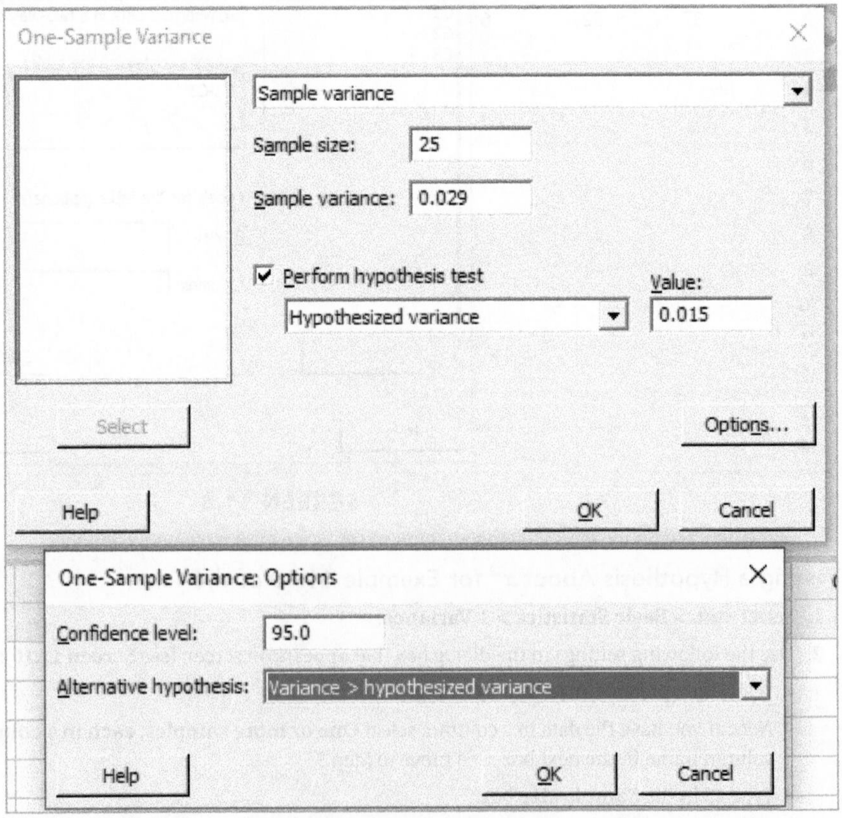

SCREEN 11.10

3. Select **Options** and use the following settings in the new dialog box that appears on screen:
   - Type 95 in the **Confidence level** box. (see **Screen** 11.10.)
   - Select Variance > hypothesized variance from the **Alternative hypothesis** drop-down menu. (see **Screen** 11.10.)
4. Click **OK** in both dialog boxes.
5. The output, including the test statistic and *p*-value, will be displayed in the Session window. The output also includes a confidence interval, but it is a one-sided confidence interval, which is beyond the scope of this text.

Excel

The Excel Technology Instructions feature of this text is written for Excel 2019. Some screens, menus, and functions will be slightly different in older versions of Excel. To enable some of the advanced Excel functions, you must enable the Data Analysis Add-In for Excel.

Performing a Chi-Square Goodness-of-Fit Test for Example 11.3 of the Text

1. In a new worksheet, enter the following text into the indicated cells (see **Screen 11.11**):
   - Type Observed in cell A1, Expected in cell B1, and Contribution in cell C1.
   - Type Chi-square in cell E1 and *p*-value in cell E2.
2. Enter the observed counts in cells A2-A6 and the expected counts in cells B2-B6. (See **Screen 11.11**.)
3. Enter the formula **=(A2-B2)^2/B2** in cell C2. Copy the formula from cell C2 to cells C3-C6. (**Screen 11.11** shows the results of this formula.)
4. Enter the formula **=SUM(C2:C6)** in cell F1. The output is the value of the test statistic. (**Screen 11.11** shows the results of this formula.)
5. Enter the formula **=CHISQ.DIST.RT(F1,4)** in cell F2. The output is the *p*-value as shown in **Screen 11.11**.

   Note: In this formula, 4 is the degrees of freedom. Adjust this value as needed for other tests.

Now compare the χ^2-value with the critical value of χ^2 or the *p*-value from Screen 11.11 with α and make a decision.

	A	B	C	D	E	F	G
1	Observed	Expected	Contribution		Chi-square	23.18333	
2	253	240	0.7042		p-value	0.00012	
3	197	240	7.7042				
4	204	240	5.4000				
5	279	240	6.3375				
6	267	240	3.0375				
7							

SCREEN 11.11

Performing a Chi-Square Independence/Homogeneity Test for Example 11.6 of the Text

1. In a new worksheet, enter Observed in cell A1, Expected in cell E1, and *p*-value in cell I1. (See **Screen 11.12**.)
2. Enter the observed counts in cells A2-C3 and the expected counts in cells E2-G3. (See **Screen 11.12**.)
3. Enter the formula **=CHISQ.TEST(A2:C3,E2:G3)** in cell J1. The output is the *p*-value. (See **Screen 11.12**.)

Now compare the *p*-value from Screen 11.12 with α and make a decision.

	A	B	C	D	E	F	G	H	I	J
1	Observed				Expected				p-value	0.016
2	93	70	12		105.0	59.5	10.5			
3	87	32	6		75.0	42.5	7.5			

SCREEN 11.12

Technology Assignments

TA11.1 According to the County Blood Bank, 44% of the population has blood type O, 42% has blood type A, 10% has blood type B and the remaining 4% of the population has blood type AB. In a random sample of 500 people, the number of people having blood types O, A, B, and AB are 215, 202, 55, and 28 respectively. At a 5% level of significance, test whether the sample data give reason to doubt the four blood type proportions as provided by the County Blood Bank.

TA11.2 A sample of 4000 persons aged 18 years and older produced the following two-way classification table:

	Men	**Women**
Single	531	357
Married	1375	1179
Widowed	55	195
Divorced	139	169

Test at a 1% significance level whether gender and marital status are dependent for all persons aged 18 years and older.

TA11.3 Two samples, one of 3000 students from urban high schools and another of 2000 students from rural high schools, were taken. These students were asked if they have ever smoked. The following table lists the summary of the results.

	Urban	**Rural**
Have never smoked	1448	1228
Have smoked	1552	772

Using a 5% significance level, test the null hypothesis that the proportions of urban and rural students who have smoked and who have never smoked are homogeneous.

TA11.4 Refer to Data Set IV on the Manchester Road Race that accompanies this text (see Appendix A). Select a random sample of 500 participants. Perform a hypothesis test with the null hypothesis that the gender of a participant (listed in column 6) is independent of whether the person is from Connecticut or not from Connecticut (listed in column 10). Do not include the participants who are not from USA.

TA11.5 Solve part (b) of Exercise 11.37 using technology.

CHAPTER 12

Analysis of Variance

JGI/Jamie Grill/Getty Images

Trying something new can be risky, and there can be uncertainty about the results. Suppose a school district plans to test three different methods for teaching arithmetic. After teachers implement these different methods for a semester, administrators want to know if the mean scores of students taught with these three different methods are all the same. What data will they require and how will they test for this equality of more than two means? (See Examples 12.2 and 12.3.)

Chapter 10 described the procedures that are used to test hypotheses about the difference between two population means using the normal and *t* distributions. Also described in that chapter were the hypothesis-testing procedures for the difference between two population proportions using the normal distribution. Then, Chapter 11 explained the procedures that are used to test hypotheses about the equality of more than two population proportions using the chi-square distribution.

This chapter explains how to test hypotheses about the equality of more than two population means. For example, suppose that teachers at a school have devised three different methods to teach arithmetic. They want to find out if these three methods produce different mean scores. Let μ_1, μ_2, and μ_3 be the mean scores of all students who will be taught by Methods I, II, and III, respectively. To test whether or not the three teaching methods produce the same mean, we test the null hypothesis

$$H_0: \mu_1 = \mu_2 = \mu_3 \quad \text{(All three population means are equal.)}$$

against the alternative hypothesis

$$H_1: \text{Not all three population means are equal.}$$

We use the analysis of variance procedure to perform such a test of hypothesis.

Note that the analysis of variance procedure can be used to compare two population means. However, the procedures learned in Chapter 10 are more efficient for performing tests of hypotheses about the difference between two population means; the analysis of variance procedure, to be discussed in this chapter, is used to compare three or more population means.

An *analysis of variance* test is performed using the *F* distribution. First, the *F* distribution is described in Section 12.1 of this chapter. Then, Sections 12.2 and 12.3 discuss the applications of the one-way and two-way analysis of variance procedures to perform tests of hypotheses, respectively.

12.1 The *F* Distribution

LEARNING OBJECTIVES

After completing this section, you should be able to:

- Describe the main characteristics of an *F* distribution.
- Identify the parameters of an *F* distribution.
- Determine the value of *F* for a given right-tail area and given degrees of freedom for an *F* distribution.

Like the chi-square distribution, the shape of a particular ***F* distribution**[1] curve depends on the number of degrees of freedom. However, the *F* distribution has *two* numbers of degrees of freedom: **degrees of freedom for the numerator** and **degrees of freedom for the denominator**. These two numbers representing two types of degrees of freedom are the *parameters of the F distribution*. Each combination of degrees of freedom for the numerator and for the denominator gives a different *F* distribution curve. The values of an *F* distribution are denoted by *F*, which assumes only nonnegative values. Like the normal, *t*, and chi-square distributions, the *F* distribution is a continuous distribution. The shape of an *F* distribution curve is skewed to the right, but the skewness decreases as both numbers of degrees of freedom increase.

The *F* Distribution

1. The *F distribution* is continuous and skewed to the right.
2. The *F* distribution has two numbers of degrees of freedom: *df* for the numerator and *df* for the denominator.
3. The units (the values of the *F*-variable) of an *F* distribution, denoted by *F*, are nonnegative.

For an *F* distribution, degrees of freedom for the numerator and degrees of freedom for the denominator are usually written as follows:

$$df = (8, 14)$$

First number denotes the *df* for the numerator

Second number denotes the *df* for the denominator

Figure 12.1 shows three *F* distribution curves for three sets of degrees of freedom for the numerator and for the denominator. In the figure, the first number gives the degrees of freedom associated with the numerator, and the second number gives the degrees of freedom associated with the denominator. We can observe from this figure that as both degrees of freedom increase, the peak of the curve moves to the right; that is, the skewness decreases.

Table VII in Appendix B lists the values of *F* for the *F* distribution. To read Table VII, we need to know three quantities: the degrees of freedom for the numerator, the degrees of freedom for the denominator, and an area in the right tail of an *F* distribution curve. Note that the *F* distribution table (Table VII) is read only for an area in the right tail of the *F* distribution curve. Also note that Table VII has four parts. These four parts give the *F* values for areas of .01, .025, .05, and .10, respectively, in the right tail of the *F* distribution curve. We can make the *F* distribution table for other values in the right tail. Example 12.1 illustrates how to read Table VII.

[1] The *F* distribution is named after Sir Ronald Fisher.

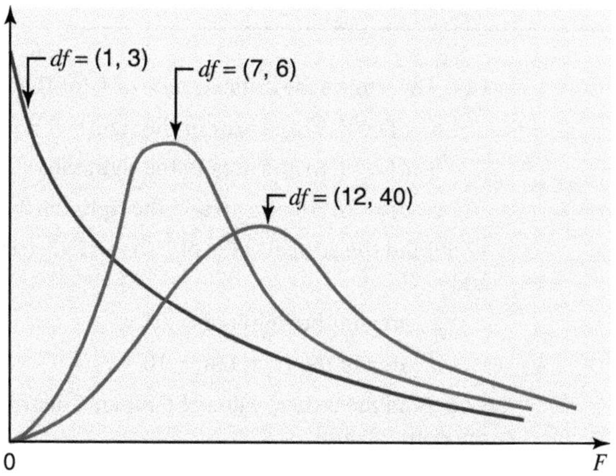

FIGURE 12.1 Three *F* distribution curves.

EXAMPLE 12.1

Reading the *F* Distribution Table

Find the *F* value for 8 degrees of freedom for the numerator, 14 degrees of freedom for the denominator, and .05 area in the right tail of the *F* distribution curve.

Solution To find the required value of *F*, we use the portion of Table VII of Appendix B that corresponds to .05 area in the right tail of the *F* distribution curve. The relevant portion of that table is shown here as Table 12.1. To find the required *F* value, we locate 8 in the row for degrees of freedom for the numerator (at the top of Table VII) and 14 in the column for degrees of freedom for the denominator (the first column on the left side in Table VII). The entry where the column for 8 and the row for 14 intersect gives the required *F* value. This value of *F* is **2.70**, as shown in **Table 12.1** and illustrated in **Figure 12.2**. The *F* value obtained from this table for a test of hypothesis is called the critical value of *F*.

TABLE 12.1 Obtaining the *F* Value From Table VII

Degrees of Freedom for the Denominator	Degrees of Freedom for the Numerator					
	1	2	...	8	...	100
1	161.5	199.5	...	238.9	...	253.0
2	18.51	19.00	...	19.37	...	19.49
.	...	...	...	...	...	...
.	...	...	...	...	...	...
14	4.60	3.74	...	2.70	...	2.19
.	...	...	...	...	...	...
.	...	...	...	...	...	...
100	3.94	3.09	...	2.03	...	1.39

The *F* value for 8 *df* for the numerator, 14 *df* for the denominator, and .05 area in the right tail

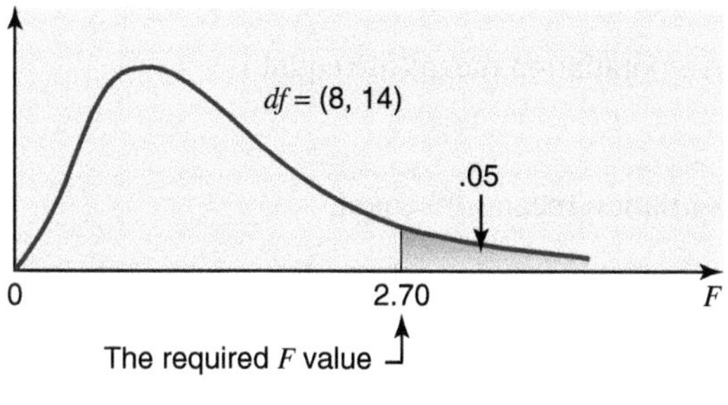

FIGURE 12.2 The value of *F* from Table VII for 8 *df* for the numerator, 14 *df* for the denominator, and .05 area in the right tail.

EXERCISES

Concepts and Procedures

12.1 Describe the main characteristics of an F distribution.

12.2 Find the critical value of F for the following:

a. $df = (3, 3)$ and area in the right tail = .05

b. $df = (3, 10)$ and area in the right tail = .05

c. $df = (3, 30)$ and area in the right tail = .05

12.3 Find the critical value of F for the following:

a. $df = (2, 6)$ and area in the right tail = .025

b. $df = (6, 6)$ and area in the right tail = .025

c. $df = (15, 6)$ and area in the right tail = .025

12.4 Determine the critical value of F for the following:

a. $df = (8, 9)$ and area in the right tail = .01

b. $df = (9, 8)$ and area in the right tail = .01

c. $df = (8, 50)$ and area in the right tail = .01

12.5 Find the critical value of F for an F distribution with $df = (7, 11)$ and

a. area in the right tail = .05

b. area in the right tail = .10

12.6 Find the critical value of F for an F distribution with .050 area in the right tail and

a. $df = (2, 13)$

b. $df = (15, 6)$

12.2 One-Way Analysis of Variance

LEARNING OBJECTIVES

After completing this section, you should be able to:

- Explain when a one-way ANOVA procedure is used to make a test of hypothesis.
- Describe the assumptions that must hold true to apply a one-way ANOVA procedure.
- Calculate between-sample and within-sample variances.
- Construct an ANOVA table and calculate an F statistic.
- Write the null and alternative hypotheses for a one-way ANOVA procedure.
- Identify the numerator and denominator degrees of freedom used in a one-way ANOVA procedure.
- Use an F distribution to illustrate the critical value, the rejection region, and the nonrejection region for a one-way ANOVA procedure.
- Apply the five-step method using a critical value approach to perform a one-way ANOVA test of hypothesis.

As mentioned in the beginning of this chapter, the analysis of variance procedure is used to test the null hypothesis that the means of three or more populations are the same against the alternative hypothesis that not all population means are the same. The analysis of variance procedure can be used to compare two population means. However, the procedures learned in Chapter 10 are more efficient for performing tests of hypotheses about the difference between two population means; the analysis of variance procedure is used to compare three or more population means.

Reconsider the example of teachers at a school who have devised three different methods to teach arithmetic. They want to find out if these three methods produce different mean scores. Let μ_1, μ_2, and μ_3 be the mean scores of all students who are taught by Methods I, II, and III, respectively. To test if the three teaching methods produce different means, we test the null hypothesis

$$H_0: \mu_1 = \mu_2 = \mu_3 \quad \text{(All three population means are equal.)}$$

against the alternative hypothesis

$$H_1: \text{Not all three population means are equal.}$$

One method to test such a hypothesis is to test the three hypotheses $H_0: \mu_1 = \mu_2$, $H_0: \mu_1 = \mu_3$, and $H_0: \mu_2 = \mu_3$ separately using the procedure discussed in Chapter 10. Besides being time consuming, such a procedure has other disadvantages. First, if we reject even one of these three hypotheses, then we must reject the null hypothesis $H_0: \mu_1 = \mu_2 = \mu_3$. Second, combining the Type I error probabilities for the three tests (one for each test) will give a very large Type I error probability for the test $H_0: \mu_1 = \mu_2 = \mu_3$. Hence, we should prefer a procedure that can test the equality of three means in one test. The **ANOVA**, short for **analysis of variance**, provides such a procedure. It is used to compare three or more population means in a single test. Note that if the null hypothesis is rejected, it does not necessarily imply that all three of the means are different or unequal. It could imply that one mean is different from the other two means, or that all three means are different, or that two means are significantly different from each other, but neither is significantly different from the third mean.

ANOVA

ANOVA is a procedure that is used to test the null hypothesis that the means of three or more populations are all equal against the alternative hypothesis that not all population means are equal.

This section discusses the **one-way ANOVA** procedure to compare the means of several populations. By using a one-way ANOVA test, we analyze only **one factor** or **variable**. For instance, in the example of testing for the equality of mean arithmetic scores of students taught by each of the three different methods, we are considering only one factor, which is the effect of different teaching methods on the scores of students. Sometimes we may analyze the effects of two factors. For example, if different teachers teach arithmetic using these three methods, we can analyze the effects of teachers and teaching methods on the scores of students. This is done by using a two-way ANOVA. The procedure under discussion in this chapter is called the analysis of variance because the test is based on the analysis of variation in the data obtained from different samples. The application of one-way ANOVA requires that the following assumptions hold true.

Assumptions of One-Way ANOVA

The following assumptions must hold true to use one-way ANOVA:

1. The populations from which the samples are drawn are approximately normally distributed.
2. The populations from which the samples are drawn have the same variance (or standard deviation).
3. The samples drawn from different populations are random and independent.

For instance, in the example about three methods of teaching arithmetic, we first assume that the scores of all students taught by each method are approximately normally distributed. Second, the means of the distributions of scores for the three teaching methods may or may not be the same, but all three distributions have the same variance, σ^2. Third, when we take samples to perform an ANOVA test, these samples are drawn independently and randomly from three different populations.

The ANOVA test is applied by calculating two estimates of the variance (σ^2) of population distributions: the **variance between samples** and the **variance within samples**. The variance between samples is also called the **mean square between samples** or **MSB**. The variance within samples is also called the **mean square within samples** or **MSW**.

The variance between samples, MSB, gives an estimate of σ^2 based on the variation among the means of samples taken from different populations. For the example of three teaching methods, MSB will be based on the values of the mean scores of three samples of students taught by three different methods. If the means of all populations under consideration are equal, the means of the respective samples will still be different, but the variation among them is expected to be small, and, consequently, the value of MSB is expected to be small. However, if the means of populations under consideration are not all equal, the variation among the means of the respective samples is expected to be large, and, consequently, the value of MSB is expected to the large.

The variance within samples, MSW, gives an estimate of σ^2 based on the variation within the data of different samples. For the example of three teaching methods, MSW will be based on the scores of individual students included in the three samples taken from three populations. The concept of MSW is similar to the concept of the pooled standard deviation, s_p, for two samples discussed in Section 10.2 of Chapter 10.

The one-way ANOVA test is always right-tailed with the rejection region in the right tail of the *F* distribution curve. The hypothesis-testing procedure using ANOVA involves the same five steps that were used in earlier chapters. The next subsection explains how to calculate the value of the test statistic *F* for an ANOVA test.

12.2.1 Calculating the Value of the Test Statistic

The value of the test statistic *F* for a test of hypothesis using ANOVA is given by the ratio of two variances, the variance between samples (MSB) and the variance within samples (MSW).

Test Statistic *F* for a One-Way ANOVA Test

The value of the *test statistic F* for an ANOVA test is calculated as

$$F = \frac{\text{Variance between samples}}{\text{Variance within samples}} \quad \text{or} \quad \frac{\text{MSB}}{\text{MSW}}$$

The calculation of MSB and MSW is explained in Example 12.2.

Example 12.2 describes the calculation of MSB, MSW, and the value of the test statistic *F*. Since the basic formulas are laborious to use, they are not presented here. We have used only the short-cut formulas to make calculations in this chapter.

EXAMPLE 12.2 Three Methods of Teaching Arithmetic

Calculating the Value of the Test Statistic *F*

Fifteen fourth-grade students were randomly assigned to three groups to experiment with three different methods of teaching arithmetic. At the end of the semester, the same test was given to all 15 students. The following table gives the scores of students in the three groups.

GeoStock/Getty Images

Method I	Method II	Method III
48	55	84
73	85	68
51	70	95
65	69	74
87	90	67

Calculate the value of the test statistic *F*. Assume that all the required assumptions mentioned in Section 12.2 hold true.

Solution In ANOVA terminology, the three methods used to teach arithmetic are called **treatments.** The table contains data on the scores of fourth-graders included in the three samples. Each sample of students is taught by a different method. Let

x = the score of a student
k = the number of different samples (or treatments)
n_i = the size of sample i
T_i = the sum of the values in sample i
n = the total number of values in all samples together $= n_1 + n_2 + n_3 + \cdots$
Σx = the sum of all values in all samples $= T_1 + T_2 + T_3 + \cdots$
Σx^2 = the sum of the squares of all values in all samples

To calculate MSB and MSW, we first compute the **between-samples sum of squares**, denoted by **SSB**, and the **within-samples sum of squares**, denoted by **SSW**. The sum of SSB and SSW is called the **total sum of squares** and is denoted by **SST**; that is,

$$\text{SST} = \text{SSB} + \text{SSW}$$

The values of SSB and SSW are calculated using the following formulas:

Between- and Within-Samples Sums of Squares

The *between-samples sum of squares*, denoted by SSB, is calculated as

$$\text{SSB} = \left(\frac{T_1^2}{n_1} + \frac{T_2^2}{n_2} + \frac{T_3^2}{n_3} + \cdots\right) - \frac{(\Sigma x)^2}{n}$$

The *within-samples sum of squares*, denoted by SSW, is calculated as

$$\text{SSW} = \Sigma x^2 - \left(\frac{T_1^2}{n_1} + \frac{T_2^2}{n_2} + \frac{T_3^2}{n_3} + \cdots\right)$$

Table 12.2 lists the scores of 15 students who were taught arithmetic by one of the three different methods; the values of T_1, T_2, and T_3; and the values of n_1, n_2, and n_3.

In Table 12.2, T_1 is obtained by adding the five scores in the first sample. Thus, $T_1 = 48 + 73 + 51 + 65 + 87 = 324$. Similarly, the sums of the values in the second and third samples give $T_2 = 369$ and $T_3 = 388$, respectively. Because there are five observations in each sample, $n_1 = n_2 = n_3 = 5$. The values of Σx and n are, respectively,

$$\Sigma x = T_1 + T_2 + T_3 = 324 + 369 + 388 = 1081$$
$$n = n_1 + n_2 + n_3 = 5 + 5 + 5 = 15$$

TABLE 12.2

Method I	Method II	Method III
48	55	84
73	85	68
51	70	95
65	69	74
87	90	67
$T_1 = 324$	$T_2 = 369$	$T_3 = 388$
$n_1 = 5$	$n_2 = 5$	$n_3 = 5$

To calculate Σx^2, we square all the scores included in all three samples and then add them. Thus,

$$\begin{aligned}\Sigma x^2 &= (48)^2 + (73)^2 + (51)^2 + (65)^2 + (87)^2 + (55)^2 + (85)^2 + (70)^2 \\ &\quad + (69)^2 + (90)^2 + (84)^2 + (68)^2 + (95)^2 + (74)^2 + (67)^2 \\ &= 80{,}709\end{aligned}$$

Substituting all the values in the formulas for SSB and SSW, we obtain the following values of SSB and SSW:

$$\text{SSB} = \left(\frac{(324)^2}{5} + \frac{(369)^2}{5} + \frac{(388)^2}{5}\right) - \frac{(1081)^2}{15} = 432.1333$$

$$\text{SSW} = 80{,}709 - \left(\frac{(324)^2}{5} + \frac{(369)^2}{5} + \frac{(388)^2}{5}\right) = 2372.8000$$

The value of SST is obtained by adding the values of SSB and SSW. Thus,

$$SST = 432.1333 + 2372.8000 = 2804.9333$$

Calculating the Values of MSB and MSW

MSB and MSW are calculated as, respectively,

$$MSB = \frac{SSB}{k-1} \quad \text{and} \quad MSW = \frac{SSW}{n-k}$$

where $k - 1$ and $n - k$ are, respectively, the *df* for the numerator and the *df* for the denominator for the *F* distribution. Remember, *k* is the number of different samples.

The variance between samples (MSB) and the variance within samples (MSW) are calculated using the formulas given in the box.

Consequently, the variance between samples is

$$MSB = \frac{SSB}{k-1} = \frac{432.1333}{3-1} = 216.0667$$

The variance within samples is

$$MSW = \frac{SSW}{n-k} = \frac{2372.8000}{15-3} = 197.7333$$

The value of the test statistic *F* is given by the ratio of MSB and MSW. Therefore,

$$F = \frac{MSB}{MSW} = \frac{216.0667}{197.7333} = \mathbf{1.09}$$

For convenience, all these calculations are often recorded in a table called the *ANOVA table.* **Table 12.3** gives the general form of an ANOVA table.

TABLE 12.3 ANOVA Table

Source of Variation	Degrees of Freedom	Sum of Squares	Mean Square	Value of the Test Statistic
Between	$k-1$	SSB	MSB	$F = \frac{MSB}{MSW}$
Within	$n-k$	SSW	MSW	
Total	$n-1$	SST		

Substituting the values of the various quantities into Table 12.3, we write the ANOVA table for our example as **Table 12.4.**

TABLE 12.4 ANOVA Table for Example 12.2

Source of Variation	Degrees of Freedom	Sum of Squares	Mean Square	Value of the Test Statistic
Between	2	432.1333	216.0667	$F = \frac{216.0667}{197.7333} = 1.09$
Within	12	2372.8000	197.7333	
Total	14	2804.9333		

12.2.2 One-Way ANOVA Test

Now suppose we want to test the null hypothesis that the mean arithmetic scores of fourth-graders are equal for three methods of instruction presented in Example 12.2 against the alternative hypothesis that the mean arithmetic scores of fourth-graders are different for at least two of the three methods of instruction. Note that in a one-way ANOVA test, the null hypothesis is that the means for all populations are equal. The alternative hypothesis is that not all population means are equal. In other words, the alternative hypothesis states that at least one of the population means is different from the others. Example 12.3 demonstrates how we use the one-way ANOVA procedure to make such a test.

EXAMPLE 12.3 Three Methods of Teaching Arithmetic

Performing a One-Way ANOVA Test: All Samples the Same Size

Reconsider Example 12.2 about the scores of 15 fourth-grade students who were randomly assigned to three groups in order to experiment with three different methods of teaching arithmetic. At a 1% significance level, can you reject the null hypothesis that the mean arithmetic score of all fourth-grade students taught by each of these three methods is the same? Assume that all the assumptions required to apply the one-way ANOVA procedure hold true.

GeoStock/Getty Images

Solution To make a test about the equality of the means of three populations, we follow our standard procedure with five steps:

STEP 1 *State the null and alternative hypotheses.*

Let μ_1, μ_2, and μ_3 be the mean arithmetic scores of all fourth-grade students who are taught, by Methods I, II, and III, respectively. The null and alternative hypotheses are

$H_0: \mu_1 = \mu_2 = \mu_3$ (The mean scores of the three methods are all equal.)

H_1: Not all three means are equal.

Note that the alternative hypothesis states that at least two population means are different.

STEP 2 *Select the distribution to use.*

Because we are comparing the means for three normally distributed populations and all of the assumptions required to apply ANOVA procedure are satisfied, we use the F distribution to make this test.

STEP 3 *Determine the rejection and nonrejection regions.*

The significance level is .01. Because a one-way ANOVA test is always right-tailed, the area in the right tail of the F distribution curve is .01, which is the rejection region shown in **Figure 12.3.**

Next we need to know the degrees of freedom for the numerator and the denominator. In our example, the students were assigned to three different methods. As mentioned earlier, these methods are called treatments. The number of treatments is denoted by k. The total number of observations in all samples taken together is denoted by n. As mentioned earlier, the number of degrees of freedom for the numerator is equal to $k - 1$ and the number of degrees of freedom for the denominator is equal to $n - k$. In our example, there are three treatments (methods of teaching) and 15 total observations (total number of students) in all three samples. Thus,

$$\text{Degrees of freedom for the numerator} = k - 1 = 3 - 1 = 2$$
$$\text{Degrees of freedom for the denominator} = n - k = 15 - 3 = 12$$

From Table VII of Appendix B, we find the critical value of F for 2 df for the numerator, 12 df for the denominator, and .01 area in the right tail of the F distribution curve. This value of F is 6.93, as shown in Figure 12.3.

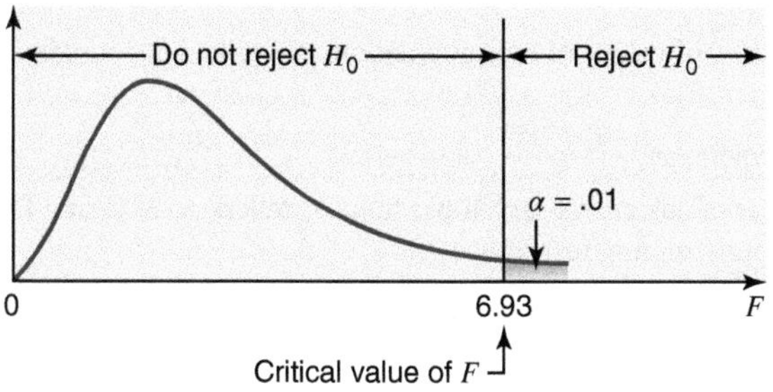

FIGURE 12.3 Critical value of F for $df = (2, 12)$ and $\alpha = .01$.

Thus, we will fail to reject H_0 if the calculated value of the test statistic F is less than 6.93, and we will reject H_0 if this value of F is 6.93 or larger.

STEP 4 *Calculate the value of the test statistic.*

We computed the value of the test statistic F for these data in Example 12.2. This value is

$$F = 1.09$$

STEP 5 *Make a decision.*

Because the value of the test statistic $F = 1.09$ is less than the critical value of $F = 6.93$, it falls in the nonrejection region. Hence, we fail to reject the null hypothesis and conclude that the sample information is not strong enough to reject the null hypothesis that the means of the three populations are equal. In other words, the three different methods of teaching arithmetic do not seem to affect the mean scores of students. The observed difference in the three mean scores in the case of our three samples most likely occurred because of sampling error.

Using technology, we find p-value = .366, which is greater than $\alpha = .01$. Hence, we fail to reject the null hypothesis.

In Example 12.3, the sample sizes were the same for all treatments. Example 12.4 describes a case in which the sample sizes are not the same for all treatments.

EXAMPLE 12.4 Bank Tellers Serving Customers

Performing a One-Way ANOVA Test: All Samples Not the Same Size

From time to time, unknown to its employees, the research department at Post Bank observes various employees for their work productivity. Recently this department wanted to check whether the four tellers at a branch of this bank serve, on average, the same number of customers per hour. The research manager observed each of the four tellers for a certain number of one-hour intervals. The following table gives the number of customers served by the four tellers during each of the observed hours.

Yin Yang/iStockphoto

Teller A	Teller B	Teller C	Teller D
19	14	11	24
21	16	14	19
26	14	21	21
24	13	13	26
18	17	16	20
	13	18	

At a 5% significance level, will you reject the null hypothesis that the mean number of customers served per hour by each of these four tellers is the same? Assume that all the assumptions required to apply the one-way ANOVA procedure hold true.

Solution To make a test about the equality of means of four populations, we follow our standard procedure with five steps:

STEP 1 *State the null and alternative hypotheses.*

Let μ_1, μ_2, μ_3, and μ_4 be the mean number of customers served per hour by tellers A, B, C, and D, respectively. The null and alternative hypotheses are, respectively,

$H_0: \mu_1 = \mu_2 = \mu_3 = \mu_4$ (The mean number of customers served per hour by each of the four tellers is the same.)

H_1: Not all four population means are equal.

STEP 2 *Select the distribution to use.*

Because we are testing for the equality of four means for four normally distributed populations and all of the assumptions required to apply the ANOVA procedure hold true, we use the *F* distribution to make the test.

STEP 3 *Determine the rejection and nonrejection regions.*

The significance level is .05, which means the area in the right tail of the *F* distribution curve is .05. In this example, there are 4 treatments (tellers) and 22 total observations in all four samples. Thus,

$$\text{Degrees of freedom for the numerator} = k - 1 = 4 - 1 = 3$$
$$\text{Degrees of freedom for the denominator} = n - k = 22 - 4 = 18$$

The critical value of F from Table VII for 3 df for the numerator, 18 df for the denominator, and .05 area in the right tail of the *F* distribution curve is 3.16. This value is shown in **Figure 12.4**.

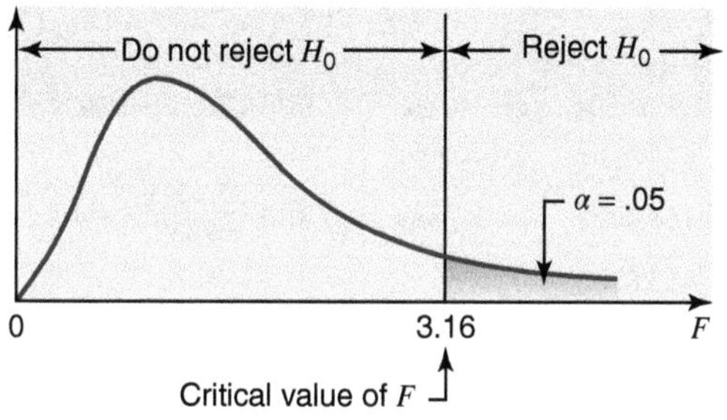

FIGURE 12.4 Critical value of F for $df = (3, 18)$ and $\alpha = .05$.

STEP 4 *Calculate the value of the test statistic.*

First we calculate SSB and SSW. **Table 12.5** lists the number of customers served by each of the four tellers during the selected hours; the values of T_1, T_2, T_3, and T_4; and the values of n_1, n_2, n_3, and n_4.

TABLE 12.5

Teller A	Teller B	Teller C	Teller D
19	14	11	24
21	16	14	19
26	14	21	21
24	13	13	26
18	17	16	20
	13	18	
$T_1 = 108$	$T_2 = 87$	$T_3 = 93$	$T_4 = 110$
$n_1 = 5$	$n_2 = 6$	$n_3 = 6$	$n_4 = 5$

The values of Σx and n are, respectively,

$$\Sigma x = T_1 + T_2 + T_3 + T_4 = 108 + 87 + 93 + 110 = 398$$
$$n = n_1 + n_2 + n_3 + n_4 = 5 + 6 + 6 + 5 = 22$$

The value of Σx^2 is calculated as follows:

$$\begin{aligned}\Sigma x^2 &= (19)^2 + (21)^2 + (26)^2 + (24)^2 + (18)^2 + (14)^2 + (16)^2 + (14)^2 \\ &\quad + (13)^2 + (17)^2 + (13)^2 + (11)^2 + (14)^2 + (21)^2 + (13)^2 \\ &\quad + (16)^2 + (18)^2 + (24)^2 + (19)^2 + (21)^2 + (26)^2 + (20)^2 \\ &= 7614\end{aligned}$$

Substituting all the values in the formulas for SSB and SSW, we obtain the following values of SSB and SSW:

$$\begin{aligned}\text{SSB} &= \left(\frac{T_1^2}{n_1} + \frac{T_2^2}{n_2} + \frac{T_3^2}{n_3} + \frac{T_4^2}{n_4}\right) - \frac{(\Sigma x)^2}{n} \\ &= \left(\frac{(108)^2}{5} + \frac{(87)^2}{6} + \frac{(93)^2}{6} + \frac{(110)^2}{5}\right) - \frac{(398)^2}{22} = 255.6182\end{aligned}$$

$$\begin{aligned}\text{SSW} &= \Sigma x^2 - \left(\frac{T_1^2}{n_1} + \frac{T_2^2}{n_2} + \frac{T_3^2}{n_3} + \frac{T_4^2}{n_4}\right) \\ &= 7614 - \left(\frac{(108)^2}{5} + \frac{(87)^2}{6} + \frac{(93)^2}{6} + \frac{(110)^2}{5}\right) = 158.2000\end{aligned}$$

Hence, the variance between samples MSB and the variance within samples MSW are, respectively,

$$\text{MSB} = \frac{\text{SSB}}{k-1} = \frac{255.6182}{4-1} = 85.2061$$

$$\text{MSW} = \frac{\text{SSW}}{n-k} = \frac{158.2000}{22-4} = 8.7889$$

The value of the test statistic F is given by the ratio of MSB and MSW, which is

$$F = \frac{\text{MSB}}{\text{MSW}} = \frac{85.2061}{8.7889} = 9.69$$

Writing the values of the various quantities in the ANOVA table, we obtain **Table 12.6**.

TABLE 12.6 ANOVA Table for Example 12.4

Source of Variation	Degrees of Freedom	Sum of Squares	Mean Square	Value of the Test Statistic
Between	3	255.6182	85.2061	
Within	18	158.2000	8.7889	$F = \frac{85.2061}{8.7889} = 9.69$
Total	21	413.8182		

STEP 5 *Make a decision.*

Because the value of the test statistic $F = 9.69$ is greater than the critical value of $F = 3.16$, it falls in the rejection region. Consequently, we reject the null hypothesis, and conclude that the sample evidence indicates that the mean number of customers served per hour by each of the four tellers is not the same. In other words, for at least one of the tellers, the mean number of customers served per hour is different from that of the other tellers.

From technology, the p-value for $df = (3, 18)$ and $F = 9.69$ is .000499, which is smaller than $\alpha = .05$. Hence, we reject the null hypothesis.

Note: What If the Sample Size Is Large and the Number of *df* Are Not in the *F* Distribution Table?

In this chapter, we used the F distribution to perform tests of hypothesis about the equality of population means for three or more populations. If we use technology to perform such tests, it does not matter how large the df (degrees of freedom) for the numerator and denominator are. However, when we use the F distribution table (Table VII in Appendix B), sometime we may not find the exact df for the numerator and/or for the denominator in this table, especially when either of these df are large. In such cases, we use the following alternative.

If the number of df is not given in the table, use the closest number of df that falls below the actual value of df. For example, if an ANOVA problem has 4 df for the numerator and 47 df for the denominator, we will use 4 df for the numerator and 40 df for the denominator to obtain the critical value of F from the table. As long as the number of df for the denominator is 3 or larger, the critical values of F become smaller as the numbers of df increase. Hence, whenever the observed value of F falls in the rejection region for a smaller number of df, it will fall in the rejection region for the larger number of df also.

EXERCISES

Concepts and Procedures

12.7 Briefly explain when a one-way ANOVA procedure is used to make a test of hypothesis.

12.8 Describe the assumptions that must hold true to apply the one-way ANOVA procedure to test hypotheses.

12.9 Consider the following data obtained for two samples selected at random from two populations that are independent and normally distributed with equal variances.

Sample I	Sample II
32	27
26	35
31	33
20	40
27	38
34	31

a. Calculate the means and standard deviations for these samples using the formulas from Chapter 3.

b. Using the procedure learned in Section 10.2 of Chapter 10, test at a 1% significance level whether the means of the populations from which these samples are drawn are different.

c. Using the one-way ANOVA procedure, test at a 1% significance level whether the means of the populations from which these samples are drawn are different.

d. Are the conclusions reached in parts (b) and (c) the same?

12.10 The following ANOVA table, based on information obtained for three samples selected from three independent populations that are normally distributed with equal variances, has a few missing values.

Source of Variation	Degrees of Freedom	Sum of Squares	Mean Square	Value of the Test Statistic
Between	4		9.28	
Within		99.56		$F =$ ____ $=$
Total	17			

a. Find the missing values and complete the ANOVA table.

b. Using $\alpha = .25$, what is your conclusion for the test with the null hypothesis that the means of the five populations are all equal against the alternative hypothesis that the means of the five populations are not all equal?

Applications

For the following exercises assume that all the assumptions required to apply the one-way ANOVA procedure hold true.

12.11 A clothing store chain is having a sale based on the use of a coupon. The company is interested in knowing whether the wording of the coupon affects the number of units of the product purchased by customers. The company created four coupons for the same product, each with different wording. Four groups of 50 customers each were selected at random. Group 1 received the first version of the coupon; Group 2 received the second version; and so on. The units of the product purchased by each customer were recorded. The following ANOVA table contains some of the values from the analysis.

Source of Variation	Degrees of Freedom	Sum of Squares	Mean Square	Value of the Test Statistic
Between				
Within		85222.189		$F =$ ____ $=$
Total		97347.192		

Assume that the four populations are approximately normally distributed with equal variances.

a. Find the missing values and complete the ANOVA table.

b. What are the appropriate null and alternative hypotheses for this analysis? Using $\alpha = .25$, what is your conclusion about the equality of population means for all four coupons?

12.12 A local "pick-your-own" farmer decided to grow blueberries. The farmer purchased and planted eight plants of each of the four different varieties of highbush blueberries. The yield (in pounds) of each plant was measured in the upcoming year to determine whether the average yields were different for at least two of the four plant varieties. The yields of these plants of the four varieties are given in the following table:

Berkeley	5.13	5.36	5.20	5.15	4.96	5.14	5.54	5.22
Duke	5.31	4.89	5.09	5.57	5.36	4.71	5.13	5.30
Jersey	5.20	4.92	5.44	5.20	5.17	5.24	5.08	5.13
Sierra	5.08	5.30	5.43	4.99	4.89	5.30	5.35	5.26

a. We are to test the null hypothesis that the mean yields for all such plants of the four varieties are the same against the alternative hypothesis that at least two of these means are different. Write the null and alternative hypotheses.

b. What are the degrees of freedom for the numerator and the denominator?

c. Calculate SSB, SSW, and SST.

d. Show the rejection and nonrejection regions on the F distribution curve for $\alpha = .01$.

e. Calculate the between-samples and within-samples variances.

f. What is the critical value of F for $\alpha = .01$?

g. What is the calculated value of the test statistic F?

h. Write the ANOVA table for this exercise.

i. Will you reject the null hypothesis stated in part (b) at a significance level of 1%?

12.13 A university employment office wants to compare the time taken by graduates with three different majors to find their first full-time job after graduation. The following table lists the time (in days) taken to find their first full-time job after graduation for a random sample of eight business majors, seven computer science majors, and six engineering majors who graduated in May 2019.

Business	Computer Science	Engineering
208	156	126
162	113	275
240	281	363
180	128	146
148	305	298
312	147	392
176	232	
292		

At a 5% significance level, can you reject the null hypothesis that the mean time taken to find their first full-time job for all May 2019 graduates in these fields is the same?

12.14 A consumer agency wanted to find out if the mean times taken by three brands of medicines to provide relief from a headache are different for one or more brands. The first drug was administered to five randomly selected patients, the second to three randomly selected patients, and the third to four randomly selected patients. The following table gives the time (in minutes) taken by each patient to get relief from a headache after taking the medicine.

Drug I	Drug II	Drug III
69	17	47
55	23	36
47	18	58
39	66	
56		

At a 5% significance level, will you reject the null hypothesis that the mean time taken to provide relief from a headache is the same for the three drugs?

12.15 A large company buys thousands of lightbulbs every year. The company is currently considering four brands of lightbulbs to choose from. Before the company decides which lightbulbs to buy, it wants to investigate if the mean lifetimes of the four types of lightbulbs are different. The company's research department randomly selected a few bulbs of each type and tested them. The following table lists the number of hours (in thousands) that each of the bulbs in each brand lasted before being burned out.

Brand I	Brand II	Brand III	Brand IV
33	28	43	36
34	31	37	34
29	26	29	19
28	37	32	37
27	30	29	41

At a 2.5% significance level, will you reject the null hypothesis that the mean lifetimes of bulbs for these four brands are equal?

12.3 Two-Way Analysis of Variance

LEARNING OBJECTIVES

After completing this section, you should be able to:

- Explain when a two-way ANOVA procedure is used to make tests of hypotheses.
- Describe the assumptions that must hold true to apply a two-way ANOVA procedure.
- Calculate between-sample and within-sample variances.
- Calculate mean square interaction and mean square error.
- Construct an ANOVA table and calculate *F* statistic related to different hypotheses.
- Write different null and alternative hypotheses for a two-way ANOVA procedure.
- Identify the numerator and denominator degrees of freedoms used in a two-way ANOVA procedure.
- Use an *F* distribution to illustrate the critical value, the rejection region, and the nonrejection region for testing different hypotheses associated with a two-way ANOVA procedure.
- Apply the five-step method using a critical value approach to perform two-way ANOVA test of hypothesis.

When two factors influence the means of several populations, the effect of the two factors can be analyzed by using a two-way ANOVA procedure. For instance, in the example of testing for the equality of mean arithmetic scores of students taught by each of the three different methods, we are analyzing only one factor, that is, the effect of different teaching methods on the scores of students.

If four different teachers teach arithmetic using these three methods, a second factor emerges—different teachers. Then the effects of teachers and teaching methods on the scores of students can be simultaneously analyzed by a two-way ANOVA. The different values or categories that each of the two factors can take are called the *levels of the corresponding factors*, and combinations of different levels of the first factor with different levels of the second factor are called the *cells*.

In the previous example, the first factor, that is, teaching method has three levels—Method I, Method II, and Method III, and the second factor, that is, the teacher has four levels—Teacher 1, Teacher 2, Teacher 3, and Teacher 4. The number of cells in the example is $3 \times 4 = 12$. The application of two-way ANOVA requires that the following assumptions hold true.

Assumptions of Two-Way ANOVA

The following assumptions must hold true to use two-way ANOVA:

1. The populations from which the samples are drawn are approximately normally distributed.
2. The populations from which the samples are drawn have the same variance (or standard deviation).
3. The samples drawn from different populations are random and independent.
4. For each cell, the sample size must be at least one.

For instance, in the hypothetical example regarding four teachers applying three methods of teaching arithmetic, we first assume that the scores of all students taught by each teacher using each method are approximately normally distributed. Second, the means of the distributions of scores for the 12 combinations of three teaching methods and four teachers may or may not be the same, but all 12 distributions have the same variance, σ^2. Third, when we take samples to perform two-way ANOVA test, these samples are drawn independently and randomly from 12 different populations. Fourth and last, for each of the 12 cells, at least one student is taught by using that combination of teaching method and teacher.

Before specifying the test statistics related to two-way ANOVA, we must specify the underlying null and alternative hypotheses which are being tested. In terms of the previous example, let us assume that μ_{ij} is the mean score of all students who will be taught using teaching method i by teacher j, where i can take values 1, 2, and 3, and j can take values 1, 2, 3, and 4. Further, assume that $\mu_{ij} = \mu + \tau_i + \beta_j$, where τ_1, τ_2, and τ_3 denote the effect of Methods I, II, and III, respectively, on the mean scores of all students, and β_1, β_2, β_3, and β_4 denote the effect of Teachers 1, 2, 3, and 4, respectively, on the mean scores of all students. μ denotes the baseline mean score of all students, that is, the hypothetical score a student is expected to get if he or she is not taught by any teacher adopting any specific teaching method.

Now, if we want to test whether the three teaching methods have the same effect, we test the null hypothesis:

$H_{\tau 0}: \tau_1 = \tau_2 = \tau_3$ (All three teaching methods have equal effects on the mean scores of students.)

against the alternative hypothesis

$H_{\tau 1}$: Not all three teaching methods have equal effects on the mean scores of students.

Alternatively, we may want to test whether the four teachers are equally effective in teaching. In that case, we test the null hypothesis.

$H_{\beta 0}: \beta_1 = \beta_2 = \beta_3 = \beta_4$ (All four teachers have equal effects on the mean scores of students.)

against the alternative hypothesis

$H_{\beta 1}$: Not all four teachers have equal effects on the mean scores of students.

Similar to the one-way ANOVA test, the two-way ANOVA test is performed by calculating different estimates of the variance (σ^2) of population distributions: the variance between samples/mean square between samples with respect to the first factor, denoted by MSB_τ, the variance between samples/mean square between samples with respect to the second factor, denoted by MSB_β, and the error variance/mean square error, denoted by MSE. For the example regarding four teachers applying three methods of teaching arithmetic, MSB_τ will be based on the values of the mean scores of the three samples of students taught by three different methods and MSB_β will be based on the values of the mean scores of four samples of students taught by four different teachers.

The calculation of MSE is explained in detail later. It is similar to that for the MSW for the one-way ANOVA procedure. If the null hypothesis $H_{\tau 0}: \tau_1 = \tau_2 = \tau_3$ is true, that is, all three teaching methods have equal effects on the mean scores of all students, then MSB_τ is expected to be small. However, if not all of τ_1, τ_2, and τ_3 are the same, then the value of MSB_τ is expected to be large. Similarly, if the other null hypothesis $H_{\beta 0}: \beta_1 = \beta_2 = \beta_3 = \beta_4$ is true, that is, all four teachers have equal effects on the mean scores of all students, then the value of MSB_β is expected to be small. However, if not all of β_1, β_2, β_3, and β_4 are same, then the value of MSB_β is expected to be large.

Similar to the one-way ANOVA test, the two-way ANOVA test is also right tailed with the rejection region in the right tail of some F distribution curve. The next subsection explains how to calculate the value of the test statistics F_τ and F_β for the two-way ANOVA test.

12.3.1 Calculating the Value of the Test Statistics

The value of the test statistic F_τ for the test of the hypothesis $H_{\tau 0}: \tau_1 = \tau_2 = \ldots = \tau_t$ using a two-way ANOVA with t levels for the first factor and b levels for the second factor is given by the ratio of MSB_τ and MSE, and the value of the other test statistic F_β for the test of the hypothesis $H_{\beta 0}: \beta_1 = \beta_2 = \ldots = \beta_b$ is given by the ratio of MSB_β and MSE.

Test Statistics F_τ and F_β for a Two-Way ANOVA Test

The value of the *test statistic* F_τ for a two-way ANOVA test is calculated as

$$F_\tau = \frac{\text{Variance between samples for } t \text{ groups formed by } t \text{ levels of the first factor}}{\text{Error variance}}$$

or $\frac{MSB_\tau}{MSE}$ and

$$F_\beta = \frac{\text{Variance between samples for } b \text{ groups formed by } b \text{ levels of the second factor}}{\text{Error variance}}$$

or $\frac{MSB_\beta}{MSE}$

The calculation of MSB_τ, MSB_β, and MSE is explained in Example 12.5.

Example 12.5 describes the calculation of MSB_τ, MSB_β, and MSE and the value of the test statistics F_τ and F_β. Only the short-cut formulas are presented here.

EXAMPLE 12.5 Four Teachers Use Three Methods to Teach Arithmetic: Each Cell Has Sample Size 1

Calculating the Value of the Test Statistics F_τ and F_β

Twelve fourth-grade students were randomly assigned to 12 groups, to experiment with three different methods of teaching arithmetic offered by each of the four different mathematics teachers. At the end of the semester, the same test was given to each of 12 students. The following table gives the scores of students in the 12 groups.

	Teacher 1	Teacher 2	Teacher 3	Teacher 4
Method I	47	72	77	81
Method II	66	62	72	74
Method III	42	57	91	82

Assume that all the required assumptions mentioned in Section 12.3 hold true and next we calculate the value of the test statistics F_τ and F_β.

Solution The table contains data on the scores of fourth-graders included in the 12 samples.

Each of the 12 samples is of size 1, and each sample of students is taught by a different combination of teaching method and teacher.

Let

x_{ij} = score of a student taught using method i and by teacher j
t = the number of levels associated with the first factor
b = the number of levels associated with the second factor
n_{ij} = the number of students taught using method i and by teacher j
n = the total number of values in all samples together = $\Sigma_{i=1}^{t}\Sigma_{j=1}^{b} n_{ij}$
$\bar{x}_{i..}$ = mean score of all students taught using method i
$\bar{x}_{.j.}$ = mean score of all students taught by teacher j
$\bar{x}_{...}$ = mean score of all n students

To calculate MSB_τ, MSB_β, and MSE, we first compute the between-samples sum of squares with respect to the first factor, that is, SSB_τ, the between-samples sum of squares with respect to the second factor, that is, SSB_β, and the error sum of squares, denoted by SSE. The sum of SSB_τ, SSB_β, and SSE is called the Total Sum of Squares and is denoted by SST; that is,

$$SST = SSB_\tau + SSB_\beta + SSE$$

The values of SSB_τ, SSB_β, and SSE are calculated using the following formulas:

Between-Samples Sum of Squares with Respect to the First Factor, Between-Samples Sum of Squares with Respect to the Second Factor, and the Error Sum of Squares

The between-samples sum of squares with respect to the first factor, SSB_τ, is calculated as

$$SSB_\tau = \sum_{i=1}^{t}\sum_{j=1}^{b}\sum_{k=1}^{n_{ij}}(\bar{x}_{i..} - \bar{x}_{...})^2$$

The between-samples sum of squares with respect to the second factor, SSB_β, is calculated as

$$SSB_\beta = \sum_{i=1}^{t}\sum_{j=1}^{b}\sum_{k=1}^{n_{ij}}(\bar{x}_{.j.} - \bar{x}_{...})^2$$

The error sum of squares, denoted by SSE, is calculated as

$$SSE = \sum_{i=1}^{t}\sum_{j=1}^{b}\sum_{k=1}^{n_{ij}}(x_{ij} - \bar{x}_{i..} - \bar{x}_{.j.} + \bar{x}_{...})^2$$

Table 12.7 lists the scores of 12 students who were taught arithmetic using one of the three different methods by four teachers and the values of $\bar{x}_{i..}$, $\bar{x}_{.j.}$, and $\bar{x}_{...}$.

In Table 12.7, $\bar{x}_{i..}$ is obtained by taking the mean of the four scores in the ith row. Thus, $\bar{x}_{1..} = (47 + 72 + 77 + 81)/4 = 277/4 = 69.25$.

Similarly, $\bar{x}_{2..} = 274/4 = 68.5$ and $\bar{x}_{3..} = 272/4 = 68$.

$\bar{x}_{.j.}$ is obtained by taking the mean of the three scores in the jth column. Thus, $\bar{x}_{.1.} = (47 + 66 + 42)/3 = 155/3 = 51.6667$. Similarly, $\bar{x}_{.2.} = 191/3 = 63.6667$, $\bar{x}_{.3.} = 240/3 = 80$, and $\bar{x}_{.4.} = 237/3 = 79$.

Because there is one observation in each of the 12 cells, so, $n_{ij} = 1$ and $n = \sum_{i=1}^{3}\sum_{j=1}^{4} n_{ij} = 12$. The value of $\bar{x}_{...}$ is given by $(47 + 72 + 77 + 81 + 66 + 62 + 72 + 74 + 42 + 57 + 91 + 82)/12 = 823/12 = 68.5833$.

TABLE 12.7

	Teacher 1	Teacher 2	Teacher 3	Teacher 4	$\bar{x}_{i..}$
Method I	47	72	77	81	69.25
Method II	66	62	72	74	68.5
Method III	42	57	91	82	68
$\bar{x}_{.j.}$	51.6667	63.6667	80	79	$\bar{x}_{...} = 68.5833$

Using the values from the table, we get

$$SSB_\tau = \sum_{i=1}^{3}\sum_{j=1}^{4}\sum_{k=1}^{1}(\bar{x}_{i..} - \bar{x}_{...})^2 = 4\sum_{i=1}^{3}(\bar{x}_{i..} - \bar{x}_{...})^2$$

$$= 4[(69.25 - 68.5833)^2 + (68.5 - 68.5833)^2 + (68 - 68.5833)^2]$$

$$SSB_\tau = 3.1667$$

Similarly, we get

$$SSB_\beta = \sum_{i=1}^{3}\sum_{j=1}^{4}\sum_{k=1}^{1}(\bar{x}_{.j.} - \bar{x}_{...})^2 = 3\sum_{j=1}^{4}(\bar{x}_{.j.} - \bar{x}_{...})^2$$

$$= 3[(51.6667-68.5833)^2 + (63.6667-68.5833)^2 + (80-68.5833)^2 + (79-68.5833)^2]$$

$$SSB_\beta = 1647.3727$$

We also calculate $SSE = \Sigma_{i=1}^{3}\Sigma_{j=1}^{4}\Sigma_{k=1}^{1}(x_{ij} - \bar{x}_{i..} - \bar{x}_{.j.} + \bar{x}_{...})^2 = \Sigma_{i=1}^{3}\Sigma_{j=1}^{4}(x_{ij} - \bar{x}_{i..} - \bar{x}_{.j.} + \bar{x}_{...})^2$ which simplifies to:

$$(47-69.25-51.6667+68.5833)^2 + (66-68.5-51.6667+68.5833)^2$$
$$+(42-68.5833)^2 + \cdots + (82-68-79+68.5833)^2 = 666.1667$$

The value of SST is obtained by adding the values of SSB_τ, SSB_β, and SSE. Thus,

$$SST = SSB_\tau + SSB_\beta + SSE = 3.1667 + 1647.3727 + 666.1667 = 2316.9167$$

Calculating the Values of MSB_τ, MSB_β, and MSE

MSB_τ, MSB_β, and MSE are calculated as follows:

$$MSB_\tau = \frac{SSB_\tau}{t-1}, \quad MSB_\beta = \frac{SSB_\beta}{b-1}, \quad \text{and} \quad MSE = \frac{SSE}{n-1-(t-1)-(b-1)} = \frac{SSE}{n-t-b+1}$$

where $t - 1$, $b - 1$, and $n - t - b + 1$ are the associated degrees of freedom. They determine the shape of the F distributions associated with two test statistics F_τ and F_β.

In Example 12.5,

$$t = 3, b = 4, \text{ and consequently } n - t - b + 1 = 12 - 3 - 4 + 1 = 6.$$

MSB_τ, MSB_β, and MSE are calculated using the formulas given in the box. Consequently,

$MSB_\tau = \frac{SSB_\tau}{t-1} = \frac{3.1667}{2} = 1.5833$, $MSB_\beta = \frac{SSB_\beta}{b-1} = \frac{1647.3727}{3} = 549.1944$, and the error variance/mean square error, denoted by $MSE = \frac{SSE}{n-t-b+1} = \frac{666.1667}{6} = 111.0278$.

The value of the test statistic F_τ is given by the ratio of MSB_τ and MSE. Similarly, the value of F_β is given by the ratio of MSB_β and MSE. Therefore,

$$F_\tau = \frac{MSB_\tau}{MSE} = \frac{1.5833}{111.0278} = .0143 \quad \text{and} \quad F_\beta = \frac{MSB_\beta}{MSE} = \frac{549.1944}{111.0278} = 4.9465$$

For convenience, all these calculations are often recorded in a table called the two-way ANOVA table. **Table 12.8** gives the general form of a two-way ANOVA table.

TABLE 12.8 Two-Way ANOVA Table

Source of Variation	Degrees of Freedom	Sum of Squares	Mean Square	Value of the Test Statistics
First factor	$t-1$	SSB_τ	$MSB_\tau = \frac{SSB_\tau}{t-1}$	$F_\tau = \frac{MSB_\tau}{MSE}$
Second factor	$b-1$	SSB_β	$MSB_\beta = \frac{SSB_\beta}{b-1}$	$F_\beta = \frac{MSB_\beta}{MSE}$
Error	$N-t-b+1$	SSE	$MSE = \frac{SSE}{n-t-b+1}$	
Total	$n-1$	SST		

Substituting the values of the various quantities into Table 12.8, we write the two-way ANOVA table for our Example 12.5 as **Table 12.9**.

TABLE 12.9 Two-Way ANOVA Table for Example 12.5

Source of Variation	Degrees of Freedom	Sum of Squares	Mean Square	Value of the Test Statistics
First factor	2	3.1667	1.5833	$F_\tau = \frac{1.5833}{111.0278} = .0143$
Second factor	3	1647.3727	549.1944	$F_\beta = \frac{549.1944}{111.0278} = 4.9465$
Error	6	666.1667	111.0278	
Total	11	2316.9167		

Regarding the two-way ANOVA procedure, we are assuming that $\mu_{ij} = \mu + \tau_i + \beta_j$, where τ_1, τ_2, and τ_3 denote the effect of different teaching methods on the mean scores of all students and β_1, β_2, β_3, and β_4 denote the effect of Teachers 1, 2, 3, and 4, respectively, on the mean scores of all students. μ denotes the baseline mean score of all students, that is, the hypothetical scores a student is expected to get if he or she is not taught by any teacher adopting any specific teaching method. Such a model that connects the mean score with the two factors via a simple sum is called an **additive model**.

One implication of the additive model is that a good teacher is good for all three methods of teaching, and a good teaching method would result in the same incremental improvement of the mean scores of the students for all four teachers. However, this assumption is unrealistic, and we can expect that a particular teacher who is tech-savvy can teach much better compared to other teachers by using one of the methods that is computer-enabled. A more realistic model that can accommodate such non-additive phenomena is given by $\mu_{ij} = \mu + \tau_i + \beta_j + \lambda_{ij}$. Here, the term λ_{ij} accommodates the non-additive component in the two-way ANOVA model and is referred to as the **interaction term**. The different values that the interaction term λ_{ij} can take are referred to as **interaction effects**. As there are $t \times b$ cells in a two-way ANOVA, there are $t \times b$ such interaction effects.

Interaction in Two-Way ANOVA

The interaction effects in a two-way ANOVA procedure inform whether the effect of the first factor on the dependent variable is the same for all values of the second factor and vice-versa. For a two-way ANOVA procedure with $t \times b$ cells, the number of such interaction effects is also $t \times b$.

Often, we intend to check whether an additive model is sufficient. It can be restated as no interaction and that can be expressed as the null hypothesis $H_{\lambda 0}$: $\lambda_{11} = \lambda_{12} = \ldots = \lambda_{34} = 0$. (All 12 interaction effects equal to zero.) The alternative hypothesis is $H_{\lambda 1}$: At least one of the 12 interaction effects is non-zero. Unfortunately, if the sample size associated with each of the $t \times b$ cells is one, $H_{\lambda 0}$ cannot be tested. For testing the null hypothesis $H_{\lambda 0}$, the sample size associated with each of the $t \times b$ cells have to be at least two.

Additional Assumption for Test of Interaction in Two-Way ANOVA

The following assumption must hold true to test the null hypothesis $H_{\lambda 0}: \lambda_{11} = \lambda_{12} = \cdots = \lambda_{34} = 0$ of no interaction effects in a two-way ANOVA.

- The sample size associated with each of the $t \times b$ cells must be two or more.

So, it is now clear that for Example 12.5, it is not possible to carry out the test for no interaction as each of the 12 cells contains exactly one value. Next, we modify Example 12.5 to create a new Example 12.6, in which we add some more values, so that each of the 12 cells contains two or more values. As a result of that, we are also able to test the null hypothesis $H_{\lambda 0}$, apart from $H_{\tau 0}$ and $H_{\beta 0}$, in Example 12.6. Note that since now more values are added, the formulas in the two-way ANOVA table and the calculations would also change.

Like the tests of $H_{\tau 0}$ and $H_{\beta 0}$, the test statistics associated with $H_{\lambda 0}$ is based on some estimate of the variance (σ^2) of population distributions: mean square interaction, denoted by MSI, and the MSE. The value of the test statistic F_λ for the test of the hypothesis $H_{\lambda 0}: \lambda_{11} = \lambda_{12} = \cdots = \lambda_{34} = 0$ is given by the ratio of MSI and MSE. Once the value of the test statistic F_λ is calculated, we can reject/not reject $H_{\lambda 0}$ based on the critical value obtained from some F distribution curve.

Test Statistics F_λ for Testing Interaction in Two-Way ANOVA Procedure

The value of the *test statistic* F_λ for a two-way ANOVA procedure with interaction is calculated as

$$F_\lambda = \frac{\text{MSI}}{\text{MSE}}$$

Example 12.6 describes the calculation of MSB_τ, MSB_β, MSI, and MSE, and the value of the test statistics F_τ, F_β, and F_λ for a two-way ANOVA procedure with interaction.

EXAMPLE 12.6 Four Teachers Use Three Methods to Teach Arithmetic: Each Cell Has Sample Size 2

Calculating the Value of the Test Statistics F_τ, F_β, and F_λ for a Two-Way ANOVA Procedure with Interaction

Twenty-four fourth-grade students were randomly assigned to 12 groups to experiment with three different methods of teaching arithmetic offered by four different mathematics teachers. Two students were taught by each of the 12 combinations of teaching methods and teachers. At the end of the semester, the same test was given to all 24 students. The following table gives the scores of students in the 12 groups.

	Teacher 1	Teacher 2	Teacher 3	Teacher 4
Method I	47, 41	72, 77	77, 81	81, 88
Method II	66, 62	62, 69	72, 79	74, 79
Method III	42, 52	57, 56	91, 85	82, 77

Assume that all the required assumptions mentioned in Section 12.3 hold true and next we calculate the value of the test statistics F_τ, F_β, and F_λ.

Solution The table contains data on the scores of fourth-graders included in the 12 samples. Each of the 12 samples is of size 2, and each sample of students is taught by a different combination of teaching method and teacher.

Let

x_{ijk} = score of the kth student taught using method i and by teacher j
$\bar{x}_{ij.}$ = mean score of all n_{ij} students taught using method i and by teacher j

The rest of the symbols used are same as those in Example 12.5.

To calculate MSB_τ, MSB_β, MSI, and MSE, we first compute the between-samples sum of squares with respect to the first factor, that is, SSB_τ, the between-samples sum of squares with respect to the second factor, that is, SSB_β, the sum of squares with respect to the interaction, denoted by SSI, and the error sum of squares, denoted by SSE. The sum of SSB_τ, SSB_β, SSI, and SSE is called the total sum of squares and is denoted by SST; that is,

$$SST = SSB_\tau + SSB_\beta + SSI + SSE$$

The values of SSB_τ, SSB_β, SSI, and SSE are calculated using the following formulas:

Between-Samples Sum of Squares with Respect to the First Factor, Between-Samples Sum of Squares with Respect to the Second Factor, and Sum of Squares with Respect to the Interaction and the Error Sum of Squares

The between-samples sum of squares with respect to the first factor, SSB_τ, is calculated as

$$SSB_\tau = \sum_{i=1}^{t}\sum_{j=1}^{b}\sum_{k=1}^{n_{ij}}(\bar{x}_{i..} - \bar{x}_{...})^2$$

The between-samples sum of squares with respect to the second factor, SSB_β, is calculated as

$$SSB_\beta = \sum_{i=1}^{t}\sum_{j=1}^{b}\sum_{k=1}^{n_{ij}}(\bar{x}_{.j.} - \bar{x}_{...})^2$$

The sum of squares with respect to the interaction, denoted by SSI, is calculated as

$$SSI = \sum_{i=1}^{t}\sum_{j=1}^{b}\sum_{k=1}^{n_{ij}}(\bar{x}_{ij.} - \bar{x}_{i..} - \bar{x}_{.j.} + \bar{x}_{...})^2$$

The error sum of squares, denoted by SSE, is calculated as

$$SSE = \sum_{i=1}^{t}\sum_{j=1}^{b}\sum_{k=1}^{n_{ij}}(x_{ijk} - \bar{x}_{ij.})^2$$

Table 12.10 lists the scores of 24 students who were taught arithmetic using one of the three different methods by four teachers and the values of $\bar{x}_{i..}$, $\bar{x}_{.j.}$, and $\bar{x}_{...}$. **Table 12.11** lists the value of $\bar{x}_{ij.}$ for each cell. In Table 12.10, $\bar{x}_{i..}$ are obtained by taking the mean of the eight scores in the ith row. Thus, $\bar{x}_{1..} = (47 + 41 + 72 + 77 + 77 + 81 + 81 + 88)/8 = 564/8 = 70.5$. Similarly, $\bar{x}_{2..} = 563/8 = 70.375$ and $\bar{x}_{3..} = 542/8 = 67.75$. $\bar{x}_{.j.}$ are obtained by taking the mean of the six scores in the jth column. Thus, $\bar{x}_{.1.} = (47 + 41 + 66 + 62 + 42 + 52)/6 = 310/6 = 51.6667$. Similarly, $\bar{x}_{.2.} = 393/6 = 65.5$, $\bar{x}_{.3.} = 485/6 = 80.8333$, and $\bar{x}_{.4.} = 481/6 = 80.1667$.

Because there are two observations in each of the 12 cells, so $n_{ij} = 2$ and $n = \Sigma_{i=1}^{3}\Sigma_{j=1}^{4} n_{ij} = 24$. The value of $\bar{x}_{...}$ is given by $(47 + 41 + 72 + 77 + 77 + 81 + 81 + 88 + 66 + 62 + 62 + 69 + 72 + 79 + 74 + 79 + 42 + 52 + 57 + 56 + 91 + 85 + 82 + 77)/24 = 1669/24 = 69.5417$.

TABLE 12.10

	Teacher 1	Teacher 2	Teacher 3	Teacher 4	$\bar{x}_{i..}$
Method I	47, 41	72, 77	77, 81	81, 88	70.5
Method II	66, 62	62, 69	72, 79	74, 79	70.375
Method III	42, 52	57, 56	91, 85	82, 77	67.75
$\bar{x}_{.j.}$	51.6667	65.5	80.8333	80.1667	$\bar{x}_{...} = 69.5417$

TABLE 12.11

$\bar{x}_{ij.}$	Teacher 1	Teacher 2	Teacher 3	Teacher 4
Method I	44	74.5	79	84.5
Method II	64	65.5	75.5	76.5
Method III	47	56.5	88	79.5

Using the values from the tables, we get

$$SSB_\tau = \sum_{i=1}^{3}\sum_{j=1}^{4}\sum_{k=1}^{2}(\bar{x}_{i..} - \bar{x}_{...})^2 = 8\sum_{i=1}^{3}(\bar{x}_{i..} - \bar{x}_{...})^2$$

$$= 8[(70.5-69.5417)^2 + (70.375-69.5417)^2 + (67.75-69.5417)^2]$$

$$SSB_\tau = 38.5833$$

Similarly,

$$SSB_\beta = \sum_{i=1}^{3}\sum_{j=1}^{4}\sum_{k=1}^{2}(\bar{x}_{.j.} - \bar{x}_{...})^2 = 6\sum_{j=1}^{4}(\bar{x}_{.j.} - \bar{x}_{...})^2$$

$$= 6[(51.6667-69.5417)^2 + (65.5-69.5417)^2 + (80.8333-69.5417)^2 + (80.1667-69.5417)^2]$$

$$SSB_\beta = 3457.4583$$

We also calculate $SSI = \Sigma_{i=1}^{3}\Sigma_{j=1}^{4}\Sigma_{k=1}^{2}(\bar{x}_{ij.} - \bar{x}_{i..} - \bar{x}_{.j.} + \bar{x}_{...})^2 = 2\Sigma_{i=1}^{3}\Sigma_{j=1}^{4}(\bar{x}_{ij.} - \bar{x}_{i..} - \bar{x}_{.j.} + \bar{x}_{...})^2$ which simplifies to $2(44 - 70.5 - 51.6667 + 69.5417)^2 + (74.5 - 70.5 - 65.5 + 69.5417)^2 + \cdots + (79.5 - 67.75 - 80.1667 + 69.5417)^2 = 982.4167$.

Finally, we calculate $SSE = \Sigma_{i=1}^{3}\Sigma_{j=1}^{4}\Sigma_{k=1}^{2}(x_{ijk} - \bar{x}_{ij.})^2$ which simplifies to $(47 - 44)^2 + (41 - 44)^2 + \cdots + (77 - 79.5)^2 = 213.5$.

The value of SST is obtained by adding the values of SSB_τ, SSB_β, SSI, and SSE. Thus,

$$SST = SSB_\tau + SSB_\beta + SSI + SSE = 38.5833 + 3457.4583 + 982.4167 + 213.5 = 4691.9583.$$

Calculating the Values of MSB_τ, MSB_β, MSI, and MSE

The formulae for MSB_τ and MSB_β are same as in case of two-way ANOVA procedure without interaction.

$MSB_\tau = \dfrac{SSB_\tau}{t-1}$, $MSB_\beta = \dfrac{SSB_\beta}{b-1}$, and MSI and MSE are calculated as follows:

$$MSI = \frac{SSI}{(t-1)(b-1)} \quad \text{and} \quad MSE = \frac{SSE}{n-1-(t-1)-(b-1)-(t-1)(b-1)} = \frac{SSE}{n-tb}$$

where $t - 1$, $b - 1$, $(t - 1)(b - 1)$, and $n - tb$ are the associated degrees of freedoms. They determine the shape of the F distributions associated with three test statistics F_τ, F_β, and F_λ.

In Example 12.6, $t = 3$, $b = 4$, $(t - 1)(b - 1) = 6$, and consequently $n - tb = 24 - 12 = 12$. MSB_τ, MSB_β, MSI, and MSE are calculated using the formulae given in the box. Consequently,

$$MSB_\tau = \frac{SSB_\tau}{t-1} = \frac{38.5833}{2} = 19.2917, \quad MSB_\beta = \frac{SSB_\beta}{b-1} = \frac{3457.4583}{3} = 1152.4861,$$

$$MSI = \frac{SSI}{(t-1)(b-1)} = \frac{982.4167}{6} = 163.7361, \quad \text{and} \quad MSE = \frac{SSE}{n-tb} = \frac{213.5}{12} = 17.7917$$

The value of the test statistic F_τ is given by the ratio of MSB_τ and MSE. Similarly, the value of F_β is given by the ratio of MSB_β and MSE, and the value of F_λ is given by the ratio of MSI and MSE. Therefore,

$$F_\tau = \frac{MSB_\tau}{MSE} = \frac{19.2917}{17.7917} = 1.0843, \quad F_\beta = \frac{MSB_\beta}{MSE} = \frac{1152.4861}{17.7917} = 64.7767,$$

and

$$F_\lambda = \frac{MSI}{MSE} = \frac{163.7361}{17.7917} = 9.2030$$

For convenience, all these calculations are often recorded in a table called the two-way ANOVA table with interaction. **Table 12.12** gives the general form of a two-way ANOVA table with interaction.

TABLE 12.12 Two-Way ANOVA Table with Interaction

Source of Variation	Degrees of Freedom	Sum of Squares	Mean Square	Value of the Test Statistics
First factor	$t-1$	SSB_τ	$MSB_\tau = \frac{SSB_\tau}{t-1}$	$F_\tau = \frac{MSB_\tau}{MSE}$
Second factor	$b-1$	SSB_β	$MSB_\beta = \frac{SSB_\beta}{b-1}$	$F_\beta = \frac{MSB_\beta}{MSE}$
Interaction	$(t-1)(b-1)$	SSI	$MSI = \frac{SSI}{(t-1)(b-1)}$	$F_\lambda = \frac{MSI}{MSE}$
Error	$n-tb$	SSE	$MSE = \frac{SSE}{n-tb}$	
Total	$n-1$	SST		

Substituting the values of the various quantities into Table 12.12, we write the two-way ANOVA table with interaction for our Example 12.6 as **Table 12.13**.

TABLE 12.13 Two-Way ANOVA Table with Interaction for Example 12.6

Source of Variation	Degrees of Freedom	Sum of Squares	Mean Square	Value of the Test Statistics
First factor	2	38.5833	19.2917	$F_\tau = \frac{19.2917}{17.7917} = 1.0843$
Second factor	3	3457.4583	1152.4861	$F_\beta = \frac{1152.4861}{17.7917} = 64.7767$
Interaction	6	982.4167	163.7361	$F_\lambda = \frac{163.7361}{17.7917} = 9.2030$
Error	12	213.5	17.7917	
Total	23	4691.9583		

12.3.2 Two-Way ANOVA Tests

Now suppose we want to test the null hypothesis that the three methods of instruction presented in Example 12.5 have the same effects on mean arithmetic scores of fourth-graders versus the alternative hypothesis that at least two of the three methods of instruction have different effects on mean arithmetic scores of fourth-graders. Similarly, we also want to test that the four teachers are equally good at teaching, that is, the four teachers have same effects on mean arithmetic scores of fourth-graders versus the alternative hypothesis that at least two of the four teachers have different effects on mean arithmetic scores of fourth-graders.

Note that in two-way ANOVA tests, the two null hypotheses are that the different levels of factor 1 have equal effects on the means of populations and the different levels of factor 2 have equal effects on the means of populations. The two alternative hypotheses are that not all levels of factor 1 have equal effects on the means of populations and not all levels of factor 2 have equal effects on the means of populations. Example 12.7 demonstrates how we use the two-way ANOVA procedure to make such tests.

EXAMPLE 12.7 Four Teachers Use Three Methods to Teach Arithmetic: Each Cell Has Sample Size 1

Reconsider Example 12.5 about the scores of 12 fourth-grade students who were randomly assigned to 12 groups to experiment with three different methods of teaching arithmetic by four teachers. At a 5% significance level, can you reject the null hypothesis that the different levels of factor 1 have equal effects on the mean scores of the populations? Also, at a 5% significance level, can you reject the null hypothesis that the different levels of factor 2 have equal effects on the mean scores of the populations? Assume that all the assumptions required to apply the two-way ANOVA procedure hold true.

Solution To perform different tests for 12 populations associated with 12 combinations of teaching methods and teachers, we follow our standard procedure with five steps:

STEP 1 *State the two null and alternative hypotheses.*

Let us assume that μ_{ij} is the mean score of all students who will be taught using the teaching method i by teacher j, where i can take values 1, 2, and 3, and j can take values 1, 2, 3, and 4. Further assume that $\mu_{ij} = \mu + \tau_i + \beta_j$, where τ_1, τ_2, and τ_3 denote the effect of Methods I, II, and III, respectively, on the mean scores of all students, and β_1, β_2, β_3, and β_4 denote the effect of Teachers 1, 2, 3, and 4, respectively, on the mean scores of all students. μ denotes the baseline mean score of all students, that is, the hypothetical scores a student is expected to get if he or she is not taught by any teacher adopting any specific teaching method.

The null and alternative hypotheses are

$H_{\tau 0}: \tau_1 = \tau_2 = \tau_3$ (All three teaching methods have equal effects on the mean scores of students.)

$H_{\tau 1}$: Not all three teaching methods have equal effects on the mean scores of students.

And,

$H_{\beta 0}: \beta_1 = \beta_2 = \beta_3 = \beta_4$ (All four teachers have equal effects on the mean scores of students.)

$H_{\beta 1}$: Not all four teachers have equal effects on the mean scores of students.

STEP 2 *Select the distribution to use.*

Because we are comparing the means for 12 normally distributed populations associated with $3 \times 4 = 12$ cells and all the assumptions required to apply the ANOVA procedure are satisfied, we use the F distribution to make the two tests.

STEP 3 *Determine the rejection and the nonrejection regions.*

The significance level is .05. Because two-way ANOVA tests are right tailed, the area in the right tail of the F distribution curve, which is the rejection region, is .05. Next, we need to know the degrees of freedom for the numerator and the denominator. In our example, students were assigned to three different methods and four different teachers. As mentioned earlier, these methods are levels of the first factor, teaching method. Sometimes, instead of the term **levels**, the term **treatments** is used. The number of levels associated with the first factor is denoted by t. The number of levels associated with the second factor, teacher, is denoted by b. The total number of observations in all samples taken together is denoted by n. Note that associated with the two tests, there are two test statistics F_τ and F_β.

As mentioned earlier, for F_τ, the number of degrees of freedom for the numerator is equal to $t - 1$ and the number of degrees of freedom for the denominator is equal to $n - t - b + 1$. In our example, there are three methods of teaching, that is, $t = 3$ and 12 total observations (total number of students) in 12 cells. Thus,

$$\text{Degrees of freedom for the numerator} = t - 1 = 3 - 1 = 2$$
$$\text{Degrees of freedom for the denominator} = n - t - b + 1 = 12 - 3 - 4 + 1 = 6$$

From Table VII of Appendix B, we find the critical value of F for 2 df for the numerator, 6 df for the denominator, and .05 area in the right tail of the F distribution curve. This value of F is 5.14.

Thus, we will fail to reject $H_{\tau 0}$ if the calculated value of the test statistic F_τ is less than 5.14, and we will reject $H_{\tau 0}$ if this value of F_τ is 5.14 or larger.

Similarly, for F_β, the number of degrees of freedom for the numerator is equal to $b - 1$, and the number of degrees of freedom for the denominator is equal to $n - t - b + 1$. In our example, there are four teachers, hence $b = 4$. Thus,

Degrees of freedom for the numerator $= b - 1 = 4 - 1 = 3$
Degrees of freedom for the denominator $= n - t - b + 1 = 12 - 3 - 4 + 1 = 6$

From Table VII of Appendix B, we find the critical value of F for 3 *df* for the numerator, 6 *df* for the denominator, and .05 area in the right tail of the F distribution curve. This value of F is 4.76. Thus, we will fail to reject $H_{\beta 0}$ if the calculated value of the test statistic F_β is less than 4.76, and we will reject $H_{\beta 0}$ if this value of F_β is 4.76 or larger.

STEP 4 *Calculate the value of the test statistics.*

We computed the value of the test statistics F_τ and F_β for these data in Example 12.5. This gave value of $F_\tau = .0143$ and $F_\beta = 4.9465$.

STEP 5 *Make a decision.*

Because the value of the test statistic $F_\tau = .0143$ is less than the critical value of $F = 5.14$, it falls in the nonrejection region. Hence, we fail to reject the null hypothesis $H_{\tau 0}$ and conclude that the sample information is not strong enough to reject the null hypothesis $H_{\tau 0}$ that all three teaching methods have equal effects on the mean scores of all students. In other words, the three different methods of teaching arithmetic do not seem to affect the mean scores of all students. Instead of deciding based on critical value, one may like to use the *p*-value. Using technology, we find *p*-value = .9858, which is much greater than $\alpha = .05$. Hence, we fail to reject the null hypothesis.

The value of the other test statistic $F_\beta = 4.9465$ is larger than the corresponding critical value of $F = 4.76$. It falls in the rejection region. Hence, we reject the null hypothesis $H_{\beta 0}$ and conclude that the sample information is strong enough to reject the null hypothesis $H_{\beta 0}$ that all the four teachers have equal effects on the mean scores of all students. In other words, different teachers do seem to affect the mean scores of the students. If one likes to use the *p*-value for decision making using technology, one finds *p*-value = .0462. Here, *p*-value is less than $\alpha = .05$. Hence, we reject the null hypothesis.

In Example 12.5, for each of the 12 cells, the sample size is one and, hence, it is not possible to carry out the test for no interaction in Example 12.5. We have modified Example 12.5 to create a new Example 12.6 in which we added some more values, so that each of the 12 cells contains two values. As a result of that, now in Example 12.6 we are also able to test the null hypothesis $H_{\lambda 0}$ of no interaction apart from $H_{\tau 0}$ and $H_{\beta 0}$.

EXAMPLE 12.8 Four Teachers Use Three Methods to Teach Arithmetic: Each Cell Has Sample Size 2

Reconsider Example 12.6 about the scores of 24 fourth-grade students who were randomly assigned to 12 groups to experiment with three different methods of teaching arithmetic by four teachers. At a 5% significance level, can you reject the null hypothesis that the different levels of factor 1 have equal effects on the mean scores of the populations? Also, at a 5% significance level, can you reject the null hypothesis that the different levels of factor 2 have equal effects on the mean scores of the populations? As each cell contains two values, we can also test for no interaction. So, we further ask, at a 5% significance level, can you reject the null hypothesis that all the interaction effects are equal to zero? Assume that all the assumptions required to apply the two-way ANOVA procedure with interaction hold true.

Solution To perform different tests for 12 populations associated with 12 combinations of teaching methods and teachers, we follow our standard procedure with five steps:

STEP 1 *State the three null and alternative hypotheses.*

Let us assume that μ_{ij} is the mean score of all the students who will be taught using the teaching method i by the teacher j, where i can take values 1, 2, and 3, and j can take values 1, 2, 3, and 4. We assume a non-additive model $\mu_{ij} = \mu + \tau_i + \beta_j + \lambda_{ij}$ that connects μ_{ij} with τ_1, τ_2, and τ_3 that denote the effect of Methods I, II, and III, respectively, on the mean scores of all students, and $\beta_1, \beta_2, \beta_3$, and β_4 that denote the effect of Teachers 1, 2, 3, and 4, respectively, on the mean scores of all students. Here, the term λ_{ij} accommodates the non-additive component in the two-way ANOVA model.

The null and alternative hypotheses are

$H_{\tau 0}: \tau_1 = \tau_2 = \tau_3$ (All three teaching methods have equal effects on the mean scores of students.)

$H_{\tau 1}$: Not all three teaching methods have equal effects on the mean scores of students.

$H_{\beta 0}: \beta_1 = \beta_2 = \beta_3 = \beta_4$ (All four teachers have equal effects on the mean scores of students.)

$H_{\beta 1}$: Not all four teachers have equal effects on the mean scores of students and the ones associated with the interaction.

$H_{\lambda 0}: \lambda_{11} = \lambda_{12} = \ldots = \lambda_{34} = 0$ (All 12 interaction terms equal to zero.)

$H_{\lambda 1}$: At least one of the 12 interaction terms is non-zero.

STEP 2 *Select the distribution to use.*

Because we are comparing the means for 24 normally distributed populations associated with $3 \times 4 = 12$ cells and all the assumptions required to apply the two-way ANOVA procedure with interaction are satisfied, we use the *F* distribution to make the three tests.

STEP 3 *Determine the rejection and nonrejection regions.*

The significance level is .05. Because two-way ANOVA tests are right tailed, the area in the right tail of the *F* distribution curve, which is the rejection region, is .05. Next, we need to know the degrees of freedom for the numerator and the denominator. In our example, students were assigned to three different methods and four different teachers. As mentioned earlier, these methods are levels of the first factor, teaching method. The number of levels associated with the first factor is denoted by *t*. The number of levels associated with the second factor, teacher, is denoted by *b*. The total number of observations in all samples taken together is denoted by *n*. Note that associated with the three tests, now there are three test statistics F_τ, F_β, and F_λ.

As mentioned earlier, for F_τ, the number of degrees of freedom for the numerator is equal to $t - 1$ and the number of degrees of freedom for the denominator is equal to $n - tb$. Note that it is different than that of a two-way ANOVA procedure without interaction, which is equal to $n - t - b + 1$. In our example, there are three methods of teaching, hence, $t = 3$ and 24 total observations (total number of students) in 12 cells. Thus,

$$\text{Degrees of freedom for the numerator} = t - 1 = 3 - 1 = 2$$
$$\text{Degrees of freedom for the denominator} = n - tb = 24 - 3 \times 4 = 12$$

From Table VII of Appendix B, we find the critical value of *F* for 2 *df* for the numerator, 12 *df* for the denominator, and .05 area in the right tail of the *F* distribution curve. This value of *F* is 3.89. Thus, we will fail to reject $H_{\tau 0}$ if the calculated value of the test statistic F_τ is less than 3.89, and we will reject $H_{\tau 0}$ if this value of F_τ is 3.89 or larger.

Similarly, for F_β the number of degrees of freedom for the numerator is equal to $b - 1$, and the number of degrees of freedom for the denominator is equal to $n - tb$. In our example, there are four teachers, hence $b = 4$. Thus,

$$\text{Degrees of freedom for the numerator} = b - 1 = 4 - 1 = 3$$
$$\text{Degrees of freedom for the denominator} = n - tb = 24 - 3 \times 4 = 12$$

From Table VII of Appendix B, we find the critical value of *F* for 3 *df* for the numerator, 12 *df* for the denominator, and .05 area in the right tail of the *F* distribution curve. This value of *F* is 3.49. Thus, we will fail to reject $H_{\beta 0}$ if the calculated value of the test statistic F_β is less than 3.49, and we will reject $H_{\beta 0}$ if this value of F_β is 3.49 or larger.

For the hypothesis associated with the interaction, the test statistics is F_λ, the number of degrees of freedom for the numerator is equal to $(t - 1)(b - 1)$, and the number of degrees of freedom for the denominator is equal to $n - tb$. Thus,

$$\text{Degrees of freedom for the numerator} = (t - 1)(b - 1) = (3 - 1) \times (4 - 1) = 6$$
$$\text{Degrees of freedom for the denominator} = n - tb = 24 - 3 \times 4 = 12$$

From Table VII of Appendix B, we find the critical value of *F* for 6 *df* for the numerator, 12 *df* for the denominator, and .05 area in the right tail of the *F* distribution curve. This value of *F* is 3.00. Thus, we will fail to reject $H_{\lambda 0}$ if the calculated value of the test statistic F_λ is less than 3.00, and we will reject $H_{\lambda 0}$ if this value of F_λ is 3.00 or larger.

STEP 4 *Calculate the value of the test statistics.*

We computed the value of the test statistics F_τ, F_β, and F_λ for these data in Example 12.6. The obtained values are $F_\tau = 1.0843$, $F_\beta = 64.7767$, and $F_\lambda = 9.2030$.

STEP 5 *Make a decision.*

Because the value of the test statistic $F_\tau = 1.0843$ is less than the critical value of $F = 3.89$, it falls in the nonrejection region. Hence, we fail to reject the null hypothesis $H_{\tau 0}$, and conclude that the sample information is not strong enough to reject the null hypothesis $H_{\tau 0}$ that all three teaching methods have equal effects on the mean scores of all students. In other words, the three different methods of teaching arithmetic do not seem to affect the mean scores of the students. Instead of deciding based on the critical value one may like to use the p-value. Using technology, we find p-value = .3691, which is much greater than $\alpha = .05$. Hence, we fail to reject the null hypothesis. For the value of the other test statistic $F_\beta = 64.7767$ is much larger than the critical value of $F = 3.49$, it falls deep inside the rejection region. Hence, we strongly reject the null hypothesis $H_{\beta 0}$, and conclude that the sample information is strong enough to reject the null hypothesis $H_{\beta 0}$ that all four teachers have equal effects on the mean scores of all students. In other words, different teachers do seem to affect the mean scores of the students to a great extent. If one likes to use the p-value for decision making, using technology, one finds p-value = .0000, which is much less than $\alpha = .05$. Hence, we reject the null hypothesis strongly.

For the value of the test statistic F_λ associated with the interaction, $F_\lambda = 9.2030$ is larger than the critical value of $F = 3.00$, and it falls in the nonrejection region. Hence, we reject the null hypothesis $H_{\lambda 0}$, and conclude that the sample information is strong enough to reject the null hypothesis $H_{\lambda 0}$ that all 12 interaction effects on the mean scores of all students are zero. In other words, the ability to teach for different teachers does not seem to depend on the teaching methods themselves. If one likes to use the p-value for decision making, using technology, one finds p-value = .0007, which is less than $\alpha = .05$. Hence, we reject the null hypothesis of no interaction. Overall, the teaching methods do not seem to affect the mean scores of all students, whereas different teachers do. Moreover, the ability to teach for the teachers does not seem to change with adoption of different teaching methods.

Note: What If Each Cell Has More Than Two Observations But the Number of Observations Are Not Same for All the *tb* Cells?

In this chapter, we discussed a two-way ANOVA procedure, where each cell has an equal number of observations, that is, n_{ij} is the same for all i, j. Such a two-way ANOVA is called balanced. It is also possible to carry out the different tests of hypotheses in a two-way ANOVA, for which n_{ij} is not the same for all i, j. Such a two-way ANOVA is called unbalanced. The test for no interaction in an unbalanced two-way ANOVA can be carried out as long as $n_{ij} \gg 2$, for all i, j. However, unlike balanced two-way ANOVA, for an unbalanced two-way ANOVA, there is no unique way of calculating the different sum of squares. Among many different approaches, three approaches are popular among statisticians: the Type I, Type II, and Type III approaches. The Type I sum of squares is sequential in nature, that is, the findings would depend on the order of entering the two factors in the procedure. The Type II sum of squares assumes that there is no interaction between factor 1 and factor 2. The Type III sum of squares is attractive in the sense that it accommodates interaction and does not depend on the order, following which the two factors enter in the two-way ANOVA procedure. As expected, adoption of the different approaches sometimes can produce conflicting results.

EXERCISES

Concepts and Procedures

12.16 The following two-way ANOVA table is obtained while studying the impact of eating a high-protein breakfast versus low-protein breakfast on swimming performance of pupils belonging to three different age groups. The table contains a few missing values.

Source of Variation	Degrees of Freedom	Sum of Squares	Mean Square	Value of the Test Statistics
Breakfast type			20.28	$F_\tau =$ _____
Age group		90.00		$F_\beta =$ _____
Error	25			
Total		200.69		

Find the missing values and complete the two-way ANOVA table.

Application

For the following exercise, assume that all the assumptions required to apply the two-way ANOVA procedure hold true.

12.17 A study examining differences in life satisfaction between young adult, middle adult, and older adult men and women was conducted. Everyone who participated in the study completed a life satisfaction questionnaire. A high score on the test indicates a higher level of life satisfaction. Test scores are recorded below.

	Young Adult	Middle Adult	Older Adult
Male	4, 3, 2, 4, 3	7, 5, 7, 6, 7	10, 9, 8, 8,11
Female	6, 6, 7, 4, 5	8, 8, 10, 7, 9	12, 11, 12, 9, 10

a. Using $\alpha = .05$, check if life satisfaction differs among different age groups. What are the appropriate null and alternative hypotheses for this analysis?

b. Does gender has a role to play in life satisfaction?

c. Also, check for interaction effect between age group and gender using $\alpha = .05$.

Uses and Misuses of Statistics

Do Not Be Late

Imagine that working at your company requires that staff travel frequently. You want to determine if the on-time performance of any one airline is sufficiently different from that of the remaining airlines to warrant a preferred status with your company. The local airport website publishes the scheduled and actual departure and arrival times for the four airlines that service this airport. You decide to perform an ANOVA test on the mean delay times for all airline carriers at the airport. The null hypothesis here is that the mean delay times for Airlines A, B, C, and D are all the same. The results of the ANOVA test tell you not to reject the null hypothesis: All airline carriers seem to have the same mean departure and arrival delay times, so that adopting a preferred status based on the on-time performance is not warranted.

When your boss tells you to redo your analysis, you should not be surprised. The choice to study flights only at the local airport was a good one because your company should be concerned about the performance of an airline at the most convenient airport. A regional airport will have a much different on-time performance profile than a large hub airport. By mixing both arrival and departure data, however, you violated the assumption that the populations are approximately normally distributed. For arrival data, this assumption could be valid: The influence of high-altitude winds, local weather, and the fact that the arrival time is an estimate in the first place result in a distribution of arrival times around the predicted arrival times. However, departure delays are not approximately normally distributed. Because a flight does not leave before its departure time but can leave after, departure delays are skewed to the right. As the statistical methods become more sophisticated, so do the assumptions regarding the characteristics of the data. Careful attention to these assumptions is required.

Glossary

Analysis of variance (ANOVA) A statistical technique used to test whether the means of three or more populations are all equal against the alternative hypothesis that not all population means are equal.

Cell Different combinations of levels of first factors with levels of second factors.

***F* distribution** A continuous distribution that has two parameters: *df* for the numerator and *df* for the denominator.

Interaction The effect of one factor on the dependent variable is not same for different levels of the other factor.

Level The different values/categories that each of the variables/factors can take.

Mean square between samples or **MSB** A measure of the variation among the means of samples taken from different populations.

Mean square within samples or **MSW** A measure of the variation within the data of all samples taken from different populations.

One-way ANOVA The analysis of variance technique that analyzes one variable only.

SSB The sum of squares between samples. Also called the sum of squares of the factor or treatment.

SST The total sum of squares given by the sum of SSB and SSW.

SSW The sum of squares within samples. Also called the sum of squares of errors.

Two-way ANOVA The analysis of variance technique that analyzes two variables simultaneously.

Exercises

Supplementary Exercises

For the following exercises, assume that all the assumptions required to apply the one-way ANOVA procedure hold true.

12.18 The following table lists the number of violent crimes reported to police on randomly selected days for this year. The data are taken from three large cities of about the same size.

City A	City B	City C
4	3	10
7	4	14
16	2	9
5	12	6
10	9	8
6	8	16
11		

Using a 5% significance level, can you reject the null hypothesis that the mean number of violent crimes reported per day is the same for all three cities?

12.19 A farmer wants to test three brands of weight-gain diets for chickens to determine if there is a difference in the mean weight gain among these brands. He selected 15 chickens and randomly put each of them on one of these three brands of diet. The following table lists the weights (in pounds) gained by these chickens after a period of one month.

Brand A	Brand B	Brand C
1.8	.5	1.3
1.1	1.5	.9
1.6	.7	.6
.7	.8	1.8
.9	.5	.8

a. At a 1% significance level, can you reject the null hypothesis that the mean weight gain for all chickens is the same for each of these three diets?

b. If you did not reject the null hypothesis in part (a), explain the Type II error that you may have made in this case. Note that you cannot calculate the probability of committing a Type II error without additional information.

12.20 An ophthalmologist is interested in determining whether a golfer's type of vision (far-sightedness, near-sightedness, no prescription) impacts how well he or she can judge distance. Random samples of golfers from these three groups (far-sightedness, near-sightedness, no prescription) were selected, and these golfers were blindfolded and taken to the same location on a golf course. Then each of them was asked to estimate the distance from this location to the pin at the end of the hole. The data (in yards) given in the following table represent how far off the estimates (let us call these errors) of these golfers were from the actual distance. A negative value implies that the person underestimated the distance, and a positive value implies that a person overestimated the distance.

Far-sighted	−12	−8	−8	−11	−4	−12	−9	2	−5
Near-sighted	−3	−4	−8	−9	−6	−8	3	−12	−11
No prescription	−6	2	0	2	4	−3	1	−9	

At a 5% significance level, can you reject the null hypothesis that the average errors in predicting distance for all golfers of the three different vision types are the same?

12.21 A resort area has three seafood restaurants, which employ students during the summer season. The local chamber of commerce took a random sample of five servers from each restaurant and recorded the tips they received on a recent Friday night. The results (in dollars) of the survey are shown in the table. Assume that the Friday night for which the data were collected is typical of all Friday nights of the summer season.

Barzini's	Hwang's	Jack's
97	67	93
114	85	102
105	92	98
85	78	80
120	90	91

a. Would a student seeking a server's job at one of these three restaurants reject the null hypothesis that the mean tips on a Friday night are the same for all three restaurants? Use a 5% level of significance.

b. What will your decision be in part (a) if the probability of making a Type I error is zero? Explain.

12.22 A college student who has a 9 A.M. class on Monday, Wednesday, and Friday mornings wants to know if the mean time taken by students to find parking spaces just before 9 A.M. differs among these three days of the week. He randomly selects five weeks and records the time taken to find a parking space on Monday, Wednesday, and Friday of each of these five weeks. These times (in minutes) are given in the following table. Assume that this student is representative of all students who need to find a parking space just before 9 A.M. on these three days at this college.

Monday	Wednesday	Friday
6	9	3
12	12	2
15	5	10
14	14	7
10	13	5

At a 5% significance level, can you reject the null hypothesis that the mean times taken by students at this college to find parking spaces before 9 A.M. is the same for Monday, Wednesday, and Friday?

For the following exercise, assume that all the assumptions required to apply the two-way ANOVA procedure hold true.

12.23 In order to check the effect of two different fertilizers on the yield of three varieties of dragon fruits, the following table on fruit yield of 30 dragon fruit plants is considered.

	Variety 1	Variety 2	Variety 3
Fertilizer A	78, 79, 83, 77, 78	91, 94, 85, 88, 89	80, 86, 82, 79, 90
Fertilizer B	56, 61, 53, 66, 68	75, 78, 76, 69, 80	66, 62, 59, 71, 70

a. Using $\alpha = .05$, check if the two different fertilizers significantly differ in terms of fruit production.

b. Does the fruit yield vary across the three different varieties of dragon fruits?

Advanced Exercises

12.24 A billiards parlor in a small town is open just four days per week—Thursday through Sunday. Revenues vary considerably from day to day and week to week, so the owner is not sure whether some days of the week are more profitable than others. He takes random samples of five Thursdays, five Fridays, five Saturdays, and five Sundays from last year's records and lists the revenues for these 20 days. His bookkeeper finds the average revenue for each of the four samples, and then calculates Σx^2. The results are shown in the following table. The value of the Σx^2 came out to be 2,950,300.

Day	Mean Revenue ($)	Sample Size
Thursday	285	5
Friday	370	5
Saturday	412	5
Sunday	323	5

Assume that the revenues for each day of the week are approximately normally distributed and that the standard deviations are equal for all four populations. At a 2.5% level of significance, can you reject the null hypothesis that the mean revenue is the same for each of the four days of the week?

12.25 The editor of an automotive magazine has asked you to compare the mean gas mileages of city driving for three makes of compact cars. The editor has made available to you one car of each of the four makes, three drivers, and a budget sufficient to buy gas and pay the drivers for approximately 1200 miles of city driving for each car.

a. Explain how you would conduct an experiment and gather the data for a magazine article comparing the gas mileage.

b. Suppose that you wish to test whether the mean gas mileages of city driving differ among the populations of these four makes of compact cars. Outline the procedure for using your data to conduct this test. Assume that the assumptions for applying analysis of variance are satisfied.

12.26 Suppose you are performing a one-way ANOVA test with only the information given in the following table:

Source of Variation	Degrees of Freedom	Sum of Squares
Between	4	200
Within	45	3547

a. Suppose the sample sizes for all groups are equal. How many groups are there? What are the group sample sizes?

b. The *p*-value for the test of the equality of the means of all populations is calculated to be .6406. Suppose you plan to increase the sample sizes for all groups but keep them all equal. However, when you do this, the sum of squares within samples and the sum of squares between samples (magically) remain the same. What are the smallest sample sizes for groups that would make this result significant at a 5% significance level?

Self-Review Test

1. The *F* distribution is

a. continuous b. discrete c. neither

2. The *F* distribution is always

a. symmetric
b. skewed to the right
c. skewed to the left

3. The units of the *F* distribution, denoted by *F*, are always

a. nonpositive b. positive c. nonnegative

4. The one-way ANOVA test analyzes only one

a. variable b. population c. sample

5. The one-way ANOVA test is always

a. right-tailed b. left-tailed c. two-tailed

6. For a one-way ANOVA with *k* treatments and *n* observations in all samples taken together, the degrees of freedom for the numerator are

a. $k-1$ b. $n-k$ c. $n-1$

7. For a one-way ANOVA with *k* treatments and *n* observations in all samples taken together, the degrees of freedom for the denominator are

a. $k-1$ b. $n-k$ c. $n-1$

8. The ANOVA test can be applied to compare

a. two or more population means
b. more than four population means only
c. more than three population means only

9. To use an *F* distribution, the random variable must be

a. a discrete random variable
b. a continuous random variable
c. a qualitative random variable

10. The shape of the *F* distribution curve is

a. skewed to the left
b. skewed to the right
c. symmetric

11. Which of the following is not a characteristic of an *F* distribution?

a. The *F* distribution is continuous and skewed to the right.
b. The *F* distribution has two numbers of degrees of freedom: degrees of freedom for the numerator and for the denominator.
c. The *F* distribution has two parameters: the mean and the standard deviation.

12. The *F* value for 9 degrees of freedom for the numerator, 12 degrees of freedom for the denominator, and a .01 area in the right tail is

a. 5.11 b. 4.74 c. 4.39

13. We can use the analysis of variance procedure to test hypotheses about

a. the proportion of one population
b. two or more population proportions
c. two or more population means

14. Which of the following assumptions is not required to use ANOVA?

a. The populations from which the samples are drawn have the same variance.
b. All samples are of the same size.
c. The samples drawn from different populations are random and independent.

15. The marketing manager for a software company is trying to decide whether to market her company's newest product by mail, on television, or through the Internet. She selects 20 communities in which to test-market the product, advertising through one of the three media in each community. The following table shows the amounts of sales, in hundreds of dollars, for each community.

Mail	TV	Internet
7	9	16
7	12	11
10	8	12
7	14	8
9	4	9
5	7	14
4		10

Sales are normally distributed, and the standard deviations are equal for all three methods of advertising. You are to test the hypothesis that the means are the same for all three treatments using the ANOVA method. What is the variance within samples, rounded to four decimal places?

a. 12.0054 b. 8.1008 c. 9.2641

16. Your associate just took a lunch break and gave you the following incomplete ANOVA table she was trying to finish before noon. The null hypothesis for this test is

$$H_0: \mu_1 = \mu_2 = \mu_3 = \mu_4 = \mu_5$$

Source of Variation	Degrees of Freedom	Sum of Squares	Mean Square	F
Between		36.4		
Within		80.2		
Total	36			

What is the df associated with the between-group variation?

a. 5 b. 30 c. 32

17. Using ANOVA, you test the hypothesis that the means of three treatments are the same. You calculate the MSW as 16.6. The value of the test statistic F is 2.1. What is the between-sample sum of squares, rounded to two decimal places?

a. 69.72 b. 213.56 c. 67.87

18. The table below gives information on a variable for four samples selected from four normally distributed populations with equal variances.

Sample I	Sample II	Sample III	Sample IV
5	8	19	16
6	14	11	12
12	21	14	19
9	20	5	8
6	12	4	10
13			13

By using ANOVA, we wish to test the null hypothesis that the means of the four corresponding populations are equal. The significance level is 5%. The critical value of F is

a. 5.48 b. 3.16 c. 7.15

19. A researcher is investigating how gender (male or female) and parental status (have children or do not have children) influence how much money participants donate to a charity that supports children living in poverty. What type of analysis could be used to answer the question?

a. One-way ANOVA

b. Two-way ANOVA

c. One-sample t-test

20. For three races, two genders, and five in each treatment group for two-way ANOVA, the degrees of freedom for source of variation due to interaction, error, and total, respectively, are

a. (6, 25, 25)

b. (6, 20, 30)

c. (2, 24, 29)

21. In a two-way ANOVA procedure, the first factor has four levels, the second factor has three levels, and each of the 12 cells has at least one sample. If the total sample size is 14, which one of the following is true?

a. One can test the effect of factor 2, but not the effect of factor 1.

b. One can test the effect of factor 1, the effect of factor 2, and the interaction effect.

c. One can test the effect of factor 1, the effect of factor 2, but not the interaction effect.

22. In an incomplete two-way ANOVA table, the values of x, F_τ, F_τ, and F_β are

Source of Variation	Degrees of Freedom	Sum of Squares	Mean Square	F
First factor	2	290		F_τ
Second Factor	3	50		F_τ, F_β
Error	12	60		
Total	x			

a. (6, 47, 4)

b. (17, 39, 11)

c. (17, 58, 5)

23. In a two-way ANOVA table, the total sample size = 40, df for second factor = 5, df for first factor = 4, and interaction df = 20. Then, the error df is

a. 10 b. 11 c. 8

Practice Exercises

1. Briefly describe the assumptions that must hold true to apply the one-way ANOVA procedure as mentioned in this chapter.

2. A small college town has four pizza parlors that make deliveries. A student doing a research paper for her business management class decides to compare how promptly the four parlors deliver. On six randomly chosen nights, she orders a large pepperoni pizza from each establishment, then records the elapsed time until the pizza is delivered to her apartment. Assume that her apartment is approximately the same distance from the four pizza parlors. The following table shows the times (in minutes) for these deliveries. Assume that all the assumptions required to apply the one-way ANOVA procedure hold true.

Tony's	Luigi's	Angelo's	Kowalski's
20.0	22.1	22.3	23.9
24.0	27.0	26.0	24.1
18.3	20.2	24.0	25.8
22.0	32.0	30.1	29.0
20.8	26.0	28.0	25.0
19.0	24.8	25.8	24.2

a. Using a 5% significance level, test whether the mean delivery times differ among the four pizza parlors.

b. Is it a Type I error or a Type II error that may have been committed in part (a)? Explain.

Mini-Projects

Mini-Project 12.1

Are some days of the week busier than others on the New York Stock Exchange (NYSE)? Record the number of shares traded on the NYSE each day for a period of six weeks (round the number of shares to the nearest million). You will have five samples—first for shares traded on six Mondays, second for shares traded on six Tuesdays, and so forth. Assume that these days make up random samples for the respective populations. Further assume that each of the five populations from which these five samples are taken follows a normal distribution with the same variance. Test if the mean number of shares traded is the same for each of the five populations. Use a 1% significance level.

Mini-Project 12.2

Pick at least 30 students at random and divide them randomly into three groups (A, B, and C) of approximately equal size. Take the students one by one, ring a bell, and 17 seconds later ring another bell. Then ask the students to estimate the elapsed time between the first and second rings. For group A, tell each student before the experiment starts that people tend to underestimate the elapsed time. Tell each student in group B that people tend to overestimate the time. Do not make any such statement to the students in group C. Record the estimates for all students, and then conduct an appropriate hypothesis test to see if the mean estimates of elapsed time are all equal for the populations represented by these groups. Use a 5% level of significance, and assume that the three populations of elapsed time are normally distributed with equal standard deviations.

Decide for Yourself

Note: The Decide for Yourself feature is located in the book's product page on *www.wiley.com*.

Chapter 12 Technology Instructions

Note: Complete TI-84, Minitab, and Excel manuals are available in the book's product page on *www.wiley.com*.

TI-84 Color/TI-84

The TI-84 Color Technology Instructions feature of this text is written for the TI-84 Plus C color graphing calculator running the 4.0 operating system. Some screens, menus, and functions will be slightly different in older operating systems. The TI-84 and TI-84 Plus can perform all of the same functions but will not have the "Color" option referenced in some of the menus.

Performing a One-Way ANOVA Test for Example 12.3 of the Text

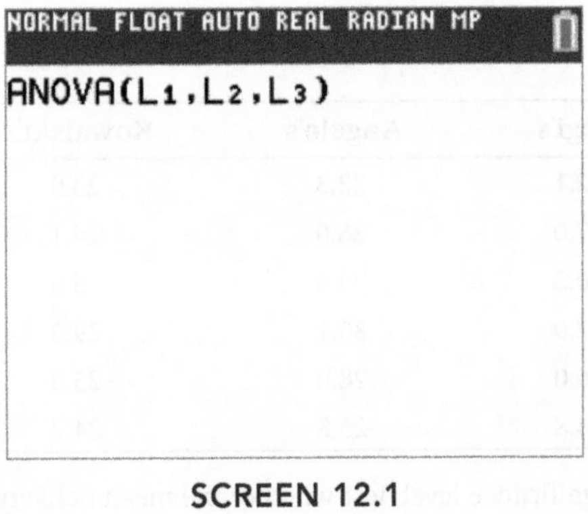

SCREEN 12.1

1. Enter the data from Example 12.3 into list1 (Method I), list2 (Method II), and list3 (Method III).
2. Select **STAT > TESTS > ANOVA**. This will paste the ANOVA function to the home screen.
3. Complete the command ANOVA(L1,L2,L3) by following these steps (see **Screen 12.1**):
   - Select **2nd > STAT > NAMES > L1.**
   - Press the comma key.
   - Select **2nd > STAT > NAMES > L2.**
   - Press the comma key.
   - Select **2nd > STAT > NAMES > L3.**
   - Press the right parenthesis key.

4. Press **ENTER.**
5. The output includes the test statistic and the *p*-value. Scroll down to see the rest of the ANOVA table. (See **Screen 12.2.**)

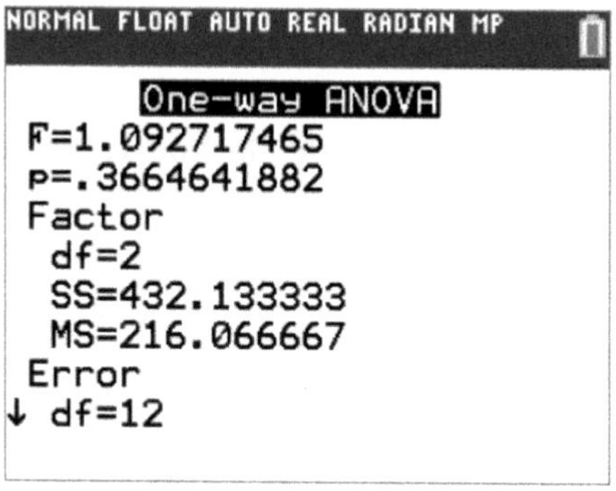

SCREEN 12.2

Now compare the *F*-value to the critical value of *F* or the *p*-value from Screen 12.2 with α and make a decision.

Minitab

The Minitab Technology Instructions in this text are written for **Minitab Professional Version 19.** Some screens, menus, and functions will be slightly different in older (or in newer) versions of Minitab. If you are using **Minitab Express**, you can download the Minitab Express Manual from the book's product page on *www.wiley.com.*

Performing a One-Way ANOVA Test for Example 12.3 of the Text

1. Enter the data from Example 12.3 into columns C1 (Method I), C2 (Method II), and C3 (Method III).
2. Select **Stat > ANOVA > One-Way.**
3. Use the following settings in the dialog box that appears on screen (see **Screen 12.3**):
   - Select **Response data are in a separate column for each factor level** from the drop-down menu.
   - Type C1-C3 in the **Responses** box.

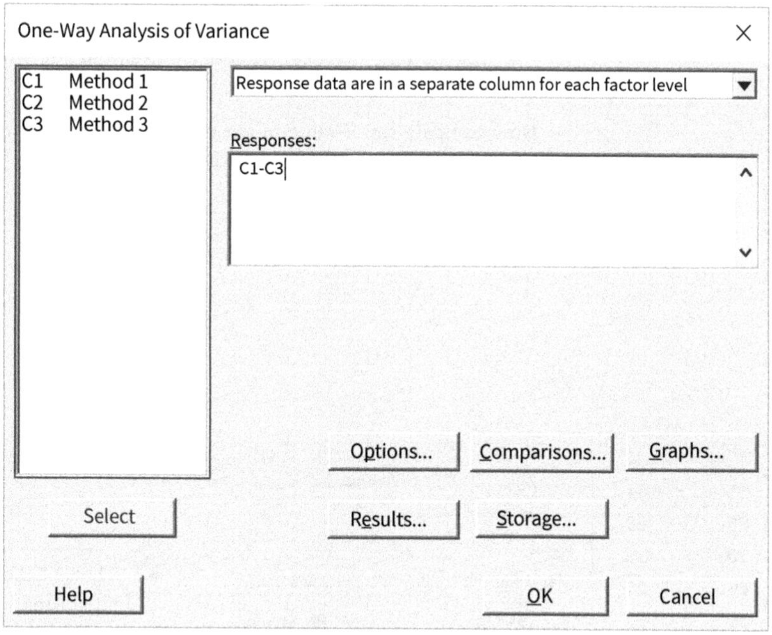

SCREEN 12.3

4. Select **Options** and use the following settings in the dialog box that appears on screen:
   - Check the **Assume equal variances** checkbox.
   - Type 99 in the **Confidence level** box.
   - Select Two-sided from the **Type of confidence interval** dropdown menu.
5. Click **OK** in both dialog boxes.
6. The output, including the test statistic and *p*-value, will be displayed in the Session window. (See **Screen 12.4.**)

   Note: By default, Minitab will show a graph called an interval plot, which is not shown here.

Now compare the *F*-value to the critical value of *F* or the *p*-value from Screen 12.4 with α and make a decision.

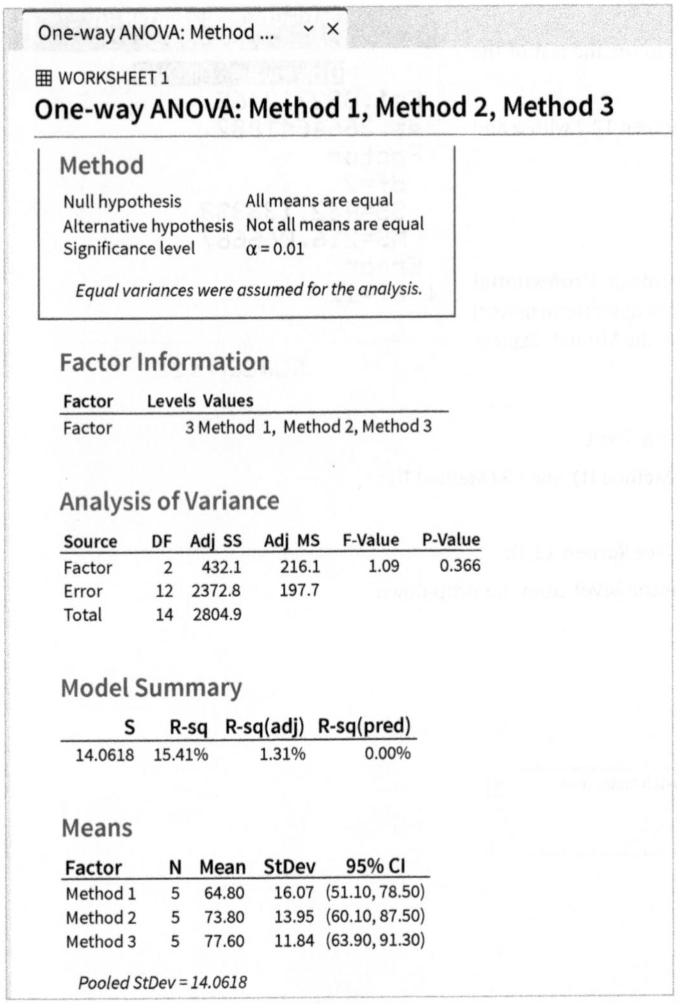

One-way ANOVA: Method ...

WORKSHEET 1

One-way ANOVA: Method 1, Method 2, Method 3

Method

Null hypothesis	All means are equal
Alternative hypothesis	Not all means are equal
Significance level	$\alpha = 0.01$

Equal variances were assumed for the analysis.

Factor Information

Factor	Levels	Values
Factor	3	Method 1, Method 2, Method 3

Analysis of Variance

Source	DF	Adj SS	Adj MS	F-Value	P-Value
Factor	2	432.1	216.1	1.09	0.366
Error	12	2372.8	197.7		
Total	14	2804.9			

Model Summary

S	R-sq	R-sq(adj)	R-sq(pred)
14.0618	15.41%	1.31%	0.00%

Means

Factor	N	Mean	StDev	95% CI
Method 1	5	64.80	16.07	(51.10, 78.50)
Method 2	5	73.80	13.95	(60.10, 87.50)
Method 3	5	77.60	11.84	(63.90, 91.30)

Pooled StDev = 14.0618

SCREEN 12.4

Excel

The Excel Technology Instructions feature of this text is written for Excel 2019. Some screens, menus, and functions will be slightly different in older versions of Excel. To enable some of the advanced Excel functions, you must enable the Data Analysis Add-In for Excel.

Performing a One-Way ANOVA Test for Example 12.3 of the Text

1. In a new worksheet, enter the data from Method I in cells A1-A5, Method II in cells B1-B5, and the data from Method III in cells C1-C5. (See **Screen 12.5**.)
2. Click **DATA** and then click **Data Analysis Tools** from the **Analysis** group.
3. Select **Anova: Single Factor** from the dialog box that appears on screen and then click OK.
4. Use the following settings in the dialog box that appears on screen (see **Screen 12.5**):
   - Type A1:C5 in the **Input Range** box.
   - Select Columns in the **Grouped by** field.
   - Type .01 in the **Alpha** box.
   - Select New Worksheet Ply from the **Output options.**
5. Click **OK.**
6. The output includes the test statistic and the *p*-value. (See Screen 12.6.)

Now compare the *F*-value to the critical value of *F* or the *p*-value from **Screen 12.6** with α and make a decision.

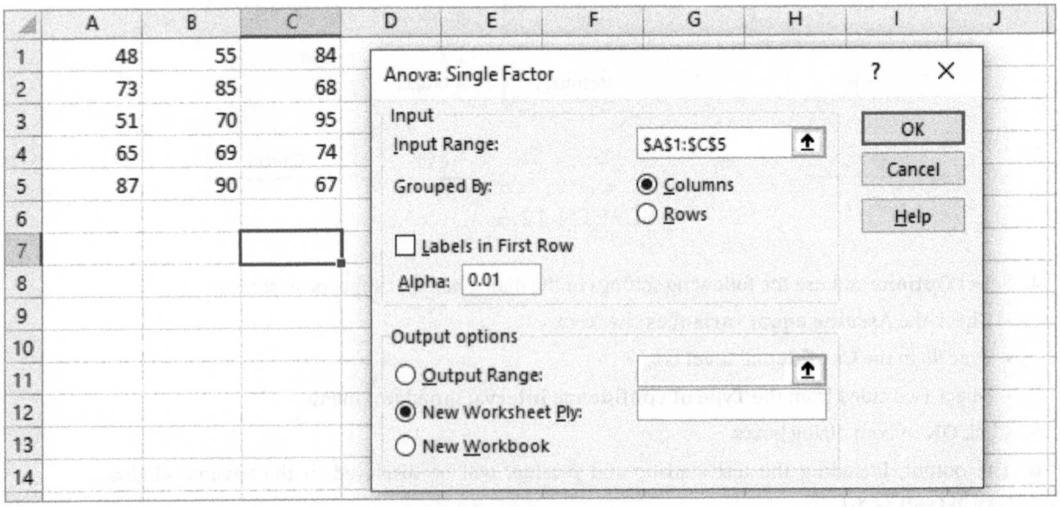

SCREEN 12.5

	A	B	C	D	E	F	G
1	Anova: Single Factor						
2							
3	SUMMARY						
4	*Groups*	*Count*	*Sum*	*Average*	*Variance*		
5	Column 1	5	324	64.8	258.2		
6	Column 2	5	369	73.8	194.7		
7	Column 3	5	388	77.6	140.3		
8							
9							
10	ANOVA						
11	*Source of Variation*	*SS*	*df*	*MS*	*F*	*P-value*	*F crit*
12	Between Groups	432.1333	2	216.0667	1.092717	0.366464	6.926608
13	Within Groups	2372.8	12	197.7333			
14							
15	Total	2804.933	14				

SCREEN 12.6

Technology Assignments

TA12.1 A consumer agency wanted to determine if the mean times taken by three brands of medicines to provide relief from a headache are different for one or more brands. The first drug was administered to six randomly selected patients, the second to four randomly selected patients, and the third to five randomly selected patients. The following table gives the time (in minutes) taken by each patient to get relief from a headache after taking the medicine.

Drug I	Drug II	Drug III
25	15	44
38	21	39
42	19	54
65	25	58
47		73
52		

At a 2.5% significance level, will you conclude that the mean time taken to provide relief from a headache is different for at least one of the three drugs?

TA12.2 A large company buys thousands of lightbulbs every year. The company is currently considering four brands of lightbulbs to choose from. Before the company decides which lightbulbs to buy, it wants to investigate whether the mean lifetimes differ among the four brands of lightbulbs. The company's research department randomly selected a few bulbs of each type and tested them. The following table lists the number of hours (in thousands) that each of the bulbs in each brand lasted before being burned out.

Brand I	Brand II	Brand III	Brand IV
23	19	23	26
24	23	27	24
19	18	25	21
26	24	26	29
22	20	23	28
23	22	21	24
25	19	27	28

At a 2.5% significance level, will you reject the null hypothesis that the mean lifetimes of bulbs for all these four brands are equal?

TA12.3 A local car dealership is interested in determining how successful their salespeople are in turning a profit when selling a car. Specifically, they are interested in the average percentage of price markups earned on various car sales. The following table lists the percentages of price markups for a random sample of car sales by three salespeople at this dealership. Note that here the markups are calculated as follows. Suppose an auto dealer pays $19,000 for a car and lists the sale price as $25,000, which gives a markup of $6000. If the car is sold for $22,000, the markup percentage earned on this sale is 50% ($3000 is half of $6000).

Ira	23.2	26.9	27.3	34.1	30.7	31.6	43.8	
Jim	19.6	41.2	60.3	34.3	52.0	23.3	39.1	44.2
Kelly	52.3	50.0	53.4	37.9	26.4	41.1	25.2	41.2

Test at a 5% significance level whether the average markup percentage earned on all car sales is different for at least one of the three salespersons.

CHAPTER 13

Simple Linear Regression

Michael Zagaris/Getty Images Sport/Getty Images

Are the heights and weights of persons related? Does a person's weight depend on his/her height? If yes, what is the change in the weight of a person, on average, for every 1 inch increase in height? What is this rate of change for National Football League players? (See Case Study 13.1.)

This chapter considers the relationship between two quantitative variables in two ways: (1) by using regression analysis and (2) by computing the correlation coefficient. By using a regression model, we can evaluate the magnitude of change in one variable due to a certain change in another variable. For example, an economist can estimate the amount of change in food expenditure due to a certain change in the income of a household by using a regression model. A sociologist may want to estimate the increase in the crime rate due to a particular increase in the unemployment rate. Besides answering these questions, a regression model also helps to predict the value of one variable for a given value of another variable. For example, by using the regression line, we can predict the (approximate) food expenditure of a household with a given income.

The correlation coefficient, on the other hand, simply tells us how strongly two variables are related. It does not provide any information about the size of the change in one variable as a result of a certain change in the other variable. For example, the correlation coefficient tells us how strongly income and food expenditure or crime rate and unemployment rate are related. In this chapter, we will discuss the linear correlation between two quantitative variables.

13.1 Simple Linear Regression

LEARNING OBJECTIVES

After completing this section, you should be able to:

- Explain the meaning of a simple linear regression.
- Explain the meaning of independent and dependent variables in a regression model.
- Explain the difference between an exact and nonexact relationship between two variables.
- Explain the difference between linear and nonlinear relationships between two variables.

- Explain the difference between a simple and a multiple regression model.
- Explain the difference between a deterministic model and a probabilistic regression model.
- Explain the reason for inclusion of a random error term in a regression model.
- Explain the term *least squares.*
- Explain the meaning of the error sum of squares in a linear regression model.
- Explain the difference between a population regression line and a sample regression line.
- Explain the difference between the true values of the slope and *y*-intercept and the estimated values of the slope and *y*-intercept.
- Construct a scatter diagram from sample data.
- Interpret the slope and *y*-intercept of a linear regression line.
- Use a linear regression model to predict a particular value of *y* for a given value of *x*.
- Use a linear regression model to estimate the mean value of *y* for a given value of *x*.

Only simple linear regression will be discussed in this chapter.[1] In the next two subsections the meaning of the words *simple* and *linear* as used in *simple linear regression* is explained.

13.1.1 Simple Regression

Let us return to the example of an economist investigating the relationship between food expenditure and income. What factors or variables does a household consider when deciding how much money it should spend on food every week or every month? Certainly, income of the household is one factor. However, many other variables also affect food expenditure. For instance, the assets owned by the household, the size of the household, the preferences and tastes of household members, and any special dietary needs of household members are some of the variables that influence a household's decision about food expenditure. These variables are called **independent** or **explanatory variables** because they all vary independently, and they explain the variation in food expenditures among different households. In other words, these variables explain why different households spend different amounts of money on food. Food expenditure is called the **dependent variable** because it depends on the independent variables. Studying the effect of two or more independent variables on a dependent variable using regression analysis is called **multiple regression.** However, if we choose only one (usually the most important) independent variable and study the effect of that single variable on a dependent variable, it is called a **simple regression.** Thus, a simple regression includes only two variables—one independent and one dependent. Note that whether we use a simple or a multiple regression analysis, there is always one and only one dependent variable. Only the number of independent variables changes in simple and multiple regressions.

Simple Regression

A regression model is a mathematical equation that describes the relationship between two or more variables. A *simple regression* model includes only two variables—one independent and one dependent. The dependent variable is the one being explained, and the independent variable is the one that explains the variation in the dependent variable.

13.1.2 Linear Regression

The relationship between two variables in a regression analysis is expressed by a mathematical equation called a **regression equation** or **model**. A regression equation, when plotted, may assume one of many possible shapes, including a straight line. A regression equation that gives a straight-line relationship between two variables is called a **linear regression model**; otherwise, the model is called a **nonlinear regression model**. In this chapter, only linear regression models are studied.

---

[1]The term *regression* was first used by Sir Francis Galton (1822–1911), who studied the relationship between the heights of children and the heights of their parents.

Linear Regression

A (simple) regression model that gives a straight-line relationship between two variables is called a *linear regression* model.

The two diagrams in **Figure 13.1** show a linear and a nonlinear relationship between the dependent variable food expenditure and the independent variable income. A linear relationship between income and food expenditure, shown in Figure 13.1*a*, indicates that as income increases, the food expenditure always increases at a constant rate. A nonlinear relationship between income and food expenditure, as depicted in Figure 13.1*b*, shows that as income increases, the food expenditure increases, although, after a point, the rate of increase in food expenditure is lower for every subsequent increase in income.

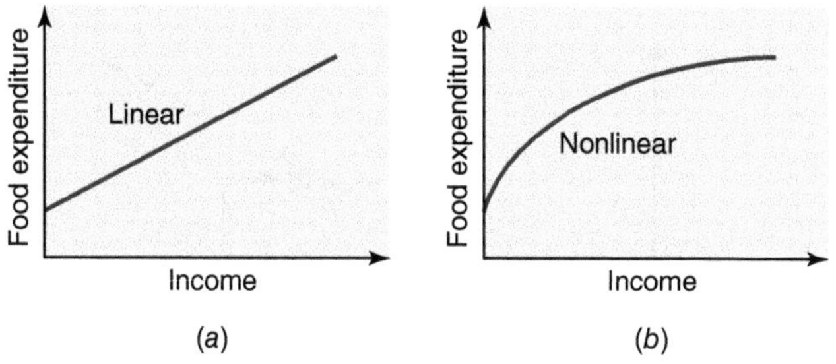

FIGURE 13.1 Relationship between food expenditure and income. (a) Linear relationship. (b) Nonlinear relationship.

The **equation of a linear relationship** between two variables x and y is written as

$$y = a + bx$$

Each set of values of a and b gives a different straight line. For instance, when $a = 50$ and $b = 5$, this equation becomes

$$y = 50 + 5x$$

To plot a straight line, we need to know two points that lie on that line. We can find two points on a line by assigning any two values to x and then calculating the corresponding values of y. For the equation $y = 50 + 5x$:

1. When $x = 0$, then $y = 50 + 5(0) = 50$.
2. When $x = 10$, then $y = 50 + 5(10) = 100$.

These two points are plotted in **Figure 13.2**. By joining these two points, we obtain the line representing the equation $y = 50 + 5x$.

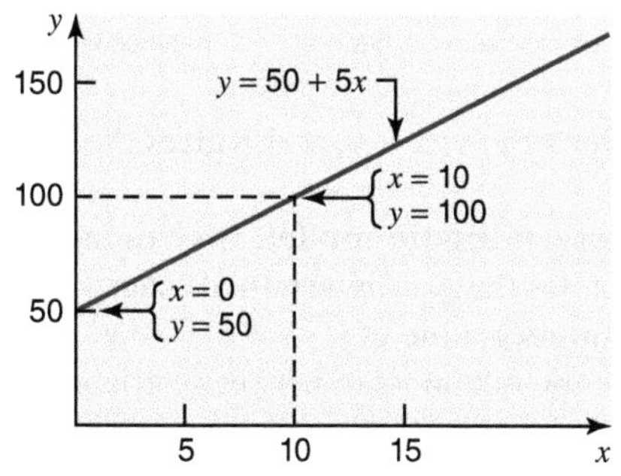

FIGURE 13.2 Plotting a linear equation.

Note that in Figure 13.2 the line intersects the y (vertical) axis at 50. Consequently, 50 is called the ***y*-intercept**. The y-intercept is given by the constant term in the equation. It is the value of y when x is zero.

In the equation $y = 50 + 5x$, 5 is called the **coefficient of** $\boldsymbol{x}$ or the **slope** of the line. It gives the amount of change in y due to a change of one unit in x. For example:

$$\text{If } x = 10, \quad \text{then} \quad y = 50 + 5(10) = 100.$$
$$\text{If } x = 11, \quad \text{then} \quad y = 50 + 5(11) = 105.$$

Hence, as x increases by 1 unit (from 10 to 11), y increases by 5 units (from 100 to 105). This is true for any value of x. Such changes in x and y are shown in **Figure 13.3**.

In general, when an equation is written in the form

$$y = a + bx$$

a gives the y-intercept and b represents the slope of the line. In other words, a represents the point where the line intersects the y-axis, and b gives the amount of change in y due to a change of one unit in x. Note that b is also called the coefficient of x.

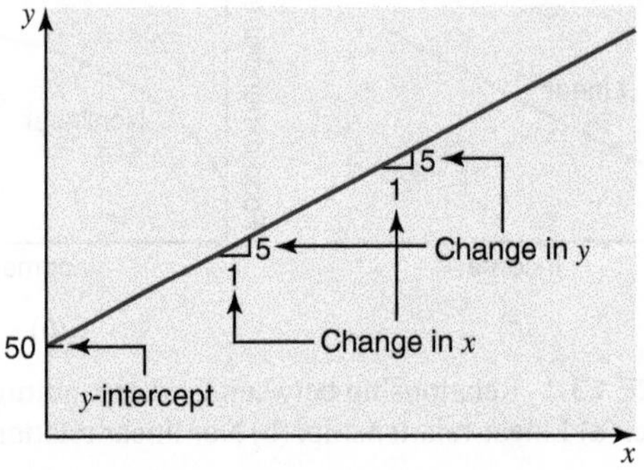

FIGURE 13.3 y-intercept and slope of a line.

13.1.3 Simple Linear Regression Model

In a regression model, the independent variable is usually denoted by x, and the dependent variable is usually denoted by y. The x variable, with its coefficient, is written on the right side of the '=' sign, whereas the y variable is written on the left side of the '=' sign. The y-intercept and the slope, which we earlier denoted by a and b, respectively, can be represented by any of the many commonly used symbols. Let us denote the y-intercept (which is also called the *constant term*) by A, and the slope (or the coefficient of the x variable) by $\boldsymbol{B}$. Then, our simple linear regression model is written as

Constant term or y-intercept → A; Slope → B

$$y = A + Bx \qquad \textbf{(1)}$$

Dependent variable → y; Independent variable → x

In model (1), A gives the value of y for $x = 0$, and B gives the change in y due to a change of one unit in x.

Model (1) is called a **deterministic model**. It gives an **exact relationship** between x and y. This model simply states that y is determined exactly by x, and for a given value of x there is one and only one (unique) value of y.

However, in many cases the relationship between variables is not exact. For instance, if y is food expenditure and x is income, then model (1) would state that food expenditure is determined by income only and that all households with the same income spend the same amount on food. As mentioned earlier, however, food expenditure is determined by many variables, only one of which is included in model (1). In reality, different households with the same income spend different amounts of money on food because of the differences in the sizes of the household, the assets they own, and their preferences and tastes. Hence, to take these variables into consideration and to make our model complete, we add another term to the right side of model (1). This term is called the **random error term**. It is denoted by ε (Greek letter *epsilon*). The complete regression model is written as

$$y = A + Bx + \varepsilon \qquad (2)$$

↑ Random error term

The regression model (2) is called a **probabilistic model** or a **statistical relationship**.

Equation of a Regression Model

In the *regression model* $y = A + Bx + \varepsilon$, A is called the y-intercept or constant term, B is the slope, and ε is the random error term. The dependent and independent variables are y and x, respectively.

The random error term ε is included in the model to represent the following two phenomena:

1. *Missing or omitted variables*. As mentioned earlier, food expenditure is affected by many variables other than income. The random error term ε is included to capture the effect of all those missing or omitted variables that have not been included in the model.
2. *Random variation*. Human behavior is unpredictable. For example, a household may have many parties during one month and spend more than usual on food during that month. The same household may spend less than usual during another month because it spent quite a bit of money to buy furniture. The variation in food expenditure for such reasons may be called random variation.

In model (2), A and B are the **population parameters**. The regression line obtained for model (2) by using the population data is called the **population regression line**. The values of A and B in the population regression line are called the **true values of the y-intercept and slope**, respectively.

However, population data are difficult to obtain. As a result, we almost always use sample data to estimate model (2). The values of the y-intercept and slope calculated from sample data on x and y are called the **estimated values of A and B** and are **denoted by a and b**, respectively. Using a and b, we write the estimated regression model as

$$\hat{y} = a + bx \qquad (3)$$

where $\hat{y}$ (read as *y hat*) is the **estimated** or **predicted value of y** for a given value of x. Equation (3) is called the **estimated regression model**; it gives the **regression of y on x**.

Estimates of A and B

In the model $\hat{y} = a + bx$, a and b, which are calculated using sample data, are called the *estimates of A and B*, respectively.

13.1.4 Scatter Diagram

Suppose we take a sample of seven households from a small city and collect information on their monthly incomes and food expenditures for the last month. The information obtained (in hundreds of dollars) is given in **Table 13.1**.

In Table 13.1, we have a pair of observations for each of the seven households. Each pair consists of one observation on income and a second on food expenditure. For example, the first household's income for the last month was \$5500 and its food expenditure was \$1400. By plotting all seven pairs of values, we obtain a **scatter diagram** or **scatterplot**. **Figure 13.4** gives the scatter diagram for the data of Table 13.1. Each dot in this diagram represents one household. A scatter diagram is helpful in detecting a relationship between two variables. For example, by looking at the scatter diagram of Figure 13.4, we can observe that there exists a strong linear relationship between food expenditure and income. If a straight line is drawn through the points, the points will be scattered closely around the line.

TABLE 13.1 Monthly Incomes and Food Expenditures of Seven Households ($00)

Income	Food Expenditure
55	14
83	24
38	13
61	16
33	9
49	15
67	17

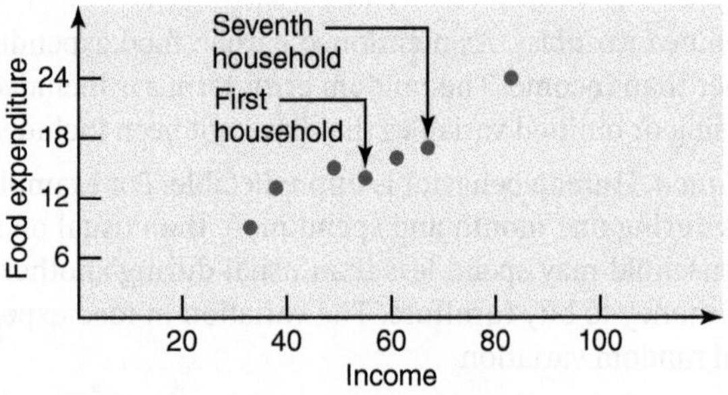

FIGURE 13.4 Scatter diagram.

Scatter Diagram

A plot of paired observations is called a *scatter diagram.*

As shown in **Figure 13.5**, a large number of straight lines can be drawn through the scatter diagram of Figure 13.4. Each of these lines will give different values for a and b of model (3).

In regression analysis, we try to find a line that best fits the points in the scatter diagram. Such a line provides a best possible description of the relationship between the dependent and independent variables. The **least squares method**, discussed in the next section, gives such a line. The line obtained by using the least squares method is called the **least squares regression line**.

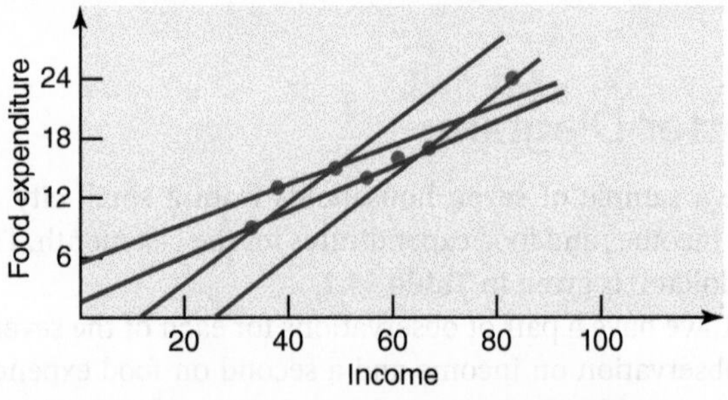

FIGURE 13.5 Scatter diagram and straight lines.

13.1.5 Least Squares Regression Line

The value of y obtained for a member from the survey is called the **observed or actual value of y.** As mentioned earlier in this section, the value of y, denoted by $\hat{y}$, obtained for a given x by using the regression line is called the **predicted value of y**. The random error ε denotes the difference between the actual value of y and the predicted value of y for population data. For example, for a given household, ε is the difference between what this household actually spent on food during the last month and what is predicted using the population regression line. The ε is also called the **residual** because it measures the surplus (positive or negative) of actual food expenditure over what is predicted by using the regression model. If we estimate model (2) by using sample data, the difference between the actual y and the predicted y based on this estimation cannot be denoted by ε. **The random error for the sample regression model is denoted by e.** Thus, e is an estimator of ε. If we estimate model (2) using sample data, then the value of e is given by

$$e = \text{Actual food expenditure} - \text{Predicted food expenditure} = y - \hat{y}$$

In **Figure 13.6**, e is the vertical distance between the actual position of a household and the point on the regression line. Note that in such a diagram, we always measure the dependent variable on the vertical axis and the independent variable on the horizontal axis.

The value of an error is positive if the point that gives the actual food expenditure is above the regression line and negative if it is below the regression line. *The sum of these errors is always zero.* In other words, the sum of the actual food expenditures for seven households included in the sample will be the same as the sum of the food expenditures predicted by the regression model. Thus,

$$\Sigma e = \Sigma(y - \hat{y}) = 0$$

Hence, to find the line that best fits the scatter of points, we cannot minimize the sum of errors. Instead, we minimize the **error sum of squares**, denoted by **SSE**, which is obtained by adding the squares of errors. Thus,

$$\text{SSE} = \Sigma e^2 = \Sigma(y - \hat{y})^2$$

The least squares method gives the values of a and b for model (3) such that the sum of squared errors (SSE) is minimum.

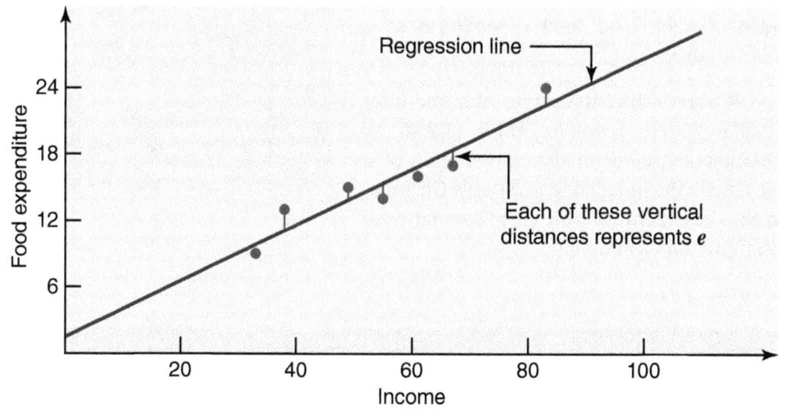

FIGURE 13.6 Regression line and random errors.

Error Sum of Squares (SSE)

The *error sum of squares*, denoted by SSE, is

$$\text{SSE} = \Sigma e^2 = \Sigma(y - \hat{y})^2$$

The values of a and b that give the minimum SSE are called the *least squares estimates* of A and B, and the regression line obtained with these estimates is called the *least squares regression line.*

The Least Squares Line

For the least squares regression line $\hat{y} = a + bx$,

$$b = \frac{SS_{xy}}{SS_{xx}} \quad \text{and} \quad a = \bar{y} - b\bar{x}$$

where

$$SS_{xy} = \Sigma xy - \frac{(\Sigma x)(\Sigma y)}{n} \quad \text{and} \quad SS_{xx} = \Sigma x^2 - \frac{(\Sigma x)^2}{n}$$

and SS stands for "sum of squares." The least squares regression line $\hat{y} = a + bx$ is also called the regression of y on x.

The least squares values of a and b are computed using the formulas given above.[2] These formulas are for estimating a sample regression line. Suppose we have access to a population data set. We can find the population regression line by using the same formulas with a little adaptation. If we have access to population data, we replace a by A, b by B, and n by N in these formulas, and use the values of Σx, Σy, Σxy, and Σx^2 calculated for population data to make the required computations. The **population regression line** is written as

$$\mu_{y|x} = A + Bx$$

where $\mu_{y|x}$ is read as "the mean value of y for a given x." When plotted on a graph, the points on this population regression line give the average values of y for the corresponding values of x. These average values of y are denoted by $\mu_{y|x}$.

Example 13.1 illustrates how to estimate a regression line for sample data.

EXAMPLE 13.1 Incomes and Food Expenditures of Households

Estimating the Least Squares Regression Line

Find the least squares regression line for the data on monthly incomes and food expenditures of the seven households given in Table 13.1. Use income as an independent variable and food expenditure as a dependent variable.

Troels Graugaard/iStockphoto

Solution We are to find the values of a and b for the regression model $\hat{y} = a + bx$. **Table 13.2** shows the calculations required for the computation of a and b. We denote the independent variable (income) by x and the dependent variable (food expenditure) by y, both in hundreds of dollars.

TABLE 13.2

Income x	Food Expenditure y	xy	x^2
55	14	770	3025
83	24	1992	6889
38	13	494	1444
61	16	976	3721
33	9	297	1089
49	15	735	2401
67	17	1139	4489
$\Sigma x = 386$	$\Sigma y = 108$	$\Sigma xy = 6403$	$\Sigma x^2 = 23{,}058$

[2]The values of SS_{xy} and SS_{xx} can also be obtained by using the following basic formulas:

$$SS_{xy} = \Sigma(x - \bar{x})(y - \bar{y}) \quad \text{and} \quad SS_{xx} = \Sigma(x - \bar{x})^2$$

However, these formulas take longer to make calculations.

The following steps are performed to compute a and b.

STEP 1 *Compute Σx, Σy, $\bar{x}$, and $\bar{y}$.*

$$\Sigma x = 386, \quad \Sigma y = 108$$
$$\bar{x} = \Sigma x/n = 386/7 = 55.1429$$
$$\bar{y} = \Sigma y/n = 108/7 = 15.4286$$

STEP 2 *Compute Σxy and Σx^2.*

To calculate Σxy, we multiply the corresponding values of x and y. Then, we sum all these products. The products of x and y are recorded in the third column of Table 13.2. To compute Σx^2, we square each of the x values and then add them. The squared values of x are listed in the fourth column of Table 13.2. From these calculations,

$$\Sigma xy = 6403 \quad \text{and} \quad \Sigma x^2 = 23{,}058$$

STEP 3 *Compute SS_{xy} and SS_{xx}.*

$$SS_{xy} = \Sigma xy - \frac{(\Sigma x)(\Sigma y)}{n} = 6403 - \frac{(386)(108)}{7} = 447.5714$$
$$SS_{xx} = \Sigma x^2 - \frac{(\Sigma x)^2}{n} = 23{,}058 - \frac{(386)^2}{7} = 1772.8571$$

STEP 4 *Compute a and b.*

$$b = \frac{SS_{xy}}{SS_{xx}} = \frac{447.5714}{1772.8571} = .2525$$
$$a = \bar{y} - b\bar{x} = 15.4286 - (.2525)(55.1429) = 1.5050$$

Thus, our estimated regression model $\hat{y} = a + bx$ is

$$\boldsymbol{\hat{y} = 1.5050 + .2525x}$$

This regression line is called the least squares regression line. It gives the *regression of food expenditure on income.*

Note that we have rounded all calculations to four decimal places. We can round the values of a and b in the regression equation to two decimal places, but we do not do this here because we will use this regression equation for prediction and estimation purposes later.

Using this estimated regression model, we can find the predicted value of y for any specific value of x. For instance, suppose we randomly select a household whose monthly income is \$6100, so that $x = 61$ (recall that x denotes income in hundreds of dollars). The predicted value of food expenditure for this household is

$$\hat{y} = 1.5050 + (.2525)(61) = \$16.9075 \text{ hundred} = \$1690.75$$

In other words, based on our regression line, we predict that a household with a monthly income of \$6100 is expected to spend \$1690.75 per month on food. This value of $\hat{y}$ can also be interpreted as a point estimator of the mean value of y for $x = 61$. Thus, we can state that, on average, all households with a monthly income of \$6100 spend about \$1690.75 per month on food.

In our data on seven households, there is one household whose income is \$6100. The actual food expenditure for that household is \$1600 (see Table 13.2). The difference between the actual and predicted values gives the error of prediction. Thus, the error of prediction for this household, which is shown in **Figure 13.7**, is

$$e = y - \hat{y} = 16 - 16.9075 = -\$.9075 \text{ hundred} = -\$90.75$$

Therefore, the error of prediction for this household is −\$90.75. The negative error indicates that the predicted value of y is greater than the actual value of y. Thus, if we use the regression model, this household's food expenditure is overestimated by \$90.75.

13.1.6 Interpretation of *a* and *b*

How do we interpret $a = 1.5050$ and $b = .2525$ obtained in Example 13.1 for the regression of food expenditure on income? A brief explanation of the y-intercept and the slope of a regression line was given in Section 13.1.2. Below we explain the meaning of a and b in more detail.

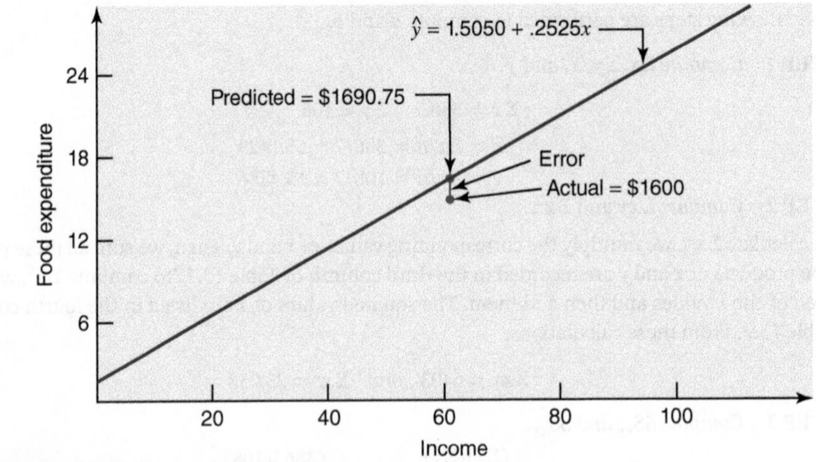

FIGURE 13.7 Error of prediction.

Interpretation of *a* Consider a household with zero income. Using the estimated regression line obtained in Example 13.1, we get the predicted value of y for $x = 0$ as

$$\hat{y} = 1.5050 + .2525(0) = \$1.5050 \text{ hundred} = \$150.50$$

Thus, we can state that a household with no income is expected to spend $150.50 per month on food. Alternatively, we can also state that the point estimate of the average monthly food expenditure for all households with zero income is $150.50. Note that here we have used $\hat{y}$ as a point estimate of $\mu_{y|x}$. Thus, $a = 150.50$ gives the predicted or mean value of y for $x = 0$ based on the regression model estimated for the sample data.

However, we should be very careful when making this interpretation of a. In our sample of seven households, the incomes vary from a minimum of $3300 to a maximum of $8300. (Note that in Table 13.1, the minimum value of x is 33 and the maximum value is 83.) Hence, our regression line is valid only for the values of x between 33 and 83. If we predict y for a value of x outside this range, the prediction usually will not hold true. Thus, since $x = 0$ is outside the range of household incomes that we have in the sample data, the prediction that a household with zero income spends $150.50 per month on food does not carry much credibility. The same is true if we try to predict y for an income greater than $8300, which is the maximum value of x in Table 13.1.

Interpretation of *b* The value of b in a regression model gives the change in the predicted value of y (dependent variable) due to a change of one unit in x (independent variable). For example, by using the regression equation obtained in Example 13.1, we see:

$$\text{When } x = 50,\ \hat{y} = 1.5050 + .2525(50) = 14.1300$$

$$\text{When } x = 51,\ \hat{y} = 1.5050 + .2525(51) = 14.3825$$

Hence, when x increased by one unit, from 50 to 51, $\hat{y}$ increased by $14.3825 - 14.1300 = .2525$, which is the value of b. Because our unit of measurement is hundreds of dollars, we can state that, on average, a $100 increase in income will result in a $25.25 increase in food expenditure. We can also state that, on average, a $1 increase in income of a household will increase the food expenditure by $.2525. Note the phrase "on average" in these statements. The regression line is seen as a measure of the mean value of y for a given value of x. If one household's income is increased by $100, that household's food expenditure may or may not increase by $25.25. However, if the incomes of all households are increased by $100 each, the average increase in their food expenditures is expected to be very close to $25.25.

Note that when b is positive, an increase in x will lead to an increase in y, and a decrease in x will lead to a decrease in y. In other words, when b is positive, the changes in x and y are in the same direction. Such a relationship between x and y is called a **positive linear relationship**. The regression line in this case slopes upward from left to right. On the other hand, if the value of b is negative, an increase in x will lead to a decrease in y, and a decrease in x will cause an increase in y. The changes in x and y in this case are in opposite directions. Such a relationship between x and y is called a **negative linear relationship**. The regression line in this case slopes downward from left to right. The two diagrams in **Figure 13.8** show these two cases.

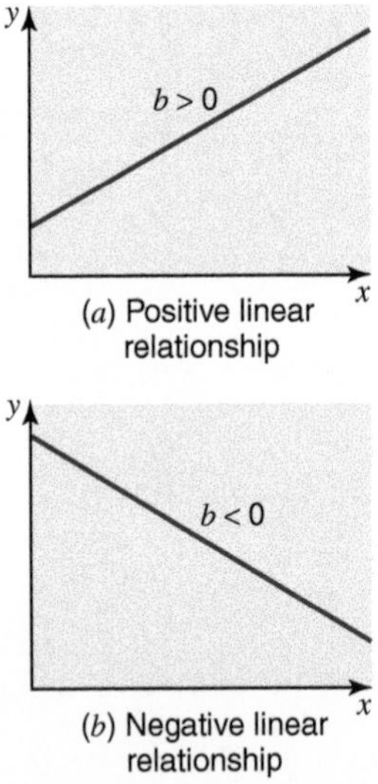

FIGURE 13.8 Positive and negative linear relationships between x and y.

For a regression model, b is computed as $b = SS_{xy}/SS_{xx}$. The value of SS_{xx} is always positive, and that of SS_{xy} can be positive or negative. Hence, the sign of b depends on the sign of SS_{xy}. If SS_{xy} is positive (as in our example on the incomes and food expenditures of seven households), then b will be positive, and if SS_{xy} is negative, then b will be negative.

Remember

Case Study 13.1 illustrates the difference between the population regression line and a sample regression line.

Case Study 13.1

Regression of Weights on Heights for NFL Players

Data Set III that accompanies this text lists many characteristics of National Football League (NFL) players who were on the rosters of all NFL teams during the 2019 season. These data comprise the population of NFL players for that season. We postulate the following simple linear regression model for these data:

$$y = A + Bx + \varepsilon$$

where y is the weight (in pounds) and x is the height (in inches) of an NFL player.

Using the population data that contain 1925 players, we obtain the following regression line:

$$\mu_{y|x} = -690.5 + 12.63x$$

This equation gives the population regression line because it is obtained by using the population data. Note that in the population regression line we write $\mu_{y|x}$ instead of $\hat{y}$. Thus, the true values of A and B are, respectively,

$$A = -690.5 \quad \text{and} \quad B = 12.63$$

The value of B indicates that for every 1-inch increase in the height of an NFL player, weight increases on average by 12.63 pounds. However, $A = -690.5$ does not make any sense. It states that the average weight of players with zero height is -690.5 pounds. (Recall from Section 13.1.6 that we must be very careful if and when we apply the regression equation to predict y for values of x outside the range of data used to find the regression line.) **Figure 13.9** gives the scatter diagram and the regression line for the heights and weights of all NFL players.

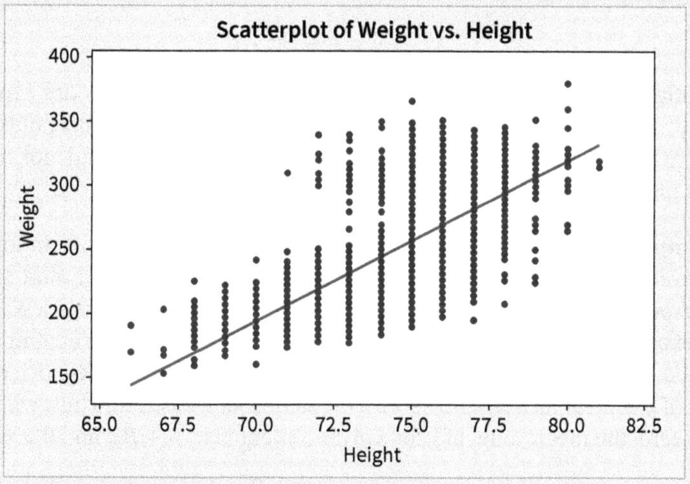

FIGURE 13.9 Scatter diagram and regression line for the data on heights and weights of all NFL players.

Next, we selected a random sample of 50 players and estimated the population regression model using these sample data. The estimated regression line from this sample is

$$\hat{y} = -703.6 + 12.78x$$

The values of a and b are

$$a = -703.6 \quad \text{and} \quad b = 12.78$$

These values of a and b give the estimates of A and B based on sample data. The scatter diagram and the regression line for the sample observations on heights and weights are given in **Figure 13.10**. Note that this figure does not show exactly 50 dots because some points/dots may be exactly the same or very close to each other.

As we can observe from Figures 13.9 and 13.10, the scatter diagrams for population and sample data both show a (positive) linear relationship between the heights and weights of NFL players, although not a very strong positive relationship.

Note that in both figures, the axes are truncated. They do not start at the zero point.

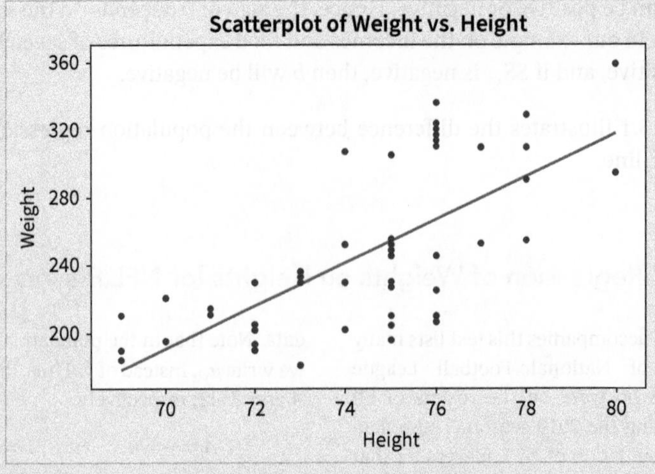

FIGURE 13.10 Scatter diagram and regression line for the data on heights and weights of 50 randomly selected NFL players.

Note that if we take another sample of 50 players and estimate the population regression line using these sample data, the new regression line will have different values of a and b, which are estimates of A and B.

13.1.7 Assumptions of the Regression Model

Like any other theory, the linear regression analysis is also based on certain assumptions. Consider the population regression model

$$y = A + Bx + \varepsilon \qquad (4)$$

Four assumptions are made about this model. These assumptions are explained next with reference to the example on incomes and food expenditures of households. Note that these assumptions are made about the population regression model and not about the sample regression model.

Assumption 1: The random error term ε has a mean equal to zero for each x. In other words, among all households with the same income, some spend more than the predicted food expenditure (and, hence, have positive errors) and others spend less than the predicted food expenditure (and, consequently, have negative errors). This assumption simply states that the sum of the positive errors is equal to the sum of the negative errors, so that the mean of errors for all households with the same income is zero. Thus, when the mean value of ε is zero, the mean value of y for a given x is equal to $A + Bx$, and it is written as

$$\mu_{y|x} = A + Bx$$

As mentioned earlier in this chapter, $\mu_{y|x}$ is read as "the mean value of y for a given value of x." When we find the values of A and B for model (4) using the population data, the points on the regression line give the average values of y, denoted by $\mu_{y|x}$, for the corresponding values of x.

Assumption 2: The errors associated with different observations are independent. According to this assumption, the errors for any two households in our example are independent. In other words, all households decide independently how much to spend on food.

Assumption 3: For any given x, the distribution of errors is approximately normal. The corollary of this assumption is that the food expenditures for all households with the same income are approximately normally distributed.

Assumption 4: The distribution of population errors for each x has the same (constant) standard deviation, which is denoted by σ_ε. This assumption indicates that the spread of points around the regression line is similar for all x values.

Figure 13.11 illustrates the meaning of the first, third, and fourth assumptions for households with incomes of \$4000 and \$7500 per month. The same assumptions hold true for any other income level. In the population of all households, there will be many households with a monthly income of \$4000. Using the population regression line, if we calculate the errors for all these households and prepare the distribution of these errors, it will look like the distribution given in Figure 13.11*a*. Its standard deviation will be σ_ε. Similarly, Figure 13.11*b* gives the distribution of errors for all those households in the population whose monthly income is \$7500. Its standard deviation is also σ_ε. Both of these distributions are identical. Note that the mean of both of these distributions is $E(\varepsilon) = 0$.

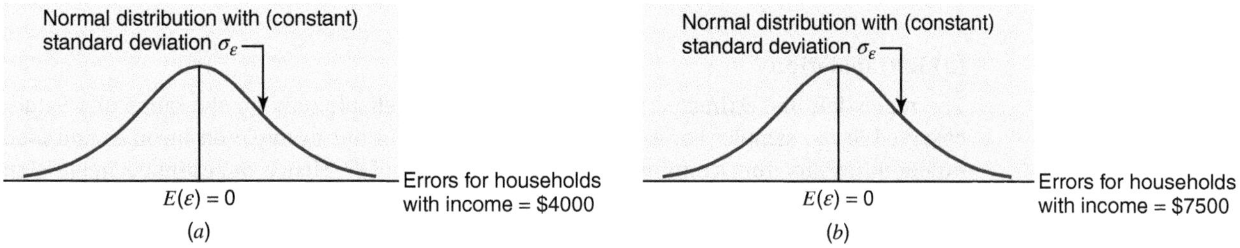

FIGURE 13.11 (*a*) Errors for households with an income of \$4000 per month. (*b*) Errors for households with an income of \$7500 per month.

Figure 13.12 shows how the distributions given in Figure 13.11 look when they are plotted on the same diagram with the population regression line. The points on the vertical line through $x = 40$ give the food expenditures for various households in the population, each of which has the same monthly income of \$4000. The same is true about the vertical line through $x = 75$ or any other vertical line for some other value of x.

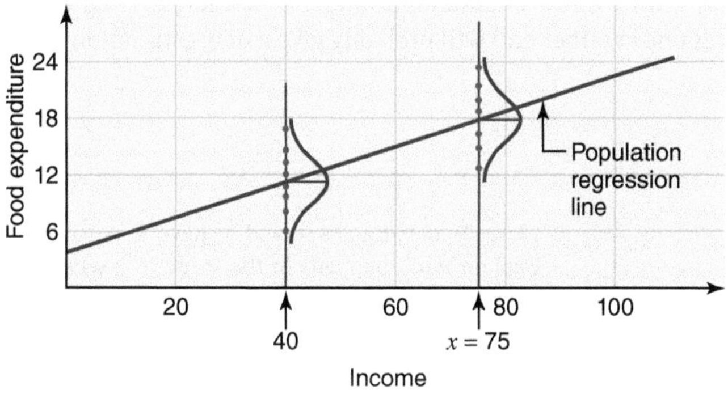

FIGURE 13.12 Distribution of errors around the population regression line.

13.1.8 Cautions in Using Regression

When carefully applied, regression is a very helpful technique for making predictions and estimations about one variable for a certain value of another variable. However, we need to be cautious when using the regression analysis, for it can give us misleading results and predictions. The following are the two most important points to remember when using regression:

(a) A Note on the Use of Simple Linear Regression

We should apply linear regression with caution. When we use simple linear regression, we assume that the relationship between two variables is described by a straight line. In the real world, the relationship between variables may not be linear. Hence, before we use a simple linear regression, it is better to construct a scatter diagram and look at the plot of the data points. We should estimate a linear regression model only if the scatter diagram indicates such a relationship. The scatter diagrams of **Figure 13.13** give two examples for which the relationship between x and y is not linear. Consequently, using linear regression in such cases would be wrong.

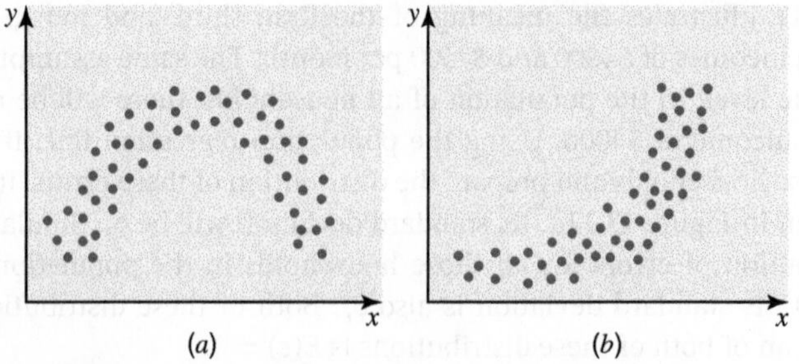

FIGURE 13.13 Nonlinear relationship between x and y.

(b) Extrapolation

The regression line estimated for the sample data is reliable only for the range of x values observed in the sample. For example, the values of x in our example on incomes and food expenditures vary from a minimum of 33 to a maximum of 83. Hence, our estimated regression line is applicable only for values of x between 33 and 83; that is, we should use this regression line to estimate the mean food expenditure or to predict the food expenditure of a single household only for income levels between \$3300 and \$8300. If we estimate or predict y for a value of x either less than 33 or greater than 83, it is called *extrapolation*. This does not mean that we should never use the regression line for extrapolation. Instead, we should interpret such predictions cautiously and not attach much importance to them.

Similarly, if the data used for the regression estimation are time-series data, the predicted values of y for periods outside the time interval used for the estimation of the regression line should be interpreted very cautiously. When using the estimated regression line for extrapolation, we are assuming that the same linear relationship between the two variables holds true for values of x outside the given range. It is possible that the relationship between the two variables may not be linear outside that range. Nonetheless, even if it is linear, adding a few more observations at either end will probably give a new estimation of the regression line.

EXERCISES

Concepts and Procedures

13.1 Explain the meaning of the words *simple* and *linear* as used in *simple linear regression*.

13.2 Explain the meaning of independent and dependent variables for a regression model.

13.3 Explain the difference between exact and nonexact relationships between two variables. Give one example of each.

13.4 Explain the difference between linear and nonlinear relationships between two variables.

13.5 Explain the difference between a simple and a multiple regression model.

13.6 Briefly explain the difference between a deterministic and a probabilistic regression model.

13.7 Why is the random error term included in a regression model?

13.8 Explain the least squares method and least squares regression line. Why are they called by these names?

13.9 Explain the meaning and concept of SSE. You may use a graph for illustration purposes.

13.10 Explain the difference between y and $\hat{y}$.

13.11 Two variables x and y have a positive linear relationship. Explain what happens to the value of y when x increases. Give one example of a positive relationship between two variables.

13.12 Two variables x and y have a negative linear relationship. Explain what happens to the value of y when x increases. Give one example of a negative relationship between two variables.

13.13 Explain the following:

a. Population regression line

b. Sample regression line

c. True values of A and B

d. Estimated values of A and B that are denoted by a and b, respectively

13.14 Briefly explain the assumptions of the population regression model.

13.15 Plot the following straight lines. Give the values of the y-intercept and slope for each of these lines and interpret them. Indicate whether each of the lines gives a positive or a negative relationship between x and y.

a. $y = 100 + 6x$ b. $y = -40 + 9x$

13.16 A population data set produced the following information:

$N = 450,\ \Sigma x = 4150,\ \Sigma y = 2500,\ \Sigma xy = 27{,}560,$
$\Sigma x^2 = 49{,}550$

Find the population regression line.

13.17 The following information is obtained from a sample data set.

$n = 15,\ \Sigma x = 75,\ \Sigma y = 625,\ \Sigma xy = 2400,\ \Sigma x^2 = 425$

Find the estimated regression line.

Applications

13.18 Bob's Pest Removal Service specializes in removing wild creatures (skunks, bats, reptiles, etc.) from private houses. He charges \$50 to go to a house plus \$30 per hour for his services. Let y be the total amount (in dollars) paid by a household using Bob's services and x the number of hours Bob spends capturing and removing the animal(s). The equation for the relationship between x and y is

$$y = 50 + 30x$$

a. Bob spent four hours removing a coyote from under Alice's house. How much will he be paid?

b. Suppose nine persons called Bob for assistance during a week. Strangely enough, each of these jobs required exactly four hours. Will each of these clients pay Bob the same amount, or do you expect each one to pay a different amount? Explain.

c. Is the relationship between x and y exact or nonexact?

13.19 A researcher took a sample of 25 electronics companies and found the following relationship between x and y:

$$\hat{y} = 3.60 - 11.75x$$

where x is the amount of money (in millions of dollars) spent on advertising by a company in 2019 and y represents the total gross sales (in millions of dollars) of that company for 2019.

a. An electronics company spent \$3 million on advertising in 2019. What are its expected gross sales for 2019?

b. Suppose four electronics companies spent \$3 million each on advertising in 2019. Do you expect these four companies to have the same actual gross sales for 2019? Explain.

c. Is the relationship between x and y exact or nonexact?

13.20 An auto manufacturing company wanted to investigate how the price of one of its car models depreciates with age. The research department at the company took a sample of eight cars of this model and collected the following information on the ages (in years) and prices (in hundreds of dollars) of these cars.

Age	8	3	6	9	2	5	6	2
Price	38	220	95	33	267	134	112	245

a. Construct a scatter diagram for these data. Does the scatter diagram exhibit a linear relationship between ages and prices of cars?

b. Find the regression line with price as a dependent variable and age as an independent variable.

c. Give a brief interpretation of the values of a and b calculated in part (b).

d. Plot the regression line on the scatter diagram of part (a) and show the errors by drawing vertical lines between scatter points and the regression line.

e. Predict the price of a 7-year-old car of this model.

f. Estimate the price of an 18-year-old car of this model. Comment on this finding.

13.21 The following table gives information on the amount of sugar (in grams) and the calorie count in one serving of a sample of nine different varieties of a beverage.

Sugar (grams)	5	11	7	14	12	9	13	10	8
Calories	328	675	455	755	825	578	685	625	555

a. Construct a scatter diagram for these data. Does the scatter diagram exhibit a linear relationship between the amount of sugar and the number of calories per serving?

b. Find the regression equation of the number of calories on the amount of sugar.

c. Give a brief interpretation of the values of a and b calculated in part (b).

d. Plot the regression line on the scatter diagram of part (a) and show the errors by drawing vertical lines between scatter points and the predictive regression line.

e. Calculate the calorie count for a beverage with 6 grams of sugar per serving.

f. Estimate the calorie count for a beverage with 18 grams of sugar per serving. Comment on this finding.

13.22 While browsing through the magazine rack at a bookstore, a statistician decides to examine the relationship between the price of a magazine and the percentage of the magazine space that contains advertisements. The data collected for eight magazines are given in the following table:

Percentage containing ads	37	43	55	49
Price (\$)	5.50	7.00	4.95	5.75
Percentage containing ads	70	28	65	32
Price (\$)	3.95	9.00	5.50	6.50

a. Construct a scatter diagram for these data. Does the scatter diagram exhibit a linear relationship between the percentage of a magazine's space containing ads and the price of the magazine?

b. Find the estimated regression equation of price on the percentage of space containing ads.

c. Give a brief interpretation of the values of a and b calculated in part (b).

d. Predict the price of a magazine with 50% of its space containing ads.

e. Estimate the price of a magazine with 99% of its space containing ads. Comment on this finding.

13.23 The following table gives the 2019 total payroll (in millions of dollars) and the percentage of games won during the 2019 season by each of the National League baseball teams.

Team	**Total Payroll (millions of dollars)**	**Percentage of Games Won**
Arizona Diamondbacks	108	52.5
Atlanta Braves	111	59.9
Chicago Cubs	208	51.9
Cincinnati Reds	129	46.3
Colorado Rockies	149	43.8
Los Angeles Dodgers	153	65.4
Miami Marlins	63	35.2
Milwaukee Brewers	130	54.9
New York Mets	162	53.1
Philadelphia Phillies	172	50.0
Pittsburgh Pirates	66	42.6
St. Louis Cardinals	150	56.2
San Diego Padres	94	43.2
San Francisco Giants	138	47.5
Washington Nationals	181	57.4

Source: www.cbssports.com

a. Find the least squares regression line with total payroll as the independent variable and percentage of games won as the dependent variable.

b. Is the equation of the regression line obtained in part (a) the population regression line? Why or why not? Do the values of the *y*-intercept and the slope of the regression line give *A* and *B* or *a* and *b*?

c. Give a brief interpretation of the values of the *y*-intercept and the slope obtained in part (a).

d. Predict the percentage of games won by a team with a total payroll of $145 million.

13.24 The following table gives the 2019 total payroll (in millions of dollars) and the percentage of games won during the 2019 season by each of the U. S. League baseball teams.

Team	Total Payroll (millions of dollars)	Percentage of Games Won
Baltimore Orioles	67	33.3
Boston Red Sox	213	51.9
Chicago White Sox	97	44.7
Cleveland Indians	107	57.4
Detroit Tigers	96	29.2
Houston Astros	177	66.0
Kansas City Royals	83	36.4
Los Angeles Angels	167	44.4
Minnesota Twins	114	62.3
New York Yankees	206	63.6
Oakland Athletics	97	59.9
Seattle Mariners	136	42.0
Tampa Bay Rays	69	59.3
Texas Rangers	102	48.1
Toronto Blue Jays	67	41.4

Source: www.cbssports.com

a. Find the least squares regression line with total payroll as the independent variable and percentage of games won as the dependent variable.

b. Is the equation of the regression line obtained in part (a) the population regression line? Why or why not? Do the values of the *y*-intercept and the slope of the regression line give *A* and *B* or *a* and *b*?

c. Give a brief interpretation of the values of the *y*-intercept and the slope obtained in part (a).

d. Predict the percentage of games won by a team with a total payroll of $145 million.

13.2 Standard Deviation of Errors and Coefficient of Determination

LEARNING OBJECTIVES

After completing this section, you should be able to:

- Explain the meaning of coefficient of determination.
- Explain the meaning of total sum of squares and regression sum of squares.
- Calculate the values of σ_ε and ρ^2 from population data.
- Calculate the values of s_e and r^2 from sample data.
- Interpret the coefficient of determination in the context of a linear regression.

In this section we discuss two concepts related to regression analysis. First we discuss the concept of the standard deviation of random errors and its calculation. Then we learn about the concept of the coefficient of determination and its calculation.

13.2.1 Standard Deviation of Errors

When we consider incomes and food expenditures, all households with the same income are expected to spend different amounts on food. Consequently, the random error ε will assume different values for these households. The standard deviation σ_ε measures the spread of these errors around the population regression line. The **standard deviation of errors** tells us how widely the errors and, hence, the values of y are spread for a given x. In Figure 13.12, which is reproduced as **Figure 13.14**, the points on the vertical line through $x = 40$ give the monthly food expenditures for all households with a monthly income of $4000. The distance of each dot from the point on the regression line gives the value of the corresponding error.

The standard deviation of errors σ_ε measures the spread of such points around the population regression line. The same is true for $x = 75$ or any other value of x.

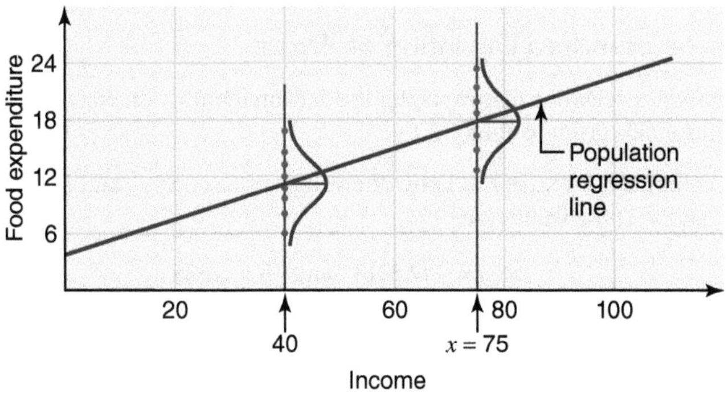

FIGURE 13.14 Spread of errors for $x = 40$ and $x = 75$.

Note that σ_ε denotes the standard deviation of errors for the population. However, usually σ_ε is unknown. In such cases, it is estimated by s_e, which is the standard deviation of errors for the sample data. The following is the basic formula to calculate s_e:

$$s_e = \sqrt{\frac{\text{SSE}}{n-2}} \quad \text{where} \quad \text{SSE} = \Sigma(y - \hat{y})^2$$

In this formula, $n - 2$ represents the **degrees of freedom** for the regression model. The reason $df = n - 2$ is that we lose one degree of freedom to calculate $\bar{x}$ and one for $\bar{y}$.

Degrees of Freedom for a Simple Linear Regression Model

The *degrees of freedom* for a simple linear regression model are

$$df = n - 2$$

For computational purposes, it is more convenient to use the following formula to calculate the standard deviation of errors s_e.

Standard Deviation of Errors

The *standard deviation of errors* is calculated as[3]

$$s_e = \sqrt{\frac{\text{SS}_{yy} - b\,\text{SS}_{xy}}{n-2}}$$

where
$$\text{SS}_{yy} = \Sigma y^2 - \frac{(\Sigma y)^2}{n}$$

The calculation of SS_{xy} was discussed earlier in this chapter.[4]

Like the value of SS_{xx}, the value of SS_{yy} is always positive.

Example 13.2 illustrates the calculation of the standard deviation of errors for the data of Table 13.1.

[3]If we have access to population data, the value of σ_ε is calculated using the formula

$$\sigma_\varepsilon = \sqrt{\frac{\text{SS}_{yy} - B\,\text{SS}_{xy}}{N}}$$

[4]The basic formula to calculate SS_{yy} is $\Sigma(y - \bar{y})^2$.

EXAMPLE 13.2 Incomes and Food Expenditures of Households

Calculating the Standard Deviation of Errors

Compute the standard deviation of errors s_e for the data on monthly incomes and food expenditures of the seven households given in Table 13.1.

Solution To compute s_e, we need to know the values of SS_{yy}, SS_{xy}, and b. In Example 13.1, we computed SS_{xy} and b. These values are

$$SS_{xy} = 447.5714 \quad \text{and} \quad b = .2525$$

To compute SS_{yy}, we calculate Σy^2 as shown in **Table 13.3**.

TABLE 13.3

Income x	Food Expenditure y	y^2
55	14	196
83	24	576
38	13	169
61	16	256
33	9	81
49	15	225
67	17	289
$\Sigma x = 386$	$\Sigma y = 108$	$\Sigma y^2 = 1792$

The value of SS_{yy} is

$$SS_{yy} = \Sigma y^2 - \frac{(\Sigma y)^2}{n} = 1792 - \frac{(108)^2}{7} = 125.7143$$

Hence, the standard deviation of errors is

$$s_e = \sqrt{\frac{SS_{yy} - b\,SS_{xy}}{n-2}} = \sqrt{\frac{125.7143 - .2525(447.5714)}{7-2}} = \mathbf{1.5939}$$

13.2.2 Coefficient of Determination

We may ask the question: How good is the regression model? In other words: How well does the independent variable explain the dependent variable in the regression model? The *coefficient of determination* is one concept that answers this question.

For a moment, assume that we possess information only on the food expenditures of households and not on their incomes. Hence, in this case, we cannot use the regression line to predict the food expenditure for any household. As we did in earlier chapters, in the absence of a regression model, we use $\bar{y}$ to estimate or predict every household's food expenditure. Consequently, the error of prediction for each household is now given by $y - \bar{y}$, which is the difference between the actual food expenditure of a household and the mean food expenditure. If we calculate such errors for all households in the sample and then square and add them, the resulting sum is called the **total sum of squares** and is denoted by **SST**. Actually SST is the same as SS_{yy} and is defined as

$$SST = SS_{yy} = \Sigma(y - \bar{y})^2$$

However, for computational purposes, SST is calculated using the following formula:

Total Sum of Squares (SST)

The *total sum of squares*, denoted by SST, is calculated as

$$SST = \Sigma y^2 - \frac{(\Sigma y)^2}{n}$$

Note that this is the same formula that we used to calculate SS_{yy}.

The value of SS_{yy}, which is 125.7143, was calculated in Example 13.2. Consequently, the value of SST is

$$SST = 125.7143$$

From Example 13.1, $\bar{y} = 15.4286$. **Figure 13.15** shows the error for each of the seven households in our sample using the scatter diagram of Figure 13.4 and $\bar{y}$.

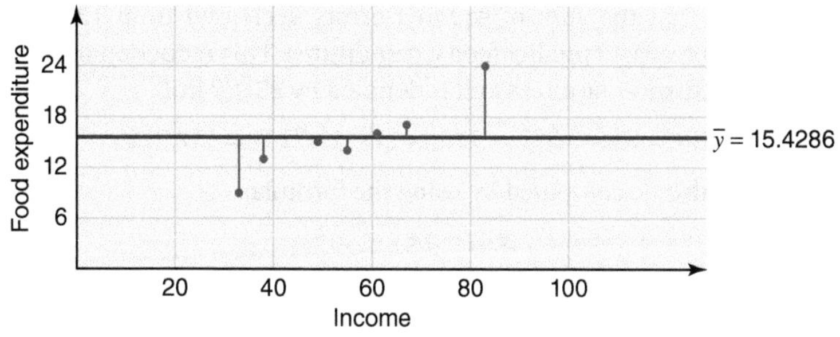

FIGURE 13.15 Total errors.

Now suppose we use the simple linear regression model to predict the food expenditure of each of the seven households in our sample. In this case, we predict each household's food expenditure by using the regression line we estimated earlier in Example 13.1, which is

$$\hat{y} = 1.5050 + .2525x$$

The predicted food expenditures, denoted by $\hat{y}$, for the seven households are listed in **Table 13.4**. Also given are the errors and error squares.

TABLE 13.4

x	y	$\hat{y} = 1.5050 + .2525x$	$e = y - \hat{y}$	$e^2 = (y - \hat{y})^2$
55	14	15.3925	−1.3925	1.9391
83	24	22.4625	1.5375	2.3639
38	13	11.1000	1.9000	3.6100
61	16	16.9075	−.9075	.8236
33	9	9.8375	−.8375	.7014
49	15	13.8775	1.1225	1.2600
67	17	18.4225	−1.4225	2.0235
				$\Sigma e^2 = \Sigma(y - \hat{y})^2 = 12.7215$

We calculate the values of $\hat{y}$ (given in the third column of Table 13.4) by substituting the values of x in the estimated regression model. For example, the value of x for the first household is 55. Substituting this value of x in the regression equation, we obtain

$$\hat{y} = 1.5050 + .2525(55) = 15.3925$$

Similarly we find the other values of $\hat{y}$. The error sum of squares SSE is given by the sum of the fifth column in Table 13.4. Thus,

$$SSE = \Sigma(y - \hat{y})^2 = 12.7215$$

The errors of prediction for the regression model for the seven households are shown in **Figure 13.16**.

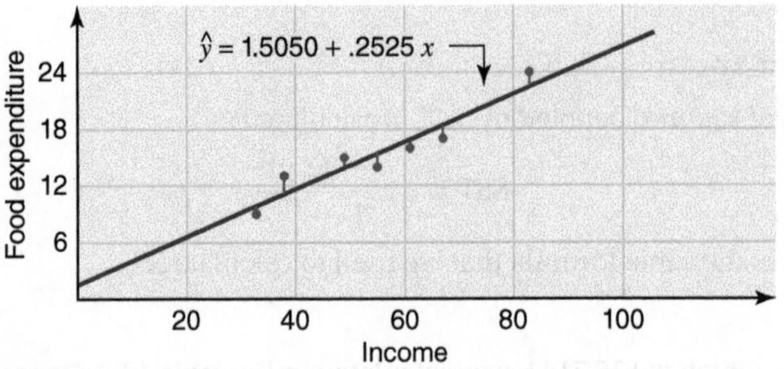

FIGURE 13.16 Errors of prediction when regression model is used.

Thus, from the foregoing calculations,

$$\text{SST} = 125.7143 \quad \text{and} \quad \text{SSE} = 12.7215$$

These values indicate that the sum of squared errors decreased from 125.7143 to 12.7215 when we used $\hat{y}$ in place of $\bar{y}$ to predict food expenditures. This reduction in squared errors is called the **regression sum of squares** and is denoted by **SSR**. Thus,

$$\text{SSR} = \text{SST} - \text{SSE} = 125.7143 - 12.7215 = 112.9928$$

The value of SSR can also be computed by using the formula

$$\text{SSR} = \Sigma(\hat{y} - \bar{y})^2$$

Regression Sum of Squares (SSR)

The *regression sum of squares*, denoted by SSR, is

$$\text{SSR} = \text{SST} - \text{SSE}$$

Thus, SSR is the portion of SST that is explained by the use of the regression model, and SSE is the portion of SST that is not explained by the use of the regression model. The sum of SSR and SSE is always equal to SST. Thus,

$$\text{SST} = \text{SSR} + \text{SSE}$$

The ratio of SSR to SST gives the **coefficient of determination**. The coefficient of determination calculated for population data is denoted by ρ^2 (ρ is the Greek letter *rho*), and the one calculated for sample data is denoted by r^2. The coefficient of determination gives the proportion of SST that is explained by the use of the regression model. The value of the coefficient of determination always lies in the range 0 to 1. The coefficient of determination can be calculated by using the formula

$$r^2 = \frac{\text{SSR}}{\text{SST}} \quad \text{or} \quad \frac{\text{SST} - \text{SSE}}{\text{SST}}$$

However, for computational purposes, the formula given next is more efficient to use to calculate the coefficient of determination.

Coefficient of Determination

The *coefficient of determination*, denoted by r^2, represents the proportion of SST that is explained by the use of the regression model. The computational formula for r^2 is[5]

$$r^2 = \frac{b\,\text{SS}_{xy}}{\text{SS}_{yy}}$$

and $$0 \le r^2 \le 1$$

[5]If we have access to population data, the value of ρ^2 is calculated using the formula

$$\rho^2 = \frac{B\,\text{SS}_{xy}}{\text{SS}_{yy}}$$

The values of SS_{xy} and SS_{yy} used here are calculated for the population data set.

Example 13.3 illustrates the calculation of the coefficient of determination for a sample data set.

EXAMPLE 13.3 Incomes and Food Expenditures of Households

Calculating the Coefficient of Determination

For the data of Table 13.1 on monthly incomes and food expenditures of seven households, calculate the coefficient of determination.

Solution From earlier calculations made in Examples 13.1 and 13.2,

$$b = .2525, \quad SS_{xy} = 447.5714, \quad \text{and} \quad SS_{yy} = 125.7143$$

Hence,

$$r^2 = \frac{b\, SS_{xy}}{SS_{yy}} = \frac{(.2525)(447.5714)}{125.7143} = .899$$

Thus, we can state that SST is reduced by approximately 89.9% (from 125.7143 to 12.7215) when we use $\hat{y}$, instead of $\bar{y}$, to predict the food expenditures of households. Note that r^2 is usually rounded to three decimal places.

The total sum of squares SST is a measure of the total variation in food expenditures, the regression sum of squares SSR is the portion of total variation explained by the regression model (or by income), and the error sum of squares SSE is the portion of total variation not explained by the regression model. Hence, for Example 13.3, we can state that we expect about 90% of the total variation in food expenditures of households to occur because of the variation in their incomes, and the remaining 10% is expected to occur due to randomness and other variables.

Usually, the higher the value of r^2, the better is the regression model. This is so because if r^2 is larger, a greater portion of the total errors is explained by the included independent variable, and a smaller portion of errors is attributed to other variables and randomness.

EXERCISES

Concepts and Procedures

13.25 What are the degrees of freedom for a simple linear regression model?

13.26 Explain the meaning of coefficient of determination.

13.27 Explain the meaning of SST and SSR. You may use graphs for illustration purposes.

13.28 A population data set produced the following information:

$$N = 275, \quad \Sigma x = 10{,}000, \quad \Sigma y = 1475, \quad \Sigma xy = 85{,}000,$$
$$\Sigma x^2 = 475{,}000, \quad \text{and} \quad \Sigma y^2 = 137{,}500$$

Find the values of σ_ε and ρ^2.

13.29 A population data set produced the following information:

$$N = 475, \quad \Sigma x = 3750, \quad \Sigma y = 2250, \quad \Sigma xy = 2750,$$
$$\Sigma x^2 = 48{,}750, \quad \text{and} \quad \Sigma y^2 = 40{,}250$$

Find the values of σ_ε and ρ^2.

13.30 The following information is obtained from a sample data set.

$$n = 15, \quad \Sigma x = 75, \quad \Sigma y = 625, \quad \Sigma xy = 2250,$$
$$\Sigma x^2 = 400, \quad \text{and} \quad \Sigma y^2 = 60{,}250$$

Find the values of s_e and r^2.

Applications

13.31 The following table gives information on the calorie count and grams of fat for eight of the many types of bagels produced and sold by Panera Bread.

Bagel	Calories	Fat (grams)
Asiago Cheese	330	6.0
Blueberry	340	1.5
Cinnamon Crunch	420	6.0
Cinnamon Swirl & Raisin	320	2.0
Everything	300	2.5
French Toast	350	4.0
Plain	290	1.5
Sesame	310	3.0

With calories as the dependent variable and fat content as the independent variable, find the following:

a. SS_{xx}, SS_{yy}, and SS_{xy}

b. Standard deviation of errors

c. SST, SSE, and SSR

d. Coefficient of determination

13.32 An auto manufacturing company wanted to investigate how the price of one of its car models depreciates with age. The research department at the company took a sample of eight cars of this model and collected the following information on the ages (in years) and prices (in hundreds of dollars) of these cars.

Age	8	3	6	9	2	5	6	2
Price	38	220	95	33	267	134	112	245

a. Calculate the standard deviation of errors.

b. Compute the coefficient of determination and give a brief interpretation of it.

13.33 The following table gives information on the amount of sugar (in grams) and the calorie count in one serving of a sample of nine different varieties of a beverage. Here calories is the dependent variable.

Sugar (grams)	5	11	7	14	12	9	13	10	8
Calories	328	675	455	755	825	578	685	625	555

a. Determine the standard deviation of errors.

b. Find the coefficient of determination and give a brief interpretation of it.

13.34 While browsing through the magazine rack at a bookstore, a statistician decides to examine the relationship between the price of a magazine and the percentage of the magazine space that contains advertisements. The data collected for eight magazines are given in the following table. Here price is the dependent variable.

Percentage Containing Ads	37	43	55	49
Price ($)	5.50	7.00	4.95	5.75
Percentage Containing Ads	70	28	65	32
Price ($)	3.95	9.00	5.50	6.50

a. Find the standard deviation of errors.

b. Compute the coefficient of determination. What percentage of the variation in price is explained by the least squares regression of price on the percentage of magazine space containing ads? What percentage of this variation is not explained?

13.35 The following table gives the 2019 total payroll (in millions of dollars) and the percentage of games won during the 2019 season by each of the National League baseball teams.

Team	**Total Payroll (millions of dollars)**	**Percentage of Games Won**
Arizona Diamondbacks	108	52.5
Atlanta Braves	111	59.9
Chicago Cubs	208	51.9
Cincinnati Reds	129	46.3
Colorado Rockies	149	43.8
Los Angeles Dodgers	153	65.4
Miami Marlins	63	35.2
Milwaukee Brewers	130	54.9
New York Mets	162	53.1
Philadelphia Phillies	172	50.0
Pittsburgh Pirates	66	42.6
St. Louis Cardinals	150	56.2
San Diego Padres	94	43.2
San Francisco Giants	138	47.5
Washington Nationals	181	57.4

Source: www.cbssports.com

a. Find the standard deviation of errors, σ_ε, with percentage of games won as the dependent variable. (Note that this data set belongs to a population.)

b. Compute the coefficient of determination, ρ^2.

13.3 Inferences About *B*

LEARNING OBJECTIVES

After completing this section, you should be able to:

- Describe the sampling distribution of the slope of a linear regression model.
- Construct a confidence interval for the true slope of a linear regression model.
- Apply the five-step procedure using a critical value approach to perform a test of hypothesis about the true slope of a linear regression model.

This section is concerned with estimation and tests of hypotheses about the population regression slope *B*. We can also make a confidence interval and test hypothesis about the *y*-intercept *A* of the population regression line. However, making inferences about *A* is beyond the scope of this text.

13.3.1 Sampling Distribution of *b*

One of the main purposes for determining a regression line is to find the true value of the slope *B* of the population regression line. However, in almost all cases, the regression line is estimated using sample data. Then, based on the sample regression line, inferences are made about the population regression line. The slope b of a sample regression line is a point estimator of the slope B of the population regression line. The different sample regression lines estimated for different samples taken from the same population will give different values of b. If only one sample is taken and the regression line for that sample is estimated, the value of b will depend on which elements are included in the sample. Thus, b is a random variable, and it possesses a probability distribution that is more commonly called its sampling distribution. The shape of the sampling distribution of b, its mean, and standard deviation are given next.

The Sampling Distribution of b and Its Mean and Standard Deviation

Because of the assumption of approximately normally distributed random errors, the sampling distribution of b is approximately normal. The mean and standard deviation of b, denoted by μ_b and σ_b, respectively, are

$$\mu_b = B \quad \text{and} \quad \sigma_b = \frac{\sigma_\varepsilon}{\sqrt{SS_{xx}}}$$

However, usually the standard deviation of population errors σ_ε is not known. Hence, the sample standard deviation of errors s_e is used to estimate σ_ε. In such a case, when σ_ε is unknown, the standard deviation of b is estimated by s_b, which is calculated as

$$s_b = \frac{s_e}{\sqrt{SS_{xx}}}$$

If σ_ε is known, the normal distribution can be used to make inferences about B. However, if σ_ε is not known, the normal distribution is replaced by the t distribution to make inferences about B.

13.3.2 Estimation of B

The value of b obtained from the sample regression line is a point estimate of the slope B of the population regression line. As mentioned in Section 13.3.1, if σ_ε is not known, the t distribution is used to make a confidence interval for B.

Confidence Interval for B

The $(1 - \alpha)100\%$ *confidence interval for B* is given by

$$b \pm t s_b$$

where

$$s_b = \frac{s_e}{\sqrt{SS_{xx}}}$$

and the value of t is obtained from the t distribution table, Table V of Appendix B, for $\alpha/2$ area in the right tail of the t distribution and $n - 2$ degrees of freedom.

Example 13.4 describes the procedure for making a confidence interval for B.

EXAMPLE 13.4 Incomes and Food Expenditures of Households

Constructing a Confidence Interval for B

Construct a 95% confidence interval for B using the data on incomes and food expenditures of seven households given in Table 13.1.

Solution From the given information and earlier calculations in Examples 13.1 and 13.2,

$$n = 7, \quad b = .2525, \quad SS_{xx} = 1772.8571, \quad \text{and} \quad s_e = 1.5939$$

The confidence level is 95%. We have

$$s_b = \frac{s_e}{\sqrt{SS_{xx}}} = \frac{1.5939}{\sqrt{1772.8571}} = .0379$$
$$df = n - 2 = 7 - 2 = 5$$
$$\alpha/2 = (1 - .95)/2 = .025$$

From the t distribution table, Table V of Appendix B, the value of t for 5 df and .025 area in the right tail of the t distribution curve is 2.571. The 95% confidence interval for B is

$$b \pm ts_b = .2525 \pm 2.571(.0379) = .2525 \pm .0974 = \mathbf{.155\ to\ .350}$$

Thus, we are 95% confident that the slope B of the population regression line is between .155 and .350.

13.3.3 Hypothesis Testing About B

Testing a hypothesis about B when the null hypothesis is $B = 0$ (that is, the slope of the population regression line is zero) is equivalent to testing that x does not determine y and that the regression line is of no use in predicting y for a given x. However, we should remember that we are testing for a linear relationship between x and y. It is possible that x may determine y nonlinearly. Hence, a nonlinear relationship may exist between x and y.

To test the hypothesis that x does not determine y linearly, we will test the null hypothesis that the slope of the population regression line is zero; that is, $B = 0$. The alternative hypothesis can be: (1) x determines y, that is, $B \neq 0$; (2) x determines y positively, that is, $B > 0$; or (3) x determines y negatively, that is, $B < 0$.

The procedure used to make a hypothesis test about B is similar to the one used in earlier chapters. It involves the same five steps. Of course, we can use the p-value approach too.

Test Statistic for b

The value of the *test statistic t for b* is calculated as

$$t = \frac{b - B}{s_b}$$

The value of B is substituted from the null hypothesis.

Example 13.5 illustrates the procedure for testing a hypothesis about B.

EXAMPLE 13.5 Incomes and Food Expenditures of Households

Conducting a Test of Hypothesis About B

Test at the 1% significance level whether the slope of the population regression line for the example on incomes and food expenditures of seven households is positive.

Solution From the given information and earlier calculations in Examples 13.1 and 13.4,

$$n = 7, \quad b = .2525, \quad \text{and} \quad s_b = .0379$$

STEP 1 *State the null and alternative hypotheses.*

We are to test whether or not the slope B of the population regression line is positive. Hence, the two hypotheses are

$$H_0: B = 0 \text{ (The slope is zero.)}$$
$$H_1: B > 0 \text{ (The slope is positive.)}$$

Note that we can also write the null hypothesis as $H_0: B \leq 0$, which states that the slope is either zero or negative.

STEP 2 *Select the distribution to use.*

Here, σ_ϵ is not known. All assumptions for the population regression model are assumed to hold true. Hence, we will use the t distribution to make the test about B.

STEP 3 *Determine the rejection and nonrejection regions.*

The significance level is .01. The > sign in the alternative hypothesis indicates that the test is right-tailed. Therefore,

$$\text{Area in the right tail of the } t \text{ distribution} = \alpha = .01$$
$$df = n - 2 = 7 - 2 = 5$$

From the t distribution table, Table V of Appendix B, the critical value of t for 5 df and .01 area in the right tail of the t distribution is 3.365, as shown in **Figure 13.17**.

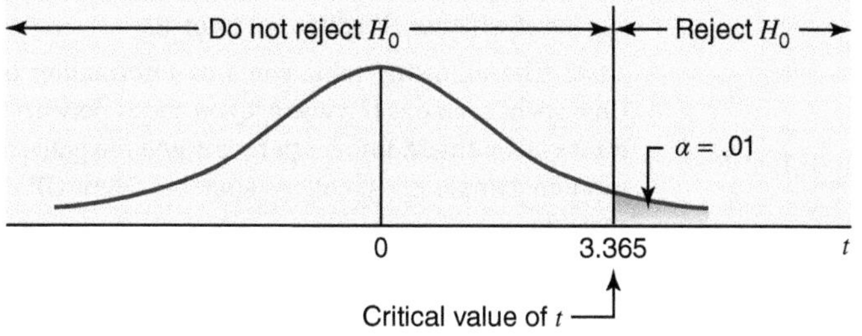

FIGURE 13.17 Rejection and nonrejection regions.

STEP 4 *Calculate the value of the test statistic.*

The value of the test statistic t for b is calculated as follows:

$$t = \frac{b - B}{s_b} = \frac{.2525 - 0}{.0379} = 6.662$$

(From H_0)

STEP 5 *Make a decision.*

The value of the test statistic $t = 6.662$ is greater than the critical value of $t = 3.365$ and falls in the rejection region. Hence, we reject the null hypothesis and based on strong evidence from the sample data we conclude that x (income) determines y (food expenditure) positively. That is, food expenditure increases with an increase in income and it decreases with a decrease in income.

Using the *p*-Value to Make a Decision

We can find the range for the p-value (as we did in Chapters 9 and 10) from the t distribution table, Table V of Appendix B, and make a decision by comparing that p-value with the significance level. For this example, $df = 5$, and the observed value of t is 6.662. From Table V (the t distribution table) in the row of $df = 5$, the largest value of t is 5.893 for which the area in the right tail of the t distribution is .001. Since our observed value of $t = 6.662$ is larger than 5.893, the p-value for $t = 6.662$ is less than .001, that is,

$$p\text{-value} < .001$$

Thus, we can state that for any α equal to or higher than .001 (the upper limit of the p-value range), we will reject the null hypothesis. For our example, $\alpha = .01$, which is larger than the p-value of .001. As a result, we reject the null hypothesis.

Using technology, we find the p-value = .000575 for this example, which is smaller than $\alpha = .01$. Hence, we reject the null hypothesis.

Note that the null hypothesis does not always have to be $B = 0$. We may test the null hypothesis that B is equal to a certain value.

A Note on Regression and Causality

The regression line does not prove causality between two variables; that is, it does not predict that a change in y is *caused* by a change in x. The information about causality is based on theory or common sense. A regression line describes only whether or not a significant linear relationship between x and y exists. Significant relationship means that we reject the null hypothesis $H_0: B = 0$ at a given significance level. The estimated regression line gives the change in y due to a change of one unit in x. Note that it does not indicate that the reason y has changed is that x has changed. In our example on incomes and food expenditures, it is economic theory and common sense, not the regression line, that tell us that food expenditure depends on income. The regression analysis simply helps determine whether or not this dependence is significant.

EXERCISES

Concepts and Procedures

13.36 Describe the mean, standard deviation, and shape of the sampling distribution of the slope b of a simple linear regression model.

13.37 The following information is obtained for a sample of 15 observations taken from a population.

$$SS_{xx} = 350, \quad s_e = 1.87, \quad \text{and} \quad \hat{y} = 15.25 + 7.50x$$

a. Make a 95% confidence interval for B.

b. Using a significance level of .025, can you conclude that B is positive?

c. Using a significance level of .01, can you conclude that B is different from zero?

d. Using a significance level of .02, test whether B is different from 6.0.

13.38 The following information is obtained for a sample of 30 observations taken from a population.

$$SS_{xx} = 275, \quad s_e = .95, \quad \text{and} \quad \hat{y} = 275 - 4.25x$$

a. Make a 95% confidence interval for B.

b. Using a significance level of .01, test whether B is negative.

c. Testing at the 5% significance level, can you conclude that B is different from zero?

d. Test if B is different from -5.0. Use $\alpha = .05$.

13.39 The following information is obtained for a sample of 78 observations taken from a population.

$$SS_{xx} = 400, \quad s_e = .90, \quad \text{and} \quad \hat{y} = 165 - 2.50x$$

a. Make a 99% confidence interval for B.

b. Using a significance level of .01, test whether B is negative.

c. Testing at the 1% significance level, can you conclude that B is different from zero?

d. Test if B is different from -2.0 . Use $\alpha = .025$.

Applications

13.40 An auto manufacturing company wanted to investigate how the price of one of its car models depreciates with age. The research department at the company took a sample of eight cars of this model and collected the following information on the ages (in years) and prices (in hundreds of dollars) of these cars.

Age	8	3	6	9	2	5	6	2
Price	38	220	95	33	267	134	112	245

a. Construct a 95% confidence interval for B.

b. Test at the 5% significance level whether B is negative.

13.41 The following table gives information on the amount of sugar (in grams) and the calorie count in one serving of a sample of nine different varieties of a beverage. Here calories is the dependent variable.

Sugar (grams)	5	11	7	14	12	9	13	10	8
Calories	328	675	455	755	825	578	685	625	555

a. Make a 95% confidence interval for B.

b. It is well known that each additional gram of carbohydrate adds 4 calories. Sugar is one type of carbohydrate. Using regression equation for the data in the table, test at the 5% significance level whether B is different from 40.

13.42 The following table contains information on the amount of time spent each day (on average) on social networks and the Internet for social or entertainment purposes and the grade point average for a random sample of 10 college students. Here GPA is the dependent variable.

Time (hours per day)	3.0	2.5	1.5	4.5	5.5
GPA	3.15	3.30	3.65	2.95	2.80
Time (hours per day)	2.0	3.5	4.0	1.0	5.0
GPA	3.50	3.08	2.88	3.40	3.05

a. Construct a 95% confidence interval for B.

b. Test at the 5% significance level whether B is negative.

13.43 While browsing through the magazine rack at a bookstore, a statistician decides to examine if the price of a magazine depends on the percentage of the magazine space that contains advertisements. The data collected for eight magazines are given in the following table:

Percentage Containing Ads	37	43	55	49
Price ($)	5.50	7.00	4.95	5.75
Percentage Containing Ads	70	28	65	32
Price ($)	3.95	9.00	5.50	6.50

a. Construct a 98% confidence interval for B.

b. Testing at the 5% significance level, can you conclude that B is different from zero?

13.44 The following table gives information on the calorie count and grams of fat for eight of the many types of bagels produced and sold by Panera Bread.

Bagel	**Calories**	**Fat (grams)**
Asiago Cheese	330	6.0
Blueberry	340	1.5
Cinnamon Crunch	420	6.0
Cinnamon Swirl & Raisin	320	2.0
Everything	300	2.5
French Toast	350	4.0
Plain	290	1.5
Sesame	310	3.0

a. Find the least squares regression line with calories as the dependent variable and fat content as the independent variable.

b. Make a 95% confidence interval for B.

c. Test at the 5% significance level whether B is different from 14.

13.4 Linear Correlation

LEARNING OBJECTIVES

After completing this section, you should be able to:

- Explain the properties of a linear correlation coefficient.
- Interpret a linear correlation coefficient in terms of the strength and direction of the linear relationship between two variables.
- Explain the relationship between the linear correlation coefficient and the slope of a linear regression model.
- Calculate the true linear correlation coefficient ρ from population data.
- Calculate the sample linear correlation coefficient r from sample data.
- Apply the five-step procedure using a critical value approach to perform a test of hypothesis about the true linear correlation coefficient.

This section describes the meaning and calculation of the linear correlation coefficient between two quantitative variables and the procedure to conduct a test of hypothesis about it.

13.4.1 Linear Correlation Coefficient

Another measure of the relationship between two variables is the correlation coefficient. This section describes the simple linear correlation, for short **linear correlation**, which measures the strength and direction of the linear association between two variables. In other words, the linear correlation coefficient measures how closely the points in a scatter diagram are spread around the regression line and indicates if the linear relationship is positive or negative. The correlation coefficient calculated for the population data is denoted by ρ (Greek letter *rho*) and the one calculated for sample data is denoted by r. (Note that the square of the correlation coefficient is equal to the coefficient of determination.)

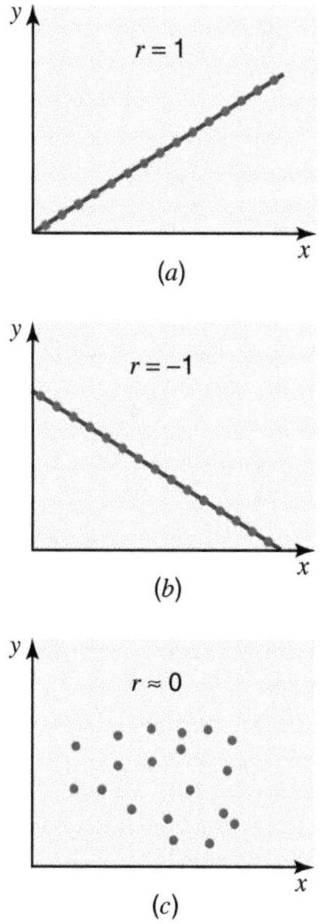

FIGURE 13.18 Linear correlation between two variables. (*a*) Perfect positive linear correlation, $r = 1$. (*b*) Perfect negative linear correlation, $r = -1$. (*c*) No linear correlation, $r \approx 0$.

Value of the Correlation Coefficient

The *value of the correlation coefficient* always lies in the range −1 to 1; that is,

$$-1 \leq \rho \leq 1 \quad \text{and} \quad -1 \leq r \leq 1$$

Although we can explain the linear correlation using the population correlation coefficient ρ, we will do so using the sample correlation coefficient r.

If $r = 1$, the two variables are said to have a **perfect positive linear relationship**. In such a case, all points in the scatter diagram lie on a straight line that slopes upward from left to right, as shown in **Figure 13.18*a***. If $r = -1$, the two variables are said to have a **perfect negative linear relationship**. In this case, all points in the scatter diagram fall on a straight line that slopes downward from left to right, as shown in **Figure 13.18*b***. If the points display a random spread all over the diagram, as shown in **Figure 13.18*c***, then there is **no evidence of a linear relationship** between the two variables, and consequently r is close to 0. Note that here r is *not* equal to zero but is very *close* to zero.

We do not usually encounter an example with perfect positive or perfect negative correlation (unless the relationship between variables is exact). What we observe in real-world problems is either a positive linear relationship with $0 < r < 1$ (that is, the correlation coefficient is greater than zero but less than 1) or a negative linear relationship with $-1 < r < 0$ (that is, the correlation coefficient is greater than −1 but less than zero).

If the correlation between two variables is positive and close to 1, we say that the variables have a **strong positive linear relationship**. If the correlation between two variables is positive but close to zero, then the variables have a **weak positive linear relationship.** In contrast, if the correlation between two variables is negative and close to −1, then the variables are said to have a **strong negative linear relationship.** If the correlation between two variables is negative but close to zero, there exists a **weak negative linear relationship** between the variables. Graphically, a strong relationship indicates that the points in the scatter diagram are very close to the regression line, and a weak relationship indicates that the points in the scatter diagram are widely spread around the regression line. These four cases are shown in **Figure 13.19*a–d*.**

The linear correlation coefficient is calculated by using the following formula. (This correlation coefficient is also called the *Pearson product moment correlation coefficient.*)

Linear Correlation Coefficient

The *simple linear correlation coefficient,* denoted by r, measures the strength and direction of the linear relationship between two variables for a sample and is calculated as[6]

$$r = \frac{SS_{xy}}{\sqrt{SS_{xx}\,SS_{yy}}}$$

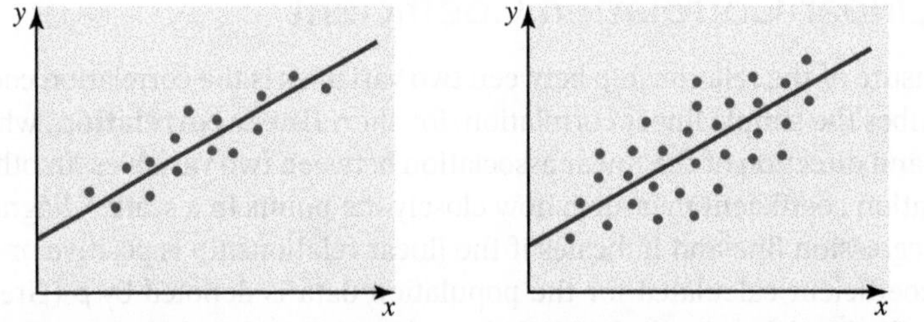

(*a*) Strong positive linear relationship (r is close to 1)

(*b*) Weak positive linear relationship (r is positive but close to zero)

(*c*) Strong negative linear relationship (r is close to −1)

(*d*) Weak negative linear relationship (r is negative and close to zero)

FIGURE 13.19 Linear correlation between two variables.

Because both SS_{xx} and SS_{yy} are always positive, the sign of the correlation coefficient r depends on the sign of SS_{xy}. If SS_{xy} is positive, then r will be positive, and if SS_{xy} is negative, then r will be negative. Another important observation to remember is that *r and b, calculated for the same sample, will always have the same sign. That is, both r and b are either positive or negative.* This is so because both r and b provide information about the direction of the relationship between x and y. Likewise, the corresponding population parameters ρ and B will always have the same sign.

Example 13.6 illustrates the calculation of the linear correlation coefficient r.

---

[6]If we have access to population data, the value of ρ is calculated using the formula

$$\rho = \frac{SS_{xy}}{\sqrt{SS_{xx}SS_{yy}}}$$

Here the values of SS_{xy}, SS_{xx}, and SS_{yy} are calculated using the population data.

EXAMPLE 13.6 Incomes and Food Expenditures of Households

Calculating the Linear Correlation Coefficient

Calculate the correlation coefficient for the example on incomes and food expenditures of seven households.

Solution From earlier calculations made in Examples 13.1 and 13.2,

$$SS_{xy} = 447.5714, \quad SS_{xx} = 1772.8571, \quad \text{and} \quad SS_{yy} = 125.7143$$

Substituting these values in the formula for r, we obtain

$$r = \frac{SS_{xy}}{\sqrt{SS_{xx}SS_{yy}}} = \frac{447.5714}{\sqrt{(1772.8571)(125.7143)}} = .948$$

Thus, the linear correlation coefficient is .948. We will round the correlation coefficient to three decimal places in this chapter.

The linear correlation coefficient simply tells us how strongly the two variables are linearly related. The correlation coefficient of .948 for incomes and food expenditures of seven households indicates that income and food expenditure are very strongly and positively associated. This correlation coefficient does not, however, provide us with any more information.

The square of the correlation coefficient gives the coefficient of determination, which was explained in Section 13.2.2. Thus, $(.948)^2$ is .90, which is the value of r^2 calculated in Example 13.3.

Sometimes the calculated value of r may indicate that the two variables are very strongly linearly related, but in reality they may not be. For example, if we calculate the correlation coefficient between the price of a haircut and the size of families in the United States using data for the last 30 years, we will find a strong negative linear association. Over time, the price of a haircut has increased and the size of families has decreased. This finding does not mean that family size and the price of a haircut are related. As a result, before we calculate the correlation coefficient, we must seek help from a theory or from common sense to postulate whether or not the two variables have a causal relationship.

Another point to note is that in a simple regression model, one of the two variables is categorized as an independent (also known as an explanatory or predictor) variable and the other is classified as a dependent (also known as a response) variable. However, no such distinction is made between the two variables when the correlation coefficient is calculated.

13.4.2 Hypothesis Testing About the Linear Correlation Coefficient

This section describes how to perform a test of hypothesis about the population correlation coefficient ρ using the sample correlation coefficient r. We use the t distribution to make this test. However, to use the t distribution, both variables should be approximately normally distributed.

Usually (although not always), the null hypothesis is that the linear correlation coefficient between the two variables is zero, that is, $\rho = 0$. The alternative hypothesis can be one of the following: (1) the linear correlation coefficient between the two variables is less than zero, that is, $\rho < 0$; (2) the linear correlation coefficient between the two variables is greater than zero, that is, $\rho > 0$; or (3) the linear correlation coefficient between the two variables is not equal to zero, that is, $\rho \neq 0$.

Test Statistic for *r*

If both variables are approximately normally distributed and the null hypothesis is $H_0: \rho = 0$, then the value of the test statistic t is calculated as

$$t = r\sqrt{\frac{n-2}{1-r^2}}$$

Here $n - 2$ gives the degrees of freedom for the test of the correlation coefficient.

Example 13.7 describes the procedure to perform a test of hypothesis about the linear correlation coefficient.

EXAMPLE 13.7 Incomes and Food Expenditures of Households

Performing a Test of Hypothesis About the Correlation Coefficient

Using a 1% level of significance and the data from Table 13.1, test whether the linear correlation coefficient between incomes and food expenditures is positive. Assume that the populations of both variables are approximately normally distributed.

Solution From Examples 13.1 and 13.6,

$$n = 7 \quad \text{and} \quad r = .948$$

Below we use the five steps to perform this test of hypothesis.

STEP 1 *State the null and alternative hypotheses.*

We are to test whether the linear correlation coefficient between incomes and food expenditures is positive. Hence, the null and alternative hypotheses are, respectively,

$H_0: \rho = 0$ (The linear correlation coefficient is zero.)
$H_1: \rho > 0$ (The linear correlation coefficient is positive.)

STEP 2 *Select the distribution to use.*

The population distributions for both variables are approximately normally distributed. Hence, we use the t distribution to perform this test about the linear correlation coefficient.

STEP 3 *Determine the rejection and nonrejection regions.*

The significance level is 1%. From the alternative hypothesis we know that the test is right-tailed. Hence,

Area in the right tail of the t distribution = .01
$df = n - 2 = 7 - 2 = 5$

From the t distribution table, Table V of Appendix B, the critical value of t is 3.365. The rejection and nonrejection regions for this test are shown in **Figure 13.20**.

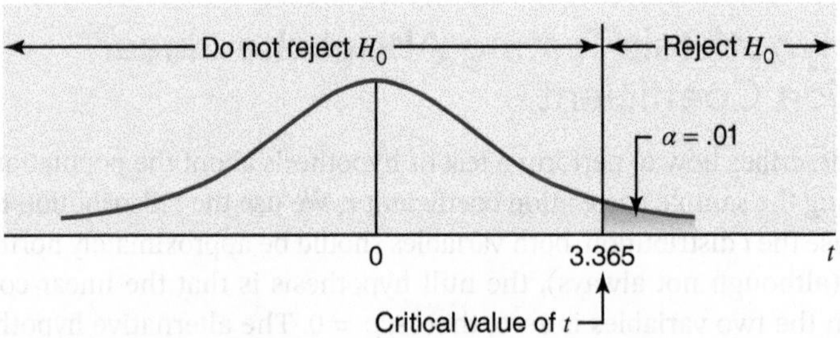

FIGURE 13.20 Rejection and nonrejection regions.

STEP 4 *Calculate the value of the test statistic.*

The value of the test statistic t for r is calculated as follows:

$$t = r\sqrt{\frac{n-2}{1-r^2}} = .948\sqrt{\frac{7-2}{1-(.948)^2}} = 6.660$$

STEP 5 *Make a decision.*

The value of the test statistic t = 6.660 is greater than the critical value of t = 3.365, and falls in the rejection region. Hence, we reject the null hypothesis and conclude that the evidence from the sample data indicates that there is a positive linear relationship between incomes and food expenditures.

Using the p-Value to Make a Decision

We can find the range for the p-value from the t distribution table (Table V of Appendix B) and make a decision by comparing that p-value with the significance level. For this example, $df = 5$, and the observed value of t is 6.660. From Table V (the t distribution table) of Appendix B in the row of $df = 5$ the largest value of t is 5.893 for which the area in the right tail of the t distribution is .001. Since our observed value of $t = 6.660$ is larger than 5.893, the p-value for $t = 6.660$ is less than .001, that is,

$$p\text{-value} < .001$$

Thus, we can state that for any α equal to or greater than .001 (the upper limit of the p-value range), we will reject the null hypothesis. For our example, $\alpha = .01$, which is greater than the p-value of .001. As a result, we reject the null hypothesis.

If we use technology, we will obtain the p-value of .000576 for $df = 5$ and $t = 6.660$. This p-value is smaller than $\alpha = .01$. Hence, we reject the null hypothesis.

EXERCISES

Concepts and Procedures

13.45 What does a linear correlation coefficient tell about the relationship between two variables? Within what range can a correlation coefficient assume a value?

13.46 What is the difference between ρ and r? Explain.

13.47 Explain each of the following concepts. You may use graphs to illustrate each concept.

a. Perfect positive linear correlation

b. Perfect negative linear correlation

c. Strong positive linear correlation

d. Strong negative linear correlation

e. Weak positive linear correlation

f. Weak negative linear correlation

g. No linear correlation

13.48 Can the values of B and ρ calculated for the same population data have different signs? Explain.

13.49 For a sample data set, the linear correlation coefficient r has a positive value. Which of the following is true about the slope b of the regression line estimated for the same sample data?

a. The value of b is positive.

b. The value of b is negative.

c. The value of b can be positive or negative.

13.50 For a sample data set, the slope b of the regression line has a negative value. Which of the following is true about the linear correlation coefficient r calculated for the same sample data?

a. The value of r is positive.

b. The value of r is negative.

c. The value of r can be positive or negative.

13.51 For a sample data set on two variables, the value of the linear correlation coefficient is (close to) zero. Does this mean that these variables are not related? Explain.

13.52 Do you expect a positive, zero, or negative linear correlation between the two variables in each of the following examples?

a. Grade of a student and hours spent on social media

b. Age of an individual and expenditure on medicine

c. Price and consumption of vegetables

d. Price of petrol and consumption of soft drinks

e. Incomes and purchases of luxury items

13.53 Do you expect a positive, zero, or negative linear correlation between the two variables in each of the following examples?

a. Blood pressure and brain stroke in individuals

b. Quality of seeds and yield of crop per acre

c. Ages and price of cars

d. Height of son and expenditure on watches

13.54 A population data set produced the following information:

$$N = 300,\ \Sigma x = 9500,\ \Sigma y = 1575,\ \Sigma xy = 86{,}050,$$
$$\Sigma x^2 = 475{,}250,\ \text{and}\ \Sigma y^2 = 137{,}575$$

Find the linear correlation coefficient ρ.

13.55 A population data set produced the following information:

$$N = 475,\ \Sigma x = 4200,\ \Sigma y = 2725,\ \Sigma xy = 25{,}250,$$
$$\Sigma x^2 = 48{,}275,\ \text{and}\ \Sigma y^2 = 38{,}975$$

Find the linear correlation coefficient ρ.

13.56 A sample data set produced the following information:

$$n = 12,\ \Sigma x = 120,\ \Sigma y = 250,\ \Sigma xy = 3575,$$
$$\Sigma x^2 = 1525,\ \text{and}\ \Sigma y^2 = 25{,}750$$

a. Calculate the linear correlation coefficient r.

b. Using a 2% significance level, can you conclude that ρ is different from zero?

Applications

13.57 An auto manufacturing company wanted to investigate how the price of one of its car models depreciates with age. The research department at the company took a sample of eight cars of this model and collected the following information on the ages (in years) and prices (in hundreds of dollars) of these cars.

Age	8	3	6	9	2	5	6	2
Price	38	220	95	33	267	134	112	245

a. Do you expect the ages and prices of cars to be positively or negatively related? Explain.

b. Calculate the linear correlation coefficient.

c. Test at a 2.5% significance level whether ρ is negative.

13.58 The following table lists the midterm and final exam scores for seven students in a statistics class.

Midterm Score	79	95	81	66	87	94	59
Final Exam Score	85	97	78	76	94	84	67

a. Do you expect the midterm and final exam scores to be positively or negatively related?

b. Plot a scatter diagram. By looking at the scatter diagram, do you expect the correlation coefficient between these two variables to be close to 0, 1, or −1?

c. Find the correlation coefficient. Is the value of r consistent with what you expected in parts (a) and (b)?

d. Using a 1% significance level, test whether the linear correlation coefficient is positive.

13.59 The following table gives information on the amount of sugar (in grams) and the calorie count in one serving of a sample of 13 different varieties of cereal.

Sugar (grams)	4	15	12	11	8	6	7	2	7	14	20	3	13
Calories	120	200	140	110	120	80	180	80	120	190	190	100	120

a. Find the correlation coefficient.

b. Test at a 1% significance level whether the linear correlation coefficient between the two variables listed in the table is positive.

13.60 The following table gives the 2019 total payroll (in millions of dollars) and the percentage of games won during the 2019 season by each of the National League baseball teams.

Team	**Total Payroll (millions of dollars)**	**Percentage of Games Won**
Arizona Diamondbacks	108	52.5
Atlanta Braves	111	59.9
Chicago Cubs	208	51.9
Cincinnati Reds	129	46.3
Colorado Rockies	149	43.8
Los Angeles Dodgers	153	65.4
Miami Marlins	63	35.2
Milwaukee Brewers	130	54.9
New York Mets	162	53.1
Philadelphia Phillies	172	50.0
Pittsburgh Pirates	66	42.6
St. Louis Cardinals	150	56.2
San Diego Padres	94	43.2
San Francisco Giants	138	47.5
Washington Nationals	181	57.4

Source: www.cbssports.com

Compute the linear correlation coefficient ρ. Does it make sense to test a hypothesis about ρ using these data? Explain.

13.61 The following table gives the 2019 total payroll (in millions of dollars) and the percentage of games won during the 2019 season by each of the U.S. League baseball teams.

Team	**Total Payroll (millions of dollars)**	**Percentage of Games Won**
Baltimore Orioles	67	33.3
Boston Red Sox	213	51.9
Chicago White Sox	97	44.7
Cleveland Indians	107	57.4
Detroit Tigers	96	29.2
Houston Astros	177	66.0
Kansas City Royals	83	36.4
Los Angeles Angels	167	44.4
Minnesota Twins	114	62.3
New York Yankees	206	63.6
Oakland Athletics	97	59.9
Seattle Mariners	136	42.0
Tampa Bay Rays	69	59.3
Texas Rangers	102	48.1
Toronto Blue Jays	67	41.4

Source: www.cbssports.com

Compute the linear correlation coefficient ρ. Does it make sense to test a hypothesis about ρ using these data? Explain.

13.5 Regression Analysis: A Complete Example

LEARNING OBJECTIVE

After completing this section, you should be able to:

- Apply the methods of this chapter to perform a complete regression analysis for a given set of sample data.

This section works out an example that includes all the topics we have discussed so far in this chapter.

EXAMPLE 13.8 Driving Experience and Monthly Auto Insurance

A Complete Example of Regression Analysis

A random sample of eight drivers selected from a small town insured with a company and having similar minimum required auto insurance policies was selected. The following table lists their driving experiences (in years) and monthly auto insurance premiums (in dollars).

Driving Experience (years)	Monthly Auto Insurance Premium ($)
5	64
2	87
12	50
9	71
15	44
6	56
25	42
16	60

Ascent/PKS Media Inc./ Getty Images

a. Does the insurance premium depend on the driving experience, or does the driving experience depend on the insurance premium? Do you expect a positive or a negative relationship between these two variables?

b. Compute SS_{xx}, SS_{yy}, and SS_{xy}.

c. Find the least squares regression line by choosing appropriate dependent and independent variables based on your answer in part (a).

d. Interpret the meaning of the values of a and b calculated in part (c).

e. Plot the scatter diagram and the regression line.

f. Calculate r and r^2, and explain what they mean.

g. Predict the monthly auto insurance premium for a driver with 10 years of driving experience.

h. Compute the standard deviation of errors.

i. Construct a 90% confidence interval for B.

j. Test at a 5% significance level whether B is negative.

k. Using $\alpha = .05$, test whether ρ is different from zero.

Solution

a. Based on theory and intuition, we expect the insurance premium to depend on driving experience. Consequently, the insurance premium is the dependent variable (variable y) and driving experience is the independent variable (variable x) in the regression model. A new driver is considered a high risk by the insurance companies, and such a driver has to pay a higher premium for auto insurance. On average, the insurance premium is expected to decrease with an increase in the years of driving experience. Therefore, we expect a negative relationship between these two variables. In other words, both the population correlation coefficient ρ and the population regression slope B are expected to be negative.

b. **Table 13.5** shows the calculation of Σx, Σy, Σxy, Σx^2, and Σy^2.

TABLE 13.5

Experience x	Premium y	xy	x^2	y^2
5	64	320	25	4096
2	87	174	4	7569
12	50	600	144	2500
9	71	639	81	5041
15	44	660	225	1936
6	56	336	36	3136
25	42	1050	625	1764
16	60	960	256	3600
$\Sigma x = 90$	$\Sigma y = 474$	$\Sigma xy = 4739$	$\Sigma x^2 = 1396$	$\Sigma y^2 = 29{,}642$

The values of $\bar{x}$ and $\bar{y}$ are

$$\bar{x} = \Sigma x/n = 90/8 = 11.25$$
$$\bar{y} = \Sigma y/n = 474/8 = 59.25$$

The values of SS_{xy}, SS_{xx}, and SS_{yy} are computed as follows:

$$SS_{xy} = \Sigma xy - \frac{(\Sigma x)(\Sigma y)}{n} = 4739 - \frac{(90)(474)}{8} = \mathbf{-593.5000}$$
$$SS_{xx} = \Sigma x^2 - \frac{(\Sigma x)^2}{n} = 1396 - \frac{(90)^2}{8} = \mathbf{383.5000}$$
$$SS_{yy} = \Sigma y^2 - \frac{(\Sigma y)^2}{n} = 29{,}642 - \frac{(474)^2}{8} = \mathbf{1557.5000}$$

c. To find the regression line, we calculate a and b as follows:

$$b = \frac{SS_{xy}}{SS_{xx}} = \frac{-593.5000}{383.5000} = -1.5476$$
$$a = \bar{y} - b\bar{x} = 59.25 - (-1.5476)(11.25) = 76.6605$$

Thus, our estimated regression line $\hat{y} = a + bx$ is

$$\hat{y} = \mathbf{76.6605 - 1.5476}x$$

d. The value of $a = 76.6605$ gives the value of $\hat{y}$ for $x = 0$; that is, it gives the predicted monthly auto insurance premium for a driver with no driving experience. However, as mentioned earlier in this chapter, we should not attach much importance to this statement because the sample contains drivers with only two or more years of experience. The value of b gives the change in $\hat{y}$ due to a change of one unit in x. Thus, $b = -1.5476$ indicates that, on average, for every extra year of driving experience, the monthly auto insurance premium decreases by \$1.55. Note that when b is negative, y decreases as x increases.

e. **Figure 13.21** shows the scatter diagram and the regression line for the data on eight auto drivers. Note that the regression line slopes downward from left to right. This result is consistent with the negative relationship we anticipated between driving experience and insurance premium.

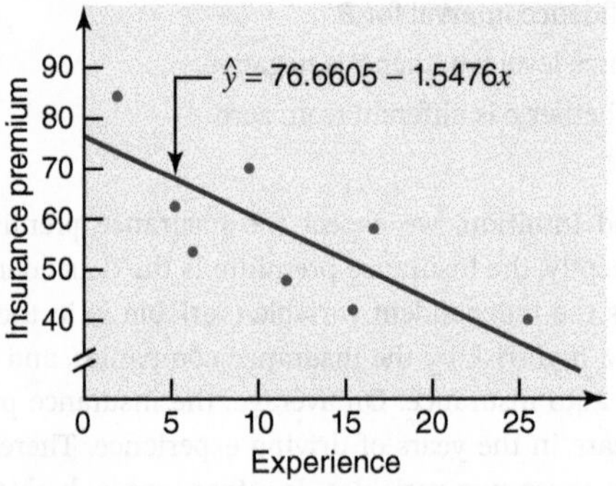

FIGURE 13.21 Scatter diagram and the regression line.

f. The values of r and r^2 are computed as follows:

$$r = \frac{SS_{xy}}{\sqrt{SS_{xx}SS_{yy}}} = \frac{-593.5000}{\sqrt{(383.5000)(1557.5000)}} = \mathbf{-.768}$$
$$r^2 = \frac{b\,SS_{xy}}{SS_{yy}} = \frac{(-1.5476)(-593.5000)}{1557.5000} = \mathbf{.590}$$

The value of $r = -.768$ indicates that the driving experience and the monthly auto insurance premium are negatively related. The linear relationship is strong but not very strong. The value of $r^2 = .590$ states that 59% of the total variation in insurance premiums is explained by years of driving experience, and 41% is not. The low value of r^2 indicates that there may be many other important variables that contribute to the determination of auto insurance premiums. For example, the premium is expected to depend on the driving record of a driver and the type and age of the car. It is also possible that the relationship between these variables is not linear.

g. Using the estimated regression line, we find the predicted value of y for $x = 10$ as

$$\hat{y} = 76.6605 - 1.5476x = 76.6605 - 1.5476(10) = \mathbf{\$61.18}$$

Thus, we expect the monthly auto insurance premium of a driver with 10 years of driving experience to be \$61.18.

h. The standard deviation of errors is

$$s_e = \sqrt{\frac{\text{SS}_{yy} - b\text{SS}_{xy}}{n-2}} = \sqrt{\frac{1557.5000 - (-1.5476)(-593.5000)}{8-2}} = \mathbf{10.3199}$$

i. To construct a 90% confidence interval for B, first we calculate the standard deviation of b:

$$s_b = \frac{s_e}{\sqrt{\text{SS}_{xx}}} = \frac{10.3199}{\sqrt{383.5000}} = .5270$$

For a 90% confidence level, the area in each tail of the t distribution is

$$\alpha/2 = (1 - .90)/2 = .05$$

The degrees of freedom are

$$df = n - 2 = 8 - 2 = 6$$

From the t distribution table, the t value for .05 area in the right tail of the t distribution and 6 df is 1.943. The 90% confidence interval for B is

$$b \pm ts_b = -1.5476 \pm 1.943(.5270)$$
$$= -1.5476 \pm 1.0240 = \mathbf{-2.57 \text{ to } -.52}$$

Thus, we can state with 90% confidence that B lies in the interval -2.57 to $-.52$. That is, on average, the monthly auto insurance premium of a driver decreases by an amount between \$.52 and \$2.57 for every extra year of driving experience.

j. We perform the following five steps to test the hypothesis about B.

STEP 1 *State the null and alternative hypotheses.*

The null and alternative hypotheses are, respectively,

$H_0: B = 0$ (B is not negative.)
$H_1: B < 0$ (B is negative.)

Note that the null hypothesis can also be written as $H_0: B \geq 0$.

STEP 2 *Select the distribution to use.*

Because σ_ε is not known, we use the t distribution to make the hypothesis test.

STEP 3 *Determine the rejection and nonrejection regions.*

The significance level is .05. The < sign in the alternative hypothesis indicates that it is a left-tailed test.

Area in the left tail of the t distribution $= \alpha = .05$
$df = n - 2 = 8 - 2 = 6$

From the t distribution table, the critical value of t for .05 area in the left tail of the t distribution and 6 df is -1.943, as shown in **Figure 13.22**.

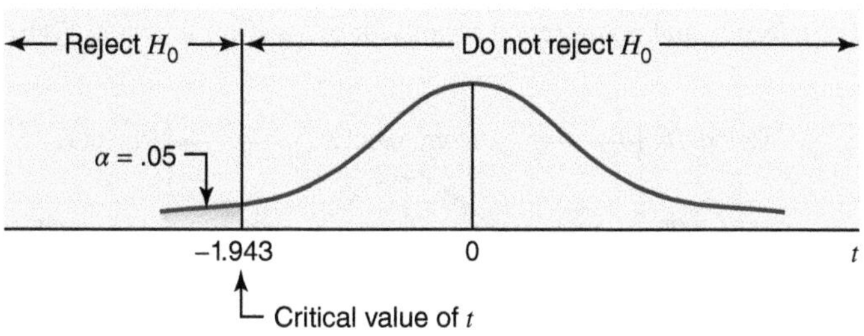

FIGURE 13.22 Rejection and nonrejection regions.

STEP 4 *Calculate the value of the test statistic.*

The value of the test statistic t for b is calculated as follows:

$$t = \frac{b - B}{s_b} = \frac{-1.5476 - 0}{.5270} = -2.937$$

(From H_0: the 0 in the numerator)

STEP 5 *Make a decision.*

The value of the test statistic $t = -2.937$ falls in the rejection region. Hence, we reject the null hypothesis and conclude that the evidence from the sample data indicates that B is negative. That is, the monthly auto insurance premium decreases with an increase in years of driving experience.

Using the p-Value to Make a Decision

We can find the range for the p-value from the t distribution table (Table V of Appendix B) and make a decision by comparing that p-value with the significance level. For this example, $df = 6$ and the observed value of t is -2.937. From Table V (the t distribution table) in the row of $df = 6$, 2.937 is between 2.447 and 3.143. The corresponding areas in the right tail of the t distribution are .025 and .01, respectively. Our test is left-tailed, however, and the observed value of t is negative. Thus, $t = -2.937$ lies between -2.447 and -3.143. The corresponding areas in the left tail of the t distribution are .025 and .01. Therefore the range of the p-value is

$$.01 < p\text{-value} < .025$$

Thus, we can state that for any α equal to or greater than .025 (the upper limit of the p-value range), we will reject the null hypothesis. For our example, $\alpha = .05$, which is greater than the upper limit of the p-value of .025. As a result, we reject the null hypothesis.

Note that if we use technology to find this p-value, we will obtain a p-value of .013. Then we can reject the null hypothesis for any $\alpha \geq .013$.

k. We perform the following five steps to test the hypothesis about the linear correlation coefficient ρ.

STEP 1 *State the null and alternative hypotheses.*

The null and alternative hypotheses are, respectively,

$H_0: \rho = 0$ (The linear correlation coefficient is zero.)
$H_1: \rho \neq 0$ (The linear correlation coefficient is different from zero.)

STEP 2 *Select the distribution to use.*

Assuming that variables x and y are approximately normally distributed, we will use the t distribution to perform this test about the linear correlation coefficient.

STEP 3 *Determine the rejection and nonrejection regions.*

The significance level is 5%. From the alternative hypothesis we know that the test is two-tailed. Hence,

$$\text{Area in each tail of the } t \text{ distribution} = .05/2 = .025$$
$$df = n - 2 = 8 - 2 = 6$$

From the t distribution table, Table V of Appendix B, the critical values of t are -2.447 and 2.447. The rejection and nonrejection regions for this test are shown in **Figure 13.23**.

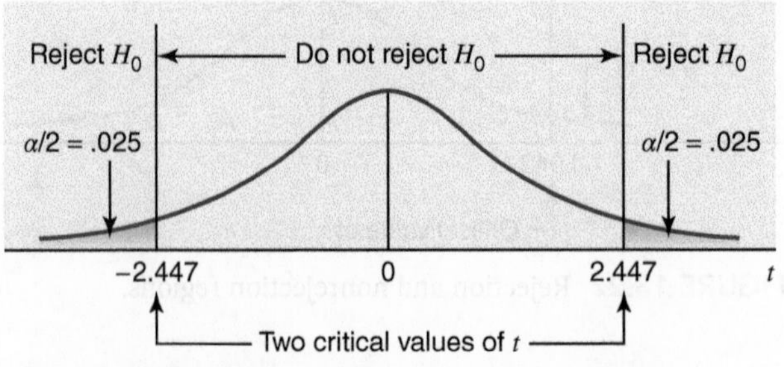

FIGURE 13.23 Rejection and nonrejection regions.

STEP 4 *Calculate the value of the test statistic.*

From part (*f*) of this example, $r = .768$. The value of the test statistic t for r is calculated as follows:

$$t = r\sqrt{\frac{n-2}{1-r^2}} = (-.768)\sqrt{\frac{8-2}{1-(-.768)^2}} = -2.937$$

STEP 5 *Make a decision.*

The value of the test statistic $t = -2.937$ falls in the rejection region. Hence, we reject the null hypothesis and conclude that the linear correlation coefficient between driving experience and auto insurance premium is different from zero.

Using the *p*-Value to Make a Decision

We can find the range for the *p*-value from the *t* distribution table and make a decision by comparing that *p*-value with the significance level. For this example, $df = 6$ and the observed value of t is -2.937. From Table V (the *t* distribution table) in the row of $df = 6$, $t = 2.937$ is between 2.447 and 3.143. The corresponding areas in the right tail of the *t* distribution curve are .025 and .01, respectively. Since the test is two tailed, the range of the *p*-value is

$$2(.01) < p\text{-value} < 2(.025) \quad \textbf{or} \quad .02 < p\text{-value} < .05$$

Thus, we can state that for any α equal to or greater than .05 (the upper limit of the *p*-value range), we will reject the null hypothesis. For our example, $\alpha = .05$, which is equal to the upper limit of the *p*-value. As a result, we reject the null hypothesis.

Using technology, we find the *p*-value = .026 for this example, which is smaller than $\alpha = .05$. Hence, we reject the null hypothesis.

EXERCISES

Applications

13.62 The recommended air pressure in a basketball is between 7 and 9 pounds per square inch (psi). When dropped from a height of 6 feet, a properly inflated basketball should bounce upward between 52 and 56 inches (http://www.bestsoccerbuys.com/balls-basketball.html). The basketball coach at a local high school purchased 10 new basketballs for the upcoming season, inflated the balls to pressures between 7 and 9 psi, and performed the *bounce test* mentioned above. The data obtained are given in the following table:

Pressure (psi)	7.8	8.1	8.3	7.4	8.9	7.2	8.6	7.5	8.1	8.5
Bounce height (inches)	54.1	54.3	55.2	53.3	55.4	52.2	55.7	54.6	54.8	55.3

a. With the pressure as an independent variable and bounce height as a dependent variable, compute SS_{xx}, SS_{yy}, and SS_{xy}.

b. Find the least squares regression line.

c. Interpret the meaning of the values of a and b calculated in part (b).

d. Calculate r and r^2 and explain what they mean.

e. Compute the standard deviation of errors.

f. Predict the bounce height of a basketball for $x = 8.0$.

g. Construct a 98% confidence interval for B.

h. Test at a 5% significance level whether B is different from zero.

i. Using $\alpha = .05$, can you conclude that ρ is different from zero?

13.63 The following table gives information on the incomes (in thousands of dollars) and charitable contributions (in hundreds of dollars) for the last year for a random sample of 10 households.

Income	Charitable Contributions
76	15
57	4
140	42
97	33
75	5
107	32
65	10
77	18
102	28
53	4

a. With income as an independent variable and charitable contributions as a dependent variable, compute SS_{xx}, SS_{yy}, and SS_{xy}.

b. Find the regression of charitable contributions on income.

c. Briefly explain the meaning of the values of a and b.

d. Calculate r and r^2 and briefly explain what they mean.

e. Compute the standard deviation of errors.

f. Construct a 99% confidence interval for B.

g. Test at a 1% significance level whether B is positive.

h. Using a 1% significance level, can you conclude that the linear correlation coefficient is different from zero?

13.64 The OYY Corporation has a large number of chain restaurants throughout a country. The research department at the company wanted to determine if the restaurants' sales depend on the mean income of households in the related areas. The company collected information on these two variables for eight restaurants randomly selected from different areas. The following table gives information on the weekly sales (in thousands of dollars) of these restaurants and the mean annual incomes (in thousands of dollars) of the households for those areas.

Sales	15	12	11	16	20	14	12	13
Income	45	42	40	47	55	47	35	38

a. Taking income as an independent variable and sales as a dependent variable, compute SS_{xx}, SS_{yy}, and SS_{xy}.

b. Find the least squares regression line.

c. Briefly explain the meaning of the values of a and b calculated in part (b).

d Calculate r and r^2 and briefly explain what they mean.

e. Compute the standard deviation of errors.

f. Construct a 95% confidence interval for B.

g. Test at a 1% significance level whether B is positive.

h. Using a 5% significance level, test whether ρ is positive.

13.65 The following table gives information on GPAs and starting salaries (rounded to the nearest thousand dollars) of 10 recent college graduates.

GPA	3.15	3.30	3.65	2.95	2.80	3.50	3.08	2.88	3.40	3.05
Starting Salary	50	55	60	45	50	60	53	70	55	65

a. With GPA as an independent variable and starting salary as a dependent variable, compute SS_{xx}, SS_{yy}, and SS_{xy}.

b. Find the least squares regression line.

c. Interpret the meaning of the values of a and b calculated in part (b).

d. Calculate r and r^2 and briefly explain what they mean.

e. Compute the standard deviation of errors.

f. Construct a 95% confidence interval for B.

g. Test at a 1% significance level whether B is different from zero.

h. Test at a 1% significance level whether ρ is positive.

13.6 Using the Regression Model

LEARNING OBJECTIVES

After completing this section, you should be able to:

- Explain the difference between estimating the mean value of *y* and predicting a particular value of *y* using a regression model.
- Construct a confidence interval for the mean value of *y* for a given value of *x*.
- Construct a prediction interval for the predicted value of *y* for a given value of *x*.

Let us return to the example on incomes and food expenditures to discuss two major uses of a regression model:

1. Estimating the mean value of y for a given value of x. For instance, we can use our food expenditure regression model to estimate the mean food expenditure of all households with a specific income (say, \$5500 per month).
2. Predicting a particular value of y for a given value of x. For instance, we can determine the expected food expenditure of a randomly selected household with a particular monthly income (say, \$5500) using our food expenditure regression model.

13.6.1 Using the Regression Model for Estimating the Mean Value of *y*

Our population regression model is

$$y = A + Bx + \varepsilon$$

As mentioned earlier in this chapter, the mean value of y for a given x is denoted by $\mu_{y|x}$, read as "the mean value of y for a given value of x." Because of the assumption that the mean value of ε is zero, the mean value of y is given by

$$\mu_{y|x} = A + Bx$$

Our objective is to estimate this mean value. The value of $\hat{y}$, obtained from the sample regression line by substituting the value of x, gives the **point estimate of $\mu_{y|x}$** for that x.

For our example on incomes and food expenditures, the estimated sample regression line (from Example 13.1) is

$$\hat{y} = 1.5050 + .2525x$$

Suppose we want to estimate the mean food expenditure for all households with a monthly income of \$5500. We will denote this population mean by $\mu_{y|x=55}$ or $\mu_{y|55}$. Note that we have written $x = 55$ and not $x = 5500$ in $\mu_{y|55}$ because the units of measurement for the data used to estimate the above regression line in Example 13.1 were hundreds of dollars. Using the regression line, we find that the point estimate of $\mu_{y|55}$ is

$$\hat{y} = 1.5050 + .2525(55) = \$15.3925 \text{ hundred} = \$1539.25$$

Thus, based on the sample regression line, the point estimate for the mean food expenditure $\mu_{y|55}$ for all households with a monthly income of \$5500 is \$1539.25 per month.

However, suppose we take a second sample of seven households from the same population and estimate the regression line for this sample. The point estimate of $\mu_{y|55}$ obtained from the regression line for the second sample is expected to be different. All possible samples of the same size taken from the same population will give different regression lines as shown in **Figure 13.24**, and, consequently, a different point estimate of $\mu_{y|x}$. Therefore, a confidence interval constructed for $\mu_{y|x}$ based on one sample will give a more reliable estimate of $\mu_{y|x}$ than will a point estimate.

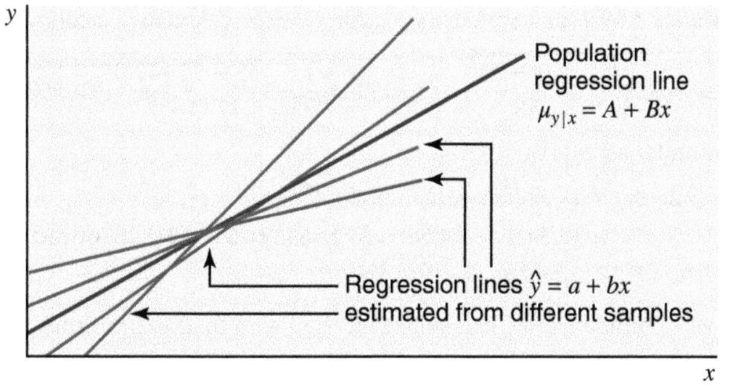

FIGURE 13.24 Population and sample regression lines.

To construct a confidence interval for $\mu_{y|x}$, we must know the mean, the standard deviation, and the shape of the sampling distribution of its point estimator $\hat{y}$.

The point estimator $\hat{y}$ of $\mu_{y|x}$ is normally distributed with a mean of $A + Bx$ and a standard deviation of

$$\sigma_{\hat{y}_m} = \sigma_\varepsilon \sqrt{\frac{1}{n} + \frac{(x_0 - \bar{x})^2}{SS_{xx}}}$$

where $\sigma_{\hat{y}_m}$ is the standard deviation of $\hat{y}$ when it is used to estimate $\mu_{y|x}$, x_0 is the value of x for which we are estimating $\mu_{y|x}$, and σ_ε is the population standard deviation of ε.

However, usually σ_ε is not known. Rather, it is estimated by the standard deviation of sample errors s_e. In this case, we replace σ_ε by s_e and $\sigma_{\hat{y}_m}$ by $s_{\hat{y}_m}$ in the foregoing expression. To make a confidence interval for $\mu_{y|x}$, we use the t distribution because σ_ε is not known.

Confidence Interval for $\mu_{y|x}$

The $(1 - \alpha)100\%$ *confidence interval for* $\mu_{y|x}$ for $x = x_0$ is

$$\hat{y} \pm t s_{\hat{y}_m}$$

where the value of t is obtained from the t distribution table for $\alpha/2$ area in the right tail of the t distribution curve and $df = n - 2$. The value of $s_{\hat{y}_m}$ is calculated as follows:

$$s_{\hat{y}_m} = s_e \sqrt{\frac{1}{n} + \frac{(x_0 - \bar{x})^2}{SS_{xx}}}$$

Example 13.9 illustrates how to make a confidence interval for the mean value of y, $\mu_{y|x}$.

EXAMPLE 13.9 Incomes and Food Expenditures of Households

Constructing a Confidence Interval for the Mean Value of y for a Given x

Refer to Example 13.1 on incomes and food expenditures. Find a 99% confidence interval for the mean food expenditure for all households with a monthly income of $5500.

Solution Using the regression line estimated in Example 13.1, we find the point estimate of the mean food expenditure for $x = 55$ as

$$\hat{y} = 1.5050 + .2525(55) = \$15.3925 \text{ hundred}$$

The confidence level is 99%. Hence, the area in each tail of the t distribution is

$$\alpha/2 = (1 - .99)/2 = .005$$

The degrees of freedom are

$$df = n - 2 = 7 - 2 = 5$$

From the t distribution table, the t value for .005 area in the right tail of the t distribution and 5 df is 4.032. From calculations in Examples 13.1 and 13.2, we know that

$$s_e = 1.5939, \quad \bar{x} = 55.1429, \quad \text{and} \quad SS_{xx} = 1772.8571$$

The standard deviation of $\hat{y}$ as an estimate of $\mu_{y|x}$ for $x = 55$ is calculated as follows:

$$s_{\hat{y}_m} = s_e\sqrt{\frac{1}{n} + \frac{(x_0 - \bar{x})^2}{SS_{xx}}} = (1.5939)\sqrt{\frac{1}{7} + \frac{(55 - 55.1429)^2}{1772.8571}} = .6025$$

Hence, the 99% confidence interval for $\mu_{y|55}$ is

$$\hat{y} \pm t s_{\hat{y}_m} = 15.3925 \pm 4.032(.6025)$$

$$= 15.3925 \pm 2.4293 = \mathbf{\$12.9632 \text{ to } \$17.8218 \text{ hundred}}$$

$$= \$1296.32 \text{ to } \$1782.18$$

Thus, with 99% confidence we can state that the mean food expenditure for all households with a monthly income of $5500 is between $1296.32 and $1782.18.

13.6.2 Using the Regression Model for Predicting a Particular Value of y

The second major use of a regression model is to predict a particular value of y for a given value of x—say, x_0. For example, we may want to predict the food expenditure of a randomly selected household with a monthly income of $5500. In this case, we are not interested in the mean food expenditure of all households with a monthly income of $5500 but in the food expenditure of one particular household with a monthly income of $5500. This predicted value of y is denoted by $\mathbf{y_p}$. Again, to predict a single value of y for $x = x_0$ from the estimated sample regression line, we use the value of **$\hat{y}$ as a point estimate of y_p**. Using the estimated regression line, we find that $\hat{y}$ for $x = 55$ is

$$\hat{y} = 1.5050 + .2525(55) = \$15.3925 \text{ hundred} = \$1539.25$$

Thus, based on our regression line, the point estimate for the food expenditure of a given household with a monthly income of $5500 is $1539.25 per month. Note that $\hat{y} = 1539.25$ is the point estimate for the mean food expenditure for all households with $x = 55$ as well as for the predicted value of food expenditure of one household with $x = 55$.

Different regression lines estimated by using different samples of seven households each taken from the same population will give different values of the point estimator for the predicted value of y for $x = 55$. Hence, a confidence interval constructed for y_p based on one sample will give a more reliable estimate of y_p than will a point estimate. The confidence interval constructed for y_p is more commonly called a **prediction interval**.

The procedure for constructing a prediction interval for y_p is similar to the one for constructing a confidence interval for $\mu_{y|x}$ except that the standard deviation of $\hat{y}$ is larger when we predict a single value of y than when we estimate $\mu_{y|x}$.

The point estimator $\hat{y}$ of y_p is normally distributed with a mean of $A + Bx$ and a standard deviation of

$$\sigma_{\hat{y}_p} = \sigma_\varepsilon \sqrt{1 + \frac{1}{n} + \frac{(x_0 - \bar{x})^2}{SS_{xx}}}$$

where $\sigma_{\hat{y}_p}$ is the standard deviation of the predicted value of y, x_0 is the value of x for which we are predicting y, and σ_ε is the population standard deviation of ε.

However, usually σ_ε is not known. In this case, we replace σ_ε by s_e and $\sigma_{\hat{y}_p}$ by $s_{\hat{y}_p}$ in the foregoing expression. To make a prediction interval for y_p, we use the t distribution when σ_ε is not known.

Prediction Interval for y_p

The $(1 - \alpha)100\%$ *prediction interval* for the predicted value of y, denoted by y_p, for $x = x_0$ is

$$\hat{y} \pm t s_{\hat{y}_p}$$

where the value of t is obtained from the t distribution table for $\alpha/2$ area in the right tail of the t distribution curve and $df = n - 2$. The value of $s_{\hat{y}_p}$ is calculated as follows:

$$s_{\hat{y}_p} = s_e \sqrt{1 + \frac{1}{n} + \frac{(x_0 - \bar{x})^2}{SS_{xx}}}$$

Example 13.10 illustrates the procedure to make a prediction interval for a particular value of y.

EXAMPLE 13.10 Incomes and Food Expenditures of Households

Making a Prediction Interval for a Particular Value of y for a Given x

Refer to Example 13.1 on incomes and food expenditures. Find a 99% prediction interval for the predicted food expenditure for a randomly selected household with a monthly income of \$5500.

Solution Using the regression line estimated in Example 13.1, we find the point estimate of the predicted food expenditure for $x = 55$:

$$\hat{y} = 1.5050 + .2525(55) = \$15.3925 \text{ hundred}$$

The area in each tail of the t distribution for a 99% confidence level is

$$\alpha/2 = (1 - .99)/2 = .005$$

The degrees of freedom are

$$df = n - 2 = 7 - 2 = 5$$

From the t distribution table, the t value for .005 area in the right tail of the t distribution curve and 5 df is 4.032. From calculations in Examples 13.1 and 13.2,

$$s_e = 1.5939, \quad \bar{x} = 55.1429, \quad \text{and} \quad SS_{xx} = 1772.8571$$

The standard deviation of $\hat{y}$ as an estimator of y_p for $x = 55$ is calculated as follows:

$$s_{\hat{y}_p} = s_e \sqrt{1 + \frac{1}{n} + \frac{(x_0 - \bar{x})^2}{SS_{xx}}}$$

$$= (1.5939) \sqrt{1 + \frac{1}{7} + \frac{(55 - 55.1429)^2}{1772.8571}} = 1.7040$$

Hence, the 99% prediction interval for y_p for $x = 55$ is

$$\hat{y} \pm t s_{\hat{y}_p} = 15.3925 \pm 4.032(1.7040)$$

$$= 15.3925 \pm 6.8705 = \mathbf{\$8.5220 \text{ to } \$22.2630 \text{ hundred}}$$

$$= \$852.20 \text{ to } \$2226.30$$

Thus, with 99% confidence we can state that the predicted food expenditure of a household with a monthly income of \$5500 is between \$852.20 and \$2226.30.

As we can observe in Example 13.10, this prediction interval is much wider than the confidence interval for the mean value of y for $x = 55$ calculated in Example 13.9, which was \$1296.32 to \$1782.18. This is always true. The prediction interval for predicting a single value of y is always larger than the confidence interval for estimating the mean value of y for a certain value of x.

EXERCISES

Concepts and Procedures

13.66 Briefly explain the difference between estimating the mean value of y and predicting a particular value of y using a regression model.

13.67 Construct a 95% confidence interval for the mean value of y and a 95% prediction interval for the predicted value of y for the following:

a. $\hat{y} = 4.75 + .75x$ for $x = 20$ given $s_e = .875$, $\bar{x} = 19.25$, $SS_{xx} = 154.25$, and $n = 15$

b. $\hat{y} = -25 + 6.25x$ for $x = 10$ given $s_e = 3.25$, $\bar{x} = 18.38$, $SS_{xx} = 402.56$, and $n = 12$

13.68 Construct a 99% confidence interval for the mean value of y and a 95% prediction interval for the predicted value of y for the following:

a. $\hat{y} = 15.15 + 2.87x$ for $x = 7$ given $s_e = 1.15$, $\bar{x} = 12.15$, $SS_{xx} = 198.57$, and $n = 10$

b. $\hat{y} = -9.3 + 3.17x$ for $x = 25$ given $s_e = 2.05$, $\bar{x} = 18.75$, $SS_{xx} = 345.57$, and $n = 10$

Applications

13.69 The following data give the experience (in years) and yearly salaries (in thousands of dollars) of nine randomly selected secretaries working at a large university.

Experience	14	3	5	6	4	9	18	5	16
Yearly Salary	49	24	30	34	29	48	53	20	48

Construct a 90% confidence interval for the mean yearly salary of all secretaries with 10 years of experience. Construct a 90% prediction interval for the yearly salary of a randomly selected secretary with 10 years of experience.

13.70 The owner of a small factory that produces work gloves is concerned about the high cost of air conditioning in the summer, but he is afraid that keeping the temperature in the factory too high will lower productivity. During the summer, he experiments with temperature settings from 68°F to 81°F and measures each day's productivity. The following table gives the temperature and the number of pairs of gloves (in hundreds) produced on each of the eight randomly selected days.

Temperature (°F)	72	71	78	75	81	77	68	76
Pairs of Gloves	37	37	32	36	33	35	39	34

Construct a 99% confidence interval for $\mu_{y|x}$ for $x = 77$ and a 99% prediction interval for y_p for $x = 77$. Here pairs of gloves is the dependent variable.

13.71 The recommended air pressure in a basketball is between 7 and 9 pounds per square inch (psi). When dropped from a height of 6 feet, a properly inflated basketball should bounce upward between 52 and 56 inches (http://www.bestsoccerbuys.com/balls-basketball.html). The basketball coach at a local high school purchased 10 new basketballs for the upcoming season, inflated the balls to pressures between 7 and 9 psi, and performed the *bounce test* mentioned above. The data obtained are given in the following table:

Pressure (psi)	7.8	8.1	8.3	7.4	8.9	7.2	8.6	7.5	8.1	8.5
Bounce Height (inches)	54.1	54.3	55.2	53.3	55.4	52.2	55.7	54.6	54.8	55.3

Construct a 99% confidence interval for the mean bounce height of all basketballs that are inflated to 8.5 psi. Construct a 99% prediction interval for the bounce height of a randomly selected basketball that is inflated to 8.5 psi. Here bounce height is the dependent variable.

13.72 The following table gives information on GPAs and annual starting salaries (rounded to the nearest thousand dollars) of 10 recent college graduates.

GPA	3.15	3.30	3.65	2.95	2.80	3.50	3.08	2.88	3.40	3.05
Starting Salary	50	55	60	45	50	60	53	70	55	65

Construct a 95% confidence interval for the mean annual starting salary of recent college graduates with a GPA of 3.33. Construct a 95% prediction interval for the annual starting salary of a randomly selected recent college graduate with a GPA of 3.33.

Uses and Misuses of Statistics

1. Processing Errors

Stuck on the far right side of the linear regression model is the Greek letter epsilon, ε. Despite its diminutive size, proper respect for the error term is critical to good linear regression modeling and analysis.

One interpretation of the error term is that it is a process. Imagine you are a chemist and you have to weigh a number of chemicals for an experiment. The balance that you use in your laboratory is very accurate—so accurate, in fact, that the shuffling of your feet, your exhaling near it, or the rumbling of trucks on the road outside can cause the reading to fluctuate. Because the value of the measurement that you take will be affected by a number of factors out of your control, you must make several measurements for each chemical, note each measurement, and then take the means and standard deviations of your samples. The distribution of measurements around a mean is the result of a random error process dependent on a number of factors out of your control; each time you use the balance, the measurement you take is the sum of the actual mass of the chemical and a "random" error. In this example, the measurements will most likely be normally distributed around the mean.

Linear regression analysis makes the same assumption about the two variables you are comparing: The value of the dependent variable is a linear function of the independent variable, plus a little bit of error that you cannot control. Unfortunately, when working with economic or survey data, you rarely can duplicate an experiment to identify the error model. As a statistician, however, you can use the errors to help you refine your model of the relationship among the variables and to guide your collection of new data. For example, if the errors are skewed to the right for moderate values of the independent variable and skewed to the left for small and large values of the independent variable, you can modify your model to account for this difference. Or you can think about other relationships among the variables that might explain this particular distribution of errors. A detailed analysis of the error in your model can be just as instructive as analysis of the slope and y-intercept of the identified model.

2. Outliers and Correlation

In Chapter 3 we learned that outliers can affect the values of some of the summary measures such as the mean, standard deviation, and range. Note that although outliers do affect many other summary measures, these three are affected substantially. Here we will see that just looking at a number that represents the correlation coefficient does not provide the entire story. A very famous data set for demonstrating this concept was created by F. J. Anscombe (Anscombe, F. J., Graphs in Statistical Analysis, *American Statistician*, 27, pp. 17–21). He created four pairs of data sets on x and y variables, each of which has a correlation of .816. To the novice, it may seem that the scatterplots for these four data sets should look virtually the same, but that may not be true. Look at the four scatterplots shown in **Figure 13.25**.

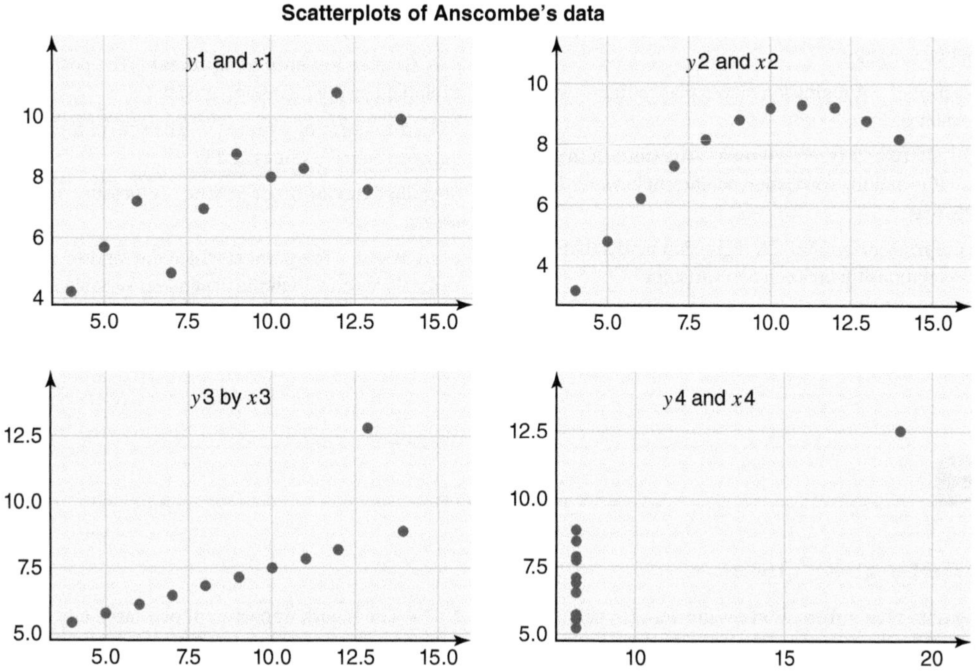

FIGURE 13.25 Four scatterplots with the same correlation coefficient.

No two of these scatterplots are even remotely close to being the same or even similar. The data used in the upper left plot are linearly associated, as are the data in the lower left plot. However the plot of $y3$ versus $x3$ contains an outlier. Without this outlier, the correlation between $x3$ and $y3$ would be 1. On the other hand, there is much more variability in the relationship between $x1$ and $y1$. As far as $x4$ and $y4$ are concerned, the strong correlation is defined by the single point in the upper right corner of the scatterplot. Without this point, there would be no variability among the $x4$ values, and the correlation would be undefined. Lastly, the scatterplot of $y2$ versus $x2$ reveals that there is an extremely well-defined relationship between these variables, but it is not linear. Being satisfied that the correlation coefficient is close to 1.0 between variables $x2$ and $y2$ implies that there is a strong linear association between the variables when actually we are fitting a line to a set of data that should be represented by another type of mathematical function.

As we have mentioned before, the process of making a graph may seem trivial, but the importance of graphs in our analysis can never be overstated.

Glossary

Coefficient of determination A measure that gives the proportion (or percentage) of the total variation in a dependent variable that is explained by a given independent variable.

Degrees of freedom for a simple linear regression model Sample size minus 2; that is, $n - 2$.

Dependent variable The variable to be predicted or explained.

Deterministic model A model in which the independent variable determines the dependent variable exactly. Such a model gives an exact relationship between two variables.

Estimated or predicted value of y The value of the dependent variable, denoted by $\hat{y}$, that is calculated for a given value of x using the estimated regression model.

Independent or explanatory variable The variable included in a model to explain the variation in the dependent variable.

Least squares estimates of A and B The values of a and b that are calculated by using the sample data.

Least squares method The method used to fit a regression line through a scatter diagram such that the error sum of squares is minimum.

Least squares regression line A regression line obtained by using the least squares method.

Linear correlation coefficient A measure of the strength and direction of the linear relationship between two variables.

Linear regression model A regression model that gives a straight-line relationship between two variables.

Multiple regression model A regression model that contains two or more independent variables.

Negative relationship between two variables The value of the slope in the regression line and the correlation coefficient between two variables are both negative.

Nonlinear (simple) regression model A regression model that does not give a straight-line relationship between two variables.

Population parameters for a simple regression model The values of A and B for the regression model $y = A + Bx + \varepsilon$ that are obtained by using population data.

Positive relationship between two variables The value of the slope in the regression line and the correlation coefficient between two variables are both positive.

Prediction interval The confidence interval for a particular value of y for a given value of x.

Probabilistic or statistical model A model in which the independent variable does not determine the dependent variable exactly.

Random error term (ε) The difference between the actual and predicted values of y.

Scatter diagram or scatterplot A plot of the paired observations of x and y.

Simple linear regression A regression model with one dependent and one independent variable that assumes a straight-line relationship.

Slope The coefficient of x in a regression model that gives the change in y for a change of one unit in x.

SSE (error sum of squares) The sum of the squared differences between the actual and predicted values of y. It is the portion of the SST that is not explained by the regression model.

SSR (regression sum of squares) The portion of the SST that is explained by the regression model.

SST (total sum of squares) The sum of the squared differences between actual y values and $\bar{y}$.

Standard deviation of errors A measure of spread for the random errors.

y-intercept The point at which the regression line intersects the vertical axis on which the dependent variable is marked. It is the value of y when x is zero.

Exercises

Supplementary Exercises

13.73 The following data give information on the ages (in years) and the number of breakdowns during the last month for a sample of eight machines at a large company.

Age (years)	11	9	8	13	15	7	10	14
Number of Breakdowns	15	10	12	18	22	13	15	20

a. Taking age as an independent variable and number of breakdowns as a dependent variable, what is your hypothesis about the sign of B in the regression line? (In other words, do you expect B to be positive or negative?)

b. Find the least squares regression line. Is the sign of b the same as you hypothesized for B in part (a)?

c. Give a brief interpretation of the values of a and b calculated in part (b).

d. Compute r and r^2 and explain what they mean.

e. Compute the standard deviation of errors.

f. Construct a 95% confidence interval for B.

g. Test at a 5% significance level whether B is positive.

h. At a 5% significance level, can you conclude that ρ is positive? Is your conclusion the same as in part (g)?

13.74 The health department of a large city has developed an air pollution index that measures the level of several air pollutants that cause respiratory distress in humans. The following table gives the pollution index (on a scale of 1 to 10, with 10 being the worst) for nine randomly selected summer days and the number of patients with acute respiratory problems admitted to the emergency rooms of the city's hospitals.

Air Pollution Index	5.1	4.7	6.7	7.3	7.7	5.4	6.3	4.1	9.5
Emergency Admissions	55	32	85	81	95	51	67	20	115

a. Taking the air pollution index as an independent variable and the number of emergency admissions as a dependent variable, do you expect B to be positive or negative in the regression model $y = A + Bx + \varepsilon$?

b. Find the least squares regression line. Is the sign of b the same as you hypothesized for B in part (a)?

c. Compute r and r^2, and explain what they mean.

d. Compute the standard deviation of errors.

e. Construct a 99% confidence interval for B.

f. Test at a 1% significance level whether B is positive.

g. Test at a 1% significance level whether ρ is positive. Is your conclusion the same as in part (f)?

13.75 The management of a supermarket wants to find if there is a relationship between the number of times a specific product is promoted on the intercom system in the store and the number of units of that product sold. To experiment, the management selected a product and promoted it on the intercom system for seven days. The following table gives the number of times this product was promoted each day and the number of units sold.

Number of Promotions per Day	Number of Units Sold per Day (hundreds)
15	11
22	22
42	30
30	26
18	17
12	15
38	23

a. With the number of promotions as an independent variable and the number of units sold as a dependent variable, what do you expect the sign of B in the regression line $y = A + Bx + \varepsilon$ will be?

b. Find the least squares regression line $\hat{y} = a + bx$. Is the sign of b the same as you hypothesized for B in part (a)?

c. Give a brief interpretation of the values of a and b calculated in part (b).

d. Compute r and r^2 and explain what they mean.

e. Predict the number of units of this product sold on a day with 35 promotions.

f. Compute the standard deviation of errors.

g. Construct a 98% confidence interval for B.

h. Testing at a 1% significance level, can you conclude that B is positive?

i. Using $\alpha = .02$, can you conclude that the correlation coefficient is different from zero?

13.76 The following table gives the average weekly retail price of a gallon of regular gasoline in New York over a 12-week period from September 23, 2019, through December 9, 2019. Consider these 12 weeks as a random sample.

Date	9/23/2019	9/30/2019	10/7/2019	10/14/2019
Price ($)	2.659	2.645	2.624	2.599
Date	10/21/2019	10/28/2019	11/4/2019	11/11/2019
Price ($)	2.587	2.596	2.596	2.610
Date	11/18/2019	11/25/2019	12/2/2019	12/9/2019
Price ($)	2.600	2.595	2.591	2.586

a. Assign a value of 0 to 9/23/2019, 1 to 9/30/2019, 2 to 10/7/2019, and so on. Call this new variable *Time*. Make a new table with the variables *Time* and *Price*.

b. With time as an independent variable and price as the dependent variable, compute SS_{xx}, SS_{yy}, and SS_{xy}.

c. Construct a scatter diagram for these data. Does the scatter diagram exhibit a negative linear relationship between time and price?

d. Find the least squares regression line $\hat{y} = a + bx$.

e. Give a brief interpretation of the values of a and b calculated in part (d).

f. Compute the correlation coefficient r.

g. Predict the average price of a gallon of regular gasoline in New York for *Time* = 26. Comment on this prediction.

13.77 The following data give information on the ages (in years) and the number of breakdowns during the last month for a sample of eight machines at a large company.

Age (years)	11	9	8	13	15	7	10	14
Number of Breakdowns	15	10	12	18	22	13	15	20

Construct a 95% confidence interval for the mean number of breakdowns per month for all machines with an age of 12 years. Find a 95% prediction interval for the number of breakdowns per month for a randomly selected machine with an age of 12 years.

13.78 The management of a supermarket wants to find if there is a relationship between the number of times a specific product is promoted on the intercom system in the store and the number of units of that product sold. To experiment, the management selected a product and promoted it on the intercom system for seven days. The following table gives the number of times this product was promoted each day and the number of units sold.

Number of Promotions per Day	Number of Units Sold per Day (hundreds)
20	12
32	15
56	18
41	17
25	10
38	13
28	10

Make a 95% confidence interval for the mean number of units of that product sold on days with 34 promotions. Construct a 95% prediction interval for the number of units of that product sold on a randomly selected day with 34 promotions.

13.79 The following table provides information on the living area (in square feet) and price (in thousands of dollars) of 10 randomly selected houses listed for sale in a city.

Living Area	3008	2032	2272	1840	2579	2583	1650	3932	2978	2176
Price	375	320	355	289	360	384	272	470	395	360

Construct a 98% confidence interval for the mean price of all houses with living areas of 2400 square feet. Construct a 98% prediction interval for the price of a randomly selected house with a living area of 2400 square feet.

Advanced Exercises

13.80 Consider the data given in the following table:

x	12	24	36	48	60	72
y	14	16	24	28	32	42

a. Find the least squares regression line and the linear correlation coefficient r.

b. Suppose that each value of y given in the table is increased by 6 and the x values remain unchanged. Would you expect r to increase, decrease, or remain the same? How do you expect the least squares regression line to change?

c. Increase each value of y given in the table by 6 and find the new least squares regression line and the correlation coefficient r. Do these results agree with your expectation in part (b)?

13.81 An economist is studying the relationship between the incomes of fathers and their sons or daughters. Let x be the annual income of a 36-year-old person and let y be the annual income of that person's father at age 36 years, adjusted for inflation. A random sample of 450 thirty-six-year-olds and their fathers yields a linear correlation coefficient of .70 between x and y. A friend of yours, who has read about this research, asks you several questions, such as: Does the positive value of the correlation coefficient suggest that the 36-year-olds tend to earn more than their fathers? Does the correlation coefficient reveal anything at all about the difference between the incomes of 36-year-olds and their fathers? If not, what other information would we need from this study? What does the correlation coefficient tell us about the relationship between the two variables in this example? Write a short note to your friend answering these questions.

13.82 Consider the formulas for calculating a prediction interval for a new (specific) value of y. For each of the changes mentioned in parts (a) through (c) below, state the effect on the width of the confidence interval (increase, decrease, or no change) and why it happens. Note that besides the change mentioned in each part, everything else such as the values of a, b, $\bar{x}$, s_e, and SS_{xx} remain unchanged.

a. The confidence level is increased.

b. The sample size is increased.

c. The value of x_0 is moved farther away from the value of $\bar{x}$.

d. What will the value of the margin of error be if x_0 equals $\bar{x}$?

13.83 Consider the following data:

x	−5	−4	−3	−2	−1	0	1	2	3	4	5
y	625	256	81	16	1	0	1	16	81	256	625

a. Calculate the correlation between x and y, and perform a hypothesis test to determine if the correlation is significantly greater than zero. Use a significance level of 10%.

b. Are you willing to conclude that there is a strong linear association between the two variables? Explain why or why not.

Self-Review Test

Multiple Choice Questions

1. A simple regression is a regression model that contains
   a. only one independent variable
   b. only one dependent variable
   c. both a and b

2. The relationship between independent and dependent variables represented by the (simple) linear regression is that of
   a. a straight line
   b. a curve
   c. both a and b

3. A deterministic regression model is a model that
   a. contains the random error term
   b. does not contain the random error term
   c. gives a nonlinear relationship

4. A probabilistic regression model is a model that
   a. contains the random error term
   b. does not contain the random error term
   c. shows an exact relationship

5. The least squares regression line minimizes the sum of
   a. errors b. squared errors c. predictions

6. The degrees of freedom for a simple regression model are
   a. $n-1$ b. $n-2$ c. $n-5$

7. Is the following statement true or false?
   The coefficient of determination gives the proportion of total squared errors (SST) that is explained by the use of the regression model.

8. Is the following statement true or false?
   The linear correlation coefficient measures the strength and direction of the linear association between two variables.

9. The value of the coefficient of determination is always in the range
   a. 0 to 1 b. −1 to 1 c. −1 to 0

10. The value of the correlation coefficient is always in the range
    a. 0 to 1 b. −1 to 1 c. −1 to 0

11. A linear regression model
    a. gives a relationship between two variables that can be described by a straight line
    b. gives a relationship between two variables that cannot be described by a line
    c. gives a relationship between three variables that can be described by a line

12. In a regression model, the slope represents the
    a. point where the y-axis intersects the x-axis
    b. change in y due to a one-unit change in x
    c. change in the independent variable due to a one-unit change in the dependent variable

13. The regression model $y = A + Bx + \varepsilon$ is

a. a probabilistic model

b. a nonlinear model

c. an exact relationship

14. In a regression model, the random error term is included to capture the effects of the

a. independent variable

b. dependent variable

c. missing variables and random variation

15. In the regression model $\hat{y} = a + bx$, a and b are the

a. estimates of the population parameters A and B

b. random variables

c. population parameters

16. In the regression model $\hat{y} = a + bx$, $\hat{y}$ is the

a. actual value of y

b. predicted value of y

c. value of y when a and b are zero

17. The random error for the sample regression model, denoted by e, is equal to the

a. predicted value of y minus the estimated value of y

b. estimated value of y minus the predicted value of y

c. actual value of y minus the predicted value of y

18. For a regression model, the error sum of squares, denoted by SSE, is equal to the sum of the

a. squares of errors

b. squares of y values

c. squares of the difference between x and y values

19. The least squares method minimizes the

a. difference between the y and x values

b. sum of the errors

c. sum of the squares of errors

20. The table shown below gives six pairs of x and y values.

x	y
16	8
20	13
12	9
10	8
13	6
9	4

The values of SS_{xx} and SS_{xy}, rounded to three decimal places, are

a. 83.333 and 50.000

b. 83.333 and 58.000

c. 74.333 and 50.000

21. The following table lists the monthly incomes (in hundreds of dollars) and the monthly rents paid (in hundreds of dollars) by a sample of six families. Let x be monthly income and y be monthly rent.

Monthly Income	Monthly Rent
24	7.0
16	4.5
19	6.5
31	113.0
12	4.5
27	8.5

Suppose we have information on the monthly incomes and the monthly rents paid by a sample of families. In the regression of monthly rent paid on the monthly income for these data, the value of a will represent

a. the point where this regression line will intersect the y-axis

b. the point where this regression line will intersect the x-axis

c. an increase in the monthly rent paid by a family due to a \$100 increase in income

22. The coefficient of determination is equal to the

a. total sum of squares divided by the regression sum of squares

b. error sum of squares divided by the regression sum of squares

c. regression sum of squares divided by the total sum of squares

23. The slope, b, of the regression line

a. is a point estimator of the slope of the population regression line

b. possesses no sampling distribution

c. will have the same value for all samples taken from the population

24. A perfect positive correlation means

a. the points in a scatter diagram lie on an upward sloping line

b. the points in a scatter diagram lie on a downward sloping line

c. r is equal to 0

25. A weak positive correlation means

a. the points in a scatter diagram are very close to an upward sloping line

b. the points in a scatter diagram are very close to a downward sloping line

c. r is positive and close to zero

26. The following table gives information on the ages (in years) and the charitable contributions (in hundreds of dollars) for the last year for a random sample of 10 individuals.

Age	Charitable Contributions
52	5
45	12
31	2
65	21
27	3
56	12
72	41
62	36
47	18
67	16

With the age as an independent variable and charitable contributions as a dependent variable, compute SS_{xx}, SS_{yy}, and SS_{xy}, rounded to one decimal place.

a. $SS_{xx} = 2048.4$

$SS_{yy} = 1568.4$

$SS_{xy} = 1383.6$

b. $SS_{xx} = 274{,}576.0$

$SS_{yy} = 27{,}556.0$

$SS_{xy} = 86{,}984.0$

c. $SS_{xx} = 524.0$

$SS_{yy} = 166.0$

$SS_{xy} = 10{,}082.0$

27. The following table lists the monthly incomes (in hundreds of dollars) and the monthly rents paid (in hundreds of dollars) by a sample of six families.

Monthly Income	Monthly Rent
24	7.0
17	4.5
19	6.5
32	11.6
10	4.5
26	8.0

What is the 99% confidence interval for the mean monthly rent of a randomly selected family with a monthly income of $2900, rounded to the nearest penny?

a. $347.50 to $1544.21

b. $458.88 to $1432.83

c. $421.84 to $1469.87

28. The following table lists the ages (in years) and the prices (in thousands of dollars) by a sample of six houses.

Age	Price
27	165
14	180
3	205
37	163
7	180
18	163

What is the 90% confidence interval for the average price of all 15-year-old homes, rounded to the nearest dollar?

a. $154,231 to $203,352

b. $155,579 to $202,004

c. $161,132 to $196,452

Practice Exercises

1. Given that $SS_{xx} = 546$ and $SS_{xy} = 1196$, the value of b in the regression of y on x, rounded to two decimal places, is ________.

2. For a data set on x and y, the value of SS_{xx} is 611, SS_{xy} is -1739, the mean of the x values is 80, and the mean of the y values is 21. The value of a in the regression of y on x, rounded to two decimal places, is ________.

3. For a sample of 18 values of x and y, $SS_{yy} = 154$, $SS_{xy} = -71$, and $b = -1.20$. The standard deviation of errors for the regression of y on x, rounded to three decimal places, is ________.

4. For a sample of nine values of x and y, $SS_{yy} = 199$, $SS_{xy} = -97$, and $b = -1.04$. The coefficient of determination for the regression of y on x, rounded to three decimal places, is ________.

5. For a data set of 10 values of x and y, the value of $SS_{xx} = 103$, the standard deviation of errors for the regression of y on x is 3.871, and the value of b is 7.17. For the purpose of conducting a hypothesis test, the null hypothesis is that the slope of the population regression line of y on x is zero and the alternative hypothesis is that the slope of this population regression line is different from zero. The significance level is 5%. The 95% confidence interval for the slope, B, of the population regression line is ________.

6. The following table lists the monthly incomes (in hundreds of dollars) and the monthly rents paid (in hundreds of dollars) by a sample of six families.

Monthly Income	Monthly Rent
24	7.0
16	4.5
19	6.5
31	12.8
11	4.5
27	8.5

The 99% confidence interval for the slope, B, of the population regression line is ________.

7. Given that $SS_{xx} = 277$, $SS_{yy} = 277$, and $SS_{xy} = 181$, what is the value of the correlation coefficient, rounded to three decimal places?

8. A population data set produced the following information:

$N = 200$, $\Sigma x = 7850$, $\Sigma y = 1100$, $\Sigma xy = 64{,}050$,
$\Sigma x^2 = 429{,}400$, and $\Sigma y^2 = 115{,}945$

Find the linear correlation coefficient ρ, rounded to three decimal places.

9. Explain why the random error term ε is added to the regression model.

10. Explain the difference between A and a and between B and b for a regression model.

11. Briefly explain the assumptions of a regression model.

12. Briefly explain the difference between the population regression line and a sample regression line.

13. The following table gives the temperatures (in degrees Fahrenheit) at 6 P.M. and the attendance (rounded to hundreds) at a minor league baseball team's night games on seven randomly selected evenings in May.

Temperature	61	70	50	65	48	75	53
Attendance	10	16	12	15	8	20	18

a. Do you think temperature depends on attendance or attendance depends on temperature?

b. With temperature as an independent variable and attendance as a dependent variable, what is your hypothesis about the sign of B in the regression model?

c. Construct a scatter diagram for these data. Does the scatter diagram exhibit a linear relationship between the two variables?

d. Find the least squares regression line. Is the sign of b the same as the one you hypothesized for B in part (b)?

e. Give a brief interpretation of the values of the y-intercept and slope calculated in part (d).

f. Compute r and r^2, and explain what they mean.

g. Predict the attendance at a night game in May for a temperature of 60°F.

h. Compute the standard deviation of errors.

i. Construct a 99% confidence interval for B.

j. Testing at a 1% significance level, can you conclude that B is positive?

k. Construct a 95% confidence interval for the mean attendance at a night game in May when the temperature is 60°F.

l. Make a 95% prediction interval for the attendance at a night game in May when the temperature is 60°F.

m. Using a 1% significance level, can you conclude that the linear correlation coefficient is positive?

Mini-Projects

Mini-Project 13.1

Using the weather sections from back issues of a local newspaper or some other source, do the following for a period of 30 or more days. For each day, record the predicted maximum temperature for the next day, and then find the actual maximum temperature in the next day's newspaper. Thus, you will have the predicted and actual maximum temperatures for 30 or more days.

a. Make a scatter diagram for your data.

b. Find the regression line with actual maximum temperature as a dependent variable and predicted maximum temperature as an independent variable.

c. Make a 99% confidence interval for the population slope.

d. Using a 1% significance level, can you conclude that the slope of the regression line is different from zero?

e. If the actual maximum temperatures were exactly the same as the predicted maximum temperatures for all days, what would the value of the correlation coefficient be?

f. Find the correlation coefficient between the predicted and actual maximum temperatures for your data.

g. Using a 1% significance level, can you conclude that the linear correlation coefficient is positive?

Mini-Project 13.2

Visit a grocery store and choose 30 different types of food items that include nutrition information on the packaging. For each food, identify the amount of fat (in grams) and the sodium content (in milligrams) per serving. Make sure that you pick a wide variety of foods in order to get a wide variety of values of these two variables. For example, selecting 30 different diet sodas would not make for an interesting analysis.

a. Calculate the linear correlation coefficient between the two variables. Do you find a positive or a negative association between the amount of fat and sodium content?

b. Create a scatterplot of these data using the amount of fat as the x variable. Does your scatterplot suggest that creating a regression line to represent these data makes sense?

c. Find a regression line for your data. If it makes sense to fit a line, interpret the values of the slope and intercept. If it does not make sense, explain why these numbers could be misleading.

Mini-Project 13.3

Refer to Data Set VII on McDonald's menu items that accompanies this text (see Appendix A). Take a random sample of 10 food items and record the grams of carbohydrates, sodium, and the number of calories for each food listed in columns 14, 12, and 3, respectively.

a. Calculate the correlation coefficient between the grams of carbohydrates and the number of calories. Do the same for the sodium and the number of calories. In each case, do you find a positive association or a negative association?

b. Create a scatterplot of these data using the carbohydrate content as the x variable and number of calories as the y variable. Does your scatterplot suggest that fitting a linear regression line to these data makes sense? Repeat the process using the sodium as the x variable and number of calories as the y variable.

c. Find the equation of the estimated regression line for each of the two models mentioned in parts (a) and (b). If it makes sense to estimate regression lines for these variables, interpret the values of the slope and intercept for each regression line. If it does not make sense, explain why the numbers could be misleading.

Decide for Yourself

Note: The Decide for Yourself feature is located in the book's product page on *www.wiley.com*.

Chapter 13 Technology Instructions

Note: Complete TI-84, Minitab, and Excel manuals are available in the book's product page on *www.wiley.com*.

TI-84 Color/TI-84

The TI-84 Color Technology Instructions feature of this text is written for the TI-84 Plus C color graphing calculator running the 4.0 operating system. Some screens, menus, and functions will be slightly different in older operating systems. The TI-84 and TI-84 Plus can perform all of the same functions but will not have the "Color" option referenced in some of the menus.

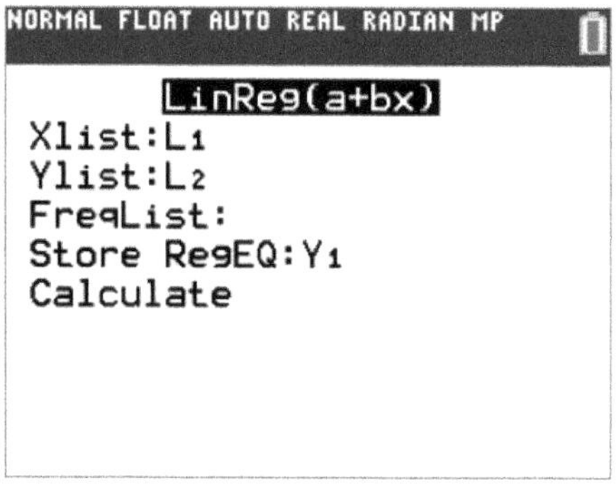

SCREEN 13.1

Finding the Regression Equation, r^2, and r for Example 13.1 of the Text

1. Enter the data from Example 13.1 into L1 (Income) and L2 (Food Expenditure).
2. Select **STAT > CALC > LinReg(a+bx)**.
3. Use the following settings in the **LinReg(a+bx)** menu (see **Screen 13.1**):
   - Enter L1 at the **Xlist** prompt.
   - Enter L2 at the **Ylist** prompt.

- Leave the **FreqList** prompt blank.
  Note: If you have another list with the frequency for each ordered pair, enter the name of that list at this prompt.
- Enter Y1 at the **Store RegEQ** prompt.
  Note: To get Y1, select **VARS > Y-VARS > Function > Y1.**

4. Highlight **Calculate** and press **ENTER.**
5. The output includes the least squares regression equation, the coefficient of determination, and the correlation coefficient. In addition, the regression equation will be stored as function Y1. (See **Screen 13.2.**)

Note: If you did not see the values of r and r^2, select **2ND > 0 > DiagnosticOn** and press **ENTER** once the function **DiagnosticOn** is pasted onto the home screen. You should see **Done** once the calculator has executed this command. This needs to be done only once.

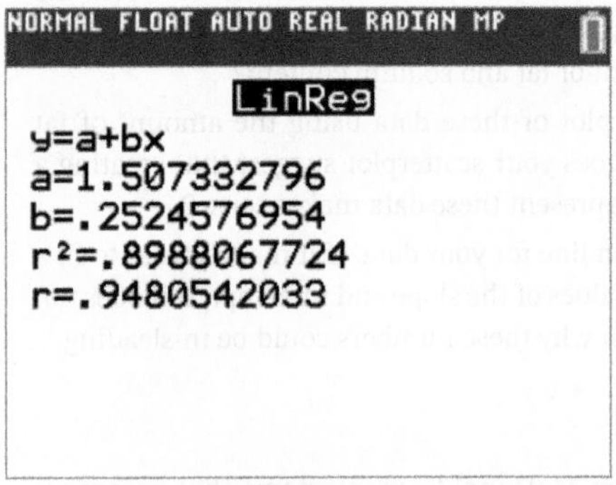

SCREEN 13.2

Constructing a Confidence Interval for Slope for Example 13.4 of the Text

1. Enter the data from Table 13.1 into L1 (Income) and L2 (Food Expenditure).
2. Select **STAT > TESTS > LinRegTInt.**
3. Use the following settings in the **LinRegTInt** menu (see **Screen 13.3**):
   - Enter L1 at the **Xlist** prompt.
   - Enter L2 at the **Ylist** prompt.
   - Type 1 at the **Freq** prompt.

   Note: If you have another list with the frequency for each ordered pair, enter the name of that list at this prompt.
   - Type .95 at the **C-Level** prompt.
   - Enter Y1 at the **RegEQ** prompt.

   Note: To get Y1, select **VARS > Y-VARS > Function > Y1.**
4. Highlight **Calculate** and press **ENTER.**
5. The output includes the confidence interval, the value of b, the degrees of freedom, the regression equation coefficients, the standard deviation of errors, the coefficient of determination, and the correlation coefficient. In addition, the regression equation will be stored as function Y1. (See **Screen 13.4.**)

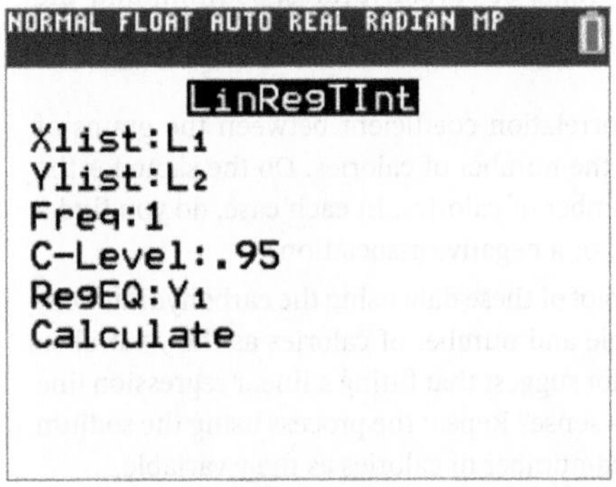

SCREEN 13.3

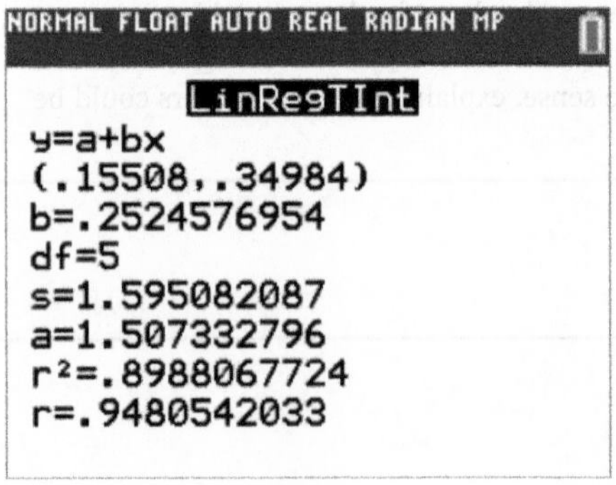

SCREEN 13.4

Testing a Hypothesis About Slope or Correlation for Example 13.5 of the Text

1. Enter the data from Table 13.1 into L1 (Income) and L2 (Food Expenditure).
2. Select **STAT > TESTS > LinRegTTest.**
3. Use the following settings in the **LinRegTTest** menu (see **Screen 13.5**):
   - Enter L1 at the **Xlist** prompt.
   - Enter L2 at the **Ylist** prompt.
   - Type 1 at the **Freq** prompt.

   Note: If you have another list with the frequency for each ordered pair, enter the name of that list at this prompt.
   - Select **>0** at the **β & ρ** prompt.
   - Enter Y1 at the **RegEQ** prompt.

   Note: To get Y1, select **VARS > Y-VARS > Function > Y1.**
4. Highlight **Calculate** and press **ENTER.**
5. The output includes the test statistic, the p-value, the degrees of freedom, the regression equation coefficients, the standard deviation of errors, the coefficient of determination, and the correlation coefficient (you must scroll down to see r and r^2). In addition, the regression equation will be stored as function Y1. (See **Screen 13.6.**)

Now compare the t-value to the critical value of t or the p-value from Screen 13.6 with α and make a decision.

Note that here we have tested both $H_0: B = 0$ vs. $H_1: B > 0$, and $H_0: \rho = 0$ vs. $H_1: \rho > 0$ for Example 13.5.

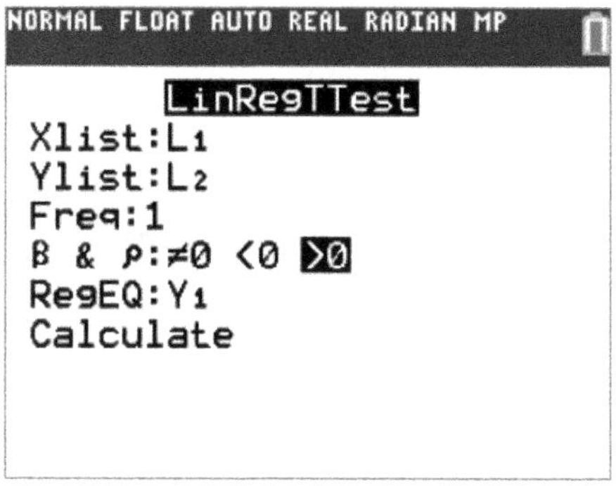

SCREEN 13.5

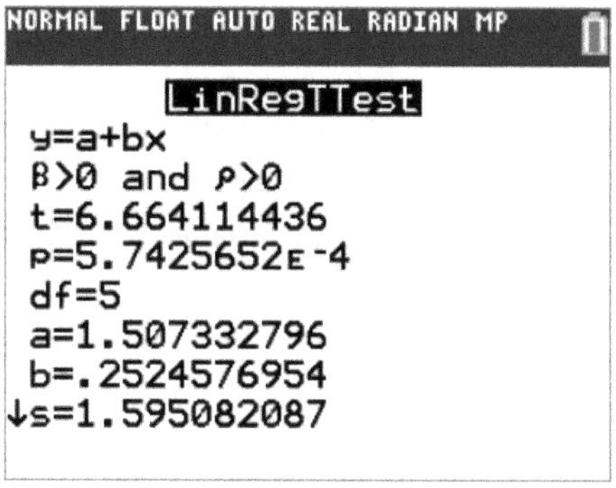

SCREEN 13.6

Minitab

The Minitab Technology Instructions in this text are written for **Minitab Professional Version 19**. Some screens, menus, and functions will be slightly different in older (or in newer) versions of Minitab. If you are using **Minitab Express**, you can download the **Minitab Express** Manual from the book's product page on *www.wiley.com*.

Finding the Regression Equation and r^2 for Example 13.1 of the Text

1. Enter the data from Example 13.1 into C1 (Income) and C2 (Food Expenditure).
2. Select **Stat > Regression > Regression > Fit Regression Model**.
3. Use the following settings in the dialog box that appears on screen:
   - Type C2 in the **Responses** box.
   - Type C1 in the **Continuous predictors** box.
   - Click on the **Options** button and enter 95 in the **Confidence level for all intervals** box. Click **OK**.
   - Click on the **Results** button and select Expanded tables from the **Display of results** drop-down menu. Click **OK**.
4. Click **OK**.
5. The output will be displayed in the Session window. The output includes the value of r^2 and the regression equation. (See **Screen 13.7**.)

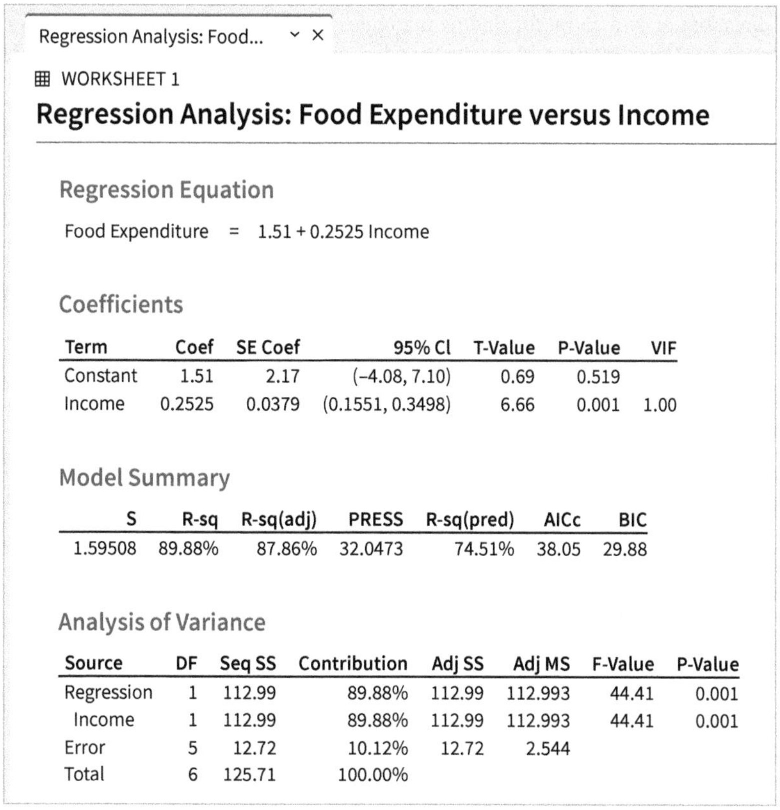
Regression Analysis: Food...

WORKSHEET 1

Regression Analysis: Food Expenditure versus Income

Regression Equation

Food Expenditure = 1.51 + 0.2525 Income

Coefficients

Term	Coef	SE Coef	95% CI	T-Value	P-Value	VIF
Constant	1.51	2.17	(−4.08, 7.10)	0.69	0.519	
Income	0.2525	0.0379	(0.1551, 0.3498)	6.66	0.001	1.00

Model Summary

S	R-sq	R-sq(adj)	PRESS	R-sq(pred)	AICc	BIC
1.59508	89.88%	87.86%	32.0473	74.51%	38.05	29.88

Analysis of Variance

Source	DF	Seq SS	Contribution	Adj SS	Adj MS	F-Value	P-Value
Regression	1	112.99	89.88%	112.99	112.993	44.41	0.001
Income	1	112.99	89.88%	112.99	112.993	44.41	0.001
Error	5	12.72	10.12%	12.72	2.544		
Total	6	125.71	100.00%				

SCREEN 13.7

Finding *r* for Example 13.6 of the Text

1. Enter the data from Table 13.1 into C1 (Income) and C2 (Food Expenditure).
2. Select **Stat > Basic Statistics > Correlation.**
3. Use the following settings in the dialog box that appears on screen:
   - Type C1 C2 in the **Variables** box. Be sure to separate C1 and C2 with a space. (See **Screen 13.8.**)
4. Click **OK.**
5. The output will be displayed in the Session window.

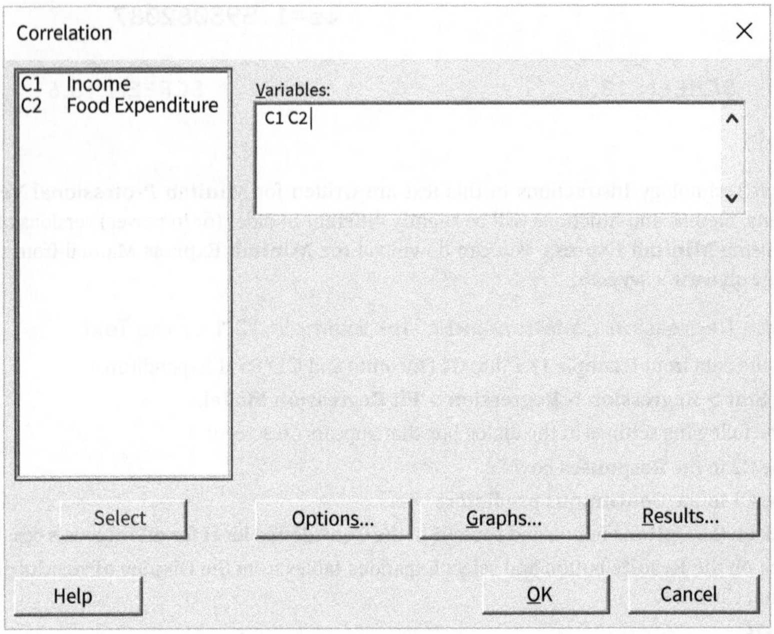

SCREEN 13.8

Constructing a Confidence Interval for Slope for Example 13.4 of the Text

Follow the directions shown above for finding the regression equation. The output includes the confidence interval in the **Coefficients** table in the **Income** row as .1551 to .3498. (See **Screen 13.7.**)

Testing a Hypothesis About Slope or Correlation for Example 13.5 of the Text

Follow the directions shown above for finding the regression equation. The output also includes the test statistic *t* (6.66) and *p*-value (.001) in the **Coefficients** table in the **Income** row. This *p*-value is for a two-sided test, so if a one-sided test is required, be sure to adjust the *p*-value accordingly. (See **Screen 13.7.**)

Now compare the *t*-value to the critical value of *t* or the *p*-value from Screen 13.7 with α and make a decision. Note that a test of $H_0: B = 0$ and $H_0: \rho = 0$ will both have the same *t* value and *p* value as long as they have the same sign in H_1.

Constructing a Prediction Interval for Slope for Example 13.10 of the Text

1. Enter the data from Table 13.1 into C1 (Income) and C2 (Food Expenditure).
2. To create a prediction interval, you must first find the regression equation (fit a regression model) as described above. If you do not fit the regression model, you will not get the correct output.
3. Select **Stat > Regression > Regression > Predict.**
4. Use the following settings in the dialog box that appears on screen (see **Screen 13.9**):
   - Select **Food Expenditure (or C2)** from the **Response** drop-down menu.
   - Select **Enter individual values** from second drop-down menu.
   - Type 55 in the list below **Income (or C1)**.
   - Click on the **Options** button and enter 99 in the **Confidence level** box. Select **Two-sided** from the **Type of Interval** drop-down menu. Click **OK.**
5. Click **OK.**
6. The output, including the prediction interval, will be displayed in the Session window.

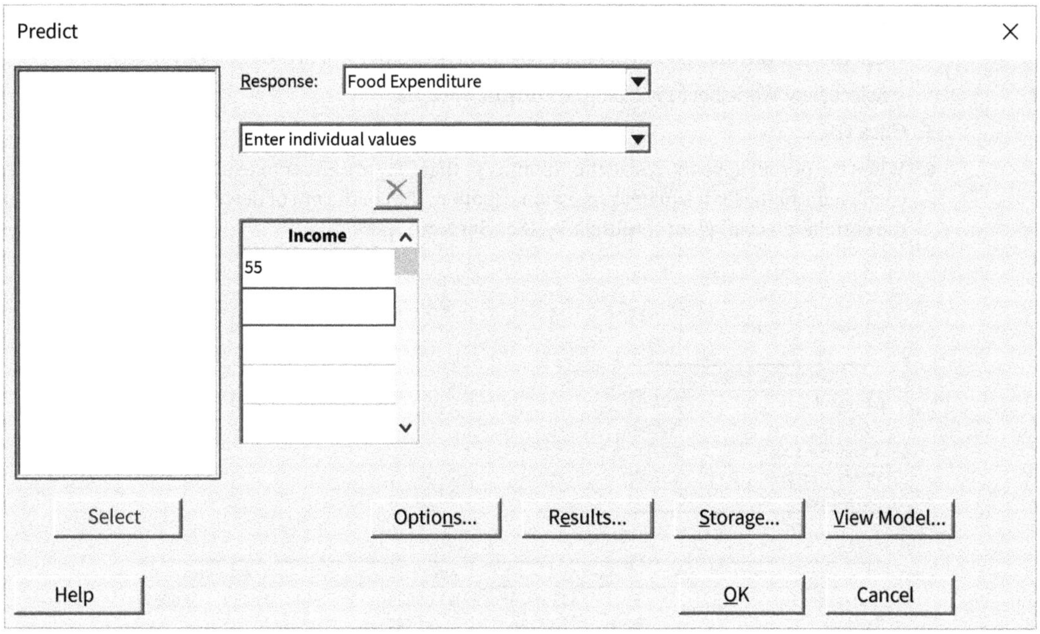

SCREEN 13.9

Excel

The Excel Technology Instructions feature of this text is written for Excel 2019. Some screens, menus, and functions will be slightly different in older versions of Excel. To enable some of the advanced Excel functions, you must enable the Data Analysis Add-In for Excel.

Finding the Regression Equation, r^2, and r for Example 13.1 of the Text

1. In a new worksheet, enter **Income** in cell A1 and **Food Expenditure** in cell B1. Then enter the data from Example 13.1 into cells A2-A8 (Income) and cells B2-B8 (Food Expenditure). (See **Screen 13.10.**)

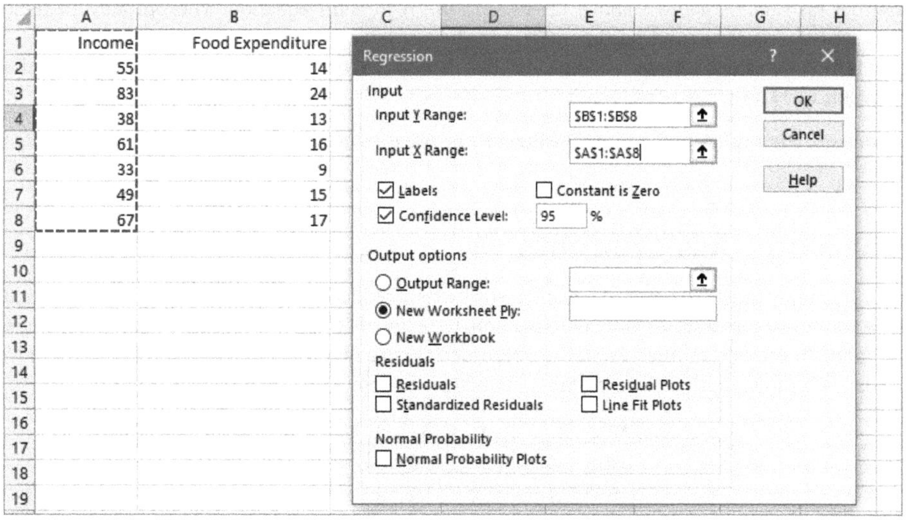

SCREEN 13.10

2. Click **DATA** and then click **Data Analysis Tools** from the **Analysis** group.
3. Select **Regression** from the dialog box that appears on screen and then click **OK**.
4. Use the following settings in the dialog box that appears on screen (see **Screen 13.10**):
   - Type B1:B8 in the **Input Y Range** box.
   - Type A1:A8 in the **Input X Range** box.

- Check the **Labels** box.
- Check the **Confidence Level** check box and type 95 in the **Confidence Level** box.
- Select New Worksheet Ply from the **Output options.**

5. Click **OK.**
6. When the output appears, resize the columns so that they are easier to read. The output includes the coefficients of the least squares regression equation, the coefficient of determination (R square), and the correlation coefficient (Multiple R). (See **Screen 13.11.**)

	A	B	C	D	E	F	G
1	SUMMARY OUTPUT						
2							
3	*Regression Statistics*						
4	Mutiple R	0.9481					
5	R Square	0.8988					
6	Adjusted R Square	0.8786					
7	Standard Error	1.5951					
8	Observations	7					
9							
10	ANOVA						
11		*df*	*SS*	*MS*	*F*	*Significance F*	
12	Regression	1	112.9929	112.9929	44.4104	0.0011	
13	Residual	5	12.7214	2.5443			
14	Total	6	125.7143				
15							
16		*Coefficients*	*Standard Error*	*t Stat*	*P-value*	*Lower 95%*	*Upper 95%*
17	Intercept	1.5073	2.1742	0.6933	0.5190	-4.0817	7.0964
18	Income	0.2525	0.0379	6.6641	0.0011	0.1551	0.3498
19							

SCREEN 13.11

Constructing a Confidence Interval for Slope for Example 13.4 of the Text

Follow the commands shown above for finding the regression equation. The output also includes the lower and upper bounds of the confidence interval in the last table of the output in the **Income** row. (See **Screen 13.11.**)

Testing a Hypothesis About Slope or Correlation for Example 13.5 of the Text

Follow the instructions shown above for finding the regression equation. The output also includes the test statistic t (6.6641) and p-value (.0011) in the last table of the output in the **Income** row. This p-value is for a two-sided test, so if a one-sided test is required, be sure to adjust the p-value accordingly. (See **Screen 13.11.**)

Now compare the t-value to the critical value of t or the p-value from Screen 13.11 with α and make a decision.

Technology Assignments

TA13.1 In a rainy coastal town in the Pacific Northwest, the local TV weatherman is often criticized for making inaccurate forecasts for daily precipitation. On each of 30 randomly selected days last winter, his precipitation forecast (x) for the next day was recorded along with the actual precipitation (y) for that day. These data are shown in the following table (in inches of rain):

x	y	x	y	x	y
1.0	.6	0	0	.4	.2
0	.1	0	.1	.2	.5
.2	0	.1	.2	.1	.1
0	0	.2	.2	0	.2
.5	.3	.1	0	.1	0
1.0	1.4	2.0	2.1	.2	.1
.5	.3	.4	.2	1.4	1.2
.1	.1	.2	.1	.5	1.0
0	.1	0	0	0	.5
2.0	.3	.3	.2	0	0

a. Construct a scatter diagram for these data.

b. Find the correlation coefficient between the two variables.

c. Find the regression line with actual precipitation as a dependent variable and predicted precipitation as an independent variable.

d. Make a 95% confidence interval for B.

e. Test at a 1% significance level whether B is positive.

f. Using a 1% significance level, can you conclude that the linear correlation coefficient is positive?

TA13.2 Refer to Data Set III on NFL players (see Appendix A). Select a random sample of 30 players from that population. Do the following for the data on heights and weights of these 30 players.

a. Construct a scatter diagram for these data. Does this scatter diagram show a linear relationship between variables?

b. Find the correlation between these two variables.

c. Find the regression line with weight as a dependent variable and height as an independent variable.

d. Make a 98% confidence interval for B.

e. Test at a 2.5% significance level whether B is positive.

f. Construct a 95% prediction interval for the weight of a randomly selected NFL player with a height of 75 inches.

TA13.3 The following data give information on the ages (in years) and the number of breakdowns during the last month for a sample of seven machines at a large company.

Age (in years)	12	7	2	8	13	9	4
Number of Breakdowns	10	5	1	4	12	7	2

a. Construct a scatter diagram for these data.

b. Find the least squares regression line with age as an independent variable and the number of breakdowns as a dependent variable.

c. Compute the correlation coefficient.

d. Construct a 99% confidence interval for B.

e. Test at a 2.5% significance level whether B is positive.

TA13.4 Refer to Data Set IV on the Manchester Road Race (see Appendix A). Take a random sample of 45 participants. Do the following for the variables on age and net time (in minutes) for this sample.

a. Construct a scatter diagram for these data, using age as the independent variable. Discuss whether it is appropriate to fit a linear regression model to these data.

b. Find the correlation coefficient for these two variables.

c. Find the equation of the regression line with age as the independent variable and net time as the dependent variable.

d. Make a 98% confidence interval for B. Explain what this interval means with regard to a person's net time for each additional year of age.

e. Test at a 1% significance level whether B is positive.

f. Test at a 1% significance level whether B is greater than .10.

CHAPTER

14

Multiple Regression

Blend Images - Hill Street Studios/Brand X Pictures/Getty Images

In Chapter 13, we discussed simple linear regression and linear correlation. A simple regression model includes one independent and one dependent variable, and it presents a very simplified scenario of real-world situations. In the real world, a dependent variable is usually influenced by a number of independent variables. For example, the sales of a company's product may be determined by the price of that product, the quality of the product, and advertising expenditure incurred by the company to promote that product. Therefore, it makes more sense to use a regression model that includes more than one independent variable. Such a model is called a multiple regression model. In this chapter, we will discuss multiple regression models.

14.1 Multiple Regression Analysis

LEARNING OBJECTIVES

After completing this section, you should be able to:

- Explain the difference between a simple linear regression model and a multiple linear regression model.
- Explain a first-order multiple regression model.
- Explain the meaning of the partial regression coefficients in a multiple linear regression model.
- Determine the degrees of freedom used to make inferences about a single parameter in a multiple linear regression model.

The simple linear regression model discussed in Chapter 13 was written as

$$y = A + Bx + \epsilon$$

This model includes one independent variable, which is denoted by x, and one dependent variable, which is denoted by y. As we know from Chapter 13, the term represented by ϵ in the above model is called the random error.

Usually a dependent variable is affected by more than one independent variable. When we include two or more independent variables in a regression model, the model is called a **multiple regression model**. Remember, whether we use a simple or a multiple regression model, there is always one and only one dependent variable.

A multiple regression model with y as a dependent variable and $x_1, x_2, x_3, \ldots, x_k$ as independent variables is written as

$$y = A + B_1x_1 + B_2x_2 + B_3x_3 + \cdots + B_kx_k + \epsilon \quad \textbf{(1)}$$

where A represents the constant term, $B_1, B_2, B_3, \ldots, B_k$ are the regression coefficients of independent variables $x_1, x_2, x_3, \ldots, x_k$, respectively, and ϵ represents the random error term. This model contains k independent variables $x_1, x_2, x_3, \ldots,$ and x_k. From model (1), it would seem that multiple regression models can only be used when the relationship between the dependent variable and each independent variable is linear. Furthermore, it also appears as if there can be no interaction between two or more of the independent variables. This is far from the truth. In the real world, a multiple regression model can be much more complex. Discussion of such models is outside the scope of this book. When each term contains a single independent variable raised to the first power as in model (1), we call it a **first-order multiple regression model**. This is the only type of multiple regression model we will discuss in this chapter.

In regression model (1), A represents the constant term, which gives the value of y when all independent variables assume zero values. The coefficients $B_1, B_2, B_3, \ldots,$ and B_k are called the **partial regression coefficients**. For example, B_1 is a partial regression coefficient of x_1. It gives the change in y due to a one-unit change in x_1 when all other independent variables included in the model are held constant. In other words, if we change x_1 by one unit but keep $x_2, x_3, \ldots,$ and x_k unchanged, then the resulting change in y is measured by B_1. Similarly, the value of B_2 gives the change in y due to a one-unit change in x_2 when all other independent variables are held constant. In model (1) above, $A, B_1, B_2, B_3, \ldots,$ and B_k are called the **true regression coefficients** or **population parameters**.

A positive value for a particular B_i in model (1) will indicate a positive relationship between y and the corresponding x_i variable. A negative value for a particular B_i in that model will indicate a negative relationship between y and the corresponding x_i variable.

Remember that in a first-order regression model such as model (1), the relationship between each x_i and y is a straight-line relationship. In model (1), $A + B_1x_1 + B_2x_2 + B_3x_3 + \cdots + B_kx_k$ is called the *deterministic portion* and ϵ is the *stochastic portion* of the model.

When we use the t distribution to make inferences about a single parameter of a multiple regression model, the **degrees of freedom** are calculated as

$$df = n - k - 1$$

where n represents the sample size and k is the number of independent variables in the model.

Multiple Regression Model

A regression model that includes two or more independent variables is called a *multiple regression model*. It is written as

$$y = A + B_1x_1 + B_2x_2 + B_3x_3 + \cdots + B_kx_k + \epsilon$$

where y is the dependent variable, $x_1, x_2, x_3, \ldots, x_k$ are the k independent variables, and ϵ is the random error term. When each of the x_i variables represents a single variable raised to the first power as in the above model, this model is referred to as a **first-order multiple regression model**. For such a model with a sample size of n and k independent variables, the degrees of freedom are

$$df = n - k - 1$$

When a multiple regression model includes only two independent variables (with $k = 2$), model (1) reduces to

$$y = A + B_1x_1 + B_2x_2 + \epsilon$$

A multiple regression model with three independent variables (with $k = 3$) is written as

$$y = A + B_1x_1 + B_2x_2 + B_3x_3 + \epsilon$$

If model (1) is estimated using sample data, which is usually the case, the estimated regression equation is written as

$$\hat{y} = a + b_1x_1 + b_2x_2 + b_3x_3 + \cdots + b_kx_k \tag{2}$$

In equation (2), a, b_1, b_2, b_3, . . ., and b_k are the sample statistics, which are the point estimators of the population parameters A, B_1, B_2, B_3, . . ., and B_k, respectively.

In model (1), y denotes the actual values of the dependent variable for members of the sample. In the estimated model (2), $\hat{y}$ denotes the predicted or estimated values of the dependent variable. The difference between any pair of y and $\hat{y}$ values gives the error of prediction. For a multiple regression model,

$$\text{SSE} = \Sigma(y - \hat{y})^2$$

where SSE stands for the error sum of squares.

As in Chapter 13, the estimated regression equation (2) is obtained by minimizing the sum of squared errors, that is,

$$\text{Minimize} \quad \Sigma(y - \hat{y})^2$$

The estimated equation (2) obtained by minimizing the sum of squared errors is called the **least squares regression equation.**

Usually the calculations in a multiple regression analysis are made by using statistical software packages for computers, such as MINITAB, instead of using the formulas manually. Even for a multiple regression equation with two independent variables, the formulas are complex and manual calculations are time consuming. In this chapter, we will perform the multiple regression analysis using MINITAB. The solutions obtained by using other statistical software packages such as JMP, SAS, S-Plus, or SPSS can be interpreted in the same way. The TI-84 calculator does not have a built-in procedure for the multiple regression model.

14.2 Assumptions of the Multiple Regression Model

LEARNING OBJECTIVES

After completing this section, you should be able to:

- State the four assumptions of a multiple regression model.
- Explain the meaning of multicollinearity.

Like a simple linear regression model, a multiple (linear) regression model is based on certain assumptions. The following are the major assumptions for the multiple regression model (1). Note that these assumptions are for the population regression model.

Assumption 1: The mean of the probability distribution of ϵ is zero, that is,

$$E(\epsilon) = 0$$

If we calculate errors for all measurements for a given set of values of independent variables for a population data set, the mean of these errors will be zero. In other words, while individual predictions will have some amount of errors, on average our predictions will be correct. Under this assumption, the mean value of y is given by the deterministic part of regression model (1). Thus,

$$E(y) = A + B_1x_1 + B_2x_2 + B_3x_3 + \cdots + B_kx_k$$

where $E(y)$ is the expected or mean value of y for the population. This mean value of y is also denoted by $\mu_{y|x_1, x_2, \ldots, x_k}$.

Assumption 2: The errors associated with different sets of values of independent variables are independent. Furthermore, these errors are approximately normally distributed and have a constant standard deviation, which is denoted by σ_ϵ.

Assumption 3: The independent variables are not linearly related. However, they can have a nonlinear relationship. When independent variables are strongly linearly related, this occurrence is referred to as **multicollinearity**. This assumption is about the nonexistence of the multicollinearity problem. For example, consider the following multiple regression model:

$$y = A + B_1x_1 + B_2x_2 + B_3x_3 + \epsilon$$

All of the following linear relationships (and other such linear relationships) between x_1, x_2, and x_3 should be invalid for this model.

$$x_1 = x_2 + 4x_3$$
$$x_2 = 5x_1 - 2x_3$$
$$x_1 = 3.5x_2$$

If any linear relationship exists among x_1, x_2, and x_3, we can substitute one variable for another, which will reduce the number of independent variables to two. However, nonlinear relationships, such as $x_1 = 4x_2^2$ and $x_2 = 2x_1 + 6x_3^2$ among x_1, x_2, and x_3, are permissible.

In practice, multicollinearity is a major issue. Examining the correlation coefficient for each pair of independent variables is a good way to determine if multicollinearity exists.

Assumption 4: There is no linear association between the random error term ϵ and each independent variable x_i.

14.3 Standard Deviation of Errors

LEARNING OBJECTIVES

After completing this section, you should be able to:

- Explain the difference between σ_ϵ and s_e in a multiple regression model.
- Explain how σ_ϵ is estimated by s_e.

The **standard deviation of errors** (also called the standard error of the estimate) for the multiple regression model (1) is denoted by σ_ϵ, and it is a measure of variation among errors. However, when sample data are used to estimate multiple regression model (1), the standard deviation of errors is denoted by s_e. The formula to calculate s_e is as follows:

$$s_e = \sqrt{\frac{\text{SSE}}{n - k - 1}} \quad \text{where} \quad \text{SSE} = \Sigma(y - \hat{y})^2$$

Note that here **SSE** is the error sum of squares. We will not use this formula to calculate s_e manually. Rather, we will obtain it from the computer solution. Note that many software packages label s_e as Root MSE, where MSE stands for mean square error.

14.4 Coefficient of Multiple Determination

LEARNING OBJECTIVES

After completing this section, you should be able to:

- Explain the difference between the coefficient of multiple determination and the adjusted coefficient of multiple determination.
- Explain the properties of the adjusted coefficient of multiple determination.

In Chapter 13, we denoted the coefficient of determination for a simple linear regression model by r^2 and defined it as the proportion of the total sum of squares **SST** that is explained by the regression model. The coefficient of determination for the multiple regression model, usually called the **coefficient of multiple determination**, is denoted by $\boldsymbol{R^2}$ and is defined as the proportion of the total sum of squares SST that is explained by the multiple regression model. It tells us how good the multiple regression model is and how well the independent variables included in the model explain the dependent variable.

Like r^2, the value of the coefficient of multiple determination R^2 always lies in the range 0 to 1, that is,

$$0 \leq R^2 \leq 1$$

Just as in the case of the simple linear regression model, SST is the total sum of squares, **SSR** is the regression sum of squares, and SSE is the error sum of squares. SST is always equal to the sum of SSE and SSR. They are calculated as follows:

$$\begin{aligned} \text{SSE} &= \Sigma e^2 = \Sigma(y - \hat{y})^2 \\ \text{SST} &= SS_{yy} = \Sigma(y - \bar{y})^2 \\ \text{SSR} &= \Sigma(\hat{y} - \bar{y})^2 \end{aligned}$$

SSR is the portion of SST that is explained by the use of the regression model, and SSE is the portion of SST that is not explained by the use of the regression model. The coefficient of multiple determination is given by the ratio of SSR and SST as follows:

$$R^2 = \frac{\text{SSR}}{\text{SST}}$$

The coefficient of multiple determination R^2 has one major shortcoming. The value of R^2 increases as we add more and more explanatory variables to the regression model (even if they do not belong in the model). Just because we can increase the value of R^2 does not imply that the regression equation with a higher value of R^2 does a better job of predicting the dependent variable. Such a value of R^2 will be misleading, and will not represent the true explanatory power of the regression model. To eliminate this shortcoming of R^2, it is preferable to use the **adjusted coefficient of multiple determination**, which is denoted by $\boldsymbol{\bar{R}^2}$. Note that $\bar{R}^2$ is the coefficient of multiple determination adjusted for degrees of freedom. The value of $\bar{R}^2$ may increase, decrease, or stay the same as we add more explanatory variables to our regression model. If a new variable added to the regression model contributes significantly to explain the variation in y, then $\bar{R}^2$ increases; otherwise it decreases. The value of $\bar{R}^2$ is calculated as follows:

$$\bar{R}^2 = 1 - (1 - R^2)\left(\frac{n-1}{n-k-1}\right) \quad \text{or} \quad 1 - \frac{\text{SSE}/(n-k-1)}{\text{SST}/(n-1)}$$

Thus, if we know R^2, we can find the value of $\bar{R}^2$. Almost all statistical software packages give the values of both R^2 and $\bar{R}^2$ for a regression model.

Another property of $\bar{R}^2$ to remember is that whereas R^2 can never be negative, $\bar{R}^2$ can be negative.

While a general rule of thumb is that a higher value of $\bar{R}^2$ implies that a specific set of independent variables does a better job of predicting a specific dependent variable, it is important to recognize that some dependent variables have a great deal more variability than others. Therefore, $\bar{R}^2 = .30$ could imply that a specific model is not a very strong model, but it could be the best possible model in a certain scenario. Many *good* financial models have values of $\bar{R}^2$ below .50.

14.5 Computer Solution of Multiple Regression

LEARNING OBJECTIVES

After completing this section, you should be able to:

- Use technology to find the summary statistics for a multiple linear regression model.
- Write an estimated multiple linear regression model.
- Interpret the meaning of the estimated regression coefficients.
- Identify the values of the standard deviation of errors, the coefficient of multiple determination, and the adjusted coefficient of multiple determination from a computer output.
- Interpret the meaning of the adjusted coefficient of multiple determination.
- Use a multiple linear regression model to predict a particular value of *y* for given values of the independent variables.
- Use a multiple linear regression model to estimate the mean value of *y* for given values of the independent variables.
- Construct a confidence interval for an individual coefficient.
- Perform a test of hypothesis about an individual coefficient.

In this section, we present an example of a multiple regression model, solve it using MINITAB, interpret the solution, and make inferences about the population parameters of the regression model.

EXAMPLE 14.1 Monthly Auto Insurance Premium as a Function of Driving Experience and Driving Violations

Using MINITAB to Find a Multiple Regression Equation

A researcher wanted to find the effect of driving experience and the number of driving violations on auto insurance premiums. A random sample of 12 drivers insured with the same company and having similar auto insurance policies was selected from a large city. **Table 14.1** lists the monthly auto insurance premiums (in dollars) paid by these drivers, their driving experiences (in years), and the numbers of driving violations committed by them during the past three years.

TABLE 14.1

Monthly Premium (dollars)	Driving Experience (years)	Number of Driving Violations (past three years)
148	5	2
76	14	0
100	6	1

TABLE 14.1 *(Continued)*

Monthly Premium (dollars)	Driving Experience (years)	Number of Driving Violations (past three years)
126	10	3
194	4	6
110	8	2
114	11	3
86	16	1
198	3	5
92	9	1
70	19	0
120	13	3

Using MINITAB, find the regression equation of monthly premiums paid by drivers on the driving experiences and the number of driving violations.

Solution Let

y = the monthly auto insurance premium (in dollars) paid by a driver

x_1 = the driving experience (in years) of a driver

x_2 = the number of driving violations committed by a driver during the past three years

We are to estimate the regression model

$$y = A + B_1x_1 + B_2x_2 + \epsilon \quad (3)$$

The first step is to enter the data of Table 14.1 into MINITAB spreadsheet as shown in **Screen 14.1**. Here, we have entered the given data in columns C1, C2, and C3 and named them Monthly Premium, Driving Experience, and Driving Violations, respectively.

↓	C1	C2	C3
	Monthly Premium	Driving Experience	Driving Violations
1	148	5	2
2	76	14	0
3	100	6	1
4	126	10	3
5	194	4	6
6	110	8	2
7	114	11	3
8	86	16	1
9	198	3	5
10	92	9	1
11	70	19	0
12	120	13	3

SCREEN 14.1

To obtain the estimated regression equation, select **Stat>Regression>Regression>Fit Regression Model**. In the dialog box you obtain, enter *Monthly Premium* in the **Response** box, and *Driving Experience* and *Driving Violations* in the **Predictors** box as shown in **Screen 14.2**. Note that you can enter the column names C1, C2, and C3 instead of variable names in these boxes. Click **OK** to obtain the output, which is shown in **Screen 14.3**.

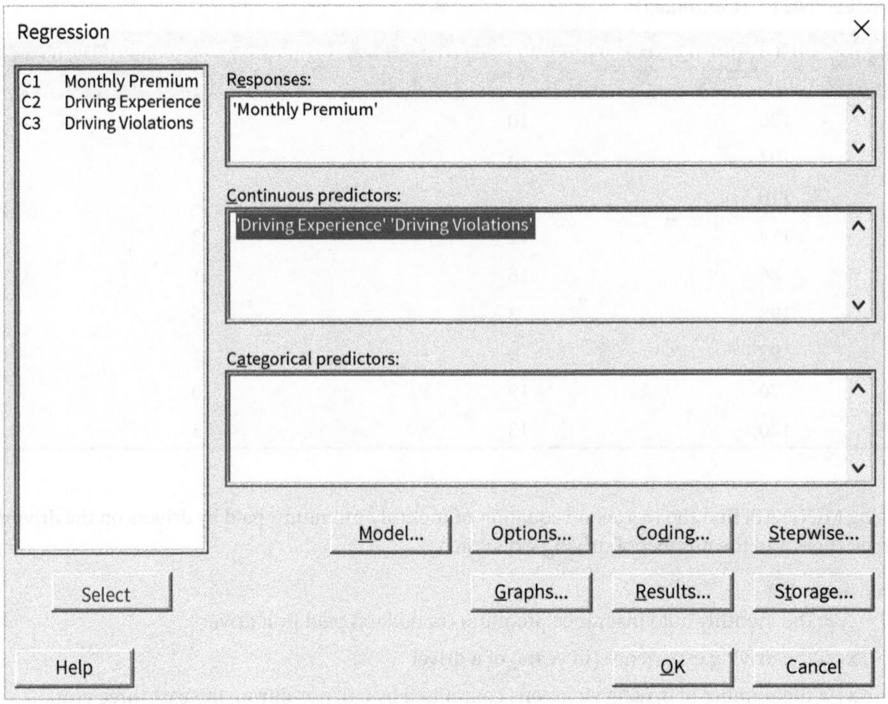

SCREEN 14.2

From the output given in Screen 14.3, the estimated regression equation is

$$\hat{y} = 110.3 - 2.747x_1 + 16.11x_2$$

Regression Analysis: Month... ×

WORKSHEET 1

Regression Analysis: Monthly Premium versus Driving Experience, Driving Violations

Regression Equation (I)

Monthly Premium = 110.3 - 2.747 Driving Experience + 16.11 Driving Violations

Coefficients (II)

Term	Coef	SE Coef	T-Value	P-Value	VIF
Constant	110.3	14.6	7.54	0.000	
Driving Experience	−2.747	0.977	−2.81	0.020	1.77
Driving Violations	16.11	2.61	6.16	0.000	1.77

Model Summary (III)

S	R-sq	R-sq(adj)	R-sq(pred)
12.1459	93.12%	91.59%	87.18%

Analysis of Variance (IV)

Source	DF	Adj SS	Adj MS	F-Value	P-Value
Regression	2	17961	8980.6	60.88	0.000
Driving Experience	1	1166	1166.5	7.91	0.020
Driving Violations	1	5603	5603.5	37.98	0.000
Error	9	1328	147.5		
Total	11	19289			

SCREEN 14.3

14.5.1 Estimated Multiple Regression Model

Example 14.2 describes, among other things, how the coefficients of the multiple regression model are interpreted.

EXAMPLE 14.2 Monthly Auto Insurance Premium

Interpreting a MINITAB Output of Multiple Regression

Refer to Example 14.1 and the MINITAB solution given in Screen 14.3.

a. Explain the meaning of the estimated regression coefficients.

b. What are the values of the standard deviation of errors, the coefficient of multiple determination, and the adjusted coefficient of multiple determination?

c. What is the predicted auto insurance premium paid per month by a driver with seven years of driving experience and three driving violations committed in the past three years?

d. What is the point estimate of the expected (or mean) auto insurance premium paid per month by all drivers with 12 years of driving experience and four driving violations committed in the past three years?

Solution

a. The section of the MINITAB solution that is marked **I** in Screen 14.3 provides the estimated regression equation as

$$\hat{y} = 110.3 - 2.747x_1 + 16.11x_2 \qquad \textbf{(4)}$$

From this equation,

$$a = 110.3, \quad b_1 = -2.747, \quad \text{and} \quad b_2 = 16.11$$

We can also read the values of these coefficients from the column labeled **Coef** in the portion of the output marked **II** in the MINITAB solution of Screen 14.3. From this column we also obtain

$$a = 110.3, \quad b_1 = -2.747, \quad \text{and} \quad b_2 = 16.11$$

Using the information in either section, we can write the estimated regression equation as

$$\hat{y} = 110.3 - 2.747x_1 + 16.11x_2 \qquad \textbf{(5)}$$

The value of $a = 110.3$ in the estimated regression equation (5) gives the value of $\hat{y}$ for $x_1 = 0$ and $x_2 = 0$. Thus, a driver with no driving experience and no driving violations committed in the past three years is expected to pay an auto insurance premium of \$110.30 per month. Again, this is the technical interpretation of a. In reality, that may not be true because none of the drivers in our sample has both zero experience *and* zero driving violations. As all of us know, some of the highest premiums are paid by teenagers just after obtaining their driver's license.

The value of $b_1 = -2.747$ in the estimated regression model gives the change in $\hat{y}$ for a one-unit change in x_1 when x_2 is held constant. Thus, we can state that a driver with one extra year of experience but the same number of driving violations is expected to pay \$2.747 (or \$2.75) less per month for the auto insurance premium. Note that because b_1 is negative, an increase in driving experience decreases the expected premium paid. In other words, y and x_1 have a negative relationship.

The value of $b_2 = 16.11$ in the estimated regression model gives the change in $\hat{y}$ for a one-unit change in x_2 when x_1 is held constant. Thus, a driver with one extra driving violation during the past three years but with the same years of driving experience is expected to pay \$16.11 more per month for the auto insurance premium.

b. The values of the standard deviation of errors, the coefficient of multiple determination, and the adjusted coefficient of multiple determination are given in part **III** of the MINITAB solution of Screen 14.3. From this part of the solution,

$$s_e = 12.1459, \quad R^2 = 93.12\%, \quad \text{and} \quad \bar{R}^2 = 91.59\%$$

Thus, the standard deviation of errors is 12.1459. The value of $R^2 = 93.12\%$ tells us that the multiple linear regression model using the years of driving experiences and the numbers of driving violations as independent variables explains 93.12% of the variation in the auto insurance premiums. The value of $\bar{R}^2 = 91.59\%$ is the value of the coefficient of multiple determination adjusted for degrees of freedom. This value states that, when adjusted for degrees of freedom, 91.59% of the variation in auto insurance premiums can be explained by the multiple linear regression model that uses the years of driving experience and the number of driving violations to predict auto insurance premiums.

c. To find the predicted auto insurance premium paid per month by a driver with seven years of driving experience and three driving violations during the past three years, we substitute $x_1 = 7$ and $x_2 = 3$ in the estimated regression model (5). Thus,

$$\hat{y} = 110.3 - 2.747x_1 + 16.11x_2$$
$$= 110.3 - 2.747\,(7) + 16.11\,(3) = \mathbf{\$139.40}$$

Note that this value of $\hat{y}$ is a point estimate of the predicted value of y, which is denoted by y_p. The concept of the predicted value of y is the same as that for a simple linear regression model discussed in Section 13.6.2 of Chapter 13.

d. To obtain the point estimate of the expected (mean) auto insurance premium paid per month by all drivers with 12 years of driving experience and four driving violations during the past three years, we substitute $x_1 = 12$ and $x_2 = 4$ in the estimated regression equation (5). Thus,

$$\hat{y} = 110.3 - 2.747x_1 + 16.11x_2$$
$$= 110.3 - 2.747\,(12) + 16.11\,(4) = \mathbf{\$141.78}$$

This value of $\hat{y}$ is a point estimate of the mean value of y, which is denoted by $E(y)$ or $\mu_{y|x_1 x_2}$. The concept of the mean value of y is the same as that for a simple linear regression model discussed in Section 13.6.1 of Chapter 13.

14.5.2 Confidence Interval for an Individual Coefficient

The values of $a, b_1, b_2, b_3, \ldots,$ and b_k obtained by estimating model (1) using sample data give the point estimates of $A, B_1, B_2, B_3, \ldots,$ and B_k, respectively, which are the population parameters. Using the values of sample statistics $a, b_1, b_2, b_3, \ldots,$ and b_k, we can make confidence intervals for the corresponding population parameters $A, B_1, B_2, B_3, \ldots,$ and B_k, respectively.

Because of the assumption that the errors are approximately normally distributed, the sampling distribution of each b_i is approximately normal with its mean equal to B_i and standard deviation equal to σ_{b_i}. For example, the sampling distribution of b_1 is approximately normal with its mean equal to B_1 and standard deviation equal to σ_{b_1}. However, usually σ_ε is not known and, hence, we cannot find σ_{b_i}. Consequently, we use s_{b_i} as an estimator of σ_{b_i} and use the t distribution to determine a confidence interval for B_i.

The formula to obtain a confidence interval for a population parameter B_i is given below. This is the same formula we used to make a confidence interval for B in Section 13.3.2 of Chapter 13. The only difference is that to make a confidence interval for a particular B_i for a multiple regression model, the degrees of freedom are $n - k - 1$.

Confidence Interval for B_i

The $(1 - \alpha) \times 100\%$ confidence interval for B_i is given by

$$b_i \pm t s_{b_i}$$

The value of t that is used in this formula is obtained from the t distribution table for $\alpha/2$ area in the right tail of the t distribution curve and $(n - k - 1)$ degrees of freedom. The values of b_i and s_{b_i} are obtained from the computer solution.

Example 14.3 describes the procedure to make a confidence interval for an individual regression coefficient B_i.

EXAMPLE 14.3 | Monthly Auto Insurance Premium

Making a Confidence Interval for an Individual Coefficient of a Multiple Regression Model

Determine a 95% confidence interval for B_1 (the coefficient of driving experience) for the multiple regression of auto insurance premium on driving experience and the number of driving violations. Use the MINITAB solution of Screen 14.3.

Solution To make a confidence interval for B_1, we use the portion marked **II** in the MINITAB solution of Screen 14.3. From that portion of the MINITAB solution,

$$b_1 = -2.747 \quad \text{and} \quad s_{b_1} = .977$$

Note that the value of the standard deviation of b_1, $s_{b_1} = .977$, is given in the column labeled **SE Coef** in part **II** of the MINITAB solution. The standard deviation of a coefficient such as b_1 is also called its standard error (SE).

The confidence level is 95%. The area in each tail of the t distribution curve is obtained as follows:

$$\text{Area in each tail of the } t \text{ distribution} = (1 - .95)/2 = .025$$

The sample size is 12, which gives $n = 12$. Because there are two independent variables, $k = 2$. Therefore,

$$\text{Degrees of freedom} = n - k - 1 = 12 - 2 - 1 = 9$$

From the t distribution table (Table V of Appendix B), the value of t for .025 area in the right tail of the t distribution curve and 9 degrees of freedom is 2.262. Then, the 95% confidence interval for B_1 is

$$\begin{aligned} b_1 \pm t s_{b_1} &= -2.747 \pm 2.262(.977) \\ &= -2.747 \pm 2.210 = \mathbf{-4.957 \text{ to } -.537} \end{aligned}$$

Thus, the 95% confidence interval for b_1 is −4.96 to −.54. That is, we can state with 95% confidence that for one extra year of driving experience, the monthly auto insurance premium changes on average by an amount between −\$4.96 and −\$.54 for all drivers. Note that since both limits of the confidence interval have negative signs, we can also state that for each extra year of driving experience, the expected monthly auto insurance premium decreases by an amount between \$.54 and \$4.96.

By applying the procedure used in Example 14.3, we can make a confidence interval for any of the coefficients (including the constant term) of a multiple regression model, such as A and B_2 in model (3). For example, the 95% confidence intervals for A and B_2, respectively, are

$$a \pm t s_a = 110.3 \pm 2.262(14.6) = 77.27 \text{ to } 143.33$$

$$b_2 \pm t s_{b_2} = 16.11 \pm 2.262(2.61) = 10.21 \text{ to } 22.01$$

14.5.3 Testing a Hypothesis about an Individual Coefficient

We can perform a test of hypothesis about any of the B_i coefficients of the regression model (1) using the same procedure that we used to make a test of hypothesis about B for a simple regression model in Section 13.3.3 of Chapter 13. The only difference is that the degrees of freedom are equal to $n - k - 1$ for a multiple regression model.

Again, because of the assumption that the errors are approximately normally distributed, the sampling distribution of each b_i is approximately normal with its mean equal to B_i and standard deviation equal to σ_{b_i}. However, usually σ_ε is not known and, hence, we cannot find σ_{b_i}. Consequently, we use s_{b_i} as an estimator of σ_{b_i}, and use the t distribution to perform the test.

Test Statistic for b_i

The value of the test statistic t for b_i is calculated as

$$t = \frac{b_i - B_i}{s_{b_i}}$$

The value of B_i is substituted from the null hypothesis. Usually, but not always, the null hypothesis is $H_0: B_i = 0$. The MINITAB solution contains this value of the t statistic.

Example 14.4 illustrates the procedure for testing a hypothesis about a single coefficient.

EXAMPLE 14.4 Monthly Auto Insurance Premium

Testing a Hypothesis about a Coefficient of a Multiple Regression Model

Using a 2.5% significance level, can you conclude that the coefficient of the number of years of driving experience in regression model (3) is negative? Use the MINITAB output obtained in Example 14.1 and shown in Screen 14.3 to perform this test.

Solution From Example 14.1, our multiple regression model (3) is

$$y = A + B_1x_1 + B_2x_2 + \epsilon$$

where y is the monthly auto insurance premium (in dollars) paid by a driver, x_1 is the driving experience (in years), and x_2 is the number of driving violations committed during the past three years. From the MINITAB solution, the estimated regression equation is

$$\hat{y} = 110.3 - 2.747x_1 + 16.11x_2$$

To conduct a test of hypothesis about B_1, we use the portion marked **II** in the MINITAB solution given in Screen 14.3. From that portion of the MINITAB solution,

$$b_1 = -2.747 \quad \text{and} \quad s_{b_1} = .977$$

Note that the value of the standard deviation of b_1, $s_{b_1} = .977$, is given in the column labeled **SE Coef** in part **II** of the MINITAB solution.

To make a test of hypothesis about B_1, we perform the following five steps:

STEP 1 *State the null and alternative hypotheses.*

We are to test whether the coefficient of the number of years of driving experience in regression model (3) is negative, that is, whether B_1 is negative. The two hypotheses are

$$H_0: B_1 = 0$$
$$H_1: B_1 < 0$$

Note that we can also write the null hypothesis as $H_0: B_1 \geq 0$, which states that the coefficient of the number of years of driving experience in the regression model (3) is either zero or positive.

STEP 2 *Select the distribution to use.*

The sample size is small ($n < 30$) and σ_ϵ is not known. The sampling distribution of b_1 is approximately normal because the errors are assumed to be approximately normally distributed. Hence, we use the t distribution to make a test of hypothesis about B_1.

STEP 3 *Determine the rejection and nonrejection regions.*

The significance level is .025. The < sign in the alternative hypothesis indicates that the test is left-tailed. Therefore, area in the left tail of the t distribution curve is $\alpha = .025$. The degrees of freedom are

$$df = n - k - 1 = 12 - 2 - 1 = 9$$

From the t distribution table (Table V in Appendix B), the critical value of t for 9 degrees of freedom and .025 area in the left tail of the t distribution curve is -2.262, as shown in **Figure 14.1**.

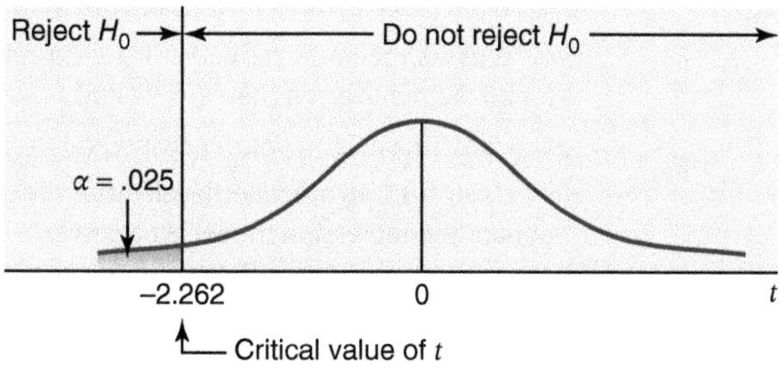

FIGURE 14.1

STEP 4 *Calculate the value of the test statistic and p-value.*

The value of the test statistic t for b_1 can be obtained from the MINITAB solution given in Screen 14.3. This value is given in the column labeled **T-Value** and the row named Driving Experience in the portion marked **II** in that MINITAB solution. Thus, the observed value of t is

$$t = \frac{b_1 - B_1}{s_{b_1}} = -2.81$$

Also, in the same portion of the MINITAB solution, the p-value for this test is given in the column labeled **P-Value** and the row named Driving Experience. This p-value is .020. However, MINITAB always gives the p-value for a two-tailed test. Because our test is one-tailed, the p-value for our test is

$$p\text{-value} = .020/2 = .010$$

STEP 5 *Make a decision.*

The value of the test statistic, $t = -2.81$, is less than the critical value of $t = -2.262$ and falls in the rejection region. Consequently, we reject the null hypothesis and conclude that the evidence from the sample data indicates that the coefficient of x_1 in regression model (3) is negative. That is, an increase in the driving experience seems to decrease the auto insurance premium.

Also the p-value for the test is .010, which is less than the significance level of $\alpha = .025$. Hence, based on this p-value also, we reject the null hypothesis and conclude that B_1 is negative.

Note that the observed value of t in Step 4 of Example 14.4 is obtained from the MINITAB solution only if the null hypothesis is $H_0: B_1 = 0$. However, if the null hypothesis is that B_1 is equal to a number other than zero, then the t value obtained from the MINITAB solution is no longer applicable. For example, suppose the null hypothesis in Example 14.4 is

$$H_0: B_1 = -2$$

and the alternative hypothesis is

$$H_1: B_1 < -2$$

In this case the observed value of t will be calculated as

$$t = \frac{b_1 - B_1}{s_{b_1}} = \frac{-2.747 - (-2)}{.977} = -.765$$

To calculate this value of t, the values of b_1 and s_{b_1} are obtained from the MINITAB solution of Screen 14.3. The value of B_1 is substituted from H_0.

EXERCISES

Concepts and Procedures

14.1 How are the coefficients of independent variables in a multiple regression model interpreted? Explain.

14.2 What are the degrees of freedom for a multiple regression model to make inferences about individual parameters?

14.3 Explain the meaning of multicollinearity in the context of multiple regression. What kinds of relationships among independent variables are permissible and which ones are not permissible in a linear multiple regression model?

14.4 Explain the meaning of the coefficient of multiple determination and the adjusted coefficient of multiple determination for a multiple regression model. What is the difference between the two?

14.5 What are the assumptions of a multiple regression model?

14.6 The following table gives data on variables y, x_1, x_2, and x_3.

y	x_1	x_2	x_3
10	19	39	75
9	27	27	65
20	33	22	48
22	37	45	32
8	15	13	81
24	51	49	36
17	29	39	43
28	61	53	19
10	15	18	73
14	22	41	59

Using MINITAB, estimate the regression model

$$y = A + B_1x_1 + B_2x_2 + B_3x_3 + \epsilon$$

Using the solution obtained, answer the following questions:

a. Write the estimated regression equation.

b. Explain the meaning of a, b_1, b_2, and b_3 obtained by estimating the given regression model.

c. What are the values of the standard deviation of errors, the coefficient of multiple determination, and the adjusted coefficient of multiple determination?

d. What is the predicted value of y for $x_1 = 42$, $x_2 = 45$, and $x_3 = 70$?

e. What is the point estimate of the expected (mean) value of y for all elements given that $x_1 = 45$, $x_2 = 28$, and $x_3 = 58$?

f. Construct a 99% confidence interval for the coefficient of x_3.

g. Using a 1% significance level, test whether or not the coefficient of x_1 is positive.

14.7 The following table gives data on variables y, x_1, and x_2.

y	x_1	x_2
26	100	55
13	52	73
19	76	61
32	110	38
11	36	92
28	121	57
27	97	59
34	151	36
15	42	62
29	112	44
28	117	49

Using MINITAB, find the regression of y on x_1 and x_2. Using the solution obtained, answer the following questions:

a. Write the estimated regression equation.

b. Explain the meaning of the estimated regression coefficients of the independent variables.

c. What are the values of the standard deviation of errors, the coefficient of multiple determination, and the adjusted coefficient of multiple determination?

d. What is the predicted value of y for $x_1 = 91$ and $x_2 = 52$?

e. What is the point estimate of the expected (mean) value of y for all elements given that $x_1 = 97$ and $x_2 = 44$?

f. Construct a 95% confidence interval for the coefficient of x_1.

g. Using a 1% significance level, test if the coefficient of x_2 in the population regression model is negative.

Applications

14.8 The salaries of workers are expected to be dependent, among other factors, on the number of years they have spent in school and their work experiences. The following table gives information on the annual salaries (in thousands of dollars) for 11 persons, the number of years each of them spent in school, and the total number of years of work experiences.

Salary	48	41	52	75	66	44	62	85	31	59	29
Schooling	15	13	12	19	16	18	15	20	13	14	11
Experience	7	11	16	9	12	3	14	6	7	8	3

Using MINITAB, find the regression of salary on schooling and experience. Using the solution obtained, answer the following questions:

a. Write the estimated regression equation.

b. Explain the meaning of the estimates of the constant term and the regression coefficients of independent variables.

c. What are the values of the standard deviation of errors, the coefficient of multiple determination, and the adjusted coefficient of multiple determination?

d. How much salary is a person with 17 years of schooling and 10 years of work experience expected to earn?

e. What is the point estimate of the expected (mean) salary for all people with 10 years of schooling and 13 years of work experience?

f. Determine a 95% confidence interval for the coefficient of schooling.

g. Using a 5% significance level, test whether or not the coefficient of experience is positive.

14.9 The CTO Corporation has a large number of chain restaurants throughout the United States. The research department at the company wanted to find if the sales of restaurants depend on the size of the population within a certain area surrounding the restaurants and the mean income of households in those areas. The company collected information on these variables for eight restaurants. The following table gives information on the weekly sales (in thousands of dollars) of these restaurants, the population (in thousands) within 15 miles of the restaurants, and the mean annual income (in thousands of dollars) of the households for those areas.

Sales	15	12	11	16	20	14	12	13
Population	35	25	18	38	45	27	30	34
Income	45	42	40	47	55	47	35	38

Using MINITAB, find the regression of (weekly) sales on population and income. Using the solution obtained, answer the following questions:

a. Write the estimated regression equation.

b. Explain the meaning of the estimates of the constant term and the regression coefficients of population and income.

c. What are the values of the standard deviation of errors, the coefficient of multiple determination, and the adjusted coefficient of multiple determination?

d. What are the predicted weekly sales for a restaurant with 40 thousand people living within a 15-mile area surrounding it and \$50 thousand mean annual income of households in that area.

e. What is the point estimate of the expected (mean) weekly sales for all restaurants with 42 thousand people living within a 15-mile area surrounding them and \$52 thousand mean annual income of households living in those areas?

f. Determine a 99% confidence interval for the coefficient of *income*.

g. Using a 5% significance level, test whether or not the coefficient of *population* is different from zero.

Uses and Misuses of Statistics

Additive versus Multiplicative Effect

A first-order multiple regression model with (quantitative) independent variables is one of the simpler types of multiple regression models. However, there are many limitations of this model. A major limitation is that the independent variables have an *additive* effect on the dependent variable. What does *additive* mean here? Suppose we have the following estimated regression equation:

$$\hat{y} = 4 + 6x_1 + 3x_2$$

From this estimated regression equation, if x_1 increases by 1 unit (with x_2 held constant), our predicted value of y increases by 6 units. If x_2 increases by 1 unit (with x_1 held constant), our predicted value of y increases by 3 units. But what happens if x_1 and x_2 both increase by 1 unit each? From this equation, our predicted value of y will increase by $6 + 3 = 9$ units. The total increase in $\hat{y}$ is simply the sum of the two increases. This change in $\hat{y}$ does not depend on the values of x_1 and x_2 prior to the increase. Since the total increase in the dependent variable is equal to the sum of the increases from the two individual parts (independent variables), we say that the effect is *additive*.

Now suppose we have the following equation:

$$\hat{y} = 4 + 6x_1 + 3x_2 + 5x_1^2x_2$$

The important difference in this case is that the increase in the value of $\hat{y}$ is no longer constant when x_1 and x_2 both increase by 1 unit each. Instead, it depends on the original values of x_1 and x_2. For example, consider the values of x_1 and x_2, and the changes in the value of $\hat{y}$ shown in the following table:

x_1	x_2	$\hat{y}$	Change in $\hat{y}$ (versus $x_1 = 2$ and $x_2 = 3$)
2	3	85	
3	3	166	81
2	4	108	23
3	4	214	129

Unlike the previous example, here the total increase in $\hat{y}$ is not equal to the sum of the increases from the individual parts. In this case, the effect is said to be *multiplicative*. It is important to recognize that the effect is multiplicative when the total increase does not equal the sum of the increases of the independent variables.

Pharmaceutical companies are always looking for multiplicative effects when creating new drugs. In many cases, a combination of two drugs might have a multiplicative effect on a certain condition. Simply stated, the two drugs provide *greater relief* when you take them together than if you take them separately so that only one drug is in your system at any time. Of course, the companies also have to look for multiplicative effects when it comes to side effects. Individual drugs may not have major side effects when taken separately, but could cause greater harm when taken together. One of the most noteworthy examples of this was the drug Fen-Phen, which was a combination of two drugs—Fenfluramine and Phentermine. Each of these two drugs had been approved for short-term (individual) control of obesity. However, the drugs used in combination became popular for long-term weight loss. Unfortunately, the combination, when associated with longtime use, resulted in severe side effects that were detailed in the following statement from the U.S. Food and Drug Administration in 1997:

> **Thanks to the reporting of health-care professionals, as of August 22, FDA has received reports of 82 cases (including Mayo's 24 cases) of cardiac valvular disease in patients—two of whom were men—on combination fenfluramine and phentermine. These reports have been from 23 different states. Severity of the cardiac valvular disease was graded as moderate or severe in over three-fourths of the cases, and two of the reports described deterioration from no detectable heart murmur to need for a valve replacement within one-and-a-half years. Sixteen of these 82 patients required surgery to repair their heart valves. At least one of these patients died following surgery to repair the valves. The agency's findings, as of July 31, are described in more detail in the current issue of *The New England Journal of Medicine*, which also carries the Mayo study. (http://www.fda.gov/cder/news/phen/fenphenupdate.htm)**

Glossary

Adjusted coefficient of multiple determination Denoted by $\bar{R}^2$, it gives the proportion of SST that is explained by the multiple regression model and is adjusted for the degrees of freedom.

Coefficient of multiple determination Denoted by R^2, it gives the proportion of SST that is explained by the multiple regression model.

First-order multiple regression model When each term in a regression model contains a single independent variable raised to the first power.

Least squares regression model The estimated regression model obtained by minimizing the sum of squared errors.

Multicollinearity When two or more independent variables in a regression model have a strong linear relationship.

Multiple regression model A regression model that contains two or more independent variables.

Partial regression coefficients The coefficients of independent variables in a multiple regression model are called the partial regression coefficients because each of them gives the effect of the corresponding independent variable on the dependent variable when all other independent variables are held constant.

SSE (error sum of squares) The sum of the squared differences between the actual and predicted values of y. It is the portion of SST that is not explained by the regression model.

SSR (regression sum of squares) The portion of SST that is explained by the regression model.

SST (total sum of squares) The sum of squared differences between actual y values and $\bar{y}$.

Standard deviation of errors Also called the *standard deviation of estimate*, it is a measure of the variation among errors.

Self-Review Test

Multiple Choice Questions

1. When using the t distribution to make inferences about a single parameter, the degrees of freedom for a multiple regression model with k independent variables and a sample size of n are equal to

a. $n+k-1$ b. $n-k+1$ c. $n-k-1$

2. The value of R^2 is always in the range

a. zero to 1 b. −1 to 1 c. −1 to zero

3. The value of $\bar{R}^2$

a. is always positive

b. is always nonnegative

c. can be positive, zero, or negative

4. A multiple regression model contains

a. one independent variable and more than one dependent variable

b. more than one independent variable and one dependent variable

c. one independent and one dependent variable

5. In a first-order multiple regression model

a. each term contains two independent variables

b. each term contains two or more independent variables raised to the first power

c. each term contains a single independent variable raised to the first power

6. A first-order multiple regression model with a sample size of 45 and eight independent variables has ________ degrees of freedom.

a. 7 b. 36 c. 44

7. A multiple regression model is written as $y = 2.5 + 1.6x_1 - .5x_2 - 2.7x_3 + 1.3x_4$. The change in y due to a one-unit change in x_1, when all other independent variables included in the model are held constant, is a(n) ________ (increase / decrease) of ________. (Separate answers with a comma.)

a. increase, 1.6

b. decrease, 1.6

c. increase, 2.7

8. The estimated multiple regression equation is

a. $E(y) = A + B_1x_1 + B_2x_2 + B_3x_3 + \cdots + B_kx_k$

b. $y = a + b_1x_1 + b_2x_2 + b_3x_3 + \cdots + b_kx_k$

c. $\hat{y} = a + b_1x_1 + b_2x_2 + b_3x_3 + \cdots + b_kx_k$

9. Which of the following is *not* an assumption for the multiple regression model?

a. There is no linear association between the random error term ϵ and each independent variable x_i.

b. The independent linear variables are linearly related.

c. The errors associated with different sets of values of independent variables are independent.

10. When independent variables are strongly linearly related, this occurrence is referred to as

a. linear correlation

b. multicollinearity

c. exact relationship

11. Which phrase does not apply to σ_ϵ?

a. Standard deviation of errors

b. Standard error of the errors

c. Standard error of the estimate

12. Which phrase does not apply to R^2?

a. Coefficient of multiple determination.

b. The proportion of the total sum of squares SST that is explained by the multiple regression model.

c. The proportion of the error sum of squares SSE that is explained by the multiple regression model.

13. The coefficient of multiple determination is given by the ratio of

a. SSE and SST

b. SST and SSR

c. SSR and SST

14. The major shortcoming of R^2 is that the value of R^2 ________ as more and more explanatory variables are added to the regression model.

a. decreases

b. increases

c. stays the same

15. Which phrase does not apply to $\bar{R}^2$?

a. The value of $\bar{R}^2$ may increase, decrease, or stay the same as more explanatory variables are added to the regression model.

b. Adjusted coefficient of multiple determination.

c. The value of $\bar{R}^2$ can never be negative.

Practice Exercises

1. What is the difference between the population multiple regression model and the estimated multiple regression model?

2. Why are the regression coefficients in a multiple regression model called the partial regression coefficients?

3. What is the difference between R^2 and $\bar{R}^2$? Explain.

4. A real estate expert wanted to find the relationship between the sale price of houses and various characteristics of the houses. She collected data on four variables, recorded in the table, for 13 houses that were sold recently. The four variables are

Price = Sale price of a house in thousands of dollars

Lot size = Size of the lot in acres

Living area = Living area in square feet

Age = Age of a house in years

Price	Lot Size	Living Area	Age
455	1.4	2500	8
278	.9	2250	12
463	1.8	2900	5
327	.7	1800	9
505	2.6	3200	4
264	1.2	2400	28
445	2.1	2700	9
346	1.1	2050	13
487	2.8	2850	7
289	1.6	2400	16
434	3.2	2600	5
411	1.7	2300	8
223	.5	1700	19

Using MINITAB, find the regression of price on lot size, living area, and age. Using the solution obtained, answer the following questions:

a. Indicate whether you expect a positive or a negative relationship between the dependent variable and each of the independent variables.

b. Write the estimated regression equation. Are the signs of the coefficients of independent variables obtained in the solution consistent with your expectations of part (a)?

c. Explain the meaning of the estimated regression coefficients of all independent variables.

d. What are the values of the standard deviation of errors, the coefficient of multiple determination, and the adjusted coefficient of multiple determination?

e. What is the predicted sale price of a house that has a lot size of 2.5 acres, a living area of 3000 square feet, and is 14 years old?

f. What is the point estimate of the mean sale price of all houses that have a lot size of 2.2 acres, a living area of 2500 square feet, and are 7 years old?

g. Determine a 99% confidence interval for each of the coefficients of the independent variables.

h. Construct a 98% confidence interval for the constant term in the population regression model.

i. Using a 1% significance level, test whether or not the coefficient of *lot size* is positive.

j. At a 2.5% significance level, test if the coefficient of *living area* is positive.

k. At a 5% significance level, test if the coefficient of *age* is negative.

Mini-Project

Mini-Project 14.1

Refer to Data Set VII on McDonald's menu (see Appendix A) and given in the book's product page on *www.wiley.com*. Use MINITAB to estimate the following regression model for that data set.

$$y = A + B_1x_1 + B_2x_2 + B_3x_3 + \epsilon$$

where y = calories

x_1 = fat (measured in grams)

x_2 = carbohydrate (measured in grams)

and x_3 = protein (measured in grams)

Now research on the Internet or in a book to find the number of calories in one gram of fat, one gram of carbohydrate, and one gram of protein.

a. Based on the information you obtain, write what the estimated regression equation should be.

b. Are the differences between your expectation in part (a) and the regression equation that you obtained from MINTAB small or large?

c. Since each gram of fat is worth a specific number of calories, and the same is true for a gram of carbohydrate, and for a gram of protein, one would expect that the predicted and observed values of y would be the same for each food item, but that is not the case. The quantities of fat, carbohydrates, and protein are reported in whole numbers. Explain why this causes the differences discussed in part (b).

Decide for Yourself

Note: The Decide for Yourself feature is located in the book's product page on *www.wiley.com*.

Chapter 14 Technology Instructions

Excel

The Excel Technology Instructions feature of this text is written for Excel 2019. Some screens, menus, and functions will be slightly different in older versions of Excel. To enable some of the advanced Excel functions, you must enable the Data Analysis Add-In for Excel

Finding the Multiple Linear Regression Equation and $\overline{R}^2$

Find the Multiple Linear Regression Equation and the Adjusted Coefficient of Multiple Determination ($\overline{R}^2$) for Example 14.1 of the Text

1. In a new worksheet, enter **Monthly Premium** in cell A1, **Driving Experience** in cell B1, and **Driving Violations** in cell C1. Then, enter the data from Example 14.1 into cells A2–A13 (Monthly Premium), cells B2–B13 (Driving Experience), and cells C2–C13 (Driving Violations). (See **Screen 14.4.**)
2. Click **DATA** and then click **Data Analysis Tools** from the **Analysis** group.
3. Select **Regression** from the dialog box that appears on screen and then click **ok.**
4. Use the following settings in the dialog box that appears on screen (see **Screen 14.4**):
   - Type A1:A13 in the **Input Y Range** box.
   - Type B1:C13 in the **Input X Range** box.
   - Check the **Labels** box.
   - Check the Confidence Level check box and the type 95 in the **Confidence Level** box.
   - Select New Workbook Ply from the **Output options.**
5. Click **OK.**
6. When the output appears, resize the columns so that they are easier to read. The output includes the coefficients of the multiple linear regression equation and the adjusted coefficient of multiple determination (Adjusted R square). (See **Screen 14.5.**)

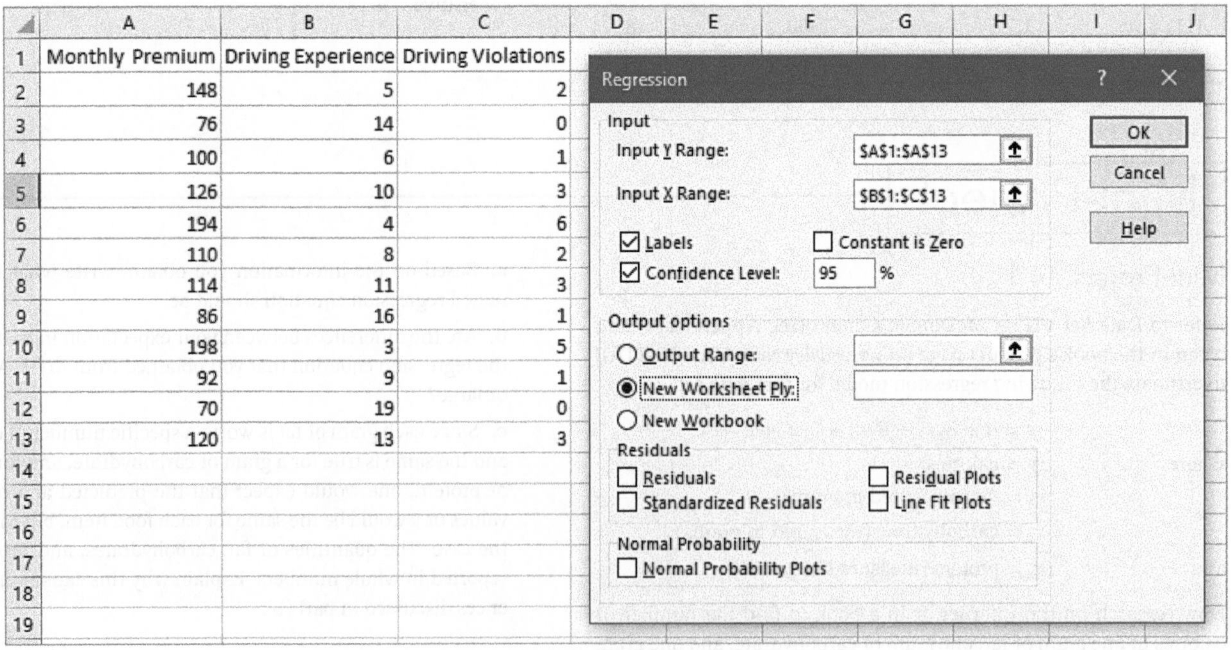

SCREEN 14.4

Constructing a Confidence Interval for B_1 (the Coefficient of Driving Experience) for Example 14.3 from the Text

Follow the directions shown above for finding the regression equation. The output also includes the lower and upper bounds of the confidence interval for B_1 (the coefficient of driving experience) in the last two entries of the **Driving Experience** row. (See **Screen 14.5.**)

	A	B	C	D	E	F	G
1	*Regression Statistics*						
2	Multiple R	0.965					
3	R Square	0.931					
4	Adjusted R Square	0.916					
5	Standard Error	12.146					
6	Observations	12					
7							
8	ANOVA						
9		*df*	*SS*	*MS*	*F*	*Significance F*	
10	Regression	2	17961.29	8980.64	60.88	0.0000	
11	Residual	9	1327.71	147.52			
12	Total	11	19289.00				
13							
14		*Coefficients*	*Standard Error*	*t Stat*	*P-value*	*Lower 95%*	*Upper 95%*
15	Intercept	110.276	14.619	7.544	0.000	77.206	143.346
16	Driving Experience	-2.747	0.977	-2.812	0.020	-4.957	-0.537
17	Driving Violations	16.106	2.613	6.163	0.000	10.194	22.018

SCREEN 14.5

Testing a Hypothesis about B_1 (the Coefficient of Driving Experience) for Example 14.4 from the Text

Follow the directions shown above for finding the regression equation. The output also includes the test statistic and *p*-value for a test of $H_0: \boldsymbol{B}_1 = 0$ against $H_1: \boldsymbol{B}_1 \neq 0$ in the 4th and 5th column of the **Driving Experience** row. The *p*-value is for a two-sided test, so if a one-sided test is required, be sure to adjust the *p*-value accordingly. (See **Screen 14.5**.)

Now compare the *t*-value to the critical value of *t* or the *p*-value from Screen 14.5 with α and make a decision.

APPENDIX A

Explanation of Data Sets

This textbook is accompanied by 12 large data sets that can be used for statistical analysis using technology. These data sets are:

Data Set I	Home Prices in Wilmington, NC
Data Set II	Data on States
Data Set III	NFL Data, 2019
Data Set IV	Manchester Road Race Data
Data Set V	Data on Movies, 2018
Data Set VI	Standard & Poor's 500 Index Data
Data Set VII	McDonald's Data, August 2019
Data Set VIII	Data on NFL Kickers, 2019
Data Set IX	Subway Menu Items Data, August 2019
Data Set X	Major League Baseball Data, 2019
Data Set XI	Data on Airline Flights
Data Set XII	Simulated Data

These data sets are available for this text in the book's product page on *www.wiley.com*. These data sets can be downloaded from this website. The website contains the following files:

1. HOMEPRICES (This file contains Data Set I)
2. STATEDATA (This file contains Data Set II)
3. NFL (This file contains Data Set III)
4. ROADRACE (This file contains Data Set IV)
5. MOVIEDATA (This file contains Data Set V)
6. S&PDATA (This file contains Data Set VI)
7. MCDONALDSDATA (This file contains Data Set VII)
8. NFLKICKERS (This file contains Data Set VIII)
9. SUBWAYMENU (This file contains Data Set IX)
10. BASEBALL (This file contains Data Set X)
11. AIRLINEFLIGHTS (This file contains Data Set XI)
12. SIMULATED (This file contains Data Set XII)

The following are the explanations of these data sets.

Data Set I: Home Prices in Wilmington, NC

This data set contains information on 801 homes sold in the Wilmington, NC, area between August 25, 2017 and July 17, 2019. This data set has 17 columns that contain the following information:

C1 The effective date that the home went on the market
C2 The date on which the home was under sale contract
C3 The number of days that the home was on the market
C4 Listing price of the home
C5 Sale price of the home
C6 Lot size in square feet
C7 Year that the home was built
C8 Number of stories
C9 Number of rooms
C10 Number of bedrooms
C11 Number of full bathrooms
C12 Number of half bathrooms
C13 Whether or not the home belongs to a home owners association
C14 The elementary school in the school district of the home
C15 The middle school in the school district of the home
C16 The high school in the school district of the home
C17 Whether or not the home is within the Wilmington city limits

Data Set II: Data on States, 2017

This data set contains information on different variables for all 50 states of the United States. These data have been obtained from various websites such as the websites of the Census Bureau, Insurance Institute for Highway Safety, Bureau of Labor Statistics, and National Center for Education Statistics. This data set has 14 columns that contain the following variables:

C1 Name of the state
C2 Total population, 2017
C3 Area of the state in square miles
C4 Median household income in dollars, 2017
C5 Total miles traveled by vehicles in millions of miles, 2017
C6 Traffic fatalities, 2017
C7 Average starting salary (in dollars) of teachers teaching in the K–12 public school system, 2017–2018
C8 Total revenue of the state in thousands of dollars, 2017
C9 Total state expenditures on education in thousands of dollars, 2017
C10 Total state expenditure on public welfare in thousands of dollars, 2017
C11 Total state expenditure on hospitals in thousands of dollars, 2017
C12 Total state expenditure on health in thousands of dollars, 2017
C13 Total state expenditure on highways in thousands of dollars, 2017
C14 Total state expenditure on police protection in thousands of dollars, 2017

Data Set III: NFL Data, 2019

This data set contains information on NFL (National Football League) players for the 2019 season as of September 14, 2019. These data were obtained from the website https://www.pro-football-reference.com/teams/. This data set has nine columns that contain the following variables:

C1 Name of the player

C2 Name of the team

C3 The position of the player

C4 Height of the player in inches

C5 Weight of the player in pounds

C6 Age of the player as of September 14, 2019

C7 Number of years in the NFL prior to the 2019 season

C8 College/university that the player attended

C9 Status of the player as of September 14, 2019

Data Set IV: Manchester Road Race Data, 2019

This data set contains information on 9956 runners who participated in and completed the 83rd annual Manchester Road Race held on November 28, 2019 (Thanksgiving Day) in Manchester, Connecticut. Eight of the 9956 runners have missing information on one or more variables. This road race is held every year on Thanksgiving Day. The length of the race is 4.748 miles. This data set was obtained from the website www.manchesterroadrace.com. This data set has 10 columns that contain the following variables:

C1 Overall place in which the runner finished

C2 Division in which the runner competed, classified by gender and age

C3 Net time to complete the race in minutes

C4 Average pace of the runner in minutes per mile

C5 Age of the runner

C6 Gender (M/F)

C7 City of residence

C8 State of residence

C9 Country of residence

C10 Whether the runner is a Connecticut resident (Yes/No)

Data Set V: Data on Movies, 2018

This data set contains information on the top 200 films of 2018 in terms of total gross domestic box office earnings until the movie's closing date in first-run theaters. Earnings do not include additional sources of revenue, such as home entertainment sales and rentals, television rights, and product placement fees. This data set was obtained from the website www.boxofficemojo.com. This data set has eight columns that contain the following variables:

C1 Movie title

C2 Studio that produced the movie

C3 Gross earnings in dollars from box office sales in the United States and Canada from the release date until the closing date of the movie

C4 Number of theater locations at which the movie was shown during the length of its release

C5 Gross earnings in dollars from box office sales in the United States and Canada during the opening weekend of the movie

C6 Number of theater locations at which the movie was shown during the opening weekend of the movie

C7 Genre of the movie as reported by IMDB, the Internet Movie Data Base

C8 Length of the film in minutes

Data Set VI: Standard & Poor's 500 Index Data

This data set contains information on the 500 stocks in the Standard & Poor's 500 index at the close of the financial markets on Monday, May 20, 2019. This data set was obtained from the website www.barchart.com. This data set has eight columns that contain the following variables:

C1 Stock symbol

C2 Company name

C3 Closing stock price for the day in dollars

C4 Difference between the closing stock price for the current day and the closing price for the previous day in dollars

C5 Difference between the closing stock price for the current day and the closing price for the previous day as a percentage of the previous day closing price

C6 Highest trade price for the day in dollars

C7 Lowest trade price for the day in dollars

C8 Total number (volume) of contracts traded for the day

Data Set VII: McDonald's Menu Data

This data set contains nutritional information on 670 menu items available at McDonald's restaurants and is obtained from the website www.nutrition.mcdonalds.com in August 2019. This data set has 23 columns that contain the following variables:

C1 Menu item

C2 Menu category

C3 Total calories

C4 Calories from fat

C5 Total fat in grams

C6 Fat, % daily value

C7 Saturated fat in grams

C8 Saturated fat, % daily value

C9 Trans fat in grams

C10 Cholesterol in milligrams

C11 Cholesterol, % daily value

C12 Sodium in milligrams

C13 Sodium, % daily value

C14 Carbohydrates in grams

C15 Carbohydrates, % daily value
C16 Dietary fiber in grams
C17 Dietary fiber, % daily value
C18 Sugars in grams
C19 Protein in grams
C20 Vitamin A, % daily value
C21 Vitamin C, % daily value
C22 Calcium, % daily value
C23 Iron, % daily value

Data Set VIII: Data on NFL Kickers, 2019

This data set contains information on the performance of 31 placekickers who had an average of at least one attempt per game during the 2019 NFL (National Football League) regular season. This data set was obtained from the website www.nfl.com. This data set has nine columns that contain the following variables:

C1 Name of the player
C2 Name of the team
C3 Number of field goals made during the season
C4 Number of field goals attempted during the season
C5 Field goal completion percentage during the season
C6 Longest field goal made during the season in yards
C7 Number of extra points made during the season
C8 Number of extra points attempted during the season
C9 Extra points completion percentage during the season

Data Set IX: Subway Menu Data

This data set contains nutritional information on 67 menu items at Subway restaurants. Breads, condiments, sides, and beverages are not included. This data set was obtained from the website www.subway.com in August 2019 and has 19 columns that contain the following variables:

C1 Menu item
C2 Menu category
C3 Whether the item is a breakfast item (Yes/No)
C4 Serving size in grams
C5 Total calories
C6 Calories from fat
C7 Total fat in grams
C8 Saturated fat in grams
C9 Trans fat in grams
C10 Cholesterol in milligrams
C11 Sodium in milligrams
C12 Carbohydrates in grams

C13 Dietary fiber in grams

C14 Sugars in grams

C15 Protein in grams

C16 Vitamin A, % daily value

C17 Vitamin C, % daily value

C18 Calcium, % daily value

C19 Iron, % daily value

Data Set X: Major League Baseball Data, 2019

This data set contains information on the offensive performance of all 442 Major League Baseball players who had at least 100 visits to the plate during the 2019 regular season. This data set, obtained from the website mlb.com, has 20 columns that contain the following variables:

C1 Name of the player

C2 Name of the team

C3 Fielding position played by the player

C4 Number of games in which the player appeared (G)

C5 Number of official at bats by a batter, with official at bat defined as the number of visits to the plate minus sacrifices, walks, and hits by pitches (AB)

C6 Number of times a base runner safely reached home plate (R)

C7 Number of times a batter hit the ball and reached base safely without the aid of an error or fielder's choice (H)

C8 Number of times a batter hit the ball and reached second base without the aid of an error or fielder's choice (2B)

C9 Number of times a batter hit the ball and reached third base without the aid of an error or fielder's choice (3B)

C10 Number of times a batter hit the ball and reached home plate, scoring a run either by hitting the ball out of play in fair territory or without the aid of an error or fielder's choice (HR)

C11 Number of runs scored safely due to a batter hitting a ball or drawing a base on balls (RBI)

C12 Number of walks by a batter (BB)

C13 Number of strikeouts by a batter (SO)

C14 Number of times the player stole a base (SB)

C15 Number of times the player was thrown out attempting to steal a base (CS)

C16 Average number of hits by a batter defined by hits divided by at bats (AVG)

C17 On-base percentage, defined as the number of times each batter reached base by hit, walk, or hit by a pitch, divided by plate appearances including at bats, walks, hit by a pitch, and sacrifice flies (OBP)

C18 Slugging percentage, defined as total bases divided by at bats (SLG)

C19 Number of times a batter was hit by a pitch (HBP)

C20 Number of times a runner tagged up and scored after a batter's fly out (SF)

Data Set XI: Data on Airline Flights

This data set contains information on airline flights departing Newark (New Jersey) Liberty International Airport during February 2019. No arriving flights are included in these data, and flights with missing data were removed, leaving 9790 flights in the data set. This data set, obtained from the website www.transtats.bts.gov, has 17 columns that contain the following variables:

C1 Carrier Code

C2 Airline carrier of the flight

C3 Date of the flight (MM/DD/YYYY)

C4 Flight number

C5 Destination airport

C6 Indicates whether the flight arrived, was cancelled, or was diverted

C7 Destination city

C8 Distance between airports in miles

C9 Scheduled departure time

C10 Actual departure time

C11 Scheduled elapsed time in minutes

C12 Actual elapsed time in minutes

C13 Departure delay in minutes

C14 Scheduled arrival time

C15 Actual arrival time

C16 Total delay in minutes; a negative value indicates an early arrival

C17 Whether the arrival was early or within 15 minutes of the scheduled arrival time

Data Set XII: Simulated Data

This data set contains four columns of simulated data from four different probability distributions. There are 1000 observations in each of the four columns, and these columns contain the following information:

C1 Simulated data from probability distribution 1

C2 Simulated data from probability distribution 2

C3 Simulated data from probability distribution 3

C4 Simulated data from probability distribution 4

APPENDIX

B

Statistical Tables

TABLE I Table of Binomial Probabilities

		p										
n	*x*	.05	.10	.20	.30	.40	.50	.60	.70	.80	.90	.95
1	0	.9500	.9000	.8000	.7000	.6000	.5000	.4000	.3000	.2000	.1000	.0500
	1	.0500	.1000	.2000	.3000	.4000	.5000	.6000	.7000	.8000	.9000	.9500
2	0	.9025	.8100	.6400	.4900	.3600	.2500	.1600	.0900	.0400	.0100	.0025
	1	.0950	.1800	.3200	.4200	.4800	.5000	.4800	.4200	.3200	.1800	.0950
	2	.0025	.0100	.0400	.0900	.1600	.2500	.3600	.4900	.6400	.8100	.9025
3	0	.8574	.7290	.5120	.3430	.2160	.1250	.0640	.0270	.0080	.0010	.0001
	1	.1354	.2430	.3840	.4410	.4320	.3750	.2880	.1890	.0960	.0270	.0071
	2	.0071	.0270	.0960	.1890	.2880	.3750	.4320	.4410	.3840	.2430	.1354
	3	.0001	.0010	.0080	.0270	.0640	.1250	.2160	.3430	.5120	.7290	.8574
4	0	.8145	.6561	.4096	.2401	.1296	.0625	.0256	.0081	.0016	.0001	.0000
	1	.1715	.2916	.4096	.4116	.3456	.2500	.1536	.0756	.0256	.0036	.0005
	2	.0135	.0486	.1536	.2646	.3456	.3750	.3456	.2646	.1536	.0486	.0135
	3	.0005	.0036	.0256	.0756	.1536	.2500	.3456	.4116	.4096	.2916	.1715
	4	.0000	.0001	.0016	.0081	.0256	.0625	.1296	.2401	.4096	.6561	.8145
5	0	.7738	.5905	.3277	.1681	.0778	.0312	.0102	.0024	.0003	.0000	.0000
	1	.2036	.3280	.4096	.3602	.2592	.1562	.0768	.0284	.0064	.0005	.0000
	2	.0214	.0729	.2048	.3087	.3456	.3125	.2304	.1323	.0512	.0081	.0011
	3	.0011	.0081	.0512	.1323	.2304	.3125	.3456	.3087	.2048	.0729	.0214
	4	.0000	.0004	.0064	.0283	.0768	.1562	.2592	.3601	.4096	.3281	.2036
	5	.0000	.0000	.0003	.0024	.0102	.0312	.0778	.1681	.3277	.5905	.7738
6	0	.7351	.5314	.2621	.1176	.0467	.0156	.0041	.0007	.0001	.0000	.0000
	1	.2321	.3543	.3932	.3025	.1866	.0937	.0369	.0102	.0015	.0001	.0000
	2	.0305	.0984	.2458	.3241	.3110	.2344	.1382	.0595	.0154	.0012	.0001
	3	.0021	.0146	.0819	.1852	.2765	.3125	.2765	.1852	.0819	.0146	.0021
	4	.0001	.0012	.0154	.0595	.1382	.2344	.3110	.3241	.2458	.0984	.0305
	5	.0000	.0001	.0015	.0102	.0369	.0937	.1866	.3025	.3932	.3543	.2321
	6	.0000	.0000	.0001	.0007	.0041	.0156	.0467	.1176	.2621	.5314	.7351
7	0	.6983	.4783	.2097	.0824	.0280	.0078	.0016	.0002	.0000	.0000	.0000
	1	.2573	.3720	.3670	.2471	.1306	.0547	.0172	.0036	.0004	.0000	.0000
	2	.0406	.1240	.2753	.3177	.2613	.1641	.0774	.0250	.0043	.0002	.0000
	3	.0036	.0230	.1147	.2269	.2903	.2734	.1935	.0972	.0287	.0026	.0002
	4	.0002	.0026	.0287	.0972	.1935	.2734	.2903	.2269	.1147	.0230	.0036
	5	.0000	.0002	.0043	.0250	.0774	.1641	.2613	.3177	.2753	.1240	.0406
	6	.0000	.0000	.0004	.0036	.0172	.0547	.1306	.2471	.3670	.3720	.2573
	7	.0000	.0000	.0000	.0002	.0016	.0078	.0280	.0824	.2097	.4783	.6983
8	0	.6634	.4305	.1678	.0576	.0168	.0039	.0007	.0001	.0000	.0000	.0000
	1	.2793	.3826	.3355	.1977	.0896	.0312	.0079	.0012	.0001	.0000	.0000
	2	.0515	.1488	.2936	.2965	.2090	.1094	.0413	.0100	.0011	.0000	.0000
	3	.0054	.0331	.1468	.2541	.2787	.2187	.1239	.0467	.0092	.0004	.0000

TABLE I *(Continued)*

		p										
n	*x*	.05	.10	.20	.30	.40	.50	.60	.70	.80	.90	.95
	4	.0004	.0046	.0459	.1361	.2322	.2734	.2322	.1361	.0459	.0046	.0004
	5	.0000	.0004	.0092	.0467	.1239	.2187	.2787	.2541	.1468	.0331	.0054
	6	.0000	.0000	.0011	.0100	.0413	.1094	.2090	.2965	.2936	.1488	.0515
	7	.0000	.0000	.0001	.0012	.0079	.0312	.0896	.1977	.3355	.3826	.2793
	8	.0000	.0000	.0000	.0001	.0007	.0039	.0168	.0576	.1678	.4305	.6634
9	0	.6302	.3874	.1342	.0404	.0101	.0020	.0003	.0000	.0000	.0000	.0000
	1	.2985	.3874	.3020	.1556	.0605	.0176	.0035	.0004	.0000	.0000	.0000
	2	.0629	.1722	.3020	.2668	.1612	.0703	.0212	.0039	.0003	.0000	.0000
	3	.0077	.0446	.1762	.2668	.2508	.1641	.0743	.0210	.0028	.0001	.0000
	4	.0006	.0074	.0661	.1715	.2508	.2461	.1672	.0735	.0165	.0008	.0000
	5	.0000	.0008	.0165	.0735	.1672	.2461	.2508	.1715	.0661	.0074	.0006
	6	.0000	.0001	.0028	.0210	.0743	.1641	.2508	.2668	.1762	.0446	.0077
	7	.0000	.0000	.0003	.0039	.0212	.0703	.1612	.2668	.3020	.1722	.0629
	8	.0000	.0000	.0000	.0004	.0035	.0176	.0605	.1556	.3020	.3874	.2985
	9	.0000	.0000	.0000	.0000	.0003	.0020	.0101	.0404	.1342	.3874	.6302
10	0	.5987	.3487	.1074	.0282	.0060	.0010	.0001	.0000	.0000	.0000	.0000
	1	.3151	.3874	.2684	.1211	.0403	.0098	.0016	.0001	.0000	.0000	.0000
	2	.0746	.1937	.3020	.2335	.1209	.0439	.0106	.0014	.0001	.0000	.0000
	3	.0105	.0574	.2013	.2668	.2150	.1172	.0425	.0090	.0008	.0000	.0000
	4	.0010	.0112	.0881	.2001	.2508	.2051	.1115	.0368	.0055	.0001	.0000
	5	.0001	.0015	.0264	.1029	.2007	.2461	.2007	.1029	.0264	.0015	.0001
	6	.0000	.0001	.0055	.0368	.1115	.2051	.2508	.2001	.0881	.0112	.0010
	7	.0000	.0000	.0008	.0090	.0425	.1172	.2150	.2668	.2013	.0574	.0105
	8	.0000	.0000	.0001	.0014	.0106	.0439	.1209	.2335	.3020	.1937	.0746
	9	.0000	.0000	.0000	.0001	.0016	.0098	.0403	.1211	.2684	.3874	.3151
	10	.0000	.0000	.0000	.0000	.0001	.0010	.0060	.0282	.1074	.3487	.5987
11	0	.5688	.3138	.0859	.0198	.0036	.0005	.0000	.0000	.0000	.0000	.0000
	1	.3293	.3835	.2362	.0932	.0266	.0054	.0007	.0000	.0000	.0000	.0000
	2	.0867	.2131	.2953	.1998	.0887	.0269	.0052	.0005	.0000	.0000	.0000
	3	.0137	.0710	.2215	.2568	.1774	.0806	.0234	.0037	.0002	.0000	.0000
	4	.0014	.0158	.1107	.2201	.2365	.1611	.0701	.0173	.0017	.0000	.0000
	5	.0001	.0025	.0388	.1321	.2207	.2256	.1471	.0566	.0097	.0003	.0000
	6	.0000	.0003	.0097	.0566	.1471	.2256	.2207	.1321	.0388	.0025	.0001
	7	.0000	.0000	.0017	.0173	.0701	.1611	.2365	.2201	.1107	.0158	.0014
	8	.0000	.0000	.0002	.0037	.0234	.0806	.1774	.2568	.2215	.0710	.0137
	9	.0000	.0000	.0000	.0005	.0052	.0269	.0887	.1998	.2953	.2131	.0867
	10	.0000	.0000	.0000	.0000	.0007	.0054	.0266	.0932	.2362	.3835	.3293
	11	.0000	.0000	.0000	.0000	.0000	.0005	.0036	.0198	.0859	.3138	.5688
12	0	.5404	.2824	.0687	.0138	.0022	.0002	.0000	.0000	.0000	.0000	.0000
	1	.3413	.3766	.2062	.0712	.0174	.0029	.0003	.0000	.0000	.0000	.0000

(Continued)

TABLE I (*Continued*)

		p										
n	*x*	.05	.10	.20	.30	.40	.50	.60	.70	.80	.90	.95
	2	.0988	.2301	.2835	.1678	.0639	.0161	.0025	.0002	.0000	.0000	.0000
	3	.0173	.0852	.2362	.2397	.1419	.0537	.0125	.0015	.0001	.0000	.0000
	4	.0021	.0213	.1329	.2311	.2128	.1208	.0420	.0078	.0005	.0000	.0000
	5	.0002	.0038	.0532	.1585	.2270	.1934	.1009	.0291	.0033	.0000	.0000
	6	.0000	.0005	.0155	.0792	.1766	.2256	.1766	.0792	.0155	.0005	.0000
	7	.0000	.0000	.0033	.0291	.1009	.1934	.2270	.1585	.0532	.0038	.0002
	8	.0000	.0000	.0005	.0078	.0420	.1208	.2128	.2311	.1329	.0213	.0021
	9	.0000	.0000	.0001	.0015	.0125	.0537	.1419	.2397	.2362	.0852	.0173
	10	.0000	.0000	.0000	.0002	.0025	.0161	.0639	.1678	.2835	.2301	.0988
	11	.0000	.0000	.0000	.0000	.0003	.0029	.0174	.0712	.2062	.3766	.3413
	12	.0000	.0000	.0000	.0000	.0000	.0002	.0022	.0138	.0687	.2824	.5404
13	0	.5133	.2542	.0550	.0097	.0013	.0001	.0000	.0000	.0000	.0000	.0000
	1	.3512	.3672	.1787	.0540	.0113	.0016	.0001	.0000	.0000	.0000	.0000
	2	.1109	.2448	.2680	.1388	.0453	.0095	.0012	.0001	.0000	.0000	.0000
	3	.0214	.0997	.2457	.2181	.1107	.0349	.0065	.0006	.0000	.0000	.0000
	4	.0028	.0277	.1535	.2337	.1845	.0873	.0243	.0034	.0001	.0000	.0000
	5	.0003	.0055	.0691	.1803	.2214	.1571	.0656	.0142	.0011	.0000	.0000
	6	.0000	.0008	.0230	.1030	.1968	.2095	.1312	.0442	.0058	.0001	.0000
	7	.0000	.0001	.0058	.0442	.1312	.2095	.1968	.1030	.0230	.0008	.0000
	8	.0000	.0000	.0011	.0142	.0656	.1571	.2214	.1803	.0691	.0055	.0003
	9	.0000	.0000	.0001	.0034	.0243	.0873	.1845	.2337	.1535	.0277	.0028
	10	.0000	.0000	.0000	.0006	.0065	.0349	.1107	.2181	.2457	.0997	.0214
	11	.0000	.0000	.0000	.0001	.0012	.0095	.0453	.1388	.2680	.2448	.1109
	12	.0000	.0000	.0000	.0000	.0001	.0016	.0113	.0540	.1787	.3672	.3512
	13	.0000	.0000	.0000	.0000	.0000	.0001	.0013	.0097	.0550	.2542	.5133
14	0	.4877	.2288	.0440	.0068	.0008	.0001	.0000	.0000	.0000	.0000	.0000
	1	.3593	.3559	.1539	.0407	.0073	.0009	.0001	.0000	.0000	.0000	.0000
	2	.1229	.2570	.2501	.1134	.0317	.0056	.0005	.0000	.0000	.0000	.0000
	3	.0259	.1142	.2501	.1943	.0845	.0222	.0033	.0002	.0000	.0000	.0000
	4	.0037	.0349	.1720	.2290	.1549	.0611	.0136	.0014	.0000	.0000	.0000
	5	.0004	.0078	.0860	.1963	.2066	.1222	.0408	.0066	.0003	.0000	.0000
	6	.0000	.0013	.0322	.1262	.2066	.1833	.0918	.0232	.0020	.0000	.0000
	7	.0000	.0002	.0092	.0618	.1574	.2095	.1574	.0618	.0092	.0002	.0000
	8	.0000	.0000	.0020	.0232	.0918	.1833	.2066	.1262	.0322	.0013	.0000
	9	.0000	.0000	.0003	.0066	.0408	.1222	.2066	.1963	.0860	.0078	.0004
	10	.0000	.0000	.0000	.0014	.0136	.0611	.1549	.2290	.1720	.0349	.0037
	11	.0000	.0000	.0000	.0002	.0033	.0222	.0845	.1943	.2501	.1142	.0259
	12	.0000	.0000	.0000	.0000	.0005	.0056	.0317	.1134	.2501	.2570	.1229
	13	.0000	.0000	.0000	.0000	.0001	.0009	.0073	.0407	.1539	.3559	.3593
	14	.0000	.0000	.0000	.0000	.0000	.0001	.0008	.0068	.0440	.2288	.4877

TABLE I *(Continued)*

		p										
n	*x*	.05	.10	.20	.30	.40	.50	.60	.70	.80	.90	.95
15	0	.4633	.2059	.0352	.0047	.0005	.0000	.0000	.0000	.0000	.0000	.0000
	1	.3658	.3432	.1319	.0305	.0047	.0005	.0000	.0000	.0000	.0000	.0000
	2	.1348	.2669	.2309	.0916	.0219	.0032	.0003	.0000	.0000	.0000	.0000
	3	.0307	.1285	.2501	.1700	.0634	.0139	.0016	.0001	.0000	.0000	.0000
	4	.0049	.0428	.1876	.2186	.1268	.0417	.0074	.0006	.0000	.0000	.0000
	5	.0006	.0105	.1032	.2061	.1859	.0916	.0245	.0030	.0001	.0000	.0000
	6	.0000	.0019	.0430	.1472	.2066	.1527	.0612	.0116	.0007	.0000	.0000
	7	.0000	.0003	.0138	.0811	.1771	.1964	.1181	.0348	.0035	.0000	.0000
	8	.0000	.0000	.0035	.0348	.1181	.1964	.1771	.0811	.0138	.0003	.0000
	9	.0000	.0000	.0007	.0116	.0612	.1527	.2066	.1472	.0430	.0019	.0000
	10	.0000	.0000	.0001	.0030	.0245	.0916	.1859	.2061	.1032	.0105	.0006
	11	.0000	.0000	.0000	.0006	.0074	.0417	.1268	.2186	.1876	.0428	.0049
	12	.0000	.0000	.0000	.0001	.0016	.0139	.0634	.1700	.2501	.1285	.0307
	13	.0000	.0000	.0000	.0000	.0003	.0032	.0219	.0916	.2309	.2669	.1348
	14	.0000	.0000	.0000	.0000	.0000	.0005	.0047	.0305	.1319	.3432	.3658
	15	.0000	.0000	.0000	.0000	.0000	.0000	.0005	.0047	.0352	.2059	.4633
16	0	.4401	.1853	.0281	.0033	.0003	.0000	.0000	.0000	.0000	.0000	.0000
	1	.3706	.3294	.1126	.0228	.0030	.0002	.0000	.0000	.0000	.0000	.0000
	2	.1463	.2745	.2111	.0732	.0150	.0018	.0001	.0000	.0000	.0000	.0000
	3	.0359	.1423	.2463	.1465	.0468	.0085	.0008	.0000	.0000	.0000	.0000
	4	.0061	.0514	.2001	.2040	.1014	.0278	.0040	.0002	.0000	.0000	.0000
	5	.0008	.0137	.1201	.2099	.1623	.0667	.0142	.0013	.0000	.0000	.0000
	6	.0001	.0028	.0550	.1649	.1983	.1222	.0392	.0056	.0002	.0000	.0000
	7	.0000	.0004	.0197	.1010	.1889	.1746	.0840	.0185	.0012	.0000	.0000
	8	.0000	.0001	.0055	.0487	.1417	.1964	.1417	.0487	.0055	.0001	.0000
	9	.0000	.0000	.0012	.0185	.0840	.1746	.1889	.1010	.0197	.0004	.0000
	10	.0000	.0000	.0002	.0056	.0392	.1222	.1983	.1649	.0550	.0028	.0001
	11	.0000	.0000	.0000	.0013	.0142	.0666	.1623	.2099	.1201	.0137	.0008
	12	.0000	.0000	.0000	.0002	.0040	.0278	.1014	.2040	.2001	.0514	.0061
	13	.0000	.0000	.0000	.0000	.0008	.0085	.0468	.1465	.2463	.1423	.0359
	14	.0000	.0000	.0000	.0000	.0001	.0018	.0150	.0732	.2111	.2745	.1463
	15	.0000	.0000	.0000	.0000	.0000	.0002	.0030	.0228	.1126	.3294	.3706
	16	.0000	.0000	.0000	.0000	.0000	.0000	.0003	.0033	.0281	.1853	.4401
17	0	.4181	.1668	.0225	.0023	.0002	.0000	.0000	.0000	.0000	.0000	.0000
	1	.3741	.3150	.0957	.0169	.0019	.0001	.0000	.0000	.0000	.0000	.0000
	2	.1575	.2800	.1914	.0581	.0102	.0010	.0001	.0000	.0000	.0000	.0000
	3	.0415	.1556	.2393	.1245	.0341	.0052	.0004	.0000	.0000	.0000	.0000
	4	.0076	.0605	.2093	.1868	.0796	.0182	.0021	.0001	.0000	.0000	.0000
	5	.0010	.0175	.1361	.2081	.1379	.0472	.0081	.0006	.0000	.0000	.0000
	6	.0001	.0039	.0680	.1784	.1839	.0944	.0242	.0026	.0001	.0000	.0000
	7	.0000	.0007	.0267	.1201	.1927	.1484	.0571	.0095	.0004	.0000	.0000
	8	.0000	.0001	.0084	.0644	.1606	.1855	.1070	.0276	.0021	.0000	.0000

(Continued)

TABLE I (*Continued*)

		p										
n	*x*	.05	.10	.20	.30	.40	.50	.60	.70	.80	.90	.95
	9	.0000	.0000	.0021	.0276	.1070	.1855	.1606	.0644	.0084	.0001	.0000
	10	.0000	.0000	.0004	.0095	.0571	.1484	.1927	.1201	.0267	.0007	.0000
	11	.0000	.0000	.0001	.0026	.0242	.0944	.1839	.1784	.0680	.0039	.0001
	12	.0000	.0000	.0000	.0006	.0081	.0472	.1379	.2081	.1361	.0175	.0010
	13	.0000	.0000	.0000	.0001	.0021	.0182	.0796	.1868	.2093	.0605	.0076
	14	.0000	.0000	.0000	.0000	.0004	.0052	.0341	.1245	.2393	.1556	.0415
	15	.0000	.0000	.0000	.0000	.0001	.0010	.0102	.0581	.1914	.2800	.1575
	16	.0000	.0000	.0000	.0000	.0000	.0001	.0019	.0169	.0957	.3150	.3741
	17	.0000	.0000	.0000	.0000	.0000	.0000	.0002	.0023	.0225	.1668	.4181
18	0	.3972	.1501	.0180	.0016	.0001	.0000	.0000	.0000	.0000	.0000	.0000
	1	.3763	.3002	.0811	.0126	.0012	.0001	.0000	.0000	.0000	.0000	.0000
	2	.1683	.2835	.1723	.0458	.0069	.0006	.0000	.0000	.0000	.0000	.0000
	3	.0473	.1680	.2297	.1046	.0246	.0031	.0002	.0000	.0000	.0000	.0000
	4	.0093	.0700	.2153	.1681	.0614	.0117	.0011	.0000	.0000	.0000	.0000
	5	.0014	.0218	.1507	.2017	.1146	.0327	.0045	.0002	.0000	.0000	.0000
	6	.0002	.0052	.0816	.1873	.1655	.0708	.0145	.0012	.0000	.0000	.0000
	7	.0000	.0010	.0350	.1376	.1892	.1214	.0374	.0046	.0001	.0000	.0000
	8	.0000	.0002	.0120	.0811	.1734	.1669	.0771	.0149	.0008	.0000	.0000
	9	.0000	.0000	.0033	.0386	.1284	.1855	.1284	.0386	.0033	.0000	.0000
	10	.0000	.0000	.0008	.0149	.0771	.1669	.1734	.0811	.0120	.0002	.0000
	11	.0000	.0000	.0001	.0046	.0374	.1214	.1892	.1376	.0350	.0010	.0000
	12	.0000	.0000	.0000	.0012	.0145	.0708	.1655	.1873	.0816	.0052	.0002
	13	.0000	.0000	.0000	.0002	.0045	.0327	.1146	.2017	.1507	.0218	.0014
	14	.0000	.0000	.0000	.0000	.0011	.0117	.0614	.1681	.2153	.0700	.0093
	15	.0000	.0000	.0000	.0000	.0002	.0031	.0246	.1046	.2297	.1680	.0473
	16	.0000	.0000	.0000	.0000	.0000	.0006	.0069	.0458	.1723	.2835	.1683
	17	.0000	.0000	.0000	.0000	.0000	.0001	.0012	.0126	.0811	.3002	.3763
	18	.0000	.0000	.0000	.0000	.0000	.0000	.0001	.0016	.0180	.1501	.3972
19	0	.3774	.1351	.0144	.0011	.0001	.0000	.0000	.0000	.0000	.0000	.0000
	1	.3774	.2852	.0685	.0093	.0008	.0000	.0000	.0000	.0000	.0000	.0000
	2	.1787	.2852	.1540	.0358	.0046	.0003	.0000	.0000	.0000	.0000	.0000
	3	.0533	.1796	.2182	.0869	.0175	.0018	.0001	.0000	.0000	.0000	.0000
	4	.0112	.0798	.2182	.1491	.0467	.0074	.0005	.0000	.0000	.0000	.0000
	5	.0018	.0266	.1636	.1916	.0933	.0222	.0024	.0001	.0000	.0000	.0000
	6	.0002	.0069	.0955	.1916	.1451	.0518	.0085	.0005	.0000	.0000	.0000
	7	.0000	.0014	.0443	.1525	.1797	.0961	.0237	.0022	.0000	.0000	.0000
	8	.0000	.0002	.0166	.0981	.1797	.1442	.0532	.0077	.0003	.0000	.0000
	9	.0000	.0000	.0051	.0514	.1464	.1762	.0976	.0220	.0013	.0000	.0000
	10	.0000	.0000	.0013	.0220	.0976	.1762	.1464	.0514	.0051	.0000	.0000
	11	.0000	.0000	.0003	.0077	.0532	.1442	.1797	.0981	.0166	.0002	.0000

TABLE I *(Continued)*

		p										
n	*x*	.05	.10	.20	.30	.40	.50	.60	.70	.80	.90	.95
	12	.0000	.0000	.0000	.0022	.0237	.0961	.1797	.1525	.0443	.0014	.0000
	13	.0000	.0000	.0000	.0005	.0085	.0518	.1451	.1916	.0955	.0069	.0002
	14	.0000	.0000	.0000	.0001	.0024	.0222	.0933	.1916	.1636	.0266	.0018
	15	.0000	.0000	.0000	.0000	.0005	.0074	.0467	.1491	.2182	.0798	.0112
	16	.0000	.0000	.0000	.0000	.0001	.0018	.0175	.0869	.2182	.1796	.0533
	17	.0000	.0000	.0000	.0000	.0000	.0003	.0046	.0358	.1540	.2852	.1787
	18	.0000	.0000	.0000	.0000	.0000	.0000	.0008	.0093	.0685	.2852	.3774
	19	.0000	.0000	.0000	.0000	.0000	.0000	.0001	.0011	.0144	.1351	.3774
20	0	.3585	.1216	.0115	.0008	.0000	.0000	.0000	.0000	.0000	.0000	.0000
	1	.3774	.2702	.0576	.0068	.0005	.0000	.0000	.0000	.0000	.0000	.0000
	2	.1887	.2852	.1369	.0278	.0031	.0002	.0000	.0000	.0000	.0000	.0000
	3	.0596	.1901	.2054	.0716	.0123	.0011	.0000	.0000	.0000	.0000	.0000
	4	.0133	.0898	.2182	.1304	.0350	.0046	.0003	.0000	.0000	.0000	.0000
	5	.0022	.0319	.1746	.1789	.0746	.0148	.0013	.0000	.0000	.0000	.0000
	6	.0003	.0089	.1091	.1916	.1244	.0370	.0049	.0002	.0000	.0000	.0000
	7	.0000	.0020	.0545	.1643	.1659	.0739	.0146	.0010	.0000	.0000	.0000
	8	.0000	.0004	.0222	.1144	.1797	.1201	.0355	.0039	.0001	.0000	.0000
	9	.0000	.0001	.0074	.0654	.1597	.1602	.0710	.0120	.0005	.0000	.0000
	10	.0000	.0000	.0020	.0308	.1171	.1762	.1171	.0308	.0020	.0000	.0000
	11	.0000	.0000	.0005	.0120	.0710	.1602	.1597	.0654	.0074	.0001	.0000
	12	.0000	.0000	.0001	.0039	.0355	.1201	.1797	.1144	.0222	.0004	.0000
	13	.0000	.0000	.0000	.0010	.0146	.0739	.1659	.1643	.0545	.0020	.0000
	14	.0000	.0000	.0000	.0002	.0049	.0370	.1244	.1916	.1091	.0089	.0003
	15	.0000	.0000	.0000	.0000	.0013	.0148	.0746	.1789	.1746	.0319	.0022
	16	.0000	.0000	.0000	.0000	.0003	.0046	.0350	.1304	.2182	.0898	.0133
	17	.0000	.0000	.0000	.0000	.0000	.0011	.0123	.0716	.2054	.1901	.0596
	18	.0000	.0000	.0000	.0000	.0000	.0002	.0031	.0278	.1369	.2852	.1887
	19	.0000	.0000	.0000	.0000	.0000	.0000	.0005	.0068	.0576	.2702	.3774
	20	.0000	.0000	.0000	.0000	.0000	.0000	.0000	.0008	.0115	.1216	.3585
21	0	.3406	.1094	.0092	.0006	.0000	.0000	.0000	.0000	.0000	.0000	.0000
	1	.3764	.2553	.0484	.0050	.0003	.0000	.0000	.0000	.0000	.0000	.0000
	2	.1981	.2837	.1211	.0215	.0020	.0001	.0000	.0000	.0000	.0000	.0000
	3	.0660	.1996	.1917	.0585	.0086	.0006	.0000	.0000	.0000	.0000	.0000
	4	.0156	.0998	.2156	.1128	.0259	.0029	.0001	.0000	.0000	.0000	.0000
	5	.0028	.0377	.1833	.1643	.0588	.0097	.0007	.0000	.0000	.0000	.0000
	6	.0004	.0112	.1222	.1878	.1045	.0259	.0027	.0001	.0000	.0000	.0000
	7	.0000	.0027	.0655	.1725	.1493	.0554	.0087	.0005	.0000	.0000	.0000
	8	.0000	.0005	.0286	.1294	.1742	.0970	.0229	.0019	.0000	.0000	.0000
	9	.0000	.0001	.0103	.0801	.1677	.1402	.0497	.0063	.0002	.0000	.0000
	10	.0000	.0000	.0031	.0412	.1342	.1682	.0895	.0176	.0008	.0000	.0000

(Continued)

TABLE I (*Continued*)

		p										
n	x	.05	.10	.20	.30	.40	.50	.60	.70	.80	.90	.95
	11	.0000	.0000	.0008	.0176	.0895	.1682	.1342	.0412	.0031	.0000	.0000
	12	.0000	.0000	.0002	.0063	.0497	.1402	.1677	.0801	.0103	.0001	.0000
	13	.0000	.0000	.0000	.0019	.0229	.0970	.1742	.1294	.0286	.0005	.0000
	14	.0000	.0000	.0000	.0005	.0087	.0554	.1493	.1725	.0655	.0027	.0000
	15	.0000	.0000	.0000	.0001	.0027	.0259	.1045	.1878	.1222	.0112	.0004
	16	.0000	.0000	.0000	.0000	.0007	.0097	.0588	.1643	.1833	.0377	.0028
	17	.0000	.0000	.0000	.0000	.0001	.0029	.0259	.1128	.2156	.0998	.0156
	18	.0000	.0000	.0000	.0000	.0000	.0006	.0086	.0585	.1917	.1996	.0660
	19	.0000	.0000	.0000	.0000	.0000	.0001	.0020	.0215	.1211	.2837	.1981
	20	.0000	.0000	.0000	.0000	.0000	.0000	.0003	.0050	.0484	.2553	.3764
	21	.0000	.0000	.0000	.0000	.0000	.0000	.0000	.0006	.0092	.1094	.3406
22	0	.3235	.0985	.0074	.0004	.0000	.0000	.0000	.0000	.0000	.0000	.0000
	1	.3746	.2407	.0406	.0037	.0002	.0000	.0000	.0000	.0000	.0000	.0000
	2	.2070	.2808	.1065	.0166	.0014	.0001	.0000	.0000	.0000	.0000	.0000
	3	.0726	.2080	.1775	.0474	.0060	.0004	.0000	.0000	.0000	.0000	.0000
	4	.0182	.1098	.2108	.0965	.0190	.0017	.0001	.0000	.0000	.0000	.0000
	5	.0034	.0439	.1898	.1489	.0456	.0063	.0004	.0000	.0000	.0000	.0000
	6	.0005	.0138	.1344	.1808	.0862	.0178	.0015	.0000	.0000	.0000	.0000
	7	.0001	.0035	.0768	.1771	.1314	.0407	.0051	.0002	.0000	.0000	.0000
	8	.0000	.0007	.0360	.1423	.1642	.0762	.0144	.0009	.0000	.0000	.0000
	9	.0000	.0001	.0140	.0949	.1703	.1186	.0336	.0032	.0001	.0000	.0000
	10	.0000	.0000	.0046	.0529	.1476	.1542	.0656	.0097	.0003	.0000	.0000
	11	.0000	.0000	.0012	.0247	.1073	.1682	.1073	.0247	.0012	.0000	.0000
	12	.0000	.0000	.0003	.0097	.0656	.1542	.1476	.0529	.0046	.0000	.0000
	13	.0000	.0000	.0001	.0032	.0336	.1186	.1703	.0949	.0140	.0001	.0000
	14	.0000	.0000	.0000	.0009	.0144	.0762	.1642	.1423	.0360	.0007	.0000
	15	.0000	.0000	.0000	.0002	.0051	.0407	.1314	.1771	.0768	.0035	.0001
	16	.0000	.0000	.0000	.0000	.0015	.0178	.0862	.1808	.1344	.0138	.0005
	17	.0000	.0000	.0000	.0000	.0004	.0063	.0456	.1489	.1898	.0439	.0034
	18	.0000	.0000	.0000	.0000	.0001	.0017	.0190	.0965	.2108	.1098	.0182
	19	.0000	.0000	.0000	.0000	.0000	.0004	.0060	.0474	.1775	.2080	.0726
	20	.0000	.0000	.0000	.0000	.0000	.0001	.0014	.0166	.1065	.2808	.2070
	21	.0000	.0000	.0000	.0000	.0000	.0000	.0002	.0037	.0406	.2407	.3746
	22	.0000	.0000	.0000	.0000	.0000	.0000	.0000	.0004	.0074	.0985	.3235
23	0	.3074	.0886	.0059	.0003	.0000	.0000	.0000	.0000	.0000	.0000	.0000
	1	.3721	.2265	.0339	.0027	.0001	.0000	.0000	.0000	.0000	.0000	.0000
	2	.2154	.2768	.0933	.0127	.0009	.0000	.0000	.0000	.0000	.0000	.0000
	3	.0794	.2153	.1633	.0382	.0041	.0002	.0000	.0000	.0000	.0000	.0000
	4	.0209	.1196	.2042	.0818	.0138	.0011	.0000	.0000	.0000	.0000	.0000
	5	.0042	.0505	.1940	.1332	.0350	.0040	.0002	.0000	.0000	.0000	.0000

TABLE I *(Continued)*

		p										
n	*x*	.05	.10	.20	.30	.40	.50	.60	.70	.80	.90	.95
	6	.0007	.0168	.1455	.1712	.0700	.0120	.0008	.0000	.0000	.0000	.0000
	7	.0001	.0045	.0883	.1782	.1133	.0292	.0029	.0001	.0000	.0000	.0000
	8	.0000	.0010	.0442	.1527	.1511	.0584	.0088	.0004	.0000	.0000	.0000
	9	.0000	.0002	.0184	.1091	.1679	.0974	.0221	.0016	.0000	.0000	.0000
	10	.0000	.0000	.0064	.0655	.1567	.1364	.0464	.0052	.0001	.0000	.0000
	11	.0000	.0000	.0019	.0332	.1234	.1612	.0823	.0142	.0005	.0000	.0000
	12	.0000	.0000	.0005	.0142	.0823	.1612	.1234	.0332	.0019	.0000	.0000
	13	.0000	.0000	.0001	.0052	.0464	.1364	.1567	.0655	.0064	.0000	.0000
	14	.0000	.0000	.0000	.0016	.0221	.0974	.1679	.1091	.0184	.0002	.0000
	15	.0000	.0000	.0000	.0004	.0088	.0584	.1511	.1527	.0442	.0010	.0000
	16	.0000	.0000	.0000	.0001	.0029	.0292	.1133	.1782	.0883	.0045	.0001
	17	.0000	.0000	.0000	.0000	.0008	.0120	.0700	.1712	.1455	.0168	.0007
	18	.0000	.0000	.0000	.0000	.0002	.0040	.0350	.1332	.1940	.0505	.0042
	19	.0000	.0000	.0000	.0000	.0000	.0011	.0138	.0818	.2042	.1196	.0209
	20	.0000	.0000	.0000	.0000	.0000	.0002	.0041	.0382	.1633	.2153	.0794
	21	.0000	.0000	.0000	.0000	.0000	.0000	.0009	.0127	.0933	.2768	.2154
	22	.0000	.0000	.0000	.0000	.0000	.0000	.0001	.0027	.0339	.2265	.3721
	23	.0000	.0000	.0000	.0000	.0000	.0000	.0000	.0003	.0059	.0886	.3074
24	0	.2920	.0798	.0047	.0002	.0000	.0000	.0000	.0000	.0000	.0000	.0000
	1	.3688	.2127	.0283	.0020	.0001	.0000	.0000	.0000	.0000	.0000	.0000
	2	.2232	.2718	.0815	.0097	.0006	.0000	.0000	.0000	.0000	.0000	.0000
	3	.0862	.2215	.1493	.0305	.0028	.0001	.0000	.0000	.0000	.0000	.0000
	4	.0238	.1292	.1960	.0687	.0099	.0006	.0000	.0000	.0000	.0000	.0000
	5	.0050	.0574	.1960	.1177	.0265	.0025	.0001	.0000	.0000	.0000	.0000
	6	.0008	.0202	.1552	.1598	.0560	.0080	.0004	.0000	.0000	.0000	.0000
	7	.0001	.0058	.0998	.1761	.0960	.0206	.0017	.0000	.0000	.0000	.0000
	8	.0000	.0014	.0530	.1604	.1360	.0438	.0053	.0002	.0000	.0000	.0000
	9	.0000	.0003	.0236	.1222	.1612	.0779	.0141	.0008	.0000	.0000	.0000
	10	.0000	.0000	.0088	.0785	.1612	.1169	.0318	.0026	.0000	.0000	.0000
	11	.0000	.0000	.0028	.0428	.1367	.1488	.0608	.0079	.0002	.0000	.0000
	12	.0000	.0000	.0008	.0199	.0988	.1612	.0988	.0199	.0008	.0000	.0000
	13	.0000	.0000	.0002	.0079	.0608	.1488	.1367	.0428	.0028	.0000	.0000
	14	.0000	.0000	.0000	.0026	.0318	.1169	.1612	.0785	.0088	.0000	.0000
	15	.0000	.0000	.0000	.0008	.0141	.0779	.1612	.1222	.0236	.0003	.0000
	16	.0000	.0000	.0000	.0002	.0053	.0438	.1360	.1604	.0530	.0014	.0000
	17	.0000	.0000	.0000	.0000	.0017	.0206	.0960	.1761	.0998	.0058	.0001
	18	.0000	.0000	.0000	.0000	.0004	.0080	.0560	.1598	.1552	.0202	.0008
	19	.0000	.0000	.0000	.0000	.0001	.0025	.0265	.1177	.1960	.0574	.0050
	20	.0000	.0000	.0000	.0000	.0000	.0006	.0099	.0687	.1960	.1292	.0238
	21	.0000	.0000	.0000	.0000	.0000	.0001	.0028	.0305	.1493	.2215	.0862

(Continued)

TABLE I (*Continued*)

		p										
n	*x*	.05	.10	.20	.30	.40	.50	.60	.70	.80	.90	.95
	22	.0000	.0000	.0000	.0000	.0000	.0000	.0006	.0097	.0815	.2718	.2232
	23	.0000	.0000	.0000	.0000	.0000	.0000	.0001	.0020	.0283	.2127	.3688
	24	.0000	.0000	.0000	.0000	.0000	.0000	.0000	.0002	.0047	.0798	.2920
25	0	.2774	.0718	.0038	.0001	.0000	.0000	.0000	.0000	.0000	.0000	.0000
	1	.3650	.1994	.0236	.0014	.0000	.0000	.0000	.0000	.0000	.0000	.0000
	2	.2305	.2659	.0708	.0074	.0004	.0000	.0000	.0000	.0000	.0000	.0000
	3	.0930	.2265	.1358	.0243	.0019	.0001	.0000	.0000	.0000	.0000	.0000
	4	.0269	.1384	.1867	.0572	.0071	.0004	.0000	.0000	.0000	.0000	.0000
	5	.0060	.0646	.1960	.1030	.0199	.0016	.0000	.0000	.0000	.0000	.0000
	6	.0010	.0239	.1633	.1472	.0442	.0053	.0002	.0000	.0000	.0000	.0000
	7	.0001	.0072	.1108	.1712	.0800	.0143	.0009	.0000	.0000	.0000	.0000
	8	.0000	.0018	.0623	.1651	.1200	.0322	.0031	.0001	.0000	.0000	.0000
	9	.0000	.0004	.0294	.1336	.1511	.0609	.0088	.0004	.0000	.0000	.0000
	10	.0000	.0001	.0118	.0916	.1612	.0974	.0212	.0013	.0000	.0000	.0000
	11	.0000	.0000	.0040	.0536	.1465	.1328	.0434	.0042	.0001	.0000	.0000
	12	.0000	.0000	.0012	.0268	.1140	.1550	.0760	.0115	.0003	.0000	.0000
	13	.0000	.0000	.0003	.0115	.0760	.1550	.1140	.0268	.0012	.0000	.0000
	14	.0000	.0000	.0001	.0042	.0434	.1328	.1465	.0536	.0040	.0000	.0000
	15	.0000	.0000	.0000	.0013	.0212	.0974	.1612	.0916	.0118	.0001	.0000
	16	.0000	.0000	.0000	.0004	.0088	.0609	.1511	.1336	.0294	.0004	.0000
	17	.0000	.0000	.0000	.0001	.0031	.0322	.1200	.1651	.0623	.0018	.0000
	18	.0000	.0000	.0000	.0000	.0009	.0143	.0800	.1712	.1108	.0072	.0001
	19	.0000	.0000	.0000	.0000	.0002	.0053	.0442	.1472	.1633	.0239	.0010
	20	.0000	.0000	.0000	.0000	.0000	.0016	.0199	.1030	.1960	.0646	.0060
	21	.0000	.0000	.0000	.0000	.0000	.0004	.0071	.0572	.1867	.1384	.0269
	22	.0000	.0000	.0000	.0000	.0000	.0001	.0019	.0243	.1358	.2265	.0930
	23	.0000	.0000	.0000	.0000	.0000	.0000	.0004	.0074	.0708	.2659	.2305
	24	.0000	.0000	.0000	.0000	.0000	.0000	.0000	.0014	.0236	.1994	.3650
	25	.0000	.0000	.0000	.0000	.0000	.0000	.0000	.0001	.0038	.0718	.2774

TABLE II Values of $e^{-\lambda}$

λ	$e^{-\lambda}$	λ	$e^{-\lambda}$
0.0	1.00000000	3.9	.02024191
0.1	.90483742	4.0	.01831564
0.2	.81873075	4.1	.01657268
0.3	.74081822	4.2	.01499558
0.4	.67032005	4.3	.01356856
0.5	.60653066	4.4	.01227734
0.6	.54881164	4.5	.01110900
0.7	.49658530	4.6	.01005184
0.8	.44932896	4.7	.00909528
0.9	.40656966	4.8	.00822975
1.0	.36787944	4.9	.00744658
1.1	.33287108	5.0	.00673795
1.2	.30119421	5.1	.00609675
1.3	.27253179	5.2	.00551656
1.4	.24659696	5.3	.00499159
1.5	.22313016	5.4	.00451658
1.6	.20189652	5.5	.00408677
1.7	.18268352	5.6	.00369786
1.8	.16529889	5.7	.00334597
1.9	.14956862	5.8	.00302755
2.0	.13533528	5.9	.00273944
2.1	.12245643	6.0	.00247875
2.2	.11080316	6.1	.00224287
2.3	.10025884	6.2	.00202943
2.4	.09071795	6.3	.00183630
2.5	.08208500	6.4	.00166156
2.6	.07427358	6.5	.00150344
2.7	.06720551	6.6	.00136037
2.8	.06081006	6.7	.00123091
2.9	.05502322	6.8	.00111378
3.0	.04978707	6.9	.00100779
3.1	.04504920	7.0	.00091188
3.2	.04076220	7.1	.00082510
3.3	.03688317	7.2	.00074659
3.4	.03337327	7.3	.00067554
3.5	.03019738	7.4	.00061125
3.6	.02732372	7.5	.00055308
3.7	.02472353	7.6	.00050045
3.8	.02237077	7.7	.00045283
7.8	.00040973	9.5	.00007485

(Continued)

TABLE II (*Continued*)

λ	$e^{-\lambda}$	λ	$e^{-\lambda}$
7.9	.00037074	9.6	.00006773
8.0	.00033546	9.7	.00006128
8.1	.00030354	9.8	.00005545
8.2	.00027465	9.9	.00005017
8.3	.00024852	10.0	.00004540
8.4	.00022487	11.0	.00001670
8.5	.00020347	12.0	.00000614
8.6	.00018411	13.0	.00000226
8.7	.00016659	14.0	.00000083
8.8	.00015073	15.0	.00000031
8.9	.00013639	16.0	.00000011
9.0	.00012341	17.0	.00000004
9.1	.00011167	18.0	.000000015
9.2	.00010104	19.0	.000000006
9.3	.00009142	20.0	.000000002
9.4	.00008272		

TABLE III Table of Poisson Probabilities

	λ									
x	0.1	0.2	0.3	0.4	0.5	0.6	0.7	0.8	0.9	1.0
0	.9048	.8187	.7408	.6703	.6065	.5488	.4966	.4493	.4066	.3679
1	.0905	.1637	.2222	.2681	.3033	.3293	.3476	.3595	.3659	.3679
2	.0045	.0164	.0333	.0536	.0758	.0988	.1217	.1438	.1647	.1839
3	.0002	.0011	.0033	.0072	.0126	.0198	.0284	.0383	.0494	.0613
4	.0000	.0001	.0003	.0007	.0016	.0030	.0050	.0077	.0111	.0153
5	.0000	.0000	.0000	.0001	.0002	.0004	.0007	.0012	.0020	.0031
6	.0000	.0000	.0000	.0000	.0000	.0000	.0001	.0002	.0003	.0005
7	.0000	.0000	.0000	.0000	.0000	.0000	.0000	.0000	.0000	.0001

	λ									
x	1.1	1.2	1.3	1.4	1.5	1.6	1.7	1.8	1.9	2.0
0	.3329	.3012	.2725	.2466	.2231	.2019	.1827	.1653	.1496	.1353
1	.3662	.3614	.3543	.3452	.3347	.3230	.3106	.2975	.2842	.2707
2	.2014	.2169	.2303	.2417	.2510	.2584	.2640	.2678	.2700	.2707
3	.0738	.0867	.0998	.1128	.1255	.1378	.1496	.1607	.1710	.1804
4	.0203	.0260	.0324	.0395	.0471	.0551	.0636	.0723	.0812	.0902
5	.0045	.0062	.0084	.0111	.0141	.0176	.0216	.0260	.0309	.0361
6	.0008	.0012	.0018	.0026	.0035	.0047	.0061	.0078	.0098	.0120
7	.0001	.0002	.0003	.0005	.0008	.0011	.0015	.0020	.0027	.0034
8	.0000	.0000	.0001	.0001	.0001	.0002	.0003	.0005	.0006	.0009
9	.0000	.0000	.0000	.0000	.0000	.0000	.0001	.0001	.0001	.0002

	λ									
x	2.1	2.2	2.3	2.4	2.5	2.6	2.7	2.8	2.9	3.0
0	.1225	.1108	.1003	.0907	.0821	.0743	.0672	.0608	.0550	.0498
1	.2572	.2438	.2306	.2177	.2052	.1931	.1815	.1703	.1596	.1494
2	.2700	.2681	.2652	.2613	.2565	.2510	.2450	.2384	.2314	.2240
3	.1890	.1966	.2033	.2090	.2138	.2176	.2205	.2225	.2237	.2240
4	.0992	.1082	.1169	.1254	.1336	.1414	.1488	.1557	.1622	.1680
5	.0417	.0476	.0538	.0602	.0668	.0735	.0804	.0872	.0940	.1008
6	.0146	.0174	.0206	.0241	.0278	.0319	.0362	.0407	.0455	.0504
7	.0044	.0055	.0068	.0083	.0099	.0118	.0139	.0163	.0188	.0216
8	.0011	.0015	.0019	.0025	.0031	.0038	.0047	.0057	.0068	.0081
9	.0003	.0004	.0005	.0007	.0009	.0011	.0014	.0018	.0022	.0027

(Continued)

TABLE III (*Continued*)

x	λ 2.1	2.2	2.3	2.4	2.5	2.6	2.7	2.8	2.9	3.0
10	.0001	.0001	.0001	.0002	.0002	.0003	.0004	.0005	.0006	.0008
11	.0000	.0000	.0000	.0000	.0000	.0001	.0001	.0001	.0002	.0002
12	.0000	.0000	.0000	.0000	.0000	.0000	.0000	.0000	.0000	.0001

x	λ 3.1	3.2	3.3	3.4	3.5	3.6	3.7	3.8	3.9	4.0
0	.0450	.0408	.0369	.0334	.0302	.0273	.0247	.0224	.0202	.0183
1	.1397	.1304	.1217	.1135	.1057	.0984	.0915	.0850	.0789	.0733
2	.2165	.2087	.2008	.1929	.1850	.1771	.1692	.1615	.1539	.1465
3	.2237	.2226	.2209	.2186	.2158	.2125	.2087	.2046	.2001	.1954
4	.1733	.1781	.1823	.1858	.1888	.1912	.1931	.1944	.1951	.1954
5	.1075	.1140	.1203	.1264	.1322	.1377	.1429	.1477	.1522	.1563
6	.0555	.0608	.0662	.0716	.0771	.0826	.0881	.0936	.0989	.1042
7	.0246	.0278	.0312	.0348	.0385	.0425	.0466	.0508	.0551	.0595
8	.0095	.0111	.0129	.0148	.0169	.0191	.0215	.0241	.0269	.0298
9	.0033	.0040	.0047	.0056	.0066	.0076	.0089	.0102	.0116	.0132
10	.0010	.0013	.0016	.0019	.0023	.0028	.0033	.0039	.0045	.0053
11	.0003	.0004	.0005	.0006	.0007	.0009	.0011	.0013	.0016	.0019
12	.0001	.0001	.0001	.0002	.0002	.0003	.0003	.0004	.0005	.0006
13	.0000	.0000	.0000	.0000	.0001	.0001	.0001	.0001	.0002	.0002
14	.0000	.0000	.0000	.0000	.0000	.0000	.0000	.0000	.0000	.0001

x	λ 4.1	4.2	4.3	4.4	4.5	4.6	4.7	4.8	4.9	5.0
0	.0166	.0150	.0136	.0123	.0111	.0101	.0091	.0082	.0074	.0067
1	.0679	.0630	.0583	.0540	.0500	.0462	.0427	.0395	.0365	.0337
2	.1393	.1323	.1254	.1188	.1125	.1063	.1005	.0948	.0894	.0842
3	.1904	.1852	.1798	.1743	.1687	.1631	.1574	.1517	.1460	.1404
4	.1951	.1944	.1933	.1917	.1898	.1875	.1849	.1820	.1789	.1755
5	.1600	.1633	.1662	.1687	.1708	.1725	.1738	.1747	.1753	.1755
6	.1093	.1143	.1191	.1237	.1281	.1323	.1362	.1398	.1432	.1462
7	.0640	.0686	.0732	.0778	.0824	.0869	.0914	.0959	.1002	.1044
8	.0328	.0360	.0393	.0428	.0463	.0500	.0537	.0575	.0614	.0653
9	.0150	.0168	.0188	.0209	.0232	.0255	.0281	.0307	.0334	.0363
10	.0061	.0071	.0081	.0092	.0104	.0118	.0132	.0147	.0164	.0181
11	.0023	.0027	.0032	.0037	.0043	.0049	.0056	.0064	.0073	.0082
12	.0008	.0009	.0011	.0014	.0016	.0019	.0022	.0026	.0030	.0034
13	.0002	.0003	.0004	.0005	.0006	.0007	.0008	.0009	.0011	.0013
14	.0001	.0001	.0001	.0001	.0002	.0002	.0003	.0003	.0004	.0005
15	.0000	.0000	.0000	.0000	.0001	.0001	.0001	.0001	.0001	.0002

TABLE III (*Continued*)

	λ									
x	5.1	5.2	5.3	5.4	5.5	5.6	5.7	5.8	5.9	6.0
0	.0061	.0055	.0050	.0045	.0041	.0037	.0033	.0030	.0027	.0025
1	.0311	.0287	.0265	.0244	.0225	.0207	.0191	.0176	.0162	.0149
2	.0793	.0746	.0701	.0659	.0618	.0580	.0544	.0509	.0477	.0446
3	.1348	.1293	.1239	.1185	.1133	.1082	.1033	.0985	.0938	.0892
4	.1719	.1681	.1641	.1600	.1558	.1515	.1472	.1428	.1383	.1339
5	.1753	.1748	.1740	.1728	.1714	.1697	.1678	.1656	.1632	.1606
6	.1490	.1515	.1537	.1555	.1571	.1584	.1594	.1601	.1605	.1606
7	.1086	.1125	.1163	.1200	.1234	.1267	.1298	.1326	.1353	.1377
8	.0692	.0731	.0771	.0810	.0849	.0887	.0925	.0962	.0998	.1033
9	.0392	.0423	.0454	.0486	.0519	.0552	.0586	.0620	.0654	.0688
10	.0200	.0220	.0241	.0262	.0285	.0309	.0334	.0359	.0386	.0413
11	.0093	.0104	.0116	.0129	.0143	.0157	.0173	.0190	.0207	.0225
12	.0039	.0045	.0051	.0058	.0065	.0073	.0082	.0092	.0102	.0113
13	.0015	.0018	.0021	.0024	.0028	.0032	.0036	.0041	.0046	.0052
14	.0006	.0007	.0008	.0009	.0011	.0013	.0015	.0017	.0019	.0022
15	.0002	.0002	.0003	.0003	.0004	.0005	.0006	.0007	.0008	.0009
16	.0001	.0001	.0001	.0001	.0001	.0002	.0002	.0002	.0003	.0003
17	.0000	.0000	.0000	.0000	.0000	.0001	.0001	.0001	.0001	.0001

	λ									
x	6.1	6.2	6.3	6.4	6.5	6.6	6.7	6.8	6.9	7.0
0	.0022	.0020	.0018	.0017	.0015	.0014	.0012	.0011	.0010	.0009
1	.0137	.0126	.0116	.0106	.0098	.0090	.0082	.0076	.0070	.0064
2	.0417	.0390	.0364	.0340	.0318	.0296	.0276	.0258	.0240	.0223
3	.0848	.0806	.0765	.0726	.0688	.0652	.0617	.0584	.0552	.0521
4	.1294	.1249	.1205	.1162	.1118	.1076	.1034	.0992	.0952	.0912
5	.1579	.1549	.1519	.1487	.1454	.1420	.1385	.1349	.1314	.1277
6	.1605	.1601	.1595	.1586	.1575	.1562	.1546	.1529	.1511	.1490
7	.1399	.1418	.1435	.1450	.1462	.1472	.1480	.1486	.1489	.1490
8	.1066	.1099	.1130	.1160	.1188	.1215	.1240	.1263	.1284	.1304
9	.0723	.0757	.0791	.0825	.0858	.0891	.0923	.0954	.0985	.1014
10	.0441	.0469	.0498	.0528	.0558	.0588	.0618	.0649	.0679	.0710
11	.0244	.0265	.0285	.0307	.0330	.0353	.0377	.0401	.0426	.0452
12	.0124	.0137	.0150	.0164	.0179	.0194	.0210	.0227	.0245	.0263
13	.0058	.0065	.0073	.0081	.0089	.0099	.0108	.0119	.0130	.0142
14	.0025	.0029	.0033	.0037	.0041	.0046	.0052	.0058	.0064	.0071
15	.0010	.0012	.0014	.0016	.0018	.0020	.0023	.0026	.0029	.0033
16	.0004	.0005	.0005	.0006	.0007	.0008	.0010	.0011	.0013	.0014
17	.0001	.0002	.0002	.0002	.0003	.0003	.0004	.0004	.0005	.0006
18	.0000	.0001	.0001	.0001	.0001	.0001	.0001	.0002	.0002	.0002
19	.0000	.0000	.0000	.0000	.0000	.0000	.0001	.0001	.0001	.0001

(*Continued*)

TABLE III (*Continued*)

	λ									
x	7.1	7.2	7.3	7.4	7.5	7.6	7.7	7.8	7.9	8.0
0	.0008	.0007	.0007	.0006	.0006	.0005	.0005	.0004	.0004	.0003
1	.0059	.0054	.0049	.0045	.0041	.0038	.0035	.0032	.0029	.0027
2	.0208	.0194	.0180	.0167	.0156	.0145	.0134	.0125	.0116	.0107
3	.0492	.0464	.0438	.0413	.0389	.0366	.0345	.0324	.0305	.0286
4	.0874	.0836	.0799	.0764	.0729	.0696	.0663	.0632	.0602	.0573
5	.1241	.1204	.1167	.1130	.1094	.1057	.1021	.0986	.0951	.0916
6	.1468	.1445	.1420	.1394	.1367	.1339	.1311	.1282	.1252	.1221
7	.1489	.1486	.1481	.1474	.1465	.1454	.1442	.1428	.1413	.1396
8	.1321	.1337	.1351	.1363	.1373	.1381	.1388	.1392	.1395	.1396
9	.1042	.1070	.1096	.1121	.1144	.1167	.1187	.1207	.1224	.1241
10	.0740	.0770	.0800	.0829	.0858	.0887	.0914	.0941	.0967	.0993
11	.0478	.0504	.0531	.0558	.0585	.0613	.0640	.0667	.0695	.0722
12	.0283	.0303	.0323	.0344	.0366	.0388	.0411	.0434	.0457	.0481
13	.0154	.0168	.0181	.0196	.0211	.0227	.0243	.0260	.0278	.0296
14	.0078	.0086	.0095	.0104	.0113	.0123	.0134	.0145	.0157	.0169
15	.0037	.0041	.0046	.0051	.0057	.0062	.0069	.0075	.0083	.0090
16	.0016	.0019	.0021	.0024	.0026	.0030	.0033	.0037	.0041	.0045
17	.0007	.0008	.0009	.0010	.0012	.0013	.0015	.0017	.0019	.0021
18	.0003	.0003	.0004	.0004	.0005	.0006	.0006	.0007	.0008	.0009
19	.0001	.0001	.0001	.0002	.0002	.0002	.0003	.0003	.0003	.0004
20	.0000	.0000	.0001	.0001	.0001	.0001	.0001	.0001	.0001	.0002
21	.0000	.0000	.0000	.0000	.0000	.0000	.0000	.0000	.0001	.0001

	λ									
x	8.1	8.2	8.3	8.4	8.5	8.6	8.7	8.8	8.9	9.0
0	.0003	.0003	.0002	.0002	.0002	.0002	.0002	.0002	.0001	.0001
1	.0025	.0023	.0021	.0019	.0017	.0016	.0014	.0013	.0012	.0011
2	.0100	.0092	.0086	.0079	.0074	.0068	.0063	.0058	.0054	.0050
3	.0269	.0252	.0237	.0222	.0208	.0195	.0183	.0171	.0160	.0150
4	.0544	.0517	.0491	.0466	.0443	.0420	.0398	.0377	.0357	.0337
5	.0882	.0849	.0816	.0784	.0752	.0722	.0692	.0663	.0635	.0607
6	.1191	.1160	.1128	.1097	.1066	.1034	.1003	.0972	.0941	.0911
7	.1378	.1358	.1338	.1317	.1294	.1271	.1247	.1222	.1197	.1171
8	.1395	.1392	.1388	.1382	.1375	.1366	.1356	.1344	.1332	.1318
9	.1255	.1269	.1280	.1290	.1299	.1306	.1311	.1315	.1317	.1318
10	.1017	.1040	.1063	.1084	.1104	.1123	.1140	.1157	.1172	.1186
11	.0749	.0775	.0802	.0828	.0853	.0878	.0902	.0925	.0948	.0970
12	.0505	.0530	.0555	.0579	.0604	.0629	.0654	.0679	.0703	.0728

TABLE III *(Continued)*

	λ									
x	8.1	8.2	8.3	8.4	8.5	8.6	8.7	8.8	8.9	9.0
13	.0315	.0334	.0354	.0374	.0395	.0416	.0438	.0459	.0481	.0504
14	.0182	.0196	.0210	.0225	.0240	.0256	.0272	.0289	.0306	.0324
15	.0098	.0107	.0116	.0126	.0136	.0147	.0158	.0169	.0182	.0194
16	.0050	.0055	.0060	.0066	.0072	.0079	.0086	.0093	.0101	.0109
17	.0024	.0026	.0029	.0033	.0036	.0040	.0044	.0048	.0053	.0058
18	.0011	.0012	.0014	.0015	.0017	.0019	.0021	.0024	.0026	.0029
19	.0005	.0005	.0006	.0007	.0008	.0009	.0010	.0011	.0012	.0014
20	.0002	.0002	.0002	.0003	.0003	.0004	.0004	.0005	.0005	.0006
21	.0001	.0001	.0001	.0001	.0001	.0002	.0002	.0002	.0002	.0003
22	.0000	.0000	.0000	.0000	.0001	.0001	.0001	.0001	.0001	.0001

	λ									
x	9.1	9.2	9.3	9.4	9.5	9.6	9.7	9.8	9.9	10
0	.0001	.0001	.0001	.0001	.0001	.0001	.0001	.0001	.0001	.0000
1	.0010	.0009	.0009	.0008	.0007	.0007	.0006	.0005	.0005	.0005
2	.0046	.0043	.0040	.0037	.0034	.0031	.0029	.0027	.0025	.0023
3	.0140	.0131	.0123	.0115	.0107	.0100	.0093	.0087	.0081	.0076
4	.0319	.0302	.0285	.0269	.0254	.0240	.0226	.0213	.0201	.0189
5	.0581	.0555	.0530	.0506	.0483	.0460	.0439	.0418	.0398	.0378
6	.0881	.0851	.0822	.0793	.0764	.0736	.0709	.0682	.0656	.0631
7	.1145	.1118	.1091	.1064	.1037	.1010	.0982	.0955	.0928	.0901
8	.1302	.1286	.1269	.1251	.1232	.1212	.1191	.1170	.1148	.1126
9	.1317	.1315	.1311	.1306	.1300	.1293	.1284	.1274	.1263	.1251
10	.1198	.1209	.1219	.1228	.1235	.1241	.1245	.1249	.1250	.1251
11	.0991	.1012	.1031	.1049	.1067	.1083	.1098	.1112	.1125	.1137
12	.0752	.0776	.0799	.0822	.0844	.0866	.0888	.0908	.0928	.0948
13	.0526	.0549	.0572	.0594	.0617	.0640	.0662	.0685	.0707	.0729
14	.0342	.0361	.0380	.0399	.0419	.0439	.0459	.0479	.0500	.0521
15	.0208	.0221	.0235	.0250	.0265	.0281	.0297	.0313	.0330	.0347
16	.0118	.0127	.0137	.0147	.0157	.0168	.0180	.0192	.0204	.0217
17	.0063	.0069	.0075	.0081	.0088	.0095	.0103	.0111	.0119	.0128
18	.0032	.0035	.0039	.0042	.0046	.0051	.0055	.0060	.0065	.0071
19	.0015	.0017	.0019	.0021	.0023	.0026	.0028	.0031	.0034	.0037
20	.0007	.0008	.0009	.0010	.0011	.0012	.0014	.0015	.0017	.0019
21	.0003	.0003	.0004	.0004	.0005	.0006	.0006	.0007	.0008	.0009
22	.0001	.0001	.0002	.0002	.0002	.0002	.0003	.0003	.0004	.0004
23	.0000	.0001	.0001	.0001	.0001	.0001	.0001	.0001	.0002	.0002
24	.0000	.0000	.0000	.0000	.0000	.0000	.0000	.0001	.0001	.0001

(Continued)

TABLE III (*Continued*)

	λ									
x	11	12	13	14	15	16	17	18	19	20
0	.0000	.0000	.0000	.0000	.0000	.0000	.0000	.0000	.0000	.0000
1	.0002	.0001	.0000	.0000	.0000	.0000	.0000	.0000	.0000	.0000
2	.0010	.0004	.0002	.0001	.0000	.0000	.0000	.0000	.0000	.0000
3	.0037	.0018	.0008	.0004	.0002	.0001	.0000	.0000	.0000	.0000
4	.0102	.0053	.0027	.0013	.0006	.0003	.0001	.0001	.0000	.0000
5	.0224	.0127	.0070	.0037	.0019	.0010	.0005	.0002	.0001	.0001
6	.0411	.0255	.0152	.0087	.0048	.0026	.0014	.0007	.0004	.0002
7	.0646	.0437	.0281	.0174	.0104	.0060	.0034	.0019	.0010	.0005
8	.0888	.0655	.0457	.0304	.0194	.0120	.0072	.0042	.0024	.0013
9	.1085	.0874	.0661	.0473	.0324	.0213	.0135	.0083	.0050	.0029
10	.1194	.1048	.0859	.0663	.0486	.0341	.0230	.0150	.0095	.0058
11	.1194	.1144	.1015	.0844	.0663	.0496	.0355	.0245	.0164	.0106
12	.1094	.1144	.1099	.0984	.0829	.0661	.0504	.0368	.0259	.0176
13	.0926	.1056	.1099	.1060	.0956	.0814	.0658	.0509	.0378	.0271
14	.0728	.0905	.1021	.1060	.1024	.0930	.0800	.0655	.0514	.0387
15	.0534	.0724	.0885	.0989	.1024	.0992	.0906	.0786	.0650	.0516
16	.0367	.0543	.0719	.0866	.0960	.0992	.0963	.0884	.0772	.0646
17	.0237	.0383	.0550	.0713	.0847	.0934	.0963	.0936	.0863	.0760
18	.0145	.0255	.0397	.0554	.0706	.0830	.0909	.0936	.0911	.0844
19	.0084	.0161	.0272	.0409	.0557	.0699	.0814	.0887	.0911	.0888
20	.0046	.0097	.0177	.0286	.0418	.0559	.0692	.0798	.0866	.0888
21	.0024	.0055	.0109	.0191	.0299	.0426	.0560	.0684	.0783	.0846
22	.0012	.0030	.0065	.0121	.0204	.0310	.0433	.0560	.0676	.0769
23	.0006	.0016	.0037	.0074	.0133	.0216	.0320	.0438	.0559	.0669
24	.0003	.0008	.0020	.0043	.0083	.0144	.0226	.0328	.0442	.0557
25	.0001	.0004	.0010	.0024	.0050	.0092	.0154	.0237	.0336	.0446
26	.0000	.0002	.0005	.0013	.0029	.0057	.0101	.0164	.0246	.0343
27	.0000	.0001	.0002	.0007	.0016	.0034	.0063	.0109	.0173	.0254
28	.0000	.0000	.0001	.0003	.0009	.0019	.0038	.0070	.0117	.0181
29	.0000	.0000	.0001	.0002	.0004	.0011	.0023	.0044	.0077	.0125
30	.0000	.0000	.0000	.0001	.0002	.0006	.0013	.0026	.0049	.0083
31	.0000	.0000	.0000	.0000	.0001	.0003	.0007	.0015	.0030	.0054
32	.0000	.0000	.0000	.0000	.0001	.0001	.0004	.0009	.0018	.0034
33	.0000	.0000	.0000	.0000	.0000	.0001	.0002	.0005	.0010	.0020
34	.0000	.0000	.0000	.0000	.0000	.0000	.0001	.0002	.0006	.0012
35	.0000	.0000	.0000	.0000	.0000	.0000	.0000	.0001	.0003	.0007
36	.0000	.0000	.0000	.0000	.0000	.0000	.0000	.0001	.0002	.0004
37	.0000	.0000	.0000	.0000	.0000	.0000	.0000	.0000	.0001	.0002
38	.0000	.0000	.0000	.0000	.0000	.0000	.0000	.0000	.0000	.0001
39	.0000	.0000	.0000	.0000	.0000	.0000	.0000	.0000	.0000	.0001

TABLE IV Standard Normal Distribution Table

The entries in the table on this page give the cumulative area under the standard normal curve to the left of *z* with the values of *z* equal to 0 or negative.

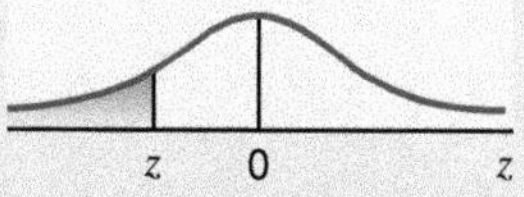

z	.00	.01	.02	.03	.04	.05	.06	.07	.08	.09
−3.4	.0003	.0003	.0003	.0003	.0003	.0003	.0003	.0003	.0003	.0002
−3.3	.0005	.0005	.0005	.0004	.0004	.0004	.0004	.0004	.0004	.0003
−3.2	.0007	.0007	.0006	.0006	.0006	.0006	.0006	.0005	.0005	.0005
−3.1	.0010	.0009	.0009	.0009	.0008	.0008	.0008	.0008	.0007	.0007
−3.0	.0013	.0013	.0013	.0012	.0012	.0011	.0011	.0011	.0010	.0010
−2.9	.0019	.0018	.0018	.0017	.0016	.0016	.0015	.0015	.0014	.0014
−2.8	.0026	.0025	.0024	.0023	.0023	.0022	.0021	.0021	.0020	.0019
−2.7	.0035	.0034	.0033	.0032	.0031	.0030	.0029	.0028	.0027	.0026
−2.6	.0047	.0045	.0044	.0043	.0041	.0040	.0039	.0038	.0037	.0036
−2.5	.0062	.0060	.0059	.0057	.0055	.0054	.0052	.0051	.0049	.0048
−2.4	.0082	.0080	.0078	.0075	.0073	.0071	.0069	.0068	.0066	.0064
−2.3	.0107	.0104	.0102	.0099	.0096	.0094	.0091	.0089	.0087	.0084
−2.2	.0139	.0136	.0132	.0129	.0125	.0122	.0119	.0116	.0113	.0110
−2.1	.0179	.0174	.0170	.0166	.0162	.0158	.0154	.0150	.0146	.0143
−2.0	.0228	.0222	.0217	.0212	.0207	.0202	.0197	.0192	.0188	.0183
−1.9	.0287	.0281	.0274	.0268	.0262	.0256	.0250	.0244	.0239	.0233
−1.8	.0359	.0351	.0344	.0336	.0329	.0322	.0314	.0307	.0301	.0294
−1.7	.0446	.0436	.0427	.0418	.0409	.0401	.0392	.0384	.0375	.0367
−1.6	.0548	.0537	.0526	.0516	.0505	.0495	.0485	.0475	.0465	.0455
−1.5	.0668	.0655	.0643	.0630	.0618	.0606	.0594	.0582	.0571	.0559
−1.4	.0808	.0793	.0778	.0764	.0749	.0735	.0721	.0708	.0694	.0681
−1.3	.0968	.0951	.0934	.0918	.0901	.0885	.0869	.0853	.0838	.0823
−1.2	.1151	.1131	.1112	.1093	.1075	.1056	.1038	.1020	.1003	.0985
−1.1	.1357	.1335	.1314	.1292	.1271	.1251	.1230	.1210	.1190	.1170
−1.0	.1587	.1562	.1539	.1515	.1492	.1469	.1446	.1423	.1401	.1379
−0.9	.1841	.1814	.1788	.1762	.1736	.1711	.1685	.1660	.1635	.1611
−0.8	.2119	.2090	.2061	.2033	.2005	.1977	.1949	.1922	.1894	.1867
−0.7	.2420	.2389	.2358	.2327	.2296	.2266	.2236	.2206	.2177	.2148
−0.6	.2743	.2709	.2676	.2643	.2611	.2578	.2546	.2514	.2483	.2451
−0.5	.3085	.3050	.3015	.2981	.2946	.2912	.2877	.2843	.2810	.2776
−0.4	.3446	.3409	.3372	.3336	.3300	.3264	.3228	.3192	.3156	.3121
−0.3	.3821	.3783	.3745	.3707	.3669	.3632	.3594	.3557	.3520	.3483
−0.2	.4207	.4168	.4129	.4090	.4052	.4013	.3974	.3936	.3897	.3859
−0.1	.4602	.4562	.4522	.4483	.4443	.4404	.4364	.4325	.4286	.4247
−0.0	.5000	.4960	.4920	.4880	.4840	.4801	.4761	.4721	.4681	.4641

(Continued)

TABLE IV *(Continued)*

The entries in the table on this page give the cumulative area under the standard normal curve to the left of *z* with the values of *z* equal to 0 or positive.

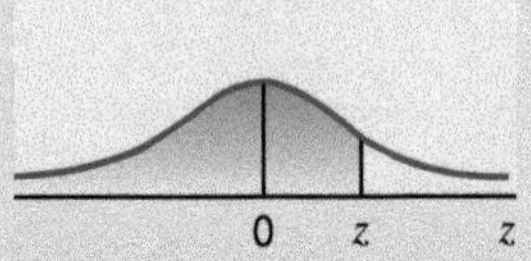

z	.00	.01	.02	.03	.04	.05	.06	.07	.08	.09
0.0	.5000	.5040	.5080	.5120	.5160	.5199	.5239	.5279	.5319	.5359
0.1	.5398	.5438	.5478	.5517	.5557	.5596	.5636	.5675	.5714	.5753
0.2	.5793	.5832	.5871	.5910	.5948	.5987	.6026	.6064	.6103	.6141
0.3	.6179	.6217	.6255	.6293	.6331	.6368	.6406	.6443	.6480	.6517
0.4	.6554	.6591	.6628	.6664	.6700	.6736	.6772	.6808	.6844	.6879
0.5	.6915	.6950	.6985	.7019	.7054	.7088	.7123	.7157	.7190	.7224
0.6	.7257	.7291	.7324	.7357	.7389	.7422	.7454	.7486	.7517	.7549
0.7	.7580	.7611	.7642	.7673	.7704	.7734	.7764	.7794	.7823	.7852
0.8	.7881	.7910	.7939	.7967	.7995	.8023	.8051	.8078	.8106	.8133
0.9	.8159	.8186	.8212	.8238	.8264	.8289	.8315	.8340	.8365	.8389
1.0	.8413	.8438	.8461	.8485	.8508	.8531	.8554	.8577	.8599	.8621
1.1	.8643	.8665	.8686	.8708	.8729	.8749	.8770	.8790	.8810	.8830
1.2	.8849	.8869	.8888	.8907	.8925	.8944	.8962	.8980	.8997	.9015
1.3	.9032	.9049	.9066	.9082	.9099	.9115	.9131	.9147	.9162	.9177
1.4	.9192	.9207	.9222	.9236	.9251	.9265	.9279	.9292	.9306	.9319
1.5	.9332	.9345	.9357	.9370	.9382	.9394	.9406	.9418	.9429	.9441
1.6	.9452	.9463	.9474	.9484	.9495	.9505	.9515	.9525	.9535	.9545
1.7	.9554	.9564	.9573	.9582	.9591	.9599	.9608	.9616	.9625	.9633
1.8	.9641	.9649	.9656	.9664	.9671	.9678	.9686	.9693	.9699	.9706
1.9	.9713	.9719	.9726	.9732	.9738	.9744	.9750	.9756	.9761	.9767
2.0	.9772	.9778	.9783	.9788	.9793	.9798	.9803	.9808	.9812	.9817
2.1	.9821	.9826	.9830	.9834	.9838	.9842	.9846	.9850	.9854	.9857
2.2	.9861	.9864	.9868	.9871	.9875	.9878	.9881	.9884	.9887	.9890
2.3	.9893	.9896	.9898	.9901	.9904	.9906	.9909	.9911	.9913	.9916
2.4	.9918	.9920	.9922	.9925	.9927	.9929	.9931	.9932	.9934	.9936
2.5	.9938	.9940	.9941	.9943	.9945	.9946	.9948	.9949	.9951	.9952
2.6	.9953	.9955	.9956	.9957	.9959	.9960	.9961	.9962	.9963	.9964
2.7	.9965	.9966	.9967	.9968	.9969	.9970	.9971	.9972	.9973	.9974
2.8	.9974	.9975	.9976	.9977	.9977	.9978	.9979	.9979	.9980	.9981
2.9	.9981	.9982	.9982	.9983	.9984	.9984	.9985	.9985	.9986	.9986
3.0	.9987	.9987	.9987	.9988	.9988	.9989	.9989	.9989	.9990	.9990
3.1	.9990	.9991	.9991	.9991	.9992	.9992	.9992	.9992	.9993	.9993
3.2	.9993	.9993	.9994	.9994	.9994	.9994	.9994	.9995	.9995	.9995
3.3	.9995	.9995	.9995	.9996	.9996	.9996	.9996	.9996	.9996	.9997
3.4	.9997	.9997	.9997	.9997	.9997	.9997	.9997	.9997	.9997	.9998

TABLE V The *t* Distribution Table

The entries in this table give the critical values of *t* for the specified number of degrees of freedom and areas in the right tail.

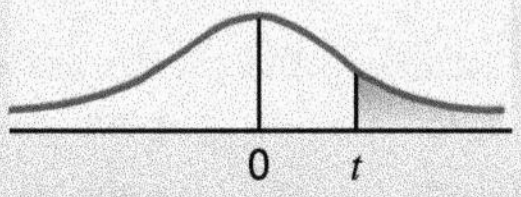

	Area in the Right Tail Under the *t* Distribution Curve					
df	.10	.05	.025	.01	.005	.001
1	3.078	6.314	12.706	31.821	63.657	318.309
2	1.886	2.920	4.303	6.965	9.925	22.327
3	1.638	2.353	3.182	4.541	5.841	10.215
4	1.533	2.132	2.776	3.747	4.604	7.173
5	1.476	2.015	2.571	3.365	4.032	5.893
6	1.440	1.943	2.447	3.143	3.707	5.208
7	1.415	1.895	2.365	2.998	3.499	4.785
8	1.397	1.860	2.306	2.896	3.355	4.501
9	1.383	1.833	2.262	2.821	3.250	4.297
10	1.372	1.812	2.228	2.764	3.169	4.144
11	1.363	1.796	2.201	2.718	3.106	4.025
12	1.356	1.782	2.179	2.681	3.055	3.930
13	1.350	1.771	2.160	2.650	3.012	3.852
14	1.345	1.761	2.145	2.624	2.977	3.787
15	1.341	1.753	2.131	2.602	2.947	3.733
16	1.337	1.746	2.120	2.583	2.921	3.686
17	1.333	1.740	2.110	2.567	2.898	3.646
18	1.330	1.734	2.101	2.552	2.878	3.610
19	1.328	1.729	2.093	2.539	2.861	3.579
20	1.325	1.725	2.086	2.528	2.845	3.552
21	1.323	1.721	2.080	2.518	2.831	3.527
22	1.321	1.717	2.074	2.508	2.819	3.505
23	1.319	1.714	2.069	2.500	2.807	3.485
24	1.318	1.711	2.064	2.492	2.797	3.467
25	1.316	1.708	2.060	2.485	2.787	3.450
26	1.315	1.706	2.056	2.479	2.779	3.435
27	1.314	1.703	2.052	2.473	2.771	3.421
28	1.313	1.701	2.048	2.467	2.763	3.408
29	1.311	1.699	2.045	2.462	2.756	3.396
30	1.310	1.697	2.042	2.457	2.750	3.385
31	1.309	1.696	2.040	2.453	2.744	3.375
32	1.309	1.694	2.037	2.449	2.738	3.365
33	1.308	1.692	2.035	2.445	2.733	3.356
34	1.307	1.691	2.032	2.441	2.728	3.348
35	1.306	1.690	2.030	2.438	2.724	3.340

(Continued)

TABLE V *(Continued)*

	Area in the Right Tail Under the *t* Distribution Curve					
df	.10	.05	.025	.01	.005	.001
36	1.306	1.688	2.028	2.434	2.719	3.333
37	1.305	1.687	2.026	2.431	2.715	3.326
38	1.304	1.686	2.024	2.429	2.712	3.319
39	1.304	1.685	2.023	2.426	2.708	3.313
40	1.303	1.684	2.021	2.423	2.704	3.307
41	1.303	1.683	2.020	2.421	2.701	3.301
42	1.302	1.682	2.018	2.418	2.698	3.296
43	1.302	1.681	2.017	2.416	2.695	3.291
44	1.301	1.680	2.015	2.414	2.692	3.286
45	1.301	1.679	2.014	2.412	2.690	3.281
46	1.300	1.679	2.013	2.410	2.687	3.277
47	1.300	1.678	2.012	2.408	2.685	3.273
48	1.299	1.677	2.011	2.407	2.682	3.269
49	1.299	1.677	2.010	2.405	2.680	3.265
50	1.299	1.676	2.009	2.403	2.678	3.261
51	1.298	1.675	2.008	2.402	2.676	3.258
52	1.298	1.675	2.007	2.400	2.674	3.255
53	1.298	1.674	2.006	2.399	2.672	3.251
54	1.297	1.674	2.005	2.397	2.670	3.248
55	1.297	1.673	2.004	2.396	2.668	3.245
56	1.297	1.673	2.003	2.395	2.667	3.242
57	1.297	1.672	2.002	2.394	2.665	3.239
58	1.296	1.672	2.002	2.392	2.663	3.237
59	1.296	1.671	2.001	2.391	2.662	3.234
60	1.296	1.671	2.000	2.390	2.660	3.232
61	1.296	1.670	2.000	2.389	2.659	3.229
62	1.295	1.670	1.999	2.388	2.657	3.227
63	1.295	1.669	1.998	2.387	2.656	3.225
64	1.295	1.669	1.998	2.386	2.655	3.223
65	1.295	1.669	1.997	2.385	2.654	3.220
66	1.295	1.668	1.997	2.384	2.652	3.218
67	1.294	1.668	1.996	2.383	2.651	3.216
68	1.294	1.668	1.995	2.382	2.650	3.214
69	1.294	1.667	1.995	2.382	2.649	3.213
70	1.294	1.667	1.994	2.381	2.648	3.211
71	1.294	1.667	1.994	2.380	2.647	3.209
72	1.293	1.666	1.993	2.379	2.646	3.207
73	1.293	1.666	1.993	2.379	2.645	3.206
74	1.293	1.666	1.993	2.378	2.644	3.204
75	1.293	1.665	1.992	2.377	2.643	3.202
∞	1.282	1.645	1.960	2.326	2.576	3.090

TABLE VI Chi-Square Distribution Table

The entries in this table give the critical values of χ^2 for the specified number of degrees of freedom and areas in the right tail.

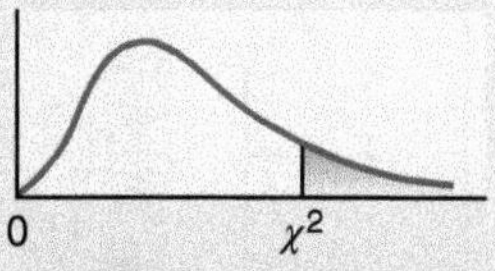

	Area in the Right Tail Under the Chi-Square Distribution Curve									
df	.995	.990	.975	.950	.900	.100	.050	.025	.010	.005
1	0.000	0.000	0.001	0.004	0.016	2.706	3.841	5.024	6.635	7.879
2	0.010	0.020	0.051	0.103	0.211	4.605	5.991	7.378	9.210	10.597
3	0.072	0.115	0.216	0.352	0.584	6.251	7.815	9.348	11.345	12.838
4	0.207	0.297	0.484	0.711	1.064	7.779	9.488	11.143	13.277	14.860
5	0.412	0.554	0.831	1.145	1.610	9.236	11.070	12.833	15.086	16.750
6	0.676	0.872	1.237	1.635	2.204	10.645	12.592	14.449	16.812	18.548
7	0.989	1.239	1.690	2.167	2.833	12.017	14.067	16.013	18.475	20.278
8	1.344	1.646	2.180	2.733	3.490	13.362	15.507	17.535	20.090	21.955
9	1.735	2.088	2.700	3.325	4.168	14.684	16.919	19.023	21.666	23.589
10	2.156	2.558	3.247	3.940	4.865	15.987	18.307	20.483	23.209	25.188
11	2.603	3.053	3.816	4.575	5.578	17.275	19.675	21.920	24.725	26.757
12	3.074	3.571	4.404	5.226	6.304	18.549	21.026	23.337	26.217	28.300
13	3.565	4.107	5.009	5.892	7.042	19.812	22.362	24.736	27.688	29.819
14	4.075	4.660	5.629	6.571	7.790	21.064	23.685	26.119	29.141	31.319
15	4.601	5.229	6.262	7.261	8.547	22.307	24.996	27.488	30.578	32.801
16	5.142	5.812	6.908	7.962	9.312	23.542	26.296	28.845	32.000	34.267
17	5.697	6.408	7.564	8.672	10.085	24.769	27.587	30.191	33.409	35.718
18	6.265	7.015	8.231	9.390	10.865	25.989	28.869	31.526	34.805	37.156
19	6.844	7.633	8.907	10.117	11.651	27.204	30.144	32.852	36.191	38.582
20	7.434	8.260	9.591	10.851	12.443	28.412	31.410	34.170	37.566	39.997
21	8.034	8.897	10.283	11.591	13.240	29.615	32.671	35.479	38.932	41.401
22	8.643	9.542	10.982	12.338	14.041	30.813	33.924	36.781	40.289	42.796
23	9.260	10.196	11.689	13.091	14.848	32.007	35.172	38.076	41.638	44.181
24	9.886	10.856	12.401	13.848	15.659	33.196	36.415	39.364	42.980	45.559
25	10.520	11.524	13.120	14.611	16.473	34.382	37.652	40.646	44.314	46.928
26	11.160	12.198	13.844	15.379	17.292	35.563	38.885	41.923	45.642	48.290
27	11.808	12.879	14.573	16.151	18.114	36.741	40.113	43.195	46.963	49.645
28	12.461	13.565	15.308	16.928	18.939	37.916	41.337	44.461	48.278	50.993
29	13.121	14.256	16.047	17.708	19.768	39.087	42.557	45.722	49.588	52.336
30	13.787	14.953	16.791	18.493	20.599	40.256	43.773	46.979	50.892	53.672
40	20.707	22.164	24.433	26.509	29.051	51.805	55.758	59.342	63.691	66.766
50	27.991	29.707	32.357	34.764	37.689	63.167	67.505	71.420	76.154	79.490
60	35.534	37.485	40.482	43.188	46.459	74.397	79.082	83.298	88.379	91.952
70	43.275	45.442	48.758	51.739	55.329	85.527	90.531	95.023	100.425	104.215
80	51.172	53.540	57.153	60.391	64.278	96.578	101.879	106.629	112.329	116.321
90	59.196	61.754	65.647	69.126	73.291	107.565	113.145	118.136	124.116	128.299
100	67.328	70.065	74.222	77.929	82.358	118.498	124.342	129.561	135.807	140.169

TABLE VII The *F* Distribution Table

The entries in the table on this page give the critical values of *F* for .01 area in the right tail under the *F* distribution curve and specified degrees of freedom for the numerator and denominator.

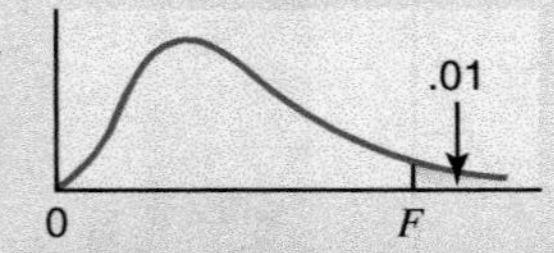

Degrees of Freedom for the Denominator	Degrees of Freedom for the Numerator																		
	1	2	3	4	5	6	7	8	9	10	11	12	15	20	25	30	40	50	100
1	4052	5000	5403	5625	5764	5859	5928	5981	6022	6056	6083	6106	6157	6209	6240	6261	6287	6303	6334
2	98.50	99.00	99.17	99.25	99.30	99.33	99.36	99.37	99.39	99.40	99.41	99.42	99.43	99.45	99.46	99.47	99.47	99.48	99.49
3	34.12	30.82	29.46	28.71	28.24	27.91	27.67	27.49	27.35	27.23	27.13	27.05	26.87	26.69	26.58	26.50	26.41	26.35	26.24
4	21.20	18.00	16.69	15.98	15.52	15.21	14.98	14.80	14.66	14.55	14.45	14.37	14.20	14.02	13.91	13.84	13.75	13.69	13.58
5	16.26	13.27	12.06	11.39	10.97	10.67	10.46	10.29	10.16	10.05	9.96	9.89	9.72	9.55	9.45	9.38	9.29	9.24	9.13
6	13.75	10.92	9.78	9.15	8.75	8.47	8.26	8.10	7.98	7.87	7.79	7.72	7.56	7.40	7.30	7.23	7.14	7.09	6.99
7	12.25	9.55	8.45	7.85	7.46	7.19	6.99	6.84	6.72	6.62	6.54	6.47	6.31	6.16	6.06	5.99	5.91	5.86	5.75
8	11.26	8.65	7.59	7.01	6.63	6.37	6.18	6.03	5.91	5.81	5.73	5.67	5.52	5.36	5.26	5.20	5.12	5.07	4.96
9	10.56	8.02	6.99	6.42	6.06	5.80	5.61	5.47	5.35	5.26	5.18	5.11	4.96	4.81	4.71	4.65	4.57	4.52	4.41
10	10.04	7.56	6.55	5.99	5.64	5.39	5.20	5.06	4.94	4.85	4.77	4.71	4.56	4.41	4.31	4.25	4.17	4.12	4.01
11	9.65	7.21	6.22	5.67	5.32	5.07	4.89	4.74	4.63	4.54	4.46	4.40	4.25	4.10	4.01	3.94	3.86	3.81	3.71
12	9.33	6.93	5.95	5.41	5.06	4.82	4.64	4.50	4.39	4.30	4.22	4.16	4.01	3.86	3.76	3.70	3.62	3.57	3.47
13	9.07	6.70	5.74	5.21	4.86	4.62	4.44	4.30	4.19	4.10	4.02	3.96	3.82	3.66	3.57	3.51	3.43	3.38	3.27
14	8.86	6.51	5.56	5.04	4.69	4.46	4.28	4.14	4.03	3.94	3.86	3.80	3.66	3.51	3.41	3.35	3.27	3.22	3.11
15	8.68	6.36	5.42	4.89	4.56	4.32	4.14	4.00	3.89	3.80	3.73	3.67	3.52	3.37	3.28	3.21	3.13	3.08	2.98
16	8.53	6.23	5.29	4.77	4.44	4.20	4.03	3.89	3.78	3.69	3.62	3.55	3.41	3.26	3.16	3.10	3.02	2.97	2.86
17	8.40	6.11	5.18	4.67	4.34	4.10	3.93	3.79	3.68	3.59	3.52	3.46	3.31	3.16	3.07	3.00	2.92	2.87	2.76
18	8.29	6.01	5.09	4.58	4.25	4.01	3.84	3.71	3.60	3.51	3.43	3.37	3.23	3.08	2.98	2.92	2.84	2.78	2.68
19	8.18	5.93	5.01	4.50	4.17	3.94	3.77	3.63	3.52	3.43	3.36	3.30	3.15	3.00	2.91	2.84	2.76	2.71	2.60
20	8.10	5.85	4.94	4.43	4.10	3.87	3.70	3.56	3.46	3.37	3.29	3.23	3.09	2.94	2.84	2.78	2.69	2.64	2.54
21	8.02	5.78	4.87	4.37	4.04	3.81	3.64	3.51	3.40	3.31	3.24	3.17	3.03	2.88	2.79	2.72	2.64	2.58	2.48
22	7.95	5.72	4.82	4.31	3.99	3.76	3.59	3.45	3.35	3.26	3.18	3.12	2.98	2.83	2.73	2.67	2.58	2.53	2.42
23	7.88	5.66	4.76	4.26	3.94	3.71	3.54	3.41	3.30	3.21	3.14	3.07	2.93	2.78	2.69	2.62	2.54	2.48	2.37
24	7.82	5.61	4.72	4.22	3.90	3.67	3.50	3.36	3.26	3.17	3.09	3.03	2.89	2.74	2.64	2.58	2.49	2.44	2.33
25	7.77	5.57	4.68	4.18	3.85	3.63	3.46	3.32	3.22	3.13	3.06	2.99	2.85	2.70	2.60	2.54	2.45	2.40	2.29
30	7.56	5.39	4.51	4.02	3.70	3.47	3.30	3.17	3.07	2.98	2.91	2.84	2.70	2.55	2.45	2.39	2.30	2.25	2.13
40	7.31	5.18	4.31	3.83	3.51	3.29	3.12	2.99	2.89	2.80	2.73	2.66	2.52	2.37	2.27	2.20	2.11	2.06	1.94
50	7.17	5.06	4.20	3.72	3.41	3.19	3.02	2.89	2.78	2.70	2.63	2.56	2.42	2.27	2.17	2.10	2.01	1.95	1.82
100	6.90	4.82	3.98	3.51	3.21	2.99	2.82	2.69	2.59	2.50	2.43	2.37	2.22	2.07	1.97	1.89	1.80	1.74	1.60

TABLE VII (*Continued*)

The entries in the table on this page give the critical values of *F* for .025 area in the right tail under the *F* distribution curve and specified degrees of freedom for the numerator and denominator.

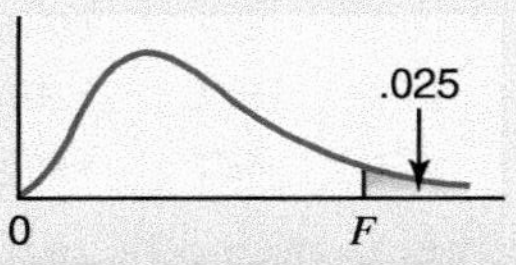

Degrees of Freedom for the Denominator	Degrees of Freedom for the Numerator																		
	1	2	3	4	5	6	7	8	9	10	11	12	15	20	25	30	40	50	100
1	647.8	799.5	864.2	899.6	921.8	937.1	948.2	956.7	963.3	968.6	973.0	976.7	984.9	993.1	998.1	1001	1006	1008	1013
2	38.51	39.00	39.17	39.25	39.30	39.33	39.36	39.37	39.39	39.40	39.41	39.41	39.43	39.45	39.46	39.46	39.47	39.48	39.49
3	17.44	16.04	15.44	15.10	14.88	14.73	14.62	14.54	14.47	14.42	14.37	14.34	14.25	14.17	14.12	14.08	14.04	14.01	13.96
4	12.22	10.65	9.98	9.61	6.36	9.20	9.07	8.98	8.90	8.84	8.79	8.75	8.66	8.56	8.50	8.46	8.41	8.38	8.32
5	10.01	8.43	7.76	7.39	7.15	6.98	6.85	6.76	6.68	6.62	6.57	6.52	6.43	6.33	6.27	6.23	6.18	6.14	6.08
6	8.81	7.26	6.60	6.23	5.99	5.82	5.70	5.60	5.52	5.46	5.41	5.37	5.27	5.17	5.11	5.07	5.01	4.98	4.92
7	8.07	6.54	5.89	5.52	5.29	5.12	4.99	4.90	4.82	4.76	4.71	4.67	4.57	4.47	4.40	4.36	4.31	4.28	4.21
8	7.57	6.06	5.42	5.05	4.82	4.65	4.53	4.43	4.36	4.30	4.24	4.20	4.10	4.00	3.94	3.89	3.84	3.81	3.74
9	7.21	5.72	5.08	4.72	4.48	4.32	4.20	4.10	4.03	3.96	3.91	3.87	3.77	3.67	3.60	3.56	3.51	3.47	3.40
10	6.94	5.46	4.83	4.47	4.24	4.07	3.95	3.85	3.78	3.72	3.66	3.62	3.52	3.42	3.35	3.31	3.26	3.22	3.15
11	6.72	5.26	4.63	4.28	4.04	3.88	3.76	3.66	3.59	3.53	3.47	3.43	3.33	3.23	3.16	3.12	3.06	3.03	2.96
12	6.55	5.10	4.47	4.12	3.89	3.73	3.61	3.51	3.44	3.37	3.32	3.28	3.18	3.07	3.01	2.96	2.91	2.87	2.80
13	6.41	4.97	4.35	4.00	3.77	3.60	3.48	3.39	3.31	3.25	3.20	3.15	3.05	2.95	2.88	2.84	2.78	2.74	2.67
14	6.30	4.86	4.24	3.89	3.66	3.50	3.38	3.29	3.21	3.15	3.09	3.05	2.95	2.84	2.78	2.73	2.67	2.64	2.56
15	6.20	4.77	4.15	3.80	3.58	3.41	3.29	3.20	3.12	3.06	3.01	2.96	2.86	2.76	2.69	2.64	2.59	2.55	2.47
16	6.12	4.69	4.08	3.73	3.50	3.34	3.22	3.12	3.05	2.99	2.93	2.89	2.79	2.68	2.61	2.57	2.51	2.47	2.40
17	6.04	4.62	4.01	3.66	3.44	3.28	3.16	3.06	2.98	2.92	2.87	2.82	2.72	2.62	2.55	2.50	2.44	2.41	2.33
18	5.98	4.56	3.95	3.61	3.38	3.22	3.10	3.01	2.93	2.87	2.81	2.77	2.67	2.56	2.49	2.44	2.38	2.35	2.27
19	5.92	4.51	3.90	3.56	3.33	3.17	3.05	2.96	2.88	2.82	2.76	2.72	2.62	2.51	2.44	2.39	2.33	2.30	2.22
20	5.87	4.46	3.86	3.51	3.29	3.13	3.01	2.91	2.84	2.77	2.72	2.68	2.57	2.46	2.40	2.35	2.29	2.25	2.17
21	5.83	4.42	3.82	3.48	3.25	3.09	2.97	2.87	2.80	2.73	2.68	2.64	2.53	2.42	2.36	2.31	2.25	2.21	2.13
22	5.79	4.38	3.78	3.44	3.22	3.05	2.93	2.84	2.76	2.70	2.65	2.60	2.50	2.39	2.32	2.27	2.21	2.17	2.09
23	5.75	4.35	3.75	3.41	3.18	3.02	2.90	2.81	2.73	2.67	2.62	2.57	2.47	2.36	2.29	2.24	2.18	2.14	2.06
24	5.72	4.32	3.72	3.38	3.15	2.99	2.87	2.78	2.70	2.64	2.59	2.54	2.44	2.33	2.26	2.21	2.15	2.11	2.02
25	5.69	4.29	3.69	3.35	3.13	2.97	2.85	2.75	2.68	2.61	2.56	2.51	2.41	2.30	2.23	2.18	2.12	2.08	2.00
30	5.57	4.18	3.59	3.25	3.03	2.87	2.75	2.65	2.57	2.51	2.46	2.41	2.31	2.20	2.12	2.07	2.01	1.97	1.88
40	5.42	4.05	3.46	3.13	2.90	2.74	2.62	2.53	2.45	2.39	2.33	2.29	2.18	2.07	1.99	1.94	1.88	1.83	1.74
50	5.34	3.97	3.39	3.05	2.83	2.67	2.55	2.46	2.38	2.32	2.26	2.22	2.11	1.99	1.92	1.87	1.80	1.75	1.66
100	5.18	3.83	3.25	2.92	2.70	2.54	2.42	2.32	2.24	2.18	2.12	2.08	1.97	1.85	1.77	1.71	1.64	1.59	1.48

(*Continued*)

TABLE VII (*Continued*)

The entries in the table on this page give the critical values of *F* for .05 area in the right tail under the *F* distribution curve and specified degrees of freedom for the numerator and denominator.

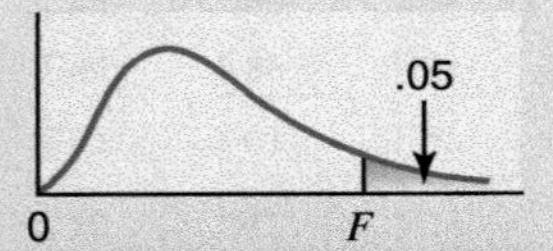

Degrees of Freedom for the Denominator	Degrees of Freedom for the Numerator																		
	1	2	3	4	5	6	7	8	9	10	11	12	15	20	25	30	40	50	100
1	161.5	199.5	215.7	224.6	230.2	234.0	236.8	238.9	240.5	241.9	243.0	243.9	246.0	248.0	249.3	250.1	251.1	251.8	253.0
2	18.51	19.00	19.16	19.25	19.30	19.33	19.35	19.37	19.38	19.40	19.40	19.41	19.43	19.45	19.46	19.46	19.47	19.48	19.49
3	10.13	9.55	9.28	9.12	9.01	8.94	8.89	8.85	8.81	8.79	8.76	8.74	8.70	8.66	8.63	8.62	8.59	8.58	8.55
4	7.71	6.94	6.59	6.39	6.26	6.16	6.09	6.04	6.00	5.96	5.94	5.91	5.86	5.80	5.77	5.75	5.72	5.70	5.66
5	6.61	5.79	5.41	5.19	5.05	4.95	4.88	4.82	4.77	4.74	4.70	4.68	4.62	4.56	4.52	4.50	4.46	4.44	4.41
6	5.99	5.14	4.76	4.53	4.39	4.28	4.21	4.15	4.10	4.06	4.03	4.00	3.94	3.87	3.83	3.81	3.77	3.75	3.71
7	5.59	4.74	4.35	4.12	3.97	3.87	3.79	3.73	3.68	3.64	3.60	3.57	3.51	3.44	3.40	3.38	3.34	3.32	3.27
8	5.32	4.46	4.07	3.84	3.69	3.58	3.50	3.44	3.39	3.35	3.31	3.28	3.22	3.15	3.11	3.08	3.04	3.02	2.97
9	5.12	4.26	3.86	3.63	3.48	3.37	3.29	3.23	3.18	3.14	3.10	3.07	3.01	2.94	2.89	2.86	2.83	2.80	2.76
10	4.96	4.10	3.71	3.48	3.33	3.22	3.14	3.07	3.02	2.98	2.94	2.91	2.85	2.77	2.73	2.70	2.66	2.64	2.59
11	4.84	3.98	3.59	3.36	3.20	3.09	3.01	2.95	2.90	2.85	2.82	2.79	2.72	2.65	2.60	2.57	2.53	2.51	2.46
12	4.75	3.89	3.49	3.26	3.11	3.00	2.91	2.85	2.80	2.75	2.72	2.69	2.62	2.54	2.50	2.47	2.43	2.40	2.35
13	4.67	3.81	3.41	3.18	3.03	2.92	2.83	2.77	2.71	2.67	2.63	2.60	2.53	2.46	2.41	2.38	2.34	2.31	2.26
14	4.60	3.74	3.34	3.11	2.96	2.85	2.76	2.70	2.65	2.60	2.57	2.53	2.46	2.39	2.34	2.31	2.27	2.24	2.19
15	4.54	3.68	3.29	3.06	2.90	2.79	2.71	2.64	2.59	2.54	2.51	2.48	2.40	2.33	2.28	2.25	2.20	2.18	2.12
16	4.49	3.63	3.24	3.01	2.85	2.74	2.66	2.59	2.54	2.49	2.46	2.42	2.35	2.28	2.23	2.19	2.15	2.12	2.07
17	4.45	3.59	3.20	2.96	2.81	2.70	2.61	2.55	2.49	2.45	2.41	2.38	2.31	2.23	2.18	2.15	2.10	2.08	2.02
18	4.41	3.55	3.16	2.93	2.77	2.66	2.58	2.51	2.46	2.41	2.37	2.34	2.27	2.19	2.14	2.11	2.06	2.04	1.98
19	4.38	3.52	3.13	2.90	2.74	2.63	2.54	2.48	2.42	2.38	2.34	2.31	2.23	2.16	2.11	2.07	2.03	2.00	1.94
20	4.35	3.49	3.10	2.87	2.71	2.60	2.51	2.45	2.39	2.35	2.31	2.28	2.20	2.12	2.07	2.04	1.99	1.97	1.91
21	4.32	3.47	3.07	2.84	2.68	2.57	2.49	2.42	2.37	2.32	2.28	2.25	2.18	2.10	2.05	2.01	1.96	1.94	1.88
22	4.30	3.44	3.05	2.82	2.66	2.55	2.46	2.40	2.34	2.30	2.26	2.23	2.15	2.07	2.02	1.97	1.94	1.91	1.85
23	4.28	3.42	3.03	2.80	2.64	2.53	2.44	2.37	2.32	2.27	2.24	2.20	2.13	2.05	2.00	1.96	1.91	1.88	1.82
24	4.26	3.40	3.01	2.78	2.62	2.51	2.42	2.36	2.30	2.25	2.22	2.18	2.16	2.03	1.97	1.94	1.89	1.86	1.80
25	4.24	3.39	2.99	2.76	2.60	2.49	2.40	2.34	2.28	2.24	2.20	2.16	2.09	2.01	1.96	1.92	1.87	1.84	1.78
30	4.17	3.32	2.92	2.69	2.53	2.42	2.33	2.27	2.21	2.16	2.13	2.09	2.01	1.93	1.88	1.84	1.79	1.76	1.70
40	4.08	3.23	2.84	2.61	2.45	2.34	2.25	2.18	2.12	2.08	2.04	2.00	1.92	1.84	1.78	1.74	1.69	1.66	1.59
50	4.03	3.18	2.79	2.56	2.40	2.29	2.20	2.13	2.07	2.03	1.99	1.95	1.87	1.78	1.73	1.69	1.63	1.60	1.52
100	3.94	3.09	2.70	2.46	2.31	2.19	2.10	2.03	1.97	1.93	1.89	1.85	1.77	1.68	1.62	1.57	1.52	1.48	1.39

TABLE VII *(Continued)*

The entries in the table on this page give the critical values of F for .10 area in the right tail under the F distribution curve and specified degrees of freedom for the numerator and denominator.

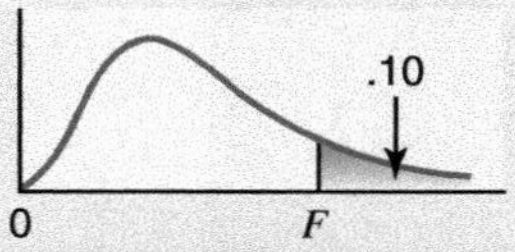

Degrees of Freedom for the Denominator	Degrees of Freedom for the Numerator																		
	1	2	3	4	5	6	7	8	9	10	11	12	15	20	25	30	40	50	100
1	39.86	49.50	53.59	55.83	57.24	58.20	58.91	59.44	59.86	60.19	60.47	60.71	61.22	61.74	62.05	62.26	62.53	62.69	63.01
2	8.53	9.00	9.16	9.24	9.29	9.33	9.35	9.37	9.38	9.39	9.40	9.41	9.42	9.44	9.45	9.46	9.47	9.47	9.48
3	5.54	5.46	5.39	5.34	5.31	5.28	5.27	5.25	5.24	5.23	5.22	5.22	5.20	5.18	5.17	5.17	5.16	5.15	5.14
4	4.54	4.32	4.19	4.11	4.05	4.01	3.98	3.95	3.94	3.92	3.91	3.90	3.87	3.84	3.83	3.82	3.80	3.80	3.78
5	4.06	3.78	3.62	3.52	3.45	3.40	3.37	3.34	3.32	3.30	3.28	3.27	3.24	3.21	3.19	3.17	3.16	3.15	3.13
6	3.78	3.46	3.29	3.18	3.11	3.05	3.01	2.98	2.96	2.94	2.92	2.90	2.87	2.84	2.81	2.80	2.78	2.77	2.75
7	3.59	3.26	3.07	2.96	2.88	2.83	2.78	2.75	2.72	2.70	2.68	2.67	2.63	2.59	2.57	2.56	2.54	2.52	2.50
8	3.46	3.11	2.92	2.81	2.73	2.67	2.62	2.59	2.56	2.54	2.52	2.50	2.46	2.42	2.40	2.38	2.36	2.35	2.32
9	3.36	3.01	2.81	2.69	2.61	2.55	2.51	2.47	2.44	2.42	2.40	2.38	2.34	2.30	2.27	2.25	2.23	2.22	2.19
10	3.29	2.92	2.73	2.61	2.52	2.46	2.41	2.38	2.35	2.32	2.30	2.28	2.24	2.20	2.17	2.16	2.13	2.12	2.09
11	3.23	2.86	2.66	2.54	2.45	2.39	2.34	2.30	2.27	2.25	2.23	2.21	2.17	2.12	2.10	2.08	2.05	2.04	2.01
12	3.18	2.81	2.61	2.48	2.39	2.33	2.28	2.24	2.21	2.19	2.17	2.15	2.10	2.06	2.03	2.01	1.99	1.97	1.94
13	3.14	2.76	2.56	2.43	2.35	2.28	2.23	2.20	2.16	2.14	2.12	2.10	2.05	2.01	1.98	1.96	1.93	1.92	1.88
14	3.10	2.73	2.52	2.39	2.31	2.24	2.19	2.15	2.12	2.10	2.07	2.05	2.01	1.96	1.93	1.91	1.89	1.87	1.83
15	3.07	2.70	2.49	2.36	2.27	2.21	2.16	2.12	2.09	2.06	2.04	2.02	1.97	1.92	1.89	1.87	1.85	1.83	1.79
16	3.05	2.67	2.46	2.33	2.24	2.18	2.13	2.09	2.06	2.03	2.01	1.99	1.94	1.89	1.86	1.84	1.81	1.79	1.76
17	3.03	2.64	2.44	2.31	2.22	2.15	2.10	2.06	2.03	2.00	1.98	1.96	1.91	1.86	1.83	1.81	1.78	1.76	1.73
18	3.01	2.62	2.42	2.29	2.20	2.13	2.08	2.04	2.00	1.98	1.95	1.93	1.89	1.84	1.80	1.78	1.75	1.74	1.70
19	2.99	2.61	2.40	2.27	2.18	2.11	2.06	2.02	1.98	1.96	1.93	1.91	1.86	1.81	1.78	1.76	1.73	1.71	1.67
20	2.97	2.59	2.38	2.25	2.16	2.09	2.04	2.00	1.96	1.94	1.91	1.89	1.84	1.79	1.76	1.74	1.71	1.69	1.65
21	2.96	2.57	2.36	2.23	2.14	2.08	2.02	1.98	1.95	1.92	1.90	1.87	1.83	1.78	1.74	1.72	1.69	1.67	1.63
22	2.95	2.56	2.35	2.22	2.13	2.06	2.01	1.97	1.93	1.90	1.88	1.86	1.81	1.76	1.73	1.70	1.67	1.65	1.61
23	2.94	2.55	2.34	2.21	2.11	2.05	1.99	1.95	1.92	1.89	1.87	1.84	1.80	1.74	1.71	1.69	1.66	1.64	1.59
24	2.93	2.54	2.33	2.19	2.10	2.04	1.98	1.94	1.91	1.88	1.85	1.83	1.78	1.73	1.70	1.67	1.64	1.62	1.58
25	2.92	2.53	2.32	2.18	2.09	2.02	1.97	1.93	1.89	1.87	1.84	1.82	1.77	1.72	1.68	1.66	1.63	1.61	1.56
30	2.88	2.49	2.28	2.14	2.05	1.98	1.93	1.88	1.85	1.82	1.79	1.77	1.72	1.67	1.63	1.61	1.57	1.55	1.51
40	2.84	2.44	2.23	2.09	2.00	1.93	1.87	1.83	1.79	1.76	1.74	1.71	1.66	1.61	1.57	1.54	1.51	1.48	1.43
50	2.81	2.41	2.20	2.06	1.97	1.90	1.84	1.80	1.76	1.73	1.70	1.68	1.63	1.57	1.53	1.50	1.46	1.44	1.39
100	2.76	2.36	2.14	2.00	1.91	1.83	1.78	1.73	1.69	1.66	1.64	1.61	1.56	1.49	1.45	1.42	1.38	1.35	1.29

TABLE VIII Critical Values of X for the Sign Test

	One tail $\alpha = .005$ Two tail $\alpha = .01$		One tail $\alpha = .01$ Two tail $\alpha = .02$		One tail $\alpha = .025$ Two tail $\alpha = .05$		One tail $\alpha = .05$ Two tail $\alpha = .10$	
n	Lower critical value	Upper critical value	Lower critical value	Upper critical value	Lower critical value	Upper critical value	Lower critical value	Upper critical value
1	—	—	—	—	—	—	—	—
2	—	—	—	—	—	—	—	—
3	—	—	—	—	—	—	—	—
4	—	—	—	—	—	—	—	—
5	—	—	—	—	—	—	0	5
6	—	—	—	—	0	6	0	6
7	—	—	0	7	0	7	0	7
8	0	8	0	8	0	8	1	7
9	0	9	0	9	1	8	1	8
10	0	10	0	10	1	9	1	9
11	0	11	1	10	1	10	2	9
12	1	11	1	11	2	10	2	10
13	1	12	1	12	2	11	3	10
14	1	13	2	12	2	12	3	11
15	2	13	2	13	3	12	3	12
16	2	14	2	14	3	13	4	12
17	2	15	3	14	4	13	4	13
18	3	15	3	15	4	14	5	13
19	3	16	4	15	4	15	5	14
20	3	17	4	16	5	15	5	15
21	4	17	4	17	5	16	6	15
22	4	18	5	17	5	17	6	16
23	4	19	5	18	6	17	7	16
24	5	19	5	19	6	18	7	17
25	5	20	6	19	7	18	7	18

Source: Adapted from D. B. Owen, *Handbook of Statistical Tables*. © 1962 by Addison-Wesley Publishing Company, Inc.

TABLE IX Critical Values of T for the Wilcoxon Signed-Rank Test

n	One-tailed $\alpha = .005$ Two-tailed $\alpha = .01$	One-tailed $\alpha = .01$ Two-tailed $\alpha = .02$	One-tailed $\alpha = .025$ Two-tailed $\alpha = .05$	One-tailed $\alpha = .05$ Two-tailed $\alpha = .10$
1	—	—	—	—
2	—	—	—	—
3	—	—	—	—
4	—	—	—	—
5	—	—	—	1
6	—	—	1	2
7	—	0	2	4
8	0	2	4	6
9	2	3	6	8
10	3	5	8	11
11	5	7	11	14
12	7	10	14	17
13	10	13	17	21
14	13	16	21	26
15	16	20	25	30

Source: From *Some Rapid Approximate Statistical Procedures*, 1964. Reprinted with permission of Lederle Pharmaceutical Division of American Cyanamid Company, Philadelphia, PA.

TABLE X Critical Values of T for the Wilcoxon Rank Sum Test

a. One-tailed $\alpha = .025$; Two-tailed $\alpha = .05$

n_2 \ n_1	3		4		5		6		7		8		9		10	
	T_L	T_U	T_L	T_U	T_L	T_U	T_L	T_U	T_L	T_U	T_L	T_U	T_L	T_U	T_L	T_U
3	5	16	6	18	6	21	7	23	7	26	8	28	8	31	9	33
4	6	18	11	25	12	28	12	32	13	35	14	38	15	41	16	44
5	6	21	12	28	18	37	19	41	20	45	21	49	22	53	24	56
6	7	23	12	32	19	41	26	52	28	56	29	61	31	65	32	70
7	7	26	13	35	20	45	28	56	37	68	39	73	41	78	43	83
8	8	28	14	38	21	49	29	61	39	73	49	87	51	93	54	98
9	8	31	15	41	22	53	31	65	41	78	51	93	63	108	66	114
10	9	33	16	44	24	56	32	70	43	83	54	98	66	114	79	131

b. One-tailed $\alpha = .05$; Two-tailed $\alpha = .10$

n_2 \ n_1	3		4		5		6		7		8		9		10	
	T_L	T_U	T_L	T_U	T_L	T_U	T_L	T_U	T_L	T_U	T_L	T_U	T_L	T_U	T_L	T_U
3	6	15	7	17	7	20	8	22	9	24	9	27	10	29	11	31
4	7	17	12	24	13	27	14	30	15	33	16	36	17	39	18	42
5	7	20	13	27	19	36	20	40	22	43	24	46	25	50	26	54
6	8	22	14	30	20	40	28	50	30	54	32	58	33	63	35	67
7	9	24	15	33	22	43	30	54	39	66	41	71	43	76	46	80
8	9	27	16	36	24	46	32	58	41	71	52	84	54	90	57	95
9	10	29	17	39	25	50	33	63	43	76	54	90	66	105	69	111
10	11	31	18	42	26	54	35	67	46	80	57	95	69	111	83	127

Source: From *Some Rapid Approximate Statistical Procedures*, 1964. Reprinted with permission of Lederle Pharmaceutical Division of American Cyanamid Company, Philadelphia, PA.

TABLE XI Critical Values for the Spearman Rho Rank Correlation Coefficient Test

	One-tailed α			
	.05	.025	.01	.005
	Two-tailed α			
n	.10	.05	.02	.01
5	±.900	—	—	—
6	±.829	±.886	±.943	—
7	±.714	±.786	±.893	±.929
8	±.643	±.738	±.833	±.881
9	±.600	±.700	±.783	±.833
10	±.564	±.648	±.745	±.794
11	±.536	±.618	±.709	±.755
12	±.503	±.587	±.678	±.727
13	±.475	±.566	±.672	±.744
14	±.456	±.544	±.645	±.714
15	±.440	±.524	±.622	±.688
16	±.425	±.506	±.601	±.665
17	±.411	±.490	±.582	±.644
18	±.399	±.475	±.564	±.625
19	±.388	±.462	±.548	±.607
20	±.377	±.450	±.534	±.591
21	±.368	±.438	±.520	±.576
22	±.359	±.428	±.508	±.562
23	±.351	±.418	±.496	±.549
24	±.343	±.409	±.485	±.537
25	±.336	±.400	±.475	±.526
26	±.329	±.392	±.465	±.515
27	±.323	±.384	±.456	±.505
28	±.317	±.377	±.448	±.496
29	±.311	±.370	±.440	±.487
30	±.305	±.364	±.432	±.478

TABLE XII Critical Values for a Two-Tailed Runs Test with $\alpha = .05$

n_2 \ n_1	5	6	7	8	9	10	11	12	13	14	15
2	—	—	—	—	—	—	—	2	2	2	2
								6	6	6	6
3	—	2	2	2	2	2	2	2	2	2	3
		8	8	8	8	8	8	8	8	8	8
4	2	2	2	3	3	3	3	3	3	3	3
	9	9	10	10	10	10	10	10	10	10	10
5	2	3	3	3	3	3	4	4	4	4	4
	10	10	11	11	12	12	12	12	12	12	12
6	3	3	3	3	4	4	4	4	5	5	5
	10	11	12	12	13	13	13	13	14	14	14
7	3	3	3	4	4	5	5	5	5	5	6
	11	12	13	13	14	14	14	14	15	15	15
8	3	3	4	4	5	5	5	6	6	6	6
	11	12	13	14	14	15	15	16	16	16	16
9	3	4	4	5	5	5	6	6	6	7	7
	12	13	14	14	15	16	16	16	17	17	18
10	3	4	5	5	5	6	6	7	7	7	7
	12	13	14	15	16	16	17	17	18	18	18
11	4	4	5	5	6	6	7	7	7	8	8
	12	13	14	15	16	17	17	18	19	19	19
12	4	4	5	6	6	7	7	7	8	8	8
	12	13	14	16	16	17	18	19	19	20	20
13	4	5	5	6	6	7	7	8	8	9	9
	12	14	15	16	17	18	19	19	20	20	21
14	4	5	5	6	7	7	8	8	9	9	9
	12	14	15	16	17	18	19	20	20	21	22
15	4	5	6	6	7	7	8	8	9	9	10
	12	14	15	16	18	18	19	20	21	22	22

Source: Adapted from Frieda S. Swed and C. Eisenhart, "Tables for Testing Randomness of Grouping in a Sequence of Alternatives," *The Annals of Statistics* 14(1943).

APPENDIX

C

Lists of Formulas

Chapter 2 • Organizing and Graphing Data

- Relative frequency of a class $= \dfrac{f}{\Sigma f}$
- Percentage of a class = (Relative frequency) × 100%
- Class midpoint or mark $= \dfrac{\text{(Upper limit + Lower limit)}}{2}$
- Class width = Upper boundary − Lower boundary
- Cumulative relative frequency $= \dfrac{\text{Cumulative frequency}}{\text{Total observations in the data set}}$
- Cumulative percentage = (Cumulative relative frequency) × 100%

Chapter 3 • Numerical Descriptive Measures

- Mean for ungrouped data: $\mu = \dfrac{\Sigma x}{N}$ and $\bar{x} = \dfrac{\Sigma x}{n}$
- Mean for grouped data: $\mu = \dfrac{\Sigma mf}{N}$ and $\bar{x} = \dfrac{\Sigma mf}{n}$

  where m is the midpoint and f is the frequency of a class
- Weighted mean for ungrouped data $= \dfrac{\Sigma xw}{\Sigma w}$
- k% Trimmed mean = Mean of the values after dropping k% of the values from each end of a ranked data set
- Geometric mean $= \sqrt[n]{x_1\, x_2\, x_3 \ldots x_n}$
- Harmonic mean $= \dfrac{n}{\left(\dfrac{1}{x_1} + \dfrac{1}{x_2} + \dfrac{1}{x_3} + \cdots + \dfrac{1}{x_n}\right)}$
- Median for ungrouped data = Value of the middle term in a ranked data set
- Range = Largest value − Smallest value
- Variance for ungrouped data:

$$\sigma^2 = \frac{\Sigma x^2 - \left(\frac{(\Sigma x)^2}{N}\right)}{N} \quad \text{and} \quad s^2 = \frac{\Sigma x^2 - \left(\frac{(\Sigma x)^2}{n}\right)}{n-1}$$

  where σ^2 is the population variance and s^2 is the sample variance
- Standard deviation for ungrouped data:

$$\sigma = \sqrt{\frac{\Sigma x^2 - \left(\frac{(\Sigma x)^2}{N}\right)}{N}} \quad \text{and} \quad s = \sqrt{\frac{\Sigma x^2 - \left(\frac{(\Sigma x)^2}{n}\right)}{n-1}}$$

  where σ and s are the population and sample standard deviations, respectively
- Coefficient of variation $= \dfrac{\sigma}{\mu} \times 100\%$ or $\dfrac{s}{\bar{x}} \times 100\%$
- Variance for grouped data:

$$\sigma^2 = \frac{\Sigma m^2 f - \left(\frac{(\Sigma mf)^2}{N}\right)}{N} \quad \text{and} \quad s^2 = \frac{\Sigma m^2 f - \left(\frac{(\Sigma mf)^2}{n}\right)}{n-1}$$

- Standard deviation for grouped data:

$$\sigma = \sqrt{\frac{\Sigma m^2 f - \left(\frac{(\Sigma mf)^2}{N}\right)}{N}} \quad \text{and} \quad s = \sqrt{\frac{\Sigma m^2 f - \left(\frac{(\Sigma mf)^2}{n}\right)}{n-1}}$$

- Chebyshev's theorem:

  For any number k greater than 1, at least $\left(1 - \dfrac{1}{k^2}\right)$ of the values for any distribution lie within k standard deviations of the mean.
- Empirical rule:

  For a specific bell-shaped distribution, about 68% of the observations fall in the interval $(\mu - \sigma)$ to $(\mu + \sigma)$, about 95% fall in the interval $(\mu - 2\sigma)$ to $(\mu + 2\sigma)$, and about 99.7% fall in the interval $(\mu - 3\sigma)$ to $(\mu + 3\sigma)$.
- Q_1 = First quartile given by the value of the middle term among the (ranked) observations that are less than the median
- Q_2 = Second quartile given by the value of the middle term in a ranked data set
- Q_3 = Third quartile given by the value of the middle term among the (ranked) observations that are greater than the median
- Interquartile range: $\text{IQR} = Q_3 - Q_1$
- The kth percentile:

$$P_k = \text{Value of the } \left(\frac{k \times n}{100}\right) \text{th term in a ranked data set}$$

- Percentile rank of x_i

$$= \frac{\text{Number of values less than } x_i}{\text{Total number of values in the data set}} \times 100$$

Chapter 4 • Probability

- Classical probability rule for a simple event:

$$P(E_i) = \frac{1}{\text{Total number of outcomes}}$$

- Classical probability rule for a compound event:

$$P(A) = \frac{\text{Number of outcomes in } A}{\text{Total number of outcomes}}$$

- Relative frequency as an approximation of probability:

$$P(A) = \frac{f}{n}$$

- Conditional probability of an event:

$$P(A|B) = \frac{P(A \text{ and } B)}{P(B)} \quad \text{and} \quad P(B|A) = \frac{P(A \text{ and } B)}{P(A)}$$

- Condition for independence of events:

$$P(A) = P(A|B) \quad \text{and/or} \quad P(B) = P(B|A)$$

- For complementary events: $P(A) + P(\overline{A}) = 1$
- Multiplication rule for dependent events:

$$P(A \text{ and } B) = P(A)\,P(B|A)$$

- Multiplication rule for independent events:

$$P(A \text{ and } B) = P(A)\,P(B)$$

- Joint probability of two mutually exclusive events:

$$P(A \text{ and } B) = 0$$

- Addition rule for mutually nonexclusive events:

$$P(A \text{ or } B) = P(A) + P(B) - P(A \text{ and } B)$$

- Addition rule for mutually exclusive events:

$$P(A \text{ or } B) = P(A) + P(B)$$

- n factorial: $n! = n(n-1)(n-2)\ldots 3 \cdot 2 \cdot 1$
- Number of combinations of n items selected x at a time:

$${}_nC_x = \frac{n!}{x!\,(n-x)!}$$

- Number of permutations of n items selected x at a time:

$${}_nP_x = \frac{n!}{(n-x)!}$$

Chapter 5 • Discrete Random Variables and Their Probability Distributions

- Mean of a discrete random variable x: $\mu = \Sigma x P(x)$
- Standard deviation of a discrete random variable x:

$$\sigma = \sqrt{\Sigma x^2 P(x) - \mu^2}$$

- Binomial probability formula:

$$P(x) = {}_nC_x\, p^x\, q^{n-x}$$

- Mean and standard deviation of the binomial distribution:

$$\mu = np \quad \text{and} \quad \sigma = \sqrt{npq}$$

- Negative binomial probability formula:

$$P(x) = {}_{x+r-1}C_x (1-p)^x p^r, \quad x = 0, 1, 2, \ldots$$

- Mean and standard deviation of a negative binomial distribution:

$$\mu = \frac{r}{p} \quad \text{and} \quad \sigma = \sqrt{\frac{rq}{p^2}}$$

- Hypergeometric probability formula:

$$P(x) = \frac{{}_rC_x \;\; {}_{N-r}C_{n-x}}{{}_NC_n}$$

- Poisson probability formula:

$$P(x) = \frac{\lambda^x e^{-\lambda}}{x!}$$

- Mean, variance, and standard deviation of the Poisson probability distribution:

$$\mu = \lambda, \quad \sigma^2 = \lambda, \quad \text{and} \quad \sigma = \sqrt{\lambda}$$

Chapter 6 • Continuous Random Variables and the Normal Distribution

- z value for an x value: $z = \dfrac{x-\mu}{\sigma}$
- Value of x when μ, σ, and z are known: $x = \mu + z\sigma$

Chapter 7 • Sampling Distributions

- Mean of $\bar{x}$: $\mu_{\bar{x}} = \mu$
- Standard deviation of $\bar{x}$ when $\dfrac{n}{N} \le .05$:

$$\sigma_{\bar{x}} = \frac{\sigma}{\sqrt{n}}$$

- z value for $\bar{x}$: $z = \dfrac{\bar{x}-\mu}{\sigma_{\bar{x}}}$
- Population proportion: $p = \dfrac{X}{N}$
- Sample proportion: $\hat{p} = \dfrac{x}{n}$
- Mean of $\hat{p}$: $\mu_{\hat{p}} = p$
- Standard deviation of $\hat{p}$ when $\dfrac{n}{N} \le .05$:

$$\sigma_{\hat{p}} = \sqrt{\frac{pq}{n}}$$

- z value for $\hat{p}$: $z = \dfrac{\hat{p}-p}{\sigma_{\hat{p}}}$

Chapter 8 • Estimation of the Mean and Proportion

- Point estimate of μ: $\bar{x}$
- Confidence interval for μ using the normal distribution when σ is known:

$$\bar{x} \pm z\sigma_{\bar{x}} \quad \text{where} \quad \sigma_{\bar{x}} = \frac{\sigma}{\sqrt{n}}$$

- Confidence interval for μ using the t distribution when σ is not known:

$$\bar{x} \pm t s_{\bar{x}} \quad \text{where} \quad s_{\bar{x}} = \frac{s}{\sqrt{n}}$$

- Margin of error of the estimate for μ:

$$E = z\sigma_{\bar{x}} \quad \text{or} \quad ts_{\bar{x}}$$

- Determining sample size for estimating μ:

$$n = \frac{z^2\sigma^2}{E^2}$$

- Point estimate of p: $\hat{p}$
- Confidence interval for p for a large sample:

$$\hat{p} \pm zs_{\hat{p}} \quad \text{where} \quad s_{\hat{p}} = \sqrt{\frac{\hat{p}\hat{q}}{n}}$$

- Margin of error of the estimate for p:

$$E = zs_{\hat{p}} \quad \text{where} \quad s_{\hat{p}} = \sqrt{\frac{\hat{p}\hat{q}}{n}}$$

- Determining sample size for estimating p:

$$n = \frac{z^2\hat{p}\hat{q}}{E^2}$$

Chapter 9 • Hypothesis Tests About the Mean and Proportion

- Test statistic z for a test of hypothesis about μ using the normal distribution when σ is known:

$$z = \frac{\bar{x} - \mu}{\sigma_{\bar{x}}} \quad \text{where} \quad \sigma_{\bar{x}} = \frac{\sigma}{\sqrt{n}}$$

- Test statistic for a test of hypothesis about μ using the t distribution when σ is not known:

$$t = \frac{\bar{x} - \mu}{s_{\bar{x}}} \quad \text{where} \quad s_{\bar{x}} = \frac{s}{\sqrt{n}}$$

- Test statistic for a test of hypothesis about p for a large sample:

$$z = \frac{\hat{p} - p}{\sigma_{\hat{p}}} \quad \text{where} \quad \sigma_{\hat{p}} = \sqrt{\frac{pq}{n}}$$

Chapter 10 • Estimation and Hypothesis Testing: Two Populations

- Mean of the sampling distribution of $\bar{x}_1 - \bar{x}_2$:

$$\mu_{\bar{x}_1 - \bar{x}_2} = \mu_1 - \mu_2$$

- Confidence interval for $\mu_1 - \mu_2$ for two independent samples using the normal distribution when σ_1 and σ_2 are known:

$$(\bar{x}_1 - \bar{x}_2) \pm z\sigma_{\bar{x}_1 - \bar{x}_2} \quad \text{where} \quad \sigma_{\bar{x}_1 - \bar{x}_2} = \sqrt{\frac{\sigma_1^2}{n_1} + \frac{\sigma_2^2}{n_2}}$$

- Test statistic for a test of hypothesis about $\mu_1 - \mu_2$ for two independent samples using the normal distribution when σ_1 and σ_2 are known:

$$z = \frac{(\bar{x}_1 - \bar{x}_2) - (\mu_1 - \mu_2)}{\sigma_{\bar{x}_1 - \bar{x}_2}}$$

- For two independent samples taken from two populations with equal but unknown standard deviations:

  Pooled standard deviation:

$$s_p = \sqrt{\frac{(n_1 - 1)s_1^2 + (n_2 - 1)s_2^2}{n_1 + n_2 - 2}}$$

  Estimate of the standard deviation of $\bar{x}_1 - \bar{x}_2$:

$$s_{\bar{x}_1 - \bar{x}_2} = s_p\sqrt{\frac{1}{n_1} + \frac{1}{n_2}}$$

  Confidence interval for $\mu_1 - \mu_2$ using the t distribution:

$$(\bar{x}_1 - \bar{x}_2) \pm ts_{\bar{x}_1 - \bar{x}_2}$$

  Test statistic using the t distribution:

$$t = \frac{(\bar{x}_1 - \bar{x}_2) - (\mu_1 - \mu_2)}{s_{\bar{x}_1 - \bar{x}_2}}$$

- For two independent samples selected from two populations with unequal and unknown standard deviations:

  Degrees of freedom: $df = \dfrac{\left(\dfrac{s_1^2}{n_1} + \dfrac{s_2^2}{n_2}\right)^2}{\dfrac{\left(\dfrac{s_1^2}{n_1}\right)^2}{n_1 - 1} + \dfrac{\left(\dfrac{s_2^2}{n_2}\right)^2}{n_2 - 1}}$

  Estimate of the standard deviation of $\bar{x}_1 - \bar{x}_2$:

$$s_{\bar{x}_1 - \bar{x}_2} = \sqrt{\frac{s_1^2}{n_1} + \frac{s_2^2}{n_2}}$$

  Confidence interval for $\mu_1 - \mu_2$ using the t distribution:

$$(\bar{x}_1 - \bar{x}_2) \pm ts_{\bar{x}_1 - \bar{x}_2}$$

  Test statistic using the t distribution:

$$t = \frac{(\bar{x}_1 - \bar{x}_2) - (\mu_1 - \mu_2)}{s_{\bar{x}_1 - \bar{x}_2}}$$

- For two paired or matched samples:

  Sample mean for paired differences:

$$\bar{d} = \frac{\Sigma d}{n}$$

  Sample standard deviation for paired differences:

$$s_d = \sqrt{\frac{\Sigma d^2 - \frac{(\Sigma d)^2}{n}}{n - 1}}$$

  Mean and standard deviation of the sampling distribution of $\bar{d}$

$$\mu_{\bar{d}} = \mu_d \quad \text{and} \quad s_{\bar{d}} = \frac{s_d}{\sqrt{n}}$$

  Confidence interval for μ_d using the t distribution:

$$\bar{d} \pm ts_{\bar{d}} \quad \text{where} \quad s_{\bar{d}} = \frac{s_d}{\sqrt{n}}$$

  Test statistic for a test of hypothesis about μ_d using the t distribution:

$$t = \frac{\bar{d} - \mu_d}{s_{\bar{d}}}$$

- For two large and independent samples, confidence interval for $p_1 - p_2$:

$$(\hat{p}_1 - \hat{p}_2) \pm z s_{\hat{p}_1-\hat{p}_2} \quad \text{where} \quad s_{\hat{p}_1-\hat{p}_2} = \sqrt{\frac{\hat{p}_1\hat{q}_1}{n_1} + \frac{\hat{p}_2\hat{q}_2}{n_2}}$$

- For two large and independent samples, for a test of hypothesis about $p_1 - p_2$ with H_0: $p_1 - p_2 = 0$:

  Pooled sample proportion:

$$\overline{p} = \frac{x_1 + x_2}{n_1 + n_2} \quad \text{or} \quad \frac{n_1\hat{p}_1 + n_2\hat{p}_2}{n_1 + n_2}$$

  Estimate of the standard deviation of $\hat{p}_1 - \hat{p}_2$:

$$s_{\hat{p}_1-\hat{p}_2} = \sqrt{\overline{p}\,\overline{q}\left(\frac{1}{n_1} + \frac{1}{n_2}\right)}$$

  Test statistic: $z = \dfrac{(\hat{p}_1 - \hat{p}_2) - (p_1 - p_2)}{s_{\hat{p}_1-\hat{p}_2}}$

Chapter 11 • Chi-Square Tests

- Expected frequency for a category for a goodness-of-fit test:

$$E = np$$

- Degrees of freedom for a goodness-of-fit test:

$$df = k - 1 \quad \text{where } k \text{ is the number of categories}$$

- Expected frequency for a cell for an independence or homogeneity test:

$$E = \frac{(\text{Row total})(\text{Column total})}{\text{Sample size}}$$

- Degrees of freedom for a test of independence or homogeneity:

$$df = (R - 1)(C - 1)$$

  where R and C are the total number of rows and columns, respectively, in the contingency table

- Test statistic for a goodness-of-fit test and a test of independence or homogeneity:

$$\chi^2 = \Sigma\frac{(O - E)^2}{E}$$

- Confidence interval for the population variance σ^2:

$$\frac{(n-1)s^2}{\chi^2_{\alpha/2}} \quad \text{to} \quad \frac{(n-1)s^2}{\chi^2_{1-\alpha/2}}$$

- Test statistic for a test of hypothesis about σ^2:

$$\chi^2 = \frac{(n-1)s^2}{\sigma^2}$$

Chapter 12 • Analysis of Variance

Let:

k = the number of different samples (or treatments)

n_i = the size of sample i

T_i = the sum of the values in sample i

n = the number of values in all samples $= n_1 + n_2 + n_3 + \cdots$

Σx = the sum of the values in all samples $= T_1 + T_2 + T_3 + \cdots$

Σx^2 = the sum of the squares of values in all samples

- For the F distribution:

  Degrees of freedom for the numerator $= k - 1$
  Degrees of freedom for the denominator $= n - k$

- Between-samples sum of squares:

$$\text{SSB} = \left(\frac{T_1^2}{n_1} + \frac{T_2^2}{n_2} + \frac{T_3^2}{n_3} + \cdots\right) - \frac{(\Sigma x)^2}{n}$$

- Within-samples sum of squares:

$$\text{SSW} = \Sigma x^2 - \left(\frac{T_1^2}{n_1} + \frac{T_2^2}{n_2} + \frac{T_3^2}{n_3} + \cdots\right)$$

- Total sum of squares:

$$\text{SST} = \text{SSB} + \text{SSW} = \Sigma x^2 - \frac{(\Sigma x)^2}{n}$$

- Variance between samples: $\text{MSB} = \dfrac{\text{SSB}}{k - 1}$

- Variance within samples: $\text{MSW} = \dfrac{\text{SSW}}{n - k}$

- Test statistic for a one-way ANOVA test:

$$F = \frac{\text{MSB}}{\text{MSW}}$$

- Test statistic for a two-way ANOVA test:

$$F_\tau = \frac{\text{MSB}_\tau}{\text{MSE}} \quad \text{or} \quad F_\beta = \frac{\text{MSB}_\beta}{\text{MSE}}$$

- The between-samples sum of squares with respect to the first factor:

$$\text{SSB}_\tau = \sum_{i=1}^{t}\sum_{j=1}^{b}\sum_{k=1}^{n_{ij}}(\overline{x}_{i..} - \overline{x}_{...})^2$$

- The between-samples sum of squares with respect to the second factor:

$$\text{SSB}_\beta = \sum_{i=1}^{t}\sum_{j=1}^{b}\sum_{k=1}^{n_{ij}}(\overline{x}_{.j.} - \overline{x}_{...})^2$$

- The error sum of squares:

$$\text{SSE} = \sum_{i=1}^{t}\sum_{j=1}^{b}\sum_{k=1}^{n_{ij}}(x_{ij} - \overline{x}_{i..} - \overline{x}_{.j.} + \overline{x}_{...})^2$$

- Values of MSB_τ, MSB_β, and MSE:

$$\text{MSB}_\tau = \frac{\text{SSB}_\tau}{t-1}, \quad \text{MSB}_\beta = \frac{\text{SSB}_\beta}{b-1}, \quad \text{and}$$

$$\text{MSE} = \frac{\text{SSE}}{n-1-(t-1)-(b-1)} = \frac{\text{SSE}}{n-t-b+1}$$

  where $t - 1$, $b - 1$, and $n - t - b + 1$ are the associated degrees of freedoms.

- Test statistics for testing interaction in a two-way ANOVA:

$$F_{\lambda} = \frac{\text{MSI}}{\text{MSE}}$$

Chapter 13 • Simple Linear Regression

- Simple linear regression model:

$$y = A + Bx + \epsilon$$

- Estimated simple linear regression model:

$$\hat{y} = a + bx$$

- Sum of squares of xy, xx, and yy:

$$\text{SS}_{xy} = \Sigma xy - \frac{(\Sigma x)(\Sigma y)}{n}$$

$$\text{SS}_{xx} = \Sigma x^2 - \frac{(\Sigma x)^2}{n}$$

$$\text{SS}_{yy} = \Sigma y^2 - \frac{(\Sigma y)^2}{n}$$

- Least squares estimates of A and B:

$$b = \frac{\text{SS}_{xy}}{\text{SS}_{xx}} \quad \text{and} \quad a = \bar{y} - b\bar{x}$$

- Standard deviation of the sample errors:

$$s_e = \sqrt{\frac{\text{SS}_{yy} - b\,\text{SS}_{xy}}{n-2}}$$

- Error sum of squares:

$$\text{SSE} = \Sigma e^2 = \Sigma(y - \hat{y})^2$$

- Total sum of squares:

$$\text{SST} = \Sigma y^2 - \frac{(\Sigma y)^2}{n}$$

- Regression sum of squares:

$$\text{SSR} = \text{SST} - \text{SSE}$$

- Coefficient of determination:

$$r^2 = \frac{b\,\text{SS}_{xy}}{\text{SS}_{yy}}$$

- Confidence interval for B:

$$b \pm ts_b \quad \text{where} \quad s_b = \frac{s_e}{\sqrt{\text{SS}_{xx}}}$$

- Test statistic for a test of hypothesis about B:

$$t = \frac{b - B}{s_b}$$

- Linear correlation coefficient:

$$r = \frac{\text{SS}_{xy}}{\sqrt{\text{SS}_{xx}\,\text{SS}_{yy}}}$$

- Test statistic for a test of hypothesis about ρ:

$$t = r\sqrt{\frac{n-2}{1-r^2}}$$

- Confidence interval for $\mu_{y|x}$:

$$\hat{y} \pm ts_{\hat{y}_m} \quad \text{where} \quad s_{\hat{y}_m} = s_e\sqrt{\frac{1}{n} + \frac{(x_0 - \bar{x})^2}{\text{SS}_{xx}}}$$

- Prediction interval for y_p:

$$\hat{y} \pm ts_{\hat{y}_p} \quad \text{where} \quad s_{\hat{y}_p} = s_e\sqrt{1 + \frac{1}{n} + \frac{(x_0 - \bar{x})^2}{\text{SS}_{xx}}}$$

Chapter 14 • Multiple Regression

- General multiple regression model:

$$y = A + B_1x_1 + B_2x_2 + B_3x_3 + \cdots + B_kx_k + \varepsilon$$

- Estimated multiple regression model:

$$\hat{y} = a + b_1x_1 + b_2x_2 + b_3x_3 + \cdots + b_kx_k$$

- A multiple regression model with two independent variables:

$$y = A + B_1x_1 + B_2x_2 + \varepsilon$$

- Estimated multiple regression model with two independent variables:

$$\hat{y} = a + b_1x_1 + b_2x_2$$

- For a multiple regression model, degrees of freedom are:

$$df = n - k - 1$$

  where n is the sample size and k is the number of independent variables
- Error of estimate:

$$e = y - \hat{y}$$

- Error sum of squares:

$$\text{SSE} = \Sigma(y - \hat{y})^2$$

- Standard deviation of errors:

$$s_e = \sqrt{\frac{\text{SSE}}{n-k-1}}$$

- Total sum of squares:

$$\text{SST} = \Sigma(y - \bar{y})^2$$

- Regression sum of squares:

$$\text{SSR} = \Sigma(\hat{y} - \bar{y})^2$$

- Coefficient of multiple determination:

$$R^2 = \frac{\text{SSR}}{\text{SST}}$$

- Adjusted coefficient of multiple determination:

$$\bar{R}^2 = 1 - (1 - R^2)\left(\frac{n-1}{n-k-1}\right) \quad \text{or} \quad 1 - \frac{\text{SSE}/(n-k-1)}{\text{SST}/(n-1)}$$

- Confidence interval for B_i:

$b_i \pm t s_{b_i}$ where s_{b_i} is the standard deviation of b_i

- Test statistic for a test of hypothesis about B_i:

$$t = \frac{b_i - B_i}{s_{b_i}}$$

Chapter 15 • Nonparametric Methods

- The **Sign test** about proportion – large sample Case ($n > 25$): Using the normal distribution as an approximation to binomial distribution, the observed value of z is:

$$z = \frac{(X \pm .5) - \mu}{\sigma}$$

where X is the number of units in the sample that belong to the outcome referring to the proportion we are testing for.

- The **Sign test** about the median of a single population – large sample case ($n > 25$):

Using the normal distribution, the observed value of z is:

$$z = \frac{(X \pm .5) - \mu}{\sigma}$$

- The **Sign test** about the median difference between paired data – large sample case ($n > 25$):

Using the normal distribution, the observed value of z is:

$$z = \frac{(X \pm .5) - \mu}{\sigma}$$

- The **Wilcoxon signed-rank test** for two dependent samples – large sample case ($n > 15$):

The observed value of z is:

$$z = \frac{T - \mu_T}{\sigma_T}$$

where $\mu_T = \dfrac{n(n+1)}{4}$ and $\sigma_T = \sqrt{\dfrac{n(n+1)(2n+1)}{24}}$

- The **Wilcoxon rank sum test** for two independent samples – large sample case: if at least one sample size is greater than 10, use the normal distribution as an approximation to Wilcoxon rank sum test. In this case, the observed value of z is:

$$z = \frac{T - \mu_T}{\sigma_T}$$

where $\mu_T = \dfrac{n_1(n_1 + n_2 + 1)}{2}$ and

$$\sigma_T = \sqrt{\frac{n_1 n_2 (n_1 + n_2 + 1)}{12}}$$

- The **Kruskal-Wallis test:**

The observed value of H is:

$$H = \frac{12}{n(n+1)}\left(\frac{R_1^2}{n_1} + \frac{R_2^2}{n_2} + \cdots + \frac{R_k^2}{n_k}\right) - 3(n+1)$$

Where R_i = sum of the ranks for sample i
n_i = sample size for sample i
$n = n_1 + n_2 + \cdots + n_k$
k = number of samples

- The **Spearman rho rank correlation coefficient test** where ρ_s and r_s are the population and sample Spearman rho rank correlation coefficients, respectively:

The observed value of r_s:

$$r_s = 1 - \frac{6\Sigma d^2}{n(n^2 - 1)}$$

where d is the difference between each pair of ranks

- The **runs test for randomness** for a large sample case when at least one of the two samples is greater than 15 and normal distribution is used to perform the test:

$$z = \frac{R - \mu_R}{\sigma_R}$$

where $\mu_R = \dfrac{2n_1 n_2}{n_1 + n_2} + 1$ and

$$\sigma_R = \sqrt{\frac{2n_1 n_2 (2n_1 n_2 - n_1 - n_2)}{(n_1 + n_2)^2 (n_1 + n_2 - 1)}}$$

Answers to Selected Odd-Numbered Exercises and Self-Review Tests

(**Note:** Due to differences in rounding, the answers obtained by readers may differ slightly from the ones given here.)

Chapter 1

1.3 **a.** inferential **b.** descriptive

1.5
- Member: Each disease included in the table
- Variable: The number of deaths
- Measurement: The number of deaths caused by each disease
- Data set: Collection of the number of deaths from each cause of death listed in the table

1.7 **a.** A variable that can be measured numerically.

b. A variable that cannot assume a numeric value but can be classified into two or more nonnumeric categories.

c. A variable whose values are countable.

d. A variable that can assume any numerical value over a certain interval or.

e. Data collected on a quantitative variable.

f. Data collected on a qualitative variable.

1.9 **a.** continuous **b.** continuous

c. continuous **d.** discrete

1.11 **a.** Nominal scale is the simplest level of the measurement in which we categorized or labeled whereas in the ordinal scales we ranked or ordered the data based on some characteristics, and the difference between the categorized does not equal in interval in it.

b. Interval scale is used for the variable where the data labeled, ordered, and do not have a consistent interval between the categories. On the other hand ratio scale has all these properties along with the true zero point.

1.13 Yes. It possess all the properties of ratio scale.

1.15 **a.** Cross-section data

b. Cross-section data

c. Time-series data

d. Time-series data

e. Cross-section data

1.17 Consider a standard deck of 52 cards. Suppose we randomly select one card from the deck and record the value and suit. If we put this card back in the deck before we randomly select a second card, this is an example of **sampling with replacement**. If we lay the first card aside and randomly select the second card from the 51 cards remaining in the deck, this is an example of **sampling without replacement**.

1.19 **a.** A sample drawn in such a way that each member of the population has some chance of being selected in the sample is called a **random sample**.

b. A sample in which some members of the population may have no chance of being selected is called a **nonrandom sample**.

c. A **convenience sample** is a sample in which the most accessible members of the population are selected.

d. A **judgment sample** is a sample in which members of a population are selected based on the judgment and prior knowledge of an expert.

e. A **quota sample** is a sample selected in such a way that each group or subpopulation is represented in the sample in exactly the same proportion as in the target population.

1.21 Simple random sample

1.23 **a.** This is a nonrandom sample since students in the university who were not in her statistics class had no chance of being included in the sample.

b. This is a convenience sample since students in her class were the most accessible members of the population.

c. This sample suffers from selection error. The population consists of all students at the university, but the sampling frame is limited to members of her statistics class.

1.25 **a.** This is a random sample since the sampling frame is the entire class.

b. This is a simple random sample since the software package gives each sample of 20 students an equal chance of being selected.

c. There should be no systematic error since the sampling frame is the entire population, and the use of the software would give each sample of 20 students an equal chance of being selected.

1.27 **a.** This is a non-random sample. Only readers of the magazine were able to answer the survey.

b. This sample is subject to voluntary response error, since only those who feel strongly enough about the issues to complete the questionnaire will respond. It also suffers from selection error since only the magazine's readers are included in the sampling frame.

1.29 This survey is subject to response error since some teachers may be reluctant to give honest answers to an interviewer's questions.

1.31 In a **survey**, data are collected without exercising any control over the factors that may affect the characteristics of interest or the results of a survey. In an **experiment**, the researchers exercise control over some or all of these factors.

1.33 **a.** This is a designed experiment since the doctors controlled the assignment of volunteers to the treatment and control groups.

b. There is not enough information to determine if this is a double-blind study. We would need to know whether or not the doctors were aware of which women were assigned to the treatment group and which were assigned to the control (placebo) group.

1.35 **a.** This is a designed experiment since the doctors controlled the assignment of people to the treatment and control groups.

b. The study is double-blind since neither the patients nor the doctors knew who was given the aspirin and who was given the placebo.

1.37 This is a designed experiment since the researcher selected participants randomly from the entire population of families on welfare and then controlled which families received the treatment (job training) and which did not.

1.39 If the data showed that the percentage of families who got off welfare was higher in the group that received job training, the conclusion is justified. Since families were randomly assigned to treatment and control groups, the two groups should have been similar, and the difference in outcomes should be due to treatment (job training).

1.41 **a.** $\Sigma y = \$622$ **b.** $(\Sigma y)^2 = 386{,}884$

c. $\Sigma y^2 = 73{,}798$

1.43 **a.** With reference to this table, we have the following definitions

- Member: Each fast-food restaurant chain included in the table
- Variable: Number of locations
- Measurement: Number of locations for a specific fast-food restaurant chain
- Data Set: Collection of different number of locations for the fast-food restaurants listed in the table

b. The data set contains measurements for different fast-food restaurant chains for the same period of time, so it is cross-section data.

1.45 **a.** Population

b. Sample

c. Sample

d. Population

1.47 **a.** $\Sigma y = 4 + 14 + 8 + 10 + 7 = 43$

b. $(\Sigma y)^2 = 1849$

c. $\Sigma y^2 = 425$

1.49 **a.** convenience sample since it includes the most accessible members of the population

b. judgment sample since it includes the elements of the population selected based on the judgment and prior knowledge of an expert

c. random sample since it was drawn in such a way that each element of the population has some chance of being included in the sample

1.51 **a.** Since the study relies on volunteers, it may not be representative of the entire population of people suffering from compulsive behavior. Furthermore, the doctors used their own judgment to form the treatment and control groups. Thus, subjective factors may have influenced them, and the two groups may not be comparable. As a result, the effect of the medicine on compulsive behavior may be confounded with other variables. Therefore, the conclusion is not justified.

b. Although this study technically satisfies the criteria for a designed experiment (experimenters controlled the assignment of people to treatment groups) it suffers from the weaknesses of an observational study, as pointed out in part (a).

c. The study is not double-blind since the physicians knew who received the treatment.

1.53 **a.** This is a designed experiment since the doctors randomly selected patients and then randomly assigned patients to the treatment and control groups.

b. The study is double-blind since neither patients nor doctors knew who was receiving the medicine.

1.55 **a.** We would expect \$61,200 to be a biased estimate of the current mean annual income for all 5432 alumni because only 1240 of the 5432 alumni answered the income question. These 1240 are unlikely to be representative of the entire group of 5432.

b. The following types of bias are likely to be present:

Nonresponse error: Alumni with low incomes may be ashamed to respond. Thus, the 1240 who actually returned their questionnaires and answered the income question would tend to have higher than average incomes.

Response error: Some of those who answered the income question may give a value that is higher than their actual income in order to appear more successful.

c. We would expect the estimate of $61,200 to be above the current mean annual income of all 5432 alumni, for the given reasons in part (b).

Self-Review Test

Multiple Choice Questions

1. b	**2.** c	**3.** a	**4.** c	**5.** c
6. c	**7.** b	**8.** c	**9.** b	**10.** a
11. b	**12.** c	**13.** c	**14.** b	**15.** b
16. b	**17.** c			

Practice Exercises

1. a. sample without replacement
b. sample with replacement

2. a. qualitative **b.** quantitative; continuous
c. quantitative; discrete **d.** qualitative

7.
- Member: Each student included in the table
- Variable: Midterm test score
- Measurement: The midterm test score of a student
- Data Set: Collection of the midterm test scores of the students listed in the table

8. a. 33 types **b.** 1089 **c.** 231

9. a. $\Sigma x = 54$ **b.** $\Sigma y = 113.6$ **c.** $\Sigma x^2 = 472$
d. $\Sigma xy = 742.7$ **e.** $\Sigma x^2 y = 5686.7$ **f.** $(\Sigma y)^2 = 12{,}904.96$

10. a. convenience and nonrandom sample
b. not likely representative of entire population
c. convenience sample
d. Answers will vary, but one better procedure is as follows: Assign numbers 1 through 40 to the pigs, and write the numbers 1 through 40 on separate pieces of paper, put them in a hat, mix them, and then draw 10 numbers. Pick the pigs whose numbers were drawn.

11. a. no
b. voluntary response error, selection error, and response error

12. a. designed experiment **b.** yes **c.** no

13. observational study

14. randomized experiment

Chapter 2

2.3 a. & b.

Category	Frequency	Relative Frequency	Percentage
Y	18	18/40 = .450	45.0
N	16	16/40 = .400	40.0
D	6	6/40 = .150	15.0

c. 45% **d.** 55%

2.5 a. & b.

Category	Frequency	Relative Frequency	Percentage
D	9	9/30 = .300	30.0
DR	4	4/30 = .133	13.3
F	2	2/30 = .067	6.7
R	11	11/30 = .367	36.7
W	4	4/30 = .133	13.3

e. 13.3%

2.7 a.

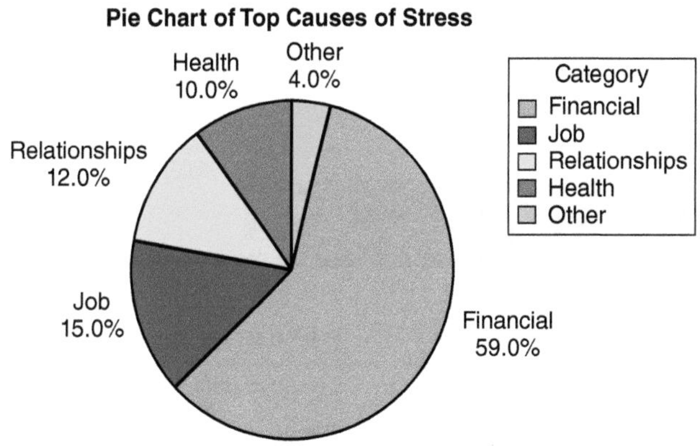

b.

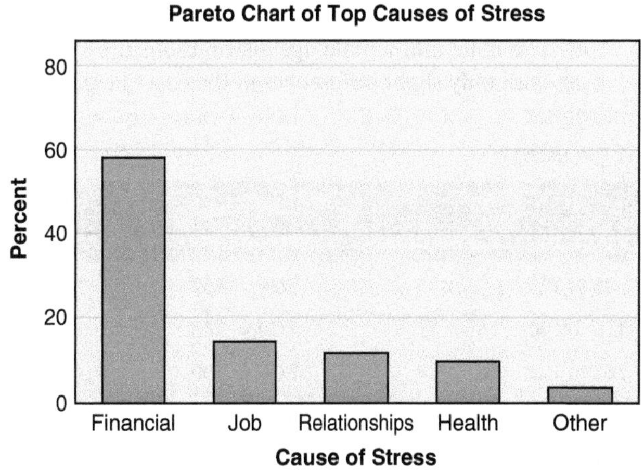

2.9 The relative frequency for a class is obtained by dividing the frequency of that class by the sum of frequencies of all classes. The percentage for a class is obtained by multiplying the relative frequency of that class by 100.

2.11 a. & b.

Number of Text Messages	Frequency	Relative Frequency	Percentage
32–37	10	10/40 = .250	25.0
38–43	9	9/40 = .225	22.5
44–49	13	13/40 = .325	32.5
50–55	6	6/40 = .150	15.0
56–61	2	2/40 = .050	5.0

d. 52.5%

2.13 a. & b.

Commuting Times	Frequency	Relative Frequency	Percentage
0 to 9	2	2/50 = .04	4
10 to 19	14	14/50 = .28	28
20 to 29	14	14/50 = .28	28
30 to 39	13	13/50 = .26	26
40 to 49	7	7/50 = .14	14

d. 40%

2.15 a. & b.

Age	Frequency	Relative Frequency	Percentage
20 to 29	8	8/30 = .267	26.7
30 to 39	9	9/30 = .300	30.0
40 to 49	6	6/30 = .200	20.0
50 to 59	6	6/30 = .200	20.0
60 to 69	1	1/30 = .033	3.30

d. 56.7%

e. The male and female age distributions are very similar, with only slight differences in the exact heights of the bars.

2.17 a. & b.

Weight	Frequency	Relative Frequency	Percentage
91 to 125	5	5/30 = .167	16.7
126 to 160	7	7/30 = .233	23.3
161 to 195	3	3/30 = .100	10.0
196 to 230	6	6/30 = .200	20.0
231 to 265	9	9/30 = .300	30.0

d. 40%

e. The weight distributions were similar, but 10% more of the females had a weight below 161 pounds than did the males. The percentage above 195 pounds was the same in both distributions, but of these more females fell in the 231 to 265 pounds range than did males.

2.19 a. & b.

Blood Glucose Level	Frequency	Relative Frequency	Percentage
75 to 89	6	6/30 = .200	20.0
90 to 104	4	4/30 = .133	13.3
105 to 119	7	7/30 = .233	23.3
120 to 134	7	7/30 = .233	23.3
135 to 149	6	6/30 = .200	20.0

d. 43.3%

2.21 a. & b.

Turnovers	Frequency	Relative Frequency	Percentage
0	5	.208	20.8
1	4	.167	16.7
2	5	.208	20.8
3	6	.250	25.0
4	4	.167	16.7

c. 15 games had two or more turnovers.

2.25 218, 245, 256, 329, 367, 383, 397, 404, 427, 433, 471, 523, 537, 551, 563, 581, 592, 622, 636, 647, 655, 678, 689, 810, 841.

2.27 a.

2	8 6 7
3	7 8 5 9 8
4	1 5 4 2 5 3 4 3 9 5
5	3 0 3 2 5 4 2 8 6 3
6	0 9 3 8 6 5 4 6 4 7 5 5 8 4 9
7	0 4 2 1 5 3 2

2	6 7 8
3	5 7 8 8 9
4	2 3 3 4 4 5 5 5 7 9
5	0 2 2 3 3 3 4 5 6 8
6	0 3 4 4 4 5 5 5 6 6 7 8 8 9 9
7	0 1 2 2 3 4 5

b.

2	
2	6 7 8
3	
3	5 7 8 8 9
4	2 3 3 4 4
4	5 5 7 9
5	0 2 2 3 3 3 4
5	5 6 8
6	0 3 4 4 4
6	5 5 5 6 6 6 7 8 8 9 9
7	0 1 2 2 3 4
7	5

c. Answers will vary, but the split stem-and-leaf display seems to better discern differences in the data in the range 30–70.

2.29

0	5 7
1	0 1 7 5 9
2	3 6 6 9 1 2
3	3 9 2
4	8 3
5	0
6	5

0	5 7
1	0 1 5 7 9
2	1 2 3 6 6 9
3	2 3 9
4	3 8
5	0
6	5

2.31 In order to prepare a **dotplot**, first we draw a horizontal line with numbers that cover the given data set. Then we place a dot above the value on the number line that represents each measurement in the data set.

2.33

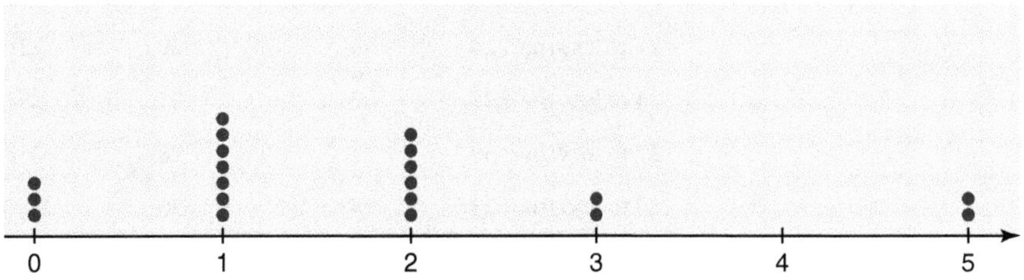

2.35

3	2 3 3 4 5 5 7 7 7 7 8 9
4	0 1 1 2 2 2 3 4 4 5 5 5 7 7 7 7 7 8 8 9
5	0 0 1 2 3 4 9
6	1

2.37

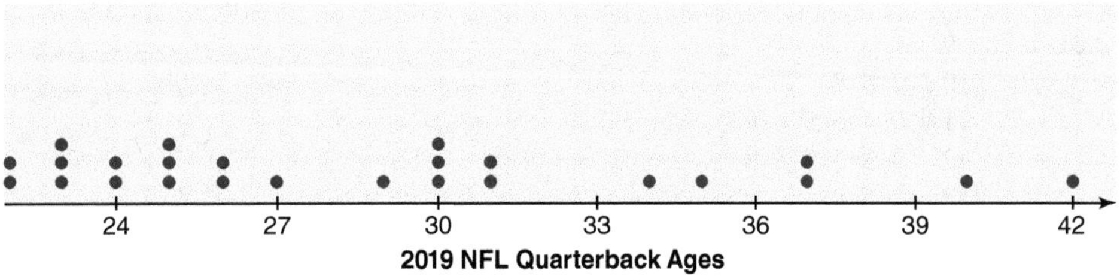

2.39 a. & b.

TV Sets Owned	Frequency	Relative Frequency	Percentage
0	1	1/40 = .025	2.5
1	14	14/40 = .350	35.0
2	14	14/40 = .350	35.0
3	8	8/40 = .200	20.0
4	3	3/40 = .075	7.5

d. 62.5%

2.41 a. & b.

Amount Spent on Video Rentals (in dollars)	Frequency	Relative Frequency	Percentage
1 – 200	25	.694	69.4
201 – 400	4	.111	11.1
401 – 600	4	.111	11.1
601 – 800	2	.056	5.6
801 – 1000	1	.028	2.8

c. 19.5%

2.43 a. & b.

Commute Length (in minutes)	Frequency	Relative Frequency	Percentage
18 to less than 21	13	13/30 = .433	43.3
21 to less than 24	9	9/30 = .300	30.0
24 to less than 27	3	3/30 = .100	10.0
27 to less than 30	1	1/30 = .033	3.3
30 to less than 33	4	4/30 = .133	13.3

c.

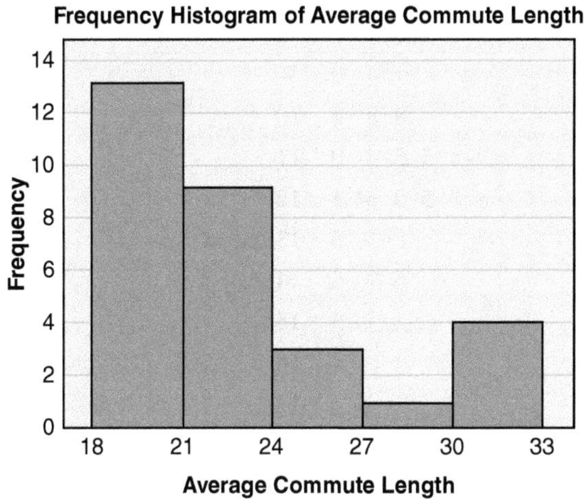

d.

Commute Length (in minutes)	Frequency	Cumulative Frequency	Cumulative Relative Frequency	Cumulative Percentage
18 to less than 21	13	13	13/30 = .433	43.3
21 to less than 24	9	22	22/30 = .733	73.7
24 to less than 27	3	25	25/30 = .833	83.3
27 to less than 30	1	26	26/30 = .867	86.7
30 to less than 33	4	30	30/30 = 1.000	100.0

2.45

2	8 4 7 7
3	4 1 8 5 2 9 3 7 0 8 4 6 0
4	4 1 7 6 1 9 5 6
5	2 3 7 0

2.49 a. 8 **b.** 5

2.51 b. i.

9	9
10	2 8 8
11	0 4 5 5 6 9
12	3 3 3 5 8 8
13	2 3 8
14	6 7 7 8
15	5 9
16	1 2 4 8
17	4 4 5 9 9 9
18	0 2 3 9
19	3 3 5
20	2 4

ii. The display shows a bimodal distribution, due to the presence of both females and males in the sample. The males tend to be heavier, so their weights are concentrated in the larger values, while the females' weights are found primarily in the smaller values.

c.

Females		Males
9	9	
8 8 2	10	
9 6 5 5 4 0	11	
8 8 5 3 3 3	12	
3	13	2 8
6 7 8	14	7
5	15	9
4	16	1 2 8
	17	4 4 5 9 9 9
	18	0 2 3 9
	19	3 3 5
	20	2 4

Self-Review Test

Multiple Choice Questions

1. b	**2.** a	**3.** a	**4.** b	**5.** a
6. a	**7.** c	**8.** a	**9.** c	**10.** b
11. b	**12.** b	**13.** c	**14.** c	**15.** c

Practice Exercises

2. a. 5 **b.** 7 **c.** 17 **d.** 13 **e.** 90 **f.** .3

4. a. & b.

Net Worth vs. $200,000	Frequency	Relative Frequency	Percentage
M	12	12/36 = .333	33.3
L	18	18/36 = .500	50.0
N	6	6/36 = .167	16.7

c. 18/36 = 50%

5. a. & b.

Monthly Expense on Gas (in dollars)	Frequency	Relative Frequency	Percentage
50 to 149	9	9/48 = .188	18.8
150 to 249	13	13/48 = .271	27.1
250 to 349	11	11/48 = .229	22.9
350 to 449	9	9/48 = .188	18.8
450 to 549	6	6/48 = .125	12.5

d. $(9 + 6)/48 = 31.25\%$

6. b. 40 dollars

d. $(6 + 8 + 7)/30 = 70\%$

7.

0	4 6 7 8
1	0 2 2 3 4 4 5 6 6 6 7 8 9
2	0 1 2 2 5 9
3	2

8. 30 33 37 42 44 46 47 49 51 53 53 56 60 67 67 71 79

Chapter 3

3.5 The mode can assume more than one value for a data set.

3.9 mean = 2.75, median = 2.50, no mode

3.11 5.7%

3.13 29.17 mph

3.15 mean = 264.58, median = 262, no mode

3.17 **a.** mean = \$27,970, median = \$25,600

b. mean = \$23,300, median = \$24,300

c. median

3.19 **a.** mean = 381.3 thousands of dollars, median = 323.5 thousands of dollars, mode = 144 thousands of dollars

b. 347.583 thousand dollars

3.21 **a.** mean = 29.85 patients, median = 29.5 patients, modes = 24, 26, 37, and 38

b. 29.86 patients

3.23 The opinion that they will not allow their children to play football

3.25 \$39.83

3.27 \$1386

3.29 88 years

3.31 5.2%

3.33 no

3.37 **a.** Yes, $\bar{x} = 118.29$

b. range = \$114, $s^2 = 1517.905$, $s = \$38.96$

3.39 **a.** range = 42 thousand dollars,
$s^2 = 154.34$,
$s = 12.42$ thousand dollars

b. 22.3%

c. sample statistics

3.41 **a.** range = 14 hours, $s^2 = 18.9333$, $s = 4.35$ hours

b. 67.97%

3.43 **a.** range = 26 minutes,
$s^2 = 75.64$, $s = 8.70$ minutes (parameters)

b. 35.66%

c. data widely spread from mean

3.45 **a.** range = 3216 thousand dollars, $s^2 = 993{,}825.51$, $s = 996.91$ thousand dollars

b. 231.95%

3.47 For yearly salaries: 10.90%
For years of experience: 13.33%

3.49 For Data Set I: $s = 3.61$, For Data Set II: $s = 10.83$ The standard deviation of the second data set is three times the standard deviation of the first data set.

3.51 $\mu = 9.15, \sigma^2 = 11.2275, s = \sqrt{11.2275} = 3.35$

3.53 $\mu = 11.24$ hours, $\sigma^2 = 36.3824$, $\sigma = 6.03$ hours

3.55 $\bar{x} = \$150, s^2 = 5730.6122, s = \$75.70$
The values in the column labeled *mf* give the approximate total amounts of electric bills for the families belonging to corresponding classes. The value $\Sigma mf =$ \$7500 is the approximate total amount of the electric bills for all 50 families included in the sample.

3.57 $\bar{x} = 36.80$ minutes, $s^2 = 597.7143$, $s = 24.45$ minutes

3.59 The empirical rule is applied to a bell–shaped distribution. According to this rule, approximately

- 68% of the observations lie within one standard deviation of the mean.
- 95% of the observations lie within two standard deviations of the mean.
- 99.7% of the observations lie within three standard deviations of the mean.

3.61 $k = 2$: at least 75% fall in (148, 312)
$k = 2.5$: at least 84% fall in (127.5, 332.5)
$k = 3$: at least 89% fall in (107, 353)

3.63 about 68% fall in (66, 98); about 95% fall in (50, 114); about 99.7% fall in (34, 130)

3.65 **a.** **i.** At least 84% of all workers have commuting times between 14 and 54 minutes.

ii. At least 75% of all workers have commuting times between 18 and 50 minutes.

b. interval is 10 to 58

3.67 **a.** About 99.7% of all workers have commuting times between 10 and 58 minutes.

b. About 95% of all workers have commuting times between 18 and 50 minutes.

3.69 To find the three quartiles:

1. Rank the given data set in increasing order.

2. Find the median using the procedure in Section 3.1.2. The median is the second quartile, Q_2.

3. The first quartile, Q_1, is the value of the middle term among the (ranked) observations that are less than Q_2.

4. The third quartile, Q_3, is the value of the middle term among the (ranked) observations that are greater that Q_2.

3.71 Given a data set of n values, to find the kth percentile (P_k):

1. Rank the given data in increasing order.

2. Calculate $kn/100$. Then, P_k is the term that is approximately ($kn/100$) in the ranking. If $kn/100$ falls between two consecutive integers a and b, it may be necessary to average the ath and bth values in the ranking to obtain P_k.

3.73 **a.** The quartiles are $Q_1 = 69$, $Q_2 = 72.5$, and $Q_3 = 76$. IQR = 7.0

b. 4.9; the 35th percentile can be approximated by the 5th term in the ranked data. So, $P_{35} = 70$.

c. Four values in the given data set are smaller than 70. Hence, the percentile rank of

$$70 = (4/14) \times 100 = 28.57\%.$$

3.75 **a.** The three quartiles are $Q_1 = 37$, $Q_2 = 42$, and $Q_3 = 48$. IQR = 11
The value 49 lies above Q_3, which means it lies in the fourth 25% group from the bottom in the ranked data set.
b. 36.4
The 91st percentile can be approximated by the 36th term in the ranked data. Therefore, $P_{91} = 54$. This means that 91% of the values in the data set are less than 54.
c. Thirteen values in the given data set are less than 40. Hence, the percentile rank of 40 = (13/40) × 100 = 32.5%. Therefore, the number of text message was 40 or higher on 67.5% of the days.

3.77 **a.** The quartiles are $Q_1 = 44.5$, $Q_2 = 57.5$, and $Q_3 = 64$. IQR = 19.5
The value 57 lies between Q_1 and Q_2 which means at least 25% of the data are smaller and at least 50% of the data are larger than 57.
b. 6; the 30th percentile is the value of the 6th term in the ranked data, which is 45. So, $P_{30} = 45$.
c. Twelve values in the given data are smaller than 61. Hence, the percentile rank of
61 = (12/20) × 100 = 60%.

3.79 median = 18.5, $Q_1 = 12$, and $Q_3 = 30.5$,
IQR = 18.5, Lower inner fence = −15.75,
Upper inner fence = 58.25
The smallest and largest values within the two inner fences are 3 and 42, respectively. There are two outliers, 62 and 75. The box-and-whisker plot is shown here.

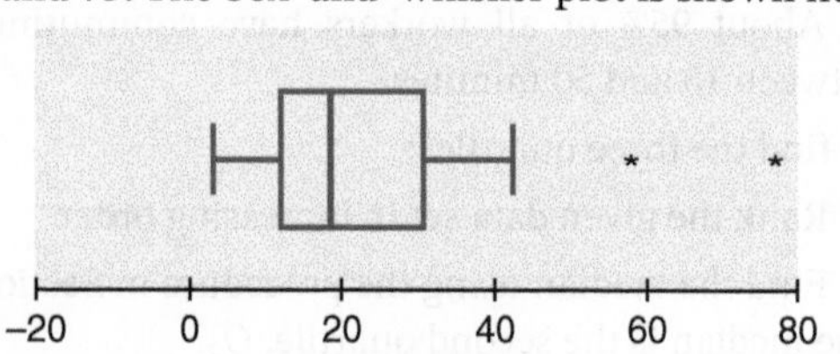

3.81 median = 136, $Q_1 = 71$, and $Q_3 = 258$,
IQR = 187, 1.5 × IQR = 280.5,
Lower inner fence = −209.5,
Upper inner fence = 538.5
The largest value exceeds the upper fence and so, is an outlier.
The data are skewed to the right (that is, toward larger values).

3.83 median = 48, $Q_1 = 45.5$, and $Q_3 = 52$,
IQR = 6.5, 1.5 × IQR = 9.75,
Lower inner fence = 35.75, Upper inner fence = 61.75
The smallest and largest values within the two inner fences are 41 and 56, respectively. There are no outliers. The data are skewed slightly to the right.

3.85 **a.** $\bar{x}$ = \$56.45 thousand, median = \$49 thousand
The modes are 27, 40, 43, and 86 since they each occur twice and all other values once.
b. \$53.94 thousand
c. range = \$67 thousand, $s^2 = 504.71$, s = \$22.47 thousand, CoV = 39.72%

3.87 \$762.33

3.89 $\bar{x}$ = 12.4 minutes, $s^2 = 20.2449$, s = 4.50 minutes (sample statistics)

3.91 **a.** **i.** At least 84% of patients had waiting times between 15 and 45 minutes.
ii. At least 89% of patients had waiting times between 12 and 48 minutes.
b. The required interval is 18 to 42 minutes.

3.93 **a.** The three quartiles are $Q_1 = 60$, $Q_2 = 76$, and $Q_3 = 97$. IQR = 37
The value 74 falls between Q_1 and Q_2, which indicates that it is at least as large as 25% of the data and no larger than 50% of the data.
b. 7; the 70th percentile occurs at the seventh term in the ranked data, which is 84. Therefore, $P_{70} = 84$. This means that about 70% of the values in the data set are smaller than or equal to 84.
c. Seven values in the given data are smaller than 97. Hence, the percentile rank of 97 = (7/10) × 100 = 70%. This means approximately 70% of the values in the data set are less than 97.

3.95 median = 83, $Q_1 = 73$, and $Q_3 = 93$,
IQR = 20, 1.5 × IQR = 30,
Lower inner fence = 43,
Upper inner fence = 123
The smallest and largest values within the two inner fences are 62 and 112, respectively. The value 135 is an outlier. The data are skewed to the right.

3.97 **a.** Jeffery would have to suggest \$60,000 be awarded to the plaintiff.
b. To prevent a juror like Jeffery from having an undue influence on the amount of damage to be awarded to the plaintiff, the jury could revise its procedure by throwing out any amounts that are outliers and then recalculate the mean, or by using the median, or by using a trimmed mean.

3.99 **a.** \$301,400
b. 18,700, mean number of trips per month per shopper = 9.35 trips
c. Mean amount spent per person per month by shoppers aged 12–17 = \$150.70.

3.101 **a.** At least 50% of the scores are within 1.41 standard deviations of the mean.
b. At least 80% of the scores are within 2.24 standard deviations of the mean, but this means that at most 10% of the scores are greater than 2.24 standard deviations above the mean.

3.103 **a.** mean = \$600.35, median = \$90, and mode = \$0
b. The mean is the largest.
c. Q_1 = \$0, Q_3 = \$272.50, IQR = \$272.50, 1.5 × IQR = \$408.75,
Lower inner fence is −408.75,
Upper inner Fence is 681.25

The largest and smallest values within the two inner fences are 0 and 501, respectively. There are three outliers at 1127, 3709 and 14,589.

The data are strongly skewed to the right.

3.105 a. 275 pieces of luggage **b.** 96

3.107 a. $\bar{x}$ = 20.80 thousand miles, median = 15 thousand miles, and mode = 15 thousand miles

b. range = 59 thousand miles, s^2 = 249.03, s = 15.78 thousand miles

c. Q_1 = 10 thousand miles and Q_3 = 26 thousand miles

d. IQR = 16 thousand miles

Since the interquartile range is based on the middle 50% of the observations it is not affected by outliers. The standard deviation, however, is strongly affected by outliers. Thus, the interquartile range is preferable in applications in which a measure of variation is required that is unaffected by extreme values.

Self-Review Test

Multiple Choice Questions

1. b	**2.** a	**3.** c	**4.** c	**5.** b
6. b	**7.** a	**8.** a	**9.** b	**10.** a
11. b	**12.** c	**13.** a	**14.** a	**15.** a
16. b	**17.** c	**18.** b	**19.** c	**20.** b
21. a	**22.** c	**23.** b	**24.** c	**25.** c
26. b	**27.** a			

Practice Exercises

1. a. $\bar{x}$ = 21 times, median = 13.5 times

The modes are 5, 8, and 14.

b. 15.6875 times

c. range = 90 times,

s^2 = 534.63, s = 23.12 times

d. coefficient of variation = 110.10%

e. sample statistics

2. $1657.914

3. mean of all scores = 75.43 points

If we drop the outlier (22), then

mean = 84.33 points.

4. Range of all scores = 73 points. If we drop the outlier (22), then range = 22 points.

5. The value of the standard deviation is zero when all the values in a data set are the same.

6. a. The frequency column gives the number of weeks for which the number of computers sold was in the corresponding class.

b. For the given data: n = 25, Σmf = 486.50, and $\Sigma m^2 f$ = 10,524.25,

$\bar{x}$ = 19.46 computers,

s^2 = 44.0400,

s = 6.64 computers

7. a. i. About 75% of the members spent between 59.4 and 124.2 minutes at the health club.

ii. About 84% of the members spent between 51.3 and 132.3 minutes at the health club.

b. 43.2 minutes to 140.4 minutes.

8. a. i. About 68% of the cars are 5.1 to 9.5 years old.

ii. About 99.7% of the cars are .7 to 13.9 years.

b. 2.9 to 11.7 years.

9. a. median = 57, Q_1 = 52, and Q_3 = 66, IQR = 14

The value 54 lies between Q_1 and the median, so it is in the second 25% group from the bottom of the ranked data set. This means that 25% of the data is less than 54 and that at least 50% of the data is larger than 54.

b. 10.8; the 60th percentile can be approximated by the 11th term in the ranked data. So, P_{60} = 61. This means that 60% of the values in the data set are less than 61.

c. Twelve values in the given data set are less than 64. Hence, the percentile rank of 64 = (12/18) × 100 = 66.7% or about 67%.

10. The data is skewed to the right.

11. $n_1 = 15$, $n_2 = 20$, $\bar{x}_1$ = \$1035, $\bar{x}_2$ = \$1090, $\bar{x}$ = \$1066.43

12. 3.17

13. a. For Data Set I: $\bar{x}$ = 19.75

For Data Set II: $\bar{x}$ = 16.75

b. For Data Set I: Σx = 79, Σx^2 = 1945, and n = 4; s = 11.32

c. For Data Set II: Σx = 67, Σx^2 = 1507, and n = 4; s = 11.32

Chapter 4

4.3 The experiment of selecting two items from the box without replacement has the following six possible outcomes: *AB, AC, BA, BC, CA, CB*. The sample space is written as *S* = {*AB, AC, BA, BC, CA, CB*}.

4.5 Let *I* = person is computer literate and *N* = person is computer illiterate. The experiment has four outcomes: *II, IN, NI,* and *NN.*

4.7 Let *G* = the selected part is good and *D* = the selected part is defective. The four outcomes for this experiment are: *GG, GD, DG,* and *DD*.

4.9 a. {*IN, NI*}; a compound event

b. {*II, IN, NI*}; a compound event

c. {*NN, NI, IN*}; a compound event

d. {*IN*}; a simple event

4.11. 1. The probability of an event always lies in the range zero to 1, that is:

$0 \le P(E_i) \le 1$ and $0 \le P(A) \le 1$

2. The sum of the probabilities of all simple events for an experiment is always 1, that is:

$\Sigma P(E_i) = P(E_1) + P(E_2) + P(E_3) + \cdots = 1$

4.15 The values −.09, 1.42, 9/4, and −1/4 cannot be probabilities of events because the probability of an event can never be less than zero or greater than one.

4.17 subjective probability

4.19 a. .48 **b.** .52

4.21 .66

4.23 4/6 classical approach

4.25 *P*(company selected offers free health fitness center on the company premises) = 130/400 = .325.

Number of companies that do not offer free health fitness center on the company premises = 400 − 130 = 270.

P(company selected does not offer free health fitness center on the company premises) = 270/400 = .675.

Yes, the sum of the probabilities is 1.0 because this experiment has only these two outcomes.

4.27 a. .086 **b.** .506

4.33 a. Events *A* and *B* are mutually exclusive.

b. $P(A) = 4/10$ and $P(A|B) = 0$. *A* and *B* are dependent.

c. $\overline{A} = \{1, 2, 5, 7, 8, 10\}$; $P(\overline{A}) = 6/10 = .600$

$\overline{B} = \{3, 4, 6, 7, 8, 9, 10\}$; $P(\overline{B}) = 7/10 = .700$

4.35 a. i. .0400 **ii.** .750

iii. .800 **iv.** .520

b. The events "man" and "yes" are not mutually exclusive. The events "yes" and "no" are mutually exclusive.

c. *P*(woman) = 200/500 = .400 and *P*(woman | yes) = 135/375 = .360. Since these two probabilities are not equal, the events "woman" and "yes" are not independent.

4.37 dependent

4.39 a. mutually exclusive, dependent **b.** .75

4.41 The complementary events are A = received job offer and B = did not receive job offer. P(A) = .3 and P(B) = .7

4.43 The **intersection** of two events is the collection of all the outcomes that are common to both events.

4.45 If *A* and *B* are dependent, then $P(A \text{ and } B) = P(A)P(B|A)$. If *A* and *B* are independent events, then $P(A \text{ and } B) = P(A)P(B)$.

4.47 a. .100 **b.** .234

4.49 a. .0230 **b.** .0511

4.51 .900

4.53 a. i. .270 **ii.** .120

4.55 a. i. .0483 **ii.** .2139

b. 0

4.57 .4667

4.59 .2667

4.61 a. .064 **b.** .216

4.63 .5278

4.65 .400

4.69 .75

4.71 a. .52 **b.** .67

4.73 a. .800 **b.** .775 **c.** 1.0

4.75 .65 **4.77** .72

4.79 .70 **4.81** .4816

4.83 1296

4.85 ${}_5C_3 = 10$ ${}_7C_4 = 35$ ${}_9C_3 = 84$ ${}_4C_0 = 1$ ${}_3C_3 = 1$

${}_6P_2 = 30$ ${}_8P_4 = 1680$

4.87 420

4.89 ${}_{25}C_3 = 2300$; ${}_{25}P_3 = 13{,}800$

4.91 ${}_{20}C_6 = 38{,}760$; ${}_{20}P_6 = 27{,}907{,}200$

4.93 ${}_{20}C_9 = 167{,}960$

4.95 a. i. .3640

ii. .4880

iii. .4297

iv. .6541

v. .2920

vi. .7080

b. $P(F) = 128/250 = .5120$ and $P(F|A) = 55/159 = .3459$. Since these two probabilities are not equal, the events "female" and "prefers watching sports" are dependent. The events "female" and "prefers watching sports" are not mutually exclusive because they can occur together.

4.97 a. i. .700 **ii.** .6296

iii. .275 **iv.** .725

4.99 .3996

4.101 .0045

4.103 a. 17,576,000 possible different license places.

b. 5200 license plates which fit the description.

4.105 a. .5000

b. .3333

4.107 a. .5000

b. .5000

c. .7368

4.109 a. .8667 **b.** .1333

4.111 a. .8851 **b.** .0035

Self-Review Test

Multiple Choice Questions

1. a **2.** b **3.** c **4.** a **5.** a

6. b **7.** c **8.** b **9.** b **10.** c

11. b **12.** c **13.** b **14.** c **15.** a

16. a **17.** b **18.** c **19.** c **20.** a

21. b **22.** b **23.** b **24.** c **25.** c

Practice Exercises

1. 120

2. a. .3333 **b.** .6667

3. **a.** P(out-of-state) = 125/200 = .6250 and P(out of state | female) = 70/110 = .6364. Since these two probabilities are not equal, the two events are dependent. Events "female" and "out of state" are not mutually exclusive because they can occur together.

b. **i.** .4500 **ii.** .6364

4. .825 **5.** .3894 **6.** .4225 **7.** .600

8. **a.** .279 **b.** .829

9. **a.** **i.** .3577 **ii.** .4047 **iii.** .2352 **iv.** .5593

b. $P(W)$ = 257/506 = .5079 and $P(W|Y)$ = 104/181 = .5746. Since these two probabilities are not equal, the events "woman" and "yes" are dependent. The events "woman" and "yes" are not mutually exclusive because they can occur together.

Chapter 5

5.3 discrete

5.5 **1.** The probability assigned to each value of a random variable x lies in the range 0 to 1; that is, $0 \le P(x) \le 1$ for each x.

2. The sum of the probabilities assigned to all possible values of x is equal to 1; that is, $\Sigma P(x) = 1$.

5.7 **a.** .15 **b.** .75 **c.** .10 **d.** .30 **e.** .55

5.9 **a.**

x	$P(x)$
1	15/35 = .43
2	8/35 = .23
3	6/35 = .17
4	4/35 = .11
5	2/35 = .06

b. **i.** .17 **ii.** .34 **iii.** .51 **iv.** .83

5.11 $P(x = 0) = P(II) = .7039$,
$P(x = 1) = P(NI) + P(IN) = .1351 + .1351 = .2702$,
$P(x = 2) = P(NN) = .0259$

x	$P(x)$
0	.7039
1	.2702
2	.0259

5.13 $P(x = 0) = P(UU) = .4225$,
$P(x = 1) = P(HU) + P(UH) = .4550$,
$P(x = 2) = P(HH) = .1225$

x	$P(x)$
0	.4225
1	.4550
2	.1225

5.17 $\mu = \Sigma x P(x) = .84$ error

$\sigma = \sqrt{\Sigma x^2 P(x) - \mu^2} = \sqrt{2.24 - (.84)^2} = 1.24$ error

5.19 $\mu = \Sigma x P(x) = 1.58$ patients

$\sigma = \sqrt{\Sigma x^2 P(x) - \mu^2} = \sqrt{4.02 - (1.58)^2}$
$= 1.23$ patients

5.21 $\mu = \Sigma x P(x) = 1.33$ magazines

$\sigma = \sqrt{\Sigma x^2 P(x) - \mu^2} = \sqrt{3.89 - (1.33)^2}$
$= 1.456$ magazines

5.23 The contractor is expected to make an average of \$3.9 million profit with a standard deviation of \$3.015 million.

5.25 $\mu = \Sigma x P(x) = .600$ athlete

$\sigma = \sqrt{\Sigma x^2 P(x) - \mu^2} = \sqrt{.7578 - (.600)^2}$
$= .631$ athlete

5.27 The parameters of the binomial distribution are n and p, which stand for the total number of trials and the probability of success, respectively.

5.29 **a.** This is an example of a binomial experiment because it satisfies all four conditions of a binomial experiment.

b. This is not a binomial experiment because the draws are not independent since the selections are made without replacement and, hence, the probabilities of drawing a red and a blue ball change with every selection.

c. This is an example of a binomial experiment because it satisfies all four conditions of a binomial experiment.

5.31 **a.**

x	$P(x)$
0	.824
1	.2471
2	.3177
3	.2269
4	.0972
5	.0250
6	.0036
7	.002

b. $\mu = np = (7)(.30) = 2.100$

$\sigma = \sqrt{npq} = \sqrt{(7)(.30)(.70)} = 1.212$

5.33 **a.** 0, 1, 2, 3, 4, 5, 6, 7, 8, 9, 10 **b.** .2001

5.35 a. .7032 **b.** .2921 **c.** .7215

5.37 a. .1575 **b.** .0187

5.39 a.

x	$P(x)$
0	.0000
1	.0004
2	.0043
3	.0287
4	.1147
5	.2753
6	.3670
7	.2097

$\mu = np = (7)(.80) = 5.6$ customers

$\sigma = \sqrt{npq} = \sqrt{(7)(.80)(.20)} = 1.058$ customers

b. .1147

5.41 a. .0527 **b.** .0507

5.43 .0048

5.47 a. .50 **b.** .033 **c.** .9667

5.49 a. .1078 **b.** .5147 **c.** .8628

5.53 a. .09158 **b.** .2157

5.55 .1185

5.57 a. .1162

b. **i.** .6625 **ii.** .1699

iii. .4941

5.59 a. .0271

b. i. .1350 **ii.** .0021

5.61 a. .5488

b.

x	$P(x)$
0	.4493
1	.3293
2	.0988
3	.0198
4	.0030
5	.0004

c. $\mu = \lambda = .6, \sigma^2 = \lambda = .6$, and $\sigma = \sqrt{\lambda} = \sqrt{.6} = .7746$

5.63 a. .0446

b. **i.** .0021

ii. .4542

iii. .0218

5.65 $\mu = \Sigma xP(x) = 3.18$ accidents

$\sigma = \sqrt{\Sigma x^2P(x) - \mu^2} = \sqrt{12.84 - (3.18)^2}$

$= 1.6516$ accidents

The mean 3.18 indicates the average number of accidents that occurred on a busy route on a given Sunday.

5.67 a. $\approx$.0000 **b.** .0351 **c.** .7214

5.69 a. .9246 **b.** .0754

5.71 a. .4885 **b.** .2418 **c.** .031

5.73 a. .0765

b. **i.** .0278 **ii.** .1263 **iii.** .5448

5.75 a. .0867

b.

x	$P(x)$
0	.3012
1	.3614
2	.2169
3	.0867
4	.0260
5	.0062
6	.0012
7	.0002

5.77 a. $1,600,000. Yes, the owner will be willing to take the risk as the expected net profit is above $500,000.

b. .3214

5.79 c. .3214 **d.** three

5.81 a. 35 **b.** 10 **c.** .2857

5.83 a. .0142 **b.** .2378

c. The rate of incoming calls is more likely to be 30 per hour.

d. You should advise the manager to hire a second operator since the rate of incoming calls is likely to be higher than 15 calls per hour.

Self-Review Test

Multiple Choice Questions

1. a **2.** b **3.** b **4.** a **5.** b
6. a **7.** c **8.** a **9.** a **10.** c
11. a **12.** b **13.** c **14.** b **15.** a
16. c **17.** b **18.** b

Practice Exercises

1. .67 **2.** 3.40 **3.** 1.555 **4.** .2241
5. .7216 **6.** .0120 **7.** .0166

9. The probability distribution table.

13. $\mu = \Sigma xP(x) = 2.04$ homes

$\sigma = \sqrt{\Sigma x^2P(x) - \mu^2} = \sqrt{6.26 - (2.04)^2} = 1.449$ homes

The four real estate agents sell an average of 2.04 homes per week.

14. a. **i.** .2128

ii. .8418

iii. .0153

b. $\mu = np = 12(.60) = 7.2$ adults

$\sigma = \sqrt{npq} = \sqrt{12\ (.60)\ (.40)} = 1.697$ adults

15. a. .4525 **b.** .0646 **c.** .0666

16. a. i. .0521

ii. .2203

iii. .2013

b.

x	$P(x)$	x	$P(x)$	x	$P(x)$
0	.0000	9	.1251	17	.0128
1	.0005	10	.1251	18	.0071
2	.0023	11	.1137	19	.0037
3	.0076	12	.0948	20	.0019
4	.0189	13	.0729	21	.0009
5	.0378	14	.0521	22	.0004
6	.0631	15	.0347	23	.0002
7	.0901	16	.0217	24	.0001
8	.1126				

Chapter 6

6.11 .6826

6.13 a. .4678 **b.** .4236 **c.** .1749 **d.** .3382 **e.** .8941

6.15 a. .0750 **b.** .0500 **c.** .5987 **d.** .7019

6.17 a. .0918 **b.** .3560 **c.** .1046 **d.** .9599

6.19 a. .25 **b.** −2.50 **c.** 1.25 **d.** −1.25

6.21 a. .0132 **b.** .2743 **c.** .1993 **d.** .0478

6.23 a. .4931 **b.** .1586

6.25 a. .0764 **b.** .0062

6.27 a. .8944 **b.** .0062

6.29 a. .8264 **b.** .1283

6.31 a. .1562 **b.** .0764

6.33 a. .0188

b. 0 approximately

Although it is possible that a given car may take more than 25 minutes for oil and lube service, the probability is almost zero.

6.35 .0232 or 2.32%

6.37 a. $z = 1.96$ **b.** $z = -1.65$ approximately

c. $z = -3.09$ approximately

d. $z = 2.33$ approximately

6.39 a. 208.50 **b.** 241.25 **c.** 178.50

d. 145.75 **e.** 158.25 **f.** 251.25

6.41 a required minimum purchase of $121 would meet the condition.

6.43 approximately $94

6.45 The normal distribution may be used as an approximation to a binomial distribution when both $np > 5$ and $nq > 5$.

6.47 a. $\mu = 50$ and $\sigma = 5.0$

b. .5398

c. .4315

6.49 .7146

6.51 a. .0332 **b.** 0 **c.** .9664

6.53 a. .1469 **b.** .8531

6.55 a. .1308 **b.** .0475

c. 0 approximately

Although it is possible for a carton to contain less than 15.20 ounces, the probability of this is very close to zero.

6.57 Jenn must leave by approximately 7:50 a.m. 40 minutes before she is due to arrive at work.

6.59 a. .0454 **b.** .0838 **c.** .8861 **d.** .2477

6.61 approximately 16.23 ounces

6.63 .0091

6.65 Ashley should leave home at about 8:10 a.m. in order to arrive at work by 9 a.m. 95% of the time.

6.67 b. The gambler has a better chance of coming out ahead with the single-number bet.

6.69 $P(-.74 \le x \le 1.34) = .9452 > .68$

$P(-1.28 \le x \le 1.88) = .9964 > .95$

$P(-1.83 \le x \le 2.43) = 1 - 0 \approx 1 > .997$

Self-Review Test

Multiple Choice Questions

1. a **2.** d **3.** b **4.** a **5.** c
6. b **7.** b **8.** a **9.** c **10.** b
11. b **12.** a

Practice Exercises

1. .1255 **2.** .0401 **3.** .2915 **4.** 3.72
5. .2072 **6.** .4596 **7.** .2381 **8.** .4306
9. .23 **10.** .7355 **11.** .0100 **12.** .5646
13. .00125 **14.** .8936 **15.** 1.27 **16.** 375.44
17. .5318

18. a. .1878 **b.** .9304 **c.** .0985 **d.** .7704

19. a. $z = -1.28$ approximately **b.** $z = .61$

c. $z = 1.65$ approximately **d.** $z = -1.07$ approximately

20. a. .5516 **b.** .0336 **c.** .8925 **d.** .1950

21. a. 139.4 minutes **b.** 168.4 minutes

22. a. i. .0318 **ii.** .9453 **iii.** .9099
iv. .0268 **v.** .4632

b. .6736

c. .3986

Chapter 7

7.5

x	$P(x)$	$xP(x)$	x^2	$x^2P(x)$
70	.20	14.00	4900	980.000
78	.20	15.60	6084	1216.80
80	.40	32.00	6400	2560.00
95	.20	19.00	9025	1805.00
		$\Sigma xP(x) = 80.60$		$\Sigma x^2P(x) = 6561.80$

$\mu = 80.60$
$\sigma = 8.09$

7.7 **a.** Mean of $\bar{x} = \mu_{\bar{x}} = \mu$
b. Standard deviation of $\bar{x} = \sigma_{\bar{x}} = \sigma/\sqrt{n}$ where σ = population standard deviation and n = sample size.

7.11 **a.** $\mu_{\bar{x}} = \mu = 100$ and $\sigma_{\bar{x}} = 4.47$
b. $\mu_{\bar{x}} = \mu = 100$ and $\sigma_{\bar{x}} = 2.828$

7.13 **a.** 14 **b.** 36

7.15 $\mu = \$482{,}000$, $\sigma = \$90{,}000$, and $n = 200$
$\mu_{\bar{x}} = \mu = \$482{,}000$ and $\sigma_{\bar{x}} = \sigma/\sqrt{n} \approx \$6363.96$

7.17 **a.** $\Sigma \bar{x}P(\bar{x}) = 80.60$
b. $\sigma_{\bar{x}} = \sqrt{\Sigma \bar{x}^2 P(\bar{x}) - \mu_{\bar{x}}^2} = 3.302$
c. $\sigma/\sqrt{n} = 4.67$ is not equal to $\sigma_{\bar{x}} = 3.30$ in this case because $n/N = 3/5 = .60 > .05$.
d. $\sigma_{\bar{x}} = \dfrac{\sigma}{\sqrt{n}}\sqrt{\dfrac{N-n}{N-1}} = 3.302$

7.21 **a.** Approximately normal because $n \geq 30$ and the central limit theorem applies.
b. Central limit theorem applies as $n \geq 30$.
c. Central limit theorem does not apply as $n < 30$, slightly skew to the right.

7.23 $\mu_{\bar{x}} = \mu = 8.4$ minutes and $\sigma_{\bar{x}} = \sigma/\sqrt{n} = .450$ minute
The sampling distribution of $\bar{x}$ is normal because the population is normally distributed.

7.25 $\mu_{\bar{x}} = \mu = 150$ pounds
$\sigma_{\bar{x}} = \sigma/\sqrt{n} = 3.536$
The sampling distribution of $\bar{x}$ is approximately normal because the sample size is large ($n \geq 30$).

7.27 $\mu_{\bar{x}} = \mu = 170.4$ minutes and $\sigma_{\bar{x}} = \sigma/\sqrt{n} = 1$ minute
The sampling distribution of $\bar{x}$ is approximately normal.

7.29 .9995

7.31 **a.** .7299 **b.** .4733

7.33 **a.** 0 **b.** .9818

7.35 **a.** .0808 **b.** .0485 **c.** .7616

7.37 **a.** .2696 **b.** .0548 **c.** .0082

7.39 **a.** .0764 **b.** .1075 **c.** .595

7.41 .0124

7.43 .13

7.45 **a.** $\mu_{\hat{p}} = p$
b. $\sigma_{\hat{p}} = \sqrt{pq/n}$
c. The sampling distribution of $\hat{p}$ is approximately normal if $np > 5$ and $nq > 5$.

7.47 Sampling error $= \hat{p} - p = .05$.

7.49 The sample proportion $\hat{p}$ is a consistent estimator of p, since $\sigma_{\hat{p}}$ decreases as the sample size increases.

7.51 **a.** $n = 300$, $\mu_{\hat{p}} = p = .34$, and $\sigma_{\hat{p}} = \sqrt{pq/n} = .027$
b. $n = 850$, $\mu_{\hat{p}} = p = .34$, and $\sigma_{\hat{p}} = \sqrt{pq/n} = .016$

7.53 A sample is considered large enough to apply the central limit theorem if $np > 5$ and $nq > 5$.

7.55 **a.** 13.5, the central limit theorem applies
b. 46.75, the central limit theorem applies
c. 48.75, the central limit theorem applies
d. The central limit theorem does not apply.

7.57 $p = .51$, $q = 1 - p = 1 - .51 = .49$, and $n = 200$
$\mu_{\hat{p}} = p = .51$, and $\sigma_{\hat{p}} = \sqrt{pq/n} = .0353$
$np = (200)(.51) = 102$ and $nq = (200)(.49) = 98$
Since $np > 5$ and $nq > 5$, the sampling distribution of $\hat{p}$ is approximately normal.

7.59 $p = .55$, $q = 1 - p = 1 - .55 = .45$, and $n = 900$
$\mu_{\hat{p}} = p = .55$, and $\sigma_{\hat{p}} = \sqrt{pq/n} = .0166$
$np = (900)(.55) = 495$ and $nq = (900)(.45) = 405$
Since $np > 5$ and $nq > 5$, the sampling distribution of $\hat{p}$ is approximately normal.

7.61 .9876

7.63 **a.** .2592 **b.** .1515

7.65 .1515

7.67 $\mu = 750$ hours, $\sigma = 55$ hours, and $n = 25$
$\mu_{\bar{x}} = \mu = 750$ hours and $\sigma_{\bar{x}} = \sigma/\sqrt{n} = 11$ hours
The sampling distribution of $\bar{x}$ is normal because the population is normally distributed.

7.69 **a.** .0582 **b.** .8325
c. .9991 **d.** .0045

7.71 $p = .70$, $q = 1 - p = 1 - .70 = .30$, and $n = 400$
$\mu_{\hat{p}} = p = .70$, and $\sigma_{\hat{p}} = \sqrt{pq/n} = .02291288$
$np = (400)(.70) = 280 > 5$, $nq = (400)(.30) = 120 > 5$
Since np and nq are both greater than 5, the sampling distribution of $\hat{p}$ is approximately normal.

7.73 **a.** .8118 **b.** .0177 **c.** .0244

7.75 **a.** .7939
b. The reporter should take a sample of at least 66 voters.

Self-Review Test

Multiple Choice Questions

1. b	**2.** b	**3.** a	**4.** a	**5.** b
6. b	**7.** c	**8.** a	**9.** a	**10.** a
11. c	**12.** a	**13.** c	**14.** b	**15.** c
16. c	**17.** a	**18.** b	**19.** b	**20.** b
21. a	**22.** a	**23.** a	**24.** b	**25.** b

Practice Exercises

1. 5.00 **2.** 7.5 pounds **3.** .1423 **4.** .4227

5. −1.40 **6.** .387 **7.** .677 **8.** .0480

9. .0203 **10.** 325 **11.** 16

12. .4789 **13.** .6029

15. a. $\mu_{\bar{x}} = \mu = \$6015$ and $\sigma_{\bar{x}} = \sigma/\sqrt{n} = \$156.53$

b. $\mu_{\bar{x}} = \mu = \$6015$ and $\sigma_{\bar{x}} = \sigma/\sqrt{n} = \$70$

c. $\mu_{\bar{x}} = \mu = \$6015$ and $\sigma_{\bar{x}} = \sigma/\sqrt{n} = \$24.75$

In all cases the sampling distribution of $\bar{x}$ is approximately normal because the population has an approximate normal distribution, and the approximation improves (in that the variance gets smaller) as n gets larger.

16. a. \$4472.14

Since the population distribution is skewed to the right, and $n < 30$, the shape of the sampling distribution of is also skewed to the right.

b. \$2000

The sampling distribution of $\bar{x}$ is approximately normal.

c. \$707.11

The sampling distribution of $\bar{x}$ is approximately normal.

17. a. .1143 **b.** .7698 **c.** .1469 **d.** .2006

e. .1711 **f.** .7422 **g.** .0735 **h.** .6820

18. a. i. .1203 **ii.** .1335 **iii.** .7486

b. .9736 **c.** .0013

19. The sampling distribution of $\hat{p}$ is approximately normal for a, b, and c.

20. a. i. .2358 **ii.** .9078 **iii.** .0158

iv. .9078 **v.** .9236 **vi.** .9236

b. .9266 **c.** .9684 **d.** .0764

Chapter 8

8.11 a. $\bar{x} = 20.4$ **b.** 19.07 to 21.73 **c.** 1.33

8.13 a. 77.84 to 85.96

b. 78.27 to 85.53

c. 78.65 to 85.15

d. Yes, the width of the confidence intervals decreases as the sample size increases. This occurs because the standard deviation of the sample mean decreases as the sample size increases.

8.15 a. 100 **b.** 75

8.17 a. 136 **b.** 114 **c.** 65

8.19 3294.96 to 3536.44 gallons

8.21 The machine does not need an adjustment.

8.23 167

8.25 72

8.33 a. $df = n - 1 = 18 - 1 = 17$ and $t = -1.333$

b. $df = n - 1 = 22 - 1 = 21$ and $t = 2.080$

c. $t = 3.281$

d. $t = -2.715$

8.35 a. 2.080 **b.** 1.671 **c.** 2.807

8.37 a. $\bar{x} = 44.10$ **b.** 38.62 to 49.58 **c.** 5.48

8.39 a. 53.93 to 56.71

b. 56.16 to 58.64

c. 54.95 to 57.55

d. The confidence intervals of parts (a) and (c) cover μ but the confidence interval of part (b) does not.

8.41 \$52,026.14 to \$57,973.86

8.43 a. 25.36 to 27.44 mpg

8.45 69.50 to 77.90 mph

8.47 a. $\bar{x} = (\Sigma x)/n = 51.89$ **b.** 48.17 to 55.61

8.53 a. The sample size is large enough.

b. The sample size is large enough.

c. The sample size is large enough.

d. The sample size is not large enough.

8.55 a. .583 to .677 **b.** .542 to .638 **c.** .624 to .716

d. The confidence intervals of parts (a) and (b) cover p, but the confidence interval of part (b) does not.

8.57 a. .615 to .845 **b.** .683 to .777 **c.** .700 to .760

d. Yes, the width of the confidence intervals decreases as the sample size increases. This occurs because increasing the sample size decreases the standard deviation of the sample proportion.

8.59 a. 380 **b.** 671

8.61 a. $\hat{p} = .67$

b. .642 to .698 $E = zs_{\hat{p}} = .028$

8.63 a. $\hat{p} = .40$, so the point estimate for the population percentage is 40%.

b. .069

8.65 a. .517 to .583

b. The sample proportion of .55 is an estimate of p based on a random sample. Because of sampling error, this estimate might differ from the true proportion p, so we make an interval estimate to allow for this uncertainty and sampling error.

8.67 a. $\hat{p} = .333$ **b.** .085 to .581

The corresponding interval for the population percentage is 8.5% to 58.1%.

8.69 1849

8.73 The machine needs an adjustment.

8.75 39.32 to 71.86 minutes

8.77 9.59 to 9.91 hours

8.79 .061 to .565. The corresponding interval for the population percentage is 6.1% to 56.5%.

8.81 .2454 to .6346. The corresponding interval for the population percentage is 24.54% to 63.46%.

8.83 221

8.85 159

8.91 1. While σ is constant, the sample standard deviation, s, is not. The sample standard deviation will change with each sample.

2. If σ is unknown, the confidence interval is calculated with a t value. This value will change as the sample size changes because the degrees of freedom will change.

Self-Review Test

Multiple Choice Questions

1. b	**2.** a	**3.** a	**4.** c	**5.** b
6. a	**7.** c	**8.** c	**9.** c	**10.** c
11. b	**12.** b	**13.** b	**14.** a	**15.** a
16. c	**17.** a	**18.** a	**19.** a	**20.** a

Practice Exercises

1. a. population parameter, sample statistic
b. sample statistic, population parameter
c. sample statistic, population parameter

2. 138 **3.** 754 **4.** 1182

5. a. $\bar{x} = \$159{,}000$
b. \$147,390 to \$170,610 $E = z\sigma_{\bar{x}} = 2.58(4500) = \$11{,}610$

6. \$571,283.30 to \$649,566.70

7. a. $\hat{p} = .37$ **b.** .343 to .397

8. 83

9. 273

10. 229

11. The width of the confidence interval can be reduced by:
1. Lowering the confidence level
2. Increasing the sample size

The second alternative is better because lowering the confidence level results in a less reliable estimate for μ.

Chapter 9

9.5 The null hypothesis is initially assumed to be true.

9.7 a. $H_0: \mu = 18$ hours, $H_1: \mu \neq 18$ hours, two-tailed test
b. $H_0: \mu = 15$ hours, $H_1: \mu > 15$ hours, right-tailed test
c. $H_0: \mu = 5$ years, $H_1: \mu \neq 5$ years, two-tailed test
d. $H_0: \mu = \$800$, $H_1: \mu < \$800$, left-tailed test
e. $H_0: \mu = 10$ minutes, $H_1: \mu > 10$ minutes, right-tailed test

9.11 Rejecting H_0 is equivalent to stating that the evidence from the sample is strong enough to claim that H_1 is true.

9.15 a. p-value = .0166
b. For $\alpha = .01$, do not reject H_0 since $.0166 > .01$.
c. For $\alpha = .025$, reject H_0 since $.0166 < .025$.

9.17 If H_0 is rejected, the difference between the observed value of $\bar{x}$ and the hypothesized value of μ is "statistically significant."

9.19 a. .03 **b.** .06 **c.** .02 **d.** .005

9.21 a. rejection region $z \leq -2.58$ or rejection region $z \geq 2.58$
b. rejection region $z \geq 2.33$
c. rejection region $z \leq -2.33$

9.23 a. Reject H_0 since $-2.67 < -1.96$.
b. Do not reject H_0 since $-1.00 > -1.96$.
c. Comparing parts (a) and (b) shows that two samples selected from the same population can lead to opposite conclusions of the same test of hypothesis.

9.25 a. Do not reject H_0 since $.0901 > .025$.
b. Do not reject H_0 since $-1.34 > -1.96$.
For both parts (a) and (b), conclude the mean life of these batteries is not less than 45 months.

9.27 a. For $\alpha = .01$, do not reject H_0 since $.0139 > .01$.
For $\alpha = .025$, reject H_0 since $.0139 < .025$.
b. For $\alpha = .01$, do not reject H_0 since $-2.20 > -2.33$.
For $\alpha = .025$, reject H_0 since $-2.20 < -1.96$.
For both parts (a) and (b), conclude the salary was less than \$55,280 at $\alpha = .01$, but not at $\alpha = .025$.

9.29 a. For $\alpha = .02$, do not reject H_0 since $.0548 > .02$.
For $\alpha = .10$, reject H_0 since $.0548 < .10$.
b. For $\alpha = .02$, do not reject H_0 since $1.92 < 2.33$.
For $\alpha = .10$, reject H_0 since $1.92 > 1.65$.
For both parts (a) and (b), the inspector will not stop this machine at $\alpha = .02$, but will stop this machine and adjust it at $\alpha = .10$.

9.31 a. For $\alpha = .05$, reject H_0 since $.0207 < .05$.
For $\alpha = .01$, do not reject H_0 since $.01 < .0207$
b. For $\alpha = .05$, reject H_0 since $2.04 > 1.65$.
For $\alpha = .01$, do not reject H_0 since $2.04 < 2.33$.
Conclude that for parts (a) and (b), the current average checking account fee is more than 49.60 per month if the maximum probability of Type 1 error is .05.
If $\alpha = .01$, conclude that the mean checking account fee is not more than \$9.60 per month.
If $\alpha = 0$, there is no rejection region, so do not reject H_0. Conclude that the average checking account fee is not more than \$9.60 per month.

9.35 a. $df = n - 1 = 15 - 1 = 14$
rejection region $t \leq -2.977$ or rejection region $t \geq 2.977$
b. $df = n - 1 = 25 - 1 = 24$
rejection region $t \leq -2.797$
c. $df = n - 1 = 22 - 1 = 21$
rejection region $t \geq 2.080$

9.37 a. <u>p-value approach</u>
$.05 < p\text{-value} < .10$.
<u>critical value approach</u>
For $\alpha = .05$ with $df = 15$, the critical value of t is -1.753.
b. <u>p-value approach</u>
$.10 < p\text{-value} < .20$.
<u>critical value approach</u>
For $\alpha = .05$ with $df = 15$, the critical values of t are -2.131 and 2.131.

9.39 a. Reject H_0 since $4.800 > 1.998$.
b. Do not reject H_0 since $1.143 < 1.998$.

c. Comparing parts (a) and (b) shows that two samples selected from the same population can lead to opposite conclusions on the same test of hypothesis.

9.41 Reject H_0 since $2.4 > 1.645$.

Conclude that the current average cost of a wedding in America is greater than $33,931.

p-value approach
Reject H_0 since p-value $< .05$.
The conclusion is the same.

9.43 Reject H_0 since p-value $< .05$.

Conclude that the manufacturer's claim is false since the hypothesis test leads to the conclusion that the number of calories in the 12-ounce cans is greater than 30.

9.45 a. Reject H_0 since $3.913 > 2.539$. Conclude that the mean drying time is longer than 60 minutes.

b. The Type I error would occur if the mean drying time for these paints was 60 minutes or less, but we concluded the mean drying time is more than 60 minutes. The probability of making such an error is $\alpha = .01$.

9.47 Do not reject H_0 since $1.369 < 1.796$.
Conclude that the average amount of time U. S. adults spent on leisure activities has not changed.

9.49 p-value approach
Reject H_0 since p-value $< .01$.
critical value approach
Reject H_0 since $2.807 > 2.326$.
Conclude that the current mean credit card debt for households with credit cards is more than $15,900.

9.53 a. Since $np > 5$ and $nq > 5$, the sample size is large enough to use the normal distribution.

b. Since $np > 5$ and $nq > 5$, the sample size is large enough to use the normal distribution.

c. Since $np < 5$, the sample size is not large enough to use the normal distribution.

d. Since $np > 5$ and $nq > 5$, the sample size is large enough to use the normal distribution.

9.55 a. observed value: $-.9132$ critical value: 1.65

b. observed value: $-.9132$ critical value: 1.96

9.57 a. rejection region $z \le -1.65$

b. rejection region $z \le -1.96$ or rejection region $z \ge 1.96$

c. rejection region $z \ge 1.65$

9.59 a. Do not reject H_0 since $-1.21 > -1.96$.

b. Reject H_0 since $-2.41 < -1.96$.

9.61 p-value approach
Reject H_0 since $.0013 < .05$.

critical value approach
Reject H_0 since $1.65 < 3.02$.

Conclude that the proportion of adults who prefer generic is higher than 18%.

9.63 p-value approach
Reject H_0 since $0 < .02$.

critical value approach
Reject H_0 since $3.92 > 2.05$.

Conclude that the current percentage of all college students who say their college experience exposed them to diverse opinions, cultures, and values is higher than 88%.

9.65 p-value approach
Do not reject H_0 since $.2177 > .025$.

critical value approach
Do not reject H_0 since $.78 < 1.96$.

Conclude that it is not the case that more than 12% of all residents ages 18 and over in this city would rate their current financial situation as excellent.

9.67 a. The machine should be stopped to make necessary adjustments.

b. We conclude that the machine should not be stopped.

c. Our decision in part (b) differs from that of part (a). Conclude that the machine should not be stopped.

9.71 a. Reject H_0 since $2.55 > 2.33$.

b. $\alpha = .02$.

c. For $\alpha = .025$, reject H_0 since $.0108 < .025$.
For $\alpha = .01$, do not reject H_0 since $.0108 > .01$.

9.73 a. Reject H_0 since $.0026 < .025$.

b. Reject H_0 since $-2.80 < -1.96$.
Conclude that the mean consumption of water per household has decreased.

9.75 a. Reject H_0 since $2.097 > 1.677$.
Conclude that the mean living area of all single-family homes in this county exceeds 2400 square feet, and the realtor's claim is false.

b. If $\alpha = .01$, the critical value of t is 2.405. Since the value of the test statistic is 2.097, we do not reject H_0.

c. Thus, our decision in part (b) differs from that of part (a); we conclude that the mean living area of all single-family homes in this county is at most 2400 square feet, and the realtor's claim is true.

9.77 Do not reject H_0 since $-2.947 < 2.083 < 2.947$.
Conclude that the mean waiting time at the emergency ward is not different than 25 minutes.

9.79 Do not reject H_0 since $1.682 < 2.718$. Conclude that the mean time it takes to learn how to use this software is not more than two hours and the company's claim is true.

9.81 a. Reject H_0 since $-2.68 < -2.33$.
Conclude that the current percentage of American adults who hold the mentioned opinion is less than 75%.

b. A Type I error would occur if the current percentage of American adults who hold the mentioned opinion is at least 75%, but we concluded this percentage is less than 75%.

c. For $\alpha = .01$, reject H_0 since $.0037 < .01$. The conclusion is the same as in part (a).

9.83 Do not reject H_0 since $-1.44 > -2.054$
Conclude that the percentage of Americans age 50–64 who own a smartphone is not less than 81%.

9.85 Reject H_0 and shut down the machine if the number of defectives in a sample of 130 parts is 9 or more.

9.87 **a.** Let p be the proportion of all people receiving the new vaccine who contract the disease within a year. Then the appropriate hypotheses are $H_0: p = .30$ and $H_1: p < .30$.

b. $z = -2.95$ and
$\alpha = P(x < 16.5 \mid p = .30) = P(z < -2.95) = .0016$

c. $\alpha = P(x < 3 \mid p = .30) = .0354$

9.89 **a.** One of the p-values is $.4546/2 = .2273$. The other p-value is the complement, $1 - .2273 = .7727$.

b. Since the value of the test statistic is negative, the p-value of .2273 is for the left-tailed test, $H_1: \mu < 15$. The p-value for the right-tailed test, $H_1: \mu > 15$, is .7727.

Self-Review Test

Multiple Choice Questions

1. a	2. b	3. a	4. b	5. a
6. a	7. a	8. b	9. c	10. a
11. c	12. b	13. c	14. a	15. b
16. b	17. a	18. c	19. b	20. c
21. a	22. c	23. b	24. c	25. b
26. a	27. c	28. c	29. b	30. a

Practice Exercises

1. **a.** Do not reject H_0. Conclude that the average deductible is \$1655.

b. Reject H_0. Conclude that the average deductible is greater than \$1655.

c. In part (a), a Type I error would occur if we reject H_0 when it is true, that is, to conclude that the average deductible is not \$1655, when it is. However, in part (a), we do not make a Type I error because we do not reject H_0. It is possible that we have made a Type II error instead. A Type I error in part (b) is to conclude that the average deductible is more than \$1655 when it isn't.

d. For $\alpha = .01$, do not reject H_0 since $.01 < .025$.

e. For $\alpha = .025$, reject H_0 since $.0125 < .25$.

2. **a.** Reject H_0 since $-3.000 < -2.438$. Conclude that the mean duration of nine-inning games has decreased after the meeting.

b. A Type I error would be to conclude that the mean durations of games have decreased after the meeting when they are actually equal to the duration of games before the meeting. The maximum probability of making such an error is $\alpha = .01$.

c. Do not reject H_0.

d. Reject H_0.

3. **a.** Reject H_0. Conclude that the mean time it takes to write a textbook is less than 31 months.

b. A Type I error would be to conclude that the editor's claim is false when it is actually true. The maximum probability of making such an error is $\alpha = .025$.

c. Do not reject H_0.

4. **a.** Reject H_0 since $-3.16 < -1.65$. Conclude that the percentage of people who have a will is less than 50%.

b. A Type I error would be to conclude that the percentage of adults with wills was less than 50% when it is actually 50%.

c. Do not reject H_0.

d. Reject H_0.

Chapter 10

10.3 **a.** 3.49

b. 1.38 to 5.60.
$E = z\sigma_{\bar{x}_1 - \bar{x}_2} = 2.11$

10.5 Do not reject H_0 since $1.66 < 1.96$.

10.7 **a.** 9 hours

b. 1.65 to 16.35 hours

c. <u>p-value approach</u>
Reject H_0 since $.0078 < .02$.
<u>critical value approach</u>
Reject H_0 since $2.66 > 2.33$.
Conclude that the claim that the mean elapsed times for repellent A and B are different is true.

10.9 **a.** −\$1024.54 to −\$75.46

b. Reject H_0 since $-2.99 < -2.58$.
Conclude that such mean repair costs are different for these two types of cars.

c. If $\alpha = 0$, there can be no rejection region, and we cannot reject H_0. Therefore, the decision would be "do not reject H_0."

10.11 **a.** −6.87 to .87 calories

b. Do not reject H_0 since $-1.81 > -2.33$. Maine Mountain Dairy's claim is false.

c. For $\alpha = .05$, reject H_0 since $.0351 < .05$.
For $\alpha = .025$, do not reject H_0 since $.0351 > .025$.

10.13 **a.** −3.93 **b.** −6.34 to −1.52

10.15 Reject H_0 since $10.0645 > 2.58$.

10.17 **a.** 2.96

b. −6.63 to 12.55

c. Do not reject H_0 since $.7768 < 2.518$.

10.19 **a.** −46.80 to −7.20 miles

b. Reject H_0 since $-2.673 < -2.326$.

Conclude that the mean distance between oil changes is less for all luxury cars than that for all compact lower price cars.

10.21 **a.** 46.86 to 63.14 seconds

b. Reject H_0 since $19.45 > 2.552$.
Conclude that the mean time taken by the Piranha to shred 100 sheets is greater than that for the Crocodile.

c. If $\alpha = 0$, there can be no rejection region, and we cannot reject H_0. Therefore, the decision would be "do not reject H_0."

10.23 **a.** $\bar{x}_1 - \bar{x}_2 = \$12.22$

b. \$7.98 to \$16.46

c. critical value approach
Reject H_0 since $6.71 > 2.326$

p-value approach
Reject H_0 since $0 < .01$.
Conclude that average annual energy cost is higher for refrigerator models with side-by-side freezers than for models with top freezers.

10.25 .0102 to .1238

10.27 Reject H_0 since $3.1308 > 2.657$.

10.29 Reject H_0 since $3.1308 > 1.999$.

10.31 a. 4.02 to 7.98 mph

b. Reject H_0 since $7.34 > 2.418$.
Conclude that the mean speed of cars driven by all men drivers on this highway is higher than that of cars driven by all women drivers.

c. 3.82 to 8.18 mph
Reject H_0 since $6.87 > 2.500$.
Conclude that the mean speed of cars driven by all men drivers on this highway is higher than that of cars driven by all women drivers.

Because the standard error and critical values are both larger, the confidence interval is wider. The conclusion for the hypothesis test did not change.

10.33 a. $-.61$ to $-.39$

b. Reject H_0 since $-10.16 < -2.58$.
Conclude that the mean satisfaction indexes for all customers for the two supermarkets are different.

c. Reject H_0 since $-10.10 < -2.58$.
Conclude that the mean satisfaction indexes for all customers for the two supermarkets are different. Because the standard error only changed slightly and the critical value is the same, the confidence interval is very slightly wider. The conclusion for the hypothesis test did not change.

10.37 a. Reject H_0 since $12.551 > 2.060$.

b. Reject H_0 since $7.252 > 2.624$.

c. Reject H_0 since $-14.389 < -1.328$.

10.39 a. -2.51 to 4.71 taps

b. Do not reject H_0 since $.6897 < 2.262$.
Conclude that the average number of taps is not different for right and left index fingers.

10.41 a. -1.02 to 1.52 ppm

b. Do not reject H_0 since $.412 < 2.093$.
Conclude that the average paired difference for concentration levels is not different from zero.

10.43 Two samples are considered large enough for the sampling distribution of the difference between two sample proportions to be approximately normal when n_1p_1, n_1q_1, n_2p_2, and n_2q_2 are all greater than 5.

10.45 $-.095$ to .175

10.47 Do not reject H_0 since $-1.96 < .7670 < 1.96$.

10.49 a. $\hat{p}_1 - \hat{p}_2 = -.04$

b. $-.086$ to .006

c. H_0: $p_1 - p_2 = 0$; H_1: $p_1 - p_2 < 0$

d. $z = -2.02$

e. Do not reject H_0 since $-2.02 > -2.33$.

10.51 a. $\hat{p}_1 - \hat{p}_2 = .024$

b. $-.020$ to .068

c. p-value approach
Do not reject H_0 since $.2758 > .05$.

critical value approach
Do not reject H_0 since $1.09 < 1.96$.
Conclude that the proportions of cars that fail the emissions test at the two stations are not different.

10.53 a. $\hat{p}_1 - \hat{p}_2 = .10$

b. .018 to .182

c. Reject H_0 since $3.04 > 2.58$.
Conclude that the proportions of all men and all women who play the lottery often are different.

10.55 a. $-.013$ to .093

b. Do not reject H_0 since $1.75 < 2.33$.
Conclude that the proportion of all sales for which at least one item is returned is not higher for Store A than for Store B.

10.57 a. -131.30 to -120.70

b. Reject H_0 since $-46.58 < -1.96$.
Conclude that the mean weekly earnings of female workers who are not union members are less than those of female workers who are union members.

10.59 a. Do not reject H_0 since $-1.058 > -2.009$.
Conclude that the mean GPAs of all male and all female college students who actively participate in sports are not different.

b. $-.31$ to .07

10.61 a. \$791.81 to \$2728.19

b. Reject H_0 since $4.06 > 2.33$.
Conclude that the average annual earnings of financial analysts is higher than that of engineers

10.63 a. -8.35 to -1.89 cards

b. Reject H_0 since $-3.11 < -1.65$.
Conclude that the average number Christmas cards mailed out by all households in Town A is different from the corresponding average for Town B.

10.65 a. -9.54 to $-.24$

b. Do not reject H_0 since $-2.425 > -2.896$.
Conclude that the course does not make any statistically significant improvement in the memory of all students.

10.67 a. $-.11$ to .12

b. p-value approach
Do not reject H_0 since $.01 < .8982$.

critical value approach
Do not reject H_0 since $-.28 > -.258$.
Conclude there is no evidence of a difference in the proportions of men and women who would develop

a mild allergic reaction within four hours of taking this drug

10.69 .2611

10.71 Nine cars should be tested.

10.73 **a.** Use a sample size of 36 for each class.

b. The sample means must differ by at least 5.89 points in order to conclude that the two population means are different.

Self-Review Test

Multiple Choice Questions

1. a	**2.** b	**3.** c	**4.** a	**5.** c
6. a	**7.** c	**8.** a	**9.** c	**10.** a
11. b	**12.** a	**13.** c	**14.** b	**15.** c
16. a				

Practice Exercises

2. **a.** 1.62 to 2.78

b. Reject H_0 since 9.86 > 1.96.
Conclude that the mean stress score of all executives is higher than that of all professors.

3. **a.** −2.72 to −1.88 hours

b. Reject H_0 since −10.997 < −2.416.
Conclude that the mean time for all wide receivers to run 40 yards is lower than the mean time required for all quarterbacks.

4. **a.** .55 to .35 hours

b. Reject H_0 since −8.72 < −2.416.
Conclude that the mean time for all wide receivers to run 40 yards is lower than the mean time required for all quarterbacks.

5. **a.** −$53.60 to $186.18

b. Do not reject H_0 since 2.050 < 2.447.
Conclude that the mean estimate for repair costs at Zeke's is not different than the mean estimate at Elmer's.

6. **a.** −.052 to .092

b. Do not reject H_0 since .60 < 2.58.
Conclude that the proportion of all male voters who voted in the last presidential election is not different from that of all female voters.

Chapter 11

11.3 $\chi^2 = 9.312$

11.5 **a.** $\chi^2 = 4.404$

b. $\chi^2 = 3.074$

11.11 Do not reject H_0 since 11 < 11.070.
Conclude that the die is fair.

11.13 Do not reject H_0 since 10.3348 < 11.345.
Conclude that the distribution of payment methods has not changed.

11.15 Do not reject H_0 since 6.534 < 9.348.
Conclude that there is no evidence against the null hypothesis that this is a random sample.

11.17 Do not reject H_0 since 8.888 < 9.210.
Conclude that the distribution of defects is the same as when the process is working properly and the process does not need an adjustment.

11.21 **a.** H_0: Row and columns are independent, H_1: Rows and columns are dependent.

c. For $\alpha = .01$, the critical value of χ^2 is 13.277.

e. Reject H_0 since 16.617 > 13.277. Conclude that the rows and columns are not independent.

11.23 Do not reject H_0 since 1.814 < 6.635.
Conclude that gender and brand preference are independent.

11.25 Do not reject H_0 since 4.059 < 9.210.
Conclude that gender and opinions are not related.

11.27 Do not reject H_0 since .5663 < 3.841.
Conclude that the distribution of good and defective parts is the same for both subsidiaries.

11.29 Reject H_0 since 25.165 > 14.449.
Conclude that the grade distributions are not homogeneous for the three professors.

11.31 **a.** 21.7631 to 153.3333

b. 24.6563 to 114.4075

c. 29.1763 to 83.1308
As the sample size increases, the confidence interval for σ^2 decreases in width.

11.33 **a.** $H_0: \sigma^2 = 1.75, H_1: \sigma^2 > 1.75$

b. 34.170

c. 22.514

d. Do not reject H_0 since 22.514 < 34.170.

11.35 **a.** 1.221 to 1.814.

b. Do not reject H_0 since 53.250 < 71.420.
Conclude that the population variance is not greater than 2.0 square micrometers.

11.37 **a.** 52.338 to 112.356

b. Do not reject H_0 since 12.401 < 29.714 < 39.364.
Conclude that the population variance is not different from 4200 square hours.

11.39 Do not reject H_0 since 9.052 < 9.348.
Conclude that there is no evidence the current distribution of responses differs from the 2019 distribution.

11.41 Reject H_0 since 35.44 > 11.345.
Conclude that the proportions of all allergic persons are not the same over the four seasons.

11.43 Reject H_0 since 11.014 > 9.348.
Conclude that the percentages of investors favoring the four choices are not equal.

11.45 Do not reject H_0 since 10.181 < 16.812.
Conclude that the percentages of people with different opinions are similar for all four regions.

11.47 **a.** 1.846 to 4.899

b. 2.887 to 5.767

11.49 Do not reject H_0 since $24.727 < 28.845$.
Conclude that the population variance is not greater than 1.1.

11.51 **a.** Do not reject H_0 since $29.920 < 33.924$.
Conclude that the population variance is not greater than .025 square millimeter, and the machine does not need an adjustment.

b. .0203 to .0681

11.53 **a.** 20.15553571

b. 2.763 to 10.671

c. Do not reject H_0 since $7.054 < 14.067$.
Conclude that the population variance is not larger than 20 square pounds.

11.55 **a.** From the normal distribution table, area to the right of $z = 1.64$ is $1 - .9495 = .0505$. $p\text{-value} = 2(.0505) = .1010$

b. $p\text{-value} = .1000$

c. $z^2 = (1.64)^2 = 2.6896$ is approximately equal to $\chi^2 = 2.713$. The p-values are the same (the difference is due to rounding), which is always true since the tests in parts (a) and (b) are equivalent.

11.57 Reject H_0 since $215.568 > 16.919$.
Conclude that a normal distribution is not an appropriate model for these data.

Self-Review Test

Multiple Choice Questions

1. b	**2.** a	**3.** c	**4.** a	**5.** b
6. b	**7.** c	**8.** b	**9.** a	**10.** b
11. c	**12.** b	**13.** a	**14.** a	**15.** a
16. b	**17.** c	**18.** c	**19.** a	**20.** c
21. c	**22.** c	**23.** a	**24.** a	

Practice Exercises

1. Do not reject H_0 since $8.70 < 11.345$.
Conclude that there is no evidence birthdays are distributed unequally across the four seasons.

2. Reject H_0 since $31.187 > 11.345$.
Conclude that educational level and ever being divorced are dependent.

3. Reject H_0 since $82.450 > 9.488$.
Conclude that the percentages of people who play the lottery often, sometimes, and never are not the same for each income group.

4. **a.** .486 to 1.154

b. Reject H_0 since $36.480 > 36.191$.
Conclude that the population variance exceeds .25 square ounce.

Chapter 12

12.3 **a.** $F = 7.26$ **b.** $F = 5.82$ **c.** $F = 5.27$

12.5 **a.** $F = 3.01$ **b.** $F = 2.34$

12.9 **a.** For Sample I: $n_1 = 6, \Sigma x = 170,$
$\Sigma x^2 = 4946$
$\bar{x}_1 = 28.333$
$s_1 = 5.08592830$
For Sample II: $n_2 = 6, \Sigma x = 204,$
$\Sigma x^2 = 7048$
$\bar{x}_2 = 34.000$
$s_2 = 4.73286383$

b. Do not reject H_0 since $-1.998 > -3.169$.

c. Do not reject H_0 since $3.99 < 10.04$.

d. For both parts (b) and (c), conclude that the population means are equal.

12.11 **b.** Reject H_0 since $9.2953 > 3.25 > 3.1834$.
Sample evidence indicates that the means of the four groups are not equal.

12.13 Do not reject H_0 since $1.30 < 3.55$.
Conclude that the mean time taken to find their first job for all 2019 graduates in these three fields is the same.

12.15 Do not reject H_0 since $.5838 < 4.08$.
Conclude that the mean life of bulbs is same for all four brands.

12.17 **a.** Reject $H_{\beta 0}$ since $62.83 > 3.40$.
Sample evidence indicates that life satisfaction is not same across different age groups.

b. Reject $H_{\tau 0}$ since $24 > 4.26$.
Sample evidence indicates that life satisfaction is not same for males and females. So, gender plays a role in life satisfaction.

c. Do not reject $H_{\lambda 0}$ since $.32 < 3.40$.
Sample evidence indicates that there is no interaction between age and gender in terms of life satisfaction.

12.19 **a.** Do not reject H_0 since $1.172 < 6.93$.
Conclude that there is no significant difference in the mean weight gained by the chickens among the three diets.

b. By not rejecting H_0, we may have made a Type II error by concluding that the mean weight gained by all chickens is the same for each of the three diets when in fact it is not.

12.21 **a.** Reject H_0 since $4.89 > 3.89$.
Conclude that the mean tips for the three restaurants are not equal.

b. If $\alpha = 0$, there can be no rejection region, and we cannot reject H_0. Therefore, the decision would be "do not reject H_0."

12.23 **a.** Reject $H_{\tau 0}$ since $101.71 > 4.26$.
Sample evidence indicates that the two different fertilizer has different effect on mean number of dragon fruit yield.

b. Reject $H_{\beta 0}$ since 20.01 > 3.40.
Sample evidence indicates that dragon fruit yield is not same across different plant varieties.

12.25 a. Lay out a route for a test run in a typical city. Let each driver drive each car several times on this route. Make sure that the length of the route is selected in such a way that the total of all tests for each car does not exceed 1200 miles. Then, calculate the gas mileage for each of these test runs. Next, compute the mean gas mileage for each of the four cars and compare them in the article.

b. Using the ANOVA procedure and the data collected in part (a) for four cars, test the null hypothesis that the mean gas mileage is the same for all four cars against the alternative hypothesis that all four means are not equal. Select your own significance level. Then, write a report explaining and interpreting these results.

Self-Review Test

Multiple Choice Questions

1. a	**2.** b	**3.** c	**4.** a	**5.** a
6. a	**7.** b	**8.** a	**9.** b	**10.** b
11. c	**12.** c	**13.** c	**14.** b	**15.** b
16. c	**17.** a	**18.** b	**19.** b	**20.** c
21. c	**22.** c	**23.** a		

Practice Exercises

2. a. Reject H_0 since 4.46 > 3.10.
Conclude that the mean prices for all four pizza parlors are not the same.
b. By rejecting H_0, we may have committed a Type I error.

Chapter 13

13.15 a. The *y*-intercept is 100. The slope is 6. There is a positive relationship between *x* and *y*.

b. The *y*-intercept is −40. The slope is 9. There is a positive relationship between *x* and *y*.

13.17 $\hat{y} = 114.167 - 14.5x$

13.19 a. $38.85 million

b. The actual gross sales would differ due to the influence of variables not included in the model and random variation.

c. The relationship is nonexact.

13.21 Let x = sugar (in grams) and y = calories.

a., d.
The scatter diagram exhibits a linear relationship between sugar and calorie count.

b. $\hat{y} = 132.993 + 48.135x$

e. 421.803

f. 999.423

13.23 Let x = total payroll (in millions of dollars) and y = percentage of games won.

a. $\mu_{y|x} = 36.8558 + .1028x$

b. The regression line obtained in part (a) is the population regression line because the data are given for all National League baseball teams. The values of the *y*-intercept and slope obtained in part (a) are those of *A* and *B*.

c. The value of $A = 36.8556$ represents the percentage of games won by a team with a total payroll of zero dollars. The value of $B = .1028$ means that, on average, the percentage of games won increases by .1028 for every $1 million increase in payroll of a National League baseball team.

d. 51.76%

13.25 For a simple linear regression model, $df = n - 2$.

13.27 SST is the sum of squared differences between the actual y values and $\bar{y}$, $SST = \Sigma(y - \bar{y})^2$. **SSR** is the portion of SST that is explained by the use of the regression model.

13.29 $\sigma_\varepsilon = 6.125$
$\rho^2 = .398$

13.31 Let x = fat (in grams) and y = calories.

a. $SS_{yy} = 11{,}550$
$SS_{xx} = 23.9688$
$SS_{xy} = 353.75$

b. $b = 14.76$
$s_e = 32.477$

c. $SST = SS_{yy} = 11{,}550$, $SSE = \Sigma e^2 = 6329.074$
$SSR = SST - SSE = 11{,}550 - 6329.074 = 5220.926$

d. $r^2 = .45$

13.33 Let x = sugar (in grams) and y = calories.

a. $s_e = 60.117$

b. $r^2 = .863$

Thus, 86.3% of the total squared errors (SST) is explained by the regression model.

13.35 Let x = total payroll (in millions of dollars) and y = percentage of games won.

a. $\sigma_e = 6.3123$

b. $\rho^2 = .292$

13.37 a. 7.2841 to 7.7159

b. Reject H_0 since 75.0751 > 2.1604.
Conclude that *B* is positive.

c. Reject H_0 since 75.0751 > 3.0123.
Conclude that *B* is different from zero.

d. Reject H_0 since 15.0150 > 2.6503.
Conclude that *B* is different from 6.0.

13.39 a. −2.616 to −2.384

b. Reject H_0 since −55.56 < −2.33.
Conclude that *B* is negative.

c. Reject H_0 since $-55.56 < -2.58$.
Conclude that B is different from zero.

d. Reject H_0 since $-11.11 < -1.96$.
Conclude that B is less than -2.0.

13.41 a. 31.008 to 65.262

b. Do not reject H_0 since 1.123 is between -2.3646 and 2.3646. Conclude that B is not different from 40.

13.43 a. $-.1569$ to $-.0017$

b. Reject H_0 since $-3.211 < -2.447$.
Conclude that B is different from zero.

13.45 The **linear correlation coefficient** measures the strength of the linear association between two variables. Its value always lies in the range -1 to 1.

13.49 a

13.51 The linear correlation coefficient r measures only linear relationships. Thus, r may be zero and the variables might still have a nonlinear relationship.

13.53 a. positive b. positive
c. negative d. zero

13.55 $\rho = .072$

13.57 a. We expect the ages and prices of cars to be negatively related because, on average, the older a car is, the less prospective buyers are willing to pay.

b. $r = -.988$

c. Reject H_0 since $-15.669 < -2.447$.
Conclude that ρ is negative.

13.59 a. $r = .7371$

b. Reject H_0 since $3.6176 > 2.718$.
Conclude that ρ is positive.

13.61 $\rho = .4348$
It does not make sense to make a confidence interval and to test a hypothesis here because the data is from a population.

13.63 a. $SS_{xx} = 6394.9$
$SS_{yy} = 1718.9$
$SS_{xy} = 3136.1$

b. $\hat{y} = -22.5355 + .4904x$

d. $r^2 = .8947$
The value of $r = .95$ indicates that the two variables have a very strong positive linear correlation. The value of $r^2 = .89$ means that 89% of the total squared errors (SST) is explained by the regression model.

e. $s_e = 4.7557$

f. .2909 to .6899

g. Reject H_0 since $8.242 > 2.896$.
Conclude that B is positive.

h. Reject H_0 since $8.254 > 3.355$.
Conclude that ρ is different from zero.

13.65 a. $SS_{xx} = .7002$
$SS_{yy} = 512.1$
$SS_{xy} = 2.752$

b. $\hat{y} = 43.818 + 3.93x$

d. The value of $r = .145$ indicates that the two variables have a very strong positive linear correlation. The value of $r^2 = .0211$ means that approximately 2.11% of the total squared errors (SST) is explained by the regression model.

e. $s_e = 7.916$

f. -17.885 to 25.745

g. Do not reject H_0 since $.4154 < 2.8964$.
Conclude that B is not different from zero.

h. Do not reject H_0 since $.4154 < 1.8595$.
Conclude that ρ is not positive.

13.67 a. The 95% confidence interval for $\mu_{y|20}$ is 19.249 to 20.251.
The 95% prediction interval for y_p for $x = 20$ is 17.795 to 21.705.

b. The 95% confidence interval for $\mu_{y|10}$ is 33.823 to 41.117.
The 95% prediction interval for y_p for $x = 10$ is 29.379 to 45.621.

13.69 The 90% confidence interval for μ_p for $x = 10$ is \$35,690 to \$43,116.
The 90% prediction interval for μ_p for $x = 10$ is \$27,882 to \$50,924.

13.71 The 99% confidence interval for $\mu_{y|8.5}$ is 54.5016 to 56.0356.
The 99% prediction interval for y_p for $x = 8.5$ is 53.2834 to 57.2538.

13.73 a. As the age of a machine increases, the number of breakdowns is expected to increase. Hence, we expect a positive relationship between these two variables. Consequently, B is expected to be positive.

b. $\hat{y} = 1.471 + 1.301x$
The sign of $b = 1.301$ is positive, which is consistent with what we expected.

c. The value of $a = 1.471$ represents the number of breakdowns per month for a new machine (age = 0). The value of $b = 1.301$ means that the average number of breakdowns per month increases by about 1.301 for every one year increase in the age of such a machine.

d. The value of $r = .9198$ indicates that the two variables have a very strong positive correlation. The value of $r^2 = .846$ means that approximately 84.6% of the total squared errors (SST) is explained by the regression model.

e. $s_e = 1.739$

f. .7472 to 1.875

g. Reject H_0 since $5.741 > 1.943$.
Conclude that B is positive.

h. Reject H_0 since $5.742 > 1.943$.
Conclude that ρ is positive. The conclusion is the same as that of part (g) (reject H_0).

13.75 **a.** We expect B to be positive.

b. $\hat{y} = 7.8304 + .5039x$
The sign of $b = .5039$ is positive, which is consistent with what we expected.

c. The value of $a = 7.8304$ represents the number of units (in hundreds) sold if there are no promotions. The value of $b = .5039$ means that the sales are expected to increase by about 50 units per day for each additional promotion.

d. The value of $r = .8861$ indicates that the two variables have a strong positive correlation. The value of $r^2 = .7853$ means that approximately 78% of the total squared errors (SST) is explained by the regression model.

e. We expect sales of about 2547 units in a day with 35 promotions.

f. $s_e = 3.3527$

g. .1073 to .9004

h. Reject H_0 since $4.278 > 3.365$.
Conclude that B is positive.

i. Reject H_0 since $4.273 > 3.365$.
Conclude that ρ is different from zero.

13.77 The 95% confidence interval for $\mu_{y|12}$ is 15.454 to 18.712. The 95% prediction interval for y_p for $x = 12$ is 12.532 to 21.646.

13.79 The 98% confidence interval for $\mu_{y|2400}$ is 333.0539 to 366.2381.
The 98% prediction interval for y_p for $x = 2400$ is 295.2779 to 404.0141.

13.83 **a.** Do not reject H_0 since $0 < 1.383029$.
Conclude that ρ is not different from zero.

b. Utilizing $r = 0$, we would conclude that there is no linear relation between two variables; in fact, a nonlinear relationship is a perfect fit for the data.

Self-Review Test

Multiple Choice Questions

1. c	2. a	3. b	4. a	5. b
6. b	7. True	8. True	9. a	10. b
11. a	12. b	13. a	14. c	15. a
16. b	17. c	18. a	19. c	20. a
21. a	22. c	23. a	24. a	25. c
26. a	27. a	28. a		

Practice Exercises

1. 2.19	2. 139.77	3. 2.074	4. .507
5. 6.29 to 8.05	6. −.01 to .77	7. .849	8. .181

Chapter 14

14.1 The **coefficients of independent variables** in a multiple regression model are interpreted as the change in y for a one-unit change in the corresponding independent variable when all other independent variables are held constant.

14.3 **Multicollinearity** is when independent variables are highly linearly related. The independent variables can have a non-linear relationship but cannot be linearly related.

14.7 **a.** $\hat{y} = 19.04 + .1051x_1 - .1592x_2$

b. The value of $a = 19.04$ gives the value for $\hat{y}$ when $x_1 = 0$ and $x_2 = 0$. However, since $x_1 = 0$ and $x_2 = 0$ do not occur together in the sample data, the estimate is invalid.

c. $s_e = 1.89576$, $R^2 = 95.43\%$, and $\bar{R}^2 = 94.29\%$

d. 24.4207

e. 26.5949

f. .0779 to .222

g. Reject H_0 since $-2.233 < 1.860$.
Conclude that B_2 is not negative.

14.9 **a.** $\hat{y} = -3.06 + .1931x_1 + .2545x_2$

b. The value of $a = -3.06$ gives the expected weekly sales for restaurants in areas with zero population and a mean annual household income of \$0. However, since the sample data does not include any restaurants in areas with zero population and a mean annual household income of \$0, the estimate is invalid.

c. $s_e = .39105$, $R^2 = 98.70\%$, and $\bar{R}^2 = 98.18\%$

d. \$17,389.

e. \$18,284.

f. .186 to .323

g. Reject H_0 since $8.698 > 2.570$.
Conclude that B_1 is different from zero.

Self-Review Test

Multiple Choice Questions

1. c	2. a	3. c	4. b	5. c
6. b	7. a	8. c	9. b	10. b
11. b	12. c	13. c	14. b	15. c

Practice Exercises

1. A regression line obtained by using population data is called **the population multiple regression model**. The **estimated multiple regression model** is obtained from sample data.

2. The regression coefficients in a multiple regression model are called the **partial regression coefficients.**

3. R^2 is the proportion of the total sum of squares (SST) that is explained by the multiple regression model. $\bar{R}^2$ is the coefficient of multiple determination adjusted for degrees of freedom.

4. **a.** We would expect the relationship between sale price and lot size to be positive, the relationship between sale price and living area to be positive, and the relationship between sale price and age to be negative.

b. $\hat{y} = 200.2 + 11.9x_1 + .0995x_2 - 7.55x_3$

The signs of the coefficients of the independent variables obtained in the solution are consistent with the expectations in part (a).

c. The value of a = 200.2 gives the expected sale price of a house for a lot size of zero and living area of zero at age zero. However, since $x_1 = 0, x_2 = 0$, and $x_3 = 0$ do not occur together in the sample data, the estimate is invalid.

d. $s_e = 37.7616$, $R^2 = 88.16\%$, and $\bar{R}^2 = 84.22\%$

e. $422,750

f. $422,280

g. The 99% confidence interval for B_1 is −65.125 to 88.925. The 99% confidence interval for B_2 is −.040 to .239. The 99% confidence interval for B_3 is −14.018 to −1.082.

h. −51.151 to 451.551

i. Do not reject H_0 since $.502 < 2.821$. Conclude that B_1 is not positive.

j. Reject H_0 since $2.319 > 2.262$. Conclude that B_2 is positive.

k. Reject H_0 since $-3.794 < -1.833$. Conclude that B_3 is negative.

Index

Note: Page numbers in *italics* refer to figures, **bolded** refer to tables, with asterisks(*) refer to the online-only Chapter 15.

N